D1758909

The Oxford Handbook of Applied Bayesian Analysis

The Oxford Handbook of Applied Bayesian Analysis

Edited by

Anthony O'Hagan
Mike West

OXFORD
UNIVERSITY PRESS

This book has been printed digitally and produced in a standard specification in order to ensure its continuing availability

OXFORD
UNIVERSITY PRESS

Great Clarendon Street, Oxford OX2 6DP
United Kingdom

Oxford University Press is a department of the University of Oxford.
It furthers the University's objective of excellence in research, scholarship,
and education by publishing worldwide.
Oxford is a registered trade mark of Oxford University Press in the UK
and in certain other countries

Reprinted 2013

British Library Cataloguing in Publication Data
Data available

Library of Congress Cataloging in Publication Data
Data available

ISBN 978-0-19-954890-3

Contents

Part IV Policy, Political and Social Sciences

Preface

A Bayesian 21st century

The diversity of applications of modern Bayesian analysis at the start of the 21st century is simply enormous. From basic biology to frontier information technology, the applications of highly structured stochastic models of increasing realism – often with high-dimensional parameters and latent variables, multiple layers of hierarchically structured random effects, and nonparametric components – are increasingly routine. Much of the impetus behind this growth and success of applied Bayesian methods over the last 20 years has come from access to the increasingly rich array of advanced computational strategies for Bayesian analysis; this has led to increasing adoption of Bayesian methods from heavily practical and pragmatic perspectives.

Coupled with this evolution in the nature of applied statistical work to a model-based, computational perspective is change in statistical scientific thought at a more fundamental level. As researchers become increasing involved in more complex stochastic model building enabled by advanced Bayesian computational methods, they also become more and more exposed to the inherent logic and directness of Bayesian model building. Scientifically relevant, highly structured stochastic models are often simply naturally developed from Bayesian formalisms and have overt Bayesian components. Hierarchical models with layers of random effects, random processes in temporal or spatial systems, and large-scale latent variables models of many flavours are just a few generic examples of nowadays standard stochastic structures in wide application, and that are all inherently Bayesian models. Much of the rapid growth in adoption of Bayesian methods from pragmatic viewpoints is engendering deeper, foundational change in scientific philosophy towards a more holistically Bayesian perspective. And this, in turn, has important implications for the core of the discipline; bringing Bayesian methods of stochastic modelling centre-stage – with models of increasing complexity and structure for reasons of increased realism – is inevitably re-energizing the core of the discipline, presenting new conceptual and theoretical challenges to Bayesian researchers as applied problems scale in dimension and complexity.

The Handbook

The Handbook of Applied Bayesian Analysis is a showcase of contemporary Bayesian analysis in important and challenging applied problems, bringing together chapters contributed by leading researchers and practitioners in interdisciplinary Bayesian analysis. Each chapter presents authoritative discussions of currently topical application areas together with key aspects of Bayesian analysis in these areas, and takes the reader to the cutting edge of research in that topic. Importantly, each chapter is built around the application, and represents personal interests, experiences and views of the authors speaking from deep and detailed expertise and engagement in the applied problem area.

Each chapter of the *Handbook* involves a concise review of the application area, describes the problem contexts and goals, discusses aspects of the data and overall statistical issues, and develops detailed analysis with relevant Bayesian models and methods. Discussion generally contacts current frontiers of research in each application, with authors presenting their own perspectives, their own latest thinking, and highlighting their own research in both the application and in related and relevant Bayesian methodology used in their application. Each chapter also includes references linking to core publications in the applied field as well as relevant models and computational methods, and some also provide access to data, software and additional material related to the study of the chapter.

Importantly, each chapter contains appendix material that adds further foundational and supporting discussion of two flavours: material on the basic statistical models and methods, with background and key references, for readers interested in going further into methodological aspects, and more traditional appendix material representing additional technical developments in the specific application. Collectively, the appendices are an important component and distinctive feature of the *Handbook*, as they reflect a broad selection of models and computational tools used widely in applied Bayesian analysis across the diverse range of applied contexts represented.

Chapter outlines

Chapters are grouped by broad field of application, namely

- Biomedical and Health Sciences
- Industry, Economics and Finance
- Environment and Ecology
- Policy, Political and Social Sciences
- Natural and Engineering Sciences

Inevitably selective in terms of broad fields as well as specific application contexts within each broad area, the chapters nevertheless represent topical, challenging and statistically illuminating studies in each case. Chapters within each area are as follows.

Biomedical and Health Sciences

Dunson discusses an epidemiological study involving pregnancy outcomes. This chapter showcases Bayesian analysis in epidemiological studies that collect continuous health outcomes data, and in which the scientific and clinical interest typically focuses on the relationships between exposures and risks of an abnormal response, corresponding to an observation in the tails of the distribution. As there is minimal interest in relationships between exposures and the centre of the response distribution in such studies, traditional regression models are inadequate. For this reason, epidemiologists typically categorize both the outcome and the predictors, with the resulting inferences very sensitive to this categorization. Bayesian analysis using density regression, mixtures and nonparametric models, as developed and applied in this pregnancy outcome study, avoid and overcome these challenges.

Green, Mardia, Nyirongo and Ruffieux discuss the alignment of biomolecules. This chapter showcases Bayesian methods for shape analysis to assist with understanding the three-dimensional structure of protein molecules, which is one of the major unsolved biological challenges. This chapter addresses the problem of matching instances of the same structure in the CoMFA (Comparative Molecular Field Analysis) database of steroid molecules, where the three-dimensional coordinates of all the atoms in each molecule are stored. The matching problem is challenging because two instances of the same three-dimensional structure in such a database can have very different sets of coordinates, due not just to noisy measurements but also to rotation, translation and scaling. The authors present an efficient Bayesian methodology to identify, given two or more biomolecules represented by the coordinates of their atoms, subsets of those atoms which match within measurement error, after allowing for appropriate geometrical transformations to align the biomolecules.

Cheng and Madigan discuss a study of pharmaceutical testing from multiple clinical trials concerned with side-effects and adverse events among patients treated with a popular pain-relieving drug. This chapter showcases the development of sensitive Bayesian analysis of clinical trials studies involving problems of missing data and particularly non-ignorable dropout of patients from studies, as well as sequential methods and meta-analysis of multiple studies. The study concerns Vioxx, an anti-inflammatory drug that was licensed for use in the

USA by the FDA in 1999, and then withdrawn from the market in 2004 due to cardiovascular safety concerns. Merck, the manufacturer of Vioxx, conducted many clinical trials both before and after 1999. In part to avoid potential future scenarios like Vioxx, analyses of the data from these multiple clinical trials are of considerable importance and interest. The study raises multiple, challenging statistical issues and questions requiring sensitive evaluation, and the chapter highlights the utility of Bayesian analysis in addressing these challenges.

Oakley and Clough discuss uncertainty in a mechanistic model that has been used to conduct a risk assessment of contamination of farm-pasteurized milk with the bacterium Vero-cytotoxigenic *E. coli* (VTEC) O157. This chapter showcases Bayesian methods for analysing uncertainties in complex computer models. The VTEC model has uncertain input parameters, and so outputs from the model used to inform the risk assessment are also uncertain. The question then arises of how to reduce output uncertainty most efficiently. The authors conduct a variance-based sensitivity analysis to identify the most important uncertain model inputs, and so prioritize what further research would be needed to best reduce model output uncertainty.

Schmidt, Hoeting, Pereira and Vieira discuss temporal prediction and spatial interpolation for out-breaks of malaria over time for municipalities in the state of Amazonas, Brazil. This chapter showcases Bayesian spatial-temporal modelling for epidemiological discrete count data. Malaria is a world-wide public health problem with 40% of the population of the world at risk of acquiring the disease. It is estimated that there are over 500 million clinical cases of malaria each year world-wide. This work falls in the area of disease mapping, where data on aggregate incidence of some disease is available for various administrative areas, but the data for Amazonas are incomplete, covering only a subset of the municipalities. Furthermore, the temporal aspect is important because malaria incidence is not constant over time. A free-form spatial covariance structure is adopted which allows for the estimation of unobserved municipalities to draw on observations in neighbouring areas, but without making strong assumptions about the nature of spatial relations. A multivariate dynamic linear model controls the temporal effects and facilitates the forecasting of future malaria incidence.

Merl, Lucas, Nevins, Shen and West discuss a study in cancer genomics. This chapter showcases the application of Bayesian concepts and methods in an overall strategy for linking the results of *in vitro* laboratory studies of gene expression to *in vivo* human observation studies. The basic problem of translating inferences across contexts constitutes a generic, critical, and growing challenge in modern biology, which typically moves from laboratory experiments with cultured cells, to animal model experiments, to human outcome studies and clinical trials. The study described here concerns this problem in the context of the genomics of several oncogene pathways that are fundamental to many human cancers. The application involves Bayesian sparse multivariate regression and sparse latent factor models for large-scale multivariate data, and details the use of such models to define and relate statistical signatures of biological phenomena between contexts. In addition, the study requires linking the resulting, model-based inferences to known biology; this is achieved using Bayesian methods for mapping summary inferences to databases of biological pathways. The study includes detailed discussion of biological interpretations of experimentally defined gene expression signatures and their elaborated subsignature representations emerging from Bayesian factor analysis of *in vivo* data, model-generated leads to design new biological experiments based on some of the findings, and contextual discussions of connections to clinical cancer profiling and prognosis.

Henderson, Boys, Proctor and Wilkinson discuss oscillations observed in the levels of two proteins, p53 and Mdm2, in single living cancer cells. This chapter showcases Bayesian methods in systems biology using genuine prior information and MCMC computation. The p53 tumour suppressor protein plays a major role in cancer. It has been described as the 'guardian of the genome', blocking cell cycle progression to allow the repair of damaged DNA. An increase of p53 due to stress causes an increase in the level of Mdm2 which in turn inhibits p53. From observations of levels of these two proteins in individual cancer cells, the objective is to learn about the rate parameters that control this feedback loop. By examining several cells, it is hoped to understand why they do not oscillate in phase. However, the modelling of the complex reactions within the cell makes this exercise highly computationally intensive. The authors develop a Bayesian approximation to the discrete time transition probabilities in the underlying continuous time stochastic model. Prior information about the unknown rate parameters is incorporated based on experimental values in the literature and they apply sophisticated MCMC methods to compute posterior distributions.

Dawid, Mortera and Vicard discuss the problem of evaluating the probability of a putative father being the real father of a child, based on his DNA profile and those of the mother and child. The chapter is a showcase for careful probabilistic reasoning. In recent years there has been heavy media coverage of DNA profiling for criminal identification, but the technique has also been useful in illuminating a number of complex genetic problems, including cases of disputed paternity. The paternity problem is complicated by the possibility of genetic mutation: the putative father could be the real father, yet a mutation in the child's DNA could seem to imply that he is not. The probability of paternity now depends strongly on the rate of mutation. On the other hand, estimates of mutation rates are themselves very sensitive to assumptions about paternity.

Using Austrian-German casework data, the authors present a meticulous study of this problem, constructing and analysing a model to handle paternity and mutation jointly.

Industry, Economics and Finance

Popova, Morton, Damien and Hanson discuss a study in Bayesian analysis and decision making in the maintenance and reliability of nuclear power plants. The chapter showcases Bayesian parametric and semiparametric methodology applied to the failure times of components that belong to an auxiliary feedwater system. This system supplies cooling water during an emergency operation or to an electro-hydraulic control system, used for the control of the main electrical generating steam turbine. The parametric models produce estimates of the hazard functions that are compared to the output from a mixture of Polya trees model. The statistical output is used as the most critical input in a stochastic optimization model which finds the optimal replacement time for a system that randomly fails over a finite horizon. The chapter also discusses decision analysis, using the model in defining strategies that minimize expected total and discounted cost of nuclear plant maintenance.

Cumming and Goldstein discuss analysis of the Gullfaks oil field using a reservoir simulation model run at two different levels of complexity. This chapter showcases Bayes linear methods to address highly complex problems for which the full Bayesian analysis may be computationally intractable. A simulator of a hydrocarbon reservoir represents properties of the reservoir on a three-dimensional grid. The finer this grid is, the more accurately the simulator is expected to predict the real reservoir behaviour, but finer resolution also implies rapidly escalating computation times. Observed behaviour of the reservoir can in principle be used to learn about values of parameters in the simulator, but this Bayesian calibration demands that the simulator can be run many times at different values of these parameters in order to search for regions of parameter space in which acceptable matches are found to the observed data. The authors employ many runs of the simulator at low resolution to augment a few runs of the fine simulator. Their approach involves careful modelling of the relationship between the two versions of the simulator, as well as how the fine simulator relates to reality.

Pievatolo and Ruggeri discuss a study in Bayesian reliability analysis concerning underground train door failures in a European underground system over a period of nine years. The chapter showcases development and application of Bayesian stochastic process models in a reliability context. Facing questions about relevant reliability 'time' scales, the authors develop a novel bivariate Poisson process as a natural way to extend the usual Poisson models for the occurrence of failures in repairable systems; the bivariate model uses both calendar time and kilometres driven by trains as metrics. An important consequence of this choice is that seasonal effects are easily incorporated into the model. The Bayesian models and MCMC methods developed lead to predictive distributions for failures and address key practical questions of how to assess reliability before warranty expiration, combining the data from several trains. This study also clarifies the advantages and disadvantages of using Poisson process models for repairable systems with a number of different failure modes.

Ferreira, Bertolde and Holan discuss an economic study of agricultural production in Espírito Santo State, Brazil, from 1990 to 2005. The chapter showcases the use of Bayesian multiscale spatio-temporal models that uses the natural geopolitical division of Espírito Santo State at levels of macroregions, microregions, and counties. The models involve multiscale latent parameters that evolve through time over a period of several years of the economic study. The analysis sheds light on the similarities and differences in agricultural production between regions within each scale of resolution, and on the temporal changes in relative agricultural importance of those regions as explicitly described by the evolution of the estimated multiscale components. The study involves a number of other questions relevant to the underlying spatio-temporal agricultural production process at each level of resolution, and builds on advanced Markov chain Monte Carlo methods for multivariate dynamic models integrated in an overall, highly-structured multiscale spatio-temporal framework.

Lopes and Polson discuss financial time series at the time of the 2007–08 credit crisis, showcasing the ability of Bayesian modelling and inference to represent a period of financial instability and to identify the underlying mechanisms. The authors consider several forms of model that exhibit stochastic volatility so as to capture the rapidly changing behaviour of the financial indicators at that time. Using Bayesian sequential model choice techniques, they show how the evidence accumulates over time that the pure stochastic volatility model is inferior to a model with jumps. Their work has implications for analysis and prediction in times of unusual market behaviour.

Quintana, Carvalho, Scott and Costigliola discuss studies in applications of the Bayesian approach to risk modelling regarding speculative trading strategies in financial futures markets. The chapter showcases applied Bayesian thinking in the context of financial investment management, highlighting the corresponding concepts of betting and investing, prices and expectations, and coherence and arbitrage-free pricing. Covering central applied methods and tools of Bayesian decision analysis and speculation in portfolio studies, risk modelling, dynamic linear models and Bayesian forecasting, and highly structured Bayesian graphical modelling approaches for multivariate, time-varying covariance matrices in multivariate dynamic models, the chapter develops studies of investment strategies and returns in futures markets over a period between 1990 and 2008 based on portfolios of currency exchange rates, government bonds and stock market indices.

Fernández-Villaverde, Guerrón-Quintana and Rubio-Ramiréz discuss macroeconomic studies of the dynamics of the US economy over the last 50 years using Bayesian analysis of dynamic stochastic equilibrium models. This chapter is a showcase of modern, model-based Bayesian analysis in mainstream economics studies, and an approach increasingly referred to as the *new macroeconometrics*. The authors formulate and estimate a benchmark dynamic stochastic equilibrium model that captures much of the time-varying structure in the US macroeconomy over these years, and describe its application in policy analysis for public institutions – such as central banks – and private organizations and businesses. Application involves likelihood evaluations that are enabled using Bayesian sequential Monte Carlo and MCMC methods. The study discusses critical questions of the roles of priors and pre-sample information, documents a range of real and nominal rigidities in the US economy and discusses the increasingly central roles of such Bayesian approaches in this context as well as frontier research issues.

Environment and Ecology

Challenor, McNeall and Gattiker discusses the potential collapse of the meridional overturning circulation in the Atlantic Ocean. This chapter showcases Bayesian methods for analysing uncertainty in complex models, and in particular for quantifying the risk of extreme outcomes. While climate science has concentrated on predictions of global warming, there are possible scenarios which, although with low probability, would have high impact. One such event is the collapse of the ocean circulation that currently ensures that Western Europe enjoys a warmer climate than, for instance, similar latitudes in Western North America. Collapse of the meridional overturning circulation (MOC) is predicted by the GENIE-1 climate model for some values of the model inputs, but the actual values of these inputs are unknown. A single run of GENIE-1 takes several hours, and the authors use Bayesian emulation to estimate the probability of MOC collapse based on a limited number of model runs, and to incorporate data comprising a sparse time series of five measurements of the MOC from 1957 to 2004.

Clark, Bell, Dietze, Hersh, Ibanez, LaDeau, McMahon, Metcalf, Moran, Pangle and Wolosin discuss demography of plant populations, showcasing applied Bayesian analysis and methods that allow for synthesis of information from multiple sources to estimate the demographic rates of trees and how they respond to environmental variation. Data come from individual (tree) measurements over a period of 18 years, including diameter, crown area, maturation status, and survival, and from seed traps, which provide indirect information on fecundity. Different observations are available for different years and trees. The multiple data sets are synthesized with a process model where each individual is represented by a multivariate state-space submodel for both continuous (fecundity potential, growth rate, mortality risk, maturation probability) and discrete states (maturation status). Each year, state variables respond to a dynamic environment. Results provide unprecedented detail on the ways in which demographic rates relate to one another, within individuals over time, among individuals, and among species. The chapter also describes how results of these Bayesian methods are being used to assess how forests can respond to changing climate.

Gelfand and Sahu discuss environmental studies that aim to combine monitoring data and computer model outputs in assessing environmental exposure. This chapter showcases Bayesian data fusion methods using spatial Gaussian process models in studies of weekly deposition data from multiple US sites monitored by the US National Atmospheric Deposition Program. Environmental exposure community numerical models are now widely available for a number of air pollutants. Based on inputs from a number of factors such as meteorological conditions, land usage, and power station emission volumes, all of which are responsible for producing air pollution, and some predictions of spatial surfaces for current, past, and future time periods, these models provide output exposures at various spatial and temporal resolutions. For large spatial regions such as the entire United States, the spatial coverage of the available network monitoring stations can never match the coverage at which the computer models produce their output. However, the monitoring data will be more accurate than the computer model output since, up to measurement error, they provide the actual true levels: observations from the realization of the pollution process surface at that time. It is important to combine these two sets of information to make inference regarding pollution exposure, and this study represents best-Bayesian practices in addressing this problem.

Choy, Murray, James and Mengersen discuss eliciting knowledge from ecological experts about the habitat of the Australian brush-tailed rock-wallaby. This chapter is a showcase of techniques for eliciting expert judgement about complex uncertainties. The rock-wallaby is an endangered species, and in order to map where it is likely to be found, it is essential to use expert judgement about how the various environmental factors (such as geology, land cover and elevation) influence the probability of rock-wallabies being present at a site. The authors employ an indirect elicitation method in which the experts are presented with descriptions of some specific sites and asked for their probabilities. The relationship between probability of occurrence and the environmental variables is then inferred and used to predict the rock-wallaby's likely habitats throughout the region of interest.

Tebaldi and Smith discuss studies in characterizing the uncertainty of climate change projections, showcasing Bayesian methods for integration and comparison of predictions from multiple models and groups. The chapter describes a suite of customised Bayesian hierarchical models that synthesize ensembles of climate model simulations, with the aim of reconciling different future projections of climate change, while characterizing their uncertainty in a rigorous fashion. Posterior distributions of future temperature and/or precipitation changes at regional scales are obtained, accounting for many peculiar data characteristics, such as systematic biases, model-specific precisions, region-specific effects, changes in trend with increasing rates of greenhouse gas emissions, and others. The chapter expands on many important issues characterizing model experiments and their collection into multimodel ensembles, and addresses the need of 'impact research', by proposing posterior predictive distributions as a representation of probabilistic projections. In addition, the calculation of the posterior predictive distribution for a new set of model data allows a rigorous cross-validation approach to assess and, in this study, confirm the reasonableness of the Bayesian modelling assumptions.

Policy, Political and Social Sciences

Carvalho and Rickershauser discuss a study of temporal volatility and information flows in political campaigns, showcasing Bayesian analysis in evaluation of information impact on vote sentiment and behaviour in highly publicized campaigns. The core application is to the 2004 US presidential campaign. The study builds a measure of information flow based on the returns and volume of the 'Bush wins the popular vote in 2004' futures contract on the tradesports/intrade prediction market. This measure links events to information level, providing a direct way to evaluate its impact in the election. Among the findings are that information flows increased as a result of the televised debates, Kerry's acceptance speech at the Democratic convention, and national security-related stories such as the report that explosives vanished in Iraq under the US's watch, the CBS story about Bush's National Guard service and the subsequent retraction, and the release of the bin Laden tape a few days before the election. Contrary to popular accounts of the election, ads attacking Kerry's military service aired by the Swift Boat Veterans for Truth in August apparently contributed only a limited amount of information to the campaign. This political science application develops novel hidden state-space models of volatility in information flows and model fitting and evaluation using Bayesian MCMC methods for nonlinear state-space models.

Gamerman, Soares and Gonçalves discuss whether cultural differences may affect the performance of students from different countries in the various test items which make up the international PISA test of mathematics ability. This chapter showcases a Bayesian model that incorporates this kind of differential item functioning (DIF) and the role of prior information. The PISA tests in mathematics and other subjects are widely used to compare the educational attainment of 15-year old students in different countries; in 2009, 67 countries have taken part from around the world. DIF is a significant issue with the potential to compromise such comparisons between countries, and substantial DIF may remain in the administered test despite preliminary screening of candidate test items. The authors seek to discover the extent of DIF remaining in the mathematics test of 2003. They employ a hierarchical three-parameter logistic model for the probability of a correct response on an individual item, where the three parameters control the difficulty of the item, its discriminating power and its guessability, and their model allows for different kinds of DIF where any of these parameters may vary between countries. The authors' Bayesian model avoids identifiability problems faced by competing approaches and requires weaker hypotheses due, especially, to the important role played by the prior distributions.

Heiner, Kennedy and O'Hagan discuss auditing of the operation of the food stamps welfare scheme in the state of New York, USA, highlighting the power of Bayesian methods in analysing data that evolve over time. Auditors examine a sample of individual awards of food stamps to see if the value awarded is correct according to the rules of the scheme. The food stamps program is a federal scheme, and if a state is found to have too large an error rate in administering it the federal government can impose large financial penalties. In New York state, the program is administered by individual counties, and sizes of audit samples in small counties can be so small that only one or two errors are found in any given year. The authors propose a model that includes a nonparametric component for the error magnitudes (taints), a hierarchical model for overall error rates across counties and parameters controlling the variation of rates from one year to the next, including an overall trend in error rates. The model allows in particular for estimation of rates in small counties to be smoothed across counties and through time.

Rubin, Wang, Yin and Zell discuss a study in estimating the effects of 'treating hospital type' on cancer survival, using administrative data from central and northern Sweden via the Karolinska Institute in Stockholm. The chapter represents a showcase in application of Bayesian causal inference, in particular using the posterior predictive approach of the 'Rubin causal model' and methods of principal stratification. The central applied question, inferring which type of hospital (e.g. large patient volume versus small volume) is superior for treating certain serious conditions, is a difficult and important problem in institutional assessment and comparisons in a medical context. Ideal data from randomized experiments are simply not available, leading to reliance on observational data. The study involves questions of which factors may reasonably be considered ignorable in the context of covariates available, and non-compliance complications due to transfers between hospital types for treatment, and showcases Bayesian causal modelling utilizing simulation-based imputation techniques.

Natural and Engineering Sciences

Cemgil, Godsill, Peeling and Whiteley discuss musical audio signal analysis in the context of an application to multipitch audio and determining a musical 'score' representation that includes pitch and time duration summary for a musical extract (the so-called 'piano-roll' representation of music). This chapter showcases applied Bayesian analysis in audio signal processing in real environments where acoustical conditions and sound sources are highly variable, yet audio signals possess strong statistical structure. There is typically much prior information about underlying structures and the detail of the recorded acoustical waveform (physical mechanisms by which sounds are generated, cognitive processes by which sounds are perceived by the human auditory system, mechanisms by which high-level sound structures are compiled). A range of Bayesian hierarchical models – involving both time and frequency domain dynamic models, and methods of fitting using simulation-based and variational approximations – are developed in this chapter. The resulting models possess complex statistical structure and so highly adaptive and powerful computational techniques are needed to perform inference, as this study exemplifies.

Higdon, Heitmann, Nakhleh and Habib discuss perhaps the grandest of all problems, the nature and evolution of the universe. This chapter showcases techniques for emulating complex computer models with many inputs and outputs. The Λ-cold dark matter model is the simplest cosmological model in agreement with the cosmic microwave background and large scale structure measurements. This model is determined by a small number of parameters which control the composition, expansion and fluctuations of the universe, and the objective of this study is to learn about the values of these parameters using measurements from the Sloan Digital Sky Survey (SDSS). Model outputs include a dark matter spectrum for the universe and a temperature spectrum for the cosmic microwave background. A key component of the Bayesian analysis is to find a parsimonious representation of such high-dimensional output. Another is innovative modelling to combine the evidence from data on both spectra to find which model input parameters influence the output appreciably, and to learn about those parameters.

Liang, Jordan and Klein discuss the use of probabilistic context-free grammars in natural language processing, involving a large-scale natural language parsing task. The chapter is a showcase of detailed, highly-structured Bayesian modelling in which model dimension and complexity responds naturally to observed data, building on the adaptive nature of the underlying nonparametric Bayesian models developed by the authors. The framework involves structured hierarchical Dirichlet process modelling and customized model fitting via variational methods, to address the core problem of identifying appropriate levels of model complexity in using probabilistic context-free grammars as important components in the modelling of syntax in natural language processing. Detailed development and evaluation in experiments with a synthetic grammar induction task complement the application to a large-scale natural language parsing study on data from the *Wall Street Journal* portion of the Penn Treebank, a large data set used in the natural language processing community for evaluating parsers.

Lee, Taddy, Gramacy and Gray discuss the development of circuit devices, bipolar junction transistors, which are used to amplify electrical current, and showcases the use of a flexible kind of emulator based on a treed Gaussian process. To aid with the design of the circuit device, a computer model predicts its peak output as a function of the input dosage and a number of design parameters. The peak output response can jump sharply with only small changes in dosage, a feature that the treed Gaussian process emulator is able to capture. The methodology also involves a novel sequential design procedure to generate data to fit the emulator, and performs sensitivity analysis and both calibration and validation using experimental data.

Prado discusses a study of experimental data involving large-scale EEG time series generated on individuals subject to tasks inducing cognitive fatigue, with the eventual goals of models able to predict cognitive fatigue based on non-invasive scalp monitoring of real-time EEG fluctuations. The chapter showcases the development and application of structured, multivariate Bayesian dynamic models for analysis of time-varying, non-stationary and erratic (brain wave) time series. Novel time-varying autoregressive and regime switching models, incorporating substantively relevant prior information via structured priors and fitted using novel, customized Bayesian computational methods, are described. The applied study involves an experimental subject asked to perform simple arithmetic operations for a period of three hours. Prior to the experiment, the subject was confirmed to be alert. After the experiment ended, the subject was fatigued, as determined by measures of performance and post-task mood. The study shows how the Bayesian analysis is used to assist practitioners in real time detection of cognitive fatigue.

Invitation

We expect the *Handbook* to be of broad interest to researchers and expert practitioners, as well as to advanced students in statistical science and related disciplines. We believe the important and challenging studies represented across a diverse ranges of applications areas, involving cutting-edge statistical thinking and a broad array of Bayesian model-based and computational methodologies, will also enthuse young researchers and non-statistical readers, and that the chapters exemplify and promote cross-fertilization in advanced statistical thinking across multiple application areas. The *Handbook* will also serve as a reference resource for researchers across these fields as well as within statistical science, and we invite you to use it broadly in support of education and teaching, as well as in disciplinary and interdisciplinary research.

Tony O'Hagan and Mike West
2009

List of Contributors

Dave Bell Nicholas School of the Environment, Duke University, Durham, NC 27708, USA

Adelmo I. Bertolde Departamento de Estatística, Universidade Federal do Espírito Santo, CCE - UFES, Av. Fernando Ferrari, s/n, Vitoria - ES - CEP: 29065-900, Brazil

Richard J. Boys School of Mathematics and Statistics, Newcastle University, Newcastle upon Tyne, NE1 7RU, UK

Carlos M. Carvalho The University of Chicago Booth School of Business, 5807 South Woodlawn Avenue, Chicago, IL 60637, USA

A. Taylan Cemgil Department of Computer Engineering, Boğaziçi University, 34342 Bebek, Istanbul, Turkey

Peter Challenor National Oceanography Centre, Southampton, SO14 3ZH, UK

Jerry Cheng Department of Statistics, 501 Hill Center, Busch Campus, Rutgers, The State University of New Jersey, 110 Frelinghuysen Road, Piscataway, NJ 08854-8019, USA

Samantha Low Choy School of Mathematical Sciences, Queensland University of Technology, Brisbane, Australia

James S. Clark Nicholas School of the Environment, Duke University, Durham, NC 27708, USA, *and*, Department of Biology, Duke University, Durham, NC 27708, USA

Helen E. Clough National Centre for Zoonosis Research, University of Liverpool, Leahurst, Chester High Road, Neston, CH64 7TE, UK

Thomas Costigliola Research and Technology, BEST, LLC, Riverview Historical Plaza II, 33–41 Newark Street, PH, Hoboken, NJ 07030, USA

Jonathan A. Cumming Department of Mathematical Statistics, Durham University, Durham, UK

Paul Damien McCombs School of Business, IROM, The University of Texas at Austin, Austin, TX 78712, USA

A. Philip Dawid Statistical Laboratory, Centre for Mathematical Sciences, Wilberforce Road, Cambridge, CB3 0WB, UK

Michael Dietze Department of Biology, Duke University, Durham, NC 27708, USA, *and* Department of Plant Biology, University of Illinois, Champaign-Urbana, Illinois, USA

David B. Dunson Department of Statistical Science, Duke University, Durham, NC 27705, USA

Jesús Fernández-Villaverde Department of Economics, University of Pennsylvania, 160 McNeil Building, 3718 Locust Walk Philadelphia, PA 19004, USA

Marco A.R. Ferreira Department of Statistics, University of Missouri – Columbia, Columbia, MO 65211-6100, USA

Dani Gamerman Instituto de Matemática, Universidade Federal do Rio de Janeiro, Brazil

James Gattiker Statistical Sciences Group, PO Box 1663, MS F600, Los Alamos, NM 87545, USA

Alan E. Gelfand Department of Statistical Science, Duke University, Durham, NC 27708, USA

Simon J. Godsill Signal Processing and Communications Laboratories, Department of Engineering, Trumpington Street, University of Cambridge, Cambridge, CB2 1PX, UK

Michael Goldstein Department of Mathematical Statistics, Durham University, Durham, UK

Flávio B. Gonçalves Department of Statistics, University of Warwick, UK

Peter J. Green School of Mathematics, University of Bristol, Bristol, BS8 1TW, UK

Robert B. Gramacy Statistical Laboratory, University of Cambridge, Wilberforce Road, Cambridge, CB3 0WB, UK

Genetha A. Gray Technical Staff Member, Sandia National Laboratories, P.O. Box 969, MS 9159, Livermore, CA 94551, USA

Pablo Guerrón-Quintana Economist, Federal Reserve Bank of Philadelphia, Ten Independence Mall, Philadelphia, PA 19106, USA

Salman Habib Nuclear & Particle Physics, Astrophysics, and Cosmology, Los Alamos National Laboratory, PO Box 1663, MS B285, Los Alamos, NM 87545, USA

Tim Hanson Division of Biostatistics, School of Public Health, University of Minnesota, Minneapolis, MN 55455, USA

Karl W. Heiner The School of Business, State University of New York at New Paltz, 1 Hawk Drive, New Paltz, New York, 12561, USA

Katrin Heitmann Space Science and Applications, Los Alamos National Laboratory, PO Box 1663, MS D466, Los Alamos, NM 87545, USA

Daniel A. Henderson School of Mathematics and Statistics, Newcastle University, Newcastle upon Tyne, NE1 7RU, UK

Michelle Hersh Department of Biology, Duke University, Durham, NC 27708, USA

Dave Higdon Statistical Sciences, Los Alamos National Laboratory, PO Box 1663, MS F600, Los Alamos, NM 87545, USA

Jennifer A. Hoeting Department of Statistics, Colorado State University, Fort Collins, CO 80523-1877, USA

Scott H. Holan Department of Statistics, University of Missouri – Columbia, Columbia, MO 65211-6100, USA

Ines Ibanez Department of Biology, Duke University, Durham, NC 27708, USA, *and* School of Natural Resources and Environment, University of Michigan, 440 Church Street, Ann Arbor, MI 48109, USA

Allan James High Performance Computing and Research Support Group, Queensland University of Technology, Brisbane, Australia

Michael I. Jordan Computer Science Division, EECS Department, University of California at Berkeley, Berkeley, CA 94720, USA, *and*, Department of Statistics, University of California at Berkeley, Berkeley, CA 94720, USA

Marc C. Kennedy The Food and Environment Research Agency, Sand Hutton, York, YO41 1LZ, UK

Dan Klein Computer Science Division, EECS Department, University of California at Berkeley, Berkeley, CA 94720, USA

Shannon L. LaDeau Department of Biology, Duke University, Durham, NC 27708, USA *and* Carey Institute of Ecosystem Studies, Milbrook, New York, USA

Herbert K.H. Lee Department of Applied Mathematics and Statistics, University of California, Santa Cruz, 1156 High Street, MS: SOE2, Santa Cruz, CA 95064, USA

Percy Liang Computer Science Division, EECS Department, University of California at Berkeley, Berkeley, CA 94720, USA

Hedibert F. Lopes The University of Chicago Booth School of Business, 5807 South Woodlawn Avenue, Chicago, IL 60637, USA

Joseph E. Lucas Institute for Genome Sciences and Policy, Duke University Medical Center, Durham, NC 27710, USA

David Madigan Department of Statistics, Columbia University, 1255 Amsterdam Ave, New York, NY 10027, USA

Kanti V. Mardia Department of Statistics, University of Leeds, Leeds, LS2 9JT, UK

Sean McMahon Nicholas School of the Environment, Duke University, Durham, NC 27708, USA

Doug McNeall Meteorological Office Hadley Centre, Fitzroy Road, Exeter, UK

Kerrie Mengersen School of Mathematical Sciences, Queensland University of Technology, Brisbane, Australia

Dan Merl Department of Statistical Science, Duke University, Durham, NC 27708, USA

Jessica Metcalf Nicholas School of the Environment, Duke University, Durham, NC 27708, USA

Emily Moran Department of Biology, Duke University, Durham, NC 27708, USA

Julia Mortera Dipartimento di Economia, Università Roma Tre, Via Silvio D'Amico 77, 00145 Roma, Italy

David Morton ORIE, Department of Mechanical Engineering, University of Texas in Austin, Austin, TX 78712, USA

Justine Murray The Ecology Centre, School of Integrative Biology, The University of Queensland, St Lucia, Australia

Charles Nakhleh Pulsed Power Sciences Center, Sandia National Laboratories, PO Box 5800, Albuquerque, NM 87185-1186, USA

Joseph R. Nevins Department of Molecular Genetics and Microbiology, Institute for Genome Sciences and Policy, Duke University Medical Center, Durham, NC 27710, USA

Vysaul B. Nyirongo MLW Clinical Research Programme, College of Medicine, University of Malawi, Blantyre 3, Malawi

Jeremy E. Oakley School of Mathematics and Statistics, University of Sheffield, The Hicks Building, Hounsfield Road, Sheffield, S3 7RH, UK

Anthony O'Hagan Department of Probability and Statistics, University of Sheffield, The Hicks Building, Hounsfield Road, Sheffield, S3 7RH, UK

Luke Pangle Nicholas School of the Environment, Duke University, Durham, NC 27708, USA

Paul Peeling Signal Processing and Communications Laboratories, Department of Engineering, Trumpington Street, University of Cambridge, Cambridge, CB2 1PX, UK

João Batista M. Pereira Departamento de Métodos Estatísticos, Universidade Federal do Rio de Janeiro, Caixa Postal 68530, Rio de Janeiro, Brazil

Antonio Pievatolo CNR IMATI, Via Bassini 15, I-20133 Milano, Italy

Nicholas G. Polson The University of Chicago Booth School of Business, 5807 South Woodlawn Avenue, Chicago, IL 60637, USA

Elmira Popova ORIE, Department of Mechanical Engineering, University of Texas in Austin, Austin, TX 78712, USA

Raquel Prado Department of Applied Mathematics and Statistics, Baskin School of Engineering, University of California, Santa Cruz, 1156 High Street, Santa Cruz, CA 95064, USA

Carole J. Proctor Institute for Ageing and Health, Newcastle University, Newcastle upon Tyne, NE4 6BE, UK

José M. Quintana President, BEST, LLC, Riverview Historical Plaza II, 33–41 Newark Street, PH, Hoboken, NJ 07030, USA

Jill Rickershauser Department of Government, American University, Washington, DC 20016, USA

Donald B. Rubin Department of Statistics, Harvard University, Cambridge, Massachusetts, USA

Juan F. Rubio-Ramírez 213 Social Sciences, Duke University, Durham, NC 27708, USA

Yann Ruffieux FSB IMA STAT Station 8, EPFL - Swiss Federal Institute of Technology, CH - 1015 Ecublens, Switzerland

Fabrizio Ruggeri CNR IMATI, Via Bassini 15, I-20133 Milano, Italy

Sujit K. Sahu School of Mathematics, Southampton Statistical Sciences Research Institute, University of Southampton, Southampton, UK

Alexandra M. Schmidt Departamento de Métodos Estatísticos, Universidade Federal do Rio de Janeiro, Caixa Postal 68530, Rio de Janeiro, Brazil

James Scott Department of Statistical Science, Duke University, Durham, NC 27708, USA

Haige Shen Novartis Research, New Jersey, USA

Richard L. Smith Department of Statistics and Operations research, University of North Carolina, Chapel Hill, NC 27599-3621, USA

Tufi M. Soares Departamento de Estatística, Universidade Federal de Juiz de Fora, Brazil

Matthew Taddy Booth School of Business, The University of Chicago, 5807 South Woodlawn Avenue, Chicago, IL 60637, USA

Claudia Tebaldi Climate Central, Princeton, NJ, USA

Paola Vicard Dipartimento di Economia, Università Roma Tre, Via Silvio D'Amico 77, 00145 Roma, Italy

Pedro P. Vieira Fundação de Medicina Tropical do Amazonas, Gerência de Malária, Av.Pedro Teixeira n.25, Dom Pedro I, 69040-000 – Manaus, AM – Brazil

Xiaoqin Wang Department of Mathematics, Natural and Computer Sciences, University College of Gävle, 801 76, Gävle, Sweden

Mike West Department of Statistical Science, Duke University, Durham, NC 27708, USA

Nick Whiteley Signal Processing and Communications Laboratories, Department of Engineering, Trumpington Street, University of Cambridge, Cambridge, CB2 1PX, UK

Darren J. Wilkinson School of Mathematics and Statistics, Newcastle University, Newcastle upon Tyne, NE1 7RU, UK

Mike Wolosin Department of Biology, Duke University, Durham, NC 27708, USA

Li Yin Department of Medical Epidemiology and Biostatistics, Karolinska Institute, Box 281, SE-171 77, Stockholm, Sweden

Elizabeth R. Zell Division of Bacterial Diseases, National Center for Immunization and Respiratory Diseases, Centers for Disease Control and Prevention, 1600 Clifton Road, Atlanta, GA 30333, USA

PART I
Biomedical and Health Sciences

·1·

Flexible Bayes regression of epidemiologic data

David B. Dunson

1.1 Introduction

Epidemiology is focused on the study of relationships between exposures and the risk of adverse health outcomes or diseases. A better understanding of the factors leading to disease is fundamental in developing better strategies for disease prevention and treatment. Hence, the findings from epidemiology studies have the potential to have a fundamental impact on public health. However, this potential is often not fully realized due to distrust among the general public and physicians about the reliability of conclusions drawn from epidemiology studies. This distrust stems in part from inconsistencies in findings from different studies.

Although statistics cannot solve certain problem in epidemiology, such as the occurrence of unmeasured confounders, many of the problems with lack of reproducibility stem from the overly simplistic statistical analyses that are routinely used. In particular, it is standard practice to reduce a potentially complex health outcome to a simple 0/1 indicator of disease and then apply a logistic regression model. By also categorizing exposures, one can then obtain exposure group-specific odds ratios, which form the primary basis for inference on exposure – disease relationships. Such an approach leads to transparent analyses, with results easily interpretable by clinicians having minimal expertise in statistics.

However, there are important limitations to this paradigm, which can lead to misinterpretations of exposure – disease relationships. The first is the obvious lack of efficiency that results from discarding data. For example, if one has a continuous health response and a continuous exposure, then categorizing both the response and exposure can lead to a reduction in power to detect an association. In addition, the results of the analysis will clearly be very sensitive to the number of categories chosen and the cutpoints for defining these categories (Boucher *et al.*, 1998). For continuous exposures, an obvious alternative to categorization is to use splines to estimate the unknown dose-response curve (Greenland, 1995). However, for continuous responses, it is not clear how to best assess risk as a function of exposures and other factors.

This chapter focuses on flexible Bayesian methods for addressing this problem motivated by pregnancy outcome data on birth weight and gestational age at delivery. As argued by Wilcox (2001), birth weight is not particularly meaningful in itself as a measure of health of the baby. However, there is often interest in assessing factors predictive of intra-uterine growth restriction (IUGR). IUGR is best studied using longitudinal ultrasound data to assess fetal growth over time as in Slaughter, Herring and Thorp (2009). However, such data are typically not available, so it is most common to study IUGR using small for gestational age (SGA) as a surrogate. SGA is defined as a 0/1 indicator that the baby is below the 10th percentile of the population distribution of birth weight stratified on gestational age at delivery. One is also concerned about large-for-gestational age (LGA) babies, which are more likely to be born by Cesarian delivery or with a low Apgar score (Nohr *et al.*, 2008). The standard approach analyses risk of SGA, LGA and preterm birth, defined as a delivery prior to 37 weeks completed gestation, in separate logistic regression analyses.

A natural question that arises is whether one can obtain a coherent picture of the impact of an exposure on pregnancy outcomes from such analyses. Although SGA is meant to provide a surrogate of growth restriction, which is adjusted for gestational age at delivery, it is not biologically plausible to assume that fetal growth and the biological process of initiating delivery are independent. Hence, it seems much more natural to consider birth weight and gestational age at delivery as a bivariate response, while avoiding the loss of information that accompanies categorization (Gage, 2003). Even if the focus is only on assessing predictors of risk of premature delivery, the cutoff of 37 weeks eliminates the possibility of assessing risk on a finer scale. In particular, babies born near the 37 week cutoff experience limited short and long term morbidity compared with babies born earlier.

Figure 1.1 shows data on gestational age at delivery and birth weight for $n = 2313$ pregnancies from the Longnecker *et al.* (2001) substudy of the US Collaborative Perinatal Project. The vertical dashed line at 37 weeks shows the cutoff used to define preterm births, while the solid lines show the cutoff for SGA depending on gestational age at delivery for male and female babies. Interest focuses on assessing how predictors, such as the level of exposure to chemicals in the environment (DDT, PCBs, etc.), impact the risk of adverse pregnancy outcomes, which correspond to lower values of birth weight and gestational age at delivery. As reducing the data to binary indicators has clear disadvantages, I propose to instead let the response for pregnancy i correspond to $y_i = (y_{i1}, y_{i2})'$, with y_{i1} = gestational age at delivery and y_{i2} = birth weight. Although this seems very natural, it does lead to some complications in that

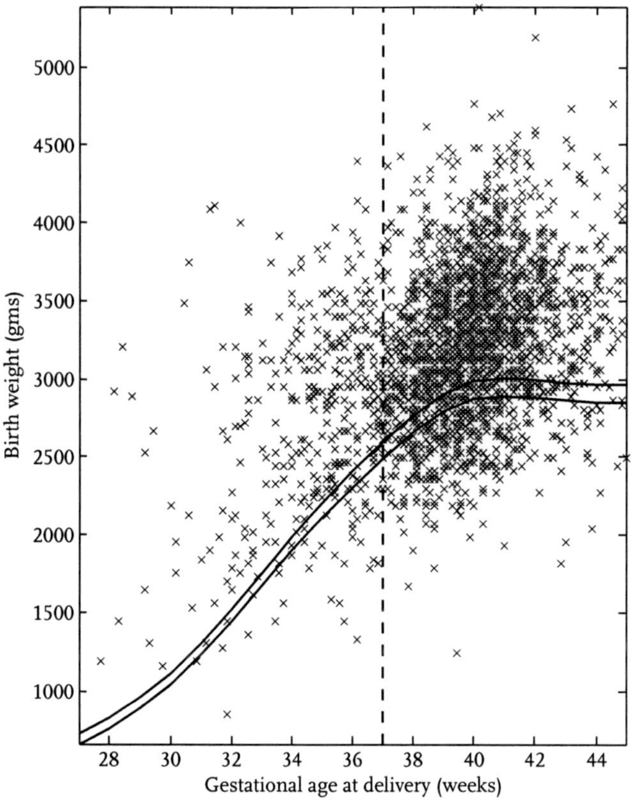

Fig. 1.1 Data on gestational age at delivery and birth weight for the Longnecker *et al.* (2001) substudy of the Collaborative Perinatal Project.

standard parametric models for bivariate continuous data, such as the bivariate normal or t-distributions, provide a poor to the data. This is clear in examining Figure 1.2, which shows an estimate of the marginal density of y_{i1}. The density is left skewed and standard transformations fail to produce a Gaussian shape.

As adverse health outcomes correspond to values in the tails of the response distribution, the main interest is in studying how predictors impact the distributional tails. For example, in studying the impact of DDE levels in maternal serum on risk of premature delivery using the Longnecker *et al.* (2001) data, we would like to assess how the left tail of the distribution in Figure 1.2 changes with dose of DDE and other predictors. Potentially, this interest can be addressed using quantile regression. However, this would necessitate choosing a particular percentile of the distribution that is of primary interest, which is in some sense as unappealing as categorization. As an alternative, one can allow the conditional distribution of y_i given predictors $x_i = (x_{i1}, \ldots, x_{ip})'$ to be unknown and changing flexibly over the predictor space.

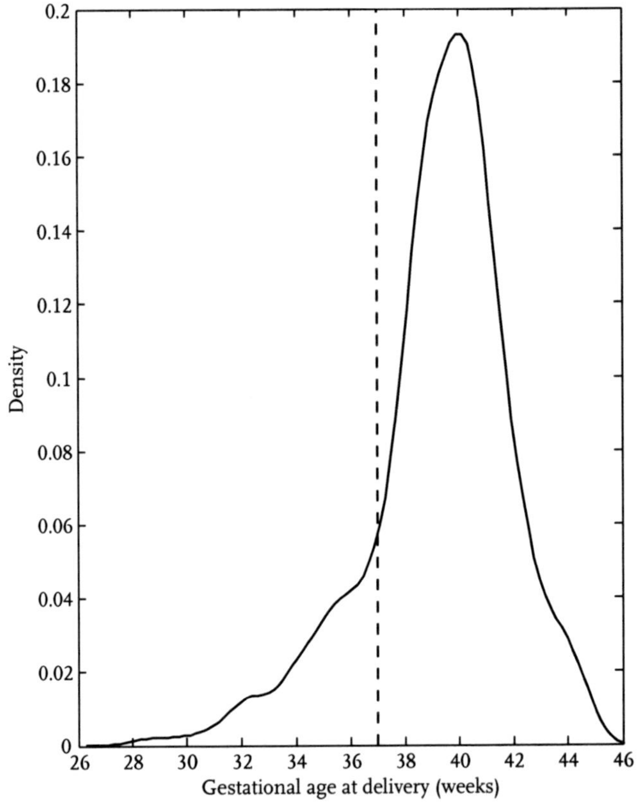

Fig. 1.2 Frequentist kernel estimate of the density of gestational age at delivery for the Longnecker *et al.* (2001) data. The vertical dashed line is the cutoff for defining premature delivery.

As further motivation, Figure 1.3 plots frequentist kernel estimates of the conditional distribution of y_{i1} given x_{i1} = maternal serum level of DDE, with x_{i1} categorized into quintiles and estimates then obtained separately for each category. As categorized DDE increases, there is an increasingly heavy left tail but the right side of the distribution remains essentially unchanged. This type of pattern is not surprising given the biological constraints limiting the possibility of a very long gestational length. I expect that similar constraints occur for many other continuous health outcomes. For such data, standard regression approaches which rely on modelling predictor effects on the mean response are completely inappropriate. Instead, what is needed is flexible methods for density regression, which allow the entire density to change flexibly with predictors, while allowing variable selection and hypothesis testing. This chapter focuses on applying such methods to the pregnancy outcome application.

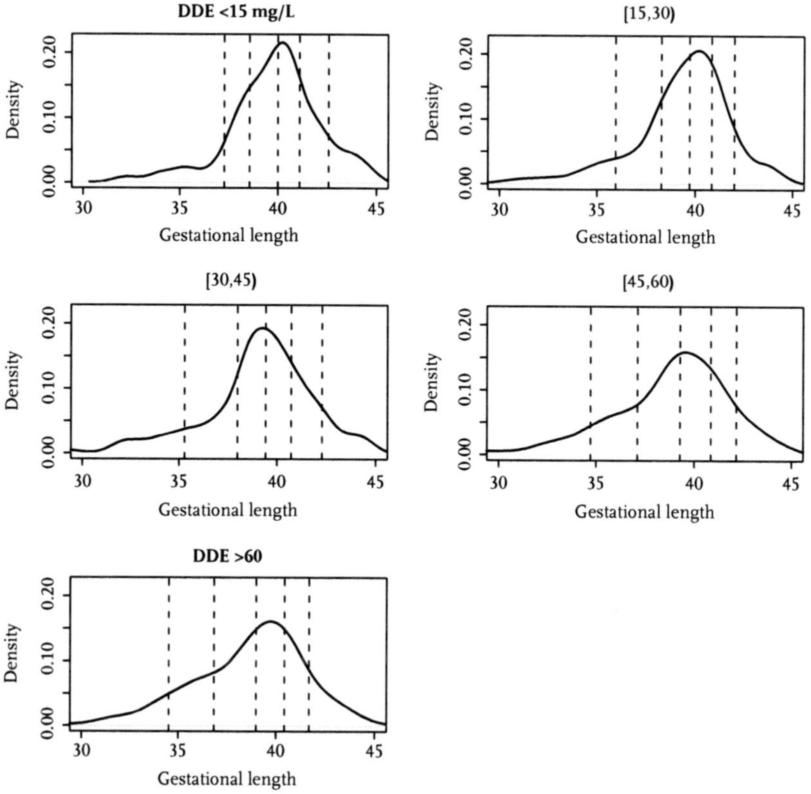

Fig. 1.3 Frequentist kernel estimate of the density of gestational age at delivery conditional on categorized DDE level. The vertical dashed lines correspond to quantiles of the empirical distribution in each group.

1.2 Mixture models

1.2.1 Finite mixture models

As a starting point, it is useful to consider finite mixture models for characterizing the conditional distribution of y_i given x_i. One possibility is to use a latent class model,

$$f(y \mid x) = \sum_{h=1}^{k} \pi_h(x)\, N_2\big(y; \mu_h, \Sigma_h\big), \tag{1.1}$$

where $h = 1, \ldots, k$ indexes mixture components, $\pi_h(x)$ is the probability of allocation to component h conditionally on the predictors x, and the distribution of $y = (y_1, y_2)'$ among pregnancies in latent class h corresponds to the bivariate normal distribution, $N(\mu_h, \Sigma_h)$. Note that model (1.1) relies on categorizing individuals into groups based on their pregnancy outcomes, with the probability of allocation to each group dependent on predictors. Hence, there are some clear similarities with standard epidemiologic analysis approaches. However, the fundamental difference is that instead of predefining hard thresholds on the ys, the latent class model adaptively allocates individuals into groups probabilistically. There are no hard thresholds and the allocation to latent classes relies on the data at hand without discarding information.

For example, in order to fit the data in Figure 1.3, one could potentially use a mixture of four normal distributions, with components corresponding to full term births, preterm births near the 37 week cutoff, early preterm births, and late term births. The need for this final component may be reduced in more recent data sets having ultrasound to accurately date the pregnancy, as many of the values in the right tail are likely due to measurement error in pregnancy dating. In order to complete a specification of the model, we can choose a regression model for the probability weights, $\pi_h(x)$. In order to aid interpretation for epidemiologists, the polytomous logistic regression model provides a convenient choice,

$$\pi_h(x) = \frac{\lambda_h \exp(x'\beta_h)}{\sum_{l=1}^{k} \lambda_l \exp(x'\beta_l)}, \tag{1.2}$$

where $\lambda = (\lambda_1, \ldots, \lambda_k)'$ are parameters characterizing the baseline probabilities of allocation to each class, β_h are parameters characterizing the effect of predictors on the probability of allocation to class h, for $h = 1, \ldots, k$, and $\lambda_1 = 1, \beta_1 = 0$ for identifiability.

As there is nothing in the model distinguishing classes $1, \ldots, k$, we face the well-known label ambiguity problem, which is omnipresent in mixture models and clustering (Stephens, 2000; Jasra, Holmes and Stephens, 2005). Frequentist analyses of model (1.1)–(1.2) can be implemented using the EM algorithm for maximum likelihood estimation. The EM algorithm will converge to a local mode, which corresponds to one possible configuration of the label indices on classes $1, \ldots, k$. As all other configurations lead to the same maximum likelihood estimate, the EM algorithm is in some sense not sensitive to the label switching problem. Each run will produce the same maximum likelihood estimates after convergence and relabelling. However, the EM algorithm just produces a point estimate, while a Bayesian analysis implemented with MCMC can allow a full probabilistic treatment of uncertainty in estimation of $f(y|x)$. Properly accounting for uncertainty is important in conducting valid inferences about predictor effects, which is the primary goal of epidemiologic analyses.

In MCMC analyses, label ambiguity will lead to a multimodal posterior and MCMC algorithms can experience a label switching problem. For example, for the first several 100 iterations, classes 1–4 may correspond to full-term, late preterm, early preterm and late term, but then the labels may switch (corresponding to a different mode in the posterior) so that classes 1–4 correspond to late term, full term, early preterm and late preterm. During an MCMC run, there may be many such relabelings, so that it becomes very difficult to summarize the results from the MCMC output in a coherent manner. Certainly, posterior means and credible intervals for component-specific parameters (e.g. λ_2, β_2) are meaningless. Several solutions have been proposed to this problem: (1) place restrictions on the parameters to avoid label ambiguity; (2) apply a post-processing relabelling algorithm to the MCMC output in order to align draws from the same component, so that one can calculate posterior summaries for component-specific parameters after realignment (Stephens, 2000; Jasra *et al.*, 2005); and (3) ignoring the label ambiguity problem, while avoiding mixture component-specific inferences.

I start by commenting on approach (1), which is by far the most widely used in the literature. In order to implement this strategy for model (1.1), the most common approach would be to place an order restriction on the means. For example, letting $\mu_h = (\mu_{h1}, \mu_{h2})'$, a possible identifying restriction would let $\mu_{11} < \mu_{21} < \cdots < \mu_{k1}$. Hence, the components would be restricted in advance to be ordered by length of gestation. Although this seems reasonable on the surface, there are a number of pitfalls making this approach unreliable. The most notable is that it does not solve the label ambiguity problem unless the components are very well separated. For example, suppose that μ_{11} and μ_{21} are close together, with the difference between classes 1 and 2 occurring in Σ_1 and Σ_2 and/or μ_{21} and μ_{22}. Then, the constraint has accomplished nothing. Such problems are well motivated in Jasra *et al.* (2005), who advocate in favour of strategy (2). In my experience, such relabelling approaches tend to work well, though they add substantially to the computational burden.

My own view is that latent class modelling and clustering is most appropriately used as a tool to obtain a flexible model, and one should avoid interpreting the clusters obtained as biologically meaningful. It is well known that clusters are entirely sensitive to parametric assumptions, with such assumptions seldom if ever justifiable by substantive theory in an application area (Bauer, 2007). That said, in some cases, clustering is useful as a hypothesis generating tool, to simplify interpretations of complex data and as a dimensionality reduction device. Even if there is no reason to suppose that biologically meaningful clusters exist in the pregnancy outcome data, one can still use model (1.1)–(1.2) in conducting inferences about changes in the distribution of the pregnancy outcomes with predictors. Such inferences are unaffected by label switching. For example, suppose that one wants to obtain an estimate of $f(y|x)$. A Gibbs sampling algorithm can be implemented for model (1.1)–(1.2), relying on data augmentation to obtain conjugate full conditional posterior distributions without any order constraints on the parameters. From this Gibbs sampler, one can obtain samples from the posterior of $f(y|x)$. Although the labels on clusters underlying each sample will vary, this has no effect on the value of $f(y|x)$, and one can calculate posterior means and credible intervals without complication.

Model (1.1) is a type of finite mixture model, and a fundamental problem that arises in applications of such models is choice of the number of mixture components, k. The most common strategy in the literature relies on fitting the mixture model for different choices of k, using the EM algorithm to obtain maximum likelihood estimates. The BIC is then used to choose the 'best' k. There are two problems with this approach. First, the BIC was originally developed by Schwarz (1978) based on a Laplace approximation to the marginal likelihood. His justification does not hold for finite mixture models, so that BIC lacks theoretical grounding, though it does seem to work well in practice in many cases (Fraley and Raftery, 1998). The second problem is that the approach of fitting models for many different choices of k and then basing inferences on the final selected model ignores uncertainty in selection of k. In my experience, such uncertainty is often substantial, with the data not providing compelling evidence in favour of any single k. Hence, it seems more appropriate to use Bayesian model averaging to allow for uncertainty in k. This can potentially be accomplished using reversible jump Markov chain Monte Carlo (RJMCMC) (Green, 1995). However, RJMCMC tends to be quite difficult to implement efficiently, so I instead advocate bypassing the model selection problem through the use of a nonparametric Bayes approach.

1.2.2 Nonparametric Bayes

Nonparametric Bayes methods allow for uncertainty in distributional assumptions within Bayesian hierarchical models. For example, initially excluding predictors from consideration, suppose that interest focuses on estimating the joint density of gestational age at delivery and birth weight. Then, one possibility is to use a Dirichlet process mixture (DPM) of normals (Lo, 1984; Escobar and West, 1995), which can be expressed in hierarchical form as

$$\begin{aligned} y_i &\sim \mathrm{N}_2\left(\mu_i^*, \Sigma_i^*\right) \\ \left(\mu_i^*, \Sigma_i^*\right) &\sim P, \qquad P \sim DP(\alpha P_0), \end{aligned} \tag{1.3}$$

where P is an unknown mixture distribution, which is assigned a Dirichlet process prior (Ferguson, 1973; 1974), parametrized in terms of a precision α and base distribution P_0. This model can accurately characterize any smooth bivariate density.

To obtain additional insight and relate the DPM model to the latent class model in (1.1), one can use the stick-breaking representation of the DP

(Sethuraman, 1994), which implies that

$$f(y) = \sum_{h=1}^{\infty} \pi_h N_2(y; \mu_h, \Sigma_h), \quad (\mu_h, \Sigma_h) \sim P_0,$$

$$\pi_h = V_h \prod_{l<h}(1 - V_l), \quad V_h \sim \text{Beta}(1, \alpha). \tag{1.4}$$

Hence, the DPM of normals model is equivalent to (1.1) except with predictors excluded, the number of mixture components (classes) equal to $k = \infty$, the component-specific parameters sampled iid from P_0, and a stick-breaking prior placed on the component weights. This stick-breaking prior, which is described in the second line of (1.4) favors allocating non-neglible weight only to the first few components for small α. This leads to a sparse representation in which only a few components are occupied by the subjects in a given sample, though infinitely many components are available, so that the model can be made more complex as subjects are added.

There is a wide variety of easy-to-implement MCMC and fast approximation algorithms available for fitting of DPMs. Refer to Dunson (2008) for a recent review of applications of nonparametric Bayes and Dirichlet process mixtures to biostatistical applications. In order for such approaches to be useful in epidemiologic applications, it is necessary to incorporate predictors. For example, a natural extension of (1.4) would let

$$f(y \mid x) = \sum_{h=1}^{\infty} \pi_h(x) N_2(y; \mu_h x, \Sigma_h), \quad (\mu_h, \Sigma_h) \sim P_0, \tag{1.5}$$

where μ_h is a $2 \times p$ matrix of coefficients specific to mixture component h, for $h = 1, \ldots, \infty$. This model is a type of bivariate infinite mixtures of experts model, which generalizes parametric mixtures of experts models that assume a finite k and focus on univariate responses (Jordan and Jacobs, 1994; Peng, Jacobs and Tanner, 1996; and Jacobs, Peng and Tanner, 1997). By mixing linear regression models with predictor-dependent weights, we allow for a sparser characterizations of complex data than possible using the framework of (1.1) with $k = \infty$. In particular, we find that it is possible to use many fewer mixture components by mixing normal linear regressions instead of normals without a regression component.

In order to complete a specification of model (1.5), it is necessary to choose a model for the unknown weight function $\pi_h(x)$. Ideally, this would be done in a highly flexible manner, while favouring a sparse representation with a few dominate components having relatively high weights. There are several possibilities that have been proposed in the literature, including the order-based dependent Dirichlet process (Griffin and Steel, 2006), the kernel stick-breaking process (KSBP) (Dunson and Park, 2008) and the local Dirichlet process (Chung

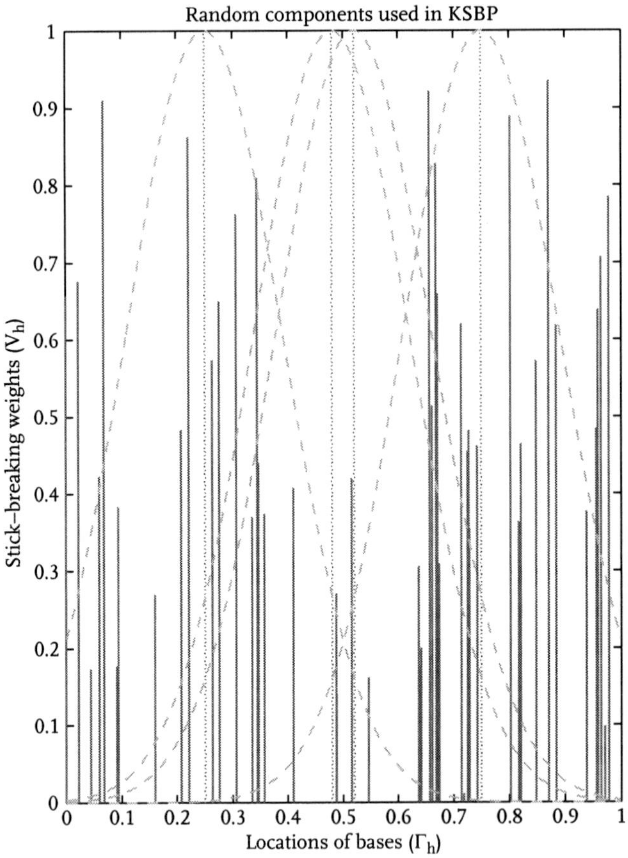

Fig. 1.4 Random components used in the KSBP. The red dotted lines represent predictor locations of interest, the blue lines correspond to weights V_h at locations Γ_h, and the green dashed line shows the kernel, K.

and Dunson, 2009a). Here, I focus on the KSBP, since it provides a relatively simple generalization of the DP stick-breaking process in (1.4), while leading to straightforward posterior computation and inferences.

1.2.3 Kernel stick-breaking process

Under the KSBP, the probability of allocation to component h conditionally on predictors x is

$$\pi_h(x) = V_h K(x, \Gamma_h) \prod_{l<h} \{1 - V_l K(x, \Gamma_l)\}, \quad h = 1, \ldots, \infty, \tag{1.6}$$

where $\Gamma_l \sim H$ is a random basis location and $V_l \sim \text{beta}(1, \alpha)$ is a random weight on that location, for $h = 1, \ldots, \infty$. Figure 1.4 plots the first 50 random components for a single draw from the KSBP in a simple case in which there is a single predictor $x \in [0, 1]$. The red vertical dotted lines correspond to predictor values of interest, which are set to $x = 0.25, 0.48, 0.52, 0.75$ for illustration. For example, consider calculating $\pi_h(0.25)$ using expression (1.6) applied to the components shown in Figure 1.4. Clearly, the locations Γ_h receiving highest weight will correspond to those that receive higher weights V_h, particularly if they are close to $x = 0.25$, so are not downweighted too much by the kernel (shown with a green dashed line), and if the index h is small, so that too much of the probability stick is not already used up by other locations.

The resulting values of $\pi_h(x)$ for $x = 0.25, 0.48, 0.52, 0.75$ are shown in Figure 1.5. Note that the probability weight, $\pi_h(x)$, at location Γ_h is the weight on $N_2(\mu_h(1, x)', \Sigma_h)$ in the mixture of bivariate normals shown in (1.5). Hence, it is appealing to have similar weights for x values that are close together, while allowing different weights for xs that are far apart. Figure 1.5 illustrates this behaviour, as the weights for $x = 0.48$ and $x = 0.52$ are close together, with both predictor values having the same two dominant mixture components. The tendency for the mixture distributions to change smoothly with predictors induces smoothness in modelling of the unknown conditional response distributions, with small changes in the predictors implying small changes in the response density.

Theoretical properties of the KSBP and an efficient MCMC algorithm for posterior computation are presented in Dunson and Park (2008). Here, I focus on the pregnancy outcome application. Note that the MCMC algorithm produces draws from the posterior distribution for $f(y|x)$ for any x value of interest. Such draws can be summarized in a broad variety of ways. For example, one can obtain nonlinear dose response curves characterizing the change in mean gestational age at delivery and birth weight with increasing dose of DDE adjusting for the other predictors. One can also obtain quantile response curves characterizing the change in a percentile of the response distribution with DDE. The next section illustrates some of the possibilities focusing on gestational age at delivery data from the Longnecker *et al.* (2001) study. The same type of approach can be applied directly in the bivariate case to jointly model gestatonal age at delivery and birth weight according to DDE exposure and other predictors.

1.3 Density regression for pregnancy outcomes

I focus on gestational age at delivery, DDE and age data from the Longnecker *et al.* (2001) study. DDE is the most persistent metabolite of the pesticide DDT. Although DDT is no longer in use in much of the developed world, including the United States, DDT is still used broadly in some parts of the world due to its effectiveness against malaria transmitting mosquitoes. Decisions of whether or not to continue use of DDT must necessarily weigh this benefit against the

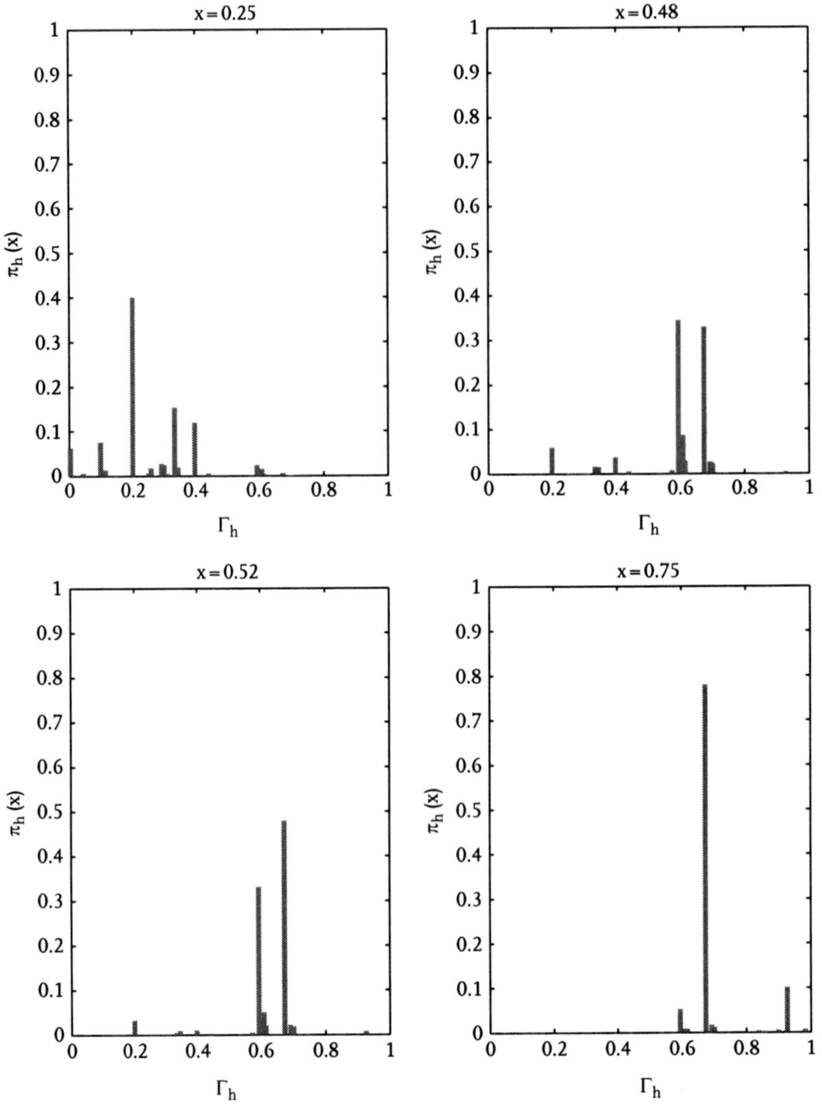

Fig. 1.5 Probability weights, $\pi_h(x)$, on each of the mixture components for different choices of x including $x = 0.25, 0.48, 0.52, 0.75$. These realizations correspond to one draw from the kernel stick-breaking process.

increasing evidence of adverse health effects. Here, the interest is in studying how risk of premature delivery is impacted by maternal exposure to DDT. DDT exposure potentially has a long term impact, as metabolites of DDT are lipophilic and hence bioaccumulate in maternal fat, with body burden often increasing over time. Premature delivery is a major public health problem, being a common condition (approximately 10–13% of pregnancies are delivered prior to 37 weeks completed gestation), which can result in substantial morbidity. Risks of Infant mortality and morbidity increase substantially for early preterm births, so it is of major public health and clinical interest to separate risk of close to full term births (close to the 37 week cutoff) from risk of early preterm births.

As shown in Figure 1.1, data are available for $n = 2313$ pregnancies after excluding pregnancies having gestational lengths longer than 45 weeks. These values were considered unreliable and to likely arise due to measurement error in pregnancy dating. Letting y_i = gestational age at delivery for pregnancy i and $x_i = (\text{DDE}_i, \text{age}_i)'$, with DDE_i dose level in mg/L measured in maternal serum during pregnancy, I focus on the following flexible mixture model:

$$f(y_i \mid x_i) = \sum_{h=1}^{\infty} \pi_h(x_i) N\big(y_i; \beta_{h1} + \beta_{h2}\text{DDE}_i + \beta_{h3}\text{age}_i, \tau^{-1}\big), \tag{1.7}$$

where the predictor-dependent mixture distributions are assumed unknown through the use of a kernel stick-breaking process prior, with a Gaussian kernel, $K(x, \Gamma_h) = \exp\{-\psi(x - \Gamma_h)^2\}$. I choose gamma hyperpriors for the parameters α and ψ to allow the data to inform about their choice. The results were found to be robust to the choice of kernel as long as hyperpriors were chosen for the kernel precision. For example, I also tried an exponential kernel, and obtained essentially indistinguishable results.

The MCMC algorithm was run for 30,000 iterations with an 8000 iteration burn-in. The convergence and mixing was good based on examination of trace plots of the conditional densities at different points and of the hyperparameter values. Figure 1.6 shows the data on DDE versus gestational age at delivery. The vertical dashed lines are quintiles of the empirical distribution of DDE. Following standard practice in epidemiologic analyses, Longnecker *et al.* (2001) used these quintiles as thresholds in categorizing DDE prior to conducting a logistic regression analysis, which relied on a 0/1 indicator of premature delivery using the standard 37 week cutoff of gestational age at delivery. Here, I instead avoid categorization of either the predictors or the response. The solid curve is the estimated expectation of y_i conditionally on DDE_i holding age_i fixed at the mean value. The curve is close to linear across much of the data range, which illustrates an appealing property of the analysis, which is a tendency to collapse on a simple submodel when that model holds approximately. In this case, the simple submodel is a linear regression model, with deviations allowed through local mixing.

A simple frequentist alternative to our approach, which also avoids categorization, lets

$$\text{E}(y_i \mid x_i) = \mu + g_1(x_{i1}) + g_2(x_{i2}) + \epsilon_i, \quad \epsilon_i \sim N\left(0, \sigma^2\right), \tag{1.8}$$

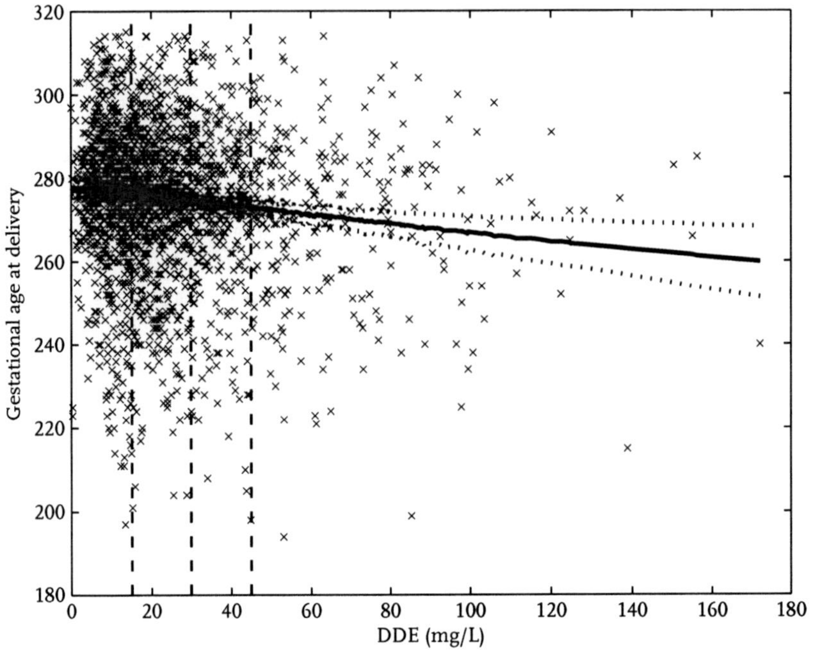

Fig. 1.6 Estimates of the expected gestational age at delivery conditionally on DDE. Gestational age at delivery and DDE data points are shown as $\times$. Solid lines are posterior means and dashed lines are 99% pointwise credible intervals.

with $g_1(\cdot)$ and $g_2(\cdot)$ unknown functions. This additive model can be fitted easily in standard software packages, such as R, and produces a similar curve to that shown in Figure 1.6, though the frequentist curve estimate is somewhat bumpier, particularly in sparse data regions. Even though the frequentist analysis makes an invalid assumption that the residuals are normally distributed with constant variance, mean estimates tend to be robust to violations of this assumption due to the central limit theorem, as the sample size is not small in this application.

A natural question is then what is gained practically by the flexible Bayesian analysis in this application? The answer is that the interest really does not focus on the mean response, but instead focuses on the tails of the distribution corresponding to adverse health responses. Applying a flexible mean regression does not tell us how the risk of premature delivery changes with level of exposure to DDE. Figure 1.7 shows the estimated conditional densities of gestational age at delivery for a range of values of DDE. It seems clear from this plot that there is an increasingly heavy left tail as DDE dose increases. Note that the estimates tell a similar story to the estimates in Figure 1.3, but categorization is avoided, so a unique density estimate is obtained for every DDE level. In

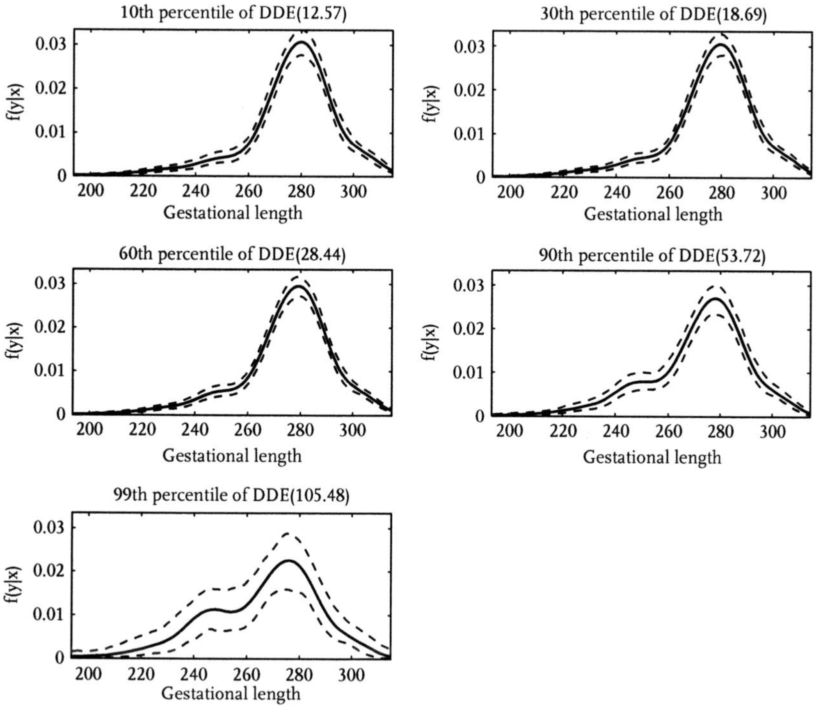

Fig. 1.7 Estimated conditional density of gestational age at delivery given DDE for a range of DDE values. Solid lines are posterior means and dashed lines are 99% pointwise credible intervals.

addition, by borrowing information across different DDE values in a flexible manner, smoother and more realistic density estimates are produced. At the higher DDE values, there is limited data available, so the width of the credible intervals increase substantially.

As it is hard to gauge subtle changes in risk of prematurity from examination of side-by-side conditional density plots, I also provide plots of the risk of falling below various thresholds for defining preterm birth in Figure 1.8. Note that unlike typical procedures which collapse the data into binary indicators and conduct separate analyses for each choice of threshold, these estimates are all based on one coherent underlying model for the conditional response distribution. The use of such a coherent model can lead to substantial efficiency gains relative to separate binary response analyses. Such efficiency gains are very important practically in attempting to estimate risk of falling within the extreme tails of a distribution, since even if the sample size is large overall, one may have limited data in the tails.

Figure 1.8 tells an interesting and disturbing story. First consider the plot in the lower left panel, which shows the risk of having a delivery prior to 37 weeks of completed gestation as a function of DDE dose. Even at low doses of

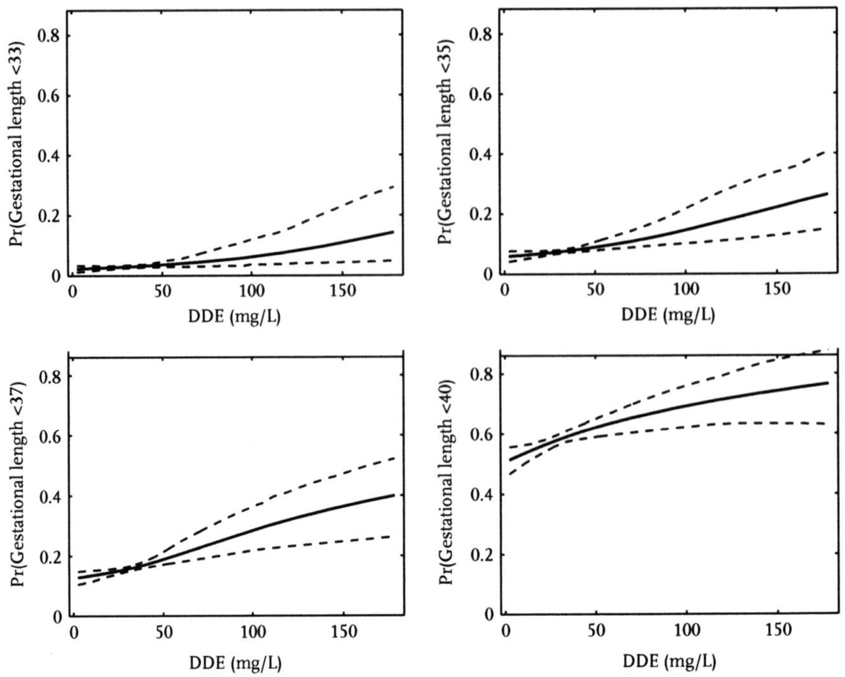

Fig. 1.8 Dose response curves for the probability of observing a gestational age at delivery below 33, 35 37 or 40 weeks. Solid lines are posterior means and dashed lines are 99% pointwise credible intervals.

DDE, this curve has a positive slope, suggesting an increasing trend in risk of premature delivery with increasing level of exposure. The result is consistent with the Longnecker *et al.* (2001) frequentist logistic regression analysis, but their analysis followed standard practice and only considered the 37 week cutoff for defining preterm birth. From Figure 1.8, there is also an increasing dose response trend for the 33 and 35 week cutoffs. These results are of great clinical and public health importance, since neonatal mortality and morbidity rates for babies born at 34 or 35 weeks are substantially higher than those for a baby born at 36 weeks (McIntire and Leveno, 2008). Hence, variability in gestational length even among late preterm births is an important determinant of risk of subsequent adverse outcomes, which should not be ignored in the analysis by focusing entirely on the 37 week cutoff. Although data become increasingly sparse as we move further into the left tail of the distribution and attempt to assess risk of early preterm births, the Bayesian approach can partly address problems with data sparsity through borrowing of information. That said, credible interval widths are wide for the dose response curve for cutoffs much below 33 weeks, since the data are so sparse in this region, as is apparent from Figure 1.6.

1.4 Discussion

This chapter has described a Bayesian approach for flexible analysis of epidemiologic data, motivated by the clear limitations of current standard approaches. In particular, it is quite common to be interested in relationships between a continuous exposure and multiple continuous health responses. As logistic regression is the default analysis strategy in epidemiology, it is very often the case that continuous responses and exposures are categorized prior to analysis. Although this approach has the advantage of resulting in simple summaries of the relationships between exposures and risk of an adverse health response, there are a number of important disadvantages, which have been highlighted in this chapter through application to studies of pregnancy outcomes. In this and many other biomedical applications, individuals with adverse responses tend to fall in the tails of the distribution. Hence, to simplify the analysis, it is tempting to choose a threshold and categorize the response as extreme/not extreme. For example, with very few exceptions, the thousands of articles in the epidemiologic literature on premature delivery use less than 37 weeks of completed gestation as the threshold, and then discard finer scale information on gestational age at delivery in conducting inferences. The main concern with premature delivery is the increased risk of subsequent adverse health outcomes in the developing child, and risk and severity of these outcomes can increase substantially with each week change within the ≤ 37 week interval. Hence, it is clearly most appropriate to assess how the entire tail of the gestational age at delivery changes with exposures.

Another important issue is how to deal with multivariate responses, such as gestational age at delivery and birth weight in the pregnancy application. Again, in epidemiology almost all analyses rely on simplifying the data, so that a univariate logistic regression can be applied. In particular, risk of premature delivery is analysed separately from risk of small-for-gestational age (SGA), which is defined as a birth that falls within the lower 10th percentile of the birth weight distribution stratified on gestational age at delivery. SGA is arguably not a meaningful summary of growth restriction, and it is difficult to interpret results from analyses assessing relationships between exposures and SGA. The conceptual problem is that SGA represents an attempt to adjust for gestational age at delivery in analyses of birth weight. However, gestational age at delivery and birth weight are so tied together that they are much more appropriately viewed as a bivariate outcome. By using the flexible mixture models proposed in this article, one can jointly model, while avoiding parametric assumptions, as normality is typically violated. If one wants to adjust for gestational length in analysing birth weight, this can be done coherently by estimating the birth weight density profile. Such approaches should be broadly useful in epidemiology.

Appendix

A. Broader context and background

In this section, we provide additional background and discussion on nonparametric Bayes mixture models. Chapters 6 and 27 present a review of some properties and applications of Dirichlet process priors. Limiting overlap, here I focus specifically on the case in which the focus is on modelling of a collection of unknown distributions indexed by $x \in \mathcal{X}$, with x corresponding to predictors, time or spatial location. A key development was the dependent Dirichlet process (DDP) prior (MacEachern, 1999), which lets

$$P_x = \sum_{h=1}^{\infty} \pi_h(x)\delta_{\theta_h(x)}, \qquad \theta_h \sim P_0, \tag{1.9}$$

where $\pi_h(x) = V_h(x)\prod_{l<h}\{1 - V_l(x)\}$, for $h = 1, \ldots, \infty$, $\theta_h(x) \sim P_{0x}$ and $V_h(x) \sim \text{Beta}(1, \alpha)$ independently for $h = 1, \ldots, \infty$, with P_{0x} the marginal of P_0 at location x. Here, the stick-breaking weights and atoms at a given location x are mutually independent, but dependence is incorporated across locations.

The DDP defines a prior for the collection $\{P_x, x \in \mathcal{X}\}$, with $P_x \sim DP(\alpha P_{0x})$ marginally at each x, while allowing dependence in P_x and $P_{x'}$. Because realizations from a Dirichlet process are almost surely discrete, the DDP is not directly useful for modeling changes in a response distribution with x. However, one can define a DDP mixture model by letting $f(y|x) = \int K(y; \theta)dP_x(\theta)$, with $K()$ a kernel (e.g. Gaussian). In practice, it is not immediately obvious how to model dependence of the stick-breaking random variables $\{V_h(x)\}$ while maintaining the necessary marginal property $V_h(x) \sim \text{Beta}(1, \alpha)$ for all x. Hence, most of the applications of DDP mixture models have relied on the fixed-π specification that lets $\pi_h(x) = \pi_h$, while allowing the atoms to vary flexibly (De Iorio *et al.*, 2004; Gelfand *et al.*, 2005; among others).

For continuous $\mathcal{X}$, a natural approach is to let $\theta_h \sim \text{GP}(\mu, C)$, with $\text{GP}(\mu, C)$ denoting a Gaussian process with mean function μ and covariance function C. For example, suppose that $x \in [0, 1]$ is a continuous predictor, with $f(y|x) = \int N(y; \mu, \sigma^2)dP_x(\mu)$ and $\{P_x, x \in \mathcal{X}\}$ assigned a fixed-π DDP prior with base measure $\text{GP}(\mu, C)$. Hence, the prior distribution for $f(y|x)$ is marginally a Dirichlet process location mixture of normals, which is highly flexible. As x varies, the probability weights on the different mixture components remain constant, but the locations of the components vary according to a Gaussian process. If one chooses a standard covariance function, such as squared exponential, one then induces smooth changes in the conditional densities with x. This favours similarity in $f(y|x)$ and $f(y|x')$ when x and x' are close together. Instead of a Gaussian process, one can potentially use a spline-based model for the atoms, which was implemented by Wang and Dunson (2007) with a focus on modeling of stochastically ordered densities and hypothesis testing.

Even in the fixed-π case, DDP mixtures are extremely flexible. However, in many settings one could more parsimoniously characterize changes in the conditional distribution $f(y|x)$ with x by allowing the weights in the mixture to vary with x. The kernel stick-breaking process described in Section 1.2.3 provides one approach to this problem. Another possibility is to induce a model for the conditional distribution $f(y|x)$ through a Dirichlet process mixture model for the joint distribution of y and x (Müller *et al.* (1996). This approach is both simple and highly flexible, leading to a predictor-dependent mixture of linear regressions when y and x are continuous and modeled using a DPM of Gaussians. However, there are two major problems. Firstly, the approach is only appropriate if x can reasonably be considered as a random variable, though one can potentially consider the joint model for y and x as an auxiliary model that is only defined to induce a coherent model on the conditional distribution. Second, there is a pitfall that often arises due to the structure of the Dirichlet process mixture model, which allocates subjects to clusters based on both y and x, while favouring few clusters. When there are more than a few predictors, it tends to be the case that the x component of the likelihood dominates, so that clusters are introduced primarily to improve fit of the predictor component. There is relatively little gain in introducing clusters that improve prediction of y given x. Hence, we have observed poor predictive performance for this approach in a variety of settings, with the performance particularly bad when y is categorical and predictors are continuous and non-Gaussian (unpublished work with Richard Hahn and Deepak Agarwal).

One possibility for addressing the second problem is to use a more flexible prior than the Dirichlet process in order to allow cluster allocation to differ between the y and x components. This can be accomplished using the local partition process (Dunson, 2008), but this approach has not yet been applied in the setting of conditional distribution modeling in regression. Another possibility is to consider x as fixed, and modify the Dirichlet process to allow the weights to vary with x. The order-based DDP (Griffin and Steel, 2006) accomplishes this by using a common set of stick-breaking variables in the Sethuraman (1994) specification of the Dirichlet process, but with the ordering in these variables dependent on x. The local Dirichlet process (Chung and Dunson, 2009a) is a simpler alternative. In this approach, one places an atom $\theta_h \sim P_0$ and stick-breaking weight $V_h \sim \text{Beta}(1, \alpha)$ at location Γ_h for $h = 1, \ldots, \infty$. Then, we let $P_x = \sum_{h \in \mathcal{L}_x} V_h \prod_{l<h}(1 - V_l)\delta_{\theta_h}$, with $\mathcal{L}_x$ denoting the subset of locations that fall within a neighborhood of location x. One obtains $P_x \sim DP(\alpha P_0)$ marginally at each location, while inducing dependence in P_x and $P_{x'}$ through including shared stick-breaking weights and atoms within the region of overlap in the neighborhoods around x and x'.

In complex and high-dimensional settings, there can be some difficulties implementing posterior computation in models allowing the mixture weights to vary with x due to lack of conjugacy. One trick that is often used to induce conjugacy in categorical data analysis is data augmentation with underlying normal variables (Albert and Chib, 1993; 2001). Suppose we have a finite mixture model with k levels and with predictor dependent weights. Then, letting $z_i = h$ denote that subject i is allocated to component h, one could use a kernel continuation ratio probit model with

$$\Pr(z_i = h \mid z_i \geq h, x_i) = \Phi\left(\alpha_h + \sum_{l=1}^{p} \beta_{hl}\eta_l(x_i)\right) = \Phi\left(w_i'\theta_h\right), \quad h = 1, \ldots, k-1,$$

where $\eta_1, \ldots, \eta_p$ are prespecified basis functions (e.g. kernels located on a fixed grid). This model can be equivalently expressed in augmented form as:

$$z_i = \arg\min_h \{z_{ih} > 0\}, \quad z_{ih} = 1\left(z_{ih}^* > 0\right),$$
$$z_{ih}^* \sim N\left(w_i'\theta_h, 1\right), \; h = 1, \ldots, k-1, \; z_{ik} = 1.$$

In particular, the allocation of individual i to a mixture component is characterized through a discrete Gaussian process in which we repeatedly sample latent normal variables until obtaining a positive observation. Using this latent variable formulation and specifying Gaussian or mixture of Gaussian priors for the coefficients θ_h, a straightforward Gibbs sampler can be implemented for posterior computation. It is also possible in such a specification to conduct variable or basis selection by using a mixture prior with a mass at zero for the coefficients θ_h. By allowing $k \to \infty$, we obtain a nonparametric Bayesian specification which avoids the need to choose a finite number of mixture components. This process was proposed by Chung and Dunson (2009b) and is referred to as the probit stick-breaking process (PSBP).

B. Posterior computation

One of the hurdles to overcome in implementing nonparametric Bayes analyses is the need to conduct posterior computation for a model with infinitely-many parameters. This initially seems to be an impossibility that cannot be overcome. However, there is a rich literature proposing a variety of simple and efficient computational algorithms that bypass the infinite dimensionality issue in a variety of ways. The most commonly used approach relies on marginalizing out the infinitely many parameters in the unknown distribution P to obtain a prior for the finitely many realizations from P represented in the sample. This strategy is referred to as collapsed, marginal or Polya urn Gibbs sampling, with MacEachern and Müller (1998) providing a good reference in the Dirichlet process case. Although this approach is quite convenient to implement, there are two disadvantages. Firstly, one looses the ability to do inferences on functionals of P, which may be of interest in certain cases. Second, such approaches typically cannot be applied in more complex settings involving generalizations of DP priors or alternatives, because in such settings it is often not possible to obtain simple expressions for conditional distributions of the realizations marginalizing out P.

Ishwaran and James (2001) proposed an alternative approach, which is referred to as conditional or blocked Gibbs sampling. Their approach, which was designed for stick-breaking priors including the Dirichlet process and many other cases, is based on the approximation:

$$P = \sum_{h=1}^{\infty} V_h \prod_{l<h}(1 - V_l)\delta_{\theta_h} \approx \sum_{h=1}^{N-1} V_h \prod_{l<h}(1 - V_l)\delta_{\theta_h} + \prod_{l<N}(1 - V_l)\delta_{\theta_N}.$$

Truncation approximations to stick-breaking representations of the Dirichlet process were previously proposed by Muliere and Tardella (1998). Using the truncated version, one can apply a standard Gibbs sampler for posterior computation, which alternates between (1) allocating each individual to one of the N clusters by sampling from multinomial conditional distributions; (2) updating the stick-breaking weights $V_1, \ldots, V_{N-1}$ from conditionally conjugate Beta distributions; and (3) updating the atom θ_h by sampling from the conditional distribution obtained by updating the prior P_0 with the likelihood for those subjects allocated to cluster h, for $h = 1, \ldots, N$. These steps are no more difficult to implement than Gibbs samplers for finite mixtures. However, unlike in the finite mixture case, N can be viewed as an upper bound on the number of clusters, with the number of occupied components varying across the MCMC samples.

Both the marginal and conditional Gibbs samplers can experience mixing problems in which the samples tend to remain for long periods in local regions of the space of configurations of subjects to clusters. This is not a problem unique to nonparametric Bayes models, but also occurs in finite mixture models due to multimodality in the posterior distribution. There has been a broad variety of approaches proposed in the literature for improve mixing in the presence of multimodality. In the setting of DPMs, split-merge algorithms were proposed by Jain and Neal (2004). Evolutionary Monte Carlo is another possibility (Goswami *et al.*, 2007). Algorithms have also been proposed for avoiding the need for truncation in conducting conditional Gibbs sampling, with Walker (2007) proposing a slice sampler and Papaspiliopoulos and Roberts (2008) developing a retrospective sampler, which also includes label switching moves to accelerate mixing. I have observed a tendency of the slice sampler to be slow mixing, while retrospective sampling is somewhat complicated to implement. Papaspiliopoulos (2008) recently proposed an exact block Gibbs sampler, which combines the two approaches, leading to a simple and efficient algorithm.

Acknowledgements

This work was supported in part by the Intramural Research Program of the NIH, National Institute of Environmental Health Sciences.

References

Albert, J.H. and Chib, S. (1993). Bayesian-analysis of binary and polychotomous response data. *Journal of the American Statistical Association*, **88**, 669–679.

Albert, J.H. and Chib, S. (2001). Sequential ordinal modeling with applications to survival data. *Biometrics*, **57**, 829–836.

Bauer, D.J. (2007). Observations on the use of growth mixture models in psychological research. *Multivariate Behavioral Research*, **42**, 757–786.

Boucher, K.M., Slattery, M.L., Berry, T.D., Quesenberry, C. and Anderson, K. (1998). Statistical methods in epidemiology: A comparison of statistical methods to analyze dose-response and trend analysis in epidemiologic studies. *Journal of Clinical Epidemiology*, **51**, 1223–1233.

Chung, Y. and Dunson, D.B. (2009*a*). The local Dirichlet process. *Annals of the Institute of Statistical Mathematics*.

Chung, Y. and Dunson, D.B. (2009*b*). Nonparametric Bayes conditional distribution modeling with variable selection. *Journal of the American Statistical Association*, **104**, 1646–1660.

De Iorio, M., Müller, P., Rosner, G.L. and MacEachern, S.N. (2004). An ANOVA model for dependent random measures. *Journal of the American Statistical Association*, **99**, 205–215.

Dunson, D.B. (2008). Nonparametric Bayes applications to biostatistics. *Bayesian Nonparametrics*, (ed. N. Hjort, C. Holmes, P. Müller and S. Walker), Chapter 7. Cambridge University Press, Cambridge.

Dunson, D.B. (2009). Nonparametric Bayes local partition models for random effects. *Biometrika*, **96**, 249–262.

Dunson, D.B. and Park, J.-H. (2008). Kernel stick-breaking processes. *Biometrika*, **95**, 307–323.

Escobar, M.D. and West, M. (1995). Bayesian density estimation and inference using mixtures. *Journal of the American Statistical Association*, **90**, 577–588.

Ferguson, T.S. (1973). A Bayesian analysis of some nonparametric problems. *Annals of Statistics*, **1**, 209–230.

Ferguson, T.S. (1974). Prior distributions on spaces of probability measures. *Annals of Statistics*, **2**, 615–629.

Fraley, C. and Raftery, A.E. (1998). How many clusters? Which clustering method? Answers via model-based cluster analysis. *Computer Journal*, **41**, 578–588.

Gage, T.B. (2003). Classification of births by birth weight and gestational age: An application of multivariate mixture models. *Annals of Human Biology*, **30**, 589–604.

Gelfand, A.E., Kottas, A. and MacEachern, S.N. (2005). Bayesian nonparametric spatial modeling with Dirichlet process mixing. *Journal of the American Statistical Association*, **100**, 1021–1035.

Goswami, G., Liu, J.S. and Wong, W.H. (2007). Evolutionary Monte Carlo methods for clustering. *Journal of Computational and Graphical Statistics*, **16**, 855–876.

Green, P.J. (1995). Reversible jump Markov chain Monte Carlo computation and Bayesian model determination. *Biometrika*, **82**, 711–732.

Griffin, J.E. and Steel, M.F.J. (2006). Order-based dependent Dirichlet process. *Journal of the American Statistical Association*, **101**, 179–194.

Greenland, S. (1995). Dose-response and trend analysis in epidemiology – alternatives to categorical analysis. *Epidemiology*, **6**, 356–365.

Jacobs, R.A., Peng, F.C. and Tanner, M.A. (1997). A Bayesian approach to model selection in hierarchical mixtures-of-experts architectures. *Neural Networks*, **10**, 231–341.

Jain, S. and Neal, R.M. (2004). A split-merge Markov chain Monte Carlo procedure for the Dirichlet process mixture model. *Journal of Computational and Graphical Statistics*, **13**, 158–182.

Jasra, A., Holmes, C.C. and Stephens, D.A. (2005). Markov chain Monte Carlo methods and the label switching problem in Bayesian mixture modeling. *Statistical Science*, **20**, 50–67.

Jordan, MI. and Jacobs, R.A. (1994). Hierarchical mixtures of experts and the EM algorithm. *Neural Computation*, **6**, 181–214.

Lo, A.Y. (1984). On a class of Bayesian nonparametric estimates. 1. Density estimates. *Annals of Statistics*, **12**, 351–357.

Longnecker, M.P., Klebanoff, M.A., Zhou, H.B. and Brock, J.W. (2001). Association between maternal serum concentration of the DDT metabolite DDE and preterm and small-for-gestational-age babies at birth. *Lancet*, **358**, 110–114.

MacEachern, S.N. (1999). Dependent nonparametric process. *ASA Proceeding of the Section on Bayesian Statistical Science*, American Statistical Association, Alexandria, VA.

McIntire, D.D. and Leveno, K.J. (2008). Neonatal mortality and morbidity rates in late preterm births compares with births at term. *Obstetrics and Gynecology*, **111**, 35–41.

Muliere, P. and Tardella, L. (1998). Approximating distributions of random functionals of Ferguson–Dirichlet priors. *Canadian Journal of Statistics*, **2**, 283–297.

Müller, P., Erkanli, A. and West, M. (1996). Bayesian curve fitting using multivariate normal mixtures. *Biometrika*, **83**, 67–79.

Nohr, E.A., Vaeth, M., Baker, J.L., Sorensen, T.I.A., Olsen, J. and Rasmussen, K.M. (2008). Combined associations of prepregnancy body mass index and gestational weight gain with the outcome of pregnancy. *American Journal of Clinical Nutrition*, **87**, 1750–1759.

Papaspiliopoulos, O. (2008). A note on posterior sampling from Dirichlet process mixture models. *Working Paper*, **08–20**, Centre for Research in Statistical Methodology, University Warwick, Coventry, UK.

Papaspiliopoulos, O. and Roberts, G.O. (2008). Retrospective Markov chain Monte Carlo methods for Dirichlet process hierarchical models. *Biometrika*, **95**, 169–186.

Peng, F.C., Jacobs, R.A. and Tanner, M.A. (1996). Bayesian inference in mixtures-of-experts and hierarchical mixtures-of-experts models with an application to speech recognition. *Journal of the American Statistical Association*, **91**, 953–960.

Schwarz, G.E. (1978). Estimating the dimension of a model. *Annals of Statistics*, **6**, 461–464.

Sethuraman, J. (1994). A constructive definition of Dirichlet priors. *Statistica Sinica*, **4**, 639–650.

Slaughter, J.C., Herring, A.H. and Thorp, J.M. (2009). A Bayesian latent variable mixture model for longitudinal fetal growth. *Biometrics*, **65**, 1233–1242.

Stephens, M. (2000). Dealing with label switching in mixture models. *Journal of the Royal Statistical Society B*, **62**, 795–809.

Walker, S.G. (2007). Sampling the Dirichlet mixture model with slices. *Communications in Statistics – Simulation and Computation*, **36**, 45–54.

Wang, L. and Dunson, D.B. (2007). Bayesian isotonic density regression. Discussion Paper, Department of Statistical Science, Duke University.

Wilcox, A.J. (2001). On the importance – and the unimportance – of birthweight. *International Journal of Epidemiology*, **30**, 1233–1241.

·2·

Bayesian modelling for matching and alignment of biomolecules

Peter J. Green, Kanti V. Mardia, Vysaul B. Nyirongo and Yann Ruffieux

2.1 Introduction

Technological advances in molecular biology over the last 15–20 years have generated many datasets – often consisting of large volumes of data on protein or nucleotide sequences and structure – that require new approaches to their statistical analysis. In consequence, some of the most active areas of research in statistics at present are aimed at such bioinformatics applications.

It is well known that proteins are the work-horses of all living systems. A protein is a sequence of amino acids, of which there are 20 types. The sequence folds into a three-dimensional structure. We can describe the shape of this structure in terms of a main chain and side chains (for examples, see Branden and Tooze, 1999; Lesk, 2000). Three atoms of each amino acid, denoted N, C_α and C', are in the main chain or backbone. The other part of the amino acid that is attached to the C_α atom is called the residue, and forms a side chain (that is, there is one side chain or residue for each amino acid in the protein).

One of the major unsolved biological challenges is the protein folding problem: how does the amino acid sequence fold into a three-dimensional protein? In particular, how can the three-dimensional protein structure (and its function) be predicted from the amino acid sequence? These are key questions, since both shape and chemistry are important in understanding a protein's function.

2.1.1 Protein data and alignment problems

One particular task where statistical modelling and inference can contribute to scientific understanding of protein structure is that of matching and alignment of two or more proteins. This chapter addresses the analysis of three data sets of the following nature.

The three-dimensional structure of a protein is important for it to perform its function. In chemoinformatics, it is a common assumption that structurally similar molecules have similar activities; in consequence, protein structure similarity can be used to infer the unknown function of a candidate protein (Leach, 2003). In drug design for example, a subject of prime interest is the local interaction between a small molecule (the ligand) and a given protein receptor. If the geometrical structure of the receptor is known, then established methods such as docking can be applied in order to specify the protein – ligand interaction. However, in most cases this structure is unknown, meaning the drug designer must rely on a study of the similarity (or diversity) in available ligands.

The alignment of the molecules is an important first step towards such a study. Thus one of the problems in protein structural bioinformatics is matching and aligning three-dimensional protein structures or related configurations, e.g. active sites, ligands, substrates and steroid molecules. To consider the matching between proteins, one normally considers the C_α atoms. A related application but with different aims is the matching of two-dimensional protein gels.

Given two or more proteins represented by configurations of points in space representing locations of particular atoms of the proteins, the generic task of matching and alignment is to discover subsets of these configurations that coincide, after allowing for measurement error and unknown geometrical transformations of the proteins. These applications require algorithms for matching, as well as statistics and distributions of measures for quantifying quality of matching and alignment.

Statistical shape analysis potentially has something to offer in solving matching and alignment problems; the field of labelled shape analysis with labelled points is well developed (see Dryden and Mardia, 1998, also Appendix A) but unlabelled shape analysis is still in its infancy. The methodology we have developed for matching and alignment is a contribution to shape analysis for unlabelled and partially labelled data.

Pairwise matching of active sites data sets An active site is a local three-dimensional arrangement of atoms in a protein that are involved in a specific function e.g. binding a ligand (and so known as a binding site). Atoms in active sites are from amino acids that are close to each other in three-dimensional space but do not necessarily follow closely in sequence order.

We consider two datasets, analysed with different objectives. We have configurations of C_α atoms in two functional sites shown in Figure 2.1 from $17-\beta$ hydroxysteroid-dehydrogenase and carbonyl reductase proteins (two dimensional view of the data, given at `http://www.stats.bris.ac.uk/~peter/Align/index.html`). These functional sites are related but which and how many atoms correspond are unknown. Our aim would be to find matching atoms and align these configurations.

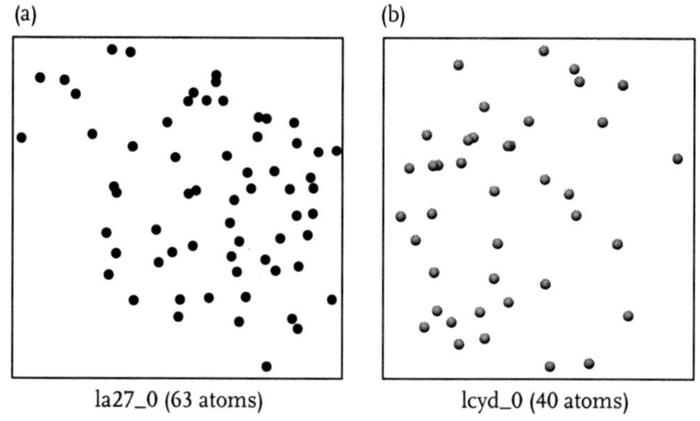

Fig. 2.1 C_α atoms in functional sites of 17 – β hydroxysteroid-dehydrogenase (1a27_0) and carbonyl reductase (1cyd_0) represented as spheres in RasMol.

We also consider matches of protein active sites from SITESDB (cf. Gold, 2003). An alcohol dehydrogenase active site (1hdx_1) is matched against NADP-binding sites of a quinone oxidoreductase protein (1o8c_1); this is a small example of a query used against a large database.

Matching multiple configurations of steroid molecules The CoMFA (Comparative Molecular Field Analysis) database is a set of steroid molecules which has become a benchmark for testing various 3D quantitative structure – activity relationship (QSAR) methods (see Coats, 1998). This database contains the three-dimensional coordinates for the atoms in each of the 31 molecules, in addition to information such as atom type and partial charge. The geometrical similarity between the molecules makes this database an ideal test-bed for our multiple alignment methods. The CoMFA steroids can be accessed from `http://www2.ccc.uni-erlangen.de/services/steroids/`.

2.1.2 Geometrical transformations

Matching and alignment is conducted generally assuming that a geometrical transformation has to be applied to each of the point configurations to bring them all onto a standard orientation and scale in which points to be matched are simply those closest together. These transformations are usually considered to be unknown and have to be inferred from the data. In the Bayesian formulations we follow below, modelling the transformations amounts to deciding on a space of transformations that are appropriate for the problem, and a prior distribution over that space; these choices will depend on understanding of both the physical processes that led to any variation in the geometry of the different configurations, and of the observational process by which the point coordinates are recorded. In the two applications presented in this chapter, we assume the space of 'rigid-body' transformations, that is, rotation and translation, and place a uniform uninformative prior on the rotation. This would be appropriate, for example, if there was no systematic distortions between the configurations, but they are presented for measurement in an arbitrary orientation. Non-trivial priors on rigid-body or other affine transformations are handled by similar methodology. For other problems, other spaces of transformation will be necessary – for example, aligning protein gels would usually require smooth 'warping' rather than any affine transformation. Methodology for such situations is still under development.

2.1.3 Structure and scope

Green and Mardia (2006) proposed a Bayesian hierarchical model allowing inference about the matching and alignment of two configurations of points; in Section 2.2 this is described as it would be used for pairwise matching of configurations of active sites. Ruffieux and Green (2009) developed the Green and Mardia (2006) model to handle the simultaneous matching of multiple configurations (see also Marín and Nieto, 2008), and this is reviewed in Section 2.3. Analysis of the three data sets introduced above is presented in Section 2.4. We end the chapter with discussion of conclusions and future directions for research. There are various appendices including Appendix A reviewing labelled shape analysis, Appendix B.1 on model formulation, Appendix B.2 on MCMC implementation, and Appendix B.3 on web data sources.

2.2 A Bayesian hierarchical model for pairwise matching

We have two point configurations, $X^{(1)} = \{x_j, j = 1, 2, \ldots, m\}$ and $X^{(2)} = \{y_k, k = 1, 2, \ldots, n\}$, in d-dimensional space $\mathcal{R}^d$. The points are labelled for identification, but arbitrarily. In our applications the points are C_α atoms.

A latent point process model The key basis for our model for the configurations is that both point sets are regarded as noisy observations on subsets of a set of unobserved true locations $\{\mu_i\}$, where we do not know the mappings from j and k to i. There may be a geometrical transformation between the x-space and the y-space, which may also be unknown. The objective is to make model-based inference about these mappings, and in particular make probability statements about matching: which pairs (j, k) correspond to the same true location?

We will assume the geometrical transformation between the x-space and the y-space to be affine, and denote it by $x = \mathcal{A}y = Ay + \tau$. Later we will restrict A to be a rotation matrix, so that this is a rigid-body transformation. We regard the true locations $\{\mu_i\}$ as being in x-space, without loss of generality.

The mappings between the indexing of the $\{\mu_i\}$ and that of the data $\{x_j\}$ and $\{y_k\}$ are captured by indexing arrays $\{\xi_j\}$ and $\{\eta_k\}$; to be specific we assume that

$$x_j = \mu_{\xi_j} + \varepsilon_{1j}, \tag{2.1}$$

for $j = 1, 2, \ldots, m$, where $\{\varepsilon_{1j}\}$ have probability density f_1, and

$$Ay_k + \tau = \mu_{\eta_k} + \varepsilon_{2k}, \tag{2.2}$$

for $k = 1, 2, \ldots, n$, where $\{\varepsilon_{2k}\}$ have density f_2. All $\{\varepsilon_{1j}\}$ and $\{\varepsilon_{2k}\}$ are independent of each other, and independent of the $\{\mu_i\}$. We take f_1 and f_2 to be normal but the method generalises to any f_1 and f_2.

Multiple matches are excluded, and thus each hidden point μ_i is observed at most once in each of the x and y configurations; equivalently, the ξ_j are distinct, as are the η_k. The label i is not used in our subsequent development, and all that is needed is the matching matrix M, defined by $M_{jk} = 1$ if $\xi_j = \eta_k$ otherwise 0. This structure is a latent variable in our model, and its distribution is derived from the latent point process model as follows.

Prior for M. Suppose that the set of true locations $\{\mu_i\}$ forms a homogeneous Poisson process with rate λ over a region $V \subset \mathcal{R}^d$ of volume v, and that N points are realized in this region. Some of these give rise to both x and y points, some to points of one kind and not the other, and some are not observed at all. We suppose that these four possibilities occur independently for each realised point, with probabilities parametrized so that with probabilities $(1 - p_x - p_y - \rho p_x p_y, p_x, p_y, \rho p_x p_y)$ we observe neither, x alone, y alone, or both x and y, respectively. The parameter ρ is a certain measure of the tendency a priori for points to be matched: the prior probability distribution of L conditional on m and n is proportional to

$$p(L) \propto \frac{(\rho/\lambda v)^L}{(m-L)!(n-L)!L!}, \tag{2.3}$$

for $L = 0, 1, \ldots, \min\{m, n\}$. The normalising constant here is the reciprocal of $H\{m, n, \rho/(\lambda v)\}$, where H can be written in terms of the confluent hypergeometric function

$$H(m, n, d) = \frac{d^m}{m!(n-m)!}\, {}_1F_1(-m, n-m+1, -1/d),$$

assuming without loss of generality that $n > m$ (see Abramowitz and Stegun, 1970, p. 504).

Assuming that M is a priori uniform conditional on L, we have

$$p(M) = p(L)p(M|L) = \frac{(\rho/\lambda v)^L}{\sum_{\ell=0}^{\min\{m,n\}} \ell! \binom{m}{\ell}\binom{n}{\ell}(\rho/\lambda v)^\ell}.$$

One application of this distribution of L is a similarity index (Davies *et al.*, 2007).

The likelihood Let

$$x_j \sim N_d\left(\mu_{\xi_j}, \sigma_x^2 I\right) \qquad \text{and} \qquad Ay_k + \tau \sim N_d\left(\mu_{\eta_k}, \sigma_y^2 I\right),$$

with $\sigma_x = \sigma_y = \sigma$, say. On integrating out the μs the joint model for parameters, latent variables and observables is (Green and Mardia, 2006)

$$p(M, A, \tau, \sigma, x, y) \propto |A|^n p(A) p(\tau) p(\sigma) \prod_{j,k:M_{jk}=1} \left(\frac{\kappa\phi\left\{(x_j - Ay_k - \tau)/\sigma\sqrt{2}\right\}}{(\sigma\sqrt{2})^d} \right), \tag{2.4}$$

where ϕ is the standard normal density in $\mathcal{R}^d$. Here $|A|$ is the Jacobian of transformation from x-space into y-space; $p(A)$, $p(\tau)$ and $p(\sigma)$ denote prior distributions for A, τ and σ; d is the dimension of the configurations i.e. $d = 2$ for two-dimensional configurations, e.g. protein gels and $d = 3$ for three-dimensional configurations, e.g. active sites; $\kappa = \rho/\lambda$ measures the tendency a priori for points to be matched and can be a function of concomitant information, e.g. amino acid types in matching protein structures. See the directed acyclic graph, Figure 2.2a, for a graphical representation of the model. Green and Mardia (2006) give a generalization where ϕ can be replaced by a more general density depending on f_1 and f_2. Details on priors are given below.

There is a connection between maximizing the joint posterior derived from equation (2.4) and minimizing root mean square deviations (RMSD), defined by

$$\text{RMSD}^2 = Q/L, \text{ where } Q = \sum_{j,k} M_{jk}||x_j - Ay_k - \tau||^2, \tag{2.5}$$

and $L = \sum_{j,k} M_{jk}$ denotes the number of matches. RMSD is the focus of study in combinatorial algorithms for matching. In the Bayesian formulation the log likelihood (with uniform priors) is proportional to

$$\textit{const.} - 2L \log \sigma + L \log \rho - \frac{1}{2} \frac{Q}{\sigma^2\sqrt{2}}.$$

The maximum likelihood estimate of σ for a given matching matrix M is the same as the $RMSD$ which is the least squares estimate. $RMSD$ is a measure commonly used in bioinformatics, although joint uncertainty in $RMSD$ and the matrix M is difficult to appreciate except in the Bayesian formulation.

Prior distributions for continuous variables For the continuous variables τ, σ^{-2} and A we use conditionally conjugate priors so

$$\tau \sim N_d\left(\mu_\tau, \sigma_\tau^2 I\right), \qquad \sigma^{-2} \sim \Gamma(\alpha, \beta), \qquad A \sim \text{Fisher}(F).$$

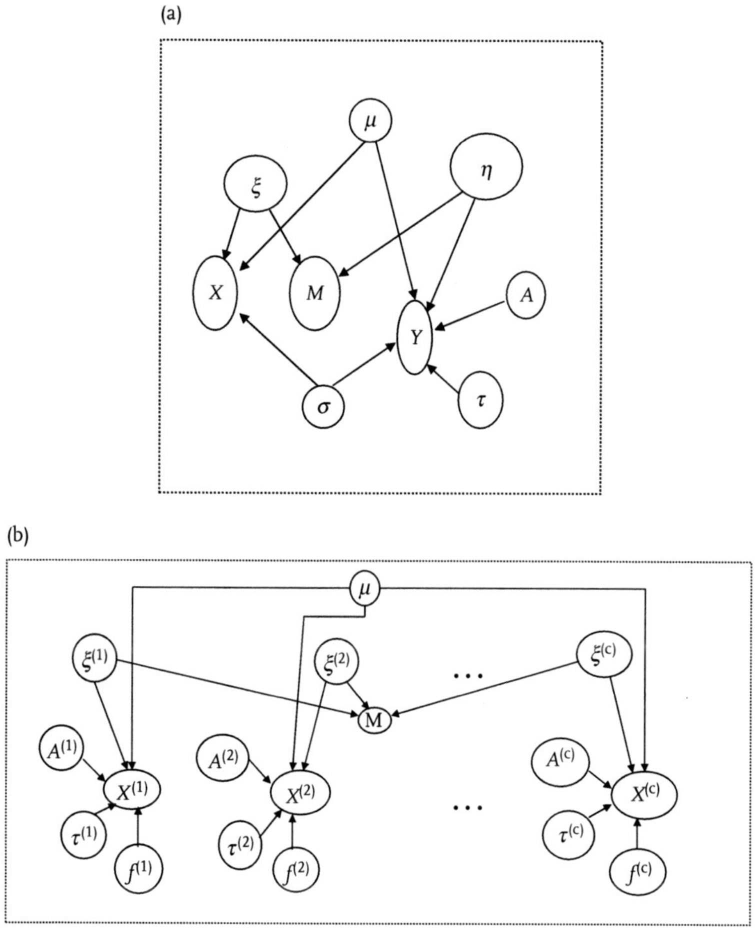

Fig. 2.2 Directed acyclic graph representing the model for matching (a) two configurations, (b) multiple configurations. The graph shows all data and parameters treated as variables.

Here, Fisher(F) denotes the matrix Fisher distribution (see, for example, Mardia and Jupp, 2000, p. 289). For $d = 2$, A has a von Mises distribution. For $d = 3$, it is useful to express A in terms of the Eulerian angles. Some efficient methods to simulate A are given in Green and Mardia (2006). If the point configurations are presented in arbitrary orientations, it is appropriate to assume a uniform distribution on A, that is, $F = 0$, and this is usually adequate.

2.3 Alignment of multiple configurations

In this section we consider a hierarchical model for matching configurations $X^{(1)}, X^{(2)}, \ldots, X^{(C)}$ simultaneously.

Multiconfiguration model The pairwise model presented above can be readily extended to the multiconfiguration context. Suppose we have C point configurations $X^{(1)}, X^{(2)}, \ldots, X^{(C)}$, such that $X^{(c)} = \{x_j^{(c)}, j = 1, 2, \ldots, n_c\}$, where $x_j^{(c)} \in \mathcal{R}^d$ and n_c is the number of points in configuration $X^{(c)}$. As in the pairwise case we assume the existence of a set of 'hidden' points $\mu = \{\mu_i\} \subset \mathcal{R}^d$ underlying the observations. Our multiple-configuration model is thus:

$$\mathcal{A}^{(c)} x_j^{(c)} = \mu_{\xi_j^{(c)}} + \varepsilon_j^{(c)}, \quad \text{for } j = 1, 2, \ldots, n_c, \quad c = 1, 2, \ldots, C. \tag{2.6}$$

The unknown transformation $\mathcal{A}^{(c)}$ brings the configuration $X^{(c)}$ back into the same frame as the μ-points, and $\xi^{(c)}$ is a labelling array linking each point in configuration $X^{(c)}$ to its underlying μ-point. As in the previous section the elements within each labelling array are assumed to be distinct. In this context a match can be seen as a set of points $(x_{j_1}^{(i_1)}, x_{j_2}^{(i_2)}, \ldots, x_{j_k}^{(i_k)})$ such that $\xi_{j_1}^{(i_1)} = \xi_{j_2}^{(i_2)} = \cdots = \xi_{j_k}^{(i_k)}$.

Thus matches may now involve more than two points at once. These are stored in a structure $\mathcal{M}$. This parameter plays the same role as the matrix M from Section 2.2, in that it contains the relevant information on the matches. We choose to categorize the matches according to their 'type'. Consider a generic set $I \subset \{1, 2, \ldots, C\}$ of configuration indices, with $I \neq \emptyset$. This set corresponds to a type of match: for example if $C = 3$, then $I = \{2, 3\}$ refers to a match involving a point from the $X^{(2)}$ configuration and a point from the $X^{(3)}$ configuration but none from the $X^{(1)}$ configuration. We call *I-match* a match involving *exactly* the configurations whose index is included in I.

The likelihood Write S_I as the set of I-matches contained in $\mathcal{M}$. The elements of S_I are written as index arrays of the form $(j_1, j_2, \ldots, j_{|I|})$, with the convention that $\{x_{j_1}^{(i_1)}, x_{j_2}^{(i_2)}, \ldots, x_{j_{|I|}}^{(i_{|I|})}\}$ is the corresponding set of matched points. Let

$$\mathcal{A}^{(c)} x_j^{(c)} \sim N_d\left(\mu_{\xi_j^{(c)}}, \sigma^2 I\right),$$

for $c = 1, 2 \ldots, C$ and $j = 1, 2, \ldots, n_c$. Supposing $\mathcal{A}^{(c)}$ is made up of a transformation matrix $A^{(c)}$ and a translation vector $\tau^{(c)}$, our posterior model has the form

$$\begin{aligned} p(\mathcal{A}, \mathcal{M} \mid X) \propto & \prod_{c=1}^{C} \left\{ p(A^{(c)}) p(\tau^{(c)}) \left| A^{(c)} \right|^{n_c} \right\} \\ & \times \prod_{I} \prod_{(j_1, \ldots, j_{|I|}) \in S_I} \kappa_I |I|^{-d/2} (2\pi\sigma^2)^{-d(|I|-1)/2} \\ & \times \exp\left\{ -\frac{1}{2\sigma^2} \gamma_{\mathcal{A}} \left(x_{j_1}^{(i_1)}, x_{j_2}^{(i_2)}, \ldots, x_{j_{|I|}}^{(i_{|I|})} \right) \right\}, \end{aligned} \tag{2.7}$$

where $\mathcal{A} = \left(\mathcal{A}^{(1)}, \mathcal{A}^{(2)}, \ldots \mathcal{A}^{(C)}\right)$ and $X = \left(X^{(1)}, X^{(2)}, \ldots, X^{(C)}\right)$, and

$$\gamma_{\mathcal{A}}\left(x_{j_1}^{(i_1)}, x_{j_2}^{(i_2)}, \ldots, x_{j_{|I|}}^{(i_{|I|})}\right) = \sum_{k=1}^{|I|} (y_k - \overline{y})^2$$

where

$$y_k = A^{(i_k)} x_{j_k}^{(i_k)} + \tau^{(i_k)} \quad \text{and} \quad \overline{y} = \sum_{k=1}^{|I|} y_k / |I|.$$

If we set $C = 2$, $A^{(1)} = I$ and $\tau^{(1)} = 0$, this posterior distribution is equivalent to the one given in the previous section.

Note that we have implicitly defined a prior distribution for $\mathcal{M}$ in the expression for the posterior distribution: this prior is described explicitly below.

Prior distributions Prior distributions for the $\tau^{(c)}$, $A^{(c)}$, and σ^2 are identical to the ones for the parameters in the pairwise model. In particular we set, for $c = 1, 2, \ldots, C$,

$$\tau^{(c)} \sim N_d\left(\mu^{(c)}, \sigma_c^2 I\right), \qquad A^{(c)} \sim \text{Fisher}(F_c),$$

and $\sigma^{-2} \sim \Gamma(\alpha, \beta)$.

To construct a prior for the matches $\mathcal{M}$, we again assume that the μ-points follow a Poisson process with constant rate λ over a region of volume v. Each point in the process gives rise to a certain number of observations, or none at all. For I as defined previously, let q_I be the probability that a given hidden location generates an I-match. We impose the following parametrization:

$$q_I = \rho_I \cdot \prod_{c \in I} q_{\{c\}},$$

where $\rho_I = 1$ if $|I| = 1$. Define L_I as the number of I-matches contained in $\mathcal{M}$, and assume the conditional distribution of $\mathcal{M}$ given the L_Is is uniform. After some combinatorial work we find that the prior distribution for the matches can be expressed as

$$p(\mathcal{M}) \propto \prod_I \left(\frac{\kappa_I}{v^{|I|-1}}\right)^{L_I},$$

where $\kappa_I = \rho_I / \lambda^{|I|-1}$. It is easy to see that this is simply a generalization of the prior distribution for the matching matrix M.

Identifiability issue To preserve symmetry between configurations, we only consider the case where $A^{(c)}$ are uniformly distributed a priori. It is then true that the relative rotations $(A^{(c_1)})' \cdot A^{(c_2)}$ are uniform and independent for $c_2 \neq c_1$ and fixed c_1. So without loss of generality, we can now impose the identifying constraint that $\mathcal{A}^{(1)}$ be fixed as the identity transformation. This is the same as saying that the first data configuration lies in the same frame as the hidden point locations, as was the case in the pairwise model.

2.4 Data analysis

2.4.1 Active sites and Bayesian refinement

2.4.1.1 *Two sites*

We now illustrate our method applied to two active sites as described in Section 2.1. A sampler described in Appendix B.2 was run for 100,000 sweeps (including 20,000 iterations for burn-in period) to match $17-\beta$ hydroxysteroid-dehydrogenase and carbonyl reductase active sites described in Section 2.1.1. Prior and hyperprior settings were $\alpha = 1$, $\beta = 2$, $\mu_\tau = (0, 0, 0)'$, $\sigma_\tau = 20$, $\lambda/\rho = 0.0005$, $F = 0$. We match 34 points as shown in Figure 2.3a with $RMSD = 0.949$Å. The algorithm was also used with restriction to match only points representing same type of amino acid. With the restriction on type of points (concomitant information), 15 matches shown in Figure 2.3b are made with $RMSD = 0.727$ Å.

2.4.1.2 *Queries in a database*

We now discuss how we could use an initial solution from other software, e.g. the graph theoretic technique (Gold 2003), derived from a physico-chemistry view point, and use our algorithm to refine the solution. In this context, the data comparisons are substantial since they involve comparing a query with other family members. We compare here with the graph theoretic approach that requires adjusting the matching distance threshold a priori according to noise in atomic positions, which is difficult to predetermine in bioinformatics applications involving matching configurations in a database with varying crystallographic precision. Furthermore, the graph method is unable to identify alternative but sometimes important solutions in the neighbourhood of the distance based solution because of strict distance thresholds. On the other hand, the graph theoretic approach is very fast, robust and can quickly give corresponding points for small configurations from which we can get initial estimates for rotation and translation. We illustrate here how our approach finds more biologically interesting and statistically significant matches between functional sites (Mardia *et al.*, 2007*a*).

We model graph theoretic matches of protein active sites from SITESDB (cf. Gold, 2003). An alcohol dehydrogenase active site (1hdx_1) is matched against NADP-binding sites of a quinone oxidoreductase protein (1o8c_1). Figure 2.4 gives matched amino acids by the graph method and refined by the Bayesian algorithm (see B.2).

(a)

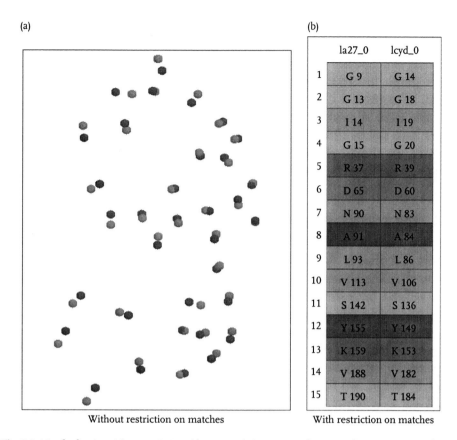

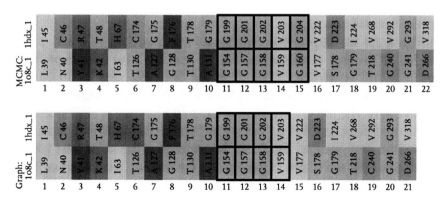

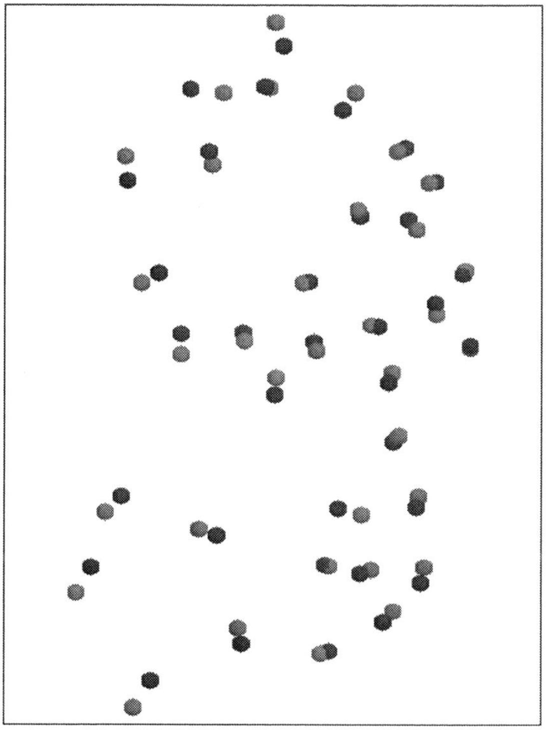

Without restriction on matches

(b)

	la27_0	lcyd_0
1	G 9	G 14
2	G 13	G 18
3	I 14	I 19
4	G 15	G 20
5	R 37	R 39
6	D 65	D 60
7	N 90	N 83
8	A 91	A 84
9	L 93	L 86
10	V 113	V 106
11	S 142	S 136
12	Y 155	Y 149
13	K 159	K 153
14	V 188	V 182
15	T 190	T 184

With restriction on matches

Fig. 2.3 Matched points (C_α atoms): (a) without restriction on matches according to amino acid type; (b) matching rotation and same types of amino acids.

MCMC:	1	2	3	4	5	6	7	8	9	10	11	12	13	14	15	16	17	18	19	20	21	22
1hdx_1	I 45	C 46	R 47	T 48	H 67	C 174	G 175	F 176	T 178	G 179	**G 199**	**G 201**	**G 202**	**V 203**	**G 204**	V 222	D 223	I 224	V 268	V 292	G 293	V 318
1o8c_1	L 39	N 40	Y 41	K 42	I 63	T 126	A 127	G 128	T 130	A 131	**G 154**	**G 157**	**G 158**	**V 159**	**G 160**	V 177	S 178	G 179	T 218	G 240	G 241	D 266

Graph:	1	2	3	4	5	6	7	8	9	10	11	12	13	14	15	16	17	18	19	20	21
1hdx_1	I 45	C 46	R 47	T 48	H 67	C 174	G 175	F 176	T 178	G 179	**G 199**	**G 201**	**G 202**	**V 203**	V 222	D 223	I 224	V 268	V 292	G 293	V 318
1o8c_1	L 39	N 40	Y 41	K 42	I 63	T 126	A 127	G 128	T 130	A 131	**G 154**	**G 157**	**G 158**	**V 159**	V 177	S 178	G 179	T 218	C 240	G 241	D 266

Fig. 2.4 Corresponding amino acids between the NAD-binding site of alcohol dehydrogenase (1hdx_1) and NADP-binding site of quinone oxidoreductase (1o8c_1) before and after MCMC refinement step. Amino acids with bold borders are part of the dinucleotide binding motif GL-GGVG.

The Markov chain Monte Carlo (MCMC) refinement step produced improvements with obvious biochemical relevance. These proteins share a well known glycine rich motif (GXGXXG) in the binding site. Before the MCMC refinement step, only two glycines in dinucleotide binding motif GLGGVG were matched by the graph theoretic approach and this increased to three glycines after MCMC refinement.

In this kind of context, only short MCMC runs may be possible, and we cannot have full confidence that the whole posterior space is being sampled. However, we should explore the mode containing the graph-theoretic initial solution, possibly refine that solution, and get an idea of uncertainty.

2.4.2 Aligning multiple steroid molecules

Here we attempt to align $C = 5$ configurations simultaneously, using the method described in Section 2.3. An MCMC algorithm (see Appendix B.2) is used to simulate a random sample from the distribution (2.7); this sample is then used as a basis for inference.

We select five molecules from the CoMFA database (see Section 2.1.1); these are aldosterone, cortisone, prednisolone, 11-deoxycorticosterone, and 17a-hydroxyprogesterone. All of these molecules contain 54 atoms. We set the following hyperparameter values: $\alpha = 1, \beta = 0.1$ and $\mu^{(c)} = 0, \sigma_c^2 = 10$, for $c = 2, 3, 4, 5$, and $F_c = 0$ for all c. For the match prior parameters we set $\kappa_I = 1$ for $|I| = 1$, $\kappa_I = 14$ for $|I| = 2$, $\kappa_I = 289$ for $|I| = 3$, $\kappa_I = 12,056$ for $|I| = 4$ and $\kappa_I = 15,070,409$ for $|I| = 5$. These values were determined by making initial 'guesses' on the number of matches of each type, and adapting the prior distribution (2.7) accordingly. As in the pairwise case we obtain estimates for the rotations, translations, and matches. In particular, the matches are ranked according to their frequency in the posterior distribution, and we choose to select the k most frequent, say.

In Figure 2.5 we display the time series traces of $L_{\{1,2,3,4,5\}}$ and $L_{\{2,3,4,5\}}$ in the MCMC output. Here we find that 56 matches have sample probability higher than 0.5. In Figure 2.6 we align the five molecules by applying the estimated transformations to each configuration. It is interesting to note that in the latter figure, the top right portion of the first molecule is slightly detached from the other four, and indeed the MCMC output suggests that those points from the first configuration should not be matched to the other four. However when aligning the molecules in pairs using the method from Section 2.2, the inference tends to favour matching these points, even if we set the match hyperparameters to be biased against matching. This confirms that inclusion of two or three additional configurations may have a positive impact on the alignment inference. One might understand this as a 'borrowing of strength' of sorts: further configurations provide further information on the number and

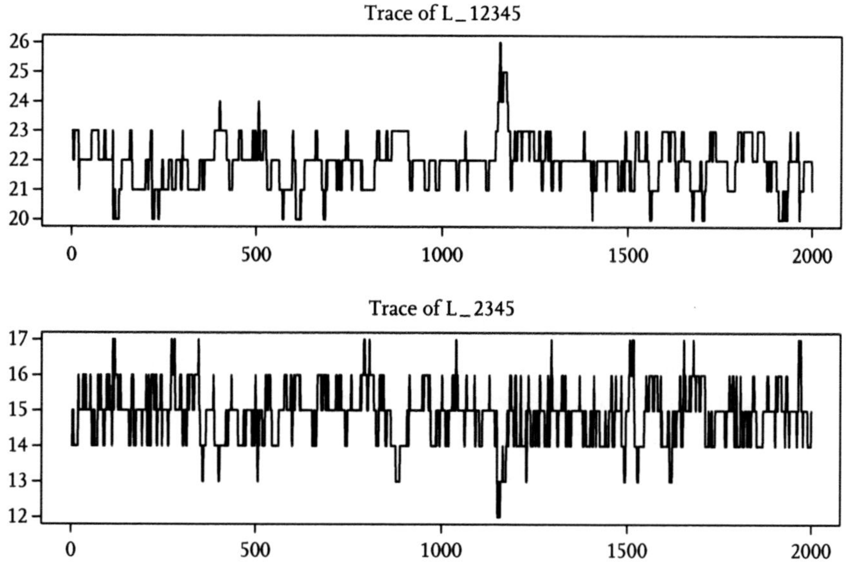

Fig. 2.5 Time series traces of the number of matches involving all configurations (top), and involving all configurations except the first (bottom). Taken from a thinned sample of 2000 after burn-in.

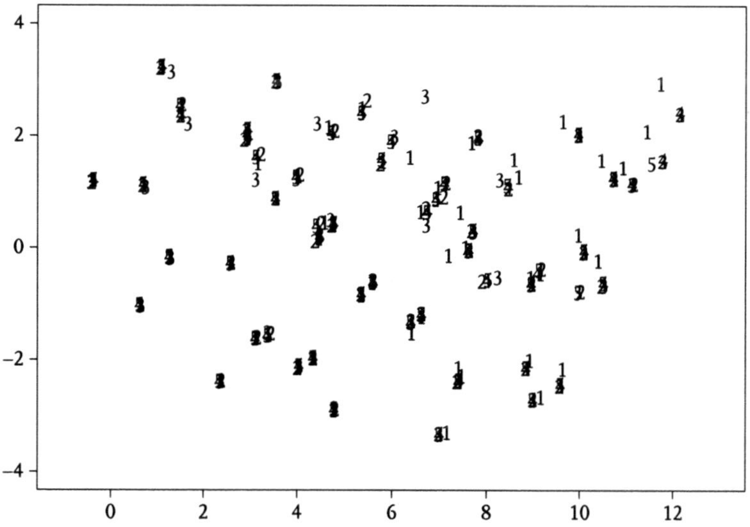

Fig. 2.6 Multiple alignment of the five steroid molecules: the full transformations are estimated from a MCMC subsample of size 2000, and are filtered out from the data. The points are projected onto the first two canonical axes, and are labelled according to the number of the configuration they belong to.

location of implied μ-points, information which can in turn be exploited in the alignment of the initial configurations. Clearly there is no way to take advantage of this information if the molecules are aligned independently in pairs.

2.5 Further discussion

2.5.1 Advantages of the Bayesian modelling approach

Some advantages of our Bayesian approach to these problems are:

1. Simultaneous inference about both discrete and continuous variables.
2. The full Bayesian posterior 'tool kit' is available for inference.
3. It allows in a natural way any prior information
4. The MCMC may be too slow in some application but it has a role to play as a gold standard against heuristic approaches.
5. The MCMC implementation provides a greater chance to escape local modes compared to optimization methods.

Wilkinson (2007*b*) has given a review of Bayesian methods in bioinformatics and computational system biology more generally, citing some of these points; in particular he has pointed out why bioinformatics and computational system biology are undergoing a Bayesian revolution similar to that already seen in genetics.

2.5.2 Outstanding issues

We expect that further work, by the authors and others, will be directed to refinement of the methodology described here.

2.5.2.1 Modelling

Spherical normality of the errors ε in (2.1), (2.2) and (2.6) was assumed for simplicity, and it may be necessary to relax this assumption. These errors represent both measurement error in recording the data, and 'model errors', small variations between the molecules in the locations of the atoms. There is an interplay between the sphericity assumption and the modelling of the geometrical transformations $\mathcal{A}$, so care is needed here, but it is straightforward to replace normality by a heavy-tailed alternative.

Further study is needed on setting hyperparameters and sensitivity to these choices. The analysis seems to be most sensitive to the parameter κ but otherwise rather robust.

As mentioned before, particular applications of matching and alignment demand more subtle modelling of the geometrical transformations $\mathcal{A}$, with the extension to non-parametric warping being most pressing.

Finally, there are interesting modelling issues concerned with using sequence information to influence the inference on matching and alignment. The most promising direction within our modelling paradigm involves the use of non-uniform 'gap' priors on the matching matrix M, encouraging or requiring matchings that respect the sequence order.

2.5.2.2 Computation

Design of MCMC samplers to deal comprehensively with problems of multimodality in the posterior distribution is the major challenge here. One can expect that generic techniques such as simulated tempering will have a part to play. We also expect further work on inferential methods that are perhaps not fully Bayesian, but are computationally faster, including the use of fast initialization methods, such as

(a) starting from the solution from three-dimensional deterministic methods such as graph theoretic, CE, geometric hashing and others;
(b) for full protein alignment of structure, using well-established sequence alignment software, e.g. BLAST as the starting point.

2.5.3 Alternative approaches

2.5.3.1 EM approach

The interplay between matching (that is, allocation), and parameter uncertainty has something in common with mixture estimation. This might suggest considering maximization of the posterior by using the EM algorithm, which could of course in principle be applied either to maximum likelihood estimation or to maximum a posteriori estimation. For the EM formulation, the 'missing data' are the matches.

The 'expectations of missing values' are just probabilities of matching. These are only tractable if we drop the assumption that a point can only be matched with at most one other point; that is, that $\sum_j M_{jk} \leq 1$ for all k, $\sum_k M_{jk} \leq 1$ for all j. We then get 'soft matching'.

EM allows us to study only certain aspects of an approximate version of our model, and is not trivial numerically. Obtaining the complete posterior by Markov chain Monte Carlo sampling gives much greater freedom in inference. For a direct EM approach see Kent *et al.* (2004).

2.5.3.2 Procrustes type approaches

Dryden *et al.* (2007) and Schmidler (2007) use a MAP estimator for M after estimating 'nuisance parameters' (A, τ, σ^2) from Procrustes registration. Schmidler (2007) has provided a fast algorithm using geometric hashing. Thus these borrow strength from labelled shape analysis. The Green and Mardia (2006) procedure also allows informative priors so the procedure is very general.

Dryden (2007) has given some initial comparisons and in particular their MAP approach often get stuck in local modes. For small variability, both approaches lead to similar results. Wilkinson (2007*a*) has touched many important problems in Bayesian alignment in particular the uniform prior would be strongly biased towards larger values of $L = |M|$. Schmidler (2007) and Dryden *et al.* (2007) effectively assume that Procrustes alignment is 'correct' and does not reflect uncertainty in geometric alignment. By integrating out the geometrical transformation as in Green and Mardia (2006) then alignment uncertainty will be propagated correctly without suffering significant computation penalty. Mardia (2007*a*) has raised a few general issues about matching in the discussion.

2.5.4 Future directions

Bayesian approaches are particularly promising in tackling problems in bioinformatics, with their inherent statistical problems of multiple testing, large parameter spaces for optimization and model or parameter non-identifiability due to high dimensionality. One of the major statistical tasks is to build simulation models of realistic proteins by incorporating local and long-range interactions between amino acids. A protein can be uniquely determined by a set of conformational angles so directional statistics (see Mardia and Jupp, 2000) plays a key role as well as shape analysis. Boomsma *et al.* (2008) have given a dynamical Bayesian network with angular distributions and amino acid sequences as its nodes for protein (local) structure prediction. This solves one of the two major bottle necks for protein based nanotechnology namely generating native(natural) protein-like structures. Another is an appropriate energy function to make it compact. The angular distributions used a priori adequately describe Ramachandran plots of the dihedral angles of the backbone (see Mardia, Taylor and Subramaniam, 2007*b*; Mardia *et al.*, 2008). Hamelryck *et al.* (2006) have given a method of simulating realistic protein conformation for C_α trace focusing on mimicking secondary structure while Mardia and Nyirongo (2008) focus on global properties, e.g. compactness and globularity.

The word homology is used in a technical sense in biology, especially in discussion of protein sequences, which are said to be homologous if they have been derived from a common ancestor. Homology implies an evolutionary relationship and is distinct from similarity. In this chapter, alignment focuses only on similarity.

To sum up, there are real challenges for statisticians in the understanding of protein structure. Similar remarks apply to understand the RNA structure (see, for example Frellsen *et al.*, 2008). All this might need is holistic statistics which implies a shift of paradigm by statisticians (Green, 2003; Mardia and

Gilks, 2005; Mardia, 2007*b*, 2008). However, protein bioinformatics is a subset of very large area of bioinformatics which has many challenging problems (see Gilks, 2004; Mardia, 2005; Wilkinson, 2007*b*).

Appendix

A. Broader context and background

Shape analysis

Advances in data acquisition technology have led to the routine collection of geometrical information, and the study of the shape of objects has been increasingly important. With modern technology, locating points on objects is now often straightforward. Such points, typically on the outline or surface of the objects, can be loosely described as landmarks, and in this discussion, and indeed throughout the chapter, we treat objects as being represented by their landmarks, regarded as points in a Euclidean space, usually $\mathcal{R}^2$ or $\mathcal{R}^3$.

What do we mean by 'shape'? The word is very commonly used in everyday language, usually referring to the appearance of an object. Mathematically, shape is all the geometrical information that remains when certain transformations are filtered out, that is, a point in shape space represents an equivalence class of objects, equivalent under transformations of the given kind. We are typically concerned with transformations such as translation and rotation, sometimes with uniform scale change and/or reflection, and less commonly with unequal scale change, and hence affine transformation, or even non-affine transformations such as nonparametric warping.

When we use the term *rigid shape analysis* we refer to the most important case in applications to bioinformatics, where the transformations in question are translations and rotations, that is, rigid-body motions. This case might more formally be termed *size and shape analysis*, or *the analysis of form*. The more common notion of shape in morphometrics is *similarity shape*, where uniform scale change is also allowed. In *reflection shape analysis*, the equivalence class also allows reflections.

To visualize the distinctions, consider right-angled triangles $\triangle ABC$ with sides in the ratio $AB : BC : CA = 3 : 4 : 5$. In reflection shape space, all such triangles are equivalent. For equivalence in similarity shape space, their vertices must be ordered in the same sense (clockwise or anticlockwise), and in rigid shape space, they must also have the same size.

Most theory and practice to date is concerned with the case of *labelled* shape analysis, where the landmarks defining an object are uniquely identified, so are regarded mathematically as an ordered set (or stacked as a matrix). But increasingly we see applications, including that dealt with in the present chapter, in which the landmarks are either not identified at all (the *unlabelled* case) or identification is incomplete, so that two different landmarks can have the same label (which we might call the *partially labelled* case). In the unlabelled case, the landmarks form an *unordered* set. To return to our 3 : 4 : 5 triangle, in unlabelled rigid shape analysis, the triangles with $AB = 3$, $BC = 4$, $CA = 5$ and with $AB = 4$, $BC = 5$, $CA = 3$ are equivalent, because for example the vertices identified with A in the two figures are not associated.

The foundation for similarity space analysis was laid by Kendall (1984). For mathematical representation, we can construct a shape space with an appropriate metric. The metric is a Procrustes distance for the Kendall shape space. For the form space (see Le, 1988) the appropriate distance is the $RMSD$ (equation 2.5). Note that in practice specific coordinate representations of shape have been useful, namely Bookstein coordinates (Bookstein, 1986) and Procrustes tangent coordinates (Kent, 1994). For further details, see for example, Dryden and Mardia (1998).

Alignment and matching problems such as those considered in this chapter extend unlabelled and partially labelled shape analysis to embrace settings where the point configurations are supersets of those that can be matched, so that the analysis includes an element of selection of points to be matched as well as inference about the geometrical transforations involved.

B.1 Model formulation and inference

Pairwise model

Using concomitant information When the points in each configuration are 'coloured', with the interpretation that like-coloured points are more likely to be matched than unlike-coloured ones, it is appropriate to use a modified likelihood that allows us to exploit such information. Let the colours for the x and y points be $\{r_j^{\mathrm{x}}, j = 1, 2, \ldots, m\}$ and $\{r_k^{\mathrm{y}}, k = 1, 2, \ldots, n\}$ respectively. The hidden-point model is augmented to generate the point colours, as follows. Independently for each hidden point, with probability $(1 - p_{\mathrm{x}} - p_{\mathrm{y}} - \rho p_{\mathrm{x}} p_{\mathrm{y}})$ we observe neither x nor y point, as before. With probabilities $p_{\mathrm{x}}\pi_r^{\mathrm{x}}$ and $p_{\mathrm{y}}\pi_r^{\mathrm{y}}$, respectively, we observe only an x or y point, with colour r from an appropriate finite set. With probability

$$\rho p_{\mathrm{x}} p_{\mathrm{y}} \pi_r^{\mathrm{x}} \pi_s^{\mathrm{y}} \exp(\gamma I[r = s] + \delta I[r \neq s]),$$

where $I[\cdot]$ is an indicator function, we observe an x point coloured r and a y point coloured s. Our original likelihood is equivalent to the case $\gamma = \delta = 0$, where colours are independent and so carry no information about matching.

If γ and δ increase, then matches are more probable, a posteriori, and, if $\gamma > \delta$, matches between like-coloured points are more likely than those between unlike-coloured ones. The case $\delta \to -\infty$ allows the prohibition of matches between unlike-coloured points, a feature that might be adapted to other contexts such as the matching of shapes with given landmarks.

In implementation of this modified likelihood, the Markov chain Monte Carlo acceptance ratios derived for M can be easily modified.

Other, more complicated, colouring distributions where the log probability can be expressed linearly in entries of M can be handled similarly.

Continuous concomitant information can be incorporated in our statistical models such as incorporating van der Waal radii. Such models will be very similar in character as above with obvious modifications, e.g. by using pairwise interaction potentials instead of indicator functions.

Loss functions and a point estimate of M

The output from the Markov chain Monte Carlo sampler derived above, once equilibrated, is a sample from the posterior distribution. As always with sample-based computation, this provides an extremely flexible basis for reporting aspects of the full joint posterior that are of interest.

We consider loss functions $L(M, \widehat{M})$ that penalize different kinds of error and do so cumulatively. The simplest of these are additive over pairs (j, k). Suppose that the loss when $M_{jk} = a$ and $\widehat{M}_{jk} = b$, for $a, b = 0, 1$, is ℓ_{ab}; we set $\ell_{00} = \ell_{11} = 0$. For example, ℓ_{01} is the loss associated with declaring a match between x_j and y_k when there is really none, that is, a 'false positive'. Then

$$E\{L(M, \widehat{M})|x, y\} = -(\ell_{10} + \ell_{01}) \sum_{j,k:\widehat{M}_{jk}=1} (p_{jk} - K),$$

where

$$K = \ell_{01}/(\ell_{10} + \ell_{01}),$$

and $p_{jk} = \text{pr}(M_{jk} = 1|x, y)$ is the posterior probability that (j, k) is a match, which is estimated from a Markov chain Monte Carlo run by the empirical frequency of this match. Thus, provided that $\ell_{10} + \ell_{01} > 0$ and $\ell_{01} > 0$, as is natural, the optimal estimate is that maximizing the sum of marginal posterior probabilities of the declared matches $\sum_{j,k:\widehat{M}_{jk}=1} p_{jk}$, penalized by a multiple K times the number of matches. The optimal match therefore depends only through the cost ratio K. If false positive and false negative matches are equally undesirable, one can simply choose $K = 0.5$.

Computation of the optimal match $\widehat{M}$ would be trivial but for the constraint that there can be at most one positive entry in each row and column of the array. This weighted bipartite matching problem is equivalent to a mathematical programming assignment problem, and can be solved by special-purpose or general LP methods; (see Burkard and Cela, 1999).

For problems of modest size, the optimal match can be found by informal heuristic methods. These may not even be necessary, especially if K is not too small. In particular, it is immediate that, if the set of all (j, k) pairs for which $p_{jk} > K$ includes no duplicated j or k value, the optimal $\widehat{M}$ consists of precisely these pairs. For aligning large size of protein chains *lpSolve* needs to be replaced by *linearass* (Jonker and Volgenant, 1987).

B.2 Model implementation

Sampling the posterior distribution for pairwise alignment

It is straightforward to update conditionally continuous variables. For updating M conditionally, we need some new ways.

The matching matrix M is updated in detailed balance using Metropolis – Hastings moves that only propose changes to a few entries: the number of matches $L = \sum_{j,k} M_{jk}$ can only increase or decrease by 1 at a time, or stay the same. The possible changes are as follows:

(a) adding a match, which changes one entry M_{jk} from 0 to 1;
(b) deleting a match, which changes one entry M_{jk} from 1 to 0;
(c) switching a match, which simultaneously changes one entry from 0 to 1 and another in the same row or column from 1 to 0.

The proposal proceeds as follows. First a uniform random choice is made from all the $m+n$ data points $x_1, x_2, \ldots, x_m, y_1, y_2, \ldots, y_n$. Suppose without loss of generality, by the symmetry of the set-up, that an x is chosen, say x_j. There are two possibilities: either x_j is currently matched, in that there is some k such that $M_{jk} = 1$, or not, in that there is no such k. This depends on a sampler parameter $p^\star$; if x_j is matched to y_k, with probability $p^\star$ it is proposed deleting the match, and with probability $1 - p^\star$ we propose switching it from y_k to $y_{k'}$, where k' is drawn uniformly at random from the currently unmatched y points. On the other hand, if x_j is not currently matched, it is proposed adding a match between x_j and a y_k, where again k is drawn uniformly at random from the currently unmatched y points.

The acceptance probabilities for these three possibilities are easily derived (see Green and Mardia, 2006)

Note that this procedure bypasses the reversible jump.

Multimodality

The issue of multimodality is a challenging issue, as we point out in Green and Mardia (2006). The MCMC samplers used here are very simple (but adequate for the presented examples), and there is a vast literature on more powerful methods that we have not yet brought to bear; this is one of the areas that needs exploring. See also discussion by Wilkinson (2007*a*).

Sampling the posterior distribution for multiple alignment

With our conditionally conjugate priors, we can update the parameters $\tau^{(c)}$, $A^{(c)}$, and σ^2 using a Gibbs move, as in the pairwise implementation. Generalizing the updating of the matches to the multiconfiguration context is less obvious. Write

$$\mathcal{M} = \left\{\left(t_1^1, t_2^1, \ldots, t_C^1\right), \left(t_1^2, t_2^2, \ldots, t_C^2\right), \ldots, \left(t_1^K, t_2^K, \ldots, t_C^K\right)\right\}.$$

Each C-tuple $(t_1^k, t_2^k \ldots, t_C^k)$ represents a match, t_c^k being the index of the point from the $x^{(c)}$ configuration involved in the match. If a given configuration is not involved in the match, a '$-$' flag is inserted at the appropriate position. For instance, if $C = 3$ the 3-tuple $(2, 4, 1)$ refers to a match between $x_2^{(1)}$, $x_4^{(2)}$ and $x_1^{(3)}$, while $(-, 2, 1)$ is a match between $x_2^{(2)}$ and $x_1^{(3)}$, with no $x^{(1)}$ – point involved. We also include unmatched points in this list: $(1, -, -)$ indicates that $x_1^{(1)}$ is unmatched, for example.

Suppose that $\mathcal{M}$ is the current list of matches in the MCMC algorithm. We define a jump proposal as follows:

- with probability q we choose to *split* a C-tuple; in this case we draw an element uniformly at random in the list $\mathcal{M}$.
  - If the C-tuple drawn corresponds to an unmatched point, we do nothing;
  - otherwise we split it into two C-tuples at random; for instance $(2, 3, 1)$ can be split into $(2, -, -)$ and $(-, 3, 1)$.
- With probability $1 - q$ we choose to *merge* two C-tuples; in this case we select two distinct elements uniformly at random from $\mathcal{M}$.
  - If the two C-tuples drawn contain a common configuration, e.g. $(j_1, k, -)$ and $(j_2, -, -)$, then we do nothing;
  - otherwise we merge the C-tuples, for example $(j, k, -)$ and $(-, -, l)$ become (j, k, l), while $(-, k, -)$ and $(-, -, l)$ become $(-, k, l)$.

This defines a Metropolis – Hastings jump, and its acceptance probability can be readily worked out from (2.7).

B.3 Other data sources

- PDB databank: This is a web resource for protein structure data (`http://www.rcsb.org/pdb/`). The RCSB PDB also provides a variety of tools and resources for studying structures of biological macromolecules and their relationship to sequence, function and disease.

  - PDBsum is a value added tool for understanding protein structure. Another important tool is SWISS-MODEL server which uses *HMM* for homology modelling to identify structural homologs of a protein sequence.
  - There are also tools for displaying three-dimensional structures, e.g. RasMol, JMol, KiNG, WebMol, MBT SimpleView, MBT Protein Workshop and QuickPDB (JMol is a new version of RasMol but is JAVA based).

- SitesBase: Resource for data on active sites. The database holds precompiled information about structural similarities between known ligand binding sites found in the Protein Data Bank. These similarities can be analysed in molecular recognition applications and protein structure – function relationships. `http://www.modelling.leeds.ac.uk/sb`.
- NPACI/NBCR resource: A database and tools for three-dimensional protein structure comparison and alignment using the Combinatorial Extension (CE) algorithm. `http://cl.sdsc.edu/ce.html`.
- The Dali database: This database is based on exhaustive, all-against-all three-dimensional structure comparison of protein structures in the Protein Data Bank. Alignments are automatically maintained and regularly updated using the Dali search engine. `http://ekhidna.biocenter.helsinki.fi/dali_server/`
- SCOP: Aims to provide a detailed and comprehensive description of the structural and evolutionary relationships between all proteins whose structure is known. As such, it provides a broad survey of all known protein folds, detailed information about the close relatives of any particular protein, and a framework for future research and classification. `http://scop.mrc-lmb.cam.ac.uk/scop/`
- CATH: Describes the gross orientation of secondary structures, independent of connectivities and is assigned manually. The topology level clusters structures into fold groups according to their topological connections and numbers of secondary structures. The homologous superfamilies cluster proteins with highly similar structures and functions. The assignments of structures to fold groups and homologous superfamilies are made by sequence and structure comparisons. `http://www.cathdb.info/`
- BLAST: Basic Local Alignment Search Tool, or BLAST, is an algorithm for comparing primary biological sequence information, such as the amino-acid sequences of different proteins. A BLAST search enables a researcher to compare a query sequence with a library or database of sequences, and identify library sequences that resemble the query sequence above a certain threshold. `http://blast.ncbi.nlm.nih.gov/Blast.cgi`

Acknowledgements

V.B. Nyirongo acknowledges funding from the School of Mathematics, University of Leeds as a visiting research fellow during the period in which part of this chapter was drafted.

References

Abramowitz, M. and Stegun, I.A. (1970). *Handbook of Mathematical Functions.* Dover, New York.

Artymiuk, P.J., Poirrette, A.R., Grindley, H.M., Rice, D.W. and Willett, P. (1994). A graph-theoretic approach to the identification of three-dimensional patterns of amino acid side-chains in protein structures. *Journal of Molecular Biology*, **243**, 327–44.

Berkelaar, M. (1996). lpsolve – simplex-based code for linear and integer programming. `http://www.cs.sunysb.edu/~algorith/implement/lpsolve/implement.shtml`.

Bookstein, F.L. (1986). Size and shape spaces for landmark data in two dimensions. *Statistical Science*, **1**, 181–242.

Boomsma, W., Mardia, K.V., Taylor, C.C., Perkinghoff-Borg, J., Krogh, A. and Hamelryck, T. (2008). A generative, probabilistic model of local protein structure. *Proceedings of the National Academy of Science*, **105**, 8932–8937.

Branden, C. and Tooze, J. (1999). *Introduction to Protein Structure* (2nd edn). Garland Publishing, New York.

Burkard, R.E. and Cela, E. (1999). Linear assignment problems and extensions. *Handbook of Combinatorial Optimization*, Vol. 4, In (ed. P. Pardalos and D.-Z. Du), pp. 75–149. Kluwer Academic Press, Boston.

Coats, E.A. (1998). The CoMFA steroid database as a benchmark dataset for development of 3D QSAR methods. *Perspectives in Drug Discovery and Design*, **12–14**, 199–213.

Davies, J.R., Jackson, R.M., Mardia, K.V. and Taylor, C.C. (2007). The Poisson index: A new probabilistic model for protein-ligand binding site similarity. *Bioinformatics*, **23**, 3001–3008.

Dryden, I.L. (2007). Discussion to Schmidler (2007), pp. 17–18. Listed here.

Dryden, I.L., Hirst, J.D. and Melville, J.L. (2007). Statistical analysis of unlabelled point sets: comparing molecules in chemoinformatics. *Biometrics*, **63**, 237–251.

Dryden, I.L., and Mardia, K.V. (1998). *Statistical Shape Analysis.* John Wiley, Chichester.

Frellsen, J., Moltke, I., Thiim, M., Mardia, K.V., Ferkinghoff-Borg, J. and Hamelryck, T. (2009). A probabilistic model of RNA conformational space. *PLoS Computational Biology*, **5**, e1000406. Published online 2009 June 19. DOI: 10.1371/journal.pcbi.1000406.

Gilks, W. (2004). Bioinformatics: new science – new statistics. *Significance*, **1**, 7–9.

Gold, N.D. (2003). Computational approaches to similarity searching in a functional site database for protein function prediction. Ph.D thesis, Leeds University, School of Biochemistry and Microbiology.

Green, P.J. (2003). Diversities of gifts, but the same spirit. *The Statistician*, **52**, 423–438.

Green, P.J. and Mardia, K.V. (2006). Bayesian alignment using hierarchical models, with applications in protein bioinformatics. *Biometrika*, **93**, 235–254.

Hamelryck, T., Kent, J.T. and Krogh, A. (2006). Sampling realistic protein conformations using local structural bias. *Computational Biology*, **2**, 1121–1133.

Holm, L. and Sander, C. (1993). Protein structure comparison by alignment of distance matrices. *Journal of Molecular Biology*, **233**, 123–138.

Horgan, G.W., Creasey, A. and Fenton, B. (1992). Superimposing two dimensional gels to study genetic variation in malaria parasites. *Electrophoresis*, **13**, 871–875.

Jonker, R. and Volgenant, A.A. (1987). Shortest augmenting path algorithm for dense and spare-linear assignment problems. *Computing*, **38**, 325–340.

Kendall, D.G. (1984). Shape manifolds, Procrustean metrics and complex projective shapes. *Bulletin of London Mathematical Society*, **16**, 81–121.

Kent, J.T. (1994). The complex Bingham distribution and shape analysis. *Journal of the Royal Statistical Society, Series B*, **56**, 285–299.

Kent, J.T., Mardia, K.V. and Taylor, C.C. (2004). Matching problems for unlabelled configurations. In *Bioinformatics, Images, and Wavelets*, (ed. R. Aykroyd, S. Barber and K. Mardia), pp. 33–36. Leeds University Press, Leeds.

Le, H.L. (1988). Shape theory in flat and curved spaces, and shape densities with uniform generators. Ph.D. thesis, University of Cambridge, Cambridge.

Leach, A.R. and Gillet, V.J. (2003). *An Introduction to Chemoinformatics*. Kluwer Academic Press, London.

Lesk, A.M. (2000). *Introduction to Protein Architecture*. Oxford University Press, Oxford.

Mardia, K.V. (2005). A vision of statistical bioinformatics. In LASR2005 Proceedings, (ed. S. Barber, P.D. Baxter, K.V. Mardia and R.E. Walls), pp. 9–20. Leeds University Press, Leeds.

Mardia, K.V. (2007*a*). Discussion to Schmidler (2007), p. 18. Listed here.

Mardia, K.V. (2007*b*). On some recent advancements in applied shape analysis and directional statistics. In *Systems Biology & Statistical Bioinformatics*, (ed. S. Barber, P.D. Baxter and K.V. Mardia), pp. 9–17. Leeds University Press, Leeds.

Mardia, K.V. (2008). Holistic statistics and contemporary life sciences. In LASR Proceedings, *The Art and Science of Statistical Bioinformatics*, (ed. S.Barker, P.D. Baxter, A.Gusnanto and K.V. Mardia), pp. 19–17. Leeds University Press, Leeds.

Mardia, K.V. and Gilks, W. (2005). Meeting the statistical needs of 21st-century science. *Significance*, **2**, 162–165.

Mardia, K.V., Hughes, G., Taylor, C.C. and Singh, H. (2008). Multivariate von mises distribution with applications to bioinformatics. *Canadian Journal of Statistics*, **36**, 99–109.

Mardia, K.V. and Nyirongo, V.B. (2008). Simulating virtual protein C_α traces with applications. *Journal of Computational Biology*, **15**, 1221–1236.

Mardia, K.V. and Jupp, P.E. (2000). *Directional Statistics*. John Wiley, Chichester.

Mardia, K.V., Nyirongo, V.B., Green, P.J., Gold, N.D. and Westhead, D.R. (2007*a*). Bayesian refinement of protein functional site matching. *BMC Bioinformatics*, **8**, 257.

Mardia, K.V., Taylor, C.C. and Subramaniam, G.K. (2007*b*). Protein bioinformatics and mixtures of bivariate von Mises distributions for angular data. *Biometrics*, **63**, 505–512.

Marín, J.M. and Nieto, C. (2008). Spatial matching of multiple configurations of points with a bioinformatics application. *Communications in Statistics – Theory and Methods*, **37**, 1977–1995.

Ruffieux, Y. and Green, P.J. (2009). Alignment of multiple configurations using hierarchical models. *Journal of Computational and Graphical Statistics*, **18**, 756–773.

Schmidler, S.C. (2007). Fast Bayesian shape matching using geometric algorithms. In *Bayesian Statistics 8* (ed. J. Bernardo, M. Bayarri, J. Berger, A. Dawid, D. Heckerman, A. Smith and W. M.), pp. 1–20. Oxford University Press, Leeds.

Wilkinson, D.J. (2007*a*). Discussion to Schmidler (2007), pp. 13–17.

Wilkinson, D.J. (2007*b*). Bayesian methods in bioinformatics and computational systems biology. *Briefings in Bioinformatics*, **8**, 109–116.

·3·

Bayesian approaches to aspects of the Vioxx trials: Non-ignorable dropout and sequential meta-analysis

Jerry Cheng and David Madigan

3.1 Introduction

Vioxx is a COX-2 selective, non-steroidal anti-inflammatory drug (NSAID). The FDA approved Vioxx in May 1999 for the relief of the signs and symptoms of osteoarthritis, the management of acute pain in adults, and for the treatment of menstrual symptoms. The COX-2 class of drugs offered the hope of lower rates of gastrointestinal adverse effects as compared with standard NSAIDs like naproxen and ibuprofen. Instead, studies would eventually show that Vioxx causes an array of cardiovascular side-effects such as myocardial infarction, stroke, and unstable angina, leading to the withdrawal of Vioxx from the market.

The potential for COX-2 inhibitors to cause cardiovascular adverse events was established early in their development. For example, in December 1997, the 'International Consensus Meeting on the Mode of Action of COX-2 Inhibition' considered adverse events. The executive summary stated, 'although gastric intolerance was the most obvious [adverse event], renal and cardiac toxicity were important and a probable link between these two phenomena had been established. It is therefore important to monitor cardiac side effects with selective COX-2 inhibitors.' By early 1998, Merck scientists were also aware of potential cardiotoxicity. Consequently, in 1998 Merck established and thereafter continued to use a standard operating procedure (SOP) that systematically monitored and adjudicated a wide array of cardiovascular events.

Several plausible mechanisms exist by which Vioxx could cause a variety of cardiovascular thrombotic (CVT) adverse events. Some mechanisms operate on a short time scale while others may concern permanent damage that could lead to adverse events after Vioxx treatment has concluded (Antman *et al.*, 2007). Hence a complete characterization of Vioxx's cardiovascular risk profile should include an array of thrombotic events, on- and off-drug.

Table 3.1 A subset of the placebo-controlled Vioxx trials. The 'RA' trials enrolled patients with rheumatoid arthritis. The 'OA' trials enrolled patients with osteoarthritis. For both RA and OA, Vioxx was used as a pain killer. The 'ALZ' patients enrolled patients with cognitive impairment in the hope that Vioxx would slow progression to Alzheimer's disease. The APPROVe study led to the withdrawal of Vioxx and enrolled patients with colon polyps. 'LPO' is the date when the last patient concluded the trial. 'PYR' is patient years at risk. Events are investigator reported cardiovascular thrombotic events.

Study block	Study No.	LPO	Vioxx		Placebo	
			Events	PYR	Events	PYR
RA	017	5/21/1997	1	8	0	7
	068	9/10/1998	1	49	0	24
	096	7/21/2000	4	97	0	58
	097	6/6/2000	0	137	0	62
	098	7/6/2000	0	11	1	12
	103	7/6/2000	0	44	0	45
	All RA		6	345	1	208
OA	010	2/8/1996	2	16	0	7
	029	2/5/1997	3	46	0	16
	033	11/18/1997	1	66	1	9
	040	1/1/1998	2	72	0	11
	044	2/18/1998	3	154	0	52
	045	2/18/1998	2	157	3	61
	058	4/1/1998	1	21	0	6
	083	2/9/2000	0	21	0	21
	085	3/3/1999	1	61	0	28
	090	5/17/1999	5	56	0	27
	112	9/8/2000	0	104	0	15
	116	6/22/2000	1	54	0	15
	136	2/5/2002	1	95	1	201
	219	11/28/2003	0	18	1	8
	220	11/24/2003	0	18	0	8
	All OA		22	959	6	485
OA/RA	All OA/RA		28	1304	7	692
ALZ	078	4/23/2003	75	1623	52	1762
	091	11/30/2000	13	369	14	381
	126	5/30/2001	11	193	7	197
	All ALZ		99	2185	73	2341
APPROVe	122	9/8/2004	75	5700	49	5828

Except for a collection of smaller and shorter trials, and two larger trials that were terminated early when Vioxx was withdrawn from the market, Table 3.1 provides a listing of the placebo-controlled trials that Merck conducted. The earliest of these trials concluded in 1996 while the latest concluded in 2004. Various publications reported the results of some of these trials individually and several meta-analysis of subsets of the trials exist (e.g. Konstam *et al.*, 2001,

Reicin *et al.*, 2002, and Weir *et al.*, 2003). However, as the trials progressed, no 'big picture' analyses were performed. We contend that large-scale drug development programmes such as that associated with Vioxx, should monitor safety on a sequential basis as data accumulate, taking into account *all* available data. Section 3.2 below provides a somewhat simplistic Bayesian approach that attempts to provide such an analysis.

Many of the Vioxx trials were longitudinal in nature, and followed participants over time to collect treatment-related measurements. Participants in such trials can deviate from the prescribed protocol in a variety of ways, either by ceasing to be evaluated (drop-out) or by failing to comply with their assigned treatments (non-compliance). Below we consider drop-out specifically. Participants that drop out provide no subsequent measurements thereby creating a missing data problem. The statistical literature provides a wide variety of methods for handling missing data – see, for example, Little and Rubin (2002) and Schafer (1997). Common simple approaches such as complete case analysis or last observation carried forward can lead to biased estimates and an underestimation of uncertainty. Model-based approaches, like those we pursue below, can present significant statistical and computational challenges, but can also provide more useful inferences. The general approach that we adopt closely follows that of Carpenter *et al.* (2002).

The outline of the paper is as follows. Section 3.2 describes an approach to sequential meta analysis in the context of Vioxx. Section 3.3 considers dropouts in the key APPROVe study and presents a Bayesian approach to handling dropout. We conclude with some discussion.

3.2 Sequential meta-analysis

The Vioxx development program included dozens of clinical trials, some focused on arthritis pain and others studying potential new indications for Vioxx. Some clinical trials compared Vioxx with other NSAIDS but, to simplify the interpretation we mostly consider placebo-controlled trials. Our general approach follows Spiegelhalter *et al.* (1994) in two key regards:

1. Rather than focusing on a single prior, we present analyses with a 'family of priors'; and
2. We adopt a simple Gaussian data model, thereby circumventing within-trial analytical issues.

We couple these with a simple graphical device to illustrate the accumulating evidence over time.

Following Spiegelhalter *et al.*, we propose a 'skeptical prior' that represents a priori skepticism that Vioxx causes cardiovascular thrombotic (CVT) adverse events.[1] We also propose a 'cautious prior' that represents a prior belief that cardiotoxicity is not so implausible. We define two thresholds to help specify these priors. The upper threshold, δ_U, is the hazard ratio above which most reasonable people would agree that Vioxx should not be on the market. We consulted with several cardiologists and settled on $\delta_U = 1.75$. This reflects the belief that anything close to a doubling of CVT risk would certainly be unacceptable. The lower threshold, δ_L, is roughly the value below which reasonable people would agree that it might be appropriate to have Vioxx on the market, and where informed patients could consider the risks and benefits on an individual basis. We chose $\delta_L = 1.1$. In between these two values, reasonable people could disagree.

We then constructed Gaussian prior distributions around these thresholds. Specifically, the skeptical prior is centered on zero with just a 5% prior probability that the true hazard ratio δ exceeds δ_U (i.e. skeptical that Vioxx is dangerous). The cautious prior is centred on δ_U with a 5% probability that δ is less than δ_L. This is cautious insofar as it reflects a prior 50% probability that the true Vioxx hazard ratio exceeds δ_U. For studies with at least 10 events per study arm, we estimated the hazard ratio using a Cox model with treatment (Vioxx or placebo) as the sole covariate. For studies with fewer events we used a simple binomial-based method (Rothman and Greenland, 1998, p. 249). We then assumed a Gaussian likelihood with mean given by the estimated hazard ratio and standard deviation equal to the standard error of the estimate.

Figure 3.1 shows the results for the pooled osteoarthritis (OA) and rheumatoid arthritis (RA) studies. Under the skeptical prior, the prior probability that the true hazard ratio exceeds δ_U is 0.05 whereas the posterior probability is 0.21. Under the cautious prior, the prior probability that the true hazard ratio exceeds δ_U is 0.5 whereas the posterior probability is 0.64. So, the OA and RA trials with modest numbers of patient years at risk, suggest a non-trivial posterior probability that $\delta > \delta_U$ under either prior. Under either prior, the posterior probability that $\delta > \delta_L$ exceeds 90%.

Figure 3.2 shows the results for the APPROVe trial. The results of the placebo-controlled APPROVe trial became available in September 2004 and led to the withdrawal of Vioxx. Under the skeptical prior, the prior probability that the true hazard ratio exceeds δ_U is again 0.05 whereas the posterior probability is 0.09. Under the cautious prior, the prior probability that the true hazard ratio exceeds δ_U is again 0.5 whereas the posterior probability is 0.29. So,

[1] The specific events are acute myocardial infarction, unstable angina pectoris, sudden and/or unexplained death, resuscitated cardiac arrest, cardiac thrombus, pulmonary embolism, peripheral arterial thrombosis, peripheral venous thrombosis, ischemic cerebrovascular stroke, stroke (unknown mechanism), cerebrovascular venous thrombosis, and transient ischemic attack.

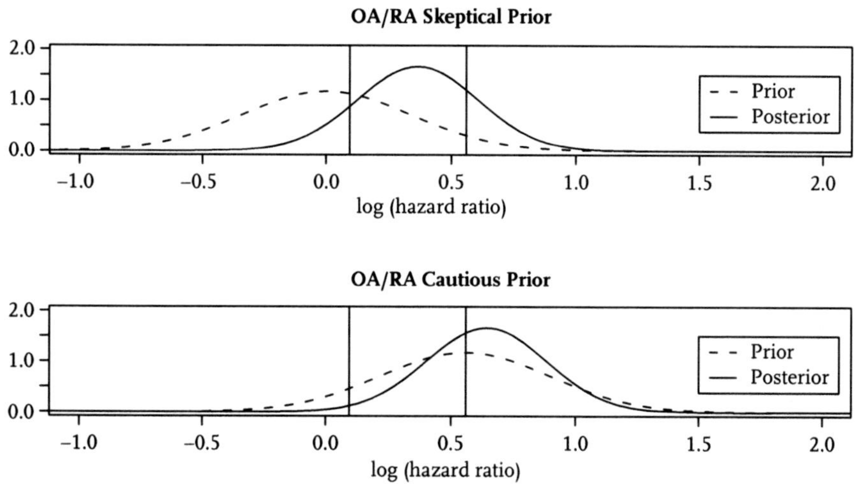

Fig. 3.1 Prior and posterior densities for the log hazard ratio of CVT events in the pooled OA and RA studies.

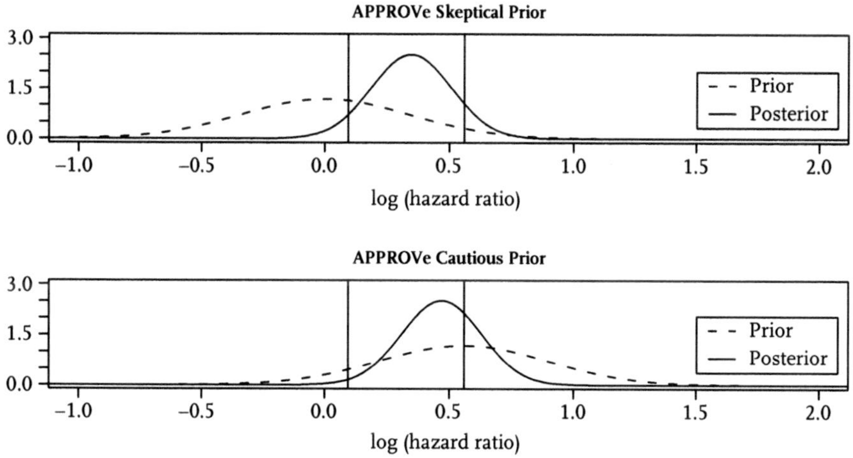

Fig. 3.2 Prior and posterior densities for the log hazard ratio of CVT events in APPROVe.

the APPROVe trial provides weaker evidence that δ exceeds δ_U although the posterior probability that $\delta > \delta_L$ under either prior exceeds 90%.

Figure 3.3 shows the results for the VIGOR trial. VIGOR was an 8000-person year-long study in rheumatoid arthritis patients, comparing Vioxx with Naproxen (Aleve), a widely-used NSAID. In this case, even the skeptical prior leads to an 80% posterior probability that $\delta > \delta_U$.

Figure 3.4 shows a sequential meta-analysis of all placebo-controlled trials. The figure shows the prior and posterior probability that $\delta > \delta_L$ under the skeptical and cautious priors as well as three further priors:

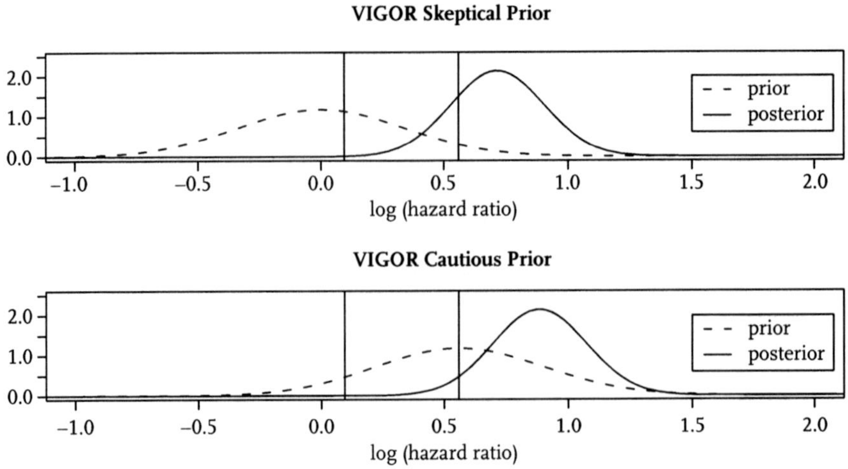

Fig. 3.3 Prior and posterior densities for the log hazard ratio of CVT events in VIGOR.

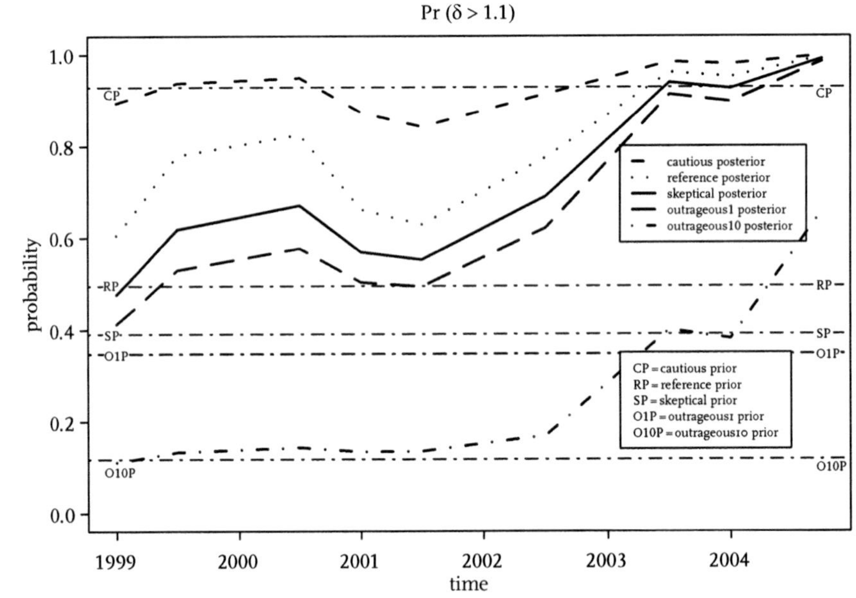

Fig. 3.4 Sequential prior and posterior $\Pr(\delta > \delta_L)$ for all Vioxx placebo-controlled trials.

- *Reference prior.* Gaussian with mean equal to 1 and variance 10^6.
- *Outrageous10 prior.* Gaussian with mean equal to 1 and a 10% probability that $\delta > \delta_L$.
- *Outrageous1 prior.* Gaussian with mean equal to 1 and a 1% probability that $\delta > \delta_U$.

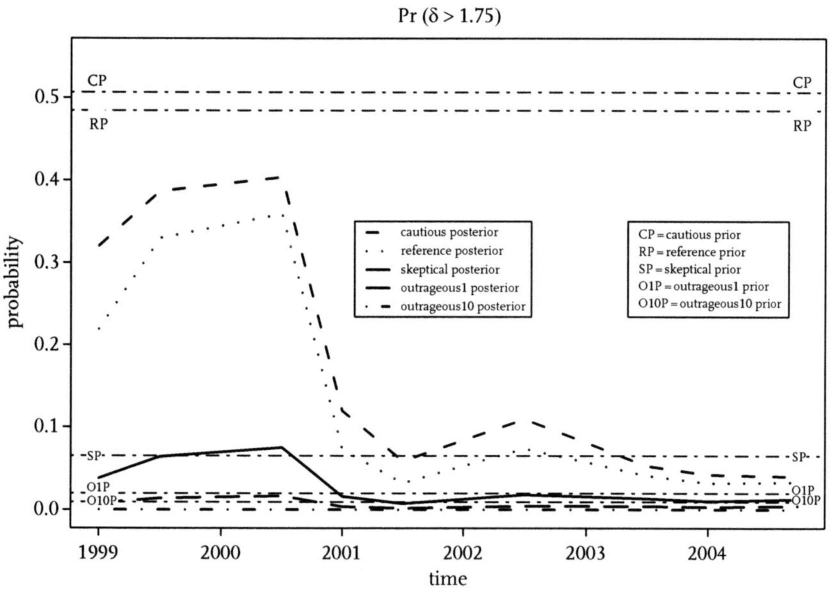

Fig. 3.5 Sequential prior and posterior $\Pr(\delta > \delta_U)$ for all Vioxx placebo-controlled trials.

We computed the posterior probabilities every six months starting January 1999 and concluding in the middle of 2004.

In the middle of 2000, some four years before Vioxx was withdrawn from the market, the posterior probability that the hazard ratio was greater than δ_L exceeded 50% under all priors except the 'outrageous10' prior. In our view, the two outrageous priors represent unreasonable prior beliefs in light of the available evidence at the outset. A striking feature of this analysis is the convergence towards $\Pr(\delta > \delta_L) = 1$ under four of the five priors.

Figure 3.5 shows the posterior probability that $\delta > \delta_U$ over time.

Clearly, in this analysis, the data from the placebo-controlled trials provide minimal support for a value of δ as large as 1.75. Again, while the different priors provide different posterior estimates at the outset, they have largely converged by 2004.

This analysis has a number of weaknesses. We excluded some large trials such as VIGOR that failed to include a placebo arm. The Gaussian data model greatly simplifies the analysis and Spiegelhalter *et al.* argue that in many situations the approximation is justified. Nonetheless, a more sophisticated approach that accounts for study-specific data models would be more satisfactory. Our analysis implicitly adopts a fixed effects approach. A random effects approach would allow the true hazard ratio to vary across studies (and possibly across time) and may be more reasonable. Beyond clinical trials, a number of other data sources contain potentially valuable information about the safety of Vioxx. For example, the US FDA's adverse event reporting system receives hundreds of thousands of drug adverse event reports each year. Medical claims databases, electronic medical record systems, and government databases also could prove useful. In principle these could also be incorporated into the meta-analysis.

Currently, safety analysis in the drug development process proceeds in a somewhat ad hoc fashion with analyses of individual trials and meta-analyses of selected groups of trials. The Bayesian paradigm provides a natural approach to monitoring safety over the entire life a drug, incorporating multiple data sources.

3.3 Non-ignorable dropout

The APPROVe (Adenomatous Polyp Prevention on Vioxx) study was a multi-centre, randomized, placebo-controlled, double-blind study to determine the effect of 156 weeks of treatment with Vioxx (25 mg daily) on the recurrence of neoplastic polyps of the large bowel. The trial enrolled 2600 patients patients with a history of colorectal adenoma. We exclude 13 patients that inadvertently took the wrong dose of Vioxx. Of the remaining 2587 patients 728 dropped out of the study. Of these patients, 410 had been randomly assigned to Vioxx whereas 318 were assigned to placebo.

Upon enrollment, the trial gathered an array of demographic and medical data from each patient. Patients then paid office visits periodically, though not at rigidly fixed intervals. Those visits captured vital sign measurements, laboratory testing results, as well as information on potential adverse events. In what follows, and to highlight particular methodological nuances, we focus on two particular safety end-points, systolic blood pressure and confirmed thrombotic cardiovascular adverse events.

The possibility that dropout could be causally related to either or both of our safety end-points seems well founded. Increased blood pressure might well lead a subject to drop out of the trial, as could the precursors of a serious thrombotic event. If, as is now believed, Vioxx causes these events, the Vioxx arm of the trial could be deprived of the very patients most likely to suffer these adverse events. All published analyses of the APPROVe trial (as well as the analysis in Section 3.2 above) assume dropout is unrelated to the safety endpoints of interest (i.e. ignorable) and hence could plausibly underestimate the effects of interest.

3.3.1 Notation

We consider a randomized longitudinal clinical trial including N patients over J time periods. For subject i at time period j, let $y_{i,j}$ denote the primary response, $i = 1, \ldots, N$, $j = 1, \ldots, J$. In what follows, $y_{i,j}$ can be real valued, $y_{i,j} \in \Re$ (e.g.

systolic blood pressure), or binary $y_{i,j} \in \{0, 1\}$ (e.g. confirmed cardiovascular adverse events). Let Y denote the $N \times J$ matrix of primary responses for all the subjects at all time periods and let $y_i = Y_{1:J,i}$ denote column i of Y.

We assume that there are C covariates such as age, sex, diabetes status, treatment type, etc. Let X denote the $C \times N$ covariate matrix for all subjects. $x_i = X_{1:C,i}$ denotes the covariates for subject i.

Let M denote the $N \times J$ matrix for the dropout status for all the subjects at all time periods. The (i, j) element of M takes a value of one when the subject i drops out at time period j, and zero otherwise. We will only consider monotone missingness in this chapter – once a patient drops out they cannot return to the trial. So, the column vector $m_i = M_{1:J,i}$ comprises a set of 0's possibly followed by a set of 1's. This vector represents the dropout status for subject i for period 1 through period J. We use the terms missingness and dropout interchangeably.

For simplicity in notation, we drop the index i and j when there is no confusion, and use only one covariate X though there might be several in practice. The primary response variable is split according to whether its value is recorded or missing with a shorthand notation: $y = \{y^{\text{obs}}, y^{\text{miss}}\}$. Equations (3.1) and (3.2) describe the relationship between Y and M for subject i:

$$y_i^{\text{obs}} = \{y_{i,j} : M_{i,j} = 0, \quad j \in \{i, 2, \ldots, J\}\} \tag{3.1}$$

$$y_i^{\text{miss}} = \{y_{i,j} : M_{i,j} = 1, \quad j \in \{1, 2, \ldots, J\}\} \tag{3.2}$$

Note that $|y_i^{\text{obs}}| + |y_i^{\text{miss}}| = J$. $(y^{\text{obs}}, y^{\text{miss}}, M, X)$ represents the 'complete data'.

3.3.2 Joint model for response and missingness

We start with the full likelihood function of the complete response y and M given covariates X. Let Ω denote the parameter space for the joint probability model. We factor Ω into subspaces that dissect the full likelihood function. Let $\Omega_R \subset \Omega$ relate the response Y to the covariates X, $\Omega_{MAR} \subset \Omega$ relate M to Y_{obs} and X, and $\Omega_{NIM} \subset \Omega$ relate M to Y^{miss},

$$\Omega = \Omega_R \cup \Omega_{MAR} \cup \Omega_{NIM}. \tag{3.3}$$

We consider parameter vectors $\beta \in \Omega_R$, $\alpha \in \Omega_{MAR}$, and $\delta \in \Omega_{NIM}$ where β is of primary interest.

The *complete data likelihood function* is then:

$$p(y, M|X, \alpha, \beta, \delta) = p(y^{\text{obs}}, y^{\text{miss}}, M|X, \alpha, \beta, \delta). \tag{3.4}$$

Since we consider that subjects in the trial are independent, the joint likelihood in equation (3.4) factors as:

$$p(y^{obs}, y^{miss}, M|X, \alpha, \beta, \delta) = \prod_{1\leq i\leq N} p\left(y_i^{obs}, y_i^{miss}, m_i|x_i, \alpha, \beta, \delta\right). \tag{3.5}$$

We can further factorize the ith component in equation (3.5) into the marginal model of y_i given x_i and the conditional model of m_i given y_i and x_i:

$$p\left(y_i^{obs}, y_i^{miss}, m_i|x_i, \alpha, \beta, \delta\right) = p\left(m_i|y_i^{obs}, y_i^{miss}, x_i, \alpha, \delta\right) p\left(y_i^{obs}, y_i^{miss}|x_i, \beta\right). \tag{3.6}$$

Equation (3.6) represents a so-called *selection model*, in which dropout depends on both observed and unobserved response variables as well as covariates. The observed data likelihood is given by:

$$p(y^{obs}, M|X, \alpha, \beta, \delta) = \int p(y^{obs}, y^{miss}, M|X, \alpha, \beta, \delta)dy^{miss}. \tag{3.7}$$

Coupled with the selection model (3.6), this becomes:

$$p(y^{obs}, M|X, \alpha, \beta, \delta) = \prod_{1\leq i\leq N} \int p\left(m_i|y_i^{obs}, y_i^{miss}, x_i, \alpha, \delta\right) p\left(y_i^{obs}, y_i^{miss}|x_i, \beta\right) dy_i^{miss}. \tag{3.8}$$

Before we discuss inference using (3.8), we first consider different standard simplifications of $p\left(m_i|y_i^{obs}, y_i^{miss}, x_i, \alpha, \delta\right)$. Little and Rubin (1987) considered three general categories for the *missing data mechanism*:

1. *Missing completely at random (MCAR)* where missingness does not depend on the response, either y^{obs} or y^{miss}:
$$p(M|y^{obs}, y^{miss}, X, \alpha, \delta) = p(M|X, \alpha) \tag{3.9}$$
2. *Missing at random (MAR)* where missingness does not depend on the unobserved response Y_{miss}:
$$p(M|y^{obs}, y^{miss}, X, \alpha, \delta) = p(M|y^{obs}, X, \alpha) \tag{3.10}$$
3. *Non-ignorable (NIM)* where missingness depends on y^{miss} and possibly on y^{obs} and covariates.

In the *NIM* scenario, the parameter δ plays a key role, defining, as it does, the relationship between dropout and missing values. Unfortunately, the data provide little or no information about δ, and checking the sensitivity of the ultimate inferences to different choices of δ is central to our approach. In APPROVe, as we discussed above, an assumption of *MAR* or *MCAR* may not be tenable.

3.3.2.1 *Inference under MCAR and MAR*

Under the assumptions of *MCAR* or *MAR*, we can simplify the observed likelihood for ith subject as the product of the likelihood of the observed response and that of the missing data mechanism:

$$p\left(y_i^{\text{obs}}, m_i | x_i, \alpha, \beta, \delta\right) = p(m_i | x_i, \alpha) p\left(y_i^{\text{obs}} | x_i, \beta\right). \tag{3.11}$$

If $\Omega_R \cap \Omega_{MAR} = 0$ and $\Omega_R \cap \Omega_{NIM} = 0$, in other words β is distinct from α and δ, inference about β depends only on the $p\left(y_i^{\text{obs}} | x_i, \beta\right)$. We refer the reader to the appendix for further details.

3.3.2.2 *Inference under NIM*

Under *NIM* no simplification of the likelihood is possible and the integration in (3.8) cannot be accomplished in closed form. Our Bayesian approach specifies prior distributions for the various parameters and uses MCMC to draw posterior inferences. Since the *NIM* model involves the non-ignorable drop-out parameter δ, its behaviour cannot be accessed directly from the data at hand. In a sensitivity study, and following Carpenter *et al.* (2002) we vary the value of δ and hence assess sensitivity to departures from ignorability.

3.3.3 A simulation study

Before turning to the APPROVe data, we conduct a small simulation study to shed some light on the robustness of the inferences about the primary parameters of interest to mismatches between the true value of δ and the assumed value of δ in a sensitivity analysis.

We set the number of subjects N to 800. For subject i, we generate data as follows:

- Step 1: With equal probability, generate covariate x_i to take value of 1 or 0, where 1 indicates the assignment of a hypothetical drug and 0 placebo.
- Step 2: Generate the response variable y_i via a simple regression model:

$$y_i = \beta x_i + \epsilon_i \tag{3.12}$$

  where $\beta = 2$, $\epsilon_i \sim N(0, \sigma^2)$ and $\sigma = 1$, $i = 1, \ldots, 800$.
- Step 3: Generate the missing data indicator m_i via a logistic regression model:

$$logit(p(m_i = 1)) = \alpha_0 + \alpha_1 x_i + \delta y_i \tag{3.13}$$

  where $\alpha_0 = 1$, $\alpha_1 = 1$, δ varies from -1 to 1 with increment of 0.2.
- Step 4: With the missing data indicator m_i controlled by δ from the previous step, reset the response to be missing ($y_i = NA$) when $m_i = 1$.

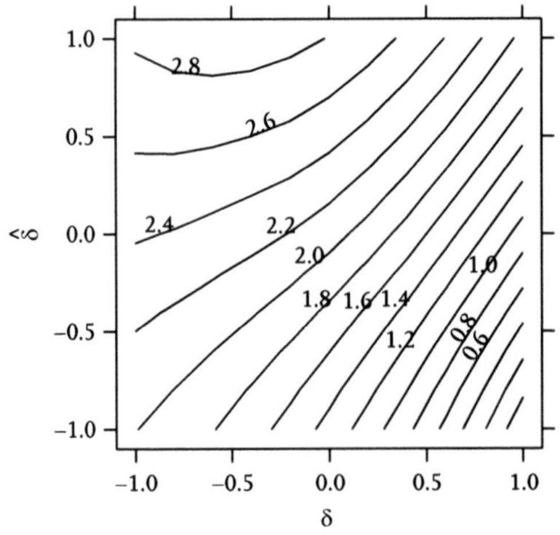

Fig. 3.6 Smoothed contour plot of the estimated treatment effect. The horizontal axis is δ and the vertical axis is $\hat{\delta}$.

For each preset value of δ, we have generated a set of data: (y, x, m). We fit the data jointly with models (3.12) and (3.13) to estimate β, α_0 and α_1. In the dropout model (3.13), we vary the assumed non-ignorable parameter $\hat{\delta}$ from -1 to 1 with an increment of 0.2.

With each combination of $(\delta,\hat{\delta})$ and diffuse priors, we draw 10,000 MCMC iterations discarding the first 200 to obtain the Bayesian estimate (posterior mean) of β – the 'treatment effect' ($\hat{\beta}$ hereafter). The MCMC convergence appears satisfactory. After smoothing via a two-dimensional loess algorithm, Figure 3.6 shows the the contours of $\hat{\beta}$ as a function of $(\delta, \hat{\delta})$.

The contour plot shows a striking pattern: the upper left corner overestimates the true value of 2 while the lower right corner underestimates the true value. Along the diagonal, where $\hat{\delta}$ is close to the true δ, the estimates are close to the true value of 2.

The treatment effect β represents the expected change of the response between the trial drug and the placebo. When the underlying non-ignorable dropout parameter δ is positive, higher valued responses are more likely to be missing. If we do not acknowledge this and estimate the treatment effect only using the observed data, $\hat{\beta}$ will be smaller, i.e. biased downward. If we misjudge the dropout mechanism and use negative values of $\hat{\delta}$ in the modeling, we will impute responses with smaller values. As a result, we will underestimate the treatment effect – this is what happens in the lower right corner of the plot. Similarly, we will overestimate the treatment effect when the true δ is negative but we assume a positive $\hat{\delta}$ as can be seen in the upper left corner of the plot.

From the plot, we can also observe that when the missing mechanism is indeed ignorable (i.e. $\delta = 0$), we will get biased estimates if we set $\hat{\delta}$ to be either positive or negative. Meanwhile when the missing data mechanism is truly *NIM*, the choice of $\hat{\delta}$ dictates the bias of the estimate.

3.3.4 Analysis of the Vioxx data

We divide the three year trial span into twelve 90-day periods. Since most patients did not re-enter the study once they dropped out, we assume a monotone missingness pattern. We set the values for M as follows:

- For systolic blood pressure: find the last period that has an observed blood pressure for a subject. Then the remaining periods have $m = 1$ (missing) and the earlier periods have $m = 0$.
- For confirmed cardiovascular events: set $m = 1$ (missing) for periods after all patient contact ceases and set $m = 0$ for other periods.

The response y is defined as follows:

- For systolic blood pressure: in the quarters with $m = 0$, y is the average systolic blood measurement over the period (most periods have only one measurement).
- For confirmed cardiovascular events: in the quarters with $m = 0$, set $y = 1$ if there is at least one confirmed event; set $y = 0$ otherwise.

We include two covariates namely treatment type (Tr) and an indicator of high cardiovascular disease risk (HC). Both covariates are binary: Tr=1 for assignment of vioxx, 0 for placebo; $HC = 1$ if the subject has high cardiovascular disease risk, 0 otherwise.

3.3.4.1 *Non-longitudinal analysis*

We begin with a non-longitudinal analysis focusing on the treatment effects at the end of the trial with the selection model (3.6). For subject i ($i = 1, \ldots, 2584$), the response model is:

- *For systolic blood pressure:*

$$y_i = \beta^0 + \beta^{tr} Tr_i + \beta^h HC_i + \epsilon_i \tag{3.14}$$

  where $\epsilon_i \sim N(0, \sigma^2)$, y_i is the systolic blood pressure measured between 990 and 1080 days – a 90 day window before the end of the third year – if such a measurement is available.

- *For confirmed cardiovascular events:*

$$logit(p(y_i = 1)) = \beta^0 + \beta^{tr} Tr_i + \beta^h HC_i \tag{3.15}$$

  where y_i is the indicator from day 1 to day 1080.

Table 3.2 Bayesian estimate of treatment effects at the end of the third year under various assumed values of the non-ignorable dropout parameter.

Scenario	Systolic blood pressure		Confirmed cardio AE	
	Estimated mean	Estimated SD	Estimated mean	Estimated SD
$\delta = 4$	0.475	0.135	0.670	0.281
$\delta = 0$	0.300	0.124	0.669	0.280
$\delta = -4$	0.213	0.126	0.669	0.280

The dropout model for m is:

$$logit(p(m_i = 1)) = \alpha^0 + \alpha^{tr} Tr_i + \alpha^h HC_i + \delta Y_i. \tag{3.16}$$

As in Section (3.3.2), we call δ in (3.16) the non-ignorable dropout parameter since it relates the dropout probability to the response which might be missing.

As in the simulation study, we use WinBUGS to perform $MCMC$ with 20,000 iterations discarding the first 1000. $MCMC$ convergence appears to be satisfactory. Table 3.2 lists posterior means and standard deviations for the treatment effects for the two endpoints with various candidate values for δ.

From the table, the results under the $MCAR$ ($\delta = 0$) and NIM ($\delta \neq 0$) are quite different for systolic blood pressure. This indicates that the estimated treatment effect is sensitive to the underlying non-ignorable missing data assumption. Moreover $\hat{\beta^{tr}}$ decreases as δ decreases. Higher values of δ indicate that the unmeasured values are indeed larger.

On the other hand, the results for $\hat{\beta^{tr}}$ for confirmed cardiovascular events remain little changed under different choices of δ, so sensitivity to non-ignorability for this endpoint appears to be modest in scale.

3.3.4.2 Longitudinal study

We expand the single time-point study from the previous section to the longitudinal setting. The total number of periods is 12 and each period has 90 days.

We specify here the two likelihood functions in the selection model (3.6). The response model for patient i ($i = 1, \ldots, 2584$) at period j ($j = 1, \ldots, 12$) is:

- *for systolic blood pressure*:

$$y_{i,j} = \beta_j^0 + \beta_j^{tr} Tr_i + \beta_j^h HC_i + \beta_j^{tm} j + \epsilon_{i,j} \tag{3.17}$$

where $\epsilon_{i,j} \sim N(0, \sigma^2)$, and

- *for confirmed cardiovascular events*:

$$logit(p(y_{i,j} = 1)) = \beta_j^0 + \beta_j^{tr} Tr_i + \beta_j^h HC_i + \beta_j^{tm} j. \tag{3.18}$$

We propose a model for dropout status $m_{i,j}$ that includes the end-point observation from both current period and the previous period. Equation (3.19) shows the model for the first period:

$$logit(p(m_{i,1} = 1)) = \alpha_1^0 + \alpha_1^{tr} Tr_i + \alpha_1^h HC_i + \alpha_1^{tm} + \delta y_{i,1}. \quad (3.19)$$

For the other periods ($j = 2, \ldots, 12$), we add an additional term $y_{i,j-1}$, i.e. the response from the previous period:

$$logit(p(m_{i,j} = 1)) = \alpha_j^0 + \alpha_j^{tr} Tr_i + \alpha_j^h HC_i + \alpha_j^{tm} j + \alpha^{\text{prev}} y_{i,j-1} + \delta y_{i,j}. \quad (3.20)$$

Again δ in (3.19) and (3.20) is the non-ignorable dropout parameter. In (3.20) we make the missing data mechanism an *MCAR* when $\delta = 0$ and $\alpha^{\text{prev}} = 0$, *MAR* when both $\delta = 0$ and α^{prev} are non-zero, and an *NIM* when δ is non-zero.

Following the inference under *NIM* in Section 3.3.2.2, we combine the likelihood functions in Equations (3.17)–(3.20) with diffuse (but proper) prior distributions. For the priors of the same group of parameters for different time periods, we adopt a hierarchical approach under which the parameters come from the same hyper-prior distribution thereby smoothing the values over time. For $j = 1, 2. \ldots 12$,

$$\begin{aligned}
\beta_j^0 &\sim N\left(\mu_\beta^0, \left(\sigma_\beta^0\right)^2\right) \\
\beta_j^{tr} &\sim N\left(\mu_\beta^{tr}, \left(\sigma_\beta^{tr}\right)^2\right) \\
\beta_j^h &\sim N\left(\mu_\beta^h, \left(\sigma_\beta^h\right)^2\right) \\
\beta_j^{tm} &\sim N\left(\mu_\beta^{tm}, \left(\sigma_\beta^{tm}\right)^2\right) \\
\alpha_j^0 &\sim N\left(\mu_\alpha^0, \left(\sigma_\alpha^0\right)^2\right) \\
\alpha_j^{tr} &\sim N\left(\mu_\alpha^{tr}, \left(\sigma_\alpha^{tr}\right)^2\right) \\
\alpha_j^h &\sim N\left(\mu_\alpha^h, \left(\sigma_\alpha^h\right)^2\right) \\
\alpha_j^{tm} &\sim N\left(\mu_\alpha^{tm}, \left(\sigma_\alpha^{tm}\right)^2\right).
\end{aligned}$$

With 200,000 iterations the *MCMC* convergence appears satisfactory. We list the posterior mean $\widehat{\mu_\beta^{tr}}$ and its standard deviation in Table 3.3 for the two end-points for various values of (δ, α^{prev}).

As in the non-longitudinal setting, the effect of Vioxx on blood pressure shows marked sensitivity to assumptions about the missing data mechanism, whereas the effect on CVT events shows less sensitivity.

Table 3.3 Bayesian estimate and its deviations under various values of the non-ignorable dropout parameter.

Scenario	Systolic blood pressure		Confirmed cardio *AE*	
	Estimated mean	Estimated SD	Estimated mean	Estimated SD
$\delta = 0,\ \alpha^{prev} = 0$	0.23	0.05	0.49	0.30
$\delta = 4$	0.32	0.05	0.53	0.22
$\delta = 0$	0.23	0.05	0.52	0.30
$\delta = -4$	0.18	0.06	0.48	0.30

3.4 Conclusion

We have highlighted two issues in drug safety where current statistical practice falls short. First, in large-scale drug development programmes, sequential analyses of *all* available safety data have the potential to uncover problems at an earlier stage of development but apparently such analyses are not performed. Second, all large human clinical trials involve noncompliance and dropout. Standard analyses either circumvent these issues with simplifying assumptions (e.g. intention to treat, see Madigan, 1999) or crude imputation methods such as last observation carried forward. Bayesian approaches to both problems have much to offer.

Appendix

A. Broader context and background

The missing competely at random (*MCAR*) assumption is implicit in many of the widely used approach to missing data such as complete case analysis. *MCAR* is possibly the most restrictive assumption one could make – missingness is unrelated to the missing value or the value of any other variable. *MCAR* (3.9) greatly simplifies the observed likelihood (3.8) for subject i:

$$\begin{aligned}
p\left(y_i^{\mathrm{obs}}, m_i|x_i, \alpha, \beta, \delta\right) &= \int p(m_i|x_i, \alpha)p\left(y_i^{\mathrm{obs}}, y_i^{\mathrm{miss}}|x_i, \beta\right)dy_i^{\mathrm{miss}} \\
&= p(m_i|x_i, \alpha)\int p\left(y_i^{\mathrm{obs}}, y_i^{\mathrm{miss}}|x_i, \beta\right)dy_i^{\mathrm{miss}} \\
&= p(m_i|x_i, \alpha)p\left(y_i^{\mathrm{obs}}|x_i, \beta\right). \qquad (3.21)
\end{aligned}$$

Missing at random (*MAR*) is slightly less restrictive. With *MAR* missingness can depend on the values of other observed variables. Under *MAR* (3.10) the observed likelihood becomes:

$$p\left(y_i^{\text{obs}}, m_i | x_i, \alpha, \beta, \delta\right) = \int p\left(m_i | y_i^{\text{obs}}, x_i, \alpha\right) p\left(y_i^{\text{obs}}, y_i^{\text{miss}} | x_i, \beta\right) dy_i^{\text{miss}}$$
$$= p\left(m_i | y_i^{\text{obs}}, x_i, \alpha\right) p\left(y_i^{\text{obs}} | x_i, \beta\right). \quad (3.22)$$

Therefore the joint observed likelihood of response and missingness can factor into a product of the likelihood of the observed response and that of the missing data mechanism. We represent this partition in a shorthanded notation as:

$$[y^{\text{obs}}, M | X, \alpha, \beta, \delta] = [M | y^{\text{obs}}, X, \alpha][y^{\text{obs}} | X, \beta]. \quad (3.23)$$

Inference about the response parameter β depends only on the likelihood from the response model $[y^{\text{obs}} | X, \beta]$.

Key references on missing data include Little and Rubin (2002) and Schafer (1997).

References

Antman, E.M., Bennett, J.S., Daugherty, A., Furberg, C., Roberts, H., and Taubert, K.A. (2007). Use of nonsteroidal antiinflammatory drugs. *Circulation*, **115**, 1634–1642.

Carpenter, J., Pocock,S., and Lamm, C.J. (2002). Coping with missing data in clinical trials: A model-based approach applied to asthma trials. *Statistics in Medicine*, **21**, 1043–1066.

Demirtas, H. and Schafter, J.L. (2003). On the performance of random-coefficient pattern-mixture models for non-ignorable drop-out. *Statistics in Medicine*, **22**, 2553–2575.

Konstam, M.A., Weir, M.R., Reicin, A., Shapiro, D., Sperling, R.S., Barr, E., and Gertz, B. (2001). Cardiovascular thrombotic events in controlled clinical trials of roefecoxib. *Circulation*, **104**, 2280–2288.

Little, R.J.A. (1994). A class of pattern-mixture models for normal missing data. *Biometrika*, **81**, 471–483.

Little, R.J.A. and Wang, Y. (1996). Pattern-mixture models for multivariate incomplete data with covariates. *Biometrika*, **52**, 98–111.

Little, R. J. A. and Rubin, D. B. (2002). *Statistical Analysis with Missing Data*. John Wiley, New York.

Madigan, D. (1999). Bayesian graphical models, intention-to-treat, and the Rubin causal model. In *Proceedings of the Seventh International Workshop on Artificial Intelligence and Statistics*, (ed. D. Heckeman and J. Whittaker), 123–132. Morgan Kaufmann Publishers, San Francisco, CA.

Reicin, A.S., Shapiro, D., Sperling, R.S., Barr, E., and Yu, Q. (2002). Comparison of cardiovascular thrombotic events in patients with osteoarthritis treated with rofecoxib versus nonselective nonsteroidal anti-inflammatory drugs. *American Journal of Cardiology*, **89**, 204–209.

Rothman, K.J. and Greenland, S. (1998). *Modern Epidemiology* (2nd edn). Lippincott Williams & Wilkins, York, PA, p. 249.

Rubin, D. B. (1976). Inference and missing data. *Biometrika*, **62**, 581–592.

Schafer, J. L. (1997). *Analysis of Incomplete Multivariate Data*. CRC Press/Chapman & Hall, London.

Spiegelhalter, D.J., Freedman, L., and Parmar, M. (1994). Bayesian analysis of randomized trials (with discussion). *Journal of the Royal Statisical Society (Series B)*, **157**, 357–416.

Spiegelhalter, D.J., Thomas, A., Best, N.G., and Gilks, W.R. (1995). *BUGS: Bayesian Inference using Gibbs Sampling, Version 0.50*. MRC Biostatistics Unit, Cambridge.

Weir, M.R., Sperling, R.S., Reicin, A., and Gertz, B.G. (2003). Selective COX-2 inhibition and cardiovascular effects: A review of the rofecoxib development program. *American Heart Journal*, **146**, 591–604.

·4·

Sensitivity analysis in microbial risk assessment: Vero-cytotoxigenic *E. coli* O157 in farm-pasteurized milk

Jeremy E. Oakley and Helen E. Clough

4.1 Introduction

In this chapter we consider the process of microbial risk assessment regarding the possible contamination of farm-pasteurized milk with the bacterium Vero-cytotoxigenic *E. coli* (VTEC) O157. A risk assessment using a mechanistic model has been conducted by Clough *et al.* (2006, 2009a), but is itself subject to uncertainty due to uncertainty in the corresponding mechanistic model inputs or parameters. Here, we use the tool of probabilistic sensitivity analysis to investigate how uncertain model inputs contribute to the uncertainty in the model outputs, with the aim of prioritizing future data collection to best reduce uncertainty in the risk assessment.

In the next two sections we give a brief introduction to microbial risk assessment and discuss the frequency and consequences of food-borne outbreaks associated with VTEC O157. The risk assessment model is reviewed in Section 4.4, and model input distributions are detailed in Section 4.5. We conduct the probabilistic sensitivity analysis in Section 4.6.

4.2 Microbial risk assessment

Microbial risk assessment (MRA) is an important approach to studying the risks from pathogens (organisms potentially harmful to humans) in the food chain, related to particular food production systems. It is common that the food production system of interest cannot be observed directly, and yet policy makers may be interested, for example, in the number of human illnesses or likely level of human exposure resulting from a particular pathogen in a given food product, or the likely effect of a particular intervention on resultant probability and burden of human illness (or infection).

MRA as defined by the CODEX Alimentarius Commission (1999), the most commonly used framework for food safety risk assessment, is a sequence of four stages: hazard identification involves the identification of the biological potentially hazardous agent; hazard characterization is the collation of all available information on the possible effects of the hazard; exposure assessment models exposure of the population under study to the hazard via the particular food production system of interest; and risk characterization is the integration of the previous three steps to produce a resultant estimate of 'risk'. MRA is based upon models, either qualitative or quantitative, which describe the production chain of interest and resultant transit of the organism of concern through the chain, in a mechanistic form.

MRA is commonly implemented via Monte Carlo (MC) simulation, using approaches such as those advocated, amongst others, by Vose (2000). MC approaches have several advantages: they permit the incorporation of uncertainty via probability distributions; the simulation-based approaches via which MC analysis is implemented are straightforward to follow and do not demand detailed knowledge of complex mathematics; and the methods can easily be implemented in packages such as @RISK, which runs as an add-on to the highly accessible Microsoft Excel.

The ease of implementation of MC methods, however, can be a disadvantage; it is possible to conduct an entire analysis on the basis of essentially arbitrary distributions describing uncertainty, and these may be structurally incorrect, and biased in that they systematically under- or over-represent the uncertainty present at various stages. It is common practice, for example, for risk analysts to use triangular distributions to represent uncertainty surrounding a particular parameter via specification of the perceived minimum, most likely, and maximum value for that parameter. However, in only considering a minimum and maximum value, it is difficult to avoid either being overly confident by excluding parameter values outside these limits, or overly pessimistic by choosing more extreme limits, which then results in too much probability near the chosen minimum and maximum.

The process of specifying probability distributions for model inputs falls naturally within the framework of Bayesian inference, as such distributions for fixed, uncertain quantities can only be describing subjective uncertainty. This is not always recognized by practitioners, and a recent review arguing the Bayesian case in MRA is given in Clough *et al.* (2009b).

4.3 Vero-cytotoxic *Escherichia coli* O157 in milk sold as pasteurized

The bacterium *Escherichia coli* (*E. coli*) is a commensal, that is, a natural inhabitant of the human intestinal system. Certain strains of *E. coli*, however, produce toxins which can cause serious illness in humans. One of these is Verocytotoxigenic *E. coli* (VTEC) O157. Common manifestations of a VTEC O157 infection are abdominal cramps and bloody diarrhoea, and the most susceptible groups for the most severe complications of a VTEC O157 infection (HUS and another condition, Thrombotic Thrombocytopaenic Purpura (TTP)) are the young, the immunocompromised, and the elderly (Ochoa and Cleary, 2003).

There have been a number of high-profile outbreaks of food-borne disease which have implicated Vero-cytotoxigenic *Escherichia coli* (VTEC) O157 as a causative agent, in the UK and worldwide. Examples in the UK include an outbreak in Wishaw, Scotland in 1996 Pennington (2000) in which more than 500 people became ill, and 20 people died; in 1999 an outbreak associated with farm-pasteurized milk resulted in 88 laboratory-confirmed cases of VTEC O157, of which 28 (32%) were admitted to hospital, and 3 (3.4%) developed the potentially life-threatening condition Haemolytic Uraemic Syndrome (HUS) (Goh *et al.*, 2002); in September 2005 the UK's second largest ever outbreak commenced in Wales, ultimately leading to illness in 158 individuals and one death (Office of the Chief Medical Officer, Welsh Assembly Government, 2005); and more recently an outbreak in a nursery in South London involved 24 cases of which one child was admitted to hospital with HUS (Health Protection Agency, 2006).

VTEC O157 can be transmitted to humans via a number of foodstuffs and environmental exposures, generally via some intermediate contamination of the vehicle of infection with the faeces of infected animals. Person-to-person transmission is also possible, and in this, combined with inherent susceptibility, in part explains why larger outbreaks are often seen in institutions such as nurseries and nursing homes. Common food products implicated in VTEC O157 outbreaks include beef (frequently undercooked hamburgers or minced beef (Shillam *et al.*, 2002)) and pasteurised milk (Goh *et al.*, 2002) and raw milk (Keene *et al.*, 1997). People can also become infected via environmental exposure, as demonstrated in an outbreak which was linked epidemiologically to the use of faecally contaminated sheep pasture at a scout camp in Scotland (Ogden *et al.*, 2002).

Historically, there have been a number of outbreaks in the UK of food-borne pathogen-related disease from milk sold as pasteurised, with three outbreaks related specifically to VTEC O157; two of these were linked to milk sold as pasteurised and one to raw milk, and in all three cases the milk was farm processed (Allen, 2002). All dairies, irrespective of size, location (on- or off-farm), and scale, are subject to the same European Union legislation European Commission (2004*a*,*b*,*c*). As a result of the greater resources which accompany large-scale off-farm dairies, however, it is easier for them to conduct their own microbiology on-site, obtaining results regularly and quickly, than it is for smaller on-farm producer-processors, who frequently must rely on external microbiological support and generally have much more limited resources and manpower at their disposal.

Two independent studies have suggested that approximately of the order of one-third of dairy herds may be infected with VTEC O157. First, Paiba *et al.* (2003, 2002) conducted an England and Wales based survey of different types of cattle herds. They considered dairy herds as a specific subgroup, and found that of 27 dairy herds randomly sampled and tested, eight were positive for VTEC O157. A second study (Kemp, 2005) found a similar result; 21 out of 63 herds sampled in an intensive dairy farming region of Cheshire were positive for VTEC O157.

4.4 A contamination assessment model

In this section we review the risk assessment model developed in Clough *et al.* (2006, 2009a). The full model describes the milk production chain from the farm through to the point of retail, using a series of linked submodels. In this chapter, we focus on one submodel which describes milk production at on-farm dairies, and is considered to be the most critical component of the overall model. The submodel output of interest is the daily level of contamination (colony forming units, or cfu, per litre) in the bulk milk tank, post-pasteurization. For simplicity we assume that the only milk processed on the farm comes from cows on the farm (i.e. no further milk is bought in for processing).

4.4.1 Model structure

We consider a herd of size 100 cows, and suppose that 80 of them will be lactating at any one time (a proportion consistent with typical practice). Within the lactating group, an unknown number of cows will be infected with VTEC O157. We define the indicator variable $I(infect_i) = 1$ if the ith cow in the lactating group is infected. An infected cow will shed VTEC O157 organisms in its faeces, with C_i defined to be the concentration measured in cfu/gramme in the faeces of the ith cow.

Each cow is milked twice per day, with the ith cow producing a total of M_i litres of milk per day. During the milking process, small quantities of animal faeces may enter the bulk tank, through various routes. For example, a cow may contaminate its own teats with either its own faeces, or those from another cow, which may then enter the milk bulk tank through the milking clusters. We define $F_{i,j}$ to be the faeces, in grammes, contributed to the bulk tank by the ith cow at the jth milking, with $j = 1, 2$. Hence contamination of the milk with VTEC O157 occurs through faeces from an infected cow entering the bulk tank.

After milking, but before pasteurisation, there is the possibility that bacterial growth will occur, and the number of organisms will increase. This may happen if the milk is not stored at the correct temperature. We define an indicator term $I(growth) = 1$ if bacterial growth occurs. If growth does occur, we suppose that the concentration of VTEC O157 in the milk multiplies by a factor of $1 + G$.

Once per day, the milk is pasteurized, i.e. heated to a temperature of $71.7\,^\circ C$ for 15 seconds. If the pasteurization process is successful, any VTEC O157 organisms will be destroyed, and the resulting concentration in the bulk tank will be 0 cfu/litre. However, there is a possibility that the process will not be successful, perhaps due to equipment failure such as incorrect thermometer calibration or a cracked heat-exchanger plate. We define an indicator function $I(fail) = 1$ if the process is unsuccessful. In the event of a failed pasteurization, we suppose that the concentration is reduced by some factor R.

Define Y to be the concentration of VTEC O157 in cfu/litre in the milk bulk tank following pasteurization. We can now write Y as

$$Y = \sum_{i=1}^{80} \left\{ I(infect_i) C_i (F_{i,1} + F_{i,2}) \right\} \{1 + I(growth)G\} I(fail) \times R \times \frac{1}{\sum_{i=1}^{80} M_i}.$$

$$(4.1)$$

We define

$$X = \left[\{ I(infect_i) \}_{i=1}^{80}, \{C_i\}_{i=1}^{80}, \{F_{i,1}\}_{i=1}^{80}, \{F_{i,2}\}_{i=1}^{80} I(growth), G, I(fail), R, \{M_i\}_{i=1}^{80} \right],$$

$$(4.2)$$

and write $Y = \eta(X)$, with X the model inputs.

Our model description so far and notation in (4.1) represent the bulk tank concentration at the end of one day at a single farm. Obviously, we would expect the concentration to vary on different days and at different farms, and a risk analysis must consider this variability. Consequently, we think of the elements of X as randomly varying quantities, that induce a distribution on Y: the population distribution of bulk tank concentrations.

4.5 Model input distributions

We consider probability distributions to represent our uncertainty about each input in the model. These distributions will need to reflect both *epistemic* and *aleatory* uncertainty. We have epistemic uncertainty (uncertainty due to lack of knowledge) about fixed population parameters, for example the mean daily milk yield, and aleatory uncertainty (uncertainty due to random variation) about individual events, for example whether pasteurisation will fail on any specific day. Distributions draw on those described presented in Clough *et al.* (2009a), although different choices have been made here in some cases.

A more general discussion of epistemic and aleatory uncertainty is given in the appendix.

For some inputs there is little or even no data to inform the distributions, and obtaining such data may be impractical. Clearly, the role of expert judgement is important here. However, expert judgement can be an expensive resource, and the necessary expertise is not always immediately available to the risk analyst, just as suitable data may not be available. In such situations, it may be necessary to prioritize obtaining expert judgement in the same way that we would prioritize future data collection. We do so by first choosing distributions that we consider to be plausible in each case, and then proceeding with the sensitivity analysis to identify the most important parameters. In effect, we have simply widened the notion of future data collection to include expert elicitation.

An important general consideration when deriving the various distributions is that the data that are available have not been obtained from studies specifically designed for the purposes of this risk analysis. Given data from a reported study, we must consider carefully whether the parameters of interest in the study correspond to the parameters in the model, or if some extra bias or uncertainty must be accounted for. A further practical difficulty to note is that data reported in the literature is not always presented or available in sufficient detail for appropriate posterior inference.

4.5.1 The number of infected cows

The most relevant data here are given in Kemp (2005), in which a cross-sectional study of infected farms was conducted and in which 30 positive animal pat samples out of 387 were observed. Denoting the number of positive samples by d_{inf}, we might then suppose that

$$\sum_{i=1}^{80} I(infect_i) | \phi_{inf} \sim Bin(80, \phi_{inf}) \tag{4.3}$$

$$d_{inf} | \phi_{inf} \sim Bin(387, \phi_{inf}), \tag{4.4}$$

and that in the absence of any other information,

$$\phi_{inf} | d_{inf} \sim Be(30, 357). \tag{4.5}$$

However, the choices of likelihood (4.3) and (4.4) are questionable. The events of infected cows (or contaminated pats) should not be treated as independent on the same farm, given the possibility of transmission of VTEC O157 between animals. It is also possible that in the study, individual cows may have been sampled more than once, and so arguably we should be reporting greater posterior uncertainty about ϕ_{inf} in any case.

Two longitudinal studies are reported in Mechie *et al.* (1997) and Robinson (2004). Mechie *et al.* (1997) visited a single farm 28 times, with the largest proportion of infected lactating cows observed to be 15/111, and no infected lactating cows observed in 13 visits. Seasonal variation was also observed. Robinson (2004) repeatedly sampled five farms over an 18 month period, and observed 29 out of 645 positive samples, with not every farm positive on every occasion.

Formally synthesizing all the available evidence would require a considerably more complex model for infections (for example, as in Turner *et al.*, 2003), but may not necessarily be advantageous for the purposes of the risk assessment. Instead, we proceed as follows. We consider the cross-sectional study only, but judge that the distribution in (4.5) may lead to overconfidence. We first choose to 'downweight' the evidence, by halving the effective sample size and setting $\phi_{inf}|d_{inf} \sim Beta(30/2, 357/2)$. Hence our expectation is unchanged, but we are now allowing for more uncertainty. We also test the sensitivity of our conclusions in Section 4.6.2 to different choices of this distribution. This will enable us to establish whether more complex modelling of infections is necessary.

4.5.2 Colony forming units (cfu) per gramme

We have one sample of 29 observations available, denoted by d_{cfu}, from the study reported in Robinson (2004). Four cows were observed to have mean counts of 2500, 250, 1400 and 1100 cfu/g respectively. A further 25 cows were observed to have counts of no more than 200 cfu/g. Precise counts are unknown at concentrations less than 200 cfu/g because this represents the lower limit of detection of the counting procedure. Below this, a highly sensitive test based upon immuno-magnetic separation can only categorize samples as positive or negative.

The observed data suggest a heavy-tailed distribution, and Clough *et al.* (2009a) found a $Weibull(\alpha, \beta)$ distribution, with α and β estimated using maximum likelihood, gave a satisfactory fit to the data, whereas other distributions such as the lognormal were not sufficiently heavy tailed. If we use fixed estimates of α and β, we cannot assess formally how our uncertainty about Y might change given better information about the distribution of cfu/g. To allow for uncertainty in α and β, we must consider a prior distribution for these two parameters.

Care is needed in the choice of prior, in particular with regard to small values of α. For fixed β, the expectation $E(C_i|\alpha, \beta)$ tends to infinity as α tends to zero, but an infinite population mean is clearly impossible here. With flat, non-informative priors there is not sufficient data to rule out the possibility of unrealistically small α, and so we must consider our prior probability of very large VTEC O157 concentrations.

To aid us, we have some additional, indirectly relevant data. In a sample of 31 calves, Zhao *et al.* (1995), reported concentrations of the order of 10^5 cfu/g in three calves, 10^4 cfu/g in eleven calves, and 10^3 cfu/g or less in the remainder. An experimental study by Cray and Moon (1995) suggests that calves shed substantially greater numbers of organisms (an order of magnitude or higher) in their faeces than adult cattle. From this we make the judgement that concentrations in adult cattle of the order of 10^6 cfu/g or more are unlikely. Specifically, we choose the following priors:

$$\alpha \sim N\left(0, 10^6\right) I\left(\alpha > 0.2\right), \tag{4.6}$$

$$\beta \sim N\left(0, 10^6\right) I\left(\beta > 0\right). \tag{4.7}$$

For the *Weibull* (α, β) distribution, we have $Pr(C_i < 10^6 | \alpha = 0.2, \beta = 2500) = 0.96$ and $Pr(C_i < 10^7 | \alpha = 0.2, \beta = 2500) = 0.995$.

We then sample from the posterior distribution $f(\alpha, \beta | d_{cfu})$ using Markov chain Monte Carlo (implemented in WinBUGS). Summary statistics from the posterior are shown in below:

	posterior mean	95% credible interval
α	0.358	(0.213, 0.574)
β	73.95	(8.528, 218.5)

We note that the 2.5th percentile for α is close to the truncation point of 0.2, and so the precise choice in our prior may be influential. We return to this issue in Section 4.6.3. The choice of 10^6 in (4.6) and (4.7) is arbitrary, but we find the posterior is insensitive to larger values.

4.5.3 Milk yields per cow

We suppose that the milk yield of each cow is normally distributed with unknown mean and variance μ_{milk} and σ^2_{milk} respectively. As in Clough *et al.* (2009a), we have data available from 30 dairy herds (NMR milk records provided by Andrew Biggs, the Vale Veterinary Centre, personal communication), with an observed mean daily yield of $d_{milk,j}$ in the jth herd. We have

$$d_{milk,j} | \mu_{milk}, \sigma^2_{milk} \sim N\left(\mu_{milk}, \sigma^2_{milk}/n_j\right),$$

where n_j is the size of the jth herd. With no expert option available, we choose an improper prior for $\mu_{milk}, \sigma^2_{milk}$: $f(\mu_{milk}, \sigma^2_{milk}) \propto \sigma^{-2}_{milk}$. This gives a normal inverse gamma posterior on $\mu_{milk}, \sigma^2_{milk}$ given milk yield data, with

$$\sigma^{-2}_{milk} | d_{milk} \sim Ga(14.5, 2.452)$$

$$\mu_{milk} | \sigma^2_{milk}, d_{milk} \sim N\left(22.178, 3.222\sigma^2_{milk}\right).$$

4.5.4 Faeces in bulk tank

No data are available here, but we have informal judgement available from an expert workshop (Pasteurized Milk Production Workshop, 2004) which was conducted to assist with the work reported in Clough *et al.* (2006). It was judged that the most likely total contamination would be approximately 2 grammes, but could vary according to the cleanliness of the farm. On a very clean farm, the total contamination may be close to 0 grammes, whereas on a dirty farm, the total contamination could be as high as 10 grammes.

We choose to model the distribution of $F_{i,j}$ (converting first to milligrammes) hierarchically. We suppose:

$$\begin{aligned}\log 1000 F_{i,j} | \mu_F, \sigma^2_{cow} &\sim N\left(\mu_F, \sigma^2_{cow}\right),\\ \mu_F | \mu_{farm}, \sigma^2_{farm} &\sim N\left(\mu_{farm}, \sigma^2_{farm}\right),\\ \mu_{farm} &\sim N(3, 1)\\ \sigma^{-2}_{farm} &\sim Ga(4.5, 0.25)\\ \sigma^{-2}_{cow} &\sim Ga(4.5, 0.01).\end{aligned}$$

Following the comments from the expert workshop, we define the parameter μ_F to be an effect representing 'cleanliness' of farm (with uncertain population mean and variance μ_{farm} and σ^2_{farm} respectively). The parameter σ^2_{cow} represents (uncertain) variability in the faeces contribution between cows within a farm. We choose these distributions such that the implied predictive distribution of $\sum_{i=1}^{80}\sum_{j=1}^{2} F_{i,j}$ gives a range representative of the initial judgements provided at the expert workshop, and reflecting our own judgement that variability is likely to be smaller within a farm than between farms. Sampling from the predictive distribution of $\sum_{i=1}^{80}\sum_{j=1}^{2} F_{i,j}$, we have a median total contamination of faeces in the bulk tank per milking of 1.7 grammes, with a central 95% interval of (0.2 g, 12.0 g).

4.5.5 Growth of VTEC O157 organisms in the bulk tank

The current EU regulations (European Commission, 2004*a*,*b*,*c*) state that the maximum storage temperature for milk pre-pasteurization is 6 °C. If this temperature is exceeded, it is possible that there will be a growth in the number of organisms. We define the proportion of occasions in which growth occurs to be ϕ_{growth}, and assume that $I(growth)|\phi_{growth} \sim Bernoulli(\phi_{growth})$. In the event that growth does occur, we are uncertain about both mean growth and variability in growth on different occasions. If the concentration increases by $100G$%, we suppose $\log G | \mu_G, \sigma^2_G \sim N(\mu_G, \sigma^2_G)$, with both μ_G and σ^2_G uncertain.

Some indirectly relevant data are available in Kauppi *et al.* (1996), which describes a study to examine the growth of VTEC O157 in milk sterilized prior to being inoculated with bacteria at a range of different temperatures, and observes that at a temperature of 6.5 °C, following a lag phase (interval between inoculation and the commencement of growth) of between two and four weeks, an average doubling time of 83 hours was recorded. At temperatures of 9.5 °C and 12 °C, doubling times of 4.6 hours and 11 hours respectively were recorded, with no noticeable lag phase. The only data relating to raw milk come from Wang *et al.* (1997), who examined milk stored at 8 °C and found an increase of between 1 and 2 logs in concentration over four days and between 2 and 3 logs over seven days.

We have no data to inform how frequently growth occurs, but we believe this to be an infrequent event. Given the observed lag phases, we note that small exceedances of the maximum allowed storage temperature may not result in any growth in the time period between milking and pasteurization.

We choose $I(growth)|\phi_{growth} \sim Bernoulli(\phi_{growth})$ with $\phi_{growth} \sim Beta(1, 99)$, so that our prior expectation is that growth occurs on 1% of occasions, with this proportion unlikely to exceed 0.05. In the event that growth does occur, we suppose that the concentration is unlikely to double before pasteurization, and set

$$\mu_G \sim N(-0.7, 0.04) \tag{4.8}$$

$$\sigma_G^{-2} \sim Ga(4.5, 0.25). \tag{4.9}$$

Sampling from the prior predictive distribution, conditional on $I\{growth\} = 1$, the median percentage increase in concentration is 50%, with a central 95% interval of (26%, 96%).

4.5.6 Pasteurization failure

Data on pasteurization failures are available from a survey of on-farm dairies, reported in Allen *et al.* (2004), in which samples of milk were tested using an alkaline phosphatase test (see Clough *et al.* 2006, 2009a for more details). Each failed test is taken as an indication of a failed pasteurization. Out of a total of 2885 samples of pasteurized milk, $d_{fail} = 115$ failures were detected. Defining ϕ_{fail} to be the population proportion of failures, and in the absence of any other information, we suppose that

$$\begin{aligned} I(fail)|\phi_{fail} &\sim Bernoulli\ (\phi_{fail}), \\ d_{fail}|\phi_{fail} &\sim Bin\ (2885, \phi_{fail}), \\ \phi_{fail}|d_{fail} &\sim Be\ (115, 2770). \end{aligned} \tag{4.10}$$

As in Section 4.5.1, the choice of likelihood can be questioned. Almost certainly, failure rates will not be equal for all farms, and in ignoring this we may be understating our uncertainty about ϕ_{fail}. Note that the predictive distribution of $I(fail)|d_{fail}$ (unconditional on ϕ_{fail}) only depends on $E(\phi_{fail}|d_{fail})$ and not

$Var(\phi_{fail}|d_{fail})$, but again, when investigating the role of ϕ_{fail} in the distribution of Y, we consider a wider range of plausible values than that implied by (4.10).

4.5.7 Reduction in concentration given pasteurization failure

We do not have any data to inform this distribution. Given a failure, we are uncertain about both the mean reduction and variability in the amount of reduction on different occasions. We suppose that on the logit scale, the reduction is normally distributed with unknown mean and variance μ_{red} and σ^2_{red} respectively. We suppose

$$\begin{aligned} \log\left(\frac{R}{1-R}\right)|\mu_{red}, \sigma^2_{red} &\sim N\left(\mu_{red}, \sigma^2_{red}\right), \\ \mu_{red} &\sim N(0, 1), \\ \sigma^{-2}_{red} &\sim Ga(4.5, 0.25). \end{aligned}$$

Sampling from the prior predictive distribution, the median reduction in concentration is 50%, with a central 95% interval of (12%, 88%).

4.6 Model output analysis

We consider an analysis of the output in two stages. At the first stage of the analysis, we investigate uncertainty about Y given our uncertainty in X. Defining

$$\theta = \left(\phi_{inf}, \alpha, \beta, \mu_{milk}, \sigma^2_{milk}, \mu_{farm}, \sigma^2_{farm}, \sigma^2_{cow}, \phi_{growth}, \mu_G, \sigma^2_G, \phi_{fail}, \mu_{red}, \sigma^2_{red}\right),$$

our distribution of $\eta(X)$ reflects both epistemic uncertainty about θ, and aleatory uncertainty about $X|\theta$. Probability statements about Y are beliefs about what we might observe on inspection of a bulk tank on any single occasion.

At the second stage of the analysis, we investigate epistemic uncertainty about the distribution of $Y|\theta$. The conditional distribution $Y|\theta$ can be interpreted as the true population distribution of bulk tank concentrations, in principle observable, but unknown as θ is unknown. Investigating how uncertainty about θ contributes to uncertainty about this population distribution will reveal where we should prioritise in reducing epistemic uncertainty through further research. Given the dependance of the analysis on prior judgement, it is crucial to understand the role of the various elements of θ in the distribution of $Y|\theta$.

4.6.1 Uncertainty about Y

We now use Monte Carlo to propagate uncertainty about X through (4.1), by sampling θ_i from $f(\theta|d)$ and X_i from $f(X|\theta_i)$ for $i = 1, \ldots, N$, and then evaluating $Y_i = \eta(X_i)$. Most elements of θ can be sampled directly, with the exception of α and β which are sampled using Markov chain Monte Carlo. Given the sample $Y_{1:N}$, we estimate $E(Y) = 0.003$ cfu/l, and $s.d(Y) = 0.34$ cfu/l. From the distributions of $I(fail|d_{fail})$ and $\sum_{i=1}^{80} I(infect_i)|d_{inf}$ we can compute $Pr(Y = 0|d_{fail}, d_{inf}) = 0.95$. Hence it will also be of interest to consider the distribution of $Y|I(fail) = 1, \sum_{i=1}^{80} I(infect_i) > 0$. Using Monte Carlo again we obtain a mean of 0.07 cfu/l, with a 95% interval of (2.871E−5 cfu/l, 0.427 cfu/l).

4.6.1.1 *Variance based sensitivity analysis*

Having measured our uncertainty about Y, we now investigate how our various sources of uncertainty contribute to our uncertainty in Y. We do so using the variance based approach to sensitivity analysis in which the importance of a particular input to model is quantified by its contribution to the output variance (see Appendix A1 for more details).

For any variable X_i, the expected reduction in the variance of Y obtained by learning the value of Y is given by

$$Var(Y) - E_{X_i}\{E(Y|X_i)\} = Var_{X_i}\{E(Y|X_i)\}, \tag{4.11}$$

so we use the term $Var_{X_i}\{E(Y|X_i)\}/Var(Y) = S_i$, known as the *main effect index* of X_i, as a measure of the 'importance' of the variable X_i. The main effect index is dependent on two factors: the uncertainty in X_i and the role of the input X_i in the function $\eta(.)$.

Similarly, we can consider learning groups of inputs at a time, for example by calculating $Var_{X_i,X_j}\{E(Y|X_i, X_j)\}/Var(Y) = S_{i,j}$. Note that we will not have $S_{i,j}$ equal to the sum of S_i and S_j if either X_i and X_j are correlated, or if there is an interaction between X_i and X_j within the function $\eta(.)$.

Related to the main effect index, is the *main effect* of X_i, defined to be $E(Y|X_i) - E(Y)$, with the variance based measure obtained by taking the the variance of the main effect of X_i. Plotting the main effect against X_i provides a useful visual tool for understanding the role of X_i in the function $\eta(.)$.

Given the simple form of (4.1), much of the computation can be done analytically, with Monte Carlo sufficiently straightforward when necessary.

We first define suitable groupings of interest of the inputs: $X_{infect} = \{I(infect_i)\}_{i=1}^{80}$, $X_{cfu} = \{C_i\}_{i=1}^{80}$, $X_{faeces} = \{F_{i,1}, F_{i,2}\}_{i=1}^{80}$ and $X_{milk} = \{M_i\}_{i=1}^{80}$. For any element in the set $\{X_{infect}, X_{cfu}, X_{faeces}, X_{milk}, I(fail), I(growth), G, R\}$, we find that the corresponding main effect makes a very small contribution to the variance, with the largest individual proportion $Var_{I(fail)}[E\{Y|I(fail)\}]/Var(Y) = 0.002$. We might expect such a result with Y the product of functions of these terms, including indicator functions that can result in $Y = 0$.

Table 4.1 Variance of main effects for various combinations of inputs.

Z	$Var_Z\{E(Y\|Z)\}/Var(Y)$
X_{infect}, X_{faeces}, $I(fail)$	0.01
X_{cfu}, X_{faeces}, $I(fail)$	0.08
X_{cfu}, X_{infect}, $I(fail)$	0.13
X_{cfu}, X_{infect}, X_{faeces}, $I(fail)$	0.84
X_{cfu}, X_{infect}, X_{faeces}, $I(fail)$, R	0.99

We now consider groups of variables at a time, to see whether a subset of variables can explain most of the variance of Y. Various group main effect variances are reported in Table 4.1.

We can see that the majority of the variance of Y can be explained through the uncertainty in X_{cfu}, X_{infect}, X_{faeces}, $I(fail)$. If we also include R, then we have accounted for 99% of the variance of Y. Hence uncertainty in the variables $I(growth)$, G, and $\{M_i\}_{i=1}^{80}$ is 'unimportant', in that it is inducing little uncertainty in Y. This is perhaps to be expected, given the $Beta(1, 99)$ prior placed on ϕ_{growth}, and the reasonably informative data about milk production. Of course, we have not used genuine expert judgement in deriving the priors for the various hyperparameters concerned, but if an expert were satisfied that the milk data were not atypical, and the priors chosen for ϕ_{growth}, μ_{growth} and σ^2_{growth} were not unreasonable, then analysis suggests that reducing uncertainty about these parameters by either collecting more data or incorporating genuine expert knowledge would not be worthwhile.

4.6.2 Uncertainty about $Y|\theta$

Recall that the distribution of $Y|\theta$ can be interpreted as the true, but unknown population distribution of bulk tank concentrations. It is possible to consider uncertainty about any summary of this distribution due to epistemic uncertainty about θ.

By considering $E(Y|\theta)$, we can first determine the respective roles of epistemic and aleatory uncertainty in our overall uncertainty about Y. We have $E_\theta\{E(Y|\theta)\} = E(Y) = 0.003$ cfu/l, and using Monte Carlo simulation, we find $Var_\theta(\{E(Y|\theta)\}) = 1.67 \times 10^{-4}$. Since $Var(Y) - E_\theta\{Var(Y|\theta\} = Var_\theta\{E(Y|\theta)\}$, we now have the expected reduction in the variance of Y obtained by removing epidemic uncertainty about θ. As $Var_\theta\{E(Y|\theta)\}/Var(Y) = 0.001$, we have the observation that uncertainty about Y is largely due to aleatory uncertainty about $X|\theta$ rather than epistemic uncertainty about θ.

We now choose to investigate the expected concentration, conditional on a pasteurization failure. If pasteurization is successful, the concentration will be zero cfu/l, and so it would be straightforward to consider the effects of uncertainty about ϕ_{fail} on a risk assessment separately. The expected concentration gives a useful indication of the likely number of contaminated cartons at the point of retail. Note that it has been hypothesized that a single microorganism has the potential to cause infection (Haas, 1983).

We define $M = E(Y|\theta, I(fail) = 1)$ to be the population mean conditional on pasteurization failure, and consider our uncertainty about M due to epistemic uncertainty about θ. From (4.1), we write

$$M = 80\phi_{infect}\beta\Gamma(1 + 1/\alpha)\frac{2}{1000}\exp\left\{\mu_{farm} + \frac{\sigma^2_{farm}}{2} + \frac{\sigma^2_{cow}}{2}\right\}$$
$$\times\phi_{growth}\left(1 + \exp\left\{\mu_{growth} + \frac{\sigma^2_{growth}}{2}\right\}\right) E\left(R|\mu_R, \sigma^2_R\right)$$
$$E\left(\frac{1}{\sum_{i=1}^{80} M_i}|\mu_{milk}, \sigma^2_{milk}\right).$$

Using Monte Carlo, we obtain $E(M) = 0.073$ and $Var(M) = 0.103$

Variance based sensitivity analysis can also be used to explore how uncertainty about individual elements of θ contribute to uncertainty in M. We again first make the groupings $\theta_{cfu} = (\alpha, \beta)$, $\theta_{faeces} = (\mu_{farm}, \sigma^2_{farm}, \sigma^2_{cow})$, $\theta_{growth} = (\phi_{growth}, \mu_{growth}, \sigma^2_{growth})$, $\theta_R = (\mu_R, \sigma^2_R)$ $\theta_{milk} = (\mu_{milk}, \sigma^2_{milk})$. Individual and group main effect variances of note are given in Table 4.2.

Hence most of our uncertainty about M can be explained through our uncertainty about $\theta_{faeces}, \theta_{cfu}, \theta_R$.

For the final part of our analysis, we consider the role of each hyperparameter in the the function (4.12). We do this by plotting individual main effects $E_{\theta_{-i}}(M|\theta_i) - E(M)$ against θ_i, where θ_{-i} denotes all elements of θ except θ_i. This is an alternative to simply fixing θ_{-i} and plotting M as θ_i varies. Fixing θ_{-i} may not be appropriate if M is a nonlinear function of θ_{-i}, or if θ is correlated with

Table 4.2 Variance of main effects for various parameters and parameter combinations.

Z	$\frac{Var_Z\{E(M\|Z)\}}{Var(M)}$
ϕ_{infect}	0.003
θ_{faeces}	0.090
θ_{cfu}	0.260
θ_G	0.000
θ_R	0.009
θ_{milk}	0.000
$\theta_{faeces}, \theta_{cfu}$	0.794
$\theta_{faeces}, \theta_{cfu}, \theta_R$	0.938

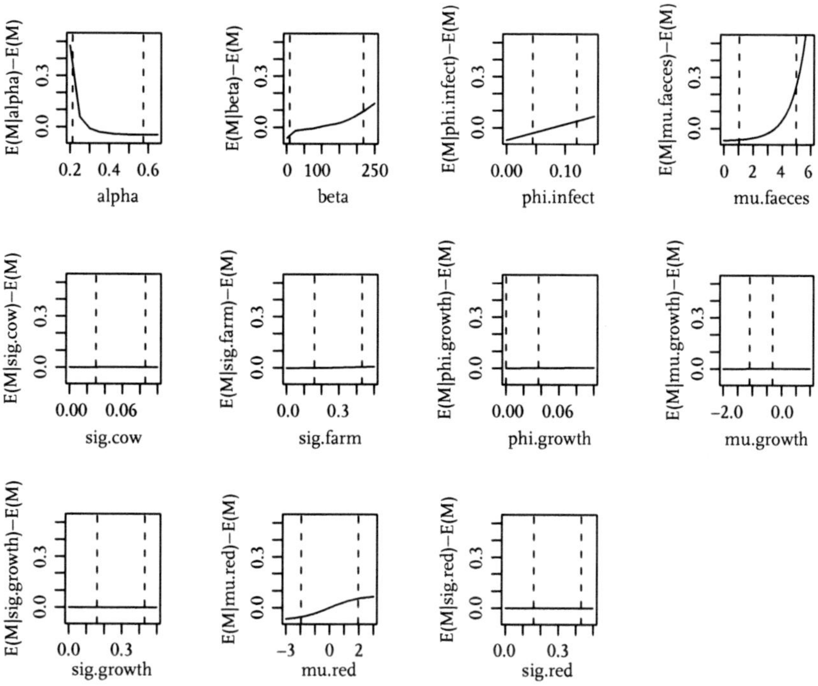

Fig. 4.1 Main effects plots. Dashed lines show 95% intervals for each parameter.

elements of θ_i. In this analysis, we have correlation between the parameters α and β, and so conditional on one parameter we must obtain a new sample from the conditional distribution of the other. The main effect plots are shown in Figure 4.1. Dashed lines indicate 95% intervals for each parameter.

Of particular interest is the main effect plot for α. We observe a rapid increase in $E(M|\alpha)$ as α decreases beyond its 2.5th percentile. This suggests that the constraint $\alpha > 0.2$ applied in the prior distribution is critical, and that obtaining either more cfu/g data, or expert knowledge about the cfu/g distribution is a priority for producing a more robust risk assessment. We also note a rapid increase in $E(M|\mu_{farm})$ for large μ_{farm}. This parameter was informed by expert judgement only, and so the plot suggests that appropriate data may be necessary. Examining the plot for ϕ_{inf} within its 95% interval, the change in the conditional expectation is relatively small. In section we noted the difficulty in formally synthesizing all the available evidence related to ϕ_{inf}. This analysis has suggested that refining the distribution of ϕ_{inf} is not a priority.

4.6.3 Sensitivity to prior distributions

We conclude the analysis by briefly investigating the sensitivity of our conclusions to particular judgements made in our choice of priors. Firstly, we consider our distribution for ϕ_{inf} and our decision to downweight the evidence from

Kemp (2005) in judging $\phi_{inf} \sim Be(30/2, 357/2)$. Repeating our analysis with alternatives such as $\phi_{inf} \sim Be(30, 357)$ or $\phi_{inf} \sim Be(30/4, 357/4)$ has very little effect on any of output means or variances of interest. Changing our expectation of ϕ_{inf} will change the expected value of Y and M, but we can be confident that under any plausible distribution for ϕ_{inf}, the importance of this parameter will be small relative to others.

The second judgement we consider is the choice of truncation point c in the prior $\alpha \sim N(0, 10^6) I(\alpha > c)$. Here, the output means and variances are less robust. For example, if we choose $c = 0.25$, we find $E(Y) = 0.002$cfu/l and $s.d(Y) = 0.07$cfu/l, in comparison with $E(Y) = 0.003$ cfu/l, and $s.d(Y) = 0.34$ cfu/l when $c = 0.2$ (although the relative contributions of the uncertain parameters/variables to the output variances do not change noticeably). This is of less practical concern, given that we have already identified cfu concentration as a key source of uncertainty requiring further research.

4.7 Further discussion

In this analysis we have investigated only one submodel of the full 'farm-to-fork' model, which predicts VTEC O157 concentrations at the point of consumption. Further sensitivity analyses would be required on other submodels to identify other potentially important sources of uncertainty in the full model. Where such models can be considered as a series of linked submodels, we believe it is highly informative to conduct sensitivity analyses on the individual submodels to provide the modeller with as much insight into input – output relationships and uncertainty as possible.

We have not specifically considered the probability of 'large' VTEC O157 concentrations in the bulk tank, with reference to some dose-response relationship. Arguably, such considerations would be premature. Given the health risks associated with infection, the dose-response relationship cannot be estimated experimentally under controlled conditions. Instead it must be inferred from observed outbreak data (see for example Teunis *et al.*, 2008), where the precise doses concerned are unknown. Doses must be estimated from mechanistic models, and so again, the importance of sensitivity analysis becomes apparent.

Appendix

A. Broader context and background

A.1 Variance-based sensitivity analysis

In this chapter we have used the variance-based approach to sensitivity analysis for identifying important model inputs. We now review some theoretical results underlying the variance-based method, and briefly discuss computational tools for more complex models. For a more detailed treatment, we refer the reader to Chan *et al.* (2000).

We consider a model represented by the function $y = \eta(x)$, with $x = x_{1:d}$, and suppose that we are interested in the value of $Y = \eta(X)$ for some uncertain value of X. The aim of a variance-based sensitivity analysis is to identify how individual elements of X contribute to the uncertainty in Y, and can be motivated by a decomposition of the variance of Y. Firstly, we can write

$$Y = \eta(X) = E(Y) + \sum_{i=1}^{d} z_i(X_i) + \sum_{i<j} z_{i,j}(X_i, X_j) + \sum_{i<j<k} z_{i,j,k}(X_i, X_j X_k) + \ldots + z_{1,2,\ldots,d}(X), \quad (4.12)$$

where

$$\begin{aligned} z_i(X_i) &= E(Y \mid X_i) - E(Y), \\ z_{i,j}(X_i, X_j) &= E(Y \mid X_i, X_j) - z_i(X_i) - z_j(X_j) - E(Y), \\ z_{i,j,k}(X_i, X_j, X_k) &= E(Y \mid X_i, Xj, X_k) - z_{i,j}(X_i, X_j) - z_{i,k}(X_i, X_k) - z_{j,k}(Xj, X_k) \\ &\quad - z_i(X_i) - z_j(X_j) - z_k(X_k) - E(Y), \end{aligned}$$

and so on. If the elements of X are independent then any pair of z terms will be uncorrelated, and we can write the variance of Y as

$$Var(Y) = \sum_{i=1}^{d} W_i + \sum_{i<j} W_{i,j} + \sum_{i<j<k} W_{i,j,k} + \ldots + W_{1,2,\ldots,d}, \quad (4.13)$$

where if p is the set of indices for the subvector $x_{(p)}$,

$$W_p = Var\{z_p(X_{(p)})\}.$$

Hence the variance of Y can be decomposed into a sum of main effect variances W_i, first order interaction variances $W_{i,j}$, and so on.

Formally, since

$$W_i = Var(Y) - E_{X_i}\{Var(Y|X_i)\}, \quad (4.14)$$

W_i gives the expected utility of learning X_i in the context of a decision problem in which the decision is to choose an estimate of Y, and the corresponding loss function for the size of the estimation error is quadratic. Hence variance-based measures are useful 'general purpose' measures for important inputs, but are not necessarily appropriate under alternative loss functions (see Oakley, 2009).

In addition to the main effect indices, various authors also consider *total effect variances*, where the total effect variance for input X_i is the sum of all the W terms in (4.13) with the subscript i. Again, we can divide the total effect variance by $Var(Y)$ to obtain the total effect index. The total effect variance can be expressed as

$$V_{T_i} = Var(Y) - E_{X_{-i}}\{Var(Y|X_i)\}, \qquad (4.15)$$

the expected variance remaining after learning all inputs except X_i. An input may have a small main effect index, but a large total effect index depending on interactions with other inputs in the function η. A small total effect index indicates that an input is genuinely 'unimportant', in that it has minimal effect on the model output.

A.1.1 Correlated inputs A variance-based analyses is complicated by the presence of correlations between the elements of X. Firstly, the decomposition in (4.13) no longer holds, as we will have correlations between the terms in (4.12). This can be seen with the simple example $Y = X_1 + X_2$, where X_1, X_2 have a bivariate normal distribution with $E(X_i) = 0$, $Var(X_i) = 1$ for $i = 1, 2$ and $Cov(X_1, X_2) = 0.5$. We have $Var(Y) = 3$, but $Var_{X_i}\{E(Y|X_i)\} = 2.25$, and so $Var(Y) < W_1 + W_2$.

Main effect indices are still useful, as their interpretation based on (4.14) still holds. The usefulness of total effect indices is less clear, as they cannot be motivated by the decomposition (4.13). We can still calculate them according to (4.15), but we cannot rely on a small total effect as an indication of an unimportant input, as it is now possible to have a total effect index smaller than a main effect index.

Note also that in the correlated input case, computation may be more demanding, due to the need to consider the conditional distribution $X_{-i}|X_i$ for different values of X_i when evaluating $Var_{X_i}\{E_{X_{-i}}(Y|X_i)\}$. In Section 4.6.2, this required repeated use of MCMC sampling to obtain $E(M|\alpha)$ and $E(M|\beta)$ for different values of α and β.

A.1.2 Computation In our risk analysis, the model was sufficiently simple so that we could evaluate main effect indices analytically, or using Monte Carlo sampling. For more complex models, simple Monte Carlo sampling can become inefficient. More efficient Monte Carlo procedures can be used, based on orthogonal arrays (Morris *et al.*, 2008), or we can first construct an emulator for the model $\eta(.)$, and then obtain sensitivity indices directly from the emulator (Oakley and O'Hagan, 2004).

A.2 *Epistemic and aleatory uncertainty*

In the treatment of uncertainty, a distinction is sometimes made between two 'types' of uncertainty: *aleatory* uncertainty, uncertainty due to random variation, and *epistemic* uncertainty, uncertainty due to a lack of knowledge. Various alternative names have been used in the literature for these two terms: for 'aleatory' these include first order, stochastic, irreducible, type A, and variability, and for 'epistemic' these include second order, subjective, reducible, type B, and simply 'uncertainty'.

From a Bayesian perspective, the distinction is somewhat irrelevant, in that the same tool, subjective probability, is used to quantify both types of uncertainty, with no need to make any special provision in either case. (Note that it is not universally accepted that subjective probability is always appropriate for quantifying epistemic uncertainty, see for example Oberkampf *et al.*, 2004). Nevertheless, the distinction is useful, both when considering how to construct appropriate probability distributions that take into account all sources of uncertainty, and in identifying possible opportunities for and benefits of reducing uncertainty through obtaining more data. Epistemic uncertainty can be reduced through the collection of more data, whereas aleatory uncertainty cannot (at least not without modifying the physical system or sampling process).

As a simple example, let X be the number of adults who currently wear glasses in a random sample of 10 adults from the UK population. In assessing $Pr(X \leq x)$, we can identity both epistemic and aleatory uncertainty. We have epistemic uncertainty about the proportion θ of all UK adults who wear glasses, and we have aleatory uncertainty about which adults (glasses-wearing or not) will be randomly selected. We can assess $Pr(X \leq x)$ by choosing a prior distribution $p(\theta)$ for θ, making the judgement $X|\theta \sim Bin(10, \theta)$, and calculating $\int_0^1 p(\theta) Pr(X \leq x|\theta) d\theta$. We could have made a direct judgement of $Pr(X \leq x)$ without considering first $p(\theta)$ and $p(X|\theta)$, but, in general, such judgements can lead to overconfidence if we do not think carefully about sources of both epistemic and aleatory uncertainty.

Acknowledgements

This work was made possible through the Environment and Human Health Programme, which is a joint three-year interdisciplinary capacity-building programme supported by NERC, EA, Defra, the MOD, MRC, The Wellcome Trust, ESRC, BBSRC, EPSRC and HPA.

References

Allen, G. (2002). Alkaline phosphatase levels in pasteurised milk and cream from on-farm dairies in North West England. Master's thesis, University of Central Lancashire.

Allen, G., Bolton, F. J., Wareing, D. R. A. and Williamson, J. K. amd Wright, P. A. (2004). Assessment of pasteurisation of milk and cream produced by on-farm dairies using a fluorimetric method for alkaline phosphatase activity. *Communicable Disease and Public Health*, **7**, 96–101.

Chan, K., Tarantola, S., Saltelli, A. and Sobol', I. M. (2000). Variance-based methods. In *Sensitivity Analysis*, (ed. A. Saltelli, K. Chan and M. Scott), John Wiley, New York.

Clough, H., Clancy, D. and French, N. (2006). Vero-cytotoxigenic *Escherichia coli* O157 in pasteurized milk: a qualitative approach to exposure assessment. *Risk Analysis*, **26**, 1291–1310.

Clough, H. E., Clancy, D. and French, N. P. (2009*a*). Quantifying exposure to Vero-cytotoxigenic *Escherichia coli* O157 in milk sold as pasteurized: a model-based approach. *International Journal of Food Microbiology*, **131**, 95–105.

Clough, H. E., Kennedy, M. C., Anderson, C. W., Barker, G., Cook, P., Hart, A., Hart, C. A., Malakar, P., Oakley, J. E., O'Hagan, A., Pieloat, A., Smid, J., Snary, E. L., Turner, J. (2009*b*). Bayesian approaches in microbial risk assessment: Past, present and future. Technical report, National Centre for Zoonosis Research, University of Liverpool.

CODEX Alimentarius Commission (1999). Principles and guidlines for the conduct of microbiological risk assessment. CAC/GL-30.

Cray, W. C. and Moon, H. W. (1995). Experimental-infection of calves and adult cattle with *Escherichia coli* O157:H7. *Applied and Environmental Microbiology*, **61**, 1586–1590.

European Commission (2004*a*). Regulation (ec) no 852/2004 of the European Parliament and of the Council on the hygiene of foodstuffs.

European Commission (2004*b*). Regulation (ec) no 853/2004 of the European Parliament and of the Council laying down specific hygiene rules for food of animal origin.

European Commission (2004*c*). Regulation (ec) no 854/2004 of the European Parliament and of the Council laying down specific rules for the organisation of official controls on products of animal origin intended for human consumption.

Goh, S., Newman, C., Knowles, M., Bolton, F. J., Hollyoak, V., Richards, S., Daley, P., Counter, D., Smith, H. R. and Keppie, N. (2002). *E.coli* O157 phage type 21/28 outbreak in North Cumbria associated with pasteurized milk. *Epidemiology and Infection*, **129**, 451–457.

Haas, C. N. (1983). Estimation of risk due to low doses of microorganisms: a comparison of alternative methodologies. *American Journal of Epidemiology*, **118**, 573–582.

Health Protection Agency (2006). Two contiguous but unconnected outbreaks of vero cytotoxin-producing E coli o157 infection in south east london. *CDR Weekly*, **16**, 2–4.

Kauppi, K. L., Tatini, S. R., Harrell, F. and Feng, P. (1996). Influence of substrate and low temperature on growth and survival of verotoxigenic *Escherichia coli*. *Food Microbiology*, **13**, 397–405.

Keene, W. E., Hedberg, K., Herriott, D. E., Hancock, D. D., McKay, R. W., Barrett, T. J. and Fleming, D. W. (1997). Prolonged outbreak of *Escherichia coli* O157:H7 infections caused by commercially distributed raw milk. *Journal of Infectious Diseases*, **176**, 815–818.

Kemp, R. (2005). The epidemiology of *E.coli O157*, non-O157 VTEC and Campylobacter spp. in a 100 km^2 dairy farming area in Northwest England. Ph.D. thesis, University of Liverpool.

Mechie, S. C., Chapman, P. A. and Siddons, C. A. (1997). A fifteen month study of Escherichia coli O157:H7 in a dairy herd. *Epidemiology and Infection*, **118**, 17–25.

Morris, M. D., Moore, L. M. and McKay, M. D. (2008). Using orthogonal arrays in the sensitivity analysis of computer models. *Technometrics*, **50**, 205–215.

Oakley, J. E. (2009). Decision-theoretic sensitivity analysis for complex computer models. *Technometrics*, **51**, 121–129.

Oakley, J. E. and O'Hagan, A. (2004). Probabilistic sensitivity of complex models: a Bayesian approach. *Journal of the Royal Statistical Society Series B*, **66**, 751–769.

Oberkampf, W. L., Helton, J. C., Joslyn, C. A., Wojtkiewicz, S. F. and Ferson, S. (2004). Challenge problems: Uncertainty in system response given uncertain parameters. *Reliability Engineering and System Safety*, **85**, 11–19.

Ochoa, T. J. and Cleary, T. G. (2003). Epidemiology and spectrum of disease of *Escherichia coli* O157. *Current Opinion in Infectious Diseases*, **16**, 259–263.

Office of the Chief Medical Officer, Welsh Assembly Government (2005). South Wales E.coli O157 outbreak – September 2005: a review commissioned by the chief medical officer.

Ogden, I. D., Hepburn, N. F., MacRae, M., Strachan, N. J. C., Fenlon, D. R., Rusbridge, S. M. and Pennington, T. H. (2002). Long-term survival of *Escherichia coli* O157 on pasture following an outbreak associated with sheep at a scout camp. *Letters in Applied Microbiology*, **34**, 100–104.

Paiba, G. A., Gibbens, J. C., Pascoe, S. J. S., Wilesmith, J. W., Kidd, S. A., Byrne, C., Ryan, J. B. M., Smith, R. R., McLaren, I. M., Futter, R. J., Kay, A. C. S., Jones, Y. E., Chappell, S. A., Willshaw, G. A. and Cheasty, T. (2002). Faecal carriage of verocytotoxin-producing *Escherichia coli* O157 in cattle and sheep at slaughter in Great Britain. *Veterinary Record*, **150**, 593–598.

Paiba, G. A., Wilesmith, J. W., Evans, S. J., Pascoe, S. J. S., Smith, R. P., Kidd, S. A., Ryan, J. B. M., McLaren, I. M., Chappell, S. A., Willshaw, G. A., Cheasty, T., French, N. P., Jones, T. W. H., Buchanan, H. F., Challoner, D. J., Colloff, A. D., Cranwell, M. P., Daniel, R. G., Davies, I. H., Duff, J. P., Hogg, R. A. T., Kirby, F. D., Millar, M. F., Monies, R. J., Nicholls, M. J. and Payne, J. H. (2003). Prevalence of faecal excretion of verocytotoxigenic *Escherichia coli* O157 in cattle in England and Wales. *Veterinary Record*, **153**, 347–353.

Pasteurized Milk Production Workshop (2004). Department of Veterinary Clinical Sciences, University of Liverpool.

Pennington, T. H. (2000). The microbiological safety of food. *Journal of Medical Microbiology*, **49**, 677–679.

Robinson, S. E. (2004). Temporal characteristics of shedding of *Escherichia coli* in UK dairy cattle. Ph.D. thesis, University of Liverpool.

Shillam, P., Woo-Ming, A., Mascola, L., Bagby, R., Lohff, C., Bidol, S., Stobierski, M. G., Carlson, C., Schaefer, L., Kightlinger, L., Seys, S., Kubota, K., Mead, P. S. and Kalluri, P. (2002). Multistate outbreak of *Escherichia coli* O157:H7 infections associated with eating ground beef – United States, June-July 2002 (Reprinted from MMWR, vol 51, pg 637–639, 2002). *Journal of the American Medical Association*, **288**, 690–691.

Teunis, P. F., Ogden, I. D. and Strachan, N. J. C. (2008). Hierarchical dose response of *E.coli* O157:H7 from human outbreaks incorporating heterogeneity in exposure. *Epidemiology and Infection*, **136**, 761–770.

Vose, D. J. (2000). *Risk Analysis: A Quantitative Guide*. John Wiley, New York.

Wang, G., Zhao, T. and Doyle, M. P. (1997). Survival and growth of *Escherichia coli* O157:H7 in unpasteurized and pasteurized milk. *Journal of Food Protection*, **60**, 610–613.

Zhao, T., Doyle, M. P., Shere, J. and Garber, L. (1995). Prevalence of enterohemorrhagic *Escherichia coli* O157:H7 in a survey of dairy herds. *Applied and Environmental Microbiology*, **61**, 1290–1293.

·5·

Mapping malaria in the Amazon rain forest: A spatio-temporal mixture model

Alexandra M. Schmidt, Jennifer A. Hoeting, João Batista M. Pereira and Pedro P. Vieira

5.1 Introduction

In this chapter we develop a spatio-temporal model for malaria outbreaks over a four year period in the state of Amazonas, Brazil. We propose a multivariate Poisson-lognormal model with random effects to capture correlation over time and space. Our goal is to predict malaria counts for unobserved municipalities and future time periods. To achieve this goal, we develop a free-form spatial covariance structure and develop methodology that allows us to 'borrow strength' from observed municipalities to predict unobserved municipalities.

There is a rich history of Bayesian models for disease counts based on a Poisson model (see Wakefield, 2007, and references therein for an overview).

Our data include observations of monthly disease counts collected over four years. To model the temporal correlation between observations, one level of our model incorporates a Bayesian dynamic linear model. Bayesian dynamic linear models have been developed and analysed for some time (West and Harrison, 1997; Migon *et al.*, 2005) and are described in more detail in the appendix.

In the Poisson model for disease counts, it is traditional to include a random effect term to allow for overdispersion in the Poisson model. For disease count modelling for observations collected over a large region, there is often support for models that account for spatial correlation among the random effects. A common model is a conditional autoregressive (CAR) model (Besag *et al.*, 1991). This formulation requires that the observations are observed on a spatial lattice which may be regular or irregular in shape. The CAR model accounts for spatial correlation for site i through a weighted average of the random effects for sites that are neighbours of i. Weights are determined via distances between centroids or are a simple indicator variable indicating whether site j is a neighbour of site i. Another approach to modelling spatial correlation is to adopt an isotropic geostatistical model (Kelsall and Wakefield, 2002). A typical approach in geostatistical modeling is to assume a model such that the correlation between observations is a function of distance and one or more function parameters. This model can be used for lattice and point-reference data.

A number of researchers have proposed spatio-temporal models of disease counts (see Sun *et al.*, 2000, for an overview). Gelfand *et al.* (2005) develop a spatio-temporal model for continuous space and discrete time. In their approach spatial effects are nested within time. Vivar (2007) proposes a generalized dynamic linear model for areal data. A spatial structure is imposed on the covariance of the error term of the evolution equation. Fernandes *et al.* (2009) propose a zero inflated Poisson spatio-temporal model to account for excess zeros. They analyse a time series of the number of cases of dengue fever across districts of the city of Rio de Janeiro. To account for overdispersion in the Poisson mean, they consider latent effects. These effects are modelled with an independent CAR prior for every time point. Nobre *et al.* (2005) develop a spatio-temporal model for malaria counts in the Brazilian state of Pará using rainfall gauge amounts as predictors. They develop a Poisson model which considers spatio-temporal latent effects following a CAR model assumed to assess spatial correlation within each time point and a dynamic linear model component to model correlation of the regression covariates over time. Nobre *et al.* (2005) also propose methodology to address the spatial misalignment issue between the observation of rainfall at gauges located throughout the state and the malaria counts observed at the county-level.

In the disease mapping literature, the region of interest is typically divided into a finite number of disjoint locations and with observations for every region. Our situation is different as we observe a complete time series for a set of regions, but only a subset of the regions that make up an entire Brazilian state. As will be described below, this data structure impacts our modelling in a number of ways. Here our goal is to assess the spatial correlation among the municipalities to assist with estimation of malaria counts over time for municipalities for which data were not collected. Successful estimation of malaria counts for unobserved regions will assist the Brazilian government in their efforts to monitor and control malaria. While we want to borrow strength from the observed municipalities to estimate malaria counts where it was unobserved, a challenge in this case is that the municipalities are scattered throughout the state. We have adopted a flexible free-form covariance model which will capture the spatial correlation, if any, among the malaria counts observed over time. The proposed structure is unique in that it is not a distance- or neighbourhood-based covariance model. Instead, we allow for correlation among all spatial locations to be estimated freely. This is similar to the approach of Le and Zidek (1992) who adopted a similar covariance structure for normally distributed data.

5.2 Motivation

Malaria is a world-wide public health problem with 40% of the population of the world at risk of acquiring the disease. It is estimated that there are over 500 million clinical cases of malaria each year world-wide. It has been shown that malaria outbreaks can cause a downward spiral into poverty for families and communities affected by the disease, and the disease can significantly decrease a country's annual economic growth. In some countries an astonishing 40% of health expenditures can be attributed to malaria treatment (World Health Organization, 2007). While the majority of the burden of the disease occurs in Sub-saharan Africa, the disease is still a major threat to human health in Brazil. Across the Brazilian Amazon basin, more than 500,000 people are infected by malaria every year, despite considerable national and international control efforts. Continued progress in prevention, treatment and the development of innovative tools for the control of malaria is required (Health Ministry of Brazil, 2006).

In this chapter we study outbreaks of malaria over time for twelve municipalities in the state of Amazonas, Brazil. The state has 62 municipalities and around 3.3 million habitants. The observed data are the number of cases of malaria from January 1999 until December 2002 obtained from the Brazilian Epidemiological Surveillance System (SIVEP)[1] for 12 municipalities. The observed time series for each of these municipalities are shown in Figure 5.1. The goal of this work is to determine whether we can model outbreaks within each municipality over time and also whether the pattern of outbreaks between the municipalities can lead to improved models for outbreaks in any one city. Moreover, our model provides sensible ways of estimating the relative risk of malaria for municipalities which were not observed.

Figure 5.2 shows the state of Amazonas. The shaded areas correspond to the municipalities for which we have malaria counts. It is clear that these municipalities do not share common boundaries; moreover, they are quite far apart. Therefore it does not seem reasonable to use standard models based on spatially-explicit conditional autoregression (CAR) or geostatistical latent effects to fit a joint model for these time series (Banerjee *et al.*, 2004). In the next section we propose a spatio-temporal model which provides a flexible covariance structure among these time series.

In the disease mapping literature it is common practice to assume that the number of cases of a disease in region i and time t, y_{ti}, are independent realizations from a Poisson distribution with mean $\mu_{ti}e_{ti}$, that is

$$y_{ti} \mid \mu_{ti}, e_{ti} \sim Po(\mu_{ti}e_{ti}).$$

[1] www.saude.gov.br/sivep_malaria.

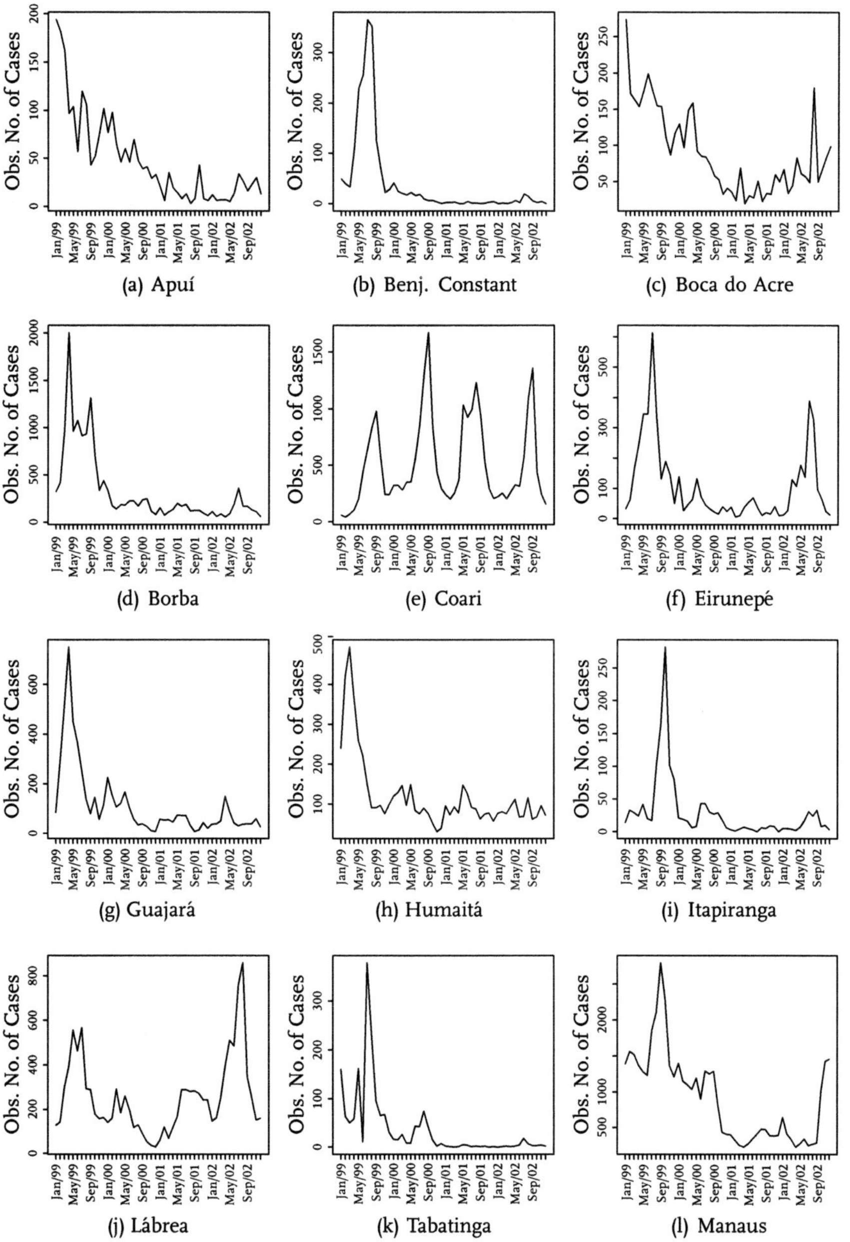

Fig. 5.1 Observed time series of the number of cases of malaria for each of the 12 municipalities, from January 1999 until December 2002.

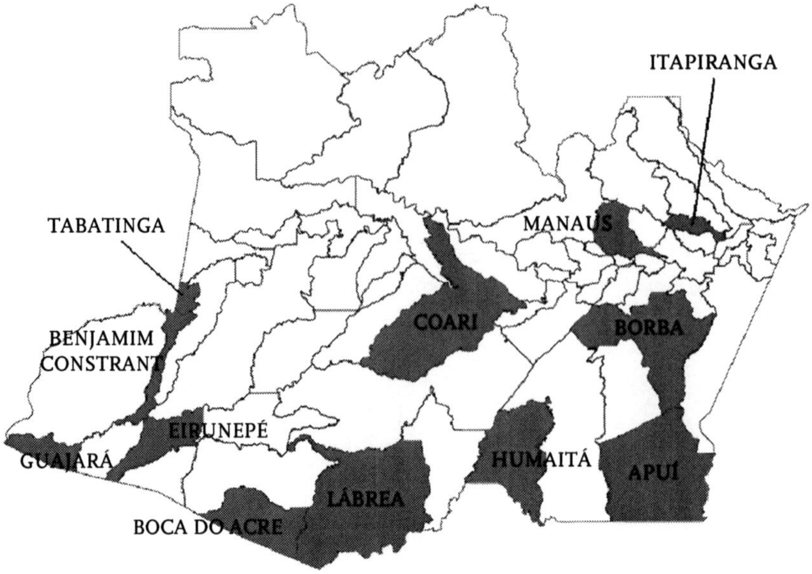

Fig. 5.2 The Amazon state and the municipalities for which we have malaria counts.

Here e_{ti} is the expected number of cases and μ_{ti} is the relative risk of the disease in region $i = 1, \ldots, n$ at time $t = 1, \ldots, T$. The focus of the analysis is to estimate μ_{ti}, but we must also calculate e_{ti}. Typically, e_{ti} is assumed fixed and a known function of pop_i, the population at risk for the disease in location i (Banerjee *et al.*, 2004). A standard approach is to assume that

$$e_{ti} = pop_i \frac{\sum_{i=1}^{n} y_{ti}}{\sum_{i=1}^{n} pop_i}, \tag{5.1}$$

therefore population effects are eliminated by including the expected number of cases in each region and time, as an offset in the analysis. Banerjee *et al.* (2004) note that this process is called internal standardization, since it centres the data but uses only the observed data to do so. Other standardization approaches are available, see Banerjee *et al.* (2004, p. 158) for details.

Based on independent realizations of the Poisson distribution, the maximum likelihood estimate of μ_{ti}(i.e. $\hat{\mu}_{ti} = y_{ti}/e_{ti}$) is commonly referred to as the standardized mortality ratio (SMR). Figure 5.3 shows the mean and variance of the SMRs across time (panels a and b) for each of the municipalities and across locations (panels c and d) for each time t. It is clear that the variances vary across municipalities, moreover, in both dimensions, spatial and temporal, we observe variances much greater than the mean. Although we do not present the results here, the estimated autocorrelation function for the SMR computed for each municipality shows temporal structure for most municipalities.

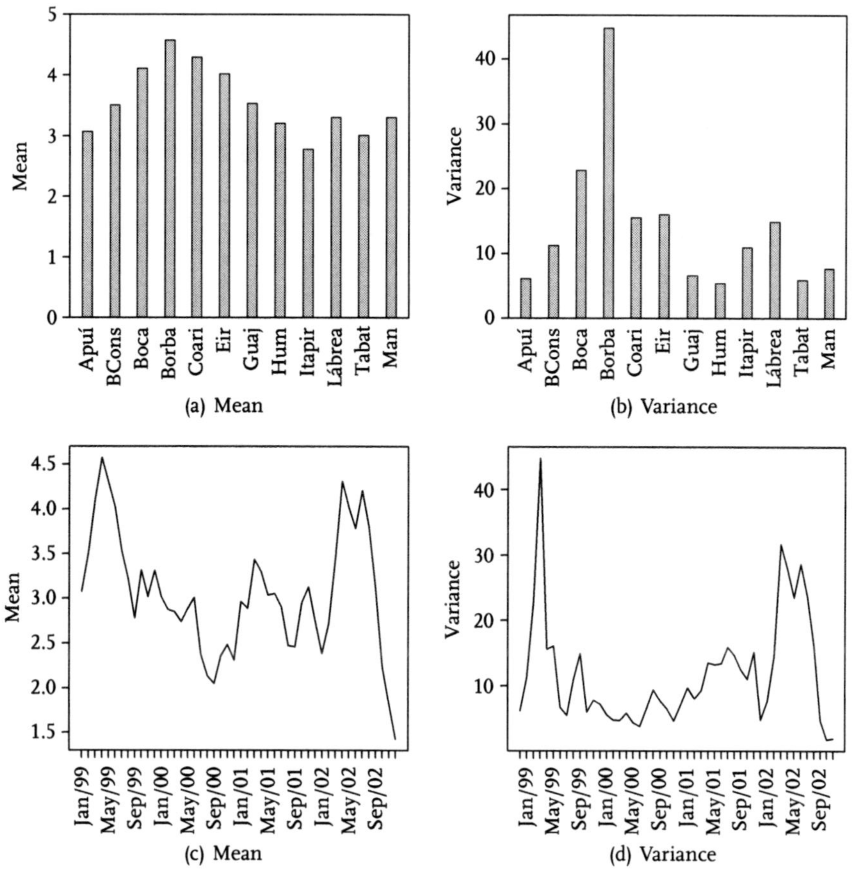

Fig. 5.3 Estimated mean and variance of the SMR of malaria, across time for each municipality (a and b), and across municipality for each time (c and d).

The overdispersion shown in Figure 5.3 is typically modelled using correlated latent structure in $\log \mu_{ti}$. Aiming to investigate if there is any spatial structure remaining in the logarithm of the SMR, we consider the sample correlation matrix obtained from $\log \text{SMR}_{ti}$ at $n = 12$ municipalities over T time points. Figure 5.4 presents the distance among centroids of the municipalities versus these estimated correlations. It is clear that for a fixed distance there is a wide range of correlations between sites. Also, some sites are negatively correlated at short distances. This suggests that distance-based spatial correlation structures might not be able to capture this complex covariance structure among the $\log \text{SMR}_{ti}$.

Generally speaking, this exploratory data analysis suggests that we should consider models which allow for different variances across municipalities, as well as overdispersion and spatio-temporal structures. Our proposed model is given in the next section.

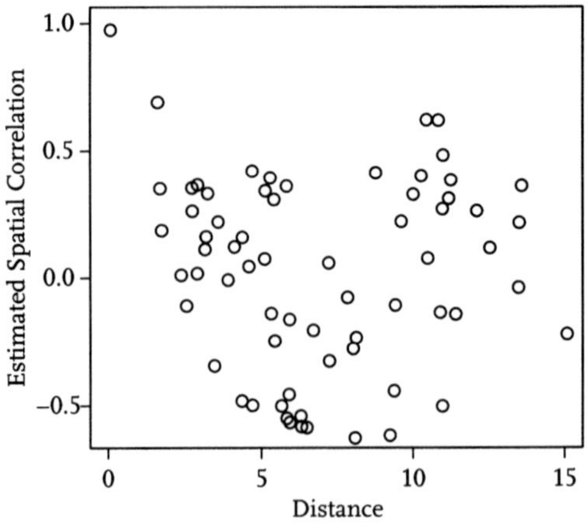

Fig. 5.4 Estimated correlation structure among the time series of the logarithm of the SMR versus the distance between centroids of the municipalities.

5.3 A multivariate Poisson – lognormal model

5.3.1 Model

Let y_{ti} be the number of cases of malaria at month t and location i, $t = 1, \cdots, T$ and $i = 1, \cdots, n$. Assume that, conditional on some parameters, each y_{ti} represents independent realizations from a Poisson distribution, that is

$$\begin{aligned} y_{ti} \mid \mu_{ti}, e_{ti} &\sim Po(\mu_{ti}\, e_{ti}), \\ \log \mu_t &= F_t\theta_t + \delta_t, \quad \delta_t \sim N_n(0, \Sigma), \\ \theta_t &= G\theta_{t-1} + \omega_t, \quad \omega_t \sim N_p(0, W), \end{aligned} \tag{5.2}$$

where $\mu_t = (\mu_{t1}, \ldots, \mu_{tn})'$, $\delta_t = (\delta_{t1}, \cdots, \delta_{tn})'$, θ_t is a vector of dimension p representing common temporal structures across locations, and $\omega_t = (\omega_{t1}, \cdots, \omega_{tp})'$. The components of $n \times p$ matrix F_t might comprise covariates, a common level, a trend, and/or seasonal components, which may vary by location and across time. The random effect terms account for any variability remaining after accounting for the covariates. We assume that δ_{ti}s are independent across time, but spatially related, for each time t. The spatial structure is described through Σ, the covariance among the elements of δ_t. We describe our treatment of Σ in Section 5.3.1.1. The $p \times p$ matrix G is a known matrix whose form will reflect the proposed structure on the θ_t parameters (see West and Harrison, 1997, for examples). We assume the components of ω_t to be independent, therefore $W = diag(W_1, \cdots, W_p)'$.

If we let $a_t = F_t\theta_t + \delta_t$, where $a_t = (a_{t1}, \cdots, a_{tn})'$, then since $\delta_t \sim N(0, \Sigma)$ it follows that $a_t \sim N(F_t\theta_t, \Sigma)$ or, equivalently, $\log \mu_t \sim N(F_t\theta_t, \Sigma)$. Therefore, our model can be described as a multivariate Poisson-log normal distribution mixture as presented in Aitchison and Ho (1989). Chib and Winkelmann (2001) discuss the fitting of this model for correlated count data via Markov chain Monte Carlo methods. Our proposed model can be seen as a generalization, to account for spatio-temporal structure, of that proposed by Chib and Winkelmann (2001).

The second level of hierarchy of the model in (5.2) is a multivariate dynamic linear model (West and Harrison, 1997, and the appendix). The relative risks μ_{ti}s are decomposed into the sum of two components, one which captures common temporal structure and another which captures local effects. Therefore this model can also be seen as a general linear model extension of the hierarchical dynamic model as proposed by Gamerman and Migon (1993).

It is useful to consider the properties of the model in (5.2). Using the properties of conditional expectations, it can be shown that

$$\begin{aligned}
E(y_{ti} \mid e_{ti}) &= e_{ti} \exp([F_t]_i \theta_t + 0.5\sigma_{ii}) = \beta_{ti}, \\
V(y_{ti} \mid e_{ti}) &= \beta_{ti} + \beta_{ti}^2\{\exp(\sigma_{ii}) - 1\}, \\
Cov(y_{ti}, y_{tj} \mid e_{ti}, e_{tj}) &= \beta_{ti}\beta_{tj}\{\exp(\sigma_{ij}) - 1\}, \qquad (5.3)\\
Corr((y_{ti}, y_{tj} \mid e_{ti}, e_{tj}) &= \frac{\exp(\sigma_{ij}) - 1}{\left[\left\{\exp(\sigma_{ii}) - 1 + \beta_{ti}^{-1}\right\}\left\{\exp(\sigma_{jj}) - 1 + \beta_{tj}^{-1}\right\}\right]^{1/2}},
\end{aligned}$$

with $[\Sigma]_{ii} = \sigma_{ii}$ and $[F_t]_i$ represents the i-th row of F_t.

5.3.1.1 *Specifying the spatial covariance structure, Σ*

A number of models are available for the spatial covariance structure. We consider three stationary models for the covariance parameter, Σ. The third structure, a free-form covariance matrix, is the proposed structure for this application.

The first covariance structure is an independent model, with separate variances for each location. The second structure is a geostatistical structure where the covariance is based on the Euclidean distance between the centroids (Kelsall and Wakefield, 2002). For example, one could consider $\Sigma_{ij} = \sigma^2\rho(d_{ij};\phi)$ where $\rho(d;\phi)$ is a valid correlation function depending on the Euclidean distance between centroids and some parameter vector ϕ. Similarly, for areal spatial data, CAR models are often the covariance structure of choice. However, note that the CAR approach does not result in a proper prior distribution for the latent effects δ_t in (5.2).

In contrast to these standard approaches, we propose a more flexible free-form covariance structure. Note that Figure 5.4 suggests that the spatial correlation structure present in the data is not isotropic. That is, for a fixed distance, the observed correlation varies significantly. Instead of the more restricting geostatistical structure, we propose that Σ follows a free-form covariance structure. More specifically we assume an inverse Wishart prior for Σ with scale matrix

Ψ and m degrees of freedom, with $m > n+1$, such that the prior mean is finite. This prior guarantees that Σ is positive definite. Moreover, it gives flexibility and allows the model to reflect the correlation structure in the data, without imposing any specific structure. One advantage of this approach is that we allow the random effect of each municipality to have its own variance, which seems to be reasonable for this dataset, as shown in panel (b) of Figure 5.3. In contrast to a CAR prior structure for Σ, our model provides an explicit structure for the covariance between the counts for any two municipalities, as shown in equation (5.3). For the free-form covariance structure the diagonal elements of Σ do represent the variance of the marginal distributions of the latent effects, avoiding a drawback of the CAR model. An advantage of the free-form covariance structure over the geostatistical approach is that it allows for both negative and positive correlations.

Care must be taken when choosing the hyperparameters for the inverse Wishart prior for Σ. It is difficult to obtain information about the covariance matrix Σ as, for each time t, the elements represent the covariance among the latent effects δ_{ti}s. While the free-form structure is very flexible, we can consider several forms for the scale matrix Ψ. One approach is to assume that elements are uncorrelated a priori. To specifically allow for spatial correlation, another approach is to assume a distance-based form for Ψ whose parameters are based on the practical spatial range. One possibility is to assume that Ψ follows an exponential correlation function with some fixed variance, and fix the decay parameter such that the associated range is one-half the maximum interlocation distance. Such an approach is common practice in the geostatistical literature (Banerjee *et al.*, 2004). This model allows more flexibility than a standard geostatistical model because the free-form covariance structure allows negative elements. Similarly, one could use the sample covariance matrix obtained from the multiple time series. However, this approach uses the data twice (both for the prior and posterior distribution). In either case, we can account for our lack of knowledge of this matrix by maintaining the degrees of freedom parameter m to be small. Indeed, to represent the lack of available information about Σ we suggest fixing m as small as possible, according to the dimension of the problem. Nadadur *et al.* (2005) call the idea of considering a structure on the form of Ψ a correlation filter and consider several different correlation filters.

5.3.2 Bayesian Inference

5.3.2.1 Prior distributions and inference procedure

From a Bayesian point of view, model specification is complete after assigning the prior distribution to the unknowns. Let Θ be the parameter vector, that is $\Theta = (\alpha_1, \ldots, \alpha_T, \theta_0, \theta_1, \ldots, \theta_T, W_1, \ldots, W_p, \Sigma, \Psi, m)$. We assume the joint prior distribution

$$p(\Theta) = \prod_{t=1}^{T} \{p(\alpha_t \mid \theta_t, \Sigma)p(\theta_t \mid \theta_{t-1}, W)\} \; p(\theta_0 \mid D_0) \prod_{i=1}^{p} p(W_i) \, p(\Sigma \mid \Psi, m).$$

Following standard practice, we assume that θ_0 follows a normal distribution with zero mean and some large variance, C_0. As we have no prior information about the variances of the evolutions of θ_t, and we do not expect them to vary a lot, we assign for the components of W an inverse gamma prior with infinite variance and scale parameter fixed at some small value. The prior distribution for Σ depends on the covariance structure, as described in Section 5.3.1.1.

5.3.2.2 Posterior distribution

Assuming we observe $y_{1:T,1:n} = (y_{11}, \ldots, y_{1n}, \ldots, y_{T1}, \ldots, y_{Tn})'$ and conditional on Θ, for each time t and municipality i, we have independent realizations from a Poisson distribution with mean $\mu_{ti}e_{ti}$; therefore, the likelihood is proportional to

$$p(y_{1:T,1:n} \mid \Theta) \propto \prod_{t=1}^{T} \prod_{i=1}^{n} \exp\left(-\mu_{ti}e_{ti}\right) \mu_{ti}^{y_{ti}}.$$

Following the Bayesian paradigm, the posterior distribution is proportional to the prior distribution times the likelihood function. It is straightforward to see that the posterior distribution does not produce the kernel of any known distribution. Therefore we make use of Markov chain Monte Carlo (MCMC) methods (Gamerman and Lopes, 2006; Givens and Hoeting, 2005) to obtain samples from the posterior distribution of Θ. Due to the temporal structure of the parameters, we have to make use of sampling methods which are efficient in spanning the parameter space. More specifically, we propose a Gibbs sampler which uses the Metropolis – Hastings algorithm for some steps. Our sampling scheme takes advantage of the multivariate dynamic linear model structure to obtain samples for θ_t. In other words, we reparameterize the model using α_t. For each iteration of the algorithm, we sample α_t, $t = 1, \cdots, T$ using a Metropolis – Hastings step based on the proposal suggested by Gamerman (1997). Now, given α_t, the likelihood does not depend on θ_t, so we have

$$\begin{aligned} \alpha_t &= F_t\theta_t + \delta_t, \quad \delta_t \sim N(0, \Sigma), \\ \theta_t &= G\theta_{t-1} + \omega_t, \quad \omega_t \sim N(0, W), \end{aligned}$$

which is a multivariate dynamic linear model, and we make use of the forward filtering, backward sampling algorithm (FFBS) as proposed by Frühwirth-Schnater (1994) (and described in the appendix), to obtain samples of θ_t. The initial parameter θ_0 follows a normal posterior full conditional distribution.

The covariance matrix Σ follows an inverse Wishart posterior full conditional distribution and is easy to sample from. Each element of W follows an inverse gamma posterior full conditional distribution, which is also easy to sample from.

5.3.3 Predictions for unobserved locations and future time periods

5.3.3.1 Spatial interpolation: Estimating the relative risk for unobserved municipalities

The model in (5.2) allows us to estimate the number of cases for municipalities for which we do not have information. This may be of interest, for example, to help allocate funds for disease prevention. Nobre *et al.* (2005) faced a similar situation and assumed that the observations were missing. They used their proposed model to estimate the counts for municipalities which were not observed. However, they assume an independent CAR prior for each time t for the latent spatial effects. In other words, these effects, for the unobserved municipalities, were not assumed to be generated from the same underlying spatial process (see Banerjee *et al.*, 2004, p. 82).

Le and Zidek (1992) propose an alternative model for kriging, based on the properties of conjugacy between the multivariate normal and the inverse Wishart distributions. The goal is to obtain an estimate of the covariance between the observed and non-observed municipalities. We can adopt a similar approach, adapted for our multivariate Poisson-lognormal model.

Following Le and Zidek (1992), define $y_t = (y_t^1, y_t^2)'$, where the $u \times 1$-vector y_t^1 contains the elements for the unmonitored municipalities, and the $g \times 1$-vector y_t^2 contains the data from monitored sites, the observed counts for the municipalities for which malaria data were collected. Here, $n = u + g$. Again we assume $\delta_t \sim N(0, \Sigma)$. If we define a similar partition for $\delta_t = (\delta_t^1, \delta_t^2)$, and $\alpha_t = (\alpha_t^1, \alpha_t^2)$, then

$$\Sigma = \begin{pmatrix} \Sigma_{11} & \Sigma_{12} \\ \Sigma_{21} & \Sigma_{22} \end{pmatrix}, \tag{5.4}$$

where Σ_{11}, a $u \times u$ matrix, represents the covariance among the unmonitored municipalities, Σ_{22}, $g \times g$, the matrix whose elements are the covariances among monitored municipalities, and Σ_{12}, is $u \times g$ with the covariance between the monitored and unmonitored municipalities. Now, in the MCMC sampling scheme described in Section 5.3.2.1 we have to include the parameters related to the unmonitored locations. For α_t^1 we have a normal posterior full conditional distribution, as the likelihood does not bring any information about the unmonitored locations. More specifically, $\alpha_t^1 \mid \alpha_t^2, \Sigma, y_{1:T,1:n} \sim N(\mu_t^{1|2}, \Sigma^{1|2})$, with $\mu_t^{1|2} = F_t^1\theta_t^1 + \Sigma_{12}\Sigma_{22}^{-1}(\alpha_t^2 - F_t^2\theta_t^2)$, where F_t^2 is the $g \times p$ matrix of covariates for the unmonitored sites and θ_t^2 is the corresponding $p \times 1$ vector of parameters, and $\Sigma^{1|2} = \Sigma_{11} - \Sigma_{12}\Sigma_{22}^{-1}\Sigma_{21}$. And for y_t^1 we have a Poisson posterior full conditional distribution.

However, following Le and Zidek (1992), if there is any information about how the monitored municipalities are related to the unmonitored sites, we can use this information in assigning the prior distribution for Σ in (5.4). Following the Bartlett decomposition, we can assume

$$\Sigma = \begin{pmatrix} \Sigma_{11} & \Sigma_{12} \\ \Sigma_{21} & \Sigma_{22} \end{pmatrix} = \begin{pmatrix} \Sigma_{1|2} + \tau\Sigma_{22}\tau' & \tau\Sigma_{22} \\ \Sigma_{22}\tau' & \Sigma_{22} \end{pmatrix},$$

where $\Sigma_{1|2} = \Sigma_{11} - \Sigma_{12}\Sigma_{22}\Sigma_{21}$ and $\tau = \Sigma_{12}\Sigma_{22}^{-1}$. Then, instead of assigning an inverse Wishart distribution to the entire covariance matrix Σ as discussed above, we assign prior distributions to the elements of the Bartlett decomposition. We assume that a priori,

$$\begin{aligned} \Sigma_{22} \mid \Psi, m &\sim IW(m - u, \Psi_{22}), \\ \Sigma_{1|2} \mid \Psi_1, m &\sim IW(m, \Psi_{1|2}), \\ \tau \mid \Sigma_{1|2}, \Psi &\sim N\left(\eta, \Sigma_{1|2} \otimes \Psi_{22}^{-1}\right), \end{aligned}$$

with Ψ following the same partition as that of Σ, so $\eta = \Psi_{12}\Psi_{22}^{-1}$. As noted by Le and Zidek (1992), $\Sigma_{1|2}$ represents the residual covariance of δ_t^1 after optimal linear prediction based on δ_t^2, and τ represents the slope of the optimal linear predictor of δ_t^1 based on δ_t^2. Based on the interpretation of τ and $\Sigma_{1|2}$, and aiming to decrease the uncertainty about the spatial interpolation, one can use the information on the estimated relative risk to assign the values for η and $\Psi_{1|2}$.

In this case we can compute the posterior full conditional distributions, which will be given by

$$\begin{aligned} \Sigma_{22} \mid \delta, \Psi_{22}, m, y_{1:T,1:n} &\sim IW(m + T - u, \hat{\Psi}_{22}), \\ \Sigma_{1|2} \mid \Psi_1, m, y_{1:T,1:n} &\sim IW(m, \Psi_{1|2}), \qquad (5.5) \\ \tau \mid y_{1:T,1:n}, \Sigma_{1|2}, \Psi &\sim N\left(\eta, \Sigma_{1|2} \otimes \Psi_{22}^{-1}\right), \\ \alpha_t^1 \mid \alpha_t^2, \Sigma, y_{1:T,1:n} &\sim N(\mu^{1|2}, \Sigma^{1|2}), \qquad (5.6) \end{aligned}$$

where $\hat{\Psi}_{22} = \Psi_{22} + S_2$, and $S_2 = \sum_{t=1}^{T}(\alpha_t^2 - F_t^2\theta_t^2)(\alpha_t^2 - F_t^2\theta_t^2)'$. Notice that since $\Sigma_{1|2}$ and τ are not in the likelihood, their posterior distributions are proportional to their prior distributions.

5.3.3.2 Temporal prediction

Another aim of this work is to obtain temporal predictions for future time periods. As mentioned previously, it is common practice in the disease mapping literature, to assume the latent effects δ_{ti} to follow a CAR or geostatistical model. However, in the case of a CAR prior, due to the independence across time, it is not possible to perform temporal predictions. This is because the posterior distribution of δ_{ti}, for times which were not observed for any municipality, would be proportional to its prior, resulting in this case on an improper posterior distribution. Here, due to the propriety of our prior distribution, we do not have this problem.

Following equation (5.2), the posterior predictive distribution for predictions k steps ahead, is given by

$$\begin{aligned} p(y_{T+k} \mid y_{1:T,1:n}) &= \int_{\Theta} p(y_{T+k} \mid y_{1:T,1:n}, \Theta) p(\alpha_{T+k} \mid \theta_{T+k}, \Sigma) \\ &\qquad p(\theta_{T+k} \mid \theta_{T+k-1}, W)\, p(\Theta \mid y_{1:T,1:n}) d\Theta \\ &\approx \frac{1}{L} \sum_{l=1}^{L} Po\left(\mu_{T+k}^{(l)} e_{T+k}\right), \end{aligned}$$

with $\log \mu_{T+k}^{(l)} = \alpha_{T+k}^{(l)}$, where the superscript (l) represents the lth MCMC sample from the posterior of Θ. Note that $\theta_{T+k}^{(l)}$, $k > 0$ is obtained by propagating the samples from the posterior through the evolution equations for θ_t. To compute k-th step ahead value for the expected number of cases, we use only the observed data, so $e_{T+k} = e_T$.

5.4 Results

Before fitting a joint model for these time series we fitted a separate dynamic linear model to each of the $n = 12$ time series to investigate whether we should consider models with trend and/or seasonality. For most of the municipalities, the simplest level-only model fitted the data best in terms of mean square error and mean absolute error. Therefore, in (5.2), we assume $F_t = 1_n$, with 1_n denoting the $n \times 1$ vector of ones, and $G = 1$. Our proposed model was fitted to the time series shown in Figure 5.1.

For all the results shown in this section, we let the MCMC procedure run for 250, 000 iterations, used a burn-in of 50,000 and stored every 100th iteration. Convergence was investigated using two independent chains starting from very different values.

5.4.1 Covariance structures

We considered three different structures for Σ, the covariance matrix among the elements of δ_t. We aim to compare the gain in assuming a free-form covariance structure which allows for spatial correlation among the municipalities to two other structures: (1) a distance-based covariance matrix which imposes the correlation as a function of the Euclidean distance between centroids; and (2) a diagonal covariance matrix which enforces independence between the latent effects of the municipalities. Therefore we fitted three models with the following structures for Σ:

Free-form covariance matrix: Σ is assumed to have an inverse Wishart prior distribution, with 20 degrees of freedom. As proposed in Section 5.3.1.1 we assume that Ψ follows a geostatistical model using an exponential correlation function with variance equal to 1. We fixed the decay parameter such that the associated range is one-half the maximum interlocation distance. (We also considered an independent structure for Ψ but this did not lead to any significant differences in the results.)

Geostatistical covariance matrix: Σ is assumed to follow an exponential correlation function, i.e. $\Sigma_{ij} = \sigma^2 \exp(-(1/\phi)\, d_{ij})$, where d_{ij} is the Euclidean distance between the centroids of municipalities i and j, $i, j = 1, \ldots, 12$. In this case, we assume a priori that σ^2 follows an inverse gamma prior distribution with mean 1 and infinite variance, whereas ϕ is assumed to follow a gamma prior with mean based on the practical range, as suggested in Section 5.3.1.1, and some large variance.

Independent covariance matrix: Σ is a diagonal matrix, with elements $\Sigma = \text{diag}(\sigma_1^2, \ldots, \sigma_n^2)$, and σ_i^2 is assumed to have an inverse gamma prior distribution with mean 1 and infinite variance, for $i = 1, \ldots, n$.

It is worth noting that when we assume an independent model for the latent effects, we are imposing that the counts of different municipalities are uncorrelated, as is clear from (5.3). In this case, the joint model produces estimates of a common temporal level θ_t. This is true because for each time t, the likelihood is based on g replicates of the number of counts. On the other hand, when we assume a full covariance matrix, although the latent effects δ_{ti}s are assumed independent across time, at each time t, we borrow strength from other locations in the estimation.

5.4.2 Model comparison

As discussed in Section 5.3.3.2 our model provides temporal forecasts for the number of cases, as well as, for the relative risk for any observed municipality for times $T+k$, $k = 1, \ldots, K$. We have fitted all three models leaving $K = 8$ observations out of the sample and obtained a sample from the predictive distribution for all 12 municipalities from May 2002 until December 2002. The panels in Figure 5.5 show the posterior summary of realizations from the predictive distribution of these temporal forecasts for two municipalities, Apuí and Humaitá. For both municipalities, the observed values (hollow circles) fall within the 95% posterior credible intervals. As expected, as time passes, the uncertainty about the predictions increases. However, it is clear that the increase is greater when the independent covariance matrix model is assumed.

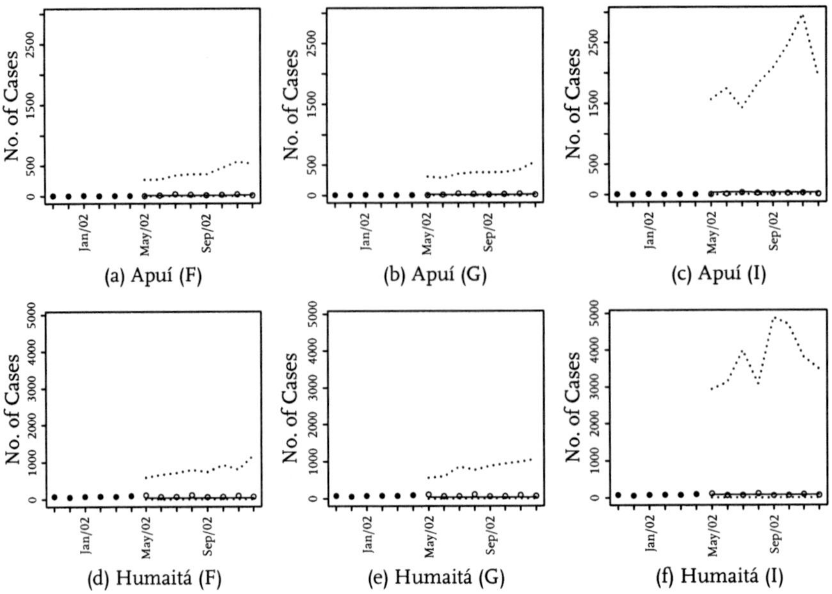

Fig. 5.5 Summary of the temporal predictive distribution ($K = 8$ steps ahead) of the number of cases of malaria for municipalities Apuí and Humaitá based on a free-form (F), geostatistical (G), and on a independent (I) covariance model. Solid line: posterior mean, and dotted lines: limits of the 95% posterior credible interval, solid circles: observed, hollow circles: omitted.

To compare how the three fitted models perform in terms of temporal predictions, we used the criteria based on logarithmic scoring rule proposed by Gneiting *et al.* (2007) and described by Gschlößl and Czado (2007). The model providing the highest log score is preferred. The panels of Figure 5.6 show the mean of the log-score for each municipality. The log-score indicates that the predictions from the free-form model for all municipalities, except Manaus, is preferred.

To investigate how the models perform in terms of spatial interpolation, we omitted one city out of the sample at a time and predicted the number of cases of malaria for the observed period. Figure 5.7 presents the summary of the predictive distribution of malaria counts for municipalities Guajará, Itapiranga, and Humaitá when they were omitted from the inference procedure, for the free-form, geostatistical, and independent models. As expected the uncertainty about the predictions is quite high, but for most of the periods and municipalities, the observed values fall within the 95% posterior credible interval. Similar to the results for the temporal forecasts, the log-scores for the free-form covariance model is considerably higher for all three cities.

As is clear from the model in (5.2) we are decomposing the relative risk of malaria in one municipality as the sum of two components: an overall temporal

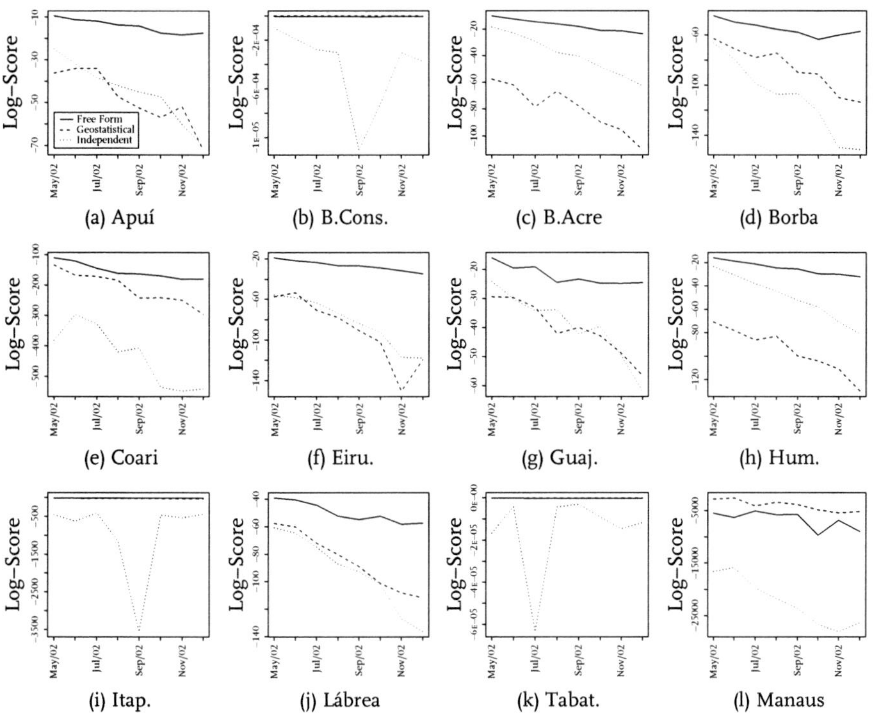

Fig. 5.6 Mean of the log-score for each time $T + k$, $k = 1, \ldots, 8$. The solid line is the mean of log score for the free-form covariance model, the dashed line is for the geostatistical structure, and the dotted line is for the independent model. In Benjamin Constant, Itapiranga and Tabatinga, the free-form and geostatistical models give very similar log score values.

level, θ_t, and a local temporal-municipal random effect, δ_{ti}. Figure 5.8 shows the summary of the posterior distribution of $\exp(\theta_t)$ from the three models, the free form spatial covariance model (panel a), the geostatistical model (panel b), and the independent covariance model (panel c). All models estimate a smoothly decaying trend in overall risk across months. However, the independent covariance model provides higher estimates of $\exp(\theta_t)$, with a clearer seasonal pattern, having higher values from December through February. All models agree that the overall relative risk is quite high, being greater than 1 for most of the months.

The panels in Figure 5.9 present the posterior summary of the relative risk for each of the municipalities under the free-form covariance model. Recall that we have a sample from the posterior distribution of each α_{ti}, therefore, we can obtain samples from $\exp(\alpha_{ti})$, for each $t = 1, \ldots, T$ and $i = 1, \ldots, N$. From these panels it is clear that the relative risk decreases over time for some municipalities, and increases for others. Manaus, the municipality with the biggest population (over 1.4 million), has a relative risk less than one, whereas Guajará,

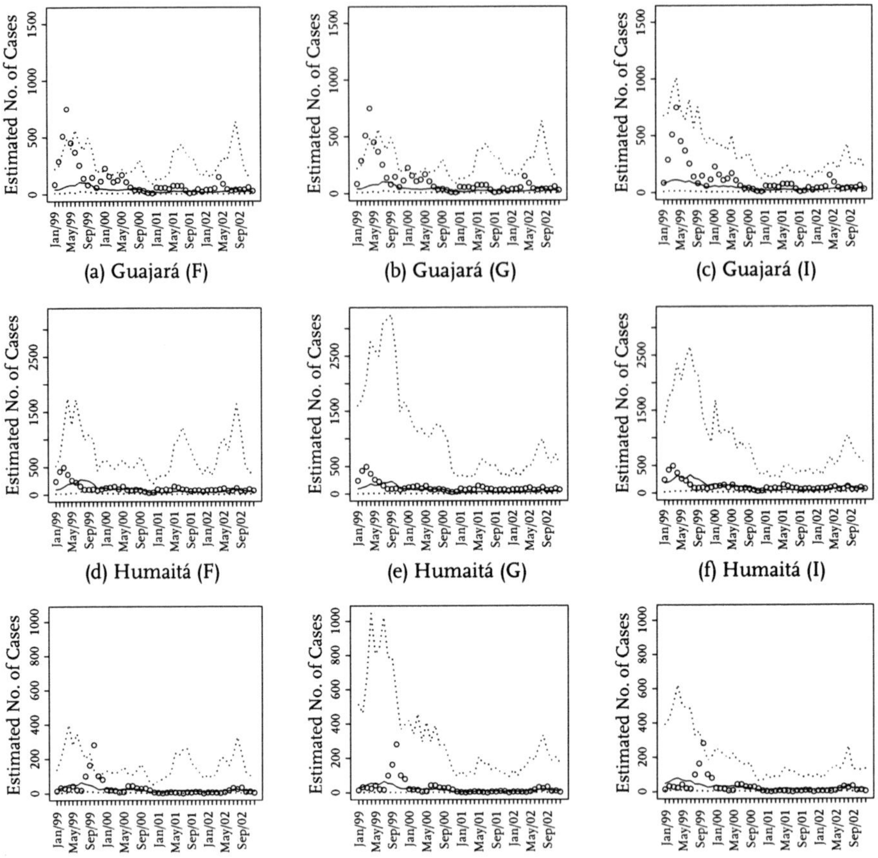

(a) Guajará (F) (b) Guajará (G) (c) Guajará (I)

(d) Humaitá (F) (e) Humaitá (G) (f) Humaitá (I)

(g) Itapiranga (F) (h) Itapiranga (G) (i) Itapiranga (I)

Fig. 5.7 Posterior summary of the predicted values when each of the municipalities are left out from the dataset, both under the free-form (F), geostatistical (G) and independent (I) models.

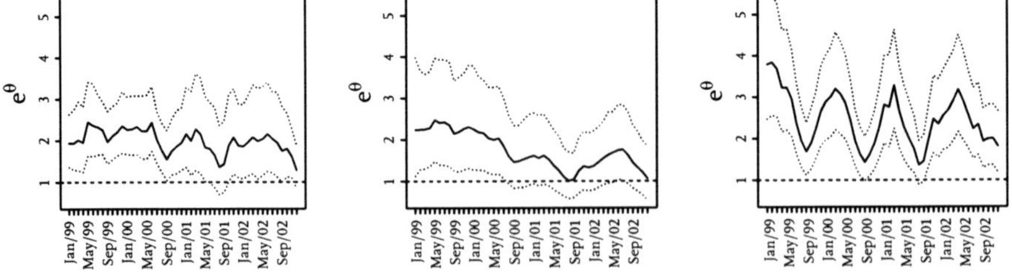

(a) Free-form (b) Geostatistical (c) Independent

Fig. 5.8 Summary of the posterior distribution (solid line: posterior mean, and dotted lines: limits of the 95% posterior credible interval) of the overall relative risk of malaria ($\exp(\theta_t)$), from January 1999 until December 2002, for the (a) free-form, (b) geostatistical, and (c) independent models.

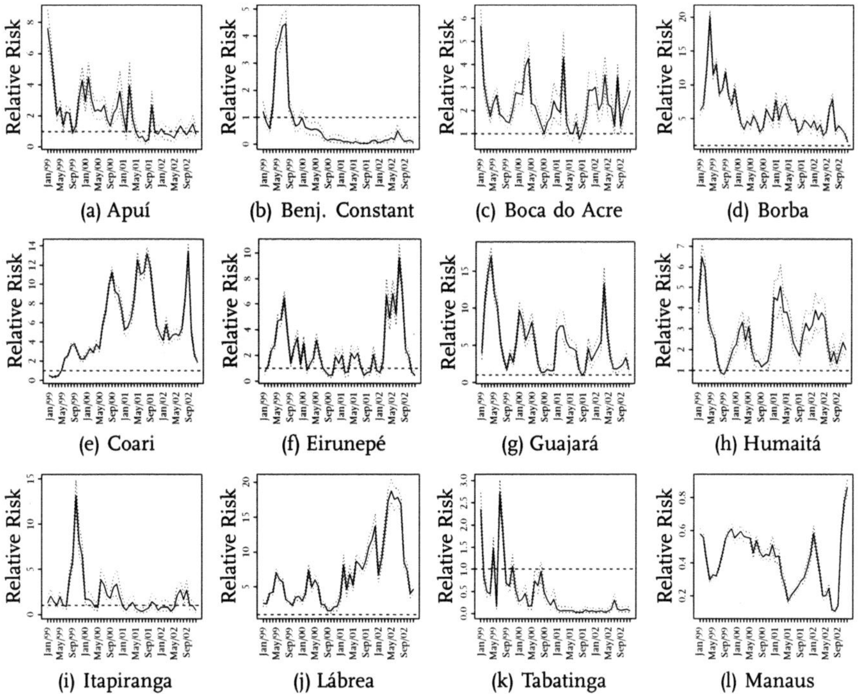

Fig. 5.9 Summary of the posterior distribution of the relative risk ($\exp(a_{ti})$) of malaria, under the free-form covariance model, for each of the 12 municipalities, from January 1999 until December 2002.

whose population in 2000 was 13,358 inhabitants, has very high relative risk, with the highest values during January and May.

Figure 5.10 shows the posterior summary of the diagonal elements of Σ under the free-form and independent models. The posterior distribution seems to give support for having different elements in the diagonal of Σ. Also, under both models the municipalities Benjamin Constant and Tabatinga produce much higher values of the variances when compared to the other municipalities. From Figure 5.1 we notice that the time series of these two municipalities start with quite high values and decrease significantly towards the end of the period.

One of the benefits of using the proposed model is that we are able to estimate the correlation among the elements of δ_t. The prior assigned to Σ allows the data to determine the shape of the correlation structure among δ_ts. Figure 5.11 shows the posterior median of the correlation among the municipalities. The pattern suggests that the simpler independence and geostatistical spatial correlation models are not appropriate as many municipalities exhibit significant correlations and a number of these are negative. It is also clear that the estimated correlation has no clear relationship with the distance between

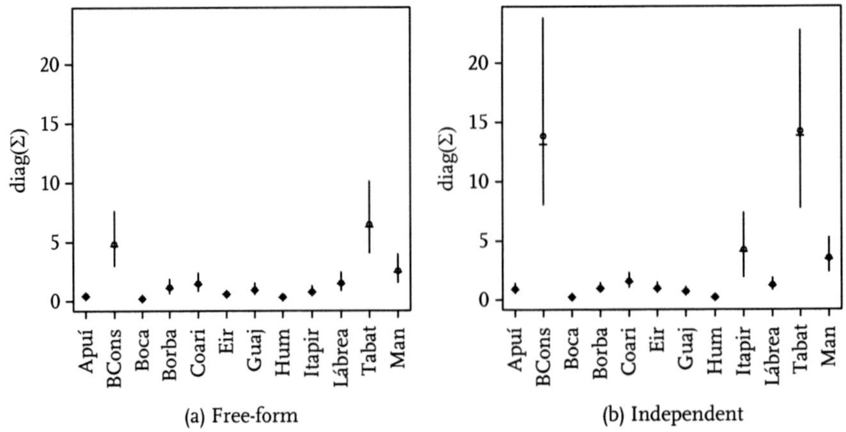

Fig. 5.10 Summary of the posterior distribution (○: posterior mean, – posterior median, and limits of the 95% CI) of the diagonal elements of the covariance matrix Σ, under the free-form and the independent models.

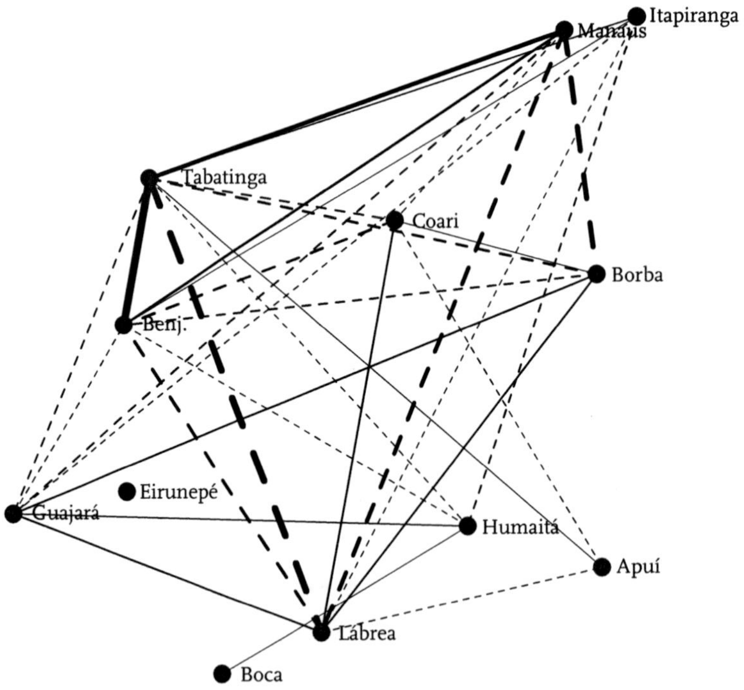

Fig. 5.11 Estimated correlation structure of δ_t (based on the posterior median of elements of Σ). Solid lines link municipalities that are positively correlated and dotted lines link negatively correlated municipalities. The thickness of each line is proportional to the strength of correlation. Only correlations that are significantly different from 0 are included.

centroids. For example, Eirunepé is close to Guajará but these municipalities are not correlated. These results thus support the conclusion that the free-form covariance structure is appropriate for this application.

5.5 Further discussion

The model in (5.2) allows for local covariates over space and time in the $n \times p$ matrix F_t. We also examined a covariate for rainfall over the entire region, but these were not informative. If additional rain gauge data from throughout the state of Amazonas or other relevant data to predict malaria outbreaks were available, the model should provide improved predictions of malaria counts, particularly for locations where malaria counts are not observed.

Other extensions of the model that could be considered include the modelling of e_{ti} in (5.1). We followed the traditional approach of internal standardization by using the observed data to specify this component. Care must be taken with the computation of this term when we aim to interpolate the number of cases in space or predict in time. Here, when we removed municipalities from the sample, we computed the expected values based solely on the set of observed locations. However, this approach likely underestimates the expected values. A reasonable approach would be to use some demographic model to estimate how these populations vary over space and time and incorporate this information into the model. In future work we plan to investigate models which consider this. It also might be useful to investigate whether a model for the covariance among random effects, Σ in (5.2), should evolve over time. We could follow the proposals described in Liu (2000) and Sansó *et al.* (2008).

If data on other diseases were available, a multivariate Poisson model may lead to new insights into the set of diseases under consideration. For example, the Brazilian government may be interested in studying the covariation between the mosquito-borne diseases of malaria and dengue fever (although these two diseases are spread by different species of mosquitoes). Meligkotsidou (2007) propose one approach to model multivariate Poisson counts in this manner, but they do not consider spatial correlation.

Appendix

A. Broader context and background

A.1 Dynamic linear models

A dynamic linear model (DLM) is a parametric model where both the observations and the parameters are assumed to follow some probability distribution and the regression parameter vector evolves over time. The class of dynamic linear models is very rich and includes standard regression models and many time series models as special cases. There is a long history of Bayesian analysis of dynamic linear models. West and Harrison (1997) is a monograph on Bayesian DLMs and Migon *et al.* (2005) gives an updated review of the subject; this appendix follows both references closely.

Let y_t be an $r \times 1$ vector of responses at time $t = 1, \ldots, T$, then a multivariate dynamic linear model can be written as

$$y_t = F_t' \theta_t + \epsilon_t, \quad \epsilon_t \sim N(0, V_t), \tag{5.7}$$

$$\theta_t = G_t \theta_{t-1} + \omega_t, \quad \omega_t \sim N(0, W_t), \tag{5.8}$$

where F_t is a $p \times r$ matrix of explanatory variables, θ_t is a p-vector of parameters, G_t is a known $p \times p$ matrix describing the parameter evolution, V_t is the $r \times r$ matrix for the variance of the observation errors, and W_t is a $p \times p$ matrix for the variance of the p-vector of parameters. Equation (5.7) is commonly referred to as the observational equation and (5.8) is called the system equation. Note that the system model is a Markovian linear model. Many applications of DLMs are for the univariate case when $r = 1$.

In a Bayesian DLM the posterior distribution of any parameters at time t is based on all available information up to time t. One aspect of dynamic linear modelling that differs from independent-assumption Bayesian modelling is that the prior distribution for the parameters at time t must be evolved over time. Thus for DLMs inference is achieved via sequential updating in three steps: evolution of the parameters, prediction of new observations, and updating of the posterior distribution of the parameters. Let D_t be the set of information available at time t. Thus D_t includes the observation vector y_t, covariates observed at time t (if any), and all previous information D_{t-1}. Define an initial prior distribution at time $t = 0$ given by $\theta_0 | D_0 \sim N(m_0, C_0)$, and the posterior distribution at time $t - 1$ given by $\theta_{t-1} | D_{t-1} \sim N(m_{t-1}, C_{t-1})$ where m_{t-1} and C_{t-1} are defined below. It can be shown that these steps are given by (West and Harrison, 1997, page 582–4):

1. Evolution: The prior distribution at time t is given by $\theta_t | D_{t-1} \sim N(a_t, R_t)$, where $a_t = G_t m_{t-1}$ and $R_t = G_t C_{t-1} G_t' + W_t$.
2. Prediction: The one-step forecast at time t is given by $y_t | D_{t-1} \sim N(f_t, Q_t)$, where $f_t = F_t' a_t$ and $Q_t = F_t' R_t F_t + V_t$.
3. Updating: The posterior distribution at time t is given by

$$\theta_t | D_t \sim N(m_t, C_t), \tag{5.9}$$

with $m_t = a_t + A_t e_t$ and $C_t = R_t - A_t Q_t A_t'$ where $A_t = R_t F_t Q_t^{-1}$ and $e_t = y_t - f_t$ (the one-step forecast error).

Two other functions of interest in DLMs are the k-step ahead distributions

$$\theta_{t+k} \mid D_t \sim N(a_t(k), R_t(k))$$

and

$$y_{t+k} \mid D_t \sim N(f_t(k), Q_t(k)),$$

where $f_t(k) = F_t' a_t(k)$, $Q_t(k) = F_t' R_t(k) F_t + V_{t+k}$, $a_t(k) = G_{t+k} a_t(k-1)$, $R_t(k) = G_{t+k} R_t(k-1) G_{t+k}' + W_{t+k}$, $a_t(0) = m_t$, and $R_t(0) = C_t$.

The typical joint posterior of interest for DLMs is given by

$$p(\theta_1, \ldots, \theta_T, V, W \mid D_T) \propto \prod_{i=1}^{T} p(y_t \mid \theta_t, V) \prod_{i=1}^{T} p(\theta_t \mid \theta_{t-1}, W) p(\theta_0 \mid m_0, C_0) p(V, W).$$

Here $p(y_t \mid \theta_t, V)$ is given by (5.7), $p(\theta_t \mid \theta_{t-1}, W)$ is given by (5.8), and $p(\theta_0)$ and $p(V, W)$ are prior distributions.

A.2 Model specification

The class of DLMs is very broad including steady, smooth trend, seasonal components, regression components, and other model forms. A simple first-order model for univariate y_t is given by

$$y_t = \theta_t + \epsilon_t, \quad \epsilon_t \sim N\left(0, \sigma_t^2\right), \tag{5.10}$$

$$\theta_t = \theta_{t-1} + \omega_t, \quad \omega_t \sim N(0, W_t), \tag{5.11}$$

where θ_t is scalar. This model can be thought of as smooth function of the time trend with Taylor series representation

$$\theta_{t+\delta t} = \theta_t + \text{higher order terms.}$$

This model is sometimes called the steady model, although note that it includes a linear trend over time.

The steady model is too simplistic for many applications. The general approach to model building for DLMs is to construct simpler sub-DLM models to describe structure, such as a slowly varying trend in the parameters over time. For example, a local linear (second-order polynomial) trend model can be thought of as a second-order Taylor series approximation of a smooth function with a parameter $\theta_{1,t-1}$ representing the level of the series and $\theta_{2,t-1}$ representing incremental growth in the level of the series. The univariate y_t model is given by

$$y_t = \theta_{1,t} + \epsilon_t, \quad \epsilon_t \sim N\left(0, \sigma_t^2\right), \tag{5.12}$$

$$\theta_{1,t} = \theta_{1,t-1} + \theta_{2,t-1} + \omega_{1,t}, \tag{5.13}$$

$$\theta_{2,t} = \theta_{2,t-1} + \omega_{2,t}, \quad (\omega_{1,t}, \omega_{2,t})' \sim N(0, W_t), \tag{5.14}$$

where W_t is a 2×2 matrix of variances of the parameters (Gamerman and Lopes, 2006, p. 64). This model is a special case of the general DLM in (5.7) and (5.8) where F_t and G_t are given by $F_t = \begin{pmatrix} 1 \\ 0 \end{pmatrix}$ and $G_t = \begin{pmatrix} 1 & 1 \\ 0 & 1 \end{pmatrix}$.

Many time series behave in a cyclical nature, often paralleling the seasons of nature. A large variety of seasonal models have been proposed (West and Harrison, 1997, Chapter 8). A commonly used model is a time varying sinusoidal model with $F_t = \begin{pmatrix} 1 \\ 0 \end{pmatrix}$ and $G_t = \begin{pmatrix} \cos(\omega) & \sin(\omega) \\ -\sin(\omega) & \cos(\omega) \end{pmatrix}$ where $\omega = 2\pi/p$ is the frequency and p is the period West (1997). This model is sometimes referred to as a harmonic component DLM. For a monthly seasonal pattern $p = 12$ thus allowing for a sine wave which varies in amplitude but with fixed periods.

Regression-type models allow for explanatory covariates as columns of F_t. For time series modeling, inclusion of lagged predictors is often appropriate so $F_t = (1, X_t, X_{t-1}, \ldots, X_{t-k})'$. Short-term forecasts based on these models can be quite accurate. However, when the regression coefficients θ_t demonstrate significant variation in time, then medium to long-term forecasts will often be poor in location and have high variance. Another form of regression-type model is the autoregressive, moving average (ARMA) model. This form assumes stationarity so that the first and second order moments of the process are constant over time (with additional constraints on the correlation structure).

A.3 MCMC for DLMs

The posterior distributions required for Bayesian inference in DLMs are typically intractable. Migon *et al.* (2005) give an in depth review of MCMC methods to facilitate inference for DLMs. One challenge in MCMC for DLMs is that correlation among the θ_t parameters can lead to slow convergence.

The forward filtering backward smoothing (FFBS) algorithm (Carter and Kohn, 1994; Frühwirth-Schnater, 1994) is often used as part of MCMC algorithms for Bayesian DLMs. The FFBS algorithm samples θ in one block to avoid convergence problems caused by correlations among the elements of θ. To simplify notation we define the information until time t as $D_t = D_{t-1} \cap \{y_t\}$ to include the values of x_t and G_t for all t. At the jth iteration of the Gibbs sampler, the FFBS algorithm proceeds as follows:

1. Sample $V^{(j)}$ from $p\left(V|\theta^{(j-1)}, W^{(j-1)}, D_T\right)$.
2. Sample $W^{(j)}$ from $p\left(W|\theta^{(j-1)}, V^{(j)}, D_T\right)$.
3. Update θ blockwise by generating from the joint conditional distribution

$$p\left(\theta|V^{(j)}, W^{(j)}, D_T\right) = p\left(\theta_T|V^{(j)}, W^{(j)}, D_T\right)\prod_{t=1}^{T} p\left(\theta_t|\theta_{t+1}^{(j)}, V^{(j)}, W^{(j)}, D_t\right).$$

This happens in two steps:

(a) Forward filtering: Sample $p\left(\theta_T | V^{(j)}, W^{(j)}, D_T\right)$ from $N(m_t, C_t)$ given in (5.9).

(b) Backward sampling: For $t = n - 1, n - 2, \ldots, 1$ sample $p(\theta_t | \theta_{t+1}^{(j)}, V^{(j)}, W^{(j)}, D_t)$ from

$$N\left(\left[G_t' W^{-1} G_t + C_t^{-1}\right]^{-1}\left[G_t' W^{-1}\theta_{t+1} + C_t^{-1} m_t\right], \left[G_t' W^{-1} G_t + C_t^{-1}\right]^{-1}\right).$$

4. Increment j and return to step 1.

In the normal model, if V and W are assumed to be a priori independent with $V \sim$ inverse Gamma and $W \sim$ inverse Wishart, then their full conditional distributions are also inverse Gamma and inverse Wishart and thus are easy to sample from.

A.4 Non-normal dynamic linear models

Much time-series data does not follow a normal distribution; for example, count data might be better modelled by a Poisson distribution. Extensions of DLMs to non-normal and generalized linear models (GLMs) have considerably increased the usefulness of DLMs (West *et al.*, 1985). In the non-normal DLM context, a probability distribution is assumed for y_t so the model is defined similarly to (5.7) and (5.8) with $p(y_t | \mu_t)$, $g(\mu_t) = F_t'\theta_t$, $\theta_t = G\theta_{t-1} + \omega_t$, $\omega_t \sim p(\omega_t)$, where $g(.)$ is an appropriate link function.

While non-normality can sometimes be dealt with via transformation, this is not always appropriate due to the lack of interpretability of the results. Much of the work on non-normal DLMs has focused on mixtures of normal distributions (Carlin *et al.*, 1992) and exponential-family GLMs.

A challenge of non-normal DLMs is that the conditional distribution of θ_t is not known and the correlation among the θ_t parameters can lead to slow convergence of MCMC algorithms. Gamerman (1998) suggested the use of an adjusted normal dynamic linear model in order to build the proposal densities in a Metropolis – Hastings step. The proposed step is made for the disturbance terms of the model, avoiding the correlation among the state parameters. However, this algorithm requires high computational effort, since it is complicated to code and can require a long time to complete a single iteration. In contrast Durbin and Koopman (2000) propose an approximation based on importance sampling and antithetic variables and thus avoid issues with Markov chain based approximations. However there are other trade-offs with this approach such as the lack of formality in checking model adequacy or accounting for model uncertainty. Ravines *et al.* (2007) propose an efficient MCMC algorithm for dynamic GLMs. They mimic the FFBS algorithm described above. However, for dynamic GLMs, the conditional distribution in the forward filtering step described in Section A.3 is not available in closed form. These authors propose an approximation of the filtering distribution. The approximation is based on a linear approximation of the posterior distribution as if it were based on a conjugate prior as proposed by West *et al.* (1985). This leads to significant improvements in algorithm efficiency as compared to previous MCMC implementations for Bayesian dynamic GLMs.

A.5 Spatio-temporal DLMs

Spatio-temporal DLMs have been of increasing interest as Bayesian DLM modelling has matured. In a spatio-temporal DLM temporal correlation is modelled as a discrete process. Spatial correlation can be considered as either a discrete process (e.g. counties) or continuous process (e.g. positive $\Re$), depending on the data structure and modeling goals. For example, Gelfand *et al.* (2005) propose a model for continuous space but discrete time.

Research in spatio-temporal DLMs has explored many areas. Migon *et al.* (2005) give a useful overview by sorting models by each approach's treatment of the terms θ_t, F_t, G_t, V_t, and W_t in (5.7) and (5.8). The book Banerjee *et al.* (2004) summarizes a number of spatio-temporal models, including some DLMs.

Multivariate spatio-temporal DLMs such as Gamerman and Moreira (2004) allow for correlation between multivariate process. Gamerman and Moreira (2004) model multivariate spatial dependence in two ways: direct modeling of spatial correlation among the observations and through spatially varying regression coefficients, described further below. The methodology of Gelfand *et al.* (2005) relies on coregionalization (i.e. mutual spatial correlation) to model a multivariate process.

Methods for incorporating spatially and/or temporally varying regression coefficients have received considerable recent interest (Gelfand *et al.*, 2003; Paez *et al.*, 2008). In this case coefficients are allowed to vary over local regions or time periods, with different range parameters for every predictor. These models can lead to useful inference in a number of application areas including economic forecasts and ecological analysis of data observed at different scales.

Non-stationary DLMs allow for more complex spatio-temporal models. Sanso and Guenni (2000) propose a nonstationary spatio-temporal DLM based on a truncated normal distribution. Stroud *et al.* (2001) propose a nonparametric type model where the mean function at each time is expressed as a mixture of locally weighted linear regression models thus accounting for spatial variability. Dependence over time is modelled through the regression coefficients resulting in a form that generalizes the process convolution approach of Higdon (1999).

A.6 Future developments

The field of Bayesian DLMs is continuing to develop in many directions. Current research includes further understanding of the impacts of incorporating spatial dependence in DLMs, improving MCMC algorithms for DLMs of all kinds, and better models for multivariate, non-normal, and nonlinear DLMs. As our ability to collect spatially explicit data has exploded in recent years and with problems of global impact such as climate change and the global economy, we anticipate that the development of models and methodology for Bayesian DLMs will continue to be an exciting area of research.

Acknowledgments

Alexandra M. Schmidt was supported by grants from CNPq and FAPERJ, Brazil. Jennifer A. Hoeting was partially supported by National Science Foundation grant DEB-0717367. The work of João B. M. Pereira was supported by a scholarship from PIBIC/CNPq. The authors wish to thank Cynthia de Oliveira Ferreira, from the Malaria Department at FMTAM and Raul Amorim, from the Epidemiology and Public Health Department at FMTAM, for making the dataset available. Schmidt and Hoeting were supported by NSF-IGERT grant DGE-0221595.

References

Aitchison, J. and Ho, C. H. (1989). The multivariate Poisson-Log Normal distribution. *Biometrika*, **76**, 643–653.

Banerjee, S., Carlin, B. P. and Gelfand, A. E. (2004). *Hierarchical Modeling and Analysis of Spatial Data*. Chapman and Hall, New York.

Besag, J., York, J. and Mollié, A. (1991). Bayesian image restoration, with two applications on spatial statistics. *Annals of the Institute of Statistical Mathematics*, **43**, 1–59.

Carlin, B. P., Polson, M. and Stoffer, D. (1992). A Monte Carlo approach to nonnormal and nonlinear statespace modeling. *Journal of the American Statistical Association*, **87**, 493–500.

Carter, C. and Kohn, R. (1994). On Gibbs sampling for state space models. *Biometrika*, **81**, 541–553.

Chib, S. and Winkelmann, R. (2001). Markov chain Monte Carlo analysis of correlated count data. *Journal of Business and Economic Statistics*, **19**, 428–435.

Durbin, J. and Koopman, S. J. (2000). Time series analysis of non-Gaussian observations based on state space models from both classical and Bayesian perspectives. *Journal of the Royal Statistical Society: Series B*, **62**, 3–56.

Fernandes, M. V., Schmidt, A. M. and Migon, H. S. (2009). Modelling zero-inflated spatio-temporal processes. *Statistical Modelling*, **9**, 3–25.

Frühwirth-Schnater, S. (1994). Data augmentation and dynamic linear models. *Journal of Time Series Analysis*, **15**, 183–202.

Gamerman, D. (1997). Sampling from the posterior distribution in generalized linear mixed models. *Statistics and Computing*, **7**, 57–68.

Gamerman, D. (1998). Markov chain Monte Carlo for dynamic generalised linear models. *Biometrika*, **85**, 215–227.

Gamerman, D. and Lopes, H. F. (2006). *Markov Chain Monte Carlo: Stochastic Simulation for Bayesian Inference*. Chapman & Hall/CRC, London (2nd edn.)

Gamerman, D. and Migon, H. (1993). Dynamic hierarchical models. *Journal of the Royal Statistical Society, B*, **55**, 629–642.

Gamerman, D. and Moreira, A. R. B. (2004). Multivariate spatial regression models. *Journal of Multivariate Analysis*, **91**, 262–281.

Gelfand, A. E., Banerjee, S. and Gamerman, D. (2005). Spatial process modelling for univariate and multivariate dynamic spatial data. *Environmetrics*, **16**, 465–479.

Gelfand, A. E., Kim, H.-J., Sirmans, C. and Banerjee, S. (2003). Spatial modeling with spatially varying coefficient processes. *Journal of the American Statistical Association*, **98**, 387–396.

Givens, G. and Hoeting, J. A. (2005). *Computational Statistics*. John Wiley, Hoboken.

Gneiting, T., Balabdaoui, F. and Raftery, A. (2007). Probabilistic forecasts, calibration and sharpness. *Journal of the Royal Statistical Society, Series B*, **69**, 243–268.

Gschlößl, S. and Czado, C. (2007). Spatial modelling of claim frequency and claim size in non-life insurance. *Scandinavian Actuarial Journal*, **2007**, 202–225.

Health Ministry of Brazil (2006). *Situação Epidemiológica da Malária no Brasil*.

Higdon, D. (1999). A process-convolution approach to modeling temperatures in the north Atlantic Ocean. *Environmental and Ecological Statistics*, **5**, 173–190.

Le, N. and Zidek, J. (1992). Interpolation with uncertain spatial covariances: A Bayesian alternative to kriging. *Journal of Multivariate Analysis*, **43**, 351–374.

Liu, F. (2000). Bayesian time series: analysis methods using simulation based computation. Ph.D. thesis, Institute of Statistics and Decision Sciences, Duke University, Durham, North Carolina, USA.

Meligkotsidou, L. (2007). Bayesian multivariate poisson mixtures with an unknown number of components. *Statistical Computing*, **17**, 93–107.

Migon, H. S., Gamerman, D., Lopes, H. F. and Ferreira, M. A. R. (2005). Dynamic models. In *Handbook of Statistics, Volume 25: Bayesian Thinking, Modeling and Computation*, (ed. D. Dey and C. R. Rao) pp. 553–588. North Holland, The Netherlands.

Nadadur, D., Haralick, R. M. and Gustafson, D. E. (2005). Bayesian framework for noise covariance estimation using the facet model. *IEEE Transactions on Image Processing*, **14**, 1902–1917.

Nobre, A. A., Schmidt, A. M. and Lopes, H. F. (2005). Spatio-temporal models for mapping the incidence of malaria in Pará. *Environmetrics*, **16**, 291–304.

Paez, M. S., Gamermana, D., Landima, F. M. and Salazara, E. (2008). Spatially varying dynamic coefficient models. *Journal of Statistical Planning and Inference*, **138**, 1038–1058.

Ravines, R. R., Migon, H. S. and Schmidt, A. M. (2007). An efficient sampling scheme for dynamic generalized models. Technical report, No. 201, Departamento de Métodos Estatísticos, Universidade Federal do Rio de Janeiro.

Sanso, B. and Guenni, L. (2000). A nonstationary multisite model for rainfall. *Journal of the American Statistical Association*, **95**, 1089–1100.

Sansó, B., Schmidt, A. M. and Nobre, A. A. (2008). Bayesian spatio-temporal models based on discrete convolutions. *Canadian Journal of Statistics*, **36**, 239–258.

Stroud, J. R., Müller, P. and Sansó, B. (2001). Dynamic models for spatiotemporal data. *Journal of the Royal Statistical Society: Series B (Statistical Methodology)*, **63**, 673–689.

Sun, D., Tsutakawa, R. K., Kim, H. and He, Z. (2000). Spatio-temporal interaction with disease mapping. *Statistics in Medicine,* **19**, 2015–2035.

Vivar, J. C. (2007). Modelos espaço-temporais para dados de área na família exponencial. Ph.D. thesis, Instituto de Matemática, Universidade Federal do Rio de Janeiro, Rio de Janeiro, Brasil.

Wakefield, J. (2007). Disease mapping and spatial regression with count data. *Biostatistics,* **8**, 153–183.

West, M. (1997). *Encyclopedia of Statistical Sciences,* Chapter on Bayesian Forecasting. John Wiley, New York.

West, M., Harrison, J. and Migon, H. (1985). Dynamic generalized linear models and Bayesian forecasting. *Journal of the American Statistical Association,* **80**, 73–83.

West, M. and Harrison, P. (1997). *Bayesian Forecasting and Dynamic Models* (2nd edn). Springer Verlag, New York.

World Health Organization (2007). *Fact Sheet No 94.* World Health Organization Press Office, Geneva, Switzerland. `http://www.who.int/mediacentre/factsheets/fs094/en/index.html`.

·6·

Trans-study projection of genomic biomarkers in analysis of oncogene deregulation and breast cancer

Dan Merl, Joseph E. Lucas, Joseph R. Nevins, Haige Shen and Mike West

6.1 Oncogene pathway deregulation and human cancers

Many genes, when mutated or subject to other forms of deregulation of normal activity, become oncogenic; the resulting effects of lack of normal function of such a gene on other genes and proteins in 'downstream pathways' can engender patterns of cell growth and proliferation that is cancer promoting. The study here concerns the extent to which patterns of gene expression associated with experimentally induced oncogene pathway deregulation can be used to query oncogene pathway activity in real human cancers. This is often referred to as the *in vitro* to *in vivo* translation problem: results drawn from controlled intervention studies using cultured cells (*in vitro*) must be made relevant for cancer cells growing in the highly heterogeneous micro-environment of a living host (*in vivo*). We address this using Bayesian sparse factor regression analysis for model-based translation and refinement of *in vitro* generated signatures of oncogene pathway activity into the domain of human breast tumour tissue samples. The promise of this strategy is the ability to directly query the degree of functionality of relevant cellular pathways in such cancers and lead into improvements in individualized diagnosis, prognosis, and treatment (Nevins *et al.* 2003; Pittman *et al.* 2004; West *et al.* 2006). Our study here is a broad and detailed application of an overall strategy being utilized in a number of areas (Huang *et al.* 2003; Bild *et al.* 2006; Seo *et al.* 2007; Chen *et al.* 2008; Carvalho *et al.* 2008; Lucas *et al.* 2009, 2010).

6.1.1 Problem context and goals

Since the late 1990s, the increasing availability of microarray data has created new opportunities for understanding the molecular basis of cancer. Microarray data, as exemplified by the Affymetrix GeneChip platform, provides a broad snapshot of the state of gene expression in a cell or tumour sample by measuring levels of cellular messenger RNA (mRNA) for many of the 20,000+ genes in the human genome. Linking outcomes in cancer to patterns of gene expression has led to several key prognostic innovations over the past decade, including the identification of relatively small subsets of genes whose collective patterns of expression define *signatures* of clinically relevant phenotypes (Alizadeh *et al.* 2000; Golub *et al.* 1999; West *et al.* 2001; van't Veer *et al.* 2002; Huang *et al.* 2003; Miller *et al.* 2005; Bild *et al.* 2006).

Various gene expression signatures have been derived *in vitro* from controlled experiments involving cell cultures subject to some combination of interventions whose effects on gene expression are of interest; targeted over-expression of a particular gene, or manipulation of the surrounding micro-environment of the cell culture. Prior to intervention, the cultured cells are in a state of quiescence, or inactivity, thus permitting all cellular reaction to the intervention – relative to some control group – to be attributed to the intervention. Though necessary for interpretability of the experiment, cells living outside of culture are rarely in such a state, and therefore were it possible to perform the same intervention on cells living *in vivo*, the observed gene expression responses would differ as a result of interactions with other normally active cellular pathways.

This study concerns nine experimentally derived signatures, each associated with the effects of over-expression of a different oncogene known to play a role in the progression of human breast cancer. Given that chemotherapies can be designed to block cancer progression by interrupting specific components of a targeted cellular pathway, it is of interest for purposes of drug discovery and utilization to evaluate the relative activity levels of these oncogene pathway signatures in observed human breast cancer tissue samples. By then associating those pathway activity levels with clinical outcomes, such as long term survival, we will be able to identify high-priority pathways, the disruption of which may be most likely to result in improved patient prognosis.

The key to the problem of *in vitro* to *in vivo* translation of the oncogene pathway signatures lies in allowing a more flexible notion of a biological 'pathway'. Rather than attempting to evaluate the activity level of an entire oncogene pathway as it is observed in the experimental setting, we use Bayesian sparse factor analysis to simultaneously decompose and refine the original signatures in the context of human breast cancer samples. Through directed latent factor search, the original pathway signatures are mapped onto the tissue data and decomposed into smaller sub-pathway modules represented by latent factors; some factors may be common to multiple oncogene pathways, while some may represent sections of pathways uniquely linked to a particular oncogene. In this way, the original collection of intersecting oncogene pathways can be

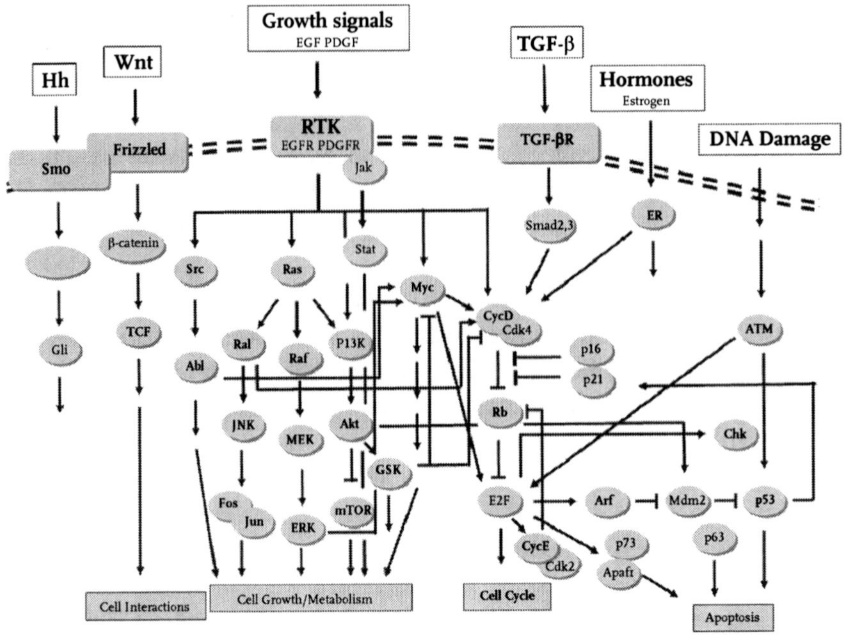

Fig. 6.1 Schematic network structure depicting interactions between oncogenes in the context of the broader cell developmental and fate network. Arrows indicate biologically defined causal influences that are activating, all of which are heavily context dependent and may be direct, indirect and reliant on activation of other pathways, and on the roles of multiple other transcription factors, binding proteins and phosphorylating agents. The edges terminating as $\perp$ indicate repressive links that are similarly well-defined and also similarly heavily context dependent.

effectively modelled in the biologically more diverse and heterogeneous human cancer data context. This approach allows inference on the activity levels of each sub-pathway, thereby aiding our search for associations between sub-pathways and clinical outcomes.

6.1.2 *In vitro* oncogene experiments

Bild *et al.* (2006) experimentally induced activation of several oncogene pathways in human mammary epithelial cell cultures (HMECs). Interventions on quiescent cell cultures were effected to artificially induce activity of the target oncogene, thereby initiating within the otherwise inactive cells that sequence of responses that collectively comprises the *oncogene pathway*. A total of nine oncogene pathways were queried in this way, representing a variety of different aspects of cancer progression (Figure 6.1). Affymetrix microarray gene expression data on the resulting perturbed cells then delivers data to measure the responses to interventions. The original study reported on the pathway signatures associated with the oncogenes human c-MYC, activated H-RAS, human c-SRC, human E2F3, and activated β-Catenin. Subsequently four additional interventions were conducted for the oncogenes P63, ATK, E2F1, and P110. Interventions were replicated across approximately nine separate cultures per oncogene. Gene expression was also measured for 10 control cultures associated with the initial five interventions, and five control cultures associated with the latter four interventions, producing a total sample size of 97 microarrays, each quantifying hybridization levels of over 50,000 oligonucleotide probes representing 20,000+ genes.

6.1.3 *In vivo* breast cancer studies

Miller *et al.* (2005) presented a collection of 251 surgically removed human breast tumour tissue samples gathered during a study conducted between the years 1987 and 1989 in Uppsala County, Sweden. Following RNA extraction and processing, gene expression measurements for each tumour sample were obtained using the Affymetrix Human U133A and U133B GeneChip microarrays. Several clinico-pathological variables were collected as part of standard surgical procedure, including age at diagnosis, tumour size, lymph node status (an indicator of metastases), and the Elston histologic grade of the tumour (a categorical rating of malignancy based on observed features of the tumour). Molecular assays were subsequently conducted to identify mutations in the estrogen receptor (ER), P53, and progesterone receptor (PgR) genes, the presence or absence of mutations in which have formed the basis of classification into several breast cancer subtypes. Patient survival histories were also monitored following surgery.

6.2 Modelling and data analysis

Section 6.2.1 concerns the initial derivation of the oncogene pathway signatures from the Bild data using sparse multivariate regression (Lucas *et al.* 2006). Section 6.2.2 describes the evaluation of the oncogene pathway signature activity in the tumor data, providing initial estimates of the activity level of each pathway in the observational data. Section 6.2.3 discusses the use of these signature scores as seeds for a targeted search of latent factor space, in order to uncover factor structure related to the original signatures. Section 6.2.4 describes exploratory analysis of the fitted latent factor model, highlighting associations between factors and several clinical variables. Section 6.2.5 takes further the *in-vitro-in-vivo* comparison by back-projecting the estimated factors to the oncogene data, allowing us to associate factors with different oncogene pathways. Section 6.2.6 details the use of shotgun stochastic search (SSS) (Hans, Dobra, and West 2007; Hans, Wang, Dobra, and West 2007) to explore connections between posterior estimates of latent factors, traditional clinical variables, and long term survival. Finally, Section 6.3 discusses biological interpretations of a few key factors emerging from the survival study using Bayesian probabilistic pathway annotation, or PROPA (Shen, 2007; Shen and West, 2010).

6.2.1 Generation of oncogene pathway signatures

Evaluation of oncogene pathway signatures from the Bild *et al.* (2006) data is based on sparse multivariate regression analysis (Lucas *et al.* 2006). A key aspect is the use of sparsity priors. Sparse regression augments a standard linear model with hierarchical point mass mixture priors on regression coefficients (West, 2003). This allows identification of the subset of differentially expressed genes that show a response to each intervention, quantified by the posterior probability of a non-zero regression coefficient for the gene $\times$ intervention effect. For each intervention, the set of genes for which this probability is sufficiently high, along with the estimated regression effects, comprise the pathway signature.

Let X^{vitro} denote the 8509×97 dimensional gene expression matrix, with elements $x_{g,i}^{\text{vitro}}$ measuring expression of gene g on array sample i (each sample i represents a different replicate in one of the nine oncogene interventions). Expression is on the $\log_2$ scale, so a unit change in x corresponds to a doubling or halving, i.e. a one-fold change in mRNA levels. Let H^{vitro} denote the 18×97 design matrix where the first 10 rows contain binary indicators associating samples with their intervention effects (nine oncogene effects and one for the second control group), and the final eight rows are covariates constructed as artifact correction factors. The latter, derived from the first eight principal components of the Affymetrix housekeeping/control probe expression levels on each sample array, are for gene-sample specific normalisation. This strategy for artifact control is demonstrated in Carvalho *et al.* (2008) and Lucas *et al.* (2006), and provides a convenient basis for model-based assimilation of data obtained through multiple experimental conditions that may have resulted in systematic errors.

The regression model for each gene $g = 1 : p$ on any sample $i = 1 : n$ is

$$x_{g,i}^{\text{vitro}} = \mu_g + \sum_{k=1}^{18} \beta_{g,k} h_{k,i}^{\text{vitro}} + \nu_{g,i},$$

or in matrix form,

$$X^{\text{vitro}} = \mu\iota' + BH + N \tag{6.1}$$

where the element μ_g of the $p \times 1$ vector μ is baseline expression of gene g, ι is the $n \times 1$ vector of ones, the elements $\beta_{g,k}$ of the $p \times 18$ matrix B are the effects of interventions and coefficients of artifact correction covariates, and the element $\nu_{g,i}$ of the $p \times n$ error matrix N represents independent residual noise for gene g on sample array i. Careful consideration is given to assignment of prior distributions. Baseline expression effects are modelled as $\mu_g \sim N(8, 100)$, based on known properties of Affymetrix RMA gene expression indices. Noise terms are modelled as $\nu_{g,i} \sim N(0, \psi_g)$ with $\psi_g \sim IG(2, 0.1)$; this reflects the view that noise standard deviations will be gene-specific and range between 0.05 and 0.7 with median values across genes near 0.2–0.25. The hierarchical sparsity prior on elements $\beta_{g,k}$ of B is

$$\beta_{g,k} \sim (1 - \pi_{g,k})\delta_0 + \pi_{g,k} N(0, \tau_k) \quad \text{with} \quad \tau_k \sim IG(1, 5),$$
$$\pi_{g,k} \sim (1 - \rho_k)\delta_0 + \rho_k Be(9, 1) \quad \text{with} \quad \rho_k \sim Be(2.5, 497.5). \qquad (6.2)$$

Here δ_0 represents a point mass at zero. Thus the effect of regressor k on gene g is non-zero and drawn from the $N(0, \tau_k)$ prior with probability $\pi_{g,k}$. The variance τ_k is regressor-specific, reflecting the fact that levels of gene activation/suppression will vary across pathways. The choice of prior for τ_k reflects expected ranges of expression changes supporting several orders of magnitude, reflecting relevant biological expectations and again based on extensive prior experience with similar data sets. The prior structure on the π and ρ terms govern the expected sparsity of the pathway signatures and reflect the prior view that approximately 99.5% of all monitored genes will show no response to a given intervention, and the pathway inclusion probability of all such genes should be exactly 0. This ensures that the data must provide overwhelming evidence of expression change for a gene to achieve high posterior probability of inclusion in the pathway.

Model fitting is acheived using Markov chain Monte Carlo (MCMC), as implemented in our BFRM software (Wang *et al.* 2007) (see Appendix B.). The analysis is based on 25,000 posterior samples following an initial burn-in of 2500. All summaries and further evaluation in this study are based on the posterior means of model parameters as reported by BFRM.

Figure 6.2 depicts the sparsity structure of the estimated oncogene pathway signatures (columns 2 through 10 of B) using a posterior probability threshold of 0.95 (i.e. $Pr(\beta_{g,k} \neq 0|X^{\text{vitro}}) > 0.95$). At this threshold, the RAS pathway involves the largest number of probes (2202), followed by the P110 pathway (1630), with the SRC pathway involving the smallest number of probes (417). The relative size of the RAS signature indicates significant opportunity for modular decomposition of this pathway via latent factor analysis, as the downstream effects of intervention on RAS clearly involve many genes and therefore, presumably, a large number of intersecting pathways.

6.2.2 Initial quantification of pathway activity by signature projection

Initial measures of activity levels of the oncogene pathways in each breast tumour sample are calculated as follows. Let $\beta_{g,k}^{\text{vitro}}$ and ψ_g^{vitro} denote the mean

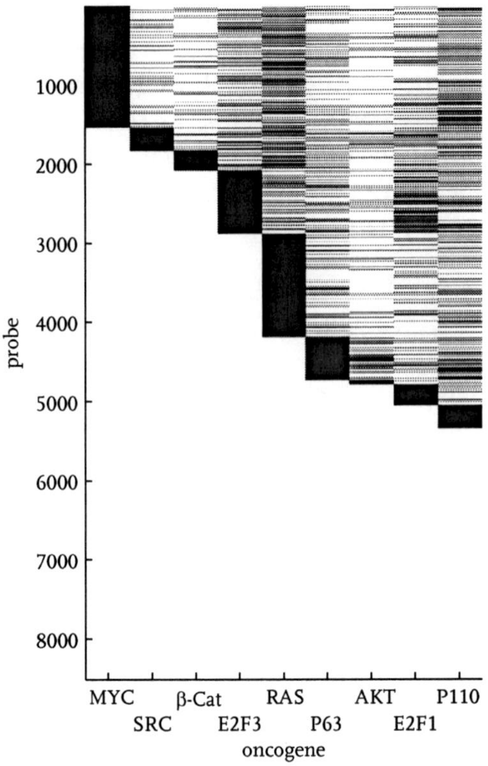

Fig. 6.2 Estimated Sparsity of oncogene pathway signatures. Black indicates genes × oncogene intervention pairs g, k such that $P(\beta_{g,k} \neq 0|X^{\text{vitro}}) > 0.95$. Genes are ordered to highlight the relative size of the pathway signature gene sets across oncogenes, and also patterns the cross-talk in terms of genes appearing responsive in multiple pathways.

posterior values of the associated model parameters in the sparse regression analysis above. Let X^{vivo} denote the matrix of RMA measures of gene expression for the tumour data. The raw *projected signature score* of pathway k in tumour i is defined, following Lucas *et al.* (2009), as $s_{k,i} = \sum_{g=1}^{p} \beta_{g,k}^{\text{vitro}} x_{g,i}^{\text{vivo}} / \psi_g^{\text{vitro}}$. In order to facilitate comparisons between signature scores and gene expression, each vector of signature scores is then transformed to have mean and variance on the same scale as the gene expression measurements in the tumor data, i.e. $s_{k,i}^{*} = m + u(s_{k,i} - m_k)/u_k$ where m_k, u_k are the mean and standard deviation of $s_{k,i}$ values over the tumour samples $i = 1:251$, m is the sample average gene expression over all genes and tumour samples, and u is the average of the set of p gene-specific standard deviations of gene expression across tumour samples.

The transformed signature scores for the tumour tissue samples are shown in Figure 6.3. Clear gradients emerge across MYC, SRC, β-Catenin, RAS, and P63 signature scores when samples are ordered by rank along the first principle

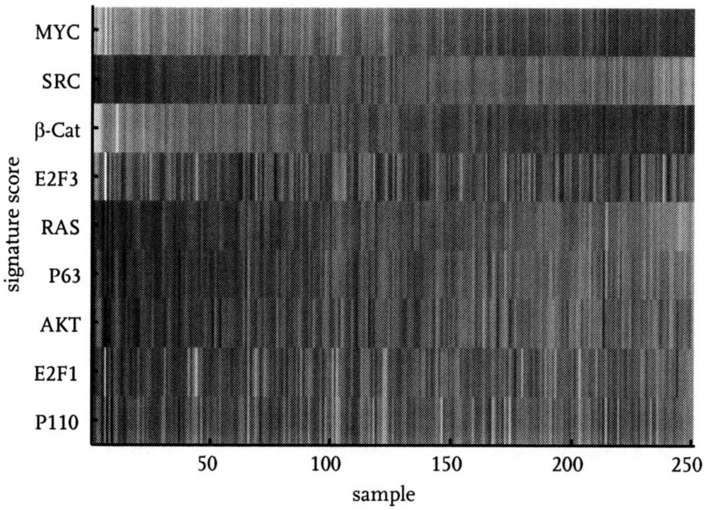

Fig. 6.3 Initial measures of oncogene pathway activity in each tumour tissue sample. Black indicates low signature scores, or pathway suppression; white indicates higher signature scores, or pathway activation. Tumor tissue samples have been reordered to emphasize dominant patterns of pathway activity, especially the correlated gradients of MYC and β-Catenin status, and SRC and RAS status.

component, indicating possible patterns of activity of these five oncogene pathways across breast cancers. High correlations exist between MYC and β-Catenin status, and between SRC and RAS status, with the former two negatively correlated with the latter. Latent factor analysis will provide the methodology for uncovering the common structure underlying these clearly interconnected pathways.

6.2.3 Latent factor models for breast tumour gene expression

The sparse latent factor extension to the sparse regression model is simply

$$x_{g,i} = \mu_g + \sum_{k=1}^{r} \beta_{g,k} h_{k,i} + \sum_{\ell=1}^{s} \alpha_{g,\ell} \lambda_{\ell,i} + \nu_{g,i} \tag{6.3}$$

where each h_i now contains any known design or covariate data of interest in the breast cancer analysis, and the $\alpha_{g,\ell}$ are coefficients that define the *loadings* of gene g on the values $\lambda_{\ell,i}$ for a set of s latent factors across the $i = 1 : n$ tumours. The value $\lambda_{\ell,i}$ is referred to as the *factor score* for factor ℓ on sample i. The general structure of the priors for μ, B, and ν terms remains as above. The prior structure on latent factor loadings $\alpha_{g,k}$ is of the same form as that for the $\beta_{g,k}$, as the same notion of expected sparsity applies; multiple factors represent the complexity of expression patterns in tumours, but genes will be only selectively related to factors as described by a sparsity prior over the factor loadings. One technical modification is that the first $s \times s$ block of the implied loading matrix is constrained to have positive diagonal values and an upper triangle of zeros; this ensures model identification.

To ease notation, the model equation (6.3) can be reexpressed as

$$x_{g,i}^{\text{vivo}*} = \mu_g + \sum_{\ell=1}^{r} a_{g,\ell}\lambda_{\ell,i} + \sum_{\ell=r+1}^{s+r} a_{g,\ell}\lambda_{\ell,i} + \nu_{g,i} \tag{6.4}$$

or in matrix form,

$$X^{\text{vivo}*} = \mu 1' + A\Lambda + N \tag{6.5}$$

where the first r columns of A are the regression coefficients associated with the first r rows of Λ, the known design effects, and the remaining columns relate to the latent factors. The only covariates used are four artifact control covariates in the first four rows of Λ, again being computed from the housekeeping genes on the tumour sample arrays.

Prior distributions are as follows. Baseline expression $\mu_g \sim N(8, 100)$, again reflecting known properties of Affymetrix expression indices, while $\nu_{g,i} \sim N(0, \psi_g)$ with $\psi_g \sim IG(2, 0.005)$ reflecting the view that residual noise should be somewhat less than that in the oncogene signature analysis due to the increased opportunity for explaining variance via an arbitrary number of latent factors. Indeed, some latent factors may reflect patterns of shared variation that are due to experimental artifacts above those captured by the artifact control covariate strategy. The sparsity prior over elements of A is

$$\begin{aligned} a_{g,\ell} &\sim (1-\pi_{g,\ell})\delta_0 + \pi_{g,\ell} N(0,\tau_\ell) \quad \text{with} \quad \tau_\ell \sim IG(2,1), \\ \pi_{g,\ell} &\sim (1-\rho_\ell)\delta_0 + \rho_\ell Be(9,1) \quad \text{with} \quad \rho_\ell \sim Be(0.2, 199.8). \end{aligned} \tag{6.6}$$

The prior is modified for $g = 1 : s$ in view of the identifying constraints that $a_{g,r+g} > 0$ and $a_{g,\ell} = 0$ when $g < \ell - r$ for $\ell = (r+2) : (r+s)$. This is done simply by taking $a_{g,r+g} \sim N(0, \tau_g)$ truncated to $a_{g,r+g} > 0$ for $g = 1 : s$, and fixing $\pi_{g,\ell} = 0$ for $g < \ell - r$, $\ell = (r+2) : (r+s)$.

The prior on τ_ℓ anticipates that variation explained by any single factor will tend to be rather less than that explained by the design effects of the oncogene study, though hedged by considerable uncertainty. The prior on inclusion probabilities now reflects an assumption that approximately 99.9% of genes will show no association with a given factor, though this assumption is enforced with less certainty than that of the 99.5% assumption made in the oncogene study.

In order to represent non-Gaussian distributions of factor scores, as well as to facilitate clustering of samples based on estimated factor profiles, the $s \times 1$ vector of latent factor scores $\Lambda_{:,i}$ on tumour sample i are modelled jointly via a

Dirichlet process mixture model (Carvalho *et al.* 2008); that is,

$$\begin{aligned} \Lambda_{:,i} &\sim F(\cdot) \quad \text{with} \quad F \sim DP(\alpha_0 G_0), \\ \alpha_0 &\sim Ga(e, f) \quad \text{and} \quad G_0 = N(0, I) \end{aligned} \tag{6.7}$$

where I is the $s \times s$ identity. The prior on the Dirichlet process precision α_0 of equation (6.7) is a standard $Ga(1, 1)$ reflecting considerable uncertainty about the complexity of structure expected in the factor distribution.

Latent factor discovery and estimation is accomplished using the same BFRM software (Wang *et al.* 2007) used for the *in vitro* oncogene analysis. For latent factor analysis, the software implements a novel evolutionary search algorithm for incrementally growing and fitting the factor model, based on specification of an initial set of genes and factors. At each step of the iterative *evolutionary search* (Carvalho *et al.* 2008), gene inclusion probabilities are calculated for all unincorporated genes, and those with highest inclusion probabilities become candidates for inclusion in the model. Increasing the set of genes within the model can then introduce additional structure in the expression data that requires expansion of the number of latent factors, and hence model refitting. Analysis terminates after a series of such steps to expand the set of genes and number of factors, subject to some specified controls on the numbers of each and guided by thresholds on gene×factor inclusion probabilities. Finally, the factor loadings and scores are estimated by MCMC for the final model (refer to Appendices A and B for references and more details on BFRM).

A key novelty of the subsequent analysis lies in applying the above evolutionary factor model in such a way as to preserve the connection between identified factors and the oncogene pathway signatures. We do this by defining the extended data matrix that has the 9×251 matrix S^* of projected signature scores as its first nine rows, viz

$$X^{\text{vivo}^*} = \begin{pmatrix} S^* \\ X^{\text{vivo}} \end{pmatrix},$$

now a $44{,}601 \times 251$ matrix. We initiate the model search with a nine-factor model incorporating only the oncogene signature scores as the initial data set. Thus, by construction the loadings of the initial nine factors correspond to the changes in gene expression of those genes most associated with changes in signature scores in the tumour tissues. Factors beyond the first nine will successively improve model fit by accounting for variation in gene expression beyond that explicitly linked with the patterns of pathway activation represented by the signature scores, thus incrementally augmenting and refining the first nine core pathway factors.

6.2.4 Exploring latent factor structure in the breast data

BFRM evolved the nine-gene, nine-factor model to include 500 genes (the maximum we allowed) in a model with 33 factors. This final model is an expanded latent factor representation of gene expression patterns that (a) link to the initial oncogene signatures, (b) evidence the greater complexity seen in the tumour data in genes defining the oncogene pathways, (c) link in numerous genes that relate to the factor representation of these initial pathways as well as (d) many other genes that reflect biological activity in intersecting pathways in the broader gene network that the initial oncogenes play roles in. Since the breast data analysis allowed exploration of over 40,000 probes on the microarray, it had the potential to identify genes that were not in the 8,509 used in the oncogene analysis; at termination, the latent factor model included 213 genes that were among the original 8,509 and an additional 287 not appearing in the *in vitro* analysis.

Figure 6.4 shows the intensity image of corrected gene expression for these 500 genes, ordered by ranking along first principle components in order to

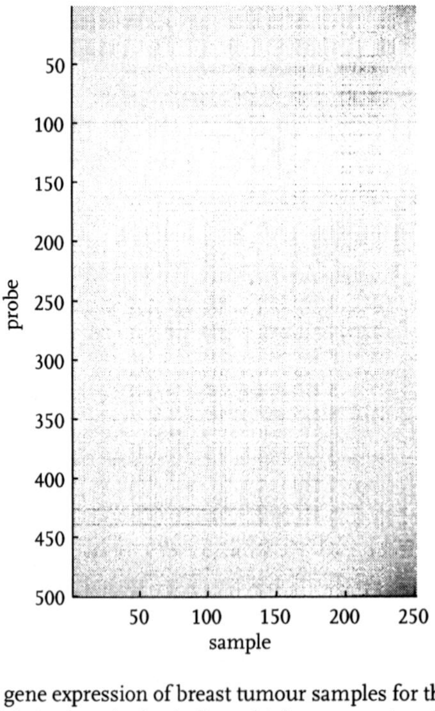

Fig. 6.4 Artifact corrected gene expression of breast tumour samples for the 500 genes incorporated into the latent factor model. Darker shades indicate higher expression than baseline. Probes and samples have been ordered by rank along first principle components in order to emphasize patterns of expression. The patterns of expression associated with the first and last 50 probes characterize a subgroup of approximately 50 tumours samples having clinical variables indicative of poor prognosis.

accentuate dominant structure. Corrected expression is calculated by subtracting baseline expression and the contribution of artifact control factors from the raw expression values, using posterior means of model parameters; i.e., $X^{vitro^*} - \mu\iota' - \alpha_{:,1:4}\Lambda_{1:4,:}$, where posterior means are used for μ, A, and Λ. A weak structure emerges, with some 100 strongly differentially expressed genes characterizing a subgroup of samples (probes 1–25 and 475–500 approximately). Inspection of clinical data on these tumours reveals that this pattern of expression defines a high risk tumour subgroup, correlated with ER negativeness, Elston grade 3, P53 positiveness, and PgR negativeness (Figure 6.5). This simply serves to demonstrate that the relatively small subset of genes incorporated into the factor model is sufficient to begin to discriminate clinically relevant tumour subtypes from one another, though the analysis was purely expression-data based and not at all trained on the clinical data. This gives initial strength to the claim that latent factors are useful for defining clinical biomarkers.

The structure of the latent factor loadings matrix gives additional insight to the factor basis of this high risk subgroup. The left frame of Figure 6.6 depicts the 'factor skeleton' identifying high probability gene-factor relationships in terms of non-zero loadings. The ordering of probes also serves to indicate blocks of genes that became incorporated upon addition of each new factor, including many that are distant from the initial nine factors. Factor 1, associated with the MYC signature, is the most heavily loaded factor, with 223 non-zero loadings. Factor 7, associated with the RAS signature, and factors 20, 22, and 28 are relatively sparsely loaded. Replotting the image of corrected gene expression with probes now ordered by factor incorporation (Figure 6.6) reveals that many of the genes defining the clinically high-risk cancer subgroup are those associated with latent factor 19.

Figure 6.7 displays estimated posterior means of the factor scores of factor 19 on tumours, coded by clinical outcomes and ordered as in Figure 6.4. The figure demonstrates an important aspect of latent factor analysis: the need for non-Gaussian modelling of latent factors. Representation of the bimodal nature of factor scores, and thus the clear separation of distinct tumour subgroups, emerges naturally under the Dirichlet process mixture prior on factor scores.

Comparing estimated factor scores for the first nine factors and their associated oncogene signature scores illustrates varying levels of 'intactness' of the original pattern of pathway activity as represented by the signature scores. High correlation, as is the case for MYC and E2F1, indicates that the patterns of MYC and E2F1 pathway activity predicted by their *in-vitro* signature scores are still largely captured by the single factors 1 and 6, respectively. The relationship between factor 4 and the SRC signature score indicates that pattern of SRC pathway activity predicted by the signature score is not evident at all in the tumour tissue samples, implying that the SRC pathway may not play a

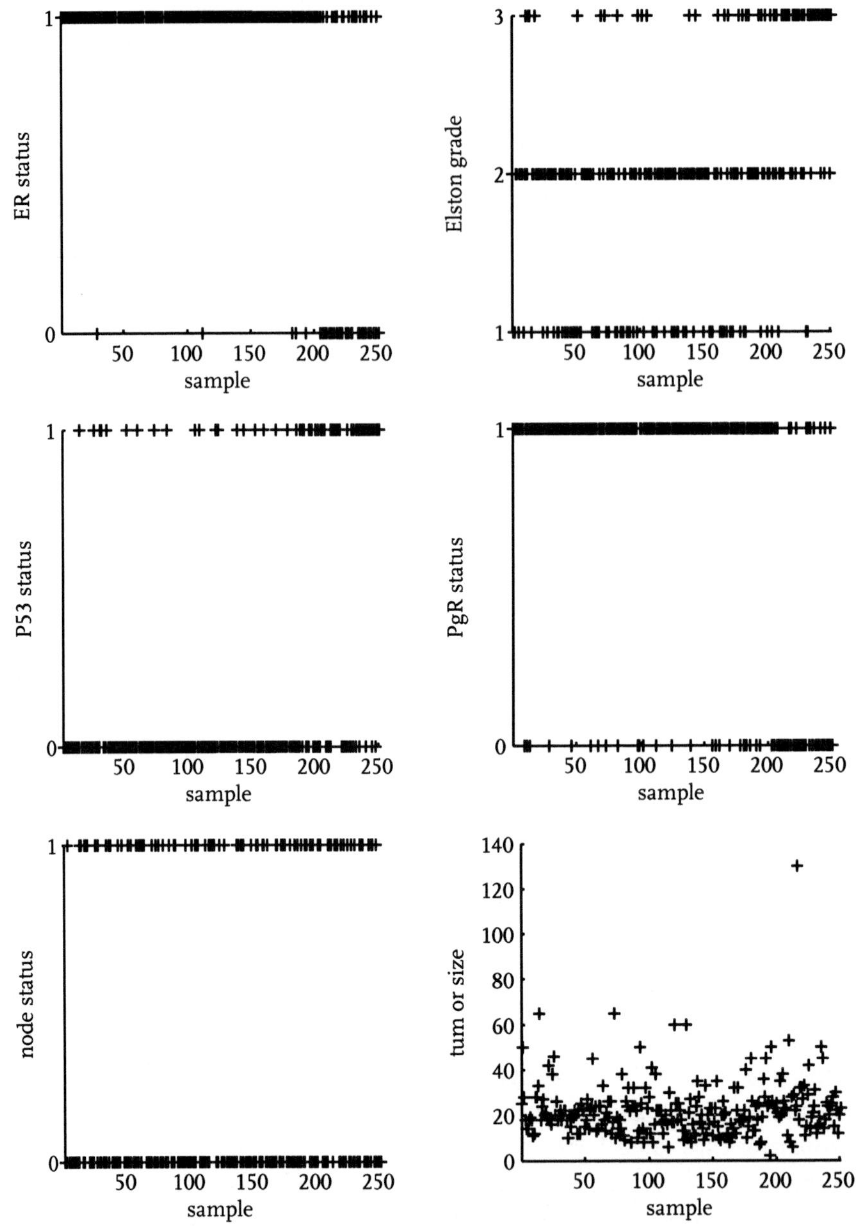

Fig. 6.5 Clinical variables associated with tumour tissue samples, with samples ordered as in Figure 6.4. The subgroup consisting of samples 200–250 (approximately) is consistently associated with ER negative, Elston grade 3, P53 positive, PgR negative tumours, all of which are primary indicators of tumour aggressiveness/malignancy. The subgroup does not have an obvious correlation with lymph node status or tumour size.

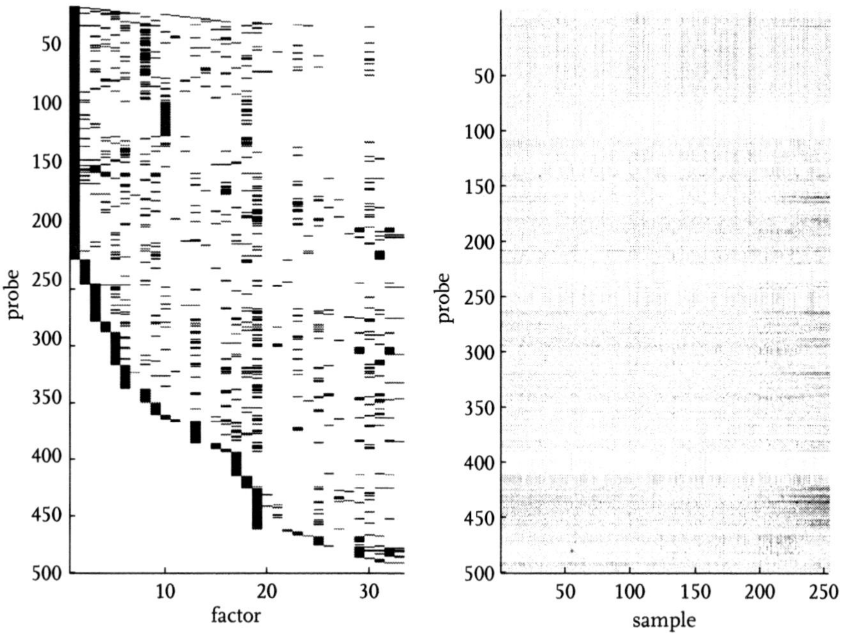

Fig. 6.6 *Left frame:* Sparsity structure of latent factors discovered during analysis of tumour tissue data. Black bars indicate genes with significant factor loadings in terms of $P(\alpha_{g,\ell} \neq 0|X^{\text{vivo}}) > 0.95$. Probes have been reordered to emphasize structure of loadings. The first nine factors are by construction associated with the MYC, β-Catenin, AKT, SRC, P63, E2F1, RAS, P110, and E2F3 signatures respectively. Note that the RAS pathway factor (factor 7) now involves significantly fewer genes than the original RAS signature. *Right frame:* Artifact corrected gene expression of breast tumour samples for the 500 genes incorporated into the latent factor model, with samples ordered as in Figure 6.4 and probes now ordered as in the factor skeleton in the left frame here. Darker shades indicate expression higher than baseline. This ordering reveals that the set of approximately 50 differentially expressed genes characterizing the high-risk tumour subgroup are some of those heavily associated with factor 19.

significant role in distinguishing these tumors. The relationship between factor 7 and the RAS signature score indicates that very little of the pattern of initial RAS pathway activity remains explained by factor 7, though the estimated scores for factor 7 show some variation (relative to that of the factor 4) implying that the RAS pathway may be active but that activity is now captured as the combined effect of multiple factors (as hypothesized above). We see this further below. We can also reflect the contrast between *in vitro* signatures and their factor counterparts by comparing estimated regression coefficients on genes defining the signature scores with the estimated loadings on the factor counterparts with similar conclusions.

6.2.5 Projection of breast tumour latent factors from *in-vivo* to *in-vitro*

We can establish more concrete connections between latent factors and oncogene pathway activity via the same strategy as was used to produce the initial

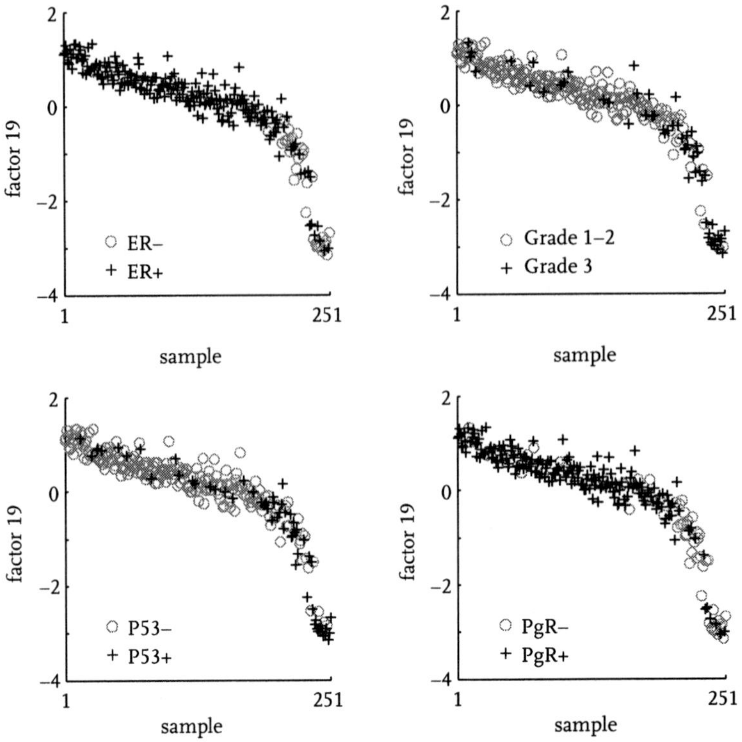

Fig. 6.7 Factor scores associated with latent factor 19 (i.e. $\Lambda_{19,:}$), coded by ER, Elston Grade, P53, and PgR status and with samples ordered as in Figure 6.4. Decreased factor 19 scores show a clear association with ER negative, Elston grade 3, P53 positive, PgR negative tumours.

signature scores, but now in reverse. This is accomplished by regarding each sample in the oncogene intervention study as a newly observed datum, and imputing approximate factor scores for the new observation. Let $x_{g,i}^{\text{vitro}}$ denote the corrected expression of gene g of sample i in the oncogene intervention data (i.e. after subtracting baseline expression and artifact terms). Write $x_{:,i}^{\text{vitro}}$ for the resulting vector on oncogene sample i, noting that we need to restrict to the 213 genes that are in common in the two analyses. Denote by A_* the 251×33 component of the *in vivo* loadings matrix related to these genes and the 33 factors, and the corresponding residual variances as the elements of the diagonal matrix Ψ_*.

The imputed factor score vectors for these *in vitro* samples are calculated, following Lucas *et al.* (2009, 2010), as approximate posterior predictive means of latent factors Λ:, i under the Dirichlet process prior; that is,

$$\Lambda_{:,i}^{\text{vivo}} = c_{i,0} f_i + \sum_{j=1}^{251} c_{i,j} \Lambda_{:,j}$$

where

$$f_i = \left(I + A_*'\Psi_* A_*\right)^{-1} A_*'\Psi_* x_{:,i}^{\text{vitro}}$$

while

$$c_{i,0} \propto \alpha_0 N\left(x_{:,i}^{\text{vitro}}|0,\ A_* A_*' + \Psi_*\right) \quad \text{and}$$
$$c_{i,j} \propto N\left(x_{:,i}^{\text{vitro}}|A_*\Lambda_{:,j},\ \Psi_*\right), \quad (j = 1 : 33),$$

and the $c_{i,j}$ sum to one over $j = 0 : 251$.

Figure 6.8 presents the imputed factor scores for a selection of factors across the complete set of observations from the oncogene study. Among the first nine factors, factors 1, 5, 6, 7, and 9, retain clear associations with their respective founding oncogene pathways. Across all 33 factors, 20 different factors, including factor 19, separate the RAS activated subgroup from the others, illustrating the degree to which the original RAS pathway signature has been dissected into constituent sub-pathways. Some oncogene subgroups, such as β-Catenin and SRC, are not distinguished by any factors, indicating these pathways may play less of a role in explaining patterns of gene expression in the tumour data than does the RAS pathway. Many of the factors identify multiple oncogene subgroups, i.e. likely intersections between pathways.

6.2.6 Factor-based prediction of clinical outcome

For over a decade, a driving interest in expression genomics has been in the potential for expression-based biomarkers of prognostic and diagnostic clinical use (West *et al.* 2001; van't Veer *et al.* 2002; Nevins *et al.* 2003; Chang *et al.* 2004; Pittman *et al.* 2004; Miller *et al.* 2005; Bild *et al.* 2006; West *et al.* 2006; Lucas *et al.* 2009), and we have already seen here that an estimated biomarker – factor 19 – strongly associates with some of the central clinical markers in regard to breast cancer recurrence risk. Identification of clinically relevant *pathways* represented by discovered factors can be explored by considering statistical models that use estimated factor scores as candidate covariates in models to predict clinical outcomes; any factors found to play significant roles in such models will warrant further investigation. We do this now in connection with survival outcomes in the breast cancer study using Weibull survival models. With t_i denoting the survival time of breast cancer patient i, the Weibull density function is $p(t_i|a, \gamma) = at_i^{a-1} \exp\left(\eta_i - t_i^a e^{\eta_i}\right)$ where $\eta_i = \gamma' y_i$ is the linear predictor based on covariate vector y_i and regression parameter vector γ, and a the Weibull index. We have previously used these models, and the Bayesian model uncertainty analysis using shotgun stochastic search (SSS) (Hans *et al.* 2007), for several applications in cancer genomics (Rich *et al.* 2005; Dressman *et al.* 2006), and now apply the approach in the breast cancer context here. Based on an overall set of candidate predictors and specified prior distributions, SSS

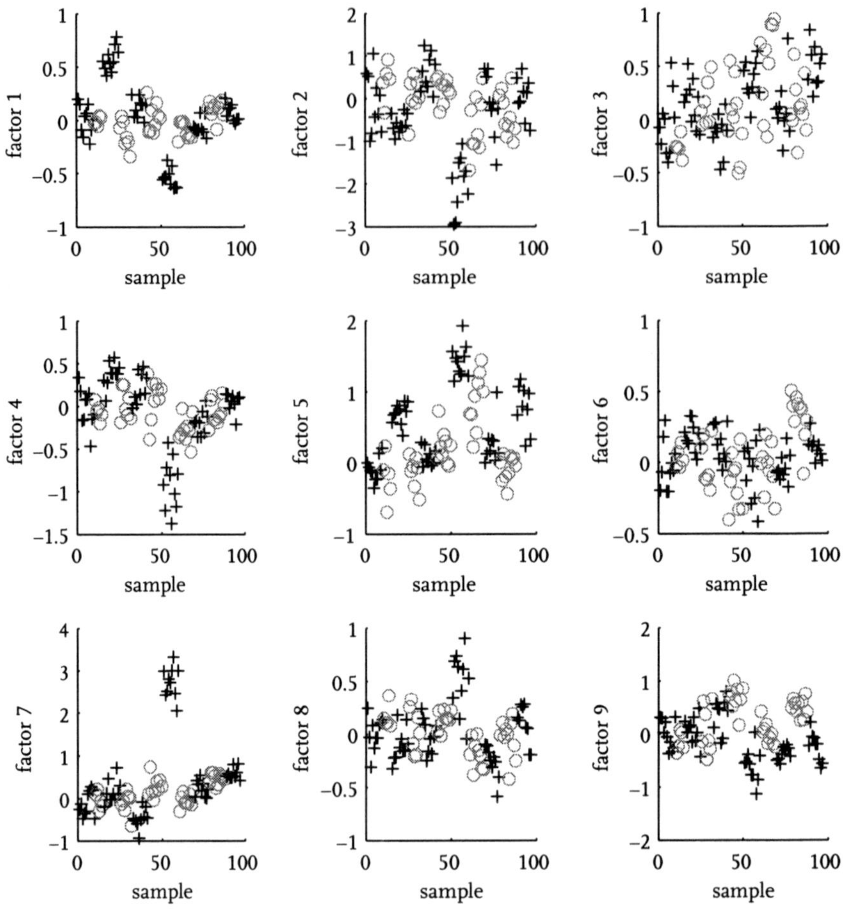

Fig. 6.8 Imputed latent factor scores for oncogene intervention data. Latent factors 1–9 are those founded by the MYC, β-Catenin, AKT, SRC, P63, E2F1, RAS, P110, E2F3 oncogene signature scores respectively. Oncogene interventions are ordered as control 1, control 2, MYC, SRC, β-Catenin, E2F3, RAS, P63, AKT, E2F1, P110, with samples from the same experimental intervention appearing consecutively, designated by the same colour/marker. Separate interventions are depicted by alternating black crosses and grey circles. Note the factor-based separation of the RAS activated experimental group achieved by factors 1, 2, 4, 5, 7, and 8.

generates a search over the space of subset regression and delivers posterior probabilities on all models visited together with posterior summaries for each model, the latter including approximate inferences on (γ, a) in any one of the set of evaluated models (see Appendix B.3 for more details and links to software).

We explored multiple Weibull survival models drawing on the 33 estimated factors (posterior means of factor scores) as possible covariates. The left panel of Figure 6.9 depicts the marginal Weibull-model inclusion probabilities for the 33 latent factors from the breast cancer survival analysis. Three factors,

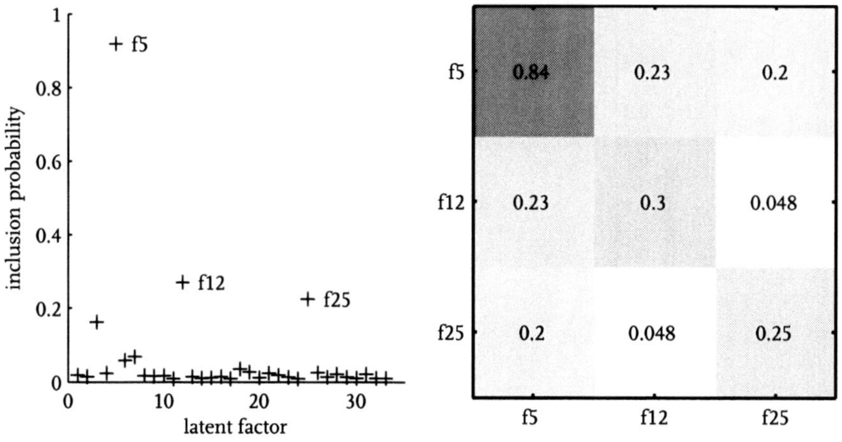

Fig. 6.9 *Left frame:* Marginal posterior inclusion probabilities for the 33 factor covariates in the Weibull model for survival time, as determined by shotgun stochastic search. *Right frame:* Probabilities of pairs of factors appearing together in the Weibull models. Darker tiles indicate higher probabilities. Factor 5 is often accompanied by either factor 12 or factor 25, but very rarely both, indicating some predictive redundancy associated with the 12,25 pair.

5, 12, and 25, appear with probabilities greater than 20%, with factor 5 being of clear interest. The right panel in Figure 6.9 displays the pairwise inclusion probabilities among factors 5, 12, and 25, and further indicates that factor 5 alone may have significant use for predicting survival. Note that factor 19, though marginally associated with clinical risk factors, does not score highly relative to these other three factors in terms of its appearance in high probability survival models.

The median survival time m for a patient with covariate vector y satisfies $m^a = \exp(-\gamma' y)\log(2)$ so that posterior inference on (a, γ) yields inference on m. From the top 1000 models visited over the course of 10,000 model-search steps we compute approximate posterior means for (a, γ) in each model as well as approximate model probabilities, and average the plug-in estimates of m over models. The resulting estimated median survival time can then be evaluated at any specified covariate vector y. In particular, we evaluate this at the covariate vectors for each of the $n = 251$ patients in the breast data set, thus generating predictions of median survival times for a future hypothetical cohort of 251 patients sharing these covariate values. Kaplan–Meier curves can provide useful visual displays for simply displaying survival data on a number of patient subgroups, and Figure 6.10 shows separate KM curves for the two groups of patients whose estimated median survival lies above/below the median value across all 251 patients. This data display clearly indicates the potential clinical value of the biological sub-pathways represented by factors 5, 12, and 25.

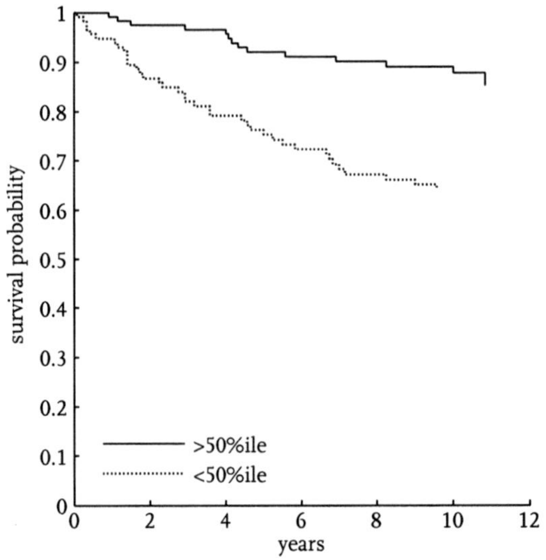

Fig. 6.10 Kaplan – Meier curves representing predicted median patient survival probabilities when patients are divided into two cohorts based on Weibull model prediction of survival time. The solid black curve represents patients for whom the posterior predictive estimate of median survival time exceeds the median of that value across the 251 patients, and the dotted curve represents the other 50% higher-risk patients.

This study was repeated with an expanded covariate set that now adds to the 33 factors the following candidates: (a) the set of nine projected oncogene pathway signature scores, and (b) the traditional clinical covariates used in breast cancer prognosis: Elston grade, ER, lymph node status, P53, PgR, tumour size and age at diagnosis. Figure 6.11 summarizes the resulting inclusion probabilities. Although the key clinical variables – lymph node status and tumour size – are identified as the top two predictors of survival time, factors 5, 12, and 25, are the third, fourth, and fifth most relevant. It is noteworthy that none of the original signature scores are among the top predictors, demonstrating that the refinement of the original signatures achieved by latent factor analysis hones the representation of biological pathway activation to deliver a refined representation of the complexity of pathway structure in the *in vivo* setting.

Kaplan–Meier curves for this expanded analysis (Figure 6.12) show greater separation between good and poor prognosis patient groups than was evident in the factor-only analysis, though this may be expected given the inclusion of clinico-pathological variables such as lymph node status that are highly indicative of poor outcome and tend to be poorly predicted by expression data.

We can directly compute estimates of the posterior predictive survival function for any future patient represented by a candidate covariate vector by

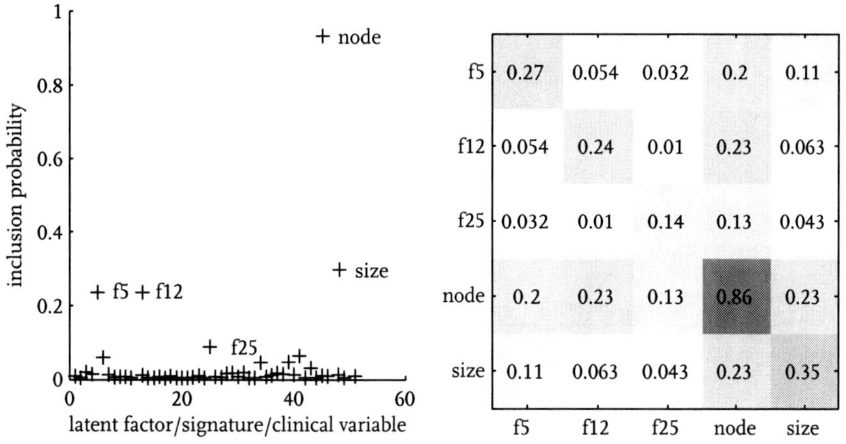

Fig. 6.11 *Left frame:* Marginal posterior inclusion probabilities for candidate covariate in the Weibull survival analysis when drawing on factors, signatures and clinical covariates. *Right frame:* Pairwise inclusion probabilities for the top 5 covariates. Again factors 12 and 25 rarely appear in conjunction. Factor 25 appears with lymph node status with substantially lower probability than do the other factors, suggesting that factor 25 can play a role as a surrogate for node status.

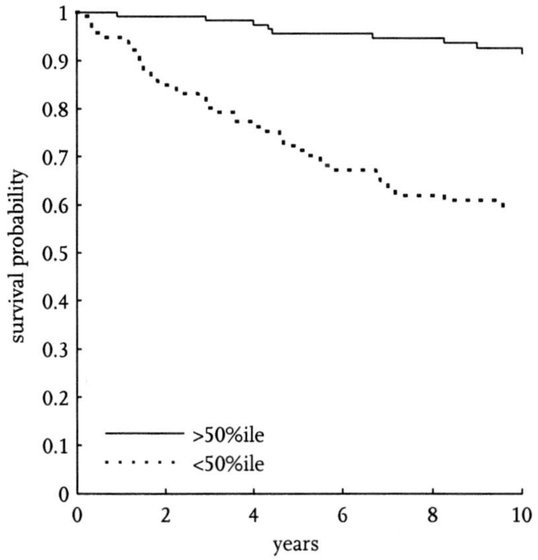

Fig. 6.12 Kaplan – Meier curves representing median patient survival from survival analysis incorporating latent factors, original signatures, and clinical variables. The solid black curve represents patients with predicted median survival time above the median over patients and the dotted curve represents the other 50% higher-risk patients. Greater stratification is achieved relative to Figure 6.10, mainly due to the use of lymph node status as a covariate.

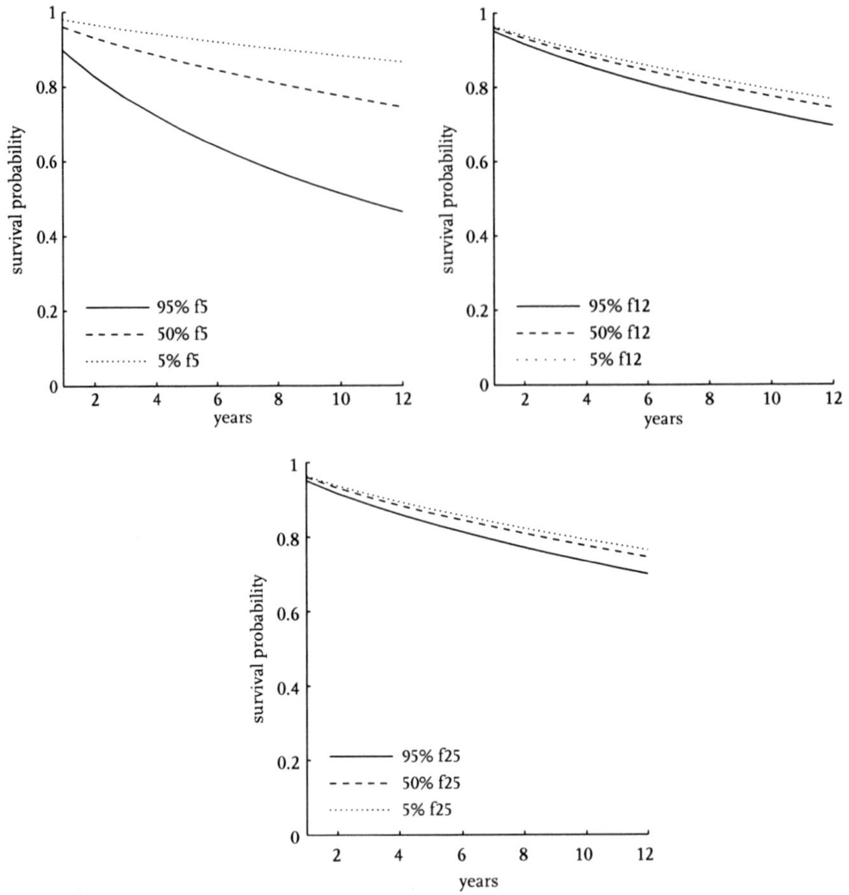

Fig. 6.13 Estimates of posterior predicted Weibull survival curves from factor-only survival analysis. The dashed lines represent survival when all factor values are fixed at their posterior median value. The solid and dotted lines in each figure represent predictions achieved by individually varying values of factors 5, 12, and 25 to values chosen as the 5th and 95th sample quantiles in the $n = 251$ patient data set. Stratification on basis of factor 5 appears to be a useful predictive diagnostic.

averaging over all models visited during SSS, thereby facilitating the the use of factor-based sub-pathway signatures for personalised prognosis. We do this using vectors y that set some covariates at their median values across the 251 samples and others at more extreme values, i.e. upper or lower 5% values with respect to the data. This enables an investigation of the predictive effect of individual covariates with the other fixed at average values. Some examples appear in Figures 6.13 and 6.14, obtained by varying the values of factors 5, 12, and 25, and then with respect to lymph node status and tumour size in Figure 6.14). In the factors-only analysis (Figure 6.13), predicted survival changes dramatically for extreme values of factor 5. In fact, the change in predicted survival associated with variation in factor 5 in the factors-only analysis is greater than

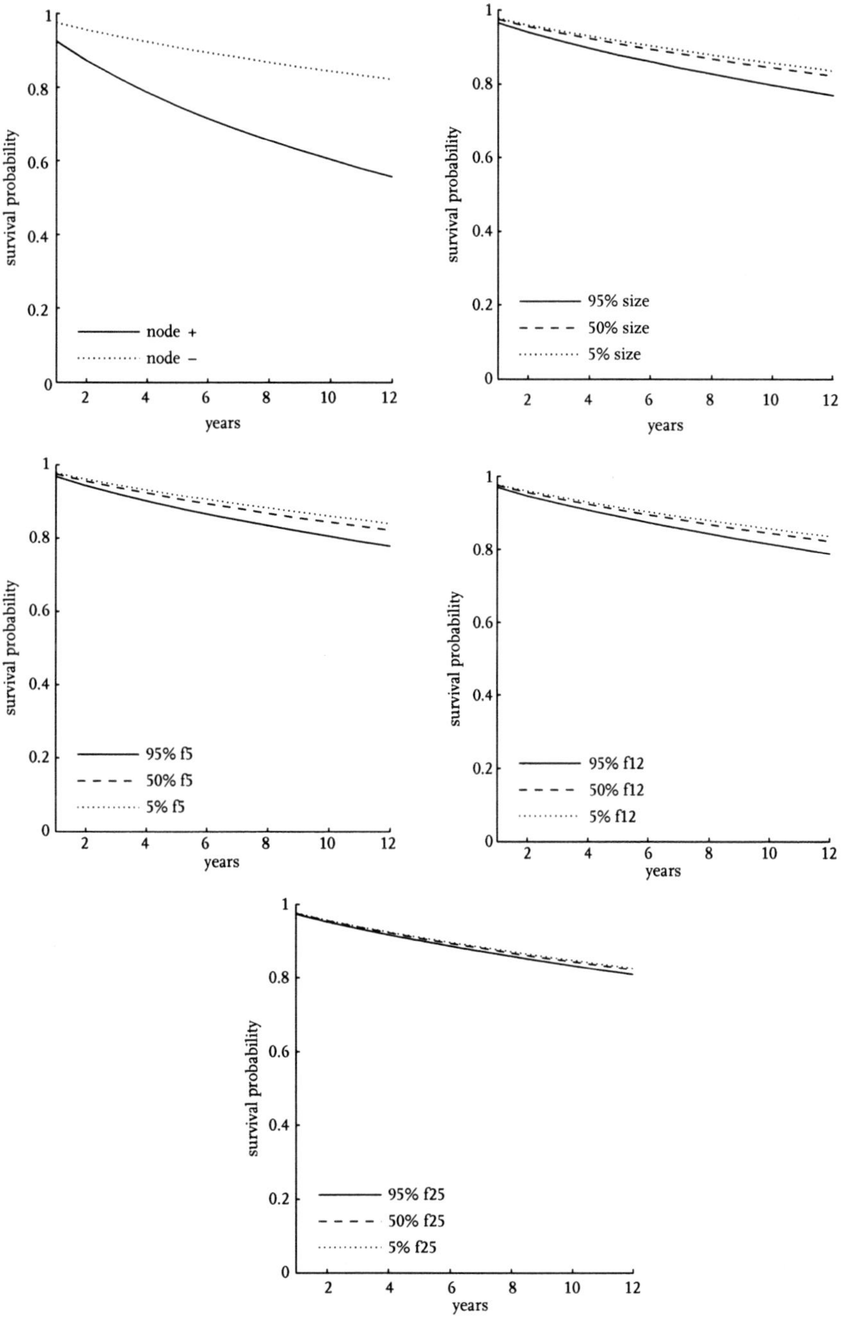

Fig. 6.14 Estimates of posterior predicted Weibull survival curves from factor/signature/clinical variable survival analysis. Dashed lines represent median covariate predictions. Solid and dotted lines represent predictions achieved by individually varying values of lymph node status, tumour size, and factors 5, 12, and 25 to their 5th and 95th quantiles. The predictive utility of factor 5 is diminished in light of lymph node status, though the separation achieved by lymph node status is comparable to that achieved by factor 5 in the Figure 6.13.

the change in predicted survival associated with variation in lymph node status in the combined factors and clinical variables analysis.

6.3 Biological evaluation and pathway annotation analysis

6.3.1 Initial discussion

In light of their obvious clinical significance, placing factors 5, 12 and 25 in their proper biological context is of primary importance. The discussion in Section 6.2.3 gave some clues as to the relationship between these factors and the oncogene pathways. Factor 5 appears to represent some intersection between the MYC, RAS, P63, and P110 pathways, and factor 12 some subcomponent of the RAS pathway, but factor 25 does not appear to associate with any of the nine *in vitro* defined oncogene pathways. We are also interested in the biological relevance of factor 19 that, although not appearing competitive in survival prediction, evidently associates with traditional clinical risk factors.

Gene-specific connections between factors and the original oncogene pathways can be explored by identifying genes with high probabilities of non-zero factor loadings and relating them to the original oncogene pathway signatures. The top 20 genes in factors 5, 12, and 25 were identified by absolute values of the estimated factor loadings. Figure 6.15 presents the probabilities of inclusion in the original oncogene signatures for each gene among the three sets of top genes. Predictably, many of the top genes in factor 5 were strongly associated with the P63 signature. This makes biological sense, given that factor 5 was founded by the P63 signature score. Top genes in factor 12 show association primarily with the RAS and P63 signatures; in particular the founder gene of factor 12 has high probability of involvement in both the RAS and P63 pathways. Very few of the top genes in factor 25 were among those included in the oncogene study, and of the four genes common to both studies, only two genes showed significant probability of involvement in oncogene pathways: probe 210761_s_at in the RAS pathway, and probe 216836_s_at in the P63 and P110 pathways. Such connections further emphasize the involvement of RAS and P63 sub-pathway activity in affecting tumour malignancy and, therefore, long-term survival.

This focuses attention on the ability of factor profiling to highlight critical pathway intersections; it turns out that these two probes identify genes well-known to play central roles in breast cancer oncogenesis. Probe 210761_s_at is a sequence within the gene GRB7, a well-known co-expression partner of the epidermal growth factor receptor ERB-B2 that is a fundamentally causative oncogene in breast cancer and one of the two major drug targets in breast cancer chemotherapy; ERB-B2 is also known as

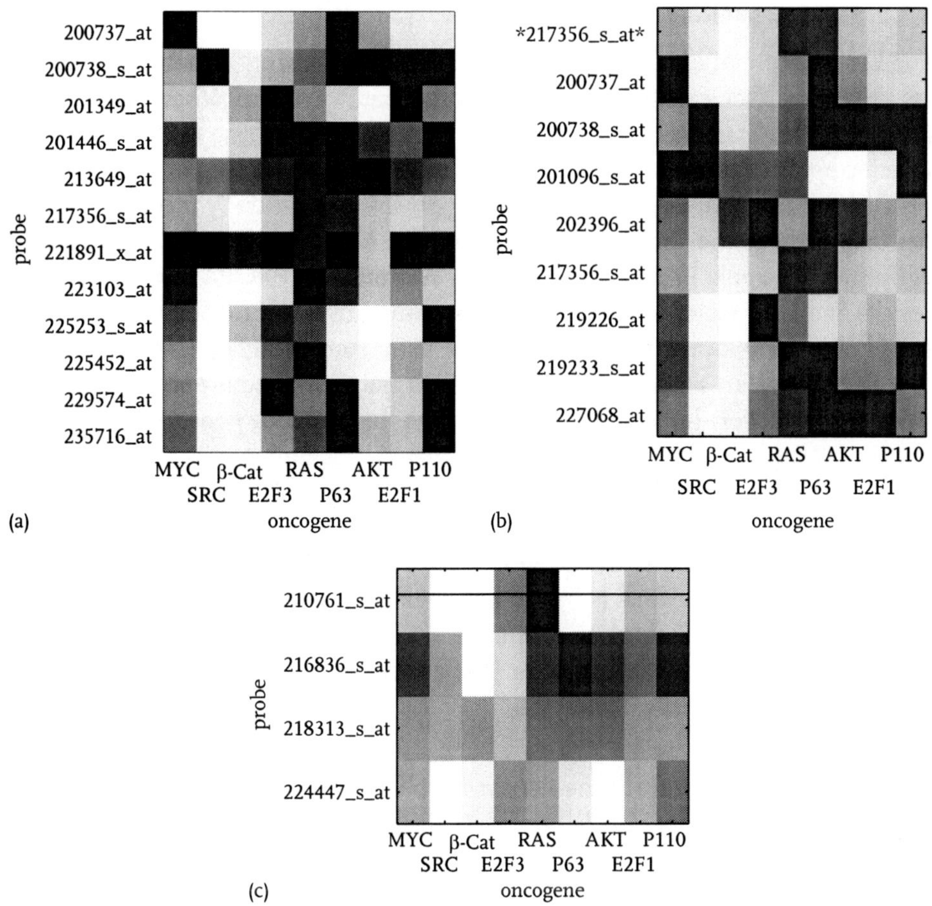

Fig. 6.15 Heatmaps showing oncogene signature inclusion probabilities $P(\beta_{g,k}^{\text{vitro}} \neq 0 | X^{\text{vitro}})$ for subsets of the top 20 genes in factors 5 (a), 12 (b), and 25 (c) also present in the oncogene signature study. Probabilities are coded by greyscale, with black designating probability 1. The top genes in each factor were determined by sorted absolute value of mean posterior factor loading. Nine of the 12 top genes in factor 5 appear in the P63 pathway signature with high probability. Several top genes in factor 12 show association with the RAS and P63 signatures, including founder probe 217356_s_at defining gene PGK1. Factor 25 shares only four genes in common with the oncogene study, and of them only two played significant roles in oncogene pathways: probe 210761_s_at in the RAS pathway is a key oncogenic marker, gene GRB7 highly correlated in expression patterns generally with ERB-B2, and probe 216836_s_at in the P63 and P110 pathways is the fundamentally causative breast cancer oncogene ERB-B2 itself.

HER-2-ν (Sørlie *et al.* 2001; Bertucci *et al.* 2004; Badache and Gonçalves 2006). Remarkably, probe 216836_s_at is in fact a probe sequence for oncogene ERB-B2 itself. Factor analysis seeded by a set of individual oncogenic signatures has identified pathway interconnections that clearly lead to ERB-B2 as represented by the clinically relevant factor 25.

6.3.2 Bayesian pathway annotation analysis

Further interpretation of factors can be developed using annotated biological databases that catalogue known cellular pathways and other versions of pathways. One standard is the Molecular Signatures Database MSigDB (2008), which contains thousands of curated lists of genes representing experimentally verified cellular pathways. Such *pathway gene lists* are naturally incomplete and typically error-prone, but nevertheless provide key representations of known pathways. We apply Probabilistic Pathway Annotation (PROPA) analysis to the results from our factor analysis in order to compare factor loadings against the set of over 1000 annotated pathway gene lists from MSigDB (2008). PROPA is a fully Bayesian analysis that provides formal assessment and rankings of pathways putatively linked to factors based on the posterior gene×factor probabilities $Pr(\alpha_{g,k} \neq 0|X^{vivo})$ and signs of estimated $\alpha_{g,k}$. Full details and examples appear in Shen (2007) and Shen and West (2010); additional comments are given here in Appendix B.4.

Application of PROPA to the gene sets comprising factors 5, 12, 19 and 25, yielded the connections below to previously annotated cellular pathways of immediate relevance in cancer. Similar evaluations can be performed on other factors; such studies, as well as other follow-on biological studies, were initiated as a result of this statistical analysis.

Factor 5. PROPA identifies the serum fibroblast cell cycle[1] pathway gene list as representing the most likely candidate biological pathway underlying factor 5. This pathway gene list relates to the gene expression programme of human cells in response to serum exposure and represents links between cancer progression and wound healing – *'cancer invasion and metastasis have been likened to wound healing gone awry'* (Chang *et al.* 2004). The 'wound-healing' expression signature that this pathway gene list is based on is a well-known and powerful predictor of risk in a number of cancers and particularly in breast cancer: a biomarker indicating increased risk of invasion, metastasis and death. This pathway very substantially dominates the other 1000+ in the PROPA analysis of factor 5, and hence we identify factor 5 as reproducing this wound-healing signature and overlaying the corresponding cell cycle pathways that are induced by serum exposure. This is consistent with our findings that factor 5 alone is the most potent predictor of survival in our breast survival studies and that it derives in our analysis from genes linked to the MYC signature, and apparently represents intersections of the MYC, RAS, P63, and P110 pathways. That many of the top genes in factor 5 were strongly associated with the P63 signature raises the potential for further studies to advance our currently limited understanding of

[1] http://www.broad.mit.edu/gsea/msigdb/cards/SERUM_FIBROBLAST_CELLCYCLE.html

the connectivities between P63 and other components of the broader cell cycle and cell growth regulatory machinery (Figure 6.1).

Factor 12. PROPA identifies a number of hypoxia-related pathways, the most highly scoring by far being the HIF-1 target pathway,[2] a set of genes regulated by hypoxia-inducible factor 1 (HIF-1) and that define pathway activities that mediate adaptive responses of cells to reduced oxygen availability. The HIF-1 pathway responses are well-known to play major roles in the pathogenesis of cancer (Semenza, 2001; Chi *et al.* 2006). Less highly scored by PROPA but of note is a well-known pathway gene list representing a breast cancer 'poor prognosis' signature (van't Veer *et al.* 2002) that consists primarily of genes regulating cell cycle, invasion, metastasis and angiogenesis. Our analysis indicates that factor 12 incorporates both hypoxia-response pathways and cancer invasion mechanisms, thereby connecting aspects of the tumour micro-environment (oxygenation) with gene pathway responses in cell regulation. Further biological investigation may be suggested to evaluate genes scoring highly in factor 12, especially in view of the fact that factor 12 arises initially as a sub-component of the RAS pathway and has many top genes that show association with the RAS signature. There are also strong connections with the P63 signature; in particular the founder gene of factor 12 has high probability of involvement in both the RAS and P63 pathways. To our knowledge, these connections of P63 with primary cancer mechanisms and pathways have not been previously explored.

Factor 19. Factor 19 is a primary ER ((o)estrogen receptor) factor; the two top PROPA pathways are expression-based ER pathways involving genes whose expression levels are either consistently positively or negatively correlated with estrogen receptor status in breast cancer.[3] This is concordant with our findings of the strong association of factor 19 with ER in the breast data set (Figure 6.5) and with the names of some of the genes that score highly on factor 19 in the BFRM analysis that can be recognised as ER related, ER targets or otherwise synergistic with ER. That an ER factor arises naturally as part of the evolutionary BFRM model refinement is no surprise given the dominance of the ER pathway in breast cancer – in terms of very many genes influenced by ER activity – and also as ER is intimately interconnected with the cell cycle and progression machinery through cyclin-D and other known pathway interactions (Figure 6.1 and the ER example and references in Carvalho *et al.* (2008)).

[2] `http://www.broad.mit.edu/gsea/msigdb/cards/HIF1_TARGETS.html`

[3] `http://www.broad.mit.edu/gsea/msigdb/cards/BRCA_ER_NEG.html` and `http://www.broad.mit.edu/gsea/msigdb/cards/BRCA_ER_POS.html`

Factor 25. The top PROPA pathway linked to factor 25 is also ER, that gene list representing genes consistently positively correlated with estrogen receptor status also identified for factor 19. The second and third most highly scoring pathways also link to ER but are very different from those defined by factor 19 and also seem to indicate a rather different biological function for the pathway activities factor 25 represents. In particular, they are pathway gene lists derived from experiments to investigate the effects on ER positive cells of drugs including Tamoxifen, the currently most-used hormonal therapy in breast cancer care, and are therefore importantly related to questions of cellular responses and resistance to Tamoxifen.[4] In an earlier study we had identified such a factor as a 'predictor of resistance to Tamoxifen' and had verified its ability to discriminate breast patients with respect to survival outcomes according to whether they were or were not resistant in the sense of high versus low values of the estimated factor (Lucas *et al.* 2009). That analysis identified this factor from a very different set of initial genes in the BFRM analysis, and that it emerges again here in an analysis initiated by the set of oncogene signatures indicates that this is clearly a key clinically relevant factor in breast cancer genomics, linking to critical underlying pathways. It is an important ER-related factor that is quite distinct from the dominant ER measures provided here by factor 19.

We have already noted that very few of the top genes in factor 25 were among those included in the oncogene study, but key among those that were ERB-B2. There is no annotated ERB-B2 pathway in the gene lists data bases, but this suggests a role for the ERB-B2 activity in mediating ER-related activity and perhaps the cellular responses to drugs including Tamoxifen. Further investigation of this together with genes most highly scoring on factor 25 may well prove informative.

Appendices

A. Broader context and background

A.1 Bayesian factor models

Factor analysis has come into play in a major way in multivariate analysis with modern, high-throughput biological data, and is evidently central to the application of this paper as well as multiple other related studies. Our work builds on a long history of Bayesian factor analysis; some of the historical picture can be explored from textual material and references in Press (1982), and more recent, broader developments relevant to this work in Lopes and West

[4] `http://www.broad.mit.edu/gsea/msigdb/cards/FRASOR_ER_UP.html` and `http://www.broad.mit.edu/gsea/msigdb/cards/BECKER_TAMOXIFEN_RESISTANT_DN.html`

(2004) and West (2003) and references therein. Write $x_i = X_{:,i}$ for column i of the $p \times n$ data matrix, the outcome vector for sample i. Similarly, write $\lambda_i = \Lambda_{:,i}$ for column i of the $k \times n$ matrix of latent factors, the k-vector of factor values on sample i. The traditional, linear, normal factor model $x_i = A\lambda_i + \nu_i$, with $\lambda_i \sim N(0, T)$ and $\nu_i \sim N(0, \Psi)$, is simultaneously a structural model that allows for inference on underlying common factors λ_i and a model for the implied variance matrix $ATA' + \Psi$.

Well-known and fundamental questions of model identification require structure and constraints that can be imposed directly on some parts of (A, T, Ψ) as well as through priors over these parameters (deterministic constraints really forming part of the global prior specification), and with consideration of the role of the dimensions k of the vector λ_i and p of x_i. Models are generic though applications involve $k < p$ as the factors are generally lower-dimensional variates that drive aspects of the association patterns among the observables in x_i. In gene expression models as in this study, p is very much larger than k so the identification problems that arise from simple numbers of parameters if k is 'too large' simply do not arise. The aim of capturing all aspects of covariation via the factors indicates a need to specify Ψ as diagonal. Lack of identification under factor rotation is typically addressed by setting $T = I$, and then additional constraints on the loadings matrix A ensure identification. With $p \gg k$, A is 'tall and skinny' and our preferred structure – favoured in applications of factor models in econometrics and finance as well as genomics – is to set the upper triangle of elements of A to zero and the diagonal elements (first k rows and columns) to be positive. One benefit of this structure is that the first k variables in x_i are then directly linked to the factors, which allows us to give 'names' to the factors and led to us referring to these variables as the 'founders' of the factors.

One positive aspect of factor modelling mentioned in the application here, and that is potentially important in all applications, is the ability of latent factors to capture *all* aspects of covariation in the data. The over-arching idea is that inferred latent factors capture scientific structure that invites interpretation and further substantive investigation. However, the adaptability of factor models means that some inferred factors may represent noise – experimental artifacts, non-Gaussian errors and outliers that show up in terms of contributions to the shared patterns of variation of subsets of variables. These noise sources have the potential to corrupt inferences, and it is a real strength of factor models that they can 'soak up' such noise contributions and isolate them in terms of one or a few factors. Interpretation of estimated models must always bear this in mind, and be aware that some estimated 'structure' may in fact be noise to identify as such rather than substantive scientific structure; more examples in Carvalho *et al.* (2008) highlight this.

A.2 Dirichlet process priors

An important aspect of the class of latent factor models applied here is the ability to represent non-Gaussian structure in gene expression patterns via the use of nonparametric Dirichlet process priors on latent factors. In many examples, it is clear that highly non-Gaussian expression patterns in observed data can arise due to substructures in the sample cohort representing natural biological variation. Nonparametric methods are measurably beneficial in such settings due to their ability to adapt to arbitrarily complicated patterns of covariation, thereby inducing robustness in posterior inferences on model parameters.

The Dirichlet process has several representations, but it is fundamentally a probability distribution on the space of probability distributions. The random measures defined by the Dirichlet process are almost surely discrete, and therefore the Dirichlet process prior is a natural choice for defining a random mixing distribution. There is a large and ever expanding literature on the use of Dirichlet process methods in Bayesian mixture modelling. From the foundational theory of Ferguson (1973) and Blackwell and MacQueen (1973), the development of computational techniques for fitting various flavors of hierarchical models involving Dirichlet processes priors (West, Müller, and Escobar 1994; Escobar and West 1995; MacEachern and Müller 1998, among many others) has established the technology as a major tool in applied Bayesian statistics.

The use of the Dirichlet process for defining a flexible distribution on the space of latent factors rests on the so-called Pólya urn representation (Blackwell and MacQueen 1973), also referred to as the Chinese Restaurant Process representation (see the appendix in Chapter 27 by Liang *et al.* in this volume). Consider the factor model component $\lambda_i \sim F(\cdot)$ where F has the Dirichlet process prior denoted by $F \sim DP(\alpha_0 F_0)$; here α is the total mass (or precision) and F_0 is the base measure that defines the pointwise prior mean, viz $E(F(\lambda)) = F_0(\lambda)$. The Pólya urn representation is based on the marginal distribution of draws $\Lambda_{1:n}$ from F and can be viewed in terms of sets of complete conditional distributions, as follows. For each $i = 1 : n$, with Λ_{-i} being the $n-1$ factor vectors apart from λ_i,

$$(\lambda_i|\Lambda_{-i}) \sim \frac{\alpha_0}{n-1+\alpha_0} F_0(\cdot) + \frac{1}{n-1+\alpha_0} \sum_{r=1, r\neq i}^{n} \delta_{\lambda_r}(\cdot)$$

where $\delta_a(\cdot)$ represents the point-mass at a. That is, the 'next' factor vector is with some probability equal to one of the already realised vectors, or with complementary probability it is a new draw from F_0. This induces a clustering on factors, with a higher probability of a smaller number of distinct values when α_0 is small. Through inference on the parameter α_0, it is possible for the data to effectively inform the number of such distinct values. The Pólya urn representation also forms the basis for the class of computational methods referred to as collapsed samplers, a common alternative to the blocked-Gibbs class of methods as described in the appendix in Chapter 27 by Liang *et al.* in this volume.

A.3 *Sparsity modelling in multivariate analysis*

In biological pathway analysis, it is inherent that multiple biological factors underlie patterns of gene expression variation, so latent factor approaches are natural – we imagine that latent factors reflect individual biological functions, i.e. variation in pathway or sub-pathway activation. This is also a motivating context for models in which A is sparse. Each biological pathway – or aspect of a complex pathway that a single factor putatively reflects, at least in part – involves a number of genes, perhaps a few to a few hundred, but not all genes; so each column of A will have many zeros. Similarly, a given gene may play roles in one or a small number of biological pathways, but will not be involved in all; so each row of A will have many zeros. The introduction of sparse models in West (2003) has led to a rich framework for large-scale multivariate analysis that is impacting applications in various fields including a broad range genomic applications.

The development of sparsity modelling in multivariate regression and analysis of variance models is basically derived from the same kinds of motivating considerations. When gene expression is the response in a designed study, many genes may show changes in expression as a result of one or more design factors, but the patterns of non-zero associations between (the very many) genes and the (relatively few) design factor variables will generally be expected to be very sparse indeed.

Technically, sparse multivariate modelling as used here – with both regression and latent factor components – builds on the standard and fundamental Bayesian approach using sparsity priors; priors over a set of regression or factor loading coefficients for a single design variable or latent factor allow some or many of the coefficients to be zero with an explicit 'point mass' at zero. The hierarchical specification of such priors allows the data to inform on the underlying base-rate of non-zero values, and then the posterior naturally shrinks inferences on individual coefficients in a way that naturally adjusts for the inherent multiple comparisons being made. In terms of the fundamental conceptual basis and technical/computational aspects, the use of sparsity priors in these multivariate models builds directly on the extensive use of such priors for regression model uncertainty and variable selection in standard Bayesian linear regression; see developments over the years from the work and references in George and McCulloch (1993), Clyde and George (2004) and Hans, Dobra, and West (2007), for example.

B. Models and computations: Bayesian latent factor regression models

B.1 MCMC in BFRM

The details of MCMC analysis of Bayesian latent factor regression models and the implementation in the BFRM software have been well documented (West 2003; Lucas *et al.* 2006; Wang *et al.* 2007; Carvalho *et al.* 2008) and illustrated in a number of related applications (Chang *et al.* 2009; Chen *et al.* 2008; Seo *et al.* 2007; Lucas *et al.* 2009, 2010; Shen and West 2010). The MCMC uses a Gibbs sampling format as fully detailed in Carvalho *et al.* (2008) and to which readers interested in implementation details should refer; the BFRM software used is fully documented with examples and is freely available.[5] In summary, the component conditional distributions sequenced through are as follows, in each case conditional on the data and all other model quantities denoted by the '$-$' in conditionings.

- Resample latent factors from conditional posteriors induced by the Dirichlet process prior. For each sample i, $p(\Lambda_{:,i}|-)$ is sampled based on the two-step configuration sampler of Dirichlet process mixture models (West *et al.* 1994; MacEachern and Müller 1998). After each sweep through the n factor vectors, there will be reduction to a smaller number of unique realized vectors induced by the inherent clustering mechanisms of the Dirichlet process model and reflecting the non-normality of the underlying latent factor distribution.
- Resample independently from the set of inverse gamma complete conditionals $p(\tau_j|-)$ for the variances of regression parameters and factor loadings.
- Resample independently from the set of inverse gamma complete conditionals $p(\psi_g|-)$ for residual variances.
- For each j in turn, resample the $\alpha_{g,j}$ parameters (denoting regression parameters and factor loadings with no loss of generality) from complete conditional posteriors under the sparsity priors. This involves a custom simulation step that samples, independently for $g = 1 : p$, the set of p distributions $p(\alpha_{g,j}, \pi_{g,j}|-)$ via composition. A detail is that with s latent factors the first s loading parameters are constrained to be non-negative, and this is incorporated in the algorithm. See Lucas *et al.* (2006) and the appendix in Carvalho *et al.* (2008).
- Resample the $\rho_{g,j}$ hyperparameters of the sparsity priors independently from implied beta complete conditionals.

[5] `http://www.stat.duke.edu/research/software/west/bfrm/`

B.2 Evolutionary factor models search

A key element of the strategy for biological pathway exploration as developed in this case study is the use of evolutionary stochastic model search to expand the model gene set and number of factors to refine an initial focus on a specified set of pathway-specific genes. This involves evolutionary search, fully detailed in Carvalho *et al.* (2008), summarized as follows:

- In any 'current' model based on p genes and k factors, compute approximate variable inclusion probabilities $Pr(\alpha_{g,k} \neq 0|X)$ for all genes g *not in the current model.*
- Rank and select genes with highest inclusion probabilities subject to exceeding a specified probability threshold. Stop if no additional genes are so significant.
- Refit the expanded model, also increasing to $k+1$ factors with an initial, random choice of the founder of the new factor (the gene listed at $g = k+1$). From this analysis identify the gene with highest estimated $Pr(\alpha_{g,k+1} \neq 0|X)$ and refit the model with that gene now at $g = k+1$.
- Cut back to fewer than $k+1$ factors by dropping any factor j for which fewer than some small pre-specified number of genes have $Pr(\alpha_{g,j} \neq 0|X)$ exceeding a high threshold. Otherwise, accept the expanded model and continue to iterate the model evolutionary search.
- Stop if the above process does not include additional genes or factors, or if the numbers exceed some pre-specified targets on the number of genes included in the model and/or the number of factors.

In our case study here, all MCMC analyses used summaries from runs of length 10,000 following 1000 discarded as burn-in. Each iteration of the evolutionary model search brought in at most 10 new genes and required a threshold of at least 0.95 on gene×factor inclusion probabilities to add such genes to the current model. In considering expanding to add a further factor, we required that at least five genes have gene×factor inclusion probabilities of at least 0.85. Model search was allowed to proceed until as many as 500 genes and 50 factors have been added to the model, with a control restricting maximum number of genes per factor to at most 30. This means that once the number of genes with non-zero loadings on a given factor reached 30, then no further genes were incorporated into the latent factor model on the basis of that factor. Note that the factor analysis of the breast cancer data terminated on reaching 500 genes, at which point the analysis had evolved from the initial nine factors to a final total of 33.

B.3 Models and computations: Bayesian Weibull survival regression models

Model search and analysis in the Weibull regression models used the SSS software (Hans *et al.* 2007) that is fully documented with examples and is freely available.[6] Under normal priors on the regression parameter vector γ and a gamma prior on the Weibull index a in any specified regression model, this analysis computes posterior modes for (a, γ) and estimates the corresponding model marginal likelihood via Laplace approximation. With a specified prior probability of covariate inclusion – treating covariates independently and each with the same prior inclusion probability – this allows for the computation of approximate posterior probabilities across any specified set of models. Shotgun stochastic search for such regression models (Hans *et al.* 2007) moves around the large space of subsets of candidate covariates visiting many possible regression models. The algorithm is designed to seek out models in regions of high posterior probability. SSS is very efficient and, in terms of large numbers of models searched, very aggressive, with the ability to rapidly explore and catalogue many models. SSS is also parallelizeable, with both serial and parallel versions available.

In the survival analysis of the case study here, SSS-based exploration of the space of Weibull survival models was allowed to proceed for 10,000 model-search iterations, and recorded the top 1000 models in terms of approximate posterior probability. The resulting approximate posterior predictive distribution for a new patient sample – the distribution required for the summary comparisons reported – is then constructed as the mixture, weighted by approximate posterior model probabilities, of the 1000 Weibull models with parameters (a, γ) set at the computed posterior modes. The prior probability of covariate inclusion, which defines an implicit penalty on model dimension, was set so that the prior expected number of covariates in the model is 3. Model search was initiated by evaluating a randomly selected model with two covariates.

B.4 Models and computation: Bayesian probabilistic pathway annotation

A biological pathway is represented by an unordered list of, typically, a small number of genes (10s to several 100s) in typical pathway data bases. In essentials, PROPA analysis of the BFRM outputs for any factor k scores each pathway gene list according to the relative enrichment of genes g with high values of $Pr(\alpha_{g,k} \neq 0 | X^{\text{vivo}})$, and also for the relative paucity of genes g with low values, in the list. Some pathway gene lists also carry information about the sign of changes in expression of genes in the list as a result of intervention on the pathway, so that the estimated signs of the $\alpha_{g,k}$ for genes in the list are also then relevant in assessing the results of PROPA.

[6] `http://www.stat.duke.edu/research/software/west/sss/`

For any one of the factors, say factor k, in the BFRM analysis of the breast data, the PROPA framework idealizes an underlying, true pathway $\mathcal{F}$ related to the factor; $\mathcal{F}$, the *factor pathway*, is simply an unknown list of genes that are changed in expression with this factor, i.e. a set of genes with non-zero loadings on the factor. The estimated values $Pr(\alpha_{g,k} \neq 0|X^{\text{vivo}})$ over all (40,000+) probes provide data for PROPA analysis to make inference on the true gene×factor associations. Write $\Pi = \{Pr(\alpha_{g,k} \neq 0|X^{\text{vivo}}),\ g = 1 : p\}$ and $A_{1:m} = (A_1, \ldots, A_m)$ for the set of annotated gene lists in the biological data base. Each A_j is a list of genes that are known to be related to an underlying biological pathway $\mathcal{A}_j$; note that A_j is typically incomplete, as there will usually be additional genes linked to $\mathcal{A}_j$ that are not yet listed in A_j; similarly, A_j may contain genes that are false positives, and may in future be removed from the list as additional biological experiments arise. PROPA addresses these issues within the analysis. PROPA represents the problem of matching the pathway $\mathcal{F}$ underlying the factor with the annotated pathways $\mathcal{A}_j$ via posterior probabilities $Pr(\mathcal{F} = \mathcal{A}_j|\Pi, A_{1:m}) \propto Pr(\mathcal{F} = \mathcal{A}_j|A_{1:m})p(\Pi|A_{1:m}, \mathcal{F} = \mathcal{A}_j)$ and focuses on the likelihood terms $p(\Pi|A_{1:m}, \mathcal{F} = \mathcal{A}_j)$ as j moves across all the pathways; these are the overall measures that underly pathway assessment, and can be applied whatever the chosen values of the priors $Pr(\mathcal{F} = \mathcal{A}_j|A_{1:m})$. Full details of the PROPA analysis, within which the terms $p(\Pi|A_{1:m}, \mathcal{F} = \mathcal{A}_j)$ are marginal likelihoods from a class of statistical models, are developed in Shen and West (2010) and rely on computations using both MCMC and variational methods, the latter for evaluation of marginal likelihoods based on the outputs of within-model MCMC. The resulting marginal likelihood values are the *pathway scores* that PROPA delivers and that are used to rank the $j = 1 : m \approx 1{,}000$ pathways on each factor.

References

Alizadeh, A., Eisen, M., Davis, R., Ma, C., Lossos, I., Rosenwald, A., Boldrick, J., Sabet, H., Tran, T., Yu, Powell, J., Yang, L., Marti, G., Moore, T., Hudson, J., Lu, L., Lewis, D., Tibshirani, R., Sherlock, G., Chan, W., Greiner, T., Weisenburger, D., Armitage, J., Warnke, R., Levy, R., Wilson, W., Grever, M., Byrd, J., Botstein, D., Brown, P. and Staudt, L. (2000). Distinct types of diffuse large B-cell lymphoma identified by gene expression profiling. *Nature*, **403**, 503–511.

Badache, A. and Gonçalves, A. (2006). The ERB-B2 signaling network as a target for breast cancer therapy. *Journal of Mammary Gland Biology and Neoplasia*, **11**, 13–25.

Bertucci, F., Borie, N., Ginestier, C., Groulet, A., Charafe-Jauffret, E., Adélaïde, J., Geneix, J., Bachelart, L., Finetti, P., Koki, A., Hermitte, F., Hassoun, J., Debono, S., Viens, P., Fert, V., Jacquemier, J. and Birnbaum, D. (2004). Identification and validation of an ERBB2 gene expression signature in breast cancers. *Oncogene*, **23**, 2564–2575.

Bild, A. H., Yao, G., Chang, J. T., Wang, Q., Potti, A., Chasse, D., Joshi, M., Harpole, D., Lancaster, J. M., Berchuck, A., Olson, J. A., Marks, J. R., Dressman, H. K., West, M. and

Nevins, J. (2006). Oncogenic pathway signatures in human cancers as a guide to targeted therapies. *Nature*, **439**, 353–357.

Blackwell, D. and MacQueen J. (1973). Ferguson distributions via Polya urn schemes. *Annals of Statistics*, **1**, 353–355.

Carvalho, C., Lucas, J., Wang, Q., Chang, J., Nevins, J. and West, M. (2008). High-dimensional sparse factor modelling – Applications in gene expression genomics. *Journal of American Statistical Association*, **103**, 1438–1456.

Chang, H., Sneddon, J., Alizadeh, A., Sood, R., West, R., Montgomery, K., Chi, J.-T., van de Rijn, M., Botstein, D., and Brown, P. (2004). Gene expression signature of fibroblast serum response predicts human cancer progression: Similarities between tumors and wounds. *PLoS Biology*, **2**, E7, `http://dx.doi.org/10.1371/journal.pbio.0020007`.

Chang, J., Carvalho, C., Mori, S., Bild, A., Wang, Q., Gatza, M., Potti, A., Febbo, P., West, M. and Nevins, J. (2009). Decomposing cellular signalling pathways into functional units: A genomic strategy. *Molecular Cell*, **34**, 104–114.

Chen, J. L.-Y., Lucas, J., Schroeder, T., Mori, S., Wu, J., Nevins, J., Dewhirst, M., West, M. and Chi, J.-T. (2008). The genomic analysis of lactic acidosis and acidosis response in human cancers. *PLoS Genetics 4*, E1000293.

Chi, J.-T., Wang, Z., Nuyten, D., Rodriguez, E., Schaner, M., Salim, A., Wang, Y., Kristensen,G., Helland, A., Borresen-Dale, A., Giaccia, A., Longaker, M., Hastie, T., Yang, G., van de Vijver, M. and Brown, P. (2006). Gene expression programs in response to hypoxia: Cell type specificity and prognostic significance in human cancers. *PLoS Medicine*, **3**, E47.

Clyde, M. and George, E. (2004). Model uncertainty. *Statistical Science*, **19**, 81–94.

Dressman, H. K., Hans, C., Bild, A., Olsen, J., Rosen, E., Marcom, P. K., Liotcheva, V., Jones, E., Vujaskovic, Z., Marks, J. R., Dewhirst, M. W., West, M., Nevins, J. R. and Blackwell, K. (2006). Gene expression profiles of multiple breast cancer phenotypes and response to neoadjuvant therapy. *Clinical Cancer Research*, **12**, 819–826.

Escobar, M. and West, M. (1995). Bayesian density estimation and inference using mixtures. *Journal of the American Statistical Association*, **90**, 577–588.

Ferguson, T. S. (1973). A Bayesian analysis of some nonparametric problems. *Annals of Statistics*, **1**, 209–230.

George, E. and McCulloch, R. (1993). Variable selection via Gibbs sampling. *Journal of the American Statistical Association*, **88**, 881–889.

Golub, T., Slonim, D., Tamayo, P., Huard, C., Gaasenbeek, M., Mesirov, J., Coller, H., Loh, M., Downing, J., Caligiuri, M., Bloomfield, C. and Lander, E. (1999). Molecular classification of cancer: Class discovery and class prediction by gene expression. *Science*, **286**, 531–537.

Hans, C., Dobra, A. and West, M. (2007). Shotgun stochastic search in regression with many predictors. *Journal of the American Statistical Association*, **102**, 507–516.

Hans, C., Wang, Q., Dobra, A. and West, M. (2007). SSS: High-dimensional Bayesian regression model search. *Bulletin of the International Society for Bayesian Analysis*, **14**, 8–9, `http://www.stat.duke.edu/research/software/west/sss/`.

Huang, E., Chen, S., Dressman, H. K., Pittman, J., Tsou, M. H., Horng, C. F., Bild, A., Iversen, E. S., Liao, M., Chen, C. M., West, M., Nevins, J. R. and Huang, A. T. (2003). Gene expression predictors of breast cancer outcomes. *The Lancet*, **361**, 1590–1596.

Huang, E., Ishida, S., Pittman, J., Dressman, H., Bild, A., D'Amico, M., Pestell, R., West, M. and Nevins, J. (2003). Gene expression phenotypic models that predict the activity of oncogenic pathways. *Nature Genetics*, **34**, 226–230.

Huang, E., West, M. and Nevins, J. R. (2003). Gene expression profiling for prediction of clinical characteristics of breast cancer. *Recent Progress in Hormone Research*, **58**, 55–73.

Lopes, H. and West, M. (2004). Bayesian model assessment in factor analysis. *Statistica Sinica*, **14**, 41–67.

Lucas, J., Carvalho, C., Wang, Q., Bild, A., Nevins, J. R. and West, M. (2006). Sparse statistical modelling in gene expression genomics. In *Bayesian Inference for Gene Expression and Proteomics*, (ed. P. Müller, K. Do, and M. Vannucci), pp. 155–176. Cambridge University Press, Cambridge.

Lucas, J. E., Carvalho, C. M., Chen, L., Chi, J.-T. and West, M. (2009). Cross-study projection of genomic markers: An evaluation in cancer genomics. PLoS ONE *4*, E4523.

Lucas, J., Carvalho, C., Merl, D. and West, M. (2010). In-vitro to In-vivo factor profiling in expression genomics. In *Bayesian Modeling in Bioinformatics* (ed. D. Dey, S. Ghosh, and B. Mallick), Taylor & Francis, London.

MacEachern, S. N. and Müller, P. (1998). Estimating mixture of Dirichlet process models. *Journal of Computational and Graphical Statistics*, **7**, 223–238.

Miller, L. D., Smeds, J., George, J., Vega, V. B., Vergara, L., Ploner, A., Pawitan, Y., Hall, P., Klaar, S., Liu, E. T. and Bergh, J. (2005). An expression signature for p53 status in human breast cancer predicts mutation status, transcriptional effects, and patient survival. *Proceedings of the National Academy of Sciences*, **102**, 13550–13555.

MSigDB (2008). *Molecular Signatures Data Base.* Broad Institute, `http://www.broad.mit.edu/gsea/msigdb/`.

Nevins, J. R., Huang, E. S., Dressman, H., Pittman, J., Huang, A. T. and West, M. (2003). Towards integrated clinico-genomic models for personalized medicine: Combining gene expression signatures and clinical factors in breast cancer outcomes prediction. *Human Molecular Genetics*, **12**, 153–157.

Pittman, J., Huang, E., Dressman, H., Horng, C. F., Cheng, S. H., Tsou, M. H., Chen, C. M., Bild, A., Iversen, E. S., Huang, A., Nevins, J. and West, M. (2004). Integrated modeling of clinical and gene expression information for personalized prediction of disease outcomes. *Proceedings of the National Academy of Sciences*, **101**, 8431–8436.

Press, S. J. (1982). *Applied Multivariate Analysis: Using Bayesian and Frequentist Methods of Inference* (2nd. edn.) Krieger, Malabar, FL.

Rich, J., Jones, B., Hans, C., Iversen, E., McClendon, R., Rasheed, A., Bigner, D., Dobra, A., Dressman, H., Nevins, J. and West, M. (2005). Gene expression profiling and genetic markers in glioblastoma survival. *Cancer Research*, **65**, 4051–4058.

Semenza, G. (2001). Hypoxia-inducible factor 1: Oxygen homeostasis and disease pathophysiology. *Trends in Molecular Medicine*, **7**, 345–50.

Seo, D. M., Goldschmidt-Clermont, P. J. and West, M. (2007). Of mice and men: Sparse statistical modelling in cardiovascular genomics. *Annals of Applied Statistics*, **1**, 152–178.

Shen, H. (2007). Bayesian analysis in cancer pathway studies and probabilistic pathway annotation. Ph.D. Thesis, Duke University, `http://www.stat.duke.edu/people/theses/ShenH.html`.

Shen, H. and West, M. (2010). Bayesian modelling for biological pathway annotation of gene expression pathway signatures. In *Frontiers of Statistical Decision Making and Bayesian Analysis* (ed. M. Chen, D. Dey, P. Mueller, D. Sun and K. Ye), Springer Verlag, New York.

Sørlie, T., Perou, C. M., Tibshirani, R., Aas, T., Geisler, S., Johnsen, H., Hastie, T., Eisen, M. B., van de Rijn, M., Jeffrey, S. S., Thorsen, T., Quist, H., Matese, J. C., Brown, P. O., Botstein, D., Lønning, P. and Børresen-Dale, A. (2001). Gene expression patterns of breast carcinomas distinguish tumor subclasses with clinical implications. *Proceedings of the National Academy of Sciences*, **98**, 10869–10874.

van't Veer, L., H. Dai, M. van de Vijver, Y. He, A. Hart, M. Mao, H. Peterse, K. van der Kooy, M. Marton, A. Witteveen, G. Schreiber, R. Kerkhoven, C. Roberts, P. Linsley, R. Bernards, and S. Friend (2002). Gene expression profiling predicts clinical outcome of breast cancer. *Nature*, **415**, 530–536.

Wang, Q., Carvalho, C., Lucas, J. and West, M. (2007). BFRM: Bayesian factor regression modelling. *Bulletin of the International Society for Bayesian Analysis*, **14**, 4–5, `http://www.stat.duke.edu/research/software/west/bfrm/`.

West, M. (2003). Bayesian factor regression models in the "large p, small n" paradigm. In *Bayesian Statistics 7* (ed. J. Bernardo, M. Bayarri, J. Berger, A. Dawid, D. Heckerman, A. Smith, and M. West), pp. 723–732. Oxford University Press.

West, M., Blanchette, C., Dressman, H., Huang, E., Ishida, S., Spang, R., Zuzan, H., Marks, J. R. and Nevins, J. R. (2001). Predicting the clinical status of human breast cancer utilizing gene expression profiles. *Proceedings of the National Academy of Sciences*, **98**, 11462–11467.

West, M., Huang, A. T., Ginsberg, G. and Nevins, J. R. (2006). Embracing the complexity of genomic data for personalized medicine. *Genome Research*, **16**, 559–566.

West, M., Müller, P. and Escobar, M. D. (1994). Hierarchical priors and mixture models, with application in regression and density estimation. In *Aspects of Uncertainty: A Tribute to D.V. Lindley* (ed. A. Smith and P. Freeman), pp. 363–386. John Wiley, London.

·7·

Linking systems biology models to data: A stochastic kinetic model of p53 oscillations

Daniel A. Henderson, Richard J. Boys, Carole J. Proctor and Darren J. Wilkinson

7.1 Introduction

7.1.1 Overview

Systems biology is an exciting new paradigm for life science research in the post-genomic era. It is a development of molecular biology in which the focus has moved from trying to understand the function of individual biomolecules (or pairs of biomolecules) to understanding how collections of biomolecules of varying types act together to accomplish the observed dynamic biological system behaviour. Systems biology involves a combination of mathematical modelling, biological experimentation and quantitative data generation. In particular, it crucially depends on the ability to adjust models in the light of experimental data. Further, there is now overwhelming evidence that intrinsic stochasticity is an important feature of intra-cellular processes. Statistical methods are therefore likely to play an increasingly important role in systems biology as models become more realistic and quantitative dynamic data becomes more routinely available (Wilkinson, 2009).

This chapter considers the assessment and refinement of a dynamic stochastic process model of the cellular response to DNA damage. The proposed model is compared to time course data on the levels of two key proteins involved in this response, captured at the level of individual cells in a human cancer cell line. The primary goal of this study is to 'calibrate' the model by finding parameters of the model (kinetic rate constants) that are most consistent with the experimental data. The model is a complex nonlinear continuous time latent stochastic process model and so Markov chain Monte Carlo (MCMC) methods are a natural way to approach the inferential analysis from a computational perspective. In addition to being computationally difficult, the problem is also conceptually hard as the data-poor scenario means that some parameters of interest are only weakly identifiable. Fortunately, the mechanistic nature of the model means that all of the parameters are clearly interpretable, and a significant level of prior information is available for many of these from biological experts or the biological literature. The problem is therefore ideally suited to a Bayesian analysis using MCMC for computation.

The model concerns oscillations observed in the levels of two proteins in single living cancer cells subsequent to gamma irradiation. The two proteins, p53 and Mdm2, appear to oscillate out of phase with one another – system behaviour typically associated with some kind of negative feedback loop. This is consistent with current biological knowledge as p53 is known to enhance the production of Mdm2, and Mdm2 is known to inhibit p53. However, these oscillations are only observed at the single cell level and are not present in data derived from cell populations. It is therefore of interest to develop a simple mechanistic model, consistent with current biological knowledge, which explains the oscillatory behaviour and also explains why it is observed only at the single cell level (Proctor and Gray, 2008). Stochasticity is the key feature required to reconcile the apparent discrepancy between the single cell and population level data, with noisy oscillations being 'averaged out' in the population level data. The stochastic process model contains several parameters whose values are uncertain. This chapter considers the problem of using time course data on levels of p53 and Mdm2 in several individual cells to improve our knowledge regarding plausible parameter values, and also to assess the extent to which the proposed stochastic model is consistent with the available data.

7.1.2 Biological background

The p53 tumour suppressor protein plays a major role in cancer as evidenced by the high incidence of TP53 gene mutations in human tumours (Hainaut and Hollstein, 2000). The TP53 gene encodes a transcription factor with target genes that are involved in DNA repair, cell cycle arrest and apoptosis. It has been described as the 'guardian of the genome' (Lane, 1992), blocking cell cycle progression to allow the repair of damaged DNA. Under normal homeostatic conditions, the cellular levels of p53 protein are kept at a low level. There is basal transcription of the p53 gene (TP53) even in unstressed cells but the protein product does not accumulate as it has a short half-life of about 15–30 minutes (Finlay, 1993) and is usually bound to Mdm2, an ubiquitin E3 ligase, which targets p53 to the proteasome for degradation (Haupt *et al.*, 1997; Clegg *et al.*, 2008). Mdm2-binding prevents the transcriptional activity of p53 (Thut *et al.*, 1997), a phenomenon that is dependent on the catalytic activity of Mdm2 (Christophorou *et al.*, 2005). Mdm2 also has a short half-life and is a substrate of its own E3 ligase activity *in vitro* (Fang *et al.*, 2000). The transcription of Mdm2 is regulated by p53 (Barak *et al.*, 1993) and so under normal conditions, both p53 and Mdm2 are kept at low levels.

It is well known that stress induces an increase in levels of p53 which in turn leads to an increase in the transcription of Mdm2 (Mendrysa and Perry, 2000). One pathway for stabilization of p53 is via the kinase ATM, which is activated by DNA damage and phosphorylates p53 close to its Mdm2 binding site, so blocking its interaction with Mdm2 (Vogelstein *et al.*, 2000). In addition, ATM phosphorylates Mdm2 which not only interferes with its ability to bind to p53 but also enhances the degradation of Mdm2 (Pereg *et al.*, 2005; Khosravi *et al.*, 1999), providing an additional route for p53 stabilization. Another mechanism for the increase in p53 levels is the activation of ARF (known as p14ARF in humans), a nucleolar protein that senses DNA damage (Khan *et al.*, 2004). Although ARF responds to DNA damage, it is better known for its response to aberrant growth signals which are triggered by oncogenes (mutated forms of normal cellular genes which when activated can induce cancer). ARF binding enhances the degradation of Mdm2, resulting in p53 stabilisation (Khan *et al.*, 2004; Zhang *et al.*, 1998). Since an increase in p53 leads to an increase in Mdm2 transcription, and Mdm2 targets p53 for degradation, p53 levels are again inhibited, providing a negative feedback loop.

Negative feedback loops have been found in several systems of interacting proteins – e.g. Hes1 in Notch signalling (Hirata *et al.*, 2002), NF-kB signalling system (Nelson *et al.*, 2004) – and have attracted the attention of mathematical modellers. In particular, models have been produced to analyse the oscillations of p53 and Mdm2 in previously published single-cell fluorescent reporter assays (Ciliberto *et al.*, 2005; Geva-Zatorsky *et al.*, 2006; Lev Bar-Or *et al.*, 2000; Ma *et al.*, 2005; Tiana *et al.*, 2007; Zhang *et al.*, 2007). The single cell assays have been very informative, revealing that increasing DNA damage results in an increased number of oscillations, but not an increased magnitude in the response (Geva-Zatorsky *et al.*, 2006; Lahav *et al.*, 2004). The data also show that there is large intercellular variation with a fraction of cells showing no response or a slowly fluctuating signal. In the cells in which oscillations were detected, there was a wide fluctuation in the amplitude (about 70%) and smaller variations in the period of the peaks (about 20%) (Geva-Zatorsky *et al.*, 2006). The oscillations in these data showed a period of about 5.5 hours with a delay of about 2 hours between p53 and Mdm2 peaks (Geva-Zatorsky *et al.*, 2006).

All previous models to date have used a deterministic approach to analyse the oscillatory behaviour. These models have used differential equations and mathematical functions requiring a fairly large number of parameters with the generation of oscillations being very dependent on the range of parameter values chosen. Geva-Zatorsky *et al.* (2006) constructed six different models and found that the simplest model, which contained one intermediary and one negative feedback loop with a delay, was unable to produce multiple oscillations and that it was necessary to either introduce a positive feedback loop or a time delay term (see Figure 6 of Geva-Zatorsky *et al.*, 2006). However, these additions were not sufficient for robustness over a wide range of parameter values. The addition of a nonlinear negative feedback loop, a linear positive feedback loop or a second negative feedback loop produced models that were able to demonstrate sustained oscillations over a wide range of parameters. As the models are deterministic, the outcome only depends on the initial conditions and so they cannot easily be used to investigate inter- and intra-cell variability. Geva-Zatorsky *et al.* (2006) incorporated some random noise in protein production in their models and found that the introduction of low-frequency noise resulted in variability in the amplitude of the oscillations as observed experimentally. Ma *et al.* (2005) also incorporated a stochastic component for the DNA damage component of their model which resulted in variability in the number of oscillations. However, for a simulated dose of 2.5 Gy, they found that the majority of cells had only one peak and that a step input of DNA damage was required to obtain sustained oscillations.

We built a mechanistic model (Proctor and Gray, 2008) within a discrete stochastic chemical kinetic framework (Wilkinson, 2006), so that the intercellular variability could be accounted for in a natural way. Our approach meant that we did not need to include complex rate laws – mass action stochastic kinetics were assumed throughout – or any forced time delay terms.

7.1.3 Construction of the stochastic kinetic model

We assume that p53 production consists of two steps: transcription to form messenger RNA (**p53_mRNA** in the model) and then translation to form protein (**p53**). Under normal conditions p53 is usually bound to Mdm2 to form a complex, Mdm2–p53 (**Mdm2_p53**). Mdm2 targets p53 to the proteasome for degradation. We assume that p53 is only transcriptionally active when not bound to Mdm2, and so the production of Mdm2 mRNA (**Mdm2_mRNA**) is dependent on the pool of unbound p53. The synthesis of **Mdm2** depends on the level of Mdm2 mRNA and so is also dependent on the level of unbound p53. Thus Mdm2 mRNA provides the intermediary link between p53 and Mdm2 to provide the necessary delay in the negative feedback loop. We also include degradation of Mdm2, Mdm2 mRNA and p53 mRNA. ATM is included in the model in two states: either inactive (**ATMI**) or active (**ATMA**). Initially all ATM is in its inactive state. After DNA damage, ATM is activated and is then able to phosphorylate both p53 and Mdm2. Phosphorylated p53 and Mdm2 are presented in the model by the species **p53_P** and **Mdm2_P** respectively. We assume that the phosphorylated proteins are unable to bind to one another and so phosphorylated p53 is not degraded. However, phosphorylation of Mdm2 leads to its enhanced degradation and **p53_P** is transcriptionally active. We also include steps for de-phosphorylation. Further details of the model are given in Proctor and Gray (2008). To carry out a 'virtual experiment', whereby the cell is subject to irradiation, the species that represents DNA damage (**damDNA**) is set to a large value for the initial period of the simulation. Damaged DNA is repaired at a rate determined by the parameter *krepair*.

The model was encoded using SBML-shorthand (Wilkinson, 2006) and then converted into the Systems Biology Markup Language (SBML) (Hucka *et al.*, 2003). SBML is a well-known modelling standard, allowing models to be shared

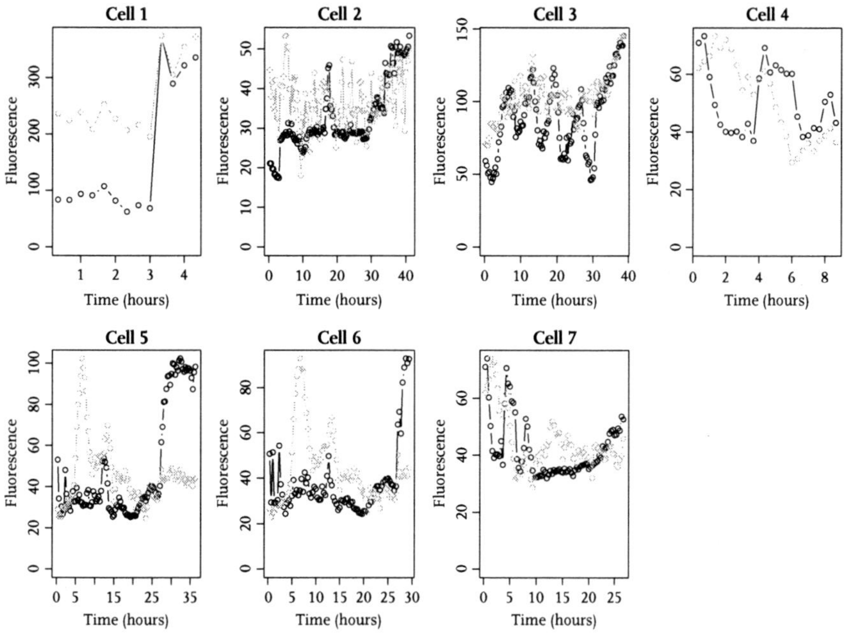

Fig. 7.1 Measured (normalized) fluorescence levels for the seven cells in movie 2; p53 (black) and Mdm2 (grey).

in a form that other researchers can use in different hardware and software environments.

7.1.4 The experiment and data

The experiments were carried out in the laboratory of Uri Alon (Department of Molecular Cell Biology, Weizmann Institute of Science, Israel). Full details of the experimental procedure can be found in Geva-Zatorsky *et al.* (2006). The cell line used was MCF7, which are human breast cancer epithelial cells. They used a clone (all cells genetically identical) which was stably transfected with p53 fused to cyan fluorescent protein (CFP) and Mdm2 fused to yellow fluorescent protein (YFP). They irradiated the cells with different doses of gamma irradiation and obtained time-lapse fluorescence microscopy movies of the cells over time periods of about 30 hours. Images were captured every 10–20 minutes. Overall they collected data for 1000 individual cells in different experiments with different doses of irradiation. We were supplied with raw data for the cells which were irradiated at a dose of 2.5 Gy (141 cells) and at 5 Gy (146 cells). The units for the raw data are relative fluorescence units and the values for Mdm2 had been normalized (by multiplying the YFP fluorescence by a constant) so that the values for p53 and Mdm2 lie in the same range; see, for example, the time-course data for the seven cells in the second movie displayed in Figure 7.1.

7.2 Stochastic kinetic model

The stochastic kinetic model describes the evolution of $k = 10$ species by using 19 reactions. Each reaction occurs at a rate governed by the numbers of molecules of the reacting species and associated rate constants. The list of reactions is

$$
\begin{array}{rcll}
\text{Mdm2_mRNA} & \longrightarrow & \text{Mdm2_mRNA} + \text{Mdm2} & rate = ksynMdm2 \times Mdm2_mRNA \\
\text{Mdm2} & \longrightarrow & \text{Sink} & rate = kdegMdm2 \times Mdm2 \times kproteff \\
\text{p53_mRNA} & \longrightarrow & \text{p53} + \text{p53_mRNA} & rate = ksynp53 \times p53_mRNA \\
\text{Mdm2_p53} & \longrightarrow & \text{Mdm2} & rate = kdegp53 \times Mdm2_p53 \times kproteff \\
\text{p53} + \text{Mdm2} & \longrightarrow & \text{Mdm2_p53} & rate = kbinMdm2p53 \times p53 \times Mdm2 \\
\text{Mdm2_p53} & \longrightarrow & \text{p53} + \text{Mdm2} & rate = krelMdm2p53 \times Mdm2_p53 \\
\text{p53} & \longrightarrow & \text{p53} + \text{Mdm2_mRNA} & rate = ksynMdm2mRNA \times p53 \\
\text{p53_P} & \longrightarrow & \text{p53_P} + \text{Mdm2_mRNA} & rate = ksynMdm2mRNA \times p53_P \\
\text{Mdm2_mRNA} & \longrightarrow & \text{Sink} & rate = kdegMdm2mRNA \times Mdm2_mRNA \\
\text{damDNA} + \text{ATMI} & \longrightarrow & \text{damDNA} + \text{ATMA} & rate = kactATM \times damDNA \times ATMI \\
\text{Mdm2_P} & \longrightarrow & \text{Sink} & rate = kdegATMMdm2 \times Mdm2_P \\
\text{ATMA} & \longrightarrow & \text{ATMI} & rate = kinactATM \times ATMA \\
\text{p53} + \text{ATMA} & \longrightarrow & \text{p53_P} + \text{ATMA} & rate = kphosp53 \times p53 \times ATMA \\
\text{p53_P} & \longrightarrow & \text{p53} & rate = kdephosp53 \times p53_P \\
\text{Mdm2} + \text{ATMA} & \longrightarrow & \text{Mdm2_P} + \text{ATMA} & rate = kphosMdm2 \times Mdm2 \times ATMA \\
\text{Mdm2_P} & \longrightarrow & \text{Mdm2} & rate = kdephosMdm2 \times Mdm2_P \\
\text{damDNA} & \longrightarrow & \text{Sink} & rate = krepair \times damDNA \\
\text{Source} & \longrightarrow & \text{p53_mRNA} & rate = ksynp53mRNA \times Source \\
\text{p53_mRNA} & \longrightarrow & \text{Sink} & rate = kdegp53mRNA \times p53_mRNA
\end{array}
$$

It will be convenient to work with the rate constants (e.g. *ksynp53mRNA*) on a log scale and so we denote the collection of the $r = 19$ 'calibration' parameters by $\theta = (\theta_1, \theta_2, \ldots, \theta_r)'$, where θ_j is the log of the jth rate constant.

Let $Y_t = (Y_{t,1}, \ldots, Y_{t,k})$ denote the state of the system at time t, where $Y_{t,j}$ is the number of molecules of species j at time t. The $k = 10$ species (and their corresponding index used in the notation) are listed in Table 7.1. Note however that these 10 species are not linearly independent due to the presence of a conservation law in the system that makes species 6 and 7 linearly related. Such conservation laws need to be preserved by inference algorithms – an example of how this is achieved is presented in Section 7.5. Also let $Y = (Y_{t_0}, Y_{t_1}, \ldots, Y_{t_n})$ denote the state of the system at the time points $(t_0, t_1, \ldots, t_n)$. The kinetic model is a Markov jump process and so the joint probability of Y factorizes as

$$
p(Y|\theta) = p(Y_{t_0}|\theta) \prod_{i=1}^{n} p(Y_{t_i}|Y_{t_{i-1}}, \theta),
$$

where we assume that the initial state Y_{t_0} is independent of the reaction constants θ, that is, $p(Y_{t_0}|\theta) = p(Y_{t_0})$.

Table 7.1 Species names and their corresponding index.

Species index	Species name
1	Mdm2
2	p53
3	Mdm2_p53
4	Mdm2_mRNA
5	p53_mRNA
6	ATMA
7	ATMI
8	p53_P
9	Mdm2_P
10	damDNA

Given full information on the process, that is, the times and types of each reaction that takes place, closed form expressions can be found for the conditional probabilities $p(Y_{t_i}|Y_{t_{i-1}}, \theta)$, and hence for the joint probability $p(Y|\theta)$. However, in the present application (as with many other practical scenarios) experimental techniques do not provide this full information, perhaps only giving the levels of some species at a limited number of time points. Here different strategies are required for analysing such partial information; see Boys *et al.* (2008) for details. The strategy we employ in this chapter is based on the fact that, for given reaction constants θ, it is possible to forward simulate realisations Y_t of the model exactly using, for example, the Gillespie algorithm (Gillespie, 1977).

7.3 Data

Suppose that the data available consist of time-course information on C cells. Specifically, the data on cell i are the scaled fluorescence measurements of two quantities, p53 and Mdm2, measured at n_i time points $t_1, \ldots, t_{n_i}$. We will sometimes refer to the p53 and Mdm2 measurements by the colours of the fluorescent proteins used, namely the 'cyan channel' and the 'yellow channel' respectively. In these data, measurements are taken every $\tau = 1200$ seconds, that is, at times $t_j = j\tau,\ j = 0, 1, 2, \ldots$ and so we simplify the notation by referring to time by its index, $j = 0, 1, 2, \ldots$. Also the measurement pair at the tth time point in the ith cell is denoted by $z_t^i = (z_{t,\mathrm{c}}^i, z_{t,\mathrm{y}}^i)'$, where the subscripts c and y refer to the channel colours.

The data on cell i is denoted by $z^i \equiv z_{1:n_i}^i = \{z_1^i, \ldots, z_{n_i}^i\}$, and the full dataset for all C cells is denoted by $z = \{z^1, \ldots, z^C\}$. Finally, an additional complication in these data is that the observed measurements of Mdm2 (yellow) have been scaled so that the maximum value is the same as the maximum observed value of p53 (cyan). This post-processing step adds an extra layer of complexity into the modelling task.

7.4 Linking the model to the data

7.4.1 Modelling the raw measurements

Let $Y_{t,c}^i$ and $Y_{t,y}^i$ denote the total amount of p53 and Mdm2, respectively, in the ith cell at the tth time point. Each of these amounts is the sum of three species counts (at time point t), namely

$$Y_{t,c}^i = Y_{t,2}^i + Y_{t,3}^i + Y_{t,8}^i \qquad \text{and} \qquad Y_{t,y}^i = Y_{t,1}^i + Y_{t,3}^i + Y_{t,9}^i.$$

Note that $Y_{t,3}^i$ is common to both $Y_{t,c}^i$ and $Y_{t,y}^i$. Let $Y_c^i = (Y_{1,c}^i, \ldots, Y_{n_i,c}^i)$ and $Y_y^i = (Y_{1,y}^i, \ldots, Y_{n_i,y}^i)$ denote the amounts (by channel) for cell i, and let the amounts over all cells be denoted by $Y_c = (Y_c^1, \ldots, Y_c^C)$, $Y_y = (Y_y^1, \ldots, Y_y^C)$ and $Y = \{Y_c, Y_y\}$.

The true fluorescence levels in the cyan and yellow channels are assumed to be directly proportional to the numbers of molecules of p53 and Mdm2, respectively, with unknown proportionality constants β_j, $j \in \{c, y\}$, that is

$$\gamma_{t,c}^i = \beta_c Y_{t,c}^i \qquad \text{and} \qquad \gamma_{t,y}^i = \beta_y Y_{t,y}^i.$$

The raw measurements of these true fluorescence levels (after adjusting for background noise), denoted by $\zeta_{t,c}^i$ and $\zeta_{t,y}^i$, are assumed to be independent and normally distributed with means $\gamma_{t,c}^i$ and $\gamma_{t,y}^i$ respectively. The measurement processes in the cyan and yellow channels are sufficiently similar that we will assume a common precision ϕ to these processes. Thus

$$\zeta_{t,c}^i | Y_{t,c}^i, \beta_c, \phi \sim N\left(\beta_c Y_{t,c}^i, \phi^{-1}\right) \qquad \text{and} \qquad \zeta_{t,y}^i | Y_{t,y}^i, \beta_y, \phi \sim N\left(\beta_y Y_{t,y}^i, \phi^{-1}\right).$$

Prior beliefs about ϕ are modelled through a gamma $Ga(a_\phi, b_\phi)$, distribution, with density

$$p(\phi) = \frac{b_\phi^{a_\phi} \phi^{a_\phi - 1} \exp(-b_\phi \phi)}{\Gamma(a_\phi)},$$

where $\Gamma(\cdot)$ denotes the gamma function. We take $a_\phi = 2$ and $b_\phi = 50$ and this reflects fairly strong beliefs that the prior precision is close to its mean of $E(\phi) = 1/25$.

7.4.2 Modelling the scaling process

Beliefs about the scaling constants β_c and β_y are modelled via independent normal distributions

$$\beta_c \sim N\left(a_{\beta_c}, b^2_{\beta_c}\right) \qquad \text{and} \qquad \beta_y \sim N\left(a_{\beta_y}, b^2_{\beta_y}\right).$$

Prior means of $a_{\beta_c} = a_{\beta_y} = 1$, representing no scaling, were adopted. The prior standard deviations b_{β_c} and b_{β_y} were chosen to be 1/3, so that the central 95% prior probability interval for β_c is approximately (0.347,1.653). Strictly speaking, these prior distributions are not consistent with the requirement that the scaling constants should take positive values. However, there is a significant benefit in choosing this form for the prior distributions, as we shall see shortly. Also, for these prior distributions, the probability of the scaling constants taking negative values is negligibly small.

7.4.3 Marginal model for the data

The scaling constants β_c and β_y are essentially nuisance parameters and their values are not of direct interest. Our choice of normal distributions for β_c and β_y allows us to marginalise them analytically from the normal models for the raw measurements, and this may result in computational benefits. This gives the marginal distributions of the raw measurements as

$$\zeta^i_{t,c}|Y^i_{t,c}, \phi \sim N\left(a_{\beta_c} Y^i_{t,c}, b^2_{\beta_c}\left(Y^i_{t,c}\right)^2 + \phi^{-1}\right) \quad \text{and}$$

$$\zeta^i_{t,y}|Y^i_{t,y}, \phi \sim N\left(a_{\beta_y} Y^i_{t,y}, b^2_{\beta_y}\left(Y^i_{t,y}\right)^2 + \phi^{-1}\right).$$

Note that the variance of the marginal measurement error distribution now depends on the signal, that is, on the true numbers of molecules.

Unfortunately the available data are not simply a collection of raw measurements: the data have been normalized so that the values for the two channels lie on the same scale. However, only data recorded for the yellow channel (Mdm2) is affected. The data recorded for the cyan channel (p53), $z^i_{t,c}$, are the raw measurements for that channel and so $z^i_{t,c} = \zeta^i_{t,c}$. However, the data recorded for the yellow channel (Mdm2), $z^i_{t,y}$, are normalized versions of the raw measurements. The normalized measurement for Mdm2 in the ith cell at time t, $z^i_{t,y}$, is obtained from the raw measurement $\zeta^i_{t,y}$ by dividing by the maximum raw Mdm2 measurement in the ith cell, $(\zeta^i_y)^{\max} = \max(\zeta^i_{1,y}, \ldots, \zeta^i_{n_i,y})$, and then multiplying by the maximum measurement in the cyan channel, $(z^i_c)^{\max} = \max(z^i_{1,c}, \ldots, z^i_{n,c})$. Thus

$$z^i_{t,y} = s\left(\zeta^i_y, z^i_c\right) \equiv \zeta^i_{t,y} \frac{\left(z^i_c\right)^{\max}}{\left(\zeta^i_y\right)^{\max}}.$$

The probability of the observed scaled measurements z_y^i is

$$p\left(z_y^i|\zeta_y^i, z_c^i\right) = \begin{cases} 1, & \text{if } s\left(\zeta_y^i, z_c^i\right) = z_y^i, \\ 0, & \text{otherwise,} \end{cases}$$

and this depends on z_c^i only through $(z_c^i)^{\max}$. The joint density of the observed data (the likelihood function) is therefore

$$\begin{aligned} p(z_c, z_y|\zeta_y, Y_c, \phi) &= p(z_c|Y_c, \phi)\, p(z_y|\zeta_y, z_c) \\ &= p(z_c|Y_c, \phi) \prod_{i=1}^{C} p(z_y^i|\zeta_y^i, z_c^i), \end{aligned}$$

where $p(z_c|Y_c, \phi)$, the joint density of the raw p53 measurements, is

$$\begin{aligned} p(z_c|Y_c, \phi) &= \prod_{i=1}^{C}\prod_{t=1}^{n_i} p\left(z_{t,c}^i|Y_{t,c}^i, \phi\right) \\ &= \prod_{i=1}^{C}\prod_{t=1}^{n_i} (2\pi)^{-1/2}\left\{b_{\beta_c}^2\left(Y_{t,c}^i\right)^2 + \phi^{-1}\right\}^{-1/2} \exp\left\{-\frac{\left(z_{t,c}^i - a_{\beta_c} Y_{t,c}^i\right)^2}{2\left\{b_{\beta_c}^2\left(Y_{t,c}^i\right)^2 + \phi^{-1}\right\}}\right\}. \end{aligned}$$

When constructing the posterior distribution (given by equation (7.2) in Section 7.5) we will also need the joint density of the raw Mdm2 measurements:

$$\begin{aligned} p(\zeta_y|Y_y, \phi) &= \prod_{i=1}^{C}\prod_{t=1}^{n_i} p\left(\zeta_{t,y}^i|Y_{t,y}^i, \phi\right) \\ &= \prod_{i=1}^{C}\prod_{t=1}^{n_i} (2\pi)^{-1/2}\left\{b_{\beta_y}^2\left(Y_{t,y}^i\right)^2 + \phi^{-1}\right\}^{-1/2} \exp\left\{-\frac{\left(\zeta_{t,y}^i - a_{\beta_y} Y_{t,y}^i\right)^2}{2\left\{b_{\beta_y}^2\left(Y_{t,y}^i\right)^2 + \phi^{-1}\right\}}\right\}. \end{aligned}$$

7.4.4 Prior specification for the model calibration parameters

Information on likely values for these parameters can be found in Proctor and Gray (2008). We found that the parameters naturally grouped into four classes, ranging from those that were known fairly accurately to those with a fair amount of uncertainty. We have fixed the parameters that were known fairly accurately to their suggested values and these are given (on a log scale) in Table 7.2. This reduces the complexity of the analysis to finding plausible values for the remaining $r^* = 7$ parameters. We have taken independent uniform prior distributions for these calibration parameters (the logged kinetic rate constants) θ, with

$$\theta_i|a_{\theta_i}, b_{\theta_i} \sim U(a_{\theta_i}, b_{\theta_i}), \qquad i = 1, \ldots, r^*$$

Table 7.2 Values (on a log scale) for known model calibration parameters.

Parameter name	Value
ksynMdm2	−7.611
kdegMdm2	−7.745
ksynp53	−5.116
kdegp53	−7.100
kbinMdm2p53	−6.764
krelMdm2p53	−11.369
ksynMdm2mRNA	−9.210
kdegMdm2mRNA	−9.210
kdegATMMdm2	−7.824
kproteff	0.000
ksynp53mRNA	−6.908
kdegp53mRNA	−9.210

and used the information in Proctor and Gray (2008) to determine reasonable values for the upper and lower limits of these distributions; see Table 7.3.

Therefore, suppressing dependence on the fixed hyperparameters a_{θ_i} and b_{θ_i}, the joint prior density is

$$p(\theta) = \prod_{i=1}^{r^*} p(\theta_i) = \prod_{i=1}^{r^*} (b_{\theta_i} - a_{\theta_i})^{-1}.$$

7.4.5 Prior specification for the initial species counts

The joint probability of the states $Y|\theta$ is given by the product

$$p(Y|\theta) = \prod_{i=1}^{C} \left\{ p\left(Y_0^i\right) \prod_{t=1}^{n_i} p\left(Y_t^i | Y_{t-1}^i, \theta\right) \right\}, \tag{7.1}$$

Table 7.3 The lower and upper limits of the uniform prior distributions for the unknown model calibration parameters (on a log scale).

Index	Parameter name	Lower limit a_{θ_i}	Upper limit b_{θ_i}
1	*kactATM*	−18.210	−0.210
2	*kinactATM*	−16.601	1.399
3	*kphosp53*	−13.601	−1.601
4	*kdephosp53*	−8.996	0.702
5	*kphosMdm2*	−7.609	2.088
6	*kdephosMdm2*	−8.996	0.702
7	*krepair*	−13.820	−7.820

in which the conditional probabilities are determined by the dynamics of the stochastic kinetic model. We model the unobserved initial state of the system in cell i using independent Poisson distributions

$$Y^i_{0,j}|\lambda_j \sim Po(\lambda_j), \qquad j = 1, \ldots, 10.$$

Note that this imposes the same prior distribution for the initial state in each of the C cells. Therefore the marginal probability of the initial state of the system for cell i is

$$p\left(Y^i_0\right) = \prod_{i=1}^{C}\prod_{j=1}^{10} p\left(Y^i_{0,j}\right) = \prod_{i=1}^{C}\prod_{j=1}^{10} \lambda_j^{Y^i_{0,j}} e^{-\lambda_j} / \Gamma\left(Y^i_{0,j} + 1\right).$$

The means of these Poisson priors (λ_j) have been chosen using information in the literature (e.g. Proctor and Gray 2008) on the most likely initial state of the system: $(\lambda_1, \ldots, \lambda_{10}) = (6, 6, 96, 11, 11, 1, 201, 1, 1, 76)$.

7.5 Posterior computation

Inferences about the values of the unknown quantities in the model are based on their joint posterior distribution, which has density proportional to the product of the joint prior density and the likelihood, that is

$$p(\theta, \phi, Y, \zeta_y|z_c, z_y) \propto p(\theta)p(\phi)p(Y|\theta)p(\zeta_y|Y, \phi)p(z_c|Y, \phi)p(z_y|\zeta_y, z_c), \qquad (7.2)$$

where all the terms on the right-hand side of (7.2) have been described previously.

Posterior computation is made difficult by the intractability of the conditional probabilities $p(Y^i_t|Y^i_{t-1}, \theta)$ from the stochastic kinetic model, which enter the Bayesian model through equation (7.1). It is possible to avoid computation of the $p(Y^i_t|Y^i_{t-1}, \theta)$ by constructing an algorithm which uses realizations from the stochastic kinetic model; see, for example, Henderson *et al.* (2009). However, such algorithms require the generation of many model realizations and so, for this approach to work well, each model realisation must be quick to simulate. Unfortunately, this is not the case for the stochastic kinetic model considered here. For example, simulating $Y^i_t|Y^i_{t-1}, \theta$ for some values of Y^i_{t-1} and θ takes a matter of milliseconds, yet for other values it can take several seconds, even on a reasonably powerful computer (2.2 GHz, 8 GB RAM) using an efficient C implementation of Gillespie's exact discrete event simulation algorithm.

One way of dealing with the slow simulation speed of the Gillespie algorithm is to use a fast approximate simulation algorithm, such as the τ-leap method; see Wilkinson (2006). Here there is a trade-off between simulation speed (and therefore speed of the inference procedure) and the exactness of the inferences made. Initial investigations with some fast approximate algorithms revealed that none were able to provide the sort of improvements in simulation speed needed for this analysis.

An alternative to using an approximate simulation algorithm is to use an exact simulation algorithm for an approximation to the stochastic kinetic model. Various authors have sought solutions along these lines. For example, Golightly and Wilkinson (2005) use an approximation based on the chemical Langevin equation (CLE; Gillespie, 2000), namely, the diffusion process that approximates most closely the Markov jump process defined by the stochastic kinetic model. These authors build on this work (in Golightly and Wilkinson 2006*a*) and demonstrate how a combination of particle filtering and MCMC methods can be used to sample from the posterior distribution of the rate constants given the CLE approximation model and data observed partially, discretely and with error.

A different strategy, and one we adopt in this chapter, is to *emulate* the stochastic kinetic model. We do this by constructing a tractable approximation to the conditional probability distribution of the stochastic kinetic model, $p(Y(t+\tau)|Y(t),\theta)$, where $Y(t)$ denotes the state of the system at any particular time t (seconds), and $Y(t+\tau)$ denotes the state of the system τ seconds in the future. For the data in this application, the time step is $\tau = 1200$ seconds. The approximate probability distribution, which we refer to as the emulator, is denoted by $p^*(\cdot|\cdot,\theta)$, and the objective is to use it in place of the probability distribution $p(\cdot|\cdot,\theta)$ in equation (7.1). The construction of emulators is commonplace in the computer models literature where complex deterministic functions are modelled via tractable stochastic processes; see Kennedy and O'Hagan (2001), O'Hagan (2006), Santner *et al.* (2003), and references therein. In this application, the function to be emulated is discrete, multivariate and stochastic. The emulator is constructed by fitting simple statistical models to output obtained by simulating the stochastic kinetic model for τ seconds from a designed collection of values of the model inputs $\{Y(t),\theta\}$. The approach to fitting the emulator that we follow is described in more detail in Appendix B.

Based on the emulator, the joint probability of the states is

$$p^*(Y|\theta) = \prod_{i=1}^{C}\left\{p\left(Y_0^i\right)\prod_{t=1}^{n_i}p^*\left(Y_t^i|Y_{t-1}^i,\theta\right)\right\},$$

and this replaces the exact probability $p(Y|\theta)$ in (7.2) to give an expression for the posterior density based on this approximation, namely

$$p^*(\theta,\phi,Y,\zeta_y|z_c,z_y) \propto p(\theta)p(\phi)p^*(Y|\theta)p(\zeta_y|Y,\phi)p(z_c|Y,\phi)p(z_y|\zeta_y,z_c). \quad (7.3)$$

Here the superscript * distinguishes probabilities (or densities) based on the emulator approximation rather than the true stochastic kinetic model. Also note that all terms in the right-hand side of (7.3) are tractable and so this formulation leads to a workable solution.

Sampling from the posterior distribution (7.3) is possible using a Metropolis – Hastings (MH) within Gibbs MCMC scheme. In the scheme, we update each set of unknown quantities in turn from their full conditional distribution by using a MH step if the full conditional distribution cannot be sampled from directly. The MCMC algorithm is constructed so that the distribution of sampled values tends to the posterior distribution as the number of iterations increases. The sampled values are then used to approximate features of the posterior distribution. For computational reasons, we find it beneficial to work with transformed values of the calibration parameters θ. Specifically we work with transformed parameters $\Lambda = (\Lambda_1, \ldots, \Lambda_{r^*})$, where $\Lambda_i = \log(\theta_i - a_{\theta_i}) - \log(b_{\theta_i} - \theta_i)$. This corresponds to a logit transformation of θ_i after it has been re-scaled to lie on the unit interval. It follows that the Λ_i have independent (standard) logistic distributions, and therefore that the joint density of Λ is

$$p(\Lambda) = \prod_{i=1}^{r^*} p(\Lambda_i) = \prod_{i=1}^{r^*} \frac{\exp(\Lambda_i)}{\{1 + \exp(\Lambda_i)\}^2}.$$

The sampled values of Λ_i are simply back transformed to give sampled values of θ_i.

In outline, one iteration of the MCMC scheme entails the following steps:

- Update $\Lambda | \cdots$ by using a MH step with a symmetric multivariate normal random walk proposal, centred at the current sampled value. The acceptance ratio for the proposed move from Λ to $\tilde{\Lambda}$ is

$$A_\Lambda = \frac{p(\tilde{\Lambda})}{p(\Lambda)} \frac{p^*(Y|\tilde{\theta})}{p^*(Y|\theta)},$$

  where, for instance, $\theta_i = \{a_{\theta_i} + b_{\theta_i} \exp(\Lambda_i)\}/\{1 + \exp(\Lambda_i)\}$.
- Update $\phi | \cdots$ by proposing a new value $\tilde{\phi}$ from a proposal distribution with density $q(\tilde{\phi}|\phi)$. The acceptance ratio for the proposed move from ϕ to $\tilde{\phi}$ is

$$A_\phi = \frac{p(\tilde{\phi})}{p(\phi)} \frac{p(\zeta_y|Y, \tilde{\phi}) p(z_c|Y, \tilde{\phi})}{p(\zeta_y|Y, \phi) p(z_c|Y, \phi)} \frac{q(\phi|\tilde{\phi})}{q(\tilde{\phi}|\phi)}.$$

- Update $Y_{\mathcal{S}\setminus\{6,7\}} | \cdots$, where $\mathcal{S} = \{1, 2, \ldots, 10\}$ as follows. Here $Y_{\mathcal{S}\setminus\{6,7\}}$ denotes the values of the states (in each cell) excluding those for species 6 (**ATMA**) and 7 (**ATMI**). Species 6 and 7 are treated separately as their sum is fixed throughout the time course.
  For cells $i = 1, \ldots, C$:
  – for $t = 0$, propose independent Poisson candidate values $\tilde{Y}^i_{t,j} | Y^i_{t,j} \sim Po(Y^i_{t,j} + a_Y)$ for $j \in \mathcal{S} \setminus \{6, 7\}$, where $a_Y > 0$ is a positive tuning

constant which is chosen to be small. Denote the proposal probability by $q(\tilde{Y}^i_{t,i}|Y^i_{t,i})$. The acceptance ratio for the proposed move from $Y^i_{t,\mathcal{S}\setminus\{6,7\}}$ to $\tilde{Y}^i_{t,\mathcal{S}\setminus\{6,7\}}$ is

$$A_{Y^i_{0,\mathcal{S}\setminus\{6,7\}}} = \frac{\prod_{j\in\mathcal{S}\setminus\{6,7\}} p\left(\tilde{Y}^i_{t,j}\right)}{\prod_{j\in\mathcal{S}\setminus\{6,7\}} p\left(Y^i_{t,j}\right)} \frac{p^*\left(Y^i_{t+1}|\tilde{Y}^i_t,\theta\right)}{p^*\left(Y^i_{t+1}|Y^i_t,\theta\right)} \frac{\prod_{j\in\mathcal{S}\setminus\{6,7\}} q\left(Y^i_{t,j}|\tilde{Y}^i_{t,j}\right)}{\prod_{j\in\mathcal{S}\setminus\{6,7\}} q\left(\tilde{Y}^i_{t,j}|Y^i_{t,j}\right)},$$

where $\tilde{Y}^i_t$ denotes the vector of proposed candidate values together with the current values for species 6 and 7.

– For $t = 1, \ldots, n-1$, propose independent Poisson candidate values $\tilde{Y}^i_{t,j}|Y^i_{t,j} \sim Po(Y^i_{t,j} + a_Y)$ for $j \in \mathcal{S}\setminus\{6,7\}$. Denote the proposal probability by $q(\tilde{Y}^i_{t,j}|Y^i_{t,j})$. The acceptance ratio for the proposed move from $Y^i_{t,\mathcal{S}\setminus\{6,7\}}$ to $\tilde{Y}^i_{t,\mathcal{S}\setminus\{6,7\}}$ is

$$A_{Y^i_{t,\mathcal{S}\setminus\{6,7\}}} = \frac{p^*\left(\tilde{Y}^i_t|Y^i_{t-1},\theta\right)}{p^*\left(Y^i_t|Y^i_{t-1},\theta\right)} \frac{p^*\left(Y^i_{t+1}|\tilde{Y}^i_t,\theta\right)}{p^*\left(Y^i_{t+1}|Y^i_t,\theta\right)} \frac{p\left(z^i_{t,c}|\tilde{Y}^i_t,\phi\right)}{p\left(z^i_{t,c}|Y^i_t,\phi\right)} \frac{p\left(\zeta^i_{t,y}|\tilde{Y}^i_t,\phi\right)}{p\left(\zeta^i_{t,y}|Y^i_t,\phi\right)} \frac{\prod_{j\in\mathcal{S}\setminus\{6,7\}} q\left(Y^i_{t,j}|\tilde{Y}^i_{t,j}\right)}{\prod_{j\in\mathcal{S}\setminus\{6,7\}} q\left(\tilde{Y}^i_{t,j}|Y^i_{t,j}\right)}.$$

– For $t = n$, propose independent Poisson candidate values $\tilde{Y}^i_{t,j}|Y^i_{t,j} \sim Po(Y^i_{t,j} + a_Y)$ for $j \in \mathcal{S}\setminus\{6,7\}$. Denote the proposal probability by $q(\tilde{Y}^i_{t,j}|Y^i_{t,j})$. The acceptance ratio for the proposed move from $Y^i_{t,\mathcal{S}\setminus\{6,7\}}$ to $\tilde{Y}^i_{t,\mathcal{S}\setminus\{6,7\}}$ is

$$A_{Y^i_{n,\mathcal{S}\setminus\{6,7\}}} = \frac{p^*\left(\tilde{Y}^i_t|Y^i_{t-1},\theta\right)}{p^*\left(Y^i_t|Y^i_{t-1},\theta\right)} \frac{p\left(z^i_{t,c}|\tilde{Y}^i_t,\phi\right) p\left(\zeta^i_{t,y}|\tilde{Y}^i_t,\phi\right)}{p\left(z^i_{t,c}|Y^i_t,\phi\right) p\left(\zeta^i_{t,y}|Y^i_t,\phi\right)} \frac{\prod_{j\in\mathcal{S}\setminus\{6,7\}} q\left(Y^i_{n,j}|\tilde{Y}^i_{t,j}\right)}{\prod_{j\in\mathcal{S}\setminus\{6,7\}} q\left(\tilde{Y}^i_{t,j}|Y^i_{t,j}\right)}.$$

- For cells $i = 1, \ldots, C$, update $Y^i_{t,\{6,7\}}|\cdots$ for $t = 0, 1, \ldots, n_i$ as follows. First, for $t = 0$ and $j = 6, 7$, propose independent Poisson candidate values $\tilde{Y}^i_{t,j}|Y^i_{t,j} \sim Po(Y^i_{t,j} + a_Y)$. Denote these proposal probabilities by $q(\tilde{Y}^i_{0,j}|Y^i_{0,j})$. This gives a proposal for the new sum of the two species in the ith cell, $\tilde{N}^i = \tilde{Y}^i_{t,6} + \tilde{Y}^i_{t,7}$. Then for $t = 1, \ldots, n_i$ propose

$$\tilde{Y}^i_{t,6} | Y^i_{t,6}, Y^i_{t,7}, \tilde{N}^i \sim Bin\left(\tilde{N}^i, \left(Y^i_{t,6} + a_{Y_6}\right) / \left(Y^i_{t,6} + Y^i_{t,7} + a_{Y_6}\right)\right)$$

where $a_{Y_6} > 0$ is a positive tuning constant which is chosen to be small. Then set $\tilde{Y}^i_{t,7} = \tilde{N}^i - \tilde{Y}^i_{t,6}$. Denote these proposal probabilities by $q(\tilde{Y}^i_{t,6} | Y^i_{t,6})$. The acceptance ratio for the proposed move from $Y^i_{\{6,7\}}$ to $\tilde{Y}^i_{\{6,7\}}$ is

$$A_{Y^i_{\{6,7\}}} = \prod_{j \in \{6,7\}} \left\{ \frac{p\left(\tilde{Y}^i_{0,j}\right) q\left(Y^i_{0,j} | \tilde{Y}^i_{0,j}\right)}{p\left(Y^i_{0,j}\right) q\left(\tilde{Y}^i_{0,6} | Y^i_{0,6}\right)} \right\} \prod_{t=1}^{n_i} \left\{ \frac{p^*\left(Y^i_t | \tilde{Y}^i_{t-1}, \theta\right) q\left(Y^i_{t,6} | \tilde{Y}^i_{t,6}\right)}{p^*\left(Y^i_t | Y^i_{t-1}, \theta\right) q\left(\tilde{Y}^i_{t,6} | Y^i_{t,6}\right)} \right\}.$$

Note that there is no contribution from the data in the above acceptance ratio since species 6 and 7 do not contribute to the total amounts of p53 or Mdm2.

- Update $\zeta_y | \cdots$ by proposing a candidate $\tilde{\zeta}_y^{\max}$ using a symmetric multivariate normal random walk centred at the current value $\zeta_y^{\max} = \{(\zeta_y^1)^{\max}, \ldots, (\zeta_y^C)^{\max}\}$, the density of which is denoted $q(\tilde{\zeta}_y^{\max} | \zeta_y^{\max})$. Then, for each cell i, set

$$\tilde{\zeta}^i_{t,y} = z^i_{t,y} \frac{\left(\tilde{\zeta}^i_y\right)^{\max}}{\left(z^i_c\right)^{\max}}, \qquad t = 1, \ldots, n_i.$$

The acceptance ratio for the proposed move from ζ_y to $\tilde{\zeta}_y$ is

$$A_{\zeta_y} = \frac{p(\tilde{\zeta}_y | Y, \phi)}{p(\zeta_y | Y, \phi)} \frac{p(z_y | \tilde{\zeta}_y, z_c)}{p(z_y | \zeta_y, z_c)} \frac{q(\zeta_y | \tilde{\zeta}_y)}{q(\tilde{\zeta}_y | \zeta_y)}.$$

The proposal ratio $q(\zeta_y | \tilde{\zeta}_y) / q(\tilde{\zeta}_y | \zeta_y) = 1$ as the proposal is symmetric. The term $p(z_y | \tilde{\zeta}_y, z_c)$ checks the validity of the proposal for ζ_y, in the sense that it is compatible with the observed z_y. Because of the form of the proposal, $\tilde{\zeta}_y$ is always compatible with z_y as we use z_y as part of the proposal. Therefore the acceptance ratio reduces to

$$A_{\zeta_y} = \frac{p(\tilde{\zeta}_y | Y, \phi)}{p(\zeta_y | Y, \phi)}.$$

7.6 Inference based on single cell data

We begin our analysis by studying the time-course information in a single cell. This cell has been chosen at random and is the third cell from the second movie. The time-course covers 38 hours and 40 minutes with data sampled every 1200 seconds, giving 116 time points; see Figure 7.1.

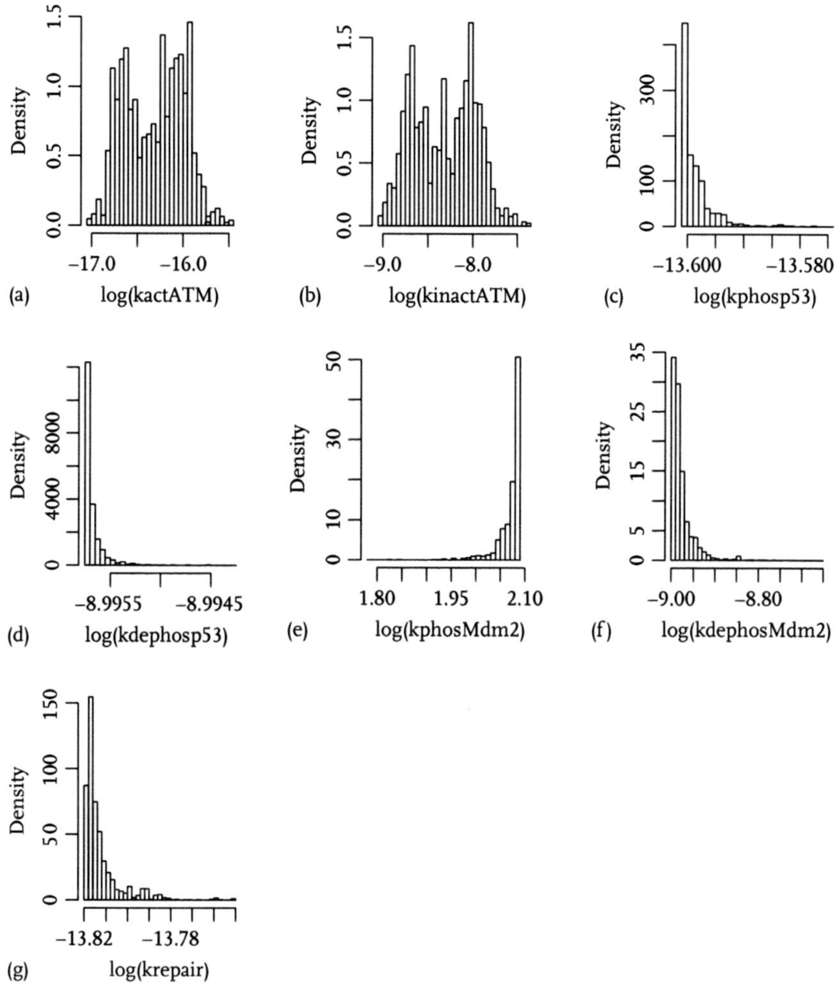

Fig. 7.2 Model calibration parameters: marginal posterior densities.

Several independent Markov chains were simulated from different starting points using the MCMC algorithm outlined in the previous section (with $C = 1$). We report here the results of one of these chains. The MCMC algorithm was run for 250,000 iterations after discarding an initial 250,000 iterates as burn-in. The output was then thinned to remove some of the high autocorrelation by taking every 25th iterate, leaving 10,000 sampled values from the posterior distribution on which to base inferences.

Figure 7.2 shows density histograms of the MCMC output for the model calibration parameters. Recall that each of these parameters has been given a constant (uniform) prior density. Estimates of posterior means and standard deviations are given in Table 7.4. Clearly, uncertainty regarding all seven model calibration parameters has reduced and the ranges of plausible values have

Table 7.4 Model calibration parameters: posterior means and standard deviations.

Index	Parameter name	Mean	Standard deviation
1	*kactATM*	−16.314	0.3269
2	*kinactATM*	−8.320	0.3496
3	*kphosp53*	−13.599	0.0030
4	*kdephosp53*	−8.996	0.0001
5	*kphosMdm2*	2.071	0.0255
6	*kdephosMdm2*	−8.977	0.0256
7	*krepair*	−13.813	0.0087

narrowed significantly. An image plot representing the posterior correlations between pairs of calibration parameters is displayed in Figure 7.3. It shows that several pairs of parameters are highly correlated and, in particular, there is a strong positive correlation between *kactATM* and *kinactATM* and a strong negative correlation between *kdephosp53* and *kphosMdm2*.

Figure 7.4 displays a histogram of the sampled values from the marginal posterior distributions of the measurement error precision ϕ and the true maximum value in the yellow (Mdm2) channel, $(\zeta_y^3)^{max}$. The posterior density

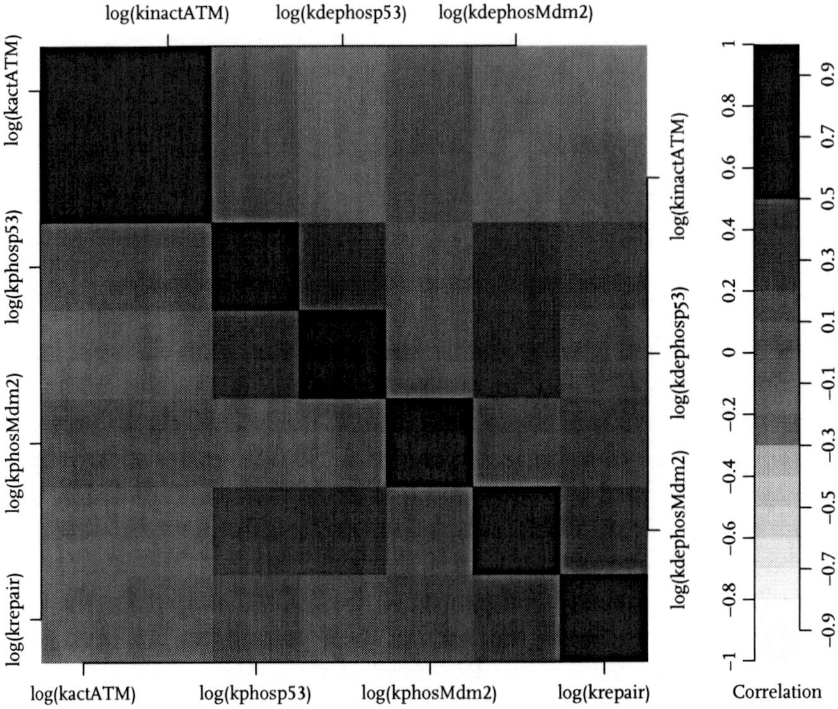

Fig. 7.3 Model calibration parameters: posterior correlations.

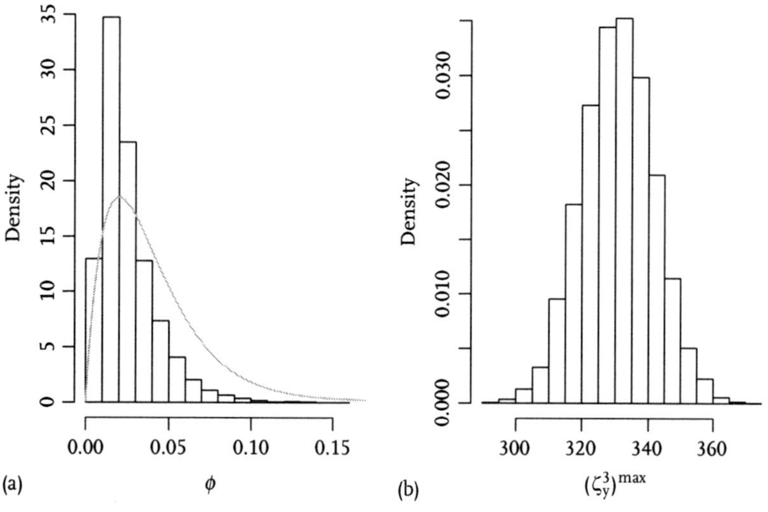

Fig. 7.4 Marginal posterior densities of the measurement error precision ϕ (left panel) and the true maximum value in the yellow (Mdm2) channel for cell 3, $(\zeta_y^3)^{max}$ (right panel). The marginal prior density for ϕ is shown by the grey curve.

for ϕ is fairly similar to its prior density, indicating that the data have not been particularly informative about likely values of ϕ. The true maximum value for Mdm2, $(\zeta_y^3)^{max}$, has posterior mean 330.8 and equal-tailed 95% credible interval (310.1, 352.7). Note that these values are considerably larger than the observed scaled maximum value of 145.0667.

We can gain some confidence in the validity of our fitted model by comparing predictive simulations from the model with the observed data z. This model validation by predictive simulations is advocated by Gelman *et al.* (2004). Figure 7.5 shows a plot of the time-course data for cell 3 (circles), together with shading representing pointwise equal-tailed 95% posterior predictive probability intervals, and a line representing the estimated posterior predictive mean. From Figure 7.5 we see that the model fits the data reasonably well. There is room for improvement in the fit of the p53 data, in terms of both the predictive mean, and the predictive variance. Despite the fact that the Mdm2 data have been normalized, we still achieve an acceptable fit to the Mdm2 data, although the predictive mean lies above the majority of the datapoints.

7.7 Inference based on multiple cells

Although it is useful to look at what is learned about the model calibration parameters and the other unknown quantities in the model from the data on a single cell, it is natural to try and use all the available experimental data (or as much of it as is feasible) in order to make inferences.

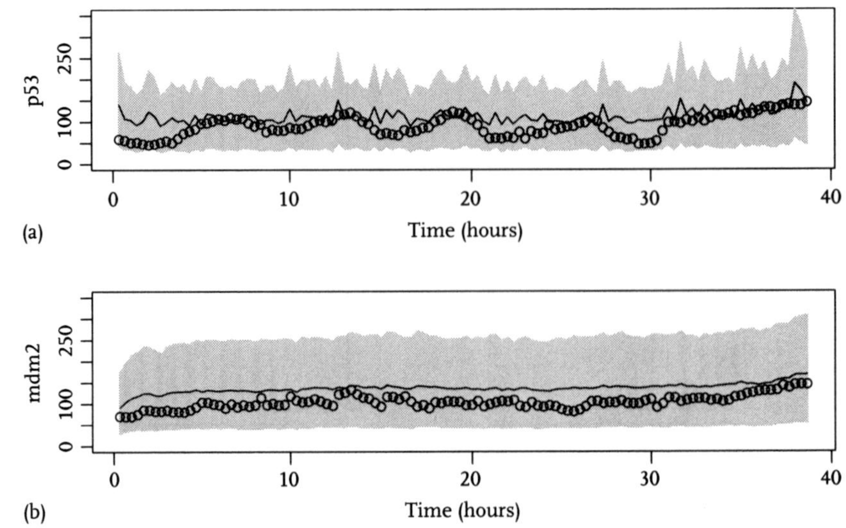

Fig. 7.5 Time-course data (circles) together with (posterior) predictive means (lines) and equal-tailed pointwise 95% predictive intervals (shading); (a) p53 fluorescence and (b) Mdm2 fluorescence.

A full Bayesian calibration of the model based on all 141 available cells is not computationally feasible at this time so here we look at all the cells in one particular movie. We chose the second movie, which consists of seven cells. The seven time-courses are all of different lengths, ranging from 4 hours 20 minutes (for cell 1) to 41 hours (for cell 2), and are shown in Figure 7.1.

Several independent Markov chains were simulated from different starting points, using the MCMC algorithm outlined in Section 7.5, with $C = 7$. We report here the results of one of these chains. The MCMC algorithm was run for 100,000 iterations after discarding an initial 150,000 iterates as burn-in. The output was thinned by taking every 25th iterate, leaving a sample of 4000 iterates on which to base inferences.

Figure 7.6 displays histograms of the sampled values of the model calibration parameters. A comparison of this figure with Figure 7.2 shows that the marginal inferences for these parameters based on all seven cells are not too dissimilar to those based only on the third cell. However these densities are less butted up against the boundaries imposed by the prior distributions than those obtained in the single cell analysis. Also the data from the additional six cells have helped to reduce further the uncertainty about these parameters. These comments are reinforced by a comparison of the values for the posterior means and standard deviations in the seven cell analysis, given in Table 7.5, with those for the single cell analysis in Table 7.4. Overall, the data have dramatically reduced uncertainty about the model calibration parameters. In comparison to their prior means, the data have been strongly suggestive that the kinetic rate constants are around two (or more) orders of magnitude smaller (on the original non-logged scale),

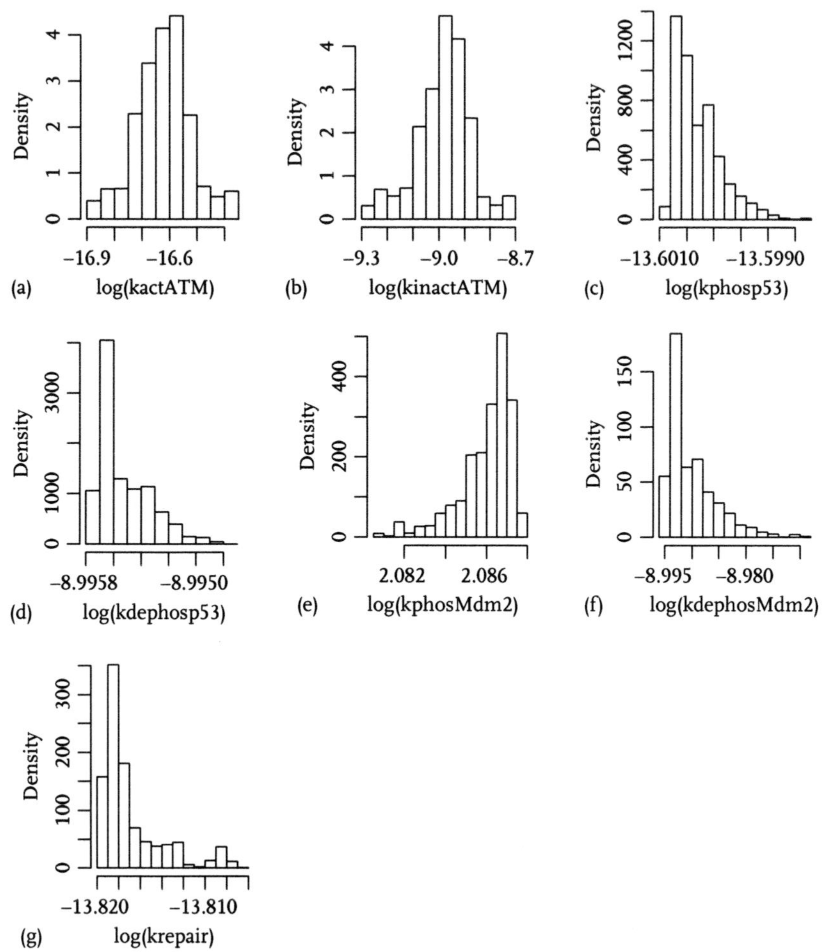

Fig. 7.6 Model calibration parameters: marginal posterior densities based on all seven cells from movie 2.

with the exception of *kphosMdm2* which is around two orders of magnitude larger.

An image plot representing posterior correlations between pairs of calibration parameters is shown in Figure 7.7. This plot reveals that incorporating the data on all seven cells has increased the already high posterior correlation between *kphosp53* and *kdephosp53*. In addition, it highlights the strong negative posterior correlation between *kdephosp53* and *kphosMdm2* that was found in the single cell analysis. Further, it now shows that there is a strong negative posterior correlation between *kphosp53* and *kphosMdm2* that was not evident in the single cell analysis.

Figure 7.8 displays histograms of sampled values from the marginal posterior distribution of the measurement error precision ϕ and the true maximum

Table 7.5 Model calibration parameters: prior and posterior means and standard deviations based on all seven cells in movie 2.

Index	Parameter name	Posterior (prior)	
		Mean	Standard deviation
1	*kactATM*	−16.621 (−9.210)	0.1015 (5.1962)
2	*kinactATM*	−8.982 (−7.601)	0.1046 (5.1962)
3	*kphosp53*	−13.600 (−7.601)	0.0004 (3.4641)
4	*kdephosp53*	−8.996 (−4.147)	0.0002 (2.8000)
5	*kphosMdm2*	2.086 (−2.761)	0.0013 (2.8000)
6	*kdephosMdm2*	−8.990 (−4.147)	0.0045 (2.8000)
7	*krepair*	−13.817 (−10.820)	0.0028 (1.7321)

values of fluorescence in the yellow channels for each of the seven cells. Our inferences about the value of the measurement error precision ϕ have changed considerably after incorporating data from multiple cells into the analysis. In particular, the data from all seven cells have been much more informative about ϕ than that from the third cell alone, suggesting that the measurement process is much less precise than expected a priori, though this could simply be a comment on an overall lack of fit of the model. Inferences on the maximum

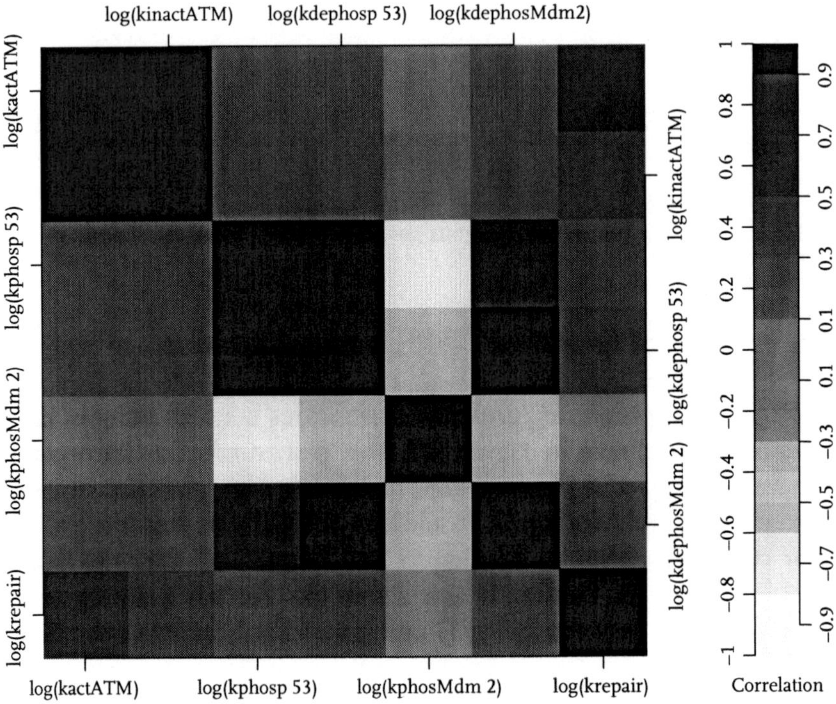

Fig. 7.7 Model calibration parameters: posterior correlations based on all seven cells from movie 2.

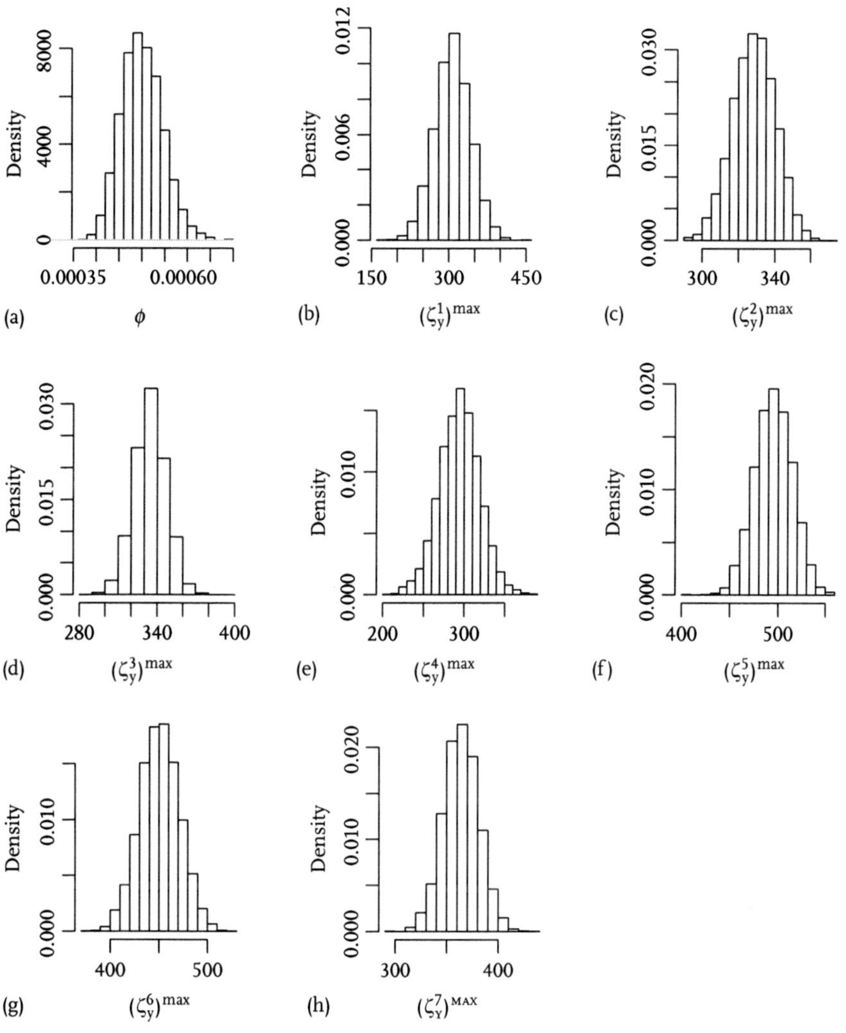

Fig. 7.8 Marginal posterior densities of (a) the measurement error precision ϕ and (b)–(h) the true maximum values in the yellow (Mdm2) channel for cells 1 to 7.

fluorescence values in the yellow (Mdm2) channels (Figure 7.8, panels b–h) confirm that they take posterior mean values, ranging from 294.1 for cell 4 to 495.3 for cell 5, which are much larger than their observed scaled maximum values.

How well does the model fit the data? Figures 7.9 and 7.10 show the data for the seven cells together with summaries of the posterior predictive distributions obtained by sampling replicate data from the calibrated model (in a similar fashion to Figure 7.5). The fit of the model to the data on cells 1 and 3 seems satisfactory, but the fit to data on the other cells shows considerable room for improvement. The predictive means are almost always larger than the observed datapoints, and there is considerable predictive uncertainty. The lack of fit

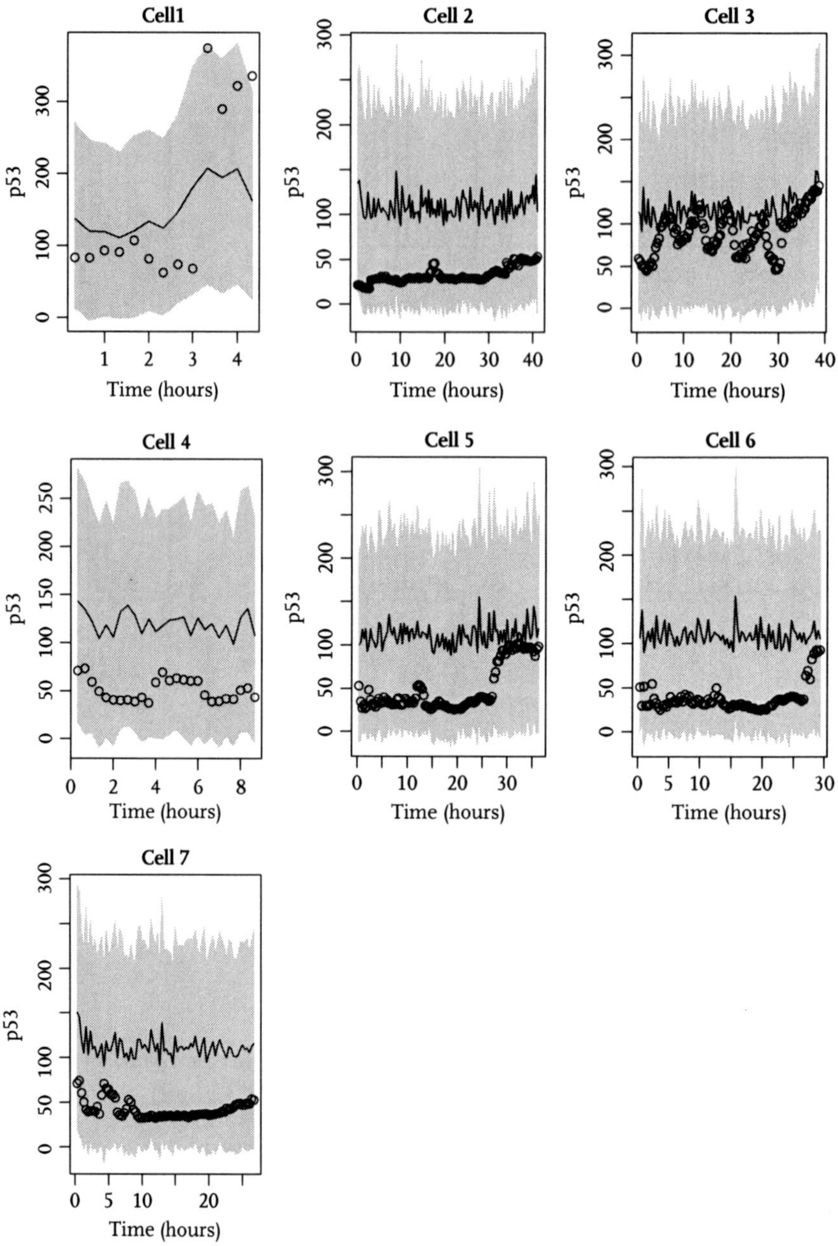

Fig. 7.9 Time-course data for p53 (circles) together with posterior predictive means (lines) and equal-tailed point-wise 95% predictive intervals (shading).

in these predictive plots goes some way to explain the preference for small values of the measurement error precision ϕ. However, the lack of fit may not be too surprising given the inferential challenges posed by the scaling and normalization of the data.

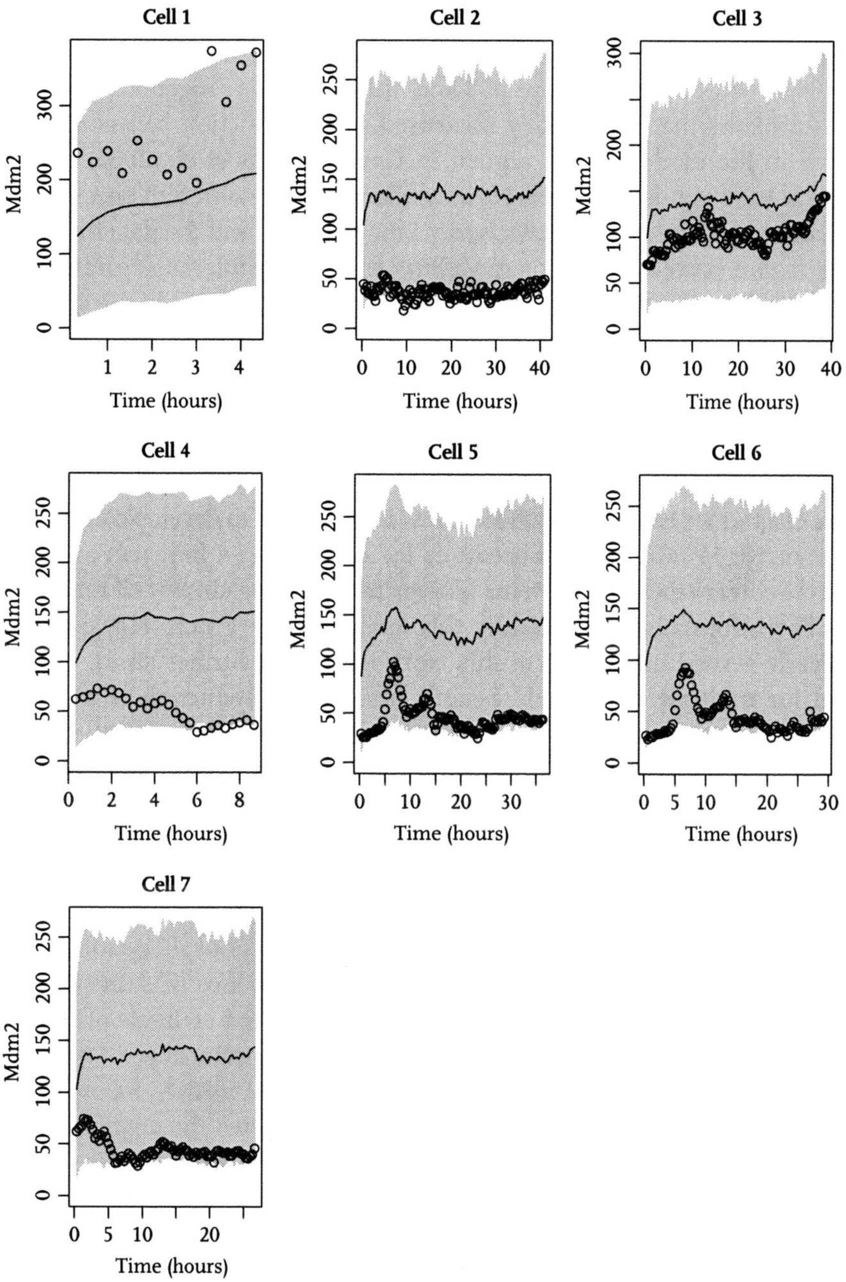

Fig. 7.10 Time-course data for Mdm2 (circles) together with posterior predictive means (lines) and equal-tailed point-wise 95% predictive intervals (shading).

Another issue that may go some way to explaining the smaller than anticipated measurement error precision (suggestive of some lack of fit) is the distinction between endogenous p53 and Mdm2 and the exogenous fluorescent fusion proteins that are actually measured. The distinction between these is ignored in the model as it is argued in Geva-Zatorsky *et al.* (2006) that they should behave similarly *in vivo*. However, ideally these different species would be modelled separately in the stochastic kinetic model and the data linked only to the fusion proteins. Explicit modelling of fusion proteins separately from the targets being reported on is currently in its infancy, and is the subject of on-going modelling work.

7.8 Further discussion

In this chapter we have demonstrated how it is possible to develop computationally intensive MCMC-based procedures for conducting a Bayesian analysis of an intra-cellular stochastic systems biology model using single-cell time course data. The information provided by this analysis is very rich. However, there are clearly several extensions of this work that merit further study. First, the model for multiple cells would benefit from the introduction of a random-effects layer in order to allow for the separation of inter- and intra-cell variation. Second, an integrated analysis allowing the comparison of competing model structures would be extremely valuable. For example, it is possible to develop an alternative (competing) model by replacing the role of ATM with ARF. In this alternative model, the species **ATMA** and **ATMI** are removed, along with the reactions involving these species. The species **p53_P** and **Mdm2_P** are also removed as ARF works by a different mechanism to phosphorylation. **ARF** is initially set to zero but its level increases in the presence of damaged DNA. ARF binds to Mdm2 with a higher affinity than p53 and so levels of unbound p53 increase. This results in an increase of p53 transcriptional activity and so it is reasonable to predict an increase in levels of Mdm2 mRNA, followed by an increase in Mdm2. Since it is known that ARF increases the degradation rate of Mdm2, this model assumes that Mdm2 which is bound to ARF is degraded at a higher rate than normal. ARF is also degraded which allows the damage signal to decline as the damaged DNA is repaired. However, this mechanism seems to play a minor role in response to irradiation compared to ATM and so it is generally believed that this model is less appropriate for the experimental data. Nevertheless, there is genuine interest in knowing whether the data provide support for the ATM-based model considered here in favour of an ARF-based model considered less plausible by some biological experts. Extension of the algorithm to allow computation of the relevant Bayes factor ought to be straightforward in principle, but is likely to be quite difficult computationally.

More generally, the problem of constructing MCMC algorithms for models of this type is currently expensive in terms of both development and computation time. It is therefore natural to seek more straightforward and more automated approaches. Sequential Monte Carlo approaches may well offer considerable advantages in this area, given the Markovian nature of the underlying processes. Such sequential algorithms have already been considered for models of this type (Golightly and Wilkinson, 2006b), but considerable work remains to be done before they can be applied routinely to this general class of problems. Moreover, the normalisation that has been applied to the data described in this chapter means that these data do not lend themselves well to analysis via sequential methods. These and other related problems are the subject of current study within the CaliBayes project (`http://www.calibayes.ncl.ac.uk`).

Appendix

A. Broader context and background

The problem considered in this chapter is a special case of the general problem of conducting inference for the parameters of a Markovian stochastic process model using time course data. Although it is possible to consider stochastic processes that are intrinsically non-Markovian, it turns out that the Markovian class is very large, covering the vast majority of models that are derived from physical considerations. The class of Markov process models that may be considered is itself very large, but can be further categorized into subclasses in various ways. It turns out that the most important attribute for classification purposes is whether the underlying stochastic process model is naturally formulated in discrete or continuous time. Time course data is typically discrete in time, but as is the case in this chapter, it is nevertheless often natural to formulate the model in continuous time. Discrete time models have been studied more widely as they are technically simpler to work with, and often quite adequate if prediction is more important than parameter inference. The class of discrete time models can be further divided depending on whether the state space of the discrete time stochastic process is discrete or continuous. If it is discrete, then the model most likely falls into the general class of *hidden Markov models* (HMMs). Bayesian inference for HMMs is a well-studied problem, with Scott (2002) providing a comprehensive review and details of computation; also see Boys and Henderson (2004) and Fearnhead (2006). Alternatively, if the state space is continuous, then the model is often referred to as a linear or nonlinear *state space* or *dynamic model*. The linear case is referred to as the *dynamic linear (state space) model* (DLM), and is studied in detail in West and Harrison (1997).

Clearly the problem considered in this chapter falls into the class of problems concerned with parameter inference for continuous time Markov process models. Here again it is helpful to subdivide this class depending on the state space of the model. If the state space is discrete and finite, then the model is a *continuous time hidden Markov model* (CTHMM), and can be tackled using techniques very similar to those used for discrete time HMMs (Fearnhead, 2008). If the state space is infinite, then inference is less straightforward (Fearnhead and Meligkotsidou, 2004). A very large class of such models (including stochastic kinetic models, and non-spatial Markovian stochastic epidemic models) is covered (in principle) by the algorithms developed in Boys *et al.* (2008), though the techniques described there do not scale well to problems of realistic size and complexity. For large and complex problems, several alternative possibilities exist. One approach is to exploit *stochastic emulators* of the process of interest, as advocated in Henderson *et al.* (2009) and this chapter. Another possibility is to exploit a combination of *sequential Monte Carlo* (Doucet *et al.*, 2001) methods and *likelihood-free MCMC* (Marjoram *et al.*, 2003). A rather different solution to the problem is to approximate the discrete state continuous time model with a continuous state model (the diffusion approximation), and then use methods for models described by stochastic differential equations (Golightly and Wilkinson, 2005).

Markovian models continuous in both time and state are typically described by stochastic differential equations. Inference for stochastic differential equation models is a rather technical topic, and problematic due to the fact that 'obvious' MCMC algorithms are subject to pathological mixing problems. An excellent introduction to the topic, describing the essential structure of basic algorithms, the inherent problems with the obvious approach, and an elegant solution for the univariate case, is given in Roberts and Stramer (2001). An effective sequential Monte Carlo algorithm for the multivariate case is described in Golightly and Wilkinson (2006*a*) and applied to realistic systems biology scenarios in Golightly and Wilkinson (2006b). An effective global MCMC algorithm is described in Golightly and Wilkinson (2008) and more generally in Golightly and Wilkinson (2009). See Wilkinson (2006) and Wilkinson (2009) for further details of stochastic process models in the context of systems biology.

B. Construction of an emulator

Our emulator is a tractable approximation to the conditional probability distribution given by $p(Y(t+\tau)|Y(t),\theta)$, where $\tau = 1200$ seconds is the time difference between each of the datapoints. It is constructed based on methodology developed in the deterministic computer models literature and successfully applied to a stochastic computer model in Henderson *et al.* (2009). In essence, we model the joint distribution by carefully chosen univariate marginal and conditional distributions, each having the form of a standard probability distribution, but whose parameters are smooth functions of the model calibration parameters and other additional inputs.

In order to construct the emulator, the stochastic kinetic model was forward simulated for $\tau = 1200$ seconds from 2000 randomly chosen state vectors $Y(t)$ and parameter values θ, and the output in the form of the values of the 10 species was recorded. The 2000 design points were generated by using an approximate maximin Latin hypercube sample. Latin hypercube sampling was popularized as a strategy for generating input points for computer experiments by McKay *et al.* (1979). A maximin Latin hypercube sample is a Latin hypercube sample which maximises the minimal distance between pairs of design points. Maximin Latin hypercube designs are described in more detail in Santner *et al.* (2003).

As the model we are approximating is stochastic rather than deterministic we ran the model independently 100 times at each of the 2000 design points. The total simulation took around three days of CPU time. This was split over 50 2.2 GHz processors, and so took less than two hours of real time.

The main complicating issue is that the conditional distribution $Y(t+\tau)|Y(t), \theta$ is 10-dimensional. A common practice when emulating multivariate outputs is to build an emulator for each output independently of all the others (although there have been recent developments on the construction of dynamic, multivariate emulators (Conti and O'Hagan, 2007; Conti *et al.*, 2007)). Independent emulators naturally ignore any correlations between the outputs and so are likely to be poorer approximations to the underlying stochastic kinetic model when such correlations exist. The approach we have used in this chapter is to construct an emulator for each univariate component of the factorisation of the joint conditional probability of the form

$$\begin{aligned} p(Y(t+\tau)|Y(t), \theta) = {} & p(Y_1(t+\tau)|Y(t), \theta) \times p(Y_2(t+\tau)|Y_1(t+\tau), Y(t), \theta) \\ & \times \cdots \times p(Y_{10}(t+\tau)|Y_1(t+\tau), \ldots, Y_9(t+\tau), Y(t), \theta). \end{aligned}$$

This reduces the task to that of fitting 10 univariate emulators to the output of the stochastic kinetic model. There is no natural ordering of the 10 species in the factorisation of the joint distribution, and so we focus on a particular ordering based on computational considerations. Because of the discrete nature of the output from the stochastic kinetic model we model the univariate component probabilities using Poisson distributions, with species 6 (**ATMA**) and 7 (**ATMI**) being modelled using a binomial distribution because their sum is constrained. By looking at the means and variances of the simulator output over the 100 replications we found that the Poisson assumption was not totally appropriate. However, since the output showed both over- and under-dispersion relative to the Poisson we decided to stick with the Poisson as a highly tractable compromise. The parameters of the Poisson or binomial distributions were assumed to be functions of the covariates (which are the species and model calibration parameters that they are conditioned on). For example, we model

$$Y_{10}(t+\tau)|Y_1(t+\tau), \ldots, Y_9(t+\tau), Y(t), \theta \\ \sim Po(\exp\{f(Y_1(t+\tau), \ldots, Y_9(t+\tau), Y(t), \theta)\}),$$

where the Poisson mean is a function of the covariates. We have found that taking the function $f(\cdot)$ to be a low-order polynomial (quadratic or cubic) in the covariates to be adequate. In particular, we have found for this particular example that no improvement in fit (to some independent validation data) was obtained by additionally allowing for code uncertainty through inclusion of a Gaussian process term in $f(\cdot)$. It would appear that allowing for stochastic variation in the output through our construction of simple probability models can account for several of the standard sources of uncertainty encountered in the analysis of complex computer models and outlined in Section 2.1 of Kennedy and O'Hagan (2001). Each component distribution in the resulting emulator is either Poisson or binomial with parameters that are deterministic functions of the appropriate set of covariates. Therefore, we have a tractable approximation to the joint distribution of interest which we denote $p^*(\cdot|\cdot, \theta)$.

Acknowledgements

We thank Uri Alon and Naama Geva-Zatorsky (Weizmann Institute of Science, Israel) for providing us with their raw experimental data, and Douglas Gray (Ottawa Health Research Institute) for advice on the biological aspects of the model construction. This work was funded by the UK Research Councils (BBSRC and EPSRC). In particular, most of the work for this chapter was directly funded by the BBSRC Bioinformatics and e-Science Initiative (BBSB16550) and the BBSRC Centre for Integrated Systems Biology of Ageing and Nutrition (BBC0082001).

References

Barak, Y., Juven, T., Haffner, R. and Oren, M. (1993). Mdm2 expression is induced by wild type-p53 activity. *Embo Journal*, **12**, 461–468.

Boys, R. J. and Henderson, D. A. (2004). A Bayesian approach to DNA sequence segmentation (with discussion). *Biometrics*, **60**, 573–588.

Boys, R. J., Wilkinson, D. J. and Kirkwood, T. B. L. (2008). Bayesian inference for a discretely observed stochastic kinetic model. *Statistics and Computing*, **18**, 125–135.

Christophorou, M. A. Martin-Zanca, D., Soucek, L., Lawlor, E., Brown-Swigart, L., Verschuren, E. W. and Evan, G. I. (2005). Temporal dissection of p53 function in vitro and in vivo. *Nature Genetics*, **37**, 718–726.

Ciliberto, A., Novak, B. and Tyson, J. J. (2005). Steady states and oscillations in the p53/Mdm2 network. *Cell Cycle*, **4**, 488–493.

Clegg, H. V., Itahana, K. and Zhang, Y. (2008). Unlocking the Mmd2-p53 loop – ubiquitin is the key. *Cell Cycle*, **7**, 1–6.

Conti, S., Gosling, J. P., Oakley, J. E. and O'Hagan, A. (2007). Gaussian process emulation of dynamic computer codes. Research Report No. 571/07, Department of Probability and Statistics, University of Sheffield.

Conti, S. and O'Hagan, A. (2007). Bayesian emulation of complex multi-output and dynamic computer models. Research Report No. 569/07, Department of Probability and Statistics, University of Sheffield.

Doucet, A., de Freitas, N. and Gordon, N. (eds) (2001). *Sequential Monte Carlo Methods in Practice*. Springer, New York.

Fang, S. Y., Jensen, J. P., Ludwig, R. L., Vousden, K. H. and Weissman, A. M. (2000). Mdm2 is a RING finger-dependent ubiquitin protein ligase for itself and p53. *Journal of Biological Chemistry*, **275**, 8945–8951.

Fearnhead, P. (2006). Exact and efficient Bayesian inference for multiple changepoint problems. *Statistics and Computing*, **16**, 203–213.

Fearnhead, P. (2008). Computational methods for complex stochastic systems: a review of some alternatives to MCMC. *Statistics and Computing*, **18**, 151–171.

Fearnhead, P. and Meligkotsidou, L. (2004). Exact filtering for partially observed continuous time models. *Journal of the Royal Statistical Society: Series B (Statistical Methodology)*, **66**, 771–789.

Finlay, C. A. (1993). The Mdm-2 oncogene can overcome wild-type p53 suppression of transformed cell growth. *Molecular Cell Biology*, **13**, 301–306.

Gelman, A., Carlin, J. B., Stern, H. S. and Rubin, D. B. (2004). *Bayesian Data Analysis* (2nd edn.). Chapman and Hall/CRC, Boca Raton, Florida.

Geva-Zatorsky, N., Rosenfeld, N., Itzkovitz, S., Milo, R., Sigal, A., Dekel, E., Yarnitzky, T., Liron, Y., Polak, P., Lahav, G. and Alon, U. (2006). Oscillations and variability in the p53 system. *Molecular Systems Biology*, **2**, 2006.0033.

Gillespie, D. T. (1977). Exact stochastic simulation of coupled chemical reactions. *The Journal of Physical Chemistry*, **81**, 2340–2361.

Gillespie, D. T. (2000). The chemical Langevin equation. *Journal of Chemical Physics*, **113**, 297–306.

Golightly, A. and Wilkinson, D. J. (2005). Bayesian inference for stochastic kinetic models using a diffusion approximation. *Biometrics*, **61**, 781–788.

Golightly, A. and Wilkinson, D. J. (2006a). Bayesian sequential inference for nonlinear multivariate diffusions. *Statistics and Computing*, **16**, 323–338.

Golightly, A. and Wilkinson, D. J. (2006b). Bayesian sequential inference for stochastic kinetic biochemical network models. *Journal of Computational Biology*, **13**, 838–851.

Golightly, A. and Wilkinson, D. J. (2008). Bayesian inference for nonlinear multivariate diffusion models observed with error. *Computational Statistics and Data Analysis*, **52**, 1674–1693.

Golightly, A. and Wilkinson, D. J. (2009). Markov chain Monte Carlo algorithms for SDE parameter estimation. In *Learning and Inference for Computational Systems Biology*, (ed. Lawrence, N. D., Girolami, M. Rattray and G. Sansuinethi), MIT Press.

Hainaut, P. and Hollstein, M. (2000). p53 and human cancer: The first ten thousand mutations. *Advances in Cancer Research*, **77**, 81–137.

Haupt, Y., Maya, R., Kazaz, A. and Oren, M. (1997). Mdm2 promotes the rapid degradation of p53. *Nature*, **387**, 296–299.

Henderson, D. A., Boys, R. J., Krishnan, K. J., Lawless, C. and Wilkinson, D. J. (2009). Bayesian emulation and calibration of a stochastic computer model of mitochondrial DNA deletions in substantia nigra neurons. *Journal of the American Statistical Association*, **104**, 76–87.

Hirata, H., Yoshiura, S., Ohtsuka, T., Bessho, Y., Harada, T., Yoshikawa, K. and Kageyama, R. (2002). Oscillatory expression of the bHLH factor Hes1 regulated by a negative feedback loop. *Science*, **298**, 840–843.

Hucka, M., Finney, A., Sauro, H. M., Bolouri, H., Doyle, J. C., Kitano, H., Arkin, A. P., Bornstein, B. J., Bray, D., Cornish-Bowden, A., Cuellar, A. A., Dronov, S., Gilles, E. D., Ginkel, M., Gor, V., Goryanin, I. I., Hedley, W. J., Hodgman, T. C., Hofmeyr, J.-H., Hunter, P. J., Juty, N. S., Kasberger, J. L., Kremling, A., Kummer, U., Novere, N. L., Loew, L. M., Lucio, D., Mendes, P., Minch, E., Mjolsness, E. D., Nakayama, Y., Nelson, M. R., Nielsen, P. F., Sakurada, T., Schaff, J. C., Shapiro, B. E., Shimizu, T. S., Spence, H. D., Stelling, J., Takahashi, K., Tomita, M., Wagner, J. and Wang, J. (2003). The Systems Biology Markup Language (SBML): a medium for representation and exchange of biochemical network models. *Bioinformatics*, **19**, 524–531.

Kennedy, M. C. and O'Hagan, A. (2001). Bayesian calibration of computer models (with discussion). *Journal of the Royal Statistical Society, Series* B, **63**, 425–464.

Khan, S., Guevara, C., Fujii, G. and Parry, D. (2004). P14ARF is a component of the p53 response following ionizing irradiation of normal human fibroblasts. *Oncogene*, **23**, 6040–6046.

Khosravi, R., Maya, R., Gottlieb, T., Oren, M., Shiloh, Y. and Shkedy, D. (1999). Rapid ATM-dependent phosphorylation of MDM2 precedes p53 accumulation in response to DNA damage. *Proceedings of the National Academy of Sciences*, **96**, 14973–14977.

Lahav, G., Rosenfeld, N., Sigal, A., Geva-Zatorsky, N., Levine, A. J., Elowitz, M. B. and Alon, U. (2004). Dynamics of the p53-Mdm2 feedback loop in individual cells. *Nature Genetics*, **36**, 147–150.

Lane, D. P. (1992). p53, guardian of the genome. *Nature*, **358**, 15–16.

Lev Bar-Or, R., Maya, R., Segel, L. A., Alon, U., Levine, A. J. and Oren, M. (2000). Generation of oscillations by the p53-Mdm2 feedback loop: A theoretical and experimental study. *Proceedings of the National Academy of Sciences*, **97**, 11250–11255.

Ma, L., Wagner, J., Rice, J. J., Hu, W. W., Levine, A. J. and Stolovitzky, G. A. (2005). A plausible model for the digital response of p53 to DNA damage. *Proceedings of the National Academy of Sciences*, **102**, 14266–14271.

Marjoram, P., Molitor, J., Plagnol, V. and Tavaré, S. (2003). Markov chain Monte Carlo without likelihoods. *Proceedings of the National Academy of Sciences, USA*, **100**, 15324–15328.

McKay, M. D., Beckman, R. J. and Conover, W. J. (1979). A comparison of three methods for selecting values of input variables in the analysis of output from a computer code. *Technometrics*, **21**, 239–245.

Mendrysa, S. M. and Perry, M. E. (2000). The p53 tumor suppressor protein does not regulate expression of its own inhibitor, MDM2, except under conditions of stress. *Molecular Cell Biology*, **20**, 2023–2030.

Nelson, D. E., Ihekwaba, A. E. C., Elliott, M., Johnson, J. R., Gibney, C. A., Foreman, B. E., Nelson, G., See, V., Horton, C. A., Spiller, D. G. and Edwards, S. W., McDowell, H. P., Unitt, J. F., Sullivan, E., Grimley, R., Benson, N., Broomhead, D., Kell, D. B. and White, M. R. (2004). Oscillations in NF-kappaB signaling control the dynamics of gene expression. *Science*, **306**, 704–708.

O'Hagan, A. (2006). Bayesian analysis of computer code outputs: a tutorial. *Reliability Engineering and System Safety*, **91**, 1290–1300.

Pereg, Y., Shkedy, D., de Graaf, P., Meulmeester, E., Edelson-Averbukh, M., Salek, M., Biton, S., Teunisse, A. F. A. S., Lehmann, W. D., Jochemsen, A. G. and Shiloh, J. (2005) Phosphorylation of Hdmx mediates its Hdm2- and ATM-dependent degradation in response to DNA damage. *Proceedings of the National Academy of Sciences*, **102**, 5056–5061.

Proctor, C. J. and Gray, D. A. (2008). Explaining oscillations and variability in the p53 – Mdm2 system. *BMC Systems Biology*, **2**, 75.

Roberts, G. O. and Stramer, O. (2001). On inference for non-linear diffusion models using Metropolis–Hastings algorithms. *Biometrika*, **88**, 603–621.

Santner, T. J., Williams, B. J. and Notz, W. I. (2003). *The Design and Analysis of Computer Experiments*. Springer, New York.

Scott, S. L. (2002). Bayesian methods for hidden Markov models: recursive computing in the 21st century. *Journal of the American Statistical Association*, **97**, 337–351.

Thut, C. J., Goodrich, J. A. and Tjian, R. (1997). Repression of p53-mediated transcription by MDM2: a dual mechanism. *Genes & Development*, **11**, 1974–1986.

Tiana, G., Krishna, S., Pigolotti, S., Jensen, M. H. and Sneppen, K. (2007). Oscillations and temporal signalling in cells. *Physical Biology*, **4**, R1–R17.

Vogelstein, B., Lane, D. and Levine, A. J. (2000). Surfing the p53 network. *Nature*, **408**, 307–310.

West, M. and Harrison, J. (1997). *Bayesian Forecasting and Dynamic Models* (2nd edn.). Springer, New York.

Wilkinson, D. J. (2006). *Stochastic Modelling for Systems Biology*. Chapman & Hall/CRC, Boca Raton, Florida.

Wilkinson, D. J. (2009). Stochastic modelling for quantitative description of heterogeneous biological systems. *Nature Reviews Genetics*, **10**, 122–133.

Zhang, L. J., Yan, S. W. and Zhuo, Y. Z. (2007). A dynamical model of DNA-damage derived p53-Mdm2 interaction. *Acta Physica Sinica*, **56**, 2442–2447.

Zhang, Y., Xiong, Y. and Yarbrough, W. G. (1998). ARF promotes MDM2 degradation and stabilizes p53: ARF-INK4a locus deletion impairs both the Rb and p53 tumor suppression pathways. *Cell*, **92**, 725–734.

·8·

Paternity testing allowing for uncertain mutation rates

A. Philip Dawid, Julia Mortera and Paola Vicard

8.1 Introduction

In 1984, Sir Alec Jeffreys and his team at Leicester University discovered 'minisatellites', stretches of DNA that are highly variable from one individual to another (Jeffreys *et al.* 1985*a*). Jeffreys quickly realised that this could be a powerful tool for identifying whether or not two biological samples originated from the same source (Jeffreys *et al.* 1985*b*), so initiating the technology of *DNA profiling*, and the new discipline of *forensic genetics*, together with the establishment of massive[1] 'intelligence databases' of DNA profiles. This technology has had a tremendous impact on crime investigation.

Although there have been changes over time to the specific DNA markers used, and to the associated instrumentation, the essential logical issues of DNA profiling persist unchanged. If two biological samples are genuinely from the same source, then they must have identical DNA, so their DNA profiles must be essentially identical (with current technology, DNA profiles can generally be measured without error, which means we should expect exact identity). However the converse is false: contrary to popular belief, different individuals can have identical DNA profiles, and so two DNA profiles from different sources might just happen to match 'by chance'. Consequently, while a failure of two samples to match can generally be taken as firm evidence of their having come from different sources, when we do find a match the evidence for a common source is logically inconclusive. Only by taking account of the probabilistic processes underlying the data can we come to a reasoned assessment of the strength of such match evidence. In particular we need to assess the frequency with which a given profile occurs in a relevant population, which is typically done by combining an assumption of independence across markers (which may be justifiable from general genetic principles) with sample-based estimates for individual markers. Forensic genetics laboratories compile 'research databases' to assist with this estimation task.

[1] For example, the UK database contains profiles from about 1% of the population.

The general logic of forensic identification (Dawid and Mortera 1996; Dawid and Mortera 1998) applies equally to non-genetic matches, such as footprints, bitemarks or glass fragments. However, the much greater discriminatory power of DNA profiling, as well as its more 'objective' basis in genetical theory and data, has tended to sideline or overturn[2] these other forms of identification. There is also one type of application that relies crucially on the genetic properties of DNA, namely when we question whether, and how, two *distinct* DNA sources might be *genetically related*. The principal question of this type, which will be the focus of this article, is that of disputed paternity. Variations include identification of remains, *e.g.* from disasters such as the twin towers or tsunami, by profiling relatives of missing individuals; and testing other relationships, such as claims made by would-be immigrants.

8.1.1 Outline

This chapter describes the construction and Bayesian analysis of a suitable model to assess paternity, when we have to take seriously the potentially misleading effect of genetic mutation.

In Section 8.2 we describe and analyse the simplest type of disputed paternity case. We introduce in Section 8.2.2 a specific disputed paternity case, in which an apparent exclusion at a single marker could indicate either non-paternity or mutation. Section 8.3 describes the features of the mutation process that need to be accounted for in the analysis. These typically need to be represented in terms of a small number of unknown parameters, by means of a suitable model: in Section 8.3.2 we describe the particular mutation model we use. Section 8.3.3 shows that, to an excellent approximation, the likelihood for the overall 'total mutation rate' τ, based on a single disputed paternity case, is a linear function.

In Section 8.4 we conduct a simple analysis of the specific disputed paternity case considered, assuming τ known, and discuss its sensitivity to the assumed value of τ. Section 8.5 then introduces the fully Bayesian approach, showing how to allow for uncertainty about τ and information about τ contained in previous data. In Section 8.6 we describe the kind of data available to make inference about τ, based on casework collected at paternity testing laboratories.

Section 8.7 is the core of our analysis, describing the computation of the likelihood function for τ based on casework data. This is applied to a particular casework data-set in Section 8.8. Finally, in Section 8.9, we conduct an analysis of the disputed paternity case under consideration, using the knowledge of τ gained from the past casework data. Section 8.10 concludes by setting this analysis in the broader context of forensic inference from DNA profiles in the presence of various complicating features.

[2] see e.g. `http://www.forensic-evidence.com/site/ID/bitemark_ID.html`

In Appendix A.1 we briefly describe the rôle of statistical reasoning in the treatment of legal evidence, while Appendix A.2 outlines the essential genetic background to DNA profiling. Some theory of the overall likelihood function for a collection of mutation rates is presented in Appendix B.

8.2 Simple paternity testing

In a simple disputed paternity case, a mother, `m`, claims that her child, `c`, was fathered by a certain man, the *putative father*, `pf`, but he denies it. In order to assess the truth of the matter, DNA profiles may be taken from all three individuals. Since the child inherits its DNA from its two parents, according to the laws of Mendelian genetics, this constellation of profiles sheds light on whether or not `pf` is the father.

We are interested in comparing two hypotheses: hypothesis P (paternity) asserts that the putative father `pf` is the true father `tf` of the disputed child, while hypothesis $\overline{P}$ (non-paternity) asserts that he is not. We could make $\overline{P}$ more specific, for example by identifying one or more alternative potential fathers, but in this article we suppose there are none such, and that, under $\overline{P}$, the true father is some unidentified alternative father `af`, who can be regarded as randomly drawn from a relevant population. Figure 8.1 gives a pictorial representation of such a case. The hypotheses P and $\overline{P}$ are represented by the states, `true` or `false`, of the 'hypothesis variable' `tf=pf?`, which determines which of `pf` and `af` is identified with `tf`, and so passes his genes to `c`.

The DNA evidence $\mathcal{E}$ in the case comprises the *DNA profiles* (each a collection of measurements on a number of *DNA markers*) for each member `c`, `m` and `pf` of the alleged family triplet. If indeed `pf` is the father of `c` (and absent complications such as mutation, measurement error, etc.), there will be certain

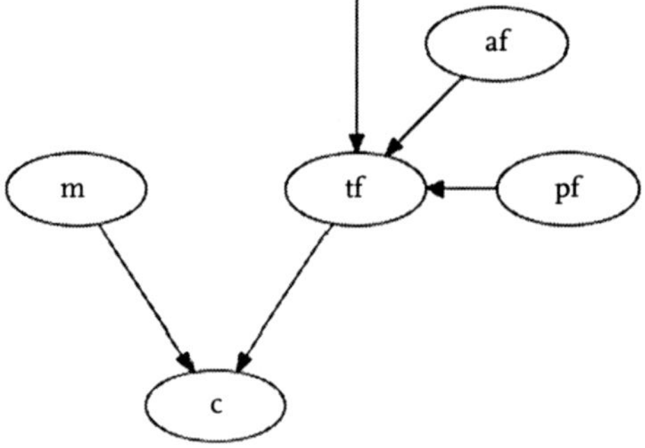

Fig. 8.1 Pedigree for simple disputed paternity.

combinations of types that are possible and others that are impossible. For example, the child may possess a type that could not have been formed by the union of `m` and `pf`, an *exclusion*. In such a case we would have to conclude that the accusation of paternity was false. Otherwise, the possibility of paternity, P, is not excluded – but neither is non-paternity, $\bar{P}$. We then need to use the genetic evidence $\mathcal{E}$ appropriately to weigh these two hypotheses in the balance.

The strength of the evidence in favour of paternity P is measured by the *paternity index*, PI, which is just the likelihood ratio for comparing the two hypotheses:

$$\text{PI} = \frac{Pr(\mathcal{E} \mid P)}{Pr(\mathcal{E} \mid \bar{P})}. \tag{8.1}$$

Under some standard assumptions (including Mendelian segregation, independent markers, known population allele frequencies, fully observed profiles and no mutation), PI can be calculated by simple and well-known algebraic formulae (Essen-Möller 1938). In particular, for independent markers the overall PI will be the product of the various marker-specific PI's, each calculated from the evidence at just that marker.

PI is often interpreted as a posterior odds on paternity, but this is only valid when the prior probability of paternity is $\frac{1}{2}$. More appropriately we should specify actual, externally assessed, prior probabilities, $Pr(P)$ and $Pr(\bar{P})$. Then posterior probabilities in the light of the evidence are obtained from Bayes's theorem, which can be expressed as:

$$\frac{Pr(P \mid \mathcal{E})}{Pr(\bar{P} \mid \mathcal{E})} = \frac{Pr(P)}{Pr(\bar{P})} \times \text{PI}. \tag{8.2}$$

8.2.1 An example

Formulae for computing PI were developed by Essen-Möller (1938), long before the introduction of DNA profiling, for use with the types of genetic evidence, *e.g.* based on blood groups, then in use. A simple example is the following.

Suppose every individual can be identified by blood tests as having genotype XX, XY, or YY, where X and Y are alternative allelic forms of a certain gene. The frequency of allele X in the population is $2/3$, and that of allele Y is $1/3$. Each genotype is formed of two genes – which however cannot be individually identified – randomly selected from the population. A child inherits one gene chosen at random (*i.e.* each with probability $1/2$) from its mother's pair, and (independently) one chosen at random from its father's pair.

Consider a specific disputed paternity case where the evidence $\mathcal{E}$ is that the mother is of type XX, the child of type XY, and the putative father of type XY. This does not result in an immediate exclusion, since under paternity the

Table 8.1 Disputed paternity case.

Marker	`cgt`	`mgt`	`pfgt`	PI formula	PI value
D3S1358	17	16/17	15/17	$1/(2\ p_{17})$	2.3256
vWA	18/19	14/18	17/18		
FGA	24/26	21.2/24	26/28	$1/(2\ p_{26})$	21.7391
TH01	6/8	6/7	8/10	$1/(2\ p_{8})$	5.9524
TPOX	8/10	8/10	8	$1/(p_8 + p_{10})$	1.692
CSF1PO	12	11/12	10/12	$1/(2\ p_{12})$	1.385
D5S818	9/11	9/11	11	$1/(p_9 + p_{11})$	2.433
D13S317	11	11/12	8/11	$1/(2\ p_{11})$	1.4749
D7S820	9/12	10/12	8/9	$1/(2\ p_{9})$	2.8249
ACTBP2	28.2/30.2	13/30.2	14/28.2	$1/(2\ p_{28.2})$	5.814
D12S391	17/19	17/19	19/21	$1/\{2\ (p_{17} + p_{19})\}$	2.2173
D8S1132	19/25	17/25	19/23	$1/(2\ p_{19})$	3.0303
Product (excluding vWA)					279272.24

evidence could be explained by the child having inherited an X from its mother (a certain event) and a Y from the putative father (an event of probability 1/2). The overall probability of this, assuming paternity, is thus 1/2. Alternatively, under non-paternity the child must still have inherited its X from its mother (again a certain event), and its Y from its true father – but since we now do not know who that is, the probability of this is just the frequency, 1/3, of Y in the general population. The paternity index is calculated as the ratio of these two probabilities: thus PI = (1/2)/(1/3) = 1.5.

8.2.2 Case analysis

The DNA evidence for a specific disputed paternity case is shown in Table 8.1. We have an alleged family triplet formed by a disputed child `c`, its mother `m`, and the putative father `pf`. Their DNA profiles, comprising their respective genotypes `cgt`, `mgt` and `pfgt` for each of a standard set of 12 forensic DNA markers, are given in columns 2–4. For each marker except vWA, the relevant paternity index formula (Essen-Möller 1938) is given in column 5, and used to compute the marker-specific PI in the last column. The allele frequencies (p_i) used are taken from Butler *et al.* (2003) and Schmid *et al.* (2005), but are not reproduced here. The overall PI based on just these 11 markers is nearly 300,000: strong evidence in favour of paternity.

However, marker vWA shows a *prima facie* exclusion: the child's allele 19 does not match any allele in `mgt` or `pfgt`, and this appears incompatible with paternity. Taken at face value, this one exclusion would totally override the opposite message from the other markers. But if there were another explanation for this apparent exclusion, we might still be able to claim paternity. One such explanation, which we shall explore in detail below, is mutation: the child inherited allele 18 from one of its two parents, but its allele 19 was the product of a mutation of one of the other parental alleles.

8.3 Mutation

If, realistically, we weaken the standard assumptions, then, although the fundamental logic leading to (8.1) and (8.2) is unaffected, computation of the correct PI can become considerably more complex. For example, to allow for the possibility of mutation during segregation, or various forms of imperfect observation of the profiles, requires considerable elaboration of the simple situation represented by Figure 8.1. Alternatively, or in addition, we might wish to admit uncertainty about some of the parameters (allele frequencies, mutation rates, ...), hitherto considered as known. In the Bayesian framework, this epistemic uncertainty is represented by probability distributions. Then both numerator and denominator of (8.1) will involve an integration over the relevant parameter distribution – which itself should previously have been updated, using a further level of Bayesian analysis, in the light of any past data that are informative about the parameters.

In this chapter we focus on the specific complication of *mutation* (Fimmers *et al.* 1992), with possibly *uncertain mutation rates.*

The markers used for forensic analysis are chosen, in part, because they are highly variable, which gives them good discriminatory power; but this variability is itself due to the fact that they have relatively high mutation rates. In the presence of non-negligible mutation, an apparent exclusion is no longer definitive evidence of non-paternity, since an alternative explanation of this finding is mutation during segregation. The potentially distorting effect of mutation needs to be properly allowed for in the analysis if misleading conclusions are to be avoided.

For the STR markers used in forensic parenthood tests, estimates of μ, the mutation rate per meiosis (single parent-child transmission), range from around 5×10^{-4} to 7×10^{-3} (Brinkmann *et al.* 1998; Henke and Henke 1999; Sajantila *et al.* 1999). However, there is considerable uncertainty and potential bias in these estimated mutation rates. The analysis of paternity allowing for mutation is sensitive to the actual mutation rates – which are only imperfectly known. In order to minimize this problem it is important to gather data from which we can obtain good mutation rate estimates. In principle this sounds straightforward, but in practice there is a further complication: the typical source of information about mutation for a forensic DNA marker is casework collected by forensic laboratories. This might include disputed paternity cases in which, say, the DNA profiles of all three parties are compatible at all but one marker. In such a case it might be decided that this was in fact a case of true paternity, but there had been a mutation event at the discrepant marker. Naïve estimates of mutation rates can be based on relative frequencies of such 'deemed mutations'. The problem is that we do not know for sure that this was a case of true paternity, and if it was not then we have been misled in inferring the existence of a mutation.

We see then that there are two possible explanations for an apparent exclusion, non-paternity and mutation, and the presence of both together complicates inference, both for individual case analysis and for parameter learning from past data. Only by formulating a complete model which incorporates both possibilities, and subjecting it to a principled statistical analysis, can we obtain both reliable mutation estimates, allowing for possible non-paternity, and reliable paternity inferences, allowing for mutation. Moreover, only a fully Bayesian analysis can properly allow for all the relevant uncertainties simultaneously.

8.3.1 The effect of mutation on paternity inference

Dawid *et al.* (2001; 2003) show how to compute a paternity index that takes appropriate account of the possibility of mutation. This analysis requires us to specify inter-allelic mutation rates $(q_{i \to j})$ for the relevant markers, where $q_{i \to j}$ denotes the probability that allele i mutates to allele j during meiosis. Data on such inter-allelic mutations are necessarily sparse. Consequently we proceed by setting up various models describing the $(q_{i \to j})$ in terms of a small set of adjustable parameters (Ohta and Kimura 1973; Valdes *et al.* 1993; Durret and Kruglyak 1999; Egeland and Mostad 2002; Dawid *et al.* 2001; Dawid *et al.* 2002) which can then be estimated more securely from data. Also, although there is some evidence that mutation rates can vary between maternal and paternal transmissions, for simplicity we shall here assume identical allele-specific mutation rates for both the maternal and paternal lines.[3] We take the *total mutation rate*, $\tau := 2\mu$, as the principal parameter of interest.

8.3.2 Mutation models

The probability that allele i is transmitted unmutated is $q_{i \to i} = 1 - \sum_{j \neq i} q_{i \to j}$, and the mutation rate per meiosis is

$$\begin{aligned} \mu &:= \sum_{(i,j):i \neq j} p_i \, q_{i \to j} \\ &= 1 - \sum_i p_i \, q_{i \to i} \,, \end{aligned}$$

where p_i denotes the population frequency of allele i.

[3] This assumption can easily be dropped, and the ratio of the paternal and maternal rates estimated along with their total (Vorkas 2005; Vicard *et al.* 2008).

A very flexible class of mutation models comprises the *scalar mutation models* (Vicard and Dawid 2004), having the general form:

$$q_{i\to j} = \lambda s_{i\to j} \quad (j \neq i),$$

where $\lambda \geq 0$ is an unknown scale parameter to be estimated. Define $s_{i\to i} := 1 - \sum_{j:j \neq i} s_{i\to j}$. By rescaling the $s_{i\to j}$ $(j \neq i)$ and reciprocally rescaling λ, if necessary, we can assume without loss of generality that $s_{i\to i} \geq 0$, so that $S := (s_{i\to j})$ is a transition matrix.

The probability of no mutation out of state i is $q_{i\to i} = 1 - \lambda(1 - s_{i\to i})$. The total mutation rate is

$$\tau = 2\mu = 2\kappa\lambda \tag{8.3}$$

where

$$\kappa := 1 - \sum_i p_i s_{i\to i}. \tag{8.4}$$

For the purposes of this chapter we shall use a particular *mixed mutation model*, which is a compromise between biological realism and mathematical tractability. It is itself constructed from two simpler models, the *proportional model* and the *single-step mutation model*, as described below.

Proportional model In this model it is assumed that, when a mutation takes place, the mutated allele is simply drawn at random from the allele frequency distribution in the population. We thus have

$$s^{\text{prop}}_{i\to j} = p_j$$
$$\kappa^{\text{prop}} = 1 - \sum p_i^2.$$

Single-step mutation model (SMM) This only allows mutations to a neighbouring allele value. It has

$$s^{\text{SMM}}_{i\to j} = \begin{cases} \frac{1}{2} & \text{if } |i-j| = 1, i \neq i_{\min} \text{ or } i_{\max} \\ 1 & \text{if } |i-j| = 1, i = i_{\min} \text{ or } i_{\max} \\ 0 & \text{otherwise} \end{cases}$$
$$\kappa^{\text{SMM}} = 1.$$

Here $i_{\min}$ and $i_{\max}$ denote the boundary values for the repertory of alleles.

Mixed model With probability h a transition follows the SMM, and with probability $(1 - h)$ the proportional model, independently for each potential mutation event. This yields

$$s^{\text{mixed}}_{i\to j} = h s^{\text{SMM}}_{i\to j} + (1-h) s^{\text{prop}}_{i\to j} \quad (i \neq j),$$
$$\kappa^{\text{mixed}} = h + (1-h)\left(1 - \sum p_i^2\right).$$

A biologically plausible value for the mixing parameter h is 0.9 (American Association of Blood Banks 2002). We use the mixed model with this value throughout.

8.3.3 Probabilities for a single triplet

Suppose we have full genotype data, f, on a certain triplet, comprising the mother's genotype `mgt` = AB, the putative father genotype `pfgt` = CD, and child's genotype `cgt` = EF. Here A, B, C, D, E, F are arbitrary allele values, and any of them could be identical.

Under paternity, P, the probability of observing this collection of genotypes is

$$\begin{aligned} Pr(f \mid P) &\propto Pr(EF \mid AB, CD; P) \\ &\propto (q_{A\to E} + q_{B\to E})(q_{C\to F} + q_{D\to F}) \\ &\quad + (q_{A\to F} + q_{B\to F})(q_{C\to E} + q_{D\to E}). \end{aligned} \tag{8.5}$$

Whittaker *et al.* (2003) use formula (8.5) to develop a likelihood analysis and comparison of various mutation models, assuming paternity in all cases.

Under non-paternity, $\bar{P}$, we have

$$\begin{aligned} Pr(f \mid \bar{P}) &\propto Pr(EF \mid AB, CD; \bar{P}) \\ &\propto 2\left\{(q_{A\to E} + q_{B\to E})\, p_F + (q_{A\to F} + q_{B\to F})\, p_E\right\}. \end{aligned} \tag{8.6}$$

In both (8.5) and (8.6) the omitted constants of proportionality are the same: in the first line, the product of the probabilities for genotypes AB and CD; in the second, $1/4$ if $E \neq F$, $1/8$ if $E = F$.

For a scalar mutation model, using (8.3), $q_{i\to j} = (\tau/2\kappa)s_{i\to j}$ $(j \neq i)$, while $q_{i\to i} = 1 - (\tau/2\kappa)(1 - s_{i\to i})$; so $q_{i\to j}$ is a known linear function of the total mutation rate τ, the constant term being 1 if $i = j$, else 0. Hence (8.5) is a quadratic function of τ, of the form

$$L_P(\tau) = K + A\tau + B\tau^2 \tag{8.7}$$

for some constants K, A, B. Moreover, the quadratic term $B\tau^2$ is entirely due to the possibility, under paternity, of two simultaneous mutations – a very rare event. Hence, to a very good approximation, we can and shall ignore it.

For an incompatible case, the constant term K is 0, and the leading term $A\tau$ then agrees with the formulae for L_P given in Table 1 of Dawid *et al.* (2001) (which likewise ignored the possibility of a double mutation).

Under non-paternity, (8.6) is exactly linear in τ, of the form

$$L_{\bar{P}}(\tau) = C + D\tau, \tag{8.8}$$

the leading term C being non-zero (assuming maternal compatibility). This leading term agrees with the formulae for $L_{\bar{P}}$ given in Table 1 of Dawid *et al.* (2001).

Since τ is very small, for many purposes we can ignore all but the leading terms in (8.7) and (8.8).

8.4 Case analysis with assumed mutation rate

In Section 8.2.2 the paternity index for the disputed paternity triplet given in Table 8.1 was computed for all markers except vWA, which shows a *prima facie* incompatibility with paternity. As for vWA, we have seen that, to an excellent approximation, the probabilities L_P and $L_{\bar{P}}$ of obtaining the case findings, under either paternity (P) or non-paternity ($\bar{P}$), are given (up to a proportionality constant) by the relevant entries in Table 1 of Dawid *et al.* (2001). For the case under discussion, the structure of the incompatibility on vWA is given by row 13 of that table. We get (under our assumption of equal paternal and maternal mutation rates):

$$L_P = \frac{1}{4}\left(q_{17\to 19} + q_{18\to 19} + q_{14\to 19} + q_{18\to 19}\right) \tag{8.9}$$

$$L_{\bar{P}} = \frac{1}{2} p_{19}. \tag{8.10}$$

For the mixed mutation model introduced in Section 8.3.1, we thus obtain

$$L_P = \frac{1}{8\kappa}\left\{h + 4\,(1-h)\,p_{19}\right\}\tau, \tag{8.11}$$

and so the paternity index PI is

$$\mathrm{PI} = \frac{A}{C}\tau = \frac{1}{4\kappa p_{19}}\left\{h + 4\,(1-h)\,p_{19}\right\}\tau. \tag{8.12}$$

We see that PI grows linearly with the total mutation rate τ.

Here we take $h = 0.9$, and use the allele frequencies for marker vWA as in Table 1 in Vicard *et al.* (2008), which yield $\kappa = 0.9807$ and $p_{19} = 0.0866$. Then $A/C = 2.75$, i.e. $\mathrm{PI} = 2.75\,\tau$.

For a range of reasonable values of $\tau = 2\mu$, from around 1×10^{-3} to 1.4×10^{-2} per generation, the paternity index for vWA ranges correspondingly between 0.00275 and 0.0385. Multiplying this by the product of the PIs calculated on all other markers, 2.79×10^5 (see Table 8.1), we find the range for the total PI (TPI) and probability of paternity (based on equal priors) as given in Table 8.2. We see that, notwithstanding an apparent exclusion on

Table 8.2 PI for marker vWA, overall TPI, and probability of paternity, corresponding to lower and upper reasonable values for τ.

	min τ	max τ
τ	0.001	0.014
PI for vWA	0.00275	0.0385
TPI	852.470	11934.576
probability of paternity	0.99882	0.99992

marker vWA, the totality of the evidence points quite strongly to paternity in this case.

8.5 Uncertain mutation rate

The above sensitivity analysis is one way of assessing the effect of mutation on the inference about paternity. An alternative, which takes proper account of any available information about τ, is to model our uncertainty by means of probability, and conduct a full Bayesian analysis. Here we show how this works out.

Suppose there is a *prima facie* exclusion on just one marker, m. Let f be the findings (genotypes of child, mother and putative father) for marker m on the case. The associated PI for this marker is sensitive to the assumed value for its total mutation rate τ; but we do not know τ perfectly, and should take proper account of this.

As we have seen in Section 8.3.3, for a typical such case we have, to a very good approximation,

$$L_P(\tau) = Pr(f \mid P, \tau) \propto A\tau \tag{8.13}$$

$$L_{\bar{P}}(\tau) = Pr(f \mid \bar{P}, \tau) \propto C \tag{8.14}$$

where A, C are simply computed positive constants. For known τ the associated paternity index is just

$$\mathrm{PI}(\tau) := (A/C)\,\tau. \tag{8.15}$$

If we do not know τ, but have data which are informative about τ, we might just compute a point estimate, e.g. the maximum likelihood estimate, and substitute this into (8.15). However this may be misleading, and it is better to perform a proper Bayesian analysis. (Note that, throughout, we consider only the mutation rates as possibly unknown; in particular, allele frequencies are supposed essentially known.)

We can express our data as `data` = (`old`, `other`), where `other` denotes the findings on other markers for the case at hand, and `old` represents whatever data we might have from other cases and sources that are informative about τ. In our analyses below, old will itself comprise findings on other disputed paternity cases.

What we eventually require is the overall posterior odds in favour of paternity in the case at hand, conditional on all the available information (f, old, other). This is

$$\frac{Pr(P \mid f, \texttt{old}, \texttt{other})}{Pr(\bar{P} \mid f, \texttt{old}, \texttt{other})} = \frac{Pr(P \mid \texttt{old})}{Pr(\bar{P} \mid \texttt{old})} \times \frac{Pr(\texttt{other} \mid P, \texttt{old})}{Pr(\texttt{other} \mid \bar{P}, \texttt{old})} \times \frac{Pr(f \mid P, \texttt{old}, \texttt{other})}{Pr(f \mid \bar{P}, \texttt{old}, \texttt{other})}. \tag{8.16}$$

The first term on the right-hand side of (8.16) is the 'prior' odds on paternity, taking into account the old data. Realistically, the old data, which do not involve the case at hand, will be totally uninformative about its paternity status, and then this term is just the 'genuine' prior odds $Pr(P)/Pr(\bar{P})$.

The second term is the likelihood ratio (overall paternity index) based on the other, non-exclusionary, markers in the case, as well as on the old data. Again the old data will typically be entirely uniformative here, and so can be ignored; while the contribution of each of the other markers is captured by its own associated PI. In the absence of an apparent exclusion in any of those markers, the possibility of mutation in them can be ignored, and the relevant PI_j for marker j calculated by means of the standard formulae of Essen-Möller (1938).

Finally, from (8.13) and (8.14) we have

$$Pr(f \mid P, \texttt{old}, \texttt{other}) = A\,E(\tau \mid \texttt{old}, \texttt{other})$$
$$Pr(f \mid \bar{P}, \texttt{old}, \texttt{other}) = C.$$

Moreover, the findings on the other markers will be essentially uninformative even about their own mutation rates, let alone that for marker m, so we can usually take

$$Pr(f \mid P, \texttt{old}, \texttt{other}) = A\,E(\tau \mid \texttt{old})$$
$$Pr(f \mid \bar{P}, \texttt{old}, \texttt{other}) = C.$$

We will thus have final posterior odds

$$\frac{Pr(P \mid f, \texttt{old}, \texttt{other})}{Pr(\bar{P} \mid f, \texttt{old}, \texttt{other})} = \frac{Pr(P)}{Pr(\bar{P})} \times \left(\prod_{j \neq m} \mathrm{PI}_j\right) \times (A/C)\,E(\tau \mid \texttt{old}). \tag{8.17}$$

In particular, the PI for marker m, properly accounting for the uncertainty and information about its unknown total mutation rate, is just $(A/C)\,E(\tau \mid \texttt{old})$, and the total paternity index TPI, based on all available information, is

$$\text{TPI} = \left(\prod_{j \neq m} \text{PI}_j\right) \times (A/C)\, E(\tau \mid \texttt{old}). \tag{8.18}$$

Hence to take proper account of the uncertainty about τ, we simply substitute in place of τ in (8.15) its posterior expectation given the old data.

In the remainder of this article we consider the implementation of the above Bayesian analysis for τ when, as is typically the case, the data `old` comprises DNA profile data from previously collected disputed paternity cases.

8.6 Paternity casework data

Estimates of the mutation rates for forensic DNA markers are themselves based on data collected at forensic laboratories, mainly consisting of the DNA profiles of putative father-mother-child triplets in disputed paternity cases. Moreover the cases used for estimation are typically not a random sample of all cases analysed at the paternity testing laboratory, but are themselves selected according to the overall probability of paternity: if this is above a certain conventional threshold (very close to 1) then the case is retained for the analysis, otherwise it is discarded. In Vicard *et al.* (2004; 2008) the effect of this preselection is discussed, and it is shown that some cases that would be discarded by this criterion can nevertheless contain relevant information about mutation: discarding them results in biased estimates.

The most informative cases are those where the child's genotype is compatible with those of the mother and of the putative father on all markers but one. Table 8.1 represents just such a case, with an incompatibility on marker vWA. Taking PI $\geq$ 0.999 as the retention criterion, this case would be considered as contributing one mutation event. But in fact paternity is not certain, and while mutation is one plausible explanation of the observed incompatibility, another possible explanation is non-paternity – in which event the case would supply essentially no information about the mutation rate. In other words, the possibility of *uncertain paternity* is a confounding influence[4] that must be taken properly into account when estimating mutation rates (Vicard and Dawid 2004; Vicard *et al.* 2004; 2008).

8.6.1 Pictorial representation

We have seen that, when using paternity casework data, we face the problem of interaction between the processes of estimating marker-specific mutation rates,

[4] Still other sources of bias affect the naïve assessment of mutation rate, for example the possibility of *hidden mutations*, that do not result in an apparent incompatibility.

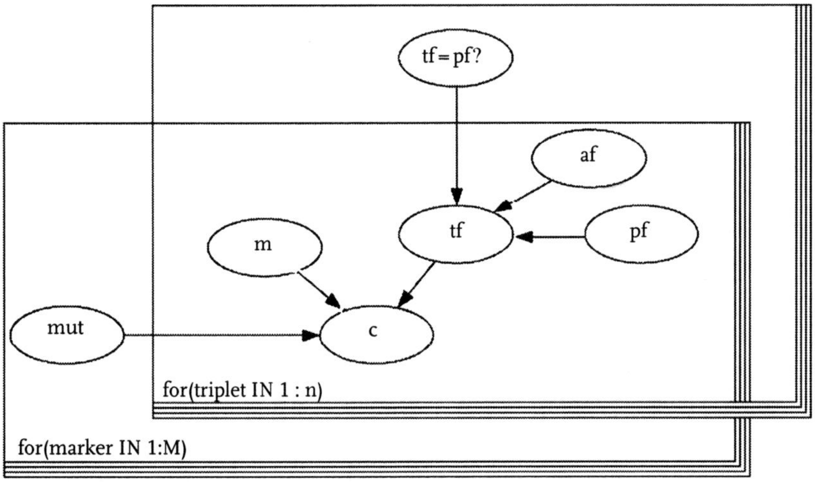

Fig. 8.2 Graph representing the two learning levels of the mutation rate estimation problem.

on the one hand, and of computing the probability of paternity for each case, on the other.

Suppose that, in a full or random subset of the casework of a forensic laboratory, M STR markers are each measured on n triplets. The estimation problem is represented in Figure 8.2, using the graphical syntax of the WINBUGS[5] software. The figure has two 'plates': one for the stack of M markers, and one for the stack of n triplets. There are two principal targets of our inference: a total mutation parameter (node `mut`) for each marker (common to all cases), and the paternity status (`tf=pf?`) for each case (common to all markers). For any combination of case and marker, we have the structure displayed in the intersection of the two plates, essentially as in Figure 8.1; the values of, and data on, these nodes will vary from one case-marker combination to another.

If we assumed all the mutation rates known, we could readily assess the paternity status for each case; conversely, if we knew the paternity status in each case we could easily estimate all the mutation rates. The complication is that, when both these inferential targets are unknown, assessment of each must take appropriate account of the uncertainty in the other.

8.7 The likelihood for the mutation rate

Here we consider the use of previous paternity casework data `old` to estimate the total mutation rate τ for a given marker m, under an assumed scalar mutation model (assuming identical mutation rates in the maternal and the paternal

[5] `http://www.mrc-bsu.cam.ac.uk/bugs/winbugs/contents.shtml`

lines). In contrast to most current practice we suppose we have unselected data, but take appropriate account of the uncertainty over paternity in the constituent cases to develop the correct likelihood function $L(\tau) \propto Pr(\texttt{old} \mid \tau)$. Once this has been obtained it can be combined with a prior density $p(\tau)$ for τ to produce the posterior density $p(\tau \mid \texttt{old})$, from which we can calculate the posterior expectation $E(\tau \mid \texttt{old})$ for insertion in (8.18), so yielding the appropriate paternity index for the case at hand.

The casework old comprises mother – putative-father – child triplets, on each of which we have information ('findings') about their DNA profiles. We suppose that these cases can be regarded as exchangeable with the new case under analysis. We shall also assume, what will typically be the case, that each of the old cases is either *compatible* (i.e. the observed genotypes could have arisen from a true family under the assumption of Mendelian inheritance without mutation) at each marker; or there is a single marker exhibiting incompatibility, which moreover could be explained by mutation in just one of the maternal or paternal lines.

8.7.1 General structure

It is shown in Appendix B that, under reasonable assumptions, the joint likelihood for the collection of mutation parameters $\tau = (\tau_m)$ across all markers can be well approximated as a product of terms, one for each τ_m. If we regard the mutation parameters for different markers as *a priori* independent, the Bayesian analysis can be effected for each parameter separately, its posterior density being proportional to its prior density multiplied by the relevant factor of the joint likelihood. We therefore concentrate on a fixed marker m, which we now drop from the notation. The overall likelihood term for its parameter τ is a product over the individual cases, the contribution from a case with findings f being

$$\ell(\tau) \propto \pi^* \; Pr(f \mid \tau, P) + (1 - \pi^*) \; Pr(f \mid \tau, \bar{P}), \tag{8.19}$$

where π^* is the probability of paternity for the case, calculated from the findings on all *other* markers, assuming no mutation.

Finally we multiply all the likelihood contributions across the different cases to obtain the overall likelihood for τ.

8.7.2 Likelihood contributions

When we have complete genotype data on a triplet, the likelihood contribution for the case is given by (8.7) if we can assume paternity, or (8.8) assuming non-paternity. In either case this is linear in τ, either exactly or to an excellent approximation. It follows that $\ell(\tau)$, calculated from (8.19), can be taken to be linear in τ, with coefficients that are readily calculated algebraically.

In Section 8.4, where we calculated the PI for the single new case at hand, the linear term in (8.8) was ignored, as negligible in comparison with the leading constant term. Here however, where we may have large numbers of compatible cases, it is important to retain it.

In certain cases we may not have full genotype information available from all three individuals in the triple. For example, we might only have a report that, for the marker concerned, the genotypes of the triple are or are not compatible with paternity. For such a case we will still have an essentially linear form for $\ell(\tau)$. Other examples of partial data arise when we only have genotypes from one of the two 'parents'. Again we will have a linear likelihood contribution.

8.7.3 Compatible cases

It is shown in Vicard *et al.* (2008) that for a case that is simply reported as showing 'full compatibility', i.e. compatibility at all markers, to a good enough approximation we can assume paternity; while for a case reported as 'locally compatible' (i.e. compatible at the marker considered, but incompatible at some other marker), we can assume non-paternity.

For such cases we can express the associated likelihood contribution as

$$\ell(\tau) \propto 1 - \alpha\tau \tag{8.20}$$

for full compatibility, and

$$\ell(\tau) \propto 1 - \frac{1}{2}\beta\tau \tag{8.21}$$

for local compatibility. With only partial data, computation of the coefficients α and β in (8.20) and (8.21) can not reasonably be effected algebraically. Vicard *et al.* (2008) (see also Vicard *et al.* 2004) describes the use of Bayesian network technology (Cowell *et al.* 1999) to determine these coefficients numerically.

For a case with genotype data on the child and just one 'parent', this will typically be a 'fully compatible pair'. The likelihood will then be of the form

$$\ell(\tau) \propto 1 - \frac{1}{2}\gamma\tau. \tag{8.22}$$

Once again we would generally need to determine the constant γ numerically, e.g. by using a Bayesian network.

8.7.4 Incompatible cases

The number of incompatible cases in the data will typically be small (it is four for our dataset). We suppose we have complete genotype data on all of these.

For an incompatible case $Pr(f \mid \tau)$ will typically be increasing in τ, and thus expressible as

$$\ell(\tau) \propto a + \tau + O(\tau^2) \tag{8.23}$$

Table 8.3 Prior probability π^* of paternity, and intercept a of the likelihood contribution $\ell(\tau) \approx a + \tau$, for each of four incompatible cases.

	Case 1	Case 2	Case 3	Case 4
π^*	0.99984	0.99959	0.99920	0.99612
a	0.00002	0.00037	0.00029	0.05228

where the intercept a, which will depend on the detailed findings for the case, determines the behaviour of the associate likelihood contribution.

Plausible values of τ typically range from 0.001 to 0.014. When the intercept $a > 0.02$, the likelihood contribution (8.23) is effectively constant in this region, and the case can thus be discarded as essentially uninformative about τ. On the other hand, when $a < 0.001$, $\ell(\tau)$ can be well approximated by τ: this is equivalent to taking paternity as confirmed. These are essentially the only two options that have typically been considered. However, in the intermediate range, $0.001 < a < 0.02$, the function $\ell(\tau)$ is not well approximated either by a constant or by τ. Then, rather than either discard the case or assume paternity, we need to compute and use its correct likelihood contribution, $\ell(\tau) \approx a + \tau$.

8.8 Data analysis for mutation rate

Professor Bernd Brinkmann, of Institut für Rechtsmedizin der Westfälischen Wilhelms-Universität, Münster, kindly provided us with partial information on Austrian-German casework data collected at his laboratory. Although this is apparently not a random sample from all of the casework, having been subject to some preselection, for illustrative purposes we analyse it as if it were. It does not include the case described in Table 8.1; for the purposes of analysing that case, it can be regarded as `old` data and used for estimating the mutation rate for vWA.

These `old` cases contain 2013 meioses in all. We have 943 fully compatible triplets (each involving two meioses); 19 triplets apparently incompatible on one or more markers other than vWA; 40 fully compatible mother – child pairs (involving a single meiosis); and 41 fully compatible putative father – child pairs. There are also four triplets that show apparent incompatibility at marker vWA. We use full genotype data for these four incompatible cases; but, not having access to the full data for the remaining cases, we have analysed these on the basis of summary information only, as described in Section 8.7.

For each of the four incompatible cases we have computed the probability π^* based on the markers other than vWA,[6] assuming a 50% initial probability of paternity. These values are shown in Table 8.3. Taking 0.999 as a typical

[6] For illustrative purposes we have used only a subset of the other markers for case 4.

Table 8.4 Coefficients α, β and γ.

α	β	γ
0.788	0.616	0.561

Table 8.5 Naïve estimate, MLE and posterior mean of τ with *Unif*(0, 1) and *Be*(0.5, 0.5) priors.

	Naïve	MLE	*Unif*(0, 1) prior	*Be*(0.5, 0.5) prior
Estimate of τ	0.0030	0.0038	0.0051	0.0044

threshold, we see that the case 4 would be discarded by the preselection process described in Section 8.6. The resulting naïve estimate of τ, defined as twice the observed fraction of meioses that are deemed to be a mutation, is then given as $2 \times 3/2011 = 0.0030$.

For the Bayesian analysis we have to compute the posterior probability density of τ, $p(\tau \mid \texttt{old}) \propto L(\tau)\, p(\tau)$, where $L(\tau)$ is the overall likelihood based on `old`, and $p(\tau)$ is the prior density. Using (8.20)–(8.23), $L(\tau)$ can be approximated by

$$(1-\alpha\tau)^{943}\left(1-\frac{1}{2}\beta\tau\right)^{19}\left(1-\frac{1}{2}\gamma\tau\right)^{81}\prod_{i=1}^{4}(a_i+\tau).$$

The intercepts a_i for the four incompatible cases, and coefficients α, β and γ, are given in Table 8.3 and Table 8.4 respectively. From Table 8.3 we see that in the first three cases paternity can be taken as confirmed, whereas the fourth case is essentially uninformative.[7]

The posterior density of τ, based on a *Unif*(0, 1) or a *Be*(0.5, 0.5) prior, is shown in Figure 8.3. The behaviour of these priors outside the range (0, 0.02) where the likelihood is non-negligible is of no consequence. Within that range, *Unif*(0, 1) is essentially non-committal, allowing the data to speak for themselves; whereas *Be*(0.5, 0.5) favours smaller over larger values. Correspondingly, using the *Be*(0.5, 0.5) prior the posterior favours smaller values of τ than the posterior based on the *Unif*(0, 1) prior.

We compute the posterior mean of τ by numerical integration. This is shown in Table 8.5, together with the naïve and the maximum likelihood estimates. We see that, compared to the Bayesian estimates, the naïve estimate exhibits considerable negative bias, and the MLE some negative bias.

[7] However, for other values of the mixing parameter h, case 4 *is* informative for mutation (Vicard *et al.* 2008).

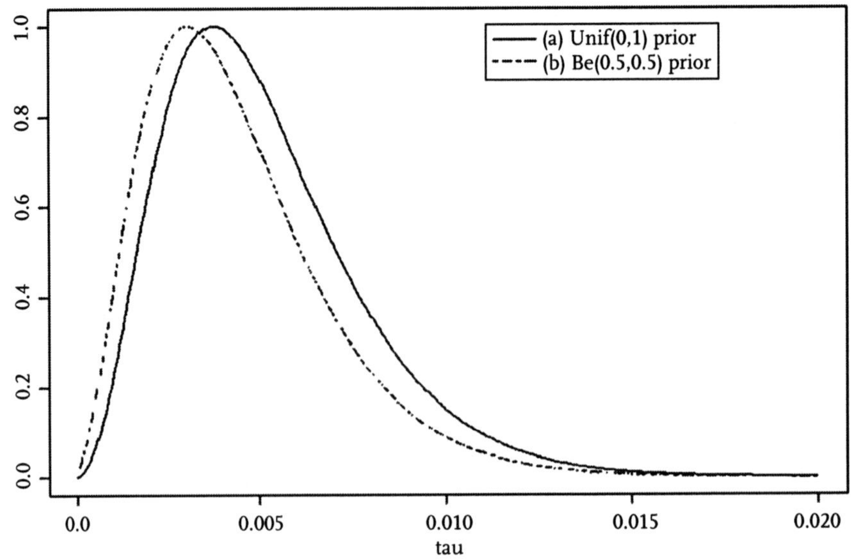

Fig. 8.3 Posterior densities (unnormalized) for τ.

8.9 Application to new case

We now apply the above statistical analysis to solve the case at hand, as described in Section 8.2.2. The PIs for markers other than vWA are given in Table 8.1. Using the values in Table 8.5, we derive by straightforward application of (8.12) the appropriate PI for vWA, and the total paternity index TPI. These are reported in Table 8.6. We see that in all cases the TPI is greater than 0.999, which strongly confirms paternity.

8.10 Further discussion

In this article we have shown how to analyse a disputed paternity case, where the DNA genotypes are compatible on all markers but one, allowing for the possibility of mutation, when the mutation rate is itself uncertain.

Table 8.6 PI for marker vWA corresponding to MLE and posterior means of τ with *Unif*(0, 1) and *Be*(0.5, 0.5) priors.

	MLE	*Unif*(0, 1) prior	*Be*(0.5, 0.5) prior
PI for vWA	0.0105	0.0139	0.0120
TPI	0.99969	0.99977	0.99973

There are other complications and uncertainties that can affect DNA paternity cases. One fairly common situation arises when we are unable to obtain profiles from all three 'family' members – typically this will be because the putative father is unavailable, or refuses to provide a DNA sample. In such a case we may be able to obtain indirect evidence by typing his relatives, e.g. parents, siblings, or undisputed children either by the same or by a different mother.

Another possible complication is that the instrumentation used to amplify and read a DNA genotype may not work perfectly, which can result in an allele that is truly present not being observed: this is called a 'null allele'. If, for example, the true genotype is heterozygous *Aa*, but the allele *a* can not be read, this could be mistaken for the homozygous genotype *AA*. A null allele can be due to sporadic failure of the measuring device (a 'missing allele'). Alternatively it can be due to intrinsic features of the biological amplification process (a 'silent allele'.) Silence of an allele is inherited in the usual way, whereas missingness is not.

We have also assumed here, unrealistically, that all founding genes have been drawn, independently, from defined 'gene pools' with known allele frequencies. Green and Mortera (2009) discuss a range of methods for dealing with uncertain allele frequencies, and/or non-independence, due, for example, to possible relatedness between founders, or subpopulation structure in a non-randomly mixing population.

For these more complicated scenarios, even with the steadying hand of statistical logic to guide our assessment of the strength and impact of the evidence, the detailed analysis can become complex and unrevealing, and to handle such problems sophisticated probabilistic modelling tools are generally required. *Bayesian networks* (Cowell *et al.* 1999), together with their associated computational methodology and technology, have been found valuable for this, particularly in their 'object-oriented' (OOBN) form. Bayesian networks for evaluating DNA evidence were introduced by Dawid *et al.* (2002); further descriptions and developments can be found in Mortera (2003); Mortera *et al.* (2003); Cowell *et al.* (2007); Dawid *et al.* (2006, 2007); Taroni *et al.* (2006). Vicard *et al.* (2004) show how to construct and apply Bayesian networks to compute the relevant likelihood for the problem considered in this chapter when only partial DNA information on the basic triplet is available.

A particularly valuable feature of Bayesian networks is their modular structure, which allows a complex problem to be broken down into simpler pieces that can then be pieced back together in different ways, so allowing us to solve a wide range of forensic queries while incorporating several complications simultaneously. Dawid *et al.* (2007) construct such generic OOBNs for paternity analysis allowing for the simultaneous possibility of unobserved actors, silent or missing alleles, and mutation. An extension to incorporate uncertain allele frequencies has also been constructed, and will be applied to mutation rate estimation in future work.

Appendix

A. Broader context and background

A.1 Statistics and the law

In recent years statistical arguments (though all too often misconceived) have played an increasingly important rôle in legal cases, both criminal and civil (Dawid 2005; 2008). Indeed, at a high level there are many similarities between the tasks faced by statistics and the law, if not in the ways they have traditionally gone about addressing them.

A.1.1 Bayesian approach Evidence presented in a case at law can be regarded as data, and the issue to be decided by the court as a hypothesis under test. There is uncertainty about the ultimate issue, the evidence, and the way in which these are related. Such uncertainty can, in principle at least, be described probabilistically. In a legal setting, where it is understood that different 'reasonable men' (and women) can reasonably hold a range of opinions, it is natural to take a subjective Bayesian interpretation of probability, regarding it as a measure of a specific individual's uncertainty about the unique event at issue, in the light of the relevant evidence.

Let $\mathcal{E}$ denote one or more items of evidence (perhaps its totality). We need to consider how this evidence affects the comparison of the hypotheses, H_0 and H_1 say, offered by either side. Thus in a criminal case with a single charge against a single defendant, the evidence might be that the defendant's DNA profile matches one found at the crime scene; hypothesis H_0, offered by the defence, is that the defendant is innocent ($\overline{G}$); the prosecution hypothesis, H_1, is that of guilt (G).

A juror needs to assess his or her conditional probability for either hypothesis, *given* the evidence: $Pr(H_0|\mathcal{E})$ and $Pr(H_1|\mathcal{E})$. However, it will typically be more reasonable to assess directly $Pr(\mathcal{E}|H_0)$ and $Pr(\mathcal{E}|H_1)$: the probability that the evidence would have arisen, under each of the competing scenarios. These can then be inserted into Bayes's theorem, expressed in the form

$$\frac{Pr(H_1|\mathcal{E})}{Pr(H_0|\mathcal{E})} = \frac{Pr(H_1)}{Pr(H_0)} \times \frac{Pr(\mathcal{E}|H_1)}{Pr(\mathcal{E}|H_0)}. \tag{8.24}$$

We can also express this in words as:

POSTERIOR ODDS = PRIOR ODDS × LIKELIHOOD RATIO.

The likelihood ratio is often constructed from frequency data and other reasonably objective ingredients, which are suitable for presentation as expert evidence. The prior odds, on the other hand, will generally contain an irreducibly subjective element, and it is not usually appropriate for experts to express their own opinions about them. Thus for the simple paternity analysis of Section 8.2, the likelihood ratio is the paternity index, PI. An expert witness could present this in court, explaining the relevant logic of Essen-Möller (1938), and arguing for the validity of its assumptions, and for estimates of allele frequencies based on suitable databases. However it would not be appropriate for the expert to opine as to the prior odds of paternity, in the absence of the DNA evidence. And in more complex cases, such as the analysis of Section 8.9, there may be necessarily subjective ingredients even in the likelihood ratio term, due for example to the need to express epistemic uncertainty about unknown parameters. To be suitable for presentation as evidence, such choices should either be justified on the basis of pre-existing data, or shown to have little effect on the conclusions. Alternatively a sensitivity analysis could be presented, varying the inputs across reasonable ranges, and leaving the final choice of value to the judge or juror.

A.1.2 Non-Bayesian arguments Notwithstanding the unarguable correctness of (8.24), it is often replaced by other, more 'intuitive', probabilistic arguments, that can be very misleading. For example, in the trial of Sally Clark for double infanticide (Dawid 2005, 2008), an expert medical witness testified that the probability that both her babies would have died from natural causes was one in 73 million. If we describe this figure as 'the probability that the babies died by innocent means' it is all too easy to misinterpret this as as the probability (on the basis of the evidence of the deaths) that Sally is innocent – such a tiny figure seeming to provide incontrovertible proof of her guilt. Mathematically, this is equivalent to misinterpreting $Pr(\mathcal{E}|\overline{G})$ as $Pr(\overline{G}|\mathcal{E})$. For obvious reasons this error is known as 'transposing the conditional', or, because it typically produces seemingly convincing evidence of guilt, 'the prosecutor's fallacy'. The prosecutor's fallacy has also been prominent in DNA identification cases, where the (generally exceedingly small) probability of the DNA match having arisen by chance, if the suspect were in fact innocent, is all too readily misinterpreted (whether through ignorance or malice) as the probability that the suspect is innocent, on the basis of the DNA match evidence.

A.2 *Genetic background*

Here we give a very brief outline of some basic genetic facts about DNA profiles: for more details see for example, Butler (2005).

For our purposes, a *gene* is simply an identified stretch of DNA, with values which are sequences of the four constituent bases, represented by the letters A,

C, G and T. A specific position on a chromosome is called a *locus* (hence there are two genes at any locus of a chromosome pair). A *DNA profile* consists of measurements on a number of *forensic markers*, which are specially selected loci on different chromosomes. Current technology uses around 8–20 *short tandem repeat* (STR) markers. Each such marker has a finite number (up to around 20) of possible values, or *alleles*, generally positive integers. For example, an allele value of 5 indicates that a certain word (e.g. CAGGTG) in the four letter alphabet is repeated exactly 5 times in the DNA sequence at that locus.

In statistical terms, a gene is represented by a random variable, whose realised value is an allele.

In a particular forensic case we may refer to the various human individuals involved in the case as 'actors'. An actor's *DNA profile* comprises a collection of *genotypes*, one for each marker. Each genotype consists of an unordered pair of genes, one inherited from the actor's father and one from the actor's mother (though one cannot distinguish which is which). When both alleles are identical the actor is *homozygous* at that marker, and only a single allele value is observed; otherwise the actor is *heterozygous*.

According to *Mendel's law of segregation*, at each marker a parent passes a copy of just one of his or her two alleles, randomly chosen, to his or her child, independently of the other parent and independently for each child.

For a number of populations, databases have been gathered from which allele frequency distributions of the various forensic markers can be estimated.

B. The joint likelihood

Here we consider the contribution, from a single case, to the joint likelihood of all mutation parameters, and show that to a good approximation it can be regarded as factorising into a number of terms, one for each marker. For simplicity we only consider cases that are fully compatible with maternity, and compatible with paternity except perhaps at a single marker – typically all cases in the data will be of this form. (More complex possibilities are considered by Vicard *et al.* 2004.) We suppose throughout that no doubt attaches to the issue of maternity, but we wish to account for the possibility that the putative father may not be the true father.

B.1 The likelihood

Suppose we have data on markers $1, \ldots, M$. For each marker m we entertain a scalar mutation model with unknown total mutation rate τ_m, any other parameters being supposed known.[8] The totality of unknown parameters is $\tau := (\tau_m : m = 1, \ldots, M)$.

[8] This requirement can be relaxed: see Vicard *et al.* (2004).

Denote by S the unobserved binary paternity status indicator: $S = P$ if the putative father is the true father, $S = \bar{P}$ if not. We denote the prior probability of paternity before taking account of any of the DNA information, $Pr(P)$, by π_0, and assume $0 < \pi_0 < 1$.

Let the findings for this case on marker m be f_m: these might give details of the genotypes of each party, or only cruder information such as compatibility/non-compatibility. We write the totality of findings on the case as $f := (f_m : m = 1, \ldots, M)$. We denote $(f_i : i \neq m)$ by $f_{\setminus m}$, etc.

We make the following reasonable assumptions, expressed in the symbolism of conditional independence (Dawid 1979):

(a) By itself, paternity status S is uninformative about mutation:

$$S \perp\!\!\!\perp \tau.$$

(b) Given τ and S, the markers behave independently:

$$\perp\!\!\!\perp \{f_m : m = 1, \ldots, M\} \mid \tau, S.$$

(c) Each mutation parameter only affects its associated marker:

$$f_m \perp\!\!\!\perp \tau_{\setminus m} \mid \tau_m, S.$$

It follows readily from assumptions (a)–(c) that the joint likelihood function $\ell(\tau) = Pr(f \mid \tau)$, based on this case, for the full parameter τ is given by:

$$\ell(\tau) = \pi_0 \prod_{m=1}^{M} Pr(f_m \mid \tau_m, P) + (1 - \pi_0) \prod_{m=1}^{M} Pr(f_m \mid \tau_m, \bar{P}). \tag{B.1}$$

B.2 Factorization

Because we are assuming compatibility with maternity, $Pr(f_m \mid \tau_m = 0, \bar{P}) > 0$ for each m; hence, since $\pi_0 < 1$, we have $\ell(0) > 0$. Applying a Taylor expansion to $\log \ell(\tau)$, we see that we can assume an approximate factorized form

$$\ell(\tau) \propto \prod_{m=1}^{M} a_m(\tau_m) \, \{1 + o(\tau)\} \tag{B.2}$$

with each $a_m(0) > 0$.

From (B.1) and (B.2), for any m we can express

$$\pi_0 \, Pr(f_{\setminus m} \mid \tau_{\setminus m}, P) \, Pr(f_m \mid \tau_m, P) + (1 - \pi_0) \, Pr(f_{\setminus m} \mid \tau_{\setminus m}, \bar{P}) \, Pr(f_m \mid \tau_m, \bar{P})$$
$$= a_m(\tau_m) \, a_{\setminus m}(\tau_{\setminus m}) \, \{1 + o(\tau)\} \tag{B.3}$$

with $a_{\backslash m}(0) > 0$. Setting $\tau_{\backslash m} = 0$ in (B.3) we see that we can take

$$\begin{aligned} a_m(\tau_m) &\propto p^*_{\backslash m}(f_m \mid \tau_m) \\ &:= \pi^*_{\backslash m}\ Pr(f_m \mid \tau_m,\ P) + \left(1 - \pi^*_{\backslash m}\right)\ Pr(f_m \mid \tau_m, \bar{P}), \end{aligned} \tag{B.4}$$

where

$$\begin{aligned} \pi^*_{\backslash m} &= \frac{\pi_0\ Pr(f_{\backslash m} \mid \tau_{\backslash m} = 0,\ P)}{\pi_0\ Pr(f_{\backslash m} \mid \tau_{\backslash m} = 0,\ P) + (1 - \pi_0)\ Pr(f_{\backslash m} \mid \tau_{\backslash m} = 0, \bar{P})} \\ &= Pr(P \mid \tau_{\backslash m} = 0,\ f_{\backslash m}), \end{aligned} \tag{B.5}$$

the posterior probability of paternity based on the findings on all markers other than m, on the assumption of the mutation rate being zero for these. That is, $p^*_{\backslash m}(f_m \mid \tau_m)$ is the marginal probability of obtaining the findings f_m, as a function of the mutation parameter τ_m, when the 'prior' probability of paternity is taken as $\pi^*_{\backslash m}$.

Whenever there is some marker j on which the findings f_j are incompatible with paternity, then, for any other marker m, $\pi^*_{\backslash m} = 0$, so that we can just take $a_m(\tau_m) = Pr(f_m \mid \tau_m, \bar{P})$, i.e. we can assume non-paternity. This is exactly equivalent to just having mother – child data. Such cases must not be ignored, since they indicate that a possible mutation in the maternal line has *not* occurred, and so constitute evidence in favour of a smaller mutation rate. When all markers other than m exhibit compatibility, then, for complete triplet data, $\pi^*_{\backslash m}$ is readily calculated by the formula of Essen-Möller (1938); for reduced findings, such as simple compatibility/non-compatibility, $\pi^*_{\backslash m}$ could be obtained by using a collection of propagations in suitable Bayesian networks, one for each other marker, to calculate likelihood ratios in favour of paternity (ignoring mutation) after entering the reduced evidence.

We have thus shown that the joint likelihood contribution $\ell(\tau)$ from a single case has the form

$$\ell(\tau) \propto \prod_{m=1}^{M} \ell_m(\tau_m) \times \left\{1 + o(\tau)\right\}, \tag{B.6}$$

with

$$\ell_m(\tau_m) := p^*_{\backslash m}(f_m \mid \tau_m). \tag{B.7}$$

B.3 *The overall likelihood*

The overall joint likelihood function $L(\tau)$, based on all cases in the data-set, is the product of all the various case-specific joint likelihood terms, $\ell(\tau)$. Assuming we have extensive data that are reasonably consistent with existing knowledge of the range of plausible values for τ, $L(\tau)$ will be negligible unless each $\tau_m < 0.01$ (roughly). Within this range the approximation which ignores the $o(1)$ term in (B.2) is likely to be excellent, so that the overall likelihood can likewise be regarded as factorising in this region; outside this range both the correct and the approximate overall likelihood are in any case negligible. Even with non-extensive or untypical data, any realistic prior distribution would serve to discount values for τ outside this region. Consequently it should be safe to proceed as if we had a factorization

$$L(\tau) \propto \prod_{m} L_m(\tau_m), \tag{B.8}$$

with each marker-specific factor $L_m(\tau_m)$ obtained as a product, over the cases in the data-set, of the terms $\ell_m(\tau_m)$, given by (B.7). For each case the factor $\ell_m(\tau_m)$ can be found by treating the findings f_m on marker m as if they were the full data available, after first updating the paternity probability from π_0 to $\pi^*_{\setminus m}$, to take account of the findings on the other markers.

If we assume that there are no a priori relations between the mutation rates across markers, inference for the different rates can thus proceed completely separately, each based on its own likelihood factor of the form L_m.

References

American Association of Blood Banks (2002). Annual report summary for testing in 2001. Online at `http://www.aabb.org/Documents/Accreditation/Parentage_Testing_Accreditation_Program/ptannrpt01.pdf`.

Anderson, T. J., Schum, D. A., and Twining, W. L. (2005). *Analysis of Evidence*, 2nd edn. Cambridge University Press, Cambridge.

Brinkmann, B., Klintschar, M., Neuhuber, F., Hühne, J., and Rolf, B. (1998). Mutation rate in human microsatellites: Influence of the structure and length of the tandem repeat. *American Journal of Human Genetics*, **62**, 1408–1415.

Butler, J. M. (2005). *Forensic DNA Typing*. Elsevier, New York.

Butler, J. M., Schoske, R., Vallone, P. M., Redman, J. W., and Kline, M. C. (2003). Allele frequencies for 15 autosomal STR loci on U.S. Caucasian, African American and Hispanic populations. *Journal of Forensic Sciences*, **48**, 908–911. `www.astm.org/JOURNALS/FORENSIC/PAGES/4412.htm`.

Cowell, R. G., Dawid, A. P., Lauritzen, S. L., and Spiegelhalter, D. J. (1999). *Probabilistic Networks and Expert Systems*. Springer, New York.

Cowell, R. G., Lauritzen, S. L., and Mortera, J. (2007). A gamma model for DNA mixture analyses. *Bayesian Analysis*, **2**, 333–348.

Dawid, A. P. (1979). Conditional independence in statistical theory (with Discussion). *Journal of the Royal Statistical Society, Series B*, **41**, 1–31.

Dawid, A. P. (2005). Probability and proof. On-line appendix to Anderson *et al.* (2005). `http://tinyurl.com/7g3bd`.

Dawid, A. P. (2008). Statistics and the law. In *Evidence* (ed. A. Bell, J. Swenson-Wright, and K. Tybjerg), pp. 119–48. Cambridge University Press, Cambridge.

Dawid, A. P. and Mortera, J. (1996). Coherent analysis of forensic identification evidence. *Journal of the Royal Statistical Society, Series B*, **58**, 425–443.

Dawid, A. P. and Mortera, J. (1998). Forensic identification with imperfect evidence. *Biometrika*, **85**, 835–849.

Dawid, A. P., Mortera, J., Dobosz, M., and Pascali, V. L. (2003). Mutations and the probabilistic approach to incompatible paternity tests. In *Progress in Forensic Genetics 9*, International Congress Series, Vol. 1239 (ed. B. Brinkmann and A. Carracedo), pp. 637–638. Elsevier Science, Amsterdam. International Society for Forensic Genetics. `doi:10.1016/S0531-5131(02)00845-2`.

Dawid, A. P., Mortera, J., and Pascali, V. L. (2001). Non-fatherhood or mutation? A probabilistic approach to parental exclusion in paternity testing. *Forensic Science International*, **124**, 55–61.

Dawid, A. P., Mortera, J., Pascali, V. L., and van Boxel, D. W. (2002). Probabilistic expert systems for forensic inference from genetic markers. *Scandinavian Journal of Statistics*, **29**, 577–595.

Dawid, A. P., Mortera, J., and Vicard, P. (2006). Representing and solving complex DNA identification cases using Bayesian networks. In *Progress in Forensic Genetics 11*, International Congress Series, Vol. 1288 (ed. A. Amorim, F. Corte-Real and N. Morling), pp. 484–491. Elsevier, Amsterdam. doi:10.1016/j.ics.2005.09.115.

Dawid, A. P., Mortera, J., and Vicard, P. (2007). Object-oriented Bayesian networks for complex forensic DNA profiling problems. *Forensic Science International*, **169**, 195–205.

Durret, R. and Kruglyak, S. (1999). A new stochastic model of microsatellite evolution. *Journal of Applied Probability*, **36**, 621–631.

Egeland, T. and Mostad, P. F. (2002). Statistical genetics and genetical statistics: A forensic perspective. *Scandinavian Journal of Statistics*, **29**, 297–308.

Essen-Möller, E. (1938). Die Beweiskraft der Ähnlichkeit im Vaterschaftsnachweis. Theoretische Grundlagen. *Mitteilungen der Anthropologischen Gesellschaft*, **68**, 9–53.

Fimmers, R., Henke, L., Henke, J., and Baur, M. (1992). How to deal with mutations in DNA testing. In *Advances in Forensic Haemogenetics 4* (ed. C. Rittner and P. M. Schneider), pp. 285–287. Springer-Verlag, Berlin.

Green, P. J. and Mortera, J. (2009). Sensitivity of inference in Bayesian networks to assumptions about founding genes. *Annals of Applied Statistics*, **3**, 731–763.

Henke, L. and Henke, J. (1999). Mutation rate in human microsatellites. *American Journal of Human Genetics*, **64**, 1473. With reply by B. Rolf and B. Brinkmann, 1473–1474.

Jeffreys, A. J., Wilson, V., and Thein, S. L. (1985*a*). Hypervariable 'minisatellite' regions in human DNA. *Nature*, **314**, 67–73.

Jeffreys, A. J., Wilson, V., and Thein, S. L. (1985*b*). Individual-specific 'fingerprints' of human DNA. *Nature*, **316**, 76–79.

Mortera, J. (2003). Analysis of DNA mixtures using Bayesian networks. In *Highly Structured Stochastic Systems* (ed. P. J. Green, N. L. Hjort, and S. Richardson), Chapter 1B, pp. 39–44. Oxford University Press, Oxord.

Mortera, J., Dawid, A. P., and Lauritzen, S. L. (2003). Probabilistic expert systems for DNA mixture profiling. *Theoretical Population Biology*, **63**, 191–205.

Ohta, T. and Kimura, M. (1973). A model of mutation appropriate to estimate the number of electrophoretically detectable alleles in a finite population. *Genetical Research*, **22**, 201–204.

Sajantila, A., Lukka, N., and Syvanen, A. C. (1999). Experimentally observed germline mutations at human micro- and minisatellite loci. *European Journal of Human Genetics*, **7**, 263–266.

Schmid, D., Anslinger, K., and Rolf, B. (2005). Allele frequencies of the ACTBP2 (=SE33), D18S51, D8S1132, D12S391, D2S1360, D3S1744, D5S2500, D7S1517, D10S2325 and D21S2055 loci in a German population sample. *Forensic Science International*, **151**, 303–305.

Taroni, F., Aitken, C., Garbolino, P., and Biedermann, A. (2006). *Bayesian Networks and Probabilistic Inference in Forensic Science*, Statistics in Practice. John Wiley, Chichester.

Valdes, A. M., Slatkin, M., and Freimer, N. B. (1993). Allele frequencies at microsatellite loci: The stepwise mutation model revisited. *Genetics*, **133**, 737–749.

Vicard, P. and Dawid, A. P. (2004). A statistical treatment of biases affecting the estimation of mutation rates. *Mutation Research*, **547**, 19–33.

Vicard, P., Dawid, A. P., Mortera, J., and Lauritzen, S. L. (2004). Estimation of mutation rates from paternity cases using a Bayesian network. Research Report 249, Department of Statistical Science, University College London.

Vicard, P., Dawid, A. P., Mortera, J., and Lauritzen, S. L. (2008). Estimating mutation rates from paternity casework. *Forensic Science International: Genetics*, **2**, 9–18.

Vorkas, S. (2005). Estimation of mutation rates using a Bayesian network. BSc Dissertation, Department of Statistical Science, University College London.

Whittaker, J. C., Harbord, R. M., Boxall, N., Mackay, I., Dawson, G., and Sibly, R. M. (2003). Likelihood-based estimation of microsatellite mutation rates. *Genetics*, **164**, 781–787.

PART II
Industry, Economics and Finance

·9·

Bayesian analysis and decisions in nuclear power plant maintenance

Elmira Popova, David Morton, Paul Damien and Tim Hanson

9.1 Introduction

It is somewhat true that in most mainstream statistical literature the transition from inference to a formal decision model is seldom explicitly considered. Since the Markov chain Monte Carlo (MCMC) revolution in Bayesian statistics, focus has generally been on the development of novel algorithmic methods to enable comprehensive inference in a variety of applications, or to tackle realistic problems which naturally fit into the Bayesian paradigm. In this chapter, the primary focus is on the formal decision or optimization model. The statistical input needed to solve the decision problem is tackled via Bayesian parametric and semiparametric models. The optimization/Bayesian models are then applied to solving an important problem in a nuclear power plant system at the South Texas Project (STP) Electric Generation Station.

STP is one of the newest and largest nuclear power plants in the US, and is an industry leader in safety, reliability and efficiency. STP has two nuclear reactors that together can produce 2500 megawatts of electric power. The reactors went online in August 1988 and June 1989, and are the sixth and fourth youngest, respectively, of more than 100 such reactors operating nationwide. STP consistently leads all US nuclear plants in the amount of electricity its reactors produce.

The STP Nuclear Operating Company (STPNOC) manages the plant for its owners, who share its energy in proportion to their ownership interests, which as of July 2004 are: Austin Energy, The City of Austin, 16%, City Public Service of San Antonio, 40%, and Texas Genco LP, 44%. All decisions that the board of directors make are of finite time since every nuclear reactor is given a *license* to operate. In the case of STP, 25 and 27 years remain on the licenses for the two reactors, respectively.

Equipment used to support production in long-lived (more than a few years) installations such as those at STP requires maintenance. While maintenance is being carried out, the associated equipment is typically out of service and the system operator may be required to reduce or completely curtail production.

In general, maintenance costs include labor, parts and overhead. In some cases, there are safety concerns associated with certain types of plant disruptions. Overhead costs include hazardous environment monitoring and mitigation, disposal fees, license fees, indirect costs associated with production loss such as wasted feed stock, and so forth.

STPNOC works actively with the Electric Power Research Institute (EPRI) to develop robust methods to plan maintenance activities and prioritize maintenance options. Existing nuclear industry guidelines, Bridges and Worledge (2002), Gann and Wells (2004), Hudson and Richards (2004), INPO (2002), recommend estimating reliability and safety performance based on evaluating changes taken one at a time, using risk importance measures supplemented by heuristics to prioritize maintenance. In our view, the nuclear industry can do better. For example, Liming *et al.* (2003) propose instead investigating 'packages' of changes, and in their study of a typical set of changes at STPNOC, the projected plant reliability and nuclear safety estimates were found to be significantly different compared to changes evaluated one at a time. STPNOC is working with EPRI to improve its preventive maintenance reliability database with more accurate probability models to aid in better quantification of preventive maintenance.

Here, we consider a single-item maintenance problem. That said, this chapter's model forms the basis for higher fidelity multi-item maintenance models that we are currently investigating. Furthermore, as we describe below, even though the model is mathematically a single-item model, in practice it could easily be applied to multiple items within an equipment class.

Equipment can undergo either preventive maintenance (PM) or corrective maintenance (CM). PM can include condition-based, age-based, or calendar-based equipment replacement or major overhaul. Also included in some strategies is equipment redesign. In all these categories of PM, the equipment is assumed to be replaced or brought back to the as-purchased condition. CM is performed when equipment has failed unexpectedly in service at a more-or-less random time (i.e. the out-of-service time is not the operator's choice as in PM).

Because such systems are generally required to meet production-run or calendar-based production goals, as well as safety goals, installation operators want assurance that the plant equipment will support these goals. On the other hand, the operator is usually constrained by a limited maintenance budget. As a consequence, operators are faced with the important problem of balancing maintenance costs against production and safety goals.

Thus, the primary decision problem with which we are confronted is interlinked, and could be stated in the form of a question: 'How should one minimize the total cost of maintaining a nuclear power plant while ensuring that its reliability meets the standards defined by the United States Department of Energy?'

There is an enormous literature on optimal maintenance policies for a single item that dates back to the early 1950s. The majority of the work covers maintenance optimization over an infinite horizon, see Valdez-Florez and Feldman (1989) for an extensive review. The problem that we address in this chapter is over a finite planning horizon, which comes from the fact that every nuclear power plant has a license to operate that expires in a finite predefined time. In addition the form of the policy is effectively predefined by the industry as a combination of preventive and corrective maintenance, as we describe below. Marquez and Heguedas (2002) present an excellent review of the more recent research on maintenance policies and solve the problem of periodic replacement in the context of a semi-Markov decision processes methodology. Su and Chang (2000) find the periodic maintenance policies that minimize the life cycle cost over a predefined finite horizon.

A review of the Bayesian approaches to maintenance intervention is presented in Wilson and Popova (1998). Chen and Popova (2000) propose two types of Bayesian policies that learn from the failure history and adapt the next maintenance point accordingly. They find that the optimal time to observe the system depends on the underlying failure distribution. A combination of Monte Carlo simulation and optimization methodologies is used to obtain the problem's solution. In Popova (2004), the optimal structure of Bayesian group-replacement policies for a parallel system of n items with exponential failure times and random failure parameter is presented. The paper shows that it is optimal to observe the system only at failure times. For the case of two items operating in parallel the exact form of the optimal policy is derived.

With respect to the form of the maintenance policy that we consider, commercial nuclear power industry practice establishes a set PM schedule for many major equipment classes. In an effort to avoid mistakenly removing from service sets of equipment that would cause production loss or safety concerns, effectively all equipment that can be safely maintained together are grouped and assigned a base calendar frequency. The grouping policy ensures PM (as well as most CM) will occur at the same time for equipment within a group. The base frequency is typically either set by the refueling schedule (equipment for which at-power maintenance is either impossible or undesirable) or calendar-based. The current thinking for PM is typified by, for example, the INPO guidance, INPO (2002), whereby a balance is sought between maintenance cost and production loss. By properly taking into account the probability of production loss, the cost of lost production, the CM cost and PM cost, the simple model described in this chapter captures industry practice. Not accounted for in the model is the probability and cost of production loss due to PM during at-power operation as well as potential critical path extension for PM during scheduled outages (such as refueling outage). This is partly justified by the fact that PM is only performed during the equipment's group assigned outage time (unlike CM when the equipment outage time is not predetermined) and because outage planning will (generally) assure PM is done off critical path. In practice, there is normally one or two major equipment activities (main turbine and generator, reactor vessel inspection, steam generator inspection, main condenser inspection) that, along with refueling, govern critical path during planned outages.

9.2 Maintenance model

The model we develop has the following constructs. The problem has a finite horizon of length L (e.g. 25 or 27 years). We consider the following (positive) costs: C_{pm} – preventive maintenance cost, C_{cm} – corrective maintenance cost, and C_d – downtime cost, which includes all lost production costs due to a disruption of power generation. Let $N(t)$ be the counting process for the number of failures in the interval $(0, t)$.

Let the random time to failure of the item from its as-new state be governed by distribution F with density f. Further assume that each failure of the item causes a loss of production (i.e. a plant trip) with probability p and in that case a downtime cost, $C_d > C_{cm}$, is instead incurred (this cost can include C_{cm}, if appropriate).

We consider the following form of a maintenance policy, which we denote (P):

(P): Bring the item to the 'as-good-as-new' state every T units of time (preventive maintenance) at a cost of C_{pm}. If it fails meanwhile then repair it to the 'as-good-as-old' state (corrective maintenance) for a cost of C_{cm} or C_d, depending on whether the failure induces a production loss.

The optimization model has a single decision variable T, which is the time between PMs, i.e. we assume constant interval lengths between PMs. The goal is to find $T \in A \subset [0, L]$ that minimizes the total expected cost, i.e.

$$\min_{T\in A} z(T) = \left[C_{pm}\lceil L/T\rceil + \left\{pC_d + (1-p)C_{cm}\right\} \lfloor L/T\rfloor E\left\{N(T)\right\} \right.$$
$$\left. + \left\{pC_d + (1-p)C_{cm}\right\} E\left\{N\left(L - T\lfloor L/T\rfloor\right)\right\}\right], \qquad (9.1)$$

where $A = \{i - \text{integer}, i \in [0, L]\}$, $\lfloor\cdot\rfloor$ is the 'floor' (round-down to nearest integer) operator, $\lceil\cdot\rceil$ is the 'ceiling' (round-up to nearest integer) operator, and $E\{N(T)\}$ is the expected number of failures, taken with respect to the failure distribution, in the interval $(0, T)$. Barlow and Hunter (1960), showed that for the above defined policy, the number of CMs in an interval of length t follows a non-homogeneous Poisson process with expected number of failures in the interval $(0, t)$,

$$E\{N(t)\} = \int_0^t q(u)du \tag{9.2}$$

where $q(u) = \frac{f(u)}{1-F(u)}$ is the associated failure rate function. First we will describe an algorithm to find the optimal T when the failure distribution is IFR (increasing failure rate). Then we model the failure rate nonparametrically.

9.3 Optimization results

The development of successful optimization algorithms in the IFR context detailed in the previous section requires a closer examination of the objective function $z(T)$. This examination will culminate in certain key conditions that will guarantee a solution to the cost minimization problem posed in equation (9.1). To this end, consider the following three propositions in which the form of the IFR function is left unspecified, that is, the theory below will hold for any arbitrary choice of an IFR function; for example, one could choose a Weibull failure rate.

The proofs of the following propositions are given in Appendix B. Also the optimization algorithm corresponding to these propositions, and which was used in the decision analysis aspect of this chapter appears in Appendix B.

Proposition 9.1 Assume that we follow maintenance policy *(P)*, and that the time between failures is a random variable with an IFR distribution, i.e. the failure rate function $q(t)$ is increasing. Then, the objective function $z(T)$ is:

(i) lower semicontinuous with discontinuities at $T = L/n$, $n = 1, 2, \ldots$, and
(ii) increasing and convex on each interval $\left(\frac{L}{n+1}, \frac{L}{n}\right)$, $n = 1, 2, \ldots$

Let $z^c(T)$ be the continuous relaxation of $z(T)$; see Appendix B.

Proposition 9.2 Let $D = \{d : d = L/n, n = 1, 2, \ldots\}$ denote the set of discontinuities of $z(T)$ (cf. Proposition 9.1). Then,

(i) $z^c(T)$ is quasiconvex on $[0, L]$.

Furthermore if $d \in D$ then

(ii) $z^c(d) = z(d)$, and
(iii) $\lim_{T \to d^-}\{z(T) - z^c(T)\} = C_{pm}$.

Proposition 9.3 Let $T_c^* \in \arg\min_{T\in[0,L]} z^c(T)$ and let n^* be the positive integer satisfying $\frac{L}{n^*+1} \leq T_c^* \leq \frac{L}{n^*}$. Then,

$$T^* \in \underset{T\in\left\{\frac{L}{n^*+1},\frac{L}{n^*}\right\}}{\arg\min}\ z^c(T)$$

solves $\min_{T\in[0,L]} z(T)$.

9.4 Data and Bayesian models

The assumption of constant failure rates with the associated exponential failure distribution and homogeneous Poisson process pervades analysis in today's nuclear power industry. Often times, the data gathered in industry are the number of failures in a given time period, which is sufficient for exponential (or constant) failure times. In general, we cannot expect to have a 'complete data set' but rather a collection of failure times and 'success' times (also known as right-censored times). Recognizing the limitations of the exponential failure rate model, the Risk Management Department at STP has implemented the procedure now recommended by the US Department of Energy, see Blanchard (1993). The constant failure rate assumption is dropped in favour of a Bayesian scheme in which the failure rate is assumed to be a gamma random variable, which is the conjugate prior for the exponential distribution, and an updating procedure is defined. This methodology requires gathering, at a local level, the number of failed items in a given month only. The mean of the posterior distribution, combining the local and industry-wide history, of the failure rate is then being used as a forecast of the frequency of failures for the upcoming month.

The hazard function associated with a gamma random variable is monotonically increasing or decreasing to unity as time tends to infinity. However, a Weibull family allows hazards to decay to zero or monotonically increase, and is therefore a common choice in reliability studies when one is uncertain a priori that the instantaneous risk of component failure becomes essentially constant after some time, as in the case of exponential and gamma random variables. Yu *et al.* (2004, 2005) apply Bayesian estimation procedures when the failure rate distribution is assumed Weibull. Thus a Weibull failure rate model is an improvement on the constant failure rate model. But parametric models such as the exponential, gamma, and Weibull models all imply that the hazard rate function is unimodal and skewed right. This has serious consequences for the estimation of the expected value in equation (9.1) if the true underlying failure rate is multi-modal with varying levels of skewness. A semiparametric approach that relaxes such stringent requirements on data while retaining the flavour of these parametric families is therefore warranted.

We first describe the nuclear power plant data to be used in the Bayesian analysis.

9.4.1 Data

The failure data analysed in this chapter are for two different systems, the auxiliary feedwater system, part of the safety system, and the electro-hydraulic control system, part of the nuclear power generation reactor.

The primary function of the auxiliary feedwater system (AF) is to supply cooling water during emergency operation. The cooling water supplied is boiled off in steam generators to remove decay heat created by products of nuclear fission. There are four pumps provided in the system, three electric motor driven and one turbine driven. Each of these pumps supplies its own steam generator although, if required, cross-connection capability is provided such that any steam generator could be cooled by any pump. In addition, isolation valves, flow control valves, and back flow prevention valves called check valves are installed to control flow through the system and to provide capability to isolate the pumping system, if needed, to control an accident. A simplified schematic diagram is shown in Figure 9.1.

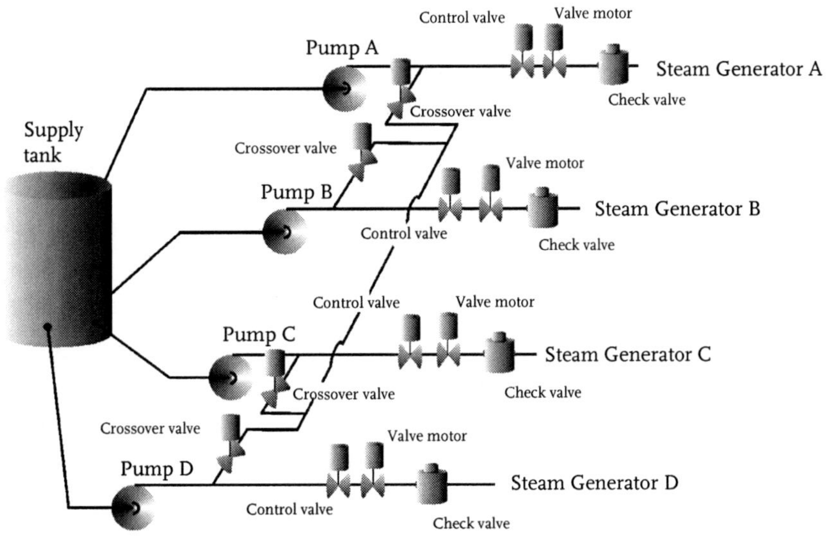

Fig. 9.1 Diagram of the auxiliary feedwater system.

We are investigating the performance of the following five groups:

- the electric motor driven pumps (six in total, three in each of two STP Units), labelled $GS1$;
- the flow control valves for all the pumps (eight in total, four in each of two STP Units), labelled $GS2$;
- the cross-connect valves (eight in total, four in each of two STP Units), labelled $GS3$;
- the accident isolation valves' motor driver (eight in total, four in each of two STP Units), labelled $GS4$;
- the check valves (eight in total, four in each of two STP Units), labelled $GS5$.

We have failure time observations for the five different groups described above starting from 1988 until May 2008. There are 56 exact observations for $GS1$, 2144 for $GS2$, 202 for $GS3$, 232 for $GS4$, and 72 for $GS5$.

The electro-hydraulic control system (EHC) consists of the high pressure electro-hydraulic fluid system and the electronic controllers, which are used for the control of the main electrical generating steam turbine as well as the steam turbines that drive the normal steam generator feed water supply pump. The high pressure steam used to turn the turbines described above is throttled for speed control and shut off with steam valves using hydraulic operators (pistons). The control functions include limiting the top speed of the turbines up to and including shut down of speeds become excessive. EHC provides high pressure hydraulic oil which is used as the motive force and the electrical signals that adjust the speed control and steam shut-off valves for the turbines. Because only one main electrical generating steam turbine is provided for each plant, production of electricity from the plant is stopped or reduced for some failures in the EHC system.

We would like to study the performance of:

- particular pressure switches that control starting and stopping of the hydraulic fluid supply pumps; the turbine shutoff warning; and temperature control of steam to the main turbine from a heat exchanger, labelled $GN1$;
- oil cleaning filters located at the outlet of the hydraulic fluid supply pumps, labelled $GN2$;
- solenoid valves (referred to as servo valves) which control the operation of the steam valves to the main electrical generating steam turbine, labelled $GN3$;
- the two hydraulic supply pumps (one is required for the system to operate), labelled $GN4$;

- solenoid valves which, when operating properly, respond to certain failures and stop the main electrical generating steam turbine as necessary to prevent damage, labelled $GN5$.

We have failure time observations for the five different groups described above starting from 1988 until May 2008. There are 221 exact and one censored observations for $GN1$, 243 exact and one censored for $GN2$, 133 exact for $GN3$, 223 exact for $GN4$, and 91 exact for $GN5$.

9.4.2 Bayesian parametric model

Here we present a Bayesian model for the failure times assuming that the lifetime distribution is Weibull with random parameters λ and β.

The sampling plan is as follows: If we have n items under observation, s of which have failed at ordered times $T_1, T_2, \ldots, T_s$, then $n - s$ have operated without failing. If there are no withdrawals then the total time on test is: $\omega = nT_s^{\beta}$, which is also the sufficient statistic for estimating λ (also known as the rescaled total time on test).

Assume that λ has an inverted gamma prior distribution with hyperparameters ν_0, μ_0 and β has uniform prior distribution with hyperparameters α_0, β_0.

We use the generic information (lognormal prior for λ) to assess the values of α_0 and β_0. First we calculate the mean and variance for the lognormal distribution with parameters:

$$M = \ln M_n - S^2/2 \tag{9.3}$$

$$S^2 = \left(\frac{\ln EF}{1.645}\right)^2 \tag{9.4}$$

where M_n is the mean and EF is the error factor as derived in Blanchard (1993). Then we compute the parameters of the Gamma prior distribution, α_0, β_0:

$$\alpha_0 = M_n/\beta_0 \tag{9.5}$$

$$\beta_0 = e^{2M}e^{S^2}(e^{S^2} - 1)/M_n. \tag{9.6}$$

We will follow the development from Martz and Waller (1982). The posterior expectations of λ and β are:

$$E[\lambda|z] = \frac{J_2}{(s + \nu_0 - 1)J_1} \tag{9.7}$$

$$E[\beta|z] = \frac{J_3}{J_1} \tag{9.8}$$

Table 9.1 Bayes estimates of the parametric Bayes model.

Group	λ	β
$GN1$	0.000152	4.763118
$GN2$	0.000448	1.994206
$GN3$	0.000294	9.980280
$GN4$	0.001594	1.006348
$GN5$	0.000298	5.931251
$GS1$	0.000313	9.910241
$GS2$	0.001499	0.348347
$GS3$	0.000932	0.655784
$GS4$	0.001667	0.723488
$GS5$	0.000308	11.109310

where z is a summary of the sample evidence, $\nu = \prod_{i=1}^{s} T_i$, $\omega_1 = nT_s + \mu_0$, $J_1 = \int_{\alpha_0}^{\beta_0} \left[\frac{\beta^s \nu^\beta}{\omega_1^{s+\nu_0}}\right] d\beta$, $J_2 = \int_{\alpha_0}^{\beta_0} \left[\frac{\beta^s \nu^\beta}{\omega_1^{s+\nu_0-1}}\right] d\beta$, and $J_3 = \int_{\alpha_0}^{\beta_0} \left[\frac{\beta^{s+1} \nu^\beta}{\omega_1^{s+\nu_0}}\right] d\beta$.

We use numerical integration to solve the above integrals.

The choice of the hyperparameters values is not random. We used the values given in the DOE database (see Blanchard, 1993) for the values of the inverted gamma parameters, and empirically assessed the parameters of the uniform prior distribution, for details see Yu *et al.* (2004). Table 9.1 shows the Bayes estimates of the model parameters for each group of data.

9.4.3 Bayesian semiparametric model

The failure data is modeled using an accelerated failure time (AFT) specification, along with the Cox proportional hazard regression model:

$$S_x(t) = S_0(e^{x'\beta}t),$$

where x is a p-dimensional vector of covariates, $S_x(\cdot)$ the corresponding reliability function, and S_0 is the 'baseline' reliability function.

The mixture of Polya trees models (MPTs) of Hanson (2006) is used to model the reliability or survival data. These models are 'centred' at the corresponding parametric models but allow significant data-driven deviations from parametric assumptions while retaining predictive power associated with parametric modeling. This point will be elaborated in Appendix A. This more flexible approach in essence assumes a nonparametric model for S_0; i.e. an MPT prior is placed on S_0. For the power plant data considered in this chapter, without loss of generality, we take the first group from each system to be represented by S_0. Then, the covariates are indicator variables corresponding to each of the remaining five groups in the data. Note that, if needed, you could easily incorporate other continuous, time-dependent covariates as well in the above formulation. Details of the MPT model are provided in Appendix A.

The MPT prior provides an intermediate choice between a strictly parametric analysis and allowing S_0 to be completely arbitrary. In some ways it provides the best of both worlds. In areas where data are sparse, such as the tails, the MPT prior places relatively more posterior mass on the underlying parametric family $\{G_\theta : \theta \in \Theta\}$. In areas where data are plentiful the posterior is more data driven; and features not allowed in the strictly parametric model, such as left-skew and multimodality, become apparent. The user-specified weight w controls how closely the posterior follows $\{G_\theta : \theta \in \Theta\}$ with larger values of w yielding inference closer to that obtained from the underlying parametric model. Hanson (2006) describes priors for w; we simply fix w to be some small value, typically $w = 1$.

Assume standard, right-censored reliability data $\mathcal{D} = \{(x_i, t_i, \delta_i)\}_{i=1}^n$, and let $\mathcal{D}_i$ denote the ith triple (x_i, t_i, δ_i). Let $T_i \sim S_{x_i}(\cdot)$. As usual, $\delta_i = 0$ indicates that t_i is a censoring time, $T_i > t_i$, and $\delta_i = 1$ denotes that t_i is a survival time, $T_i = t_i$.

Given S_0 and β, the survival function for covariates x is

$$S_x(t|\mathcal{Y}, \theta, \beta) = S_0(e^{x'\beta}t|\mathcal{Y}, \theta), \tag{9.9}$$

and the pdf is

$$f_x(t|\mathcal{Y}, \theta, \beta) = e^{x'\beta} f_0(e^{x'\beta}t|\mathcal{Y}, \theta), \tag{9.10}$$

where $S_0(t|\mathcal{Y}, \theta)$ and $f_0(t|\mathcal{Y}, \theta)$ are given by (9.16) and (9.17) in Appendix A.

Assuming uninformative censoring, the likelihood for right censored data is then given by

$$\mathcal{L}(\mathcal{Y}, \theta, \beta) = \prod_{i=1}^{n} f_{x_i}(t_i|\mathcal{Y}, \theta, \beta)^{\delta_i} S_{x_i}(t_i|\mathcal{Y}, \theta, \beta)^{1-\delta_i}. \tag{9.11}$$

The MCMC algorithm alternately samples $[\beta, \theta|\mathcal{Y}, \mathcal{D}]$ and $[\mathcal{Y}|\beta, \theta, \mathcal{D}]$. The vector $[\beta, \theta|\mathcal{Y}, \mathcal{D}]$ is efficiently sampled with a random walk Metropolis – Hastings step Tierney (1994). A proposal that has worked very well in practice is obtained from using the large sample estimated covariance matrix from fitting the parametric log-logistic model via maximum likelihood. A simple Metropolis – Hastings step for updating the components $(Y_{\epsilon 0}, Y_{\epsilon 1})$ one at a time first samples a candidate $(Y^*_{\epsilon 0}, Y^*_{\epsilon 1})$ from a Dirichlet$(mY_{\epsilon 0}, mY_{\epsilon 1})$ distribution, where

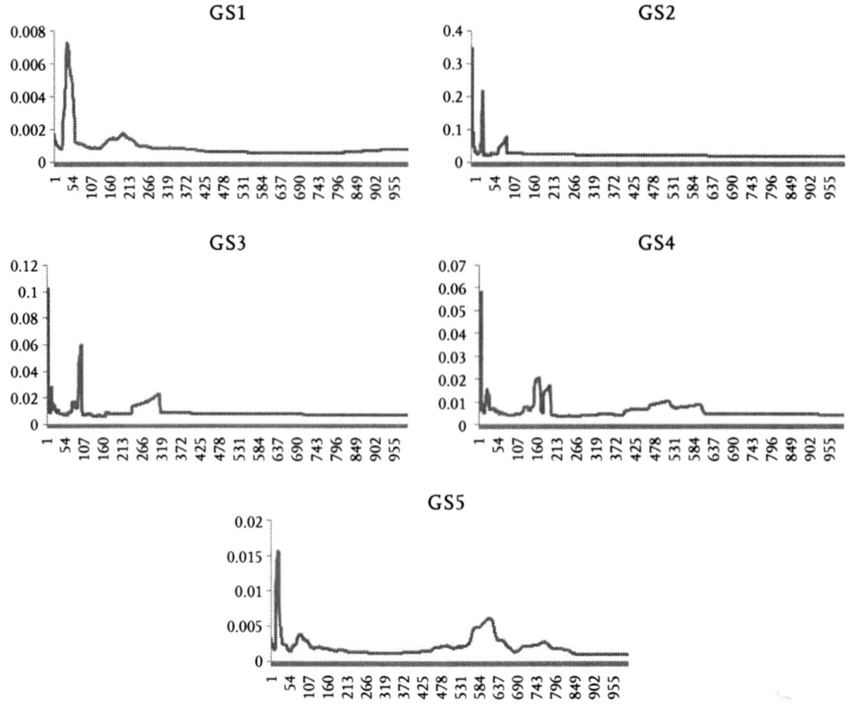

Fig. 9.2 Estimated hazard curves for the AF system using the MPT model.

$m > 0$, typically $m = 20$. This candidate is accepted as the 'new' $(Y_{\epsilon 0}, Y_{\epsilon 1})$ with probability

$$\rho = \min\left\{1, \frac{\Gamma(mY_{\epsilon 0})\Gamma(mY_{\epsilon 1})(Y_{\epsilon 0})^{mY^*_{\epsilon 0}-wj^2}(Y_{\epsilon 1})^{mY^*_{\epsilon 1}-wj^2}\mathcal{L}(\mathcal{Y}^*, \theta, \beta)}{\Gamma\left(mY^*_{\epsilon 0}\right)\Gamma\left(mY^*_{\epsilon 1}\right)\left(Y^*_{\epsilon 0}\right)^{mY_{\epsilon 0}-wj^2}\left(Y^*_{\epsilon 1}\right)^{mY_{\epsilon 1}-wj^2}\mathcal{L}(\mathcal{Y}, \theta, \beta)}\right\},$$

where j is the number of digits in the binary number $\epsilon 0$ and $\mathcal{Y}^*$ is the set $\mathcal{Y}$ with $(Y^*_{\epsilon 0}, Y^*_{\epsilon 1})$ replacing $(Y_{\epsilon 0}, Y_{\epsilon 1})$. The resulting Markov chain has mixed reasonably well for many data sets unless w is set to be very close to zero.

Based on Figures 9.2, 9.3, and 9.4 it is clear that the assumption of IFR is inappropriate. The 10 groups depict different hazard patterns that is nicely captured by the MPT model. Also, the survival probabilities are widely different. For example, for the EHC system, the group $GN3$ is by far the most reliable, whereas the group $GS1$ performs best for the AF system.

In contrast, the parametric Weibull model produces different results. Figures 9.5 and 9.6 show the corresponding hazard curves using the Bayes point estimators of the model parameters.

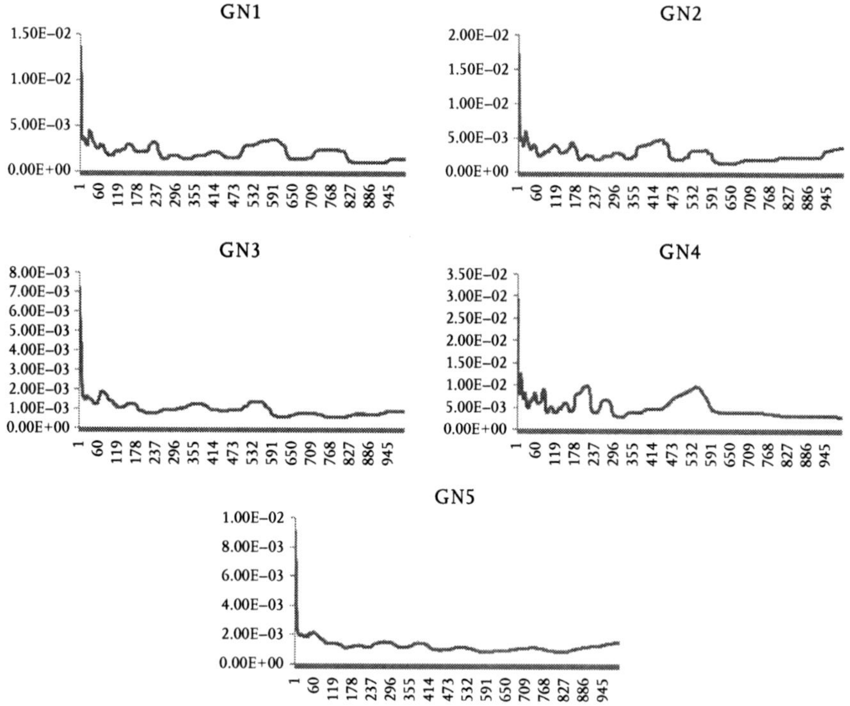

Fig. 9.3 Estimated hazard curves for the EHC system using the MPT model.

Table 9.2 shows the optimal time to perform preventive maintenance for each of the groups using the Bayesian nonparametric and the Bayesian parametric models to estimate the expected number of failures. The optimal preventive maintenance time is the same for groups GN1, GN2, GN3, GN4, GN5, GS1, and GS5. The associated total costs are different. This is due to the

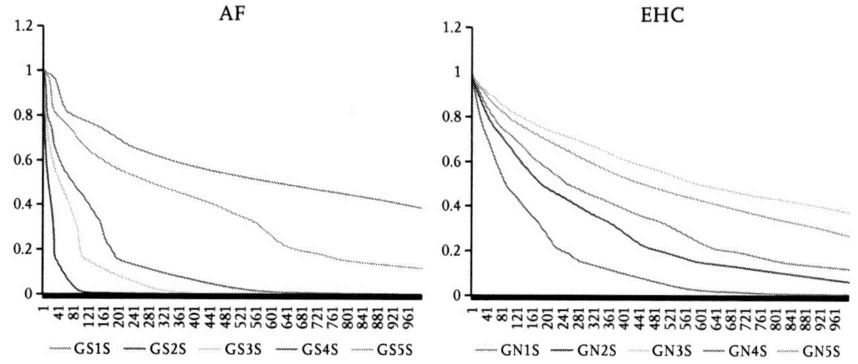

Fig. 9.4 Estimated survival curves for the AF and EHC systems using the MPT model.

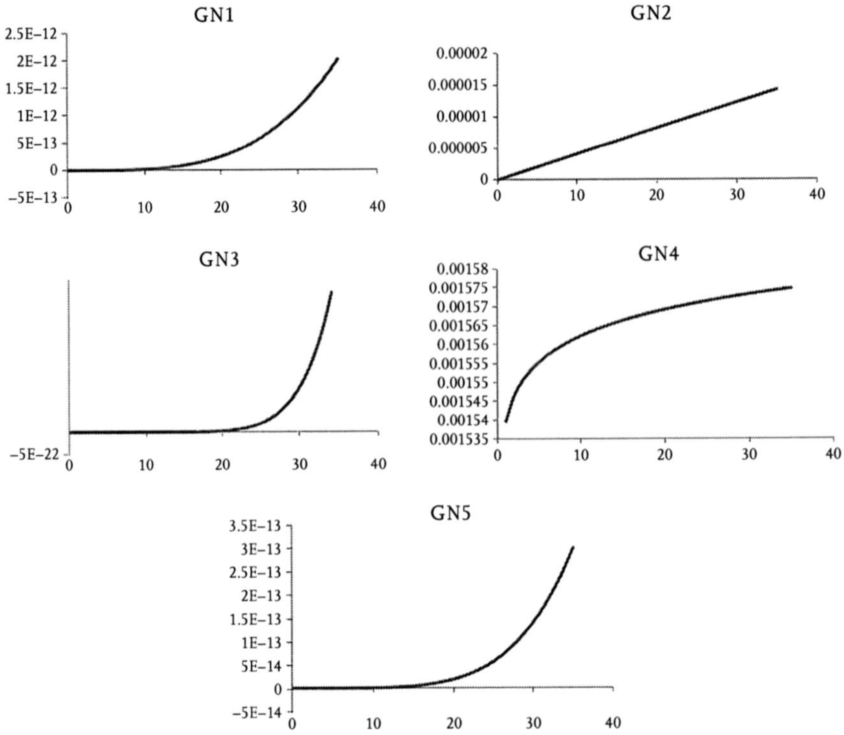

Fig. 9.5 Estimated hazard curves for the EHC system using the parametric Bayes model.

fact that the Bayesian parametric model underestimates the expected number of failures in a given interval, and the result is less total cost. Groups GS2, GS3, and GS4 have different preventive maintenance intervals. For all of them the Bayesian parametric model produced decreasing hazard curves, hence,

Table 9.2 Optimal PM time (in days) and the associated total cost using the Bayesian Nonparametric Model (BNP) and Bayesian Parametric Model (BP).

Group	Optimal PM time (BNP)	Total cost (BNP)	Optimal PM time (BP)	Total cost (BP)
GN1	501	226,996	501	104,070
GN2	501	157,019	501	64,145
GN3	501	195,147	501	120,485
GN4	501	400,072	501	263,641
GN5	501	71,669	501	40,027
GS1	501	172,614	501	104,069
GS2	501	921,880	1000	529,848
GS3	501	850,355	1000	923,097
GS4	501	460,057	1000	1,117,478
GS5	501	82,777	501	40,026

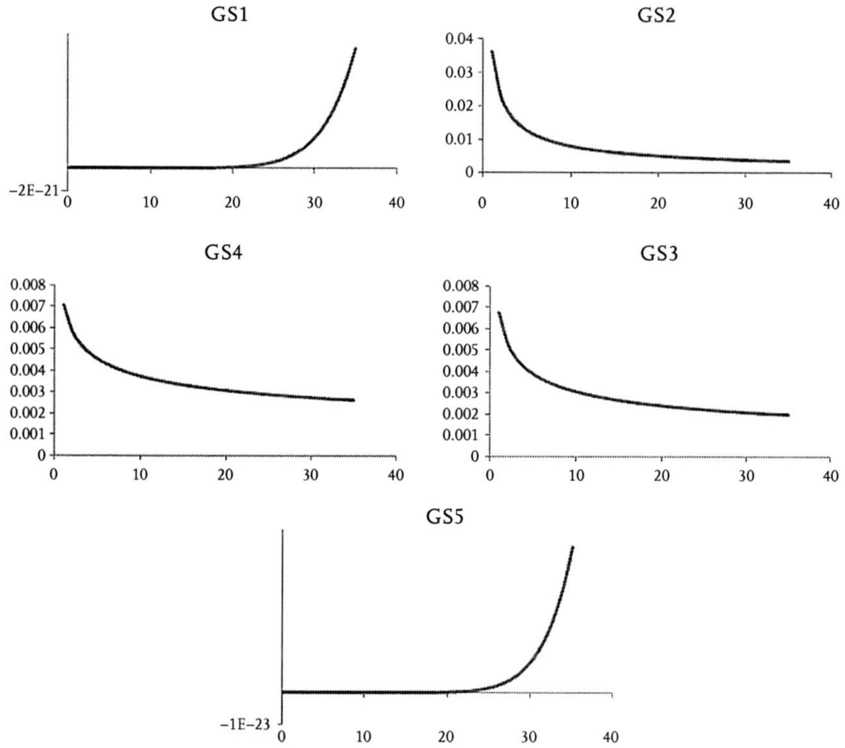

Fig. 9.6 Estimated hazard curves for the AF system using the parametric Bayes model.

no preventive maintenance is optimal to be performed before the end of the time horizon (replacement at the end of the time horizon is one of our initial assumptions.)

Appendix

A. Broader context and background

In this chapter we have introduced Bayesian decision analysis to enable engineers to reach an optimal decision in the context of maintaining nuclear power plants. Typical Bayesian analysis usually stops at inference. However, as noted by DeGroot (2004), Smith (1988), and Berger (1980), for Bayesian methods to gain acceptance, one needs to extend posterior inference to actual decisions.

In this chapter we have used Polya trees and mixtures of Polya trees to carry out inference. We provide some details of this methodology.

Extending the work of Lavine (1992, 1994), mixtures of Polya trees have been considered by Berger and Guglielmi (2001), Walker and Mallick (1999), Hanson and Johnson (2002), Hanson (2006), Hanson *et al.* (2006), and Hanson (2007).

The last four references deal with survival or reliability data. This prior is attractive in that it is easily centred at a parametric family such as the class of Weibull distributions. When sample sizes are small, posterior inference has much more of a parametric flavour due to the centring family. When sample sizes increase, the data take over the prior and features such as multimodality and left skew in the reliability distributions are better modelled.

Consider a mixture of Polya trees prior on S_0,

$$S_0|\theta \sim PT(c, \rho, G_\theta), \tag{9.12}$$

$$\theta \sim p(\theta), \tag{9.13}$$

where (9.12) is shorthand for a particular Polya tree prior; see Hanson and Johnson (2002). We briefly describe the prior but leave details to the references above.

Let J be a fixed, positive integer and let G_θ denote the family of Weibull cumulative distribution functions, $G_\theta(t) = 1 - \exp\{-(t/\lambda)^\alpha\}$ for $t \geq 0$, where $\theta = (\alpha, \lambda)'$. The distribution G_θ serves to centre the distribution of the Polya tree prior. A Polya tree prior is constructed from a set of partitions Π_θ and a family $\mathcal{A}$ of positive real numbers. Consider the family of partitions $\Pi_\theta = \{B_\theta(\epsilon) : \epsilon \in \bigcup_{l=1}^{J}\{0, 1\}^l\}$. If j is the base-10 representation of the binary number $\epsilon = \epsilon_1 \cdots \epsilon_k$ at level k, then $B_\theta(\epsilon_1 \cdots \epsilon_k)$ is defined to be the interval $(G_\theta^{-1}(j/2^k), G_\theta^{-1}((j+1)/2^k))$. For example, with $k = 3$, and $\epsilon = 000$, then $j = 0$ and $B_\theta(000) = (0, G_\theta^{-1}(1/8))$, and with $\epsilon = 010$, then $j = 2$ and $B_\theta(010) = (G_\theta^{-1}(2/8), G_\theta^{-1}(3/8))$, etc.

Note then that at each level k, the class $\{B_\theta(\epsilon) : \epsilon \in \{0, 1\}^k\}$ forms a partition of the positive reals and furthermore $B_\theta(\epsilon_1 \cdots \epsilon_k) = B_\theta(\epsilon_1 \cdots \epsilon_k 0) \bigcup B_\theta(\epsilon_1 \cdots \epsilon_k 1)$ for $k = 1, 2, \ldots, M-1$. We take the family $\mathcal{A} = \{\alpha_\epsilon : \epsilon \in \bigcup_{j=1}^{M}\{0, 1\}^j\}$ to be defined by $\alpha_{\epsilon_1 \cdots \epsilon_k} = wk^2$ for some $w > 0$; see Walker and Mallick (1999); Hanson and Johnson (2002). As w tends to zero the posterior baseline is almost entirely data-driven. As w tends to infinity we obtain a fully parametric analysis.

Given Π_θ and $\mathcal{A}$, the Polya tree prior is defined up to level J by the random vectors $\mathcal{Y} = \{(Y_{\epsilon 0}, Y_{\epsilon 1}) : \epsilon \in \bigcup_{j=0}^{M-1}\{0, 1\}^j\}$ through the product

$$S_0\{B_\theta(\epsilon_1 \cdots \epsilon_k)|\mathcal{Y}, \theta\} = \prod_{j=1}^{k} Y_{\epsilon_1 \cdots \epsilon_j}, \tag{9.14}$$

for $k = 1, 2, \ldots, M$, where we define $S_0(A)$ to be the baseline measure of any set A. The vectors $(Y_{\epsilon 0}, Y_{\epsilon 1})$ are independent Dirichlet:

$$(Y_{\epsilon 0}, Y_{\epsilon 1}) \sim \text{Dirichlet}(\alpha_{\epsilon 0}, \alpha_{\epsilon 1}), \quad \epsilon \in \bigcup_{j=0}^{M-1}\{0, 1\}^j. \tag{9.15}$$

The Polya tree parameters 'adjust' conditional probabilities, and hence the shape of the survival density f_0, relative to a parametric centring family of distributions. If the data are truly distributed G_θ observations should be on average evenly distributed among partition sets at any level j. Under the Polya tree posterior, if more observations fall into interval $B_\theta(\epsilon 0) \subset \mathbb{R}^+$ than its companion set $B_\theta(\epsilon 1)$, the conditional probability $Y_{\epsilon 0}$ of $B_\theta(\epsilon 0)$ is accordingly stochastically 'increased' relative to $Y_{\epsilon 1}$. This adaptability makes the Polya tree attractive in its flexibility, but also anchors the random S_0 firmly about the family $\{S_\theta : \theta \in \Theta\}$.

Beyond sets at the level J in Π_θ we assume $S_0|\mathcal{Y}, \theta$ follows the baseline G_θ. Hanson and Johnson (2002) show that this assumption yields predictive distributions that are the same as from a fully specified (infinite) Polya tree for large enough J; this assumption also avoids a complication involving infinite probability in the tail of the density $f_0|\mathcal{Y}, \theta$ that arises from taking $f_0|\mathcal{Y}, \theta$ to be flat on these sets.

Define the vector of probabilities $p = p(\mathcal{Y}) = (p_1, p_2, \ldots, p_{2^J})'$ as $p_{j+1} = S_0\{B_\theta(\epsilon_1 \cdots \epsilon_J)|\mathcal{Y}, \theta\} = \prod_{i=1}^{J} Y_{\epsilon_1 \cdots \epsilon_i}$ where $\epsilon_1 \cdots \epsilon_J$ is the base-2 representation of j, $j = 0, \ldots, 2^J - 1$. After simplification, the baseline survival function is,

$$S_0(t|\mathcal{Y}, \theta) = p_N\left[N - 2^J G_\theta(t)\right] + \sum_{j=N+1}^{2^J} p_j, \tag{9.16}$$

where N denotes the integer part of $2^J G_\theta(t) + 1$ and where $g_\theta(\cdot)$ is the density corresponding to G_θ. The density associated with $S_0(t|\mathcal{Y}, \theta)$ is given by

$$f_0(t|\mathcal{Y}, \theta) = \sum_{j=1}^{2^J} 2^J p_j g_\theta(t) I_{B_\theta(\epsilon_J(j-1))}(t) = 2^J p_N g_\theta(t), \tag{9.17}$$

where $\epsilon_J(i)$ is the binary representation $\epsilon_1 \cdots \epsilon_J$ of the integer i and N as is in (9.16). Note that the number of elements of $\mathcal{Y}$ may be moderate. This number is $\sum_{j=1}^{J} 2^j = 2^{J+1} - 2$. For $J = 5$, a typical level, this is 62, of which half need to be sampled in an MCMC scheme.

B. Proof of Propositions

Proposition 9.1 Assume that we follow maintenance policy (P), and that the time between failures is a random variable with an IFR distribution, i.e. the failure rate function $q(t)$ is increasing. Then, the objective function $z(T)$ is:

(i) lower semicontinuous with discontinuities at $T = L/n$, $n = 1, 2, \ldots$, and
(ii) increasing and convex on each interval $\left(\frac{L}{n+1}, \frac{L}{n}\right)$, $n = 1, 2, \ldots$.

Proof We first show (ii). Assume that $T \in (\frac{L}{n+1}, \frac{L}{n})$ for some n. Within this interval

$$z'(T) \equiv \frac{dz(T)}{dT} = n\bar{C}_{cm}\left\{q(T) - q(L - nT)\right\},$$

where $\bar{C}_{cm} = pC_d + (1-p)C_{cm}$. Here, we have used $q(t) = \frac{d}{dt}E[N(t)]$ and the fact that $\lfloor L/T \rfloor = n$ is constant within this n-th interval. That z increases on this interval follows from the fact that $z'(T)$ is positive since q is increasing and $L - nT < T$ for $T \in (\frac{L}{n+1}, \frac{L}{n})$. To show convexity, we show that $z'(T)$ is increasing. Let $T + \Delta T \in (\frac{L}{n+1}, \frac{L}{n})$. Then, $q(T + \Delta T) \geq q(T)$ and $q\{L - n(T + \Delta T)\} \leq q(L - nT)$, and hence $z'(T + \Delta T) \geq z'(T)$. This completes the proof of (ii).

The convexity result for $z(T)$ from (ii) shows that $T = L, L/2, L/3, \ldots$ are the only possible points of discontinuity in $z(T)$. The first term in $z(T)$ (i.e. the PM term) is lower semicontinuous because of the ceiling operator. The second term (i.e. $z(T)$ when $C_{pm} = 0$) is continuous in T. To show this it suffices to verify that

$$\lfloor L/T \rfloor E\left\{N(T)\right\} + E\left\{N\left(L - T\left\lfloor \frac{L}{T} \right\rfloor\right)\right\}$$

has limit $nE\{N(d)\}$ for both $T \to d^-$ and $T \to d^+$, where $d = L/n$ for any $n = 1, 2, \ldots$. This follows in a straightforward manner using the bounded convergence theorem, e.g. Cinlar (1975, pp. 33). Hence, we have (i). □

Let

$$z^c(T) = C_{pm}(L/T) + \bar{C}_{cm}(L/T)E\left\{N(T)\right\},$$

i.e. $z^c(T)$ is $z(T)$ with $\lfloor L/T \rfloor$ and $\lceil L/T \rceil$ replaced by L/T. The following proposition characterizes $z^c(T)$ and its relationship with $z(T)$.

Proposition 9.2 Let $D = \{d : d = L/n, n = 1, 2, \ldots\}$ denote the set of discontinuities of $z(T)$ (cf. Proposition 9.1). Then,

(i) $z^c(T)$ is quasiconvex on $[0, L]$.

Furthermore if $d \in D$ then

(ii) $z^c(d) = z(d)$, and

(iii) $\lim_{T \to d^-} \left\{z(T) - z^c(T)\right\} = C_{pm}$.

Proof (i) We have

$$z^c(T) = \frac{C_{pm} + \bar{C}_{cm}E\left\{N(T)\right\}}{T/L}. \tag{9.18}$$

The numerator of (9.18) is convex in T because $E\{N(T)\}$ has increasing derivative $q(T)$. As a result, $z^c(T)$ has convex level sets and is hence quasiconvex.

(ii) The result is immediate by evaluating z and z^c at points of D.

(iii) Let $d \in D$. Then, there exists a positive integer n with $d = L/n$. Note that due to the ceiling operator, we have the following result

$$\lim_{T\to d^-} \left\{ \left\lceil \frac{L}{T} \right\rceil - \frac{L}{T} \right\} = 1. \tag{9.19}$$

Based on (9.19) and the definitions of z and z^c it suffices to show

$$\lfloor L/T \rfloor E\{N(T)\} + E\left\{N\left(L - T\left\lfloor \frac{L}{T} \right\rfloor\right)\right\}$$

has limit $nE\{N(d)\}$ as $T \to d^-$. This was established in the proof of part (i) in Proposition 1.

□

Proposition 9.2 shows that at its points of discontinuity, our objective function z drops by magnitude C_{pm} to agree with z^c. Before developing the associated algorithm, the following proposition shows how to solve a simpler variant of our model in which the set A of feasible replacement intervals is replaced by $[0, L]$.

Proposition 9.3
Let

$$T_c^* \in \arg\min_{T\in[0,L]} z^c(T)$$

and let n^* be the positive integer satisfying $\frac{L}{n^*+1} \le T_c^* \le \frac{L}{n^*}$. Then,

$$T^* \in \arg\min_{T\in\left\{\frac{L}{n^*+1}, \frac{L}{n^*}\right\}} z^c(T)$$

solves $\min_{T\in[0,L]} z(T)$.

Proof Proposition 9.1 says that $z(T)$ increases on each interval $\left(\frac{L}{n+1}, \frac{L}{n}\right)$, and is lower semicontinuous. As a result, $\min_{T\in[0,L]} z(T)$ is equivalent to $\min_{T\in D} z(T)$, where $D = \{d : d = L/n, n = 1, 2, \ldots\}$ is the set of discontinuities of z. This optimization model is, in turn, equivalent to $\min_{T\in D} z^c(T)$ since, part (ii) of Proposition 9.2 establishes that $z(T) = z^c(T)$ for $T \in D$. Finally, since $z^c(T)$ is quasiconvex (see Proposition 9.2 part (i)), we know $z^c(T)$ is nondecreasing on $[T_c^*, L]$ and non-increasing on $[0, T_c^*]$. Proposition 9.3 follows. □

B.1 The optimization algorithm

Proposition 9.3 shows how to solve $\min_{T\in[0,L]} z(T)$. The key idea is to first solve the simpler problem $\min_{T\in[0,L]} z^c(T)$, which can be accomplished efficiently, e.g. via a Fibonacci search. Let T_c^* denote the optimal solution to the latter problem. Then, the optimal solution to the former problem is either the first point in D to the right of T_c^* or the first point in D to the left of T_c^*, whichever yields a smaller value of $z(T)$ (or $z^c(T)$ since they are equal on D).

As described above, our model has the optimization restricted over a finite grid of points $A \subset [0, L]$. Unfortunately, it is not true that the optimal solution is given by either the first point in A to the right of T_c^* or the first point in A to the left of T_c^*. However, the results of Proposition 9.1, part (ii) establish that within the set $A \cap (\frac{L}{n+1}, \frac{L}{n})$ we can restrict attention to the left-most point (or ignore the segment if the intersection is empty). To simplify the discussion that follows we will assume that A has been redefined using this restriction. Furthermore, given a feasible point $T_A \in A$ with cost $z(T_A)$ we can eliminate from consideration all points in A to the right of $\min\{T \in D : T \geq T_c^*, z^c(T) \geq z(T_A)\}$ and all points in A to the left of $\max\{T \in D : T \leq T_c^*, z^c(T) \geq z(T_A)\}$. This simplification follows from the fact that z increases on each of its intervals and is equal to the quasiconvex z^c on D.

Our algorithm therefore consists of:

- Step 0: Let $T_c^* \in \arg\min_{T\in[0,L]} z^c(T)$.
- Step 1: Let $T^* \in A$ be the first point in A to the right of T_c^* and let $z^* = z(T^*)$. Let $T_A = T^*$.
- Step 2: Increment T_A to be the next point in A to the right. If $z(T_A) < z^*$ then $z^* = z(T_A)$ and $T^* = T_A$. Let d be the next point to the right of T_A in D. If $z^c(d) \geq z^*$ then let T_A to be the first point in A to the left of T_c^* and proceed to the next step. Otherwise, repeat this step.
- Step 3: If $z(T_A) < z^*$ then $z^* = z(T_A)$ and $T^* = T_A$. Let d be the next point to the left of T_A in D. If $z^c(d) \geq z^*$ then output T^* and z^* and stop. Otherwise, increment T_A to be the next point in A to the left and repeat this step.

Step 0 solves the simpler continuous problem for T_c^*, as described above and Step 1 simply finds an initial candidate solution. Then, Steps 2 and 3, respectively, increment to the right and left of T_c^*, moving from the point of A in, say, $A \cap (\frac{L}{n+1}, \frac{L}{n})$ to the point (if any) in the next interval. Increments to the right and to the left stop when further points in that direction are provably suboptimal. The algorithm is ensured to terminate with an optimal solution.

Acknowledgements

The authors would like to thank the OR/IE graduate students Alexander Galenko and Dmitriy Belyi for their help in proving the propositions and computational work. Thank you to Ernie Kee for providing the data and helping with the nuclear engineering part of the paper. This research has been partially supported by the Electric Power Research Institute under contract EP-P22134/C10834, STPNOC under grant B02857, and the National Science Foundation under grants CMMI-0457558 and CMMI-0653916.

References

Barlow, R. and Hunter, L. (1960). Optimum preventive maintenance policies. *Operations Research*, **8**, 90–100.

Berger, J. O. (1980). *Statistical Decision Theory and Bayesian Analysis* (2nd edn). Springer Series in Statistics, Springer-Velag, Berlin.

Berger, J. O. and Guglielmi, A. (2001). Bayesian testing of a parametric model versus nonparametric alternatives. *Journal of the American Statistical Association*, **96**, 174–184.

Blanchard, A. (1993). Savannah river site generic data base development. Technical Report WSRC-TR 93-262, National Technical Information Service, U.S. Department of Commerce, 5285 Port Royal Road, Springfield, VA 22161, Savannah River Site, Aiken, South Carolina 29808.

Bridges, M. and Worledge, D. (2002). Reliability and risk significance for maintenance and reliability professionals at nuclear power plants. EPRI Technical Report 1007079.

Chen, T.-M. and Popova, E. (2000). Bayesian maintenance policies during a warranty period. *Communications in Statistics – Stochastic Models*, **16**, 121–142.

Cinlar, E. (1975). *Introduction to Stochastic Processes*. Prentice Hall, Englewood Cliffs.

DeGroot, M. (2004). *Optimal Statistical Decisions*. Wiley Classics Library (Originally published 1970). John Wiley, New York.

Gann, C. and Wells, J. (2004). Work Process Program. STPNOC plant procedure 0PGP03-ZA-0090, Revision 28.

Hanson, T. (2006). Inferences on mixtures of finite Polya tree models. *Journal of the American Statistical Association*, **101**, 1548–1565.

Hanson, T. (2007). Polya trees and their use in reliability and survival analysis. In *Encyclopedia of Statistics in Quality and Reliability* (ed. F. Ruggeri, R. Kenett and F. W. Faltin). John Wiley, Chichester. pp. 1385–1390.

Hanson, T., Johnson, W. and Laud, P. (2006). A semiparametric accelerated failure time model for survival data with time dependent covariates. In *Bayesian Statistics and its Applications* (ed. S. Upadhyay, U. Singh, D. Dey), Anamaya Publishers, New Delhi, pp. 254–269.

Hanson, T. and Johnson, W. O. (2002). Modeling regression error with a mixture of Polya trees. *Journal of the American Statistical Association*, **97**, 1020–1033.

Hudson, E. and Richards, D. (2004). Configuration Risk Management Program. STPNOC plant procedure 0PGP03-ZA-0091, Revision 6.

INPO (2002). Equipment Reliability Process Description. Institute of Nuclear Power Operations process description, AP 913, Revision 1.

Lavine, M. (1992). Some aspects of Polya tree distributions for statistical modeling. *Annals of Statistics*, **20**, 1222–1235.

Lavine, M. (1994). More aspects of Polya trees for statistical modeling. *Annals of Statistics*, **22**, 1161–1176.

Liming, J., Kee, E., Johnson, D. and Sun, A. (2003). Practical application of decision support metrics for nuclear power plant risk-informed asset management. In: Proceedings of the 11th International Conference on Nuclear Engineering. Tokyo, Japan.

Marquez, A. and Heguedas, A. (2002). Models for maintenance optimization: a study for repairable systems and finite time periods. *Reliability Engineering and System Safety*, **75**, 367–377.

Martz, H. F. and Waller, R. A. (1982). *Bayesian Reliability Analysis*. John Wiley, New York.

Popova, E. (2004). Basic optimality results for Bayesian group replacement policies. *Operations Research Letters*, **32**, 283–287.

Smith, J. (1988). *Decision Analysis: A Bayesian Approach*. Chapman and Hall, Boca Raton.

Su, C.-T. and Chang, C.-C. (2000). Minimization of the life cycle cost for a multistate system under periodic maintenance. *International Journal of Systems Science*, **31**, 217–227.

Tierney, L. (1994). Markov chains for exploring posterior distributions. *The Annals of Statistics*, **22**, 1701–1762.

Valdez-Florez, C. and Feldman, R. (1989). A survey of preventive maintenance models for stochastically deteriorating single-unit systems. *Naval Research Logistics*, **36**, 419–446.

Walker, S. G. and Mallick, B. K. (1999). Semiparametric accelerated life time model. *Biometrics*, **55**, 477–483.

Wilson, J. G. and Popova, E. (1998). Bayesian approaches to maintenance intervention. In: Proceedings of the Section on Bayesian Science of the American Statistical Association. pp. 278–284.

Yu, W., Popova, E., Kee, E., Sun, A. and Grantom, R. (2005). Basic factors to forecast maintenance cost for nuclear power plants. In: Proceedings of 13th International Conference on Nuclear Engineering. Beijing, China.

Yu, W., Popova, E., Kee, E., Sun, A., Richards, D. and Grantom, R. (2004). Equipment data development case study–Bayesian Weibull analysis. Technical Report, STPNOC.

·10·

Bayes linear uncertainty analysis for oil reservoirs based on multiscale computer experiments

Jonathan A. Cumming and Michael Goldstein

10.1 Introduction

Reservoir simulators are important and widely used tools for oil reservoir management. These simulators are computer implementations of high-dimensional mathematical models for reservoirs, where the model inputs are physical parameters, such as the permeability and porosity of various regions of the reservoir, the extent of potential faults, aquifer strengths and so forth. The outputs of the model, for a given choice of inputs, are observable characteristics such as pressure readings, oil and gas production levels, for the various wells in the reservoir.

Usually, we are largely uncertain as to the physical state of the reservoir, and thus we are unsure about appropriate choices of the input parameters for a reservoir model. Therefore, an uncertainty analysis for the model often proceeds by first calibrating the simulator against observed production history at the wells and then using the calibrated model to forecast future well production, and act as an information tool for the efficient management of the reservoir.

In a Bayesian analysis, all of our uncertainties are incorporated into the system forecasts. In addition to the uncertainty about the input values, there are three other basic sources of uncertainty. First, although the simulator is deterministic, an evaluation of the simulator for a single choice of parameter values can take hours or days, so that the function is unknown to us, except at the subset of values which we have chosen to evaluate. Secondly, the reservoir simulator, even at the best choice of input values, is only a model for the reservoir, and we must therefore take into account the discrepancy between the model and the reservoir. Finally, the historical data which we are calibrating against is observed with error.

This problem is typical of a very wide and important class of problems each of which may be broadly described as an uncertainty analysis for a complex physical system based on a model for the system (Sacks, Welch, Mitchell, and

Wynn 1989; Craig, Goldstein, Seheult, and Smith 1996; O'Hagan 2006). Such problems arise in almost all areas of scientific enquiry; for example climate models to study climate change, or models to explore the origins and generating principles of the universe. In all such applications, we must deal with the same four basic types of uncertainty: input uncertainty, function uncertainty, model discrepancy and observational error. A general methodology has been developed to deal with this class of problems. Our aim, in this chapter, is to provide an introduction to this methodology and to show how it may be applied for a reservoir model of realistic size and complexity. We shall therefore analyse a particular problem in reservoir description, based upon a description of our general approach to uncertainty analysis for complex models. In particular, we will highlight the value of fast approximate versions of the computer simulator for making informed prior judgements relating to the form of the full simulator. Our account is based on the use of Bayes linear methodology for simplifying the specification and analysis for complex high-dimensional problems, and so this chapter also serves as an introduction to the general principles of this approach.

10.2 Preliminaries

10.2.1 Model description

The focus of our application is a simulation of a hydrocarbon reservoir provided to us by Energy SciTech Ltd. The model is a representation of the Gullfaks oil and gas reservoir located in the North Sea, and the model is based around a three-dimensional grid of size $38 \times 87 \times 25$ where each grid cell represents a cuboid region of subterranean rock within the reservoir. Each grid cell has different specified geological properties and contains varying proportions of oil, water and gas. The reservoir also features a number of wells which, during the course of the simulation, either extract fluids from or inject fluids into the reservoir. The overall purpose of the simulation is to model changes in pressure, and the flows and changes in distribution of the different fluids throughout the reservoir, thereby giving information on the pressures and production levels at each of the wells. A simple map of the reservoir is shown in Figure 10.1.

The inputs to the computer model are a collection of scalar multipliers which adjust the magnitudes of the geological properties of each grid cell uniformly across the entire reservoir. This results in four field multipliers – one each for porosity (ϕ), x-permeability (k_x), z-permeability (k_z), and critical saturation (crw). There is no multiplier for y-permeability as the (x, y) permeabilities are treated as isotropic. In addition to these four inputs, we have multipliers for aquifer permeability (A_p) and aquifer height (A_h) giving a total of six input parameters. The input parameters and their ranges are summarised in Table 10.1.

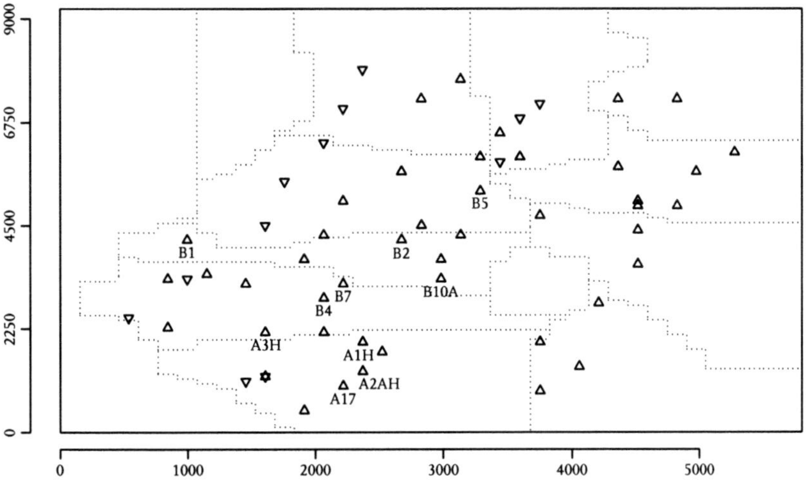

Fig. 10.1 Map of the Gullfaks reservoir grid. Production wells are marked △, injection wells are marked ▽, wells considered in our analysis are labelled, and the boundaries of different structural regions of the reservoir are indicated by dotted lines.

The outputs of the model are collections of time series of monthly values of various production quantities obtained for each well in the reservoir. The output quantities comprise monthly values of oil, water and gas production rates, oil, water and gas cumulative production totals, water-cut, gas-oil ratio, bottom-hole pressure and tubing-head pressure. For the purposes of our analysis, we shall focus exclusively on oil production rate since this is the quantity of greatest practical interest and has corresponding historical observations. In terms of the time series aspect of the output, we shall focus on a three-year window in the operation of the reservoir beginning at the start of the third year of production. We smooth these 36 monthly observations by taking four-month averages. By making these restrictions, we focus our attention on the 10 production wells which were operational throughout this period and so our outputs now consist of a collection of 10 time series with each 12 time points.

Table 10.1 The six input parameters to the hydrocarbon reservoir model.

Description	Symbol	Initial range
Porosity	ϕ	[0.5, 1.5]
x-permeability	k_x	[0.25, 6]
z-permeability	k_z	[0.1, 0.75]
Critical saturation	crw	[0.4, 1.6]
Aquifer height	A_h	[50, 500]
Aquifer permeability	A_p	[300, 3000]

For the purposes of our analysis, it will be necessary to have access to a fast approximate version of the simulator. To obtain such an approximation, we coarsened the vertical gridding of the model by a factor of 10. The evaluation time for this coarse model is between 1 and 2 minutes, compared to 1.5–3 hours for the full reservoir model.

10.2.2 Uncertainty analysis for complex physical systems

We now describe a general formulation for our approach to uncertainty analysis for complex physical systems given a computer model for that system, which is appropriate for the analysis of the reservoir model. There is a collection, x^+, of system properties. These properties influence system behaviour, as represented by a vector of system attributes, $y = (y_h, y_p)$, where y_h is a collection of historical values and y_p is a collection of values that we may wish to predict. We have an observation vector, z_h, on y_h. We write

$$z_h = y_h + e \tag{10.1}$$

and suppose that the observational error, e, is independent of y with $\mathrm{E}[e] = 0$.

Ideally, we would like to construct a deterministic computer model, $F(x) = (F_h(x), F_p(x))$, embodying the laws of nature, which satisfies $y = F(x^+)$. In practice, however, our actual model F usually simplifies the physics and approximates the solution of the resulting equations. Therefore, our uncertainty description must allow both for the possible differences between the physical value of x^+ and the best choice of inputs to the simulator, and also for the discrepancy between the model outputs, evaluated at this best choice, and the true values of the system attributes, y.

Therefore, we must make explicit our assumptions relating the computer model $F(x)$ and the physical system, y. In general, this will be problem dependent. The simplest and most common way to relate the simulator and the system is the so-called 'Best Input Approach'. We proceed as though there exists a value x^*, independent of the function F, such that the value of $F^* = F(x^*)$ summarizes all of the information that the simulator conveys about the system, in the following sense. If we define the *model discrepancy* as the difference between y and F^*, so that

$$y = F^* + \epsilon \tag{10.2}$$

then our assumption is that ϵ is independent of both F and x^*. (Here, and onwards, all probabilistic statements relate to the uncertainty judgements of the analyst.) For some models, this assumption will be justified as we can identify x^* with the true system values x^+. In other cases, this should be viewed more as a convenient simplifying assumption which we consider to be approximately true because of an approximate identification of this type. For many problems, whether this formulation is appropriate is, itself, the question of interest; for a careful discussion of the status of the best input approach, and a more general formulation of the nature of the relationship between simulators and physical systems, see Goldstein and Rougier (2008).

Given this general framework, our overall aim is to tackle previously intractable problems arising from the uncertainties inherent in imperfect computer models of highly complex physical systems using a Bayesian formulation. This involves a specification for (i) the prior probability distribution for best input x^*, (ii) a probability distribution for the computer function F, (iii) a probabilistic discrepancy measure relating $F(x^*)$ to the system y, (iv) a likelihood function relating historical data z to y. This full probabilistic description provides a formal framework to synthesise expert elicitation, historical data and a careful choice of simulator runs. From this synthesis, we aim to learn about appropriate choices for the simulator inputs and to assess, and possibly to control, the future behaviour of the system. For problems of moderate size, this approach is appropriate, practical and highly effective (Kennedy and O'Hagan, 2001; Santner, Williams, and Notz, 2003). As the scale of the problem increases, however, the full Bayes analysis becomes increasingly difficult because (i) it is difficult to give a meaningful full prior probability specification over high-dimensional spaces; (ii) the computations, for learning from data (observations and computer runs), particularly in the context of choosing informative sets of input values at which to evaluate the simulator, become technically difficult and extremely computer intensive; (iii) the likelihood surface tends to be very complicated, so that full Bayes calculations may become highly non-robust.

However, the idea of the Bayesian approach, namely capturing our expert prior judgements in stochastic form and modifying them by appropriate rules given observations, is conceptually appropriate, and indeed there is no obvious alternative. In this chapter, we therefore describe the Bayes linear approach to uncertainty analysis for complex models. The Bayes linear approach is (relatively) simple in terms of belief specification and analysis, as it is based only on mean, variance and covariance specifications. These specifications are made directly as, following de Finetti (1974, 1975), we take expectation as our primitive quantification of uncertainty. The adjusted expectation and variance for a random vector y, given random vector z, are as follows.

$$\mathrm{E}_z[y] = \mathrm{E}[y] + \mathrm{Cov}[y, z]\,\mathrm{Var}[z]^{-1}(z - \mathrm{E}[z]), \tag{10.3}$$

$$\mathrm{Var}_z[y] = \mathrm{Var}[y] - \mathrm{Cov}[y, z]\,\mathrm{Var}[z]^{-1}\,\mathrm{Cov}[z, y]. \tag{10.4}$$

(If Var[z] is not invertible, then an appropriate generalized inverse is used in the above forms.)

For the purpose of this account, we may either view Bayes linear analysis as a simple approximation to a full Bayes analysis, or as the appropriate analysis given a partial specification based on expectation. We give more details of the rationale and practicalities of the Bayes linear approach in the appendix, and for a detailed treatment, see Goldstein and Wooff (2007).

10.2.3 Overview of the analysis

The evaluation of complex computer models, such as the hydrocarbon reservoir simulation, at a given choice of input parameters is often a highly expensive undertaking both in terms of the time and the computation required. This expense typically precludes a large-scale investigation of the behaviour of the simulation with respect to its input parameters on the basis of model evaluations alone. Therefore, since the number of available model evaluations is limited by available resources there remains a substantial amount of uncertainty about the function and its behaviour, which we represent by means of an *emulator* (see Section 10.3.2.1).

For some problems, an approximate version of the original simulation may also be available. This *coarse* simulator, denoted F^c, can be evaluated in substantially less time and for substantially less cost, albeit with a consequent lower degree of accuracy. Since, both this coarse simulator and the original *accurate* simulator, F^a, are models of the same physical system, it is reasonable to expect that there will be strong qualitative and quantitative similarities between the two models. Therefore, with an appropriate belief framework to link the two simulators, we can use a large batch of evaluations of F^c to construct a detailed emulator of the coarse simulator, which we can then use to inform our beliefs about F^a and supplement the sparse collection of available full model evaluations. This is the essence of the multiscale emulation approach.

Our multiscale analysis of the hydrocarbon reservoir model proceeds in the following stages:

1. Initial model runs and screening – we perform a large batch of evaluations of $F^c(x)$ and then identify which wells are most informative and therefore most important to emulate.
2. Emulation of the coarse simulator – given the large batch of evaluations of $F^c(x)$, we now emulate each of the remaining outputs after the screening process.
3. Linking the coarse and accurate emulators – we use our emulators for $F^c(x)$ to construct an informed prior specification for the emulators of $F^a(x)$, which we then update by a small number of evaluations of $F^a(x)$.

4. History matching – using our updated emulators of $F^a(x)$, we apply the history matching techniques of Section 10.3.4.1 to identify a set $\mathcal{X}^*$ of possible values for the best model input x^*.
5. Re-focusing – we now focus on the reduced space, $\mathcal{X}^*$, identified by our history matching process. In addition to the previous outputs we now consider an additional time point 12 months after the end of our original time series, which is to be the object of our forecast. We then build our emulators in the reduced space for $F^c(x)$ and $F^a(x)$ over the original time series and the additional forecast point.
6. Forecasting – using our emulators of $F^a(x)$ within the reduced region, we forecast the 'future' time point using the methods from Section 10.3.5.1.

10.3 Uncertainty analysis for the Gullfaks reservoir

10.3.1 Initial model runs and screening

We begin by evaluating the coarse model $F^c(x)$ over a 1000-point Latin hypercube design (McKay, Beckman, and Conover, 1979; Santner *et al.*, 2003) in the input parameters. Since emulation, history matching and forecasting are computationally demanding processes, we choose to screen the collection of 120 outputs and determine an appropriate subset which will serve as the focus of our analysis. In order to identify this reduced collection, we will apply the principal variable selection methods of Cumming and Wooff (2007) to the 120×120 correlation matrix of the output vectors $\{F^c(x_i)\}$, $i = 1, \ldots, 1000$.

10.3.1.1 Methodology – Principal variables

Given a collection of outputs, $y_{1:q}$, with correlation matrix R, the principal variable selection procedure operates by assigning a score $h_i = \sum_{j=1}^{q} \mathrm{Corr}[y_i, y_j]^2$ to each output y_i. The first principal variable is then identified as the output which maximises this score. Subsequent outputs are then selected using the partial correlation given the set of identified principal variables. This allows for the choice of additional principal variables to be made having removed any effects of those variables already selected. To calculate this partial correlation, we first partition the correlation matrix into block form

$$R = \begin{pmatrix} R_{11} & R_{12} \\ R_{21} & R_{22} \end{pmatrix},$$

where R_{11} corresponds to the correlation matrix of the identified principal variables, R_{22} is the correlation matrix of the remaining variables, and R_{12} and

R_{21} are the matrices of correlations between the two groups. We then determine the partial correlation matrix $R_{22\cdot1}$ as

$$R_{22\cdot1} = R_{22} - R_{21} R_{11}^{-1} R_{12}.$$

The process continues until sufficient variables are chosen that the partial variance of each remaining output is small, or a sufficient proportion of the overall variability of the collection has been explained. In general, outputs with large values of h_i have, on average, large loadings on important principal components of the correlation matrix and thus correspond to structurally important variables.

10.3.1.2 Application and results

The outputs from the hydrocarbon reservoir model have a group structure, with groups formed by the different wells, and different time points. We intend to retain all time points at a given well to allow for a multivariate temporal treatment of the emulation. Therefore we make our reduction in the number of wells in the model output by applying a modified version of the above procedure where, rather than selecting a single output, y_i, at each stage, we instead select all 12 outputs corresponding to the well with highest total h_i score. We then continue as before, though selecting a block of 12 outputs at each stage. The results from applying this procedure to the 10 wells in the model are given in Table 10.2.

We can see from the results that there is a substantial amount of correlation among the outputs at each of the wells, as the first identified principal well accounts for 77.7% of the variation of the collection. Introducing additional wells into the collection of principal outputs only increases the amount of variation of all the outputs explained by the principal variables by a small

Table 10.2 Table of summary statistics for the selection of principal wells.

Well name	Well h_i	Cumulative % of variation
B2	4526.0	77.7
A3H	26.5	81.8
B1	19.5	84.6
B5	14.1	87.1
B10A	10.5	94.7
B7	7.0	95.2
A2AH	7.3	99.2
A17	1.1	99.7
B4	1.0	99.7
A1H	1.1	100.0

amount. On the basis of this information one could choose to retain the first four or five principal wells and capture between 87% and 95% of the variation in the collection. For simplicity, we choose to retain four of the ten wells, namely B2, A3H, B1 and B5.

10.3.2 Representing beliefs about F using emulators

10.3.2.1 Methodology – Coarse model emulation

We express our beliefs about the uncertainty in the simulator output by constructing a stochastic belief specification for the deterministic simulator, which is often referred to as an *emulator*. Our emulator for component i of the coarse simulator, $F^c(x)$, takes the following form:

$$F_i^c(x) = \sum_j \beta_{ij}^c \, g_{ij}(x) + u_i^c(x). \tag{10.5}$$

In this formulation, $\beta_i^c = (\beta_{i1}^c, \ldots, \beta_{ip_i}^c)$ are unknown scalars, $g_i(x) = (g_{i1}(x), \ldots, g_{ip_i}(x))$ are known deterministic functions of x (typically monomials), and $u_i^c(x)$ is a stochastic residual process. The component $g_i(x)^T \beta_i^c$ is a regression term which expresses the global variation in F_i^c, namely that portion of the variation in $F_i^c(x)$ which we can resolve without having to make evaluations for F_i^c at input choices which are near to x. The residual $u^c(x)$ expresses local variation, which we take to be a weakly stationary stochastic process with constant variance.

Often, we discover that most of the global variation for output component F_i^c is accounted for by a relatively small subset, $x_{[i]}$ say, of the input quantities called the *active variables*. In such cases, we may further simplify our emulator, as

$$F_i^c(x) = \sum_j \beta_{ij}^c \, g_{ij}(x_{[i]}) + u_i^c(x_{[i]}) + v_i^c(x) \tag{10.6}$$

where $u_i^c(x_{[i]})$ is now a stationary process in the $x_{[i]}$ only, and $v_i^c(x)$ is an uncorrelated 'nugget' term expressing all of the residual variation which is attributable to the inactive inputs. When variation in these residual terms is small, and the number of inactive inputs is large, this simplification enormously reduces the dimension of the computations that we must make, while usually having only a small impact on the accuracy of our results.

The emulator expresses prior uncertainty judgements about the function. In order to fit the emulator, we must choose the functions $g_{ij}(x)$, specify prior uncertainties for the coefficients β_i^c and update these by carefully chosen evaluations of the simulator, and choose an appropriate form for the local variation $u_i^c(x)$. For a full Bayesian analysis, we must make a full prior specification for each of the key uncertain quantities, $\{\beta_i^c, u_i^c(x_{[i]}), v_i^c(x)\}$, often choosing a

Gaussian form. Within the Bayes linear formulation, we need only specify the mean, variance and covariance across each of the elements and at each input value x. From the prior form (10.6), we obtain the prior mean and variance of the coarse emulator as

$$\mathrm{E}\left[F_i^c(x)\right] = g_i(x_{[i]})^T\mathrm{E}\left[\beta_i^c\right] + \mathrm{E}\left[u_i^c(x_{[i]})\right] + \mathrm{E}\left[v_i^c(x)\right], \tag{10.7}$$
$$\mathrm{Var}\left[F_i^c(x)\right] = g_i(x_{[i]})^T\mathrm{Var}\left[\beta_i^c\right]g_i(x_{[i]}) + \mathrm{Var}\left[u_i^c(x_{[i]})\right] + \mathrm{Var}\left[v_i^c(x)\right], \tag{10.8}$$

where a priori we consider $\{\beta_i^c, u_i^c(x_{[i]}), v_i^c(x)\}$ as independent. There is an extensive literature on functional emulation (Sacks *et al.*, 1989; Currin, Mitchell, Morris, and Ylvisaker, 1991; Craig, Goldstein, Rougier, and Seheult, 2001; Santner *et al.*, 2003; O'Hagan, 2006).

As the coarse simulator is quick to evaluate, emulator choice may be made solely on the basis of a very large collection of simulator evaluations. If coarse simulator evaluations had been more costly, then we would need to rely on prior information to direct the choice of evaluations and the form of the collection $\mathcal{G}_i = \cup_{i,j}\{g_{ij}(\cdot)\}$ (Craig, Goldstein, Seheult, and Smith, 1998). We may make many runs of the fast simulator, which allows us to develop a preliminary view of the form of the function, and therefore to make a preliminary choice of the function collection $\mathcal{G}_i$ and therefore to suggest an informed prior specification for the random quantities that determine the emulator for F^a. We treat the coarse simulator as our only source of prior information about $F^a(x)$. This prior specification will be updated by careful choice of evaluations of the full simulator, supported by a diagnostic analysis, for example based on looking for systematic structure in the emulator residuals.

With such a large number of evaluations of the coarse model, the emulator (10.6) can be identified and well-estimated from the data alone. For a Bayesian treatment at this stage, our prior judgements would be dominated by the large number of model evaluations. In contrast, our prior judgements will play a central role in our emulation of $F^a(x)$, as in that case the data are far more scarce.

In general, our prior beliefs about the emulator components are structured as follows. First, we must identify, for each $F_i^c(x)$, the collection of active inputs which describe the majority of global variation, their associated basis functions $\mathcal{G}$ and the coefficients β^c. Having such an ample data set allows for model selection and parameter estimation to be carried out independently for each component of F^c and to be driven solely by the information from the model runs. The residual process $u_i^c(x_{[i]})$ is a weakly stationary process in $x_{[i]}$ which represents the residual variation in the emulator that is not captured by our trend in the active variables. As such, residual values will be strongly correlated for neighbouring values of $x_{[i]}$. We therefore specify a prior covariance structure over values of $u_i^c(x_{[i]})$ which is a function of the separation of the active variables. The prior form we use is the Gaussian covariance function

$$Cov\left[u_i^c(x_{[i]}), u_i^c(x'_{[i]})\right] = \sigma_{u_i}^2 \exp\left(-\theta_i^c ||x_{[i]} - x'_{[i]}||^2\right), \qquad (10.9)$$

where $\sigma_{u_i}^2$ is the point variance at any given x, θ_i^c is a correlation length parameter which controls the strength of correlation between two separated points in the input space, and $|| \cdot ||$ is the Euclidean norm.

The nugget process $v_i^c(x)$ expresses all the remaining variation in the emulator attributable to the inactive inputs. The magnitude of the nugget process is often small and so is treated as uncorrelated random noise with $\text{Var}\left[v_i^c(x)\right] = \sigma_{v_i}^2$. We consider the point variances of these two processes to be proportions of the overall residual variance of the computer model given the emulator trend, σ_i^2, so that $\sigma_{u_i}^2 = (1 - \delta_i)\sigma_i^2$, and $\sigma_{v_i}^2 = \delta_i \sigma_i^2$, for some σ_i^2 and some typically small value of δ_i.

10.3.2.2 Application and results

We have accumulated 1000 simulator runs and identified which production wells in the reservoir are of particular interest. Prior to emulation, the design was scaled so that all inputs took the range $[-1, 1]$, and all outputs from F^c were scaled by the model runs to have mean 0 and variance 1. We now describe the emulator of component $i = (w, t)$ of the coarse simulator F^c, where w denotes the well, and t denotes the time associated with the ith output component of the computer model.

The first step in constructing the emulator is to identify, for each output component F_i^c, the subset of active inputs $x_{[i]}$ which drive the majority of global variation in F_i^c. Using the large batch of coarse model runs, we make this determination via a stepwise model search using simple linear regression. We begin by fitting each F_i^c on all linear terms in x using ordinary least squares. We then perform a stepwise delete on each regression, progressively pruning away inactive inputs until we are left with a reduced collection $x_{[i]}$ of between 3 and 5 of the original inputs. The chosen active variables for a subset of the wells of F_i^c are presented in the third column of Table 10.3. We can see from these results that the inputs ϕ and crw are active on almost all emulators for those two wells, a pattern that continues on the remaining two wells. Clearly ϕ and crw are important in explaining the global variation of F^c across the input space. Conversely, the input variable A_h appears to have no notable effect on model output.

The next stage in emulator construction is to choose the functions $g_{ij}(x_{[i]})$ for each $F_i^c(x)$ which form the basis of the emulator trend. Again, since we have an ample supply of computer evaluations we determine this collection by stepwise fitting. For each $F_i^c(x)$, we define the maximal set of basis functions to

Table 10.3 Emulation summary for wells B2 and A3H.

Well	Time	$x_{[i]}$	No. model terms	Coarse simulator R^2	Accurate simulator $\tilde{R}^2$
B2	4	ϕ, crw, A_p	9	0.886	0.951
B2	8	ϕ, crw, A_p	7	0.959	0.958
B2	12	ϕ, crw, A_p	10	0.978	0.995
B2	16	ϕ, crw, k_z	7	0.970	0.995
B2	20	ϕ, crw, k_x	11	0.967	0.986
B2	24	ϕ, crw, k_x	10	0.970	0.970
B2	28	ϕ, crw, k_x	10	0.975	0.981
B2	32	ϕ, crw, k_x	11	0.980	0.951
B2	36	ϕ, crw, k_x	11	0.983	0.967
A3H	4	ϕ, crw, A_p	9	0.962	0.824
A3H	8	ϕ, crw, k_x	7	0.937	0.960
A3H	12	ϕ, crw, k_z	10	0.952	0.939
A3H	16	ϕ, crw, k_z	7	0.976	0.828
A3H	20	ϕ, crw, k_x	11	0.971	0.993
A3H	24	ϕ, crw, k_x	10	0.964	0.899
A3H	28	ϕ, k_z, A_p	10	0.961	0.450
A3H	32	ϕ, crw, k_z	11	0.968	0.217
A3H	36	ϕ, crw, k_x	11	0.979	0.278

include an intercept with linear, quadratic, cubic and pairwise interaction terms in $x_{[i]}$. The saturated linear regression over these terms is then fitted using the coarse model runs and we again prune away any unnecessary terms via stepwise selection. For illustration, the trend and coefficients of the coarse emulator for well B1 oil production rate at time $t = 28$ are given in the first row of Table 10.4.

For each component F_i^c, we have now identified a subset of active inputs $x_{[i]}$ and a collection of p_i basis functions $g_i(x_{[i]})$ which adequately capture the majority of the global behaviour of F_i^c. The next stage is to quantify beliefs

Table 10.4 Table of mean coefficients from the emulator of oil production rate at well B1 and time 28.

Model	*Int*	ϕ	ϕ^2	ϕ^3	*crw*	*crw*^{2}	*crw*^{3}
Coarse	0.663	−0.326	−1.858	2.313	−0.219	0.064	0.079
Accurate	0.612	−0.349	−0.599	0.811	0.313	−0.331	−0.822
Refocused coarse	−0.149	2.204	−0.614	−0.858	−0.586	0.386	0.119
Refocused accurate	0.678	0.402	−0.456	−0.098	−0.057	−0.055	0.053

	$\phi \times crw$	A_p	kz	$w^c(x)$	R^2	$\tilde{R}^2$
Coarse	0.407	−0.008			0.904	
Accurate	0.072	0.111		0.112		0.952
Refocused coarse	0.206	0.044	−0.037		0.905	
Refocused accurate	−0.045	−0.034	−0.045	−0.109		0.945

about the emulator coefficients β_i^c. We fit our linear description in the selected active variables using ordinary least squares assuming uncorrelated errors to obtain appropriate estimates for these coefficients. The value of $\mathrm{E}\left[\beta_i^c\right]$ is then taken to be the estimate $\hat{\beta}_{ij}^c$ from the linear regression and $\mathrm{Var}\left[\beta_i^c\right]$ is taken to be the estimated variance of the corresponding estimates. As we have 1000 evaluations in an approximately orthogonal design, the estimation error is negligible.

The results of the stepwise selection and model fitting are given in the first five columns of Table 10.3. We can see from the R^2 values that the emulator trends are accounting for a very high proportion of the variation in the model output. We observe similar performance on the emulators of the remaining wells, with the exceptions of the emulators of well B5 at times $t = 4$ and $t = 8$, which could not be well-represented using any number or combination of basis functions. These two model outputs were therefore omitted from our subsequent analysis leaving us with a total of 34 model outputs.

The final stage is to make assessments for the values of the hyperparameters in our covariance specifications for $u_i(x_{[i]})$ and $v_i(x)$. We estimate σ_i^2 by the estimate of variance of the emulator trend residuals, and then obtain estimates for θ_i and δ_i by applying the robust variogram methods of Cressie (1991). We then use these estimates as plug-in values for θ_i, δ_i^2, and σ_i^2.

For diagnostic purposes, we then performed a further 100 evaluations of $F^c(x)$ at points reasonably well separated from the original design. For each of these 100 runs, we compared the actual model output, $F_i^c(x)$, with the predictions obtained from our coarse emulators. For all emulators, the variation of the prediction errors of the 100 new points was comparable to the residual variation of the original emulator trend, indicating that the emulators are interpolating well and are not over-fitted to the original coarse model runs. Investigation of residual plots also corroborated this result.

10.3.3 Linking the coarse and accurate emulators

10.3.3.1 Methodology – Multiscale emulation

We now develop an emulator for the accurate version of the computer model $F^a(x)$. We consider that $F^c(x)$ is sufficiently informative for $F^a(x)$ that it serves as the basis for an appropriate prior specification for this emulator. We initially restrict our emulator for component i of $F^a(x)$ to share the same set of active variables and the same basis functions as its coarse counterpart $F_i^c(x)$. Since the coarse model $F^c(x)$ is well-understood due to the considerable number of model evaluations, we consider the coarse emulator structure as known and fixed and use this as a structural basis for building the emulator of $F^a(x)$. Thus we specify a prior accurate emulator of the form

$$F_i^a(x) = g_i(x_{[i]})^T \beta_i^a + \beta_{w_i}^a w_i^c(x) + w_i^a(x), \qquad (10.10)$$

where $w_i^c(x) = u_i^c(x_{[i]}) + v_i^c(x_i)$, $w_i^a(x) = u_i^a(x_{[i]}) + v_i^a(x_i)$, and we have an identical global trend structure over the inputs albeit with different coefficients. On this accurate emulator, we also introduce some unknown multiple of the coarse emulator residuals $\beta_{w_i}^a w_i^c(x)$, and include a new residual process $w_i^a(x)$ which will absorb any structure of the accurate computer model that cannot be explained by our existing set of active variables and basis functions. Alternative methods for constructing such a multiscale emulator can be found in Kennedy and O'Hagan (2000) and Qian and Wu (2008).

As we have performed a large number of evaluations of $F^c(x)$, over a roughly orthogonal design, our estimation error from the model fitting is negligible and so we consider the β_i^c as known for each component i, and further for any x at which we have evaluated $F^c(x)$, the residuals $w_i^c(x)$ are also known. Thus we incorporate the $w_i^c(x)$ into our collection of basis functions with associated coefficient $\beta_{w_i}^a$. Absorbing $w^c(x)$ into the basis functions and $\beta_{w_i}^a$ into the coefficient vector β_i^a, we write the prior expectation and variance for the accurate simulator as

$$\mathrm{E}\left[F_i^a(x)\right] = g_i(x_{[i]})^T \mathrm{E}\left[\beta_i^a\right] + \mathrm{E}\left[w_i^a(x)\right], \tag{10.11}$$

$$\mathrm{Var}\left[F_i^a(x)\right] = g_i(x_{[i]})^T \mathrm{Var}\left[\beta_i^a\right] g_i(x_{[i]}) + \mathrm{Var}\left[w_i^a(x)\right], \tag{10.12}$$

where now $g_i(x_{[i]}) = (g_{i1}(x_{[i]}), \ldots, g_{ip_i}(x_{[i]}), w_i^c(x))$, and $\beta_i^a = (\beta_{i1}^a, \ldots, \beta_{ip_i}^a, \beta_{w_i}^a)$. We also specify the expectation and variance of the residual process $w^a(x)$ to be

$$\mathrm{E}\left[w_i^a(x)\right] = 0, \tag{10.13}$$

$$\mathrm{Cov}\left[w_i^a(x), w_i^a(x')\right] = \sigma_{u_i}^2 \exp\left(-\theta_i^c ||x_{[i]} - x'_{[i]}||^2\right) + \sigma_{v_i}^2 \mathbb{I}, \tag{10.14}$$

where the covariance function between any pair of residuals on the accurate emulator has the same prior form and hyperparameter values as that used for $u^c(x_{[i]})$ in (10.9).

We now consider the prior form of β_i^a in more detail. If we believe that each of the terms in the emulator trend corresponds to a particular qualitative physical effect, then we may expect that the magnitude of these effects will change differentially as we move from the coarse to the accurate simulator. This would suggest allowing the contribution of each $g_{ij}(x_{[i]})$ to the trend of $F^a(x)$ to be changed individually. One prior form which allows for such changes is

$$\beta_{ij}^a = \rho_{ij}\beta_{ij}^c + \gamma_{ij} \tag{10.15}$$

where ρ_{ij} is an unknown multiplier which scales the contribution of β_{ij}^c to β_{ij}^a, and γ_{ij} is a shift that can accommodate potential changes in location. We consider ρ_{ij} to be independent of γ_{ij}. In order to construct our prior form for $F_i^a(x)$, we must specify prior means, variances and covariances for ρ_{ij} and

γ_{ij}. We develop choices appropriate to the hydrocarbon reservoir model in Section 10.3.3.2.

As our prior statements about $F^a(x)$ describe our beliefs about the uncertain value of the simulator output, we can use observational data, namely the matrix $F^a(X^a)$ of evaluations of the accurate simulator over the elements of the chosen design X^a, to compare our prior expectations to what actually occurs. A simple such comparison is achieved by the *discrepancy ratio* for $F_i^a(X^a)$, the vector containing accurate simulator evaluations over X^a for the ith output component, defined as follows

$$\mathrm{Dr}(F_i^a(X^a)) = \frac{\left\{F_i^a(X^a) - \mathrm{E}\left[F_i^a(X^a)\right]\right\}^T \mathrm{Var}\left[F_i^a(X^a)\right]^{-1}\left\{F_i^a(X^a) - \mathrm{E}\left[F_i^a(X^a)\right]\right\}}{\mathbf{rk}\left\{\mathrm{Var}\left[F_i^a(X^a)\right]\right\}}, \tag{10.16}$$

which has prior expectation 1, and where $\mathbf{rk}\{\mathrm{Var}\left[F_i^a(X^a)\right]\}$ corresponds to the rank of the matrix $\mathrm{Var}\left[F_i^a(X^a)\right]$. Very large values of $\mathrm{Dr}(F_i^a(X^a))$ may suggest a mis-specification of the prior expectation or a substantial underestimation of the prior variance. Conversely, very small values of $\mathrm{Dr}(F_i^a(X^a))$ may suggest an overestimation of the variability of $F_i^a(x)$.

Given the prior emulator for $F_i^a(x)$ and the simulator evaluations $F_i^a(X^a)$, we now update our prior beliefs about $F_i^a(x)$ by the model runs via the Bayes linear adjustment formulae (10.3) and (10.4). Thus we obtain an adjusted expectation and variance for $F_i^a(x)$ given $F_i^a(X^a)$

$$\begin{aligned}\mathrm{E}_{F_i^a(X^a)}\left[F_i^a(x)\right] = \mathrm{E}\left[F_i^a(x1)\right] + \mathrm{Cov}\left[F_i^a(x), F_i^a(X^a)\right]\mathrm{Var}\left[F_i^a(X^a)\right]^{-1}\\ \left\{F_i^a(X^a) - \mathrm{E}\left[F_i^a(X^a)\right]\right\}, \end{aligned}\tag{10.17}$$

$$\begin{aligned}\mathrm{Var}_{F_i^a(X^a)}\left[F_i^a(x)\right] = \mathrm{Var}\left[F_i^a(x)\right] - \mathrm{Cov}\left[F_i^a(x), F_i^a(X^a)\right]\mathrm{Var}\left[F_i^a(X^a)\right]^{-1}\\ \mathrm{Cov}\left[F_i^a(X^a), F_i^a(x)\right]. \end{aligned}\tag{10.18}$$

The constituent elements of this update can be derived from our prior specifications for $F^a(x)$ from (10.11) and (10.12), and our belief statements made above.

10.3.3.2 Application and results

We first consider that the prior judgement that the expected values of the fine emulator coefficients are the same as those of the coarse emulator is appropriate, and so we specify expectations $\mathrm{E}\left[\rho_{ij}\right] = 1$ and $\mathrm{E}\left[\gamma_{ij}\right] = 0$. We now describe the covariance structure for the ρ_{ij} and γ_{ij} parameters. Every ρ_{ij} (and similarly γ_{ij}) is associated with a unique well w and time point t via the simulator output component $F_i^a(x)$. Additionally, every ρ_{ij} is associated with a unique regression basis function $g_{ij}(\cdot)$. Given these associations, we consider there to be two sources of correlation between the ρ_{ij} at a given well. First, for a given well w we consider there to be temporal effects correlating all $(\rho_{ij}, \rho_{i'j'})$ pairs to a degree governed by their separation in time. Secondly, we consider that there are model term effects which introduce additional correlation when both ρ_{ij} and $\rho_{i'j'}$ are multipliers for coefficients of the same basis functions, i.e. $g_{ij}(\cdot) \equiv g_{i'j'}(\cdot)$.

To express this covariance structure concisely, we extend the previous notation and write ρ_{ij} as $\rho_{(w,t,k)}$ where w and t correspond to the well and time associated with $F_i^a(x)$, and where k indexes the unique regression basis function associated with ρ_{ij}, namely the single element of the set of all basis functions $\mathcal{G} = \cup_{i,j}\{g_{ij}(\cdot)\}$. Under this notation, for a pair of multipliers $(\rho_{(w,t,k)}, \rho_{(w,t',k')})$ then $k = k'$ if and only if both are multipliers for coefficients of the same basis function, say ϕ^2, albeit on different emulators. On this basis, we write the covariance function for $\rho_{(w,t,k)}$ as

$$\text{Cov}\left[\rho_{(w,t,k)}, \rho_{(w,t',k')}\right] = \left(\sigma_{\rho 1}^2 + \sigma_{\rho 2}^2 \mathbb{I}_{k=k'}\right) R_T(t, t'), \tag{10.19}$$

where $\sigma_{\rho 1}^2$ governs the contribution of the overall temporal effect to the covariance, and $\sigma_{\rho 2}^2$ controls the magnitude of the additional model term effect, $R_T(t, t') = \exp\{-\theta_T(t - t')^2\}$ is a Gaussian correlation function over time, and $\mathbb{I}_p$ is the indicator function of the proposition p. Our covariance specification for the $\gamma_{(w,t,k)}$ takes the same form as (10.19) albeit with variances $\sigma_{\gamma 1}^2$ and $\sigma_{\gamma 2}^2$.

To complete our prior specification over $F^a(x)$, we assign $\sigma_{\rho 1}^2 = \sigma_{\gamma 1}^2 = 0.1$, and $\sigma_{\rho 2}^2 = \sigma_{\gamma 2}^2 = 0.1$ for all output components, which correspond to the belief that coefficients are weakly correlated with other coefficients on the same emulator, and that the model term effect has a similar contribution to the covariance as the temporal effect. We also assigned $\theta_T = 1/12^2$ to allow for a moderate amount of correlation across time.

We now evaluate a small batch of 20 runs of the accurate simulator. The runs were chosen by generating a large number of Latin hypercube designs and selecting that which would be most effective at reducing our uncertainty about β^a by minimising $\mathbf{tr}\{\text{Var}[\hat{\beta}^a]\}$ by least squares. Considering the simulator output for each well individually, since information on $F_w^a(x) = (F_{(w,4)}^a(x), F_{(w,8)}^a(x), \ldots, F_{(w,36)}^a(x))$ for each well w is now available in the form of the model runs $F_w^a(X^a)$ over the design X^a, we can make a diagnostic assessment of the choices made in specifying prior beliefs about F_w^a. In the case of our prior specifications for the multivariate emulators for each well, we obtain discrepancy ratio values of 0.86, 1.14, 0.67, and 1.07 suggesting our prior beliefs are broadly consistent with the behaviour observed in the data. For more detailed diagnostic methods for evaluating our prior and adjusted beliefs see Goldstein and Wooff (2007).

Using the prior emulator for $F_w^a(x)$ and the simulator evaluations $F_w^a(X^a)$, we update our beliefs about $F_w^a(x)$ by the model runs via the Bayes linear adjustment formulae (10.17), (10.18). To assess the adequacy of fit for the updated accurate emulators of the outputs updated using the 20 runs of $F^a(x)$, we calculate a version of the R^2 statistic by using the residuals obtained from the adjusted emulator trend $g_i(x_{[i]})^T \mathrm{E}_{F_i^a(X^a)}\left[\beta_i^a\right]$, denoted $\tilde{R}^2$. These are given in the final column of Table 10.3. It is clear that the majority of the accurate emulators perform well and accurately represent the fine simulator, except for the emulators at well A3H and times $t = 28, 32, 36$, which display poor fits to the fine model due to the behaviour of $F^c(x)$ at those locations being uninformative for the corresponding accurate model. For additional illustration, the coefficients for the coarse emulator and the adjusted expected coefficients of the accurate emulator of well B1 oil production rate at time $t = 28$ are given in the first two rows of Table 10.4. We can see from these values in this case that both emulators have good fits to the simulator despite the different coefficients.

10.3.4 History matching and calibration

10.3.4.1 Methodology – History matching via implausibility

Most models go through a series of iterations before they are judged to give an adequate representation of the physical system. This is the case for reservoir simulators, where a key stage in assessing simulator quality is termed *history matching*, namely identifying the set $\mathcal{X}^*$ of possible choices of x^* for the reservoir model (i.e. those choices of input geology which give a sufficiently good fit to historical observations, relative to model discrepancy and observational error). If our search reveals no possible choices for x^*, this is usually taken to indicate structural problems with the underlying model, provided that we can be reasonably confident that the set $\mathcal{X}^*$ is indeed empty. This can be difficult to determine, as the input space over which we must search may be very high dimensional, the collection of outputs over which we may need to match may be very large, and each single function evaluation may take a very long time. We now describe the method followed in Craig, Goldstein, Seheult, and Smith (1997).

History matching is based on the comparison of simulator output with historical observations. If we evaluate the simulator at a value, x, then we can judge whether x is a member of $\mathcal{X}^*$ by comparing $F(x)$ with data z. We do not expect an exact match, due to observational error and model discrepancy, and so we only require a match at some specified tolerance, often expressed in terms of the number of standard deviations between the function evaluation and the data. In practice, we cannot usually make a sufficient number of function evaluations to determine $\mathcal{X}^*$ in this way. Therefore, using the emulator, we obtain, for each x, the values $\mathrm{E}\left[F(x)\right]$ and $\mathrm{Var}\left[F(x)\right]$. We seek to rule out regions of x space for which we expect that the evaluation $F(x)$ is likely to be a very poor match to observed z.

For a particular choice of x, we may assess the potential match quality, for a single output F_i, by evaluating

$$\mathcal{I}_{(i)}(x) = \frac{|\mathrm{E}\left[F_i(x)\right] - z_i|^2}{\mathrm{Var}\left[\mathrm{E}\left[F_i(x)\right] - z_i\right]}, \tag{10.20}$$

which we term the *implausibility* that $F_i(x)$ would give an acceptable match to z_i. For given x, implausibility may be evaluated over the vector of outputs, or over selected subvectors or over a collection of individual components. In the latter case, the individual component implausibilities must be combined, for example by using

$$\mathcal{I}_M(x) = \max_i \mathcal{I}_{(i)}(x). \tag{10.21}$$

We may identify regions of x with large $\mathcal{I}_M(x)$ as implausible, i.e. unlikely to be good choice for x^*. These values are eliminated from our set of potential history matches, $\mathcal{X}^*$.

If we wish to assess the potential match of a collection of q outputs F, we use a multivariate implausibility measure analogous to (10.20) given by

$$\mathcal{I}(x) = \frac{(\mathrm{E}\left[F(x)\right] - z)^T \mathrm{Var}\left[E\left[F(x)\right] - z\right]^{-1} (\mathrm{E}\left[F(x)\right] - z)}{q}, \tag{10.22}$$

where $\mathcal{I}(x)$ is scaled to have expectation 1 if we set $x = x^*$. Unlike (10.20), the calculation of $\mathcal{I}(x)$ from (10.22) requires the specification of the full covariance structure between all components of z and F, for any pair of x values.

For comparison, a direct Bayesian approach to model calibration is described in Kennedy and O'Hagan (2001). The Bayesian calibration approach involves placing a posterior probability distribution on the 'true value' of x^*. This is meaningful to the extent that the notion of a true value for x^* is meaningful. In such cases, we may make a direct Bayesian evaluation over the reduced region $\mathcal{X}^*$, based on careful sampling and emulation within this region. If our history matching has been successful, the space over which we must calibrate will have been sufficiently reduced that calibration should be tractable and effective, provided our prior specification is sufficiently careful. As a simple approximation to this calculation, we may re-weight the values in this region by some function of our implausibility measure. The Bayes linear approach to prediction that we will describe in Section 10.3.5.1 does not need a calibration stage and so may be used directly following a successful history matching stage.

10.3.4.2 Application and results

We now use our updated emulator of $F^a(x)$ to history match the reservoir simulator. At a given well, we consider outputs corresponding to different times to be temporally correlated. Thus we apply the multivariate implausibility measure (10.22) to obtain an assessment of the potential match quality of a given input x at each well. Incorporating the definitions of z and y from (10.1) and (10.2) into the implausibility formulation, we can write the implausibility function as

$$\mathcal{I}(x) = \left\{\mathrm{E}_{F^a(x)}\left[F^a(x)\right] - z\right\}^T \left\{\mathrm{Var}_{F^a(x)}\left[F^a(x)\right] + \mathrm{Var}\left[e\right] + \mathrm{Var}\left[\epsilon\right]\right\}^{-1} \left\{\mathrm{E}_{F^a(x)}\left[F^a(x)\right] - z\right\}, \quad (10.23)$$

which is a function of the adjusted expectations and variances of our emulator for $F^a(x)$ given the model evaluations, combined with the corresponding observational data, z, and the covariances for the observational error, e, and the model discrepancy, ϵ.

We now specify our prior expectation and variance for the observational error e and the model discrepancy ϵ. We do not have any prior knowledge of biases of the simulator or the data therefore we assign $\mathrm{E}[e] = 0$ and $\mathrm{E}[\epsilon] = 0$. It is believed that our available well production history has an associated error of approximately $\pm 10\%$, therefore we assign $2 \times \mathrm{sd}(e_i) = 0.1 \times z_i$ for each emulator component $F_i^a(x)$ and we assume that there is no prior correlation between observational errors. Assessing model discrepancy is a more conceptually challenging task requiring assessment of the difference between the model evaluated at the best, but unknown, input, x^*, and the true, also unknown, value of the system. For simplicity, we assign the variance of the discrepancy to be twice that of the observational error to reflect a belief that the discrepancy has a potentially important and proportional effect. In contrast to observational errors, we introduce a relatively strong temporal correlation over the $\epsilon_i = \epsilon_{(w,t)}$ such that $\mathrm{Corr}[\epsilon_{(w,t)}, \epsilon_{(w,t')}] = \exp\{-\theta_T(t-t')^2\}$, where we assign $\theta_T = 1/36^2$ to allow the correlation to persist across all 12 time points spanning the 36-month period. We specify such a correlation over the model discrepancy since we believe that if the simulator is, for example, substantially under-estimating the system at time t, then it will be highly likely that it will also under-predict at time $t+1$.

To assess how the implausibility of input parameter choices changes with x, we construct a grid over the collection of active inputs spanning their feasible ranges and we evaluate (10.23) for each of the four selected wells, at every x point in that grid. We then have a vector of four implausibilities for every input parameter combination in the grid. To collapse these vectors into a scalar for each x, we use the maximum projection (10.21) where we maximise over the different wells to obtain a single measure $\mathcal{I}_M(x)$. This gives a conservative measure for the implausibility of a parameter choice, x, since if x is judged implausible on any one of the wells then it is deemed implausible for the collection. Thus the implausibility scores are combined in such a way that a particular input point x is only ever considered a potential match to the simulator if it is an acceptable match across all wells.

Thus we obtain a quantification for the match quality for a representative number of points throughout the possible input space. The domain of the implausibility measure is a five-dimensional cube and, as such, it is hard to visualize the implausibility structure within that space. To address this problem, we project this hypercube of implausibility values down to 2D spaces in every pair of active inputs using the method described in Craig *et al.* (1997). If we partition x such that $x = (x', x'')$, then we obtain a projection of $\hat{\mathcal{I}}(x)$ onto the subspace x' of x by calculating

$$\min_{x''} \mathcal{I}_M(x),$$

which is a function only of x'.

$\mathcal{I}_M(x)$ is a Mahalanobis distance over the four time points for each well. We produce the projections of the implausibility surface in Figure 10.2, colouring by appropriate χ^2_4 quantiles for comparison. The first plot in Figure 10.2(a) shows the approximate proportion of the implausibility space that would be excluded if we were to eliminate all points x with $\mathcal{I}_M(x)$ greater than a number of the standard deviations from the re-scaled χ^2 distribution. For example, thresholding at three standard deviations, corresponding to $\mathcal{I}_M(x) \simeq 4$, would excludes approximately 90% of of input space. The subsequent plots in Figure 10.2(b) to Figure 10.2(f) are a subset of the 2D projections of the implausibility surface onto pairs of active variables. It is clear from these plots that there are regions of low implausibility corresponding to values of ϕ less than approximately 0.8 which indicates a clear region of potential matches to our reservoir history. Higher values of ϕ are much more implausible and so would be unlikely history matches. Aside from ϕ, there appears to be little obvious structure on the remaining active variables. This is reinforced by Figure 10.2(f), which is representative of all the implausibility projections in the remaining active inputs. This plot clearly shows that there is no particular set of choices for k_x or k_z that could reasonably be excluded from consideration without making very severe restrictions of the input space. Therefore, we decide to define our region of potential matches, $\mathcal{X}^*$, by the set $\{x : \mathcal{I}_M(x) \leq 4\}$. Closer investigation revealed that this set can be well-approximated by the restriction that ϕ should be constrained to the sub-interval [0.5, 0.79].

10.3.4.3 Re-emulation of the model

Given the reduced space of input parameters, we now re-focus our analysis on this subregion with a view towards our next major task – forecasting. Our intention for the forecasting stage is, for each well, to use the

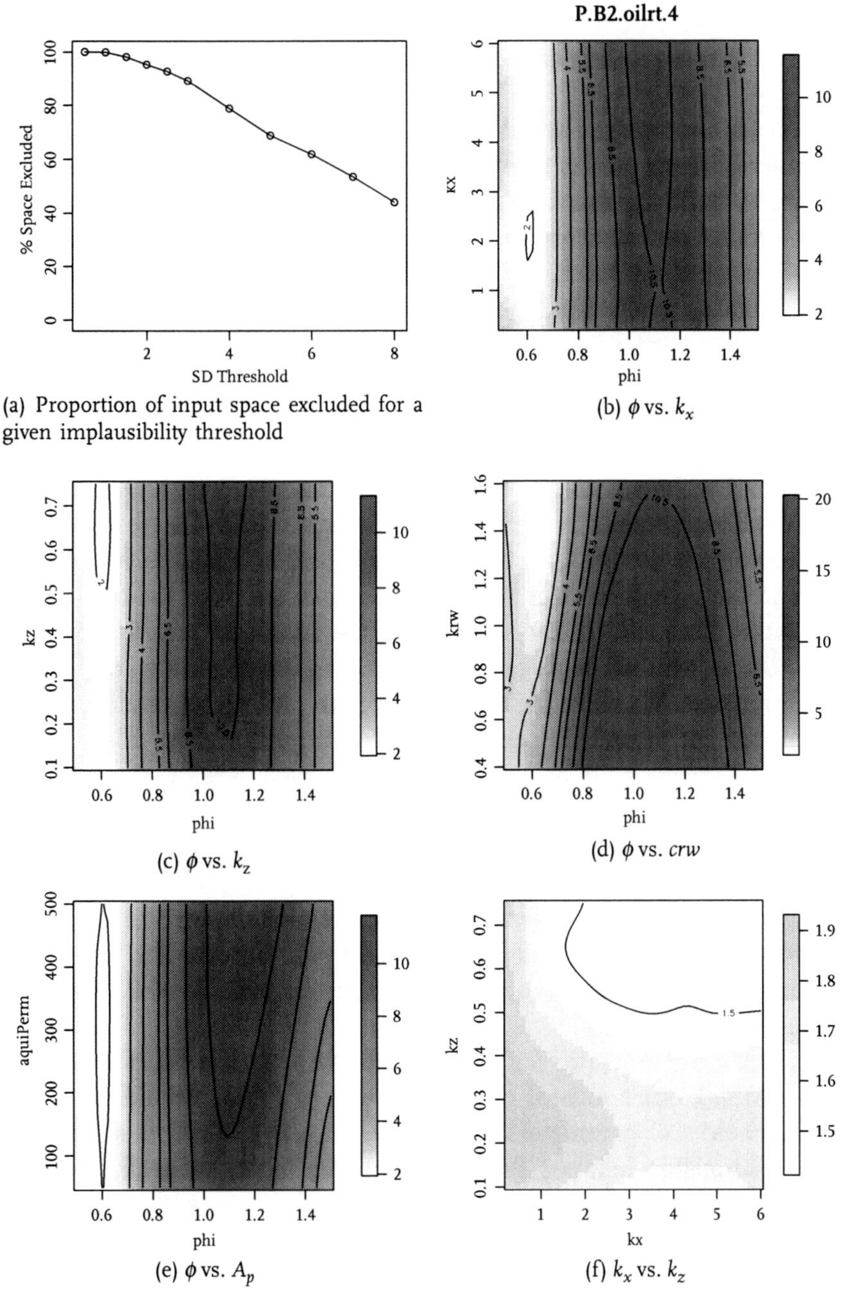

(a) Proportion of input space excluded for a given implausibility threshold

(b) ϕ vs. k_x

(c) ϕ vs. k_z

(d) ϕ vs. crw

(e) ϕ vs. A_p

(f) k_x vs. k_z

Fig. 10.2 Implausibility summary and projections for the hydrocarbon reservoir simulator.

last four time points in our existing series to forecast an additional time point located 12-months beyond the end of our original time series at $t = 48$ months. Therefore, we no longer continue investigating the behaviour of well B2 since it ceases production shortly after our original three-year emulation period.

To forecast, we require emulators for each the four historical time points as well as the additional predictive point, which we now construct over the reduced space $\mathcal{X}^*$. To build the emulators, we follow the same process as described in Section 10.3.2.2 and Section 10.3.3.1, requiring a batch of additional model runs. Of our previous batch of 1000 evaluations of $F^c(x)$, 262 were evaluated within $\mathcal{X}^*$ and so can be used at this stage. Similarly, of the 20 runs of $F^a(x)$ a total of six remain valid. These runs will be supplemented by an additional 100 evaluations of $F^c(x)$ and then by an additional 20 evaluations of $F^a(x)$.

Adopting the same strategy as Section 10.3.2.2, we construct our coarse emulators from the information contained within the large pool of model evaluations, albeit with two changes to the process. Since we have already emulated these output components in the original input space (with the exception of our predictive output), we already have some structural information in the form of the $x_{[i]}$ and the $g_i(x_{[i]})$ for each $F_i^c(x)$ that we obtained from the original emulation. Rather than completely re-executing the search for active variables and basis functions, we shall begin our searches using the original $x_{[i]}$ and $g_i(x_{[i]})$ as the starting point. We allow for the emulators to pick up any additional active variables, but not to exclude previously active inputs; and we allow for basis functions in the new $x_{[i]}$ to be both added and deleted to refine the structure of the emulator.

An emulation summary for $F^c(x)$ within the reduced region $\mathcal{X}^*$ is given in Table 10.5 for wells A3H and B1. We can see that the emulator performance is still good with high R^2 indicating that the emulators are still explaining a large proportion of the variation in model output. Observe that many of the emulators have picked up an additional active input variable when we re-focus in the reduced input space.

Considering the emulator for $F^a(x)$, we make a similar belief specification as before to link our emulator of $F^c(x)$ to that of $F^a(x)$. We choose to make the same choices of parameters $(\sigma_{\rho 1}, \sigma_{\rho 2}, \sigma_{\gamma 1}, \sigma_{\gamma 2}, \theta_T)$ to reflect a prior belief that the relationship between the two emulators in the reduced space is the similar to that over the original space. Comparing this prior with the data via the discrepancy ratio (10.16) showed that it was again reasonably consistent with $\text{Dr}(F^a(X^a))$ taking values of 2.14, 0.98, and 2.02, although perhaps we may be slightly understating our prior variance. The prior emulator was then updated by the runs of the accurate model. Looking at the final column at Table 10.5 we see that the emulator trend fits the data well, although the $\tilde{R}^2$ values appear to be decreasing over time. Example coarse and accurate emulator coefficients for

Table 10.5 Re-focused emulation summary for wells A3H and B1.

Well	Time	$x_{[i]}$	No. model terms	Coarse trend R^2	Accurate trend $\tilde{R}^2$
A3H	24	ϕ, crw, k_x, k_z	8	0.981	0.974
A3H	28	ϕ, crw, k_x, k_z, A_p	11	0.971	0.989
A3H	32	ϕ, crw, k_x, k_z	11	0.973	0.958
A3H	36	ϕ, crw, k_x	10	0.958	0.917
A3H	48	ϕ, crw, k_x	10	0.981	0.888
B1	24	ϕ, crw, k_x, A_p	11	0.894	0.982
B1	28	ϕ, crw, k_z, A_p	9	0.905	0.945
B1	32	ϕ, crw, k_x	11	0.946	0.953
B1	36	ϕ, crw, k_x	11	0.953	0.927
B1	48	ϕ, crw, k_x, A_p	11	0.941	0.880

the re-focused emulator are also given in the bottom two rows of Table 10.4, which shows more variation in the coefficients as we move from coarse to fine in the reduced space and also shows the presence of an additional active input.

10.3.5 Forecasting

10.3.5.1 Methodology – Bayes linear prediction

We wish to predict the collection y_p of future well production using the observed well history z_h. This is achieved by making an appropriate specification for the joint mean and variance of the collection (y_p, z_h), and so our prediction for y_p using the history z_h is the adjusted expectation and variance of y_p given z_h. This Bayes linear approach to forecasting is discussed extensively in Craig *et al.* (2001).

The Bayes linear forecast equations for y_p given z_h are given by

$$\mathrm{E}_{z_h}\left[y_p\right] = \mathrm{E}\left[y_p\right] + \mathrm{Cov}\left[y_p, z_h\right]\mathrm{Var}\left[z_h\right]^{-1}\left(z_h - \mathrm{E}\left[z_h\right]\right), \tag{10.24}$$

$$\mathrm{Var}_{z_h}\left[y_p\right] = \mathrm{Var}\left[y_p\right] - \mathrm{Cov}\left[y_p, z_h\right]\mathrm{Var}\left[z_h\right]^{-1}\mathrm{Cov}\left[z_h, y_p\right]. \tag{10.25}$$

Given the relations (10.1) and (10.2), we can express this forecast in terms of the 'best' simulator run $F^* = F^a(x^*)$, the model discrepancy ϵ, the observed history z_h, and the observational error e. From (10.2) we write the expectation and variance of y as $\mathrm{E}[y] = \mathrm{E}_{F^a(X^a)}[F^*] + \mathrm{E}[\epsilon]$ and $\mathrm{Var}[y] = \mathrm{Var}_{F^a(X^a)}[F^*] + \mathrm{Var}[\epsilon]$, namely the adjusted expectation and variance of the best accurate simulator run F^*, given the collection of available simulator evaluations $F^a(X^a)$, plus the model discrepancy. For simplicity of presentation, we introduce the shorthand notation $\mu^* = \mathrm{E}_{F^a(X^a)}[F^*]$, $\Sigma^* = \mathrm{Var}_{F^a(X^a)}[F^*]$, $\Sigma^\epsilon = \mathrm{Var}[\epsilon]$, and $\Sigma^e = \mathrm{Var}[e]$, and we again use the subscripts h, p to indicate the relevant subvectors and submatrices of these quantities corresponding to the historical and predictive components. We also assume $\mathrm{E}[\epsilon] = 0$ to reflect the belief that there are no systematic biases in the model known a priori. The Bayes linear forecast equations are now fully expressed as follows

$$\mathrm{E}_{z_h}\left[y_p\right] = \mu_p^* + \left(\Sigma_{ph}^* + \Sigma_{ph}^\epsilon\right)\left(\Sigma_h^* + \Sigma_h^\epsilon + \Sigma_h^e\right)^{-1}\left(z_h - \mu_h^*\right), \tag{10.26}$$

$$\mathrm{Var}_{z_h}\left[y_p\right] = \left(\Sigma_p^* + \Sigma_p^\epsilon\right) - \left(\Sigma_{ph}^* + \Sigma_{ph}^\epsilon\right)\left(\Sigma_h^* + \Sigma_h^\epsilon + \Sigma_h^e\right)^{-1}\left(\Sigma_{hp}^* + \Sigma_{hp}^\epsilon\right). \tag{10.27}$$

Given a specification for Σ^ϵ and Σ^e, we can assess the first and second order specifications $\mathrm{E}\left[F_i^a(x)\right]$, $\mathrm{Cov}\left[F_i^a(x), F_i^a(x')\right]$ from our emulator of F^a for every $x, x' \in \mathcal{X}^*$. We may therefore obtain the mean and variance of $F^* = F^a(x^*)$ by first conditioning on x^* and then integrating out with respect to an appropriate prior specification over $\mathcal{X}^*$ for x^*. Hence $\mathrm{E}_{F^a(X^a)}[F^*]$ and $\mathrm{Var}_{F^a(X^a)}[F^*]$ are calculated to be the expectation and variance (with respect to our prior belief specification about x^*) of our adjusted beliefs about $F^a(x)$ at $x = x^*$ given the model evaluations $F^a(X^a)$. Specifically, this calculation requires the computation of the expectations, variances and covariances of all $g_{ij}(x_{[i]}{}^*)$, and $w_i^a(x)$, which, in general, may require substantial numerical integration.

This analysis makes predictions without a preliminary calibration stage. Therefore, the approach is tractable even for large systems, as are search strategies to identify collections of simulator evaluations chosen to minimize adjusted forecast variance. The approach is likely to be effective when global variation outweighs local variation and the overall collection of global functional forms $g(x)$ for F_h^a and F_p^a are similar. It does not exploit the local information relevant to the predictive quantities, as represented by the residual terms $w_i^a(x)$ in the component emulators. If some quantities that we wish to predict have substantial local variation, then we may introduce a Bayes linear calibration stage before forecasting, whilst retaining tractability (Goldstein and Rougier, 2006).

10.3.5.2 Application and results

We now apply the forecasting methodology, as described in Section 10.3.5.1, to the three wells under consideration from the hydrocarbon reservoir model. The goal of this forecasting stage is to forecast the collection of future system output, y_p, using the available historical observations z_h. For a given well in the hydrocarbon reservoir, we will consider the vector of four average oil production rates at times $t = 24, 28, 32$, and 36 as historical values, and the quantity to be predicted is the corresponding production rate observed 12 months later at time $t = 48$. As we actually have observations z_p on y_p, these may act as a check on the quality of our assessments.

By history matching the hydrocarbon reservoir, we have determined a region $\mathcal{X}^*$ in which we believe it is feasible that an acceptable match x^* should lie. For our forecast, we shall consider that x^* is equally likely to be any input point contained within the region $\mathcal{X}^*$. This means that we take our prior for x^* to be uniform over $\mathcal{X}^*$. As we take the $g_{ij}(x)$ to be polynomials in x, then the expectations, variances and covariances of the $g_{ij}(x^*)$ can be found analytically from the moments of a multivariate uniform random variable which greatly simplifies the calculations of μ^* and Σ^*, the adjusted mean and variance for $F^* = F^a(x^*)$. We now refine our previously generous specification for $\mathrm{Var}[e]$. Since each output quantity is an average of four monthly values, we now take $\mathrm{Var}[e]$ to be $1/4$ its previous value to reflect the reduced uncertainty associated with the mean value.

Before we can obtain a prediction for y_p given z_h we require an appropriate belief specification for the model discrepancy ϵ, both at the past time points, and also at the future time point to be predicted. The role of the model discrepancy, ϵ, is important in forecasting as it quantifies the amount by which we allow the prediction to differ from our mean simulator prediction, $\mu^* = \mathrm{E}[F^*]$, in order to move closer to the true system value y_p. If the specified discrepancy variance is too small, then we will obtain over-confident and potentially inaccurate forecasts located in the neighbourhood of μ^*. If the discrepancy variance is too large then the forecast variances could be unfavourably large or we could over-compensate by the discrepancy and move too far from y_p. We now briefly consider the specification of $\mathrm{Var}[\epsilon]$.

Consider the plots in Figure 10.3 depicting the outputs from the reservoir model over the time period of our predictions. Observe that for time points 24 to 36 at wells B1 and B5, the mean values μ^* (indicated by the solid black circles) underestimate the observational data (indicated by the thick solid line). However at well A3H, the simulator 'overestimates' observed production. Furthermore, we observe that for well A3H the size of $|\mu^* - z|$ is a decreasing function of time, for well B1 this distance increases over time, and for well B5 $|\mu^* - z|$ appears to be roughly constant.

Given the specification for the observational error used in Section 10.3.4.2 and the observed history, we can compare $\mathrm{Var}[e_h]$ to the observed values of $(\mu^* - z_h)^2$ at each well and historical time point to obtain a simple order of magnitude data assessment for the discrepancy variance at the historical time points. To obtain our specification for $\mathrm{Var}[\epsilon]$, we took a weighted combination of prior information in the form of our belief specification for $\mathrm{Var}[\epsilon]$ from Section 10.3.4.2 and these order of magnitude numerical assessments. As the value of z_p is unknown at the prediction time $t = 48$, we must make a specification for $\mathrm{Var}\left[\epsilon_p\right]$ in the absence of any sample information. To make this assessment, we performed simple curve fitting to extrapolate the historical discrepancy variances to the forecast point $t = 48$. The resulting specification for $\mathrm{Var}[\epsilon]$ is given in Table 10.6; the correlation structure over the discrepancies is the same as in Section 10.3.4.2

Given all these specifications, we evaluate the forecast equations (10.26) and (10.27). The results of the forecasts are presented in Table10.7 alongside the corresponding prior values, and the actual observed production z_p at $t = 48$.

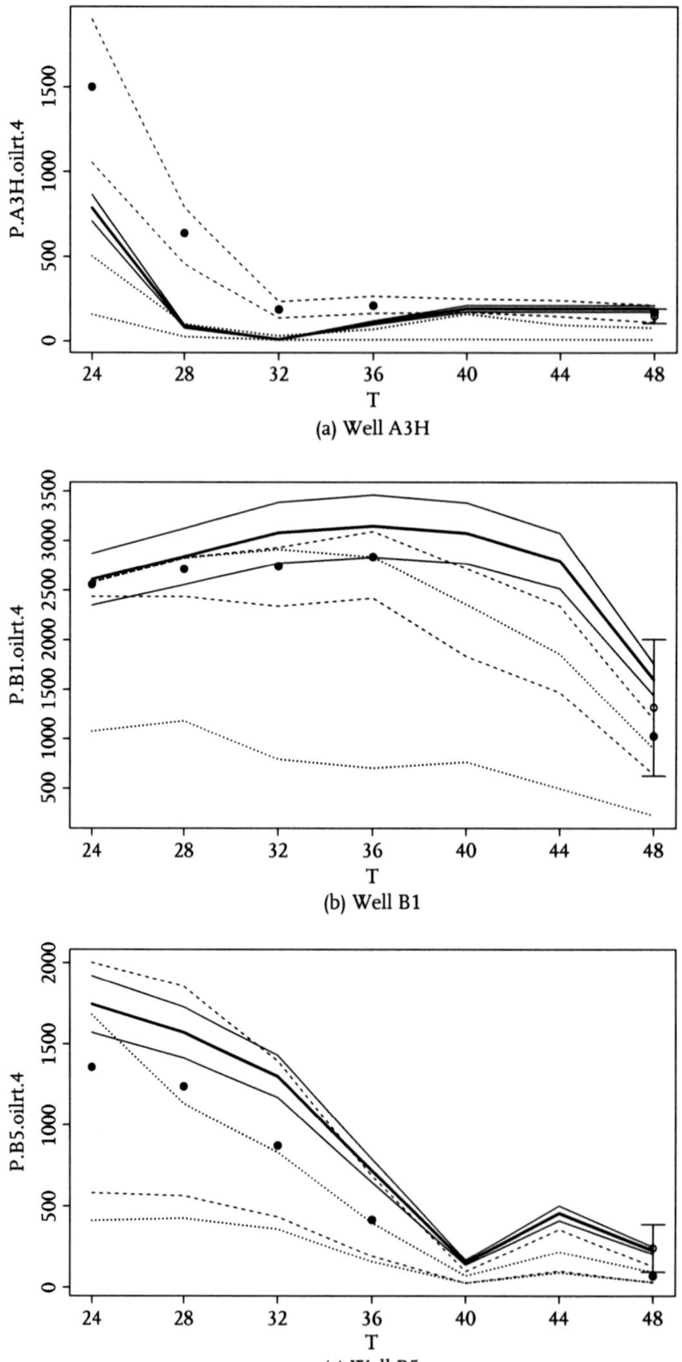

Fig. 10.3 Simulator outputs, observational data and forecasts for each well. The solid lines indicate z with error bounds of $2 \times \text{sd}(e)$. The dotted and dashed lines represent the maximum and minimum values of the runs of $F^c(x)$ and $F^a(x)$ respectively in $\mathcal{X}^*$. The solid black dots correspond to μ^*. The forecast is indicated by a hollow circle with attached error bars.

Table 10.6 Table of specified values for $2 \times \text{sd}\left(\epsilon_{(w,t)}\right)$.

$2 \times \text{sd}\left(\epsilon_{(w,t)}\right)$	$t = 24$	$t = 28$	$t = 32$	$t = 36$	$t = 48$
A3H	504.9	390.4	124.5	71.5	14.7
B1	130.5	142.0	260.4	245.1	408.0
B5	284.3	239.3	305.8	214.7	260.8

The forecasts and their errors are also shown on Figure 10.3 by a hollow circle with attached error bars.

We can interpret the prediction $\text{E}_{z_h}\left[y_p\right]$ as the forecast from the simulator for y_p which is modified in the light of the discrepancy between $\text{E}_{z_h}\left[y_h\right]$ and the observed z_h. In the case of the wells tabulated, the simulator for well A3H over-estimates the value of z_h during the period $t = 24, \ldots, 36$ resulting in a negative discrepancy and a consequent downward correction to our forecast. However, interestingly in the intervening period the simulator changes from under-estimating to over-estimating observed well production, and so on the basis of our observed history alone we under-predict the oil production rate for this well. The wells B1 and B5 behaved in a more consistent manner with a constant under-prediction of the observed data compared to the observed history being reflected by an increase to our forecast. Note that whilst some of the best runs of our computer model differ substantially from z_p, the corrections made by the model discrepancy result in all of our forecast intervals being within the measurement error of the data.

In practice, the uncertainty analysis for a reservoir model is an iterative process based on monitoring the sizes of our final forecast intervals and investigating the sensitivity of our predictions to the magnitude of the discrepancy variance and correlation, for example by repeating the forecasts using the discrepancy variance values from Table 10.6 scaled by some constant α, for various values of α and varying the degree of temporal correlation across discrepancy terms. If we obtain forecast intervals which are sufficiently narrow to be useful to reservoir engineers then the we can end our analysis at this

Table 10.7 Forecasts of y_p at $t = 48$ using z_h for three wells in the hydrocarbon reservoir.

		A3H	B1	B5
Observation	z_p	190.0	1598.6	227.0
	$2 \times \text{sd}\,(e)$	19.0	159.9	22.7
Prior	$\text{E}\left[y_p\right]$	170.2	1027.1	69.4
	$2 \times \text{sd}\left(y_p\right)$	62.9	595.1	270.3
Forecast	$\text{E}_{z_h}\left[y_p\right]$	170.3	1207.1	299.8
	$2 \times \text{sd}_{z_h}(y_p)$	39.1	348.9	167.4

stage. If, however, our prediction intervals are unhelpfully large then we return to and repeat earlier stages of the analysis. For example, introducing additional wells and other aspects of historical data into our analysis could be helpful in further reducing the size of $\mathcal{X}^*$, and allow us to refocus again and therefore reduce the uncertainties attached to our emulator, and so narrow the forecast interval.

Appendix

A. Broader context and background

The Bayes linear approach is similar in spirit to conventional Bayes analysis, but derives from a simpler system for prior specification and analysis, and so offers a practical methodology for analysing partially specified beliefs for large problems. The approach uses expectation rather than probability as the primitive for quantifying uncertainty; see De Finetti (1974, 1975). In the Bayes linear approach, we make direct prior specifications for that collection of means, variances and covariances which we are both willing and able to assess. Given two random vectors, B, D, the *adjusted* expectation for element B_i, given D, is the linear combination $a_0 + a^T D$ minimising $\mathrm{E}\left[(B_i - a_0 - a^T D)^2\right]$ over choices of a_0, a. The adjusted expectation vector, $\mathrm{E}_D[B]$ is evaluated as

$$\mathrm{E}_D[B] = \mathrm{E}[B] + \mathrm{Cov}[B, D](\mathrm{Var}[D])^{-1}(D - \mathrm{E}[D]).$$

(If $\mathrm{Var}[D]$ is not invertible, then we use an appropriate generalised inverse). The *adjusted variance matrix* for B given D, is

$$\mathrm{Var}_D[B] = \mathrm{Var}[B - \mathrm{E}_D[B]] = \mathrm{Var}[B] - \mathrm{Cov}[B, D](\mathrm{Var}[D])^{-1}\mathrm{Cov}[D, B].$$

Stone (1963), and Hartigan (1969) were among the first to discuss the role of such assessments in partial Bayes analysis. A detailed account of Bayes linear methodology is given in Goldstein and Wooff (2007), emphasizing the interpretive and diagnostic cycle of subjectivist belief analysis.

The basic approach to statistical modelling within this formalism is through second order exchangeability. An infinite sequence of vectors is second-order exchangeable if the mean, variance and covariance specification is invariant under permutation. Such sequences satisfy the second order representation theorem which states that each element of such a sequence may be decomposed as the uncorrelated sum of an underlying 'population mean' quantity and an individual residual, where the residual quantities are themselves uncorrelated with zero mean and equal variances. This is similar in spirit to de Finetti's representation theorem for fully exchangeable sequences but is sufficiently weak, in the requirements for prior specification, that it allows us to construct statistical models directly from simple collections of judgements over observable quantities.

Within the usual Bayesian view, adjusted expectation offers a simple, tractable approximation to conditional expectation, and adjusted variance is a strict upper bound for expected posterior variance, over all prior specifications consistent with the given moment structure. The approximations are exact in certain special cases, and in particular if the joint probability distribution of B, D is multivariate normal. Adjusted expectation is numerically equivalent to conditional expectation when D comprises the indicator functions for the elements of a partition, i.e. each D_i takes value one or zero and precisely one element D_i will equal one. We may therefore view adjusted expectation as a generalization of de Finetti's approach to conditional expectation based on 'called-off' quadratic penalties, where we remove the restriction that we may only condition on the indicator functions for a partition. Geometrically, we may view each individual random quantity as a vector, and construct the natural inner product space based on covariance. In this construction, the adjusted expectation of a random quantity Y, by a further collection of random quantities D, is the orthogonal projection of Y into the linear subspace spanned by the elements of D and the adjusted variance is the squared distance between Y and that subspace. This formalism extends naturally to handle infinite collections of expectation statements, for example those associated with a standard Bayesian analysis.

A more fundamental interpretation of the Bayes linear approach derives from the temporal sure preference principle, which says, informally, that if it is necessary that you will prefer a certain small random penalty A to C at some given future time, then you should not now have a strict preference for penalty C over A. A consequence of this principle is that you must judge now that your actual posterior expectation, $\mathrm{E}_T[B]$, at time T when you have observed D, satisfies the relation $\mathrm{E}_T[B] = \mathrm{E}_D[B] + R$, where R has, a priori, zero expectation and is uncorrelated with D. If D represents a partition, then $\mathrm{E}_D[B]$ is equal to the conditional expectation given D, and R has conditional expectation zero for each member of the partition. In this view, the correspondence between actual belief revisions and formal analysis based on partial prior specifications is entirely derived through stochastic relationships of this type.

Acknowledgements

This study was produced with the support of the Basic Technology initiative as part of the Managing Uncertainty for Complex Models project. We are grateful to Energy SciTech Limited for the use of the reservoir simulator software and providing the Gullfaks reservoir model.

References

Craig, P. S., Goldstein, M., Rougier, J. C., and Seheult, A. H. (2001), Bayesian forecasting for complex systems using computer simulators, *Journal of the American Statistical Association*, 96, 717–729.

Craig, P. S., Goldstein, M., Seheult, A. H., and Smith, J. A. (1996). Bayes linear strategies for history matching of hydrocarbon reservoirs. In *Bayesian Statistics 5* (eds. J. M. Bernardo, J. O., Berger, A. P., Dawid and A. F. M., Smith), Clarendon Press, Oxford, UK, pp. 69–95.

Craig, P. S., Goldstein, M., Seheult, A. H., and Smith, J. A. (1997). Pressure matching for hydrocarbon reservoirs: A case study in the use of Bayes linear strategies for large computer experiments. In *Case Studies in Bayesian Statistics* (eds. C., Gatsonis, J. S., Hodges, R. E., Kass, R., McCulloch, P., Rossi, and N. D., Singpurwalla), Springer-Verlag, New York, vol. 3, pp. 36–93.

Craig, P. S., Goldstein, M., Seheult, A. H., and Smith, J. A. (1998). Constructing partial prior specifications for models of complex physical systems. *Applied Statistics*, **47**, 37–53.

Cressie, N. (1991). *Statistics for Spatial Data*. John Wiley, New York.

Cumming, J. A. and Wooff, D. A. (2007). Dimension reduction via principal variables, *Computational Statistics & Data Analsysis*, **52**, 550–565.

Currin, C., Mitchell, T., Morris, M., and Ylvisaker, D. (1991). Bayesian prediction of deterministic functions with applications to the design and analysis of computer experiments. *Journal of the American Statistical Association*, **86**, 953–963.

De Finetti, B. (1974), *Theory of Probability*, Vol. 1, New York: John Wiley.

De Finetti, B. (1975). *Theory of Probability*. Vol. 2. John Wiley, New York.

Goldstein, M. and Rougier, J. C. (2006), Bayes linear calibrated prediction for complex systems, *Journal of the American Statistical Association*, **101**, 1132–1143.

Goldstein, M. and Rougier, J. C. (2008), Reified Bayesian modelling and inference for physical Systems. *Journal of Statistical Planning and Inference*, **139**, 1221–1239.

Goldstein, M. and Wooff, D. A. (2007). *Bayes Linear Statistics: Theory and Methods*. John Wiley, New York.

Hartigan, J. A. (1969). Linear Bayes methods. *Journal of the Royal Statistical Society, Series B*, **31**, 446–454.

Kennedy, M. C. and O'Hagan, A. (2000). Predicting the output from a complex computer code when fast approximations are available, *Biometrika*, **87**, 1–13.

Kennedy, M. C. and O'Hagan, A. (2001). Bayesian calibration of computer models. *Journal of the Royal Statistical Society, Series B*, **63**, 425–464.

McKay, M. D., Beckman, R. J., and Conover, W. J. (1979). A comparison of three methods for selecting values of input variables in the analysis of output from a computer code. *Technometrics*, **21**, 239–245.

O'Hagan, A. (2006). Bayesian analysis of computer code outputs: A tutorial. *Reliability Engineering and System Safety*, **91**, 1290–1300.

Qian, Z. and Wu, C. F. J. (2008). Bayesian hierarchical modeling for integrating low-accuracy and high-accuracy experiments. *Technometrics*, **50**, 192–204.

Sacks, J., Welch, W. J., Mitchell, T. J., and Wynn, H. P. (1989). Design and analysis of computer experiments. *Statistical Science*, **4**, 409–435.

Santner, T. J., Williams, B. J., and Notz, W. I. (2003). *The Design and Analysis of Computer Experiments*, Springer-Verlag, New York.

Stone, M. (1963). Robustness of non-ideal decision procedures. *Journal of the American Statistical Association*, **58**, 480–486.

·11·

Bayesian modelling of train door reliability

Antonio Pievatolo and Fabrizio Ruggeri

11.1 Train door reliability

We consider failure data from 40 underground trains, which were delivered to an European transportation company between November 1989 and March 1991 and all of them were put in service from 20th March 1990 to 20th July 1992. Failure monitoring ended on 31st December 1998. When a failure took place, both the odometer reading and failure date were recorded, along with the code of the failed component.

The original research, described in Pievatolo *et al.* (2003), was aimed to analyse door reliability before warranty expiration. That paper considered the failures altogether, without making any distinction among different failure modes. In the current study, we first perform exploratory data analysis to identify the most important failure modes among the seven classified ones, and, then we consider models for the failures due either to opening commands or mechanical parts.

Doors, like other train components, are to be built following some technical requirements, ensuring a minimum level of reliability, i.e. a constraint on the maximum number of failures expected within specified time (in days) and space (in kilometres) intervals.

The research is aimed to find models able to describe the failure history and to predict the number of failures in future time intervals.

Train doors can be thought of as complex repairable systems, subject to minimal repair upon failure. In fact, doors are not replaced, in general, when a failure occurs but the repair intervention is only on their failed component, either fixing or replacing it. Therefore, interventions after failures do not significantly affect the behavior and the reliability of the door. This is the typical situation described in reliability as *bad as old*, since repairs are minimal ones and they are bringing the system reliability back to its status just before the failure. Some authors prefer the expression *same as old* since reliability before failure and after repair could both actually be high. The opposite situation (*good* (*same*) *as new*) is encountered when the entire system is replaced upon failure. The former situation, corresponding to our case, is usually modelled using a

Poisson process, while the latter is in general described by a renewal process. More details on the statistical analysis of repairable systems can be found in Rigdon and Basu (2000).

The choice of a Poisson process (see Definition 1 in the appendix) is also motivated by the superposition of different causes of failures, as stated by Theorem 1 in the appendix. In our case the seven possible failure modes are the result of a coarser classification with respect to the actual data. Therefore, the Poisson process assumption is justified by pooling components in seven subsystems. Our research is thus aimed at finding an adequate Poisson process model to describe door failures. We started from a descriptive statistical analysis to identify desirable features of the model and then we entertained some models until we found the ones presented in Section 11.2.2. First of all, data are collected on a double scale (kilometres and days) and both pieces of information are relevant. Not many papers in reliability have considered failure data on two scales, one of the major exceptions being Singpurwalla and Wilson (1998).

In previous work (Pievatolo *et al.* 2003) we considered a hierarchical model to relate kilometres and days; here we are considering a bivariate intensity function for the Poisson process. Plots of kilometres versus days for each failure show a rather linear relation between them and it is reflected in the bivariate intensity function. In the same paper, combined data about all failure modes showed a 12-month periodicity; here we are introducing a periodic component in the intensity function as well, based upon findings from periodograms. Furthermore, concavity of the curve of cumulative number of failures with respect to time induces us to consider a baseline component decreasing over time in the intensity function.

Quality of data might be a problem since few failures are recorded at the same time and there are serious doubts about such coincidences. Furthermore, it is not clear if few big gaps between successive failures are actually true or due to delayed recording. Unfortunately, it is no longer possible to contact the transport company to get explanations about such cases. The limited number of dubious data should not affect the analysis significantly. We kept both big gaps and coincident failures.

Posterior distributions of the parameters of the entertained models have been obtained via MCMC simulations and they have been used mostly to make forecast on the number of future failures. The inspiring motivation for the research was the need for the company to determine the reliability level of the doors train before warranty expiration. Trains are to be built according to some technical standards and they must fulfil given reliability requirements. It is company's interest if such requirements are satisfied before the expiration of the warranty. If the transport company is able to prove the poor reliability of the trains within such a deadline, then the manufacturer will be asked to intervene and improve the reliability at its own expense; otherwise, interventions after the deadline will be the company's responsibility.

11.2 Modelling and data analysis

The door opening system is composed of many parts, and the failure tree is available at the single component level. As previously discussed, a model for minimal repair based on the Poisson process can be postulated only at higher levels, by pooling components into subsystems. The transport company identifies seven subsystems; we analyse in detail two of them, because the other ones either feature a number of failures that is too small for fitting a model, or show changes in the failure occurrence pattern that would require a deeper knowledge of the maintenance actions carried out by the company to be addressed seriously.

11.2.1 Exploratory data analysis

The failures of 40 trains are observed from 20th March 1990, when the first train was put into service, until 31st December 1998. The possible failure modes are coded as shown in Table 11.1.

The plots of the cumulative number of failures versus calendar days (reckoned from a common origin) appear in Figures 11.1 and 11.2. We make the following remarks.

1. The 40 trains were put into service gradually over a period of about 1000 calendar days (from day 1 up to day 932): this explains the convexity of the initial part of the curves.
2. Failure modes 4 and 5 occur very rarely, so there is not enough information for fitting a stochastic process model.
3. Failure modes 6 and 7 show change-points very clearly. We could model those change-points as in Ruggeri and Sivaganesan (2005), but it would

Table 11.1 Classification of failure modes and total failures per mode for all trains in nine years.

Code	Subsystem	No. of parts	Total failures
1	opening commands (electrical)	14	530
2	cables and clamps	4	33
3	mechanical parts	67	1182
4	electrical protections	12	9
5	power supply circuit	2	7
6	pneumatic gear	31	295
7	electro-valves	8	39

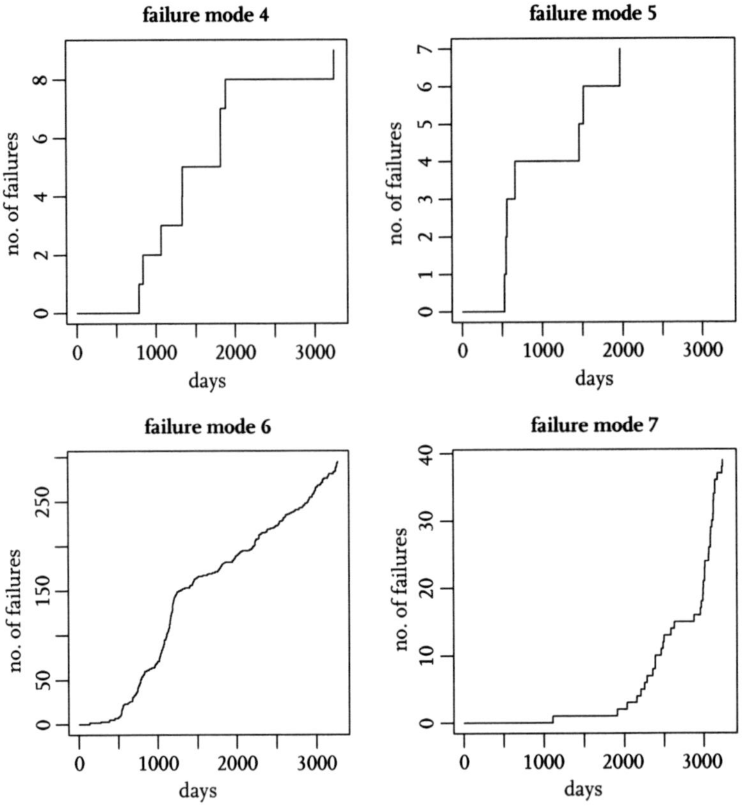

Fig. 11.1 Cumulative number of failures versus time for all trains (failure modes 4–7).

be more interesting to understand why they occurred, rather than obtain descriptive change-point models.

4. Failure modes 1, 2 and 3 display a more regular pattern. Mode 2 failures are only 0.11 per train and per year,[1] therefore we concentrate on modes 1 and 3. In general, the curves which appear to follow a regular pattern are those related to subsystems with a larger number of parts, which is an expected feature of the superposition of many point processes.

There are two reasons for considering time instead of space on the x-axis of the plots. First, the observation of the failure process stops at the end of the ninth year, whereas there is no stopping rule on the space scale. Second, it is easier to identify seasonal effects: in Figure 11.2, a cyclic component is apparent for failure mode 2. On the other hand, distance is the natural unit for measuring wear, therefore the modelling itself must consider both scales, by means of a two-dimensional Poisson process.

[1] This figure is obtained as 33/(113,748/365.25), where 113,748 is the total number of days run by all trains.

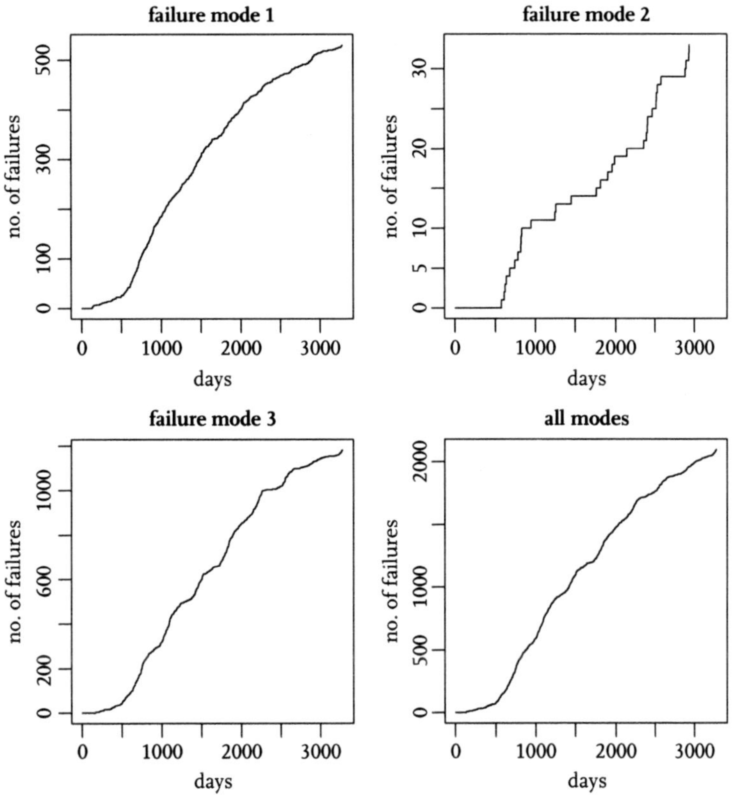

Fig. 11.2 Cumulative number of failures versus time for all trains (failure modes 1–3 and all modes).

For individual trains, we may plot kilometres versus days, using the pairs recorded at failure times. Such plots show an essentially linear relationship, even though with a smaller slope than that implied by the mission profile indicated by the company.[2]

11.2.2 A bivariate intensity function

For failure modes 1 and 3, the concavity in the plots of cumulative number of failures indicates a decreasing rate of occurrence of failures. Failures are concentrated along the line of the day-kilometer relationship. Therefore the intensity function of the two-dimensional Poisson process should be higher above the line of the day-kilometer relationship, decaying to zero as one moves away from it, and also decreasing as one moves along the line away from the origin.

We denote the intensity function as $\lambda(t, s)$, where t is time (days) and s is space (kilometres). A form for it that meets the above requirements is the

[2] Each train should have run 85,000 km per year, that is, 233 km per day. However, the actual mean daily distance is 194 km, with groupings of trains around two different mean values (see Figure 11.3).

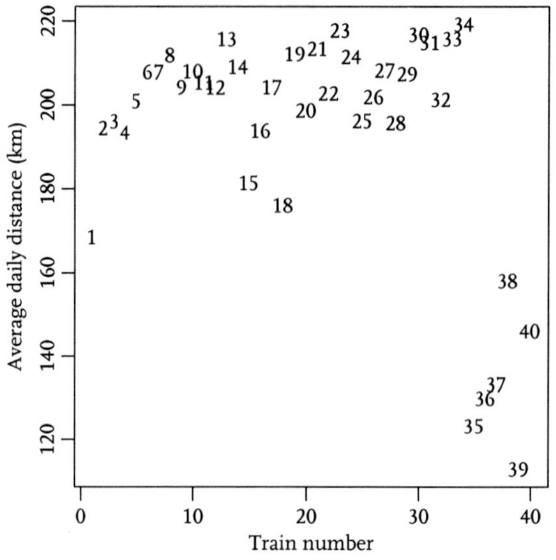

Fig. 11.3 Average daily distance (km) run from individual trains versus train number.

following, for train number i:

$$\lambda_i(t, s) = \mu \exp\left\{-\gamma(s - a_i - c_i(t - t_{0i}))^2 w(t - t_{0i})\right\}$$
$$\times \exp\left\{A \cos(\omega(t - d))\right\} \lambda_0(t - t_{0i}), \tag{11.1}$$

where t_{0i} is the date when the train was put into service, $a_i + c_i(t - t_{0i})$ is the expected distance after $(t - t_{0i})$ days in service, $w(\cdot)$ is a positive weight function, $\lambda_0(\cdot)$ is a baseline intensity function, and the exponentiated cosine is a periodic component with phase d. The remaining quantities are nonnegative unknown parameters. We observe that the baseline and the distance are a function of the number of days in service, whereas the periodic component is a function of calendar days, because a seasonal effect will be the same for all trains. The intercept and slope (a_i, c_i) are different for every train to reflect individual departures from the average (see Figure 11.3). We have included a nonzero intercept term because the relationship between days and distance is linear only after a short initial transient phase, which is accounted for by downweighting the squared differences $(s - a_i - c_i(t - t_{0i}))^2$ for small values of $(t - t_{0i})$; for large values, the weight function tends to one.

The marginal intensity function on the temporal scale is obtained after integrating $\lambda_i(t, s)$ over $[0, +\infty)$ with respect to s:

$$\lambda_i(t) = \mu \sqrt{\frac{\pi}{\gamma w(t - t_{0i})}}\, \Phi\left\{(a_i + c_i(t - t_{0i}))\sqrt{2\gamma w(t - t_{0i})}\right\}$$
$$\times \exp\left\{A \cos(\omega(t - d))\right\} \lambda_0(t - t_{0i}), \tag{11.2}$$

where Φ denotes the standard Gaussian cumulative distribution function.

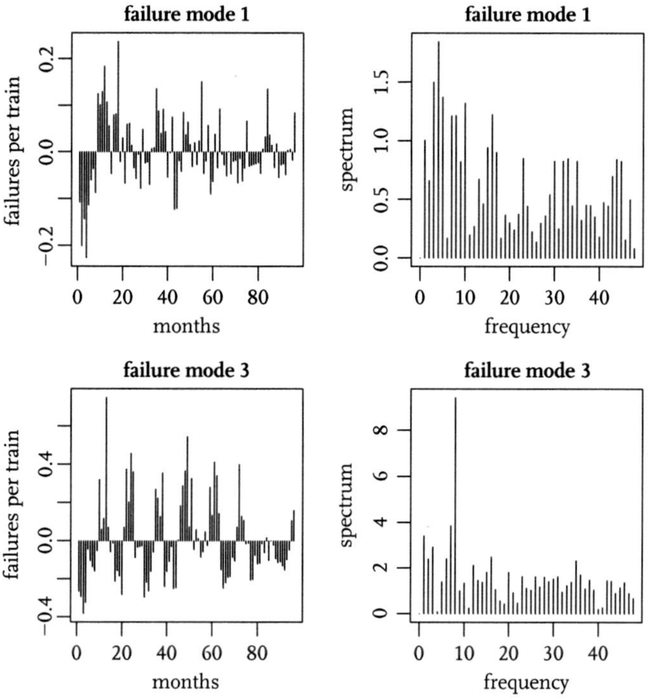

Fig. 11.4 Detrended average monthly number of failures per train and its spectrum.

The intensity function (11.1) contains several unknown parameters, that is $(\mu, \gamma, a_i, c_i, A, \omega, d)$ and those in the baseline; they will be estimated with a Bayesian model, except for ω, whose value can be found from a simple descriptive analysis of the data. Consider in fact the periodogram of the monthly time series of failures for failure modes 1 and 3 (after detrending), shown in Figure 11.4. There is no clear predominant frequency for the former, whereas a 12-month cycle is evident for the latter. Thus, we will simply omit the periodic component in the intensity function for failure mode 1, whereas we take $\omega = 2\pi/365$ for failure mode 3.

Regarding the phase d, the detrended time series of monthly failures for mode 3 has peaks during winter. There is also the additional information that the deposit of the trains is outdoor, and cold is an adverse condition for mechanical parts. To reflect this information, we take a random d uniformly distributed on days from 1 October to 31 March.

The plots of the cumulative number of failures of all trains versus days have proven useful for spotting a periodic component in the intensity function. While they show an improvement in the rate of occurrence of failures, they do not give the correct information for postulating the form of $\lambda_0(\cdot)$. They are in fact the result of the superposition of point processes with different origins and parameters, hence they do not reflect the shape of $\lambda_0(\cdot)$ only and we should examine the trains individually.

One possibility for $\lambda_0(\cdot)$ is the popular power law process, that is, $\lambda_0(u)$ behaves as u^{b-1}, up to a multiplicative constant, with $b < 1$ indicating a reliability growth. The corresponding mean value function, denoted by $\Lambda_0(u)$, is proportional to u^b. If the data are from a power law process, then the failure times can be transformed into an order statistics from the standard exponential distribution and exponentiality can be assessed via a total time on test plot. The resulting plots are quite heterogeneous, that is, the observed points follow quite closely the bisector of the unit square for some trains, but not for others. The same is obtained with a different mean value function, $\Lambda_0(u) = \ln(1 + bu)$, which seems to fit a little better overall, but does not eliminate the heterogeneity. We call *reciprocal* this mean value function because of the shape of its intensity function. Another candidate (*exponential*) that brings in a further little improvement is $\Lambda_0(u) = (1 - e^{-bt})/b$.[3] In short, there is no single intensity function that is suitable for all trains. Nonetheless, even though a function may not fit every single train perfectly, it may fit the superposed data, because different behaviors of individual trains may compensate each other. Hence we follow the route of estimating models with different baseline intensities and choose the one that fits the superposed failure data better.

11.2.3 The likelihood, the prior and parameter estimation

We can reasonably assume that the trains are conditionally independent (different assumptions are discussed in the appendix). Let $n = 40$ denote the total number of trains. Then let $a = (a_1, \ldots, a_n)$, $c = (c_1, \ldots, c_n)$ and gather all the unknown parameters in the vector $\theta = (\mu, \gamma, a, c, A, d, b)$. For the observed data, we denote by n_i the number of failures of train i and by (t_{ij}, s_{ij}) the day and dilometres of failure j for train i. Finally let T be the last calendar day when failures were observed. Then the likelihood function is

$$L(\theta) = \prod_{i=1}^{n} \left[\left\{ \prod_{j=1}^{n_i} \lambda_i(t_{ij}, s_{ij}) \right\} \exp\left(- \int_{t_{0i}}^{T} \lambda_i(t)\, dt \right) \right], \tag{11.3}$$

where the integral of $\lambda_i(t, s)$ over $[0, T] \times [0, +\infty)$ appears in the negative exponential function, using (11.2). The integral of $\lambda_i(t)$ cannot be found in closed form, therefore numerical integration will be necessary. A look at equation (11.2) shows that there will be no conjugate distribution for the parameters

[3] For the logarithmic and the exponential mean value functions we are aware of no parameter-independent transformation of the failure times leading to a standard distribution, so that one must transform with the mean value function itself, where parameter estimates must be inserted; then the transformed data follow a homogeneous Poisson process.

except for μ, so that the Markov chain Monte Carlo algorithm for the model estimation can only be Metropolis–Hastings, which we will design componentwise.

For the calculation of the logarithm of the acceptance ratios in the Metropolis–Hastings algorithms, it is convenient to break up the log-likelihood, $\ell(\theta)$, into blocks. For example, with the negative exponential baseline $\lambda_0(u)$ proportional to $\exp\{-bu\}$, we have:

$$\ell(\theta) = -\mu\sqrt{\frac{\pi}{\gamma}}\sum_{i=1}^{m} \ell_{0i}(\theta) + \left(\sum_{i=1}^{n} n_i\right)\ln\mu - \gamma\sum_{i=1}^{m} \ell_1(a_i, c_i) + A\ell_2(d) - b\ell_3, \tag{11.4}$$

where

$$\ell_{0i}(\theta) = \int_{t_{0i}}^{T} \frac{1}{\sqrt{w(t - t_{0i})}}\, \Phi\left\{(a_i + c_i(t - t_{0i}))\sqrt{2\gamma w(t - t_{0i})}\right\} \times e^{A\cos\{\omega(t-d)\} - b(t - t_{0i})}\, dt \tag{11.5}$$

$$\ell_1(a_i, c_i) = \sum_{j=1}^{n_i} [s_{ij} - \{a_i + c_i(t_{ij} - t_{0i})\}]^2 w(t_{ij} - t_{0i}) \tag{11.6}$$

$$\ell_2(d) = \sum_{i=1}^{n}\sum_{j=1}^{n_i} \cos\{\omega(t_{ij} - d)\} \tag{11.7}$$

$$\ell_3 = \sum_{i=1}^{n}\sum_{j=1}^{n_i} (t_{ij} - t_{0i}). \tag{11.8}$$

Hence, for all parameters except for μ, the logarithm of the acceptance ratio will involve the evaluation of the terms ℓ_{0i} by numerical integration, which depends on the parameter values themselves. In this situation it is highly convenient to use uniform priors on suitable ranges (that is, wide enough) for all parameters, whereas for μ we choose a vague prior.

Prior independence is assumed among all components of θ; the elements of a are taken as exchangeable, as well as the elements of c. Specifically, the a_i's are conditionally i.i.d. Gaussian with mean α and known standard deviation σ_a, with a vague prior for α; the c_i's are conditionally i.i.d. exponential with mean $1/\tau$, with a vague prior for τ. In this way it is possible to obtain the predictive distribution of failures of a new train, through the posterior distribution of the parameters common to all trains, that is (μ, γ, A, d, b), and of the train-specific intercept and slope parameters.

The update of μ, τ and α is done through their full conditionals. The full conditional of μ is Gamma with shape $\sum n_i + 1$ and rate $\sqrt{\pi/\gamma}\sum \ell_{0i}(\theta)$; the full conditional of τ is Gamma with shape $n + 1$ and rate $\sum c_i$; the full conditional of α is Gaussian with mean $\sum a_i/n$ and variance σ_a/n. For the other parameters we use a random walk with a truncated normal distribution as proposal. Then, if θ_i is a generic parameter with support $(\theta_{i0}, \theta_{i1})$, the proposal density evaluated at θ_i' is

$$q(\theta_i, \theta_i') = \frac{\frac{1}{\sigma_i}\varphi\left(\frac{\theta_i'-\theta_i}{\sigma_i}\right)}{\Phi\left(\frac{\theta_{i1}-\theta_i}{\sigma_i}\right) - \Phi\left(\frac{\theta_{i0}-\theta_i}{\sigma_i}\right)}, \qquad \theta_i' \in (\theta_{i0}, \theta_{i1}),$$

where σ_i is a fixed dispersion parameter and φ and Φ are the density and the cumulative distribution function of a standard normal, respectively. With these specifications, the logarithm of the acceptance ratio of γ' when the current state is γ, for example, is

$$-\mu\sqrt{\pi}\left\{\frac{\sum_i \ell_{0i}(\theta')}{\sqrt{\gamma'}} - \frac{\sum_i \ell_{0i}(\theta)}{\sqrt{\gamma}}\right\} - (\gamma' - \gamma)\sum_i \ell_1(a_i, c_i)$$
$$- \ln\left\{\frac{\Phi\left(\frac{\gamma_{i1}-\gamma_i'}{\sigma_\gamma}\right) - \Phi\left(\frac{\gamma_{i0}-\gamma_i'}{\sigma_\gamma}\right)}{\Phi\left(\frac{\gamma_{i1}-\gamma_i}{\sigma_\gamma}\right) - \Phi\left(\frac{\gamma_{i0}-\gamma_i}{\sigma_\gamma}\right)}\right\}$$

where θ' is the parameter vector with γ' replacing γ, $(\gamma_{i0}, \gamma_{i1})$ is the support of γ and σ_γ is the (constant) dispersion parameter of the proposal.

The numerical integration required for the ℓ_{0i} terms will not encounter convergence problems if the integrand is bounded over the domain $[t_{0i}, T]$. In the present case, we only need to ensure that $w(\cdot)$ is bounded away from zero, which is true if

$$w(u) = \frac{\sqrt{1+u}}{1+\sqrt{1+u}}, \qquad u \geq 0. \tag{11.9}$$

This weight function is bounded from below by 1/2 and tends to 1 as u goes to infinity, meeting the requirements introduced previously.

All this modelling and computation might seem a little cumbersome at first, but a short pause for reconsideration of all the steps made so far will show instead that this is a simple model indeed, if one wants to capture at least the main features of the data.

We are now ready to produce parameter estimates and predictions, but first a quick model checking is in order, mainly to choose the best fitting baseline intensity function. Then, with the chosen baseline, a comparison between predictions and observed data will provide a further model validation.

The quick model checking is done by drawing the observed cumulative number of failures of all trains for a given failure mode and an estimate of the corresponding mean value function in the same plot. For the estimate, we

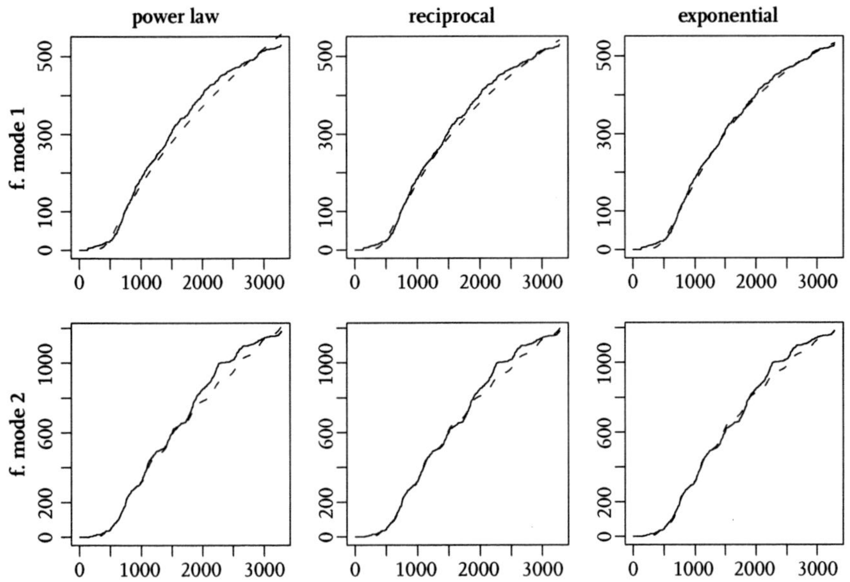

Fig. 11.5 Cumulative number of failures for all trains and estimated mean value function (dashed). First row: failure mode 1. Second row: failure mode 2. Each column is for a different baseline intensity.

approximate the posterior mean of θ from an MCMC run, let us say $\hat{\theta}$, and plot

$$\Lambda(t;\hat{\theta}) = \sum_{i=1}^{m} \int_{t_{0i}}^{t} \lambda_i(u;\hat{\theta})\,du, \qquad t = 1, \ldots, 3287$$

on a grid of days covering the nine years of observed failure data. The $\lambda_i(\cdot;\hat{\theta})$ functions are those defined in (11.2), where we have made the dependence on $\hat{\theta}$ explicit. The estimate $\Lambda(t;\hat{\theta})$ is not the optimal Bayesian estimate, because we should use the posterior mean of $\Lambda(t;\theta)$ instead. The reason for not doing so is the need of avoiding an array of numerical integrations for every iteration in the MCMC run. In Figure 11.5, we see that the best baseline intensity function among those proposed is the negative exponential.

Goodness of fit of the model could be assessed, but in a non-Bayesian fashion, using Theorem 4 in the appendix which leads to a Kolmogorov–Smirnov test with a uniform distribution as null hypothesis, as in Ogata (1988); in fact, should the estimated model be the *true* one, then adequately transformed failure times would be uniformly distributed.

11.2.4 Failure forecasting and warranty assessment

The Poisson process model is a basic probabilistic tool that the transport company can use for various purposes, such as failure forecasting or assessment of the warranty parameters.

Let us focus on failure forecasting first. Suppose we have observed the failures up to a calendar day $T_0 \leq T$, then we will be interested in the predictive distribution of the number of failures in the interval $(T_0, T_0 + u)$ for some value of u (such as $u = 365$ or $u = 730$). The predictive distribution can be approximated at all points using a single stream $\{\theta^{(j)}\}$ of parameter values generated from the MCMC sampling scheme. In fact, let us denote by N_t the number of failures of a given mode observed up to day t and let us denote by D_{T_0} the data available at day T_0. Then, for $x = 0, 1, 2, \ldots,$

$$P(N_{T_0+u} - N_{T_0} = x \mid D_{T_0}) = \int e^{-\{\Lambda(T_0+u;\theta)-\Lambda(T_0;\theta)\}} \times \frac{\{\Lambda(T_0 + u; \theta) - \Lambda(T_0; \theta)\}^x}{x!} \pi(\theta \mid D_{T_0})\, d\theta \qquad (11.10)$$

where $\pi(\cdot \mid D_{T_0})$ is the posterior density of θ. Similarly,

$$\mathrm{E}(N_{T_0+u} - N_{T_0} \mid D_{T_0}) = \int \{\Lambda(T_0 + u; \theta) - \Lambda(T_0; \theta)\}, \pi(\theta \mid D_{T_0})\, d\theta. \qquad (11.11)$$

Sample path average approximations of both integrals are then obtained from the sequence $\{\theta^{(j)}\}$.

Turning to warranty parameters assessment, we consider the situation where the assessment takes place when the first delivered train reaches the end of its warranty period. Using the failure data from all trains, the transport company will evaluate whether the warranty specifications have been met. If this is not the case, the warranty period for all trains will be extended one year beyond the expiration date of the last delivered train.[4]

Assume for example that the warranty contract specifies an upper limit on the number of failures of a given type for a train during the first two years in service. Then we can calculate the predictive probability that a new train exceeds that limit as a measure of how far it lies in the tail of the actual distribution of the number of failures. Let $N_t^{(H)}$ be the failure counting process for the new train and let $\lambda_H(t; \theta)$ be the process intensity function as defined by (11.2), but with $w(\cdot) \equiv 1$, $i \equiv H$ and parameters a_H and c_H, used to determine the time – distance relationship, set to zero and 233, respectively. Finally, let x_U denote the upper limit for the number of failures, and let $T_0 - t_{0H}$ be a two-year period starting on calendar day t_{0H}. We want to calculate

$$\Pr(N_{T_0}^{(H)} > x_U \mid D_t) = 1 - \int \sum_{x=0}^{x_U} e^{-\Lambda_H(T_0;\theta)} \frac{[\Lambda_H(T_0; \theta)]^x}{x!} \pi(\theta \mid D_t)\, d\theta, \qquad (11.12)$$

[4] In the present case, it is natural to express the warranty duration in years, because trains are expected to run a common predetermined distance every year.

Table 11.2 Forecasts of the number of failures for failure mode 1. Forecasts are presented for all trains after one, two and three years beyond the recording period, which is set on 31 December of the indicated years.

End of recording period	Forecasting horizon (years)	95% credibility interval	True value	Posterior mean
1992	1	(86, 143)	83	114
	2	(79, 140)	72	109
	3	(71, 138)	62	105
1993	1	(69, 124)	72	97
	2	(59, 121)	62	90
	3	(50, 119)	42	85
1994	1	(50, 100)	62	74
	2	(41, 95)	42	66
	3	(32, 91)	35	59
1995	1	(38, 81)	42	59
	2	(30, 74)	35	51
	3	(24, 68)	23	44
1996	1	(27, 60)	35	43
	2	(20, 52)	23	35
1997	1	(19, 46)	23	39

where D_t are the failure data available at the expiration time of the warranty of the first delivered train. If the sample path average approximation of this probability using $\{\theta^{(j)}\}$ is large, then x_U is not consistent with the observed data.

In Table 11.2 we present the forecasts of the number of failure for failure mode 1. Despite the good fit seen in Figure 11.5 for the exponential baseline, there are five out of 15 95% credibility intervals that do not contain the true observed values. This happens for all forecasting horizons when very few data are used to estimate the model (that is, up to 31 December 1992) and in two other cases for the three-year-ahead forecasts. An obvious reason for this behaviour is that the chosen baseline does not reflect the 'true' failure process closely enough, especially when the fitting is done on a subset of data. Therefore, small variations in the posterior distribution of b are magnified when summing up the 40 intensity functions to obtain the superposed process (still a Poisson process because of Theorem 2 in the appendix) for extrapolation. Nonetheless, the forecasts we have computed can be useful for year-to-year programming of the maintenance facilities, especially after enough data have come in.

Consider now the forecasts for failure mode 3 displayed in Table 11.3. Those with recording period ending in 1992, 1993 and 1994 are surprisingly good, then they become very bad in years 1995 and 1996, ending with a reasonable

Table 11.3 Forecasts of the number of failures for failure mode 3. Forecasts are presented for all trains after one, two and three years beyond the recording period, which is set on 31st December of the indicated years.

End of recording period	Forecasting horizon (years)	95% credibility interval	True value	Posterior mean
1992	1	(159, 255)	165	209
	2	(137, 255)	169	196
	3	(117, 261)	188	184
1993	1	(132, 213)	169	171
	2	(111, 205)	188	154
	3	(92, 198)	124	139
1994	1	(120, 189)	188	153
	2	(102, 177)	124	138
	3	(87, 167)	74	124
1995	1	(132, 200)	124	165
	2	(120, 193)	74	156
	3	(109, 188)	52	147
1996	1	(105, 164)	74	134
	2	(93, 153)	52	122
1997	1	(71, 117)	52	94

value in 1997. Why this happens should be clear from the shape of the curve of cumulative number of failures, where a change in slope starts to occur at around day 2200, that is, near the end of 1995.

The results of this forecasting exercise suggest that medium-term forecasts (beyond one year) can be useful but should be handled with care. After all, the Poisson process is completely determined by the distribution of the first failure time and the assumption of independent increments, thus it cannot describe things such as changes in maintenance practices, delayed recording of failures or technical upgrades.

As far as the warranty parameter assessment is concerned, this does not involve extrapolation, but only a retrospective evaluation of the consistency of a threshold for the number of failures with the observed data. In Table 11.4 we report the sample path average approximation of (11.12) for failures

Table 11.4 Sample path average approximation of probabilities in equation (11.12) with a two-year warranty.

f. mode 1	x_U	3	4	5	6	7	8	9	10	11	12	13
	prob.	0.82	0.68	0.52	0.36	0.23	0.14	0.07	0.04	0.02	0.01	0.00
f. mode 3	x_U	12	13	14	15	16	17	18	19	20	21	22
	prob.	0.47	0.36	0.26	0.18	0.12	0.08	0.05	0.03	0.01	0.01	0.00

mode 1 and 3 and different values of x_U, $T_0 = 730$ (two years) and $t_{0H} = 0$ (corresponding to 31 December in our day reckoning). We see that, for example, thresholds of 9 and 18 failures in the first two years for failure modes 1 and 3, respectively, would be warranty thresholds consistent with the observed data.

11.3 Further discussion

As discussed in the chapter, the choice of a baseline intensity function $\lambda_0(\cdot)$ is not an easy task. We have compared three of them and chose one according to the best fit between cumulative number of failures and estimated mean value function. We think the parametric assumptions could be relaxed by considering a nonparametric approach; one possible direction of research is provided by the work by Cavallo and Ruggeri (2001), who, stemming from Lo (1982), considered an extended Gamma process and treated the mean value function $\Lambda(\cdot)$ as a random measure. They centered the extended Gamma process around a parametric model and compared the two of them via Bayes factor.

Considering again a parametric model, it could be worth extending the exchangeability assumption to parameters other than the intercept and slope of the day-kilometre line, in order to account for other possible individual differences among trains.

The doors failure example illustrated in the chapter shows very clearly the importance of a continuous interaction between company employees and statisticians. We experienced some problems about the quality of data and we could not seriously analyse data about failure modes 6 and 7 showing a change-point which we could not justify at the best of our knowledge. Furthermore, company expertise would be required to elicit priors properly and to update the model in progress. Here we have in mind the knowledge about the type of repair performed upon failure and the (possible) preventive maintenance intervention. We have assumed minimal repairs which occurred almost instantaneously after the failures; these assumptions justify the choice of a nonhomogeneous Poisson process but repairs could have been different, ranging from an imperfect one to a complete replacement (or relevant intervention) of the door. We assumed the company was intervening only after failures and not before to prevent them. Probably a maintenance programme for all train doors is too expensive with respect to the cost of repairing them upon failure. The problem can be addressed by maximizing an expected utility function based upon costs due to intervention after failures and maintenance checks. Different maintenance strategies have been proposed, ranging from check at predetermined time or kilometre intervals to thorough checks only when some simple-to-check feature has deteriorated (*condition-based* maintenance). An up-to-date review of repair policies, maintenance strategies, along with warranty related issues, can be found in Ruggeri *et al.* (2007).

Appendix

A. Broader context and background

A.1 Foundations

Poisson processes are widely used to describe failures in repairable systems under *minimal* and *instantaneous repairs* which bring system reliability back to its value just before the failure (*bad as old* property).

The choice of a Poisson process to describe failures in the context of the current study is also justified by a theorem due to Grigelionis; see, e.g. Thompson (1988), p. 69. In the chapter we have considered the failure labelled as 1062; actually many different kinds of failures go under the label 1062. We can suppose that m different kinds of arrivals (or failures) are identified and that $N_t^{(i)}$ denotes the number of arrivals of type i up to time t. Consider the summary process $N_t = \sum_{i=1}^m N_t^{(i)}$ which denotes the total number of arrivals. Take $M_i(t) = E(N_t^{(i)})$ and $M(t) = \sum_{i=1}^m M_i(t)$.

Theorem 1 For a system of arrivals satisfying

1. $\lim_{m\to\infty} \min_{i=1,\ldots,m} Pr(N_t^{(i)} = 0) = 1$
2. $\lim_{m\to\infty} \sum_{i=1}^m Pr(N_t^{(i)} \geq 2) = 0,$

the necessary and sufficient condition that the summary process N_t converges to a Poisson process $\tilde{N}_t$ with $E(\tilde{N}_t) = M(t)$ is that for any fixed s and t $(s < t)$

$$\lim_{m\to\infty} \sum_{i=1}^{m} Pr(N^{(i)}(s,t) = 1) = M(t) - M(s),$$

where $N^{(i)}(s,t)$ denotes the number of arrivals of type i in the interval $(s,t]$.

The theorem allows for an approximation for *large* fixed m provided m approaches infinity in such a way that arrivals become increasingly rare.

The theorem gives a strong justification to the use of Poisson processes in describing reliability in complex repairable systems where failure causes can be many. This theorem is even more general of the superposition theorem 2 (see next) which shows that the sum of independent Poisson processes is still a Poisson process.

A.2 Models

Detailed illustration of NHPPs applied to reliability can be found in Rigdon and Basu (2000), whereas general Poisson processes are described in, e.g. Kingman (1993).

Let $N(s, t)$ denote the number of events (typically failures in reliability) in the interval $(s, t]$, whereas the notation N_t is used instead of $N(0, t)$.

Definition 1 A counting process N_t is said to be a NHPP with intensity function $\lambda(t)$ if

1. $N_0 = 0$;
2. it has independent increments;
3. $Pr(N(t, t+h) \geq 2) = o(h)$;
4. $Pr(N(t, t+h) = 1) = \lambda(t)h + o(h)$.

When the intensity $\lambda(t)$ is constant over time, then an homogeneous Poisson process (HPP) is obtained. If λ is the intensity rate of an HPP, then interarrival times between two events are i.i.d. $Exp(\lambda)$ r.v.'s. Such property implies that the HPP is also a renewal process, unlike the other NHPP's whose interarrival times are not i.i.d r.v.'s.

The intensity function of a NHPP can be interpreted as

$$\lambda(t) = \lim_{\Delta t \to 0} \frac{Pr(N(t, t+\Delta t] \geq 1)}{\Delta t}.$$

The mean value function (m.v.f.) of the NHPP is defined as the nondecreasing, nonnegative, function $M(s, t) = E(N(s, t)), 0 \leq s < t$, with $M(t) = E(N_t), t \geq 0$. Assuming that $M(t)$ is differentiable, then $\mu(t) = dM(t)/dt$ is the rate of occurrence of failures (ROCOF) for the NHPP. Property (3) in Definition 1 implies that $\mu(t) = \lambda(t)$ a.e. so that $M(y, s) = \int_y^s \lambda(t)dt$.

The NHPP owes its name to the fact that

$$Pr(N(y, s) = k) = \frac{[M(y, s)]^k}{k!} \exp\{-M(y, s)\}$$

for any integer k.

Suppose that the system is observed in the interval $(0, y]$ and failures are recorded at $0 < t_1 < \ldots < t_n \leq y$. The upper limit y could be either a predetermined endpoint or the time of the n-th (predetermined) failure. In the former case we talk about a *time truncated* experiment, whereas the latter is a *failure truncated* one. Under both experiments and the intensity function $\lambda(t) = \lambda(t; \theta)$, the likelihood function is

$$L(\theta; \underline{t}) = \prod_i^n \lambda(t_i; \theta) \exp\left\{-\int_0^y \lambda(t; \theta)dt\right\}. \quad (11.13)$$

When considering an HPP with intensity rate λ, then the likelihood (11.13) becomes $\lambda^n \exp\{-\lambda y\}$.

Many NHPP's with different intensity functions have been considered in literature; an extensive review of NHPP's considered in reliability is in McCollin (2007).

Here we concentrate on a general class of NHPP's, presented in Ruggeri and Sivaganesan (2005), whose intensity function can be represented as $\lambda(t; \alpha, \beta) = \alpha g(t, \beta)$, with $\alpha, \beta > 0$, such that their m.v.f. is $M(t; \alpha, \beta) = \alpha G(t, \beta)$, with $G(t, \beta) = \int_0^t g(u, \beta) du$. The class contains well-known processes, such as the Musa–Okumoto, the Cox – Lewis and the power law processes.

The first process, described in Musa and Okumoto (1984), has been widely used in modelling software reliability; it has intensity function $\lambda(t; \alpha, \beta) = \alpha/(t + \beta)$ and m.v.f. $M(t; \alpha, \beta) = \alpha \log(t + \beta)$. The second process, described in Cox and Lewis (1966), is such that $\lambda(t; \alpha, \beta) = \alpha \exp\{\beta t\}$ and $M(t; \alpha, \beta) = (\alpha/\beta) [\exp\{\beta t\} - 1]$. The third process, the power law process (PLP), is the mostly used NHPP in reliability, and its intensity function and m.v.f. are given, respectively, by $\lambda(t; \alpha, \beta) = \alpha\beta t^{\beta-1}$ and $M(t) = \alpha t^\beta$.

The plethora of proposed intensity functions is justified by the need of modelling different behaviours. NHPP's might differ because of many features, including monotonicity of the intensity function, asymptotic expected number of events, i.e. $\lim_{t\to\infty} M(t)$ (finite or not), and different functional form of the intensity function. Choice among different NHPP's is not an easy task, and it can be performed by using the *classical* tools of Bayesian analysis, like Bayes factors and posterior probabilities, or more heuristic ones like comparison between estimated mean value function and cumulative number of failures.

Even within the same NHPP, different choices of the parameter correspond to different physical behaviours of the system. The typical example is provided by the PLP with intensity function $\lambda(t; \alpha, \beta) = \alpha\beta t^{\beta-1}$. Reliability growth (i.e. decreasing intensity function over time) is obtained for $0 < \beta < 1$, whereas reliability decay is obtained when $\beta > 1$ and constant reliability (and an HPP) is obtained for $\beta = 1$. When considering a PLP to describe failures, it is even possible, unlike in classical statistics, to estimate $Pr(\beta < 1|data)$, i.e. the probability that the system is subject to reliability growth.

Sometimes, it is not possible to specify a unique mathematical expression for the intensity function; an evident example is provided by the typical behavior of complex systems, which are first subject to an *early mortality* phase, modelled by an intensity function which starts very high and then drops down, followed by a phase of *useful life* with constant intensity rate until the systems are subject to *aging*, when the intensity function increases dramatically. The obtained intensity function has the typical *bathtub* shape. It is difficult to express such behaviour with a unique formula; therefore different intensity functions are considered in different intervals. The issue here is not only the choice of adequate intensity functions and estimation of their parameters, but also the detection of the points at which the changes between different intensity functions occur. A thorough illustration of Bayesian methods to address this *change point* problem can be found in Ruggeri and Sivaganesan (2005). In one of their models, they consider different values of the parameter β of a PLP. In particular, at time t_i^+, $i = 1, \ldots, n$, right after a failure, the parameter β_{i-1}, identifying the process over $(t_{i-1}, t_i]$, is modified according to

$$\log \beta_i = \log a + \log \beta_{i-1} + \epsilon_i,$$

where β_i is the value of β over $(t_i, t_{i+1}]$, a is a positive constant and ϵ_i is a normally distributed random variable.

Furthermore, Ruggeri and Sivaganesan (2005) considered a reversible jump Markov chain Monte Carlo simulation method which allowed for a very flexible model, in which number and location of change points were not limited by the actual failure times.

NHPP's can be considered on more complex spaces than time intervals; a thorough illustration can be found in Kingman (1993). A simple extension was presented in Pievatolo and Ruggeri (2004) when gas escapes in a city network were considered in function of both time and space (the length of the network). Assuming an aging effect w.r.t. time and constancy over space, the intensity function was taken such that

$$\lambda(s, t; \mu, \theta) = \mu\rho(t; \theta).$$

We present two relevant results which allow for merging and splitting of NHPPs in such a way new NHPPs can be obtained.

The first result is given by the following superposition theorem (see Kingman 1993 for a proof).

Theorem 2 Consider n independent Poisson processes $N_t^{(i)}, t \geq 0$, with intensity function $\lambda_i(t)$, $i = 1, \ldots, n$. The sum process $N_t, t \geq 0$, defined as $N_t = \sum_{i=1}^n N_t^{(i)}$ is a Poisson process with intensity function $\lambda(t) = \sum_{i=1}^n \lambda_i(t)$.

Theorem 2 has been applied by Pievatolo and Ruggeri (2004) in a Bayesian framework, when considering gas escapes in a city network of steel pipelines. The city network was split according to the installation year of the pipelines and for each subnetwork a NHPP, independent from the others, was considered. Each NHPP had its intensity function $\lambda_i(t; \theta_i)$, which is nonzero only after the installation date. Using the gas escape example, we discuss four main scenarios about the choice of a prior distribution on the θ_i's.

We consider m subsystems and n_i failures (or gas escapes), $i = 1, \ldots, m$, and we denote the failure times of system i as $\underline{t}^i = (t_1^i, \ldots, t_{n_i}^i)$.

1. **[Same θ]** Gas pipes are considered to be identically built and operated, so that the parameter θ, summarizing the physical properties of the pipes, is the same for all subsystems. As mentioned earlier, the intensity functions might differ for the starting point, so that we denote them by $\lambda_i(t;\theta)$. In this case the posterior distribution is given by
$$\pi(\theta;\underline{t}) \propto \pi(\theta) \prod_{i=1}^{m} \prod_{j=1}^{n_i} \lambda_i\left(t_j^i;\theta\right) \exp\left\{-\sum_{i=1}^{m} \int_0^y \lambda_i(t;\theta)dt\right\}.$$
2. **[Independent θ's]** Subsystems can be thought independent one another, since, e.g. operating conditions and environment are completely different. In this case, each subsystem has its own intensity function with its own parameter, i.e. $\lambda_i(t;\theta_i)$, $i = 1, m$. Posterior distributions are obtained for each parameter separately:
$$\pi(\theta_i;\underline{t}^i) \propto \pi_i(\theta_i) \prod_{j=1}^{n_i} \lambda_i\left(t_j^i;\theta_i\right) \exp\left\{-\int_0^y \lambda_i(t;\theta_i)dt\right\},$$
regardless of choosing independent priors $\pi_i(\theta_i)$ or dependent ones.
3. **[Exchangeable θ's]** Similarity among parameters θ_i's of different subsystems can be modelled by assuming they are drawn from the same distribution $\pi(\theta|\omega)$ and a prior distribution $\pi(\omega)$ is specified as well. The posterior distribution becomes
$$\pi(\theta_1, \ldots, \theta_m, \omega; \underline{t}) \propto \pi(\omega) \prod_{i=1}^{m} \pi(\theta_i|\omega) \prod_{j=1}^{n_i} \lambda_i\left(t_j^i;\theta_i\right)$$
$$\times \exp\left\{-\sum_{i=1}^{m} \int_0^y \lambda_i(t;\theta_i)dt\right\}.$$
As an alternative, an empirical Bayes approach can be taken by looking at
$$\hat{\omega} = \arg\max \int \prod_{i=1}^{m} \pi(\theta_i|\omega) \prod_{j=1}^{n_i} \lambda_i\left(t_j^i;\theta_i\right)$$
$$\times \exp\left\{-\sum_{i=1}^{m} \int_0^y \lambda_i(t;\theta_i)dt\right\} d\theta_1 \ldots d\theta_m,$$
and performing the *usual* Bayesian analysis with priors $\pi(\theta_i;\hat{\omega})$.
4. **[Covariate dependent θ's]** Pipes might differ for many aspects (e.g. environment, when stray currents are massively present or huge moisture is in the ground) and covariates $\underline{X} = (X_1, \ldots, X_k)$ could be considered to study their effect on the reliability of the system. Two different approaches have

been proposed in Masini *et al.* (2006). In the former each subsystem has a set $\underline{X}^i$ of covariates and the intensity of the corresponding NHPP depends on a parameter $\theta_i = \theta(\underline{X}^i)$. As an example, the parameter α of the PLP can be modelled, when considering dichotomic covariates, as

$$\alpha = \gamma_0 \prod_{j=1}^{k} \gamma_j^{X_j^i}.$$

In this case priors on $(\gamma_0, \ldots, \gamma_k)$ are considered and the corresponding posterior distributions are used to study the influence of the covariates. The latter approach considers covariate dependent priors on the parameters θ_i of each subsystem. As an example, Gamma priors can be considered for the parameter α of the PLP such that $\alpha_i \sim Ga(a \exp\{b' \underline{X}^i\}, a)$, $i = 1, m$. In this way, the prior expected value of α_i is given by $\exp\{b' \underline{X}^i\}$, whereas a measures the strength of such assertion, influencing the prior variance.

Other dependence structures are possible among the parameters, depending on the problem at hand. In the gas escape example, an evolution of the behavior of the pipes over the installation date could be considered, as a result of, say, the improved techniques to lay the pipes in the ground. Therefore an autoregressive model $\theta_i = \rho\theta_{i-1} + \epsilon_i$, with $|\rho| < 1$ and ϵ_i a r.v., could be considered when the i-th subsystem is influenced only by the $(i-1)$-th one.

The colouring theorem (sometimes called thinning theorem) is a sort of reciprocal of the superposition theorem 2. See Kingman (1993) for a proof.

Theorem 3 Consider a Poisson process $N_t, t \geq 0$ with intensity function $\lambda(t)$ and assign each event (failure) to one of n classes with probability p_i, $i = 1, \ldots, n$. Suppose each assignment is independent of one another and the other events. The events in each class determine n independent Poisson processes $N_t^{(i)}, t \geq 0$, with intensity function $\lambda_i(t) = p_i\lambda(t)$, $i = 1, \ldots, n$.

Bayesian conjugate analysis is straightforward when N total events from an HPP, with intensity λ, are recorded in the interval $(0, T]$ along with the numbers $n_1, \ldots, n_m$ of the events assigned to m classes with the random mechanism described in Theorem 3. From the likelihood given by

$$Pr(n_1, \ldots, n_m, N) = Pr(n_1, \ldots, n_m | N) Pr(N) \propto p_1^{n_1} \ldots p_m^{n_m} \lambda^N \exp\{-\lambda T\}$$

and the priors $(p_1, \ldots, p_m) \sim Dir(\alpha_1, \ldots, \alpha_m)$ and $\lambda \sim Ga(a, b)$, it follows that the posterior distributions are given by $(p_1, \ldots, p_m)|data \sim Dir(\alpha_1 + n_1, \ldots, \alpha_m + n_m)$ and $\lambda|data \sim Ga(a + N, b + T)$.

The models presented so far are parametric ones, as most of the Bayesian literature on reliability of repairable systems. Some Bayesian nonparametric papers stems from the work by Lo (1982). Lo considered the m.v.f. $M(t)$ as the distribution function of a random measure M, i.e. the intensity measure of the process. Gamma and extended Gamma processes were considered as prior for M. When observing data from Poisson processes, Lo (1982) proved that Gamma and extended Gamma processes were conjugate priors for the Poisson processes. Bayesian nonparametric estimator of reliability and comparison of Bayesian parametric and nonparametric models via Bayes factor are illustrated in Cavallo and Ruggeri (2001).

Another important theorem is given by the following.

Theorem 4 Let $\Lambda(t)$ be a continuous nondecreasing function. Then $T_1, T_2, \ldots$ are arrival times in a Poisson process N_t with m.v.f. $\Lambda(t)$ if and only if $\Lambda(T_1), \Lambda(T_2), \ldots$ are arrival times in an HPP H_t with unit failure rate.

Once a NHPP is transformed into an HPP using Theorem 4 and considering arrival times $Y_i = \Lambda(T_i)$, it is immediate to show that the interarrival times $X_i = Y_i - Y_{i-1}$ are independent r.v.'s such that $X_i \sim Exp\ (1)$. Furthermore, the r.v.'s $U_i = \exp\{-X_i\}$ have uniform distributions on $[0, 1]$.

The use of Theorem 4 is, at least, twofold. It is possible to simulate arrival (failure, in reliability jargon) times by a NHPP with invertible m.v.f. $\Lambda(t)$, by drawing data u_i, $i = 1, \ldots, n$, from a uniform distribution, transforming them into interarrival times $x_i = -\log u_i$ of an HPP with unit intensity rate, computing its failure times $y_i = \sum_{j=1}^{i} x_i$ and, finally, getting the arrival times $t_i = \Lambda^{-1}(y_i)$.

Theorem 4 can be used also in checking goodness of fit of the estimated model. Given a NHPP with intensity function $\lambda(t;\theta)$ and m.v.f. $\Lambda(t;\theta)$, let $\hat{\theta}$ be the Bayes estimator, namely the posterior mean, of θ. Should the estimated model $\Lambda(\hat{t}) = \Lambda(t;\hat{\theta})$ be the *true* one, then the r.v.'s $U_i = \exp\{\Lambda(\hat{T}_{i-1}) - \Lambda(\hat{T}_i)\}$ are drawn from a uniform distribution $F_0(x)$ on $[0, 1]$.

Therefore, it is possible to test if the transformed failure times are uniformly distributed, considering the null hypothesis

$$H_0 : F(x) = F_0(x)\ \forall x \in \Re$$

against the alternative

$$H_1 : F(x) \neq F_0(x) \text{ for some } x \in \Re,$$

and using the two-sided Kolmogorov–Smirnov statistics, defined as

$$D_n = \sup_{x \in \Re} |\hat{F}_n(x) - F_0(x)|.$$

This (non-Bayesian) goodness of fit tests has been originally applied to earthquake data by Ogata (1988).

B. Computation

We consider two simple, but significant, cases.

Let N_t be an HPP with intensity rate λ, like in Cagno *et al.* (1999) where gas escapes were considered in cast iron pipelines, which, typically, are not subject to corrosion. A conjugate Gamma prior $Ga(a, b)$ can be considered for λ so that the posterior distribution is $Ga(a + n, b + y)$.

The prior $p(\alpha, \beta) = p(\alpha|\beta)p(\beta)$, with $\alpha|\beta \sim Ga(a, b)$, can be chosen for the general class of NHPPs with

$$\lambda(t; \alpha, \beta) = \alpha g(t, \beta), \qquad \alpha, \beta > 0. \tag{11.14}$$

Recording failures at $0 < t_1 < \ldots < t_n \leq y$ over the interval $(0, y]$, the likelihood based on the intensity function (11.14) is given by $L(\alpha, \beta; \underline{t}) = \alpha^n \prod_i^n g(t_i, \beta) \exp\{-\alpha G(y, \beta)\}$.

Gibbs sampling, with a Metropolis – Hastings step within (see Gamerman and Lopes 2006), is required to get a sample from the posterior distribution since the conditional distributions is given by

$$\alpha|\beta, \underline{t} \sim Ga(a + n, b + G(y, \beta))$$

and

$$\beta|\alpha, \underline{t} \propto p(\beta) \prod_i^n g(t_i, \beta) \exp\{-\alpha G(y, \beta)\}.$$

Bayesian estimation of parameters of the PLP is thoroughly discussed in Bar-Lev *et al.* (1992).

References

Bar-Lev, S., Lavi, I. and Reiser, B. (1992). Bayesian inference for the power law process. *Annals of the Institute of Mathematical Statistics*, **44**, 623–639.

Cagno, E., Caron, F., Mancini, M. and Ruggeri, F. (1999). Using AHP in determining the prior distributions on gas pipeline failures in a robust Bayesian approach. *Reliability Engineering and Systems Safety*, **67**, 275–284.

Cavallo, D. and Ruggeri, F. (2001). Bayesian models for failures in a gas network. In *Safety and Reliability* (ed. E. Zio, M. Demichela and N. Piccinini), pp. 1963–1970. Politecnico di Torino, Torino.

Cox, D. R. and Lewis, P. A. (1966). *Statistical Analysis of Series of Events*. Methuen, London.

Gamerman, D. and Lopes, H. F. (2006). *Markov Chain Monte Carlo – Stochastic Simulation for Bayesian Inference*, (2nd edn.). Chapman & Hall/CRC Press, Boca Raton.

Kingman, J. F. C. (1993). *Poisson Processes*. Clarendon Press, Oxford.

Lo, A. (1982). Bayesian nonparametric statistical inference for poisson point processes. *Zeitschrift für Wahrscheinlichkeitstheorie und verwandte Gebiete*, **59**, 55–66.

Masini, L., Pievatolo, A., Ruggeri, F. and Saccuman, E. (2006). On Bayesian models incorporating covariates in reliability analysis of repairable systems. In *Bayesian Statistics and its*

Applications (ed. S. K. Upadhyay, U. Singh and D. K. Dey), pp. 331–341. Anamaya, New Delhi.

McCollin, C. (2007). Intensity functions for nonhomogeneous Poisson processes. In *Encyclopedia of Statistics in Quality and Reliability* (ed. F. Ruggeri, R. Kenett and F. W. Faltin), pp. 861–868. John Wiley, Chichester.

Musa, J. D. and Okumoto, K. (1984). A logarithmic Poisson execution time model for software reliability measurement. Proceedings Seventh International Conference on Software Engineering, Orlando, Florida, pp. 230–238.

Ogata, Y. (1988). Statistical models for earthquake occurrences and residual analysis for point processes. *Journal of the American Statistical Association*, **83**, 9–27.

Pievatolo, A. and Ruggeri, F. (2004). Bayesian reliability analysis of complex repairable systems. *Applied Stochastic Models in Business and Industry*, **20**, 253–264.

Pievatolo, A., Ruggeri, F. and Argiento, R. (2003). Bayesian analysis and prediction of failures underground trains. *Quality Reliability Engineering International*, **19**, 327–336.

Rigdon, S. E. and Basu, A. P. (2000). *Statistical Methods for the Reliability of Repairable Systems.* John Wiley, New York.

Ruggeri, F., Kenett, R. and Faltin, F. (2007). *Encyclopedia of Statistics in Quality and Reliability.* John Wiley, Chichester.

Ruggeri, F. and Sivaganesan, S. (2005). On modeling change points in non-homogeneous Poisson processes. *Statistical Inference for Stochastic Processes*, **8**, 311–329.

Singpurwalla, N. and Wilson, S. (1998). Failures models indexed by two scales. *Advances in Applied Probability*, **30**, 1058–1072.

Thompson, W. A. J. (1988). *Point Process Models with Applications to Safety and Reliability.* Chapman & Hall, New York.

·12·

Analysis of economic data with multiscale spatio-temporal models

Marco A. R. Ferreira, Adelmo I. Bertolde and Scott H. Holan

12.1 Introduction

Many economic processes are naturally spatio-temporal. Often times, quantities related to these processes are available as areal data, partitioned by the geographical domain of interest, for example totals or averages of some variable of interest for each state of a country. Furthermore, these economic processes may often be considered at several different levels of spatial resolution. Here we present a general modelling approach for multiscale analysis of spatio-temporal processes with areal data observations and illustrate the utility of this approach for economic applications with an analysis of agricultural production per county in the Brazilian state of Espírito Santo from 1990 to 2005.

In practice, economic data may be considered at different levels of resolution, e.g. census tract, county, state, country, continent. Our analysis of agricultural production in Espírito Santo is based on the states geopolitical organization as depicted in Figure 12.1. Specifically, each state in Brazil is divided into counties, microregions and macroregions; counties are grouped into microregions, which are then grouped into macroregions, according to their socioeconomic similarity. Thus, our analysis considers three levels of resolution: county, microregion, and macroregion.

The county-level data are aggregated to obtain the data at the other scales of resolution. Then, at each time point, the observed data are decomposed into empirical multiscale coefficients that represent a vector of weighted differences between each subregion and its descendants. To this extent, we consider a multiscale decomposition, proposed by Kolaczyk and Huang (2001), which includes Haar wavelet decompositions (Vidakovic, 1999) as particular case. This decomposition is quite flexible and can accommodate irregular grids and heteroscedastic errors. Finally, we assume that the empirical multiscale coefficients correspond to latent multiscale coefficients that evolve through time following singular random walks.

An interesting feature of our approach is that the multiscale coefficients have an interpretation of their own; thus, the multiscale spatio-temporal framework

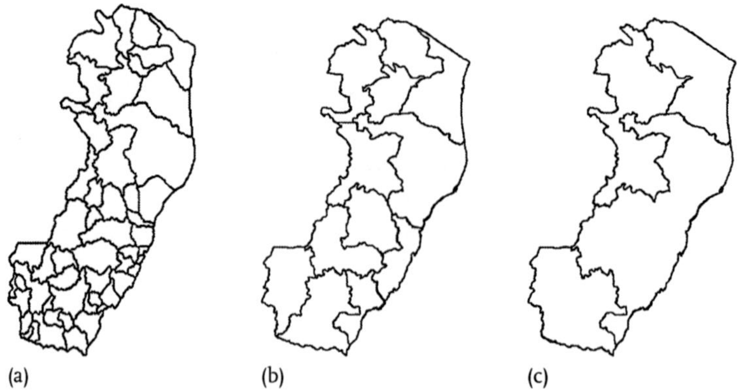

Fig. 12.1 Geopolitical organization of Espírito Santo State by (a) counties, (b) microregions, and (c) macroregions.

may offer new insight on understudied multiscale aspects of spatio-temporal observations. Moreover, the analysis of the estimated multiscale coefficients that we propose sheds light on the similarities and differences in agricultural production between regions within each scale of resolution. In addition, the temporal change in relative agricultural importance of those regions is explicitly described by the evolution of the estimated latent multiscale coefficients.

Multiscale spatio-temporal modeling has only recently been addressed in the literature, with just a hand full of papers published to date. One early attempt at multiscale spatio-temporal modeling was given by Berliner *et al.* (1999), where the authors propose a hierarchical model for turbulence and consider a wavelet decomposition for an underlying latent process, and allow the wavelet coefficients to evolve through time. Note that the multiscale decomposition that they consider is a wavelet decomposition applied to data on a regular grid. In contrast, we employ a decomposition that can be applied naturally to data on irregular grids without having to appeal to lifting schemes (Sweldens, 1996). Additionally, wavelets yield a decomposition where each node, in a given resolution level, gives rise to two children nodes (i.e. a dyadic expansion); this would lead to a two-dimensional wavelet coefficient but, as the sum of its elements is zero, only one of its elements need to be retained. Conversely, in our construction, there is no limitation on the number of children nodes at each level of resolution in the decomposition. For example, in the agricultural production application each macroregion has between two and five descendant microregions, and each microregion has between two and twelve descendant counties. As each empirical multiscale coefficient is a vector with number of elements equal to the corresponding number of descendants, and the sum of its elements is zero, the distribution of the empirical multiscale coefficient is a degenerate Gaussian distribution.

A more recent reference related to our work is Johannesson *et al.* (2007). Although the authors propose a multiscale decomposition that uses a coarse to fine construction, there are several differences between their approach and ours. First, they consider a regular grid while we consider more general data structures. Second, they estimate the relationship between the different levels of resolution subject to a mass balance constraint; conversely, we obtain the relationship between the different levels of resolution as a function of the different variances of the different counties, and use the resulting relationship in order to define our multiscale coefficients. Lastly, they assume temporal dynamics only for the aggregated coarse level. In contrast, we include dynamics not only for the aggregated coarse level, but also for the multiscale coefficients at each resolution level. The inclusion of the latter dynamics is fundamentally difficult due to the singular nature of the multiscale coefficients and it is one of the main contributions of the approach developed in this paper.

A desirable property for methods that provide estimates at different scales of resolution is that the estimates respect deterministic relationships between the different scales of resolution. For example, the sum of per county agricultural production within a given microregion is equal to the total production of that microregion. Here we assume that there is an underlying latent process that evolves through time such that the observations per county are independent conditional on the underlying process. Moreover, conditional on this underlying process we can compute the expected value of the observations at the different resolution levels; these expected values should respect the fact that the sum of per county agricultural production within a given microregion is equal to the total production of that microregion. This is achieved by requiring that the latent multiscale coefficients satisfy the same constraint as their empirical multiscale coefficients counterparts; namely, that the sum of its elements is equal to zero. Moreover, this is accomplished by assuming that the latent multiscale coefficients evolve through time according to singular random walks, with singular covariance matrix proportional to the singular covariance matrix of the empirical multiscale coefficients. As a consequence, our framework insures that deterministic relationships between different levels of resolution are automatically respected for both the observations and the latent process as well as for the estimated latent process.

Finally, we propose a Bayesian analysis for the spatio-temporal multiscale model, performed in three steps. First, we obtain a preliminary estimate of the variance for each county using a standard first-order dynamic linear model (West and Harrison, 1997). Then, based on these preliminary estimates of the county variances we compute the empirical multiscale coefficients. The original likelihood function for the county observations is then decomposed into the product of the likelihood for the coarse level observations and the likelihood for the empirical multiscale coefficients at the different levels of resolution. This allows us to implement a very efficient MCMC algorithm which combines the standard Forward Filter Backward Sampler (FFBS) (Carter and Kohn, 1994; Frühwirth-Schnatter, 1994) with a novel Singular Forward Filter Backward Sampler (SFFBS) specially tailored for the simulation of the latent multiscale coefficients.

Since the seminal work by Donoho and Johnstone (1994) on nonparametric regression using wavelets, wavelet-based multiscale methods and models have flourished. In fact, for many researchers the term *multiscale* means use of *wavelets*. However, this is far from the truth: there are many multiscale methods and models for processes defined at several scales of resolution that are not wavelet-based. Our multiscale spatio-temporal model fits better in this last category as it explores a non-wavelet-based multiscale decomposition useful for Gaussian areal data observed on an irregular grid. For this reason, the discussion of other multiscale modelling techniques is out of the scope of this chapter. For a comprehensive discussion on multiscale modelling see Ferreira and Lee (2007) and the references therein.

The remainder of this paper is organized as follows. Section 12.2 describes multiscale factorizations for spatial processes. Section 12.3 presents an exploratory multiscale data analysis and provides the motivation for multiscale spatio-temporal models. The temporal evolution of the underlying latent multiscale coefficients is introduced in Section 12.4. Section 12.5 presents a Bayesian analysis based on the multiscale decomposition of the likelihood function along with MCMC. Section 12.6 illustrates our methodology through analysis of data from our motivating example. Conclusions and possible extensions are presented in Section 12.7. For convenience of exposition the derivation of all full conditional distributions is left to the Appendix.

12.2 Multiscale factorization

At each time point we decompose the data into empirical multiscale coefficients using the spatial multiscale modelling framework of Kolaczyk and Huang (2001), where the authors consider spatial aggregations by sums. In what follows we review their approach, in the context of our motivating example, using the notation of Ferreira and Lee (2007, Chapter 9). For simplicity of notation, in this section, we omit any temporal dependence.

In our motivating example, interest lies in agricultural production observed at the county level, which we assume is the Lth level of resolution (i.e. the finest level of resolution), on a partition of a domain $S \subset \mathbb{R}^2$, denoted by $\{B_{L1}, \ldots, B_{Ln_L}\}$, with $B_{Lj} \in S$, $j = 1, \ldots, n_L$, $B_{Li} \cap B_{Lj} = \emptyset$, $i \neq j$, and $\cup_{j=1}^{n_L} B_{Lj} = S$. In our case, the region of interest S is the Brazilian state of Espírito

Santo, $L = 3$, and $B_{Lj} = B_{3j}$ refers to the jth county. Furthermore, for the jth county, let y_{Lj} and $\mu_{Lj} = E(y_{Lj})$ respectively denote agricultural production and its latent expected value.

We assume that $y_{L1}, \ldots, y_{Ln_L}$ are conditionally independent given the latent mean process $\mu_{L,1:n_L} = (\mu_{L1}, \ldots, \mu_{Ln_L})'$. This is a fairly reasonable assumption and is equivalent to assuming independent measurement errors. Note that in general the latent mean process will be correlated and this dependence will carry over to the observations $y_{L,1:n_L} = (y_{L1}, \ldots, y_{Ln_L})'$. Thus, the marginal distribution of the observations will exhibit dependence.

Interest not only lies in the latent mean process μ at the Lth scale of resolution but also in the process at aggregated coarser scales. At the lth scale of resolution, the domain S is partitioned in n_l subregions $B_{l1}, \ldots, B_{ln_l}$, $l = 1, \ldots, L$. In our case, $l = 1$ corresponds to the macroregion resolution level, whereas $l = 2$ corresponds to the microregion level. It is assumed that the partition at level $l + 1$ is a refinement of the partition at level l. That is, $B_{lj} = \cup_{(l+1,j')\in D_{lj}} B_{l+1,j'}$, where D_{lj} is the set of descendants of subregion j at level l, and $D_{lj} \cap D_{li} = \emptyset, i \neq j$. Note that the number of descendants does not need to be constant. In what follows we denote the number of descendants of a particular subregion (l, j) by m_{lj}. This construction is very similar to the construction of Basseville *et al.* (1992a,b) for multiscale models on trees.

With this notation at hand, the aggregated measurements at the lth level of resolution are recursively defined by

$$y_{lj} = \sum_{(l+1,j')\in D_{lj}} y_{l+1,j'}.$$

Analogously, the aggregated agricultural production mean process is defined by

$$\mu_{lj} = \sum_{(l+1,j')\in D_{lj}} \mu_{l+1,j'}.$$

Next, let A be a set of subregions and denote by y_A and μ_A the corresponding vectors of measurements and means, respectively. Then $y_{D_{lj}}$ represents the vector of descendants of y_{lj}. Note that y_{lj} is a deterministic function of its descendants; thus, the distribution of $y_{D_{lj}}$ conditional on y_{lj} is a degenerate Gaussian distribution with a singular covariance matrix.

In addition, let σ^2_{lj} denote the variance of y_{lj} and define $\Sigma_l = diag\left(\sigma^2_{l,1:n_l}\right)$. Assuming $y_{L1}, \ldots, y_{Ln_L}$ are conditionally independent given $\mu_{L,1:n_L}$ and $\sigma^2_{L,1:n_L}$, the variance at subregion (l, j) can be recursively computed as

$$\sigma^2_{lj} = \sum_{(l+1,j')\in D_{lj}} \sigma^2_{l+1,j'}.$$

With these assumptions and as explained in Appendix B.1, the likelihood function can be factorized as (Kolaczyk and Huang, 2001)

$$\prod_{j=1}^{n_L} p(\gamma_{Lj}|\mu_{Lj}, \sigma_{Lj}^2) = p(\gamma_{1,1:n_1}|\mu_{1,1:n_1}, \Sigma_1) \prod_{l=1}^{L-1} \prod_{j=1}^{n_l} p(\theta_{lj}^e|\gamma_{lj}, \theta_{lj}, \Omega_{lj}),$$

where $\nu_{lj} = \sigma_{D_{lj}}^2/\sigma_{lj}^2$, $\theta_{lj} = \mu_{D_{lj}} - \nu_{lj}\mu_{lj}$, and $\Omega_{lj} = \Sigma_{D_{lj}} - \sigma_{lj}^{-2}\sigma_{D_{lj}}^2\left(\sigma_{D_{lj}}^2\right)'$, $l = 1, \ldots, L-1$, $j = 1, \ldots, n_l$. Note that the scale specific parameter θ_{lj} is a vector corresponding to the difference between the mean level process at the descendants of subregion B_{lj} and the scaled mean level process at B_{lj}. Finally, $\theta_{lj}^e = \gamma_{D_{lj}} - \nu_{lj}\gamma_{lj}$ is an empirical estimate of θ_{lj}, and $\theta_{lj}^e|\gamma_{lj}, \mu_L, \sigma_L^2 \sim N(\theta_{lj}, \Omega_{lj})$.

This factorization is equivalent to a reparametrization of the model, where $\mu_{L,1:n_L}$ are replaced by the mean level process at the coarsest level $\mu_{1,1:n_1}$ and by the scale-specific multiscale parameters θ_{lj}s.

12.3 Exploratory multiscale data analysis

In this section we present an exploratory analysis of agricultural production data, from 1990 to 2005, in Espírito Santo State across several levels of resolution. The data are in inflation-corrected monetary value in the Brazilian currency which is the Real. The first column of Figure 12.2 displays time series of total per macroregion agricultural production from 1990 to 2005 in Espírito Santo State. As a result of these plots we can immediately make several observations. First, Macroregion 1 shows a rapid increase in production from 1993 to 1995 and extremely volatile behavior from 1995 to 2005. Second, Macroregion 2 presents a strong steady increase in production from 1990 to 2005. Additionally, Macroregion 3 displays a less impressive increase in production, whereas Macroregion 4 shows a striking increase in production from 1993 to 1997 and a not less surprising sharp decrease from 1997 to 2001.

The second column of Figure 12.2 displays the time series of agricultural production, for each macroregion, from 1990 to 2005 in Espírito Santo State per microregion. The disaggregation of production from Macroregion 1 shows that Microregion 2 has the largest production in almost all the years considered, and also the largest volatility. Microregions 1 and 3 have comparatively fairly low production while Microregions 4 and 5 have moderate and increasing production.

Macroregion 2 is formed by Microregions 6 and 7, and both microregions had an increase in production during the period of interest. Macroregion 3 contains Microregions 8, 9, and 10 and each of these microregions showed an increase in production, with Microregion 8 also exhibiting high volatility.

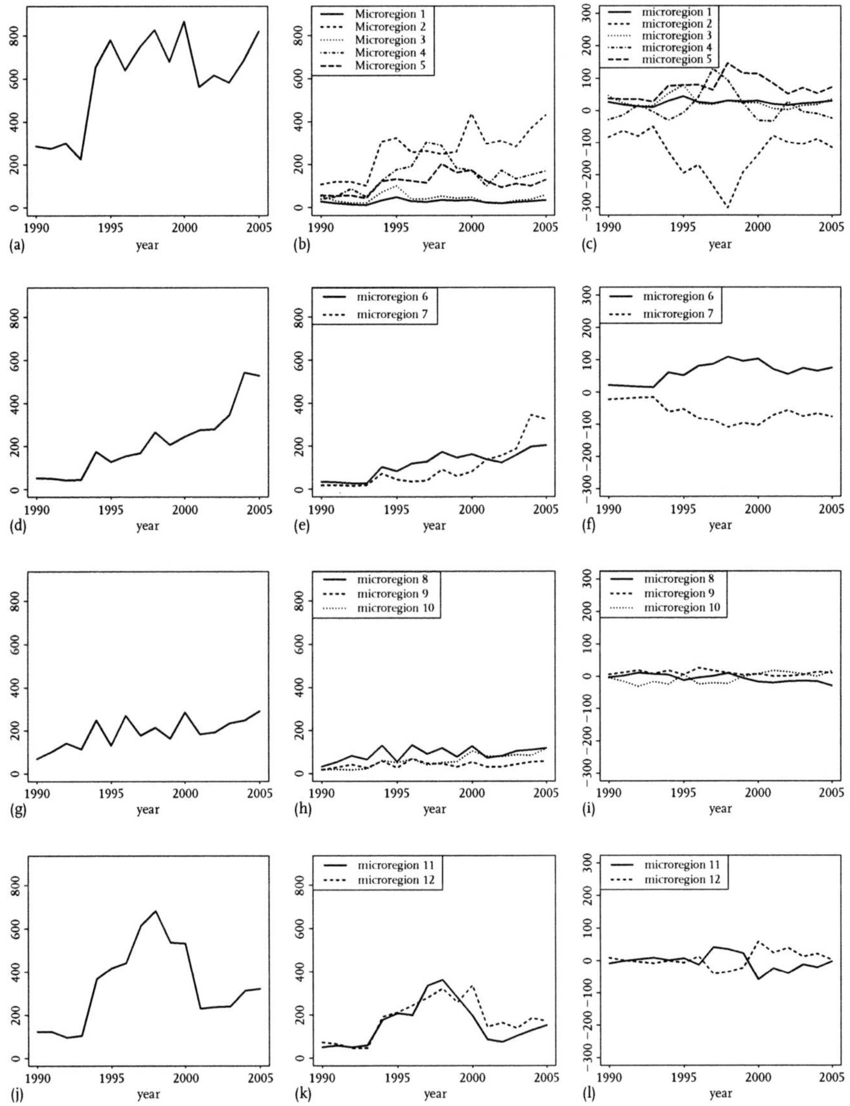

Fig. 12.2 Agriculture production per macroregion of Espírito Santo State. Macroregion 1 (a–c), 2 (d–f), 3 (g–i), and 4 (j–l). Macroregion totals (a,d,g,j); macroregion disaggregated by microregion (b,e,h,k); empirical multiscale coefficients (c,f,i,l). Note, for ease of comparison, all plots have been constructed using the same range on the agricultural production-axis (y-axis).

Finally, Macroregion 4 contains Microregions 11 and 12; both microregions exhibit similar behaviour with rapid increase from 1993 to about 1998 and then rapid decrease from 1999 to 2001.

The third column of Figure 12.2 displays the temporal evolution of the empirical multiscale coefficients, for different macroregions. Recall that, for each macroregion, the empirical multiscale coefficient, at a given time, is a vector with each element corresponding to a descendant microregion. Thus, for Macroregion 1 we have 5 time series representing the evolution of the empirical multiscale coefficient. Moreover, this figure illustrates several other interesting aspects. First, the empirical multiscale coefficient varies through time in a fairly smooth manner. Second, the coefficient corresponding to Microregion 2 is the most prominent, clearly indicating that the microregion lost relative importance from 1996 to 1998, but then regained importance within Macroregion 1. Unfortunately, due to noise, it is more difficult to evaluate what is transpiring with the other microregions contained in Macroregion 1.

Macroregions 2 and 4 have only two microregions each, and thus at each time point the respective empirical multiscale coefficient has only two elements. These two elements sum to zero, and thus through time the respective series have symmetric behaviour around zero. Finally, Macroregion 3 contains three microregions, all of which do not seem to differ significantly.

Another interesting aspect is that the empirical multiscale coefficient for all of the macroregions seem to be noisy versions of latent multiscale coefficients that vary somewhat smoothly over time. Thus, it makes sense to consider a model that focuses on the temporal evolution of the mean level process at the coarsest level and on the temporal evolution of multiscale coefficients. The multiscale coefficient parameter acts similar to a wavelet coefficient in wavelet analysis; that is, it provides the detail information necessary to recover the mean process at resolution level $l + 1$ from the mean process at resolution level l. Thus, we can think of the multiscale factorization as providing a hierarchical structure, where each level of the structure is related to a particular resolution level and its specific mean process behavior.

12.4 Dynamic multiscale modelling

12.4.1 Multiscale coefficients and dynamic priors

In order to analyse the data arising out of our application of interest and to generally increase the flexibility of the approach reviewed in Section 12.2, we introduce temporal dynamics to the spatial multiscale framework. In particular, we add a subscript t to indicate time and assume that the mean level process at the coarsest level μ_{t1} and the scale-specific parameter θ_{tlj} evolve through time according to random walks

$$\mu_{t1} = \mu_{t-1,1} + w_{t1}, \qquad w_{t1} \sim N(0, W), \tag{12.1}$$

$$\theta_{tlj} = \theta_{t-1,lj} + \omega_{tlj}, \qquad \omega_{tlj} \sim N(0, \Upsilon_{lj}). \tag{12.2}$$

Fundamental to the construction of our multiscale spatio-temporal modeling approach is the specification of the evolution covariance matrices W and Υ_{lj}.

These matrices have to be properly specified in order to preserve the special correlation structure of the elements of the scale specific parameters θ_{tlj}s and to respect the constraint $1'_{m_{lj}}\theta_{tlj} = 0$.

The model we propose also sets out a structure for the covariance matrices defined in (12.1) and (12.2). Particularly, for the covariance matrix at the coarsest level, we assume $W = diag(W_{1:n_1})$, with $W_k = \xi_k \sigma^2_{1k}$. This parametrization is extremely useful since ξ_k can be interpreted as the signal-to-noise ratio at the coarsest level and, as a result, it is easier to assign a meaningful prior to ξ_k than to W_k. Now, for the covariance matrix of the evolution innovations of the multiscale coefficients, we assume $\Upsilon_{lj} = \psi_{lj}\Omega_{lj}$. This specification both preserves the special correlation structure of the θ_{tlj}s and implies the constraint $1'_{m_{lj}}\theta_{tlj} = 0$.

12.4.2 Priors for the hyperparameters

The parameters μ_{01k} and θ_{0lj} provide the initial conditions at time $t = 0$. Let D_0 be the prior information about the system at time 0, and recursively define the accumulated information about the system at time t as $D_t = D_{t-1} \cup \{y_L\}$. Then, we assume the following conjugate priors for μ_{01k} and θ_{0lj}

$$\mu_{01k} | D_0 \sim N\big(m_{01k}, c_{01k}\sigma^2_{1k}\big)$$

and

$$\theta_{0lj} | D_0 \sim N(m_{0lj}, C_{0lj}\Omega_{lj}),$$

where m_{01k}, c_{01k}, m_{0lj}, and C_{0lj} are fixed a priori. Note that noninformative uniform priors $p(\mu_{01k}) \propto 1$ and $p(\theta_{0lj}) \propto 1(p'_{D_{lj}}\theta_{0lj} = 0)$, where $1(\cdot)$ is the indicator function and is equal to 1 if the condition is true, and is equal to 0 otherwise, can be obtained as limiting cases of the conjugate priors when $c_{01k} \to \infty$ and $C_{0lj} \to \infty$.

The signal-to-noise ratio parameters ξ_k and ψ_{lj}, described in Section 12.4.1, are most likely small. Otherwise, the components of the latent process would vary too much over time making it difficult to predict the several multiscale components of the multivariate spatio-temporal process. As a result, we expect ξ_k, $k = 1, \ldots, n_1$, and ψ_{lj}, $l = 1, \ldots, L$, $j = 1, \ldots, n_l$, to be much smaller than 1. Therefore, we assume that the prior distribution for each of ξ_k is $IG(0.5\tau_k, 0.5\kappa_k)$ with density $p(\xi_k) \propto \xi_k^{-0.5\tau_k - 1}\exp(-0.5\kappa_k/\xi_k)$, where τ_k and κ_k are fixed a priori such that there is a high probability that ξ_k is smaller than 0.3. Similarly, the prior distribution for ψ_{lj} is $IG(0.5\varrho_{lj}, 0.5\varsigma_{lj})$ where ϱ_{lj} and ς_{lj} are fixed a priori such that there is a high probability that ψ_{lj} is smaller than 0.3.

12.4.3 The multiscale spatio-temporal model

In the previous sections we have detailed an approach to multiscale spatio-temporal modeling. For the sake of clarity we summarize our multiscale spatio-temporal model.

Observation equation:

$$y_{tL,1:n_L} = \mu_{tL,1:n_L} + v_{tL,1:n_L}, \; v_{tL,1:n_L} \sim N(0, \Sigma_L), \tag{12.3}$$

where $\Sigma_L = diag\left(\sigma^2_{L,1:n_L}\right)$.

Multiscale decomposition of the observation equation:

$$y_{t1k} \mid \mu_{t1k} \sim N\left(\mu_{t1k}, \sigma^2_{1k}\right). \tag{12.4}$$

$$\theta^e_{tlj} \mid \theta_{tlj} \sim N(\theta_{tlj}, \Omega_{lj}). \tag{12.5}$$

System equations:

$$\mu_{t1k} = \mu_{t-1,1k} + w_{t1k}, \, w_{t1k} \sim N\left(0, \xi_k \sigma^2_{1k}\right).$$

$$\theta_{tlj} = \theta_{t-1,lj} + \omega_{tlj}, \quad \omega_{tlj} \sim N(0, \psi_{lj}\Omega_{lj}).$$

12.5 Estimation

12.5.1 Empirical Bayes estimation of σ^2_{tLj}, ν_{lj} and Ω_{lj}

The parameter ν_{lj} is the vector of relative volatility of the descendants of subregion (l, j), and Ω_{lj} is the singular covariance matrix of the empirical multiscale coefficient of subregion (l, j). Here we use empirical Bayes (Carlin and Louis, 2000) for the estimation of ν_{lj} and Ω_{lj}, that is, we estimate these parameters upfront and then hold them fixed at their estimates when we perform the analysis of the other parameters of the model.

The parameters ν_{lj} and Ω_{lj}, $l = 1, \ldots, L$, $j = 1, \ldots, n_l$, are deterministic functions of $\sigma^2_{tLj}, j = 1, \ldots, n_L$, the conditional variance of y_{tLj} given μ_{tLj}. A preliminary analysis of the agricultural production of each county through time indicates that $\sigma^2_{tLj}, j = 1, \ldots, n_L$, may be assumed constant over time. Henceforth, we assume that $\sigma^2_{tLj} = \sigma^2_{Lj}, t = 1, \ldots, T, j = 1, \ldots, n_L$. In order to obtain an estimate of σ^2_{Lj}, we perform a univariate time series analysis for each county using first-order dynamic linear models (West and Harrison, 1997). These analyses yield estimates $\tilde{\sigma}^2_{Lj}$, that we use for the empirical Bayes estimation of ν_{lj} and Ω_{lj}. More specifically, we estimate these parameters by

$$\tilde{\nu}_{lj} = \tilde{\sigma}^2_{D_{lj}} / \tilde{\sigma}^2_{lj},$$

$$\tilde{\Omega}_{lj} = \tilde{\Sigma}_{D_{lj}} - \tilde{\sigma}^{-2}_{lj},$$

where the estimates $\tilde{\sigma}^2_{D_{lj}}$, $\tilde{\sigma}^2_{lj}$, and $\tilde{\Sigma}_{D_{lj}}$ are directly computed from $\tilde{\sigma}^2_{Lj}$, $j = 1, \ldots, n_L$.

12.5.2 Posterior exploration

Here we present a Markov chain Monte Carlo algorithm for performing a Bayesian analysis of our multiscale multivariate spatio-temporal model. More specifically, we build a Gibbs sampler (Geman and Geman, 1984; Gelfand and Smith, 1990) to explore the posterior distribution. We block the parameters of the model as follows: $\xi_1, \ldots, \xi_{n_1}, \psi_{11}, \ldots, \psi_{1n_1}, \ldots, \psi_{L-1,1}, \ldots, \psi_{L-1,n_{L-1}}, \mu_{0:T,11}, \ldots, \mu_{0:T,1n_1}, \theta_{0:T,11}, \ldots, \theta_{0:T,1n_1}, \ldots, \theta_{0:T,L-1,1}, \ldots, \theta_{0:T,L-1,n_{L-1}}$. It can be shown that, given $\sigma^2_{L,1:n_L}$, $\xi_{1:n_1}$, and $\psi_{1,1:n_1}, \ldots, \psi_{L-1,1:n_{L-1}}$, the vectors $\mu_{0:T,11}, \ldots, \mu_{0:T,1n_1}, \theta_{0:T,11}, \ldots, \theta_{0:T,1n_1}, \ldots, \theta_{0:T,L-1,1}, \ldots, \theta_{0:T,L-1,n_{L-1}}$ are conditionally independent and normally distributed a posteriori. As a result, simulation of $\mu_{0:T,11}, \ldots, \mu_{0:T,1n_1}, \theta_{0:T,11}, \ldots, \theta_{0:T,1n_1}, \ldots, \theta_{0:T,L-1,1}, \ldots, \theta_{0:T,L-1,n_{L-1}}$ can be performed in parallel at each iteration of the MCMC algorithm. In this direction, we simulate each $\mu_{0:T,1k}$, $k = 1, \ldots, n_1$, from its full conditional distribution using the Forward Filter Backward Sampler (FFBS) (Carter and Kohn, 1994; Frühwirth-Schnatter, 1994). Simulation of each $\theta_{0:T,lj}$, $l = 1, \ldots, L-1; j = 1, \ldots, n_l$, poses a non-trivial step in the MCMC procedure because the covariance matrix of the system equation is singular. Therefore, to accommodate this simulation we describe a modified FFBS for models with singular system equation covariance matrix (see Appendix B.3).

Simulation of ξ_k and ψ_{lj} can be performed with Gibbs steps sampling directly from their full conditional distributions as we describe below. Specifically, the parameter ξ_k, $k = 1, \ldots, n_1$, is sampled from its full conditional distribution: $\xi_k|\mu_{0:T,1k}, \sigma^2_{1k}, D_T \sim IG\left(0.5\tau^*_k, 0.5\kappa^*_k\right)$, where $\tau^*_k = \tau_k + T$ and $\kappa^*_k = \kappa_k + \sigma^{-2}_{1k}\sum_{t=1}^{T}(\mu_{t1k} - \mu_{t-1,1k})^2$. Finally, the parameter ψ_{lj}, $l = 1, \ldots, L$, $j = 1, \ldots, n_l$, is sampled from its full conditional distribution: $\psi_{lj}\,|\theta_{0:T,lj}, D_T \sim IG(0.5\varrho^*_{lj}, 0.5\varsigma^*_{lj})$, where $\varrho^*_{lj} = \varrho_{lj} + T(m_{lj} - 1)$ and $\varsigma^*_{lj} = \varsigma_{lj} + \sum_{t=1}^{T}(\theta_{tlj} - \theta_{t-1,lj})'\Omega^-_{lj}(\theta_{tlj} - \theta_{t-1,lj})$, where Ω^-_{lj} is a generalized inverse of Ω_{lj}.

Simulation of $\mu_{0:T,1k}$ is performed with the usual FFBS as introduced and described in Carter and Kohn (1994) and Frühwirth-Schnatter (1994). This step is nowadays fairly standard, therefore we omit the exact equations for the sake of brevity. For a comprehensive discussion, see Gamerman and Lopes (2006).

Simulation of $\theta_{\cdot lj}$ is performed with the Singular Forward Filter Backward Sampler (SFFBS) which we describe in Appendix B.3.

12.5.3 Reconstruction of the latent mean process

One of the main interests of any multiscale analysis is the estimation of the latent agricultural production mean process at each scale of resolution. This task can be performed using the sample of the posterior distribution simulated with the MCMC algorithm described above and in Appendix B. In particular, from the gth draw from the posterior distribution, we can recursively compute the corresponding latent mean process at each level of resolution using the equation

$$\mu^{(g)}_{t,D_{lj}} = \theta^{(g)}_{tlj} + \nu_{tlj}\mu^{(g)}_{tlj},$$

proceeding from the coarsest to the finest resolution level. With these draws, we can then compute the posterior mean, standard deviation and credible intervals for the latent mean process. Finally, the results can be presented graphically via maps and dynamic movies of the estimated latent mean process over the region of interest.

12.6 Agricultural production in Espírito Santo

Espírito Santo State has a tropical humid climate, with an average annual temperature around 73 degrees Fahrenheit and rain precipitation around 1400 millimeters per year. The state borders the Atlantic Ocean to the east, and the states of Rio de Janeiro to the south, Bahia to the north and Minas Gerais to the west. Its main agriculture products are coffee, sugar cane, manioc, coconut, and tropical fruits such as banana, passion fruit, and papaya. Fluctuations in the total value of agricultural production may be caused by price changes, expansions and reductions in cultivated area, variations in weather, crop diseases, and productivity improvement.

Here we analyse the data on agricultural production in Espírito Santo with the multiscale spatio-temporal model using the estimation procedure described in Section 12.5. For this analysis, we ran the MCMC algorithm for a total of 10,000 iterations. Convergence of the MCMC is verified through trace plots of the posterior and was achieved after the first 500 iterations. Consequently, for the purposes of inference, we have conservatively discarded the first 1000 iterations as burnin.

Figure 12.3 shows the marginal posterior densities for the signal-to-noise ratio ξ_k of the evolution of the aggregated coarsest level. The posterior density for ξ_3 of Macroregion 3 is relatively closer to 0 than for the other macroregions, implying that the underlying latent process μ_{t3} varies through time with less variability than μ_{tk} for $k = 1, 2, 4$. This is reflected in Figure 12.4, which displays the posterior mean and 95% credible interval for μ_{t1k}, $k = 1, \ldots, 4$, as a function of time, along with the coarse level observations. Even though the coarse level observations are fairly noisy, we are still able to estimate a meaningful and reasonably smooth underlying latent mean process. This allows us to refine our conclusions from the exploratory data analysis of Section

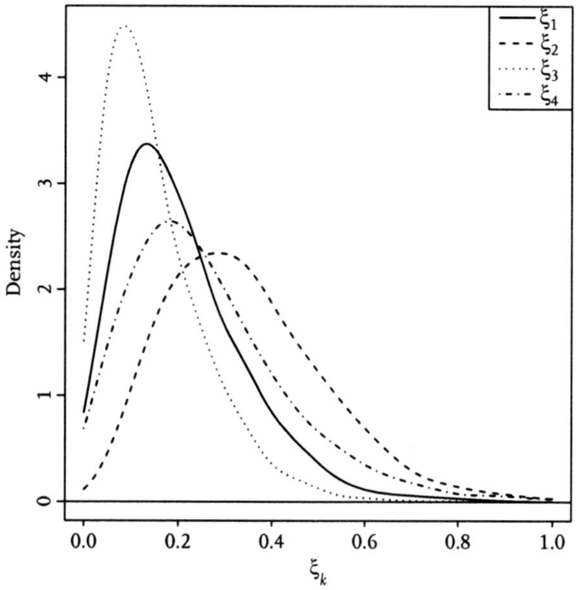

Fig. 12.3 Marginal posterior densities for the signal-to-noise ratio ξ_k of the evolution of the aggregated coarsest level.

12.3. First, Macroregion 1 shows a rapid increase in production from 1993 to about 1995 due to expansion of the planted area; after 1996 the expansion stopped and the production seems to have just varied around the same level due to price fluctuations. Second, Macroregion 2 presents a steady and relatively strong increase in production from 1990 to 2005 due to incorporation of new areas for agriculture. Additionally, Macroregion 3 displays a slow increase in production probably due to a combination of slow expansion of cultivated area and improvement in productivity. Finally, Macroregion 4 shows a striking increase in production from 1994 to 1998 and a not less surprising decrease from 1999 to 2001. This coincides with a period of economic stability in Brazil from 1994 to 1998 which abruptly ended in the beginning of 1999 with a strong devaluation of the Real and a credit squeeze. Probably many farmers in Macroregion 4 used credit to expand the planted area during the period of economic stability and were driven out of business by the 1999 Brazilian economic crisis.

Figure 12.5 shows the marginal posterior densities for the signal-to-noise ratio ψ_{1j} of the evolution of the coarsest level multiscale coefficients. The signal-to-noise ratio for Macroregion 1 is relatively large, reflecting the fact that the underlying multiscale coefficient varies significantly through time (see, for example, Figure 12.2c). The signal-to-noise ratio for Macroregion 2 is smaller than for Macroregion 1, reflecting a less volatile temporal evolution. Meanwhile, the signal-to-noise ratios for Macroregions 3 and 4 are concentrated close to 0

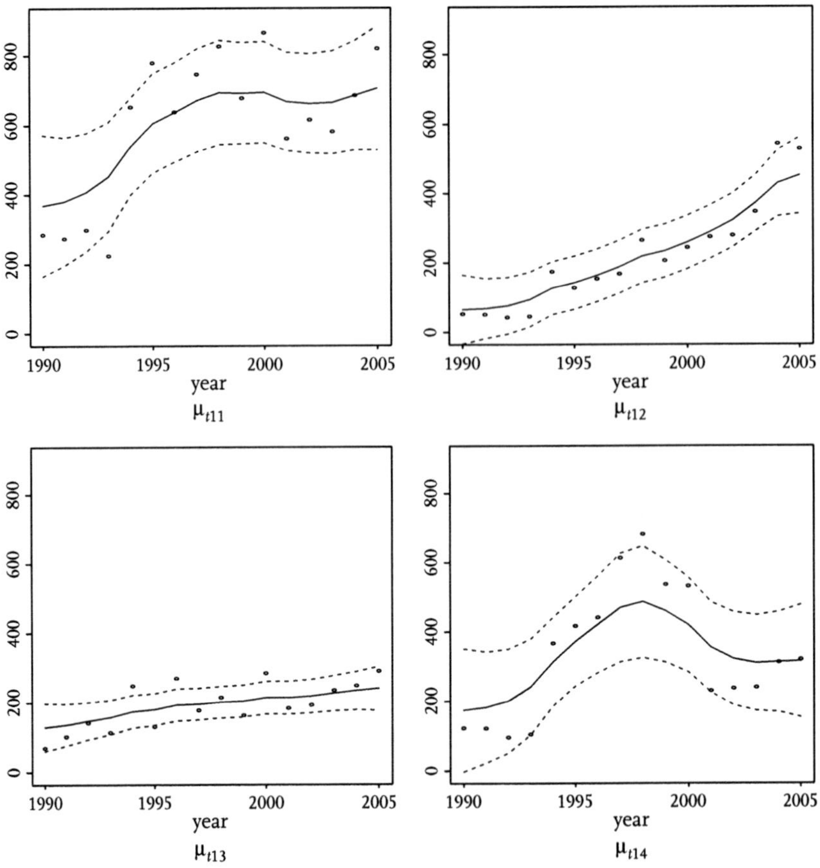

Fig. 12.4 Coarse level observations (dots), and posterior mean (solid line) and 95% credible interval (dashed lines) for μ_{t1k}, $k = 1, \ldots, 4$, through time. Note, for ease of comparison, all plots have been constructed with identical range on the y-axis.

implying that most of the variability of the empirical multiscale coefficients in Figures 12.2(i) and (l) are due to noise.

The analysis of the latent multiscale coefficient sheds light on possible changes over time in the relative importance of different subregions within the same parent region. With regard to agricultural production in Espírito Santo State, there are four multiscale coefficients that relate each macroregion with the respective microregions and 12 multiscale coefficients that relate each microregion with the respective counties. In the present application, these multiscale coefficients are vectors with lengths varying from 2 to 12. Thus, the number of graphs to be examined is quite extensive; therefore, for the sake of brevity we focus here only on a subset of these graphs. Specifically, we illustrate possible patterns that may emerge with the analyses of the multiscale coefficients for Macroregions 1, 2 and 4. Figure 12.6 presents the posterior mean

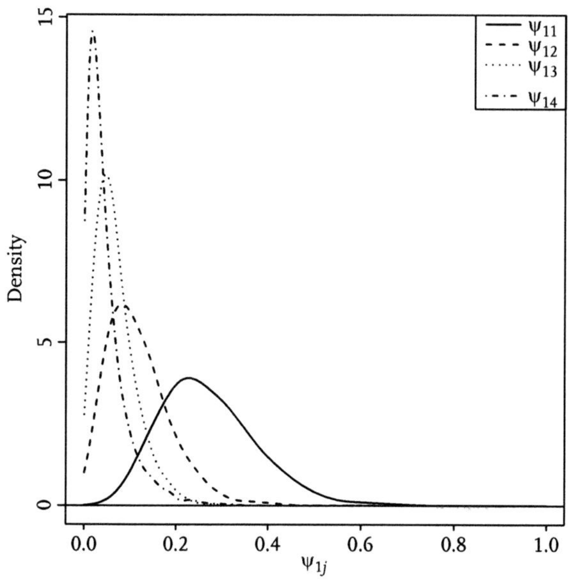

Fig. 12.5 Marginal posterior densities for the signal-to-noise ratio ψ_{1j} of the evolution of the coarsest level multiscale coefficients.

and 95% credible interval through time, as well as the empirical multiscale coefficient for each of the five components of the multiscale coefficient for Macroregion 1, $\theta_{t11} = \theta_{t11,1:5}$. From this figure we note the significant change through time in the relative importance of Microregions 2, 4, and 5, within Macroregion 1. Figure 12.7 focuses on Macroregion 2; particularly, this figure illustrates that the relative change in importance of Microregions 6 and 7 within Macroregion 2 is evident. Finally, Figure 12.8 displays posterior summaries for the multiscale coefficient for Macroregion 4; whereas the empirical multiscale coefficients are fairly noisy, the posterior mean and credible intervals seem to indicate that the relative importance of Microregions 11 and 12 has not changed significantly through time.

Figure 12.9 presents, on the logarithm scale, maps of observed agricultural production and estimated latent processes for years 1993, 1997, 2001, and 2005. Whereas the observed production varies substantially both spatially and temporally, the estimated latent process is reasonably smooth in both the spatial and temporal domains. The reason for this is because each subregion borrows information both across time and from the other subregions through its parent subregion. At the same time, differences between subregions are respected and thus the maps of estimated latent process do not exhibit the spatial over-smoothing typical of many spatial models.

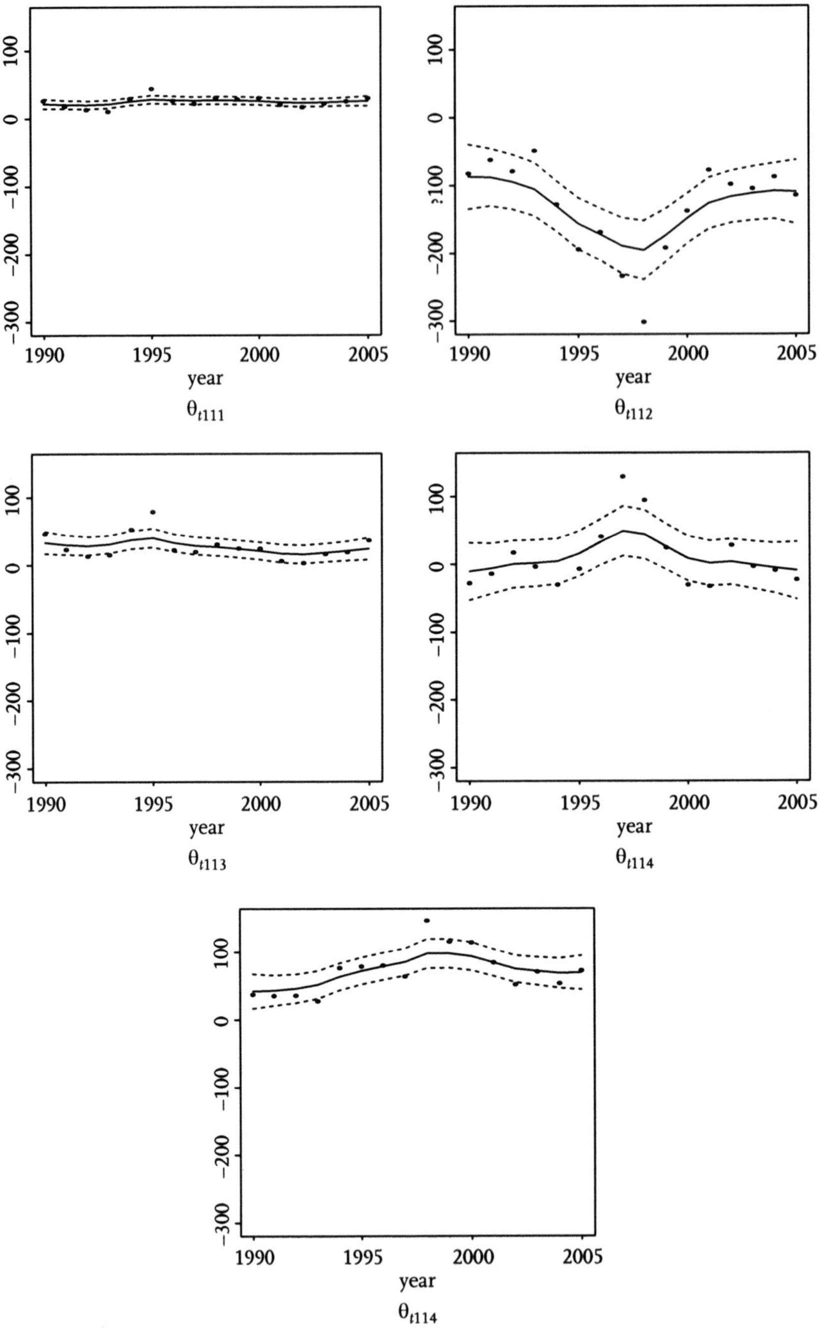

Fig. 12.6 Multiscale coefficient for Macroregion 1, $\theta_{t11,1:5} = (\theta_{t111}, \ldots, \theta_{t115})$. Empirical multiscale coefficient (dots), posterior mean (solid line) and 95% credible interval (dashed lines). Note, for ease of comparison, all plots have been constructed with identical range on the y-axis.

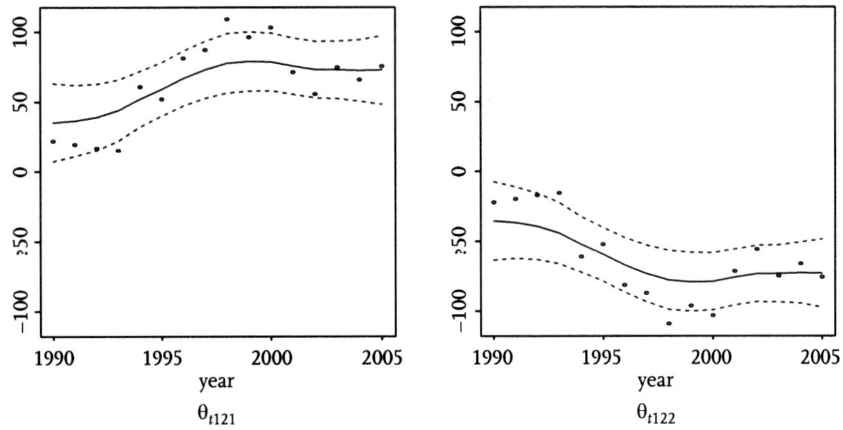

Fig. 12.7 Multiscale coefficient for Macroregion 2, $\theta_{t12,1:2} = (\theta_{t121}, \theta_{t122})$. Empirical multiscale coefficient (dots), posterior mean (solid line) and 95% credible interval (dashed lines).

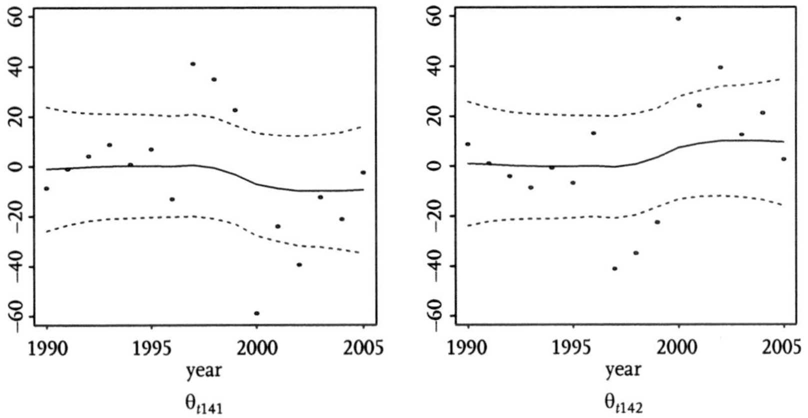

Fig. 12.8 Multiscale coefficient for Macroregion 4, $\theta_{t14,1:2} = (\theta_{t141}, \theta_{t142})$. Empirical multiscale coefficient (dots), posterior mean (solid line) and 95% credible interval (dashed lines).

12.7 Further discussion

We have presented an analysis of agricultural production in Espírito Santo State using a multiscale spatio-temporal model. This model decomposes the agricultural production data into multiscale coefficients at different scales of resolution and evolves the multiscale coefficients through time using state-space models. The resulting framework is able to produce estimates of the underlying latent production mean process that are coherent across the different scales of resolution.

A potentially important aspect of our spatio-temporal framework is the ability to effectively analyse massive economics data sets. More specifically, the

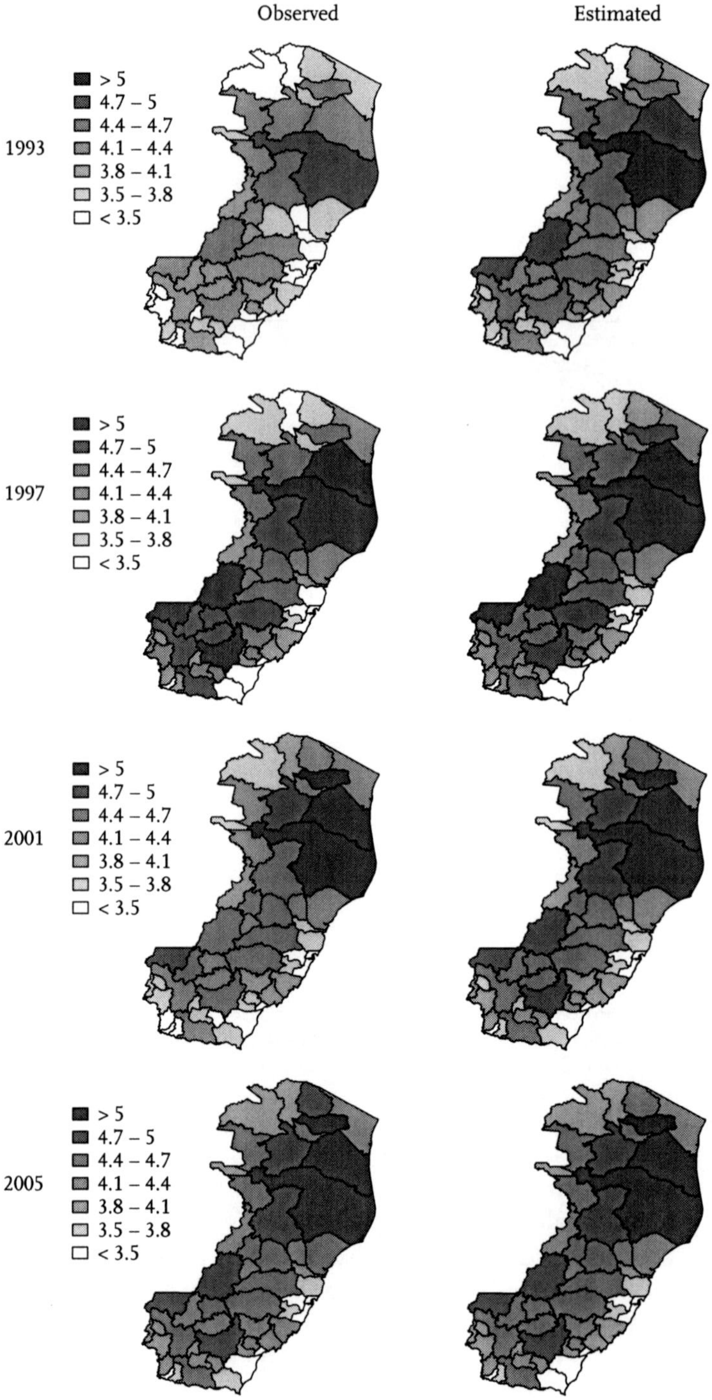

Fig. 12.9 Logarithm of observed agriculture production and logarithm estimated latent process for years 1993, 1997, 2001, and 2005.

multiscale decomposition coupled with the state-space time evolution allow an extremely efficient *divide-and-conquer* estimation algorithm. This estimation algorithm has good granularity and can, therefore, be implemented in a parallel computing environment. As a consequence, our multiscale spatio-temporal approach can be made extremely computationally efficient and thus is well designed for dealing with massive data sets. For example, it is conceptually feasible to analyse economic data sets comprising county level information for large countries through several years.

Appendix

A. Broader context and background

A.1 Multiscale models

Multiscale modelling broadly refers to models for data and processes that may be structured by scale. A broad coverage of existing statistical multiscale models is given by Ferreira and Lee (2007). Multiscale modelling includes not only the well-known wavelet multiscale decompositions (Daubechies, 1992; Mallat, 1999; Vidakovic, 1999), but many other models such as for example Gaussian models on trees (Willsky, 2002), hidden Markov models on trees (Kato *et al.*, 1996), multiscale random fields (Chapter 10, Ferreira and Lee, 2007), multiscale time series (Ferreira *et al.*, 2006), change of support models (Banerjee *et al.*, 2003), and implicit computationally linked multiscale models (Higdon *et al.*, 2002; Holloman *et al.*, 2006). Multiscale models have been successfully applied to several different scientific areas such as image segmentation (Nowak, 1999), permeability estimation (Ferreira *et al.*, 2003), agronomy (Banerjee and Johnson, 2006), single photon emission computed tomography (Holloman *et al.*, 2006), and disease mapping (Louie and Kolaczyk, 2006).

A.2 Dynamic linear models

Dynamic models, also known as state-space models, have been successfully used in many areas of science to model processes that evolve through time, as for example in economics (Azevedo *et al.*, 2006), finance (Jacquier *et al.*, 2007), ecology (Wikle, 2003), epidemiology (Knorr-Held and Richardson, 2003), and climatology (Wikle *et al.*, 2001). The books by West and Harrison (1997) and Harvey (1989) give thorough coverage of dynamic models from the Bayesian and frequentist perspectives, respectively. Migon *et al.* (2005) provide a review along with recent developments on Bayesian dynamic models.

In this chapter we consider dynamic linear models, which can be written as

$$\mathbf{y}_t = \mathbf{F}_t'\theta_t + \varepsilon_t, \qquad \varepsilon_t \sim N(\mathbf{0}, \mathbf{V}_t),$$
$$\theta_t = \mathbf{G}_t\theta_{t-1} + \omega_t, \qquad \omega_t \sim N(0, \mathbf{W}_t),$$

where the first equation is known as the observation equation and the second equation is the system equation. At time t, $\mathbf{y}_t$ is the vector of observations, $\boldsymbol{\theta}_t$ is the latent process, $\mathbf{F}_t$ relates the observations to the latent process, $\mathbf{G}_t$ describes the evolution of the latent process through time, $\mathbf{V}_t$ is the observational covariance matrix, and $\mathbf{W}_t$ is the covariance matrix of the system equation innovation. Typically, $\mathbf{F}_t$, $\mathbf{G}_t$, $\mathbf{V}_t$ and $\mathbf{W}_t$ are known up to a few hyperparameters.

If the matrices $\mathbf{F}_t$, $\mathbf{G}_t$, $\mathbf{V}_t$, and $\mathbf{W}_t$ are completely known, then the Kalman filter can be used to estimate the latent process. If $\mathbf{F}_t$, $\mathbf{G}_t$, $\mathbf{V}_t$, and/or $\mathbf{W}_t$ are functions of unknown hyperparameters, then estimation can be performed using Markov chain Monte Carlo (MCMC). See Robert and Casella (2004) and Gamerman and Lopes (2006) for detailed information on MCMC methods. In this context, each iteration of the MCMC algorithm is then divided in two blocks: simulation of the unknown hyperparameters, and simulation of the latent process. The details of the simulation of the hyperparameters is model-specific. If the matrices $\mathbf{V}_t$ and $\mathbf{W}_t$ are positive definite, the latent process can be efficiently simulated using the forward filter backward sampler (Frühwirth-Schnatter, 1994; Carter and Kohn, 1994).

B. Multiscale decomposition and computations

B.1 Multiscale decomposition

Since the joint distribution of the observations at the finest level L conditional on the mean process $\mu_{L,1:n_L}$ is multivariate Gaussian, it follows that $y_{l,1:n_l}|\mu_{l,1:n_l}, \Sigma_l \sim N(\mu_{l,1:n_l}, \Sigma_l)$. Moreover, the joint distribution of y_{lj} and $y_{D_{lj}}$ is

$$\begin{pmatrix} y_{lj} \\ y_{D_{lj}} \end{pmatrix} \Bigg| \mu_{L,1:n_L}, \sigma^2_{L,1:n_L} \sim N\left[\begin{pmatrix} \mu_{lj} \\ \mu_{D_{lj}} \end{pmatrix}, \begin{pmatrix} \sigma^2_{lj} & \left(\sigma^2_{D_{lj}}\right)' \\ \sigma^2_{D_{lj}} & \Sigma_{D_{lj}} \end{pmatrix}\right].$$

Therefore, using standard results from the theory of multivariate normal distributions (Mardia *et al.*, 1979), the conditional distribution of $y_{D_{lj}}$ given y_{lj} is

$$y_{D_{lj}} \big| y_{lj}, \mu_{L,1:n_L}, \sigma^2_{L,1:n_L} \sim N(\nu_{lj} y_{lj} + \theta_{lj}, \Omega_{lj}),$$

with

$$\begin{aligned} \nu_{lj} &= \sigma^2_{D_{lj}} / \sigma^2_{lj}, \\ \theta_{lj} &= \mu_{D_{lj}} - \nu_{lj}\mu_{lj}, \end{aligned}$$

and

$$\Omega_{lj} = \Sigma_{D_{lj}} - \sigma^{-2}_{lj}\sigma^2_{D_{lj}}\left(\sigma^2_{D_{lj}}\right)',$$

$l = 1, \ldots, L-1, j = 1, \ldots, n_l$. Note that the scale specific parameter θ_{lj} is a vector corresponding to the difference between the mean level process at the descendants of subregion B_{lj} and the scaled mean level process at B_{lj}.

Next, consider

$$\theta^e_{lj} = y_{D_{lj}} - \nu_{lj} y_{lj},$$

which is an empirical estimate of θ_{lj}. Then,

$$\theta^e_{lj} | y_{lj}, \mu_L, \sigma^2_L \sim N(\theta_{lj}, \Omega_{lj}), \tag{12.6}$$

which is a singular Gaussian distribution (Anderson, 1984, p. 33). Straightforward linear algebra shows that θ^e_{lj} is subject to the constraint $1'_{m_{lj}} \theta^e_{lj} = 0$. Moreover, this constraint is implicitly embedded in the singular Gaussian distribution in (12.6). In order to see this is true, note that

$$E\left(1'_{m_{lj}} \theta^e_{lj}\right) = 1'_{m_{lj}} \theta_{lj} = 1'_{m_{lj}} (\mu_{D_{lj}} - \nu_{lj} \mu_{lj}) = 0$$

and

$$V\left(1'_{m_{lj}} \theta^e_{lj}\right) = 1'_{m_{lj}} \Omega_{lj} 1_{m_{lj}} = 1'_{m_{lj}} \Sigma_{D_{lj}} 1_{m_{lj}} - \sigma^{-2}_{lj} 1'_{m_{lj}} \sigma^2_{D_{lj}} \left(\sigma^2_{D_{lj}}\right)' 1_{m_{lj}} = 0.$$

B.2 Derivation of full conditional distributions

Full conditional of ξ_k. The full conditional density of ξ_k, $k = 1, \ldots, n_1$, is

$$p\left(\xi_k | \mu_{0:T,1k}, \sigma^2_{1k}, D_T\right) \propto p(\xi_k | \tau_k, \kappa_k) \prod_{t=1}^{T} p\left(\mu_{t1k} | \mu_{t-1,1k}, \sigma^2_{1k}, \xi_k\right)$$
$$\propto \xi_k^{-0.5(T+\tau_k)-1} \exp\left[-\frac{1}{2\xi_k}\left\{\kappa_k + \sigma^{-2}_{1k} \sum_{t=1}^{T} (\mu_{t1k} - \mu_{t-1,1k})^2\right\}\right].$$

Therefore, $\xi_k | \mu_{0:T,1k}, \sigma^2_{1k}, D_T \sim IG\left(0.5\tau^*_k, 0.5\kappa^*_k\right)$, where $\tau^*_k = \tau_k + T$ and $\kappa^*_k = \kappa_k + \sigma^{-2}_{1k} \sum_{t=1}^{T} (\mu_{t1k} - \mu_{t-1,1k})^2$.

Full conditional of ψ_{lj}. The full conditional density of ψ_{lj}, $l = 1, \ldots, L$, $j = 1, \ldots, n_l$, is

$$p(\psi_{lj} | \theta_{0:T,lj}, \sigma^2, D_T) \propto p(\psi_{lj} | \tau, \kappa) \prod_{t=1}^{T} p(\theta_{tlj} | \theta_{t-1,lj}, \psi_{lj}, \sigma^2, D_{t-1})$$
$$\propto \psi_{lj}^{-\frac{1}{2}\left(T(m_{lj}-1)+\tau\right)-1}$$
$$\times \exp\left[-\frac{1}{2\psi_{lj}}\left\{\kappa_{lj} + \sum_{t=1}^{T} \left(\theta_{tlj} - \theta_{t-1,lj}\right)' \Omega^-_{lj} \left(\theta_{tlj} - \theta_{t-1,lj}\right)\right\}\right].$$

Therefore $\psi_{lj} | \theta_{0:T,lj}, \sigma^2, D_T \sim IG\left(0.5\tau^*_{lj}, 0.5\kappa^*_{lj}\right)$, where $\tau^*_{lj} = T(m_{lj} - 1) + \tau$ and $\kappa^*_{lj} = \kappa + \sum_{t=1}^{T} (\theta_{tlj} - \theta_{t-1,lj})' \Omega^-_{lj} (\theta_{tlj} - \theta_{t-1,lj})$.

B.3 Singular forward filter backward sampler

From the multiscale decomposition of the observation equation, we have that

$$\theta^e_{tlj} = \theta_{tlj} + v_{tlj}, \qquad v_{tlj} \sim N(0, \Omega_{lj}).$$

Moreover, from the system equations,

$$\theta_{tlj} = \theta_{t-1,lj} + \omega_{tlj}, \qquad \omega_{tlj} \sim N(0, \psi_{lj}\Omega_{lj}).$$

Then the singular forward filter backward sampler proceeds as follows.

1. Use the Kalman filter to obtain the mean and covariance matrix of $p(\theta_{1lj}|\sigma^2, \psi_{lj}, D_1), \ldots, p(\theta_{Tlj}|\sigma^2, \psi_{lj}, D_T)$:
   - posterior at $t-1$: $\theta_{t-1,lj}|D_{t-1} \sim N\left(m_{t-1,lj}, C_{t-1,lj}\Omega_{lj}\right)$;
   - prior at t: $\theta_{tlj}|D_{t-1} \sim N\left(a_{tlj}, R_{tlj}\Omega_{lj}\right)$, where $a_{tlj} = m_{t-1,lj}$ and $R_{tlj} = C_{t-1,lj} + \psi_{lj}$;
   - posterior at t: $\theta_{tlj}|D_t \sim N\left(m_{tlj}, C_{tlj}\Omega_{lj}\right)$, where $C_{tlj} = \left(1 + R^{-1}_{tlj}\right)^{-1}$ and $m_{tlj} = C_{tlj}\left(\theta^e_{tlj} + R^{-1}_{tlj}a_{tlj}\right)$.
2. Simulate θ_{Tlj} from $\theta_{Tlj}|\sigma^2, \psi_{lj}, D_T \sim N(m_{Tlj}, C_{Tlj}\Omega_{lj})$.
3. Recursively simulate θ_{tlj}, $t = T-1, \ldots, 0$, from

$$\theta_{tlj}|\theta_{(t+1):T,lj}, D_T \equiv \theta_{tlj}|\theta_{t+1,lj}, D_t \sim N(h_{tlj}, H_{tlj}\Omega_{lj}),$$

where $H_{tlj} = \left(C^{-1}_{tlj} + \psi^{-1}_{lj}\right)^{-1}$ and $h_{tlj} = H_{tlj}\left(C^{-1}_{tlj}m_{tlj} + \psi^{-1}_{lj}\theta_{t+1,lj}\right)$.

Acknowledgments

Marco A. R. Ferreira was partially supported by CNPq grant 479093/2004-0. Part of this work was performed while Adelmo I. Bertolde was a graduate student in the Statistics Program at the Federal University of Rio de Janeiro with a scholarship from CAPES, Brazil.

References

Anderson, T. W. (1984). *An Introduction to Multivariate Statistical Analysis* (2nd edn). John Wiley, New York.

Azevedo, J. V., Koopman, S. J. and Rua, A. (2006). Tracking the business cycle of the euro area: A multivariate model-based bandpass filter. *Journal of Business and Economic Statistics*, **24**, 278–290.

Banerjee, S., Carlin, B. P. and Gelfand, A. E. (2003). *Hierarchical Modeling and Analysis for Spatial Data*. Chapman & Hall, Boca Raton, Florida.

Banerjee, S. and Johnson, G. A. (2006). Coregionalized single- and multi-resolution spatially-varying growth curve modelling with application to weed growth. *Biometrics*, **61**, 617–625.

Basseville, M., Benveniste, A. and Willsky, A. S. (1992*a*). Multiscale autoregressive processes, part I: Schur-Levinson parametrizations. *IEEE Transactions on Signal Processing*, **40**, 1915–1934.

Basseville, M., Benveniste, A. and Willsky, A. S. (1992*b*). Multiscale autoregressive processes, part II: Lattice structures for whitening and modeling. *IEEE Transactions on Signal Processing*, **40**, 1935–1953.

Berliner, L. M., Wikle, C. K. and Milliff, R. F. (1999). Multiresolution wavelet analyses in hierarchical Bayesian turbulence models. In *Bayesian Inference in Wavelet Based Models*, (ed. P. Müller and B. Vidakovic), pp. 341–359. Springer-Verlag, New York.

Carlin, B. P. and Louis, T. A. (2000). *Bayes and Empirical Bayes Methods for Data Analysis*, (2nd edn.) Chapman Hall / CRC. Boca Raton, Florida.

Carter, C. K. and Kohn, R. (1994). On Gibbs sampling for state space models. *Biometrika*, **81**(3), 541–553.

Daubechies, I. (1992). *Ten Lectures on Wavelets*. SIAM: Philadelphia, PA.

Donoho, D. L. and Johnstone, I. M. (1994). Ideal spatial adaptation via wavelet shrinkage. *Biometrika*, **81**, 425–455.

Ferreira, M. A. R., West, M., Bi, Z., Lee, H. and Higdon, D. (2003). Multi-scale modelling of 1-D permeability fields. In *Bayesian Statistics 7*, (ed. J. Bernardo, M. J. Bayarri, J. O. Berger, A. P. Dawid, D. Heckerman, A. F. M. Smith, and M. West), pp. 519–527. Oxford University Press, Oxford.

Ferreira, M. A. R. and Lee, H. K. H. (2007). *Multiscale Modeling: A Bayesian Perspective*. Springer Series in Statistics. Springer, New York.

Ferreira, M. A. R., West, M. Lee, H. K. H. and Higdon, D. (2006). Multi-scale and hidden resolution time series models. *Bayesian Analysis*, **1**, 947–968.

Frühwirth-Schnatter, S. (1994). Data augmentation and dynamic linear models. *Journal of Time Series Analysis*, **15**, 183–202.

Gamerman, D. and Lopes, H. F. (2006). *Markov Chain Monte Carlo: Stochastic Simulation for Bayesian Inference*, (2nd edn.) Chapman Hall/CRC, Boca Raton, Florida.

Gelfand, A. E. and Smith, A. F. M. (1990). Sampling-based approaches to calculating marginal densities. *Journal of the American Statistical Association*, **85**, 398–409.

Geman, S. and Geman, D. (1984). Stochastic relaxation, Gibbs distributions, and the Bayesian restoration of images. *IEEE Transactions on Pattern Analysis and Machine Intelligence*, **6**, 721–741.

Harvey, A. C. (1989). *Forecasting, Structural Time Series Models and the Kalman Filter*. Cambridge University Press, Cambridge.

Higdon, D., Lee, H. and Bi, Z. (2002). A Bayesian approach to characterizing uncertainty in inverse problems using coarse and fine scale information. *IEEE Transactions on Signal Processing*, **50**, 389–399.

Holloman, C. H., Lee, H. K. H. and Higdon, D. M. (2006). Multi-resolution genetic algorithms and Markov chain Monte Carlo. *Journal of Computational and Graphical Statistics*, **15**, 861–879.

Jacquier, E., Johannes, M. and Polson, N. (2007). MCMC maximum likelihood for latent space models. *Journal of Econometrics*, **137**, 615–640.

Johannesson, G., Cressie, N. and Huang, H. (2007). Dynamic multi-resolution spatial models. *Environmental and Ecological Statistics*, **14**, 5–25.

Kato, Z., Berthod, M. and Zerubia, J. (1996). A hierarchical markov random field model and multi-temperature annealing for parallel image classification. *Graphical Models and Image Processing*, **58**, 18–37.

Knorr-Held, L. and Richardson, S. (2003). A hierarchical model for space-time surveillance data on meningococcal disease incidence. *Applied Statistics*, **52**, 169–183.

Kolaczyk, E. D. and Huang, H. (2001). Multiscale statistical models for hierarchical spatial aggregation. *Geographical Analysis*, **33**, 95–118.

Louie, M. M. and Kolaczyk, E. D. (2006). A multiscale method for disease mapping in spatial epidemiology. *Statistics in Medicine*, **25**, 1287–1308.

Mallat, S. G. (1999). *A Wavelet Tour of Signal Processing*, (2nd edn). Academic Press: San Diego.

Mardia, K. V., Kent, J. T. and Bibby, J. M. (1979). *Multivarite Analysis*. Academic Press: San Diego.

Migon, H. S., Gamerman, D., Lopes, H. F. and Ferreira, M. A. R. (2005). Bayesian dynamic models. In *Handbook of Statistics*, Volume 25, (ed. D. Dey and C. R. Rao), pp. 553–588. Elsevier, Amsterdam.

Nowak, R. D. (1999). Multiscale hidden Markov models for Bayesian image analysis. In *Bayesian Inference in Wavelet Based Models*, (ed. P. Müller and B. Vidakovic), pp. 243–265. Springer-Verlag, New York.

Robert, C. P. and Casella, G. (2004). *Monte Carlo Statistical Methods*, (2nd edn.) Springer-Verlag, New York.

Sweldens, W. (1996). The lifting scheme: a custom-design construction of biorthogonal wavelets. *Applied and Computational Harmonic Analysis*, **3**, 186–200.

Vidakovic, B. (1999). *Statistical Modeling by Wavelets*. John Wiley, New York.

West, M. and Harrison, J. (1997). *Bayesian Forecasting and Dynamic Models* (2nd ed.). Springer-Verlag, New York.

Wikle, C. K. (2003). A kernel-based spectral approach for spatiotemporal dynamic models. Technical report, Department of Statistics, University of Missouri.

Wikle, C. K., Milliff, R. F., Nychka, D. and Berliner, L. M. (2001). Spatio-temporal hierarchical Bayesian modeling: Tropical ocean surface winds. *Journal of the American Statistical Association*, **96**, 382–397.

Willsky, A. S. (2002). Multiresolution Markov models for signal and image processing. *Proceedings of the IEEE*, **90**, 1396–1458.

·13·

Extracting S&P500 and NASDAQ volatility: The credit crisis of 2007–2008

Hedibert F. Lopes and Nicholas G. Polson

13.1 Introduction

Volatility and volatility dynamics are important for understanding the behaviour of financial markets and market pricing. In this chapter we estimate volatility and volatility dynamics for daily data for the year 2007 for three market indicies: the Standard and Poor's S&P500, the NASDAQ NDX100 and the financial equity index called XLF. We study how the three series reflect the beginning of the credit crisis; the XLF index reflects the effect on financial companies that were among the earliest to be affected while the other two market indices reflect broader, economy wide effects. Three models of financial time series are estimated: a model with stochastic volatility, a model with stochastic volatility that also incorporates jumps in volatility and a Garch model. We compare our volatility estimates from these three models with subjective implied market volatility calculated from option prices, the VIX and VXN, for the S&P500, the NASDAQ NDX100, respectively. By sequentially computing marginal likelihoods or Bayes factors we can also evaluate the ability of the different models to capture the time series behavior over this turbulent period.

Volatility and volatility dynamics are central to many issues in financial markets including derivative prices, leverage ratios, credit spreads, and portfolio decisions. In times of low market volatility it is relatively straightforward to measure volatility and understand volatility dynamics. At other times, financial markets are affected by severe disruptions which may be largely isolated events like the market crash of 1987 or may be a series of events such as the Russian default and Long Term Capital Management Fund Crisis of 1998, or the current credit crisis. During such periods, apparent spikes in volatility and large movements in asset prices complicate estimation of volatility and volatility dynamics. This chapter describes how Bayesian particle filtering methods can track volatility in an online fashion under these circumstances. Our particle filtering methodology presented here provides sequential inference for financial time series in general and can also be applied to other problems such as longitudinal time series studies.

Sequential Bayesian methods based on particle filtering provide a natural solution to estimating volatility and volatility dynamics. Sequential inference is important as it provides estimates of current (spot) volatility and parameters for the evolution of the volatility dynamics given the currently available information. This includes the historical path of prices or returns available up until the current time together with market beliefs about volatility dynamics implicit in option prices. These subjective market beliefs are measured by the so-called implied option volatilities and can be compared to model-based estimated volatility after accounting for the market price of volatility risk. We describe this further at the end of this section and in the appendicies.

Our primary filtering results are from two financial asset price models, one with stochastic volatility (SV) and one with stochastic volatility and jumps in volatility (SVJ). Jumps are transient by their nature and the resulting price and volatility dynamics can be very different depending upon whether jumps are included or not. In our estimation, we will identify the posterior distribution of the latent state variables at each point in time, denoted by $L_t = (V_t, J_t, Z_t)$ corresponding to the stochastic variance, jump time and the jump size. Our approach will also examine how quickly volatility estimation procedures react to changes in underlying asset returns.

We find a number of important empirical finance results. First, we compare our filtering results from the SV model and SVJ model with a Garch(1,1) model frequently applied to financial time series. We find that in periods of market turbulence that the Garch(1,1) model does not track option implied volatility as as well the two stochastic volatility models particularly for the more volatile NDX100 index. These facts also lead to differences in market prices of volatility risk between stochastic volatility models and Garch(1,1) throughout the period under investigation. Second, we examine the reaction of volatility estimates in periods of market stress and extreme movements can be surprisingly different even though SVJ is designed to capture jump effects. An attractive feature of the particle filtering approach is that we have a model diagnostic in the form of a sequential Bayes factor. We find that while the SVJ model outperforms the SV model for the entire period under investigation, the relative superiority of the SVJ model increases with market turbulence.

The methods used here build upon a number of recent papers that develop particle filtering algorithms for SV and SVJ models. Carvalho and Lopes (2007) develop sequential particle methods for Markov switching SV models and Carvalho *et al.* (2008) provide a general particle learning algorithm for SV models. Johannes *et al.* (2008) consider a general class of continuous-time asset price diffusion models and continuous-time stochastic volatility jump models. Particle filtering has an advantage over commonly used MCMC methods

(Jacquier *et al.*, 1994, 2004, Johannes and Polson, 2009) which are computationally burdensome and inherently non-sequential.

The rest of the paper proceeds as follows. We begin by describing the applied problem and goal of the Bayesian analysis. Section 13.2 presents the estimation models. Section 13.3 discusses the simulation-based methodology for inference, including MCMC and particle filtering methods for filtering and parameter learning. Detailed analysis of our empirical findings appear in Section 13.4. Finally, Section 13.5 concludes.

13.1.1 Problem context and goals

Subjective beliefs about volatility are available from option prices in the form of implied volatilities. Model-based volatility estimates can be determined in an online fashion using historical return data and a specification for volatility dynmaics. Our goal then is to extract volatility estimates sequentially and compare them with market-based volatility measures. Our three underlying equity time series are daily data for the S&P500, the NDX100 and the XLF. We use two common market volatility indicies, the VIX and VXN indicies corresponding to the option implied volatilities of the benchmark indicies.

We begin by studying the volatility of the financial stock index, XLF. Figure 13.1 plots the XLF together with realized volatility estimates based on a simple rolling window of past returns. At the end of February, there are is a volatility spike, when the first credit crisis indicators in the form of sub-prime mortgage issues and collateral debt obligations (CDOs) came to light. The XLF stock index fell from 36 to 31 in the week at the end of February. The volatility spike on February 27th, 2007 occurred when the Dow Jones index dropped 320 points. This followed a fall of over 10% in the Chinese stock market. Our estimate of volatility, $\sqrt{V_t}$, of the XLF index moved from around 10% to nearly 30% very quickly in a matter of days.

In the next few months prices stabilized an volatility mean-reverted decaying back to the 20% range. The most dramatic change in volatility dynamics occurs just before the beginning of August and persists throughout the rest of the year. For example, the volatility spikes higher again in October back to the 40% range with the XLF dropping further from 36 at the beginning of October to 30 in December. We now describe theoretically and empirically the relationship between option prices and volatility.

13.1.2 Option prices and volatility

To study the relationship between option prices and volatility more closely we observe that a financial model describes asset price dynamics for the physical or objective measure $\mathbb{P}$ whereas derivatives are priced under $\mathbb{Q}$ the risk-neutral

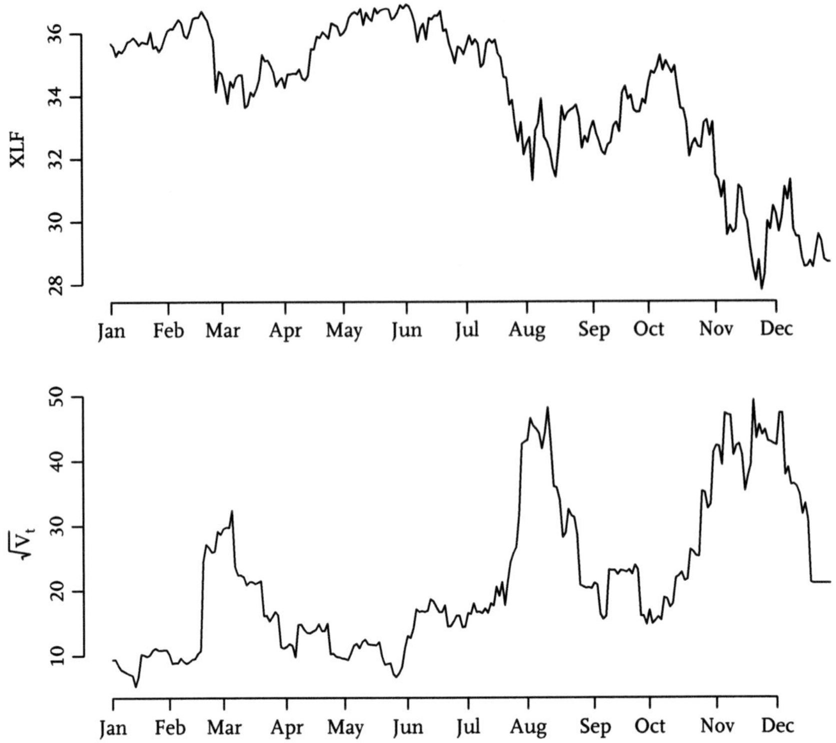

Fig. 13.1 *Year 2007*: XLF underlying price (top) and rolling window annualized standard deviations (bottom). Volatility increases just before August 2007.

dynamics. These two different probability distributions underpin financial market pricing. In the next section we describe clearly the stochastic volatility dynamics describing these two distributions. Historical returns can be used to estimate the dynamics of the physical dynamics $\mathbb{P}$ and option returns can be used to assess the risk neutral dynamics $\mathbb{Q}$. Both returns and option prices will have information concerning the latent stochastic variance V_t. This is because, derivative prices depend crucially on the current volatility state, V_t, through a pricing formula

$$C(S_t, V_t) = e^{-r(T-t)} E_t^{\mathbb{Q}} [\max(S_T - K, 0) \mid S_t, V_t],$$

where S_t is the current value of the equity index and V_t is the current (spot) volatility. Appendix B provides explicit formulas for option pricing with stochastic volatility. This is related to the implied volatilities, VIX and VXN indices. The VIX index is computed from option prices and it also serves as a basis for volatility swaps which are a tradable asset.

We term these indices option market implied volatility denoted by IV_t. Formally, they are related to the expected future cumulative volatility via

$$IV_t = E_t^{\mathbb{Q}}\left[\int_t^{t+\tau} V_s ds\right].$$

If we include jumps, then

$$IV_t^J = E_t^{\mathbb{Q}}\left[\int_t^{t+\tau} V_s ds\right] + var_t^{\mathbb{Q}}\left[\sum_{n=N_t^s+1}^{N_{t+\tau}^s} Z_n^s\right],$$

where V_s denotes the path of volatility and N_t and Z_t denote the number of jumps and the jump sizes, respectively. As expected, the option implied volatility is providing a market-based prediction of future cumulative volatility.

Most approaches for estimating volatility and jumps rely exclusively on returns, ignoring the information embedded in option prices. In principle, options are a rich source of information regarding volatility, which explains the common use of Black–Scholes implied volatility as a volatility proxy for practitioners. In contrast to this is the common academic approach of only using returns to estimate volatility.

Options are highly informative about volatility, but the information content is dependent on the model specification and $\mathbb{P}$ and $\mathbb{Q}$-measure parameter estimates. Misspecification or poorly estimated risk premia could results in directionally biased estimates of V_t using option prices. This informativeness of index options is mitigated by the fact that these options contain relatively large bid-ask spreads, as noted by Bakshi *et al.* (1997) or George and Longstaff (1993). This implies that while option prices are informative, they may be quite noisy as a measure of volatility. There is also empirical evidence for jumps in returns (Bates, 2000; Bates, 2006; Bakshi *et al.*, 1997; and Eraker, 2004) as well as volatility (see Eraker *et al.*, 2003).

Figure 13.2 compares the filtered XLF volatilities with the option implied volatilities of the broader indices. Not surprisingly, the volatilities are better tracked by the VIX than for the VXN. This is primarily due to the fact that financial companies comprised 25% of the SP500 index at the time. Even though the financial weighting in the NASDAQ is lower, it is surprising that the VXN volatility also tracks the movements in the XLF albeit not as well. This shows the influence of contagion effects of the credit crisis even to a sector of the market that has very little financial leverage. Implied volatility versus a 10-day moving average tracks very well, hence confirming the notion that implied volatility has information about future average volatility.

Over our data period the move in XLF prices was mirrored by the changes in the market option volatility indices (VIX and VXN). The sharp moves in market

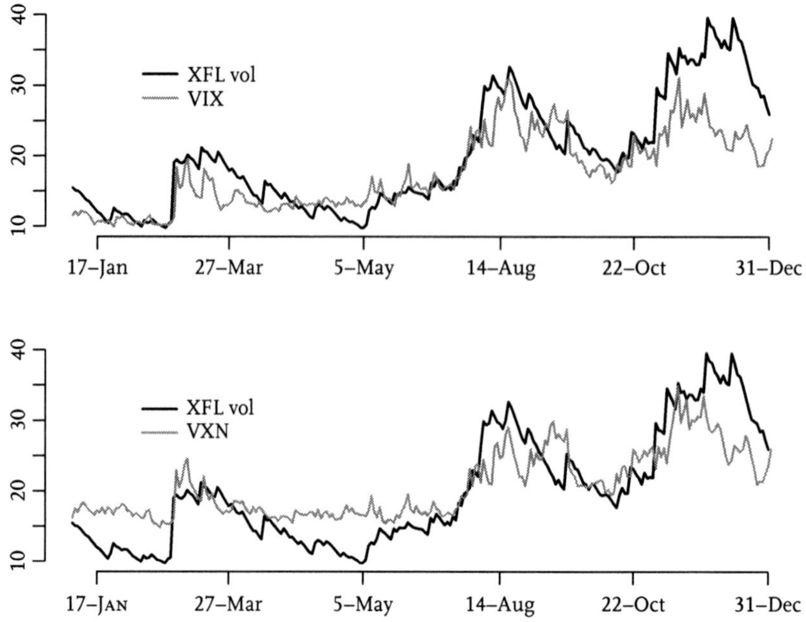

Fig. 13.2 *Year 2007*: XLF volatility versus VIX and VXN.

option implied volatility are hard to fully capture with a pure diffusive SV model even though filtered volatility estimates move quicker than smooth estimates which incorporate future and past returns. Hence we will also study a stochastic volatility model that allows for jumps. However, we also show that these moves are even harder to capture with a deterministic time-varying volatility model like Garch (Rosenberg, 1972, Engle, 1982). We will see later than these types of return movements provide large statistical evidence in favor of the stochastic volatility jump model. For the next three months volatility mean reverts back to around the 10% level before the next shock hits at the end of July. In the next section we describe more clearly our methodology and comparison with the SP500 and NDX100 stock indices.

In principle, option prices also allow us to extract information about the current volatility state although this is not the focus of the study here. It is important to realise that in going from risk-neutral to physical measures we can assess the market price of volatility risk, see Polson and Stroud (2003), Eraker (2004) and Todorov (2010) and Appendix B for further discussion. This is also related to a well-known volatility puzzle where option implied volatility is usually higher than estimated historical volatility. This is due to the fact that there exists a positive market price for taking volatility risk. Our particle methods will be able to measure this quantity.

13.2 Models

Our empirical analysis will focus on sequential volatility filtering in the 2007. Parameter estimates will be performed on a longer historical period of returns from 2002–2006. Our particle filtering approach will then be implemented for the year of 2007. We use daily return data for the Standard and Poor's SP500 stock index and the NASDAQ NDX 100 index, denoted here by SP500 and NASDAQ, respectively. We also study the behaviour of the XLF which is an equity index for the prices for US financial firms. The corresponding option volatility indices are the VIX and the VXN; again these are available on a continuous basis. We now describe the three models that we use for price dynamics: pure diffusive SV model, SVJ model and Garch(1,1) model. Then, we describe the different volatility indices and how they are related to option prices and the market price of volatility risk.

13.2.1 Stochastic volatility (SV) model

A common model of asset price dynamics results in the following two equation describing the movements of an equity index S_t and its stochastic volatility V_t,

$$\begin{aligned}\frac{dS_t}{S_t} &= \mu + \sqrt{V_t}dB_t^{\mathbb{P}}\\ d\log V_t &= \kappa_v(\theta_v - \log V_t) + \sigma_v dB_t^V\end{aligned}$$

where μ is an expected rate of return and the parameters governing the volatility evolution are $\Theta = (\kappa_v, \theta_v, \sigma_v)$. The Brownian motions $\left(B_t^{\mathbb{P}}, B_t^V\right)$ are possibly correlated giving rise to a leverage effect.

This is a pure stochastic volatility (SV) model. The probabilistic evolution $\mathbb{P}$ describes what is known as the physical dynamics as opposed to the risk-neutral dynamics $\mathbb{Q}$ which is used for pricing. To analyse this model in a sequential fashion it is common to using an Euler discretization of the above model for continuously compounded returns (see details in Appendix A). In the subsequent analysis we will use daily time scale.

Let $Y_{t+1} = \log(S_{t+1}/S_t)$ be log-returns and transform volatility to logarithms as well, $X_{t+1} = \ln V_{t+1}$, then the Euler discretization is

$$\begin{aligned}Y_{t+1} &= e^{\frac{X_{t+1}}{2}}\varepsilon_{t+1}\\ X_{t+1} &= \alpha_v + \beta_v X_t + \sigma_v \eta_{t+1}.\end{aligned}$$

Table 13.1 *Stochastic volatility (SV) model.* Mean and StDev are posterior mean and standard deviation, respectively. 2.5% and 97.5% are posterior percentiles. Time span: 1/02/2002–12/29/2006.

	Mean	StDev	2.5%	97.5%
SP500				
α_v	−0.0031	0.0029	−0.0092	0.0022
β_v	0.9949	0.0036	0.9868	1.0011
σ_v^2	0.0076	0.0026	0.0041	0.0144
NASDAQ				
α_v	−0.0003	0.0023	−0.0046	0.0045
β_v	0.9968	0.0024	0.9914	1.0011
σ_v^2	0.0046	0.0015	0.0024	0.0084
XLF				
α_v	−0.0020	0.0032	−0.0082	0.0040
β_v	0.9924	0.0042	0.9830	0.9996
σ_v^2	0.0115	0.0036	0.0064	0.0203

Here ε_{t+1} and η_{t+1} are normally distributed, serially and contemporaneously independent shocks. The parameters that govern the evolution of volatility dynamics are transformed to $\alpha = \kappa_v \theta_v$ and $\beta_v = 1 - \kappa_v$.

The model specification is completed with independent prior distributions for the components of $\Theta = \left(X_0, \alpha_v, \beta_v, \sigma_v^2\right)$, i.e. $N(X_0, V_X)$, $N(\alpha_0, V_\alpha)$, $N(\beta_0, V_\beta)$ and $IG\left(\nu/2, \nu\bar{\sigma}_v^2/2\right)$, respectively, where N and IG denote the normal distribution and the inverse Gamma distribution. Table 13.1 provides posterior summaries for parameter estimates the stochastic volatility (SV) model for all three series, SP500, NASDAQ and XLF, for based on data from January 2002 to December 2006. We assumed relatively uninformative priors and set the hyperparameters in common choices from the literature, i.e. $V_X = 10$, $V_\alpha = V_\beta = 1$ and $\nu = 3$. The hyperparameters α_0, β_0 and $\bar{\sigma}_v^2$ are simple ordinary least square estimates based on a 10-day moving window procedure. Setting these hyperparameters at $(0, 0, 1)$, for instance, produced virtually the same posterior summaries from Table 13.1. The estimates (α_0, β_0) and $\bar{\sigma}_v^2$ are used as initial values for the MCMC algorithm. Other sound initial values also produced the same posterior results. Finally, the results are similar to the ones found in Eraker *et al.* (2003), with all three series exhibiting high persistent evolution.

13.2.2 Stochastic volatility jump (SVJ) model

The stochastic volatility jump (SVJ) model includes the possibility of jumps to asset prices. Now an equity index S_t and its stochastic variance V_t

jointly solve

$$\frac{dS_t}{S_t} = \mu + \sqrt{V_t}dB_t^{\mathbb{P}} + d\left(\sum_{s=N_t}^{N_{t+1}} Z_s\right)$$

$$d\log V_t = \kappa_v(\theta_v - \log V_t) + \sigma_v dB_t^V$$

where the additional term in the equity price evolution describes the jump process. Since data are observed in discrete time it is again common to use an Euler discretization of this continuous time process (Appendix A). Specifically,

$$\begin{aligned} Y_{t+1} &= e^{\frac{X_{t+1}}{2}}\varepsilon_{t+1} + J_{t+1}Z_{t+1} \\ X_{t+1} &= \alpha_v + \beta_v X_t + \sigma_v \eta_{t+1} \\ J_{t+1} &\sim Ber(\lambda) \\ Z_{t+1} &\sim N\left(\mu_z, \sigma_z^2\right) \end{aligned}$$

with $Y_{t+1} = \log(S_{t+1}/S_t)$. The log-returns with ε_{t+1} and η_{t+1} normally distributed, serially and contemporaneously independent shocks. The parameter vector is $\Theta = (\lambda, \mu_z, \sigma_z, \alpha_v, \beta_v, \sigma_v)$.

Prior distributions are required for the initial volatility state, X_0, and for all parameters governing the dynamic of the volatilities $(\alpha_v, \beta_v, \sigma_v^2)$. See Section 13.2.1 for explicit details. For the jump specification, we use a conditionally conjugate prior structure for parameters $(\lambda, \mu_z, \sigma_z^2)$, i.e. $\lambda \sim Beta(a, b)$, $\mu_z \sim N(c, d)$ and $\sigma_z^2 \sim IG\left(\nu/2, \nu\bar{\sigma}_z^2/2\right)$, respectively. We set $c = -3$ and $d = 0.01$ and $a = 2$ and $b = 100$ such that the prior mean and standard deviation of λ are around 0.02 and 0.014. ν and $\bar{\sigma}_z^2$ are set at 20 and 0.05, respectively, such that the prior mean and standard deviation of σ_z^2 are roughly 0.05 and 0.02. These prior specifications predict around five large negative jumps per year (roughly 250 business days) whose magnitude are around an additional three percent.

This structure naturally leads to conditional posterior distributions that can be easily simulated to form a Gibbs sampler (see Eraker *et al.*, 2003). Table 13.2 provides posterior summaries for parameter estimates the stochastic volatility jump (SVJ) model for all three series, SP500, NASDAQ and XLF. Jump probability estimates are all similar at about 0.04 or 10 jumps per year. The largest estimates jump sizes are −3.72% for the XLF, −2.14% for the SP500 and −1.98% for the NASDAQ. The table also provides posterior means, standard deviations and 5% and 95% quantiles for $(\alpha_v, \beta_v, \sigma_v^2)$ and $(\lambda, \mu_z, \sigma_z^2)$.

Table 13.2 *Stochastic volatility with jump (SVJ) model.* Mean and StDev are posterior mean and standard deviation, respectively. 2.5% and 97.5% are posterior percentiles. Time span: 1/02/2002–12/29/2006.

	Mean	StDev	2.5%	97.5%
SP500				
α_v	−0.0117	0.0070	−0.0262	0.0014
β_v	0.9730	0.0084	0.9551	0.9886
σ_v^2	0.0432	0.0082	0.0302	0.0613
λ	0.0025	0.0017	0.0003	0.0066
μ_z	−2.7254	0.1025	−2.9273	−2.5230
σ_z^2	0.3809	0.2211	0.1445	0.9381
NASDAQ				
α_v	0.0079	0.0065	−0.0042	0.0215
β_v	0.9785	0.0073	0.9631	0.9916
σ_v^2	0.0390	0.0071	0.0275	0.0553
λ	0.0031	0.0021	0.0004	0.0082
μ_z	−3.9033	0.1001	−4.0906	−3.7054
σ_z^2	0.6420	0.3856	0.2445	1.6314
XLF				
α_v	−0.0044	0.0064	−0.0176	0.0083
β_v	0.9728	0.0085	0.9554	0.9880
σ_v^2	0.0472	0.0091	0.0324	0.0677
λ	0.0026	0.0018	0.0003	0.0071
μ_z	−3.2676	0.0997	−3.4593	−3.0700
σ_z^2	0.8983	0.5490	0.3467	2.2171

13.2.3 Garch model

The Garch(1,1) model is a time-varying volatility model that uses The evolution of returns and volatility is then given by

$$Y_{t+1} = \sqrt{V_{t+1}}\varepsilon_{t+1}$$
$$V_{t+1} = \alpha_v + \beta_v V_t + \gamma_v \varepsilon_t^2.$$

This leads to a time-varying volatility sequence given the residuals from the observation equation. The parameters have the usual constraints: $\alpha_v > 0$, $\beta_v > 0$ and $\gamma_v > 0$ to ensure positive variance, and $\alpha_v + \beta_v + \gamma_v < 1$ for stationarity. This model is fundamentally different from stochastic volatility, since last period's return shock ε_t^2 is used as a regressor as opposed to a stochastic volatility term $\sigma_v \eta_{t+1}$. Other differences include assessments of tail-probabilities and predictives. MCMC approaches for Bayesian inference on Garch models and several of its variants appear in Müller and Pole (1994), Bauwens and Lubrano (1998), Vontros and Politis (2000), Wago (2004) and Ausín and Galeano (2007) amongst many others.

Table 13.3 shows the parameters α_v, β_v and γ_v based on the 2002–2006 period. We then sequentially apply the model through the time period of 2007.

Table 13.3 *Garch model.* Parameter estimates. Time span: 1/02/2002–12/29/2006.

Index	α_v	β_v	γ_v	$\alpha_v + \beta_v + \gamma_v$
SP500	0.0042	0.9440	0.0501	0.9983
NADSAQ	0.0035	0.9652	0.0319	1.0006
XLF	0.0072	0.9354	0.0573	1.0000

Of particular interest is the comparison in August 2007 at the beginning of the credit crisis. We will also compare with the implied volatility series to see what the implications are for market prices of volatility risk. We find that the Garch volatility effect γ_v on the squared residual ε_t^2 is largest for the XLF at 0.057 and smallest for the NASDAQ at 0.03. This is not surprising as the credit crisis affected leveraged finance companies as opposed to technology stocks that traditionally has low debt levels.

13.3 Sequential learning via particle filtering

Given a time series of observations, $Y_{1:T} = (Y_1, \ldots, Y_T)$, the usual estimation problem is to estimate the parameters, Θ, and the unobserved states, $L_{1:T}$, from the observed data. In our case, the latent variables include (1) the volatility states, (2) the jump times, and (3) the jump sizes. In a Bayesian setting, this information is summarized by the joint posterior distribution $p(\Theta, L_{1:T}|Y_{1:T})$. In turn this joint posterior can be used to find estimates of the current state variables and parameters estimates using $\hat{L}_t = E(L_t|Y_{1:t})$ and $\hat{\Theta}_t = E(\Theta|Y_{1:t})$. Samples from this distribution are usually obtained via MCMC methods by iteratively sampling from the complete conditional distributions, $p(L_{1:T}|\Theta, Y_{1:T})$ and $p(\Theta|L_{1:T}, Y_{1:T})$. From these samples, it is straightforward to obtain smoothed estimates of the parameters and states.

Alternatively, particle filtering provides sequential estimates from the set of joint distributions $p(L_{1:t}|Y_{t:t})$. Particle filtering (PF, Gordon *et al.*, 1993) methods are a simulation-based approach to sequential Bayesian filtering. Doucet *et al.* (2001) provide a reference text for a detailed discussion of the theoretical properties and applications. We also use a variant of the PF known as the auxiliary particle filter (APF, Pitt and Shephard, 1999).

The sequential estimation procedure is implemented in what follows, where Θ is assumed known to simplify the presentation. In our application, Θ is estimated based on daily data on SP500, NASDAQ and XLF from January 2002 to December 2006. Carvalho *et al.* (2008) provides more details about the following resample-propagate scheme including parameter learning.

13.3.1 Extracting state variables $L_t = (V_t, J_t, Z_t)$

The optimal filtering problem is solved by the sequential computation of the set of posteriors $p(V_t, J_t, Z_t | Y_{1:t})$. By Bayes rule these posteriors satisfy the recursion

$$p(V_{t+1}, J_{t+1}, Z_{t+1} | Y_{1:t+1}) \propto p(Y_{t+1} | V_{t+1}, J_{t+1}, Z_{t+1}) p(V_{t+1}, J_{t+1}, Z_{t+1} | Y_{1:t})$$

where the normal likelihood $p(Y_{t+1} | V_{t+1}, J_{t+1}, Z_{t+1})$ has mean $J_{t+1} Z_{t+1}$ and variance V_{t+1}, while the prior $p(V_{t+1}, J_{t+1}, Z_{t+1} | Y_{1:t})$ is given by

$$p(V_{t+1}, J_{t+1}, Z_{t+1} | Y_{1:t}) \propto p(V_{t+1} | V_t) p(J_{t+1}, Z_{t+1}) p(V_t, J_t, Z_t | Y_{1:t}).$$

The state evolution gives rise to a normal density $p(V_{t+1} | V_t)$ with mean $\alpha_v + \beta_v V_t$ and variance σ_v^2. Finally, as the jumps as transient and conditionally i.i.d. with $Pr(J_{t+1} = 1) = \lambda$ and normal density $p(Z_{t+1} | J_{t+1} = 1)$ with mean μ_z and variance σ_z^2. It would also be straightforward to allow for the jump probability to depend on the volatility state through a conditional $p(J_{t+1} = 1 | V_t)$ (see Johannes *et al.*, 1998). These equations will form the basis of our particle filtering algorithm that we develop in what follows. Following Carvalho *et al.* (2008) we now describe our resample-propagate filtering scheme.

13.3.2 Resample-propagation filter

Let the current filtering posterior be denoted by $p(L_t | Y_{1:t})$. The next likelihood is $p(Y_{t+1} | L_{t+1})$ and the state evolution is $p(L_{t+1} | L_t)$. Bayes rule links these to the next filtering distribution through Kalman updating. This takes the form of a smoothing step and a prediction step

$$p(L_t | Y_{1:t+1}) \propto p(Y_{t+1} | L_t) p(L_t | Y_{1:t})$$
$$p(L_{t+1} | Y_{1:t+1}) = \int p(L_{t+1} | L_t) p(L_t | Y_{1:t+1}) dL_t$$

where $Y_{1:t}$ are the continuously compounded log-returns. Specifically, for extracting volatility and jumps as latent states we let $L_t = (V_t, J_t, Z_t)$. Uncertainty about these quantities is summarized via the filtered posterior distribution $p(L_t | Y_{1:t})$. Particle methods will represent this distribution as

$$p^N(L_t | Y_{1:t}) = \frac{1}{N} \sum_{i=1}^{N} \delta_{L_t^{(i)}}.$$

for particles $L_t^{(1)}, \ldots, L_t^{(N)}$. The previous filtered distribution is represented by its particle approximation and the key is how to propagate particle forward. From the updating formulas we can approximate

$$p^N(L_{t+1} | Y_{1:t+1}) = \sum_{i=1}^{N} w_t^{(i)} p\left(L_{t+1} | L_t^{(i)}, Y_{t+1}\right)$$

where weights are given by

$$w_t^{(i)} = p\left(Y_{t+1}|L_t^{(i)}\right) \Big/ \sum_{\ell=1}^{N} p\left(Y_{t+1}|L_t^{(\ell)}\right) \qquad i = 1, \ldots, N.$$

Hence after we have resampled the initial particles with weights proportional to $w_t^{(i)}$ we then propagate new particles using $p\left(L_{t+1}|L_t^{(i)}, Y_{t+1}\right)$. This then leads us to the following simulation algorithm.

Resample-propagate filter

1. *Resample:* For $i = 1, \ldots, N$, compute

$$w_t^{(i)} = p\left(Y_{t+1}|L_t^{(i)}\right) \Big/ \sum_{\ell=1}^{N} p\left(y_{t+1}|L_t^{(\ell)}\right),$$

draw

$$z(i) \sim Mult\left(N; w_t^{(1)}, \ldots, w_t^{(N)}\right),$$

and set $L_t^{(i)} = L_t^{z(i)}$ for $i = 1, \ldots N$.

2. *Propagate:* For $i = 1, \ldots, N$, draw

$$L_{t+1}^{(i)} \sim p\left(L_{t+1}|L_t^{(i)}, Y_{t+1}\right).$$

For the SV and SVJ models, it is common to include V_{t+1} into the definition of the latent state variable L_t. This is due to the nonlinearities that the volatility induces into the asset price dynamics. In the SVJ model the state variables describing the jump process (J_t, Z_t) can be marginalized out. This is due to the fact that they are independent of V_t and we have the integral decomposition

$$p(Y_{t+1}|V_{t+1}) = \int p(Y_{t+1}|V_{t+1}, J_{t+1}, Z_{t+1}) p(J_{t+1}, Z_{t+1}) dJ_{t+1} dZ_{t+1}.$$

Therefore, in practice, we propagate forward the volatility $V_{t+1} \sim p(V_{t+1}|V_t)$ and then attach it to the current particle before using the resample-propagate filter described above.

13.3.3 Sequential model choice

An important by-product of sequential Monte Carlo methods is the ability to easily compute approximate marginal predictive densities and then Bayes factors. Let $\mathcal{M}$ denote a given model, then the sequential marginal predictive for any t can be approximated via

$$p(Y_{t+1}|Y_{1:t}, \mathcal{M}) = \frac{1}{N}\sum_{i=1}^{N} p\left(Y_{t+1}|L_t^{(i)}, \mathcal{M}\right)$$

where $L_t^{(i)}$ is the particle for the latent volatility state of model $\mathcal{M}$. This approximation allows one to sequentially compute the Bayes factors (West, 1986), namely $\mathcal{BF}_{1:t}$, for competing models $\mathcal{M}_0$ and $\mathcal{M}_1$

$$\mathcal{BF}_{1:t} = \frac{p(Y_{1:t}|\mathcal{M}_1)}{p(Y_{1:t}|\mathcal{M}_0)}.$$

The Bayes factor is related to the Bayesian posterior probabilities of the models being true via the following identity

$$\frac{p(\mathcal{M}_1|Y_{1:t})}{p(\mathcal{M}_0|Y_{1:t})} = \mathcal{BF}_{1:t} \times \frac{p(\mathcal{M}_1)}{p(\mathcal{M}_0)}$$

where $p(\mathcal{M}_1)/p(\mathcal{M}_0)$ is typically set equal to one to denote a priori equal weight on either model. One advantage of this approach is that we can interpret the relative posterior model probabilities whether or not the 'true' data generating process is either on of the models under consideration.

A sequential decomposition of the joint distribution is also available and we write

$$p(Y_{1:t}|\mathcal{M}) = p(Y_1|\mathcal{M})\prod_{j=1}^{t-1} p(Y_{j+1}|Y_{1:j}, \mathcal{M})$$

Now we can use the particle approximation in an on-line fashion to compute the quantities $p(Y_{j+1}|Y_{1:j}, \mathcal{M})$. Hence we can sequentially compute the Bayes factor for the full data sequence by compounding

$$\mathcal{BF}_{1:T} = \prod_{j=1}^{T-1} \mathcal{BF}_{j:j+1}.$$

When competing models (and priors) are nested, which is the case in the SV versus SVJ case, for example, a further simplification is the implementation of the Savage–Dickey density ratio (Verdinelli and Wasserman, 1995). More specifically, let $\Theta = \left(\lambda, \mu_z, \sigma_z^{-2}\right)$, i.e. the parameters of the SVJ model (the full model or $\mathcal{M}_1$) that are not in the SV model (the reduced model or $\mathcal{M}_0$), then

$$\mathcal{BF}_{1:t} = \frac{p(\Theta = \Theta_0|Y_{1:t}, \mathcal{M}_1)}{p(\Theta = \Theta_0|\mathcal{M}_1)} \approx \frac{1}{N}\sum_{i=1}^{N} \frac{p\left(\Theta = \Theta_0|Y_{1:t}, L_{1:t}^{(i)}, \mathcal{M}_1\right)}{p(\Theta = \Theta_0|\mathcal{M}_1)}$$

where $\Theta_0 = (0, 0, 0)$. The computation of $p(\Theta = \Theta_0 | Y_{1:t}, L_{1:t}^{(i)}, \mathcal{M}_1)$ is the main challenge since

$$p\left(\Theta = \Theta_0 | Y_{1:t}, L_{1:t}^{(i)}, \mathcal{M}_1\right) \propto p\left(Y_{1:t} | L_{1:t}^{(i)}, \Theta = \Theta_0, \mathcal{M}_1\right) p\left(\Theta = \Theta_0 | L_{1:t}^{(i)}, \mathcal{M}_1\right).$$

Hence this provides a computational tractable approach for calculating Bayes factors from particle filtering output.

Figure 13.6 below provides the sequential log-Bayes factor $\mathcal{BF}(SV, SVJ)$ for comparing the pure SV model with an SVJ model. We provide the Bayes factor diagnostic for all three series, SP500, NASDAQ and XLF. Not surprisingly, the large negative shock in February 2007 leads to substantial evidence in favour of SVJ over SV. However, for the rest of the year as there are no extreme shocks to returns the evidence decays back and at the end of the period slightly favours the pure SV model. Providing these estimates within a pure MCMC framework would be computationally expensive, see, for example, the discussion in Chapter 7 of Gamerman and Lopes (2006).

13.4 Empirical results

We implement our particle filtering methodology to find volatility estimates for all the models on a daily basis in 2007. Special focus is on the period at the onset of the credit crisis, namely August 2007. We compare volatility estimates and implied prices of volatility risk for each model. Table 13.4 provides this comparison for the month of August. The volatilities are filtered and a direct comparison with the VIX and VXN for each model gives a measurement of the price of volatility risk for each model. The SVJ model tracks the movements more closely than the pure SV model. The SVJ model implies a relatively constant λ_v until the end of the month when all model imply a very small λ_v. It is also interesting to see how the sequential Bayes factor discriminates between these models. see Figure 13.6 and the subsequent discussion.

Periods of high volatility risk premia occur at the end of July for both SV and SVJ models while lower premia occur for the Garch(1,1) model. A high premia is empirically justified by subsequent increases and volatility spikes in the August period. Remember that one can interpret the risk premia as market expectations of future changes in average volatility.

This difference can be explained as follows. In the last week of July the Dow Jones index dropped from 14, 000 to 13, 000 with a sequence of negative shocks. The Garch(1,1) model therefore estimates volatility at 23% on August 1st as negative shocks feed directly into the Garch(1,1) model via $\gamma\varepsilon_t^2$. Both SV and SVJ models attribute some of their negative shocks to the stochastic volatility error term of the jump component rather than directly to $\sqrt{V_t}$. To initialize our estimate of volatility we use 17% for the SV model.

Table 13.4 SVJ, SV and Garch comparison. Columns 2–10 are annualized standard deviations.

Day	SP&500			NASDAQ			XLF				
	SVJ	SV	Garch	SVJ	SV	Garch	SVJ	SV	Garch	VIX	VXN
1	18.1	16.5	17.4	18.6	14.9	16.5	25.9	22.6	21.3	23.8	23.6
2	17.0	16.1	17.0	18.3	14.9	16.4	24.2	22.0	20.9	25.1	25.4
3	21.1	18.3	16.5	20.2	15.8	16.2	31.0	26.3	20.3	23.4	24.5
6	23.2	19.5	18.6	20.1	15.8	16.9	37.9	29.9	23.5	22.8	24.5
7	21.9	19.1	19.5	19.3	15.7	16.8	35.4	29.3	26.4	24.6	27.1
8	21.7	19.2	18.9	19.1	15.7	16.6	35.0	29.6	25.6	24.0	26.6
9	25.3	21.1	18.7	22.0	17.0	16.5	37.6	31.4	25.4	26.2	29.2
10	23.6	20.7	20.4	21.3	16.9	17.5	34.7	30.6	26.1	27.4	29.9
13	21.8	20.2	19.7	20.4	16.7	17.3	32.2	29.8	25.3	25.3	27.5
14	22.8	20.7	19.0	21.4	17.2	17.1	31.3	29.5	24.5	25.0	28.0
15	22.7	20.7	19.2	22.7	17.8	17.4	29.1	28.8	24.1	24.8	27.4
16	21.1	20.2	19.0	22.2	17.8	17.8	31.9	30.0	23.4	24.9	27.0
17	23.7	21.3	18.4	23.2	18.3	17.7	36.5	32.6	24.1	26.5	28.7
20	22.0	20.8	19.4	21.8	18.1	18.0	34.4	31.9	25.5	20.4	23.3
21	20.4	20.3	18.7	21.3	18.0	17.6	31.7	30.8	24.9	20.0	22.5
22	19.8	20.3	18.1	21.3	18.1	17.5	29.0	29.7	24.2	20.4	22.0
23	18.4	19.8	17.9	20.1	17.9	17.4	26.5	28.5	23.4	19.0	20.7
24	18.3	19.6	17.3	20.4	18.0	17.1	24.0	27.4	22.6	19.4	20.8
27	17.7	19.3	17.1	19.5	17.8	17.2	23.3	26.8	21.9	18.6	20.8
28	21.1	20.6	16.8	22.8	18.9	16.9	28.1	28.8	21.5	17.6	20.6
29	22.7	21.5	18.3	25.9	20.4	17.9	26.8	28.2	22.4	17.0	20.6
30	21.2	21.0	18.9	24.4	20.0	18.9	24.6	27.0	22.0	18.0	21.0
31	20.7	20.7	18.3	23.8	19.9	18.5	24.0	26.5	21.4	17.8	21.1

Our comparisons are summarized in more detail in Figures 13.3–13.5. They provide sequential comparisons of volatility estimates relative to the VIX and VXN indices. Specifically, Figure 13.4 compares the underlying indices S&P500 and NASDAQ with their respective implied volatility series. Figure 13.4 compares Garch(1,1) with SV and SVJ against the option implied series VIX and VXN.

Figure 13.5 provides a direct comparison between the models at the beginning of the credit crisis in August 2007. For reasons described before, the Garch(1,1) model does not react as quickly as a stochastic volatility model. It also starts the month of August at higher estimates of V_t or equivalently lower estimates of the volatility risk premia. The filtered Garch(1,1) estimates are also considerably smoother than the SV and SVJ estimates, this is because the SV model has the extra flexibility in the random variance term to adapt to large shocks.

Figure 13.5 also compares the three models with their option implied volatilities. This allows the researcher to gauge the movement in the market price of volatility risk, λ_v. It is also useful to look at the sequential Bayes factors

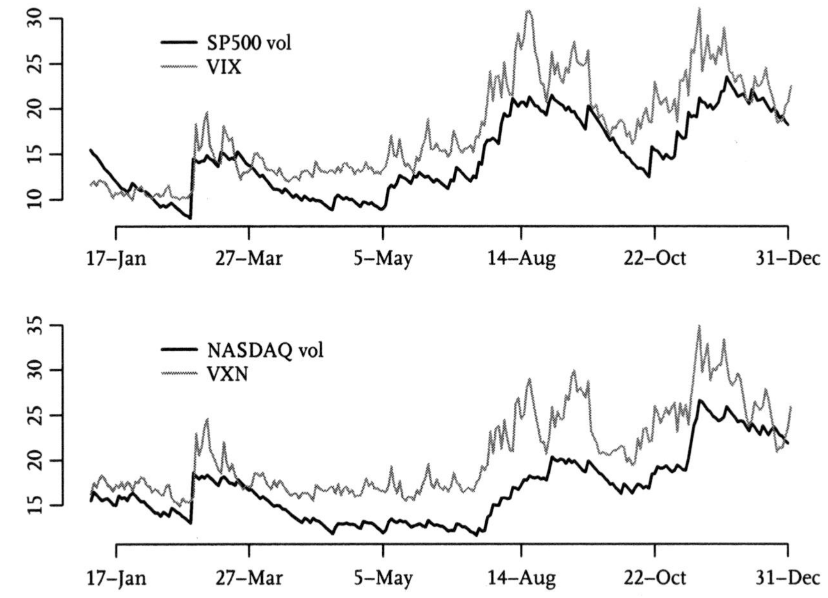

Fig. 13.3 *Year 2007*: SP500 and NASDAQ volatilities and VIX and VXN indices.

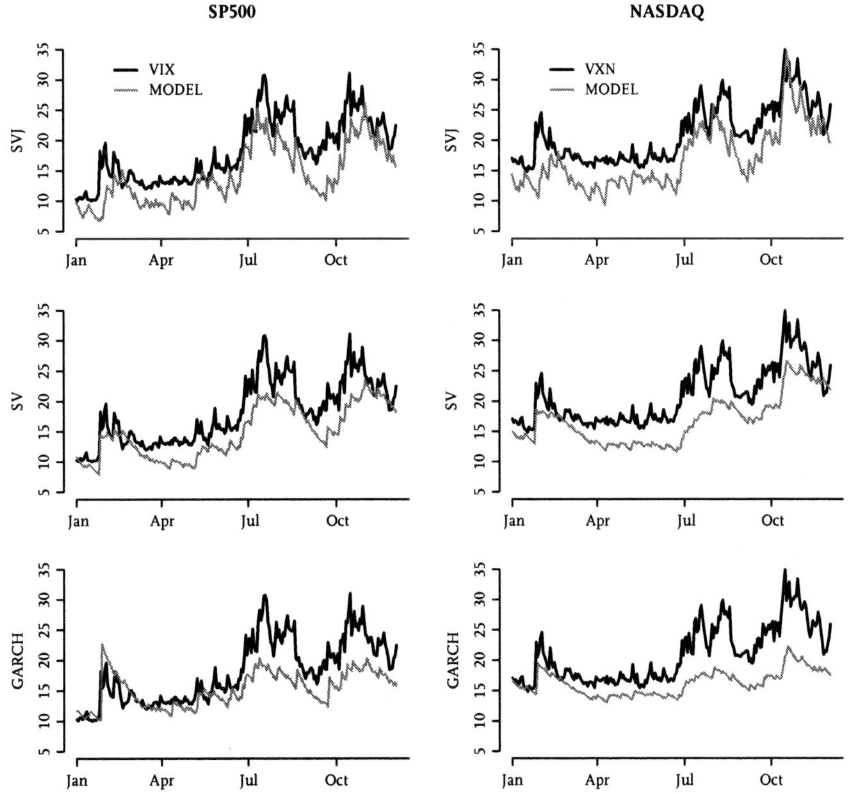

Fig. 13.4 *Year 2007*: Comparing Garch, SV and SVJ models.

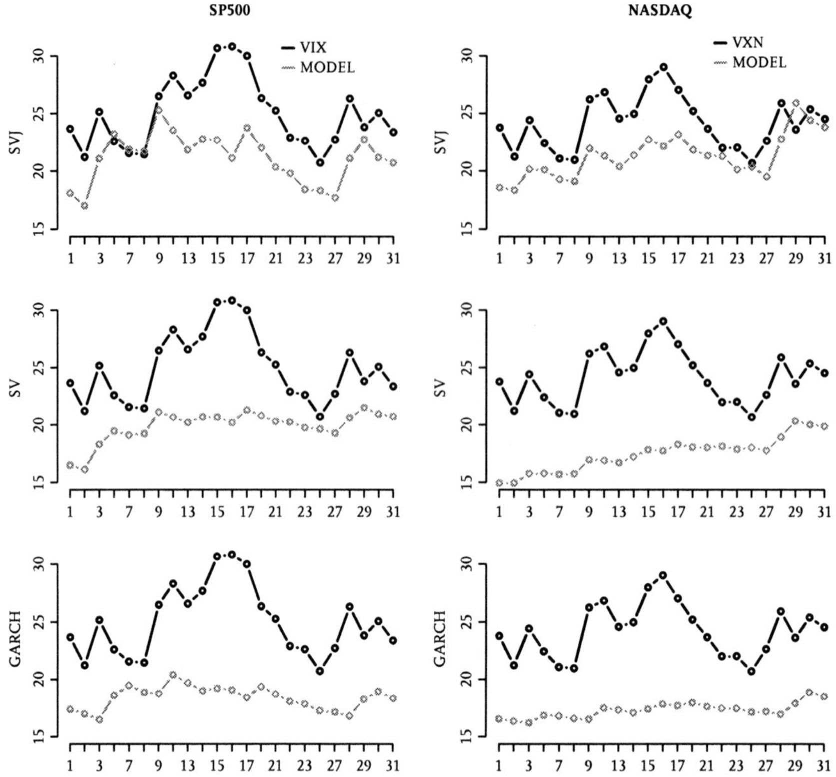

Fig. 13.5 *August 2007*: Comparing Garch, SV and SVJ models.

as a measurement of the market price of volatility risk is conditional on the model. Within the SV model, the average level of risk neutral volatility is $\theta^{\star} = \kappa\theta/(\kappa + \lambda_v)$. See Duffie *et al.* (2000) and Pan (2002) for a discuss with the SVJ model. By the end of August the initial shock has dissipated and all three models imply that there is little price of volatility risk.

Figure 13.6 shows that just before August 2007 the Bayes factor favors the stochastic volatility jump model. Hence we would expect the volatility estimates from the SVJ model to more closely track the option implied volatility VXN. This is indeed the case. The market volatility risk premia is effectively constant for this model over this data period, except at the very end of the period where the implied option volatility decay quickly and the estimated volatility does not. This is also coincident with the Bayes factor decaying back in favor of the pure SV model for the NASDAQ index. By the end of 2007, the odds favor the pure SV model over the SVJ model for the NASDAQ index.

For the XLF, most of the evidence for jumps is again contained in the February move. The sequential Bayes factor tends to lie in between the strong evidence for the SP500 and weaker evidence for the NASDAQ index. The story

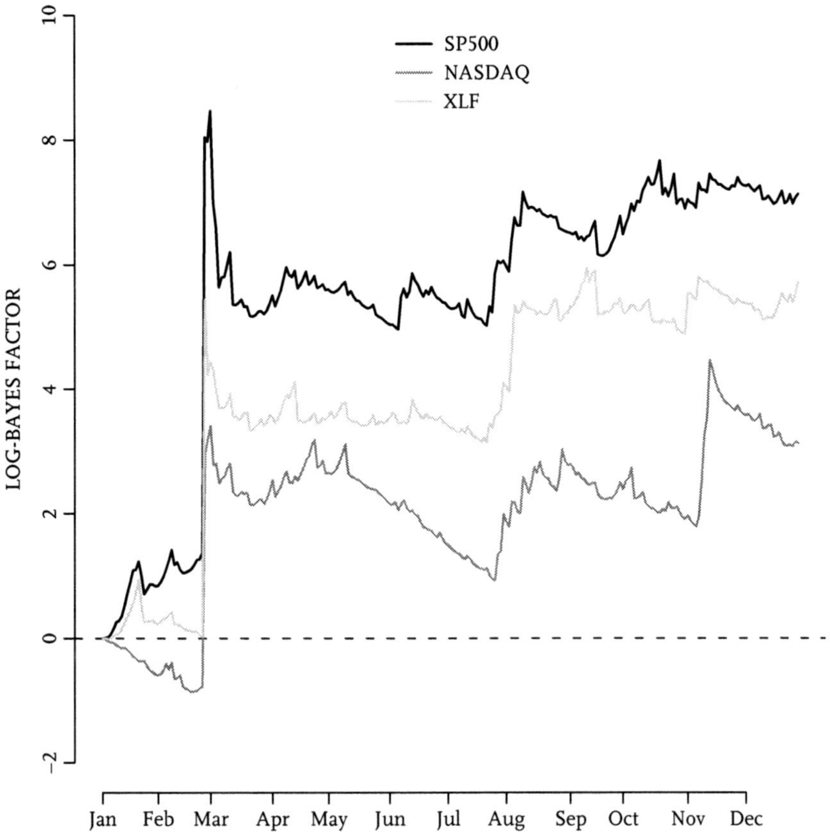

Fig. 13.6 *Year 2007*: Sequential (log) Bayes factor, $BF(M_1, M_0)$. $M_1 \equiv$ SVJ model $M_0 \equiv$ SV model.

for the SP500 is different. Figure 13.6 shows that after the February shock, the SVJ model is preferred to the SV model for the whole period. When comparing with VIX the jump model seems to track the option implied volatility with an appropriate market price of volatility risk.

13.5 Conclusions

In this chapter we proposed and implemented sequential particle filtering methods to study the US credit crisis which began in 2007. The purpose of our study was to show how model-based volatility estimates can be compared to market-based ones in a dynamic setting. Online daily volatility for S&P500, NDX100 and XLF are estimated for SV, SVJ and Garch(1,1) models. Market-based volatilities are based on the implied volatility indicies for the S&P500 and NDX100 indices, namely the VIX and VXN. We show that particle filtering methods naturally allow for online volatility, jump and sequential model comparison.

We find a number of empirical results. First, tracking volatility in turbulent periods is much harder than low volatility periods. The inclusion of the possibility of jumps can change current estimates of volatility dramatically. The pure stochastic volatility (SV) and Garch(1,1) models perform significantly worse in periods of market stress, both in terms of tracking subjective market-based volatility and relative marginal likelihoods versus the SVJ model. Second, by calculating sequential marginal likelihoods for our data period, we see that the stochastic volatility jump model is clearly preferred. Not surprisingly, this evidence accumulates mainly on a few days where the stock returns are most extreme.

Extensions include multivariate modelling of the return series possibly with the use of a factor stochastic volatility models as proposed by Aguilar and West (2000), Lopes (2000) and Lopes and Migon (2002). Lopes and Carvalho (2007) show how to perform sequential inference for this class of models. More flexible volatility dynamics can also be modeled. Recently, Carvalho *et al.* (2009) introduced a class of affine *shot-noise* continuous-time stochastic volatility model that accounts for both fast moving, rapidly mean-reverting shot-noise volatility and slow-moving square-root diffusive volatility.

Appendix

A. Broader context and background

A major advance in financial modelling was the development of continuous-time models for the evolution of stock prices and their volatility. The simplest model is a geometric Brownian motion Black – Scholes model where stock prices S_t solves

$$dS_t = S_t(\mu\, dt + \sigma\, dB_t)$$

for an instantaneous expected return μ and volatility σ. Here B_t is a standard Brownian motion with distribution $B_t \sim \mathcal{N}(0, t)$. A natural extension is the inclusion of stochastic volatility (SV). Transforming to a log-scale and replacing the volatility by $\sqrt{V_t}$ gives a model of the form

$$d \ln S_t = \mu dt + \sqrt{V_t} dB_t$$

where typically V_t solves its own stochastic evolution such as a square-root Ornstein–Uhlenbeck process.

In practice, we observe data on a discrete time scale and it is common to use a time discretisation of these models. The most common discretisation is an Euler scheme. Specifically, on a time interval $(t, t + \Delta)$ we have the evolution

$$\ln S_{t+\Delta} - \ln S_t = \mu\Delta + \sqrt{V_t}\sqrt{\Delta}\epsilon_t$$

where $\epsilon_t \sim \mathcal{N}(0, 1)$. With loss of generality we take $\Delta = 1$. Moreover, writing the log-returns $R_{t+1} = \ln(S_{t+1}/S_t)$ we we now have a dynamic linear models (DLM) with a hidden state that can be filtered from the partially observed data.

B. Informational content of returns and option prices

Option prices provide information about state variables and parameters via the pricing equation. An option price is given by

$$C(S_t, V_t, \theta) = e^{-r(T-t)}\mathbb{E}_t^{\mathbb{Q}}[\max(S_t - K, 0) \mid S_t, X_t, \Theta]$$

where the expectation is taken under $\mathbb{Q}$ the risk-neutral probability measure. here we assume that the risk-free rate r is constant. This simplifies by letting $A = \{S_T \geq K\}$ denote the event that the stock ends in the money. Then the option price is given by

$$\begin{aligned} e^{-rt}\mathbb{E}_t^{\mathbb{Q}}(\max(S_T - K, 0)) &= e^{-rt}\mathbb{E}_t^{\mathbb{Q}}(S_T\mathbb{I}_A) - e^{-rt}K\mathbb{E}(\mathbb{I}_A) \\ &= e^{-rt}\mathbb{E}_t^{\mathbb{Q}}(S_T\mathbb{I}_A) - e^{-rt}K\mathbb{P}(A). \end{aligned}$$

A common approach is to use a leverage stochastic volatility model (Heston, 1993). Here the underlying equity price S_t evolves according to a stochastic volatility model with square-root dynamics for variance V_t, and correlated errors. The logarithmic asset price and volatility follow an affine process with $Y_t = (\log S_t, V_t) \in \Re \times \Re^+$ satisfying

$$\begin{aligned} dY_t &= \left(r - \frac{1}{2}V_t\right)dt + \sqrt{V_t}dB_{1,t} \\ dV_t &= \kappa(\theta - V_t)dt + \sigma_v\sqrt{V_t}dZ_t \end{aligned}$$

where $Z_t = \rho B_{1,t} + \sqrt{1-\rho^2}B_{2,t}$. The correlation, ρ, or so-called leverage effect (Black, 1976) is important to explain the empirical fact that volatility increases faster as equity prices drop. The parameters (κ, θ) govern the speed of mean reversion and the long-run mean of volatility and σ_v measures the volatility of volatility. Under the risk-neutral measure they become $(\kappa/\kappa + \lambda_v)\theta$ and $\kappa + \lambda_v$ where λ_v is the market price of volatility risk.

From affine process theory, the discounted transform

$$\psi(u) = e^{-rt}\mathbb{E}\left(e^{uY_T} \mid Y_t\right) = e^{\alpha(t,u)+uY_t+\beta(t,u)V_t}$$

where α, β satisfy Riccati equations. We can then use transform analysis (Duffie *et al.*, 2000) and compute prices by inverting a fast fourier transform. Specifically, there exists a pair of probabilities $\mathbb{P}_j(V_t, \Theta, \Lambda)$, $j = 1, 2$, such that the call price $C\left(Y_t^S, V_t, \Theta, \Lambda\right)$ is given by:

$$C\left(Y_t^S, V_t, \Theta, \Lambda\right) = S_t\mathbb{P}_1(V_t, \Theta, \Lambda) - Ke^{-r\tau}\mathbb{P}_2(V_t, \Theta, \Lambda).$$

Specifically, $\mathbb{P}_j(V_t, \Theta, \Lambda) = Pr_j\ (\ln(S_T/K)|\, V_t, \Theta, \Lambda)$, where these probabilities can be determined by inverting a characteristic function

$$\mathbb{P}_j(V_t, \Theta, \Lambda) = \frac{1}{2} + \frac{1}{\pi}\int_0^{\infty} Re\left[\frac{e^{-i\phi\ln(K)} f_j(V_t, \Theta, \Lambda)}{i\phi}\right] d\phi$$

where

$$f_j(V_t, \Theta, \Lambda) = e^{C(\tau;\phi)+D(\tau;\phi)V_t+i\phi\ln(S_t)},$$

and the coefficients $C(\tau;\phi)$ and $D(\tau;\phi)$ are defined in Heston (1993).

$$C(\tau;\phi) = r\phi i\tau + \frac{a}{\sigma^2}\left\{(b_j - \rho\sigma\phi i + d)\tau - 2\log\left[\frac{1-ge^{d\tau}}{1-g}\right]\right\},$$

$$D(\tau;\phi) = \frac{b_j - \rho\sigma\phi i + d}{\sigma^2}\left[\frac{1-e^{d\tau}}{1-ge^{d\tau}}\right],$$

with parameters

$$g = \frac{b_j - \rho\sigma\phi i + d}{b_j - \rho\sigma\phi i - d},$$

$$d = \sqrt{(\rho\sigma\phi i - b_j)^2 - \sigma^2(2u_j\phi i - \phi^2)}.$$

The correlation and volatility of volatility parameters affect the skewness and kurtosis of the underlying return distribution and also how the correlation ρ affects the probabilities $\mathbb{P}_j(V_t, \Theta, \Lambda)$. For example, as is typically the case, when the correlation is negative then the underlying risk neutral distribution has a skewed left tail which in turn decreases the price of out-of-the-money call options while increasing the out-of-the-money put options.

Acknowledgements

We are grateful to the Editors.

References

Aguilar, O. and West, M. (2000). Bayesian dynamic factor models and variance matrix discounting for portfolio allocation. *Journal of Business and Economic Statistics*, **18**, 338–357.

Ausín, M. C. and Galeano, P. (2007). Bayesian estimation of the Gaussian mixture Garch model. *Computational Statistics & Data Analysis*, **51**, 2636–2652.

Bakshi, G., Cao, C. and Chen, Z. (1997). Empirical performance of alternative option pricing models. *Journal of Finance*, **52**, 2003–2049.

Bates, D. S. (2000). Post-'87 crash fears in S&P500 futures options. *Journal of Econometrics*, **94**, 181–238.

Bates, D. S. (2006). Maximum likelihood estimation of latent affine processes. *Review of Financial Studies*, **19**, 909–965.

Bauwens, L. and Lubrano, M. (1998). Bayesian inference on Garch models using the Gibbs sampler. *Econometrics Journal*, **1**, C23–C46.

Black, F. (1976). Studies of stock price volatility changes. In *1976 Meetings of the Business and Economics Statistics Section*, pp. 177–81. Proceedings of the American Statistical Association.

Carvalho, C., Johannes, M., Lopes, H. F. and Polson, N. (2008). Particle learning and smoothing. Working paper, University of Chicago Booth School of Business.

Carvalho, C., Johannes, M., Lopes, H. F. and Polson, N. (2009). Stochastic volatility shot noise. Working paper, University of Chicago Booth School of Business.

Carvalho, C. and Lopes, H. F. (2007). Simulation-based sequential analysis of markov switching stochastic volatility models. *Computational Statistics & Data Analysis*, **51**, 4526–4542.

Doucet, A., de Freitas, N. and Gordon, N. (2001). *Sequential Monte Carlo Methods in Practice*, Springer: New York.

Duffie, D., Pan, J. and Singleton, K. (2000). Transform analysis and asset pricing for affine jump-diffusions. *Econometrica*, **68**, 1343–1376.

Engle, R. (1982). Autoregressive conditional heteroskedasticity with estimates of the variance of U.K. inflation. *Econometrica*, **50**, 987–1008.

Eraker, B. (2004). Do equity prices and volatility jump? reconciling evidence from spot and option prices. *Journal of Finance*, **59**, 1367–1403.

Eraker, B., Johannes, M. and Polson, P. (2003). The impact of jumps in volatility and returns. *Journal of Finance*, **59**, 227–160.

Gamerman, D. and Lopes, H. F. (2006). *Markov Chain Monte Carlo: Stochastic Simulation for Bayesian Inference*. Chapman & Hall/CRC, Baton Rouge.

George, T. and Longstaff, F. (1993) Bid-ask spreads and trading activity in index options market. *Journal of Financial and Quantitative Analysis*, **28**, 381–397.

Gordon, N., Salmond, D. and Smith, A. F. M. (1993). Novel approach to nonlinear/non-Gaussian Bayesian state estimation. *IEE Proceedings*, **F,140**, 107–113.

Heston, S. (1993). A closed-form solution for options with stochastic volatility with applications to bond and currency options. *Review of Financial Studies*, **6**, 327–343.

Jacquier, E., Polson, N. G. and Rossi, P. E. (1994). Bayesian analysis of stochastic volatility models. *Journal of Business and Economic Statistics*, **20**, 69–87.

Jacquier, E., Polson, N. G. and Rossi, P. E. (2004). Bayesian analysis of stochastic volatility models with fat-tails and correlated errors. *Journal of Econometrics*, **122**, 185–212.

Johannes, M., Kumar, R. and Polson, N. G. (1998). State dependent jump models: How do US equity indices jump? Working paper, University of Chicago Booth School of Business.

Johannes, M. and Polson, N. (2009). MCMC methods for continuous-time financial econometrics. In *Handbook of Financial Econometrics*, Vol. 2 (ed. Y. Ait-Sahalia, L. Hansen), pp. 1–72. Elsevier, Oxford.

Lopes, H. F. (2000). Bayesian analysis in latent factor and longitudinal models. Ph.D. thesis, Institute of Statistics and Decision Sciences, Duke University.

Lopes, H. F. and Carvalho, C. M. (2007). Factor stochastic volatility with time varying loadings and Markov switching regimes. *Journal of Statistical Planning and Inference*, **137**, 3082–3091.

Lopes, H. F. and Migon, H. S. (2002). Comovements and contagion in emergent markets: stock indexes volatilities. In *Case Studies in Bayesian Statistics*, Vol. 6 (ed. A. Carriquiry, C. Gatsonis and A. Gelman), pp. 285–300. Springer, New York.

Müller, P. and Pole, A. (1994). Monte Carlo posterior integration in Garch models. *Sankhya, Series B*, **60**, 127–144.

Pan, J. (2002). The jump-risk premia implicit in options: evidence from an integrated time-series study. *Journal of Financial Economics*, **63**, 3–50.

Pitt, M. and Shephard, N. (1999). Filtering via simulation: auxiliary particle filter. *Journal of the American Statistical Association*, **94**, 590–599.

Polson, N. G. and Stroud, J. (2003). *Bayesian Statistics 7*, Chapter on Bayesian inference for derivative prices, pp. 641–650. Clarendon Press, Oxford.

Rosenberg, B. (1972). The behaviour of random variables with nonstationary variance and the distribution of security prices. Working Paper 11, Research Programme in Finance, Graduate School of Business Administration, Institute of Business and Economic Research, University of California, Berkeley.

Todorov, V. (2010). Variance risk premium dynamics: The role of jumps. *Review of Financial Studies*, **23**, 345–383.

Verdinelli, I. and Wasserman, L. (1995). Computing Bayes factor using a generalization of the Savage-Dickey density ratio. *Journal of the American Statistical Association*, **90**, 614–618.

Vontros, I., D. P. and Politis, D. I. (2000). Full Bayesian inference for Garch and EGarch models. *Journal of Business & Economic Statistics*, **18**, 187–198.

Wago, H. (2004). Bayesian estimation of smooth transition Garch model using Gibbs sampling. *Mathematics and Computers in Simulation*, **44**, 63–78.

West, M. (1986). Bayesian model monitoring. *Journal of the Royal Statistical Society (Series B)*, **48**, 70–78.

·14·

Futures markets, Bayesian forecasting and risk modelling

José M. Quintana, Carlos M. Carvalho, James Scott and Thomas Costigliola

14.1 Introduction

On the one hand, the media often refers (presumably in a pejorative sense) to certain trading activities, particularly those involving financial derivative instruments, as gambling. On the other hand, the origin of probability theory is traced back to a betting puzzle, on how to redistribute fairly the stakes of an interrupted game of chance, posed by the gambler Antoine Gombaud (also known as Chevalier de Méré) to Blaise Pascal. Thus, it should not be surprising that a very strong bond exists between the fields of subjective probability and derivative finance. In this chapter, we explore overlapping concepts in these fields while focusing on the application of dynamic risk modeling *à propos* the hedging activity from the speculative perspective of a hedge fund manager.

Subjective expectations are motivated as fair prices of futures contracts in Section 14.2. The futures markets are presented, in Section 14.3, as a Bayesian market maker engine that dynamically reveals rational (coherent and proficient) expectations of random quantities as prices of futures contracts. A portfolio mean-variance efficiency generalization is motivated, in Section 14.4, as a sensible quantitative trading strategy for a hedge fund manager adopting the role of a Bayesian speculator (as opposed to the role of a Bayesian market maker) to highlight the critical role of hedging to ensue attractive risk-adjusted performance. Finally, general Bayesian dynamic models and specific Bayesian dynamic linear models are presented, in Section 14.5, to entertain a method, in Section 14.6, for assessing risk models in terms of their hedging effectiveness in the context of the risk-adjusted performance of trading strategies.

14.2 Subjective expectations

Most children know how to make a sibling cut two pieces of a pie fairly: she would let her brother do it as long as he agrees that she can choose her piece first. To elicit from someone a current fair price of an item with a future settlement (payment and delivery), one can proceed in a similar way: the person is free to name any price as long as it has been agreed that one can decide afterwards how many items to buy from, or sell to, the individual. The transaction will happen in the future but at a specific place, date and price agreed at the present time (e.g. a gallon of regular gasoline in Hoboken, New Jersey on the last business day of the next month at a specified price). The individual should avoid naming a price that is too low (or too high) just in case he or she is forced to sell (or to buy) the item. Furthermore, a smart (money-seeking, risk-averse) person should name the expected future spot price of the item on the settlement date as the fair price because this value minimizes the maximum, and potentially huge, expected loss (assuming that the individual will buy, or sell, the items at the future spot price to accomplish the settlement).

The above betting scheme corresponds to de Finetti's 1931; 1974 operational foundation of Bayesian subjective probability and expectations in which the individual is forced to play the role of a liquid market maker. Interestingly, recent informal experiments performed by the first author confirm that personal fair prices (with future settlement dates), acquired by simultaneous elicitation from individuals familiar with the concept of coherence (the avoidance of becoming a sure loser in this setup), diverge. This is just a confirmation of the obvious: subjective personal fair prices (i.e. expectations) are subjective and personal; yet, this implies that a group of individually coherent people, operating independently, typically would act incoherently as an entity. It is even unfair to force a single person to act as a liquid market maker and one should never really quote fair prices because one would be potentially vulnerable unnecessarily. Only an immeasurably rich irrational individual or the ultimate genius would dare to do so for an extended period of time because, in the words of Barnard (1980), 'fanatical insistence on freedom from "incoherence" can lead to such complicatedly interrelated analyses of data as to go well beyond the capacity of our understanding.' Yet, to a great extent, fair prices (with future settlement dates) of many standardized commodities began to be quoted almost three centuries ago, and they continue to be quoted in the present futures markets. This suggests the following questions: Are the futures markets coherent? Are their expectations worthwhile? Are they beatable? Furthermore, if they are, to what degree are they beatable?

14.3 Futures markets

A futures contract represents the obligation to deliver a standardized commodity at a specified future maturity date and at a prearranged location and price.

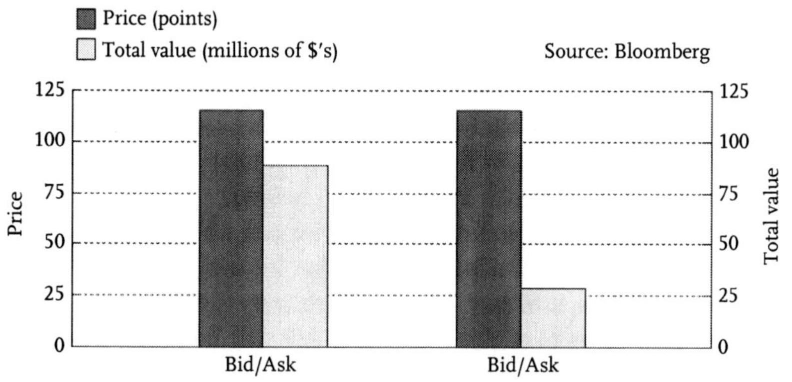

Fig. 14.1 A typical liquid market snapshot.

From an operational standpoint perfect liquid futures markets, as entities, are quoting fair prices (i.e. they are prepared to buy or sell at the quoted prices unlimited quantities of standardized commodities at future settlement dates). Perfect liquid markets, of course, do not exist; but, futures markets trade daily a variety of commodities (e.g. gallons of gasoline, S&P 500 Index units; Japanese Government Bonds, pounds of rice, shares of Google, barrels of WTI crude oil, euros vs. dollars, troy oz. of gold, etc.) with a total worth of trillions of dollars per day. Figure 14.1 shows, in random order, a snapshot of bid and ask prices of the US Treasury Bill Futures Contract and their corresponding stakes. It is apparent that the bid and ask prices are virtually indistinguishable. In addition, a $25,000,000 transaction would not have affected the bid and ask price quotes. Indeed, the spread between bid and ask prices of liquid futures markets are typically measured by a few basis points (a basis point is one percent of one percent) and the associated stakes are worth millions of dollars.

Are these liquid futures markets coherent over time? The straightforward answer is yes, and a simple financial economics explanation resembles the anthropic principle arguments: the fact that the question can be posed implies its answer (if the futures markets were incoherent they would have been an ideal fountain of wealth but only for a short while; afterwards they would have dried out and disappeared). More plausibly, as soon as traders would identify and exploit incoherencies in the markets, the trading activity would eventually restore coherent prices, since perfect liquid markets do not exist (prices are affected by supply and demand). Remarkably, although the futures markets participants might not be individually coherent, as an entity, they are acting in concert coherently by sharing between all of them two crucial pieces of information: the bid and ask prices. The ending value of a futures contract (i.e. the future spot price) can be tied to a general commodity (e.g. gallons of gasoline) but it also can be associated to an artificial item (e.g. S&P 500 index) and in that case is settled in cash. Further, in principle, the future spot price could be linked to any random variable yet to be revealed. For example, a well defined global average temperature measurement for the year 2025 could define a futures market where the global warmers and global coolers could settle their differences (actually, contracts of this kind but with near settlements are traded in the form of binary bets, which are described below).

A futures contract where the future spot price can only take two values and one (and only one) potential value is zero (also known as a binary bet) is of particular interest. Thus, a binary bet on an arbitrary event, the condition that defines a non zero value (e.g. the S&P500 Index will be up at the end of the day, Barack Obama will be the 2008 US presidential election winner, etc.), is a futures contract that pays a specified amount (e.g. \$100) if the event occurs and zero otherwise. The concept of a binary bet, from a practical perspective, was introduced in the UK early in this millennium; however, from a theoretical perspective it can be traced back to de Finetti (1931), Ramsey (1926) and arguably to Bayes (1763). Indeed, his definition 'The *probability of any event* is the ratio between the value at which an expectation depending on the happening of the event ought to be computed, and the chance of the thing expected upon its happening' written in centuries old English might be hard to comprehend at first sight but it is easily illustrated by example. Bayes' insight behind his definition of probability of an event becomes apparent if we try to compute the expected payoff of a binary contract divided by the conditional reward (\$100) according to a hypothetical market probability of its associated event,

$$\frac{E\,(\$\text{Payoff})}{\$100} = \frac{(1 - P\,(\text{Event}))\,\$0 + P\,(\text{Event})\,\$100}{\$100} = P\,(\text{Event})\,.$$

Figure 14.2 depicts a partial time series of fair price transactions from the Irish Intrade exchange of several binary contracts for the US presidential election winner (paying \$100 if, and only if, the particular candidate wins). Thus, using the connection between fair prices and expectations, the time series of market fair prices represent the probabilities in percent (according to the market participants) of their associated events. These are time series of coherent market probabilities (they do not add up to 100% because other contracts of potential contenders, such as Al Gore, with small non-zero probabilities are not shown in the graph) assimilating relevant information as it is uncovered (e.g. the January primaries results). In fairness, the markets are as coherent as they are liquid, and this particular market is not as liquid (in terms of tight bid and ask spreads and deep associated stakes) as other markets. Yet, the implication is that since liquid markets ought to be coherent then liquid markets trading binary contracts must obey the rules of probability theory associated to coherence. That is, denoting generic events by A and B and a sure event by Ω, the

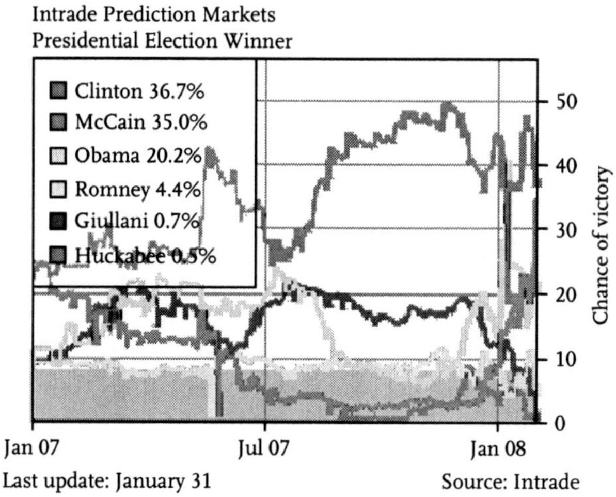

Fig. 14.2 Market implied probabilities.

market implied probabilities should comply with: $P(A)$ takes a unique value; $P(A) \geq 0$; $P(\Omega) = 1$; $P(A + B) = P(A) + P(B)$, where $A + B$ denotes the union of two inconsistent events, and $P(AB) = P(A)P(B|A)$ where $P(B|A)$ denotes a conditional binary bet on the event B that is called off if and only if A fails to happen. For completeness, a concise direct derivation of these coherent rules, following Quintana (2005), is presented in the Appendix A. Furthermore, as pointed out by de Finetti (1974), these probability rules can be alternatively derived indirectly from three basic rules regarding generic fair prices of payoffs connected to arbitrary random quantities (denoting the fair price of a contract tied to a random quantity X as $E(X)$ to emphasize that it is, or should be, its expectation): $E(X)$ takes a unique value; $E(X + Y) = E(X) + E(Y)$; and $X \leq a$ (meaning, a fixed quantity, a, is greater than or equal to any possible outcome of X) implies $E(X) \leq a$. Therefore, liquid futures markets ought to obey these rules and all their implications to remain coherent (e.g. for several random quantities, the associated fair prices' points must lie in the closed convex hull of the possible outcomes).

Are these implied market expectations worthwhile? Are the futures markets unbeatable? The supporters of the Efficient Markets Hypothesis (EMH) answer these questions affirmatively. The conjecture and its underlying argument can be traced to Bachelier's (1900) conclusion: 'the mathematical expectation of the speculator is zero'. The argument stretches convincingly the reasoning supporting the liquid markets' coherence as follows: if the futures markets were inefficient then the trading triggered by the speculators taking advantage of expected profitable opportunities would re-establish price levels that fully assimilate the relevant information and no more expected gains would remain, thus, effectively restoring the market efficiency. Empirical analyses (e.g. Fama

1970) support the EMH in a variety of setups, including binary bets (also known as prediction) markets (e.g. Carvalho and Rickershauser 2010) and the early period of the Dojima futures market (Hamori *et al.* 2001). Interestingly, the Dojima rice futures market, which can be regarded as the first Bayesian practitioner, started to quote fair prices in 1730; that is, 33 years before the work of the first Bayesian theorist was published posthumously.

The liquid futures markets' coherence and efficiency appears to be accomplishing an amazing feat. Yet, it can be explained because: first, key markets' speculators are well aware of the arbitrage-free concept (essentially the financial economics jargon for coherence) and eager to exploit any potential opportunities; second, the market consensus expectations (fair prices) integrate over a spectrum of presumably relevant pieces of information pondered by the conviction measured by the amount of money willing to be committed by each one of the participants; and third, a financial evolution process ensures that only the best fitted participants survive in the long run. Furthermore, the market coherence and efficiency ought to be dynamic; that is, coherent conditional market expectations (fair prices) are produced in real time assimilating available relevant information under all circumstances even while extreme events are happening. These liquid futures markets come across, as a whole, as an invincible juggernaut assimilating not only relevant information but also any applicable knowledge (e.g. financial economics, computing and statistical technologies among others) in its path.

14.4 Bayesian speculation

Last section ending notwithstanding, one might sense a vulnerability on any market maker due to the willingness to buy or sell at the, virtually, same quoted price. Let us adopt henceforth the offensive role of a Bayesian speculator instead of the defensive attitude of a Bayesian market maker (just as one would rather choose a piece, as opposed to cutting two pieces fairly, of a pie). An encouraging indication, from the speculator's viewpoint of a hedge fund manager, is that eighty years after the aforementioned Bachelier's (1900) pronouncement, a disarming counterargument was proposed by Grossman and Stiglitz (1980): If the markets were efficient there would not be any incentives to get, gather and process information at a certain non-zero cost, and incorporate it, by trading, into the markets; therefore, the market would not assimilate the relevant information and it would become inefficient. This leads to the EMH paradox: markets' efficiency implies markets' inefficiency, and markets' inefficiency implies markets' efficiency! As a smart (profit-seeking, risk-averse) speculator one could perceive markets incoherence and efficiency as black (when a sure profit is available) and white (when a positive expected gain is unattainable) with shades of gray in between. A general speculating strategy consists of two steps: first, form a (probabilistic) view of the markets; second, trade an advantageous book of bets based on that view.

One way a Bayesian speculator can quantitatively implement the latter step is by maximizing the expected net payoff (also known as excess return) of a book of bets $(\mathbf{w}_t'\mathbf{y}_t)$ at a certain time (t) for a certain holding period (e.g. a week) per unit of risk measured by its standard deviation (the so-called ex-ante information ratio); that is,

$$\max_{\mathbf{w}_t} \left(\frac{\underset{\mathbf{y}_t|\mathbf{x}_1,\mathbf{y}_1,\ldots,\mathbf{y}_{t-1},\mathbf{x}_t}{E} \mathbf{w}_t'\mathbf{y}_t}{\sqrt{\underset{\mathbf{y}_t|\mathbf{x}_1,\mathbf{y}_1,\ldots,\mathbf{y}_{t-1},\mathbf{x}_t}{V} \mathbf{w}_t'\mathbf{y}_t}} \right) \tag{14.1}$$

where $\mathbf{y}_t, \mathbf{x}_t, \mathbf{w}_t$ denote respectively the vector of net payoffs (the differences between buying prices and selling prices of the futures contracts), their associated vector of relevant explanatory variables (that presumably would contribute to its predictability) and the corresponding vector of trading (holding) amounts. Notice that the weights do not have to comply with the traditional investments constraints (i.e. they do not have to be non-negative, neither do they have to add up to one) since $\mathbf{w}_t$ represents a book of side bets as opposed to an allocation of investment capital (capital would still be required to support the book of bets to cover potential losses but it could be earning interest or be invested in any other liquid assets). The optimal information ratio is bounded by zero on the left hand side and by infinity on the right hand side; the former bound corresponds, from the speculator's point of view, to market efficiency whereas the latter implies an incoherent market condition (although market incoherence, in principle, could also occur in a situation with a finite information ratio). Thus, the information ratio provides quantitative means to evaluate to what degree the market expectations are beatable by measuring the speculator's expected gain per unit of risk taken in an optimal scenario.

This optimization step is equivalent to the following two-way Pareto optimization,

$$\max_{\mathbf{w}_t} \left(\left(\underset{\mathbf{y}_t|\mathbf{x}_1,\mathbf{y}_1,\ldots,\mathbf{y}_{t-1},\mathbf{x}_t}{E} \mathbf{y}_t \right) - \frac{1}{2}\lambda_t \left(\underset{\mathbf{y}_t|\mathbf{x}_1,\mathbf{y}_1,\ldots,\mathbf{y}_{t-1},\mathbf{x}_t}{V} \mathbf{y}_t \right) \right) \tag{14.2}$$

and optimal trades are given by,

$$\mathbf{w}_t^* = \frac{1}{\lambda_t} \left(\underset{\mathbf{y}_t|\mathbf{x}_1,\mathbf{y}_1,\ldots,\mathbf{y}_{t-1},\mathbf{x}_t}{V} \mathbf{y}_t \right)^{-1} \underset{\mathbf{y}_t|\mathbf{x}_1,\mathbf{y}_1,\ldots,\mathbf{y}_{t-1},\mathbf{x}_t}{E} \mathbf{y}_t \tag{14.3}$$

where λ_t is an arbitrary inverse scaling factor representing the speculator's greed-fear tradeoff. The optimization (14.1) corresponds to the Modern Portfolio Theory (MPT) objective function but the feasible set for the weights is different: in this setup there are no constraints whereas in the MPT setup the weights are constrained to lie on the simplex and the solution formula (14.3) would not apply; instead, a quadratic programming technique would be required to obtain the solution. The online optimal book (portfolio) given by the formula (14.2) turns out to be also optimal in a multiperiod utility maximization setup; this is discussed briefly in Appendix A.

It is interesting to note at this point that once a liquid futures market has emerged the subjectivity (of probabilities and expectations elicited according to de Finetti's recipe) should disappear because a smart market maker should simply quote the liquid market price and hedge accordingly since this would not only make his expected loss null but the potential loss itself would be null (quoting a different price would not necessarily imply incoherence but a quasi-incoherence in the sense that he would be willing to buy at a higher price or to sell at a lower price than the market price giving a perfect arbitrage opportunity to another speculator to make sure money). However, if the expectations of the Bayesian speculator coincides with the futures market's expectations (i.e. prices), then the expected excess return of any book would necessarily be null. To avoid this pitfall the loophole should be closed and the elicitation process should, in principle, assume that the Bayesian speculator has no access to liquid futures markets during his hypothetical role as a market maker.

14.5 Bayesian forecasting

The first step of the aforementioned trading strategy can be implemented quantitatively in the following manner: form the predictive density function of the payoffs for each holding period by a Bayesian updating process of an underlying Bayesian dynamic model. Coherency, as mentioned in the appendix, requires that conditional fair prices must correspond to the customary conditional expectations before the conditioning information is observed, but does not require that this correspondence must remain after the conditioning information is revealed. Although perceived by some as a weakness in the foundations of the Bayesian subjective approach, this feature is actually a blessing in disguise from the Bayesian speculator's perspective: on the one hand, the speculator could maintain consistency and rely routinely on the default Bayesian updating based on conditional probability and expectations since they ensure coherence for the predictive probabilistic view at each holding period, and this is an essential requirement for the strategy (otherwise, for example, one could dangerously conclude that there is a perfect arbitrage opportunity when in fact there is none); on the other hand, the Bayesian speculator is always free to consider and switch to an alternative Bayesian model at any time if a

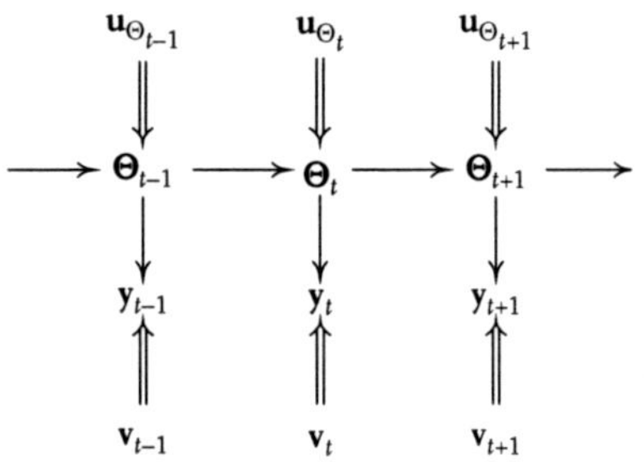

Fig. 14.3 Functional network of a generic Bayesian dynamic model.

decisive new information, knowledge or technology becomes available (only a mad speculator would blindly commit perpetually to a fixed model).

A Bayesian dynamic (also known as stochastic) model presumably would allow the trading strategy to adapt to regime changes and assimilate financial market shocks. Figure 14.3 represents the stochastic structure of a generic Bayesian dynamic model in a manner introduced by Quintana *et al.* (2003). This representation relies on the principle that a stochastic model is well defined if and only if its simulation is fully specified. Arrows denote inputs to a deterministic function at each node; double arrows denote generated random entity (vector, matrix, etc.) inputs whose distributions are fully specified (involving only known parameters) and they constitute the primary sources of randomness; and all the primary random entities are jointly independent. The Bayesian dynamic model consists of a Markovian evolution of the parameter time series Θ_t coupled with contemporaneous random observations $\mathbf{y}_t$. The basic premise is that the model structure, described by the values of the system parameters Θ_t, is changing according to a stochastic process (as opposed to traditional rigid static formulations) but there is a degree of persistence making the sampling and filtering process worthwhile.

A Dynamic Linear Model (DLM) variant of the formulations entertained by West and Harrison (1997) is of particular interest when the speculator suspects that the markets are not assimilating efficiently all the available public (relevant) information; that is when the so-called semi-strong form of the EMH (Fama 1970) might be vulnerable. This model is a particular Bayesian dynamic model where the parameters are decomposed as follows: the observation equation, at each time period, takes the form of a multivariate multiple regression that relates the excess returns (or logarithmic excess returns that are more convenient from a modelling perspective) $\mathbf{y}_t$ and associated relevant available public information $\mathbf{x}_t$ via the regression coefficients β_t with a residual variance-covariance Σ_t representing the variance-covariance of the difference of the random observations and their corresponding expectations; and the random evolution of the parameters follows a form of random walk.

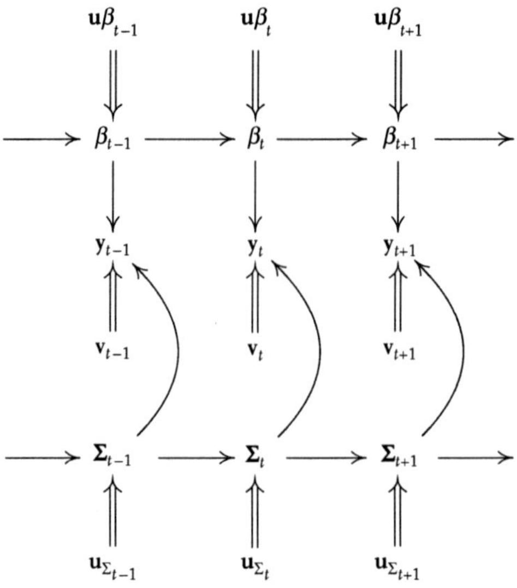

Fig. 14.4 Functional network of a specific Bayesian DLM model.

This model's functional network can be decomposed as a marginal network (for Σ_t) coupled with a conditional network (given Σ_t), or vice versa in terms of β_t, in a hierarchical fashion as is depicted in Figure 14.4. A more detailed specification of this model, including specific functional forms and random generation procedure descriptions, is presented in the Appendix B.

To what degree are the market expectations beatable? A comprehensive discussion on this subject is beyond the scope of this chapter; however, the answer to the question depends, of course, on the respondent. According to simulated and, more importantly, actual trading of strategies based on the above described specific Bayesian DLM and a variant the information ratio optimization have generated attractive risk-adjusted returns (see Quintana, 2005 and Quintana *et al.*, 2003 for details). Nevertheless, the customary disclaimer applies: PAST PERFORMANCE IS NOT A GUARANTEE OF FUTURE RESULTS and the only way to beat the markets going forward is by taking some risk of losing money (unless market incoherencies are identified and exploited).

14.6 Risk modelling

Risk reduction is more important than expected return enhancement in the sense that according to the ex-ante information ratio formula 14.1 a risk reduction of 50% is equivalent to an expected return enhancement of 100%. Someone might still try to argue pointing out that doubling the expected excess return at the same risk in some circumstances could be more appealing than halving the risk while maintaining the same expected return. However, the former book can be derived from the latter book using leverage (by doubling the bets; that is, the allocation weights). Thus, risk modeling is a crucial mission for a speculator seeking the best risk-adjusted performance (i.e. the best information ratio). In this section the focus is on the predictive ability, for hedging purposes, of the variance covariance of a book of bets $(\mathbf{w}_t'\mathbf{y}_t)$; which is the variance appearing in the denominator of expression (14.1),

$$\underset{\mathbf{y}_t|\mathbf{x}_1,\mathbf{y}_1,\ldots,\mathbf{y}_{t-1},\mathbf{x}_t}{V} \mathbf{w}_t'\mathbf{y}_t = \mathbf{w}_t'\mathbf{V}_t\mathbf{w}_t \tag{14.4}$$

where V_t is shorthand for the predictive variance of the book of bets $\underset{\mathbf{y}_t|\mathbf{x}_1,\mathbf{y}_1,\ldots,\mathbf{y}_{t-1},\mathbf{x}_t}{V} \mathbf{y}_t$.

Although risk and return are inherently connected, it is useful from a modelling perspective to analyze them as separately as possible. Because of the availability of leverage, a speculator trying to evaluate the expected return capability of a model cannot just set λ_t equal to zero in the optimization (14.2); one possible way to alleviate this difficulty is, for evaluation purposes, to force the weights to add up to one. However, because the book expectation, given a direction, is directly proportional to the square of its norm, forcing the weights to add up to one would unfairly punish the minimum length book satisfying the constraint (i.e. the long equal weighted book) and favour other books as their length increase. This drawback can be avoided simply by constraining the potential weights to have a unitary norm instead. Similarly, an ideal testing hedging task would be to find the minimum variance book lying on the hyper-circumference,

$$\min_{\mathbf{w}_t} \left(\mathbf{w}_t'\mathbf{V}_t\mathbf{w}_t\right), \ s.t.\ \mathbf{w}_t'\mathbf{w}_t = 1 \tag{14.5}$$

or equivalently, to find the minimum $\mathbf{V}_t$'s Rayleigh quotient $\left(\left(\mathbf{w}_t'\mathbf{V}_t\mathbf{w}_t\right)/\left(\mathbf{w}_t'\mathbf{w}_t\right)\right)$. The solution to both problems can be derived directly but it is a known result in extremizing quadratic forms (e.g. Noble and Daniel 1987): the optimal weights correspond to the normalized eigenvector associated to the smallest eigenvalue (which itself is the value of the minimum variance book) of the spectral decomposition of $\mathbf{V}_t$; this is the counterpart of the other better known maximum variance result where the book (portfolio) corresponding to the largest eigenvalue exhibits the most variation (as discussed by Quintana and West 1987).

Although modelling assumptions about the payoff expectations would obviously influence risk predictions given the liquid markets' efficiency those expectations should not be far away from the markets' expectations. Thus, risk models evaluation results based on markets' payoff expectations should still apply and for this reason these markets' expectations are entertained henceforth. That is, it is assumed in this section that the expectation $\underset{\mathbf{y}_t|\mathbf{x}_1,\mathbf{y}_1,\ldots,\mathbf{y}_{t-1},\mathbf{x}_t}{E} \mathbf{y}_t$ corresponds to the market's expectation (i.e. it is null).

The merits of alternative risk models, since the expected excess returns are assumed to be null, can be assessed by comparing the cumulative sum of squares of the excess returns of these extreme books over time associated to each model, or equivalently, rolling (e.g. one year) ex-post (sample) standard deviations can be used to discriminate between models. Two alternative risk models embedded in the functional network structure (Figure 14.4), for $\beta_t = \mathbf{0}$, are considered in this study. The first, and the simplest, assumes that the Σ_t stochastic evolution follows a form of random walk that allows for a conjugate analysis with Wishart probability distributions; this model corresponds to the stochastic multiple-factor model discussed by Quintana *et al.* 2003 with the constraint $\beta_t = \mathbf{0}$ which was introduced by Quintana and West 1987. The second model, entertains a graphical formulation for Σ_t that encompasses a conditional independence structure via a pattern of zeros in the corresponding precision matrix. As introduced by Carvalho and West (2007), the idea of conditioning the sequence of variance-covariance matrices on set of graphical constraints provides a parsimonious model for Σ potentially improving the empirical accuracy of book (portfolio) predictions and reducing the variation of realized returns. In this application, the set of constraints (i.e. the graph) was selected as the top model from a stochastic search procedure (Scott and Carvalho 2008) using observations prior to the period analysed here. A comprehensive description of both models appears in Appendix B.

The dataset used to evaluate the trading performance of the alternative dynamic variance-covariance models consists of weekly excess returns, from 5/25/1990 through 6/19/2008, derived from adjusted indexed futures prices provided by Bloomberg using the indexed price series. This futures dataset entails currency, government bonds and stock indices instruments typically traded by a Global Macro Hedge Fund and it is described at the end of this section. Figures 14.5–14.7 show the hedging performance of the full model versus the graphical model. The figures' left-hand sides depict the corresponding annualized one-year rolling ex-post standard deviations for each model. The right-hand sides show the ratio (full/graph) of these rolling standard deviations. The different pictures result from varying the discount factor as follows: $\delta = 0.97, 0.98, 0.99$. It is apparent from these figures that for a high discount factor the performances of the models are similar but as the values decrease the dominance of the graphical model over the full model becomes apparent. This result is consistent with the intuition that the regularization from the graphical structure is more relevant when information is scarcer (i.e. when the discount factor is smaller) and suggests that the hedging performance of the graphical model is more robust in relation to the choices of the discount factors making it very attractive.

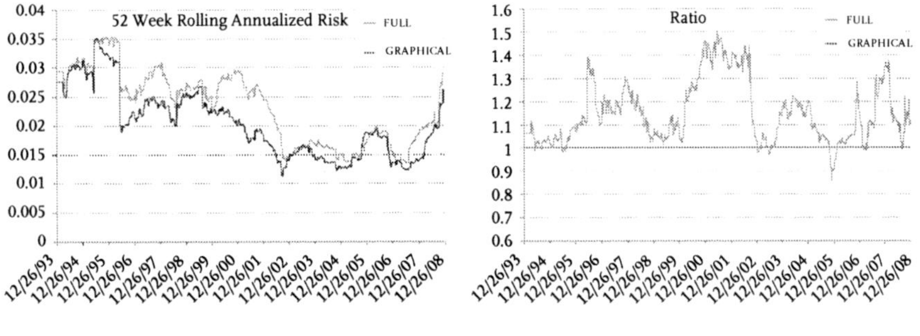

Fig. 14.5 $\delta = 0.97$.

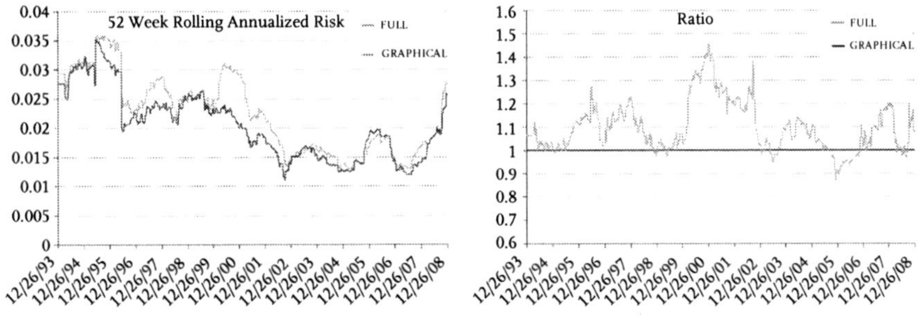

Fig. 14.6 $\delta = 0.98$.

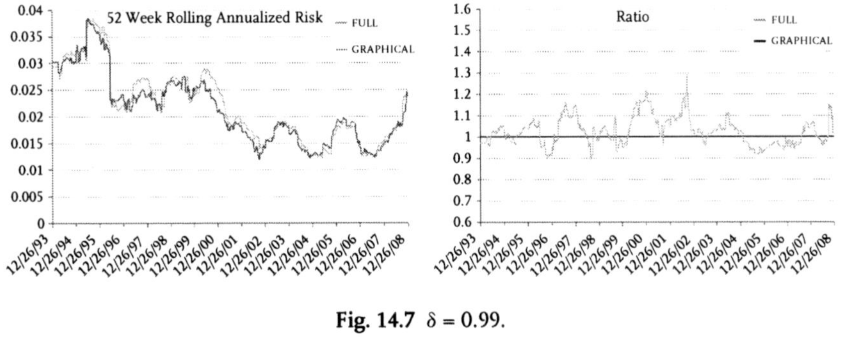

Fig. 14.7 $\delta = 0.99$.

The discussion at the beginning of this section hints that the hedging performance dominance of the graphical models over the full models should be carried into risk-adjusted performance dominance. This is confirmed, in the case at hand, by the simulated risk-adjusted performance exhibited in Figures 14.8–14.10. The figures depict the one-year rolling ex-post information ratio attained when the risk models are coupled with the expected returns formulation of the Bayesian DLM model mentioned in Section 14.5 using proprietary explanatory

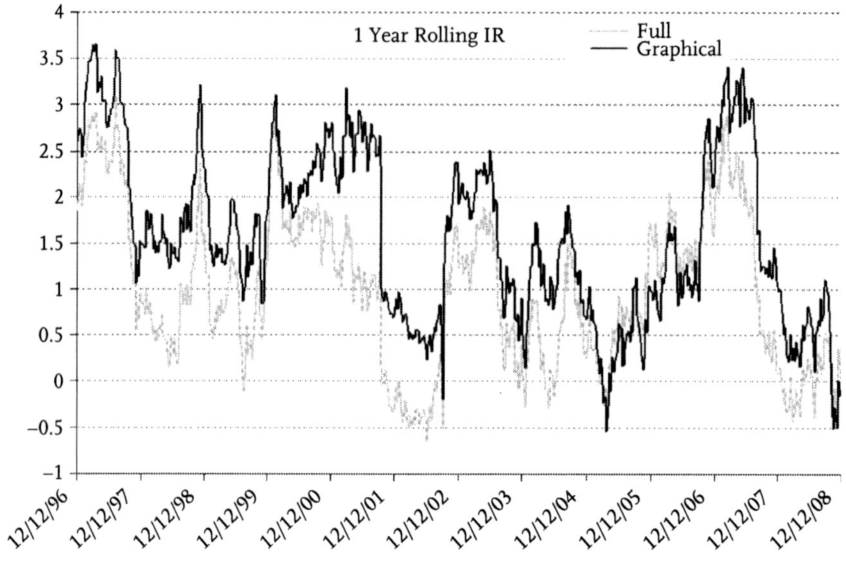

Fig. 14.8 $\delta = 0.97$.

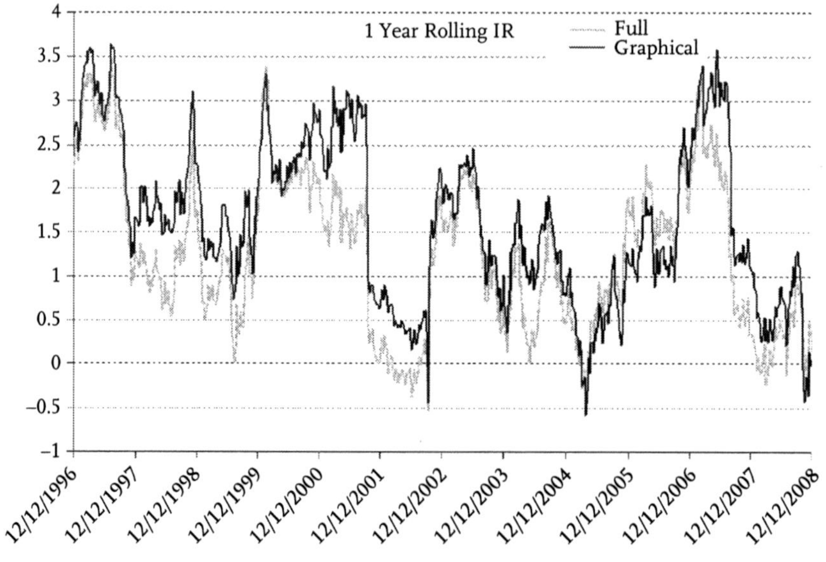

Fig. 14.9 $\delta = 0.98$.

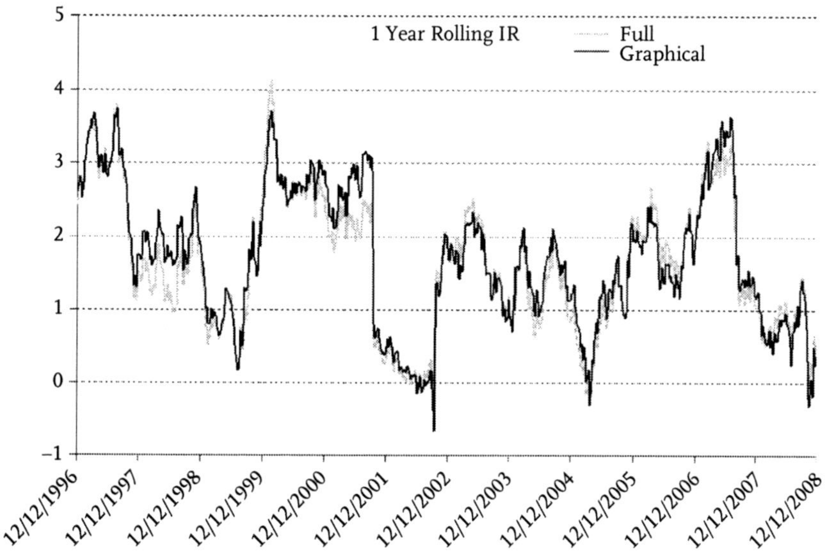

Fig. 14.10 $\delta = 0.99$.

variables that include measures of short, medium and long interest rates, liquidity and credit conditions, proxies for inflation, etc. These simulations begin after a buffer of three years used as a learning period for estimating the parameters associated to the payoffs expectations and take into account estimated transaction costs and employ a realistic production four-way (mean-variance-size-turnover) Pareto optimization variant of the simpler theoretical two-way (mean-variance) optimization discussed in Section 14.4 to promote the book's liquidity and control its transaction costs.

A more detailed description of the dataset follows to complete this section. The currencies weekly excess returns in the dataset were calculated as $y_t = C_t/C_{t-1} - 1$ where C_{t-1} and C_t are the specific indexed closing currency futures prices on consecutive Fridays (i.e. they are excess returns from a US dollar speculator's perspective); the stock indices and the government bond weekly excess returns were calculated as $y_t = (P_t/P_{t-1} - 1)C_t/C_{t-1}$ where P_{t-1} and P_t are the indexed closing futures prices and C_{t-1} and C_t are the corresponding indexed closing currency futures prices (i.e. they are currency hedged excess returns from a U.S. dollar speculator's perspective).

The Bloomberg tickers of the liquid futures contracts considered are as follows,

Currencies: AD1 Crncy, CD1 Crncy, SF1 Crncy, BP1 Crncy, EC1 Crncy, JY1 Crncy.

Government Bonds: XM1 Comdty, CN1 Comdty, FB1 Comdty, RX1 Comdty, G 1 Comdty, JB1 Comdty, ZT1 Comdty, TY1 Comdty.

Stock Indices: XP1 Index, PT1 Index, SM1 Index, GX1 Index, IB1 Index, CF1 Index, Z 1 Index, ST1 Index, TP1 Index, EO1 Index, SP1 Index, ZI1 Index.

Indexing based on price ratio was used to smooth out jumps that occur when a contract rolls into a new month (i.e. the speculator is assumed to maintain the same exposure while rolling contracts). To adjust by ratio when a contract changes, the previous indexed prices were multiplied by the ratio of the price of the new contract on the roll date to the price of the previous contract just before the roll date. The rolling schedule used for the currency assets was five days prior to the contract expiration date. The equity indices roll schedule was four days prior to the contract expiration date with the exception of the French CAC40 Equity Index (CF1 Index) – which was five days prior to expiration. The bond assets were rolled five days prior to the contract expiration with the following exceptions: CN1 Comdty – 28 days prior to expiration, G 1 Comdty – 35 days prior to expiration, TY1 Comdty – 25 days prior to expiration, XM1 Comdty and ZT1 Comdty – on the expiration date. Adjustments to imply excess returns were performed for the latter due to the Australian ad hoc quoting methods (described in its Bloomberg description page). The Deutsche mark was considered instead of the Euro prior to 1999. Finally, where futures prices were not available for the entire history considered, implied fair future prices were calculated based on suitable proxies.

14.7 Conclusions

According to subjective formulations of probability theory, Bayesians are, in principle, playing the role of altruist bookmakers in a betting scheme. Correspondingly, liquid financial futures markets act virtually as Bayesian market makers quoting fair prices that, consequently, represent objective coherent consensus expectations of the underlying commodities (including currencies, government bonds, stock indices, etc.). This amazing feat is accomplished by market participants acting in concert, albeit driven by selfish motivation, and putting into practice the Japanese proverb 'none of us is as smart as all of us'. Furthermore, the futures markets, as an entity, assimilates not only all the relevant information but also the applicable methodologies and technologies resulting in a formidable apparatus providing key economic ongoing expectations that, according to the efficient markets hypothesis advocates, cannot be improved.

Although the futures markets appear to be coherent, a Bayesian speculator driven by dynamic models predictions, which are hypothetically elicited without access to the markets to allow back subjectivity, could be able to generate attractive returns, given the risk taken, by trading in different markets as a counterpart. Regarding risk-adjusted performance, risk modeling and risk predictions are at least as important as the expected payoffs' formulations. Moreover, the hedging ability of different risk models can be assessed by comparing their success in producing normalized books (portfolios) with minimum risk; by this criterion, graphical models seem to be more robust having either better or similar hedging effectiveness in relation to the corresponding models without a conditional independence structure. Furthermore, their associated risk-adjusted performance exhibits, as expected, a similar dominance relationship.

Ultimately, resistance is futile, since the market Borg would dynamically assimilate any Bayesian speculating trading strategy that is implemented and would reward (or punish) the Bayesian speculator to the exact extent to which the futures markets' expectations improve (or deteriorate) as a result of this particular trading activity.

Appendix

A. Broader context and background

A.1 Foundations: Probabilty axioms derivation

An implication of the probability axioms (showing the steps to get sure money from an otherwise incoherent market maker) is concisely presented in Quintana (2005) as follows:

1. $P(A)$ is unique (otherwise, buy it low and sell it high).
2. $P(A) \geq 0$ (otherwise, buy it if $P(A) < 0$).
3. $P(\Omega) = 1$ (otherwise, buy it if $P(\Omega) < 1$, or sell it if $P(\Omega) > 1$; recall, Ω denotes a sure event).
4. $P(A + B) = P(A) + P(B)$ (a synthetic binary bet for $A + B$ is a binary bet for A plus a binary bet for B; $A + B$ denotes the union of two inconsistent events).
5. $P(AB) = P(A)P(B \mid A)$ (a synthetic binary bet for AB is $P(B \mid A)$ units of a binary bet on A plus a contingent binary bet on $B \mid A$ if A happens, using the contingent payoff to buy it; recall, a binary bet on $B \mid A$ is called off if A does not happen).

However, there are two caveats worth mentioning: First, Axiom 4 only implies finite additivity, and its implied generality is somewhat controversial. Second, the price $P(B \mid A)$ is quoted beforehand; but the market maker could quote a different one when and if A happens, even if this is the only extra information revealed (then again, as pointed out by Goldstein (2005), typically extra relevant information would also become simultaneously available). There is an alleviating provision, the market maker cannot tell in advance, without becoming incoherent, that her future quote will always be higher (or lower) afterwards.

A.2 Foundations: Bayesian portfolio optimization

The optimization (14.2) corresponds to the Markowitz (1959) mean-variance approach to finding optimal allocation weights in a traditional investment setup but without any of the traditional investment constraints for the weights to lie on the simplex. Although this approach, where only the first two predictive moments of the excess returns are in play, is not a panacea it seems to work well in practice. This direct predictive formulation is equivalent to solving first the optimization problem conditional on the model's unknown parameters and taking the utility expectation afterwards, as prescribed by the early work of Zelner and Chetty (1965), and naturally deals with the uncertainty inherent in the unknown modeling parameters. In contrast, there is an industry (e.g. DiBartolomeo 2003) to compensate for naive non-Bayesian alternatives that rely on plug-in estimates.

Moreover, it can be justified by the Bayesian paradigm of maximizing expected utility and it is the solution to a multiple-period decision problem even though it can be implemented myopically. This can be induced by defining the multiple-period utility as a weighted additive (where the components are the single-period utilities). In this framework, the associated stochastic dynamic programming problem is broken down into several isolated single-period optimizations and each optimal portfolio can be determined online requiring only single-step conditional predictive distributions for the payoffs. In addition, the single-period optimization (14.2) can be derived from the principle of utility maximization when the single-period utility is set as the difference of (positive) target payoff minus a square error penalty for deviating from that target. This approach and its motivations are discussed by Quintana (1992) and Quintana *et al.* (2003) and references therein. It is worth noting that, despite common Wall Street wisdom to the contrary, the speculator does not necessarily have to assume normality, nor miss potentially unlimited opportunities (due to the presence of market incoherence) as it would be the case when using a naive quadratic loss formulation.

B. Models and computations

B.1 Models

Generic Bayesian dynamic model The generic statistical model depicted by the functional network in Figure 14.3 and described by the corresponding specific functional forms for the observations and parameters random evolution,

$$\begin{aligned} \mathbf{y}_t &= f(\Theta_t, \mathbf{v}_t) \\ \Theta_t &= g\left(\Theta_{t-1}, \mathbf{u}_{\Theta_t}\right) \end{aligned} \tag{A.1}$$

coupled with independent random generation procedures for $\mathbf{v}_t$ and $\mathbf{u}_{\Theta_t}$ has a conditional probabilistic structure; this generic model specification can be defined in a traditional way by the set of stochastic Markovian parametric probability density functions $p(\Theta_t \mid \Theta_{t-1})$ and the set of observational probability density (i.e. likelihood) functions $p(\mathbf{y}_t \mid \Theta_t)$. For notational convenience we are omitting all the given available information (e.g. $\mathbf{x}_1, \mathbf{y}_1, \ldots, \mathbf{y}_{t-1}, \mathbf{x}_t$). The initial parameter, without loss of generality, can be set $\Theta_0 = 0$. The updating recurrences for the evolution, prediction and filtering are respectively:

$$\begin{aligned} p(\Theta_t) &= \int_{\Theta_{t-1}} p(\Theta_t \mid \Theta_{t-1})\, p(\Theta_{t-1})\, d\Theta_{t-1} \\ p(\mathbf{y}_t) &= \int_{\Theta_t} p(\mathbf{y} \mid \Theta_t)\, p(\Theta_t)\, d\Theta_t \\ p(\Theta_t \mid \mathbf{y}_t) &= \frac{p(\mathbf{y}_t \mid \Theta_t)\, p(\Theta_t)}{p(\mathbf{y}_t)} \end{aligned}$$

that is, the steps involve the computation of two marginal densities and one conditional density.

Specific Bayesian DLM The observational functional equations and the associated random generation procedures (given Σ_t) of the generic Bayesian DLM, depicted in Figure 14.4, correspond to a multivariate multiple regression model with a global scaling factor for the residual variance covariance and are defined as follows, $\mathbf{y}_t = \mathbf{X}_t\beta_t + C(\Sigma_t)\mathbf{v}_t$, where $\mathbf{v}_t \sim N(0, \mathbf{I})$ and $C(\Sigma_t)$ denotes the Cholesky decomposition of Σ_t (i.e. $(\mathbf{y}_t \mid \beta_t) \sim N(\mathbf{X}_t\beta_t, \Sigma_t)$.)

The aim of the evolution functional equations and the (independent) random generation procedures corresponding to the network depicted in Figure 14.4 (given Σ_t) is to provide a bridge from a posterior conjugate distribution at time $t-1$ to a prior conjugate distribution at time t. That is, to evolve from $\beta_{t-1} \sim N(\mathbf{m}_{t-1}, \mathbf{C}_{t-1})$ (i.e. $\beta_{t-1} \sim N(\mathbf{m}_{t-1}, \mathbf{C}_{t-1})$) to $\beta_t \sim N\left(\mathbf{m}_t^*, \mathbf{C}_t^*\right)$ where

$$\mathbf{m}_t^* = \mathbf{G}_t\mathbf{m}_{t-1} \tag{A.2a}$$

$$\mathbf{C}_t^* = \mathbf{C}_{t-1} + \mathbf{G}_t\mathbf{U}_t\mathbf{G}_t' \tag{A.2b}$$

This is accomplished by the following functional equations and random generation procedures, $\beta_t = \mathbf{G}_t\beta_{t-1} + C(\mathbf{U}_t)\mathbf{u}_{\beta_t}$ where $\mathbf{u}_{\beta_t} \sim N(0, I)$; $\mathbf{U}_t$ denotes a variance-covariance matrix associated to the evolution of β_t. Conversely, the evolutional functional equations and the (independent) random generation procedures for the evolution of Σ_t is chosen aiming to produce a dynamic conjugate updating procedure given β_t. The details of this evolution are found in Quintana *et al.* (2003) and can be summarized as follows: First, pre-multiply and post-multiply Σ_t by the inverse of the Cholesky decomposition of the scale parameter of its inverse Wishart distribution (this new random matrix has also a Wishart distribution but with an identity matrix as its scale parameter). Second, find the Cholesky decomposition of this new random matrix. Third, multiply the diagonal elements of the latter by suitable transformations of Beta independent random variables to discount their degrees of freedom. Finally, reverse the transformations of the first and second steps. The entire process is similar to taking a computer apart, tweaking one component, and putting it back together. The end result of choosing this form of evolution is that (given β_t) if the posterior distribution of $\Sigma_{t-1} \sim IW(d_{t-1}, \mathbf{S}_{t-1})$ is an inverse Wishart then the prior distribution of $\Sigma_t \sim IW\left(d_t^*, \mathbf{S}_t^*\right)$ where

$$\mathbf{S}_t^* = \delta_t \mathbf{S}_{t-1} \tag{A.3a}$$

$$d_t^* = \delta_t d_{t-1} \tag{A.3b}$$

and $0 < \delta_t \leq 1$ is interpreted as an information discount factor.

This provides a framework where the sequence of estimates of Σ_t keep adapting to new information while further discounting past observations.

Variance – covariance graphical model A graphical model is a probability model that characterizes the conditional independence structure of a set of random variable by a graph (Lauritzen 1996; Jones *et al.* 2005). Graphs provide a way to decompose the sample space into subsets of variables generating efficient ways to model conditional and marginal distributions locally. In high-dimensional problems, graphical model structuring is a key approach to parameter dimension reduction and, hence, to scientific parsimony and statistical efficiency when appropriate graphical structures can be identified.

In the context of a multivariate normal distribution, conditional independence restrictions are simply expressed through zeros in the off-diagonal elements of the precision matrix (inverse of the covariance matrix), establishing a parsimonious way to model covariance structures. Let a p-vector $\mathbf{y} \sim N(0, \Sigma)$ and $\Omega = \Sigma^{-1}$ with elements ω_{ij}. Write $G = (V, E)$ for the undirected graph whose vertex set V corresponds to the set of p random variables in $\mathbf{y}$, and E contains the elements (i, j) for which ω_{ij} are not equal to 0. The canonical parameter Ω belongs to $M(G)$, the set of all positive-definite symmetric matrices with elements equal to zero for all $(i, j) \notin E$.

In working with decomposable Gaussian graphical models, Dawid and Lauritzen (1993) defined a family of conjugate Markov probability distributions called *hyper-inverse Wishart*. If $\Omega \in M(G)$, the hyper-inverse Wishart

$$\Sigma \sim HIW_G(d, \mathbf{S})$$

has a degree-of-freedom parameter b and location matrix $\mathbf{S} \in M(G)$ implying that each clique $C \in \mathcal{C}$, $\Sigma_C \sim IW(d, \mathbf{S}_C)$ where $\mathbf{S}_C$ is the diagonal block of $\mathbf{S}$ corresponding to the vertices in C.

As shown by Carvalho and West (2007), graphical structuring can be incorporated in matrix normal DLMs to provide parsimonious models for the innovation variance-covariance matrix Σ. For a given decomposable graph G, take the hyper-inverse Wishart as a conjugate prior for Σ; it turns out that the closed-form, sequential updating theory of DLMs can be generalized to this richer model class.

This is also true for the case described in the previous section where Σ is time-varying. In detail, if the posterior at time $t-1$ is

$$\Sigma_{t-1} \sim HIW_G(d_{t-1}, \mathbf{S}_{t-1})$$

the prior at time t takes the form

$$\Sigma_t \sim HIW_G(\delta_t d_{t-1}, \delta_t \mathbf{S}_{t-1}).$$

This development is based on the same stochastic model of independent Beta shocks applied to the diagonal elements of the Cholesky decomposition of Σ_{t-1} presented above.

B.2 Computations

The specific DLM described above does not allow for tractable exact recursive updating formulas; however, as a result of the choice for the specific Markovian parametric evolution, given the time series Σ_t, the conventional updating recursions (West and Harrison 1997) apply for updating the hyperparameters of the normal distribution corresponding to the parameter β_t; and conversely, given the time series β_t, the conventional updating recursions (West and Harrison 1997) apply for updating the hyperparameters of the inverse Wishart distribution corresponding to the parameters Σ_t. This suggests a Gibbs sampling scheme for the implementation of the model.

References

Bachelier, L. (1900). *Théorie de la Spéculation*. Annales de l'École normale supérieure. Paris. Translated from French in *The Random Character of Stock Market Prices*. (ed. P.H. Cootner). The MIT Press, Cambridge, MA. 1964, pp. 17–78.

Barnard, G. A. (1980). Discussion of H. Akaike 'Likelihood and the Bayes procedure' and A. P. Dawid 'A Bayesian look at nuisance parameters'. In *Bayesian Statistics*, (ed. J. M. Bernardo, M. H. DeGroot, D. V. Lindley, and A. F. M. Smith), pp. 185–189. Valencia Univeristy Press, Valencia.

Bayes, T. (1763). An essay towards solving a problem in the doctorine of chances. *Philosophical Transactions of the Royal Society, London* , **53**, 370–418. Published posthumously. Electronic edition: `http://www.stat.ucla.edu/history/essay.pdf`.

Carvalho, C. and West, M. (2007). Dynamic matrix-variate graphical models. *Bayesian Analysis*, **2**, 69–96.

Carvalho, C. M. and Rickershauser, J. (2010). Volatility in prediction markets: a measure of information flow in political campaigns. In *The Handbook of Applied Bayesian Analysis*. (ed. A. O'Hagan and M. West), Oxford University Press, Oxford.

Dawid, A. P. and Lauritzen, S. L. (1993). Hyper-Markov laws in the statistical analysis of decomposable graphical models. *The Annals of Statistics*, **3**, 1272–1317.

de Finetti, B. (1931). Sul Significato Soggettivo della Probabilità. *Fundamenta Mathematicae*, **17**, 298–329. Translated into English as "On the subjective meaning of probability," in *Probabilitàe Induzione*, (ed. Paola Monari and Daniela Cocchi, 1993) Clueb, Bologna, pp. 291–321.

de Finetti, B. (1974). *Theory of Probability, a Critical Introductory Treatment*. John Wiley, New York.

DiBartolomeo, D. (2003). Portfolio optimization: The robust solution. In *Prudential Securities Quantitative Conference*. Electronic edition: `http://www.northinfo.com/documents/45.pdf`.

Fama, E. (1970). Efficient capital markets: A review of theory and empirical work. *Journal of Finance*, **25**, 383–417.

Goldstein, M. (2005). Subjective Bayesian analysis: Principles and practice. In *Case Studies in Bayesian Statistics, Workshop 8–2005*. Electronic edition `http://lib.stat.cmu.edu/bayesworkshop/2005/goldstein.pdf`.

Grossman, S. and Stiglitz, J. (1980). On the impossibility of informationally efficient markets. *American Economic Review*, **70**, 393–408.

Hamori, S., Hamori, N. and Anderson, D. A. (2001). An empirical analysis of the efficiency of the Osaka rice market during Japan's Tokugawa era. *Journal of Futures Markets*, **21**, 861–874.

Jones, B., Carvalho, C., Dobra, A., Hans, C., Carter, C. and West, M. (2005). Experiments in stochastic computation for high-dimensional graphical models. *Statistical Science*, **20**, 388–400.

Lauritzen, S. L. (1996). *Graphical Models*. Clarendon Press, Oxford.

Markowitz, H. M. (1959). *Portfolio Selection: Efficient Diversification of Investments* (2nd edn). Blackwell, Oxford.

Noble, B. and Daniel, J. (1987). *Applied Linear Algebra* (2nd edn). Prentice Hall, Englewood Cliffs.

Quintana, J. M. (1992). Optimal portfolios of forward currency contracts. In *Bayesian Statistics 4*, (ed. J. M. Bernardo, J. O. Berger, A. P. Dawid, and A. F. M. Smith), pp. 753–762. Oxford University Press, Oxford.

Quintana, J. M. (2005). Bayesian efficient trading. In *Proceedings of the Second Brazilian Conference on Statistical Modelling in Insurance and Finance*. (ed. N. Kolev and P. Morettin), Institute of Mathematics and Statistics at the University of Sao Paulo.

Quintana, J. M., Lourdes, V., Aguilar, O. and Liu, J. (2003). Global gambling. In *Bayesian Statistics 7*, (ed. J. M. Bernardo, M. J. Bayarri, J. O. Berger, A. P. Dawid, D. Heckerman, A. F. M. Smith, and M. West), pp. 349–367. Oxford University Press. (with discussion).

Quintana, J. M. and West, M. (1987). Multivariate time series analysis: New techniques applied to international exchange rate data. *The Statistician*, **36**, 275–281.

Ramsey, F. P. (1926). Truth and probability, In: Ramsey, 1931, *The Foundations of Mathematics and other Logical Essays*, R.B. Braithwaite, Ch. VII, p. 156–198, London: Kegan, Paul, Trench, Trubner & Co., New York: Harcourt, Brace and Company.

Scott, J. G. and Carvalho, C. M. (2008). Feature-inclusion stochastic search for Gaussian graphical models. *Journal of Computational and Graphical Statistics*, **17**, 790–808.

West, M. and Harrison, P. J. (1997). *Bayesian Forecasting and Dynamic Models* (2nd edn). Springer-Verlag, New York.

Zelner, A. and Chetty, V. K. (1965). Prediction and decision problems in regression models from the Bayesian point of view. *Journal of the American Statistical Association*, **60**, 608–616.

·15·

The new macroeconometrics: A Bayesian approach

Jesús Fernández-Villaverde, Pablo Guerrón-Quintana and Juan F. Rubio-Ramírez

15.1 Introduction

This chapter studies the dynamics of the US economy over the last 50 years via the Bayesian analysis of dynamic stochastic general equilibrium (DSGE) models. Our data are aggregate, quarterly economic variables and our approach combines macroeconomics (the study of aggregate economic variables like output or inflation) with econometrics (the application of formal statistical tools to economics). This is an example application in what is often called the *new macroeconometrics.*

Our application is of particular interest because modern macroeconomics is centred around the construction of DSGE models. Economists rely on DSGE models to organize and clarify their thinking, to measure the importance of economic phenomena, and to provide policy prescriptions. DSGE models start by specifying a number of economic agents, typically households, firms, a government, and often a foreign sector, and embodying them with behavioural assumptions, commonly the maximization of an objective function like utility or profits.

The inelegant name of DSGE comes from what happens next. First, economists postulate some sources of shocks to the model: shocks to productivity, to preferences, to taxes, to monetary policy, etc. Those shocks are the 'stochastic' part that will drive the model. Second, economists study how agents make their decisions over time as a response to these shocks (hence 'dynamic'). Finally, economists focus on the investigation of aggregate outcomes, thus 'general equilibrium'.

Before we discuss our application, it is important to emphasize that economists employ the word 'equilibrium' in a particular and precise technical sense, which is different from daily use or from its meaning in other sciences. We call equilibrium a situation where (a) the agents in the model follow a concrete behavioural assumption (usually, but not necessarily, maximization of either their utility or of profits) and (b) the decisions of the agents are consistent with each other (for example, the number of units of the good sold must be equal to the number of units of the good bought). Nothing in the definition of equilibrium implies that prices or allocations of goods are constant over time (they may be increasing, decreasing, or fluctuating in rather arbitrary ways), or that the economy is at a rest point, the sense in which equilibrium is commonly used in other fields, particularly in the natural sciences. Many misunderstandings about the accomplishments and potentialities of the concept of equilibrium in economics come about because of a failure to appreciate the subtle difference in language across fields.

General equilibrium has a long tradition in economics, being already implicit in the work of the French Physiocrats of the 18th century, such as Cantillon, Turgot, or Quesnay, or in Adam Smith's magnum opus, *The Wealth of Nations*. However, general equilibrium did not take a front seat in economics until 1874 when Léon Walras published a pioneering book, *Elements of Pure Economics*, in which he laid down the foundations of modern equilibrium theory.

For many decades after Walras, general equilibrium theory focused on static models. This was not out of a lack of appreciation by economists of the importance of dynamics, which were always at the core of practical concerns, but because of the absence of proper methodological instruments to investigate economic change.

Fortunately, the development of recursive techniques in applied mathematics (dynamic programming, Kalman filtering, and optimal control theory, among others) during the 1950s and 1960s provided the tools that economists required for the formal analysis of dynamics. These new methods were first applied to the search for microfoundations of macroeconomics during the 1960s, which attempted to move from ad hoc descriptions of aggregate behaviour motivated by empirical regularities to descriptions based on first principles of individual decision-making. Later, in the 1970s, recursive techniques opened the door to the rational expectations revolution, started by John Muth (1961) and Robert Lucas (1972), who highlighted the importance of specifying the expectations of the agents in the model in a way that was consistent with their behaviour.

This research culminated in 1982 with an immensely influential paper by Finn Kydland and Edward Prescott, *Time to Build and Aggregate Fluctuations*.[1] This article presented the first modern DSGE model, on this propitious occasion, one tailored to explaining the fluctuations of the US economy. Even if most of the material in the paper was already present in other articles written in the previous years by leading economists like Robert Lucas, Thomas Sargent,

[1] The impact of this work was recognized in 2004 when Kydland and Prescott received the Nobel Prize in economics for this paper and some previous work on time inconsistency in 1977.

Christopher Sims, or Neil Wallace, the genius of Kydland and Prescott was to mix the existing ingredients in such a path-breaking recipe that they redirected the attention of the profession to DSGE modelling.

Kydland and Prescott also opened several methodological discussions about how to best construct and evaluate dynamic economic models. The conversation that concerns us in this chapter was about how to take the models to the data, how to assess their behaviour, and, potentially, how to compare competing theories. Kydland and Prescott were skeptical about the abilities of formal statistics to provide a useful framework for these three tasks. Instead, they proposed to 'calibrate' DSGE models, i.e. to select parameter values by matching some moments of the data and by borrowing from microeconomic evidence, and to judge models by their ability to reproduce properties of the data that had not been employed for calibration (for example, calibration often exploits the first moments of the data to determine parameter values, while the evaluation of the model is done by looking at second moments).

In our previous paragraph, we introduced the idea of 'parameters of the model'. Before we proceed, it is worthwhile to discuss further what these parameters are. In any DSGE economy, there are functions, like the utility function, the production function, etc. that describe the preferences, technologies, and information sets of the agents. These functions are indexed by a set of parameters. One of the attractive features of DSGE models in the eyes of many economists is that these parameters carry a clear behavioral interpretation: they are directly linked to a feature of the environment about which we can tell an economic history. For instance, in the utility function, we have a discount factor, which tells us how patient the households are, and a risk aversion, which tells us how much households dislike uncertainty. These behavioural parameters (also called 'deep parameters' or, more ambiguously, 'structural parameters') are of interest because they are invariant to interventions, including shocks by nature or, more important, changes in economic policy.[2]

Kydland and Prescott's calibration became popular because, back in the early 1980s, researchers could not estimate the behavioural parameters of the models in an efficient way. The bottleneck was how to evaluate the likelihood of the model. The equilibrium dynamics of DSGE economies cannot be computed analytically. Instead, we need to resort to numerical approximations. One key issue is, then, how to go from this numerical solution to the likelihood of the model. Nowadays, most researchers apply filtering theory to accomplish this

[2] Structural parameters stand in opposition to reduced-form parameters, which are those that index empirical models estimated with a weaker link with economic theory. Many economists distrust the evaluation of economic policy undertaken with reduced parameters because they suspect that the historical relations between variables will not be invariant to the changes in economic policy that we are investigating. This insight is known as the *Lucas' critique*, after Robert Lucas, who forcefully emphasized this point in a renown 1976 article (Lucas, 1976).

goal. Three decades ago, economists were less familiar with filters and faced the speed constraints of existing computers. Only limited estimation exercises were possible and even those were performed at a considerable cost. Furthermore, as recalled by Sargent (2005), the early experiments on estimation suggested that the prototype DSGE models from the early 1980s were so far away from fitting the data that applying statistical methods to them was not worth the time and effort. Consequently, for over a decade, little work was done on the estimation of DSGE models.[3]

Again, advances in mathematical tools and in economic theory rapidly changed the landscape. From the perspective of macroeconomics, the streamlined DSGE models of the 1980s begot much richer models in the 1990s. One remarkably successful extension was the introduction of nominal and real rigidities, i.e. the conception that agents cannot immediately adjust to changes in the economic environment. In particular, many of the new DSGE models focused on studying the consequences of limitations in how frequently or how easily agents can changes prices and wages (prices and wages are 'sticky'). Since the spirit of these models seemed to capture the tradition of Keynesian economics, they quickly became known as New Keynesian DSGE models (Woodford, 2003). By capturing these nominal and real rigidities, New Keynesian DSGE models began to have a fighting chance at explaining the dynamics of the data, hence suggesting that formally estimating them could be of interest. But this desire for formal estimation would have led to naught if better statistical/numerical methods had not become widely available.

First, economists found out how to approximate the equilibrium dynamics of DSGE models much faster and more accurately. Since estimating DSGE models typically involves solving for the equilibrium dynamics thousands of times, each for a different combination of parameter values, this advance was crucial. Second, economists learned how to run the Kalman filter to evaluate the likelihood of the model implied by the linearized equilibrium dynamics. More recently, economists have also learned how to apply sequential Monte Carlo (SMC) methods to evaluate the likelihood of DSGE models where equilibrium dynamics is nonlinear and/or the shocks in the model are not Gaussian. Third, the popularization of Markov chain Monte Carlo (McMc) algorithms has facilitated the task of exploring and characterizing the likelihood of DSGE models.

[3] There was one partial exception. Lars Peter Hansen proposed the use of the Generalized Method of Moments, or GMM (Hansen, 1982), to estimate the parameters of the model by minimizing a quadratic form built from the difference between moments implied by the model and moments in the data. The GMM spread quickly, especially in finance, because of the simplicity of its implementation and because it was the generalization of well-known techniques in econometrics such as instrumental variables OLS. However, the GMM does not deliver all the powerful results implied by likelihood methods and it often has a disappointing performance in small samples.

Those advances have resulted in an explosion of the estimation of DSGE models, both at the academic level (An and Schorfheide, 2006) and, more remarkable still, at the institutional level, where a growing number of policy-making institutions are estimating DSGE models for policy analysis and forecasting (see Appendix A for a partial list). Furthermore, there is growing evidence of the good forecasting record of DSGE models, even if we compare them with the judgmental predictions that staff economists prepare within these policy-making institutions relying on real-time data (Edge, Kiley and Laforte, 2008).

One of the features of the new macroeconometrics that the readers of this volume will find exciting is that the (overwhelming?) majority of it is done from an explicitly Bayesian perspective. There are several reasons for this. First, priors are a natural device for economists. To begin with, they use them extensively in game theory or in learning models. More to the point, many economists feel that they have reasonably concrete beliefs about plausible values for most of the behavioural parameters, beliefs built perhaps from years of experience with models and their properties, perhaps from introspection (how risk adverse am I?), perhaps from well-educated economic intuition. Priors offer researchers a flexible way to introduce these beliefs as pre-sample information.

Second, DSGE models are indexed by a relatively large number of parameters (from around 10 to around 60 or 70 depending on the size of the model) while data are sparse (in the best case scenario, the US economy, we have 216 quarters of data from 1954 to 2007).[4] Under this low ratio data/parameters, the desirable small sample properties of Bayesian methods and the inclusion of pre-sample information are peculiarly attractive.

Third, and possibly as a consequence of the argument above, the likelihoods of DSGE models tend to be wild, with dozens of maxima and minima that defeat the best optimization algorithms. McMc are a robust and simple procedure to get around, or at least alleviate, these problems (so much so, that some econometricians who still prefer a classical approach to inference recommend the application of McMc algorithms by adopting a 'pseudo-Bayesian' stand; see Chernozhukov and Hong, 2003). Our own experience has been that models that we could estimate in a few days using McMC without an unusual effort turned out to be extremely difficult to estimate using maximum likelihood. Even the most pragmatic of economists, with the lowest possible interest in axiomatic foundations of inference, finds this feasibility argument rather compelling.

[4] Data before 1954 are less reliable and the structure of the economy back then was sufficiently different as to render the estimation of the same model problematic. Elsewhere (Fernández-Villaverde and Rubio-Ramírez, 2008), we have argued that a similar argument holds for data even before 1981. For most countries outside the US, the situation is much worse, since we do not have reliable quarterly data until the 1970s or even the 1980s.

Fourth, the Bayesian approach deals in a transparent way with misspecification and identification problems, which are pervasive in the estimation of DSGE models (Canova and Sala, 2006). After all, a DSGE model is a very stylized and simplified view of the economy that focuses only on the most important mechanisms at play. Hence, the model is false by construction and we need to keep this notion constantly in view, something that the Bayesian way of thinking has an easier time with than frequentist approaches.

Finally, Bayesian analysis has a direct link with decision theory. The connection is particularly relevant for DSGE models since they are used for policy analysis. Many of the relevant policy decisions require an explicit consideration of uncertainty and asymmetric risk assessments. Most economists, for example, read the empirical evidence as suggesting that the costs of a 1% deflation are considerably bigger than the costs of 1% inflation. Hence, a central bank with a price stability goal (such as the Federal Reserve System in the US or the European Central Bank) faces a radically asymmetric loss function with respect to deviations from the inflation target.

To illustrate this lengthy discussion, we will present a benchmark New Keynesian DSGE model, we will estimate it with US data, and we will discuss our results. We will close by outlining three active areas of research in the estimation of DSGE models that highlight some of the challenges that we see ahead of us for the next few years.

15.2 A benchmark new Keynesian model

Due to space constraints, we cannot present a benchmark New Keynesian DSGE model in all its details. Instead, we will just outline its main features to provide a flavour of what the model is about and what it can and cannot deliver. In particular, we will omit many discussions of why are we doing things the way we do. We will ask the reader to trust that many of our choices are not arbitrary but are the outcome of many years of discussion in the literature. The interested reader can access the whole description of the model at a complementary technical appendix posted at `www.econ.upenn.edu/~jesusfv/benchmark_DSGE.pdf`. The model we select embodies what is considered the current standard of New Keynesian macroeconomics (see Christiano, Eichenbaum and Evans, 2005) and it is extremely close to the models being employed by several central banks as inputs for policy decisions.

The agents in the model will include (a) households that consume, save, and supply labour to a labour 'packer', (b) a labour 'packer' that puts together the labour supplied by different households into an homogeneous labour unit, (c) intermediate good producers, who produce goods using capital and aggregated labour, (d) a final good producer that mixes all the intermediate goods into a final good that households consume or invest in, and (e) a government that sets monetary policy through open market operations. We present in turn each type of agents.

15.2.1 Households

There is a continuum of infinitely lived households in the economy indexed by j. The households maximize their utility function, which is separable in consumption, c_{jt}, real money balances (nominal money, mo_{jt}, divided by the aggregate price level, p_t), and hours worked, l_{jt}:

$$\mathbb{E}_0 \sum_{t=0}^{\infty} \beta^t e^{d_t} \left\{ \log \left(c_{jt} - hc_{jt-1} \right) + \upsilon \log \left(\frac{mo_{jt}}{p_t} \right) - e^{\varphi_t} \psi \frac{l_{jt}^{1+\vartheta}}{1+\vartheta} \right\}$$

where β is the discount factor, h controls habit persistence, ϑ is the inverse of the Frisch labour supply elasticity (how much labour supply changes when the wage changes while keeping consumption constant), d_t is a shock to intertemporal preference that follows the AR(1) process:

$$d_t = \rho_d d_{t-1} + e^{\sigma_d} \varepsilon_{d,t} \text{ where } \varepsilon_{d,t} \sim \mathcal{N}(0, 1),$$

and φ_t is a labour supply shock that also follows an AR(1) process:

$$\varphi_t = \rho_\varphi \varphi_{t-1} + e^{\sigma_\varphi} \varepsilon_{\varphi,t} \text{ where } \varepsilon_{\varphi,t} \sim \mathcal{N}(0, 1).$$

These two shocks, $\varepsilon_{d,t}$ and $\varepsilon_{\varphi,t}$, are equal across all agents.

Households trade on assets contingent on idiosyncratic and aggregate events. By a_{jt+1} we indicate the amount of those securities that pay one unit of consumption at time $t+1$ in event $\omega_{j,t+1,t}$ purchased by household j at time t at real price $q_{jt+1,t}$. These one-period securities are sufficient to span the whole set of financial contracts regardless of their duration. Households also hold an amount b_{jt} of government bonds that pay a nominal gross interest rate of R_t and invest a quantity x_t of the final good. Thus, the jth household's budget constraint is given by:

$$c_{jt} + x_{jt} + \frac{mo_{jt}}{p_t} + \frac{b_{jt+1}}{p_t} + \int q_{jt+1,t} a_{jt+1} d\omega_{j,t+1,t}$$
$$= w_{jt} l_{jt} + \left(r_t u_{jt} - \mu_t^{-1} \Phi \left[u_{jt} \right] \right) k_{jt-1} + \frac{mo_{jt-1}}{p_t} + R_{t-1} \frac{b_{jt}}{p_t} + a_{jt} + T_t + \mathcal{F}_t \quad (15.1)$$

where w_{jt} is the real wage, r_t the real rental price of capital, $u_{jt} > 0$ the intensity of use of capital, $\mu_t^{-1} \Phi[u_{jt}]$ is the physical cost of u_{jt} in resource terms (where $\Phi[u] = \Phi_1(u-1) + \Phi_2/2(u-1)^2$ and $\Phi_1, \Phi_2 \geq 0$), μ_t is an investment-specific technological shock that we will describe momentarily, T_t is a lump-sum transfer from the government, and $\mathcal{F}_t$ is the household share of the profits of the firms in the economy.

Given a depreciation rate δ and a quadratic adjustment cost function

$$V\left[\frac{x_t}{x_{t-1}}\right] = \frac{\kappa}{2}\left(\frac{x_t}{x_{t-1}} - \Lambda_x\right)^2,$$

where $\kappa \geq 0$ and Λ_x is the long-run growth of investment, capital evolves as:

$$k_{jt} = (1-\delta)\,k_{jt-1} + \mu_t\left(1 - V\left[\frac{x_{jt}}{x_{jt-1}}\right]\right)x_{jt}.$$

Note that in this equation there appears an investment-specific technological shock μ_t that also follows an autoregressive process:

$$\mu_t = \mu_{t-1}\exp\left(\Lambda_\mu + z_{\mu,t}\right) \text{ where } z_{\mu,t} = \sigma_\mu \varepsilon_{\mu,t} \text{ and } \varepsilon_{\mu,t} \sim \mathcal{N}(0,1).$$

Thus, the problem of the household is a Lagrangian function formed by the utility function, the law of motion for capital, and the budget constraint. The first order conditions of this problem with respect to c_{jt}, b_{jt}, u_{jt}, k_{jt}, and x_{jt} are standard (although messy), and we refer the reader to the online appendix for a detailed exposition.

The first order condition with respect to labor and wages is more involved because, as we explained in the introduction, households will face rigidities in changing their wages.[5] Each household supplies a slightly different type of labour service (for example, some households write chapters on Bayesian estimation of DSGE models and some write chapters on hierarchical models) that is aggregated by a labour 'packer' (in our example, an editor) into an homogeneous labor unit ('Bayesian research') according to the function:

$$l_t^d = \left(\int_0^1 l_{jt}^{\frac{\eta-1}{\eta}}\,dj\right)^{\frac{\eta}{\eta-1}} \tag{15.2}$$

where η controls the elasticity of substitution among different types of labour and l_t^d is the aggregate labour demand. The technical role that the labour 'packer' plays in the model is to allow for the existence of many different types of labour, and hence for the possibility of different wages, while keeping the heterogeneity of agents at a tractable level.

The labour 'packer' takes all wages w_{jt} of the differentiated labor and the wage of the homogeneous labor unit w_t as given and maximizes profits subject to the production function (15.2). Thus, the first order conditions of the labour

[5] In this type of model, households set up their wages and firms decide how much labour to hire. We could also have specified a model where wages are posted by firms and households decide how much to work. For several technical reasons, our modelling choice is easier to handle. If the reader has problems seeing a household setting its wage, she can think of the case of an individual contractor posting a wage for hour worked, or a union negotiating a contract in favour of its members.

'packer' are:

$$l_{jt} = \left(\frac{w_{jt}}{w_t}\right)^{-\eta} l_t^d \qquad \forall j. \tag{15.3}$$

We assume that there is free entry into the labour packing business. After a few substitutions, we get an expression for the wage w_t as a function of the different wages w_{jt}:

$$w_t = \left(\int_0^1 w_{jt}^{1-\eta} dj\right)^{\frac{1}{1-\eta}}.$$

The way in which households set their wages is through a device called Calvo pricing (for the economist Guillermo Calvo, who proposed this formulation in 1983). In each period, the household has a probability $1 - \theta_w$ of being able to change its wages. Otherwise, it can only partially index its wages by a fraction $\chi_w \in [0, 1]$ of past inflation. Therefore, if a household has not been able to change its wage for τ periods, its real wage is

$$\prod_{s=1}^{\tau} \frac{\Pi_{t+s-1}^{\chi_w}}{\Pi_{t+s}} w_{jt},$$

the original wage times the indexation divided by the accumulated inflation. This probability $1 - \theta_w$ represents a streamlined version of a more subtle explanation of wage rigidities (based on issues like contracts costs, Caplin and Leahy, 1991, or information limitations, Sims, 2002), which we do not flesh out entirely in the model to keep tractability.[6]

The average duration of a wage decision in this economy will be $1/(1 - \theta_w)$, although all wages change period by period because of the presence of indexation (in that sense the economy would look surprisingly flexible to a naïve observer!). Since a suitable law of large numbers holds for this economy, the probability of changing wages, $1 - \theta_w$, will also be equal to the fraction of households reoptimizing their wages in each period.

Calvo pricing implies three results. First, the form of the utility function that we selected and the presence of complete asset markets deliver the remarkable result that all households that choose their wage in this period will choose exactly the same wage, w_t^* (also, the consumption, investment, and Lagrangian multiplier λ_t of the budget constraint of all households will be the same and we drop the subindex j when no confusion arises).

Second, in every period, a fraction $1 - \theta_w$ of households set w_t^* as their wage, while the remaining fraction θ_w partially index their price by past inflation.

[6] There is, however, a discussion in the literature regarding the potential shortcomings of Calvo pricing. See, for example, Dotsey, King and Wolman (1999), for a paper that prefers to be explicit about the problem faced by agents when changing prices.

Consequently, the real wage index evolves as:

$$w_t^{1-\eta} = \theta_w \left(\frac{\Pi_{t-1}^{\chi_w}}{\Pi_t} \right)^{1-\eta} w_{t-1}^{1-\eta} + (1 - \theta_w) \, w_t^{*1-\eta}.$$

Third, after a fair amount of algebra, we can show that the evolution of w_t^* and an auxiliary variable f_t are governed by two recursive equations:

$$f_t = \frac{\eta - 1}{\eta} \left(w_t^* \right)^{1-\eta} \lambda_t w_t^{\eta} l_t^d + \beta \theta_w \mathbb{E}_t \left(\frac{\Pi_t^{\chi_w}}{\Pi_{t+1}} \right)^{1-\eta} \left(\frac{w_{t+1}^*}{w_t^*} \right)^{\eta-1} f_{t+1}$$

and:

$$f_t = \psi d_t \varphi_t \left(\frac{w_t}{w_t^*} \right)^{\eta(1+\vartheta)} \left(l_t^d \right)^{1+\vartheta} + \beta \theta_w \mathbb{E}_t \left(\frac{\Pi_t^{\chi_w}}{\Pi_{t+1}} \right)^{-\eta(1+\vartheta)} \left(\frac{w_{t+1}^*}{w_t^*} \right)^{\eta(1+\vartheta)} f_{t+1}.$$

15.2.2 The final good producer

The final good producer aggregates all intermediate goods into a final good with the following production function:

$$y_t^d = \left(\int_0^1 y_{it}^{\frac{\varepsilon-1}{\varepsilon}} di \right)^{\frac{\varepsilon}{\varepsilon-1}}. \tag{15.4}$$

where ε controls the elasticity of substitution.

The role of the final good producer is to allow for the existence of many firms with differentiated products while keeping a simple structure in the heterogeneity of products and prices. Also, the final good producer takes as given all intermediate goods prices p_{ti} and the final good price p_t, which implies a demand function for each good

$$y_{it} = \left(\frac{p_{it}}{p_t} \right)^{-\varepsilon} y_t^d \quad \forall i,$$

where y_t^d is the aggregate demand and by a zero profit condition:

$$p_t = \left(\int_0^1 p_{it}^{1-\varepsilon} di \right)^{\frac{1}{1-\varepsilon}}.$$

15.2.3 Intermediate good producers

There is a continuum of intermediate good producers. Each intermediate good producer i has access to a technology described by a production function of the form $y_{it} = A_t k_{it-1}^{\alpha} \left(l_{it}^d \right)^{1-\alpha} - \phi z_t$ where k_{it-1} is the capital rented by the firm, l_{it}^d is the amount of the homogeneous labour input rented by the firm, the parameter ϕ corresponds to the fixed cost of production, and where A_t follows a unit root process in logs, $A_t = A_{t-1} \exp(\Lambda_A + z_{A,t})$ where $z_{A,t} = \sigma_A \varepsilon_{A,t}$ and $\varepsilon_{A,t} \sim \mathcal{N}(0, 1)$.

The fixed cost ϕ is scaled by the variable $z_t = A_t^{\frac{1}{1-\alpha}} \mu_t^{\frac{\alpha}{1-\alpha}}$ to guarantee that economic profits are roughly equal to zero. The variable z_t evolves over time as $z_t = z_{t-1} \exp(\Lambda_z + z_{z,t})$ where $z_{z,t} = \frac{z_{A,t}+\alpha z_{\mu,t}}{1-\alpha}$ and $\Lambda_z = \frac{\Lambda_A + \alpha\Lambda_\mu}{1-\alpha}$. We can also prove that Λ_z is the mean growth rate of the economy.

Intermediate good producers solve a two-stage problem. First, given w_t and r_t, they rent l_{it}^d and k_{it-1} to minimize real costs of production, which implies a marginal cost of:

$$mc_t = \left(\frac{1}{1-\alpha}\right)^{1-\alpha} \left(\frac{1}{\alpha}\right)^{\alpha} \frac{w_t^{1-\alpha} r_t^{\alpha}}{A_t}.$$

This marginal cost does not depend on i: all firms receive the same shocks and rent inputs at the same price.

Second, intermediate good producers choose the price that maximizes discounted real profits under the same Calvo pricing scheme as households. The only difference is that now the probability of changing prices is given by $1 - \theta_p$ and the partial indexation by the parameter $\chi \in [0, 1]$. We will call p_t^* the price that intermediate good producers select when they are allowed to optimize their prices at time t.

Again, after a fair amount of manipulation, we find that the price index evolves as:

$$p_t^{1-\varepsilon} = \theta_p \left(\Pi_{t-1}^{\chi}\right)^{1-\varepsilon} p_{t-1}^{1-\varepsilon} + (1 - \theta_p)\, p_t^{*1-\varepsilon}$$

and that p_t^* is determined by two recursive equations in the auxiliary variable g_t^1 and g_t^2:

$$g_t^1 = \lambda_t mc_t y_t^d + \beta\theta_p \mathbb{E}_t \left(\frac{\Pi_t^{\chi}}{\Pi_{t+1}}\right)^{-\varepsilon} g_{t+1}^1$$

$$g_t^2 = \lambda_t \Pi_t^* y_t^d + \beta\theta_p \mathbb{E}_t \left(\frac{\Pi_t^{\chi}}{\Pi_{t+1}}\right)^{1-\varepsilon} \left(\frac{\Pi_t^*}{\Pi_{t+1}^*}\right) g_{t+1}^2$$

where $\varepsilon g_t^1 = (\varepsilon - 1) g_t^2$ and $\Pi_t^* = p_t^* / p_t$.

15.2.4 The government

The government plays an extremely limited role in this economy. It sets the nominal interest rates according to the following policy rule:

$$\frac{R_t}{R} = \left(\frac{R_{t-1}}{R}\right)^{\gamma_R} \left(\left(\frac{\Pi_t}{\Pi}\right)^{\gamma_\Pi} \left(\frac{\frac{y_t^d}{y_{t-1}^d}}{e^{\Lambda_z}}\right)^{\gamma_y}\right)^{1-\gamma_R} e^{m_t}. \tag{15.5}$$

The policy is implemented through open market operations that are financed with lump-sum transfers T_t to ensure that the government budget is balanced period by period. The variable Π is the target level of inflation, R the steady-state gross return of capital, and Λ_z, as mentioned above, is the average gross growth rate of y_t^d (these last two variables are determined by the model, not by the government). The term m_t is a random shock to monetary policy that follows $m_t = \sigma_m \varepsilon_{mt}$ where ε_{mt} is distributed according to $\mathcal{N}(0, 1)$.

The intuition behind the policy rules, motivated on a large body of empirical literature (Orphanides, 2002), is that a good way to describe the behavior of the monetary authority is to think about central banks as setting up the (short-term) interest rate as a function of the past interest rate, R_{t-1}, the deviation of inflation from a target, and the deviation of output growth from the long-run average growth rate.

15.2.5 Aggregation

With some additional work, we can sum up the behaviour of all agents in the economy to find expressions for the remaining aggregate variables, including aggregate demand:

$$y_t^d = c_t + x_t + \mu_t^{-1} \Phi\left[u_t\right] k_{t-1}$$

aggregate supply:

$$y_t^s = \frac{A_t \left(u_t k_{t-1}\right)^{\alpha} \left(l_t^d\right)^{1-\alpha} - \phi z_t}{v_t^p}$$

where:

$$v_t^p = \int_0^1 \left(\frac{p_{it}}{p_t}\right)^{-\varepsilon} di$$

and aggregate labour supply $l_t^d = l_t / v_t^w$ where

$$v_t^w = \int_0^1 \left(\frac{w_{jt}}{w_t}\right)^{-\eta} dj.$$

The terms v_t^p and v_t^w represent the loss of efficiency induced by price and wage dispersion. Since all households and firms are symmetrical, a social planner would like to set every price and wage at the same level. However, the nominal rigidities prevent this socially optimal solution. By the properties of the index under Calvo's pricing, v_t^p and v_t^w evolve as:

$$v_t^p = \theta_p \left(\frac{\Pi_{t-1}^{\chi}}{\Pi_t} \right)^{-\varepsilon} v_{t-1}^p + \left(1 - \theta_p\right) \Pi_t^{*-\varepsilon}$$

$$v_t^w = \theta_w \left(\frac{w_{t-1}}{w_t} \frac{\Pi_{t-1}^{\chi_w}}{\Pi_t} \right)^{-\eta} v_{t-1}^w + (1 - \theta_w) \left(\Pi_t^{w*}\right)^{-\eta}.$$

15.2.6 Equilibrium

The equilibrium of this economy is characterized by the first order conditions of the household, the first order conditions of the firms, the policy rule of the government, and the aggregate variables that we just presented, with a total of 21 equations (the web appendix has the list of all of these 21 equations). Formally, we can think about them as a set of nonlinear functional equations where the unknowns are the functions that determine the evolution over time of the variables of the model. The variables that enter as arguments of these functions are the 19 state variables that determine the situation of the system at any point in time.

A simple inspection of the system reveals that it does not have an analytical solution. Instead, we need to resort to some type of numerical approximation to solve it. The literature on how to do so is enormous, and we do not review it here (see, instead, the comprehensive textbook by Judd, 1998). In previous work (Aruoba, Fernández-Villaverde, and Rubio-Ramírez, 2006), we convinced ourselves that a nice compromise between speed and accuracy in the solution of systems of this sort can be achieved by the use of perturbation techniques.

The foundation of perturbation methods is to substitute the original system by a suitable variation of it that allows for an analytically (or at least numerically) exact solution. In our particular case, this variation of the problem is to set the standard deviations of all of the model's shocks to zero. Under this condition, the economy converges to a balanced growth path, where all the variables grow at a constant (but potentially different) rate. Then, we use this solution and a functional version of the implicit function theorem to compute the leading coefficients of the Taylor expansion of the exact solution.

Traditionally, the literature has focused on the first order approximation to the solution because it generates a linear representation of the model whose likelihood is easily evaluated using the Kalman filter. Since this chapter is a brief introduction to the new macroeconometrics, we will follow that convention. However, we could also easily compute a second order approximation. Higher order approximations have the advantage of being more accurate and of capturing precautionary behaviour among agents induced by variances.

We have found in our own research that in many problems (but not in all!), those second order terms are important.

Further elaborating this point, note that we end up working with an approximation of the solution of the model and not with the solution itself raises an interesting technical issue. Even if we were able to evaluate the likelihood implied by that solution, we would not be evaluating the likelihood of the exact solution of the model but the likelihood implied by the approximated solution of the model. Both objects may be quite different and some care is required when we proceed to perform inference (Fernández-Villaverde, Rubio-Ramírez and Santos, 2006).

However, before digging into the details of the solution method, and since we have technological progress, we need to rescale nearly all variables to make them stationary. To do so, define $\widetilde{c}_t = \frac{c_t}{z_t}$, $\widetilde{\lambda}_t = \lambda_t z_t$, $\widetilde{r}_t = r_t \mu_t$, $\widetilde{q}_t = q_t \mu_t$, $\widetilde{x}_t = \frac{x_t}{z_t}$, $\widetilde{w}_t = \frac{w_t}{z_t}$, $\widetilde{w}_t^* = \frac{w_t^*}{z_t}$, $\widetilde{k}_t = \frac{k_t}{z_t \mu_t}$, and $\widetilde{y}_t^d = \frac{y_t^d}{z_t}$. Also note that $\Lambda_c = \Lambda_x = \Lambda_w = \Lambda_{w^*} = \Lambda_{y^d} = \Lambda_z$. Furthermore, we normalize $u = 1$ in the steady state by setting $\Phi_1 = \widetilde{r}$, eliminating it as a free parameter.

For each variable var_t, we define $\widehat{var}_t = var_t - var$, as the deviation with respect to the steady state. Then, the states of the model $\overline{S}_t$ are given by:

$$\overline{S}_t = \begin{pmatrix} \widehat{\Pi}_{t-1}, \widehat{\widetilde{w}}_{t-1}, \widehat{g}^1_{t-1}, \widehat{g}^2_{t-1}, \widehat{\widetilde{k}}_{t-1}, \widehat{R}_{t-1}, \widehat{\widetilde{y}}^d_{t-1}, \widehat{\widetilde{c}}_{t-1}, \widehat{v}^p_{t-1}, \widehat{v}^w_{t-1}, \\ \widehat{\widetilde{q}}_{t-1}, \widehat{f}_{t-1}, \widehat{\widetilde{x}}_{t-1}, \widehat{\widetilde{\lambda}}_{t-1}, \widehat{\widetilde{z}}_{t-1}, z_{\mu,t-1}, \widehat{d}_{t-1}, \widehat{\varphi}_{t-1}, z_{A,t-1} \end{pmatrix}',$$

and the exogenous shocks are $\varepsilon_t = \left(\varepsilon_{\mu,t}, \varepsilon_{d,t}, \varepsilon_{\varphi,t}, \varepsilon_{A,t}, \varepsilon_{m,t}\right)'$.

From the output of the perturbation, we build the law of motion for the states:

$$\overline{S}_{t+1} = \Psi_{s1}\left(\overline{S}_t', \varepsilon_t'\right)' \tag{15.6}$$

where Ψ_{s1} is a 19×24 matrix, and the law of motion for the observables:

$$\mathbb{Y}_t = \Psi_{o1}\left(S_t', \varepsilon_t'\right)' \tag{15.7}$$

where $S_t = \left(\overline{S}_t', \overline{S}_{t-1}', \varepsilon_{t-1}'\right)$ and Ψ_{o1} is a 5×48 matrix. We include lagged states in our observation equation, because some of our data will appear in first differences. Equivalently, we could have added lagged observations to the states to accomplish the same objective. Also, note that the function (15.7) includes a constant that captures the mean of the observables as implied by the equilibrium of the model.

Our observables are the first differences of the relative price of investment, output, real wages, inflation, and the federal funds rate, or in our notation:

$$\mathbb{Y}_t = \left(\Delta \log \mu_t^{-1}, \Delta \log y_t, \Delta \log w_t, \log \Pi_t, \log R_t\right)'.$$

While the law of motion for states is unique (or at least belonging to an equivalence class of representations), the observation equation depends on what the econometrician actually observes or chooses to observe. Since little is known about how those choices affect our estimation, we have selected the time series that we find particularly informative for our purposes (see Guerrón-Quintana, 2008, for a detailed discussion).

For observables, we can only select a number of series less than or equal to the number of shocks in the model. Otherwise, the model will be stochastically singular, i.e. we could write the extra observables as *deterministic* functions of the other observables and the likelihood would be $-\infty$. In the jargon of macroeconomics, these functions are part of the equilibrium cross-equation restrictions.

A popular way around the problem of stochastic singularity has been to assume that the observables come with measurement error. This assumption delivers, by itself, one shock per observable. There are three justifications for measurement error. First, statistical agencies make mistakes. Measuring US output or wages is a daunting task, undertaken with extremely limited resources by different government agencies. Despite their remarkable efforts, statistical agencies can only provide us with an estimate of the series we need. Second, there are differences between the definitions of variables in the theory and in the data, some caused by divergent methodological choices (see, for instance, the discussion in Appendix B about how to move from nominal output into real output), some caused by the limitations of the data the statistical agency can collect. Finally, measurement errors may account for parts of the dynamic of the data that are not captured by the model.

On the negative side, including measurement error complicates identification and faces the risk that the data ends up being explained by this measurement error and not by the model itself. In our own research we have estimated DSGE models with and without measurement error, according to the circumstances of the problem. In this chapter, since we have a rich DSGE economy with many sources of uncertainty, we assume that our data come without measurement error. This makes the exercise more transparent and easier to follow.

15.2.7 The likelihood function

Equations (15.6) and (15.7) plus the definition of S_t are the state space representation of a dynamic model. It is well understood that we can use this state representation to find the likelihood $\mathcal{L}(\mathbb{Y}_{1:T}; \Psi)$ of our DSGE model, where we have stacked all parameter values in:

$$\Psi = \{\beta, h, \upsilon, \vartheta, \delta, \eta, \varepsilon, a, \phi, \theta_w, \chi_w, \theta_p, \chi_p, \Phi_2, \gamma_R, \gamma_y, \gamma_\Pi, \\ \Pi, \Lambda_\mu, \Lambda_A, \rho_d, \rho_\varphi, \sigma_\mu, \sigma_d, \sigma_A, \sigma_m, \sigma_\varphi\}.$$

and where $\mathbb{Y}_{1:T}$ is the sequence of data from period 1 to T.

To perform this evaluation of the likelihood function given some parameter values, we start by factorizing the likelihood function as:

$$\mathcal{L}\left(\mathbb{Y}_{1:T};\Psi\right)=\prod_{t=1}^{T}\mathcal{L}\left(\mathbb{Y}_t|\mathbb{Y}_{1:T-1};\Psi\right).$$

Then:

$$\mathcal{L}\left(\mathbb{Y}_{1:T};\Psi\right)=\int\mathcal{L}\left(\mathbb{Y}_1|S_0;\Psi\right)d\,S_0\prod_{t=2}^{T}\int\mathcal{L}\left(\mathbb{Y}_t|S_t;\Psi\right)p\left(S_t|\mathbb{Y}_{1:T-1};\Psi\right)d\,S_t. \tag{15.8}$$

If we know S_t, computing $\mathcal{L}\left(\mathbb{Y}_t|S_t;\Psi\right)$ is conceptually simple since, conditional on S_t, the measurement equation (15.7) is a change of variables from ε_t to $\mathbb{Y}_t$. Similarly, if we know S_0, it is easy to compute $\mathcal{L}\left(\mathbb{Y}_1|S_0;\Psi\right)$ with the help of (15.6) and (15.7). Thus, all that remains to evaluate the likelihood is to find the sequence $\{p\left(S_t|\mathbb{Y}_{1:t-1};\Psi\right)\}_{t=1}^{T}$ and the initial distribution $p\left(S_0;\Psi\right)$.

This second task is comparatively simple. There are two procedures for doing so. First, in the case where we can characterize the ergodic distribution, for example, in linearized models, we can draw directly from it (this procedure takes advantage of the stationarity of the model achieved by the rescaling of variables). In our model, we just need to set the states to zero (remember, they are expressed in differences with respect to the steady state) and specify their initial variance-covariance matrix. Second, if we cannot characterize the ergodic distribution, we can simulate the model for a large number of periods (250 for a model like ours may suffice). Under certain regularity conditions, the values of the states at period 250 are a draw from the ergodic distribution of states. Consequently, the only remaining barrier is to characterize the sequence of conditional distributions $\{p\left(S_t|\mathbb{Y}_{1:T-1};\Psi\right)\}_{t=1}^{T}$ and to compute the integrals in (15.8).

Since (a) we have performed a first order perturbation of the equilibrium equations of the problem and generated a linear state space representation, and (b) our shocks are normally distributed, the sequence $\{p\left(S_t|\mathbb{Y}_{1:T-1};\Psi\right)\}_{t=1}^{T}$ is composed of conditionally Gaussian distributions. Hence, we can use the Kalman filter, find $\{p\left(S_t|\mathbb{Y}_{1:T-1};\Psi\right)\}_{t=1}^{T}$, and evaluate the likelihood in a quick and efficient way. A version of the Kalman filter applicable to our setup is described in Appendix B. The Kalman filter is extremely efficient: it takes less than one second to run for each combination of parameter values.

However, in many cases of interest, we may want to perform a higher order perturbation or the shocks to the model may not be normally distributed. Then, the components of $\{p\left(S_t|\mathbb{Y}_{1:T-1};\Psi\right)\}_{t=1}^{T}$ would not be conditionally Gaussian. In fact, in general, it will not belong to any known parametric family of density functions. In all of those cases, we can track the desired sequence $\{p(S_t|\mathbb{Y}_{1:T-1};\Psi)\}_{t=1}^{T}$ using SMC methods (see Appendix A).

Once we have the likelihood of the model $\mathcal{L}(\mathbb{Y}_{1:T};\Psi)$, we combine it with a prior density for the parameters $p(\Psi)$ to form a posterior

$$p(\Psi|\mathbb{Y}_{1:T}) \propto \mathcal{L}(\mathbb{Y}_{1:T};\Psi)\, p(\Psi).$$

Since the posterior is also difficult to characterize (we do not even have an expression for the likelihood, only an *evaluation*), we generate draws from it using a random walk Metropolis – Hastings algorithm. The scaling matrix of the proposal density will be selected to generate the appropriate acceptance ratio of proposals (Roberts, Gelman and Gilks, 1997).

We can use the resulting empirical distribution to obtain point estimates, variances, etc., and to build the marginal likelihood that is a fundamental component in the Bayesian comparison of models (see Fernández-Villaverde and Rubio-Ramírez, 2004, for an application of the comparison of dynamic models in economics). An important advantage of having draws from the posterior is that we can also compute other objects of interest, like the posterior of the welfare cost of business cycle fluctuations or the posterior of the responses of the economy to an unanticipated shock. These objects are often of more interest to economists than the parameter themselves.

15.3 Empirical analysis

Now we are ready to take our model to the data and obtain point estimates and posterior distributions. Again, because of space limitations, our analysis should be read more as an example of the type of exercises that can be implemented than as an exhaustive investigation of the empirical properties of the model. Furthermore, once we have parameter posteriors, we can undertake many additional exercises of interest, like forecasting (where is the economy going?), counterfactual analysis (what would have happen if?), policy design (what is the best way to respond to shocks?), and welfare analysis (what is the cost of business cycle fluctuations?).

We estimate our New Keynesian model using five time series for the US economy: (1) the relative price of investment with respect to the price of consumption, (2) real output per capita growth, (3) real wages per capita growth, (4) the consumer price index growth, and (5) the federal funds rate (the interest rate at which banks lend balances at the Federal Reserve System to each other, usually overnight). Our sample goes from 1959Q1–2007Q1. Appendix B explains how we construct these series.

Before proceeding further, we specify priors for the parameters. We will have two sets of priors that will provide us with two sets of posteriors. First, we adopt flat priors for all parameters except a few that we fix at some predetermined values. These flat priors are modified only by imposing boundary constraints to make the priors proper and to rule out parameter values that are incompatible with the model (i.e. a negative value for a variance or Calvo parameters outside the unit interval). Second, we use a set of priors nearly identical to the one proposed by Smets and Wouters (2007) in a highly influential study in which they estimate a similar DSGE model to ours.

Our first exercise with flat priors is motivated by the observation that, with this prior, the posterior is proportional to the likelihood function.[7] Consequently, our Bayesian results can be interpreted as a classical exercise where the mode of the likelihood function (the point estimate under an absolute value loss function for estimation) is the maximum likelihood estimate. Moreover, a researcher who prefers alternative priors can always reweight the draws from the posterior using importance sampling to encompass his favorite priors. We do not argue that our flat priors are uninformative. After a reparameterization of the model, a flat prior may become highly curved. Instead, we have found in our research that an estimation with flat priors is a good first step to learn about the amount of information carried by the likelihood and a natural benchmark against which to compare results obtained with nonflat priors. The main disadvantage of this first exercise is that, in the absence of further information, it is extremely difficult to get the McMc to converge and characterize the posterior accurately. The number of local modes is so high that the chain wanders away for the longest time without settling into the ergodic distribution.

In comparison, our second exercise, with Smets and Wouters' (2007) priors, is closer to the exercises performed at central banks and other policy institutions. The priors bring a considerable amount of additional information that allows us to achieve much more reasonable estimates. However, the use of more aggressive priors requires a careful preparation by the researcher, and in the case of institutional models, a candid conversation with the principal regarding the assumptions about parameters she feels comfortable with.

15.3.1 Flat priors

Now we present results from our exercise with flat priors. Before reporting results, we fix some parameter values at the values reported in Table 15.1. In a Bayesian context, we can think about fixing parameter values as having Dirac priors over them.

[7] There is a small qualifier: the bounded support of some of the priors. We can eliminate this small difference by thinking about those bounds as frontiers of admissible parameter values in a classical perspective.

Table 15.1 Fixed parameters.

β	h	ψ	ϑ	δ	α	ε	η	κ	ϕ	Φ_2
0.99	0.9	9	1.35	0.015	0.30	8	8	30	0	0.001

Macroeconomists fix some parameters for different reasons. One reason is related to the amount of data: being too ambitious in estimating all the parameters in the model with limited data sometimes delivers posteriors that are too spread for useful interpretation and policy analysis. A related reason is that some parameters are poorly identified with macro data (for example, the elasticity of output with respect to capital), while, at the same time, we have good sources of information from micro data not used in the estimation. It seems, then, reasonable to set these parameters at the conventional values determined by microeconometricians. In our first exercise with flat priors, the number of parameters that we need to fix is relatively high, 11. Otherwise, the likelihood has too many local peaks, and it turns out to be extremely difficult to design a successful McMC.

We briefly describe some of these parameter values in Table 15.1. The discount factor, β, is fixed at 0.99, a number close to one, to match the low risk free rate observed in the US economy (the gross risk free rate depends on the inverse of β and the growth rate of the economy). Habit persistence, $h = 0.9$, arises because we want to match the evidence of a sluggish adjustment of consumption decisions to shocks. The parameter ϑ pins down a Frisch elasticity of 0.74, a value well within the range of the findings of microeconometrics (Browning, Hansen and Heckman, 1999). Our choice of δ, the depreciation rate, matches the capital-output ratio in the data. The elasticity of capital to output, $\alpha = 0.3$, is informed by the share of national income that goes to capital. The elasticity of substitution across intermediate goods, ε, accounts for the microeconomic evidence on the average mark-up of US firms, which is estimated to be around 10–15 percent (the mark-up in our model is approximately $\varepsilon/(\varepsilon - 1)$). By symmetry, we pick the same value for η. We take a high value for the adjustment cost κ, 30, to dampen the fluctuations of investment in the model and get them closer to the behaviour of investment in the data. We set ϕ to zero, since we do not have information on pure profits by firms. Luckily, since we do not have entry and exit of firms in the model, the parameter is nearly irrelevant for equilibrium dynamics. Finally, the parameter controlling money demand υ does not affect the dynamics of the model because the monetary authority will supply as much money as required to implement the nominal interest rate determined by the policy rule.

We summarize the posterior distribution of the parameters in Table 15.2, where we report the median of each parameter and its 5 and 95 percentile values and in Figure 15.1, where we plot the histograms of the draws of each

Table 15.2 Median estimated parameters (5 and 95 percentile in parentheses).

θ_p	χ	θ_w	χ_w	γ_R	γ_y	γ_π	Π	
0.72	0.94	0.62	0.92	0.87	0.99	2.65	1.012	
[0.68, 0.77]	[0.81, 0.99]	[0.56, 0.70]	[0.73, 0.99]	[0.83, 0.90]	[0.59, 1.73]	[2.10, 3.67]	[1.011, 1.013]	
ρ_d	ρ_φ	σ_A	σ_d	σ_φ	σ_μ	σ_e	Λ_μ	Λ_A
0.14	0.91	−4.45	−2.53	−2.29	−5.49	−5.97	3e−3	1e−4
[0.02, 0.27]	[0.86, 0.93]	[−4.61, −4.29]	[−2.64, −2.43]	[−2.59, −1.74]	[−5.57, −5.40]	[−6.09, −5.84]	[0.002, 0.004]	[0.00, 0.0003]

parameter. The results are based on a chain of 75,000 draws, initialized after a considerable amount of search for good initial starting values, and a 30 percent acceptance rate. We performed standard analysis of convergence of the chain to ensure that the results are accurate.

We highlight a few results. The Calvo parameter for prices, θ_p, is closely estimated to be around 0.72. This indicates that firms revise their pricing decisions around 3.5 quarters on average, a number that seems close to a natural benchmark of a yearly pricing cycle in many firms complemented with some firms that revise their prices more often. The indexation level, χ, of 0.94, has a higher degree of uncertainty but also suggests that firms respond quickly to inflation. High indexation generates a high level of inflation inertia, and with it, a bigger role for monetary policy. The Calvo parameter for wages, θ_w, implies wage decisions every 2.6 quarters, also with high indexation.

The parameters of the policy rule show that the monetary authority smooths to a large degree the evolution of the federal funds rate, γ_R is equal to 0.87, cares about the growth level of the economy, γ_y is equal to 0.99, and finally, is rather aggressive against inflation, γ_π is equal to 2.65. This last parameter value is important, because it implies that the monetary authority respects the 'Taylor principle' that requires nominal interest rates to rise more than inflation rises. Otherwise, the real interest rate falls, inflation rises, and we generate bad policy outcomes due to the indeterminacy of equilibria (Lubick and Schorfheide, 2004). The inflation target of 1 percent per quarter is higher than the stated goals of the Federal Reserve but consistent with the behavior of inflation in the sample.

The standard deviations of the five shocks in the economy show the importance of the preference shocks and the higher importance of the neutral technological shock in comparison with the investment-specific technological shock in accounting for fluctuations in the economy. In comparison, the posterior clearly indicated that the *average* growth of the economy is driven by investment-specific technological change, whose mean, Λ_μ, is an order of magnitude bigger than neutral technological change, Λ_A. These findings are similar to the ones

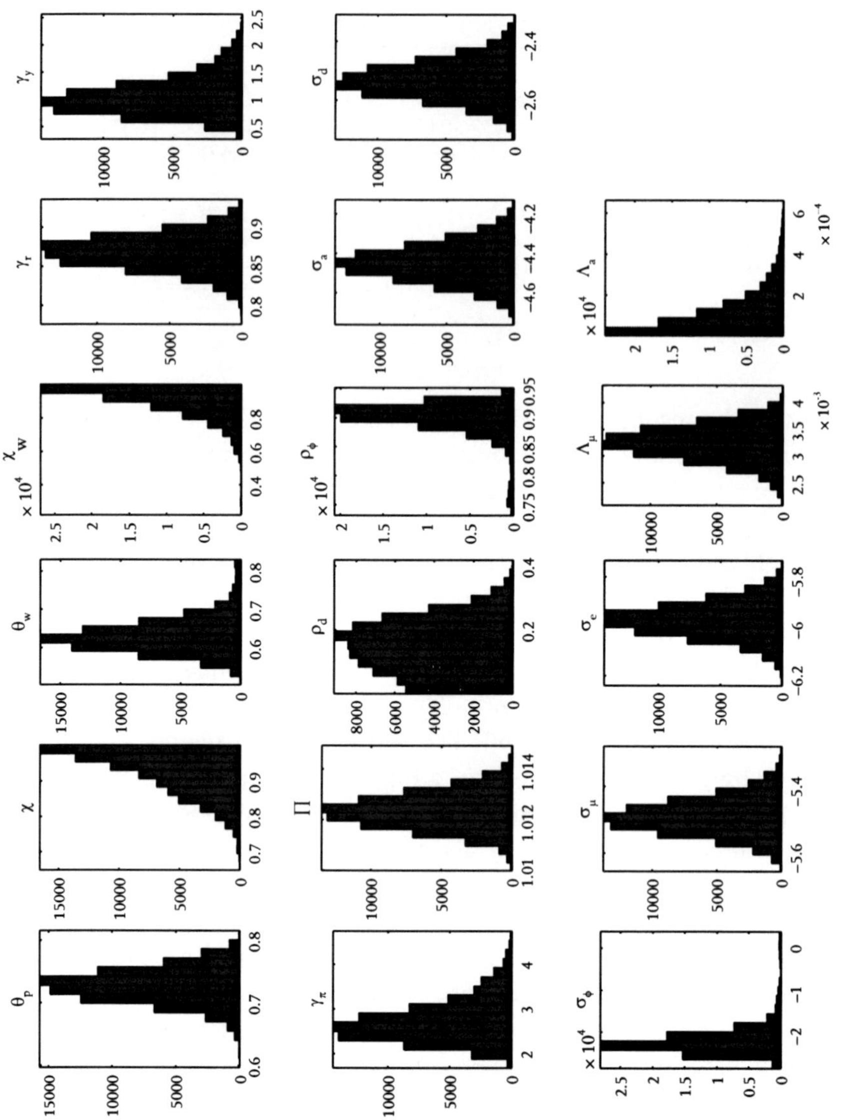

Fig. 15.1 Posterior distribution, flat priors.

obtained using a calibration approach by Greenwood, Herkowitz and Krusell (1997).

15.3.2 Smets and Wouters' (2007) priors

In our second estimation exercise, we incorporate priors for the parameters of the model based on the work of Smets and Wouters (2007). First, as we did before, we fix some parameter values. In this case, since the priors help to

Table 15.3 Fixed parameters.

δ	ε	η	ϕ	Φ_2
0.025	10	10	0	0.001

achieve identification, we can free six out of the 11 parameters fixed in the first exercise. The remaining five fixed parameters are those still difficult to identify with aggregate data and are summarized in Table 15.3. Note that, to maintain comparability with Smets and Wouters' paper, we increase the elasticities of substitution, ε and η, to 10, an increase that has a minuscule effect on the dynamics of the data.

We modify several of the priors of Smets and Wouters to induce better identification and to accommodate some of the differences between their model and ours. Instead of a detailed discussion of the reasons behind each prior, it is better to highlight that the priors are common in the literature and are centered around values that are in the middle of the usual range of estimates, using both macro and micro data, and diverse econometric methodologies. We summarize our priors in Table 15.4.

We generate 75,000 draws from the posterior using our Metropolis – Hastings, also after an exhaustive search for good initial parameters of the chain. In this case, the search is notably easier than in the first exercise, showing the usefulness of the prior in achieving identification and a good behaviour of the simulation. Table 15.5 reports the posterior medians, and the 5 and 95 percentile values of the 23 estimated parameters of the model, while Figure 15.2 plots the histograms of each parameter.

The value of the discount factor, β, goes very close to 1. This result is typical in DSGE models. The growth rate of the economy and the log utility function for consumption induce a high risk free real interest rate even without any discounting. This result is difficult to reconcile with an observed low average real interest rate in the US Hence, the likelihood wants to push β as close as possible to 1 to avoid an even worse fit of the data. We find a quite high

Table 15.4 Priors.

$100\,(\beta^{-1}-1)$	h	ψ	θ_p	χ	θ_w
$Ga(0.25, 0.1)$	$Be(0.7, 0.1)$	$N(9, 3)$	$Be\,(0.5, 0.1)$	$Be\,(0.5, 0.15)$	$Be\,(0.5, 0.1)$
χ_w	γ_R	γ_y	γ_π	$100(\Pi-1)$	ϑ
$Be\,(0.5, 0.1)$	$Be\,(0.75, 0.1)$	$N(0.12, 0.05)$	$N(1.5, 0.125)$	$Ga\,(0.95, 0.1)$	$N(1, 0.25)$
κ	α	ρ_d	ρ_φ	$\exp(\sigma_A)$	$\exp(\sigma_d)$
$N(4, 1.5)$	$N(0.3, 0.025)$	$Be\,(0.5, 0.2)$	$Be\,(0.5, 0.2)$	$IG\,(0.1, 2)$	$IG\,(0.1, 2)$
$\exp(\sigma_\varphi)$	$\exp(\sigma_\mu)$	$\exp(\sigma_e)$	$100\Lambda_\mu$	$100\Lambda_A$	
$IG\,(0.1, 2)$	$IG\,(0.1, 2)$	$IG\,(0.1, 2)$	$N(0.34, 0.1)$	$N(0.178, 0.075)$	

Table 15.5 Median estimated parameters (5 and 95 percentile in parentheses).

β	h	ψ	ϑ	κ	α
0.998	0.97	8.92	1.17	9.51	0.21
0.997,0.999]	[0.95,0.98]	[4.09,13.84]	[0.74,1.61]	[7.47,11.39]	[0.17,0.26]
θ_p	χ	θ_w	χ_w	γ_R	γ_y
0.82	0.63	0.68	0.62	0.77	0.19
0.78,0.87]	[0.46,0.79]	[0.62,0.73]	[0.44,0.79]	[0.74,0.81]	[0.13,0.27]
γ_π	Π	ρ_d	ρ_φ	σ_A	σ_d
1.29	1.010	0.12	0.93	−3.97	[−1.82,−1.11]
1.02,1.47]	[1.008,1.011]	[0.04,0.22]	[0.89,0.96]	[−4.17,−3.78]	−1.51
σ_φ	σ_μ	σ_e	Λ_μ	Λ_A	
−2.36	−5.43	−5.85	3.4e−3	2.8e−3	
−2.76,−1.74]	[−5.52,−5.35]	[−5.94,−5.74]	[0.003,0.004]	[0.002,0.004]	

level of habit, 0.97, which diverges from other findings in the literature that hover around 0.7. The Frisch elasticity of labor supply of 0.85 (1/1.17) is similar to the one we fixed in the first estimation exercise, a reassuring result since DSGE models have often been criticized for relying on implausible high Frisch elasticities.

The adjustment cost of investment, κ, is high, 9.51, although lower than the value we fixed in the first exercise. Since our specification of adjustment costs was parsimonious, this result hints at the importance of further research on the nature of investment plans by firms. The parameter α is centred around 0.2. This result, which coincides with the findings of Smets and Wouters, is lower than in other estimates, but it is hard to interpret because the presence of monopolistic competition complicates the mapping between this parameter and observed income shares in the national income and product accounts.

The Calvo parameter for price adjustment, θ_p, is 0.82 (an average five-quarter pricing cycle) and the indexation level χ is 0.63, although the posterior for this parameter is quite spread. The Calvo parameter for wage adjustment, θ_w, is 0.68 (wage decisions are revised every three quarters), while the indexation, χ_w, is 0.62. We see how the data push us considerably away from the prior. There is information to be learned from the observations. However, at the same time, the median of the posterior of the nominal rigidities parameters is relatively different from the median in the case with flat priors. This shows how more informative priors *do* have an impact on our inference, in particular in parameter values of key importance for policy analysis as θ_p and θ_w. We do not necessarily judge this impact in a negative way. If the prior is bringing useful information into the table (for example, from micro data on individual firms' price changes, as in Bils and Klenow, 2004), this is precisely what we want. Nevertheless, the researcher and the final user of the model must be aware of this fact.

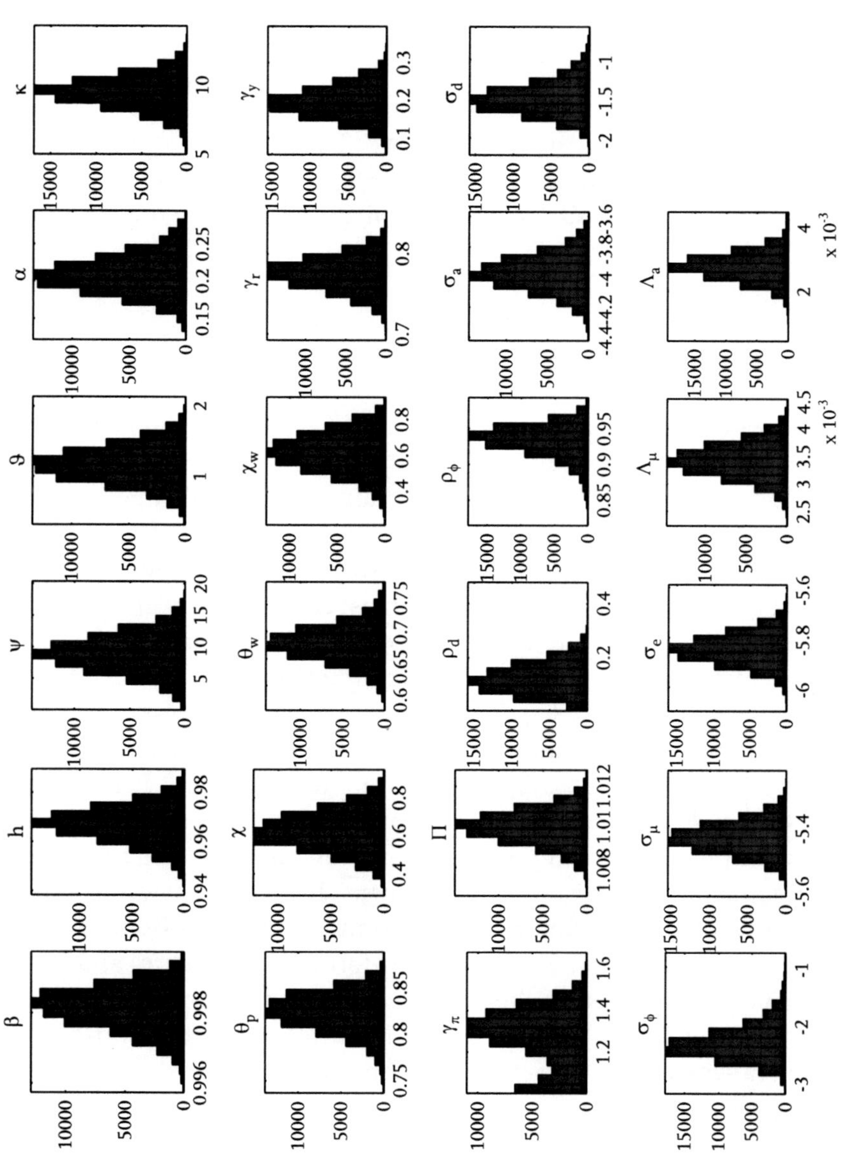

Fig. 15.2 Posterior distribution, Smets – Wouters priors.

The parameters of the policy rule $\{\gamma_R, \gamma_\Pi, \gamma_y, \Pi\}$ are standard, with the only significant difference that now the Fed is less responsive to inflation, with the coefficient $\gamma_\pi = 1.29$. However, this value still respects the Taylor principle (even the value at the 5 percentile does). The Fed smooths the interest rate over time (γ_R is estimated to be 0.79) and responds actively to inflation (γ_R is 1.25) and weakly to the output growth gap (γ_y is 0.19). We estimate that the Fed has a target for quarterly inflation of 1 percent.

The growth rates of the investment-specific technological change, Λ_μ, and of the neutral technology, Λ_A, are roughly equal. The estimated average growth rate of the economy in per capita terms, $(\Lambda_A + \alpha\Lambda_\mu) / (1 - \alpha)$ is 0.43 percent per quarter, or 1.7 percent annually, roughly the historical mean in the US in the period 1865–2007.

15.4 Lines of further research

While the previous pages offered a snapshot of where the frontier of the research is, we want to conclude by highlighting three avenues for further research. Our enumeration is only a small sampler from the large menu of items to be explored in the new macroeconometrics. We could easily write a whole chapter just signaling areas of investigation where entrepreneurial economists can spend decades and yet leave most of the territory uncharted.

First, we are particularly interested in the exploration of 'exotic preferences' that go beyond the standard expected utility function used in this chapter (Backus, Routledge and Zin, 2005). There is much evidence that expected utility functions have problems accounting for many aspects of the behaviour of economic agents, in particular those aspects determining asset prices. For example, expected utility cannot explain the time inconsistencies that we repeatedly see at the individual level (how long did you stick with your last diet?). Recently, Bansal and Yaron (2004) and Hansen and Sargent (2007) have shown that many of the asset pricing puzzles can be accounted for if we depart from expected utility. However, these authors have used little formal econometrics and we do not know which of the different alternatives to expected utility will explain the data better. Therefore, the estimation of DSGE models with 'exotic preferences' seems an area where the interaction between empirical work and theory is particularly promising.

Second, we would like to relax some of the tight parametric restrictions of the model, preferably within a Bayesian framework. DSGE models require many auxiliary parametric assumptions that are not central to the theory. Unfortunately, many of these parametric assumptions do not have a sound empirical foundation. But picking the wrong parametric form may have horrible consequences for the estimation of dynamic models. These concerns motivate the study of how to estimate DSGE models combining parametric and nonparametric components and, hence, conducting inference that is more robust to auxiliary assumptions.

Finally, a most exciting frontier is the integration of microeconomic heterogeneity within estimated DSGE models. James Heckman and many other econometricians have shown beyond reasonable doubt that individual heterogeneity is the defining feature of micro data (see Browning, Hansen and Heckman, 1999). Our macro models need to move away from the basic representative agent paradigm and include richer configurations. Of course, this raises the difficult challenge of how to effectively estimate these economies, since the computation of the equilibrium dynamics of the model is a challenge by itself. However, we are optimistic: advances in computational power and methods have allowed us to do things that were unthinkable a decade ago. We do not see any strong reasons why estimating DSGE models with heterogeneous agents will not be any different in a few years.

Appendix

A. Broader context and background

A.1 Policy-making institutions and DSGE models

We mentioned in the main text that numerous policy institutions are using estimated DSGE models as an important input for their decisions. Without being exhaustive, we mention the Federal Reserve Board (Erceg, Guerrieri and Gust, 2006), the European Central Bank (Christoffel, Coenen and Warne, 2007), the Bank of Canada (Murchison and Rennison, 2006), the Bank of England (Harrison *et al.*, 2005), the Bank of Sweden (Adolfson *et al.*, 2005), the Bank of Finland (Kilponen and Ripatti, 2006 and Kortelainen, 2002), and the Bank of Spain (Andrés, Burriel and Estrada, 2006), among several others.

A.2 Sequential Monte Carlo methods and DSGE models

The solution of DSGE models can be written in terms of a state space representation. Often, that state space representation is nonlinear and/or non-Gaussian. There are many possible reasons for this. For example, we may want to capture issues, such as asymmetries, threshold effects, big shocks, or policy regime switching that are approximated very poorly (or not at all) by a linear solution. Also, one of the key issues in macroeconomics is the time-varying volatility in time series. McConnell and Pérez-Quirós (2000) and Kim and Nelson (1998) presented extremely definitive evidence that the US economy had been more stable in the 1980s and 1990s than before. Fernández-Villaverde and Rubio-Ramírez (2007) and Justiniano and Primiceri (2008) demonstrated that DSGE models, properly augmented with non-Gaussian shocks, could account for this evidence.

Nonlinear and/or non-Gaussian state space representations complicate the filtering problem. All the relevant conditional distributions of states became non-normal and, except for a few cases, we cannot resort to analytic methods to track them. Similarly, the curse of dimensionality of numerical integration precludes the use of quadrature methods to compute the relevant integrals of filtering (those that appear in the Chapman-Kolmogorov formula and in Bayes' theorem).

Fortunately, during the 1990s, a new set of tools was developed to handle this problem. This set of tools has become to be known as sequential Monte Carlo (SMC) methods because they use simulations period by period in the sample. The interested reader can see a general introduction to SMC methods in Arulampalam *et al.* (2002), a collection of background material and applications in Doucet, de Freitas and Gordon (2001), and the first applications in macroeconomics in Fernández-Villaverde and Rubio-Ramírez (2005 and 2007). The appendix in that last paper also offers references to alternative approaches.

The simplest SMC method is the particle filter, which can easily be applied to estimate our DSGE model. This filter replaces the conditional distribution of states $\{p(S_t|\mathbb{Y}_{1:T-1};\Psi)\}_{t=1}^{T}$ by an empirical distribution of N draws $\left\{\left\{s_{t|t-1}^{i}\right\}_{i=1}^{N}\right\}_{t=1}^{T}$ from the sequence $\{p(S_t|\mathbb{Y}_{1:T-1};\Psi)\}_{t=1}^{T}$. These draws are generated by simulation. Then, by a trivial application of the law of large numbers:

$$\mathcal{L}(\mathbb{Y}_{1:T};\Psi) \simeq \frac{1}{N}\sum_{i=1}^{N}\mathcal{L}(\mathbb{Y}_1|s_{0|0}^{i};\Psi)\prod_{t=2}^{T}\frac{1}{N}\sum_{i=1}^{N}\mathcal{L}(\mathbb{Y}_t|s_{t|t-1}^{i};\Psi)$$

where the subindex tracks the conditioning set (i.e., $t|t-1$ means a draw at moment t conditional on information until $t-1$) and lower cases are realizations of a random variable.

The problem is then to draw from the conditional distributions $\{p(S_t|\mathbb{Y}_{1:T-1};\Psi)\}_{t=1}^{T}$. Rubin (1988) first noticed that such drawing can be done efficiently by an application of sequential sampling:

Proposition 15.1 Let $\left\{s_{t|t-1}^{i}\right\}_{i=1}^{N}$ be a draw from $p(S_t|\mathbb{Y}_{1:T-1};\Psi)$. Let the sequence $\left\{\widetilde{s}_t^{i}\right\}_{i=1}^{N}$ *be a draw with replacement from* $\left\{s_{t|t-1}^{i}\right\}_{i=1}^{N}$ *where the resampling probability is given by*

$$q_t^i = \frac{\mathcal{L}(\mathbb{Y}_t|s_{t|t-1}^{i};\Psi)}{\sum_{i=1}^{N}\mathcal{L}(\mathbb{Y}_t|s_{t|t-1}^{i};\Psi)}.$$

Then $\left\{\widetilde{s}_t^{i}\right\}_{i=1}^{N}$ is a draw from $p(S_t|\mathbb{Y}_{1:T};\Psi)$.

Proposition 15.1 recursively uses a draw $\{s^i_{t|t-1}\}^N_{i=1}$ from $p(S_t|\mathbb{Y}_{1:T-1};\Psi)$ to draw $\{s^i_{t|t}\}^N_{i=1}$ from $p(S_t|\mathbb{Y}_{1:T};\Psi)$. But this is nothing more than the update of our estimate of S_t to add the information on y_t that Bayes' theorem is asking for. Resampling ensures that this update is done in an efficient way.

Once we have $\{s^i_{t|t}\}^N_{i=1}$, we draw N vectors of the five exogenous innovations in the DSGE model (the two shocks to productivity, the two shocks to technology, and the monetary policy shock) from the corresponding distributions and apply the law of motion for states to generate $\{s^i_{t+1|t}\}^N_{i=1}$. This step, known as forecast, puts us back at the beginning of Proposition 15.1, but with the difference that we have moved forward one period in our conditioning, implementing in that way the Chapman – Kolmogorov equation. By going through the sample repeating these steps, we complete the evaluation of the likelihood function. Künsch (2005) provides general conditions for the consistency of this estimator of the likelihood function and for a central limit theorem to apply.

The following pseudo-code summarizes the description of the algorithm:

Step 0, Initialization: Set $t \rightsquigarrow 1$. Sample N values $\{s^i_{0|0}\}^N_{i=1}$ from $p(S_0;\Psi)$.

Step 1, Prediction: Sample N values $\{s^i_{t|t-1}\}^N_{i=1}$ using $\{s^i_{t-1|t-1}\}^N_{i=1}$, the law of motion for states and the distribution of shocks ε_t.

Step 2, Filtering: Assign to each draw $(s^i_{t|t-1})$ the weight ω^i_t in Proposition 15.1.

Step 3, Sampling: Sample N times with replacement from $\{s^i_{t|t-1}\}^N_{i=1}$ using the probabilities $\{q^i_t\}^N_{i=1}$. Call each draw $(s^i_{t|t})$. If $t < T$ set $t \rightsquigarrow t+1$ and go to step 1. Otherwise stop.

B. Model and computation

B.1 Computation of the model

A feature of the new macroeconometrics that some readers from outside economics may find less familiar is that the researcher does not specify some functional forms to be directly estimated. Instead, the economist postulates an environment with different agents, technology, preferences, information, and shocks. Then, she concentrates on investigating the equilibrium dynamics of this environment and how to map the dynamics into observables. In other words, economists are not satisfied with describing behaviour, they want to explain it from first principles. Therefore, a necessary first step is to solve for the equilibrium of the model given arbitrary parameter values. Once we have that equilibrium, we can build the associated estimating function (the likelihood, some moments, etc.) and apply data to perform inference. Thus, not only is finding the equilibrium of the model of key importance, but doing it rapidly and accurately, since we may need to do it for many different combinations of parameter values (for example, in an McMC simulation, we need to solve the model for each draw of parameter values).

As we described in the main text, our solution algorithm for the model relies on the perturbation of the equations that characterize the equilibrium of the model given some fixed parameter values. The first step of the perturbation is to take partial derivatives of these equations with respect to the states and control variables. This step, even if conceptually simple, is rather cumbersome and requires an inordinate amount of algebra.

Our favourite approach for solving this step is to write a `Mathematica` program that generates the required analytic derivatives (which are in the range of several thousands) and that writes automatic Fortran 95 code with the corresponding analytic expressions. This step is crucial because, once we have paid the fixed cost of taking the analytic derivatives (which takes several hours on a good desktop computer), solving the equilibrium dynamics for a new combination of parameter values takes less than a second, since now we only need to solve a numerical system in Fortran. The solution algorithm will be nested inside a Metropolis – Hastings. For each proposed parameter value in the chain, we will solve the model, find the equilibrium dynamics, use those dynamics to find the likelihood using the Kalman filter, and then accept or reject the proposal.

B.2 Kalman filter

The implementation of the Kalman filter to evaluate the likelihood function in a DSGE model like ours follows closely that of Stengel (1994). We start by writing the first order linear approximation to the solution of the model in a standard state space representation:

$$s_t = As_{t-1} + B\varepsilon_t \tag{15.9}$$

$$y_t = Cs_t + D\varepsilon_t \tag{15.10}$$

where s_t are the states of the model, y_t are the observables, and $\varepsilon_t \sim \mathbf{N}(0, I)$ is the vector of innovations to the model.

We introduce some definitions. Let $s_{t|t-1} = \mathbb{E}(s_t | Y_{t-1})$ and $s_{t|t} = \mathbb{E}(x_t | Y_t)$ where $Y_t = \{y_1, y_2, ..., y_t\}$. Also, we have $P_{t-1|t-1} = \mathbb{E}(s_{t-1} - s_{t-1|t-1})(s_{t-1} - s_{t-1|t-1})'$ and $P_{t|t-1} = \mathbb{E}(s_{t-1} - s_{t|t-1})(s_{t-1} - s_{t|t-1})'$. With this notation, the one-step-ahead forecast error is $\eta_t = y_t - Cs_{t|t-1}$.

We forecast the evolution of states:

$$s_{t|t-1} = As_{t-1|t-1}. \tag{15.11}$$

Since the possible presence of correlation in the innovations does not change the nature of the filter, it is still the case that

$$s_{t|t} = s_{t|t-1} + K\eta_t, \tag{15.12}$$

where K is the Kalman gain at time t. Define variance of forecast as $V_y = C P_{t|t-1} C' + DD'$.

Then, the conditional likelihood is just:

$$\log \mathcal{L} = -\frac{n}{2}\log 2\pi - \frac{1}{2}\log\det\left(V_y\right) - \frac{1}{2}\eta_t V_y^{-1}\eta_t.$$

The last step is to update our estimates of the states. Define residuals $\xi_{t|t-1} = s_t - s_{t|t-1}$ and $\xi_{t|t} = s_t - s_{t|t}$. Subtracting equation (15.11) from equation (15.9)

$$\begin{aligned} s_t - s_{t|s-1} &= A\left(s_{t-1} - s_{t-1|t-1}\right) + Bw_t, \\ \xi_{t|t-1} &= A\xi_{t-1|t-1} + Bw_t. \end{aligned} \tag{15.13}$$

Now subtract equation (15.12) from equation (15.9)

$$\begin{aligned} s_t - s_{t|t} &= s_t - s_{t|t-1} - K\left[Cx_t + Dw_t - Cx_{t|t-1}\right] \\ \xi_{t|t} &= \xi_{t|t-1} - K\left[C\xi_{t|t-1} + Dw_t\right]. \end{aligned} \tag{15.14}$$

Note $P_{t|t-1}$ can be written as

$$\begin{aligned} P_{t|t-1} &= \mathbb{E}\xi_{t|t-1}\xi'_{t|t-1}, \\ &= \mathbb{E}\left(A\xi_{t-1|t-1} + Bw_t\right)\left(A\xi_{t-1|t-1} + Bw_t\right)' \\ &= AP_{t-1|t-1}A' + BB'. \end{aligned} \tag{15.15}$$

For $P_{t|t}$ we have:

$$\begin{aligned} P_{t|t} &= \mathbb{E}\xi_{t|t}\xi'_{t|t} \\ &= (I - KC)\,P_{t|t-1}\left(I - C'K'\right) + KDD'K' - KDB' \\ &\quad - BD'K' + KCBD'K' + KDB'C'K'. \end{aligned} \tag{15.16}$$

The optimal gain minimizes $P_{t|t}$:

$$K_{opt} = \left[P_{t|t-1}C' + BD'\right]\left[V_Y + CBD' + DB'C'\right]^{-1}$$

and, consequently, the updating equations are:

$$\begin{aligned} P_{t|t} &= P_{t|t-1} - K_{opt}\left[DB' + CP_{t|t-1}\right], \\ s_{t|t} &= s_{t|t-1} + K_{opt}\eta_t \end{aligned}$$

and we close the iterations.

B.3 Construction of data

When we estimate the model, we need to make the series provided by the national and income product accounts (NIPA) consistent with the definition of variables in the theory. The main adjustment that we undertake is to express both real output and real gross investment in consumption units. Our DSGE model implies that there is a numeraire in terms of which all the other prices need to be quoted. We pick consumption as the numeraire. The NIPA, in comparison, uses an index of all prices to transform nominal GDP and investment into real values. In the presence of changing relative prices, such as the ones we have seen in the US over the last several decades with the fall in the relative price of capital, NIPA's procedure biases the valuation of different series in real terms.

We map theory into data by computing our own series of real output and real investment. To do so, we use the relative price of investment, defined as the ratio of an investment deflator and a deflator for consumption. The denominator is easily derived from the deflators of nondurable goods and services reported in the NIPA. It is more complicated to obtain the numerator because, historically, NIPA investment deflators were poorly constructed. Instead, we rely on the investment deflator computed by Fisher (2006), a series that unfortunately ends early in 2000Q4. Following Fisher's methodology, we have extended the series to 2007Q1.

For the real output per capita series, we first define nominal output as nominal consumption plus nominal gross investment. We define nominal consumption as the sum of personal consumption expenditures on nondurable goods and services. We define nominal gross investment as the sum of personal consumption expenditures on durable goods, private residential investment, and non-residential fixed investment. Per capita nominal output is equal to the ratio between our nominal output series and the civilian non-institutional population between 16 and 65. To obtain per capita values, we divide the previous series by the civilian non-institutional population between 16 and 65. Finally, real wages are defined as compensation per hour in the non-farm business sector divided by the CPI deflator.

Acknowledgement

We thank Alex Groves for his help with preparing the data used in the estimation. Beyond the usual disclaimer, we must note that any views expressed herein are those of the authors and not necessarily those of the Federal Reserve Bank of Atlanta or the Federal Reserve System. Finally, we also thank the NSF for financial support.

References

Adolfson, M., Laséen, S., Lindé, J. and Villani, M. (2005). Bayesian estimation of an open economy DSGE model with incomplete pass-through. Sveriges Riksbank. Working Paper Series 179.

An, S. and Schorfheide, F. (2006). Bayesian analysis of DSGE models. *Econometric Reviews*, **26**, 113–172.

Andrés, J., Burriel, P. and Estrada, A. (2006). BEMOD: A DSGE model for the Spanish economy and the rest of the euro area. Documento de Trabajo del Banco de España 0631.

Aruoba, S.B., Fernández-Villaverde, J. and Rubio-Ramírez, J. (2006). Comparing solution methods for dynamic equilibrium economies. *Journal of Economic Dynamics and Control*, **30**, 2477–2508.

Arulampalam, A.S., Maskell, S., Gordon, N. and Clapp, T. (2002). A tutorial on particle filters for online nonlinear/non-Gaussian Bayesian tracking. *IEEE Transactions on Signal Processing*, **50**, 174–188.

Backus, D.K., Routledge, B.R. and Zin, S.E. (2005). Exotic preferences for macroeconomists. *NBER Macroeconomics Annual*, 2004, 319–390.

Bansal, R. and Yaron, A. (2004). Risks for the long run: A potential resolution of asset pricing puzzles. *Journal of Finance*, **59**, 1481–1509.

Bils, M. and Klenow, P. (2004). Some evidence on the importance of sticky prices. *Journal of Political Economy*, **112**, 947–985.

Browning, M., Hansen, L.P. and Heckman, J.J. (1999). Micro data and general equilibrium models. In *Handbook of Macroeconomics*, (ed. J.B. Taylor and M. Woodford), vol. 1, pp. 543–633. Elsevier, Amsterdam.

Calvo, G. A. (1983). Staggered prices in a utility-maximizing framework. *Journal of Monetary Economics*, **12**, 383–398.

Canova, F. and Sala, L. (2006). Back to square one: Identification issues in DSGE models. Mimeo. Pompeu Fabra University.

Caplin, A. and Leahy, J. (1991). State-dependent pricing and the dynamics of money and output. *Quarterly Journal of Economics*, **106**, 683–708.

Chernozhukov, V. and Hong, H. (2003). A MCMC approach to classical estimation. *Journal of Econometrics*, **115**, 293–346.

Christiano, L., Eichenbaum, M. and Evans, C.L. (2005). Nominal rigidities and the dynamic effects of a shock to monetary policy. *Journal of Political Economy*, **113**, 1–45.

Christoffel, K., Coenen, G. and Warne, A. (2007). The new area-wide model of the euro area: Specification and first estimation results. *Mimeo*. European Central Bank.

Dotsey, M., King, R.G. and Wolman, A. (1999). State dependent pricing and the general equilibrium dynamics of money and output. *Quarterly Journal of Economics*, **144**, 655–690.

Doucet, A., de Freitas, N. and Gordon, N. (2001). *Sequential Monte Carlo Methods in Practice*. Springer Verlag, Berlin.

Edge, R., Kiley, M. and Laforte, J.P. (2008). A comparison of forecast performance between Federal Reserve staff forecasts, simple reduced-form models, and a DSGE model. Mimeo. Federal Reserve Board.

Erceg, C.J., Guerrieri, L. and Gust, C. (2006). SIGMA: A new open economy model for policy analysis. *International Journal of Central Banking*, **2**, 1–50.

Fernández-Villaverde, J. and Rubio-Ramírez, J. (2004). Comparing dynamic equilibrium models to data: A Bayesian approach. *Journal of Econometrics*, **123**, 153–187.

Fernández-Villaverde, J. and Rubio-Ramírez, J. (2005). Estimating dynamic equilibrium economies: Linear versus nonlinear likelihood. *Journal of Applied Econometrics*, **20**, 891–910.

Fernández-Villaverde, J. and Rubio-Ramírez, J. (2007). Estimating macroeconomic models: A likelihood approach. *Review of Economic Studies*, **74**, 1059–1087.

Fernandez-Villaverde, J. and Rubio-Ramirez, J. (2008). How structural are structural parameters? *NBER Macroeconomics Annual*, 2007, 83–137.

Fernández-Villaverde, J., Rubio-Ramírez, J. and Santos, M.S. (2006). Convergence properties of the likelihood of computed dynamic models. *Econometrica*, **74**, 93–119.

Fisher, J. (2006). The dynamic effects of neutral and investment-specific technology shocks. *Journal of Political Economy*, **114**, 413–52.

Greenwood, J., Herkowitz, Z. and Krusell, P. (1997). Long-run implications of investment-specific technological change. *American Economic Review*, **87**, 342–362.

Guerrón-Quintana, P. (2008). What you match does matter: The effects of data on DSGE estimation. Working Paper. North Carolina State University.

Judd, K.L. (1998). *Numerical Methods in Economics*. MIT Press, Cambridge, MA.

Justiniano, A. and Primiceri, G.E. (2008). The time varying volatility of macroeconomic fluctuations. *American Economic Review*, **98**, 604–641.

Hansen, L.P. (1982). Large sample properties of generalized method of moments estimators. *Econometrica*, **50**, 1029–54.

Hansen, L.P. and Sargent, T. (2007). *Robustness*. Princeton University Press, Princeton.

Harrison, R., Nikolov, K., Quinn, M., Ramsey, G., Scott, A. and Thomas, R. (2005). *The Bank of England Quarterly Model*. The Bank of England.

Kim, C. and Nelson, C.R. (1998). Has the U.S. economy become more stable? A Bayesian approach based on a Markov-switching model of the business cycle. *Review of Economics and Statistics*, **81**, 608–616.

Kilponen, J. and Ripatti, A. (2006). Introduction to AINO. Mimeo, Bank of Finland.

Kortelainen, M. (2002). EDGE: A model of the euro area with applications to monetary policy. Bank of Finland Studies E:23.

Künsch, H.R. (2005). Recursive Monte Carlo filters: Algorithms and theoretical analysis. *Annals of Statistics*, **33**, 1983–2021.

Kydland, F. and Prescott, E.C. (1982). Time to build and aggregate fluctuations. *Econometrica*, **50**, 1345–1370.

Lubick, T. and Schorfheide, F. (2004). Testing for indeterminacy: An application to U.S. monetary policy. *American Economic Review*, **94**, 190–217.

Lucas, R. (1972). Expectations and the neutrality of money. *Journal of Economic Theory*, **4**, 103–124.

Lucas, R. (1976). Econometric policy evaluation: A critique. *Carnegie-Rochester Conference Series on Public Policy*, **1**, 19–46.

McConnell, M.M. and Pérez-Quirós, G. (2000). Output fluctuations in the United States: What has changed since the early 1980's? *American Economic Review*, **90**, 1464–1476.

Murchison, S. and Rennison, A. (2006). ToTEM: The Bank of Canada's new quarterly projection model. Bank of Canada Technical Report, 97.

Muth, J. F. (1961). Rational expectations and the theory of price movements. *Econometrica*, **29**, 315–335.

Orphanides, A. (2002). Monetary policy rules and the great inflation. *American Economic Review*, **92**, 115–120.

Roberts, G., Gelman, A. and Gilks, W. (1997). Weak convergence and optimal scaling of random walk Metropolis algorthims. *Annals of Applied Probability*, **7**, 110–120.

Rubin, D.B. (1988). Using the SIR algorithm to simulate posterior distributions. In *Bayesian Statistics 3*, (ed. J.M. Bernardo, M.H. DeGroot, D.V. Lindley, and A.F.M. Smith), pp. 395–402. Oxford University Press, Oxford.

Sargent, T.J. (2005). An interview with Thomas J. Sargent by George W. Evans and Seppo Honkapohja. *Macroeconomic Dynamics*, **9**, 561–583.

Sims, C. A. (2002). Implications of rational inattention. Mimeo. Princeton University.

Smets, F. and Wouters, R. (2007). Shocks and frictions in US business cycles: A Bayesian DSGE approach. *American Economic Review*, **97**, 586–606.

Smith, A. (1776). *An Inquiry into the Nature and Cause of the Wealth of Nations.* Published by Clarendon Press (1979), Oxford, Glasgow Edition of the Works and Correspondence of Adam Smith, vol. 2.

Stengel, R.F. (1994). *Optimal Control and Estimation.* Dover Publications, New York.

Walras, L. (1874). *Elements of Pure Economics.* Republished by Harvard University Press (1954), Cambridge.

Woodford, M. (2003). *Interest and Prices: Foundations of a Theory of Monetary Policy.* Princeton University Press, Princeton.

PART III
Environment and Ecology

·16·

Assessing the probability of rare climate events

Peter Challenor, Doug McNeall, and James Gattiker

16.1 Introduction

16.1.1 The problem of rare climate events

Until recently, climate scientists have concentrated on demonstrating that global warming is a problem that needs to be addressed. This has involved predicting expected climate change conditional on some future greenhouse gas emissions scenario (e.g. IPCC 2007). The purpose of these studies has been to show the most probable consequences of anthropogenic (human induced) climate change. In addition to the widely predicted warming, there are other *possible*, albeit less likely, climate futures where dramatic changes occur. These are possible futures with 'low probability, high impact' events, which are often associated with tipping elements (Lenton *et al.*, 2008) in the Earth climate system. A tipping element is a part of the Earth system that shows some form of non-linear behaviour. A small external forcing can 'tip' the system into a different *type* of behaviour. Such a change often happens faster than the forcing (termed 'abrupt'), and may be irreversible, or at least show some hysteresis. Such events have been seen often in the past, and possible future events include; the collapse of the Greenland or West Antarctic Ice Sheets (Huybrechts and De Wolde, 1999), the loss of sea ice in the Arctic, the die-back of the tropical Amazon rainforest (Cox *et al.*, 2000) or the collapse of the thermohaline circulation in the North Atlantic (Rahmstorf and Ganopolski, 1999). All of these events, though unlikely, would have severe consequences for both the Earth system, and any society which had to cope with them. Some events would result in strong positive feedbacks to a changing climate; for example a die-back of the Amazon rainforest would hugely increase the concentration of carbon dioxide in the atmosphere, further increasing global warming. The most severe of these events could effect entire societies on a continental scale. If we informally define the *risk* of an unwelcome future climate event as the probability of its occurrence, multiplied by the damage that it would cause, it is arguable that we should devote significant resources to identify and study 'rare' events.

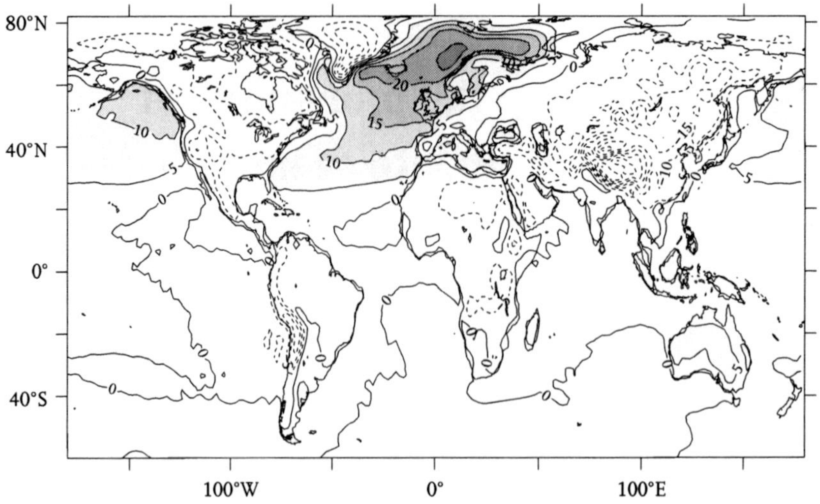

Fig. 16.1 The difference in mean winter surface air temperature from the zonal average (K).

In this paper we will look at statistical methods for estimating the probability of such rare events happening, using complex numerical models of the climate system. To illustrate the methods we will concentrate on a single such rare event – the potential collapse of the meridional overturning circulation in the Atlantic Ocean.

16.1.2 The meridional overturning circulation

A look at winter air temperature around the globe in Figure 16.1 confirms that the eastern North Atlantic region, including North West Europe, is significantly warmer than similar regions on the same latitude. This is not just the effect of the ocean; British Columbia, on the east side of the Pacific, is colder than western Europe at the same latitude. The reason is the meridional (North-South) overturning circulation in the Atlantic; the MOC. A crucial part of the MOC is driven by density differences due to heat and salt, and is known as the thermohaline circulation (THC). At high latitudes in the Atlantic the atmosphere is much colder than the ocean. This means that (i) sea ice is created by freezing and (ii) there is a strong flux of water from the ocean to the atmosphere. These processes make the water both cold – near freezing – and more saline. This cold, saline water is denser than the water beneath it, and so it sinks. To replace the surface water, less dense, warmer, fresher water is moves northward through the Atlantic. This flow of warm, fresh water at the surface of the Atlantic acts as a 'heat pump', and keeps Europe relatively warm in winter.

The dense water that has sunk to the bottom of the ocean makes the opposite, southward trip. Moving south though the Atlantic deep in the ocean it slowly warms and freshens, rising to the surface in the Southern Ocean and Pacific. This water is now the warm, fresh water that can be drawn into the Atlantic. The shallow northwards circulation, and deep southwards circulation can be thought of as a great ocean conveyor belt of heat and salt Broecker (1997). The full MOC is more complex than this simple conveyer belt; there is also a wind driven component that creates western boundary currents such as the Gulf Stream in the Atlantic, and the Kuroshio in the Pacific.

It is possible to model the thermohaline circulation simply, as two boxes representing the low and high latitude oceans, connected by 'pipes', carrying an overturning flow (Stommel, 1961). In this simple case, the so-called Stommel model, it is possible to show that there are equilibrium solutions to the system both with, and without, a thermohaline circulation. This raises questions about the possibility of the stable (equilibrium) solutions for the thermohaline circulation in the real world. Some studies have suggested that global warming would result in changes in both the heat driven (through surface heat flux), and salt driven (through increased fresh meltwater from Greenland, or enhanced rainfall) parts of the THC.

Is it feasible that the MOC is at risk of a collapse, and is global warming likely to make it happen in the near future? We can seek to answer these questions in two ways. The first is to reconstruct past climate, using palaeodata; data locked away for millennia in ice cores, deep ocean sediments and caves, for example. The second is to use our knowledge of the dynamics of the Earth system to construct models, and then simulate the future in computer experiments. Ultimately, these two activities overlap, as past climate can provide valuable out-of-sample data to test the models, and models can be the best method of reconstructing past climate.

The use of palaeoclimate data is a useful way to test our theories about how the climate system works and the mechanisms of climate change. Unfortunately there are no exact analogues for the current anthropogenic global warming so we cannot use past data to directly make predictions about what will happen in the future. However, there is good evidence that abrupt changes in the MOC system occurred in the past (Alley *et al.*, 2003). If we look at the time since the last ice age, the Holocene, we have good records of the temperature in Greenland from ice cores (Grootes *et al.*, 1993). These records show that about eight and a half thousand years ago there was a sudden cooling of the temperatures in Greenland. Other palaeodata around the North Atlantic show a similar cooling but the phenomenon was not global (Rohling and Palike, 2005). It is hypothesized that as the ice sheet retreated across North America, the ice dam holding back the huge Lake Agassiz broke, and vast amounts of fresh water flowed through Hudson Bay into the North Atlantic (Barber *et al.*,

1999). This low density, fresh water 'capped' the North Atlantic, stopping the deep convection and effectively shutting down the thermohaline circulation, causing rapid cooling. In this case the circulation recovered after a few hundred years.

To ascertain whether there is a significant risk that the thermohaline circulation could collapse in the near future (100–200 years) we need to use climate models. Studies using more complex models than the simple box model (Vellinga and Wood, 2002; Gregory *et al.*, 2005; Rahmstorf *et al.*, 2005) have shown mixed results in reproducing the nonlinear behaviour in the simple Stommel model, with the most complex models the least likely to see a collapse of the MOC.

16.2 Climate models

A climate model is primarily a tool for learning about the climate. It is an encoding of our knowledge about the way that the climate system works. The climate is a system where it is not possible to perform multiple, controlled, real-world experiments on a scale that encompasses the whole system,[1] and so we construct a climate model as a form of thought experiment.

These models can consist of several parts. A *conceptual model* of the dynamics (feedbacks, physics, chemistry, biology, and possibly, economics) of the Earth system is encoded in a *mathematical model* – usually a series of partial differential equations and equations of state. The mathematical model is then discretized and rendered as code, so that it can be solved on a computer. We can then perform experiments on this climate *simulator* within the confines of the computer, varying properties of the model at will. Some processes within the Earth system will be better understood than others; for example, geophysical fluid dynamics (e.g. Pedlosky 1992) gives us the tools to solve for the circulation of both the atmosphere and the ocean with a relatively low uncertainty. Other processes – parts of the carbon cycle, for example – are relatively unknown, and contain many simplifications or 'paramatrizations'.

16.2.1 How the climate system works

The Earth's climate is ultimately driven by energy from the Sun. In steady state, the amount of energy radiated and reflected back into space must equal the amount coming in from the Sun, otherwise the Earth would warm, or cool. Without an atmosphere, the Earth would behave as a 'grey' body, reflecting and emitting radiation according to it's albedo, and maintaining a (roughly)

[1] It is possible to regard the increase in greenhouse gas emissions as one, uncontrolled, high-risk experiment, of course.

steady temperature of −18 °C. The Earth's *actual* temperature is modified by the atmosphere. Incoming short-wave radiation is absorbed by the planet surface, and re-emitted as long-wave radiation. This long-wave energy is then 'trapped' by gas molecules in the atmosphere, and the resulting increase in average surface temperature to 14 °C is known as the 'greenhouse effect'. Every planet in the Solar System with an atmosphere is subject to a greenhouse effect, with the exact value of the steady state temperature increase determined in part by the chemical composition of the atmosphere. The gases responsible for the greenhouse effect on Earth include water vapour, carbon dioxide and methane. Changing the relative composition of the atmosphere, for example by increasing the concentration of these greenhouse gases, forces a change in the radiative balance of the Earth. Stopping the forcing, by stopping the release of greenhouse gases, for example, would lead to a recovery to a steady state albeit at a higher temperature. The radiative balance would not be achieved instantly however, because unlike the atmosphere, the ocean can store energy on long time-scales, and adjusts its temperature slowly. Thus, even if we ceased to emit any more greenhouse gases tomorrow it would take many decades for the system to reach equilibrium.

Solar radiation heats the tropics more than the poles, so that the atmosphere and ocean can be thought of as heat engines transferring heat from tropical to polar regions. As we have seen above there is also transfer of heat from the surface to the ocean depths. The simplest climate models have a single box representing the whole Earth, and balance the radiation coming in and out. These are known as energy balance models and, although they are surprisingly effective, they give us no information about the spatial distribution of climate variables caused by the transfer of energy over the surface of the planet. The most complex models are modified from atmospheric models used for weather forecasting. The resolution is often reduced, and components to model the oceans and sea ice, and in some cases the terrestrial and marine biosphere are added. These General Circulation Models (GCMs) take huge amounts of computer time on the largest computers currently available, often taking many weeks to simulate a hundred years of climate. Such models are *expensive*. In between these two extremes are the Earth system models of intermediate complexity (EMICs). These models are more complex than the simple radiation balance models but not as complex as a full GCM. They aim to represent as many of the important components of the Earth system as possible, without incurring the expense of the full GCM based models. They are solved on a geographical grid and usually resolve a number of layers in the ocean. Although mainly used for simulating climates on long time scales (thousands or tens of thousands of years) such models are fast enough to allow us to run large sets of experiments without the need of specialised computer hardware.

16.2.2 Climate scenarios

To run climate simulations into the future we need to be able to predict the important drivers, or forcings of the climate. Forcings such as those due to the variation in Earth's orbit are known with a very small uncertainty, far into the future. Much less easy to predict are those due to societal factors, such as greenhouse gas emissions, and land use changes. Ideally such forcings would be predicted by an integrated model, that not only modelled the Earth system but also the social and economic systems that interact with it. Such models are being developed but are still some years away from being usable, and the uncertainties associated with societal forcing are very large. The approach of the IPCC is to specify a range of indicative scenarios to give a possible time series of emissions through the 21st century (Nakicenovic and Swart, 2000). It is important to note that the scenario developers refuse to attach probabilities to these scenarios, and it is therefore impossible to integrate over them to obtain the 'expected' future.

16.3 Inference from climate models

16.3.1 Uncertainty in climate simulators

There are two main sources of uncertainty in our modelling. The first is *epistemic* uncertainty, deriving from our lack of knowledge about the system. The second is *aleatoric* uncertainty; this is the randomness in the natural world. The uncertainty in weather forecast for the next few days is believed to be epistemic; if we knew perfectly the state of the atmosphere and ocean at the start of our forecast, there would be little or no uncertainty in the forecast. Beyond a few days, the weather is chaotic; that is, tiny perturbations in the initial conditions lead to very different and fundamentally unpredictable, states of the system (Lorentz, 1963). The uncertainty in the weather beyond this horizon of a few days has become aleatory. The climate can be thought of as the *distribution* of weather, and so on small scales of time and space, the uncertainty is aleatory in nature. Over larger scales however, the climate is driven by the long term energy balance and dynamics, or the *boundary conditions*. The uncertainty has once again become epistemic.

The epistemic uncertainty can itself be split into two parts. There is structural uncertainty which is the uncertainty that arises because we do not know the *form* of the model. We can start to estimate this structural error by looking at data from the system we are trying to model. As an example, imagine predicting the temperature in the Northern Hemisphere in winter; an intermediate complexity model may be too simple to include seasonal cycle, so we could include a *model discrepancy* term; an offset, with associated uncertainty, that took us from the annual mean temperature to a seasonal one. We could estimate this discrepancy by comparing the model output with observations. However, it is more difficult to get at the possible model discrepancy in modelling the future, as we have no access to such calibration or validation data. If the future relationship between the annual mean temperature and the seasonal temperature changes in an unpredictable fashion, our estimates – including our uncertainty – will be wrong. We can also examine the structural uncertainty by examining multimodel ensembles (Furrer *et al.*, 2007). Here, the same climate is predicted by different versions of models, or even those built by different institutions and people. There are several problems with this. The first is that many models share a common underlying *conceptual model*, which may be wrong. Models are often built in collaboration between centres, developed from older, simpler models, or perhaps even share some of the code that performs the climate calculations. Because of this, the models are not *independent* samples from the possible 'model space', and are unlikely even to span this space. There have been attempts to define a model independence metric, but drawing inference from groups of models will continue to be an active area of research for some time.

The second form of epistemic uncertainty is parameter uncertainty. Climate models contain a number of *parametrizations*, that represent simplifications of processes within the real system. The paramatrizations are often expressed as numerical inputs[2] to the model, and are often treated as variable to some degree. The relationship between the parameters and the model output can be represented by the mapping $Y = g(X)$, where Y is the climate model output, g is a function that represents the climate model, and $X \in \mathcal{X}$ is a vector of parameters. The parameters, or inputs X can be 'tuned' to find X^*, the 'best' input configuration, so that the model reproduces some observed data well. In most situations, however, it is better to allow uncertainty in X, finding a distribution $Pr(X^*)$ rather than committing to a single best input configuration. Individual components of X may be relatively well known, for example the acceleration due to gravity. Others may be hard to measure in the real world, or very uncertain in the future. Uncertainty about these inputs propagates through the model g to give us uncertainty on the outputs Y.

16.3.2 The role of data in climate prediction

A major problem with making predictions of climate is the absence of data that are directly relevant to the predictions we are trying to make. Due to the timescales of climate, and unlike weather prediction, we have no chance of directly testing our predictions with validation data. Palaeodata are often very uncertain, and no climate of the past is a direct analogue to that of the present day or the immediate future. If we can build a climate model from first

[2] Generally, parameters are time independent inputs.

principles that also simulates known climates, we can have some confidence that it will do a good job predicting the future. There is no guarantee that this is correct, however if we have a model which cannot reproduce the known climate, we have little confidence in it and would wish to down weight or discard its predictions. The role of data is to help us *learn* which model structures and parameter values best simulate the climate we know. Learning about the model structure often occurs on an ad hoc basis, in the process of *model development*. Learning which choice of parameter values lead to a given model best reproducing climate is the process of *calibration*. We link our uncertainty in inputs to the climate data through the output of the climate model, and modify our prior distributions for the model parameters using Bayes' theorem. To find the probability of the future state being in some state (a collapsed MOC at 2100, for example), we integrate over the posterior model input space where our model output shows a corresponding collapse. To perform the integral, we must specify the model discrepancy, and include uncertainty due to the error in observations.

16.3.3 Monte Carlo methods

If the climate model were very cheap to run, an effective way of performing the integral over the model input space might be via Monte Carlo methods. Sampling randomly from the uncertainty distributions of the inputs, we run an *ensemble* of model simulations at the chosen points in input space. The output of the model can be estimated over all of the input space by taking some average of the output at the chosen points. The mean of the output $E[g(X)]$ can be evaluated by drawing samples x_i, $i = 1, \ldots, N$ from our input distributions, and then approximating

$$E[g(X)] \approx \frac{1}{N}\sum_{i=1}^{N} g(x_i). \tag{16.1}$$

As is well known the variance of this estimate reduces as

$$\frac{\sigma^2}{N} \tag{16.2}$$

where σ^2 is the variance of the function, g(.). The usefulness of such naive Monte Carlo methods is limited because a climate model is often too expensive to run a large number of times.

16.3.4 Emulators

As we cannot run the expensive climate model as many times as we need to to make inferences about the climate system, we use statistical methods developed for such complex computer codes (Kennedy and O'Hagan, 2001).

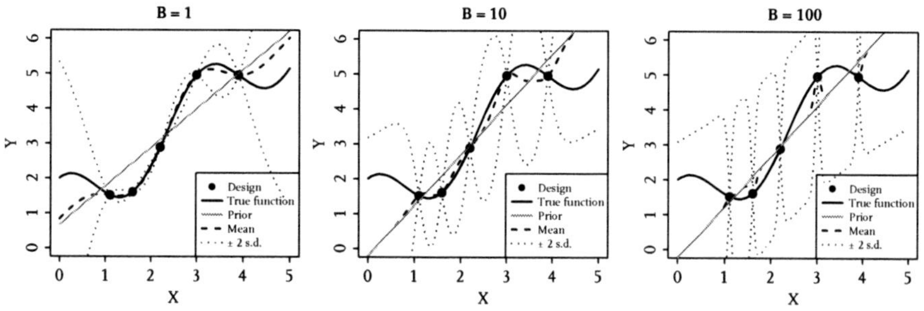

Fig. 16.2 Gaussian process emulator behaviour is dependent on hyperparameters; in this case, smoothing parameter B. The function $Y = 1 + X + \cos(2X)$ is approximated with a GP emulator trained at five design points.

These are based on the idea of an emulator; a 'cheap' statistical approximation to the model. We run the full climate model in a designed experiment and use the results to build our emulator. Any inference, by Monte Carlo sampling for example, then uses the emulator in place of the model itself. The emulator describes what (Kennedy and O'Hagan, 2001) call *code uncertainty*, that is, our uncertainty about the model output because we have a limited number of samples. The emulator describes the full posterior distribution for the output of the model at any input, given the training sample of inputs $\mathbf{D}$ and outputs Y. Because the model is deterministic, we know the output at the training sample, and so the uncertainty at those points is zero. The uncertainty between the training points expands, according to the distance from the nearest training point, and the smoothness of the response of the model. We use a Gaussian process emulator, seen in use in an example in Figure 16.2, and described in the appendix.

16.4 Application to the collapse of the meridional overturning circulation

We run an experiment to combine information from experts, with that of observed climate data, to simulate the evolution of the MOC through the 21st century. We take into account uncertainty due to the small number of expensive simulations that we can afford to run.

16.4.1 GENIE-1

We use an ensemble of simulations of the 'Grid ENabled Integrated Earth system model', GENIE-1 (see Edwards and Marsh (2005), and `http//www.genie.ac.uk` for full details). GENIE-1 is a highly computationally efficient EMIC. It is capable of completing a several thousand year integration

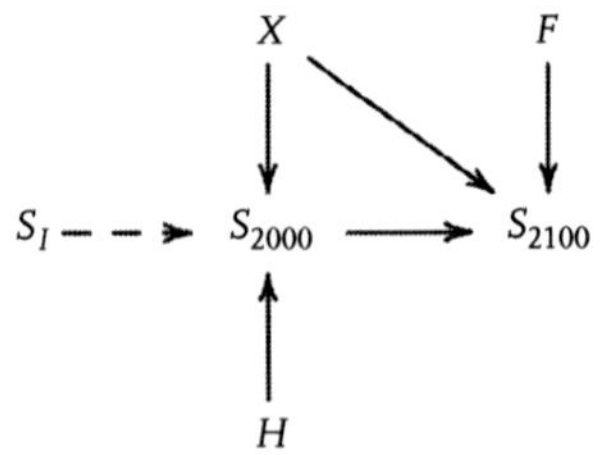

Fig. 16.3 Schematic of the experiment to simulate the MOC over the 21st century. X is a 16-dimensional vector of inputs to the model, and H is the historic CO_2 forcing for the years 1800–2000. F is a two-dimensional vector that controls forcing over the 21st century. S is the state of the system, with subscript indicating the time. We hope that spinning up the model for 4000 years removes the dependence on initial conditions S_I, so S_{2000} depends solely on the inputs X and forcing H.

in a few hours on a UNIX workstation. The model features a reduced physics (frictional geostrophic) 3D ocean circulation component coupled to a 2D energy-moisture balance (EMBM) atmosphere and a dynamic-thermodynamic sea ice model. The model grid differs from that of Edwards and Marsh (2005), in that there is increased resolution in longitude, and slightly decreased in latitude. This version of GENIE has 64 longitudes and 32 latitudes, uniform in the longitude and sin(latitude) coordinates, giving boxes of equal area in physical space. There are eight depth levels in the ocean on a uniformly logarithmically stretched grid, so that the box depth increases from 175 m to 1420 m. GENIE-1 is *deterministic*; when run with the same input parameters, it always produces the same output. A consequence of this is that stochastic processes (e.g. weather) are not represented in the model. The version we use does not include any seasonal variation.

16.4.2 Structure of the experiment

Our experiment is in two parts; the first is an initialisation, or 'spin-up' to bring the state of the model to a state representing the present day. The second part is a projection of this model state into the future. A schematic of the structure can be seen in figure 16.3.

After identifying 16 uncertain model inputs, we vary them together according to a space-filling Latin hypercube design (McKay *et al.*, 1979). The inputs are perturbed randomly and uniformly across an input space identified by one of GENIE-1's developers (Marsh) as beyond the maximum plausible region where the model is of use in understanding the climate. It is important to note that this design is not meant to represent the true uncertainty associated with the input parameters, except in a very broad sense; the design is primarily to learn as much about the behaviour of the model as possible, in a plausible input region. We use a *maximin* variant of the Latin hypercube design, by choosing the example with the largest minimum distance between pairs of points, from 10,000

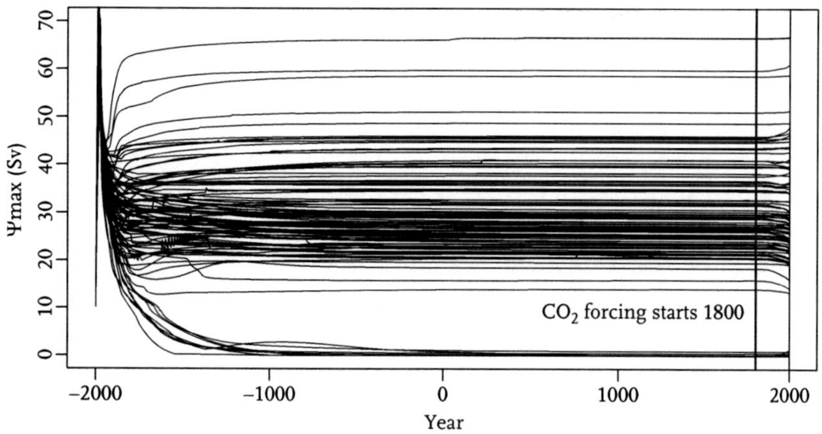

Fig. 16.4 Trajectories of the maximum meridional overturning circulation (Ψ_{max}) in the spin-up phase of an ensemble of GENIE-1.

randomly generated candidates. This type of design is a compromise, selected on a number of criteria including (i) space filling properties – important both in sampling input space $\mathcal{X}$ as efficiently as possible, and reducing numerical problems with the inversion of large matrices with nearly identical rows in the later emulation process; (ii) 'wide' boundaries, to ensure that the emulator is interpolating and not extrapolating, where it is expected to perform poorly; and (iii) ease and speed of calculation.

The spin-up of the model is important, not just as a prerequisite to our further experiment, but in its own right as an opportunity to examine the climate in the model in equilibrium or near-equilibrium. The ensemble of trajectories of the maximum value of the MOC in the North Atlantic, Ψ_{max}, throughout the spin-up phase of the experiment can be seen in Figure 16.4. Most of the trajectories follow a typical pattern; a start from the 'bland earth' value of between 10 Sv and 19 Sv is followed by a rapidly increasing overturning. After a peak, the overturning generally reduces to nearer an equilibrium value after 200 years. The MOC appears unstable at this point, as some ensemble members show a collapse in overturning circulation, continuing at a very low overturning throughout the remainder of the spin-up phase. The majority of more stable ensemble members have reached a near-equilibrium overturning after approximately 1000 years. Of the 108 original members of the ensemble design, eight collapse early in the integration, and a further 12 fail, usually due to numerical error.

We designate the model year 3800 as the calendar 1800AD. From this date the model is forced with historical atmospheric CO_2 concentrations, based initially on ice core data (Etheridge *et al.*, 1998) and then on direct atmospheric measurements from Mauna Loa (Keeling and Whorf, 2005). The forcing perturbs the MOC trajectories from equilibrium. Many of the ensemble members with less vigorous overturning exhibiting a significant and accelerating reduction. Ensemble members with a higher equilibrium overturning tend to increase in strength when forced with higher CO_2 concentrations. There are no stable ensemble members with an overturning strength between zero (or close to it) and around 10 Sv. As the ensemble is carefully designed to sample parameter space as evenly as possible, it is likely that this is an emergent property of the model, and not a sampling error.

Surviving members of the ensemble at the end of the spin-up are then run on to simulate the 21st century. Emissions of CO_2 into the atmosphere during the 21st century are varied according to the Intergovernmental Panel on Climate Change (IPCC) *Special Report on Emissions Scenarios* (SRES) (Nakicenovic and Swart, 2000). These are six indicative and self-consistent 'emissions pathways': scenarios of greenhouse gas emissions that take into account possible energy use, economic development and various other factors. These scenarios only give the anthropogenic emissions. To convert these into atmospheric concentrations we need to model the natural removal of CO_2, both by the terrestrial and oceanic ecosystems. We do this though a simple exponential decay in the concentration. The possible response of the Greenland Ice Sheet to warming in the Arctic region is accounted for via a parameter which increases freshwater input to the North Atlantic Ocean as the Greenland Ice Sheet melts with increased warming. The SRES authors are also careful not to make statements regarding the probability of any of the scenarios being more correct than any other. We therefore run our model ensemble through the 21st century using three indicative scenarios, and choose scenario A1B for illustrative diagrams.

This is our first formal experiment with this version of the model, and it is important to note that the design of experiment, and associated ensemble is exploratory in nature. Previous work looking at this problem (Challenor *et al.*, 2006) used a version of the model with a $36 \times 36 \times 8$ grid. Our work builds on this analysis, but the version of GENIE we are using behaves significantly differently from this earlier version.

16.4.3 Eliciting bounds on input parameters

The output of the model ensemble does not represent our uncertainty about the way that the true system will behave. We have further sources of information about the model output's ability to represent reality, in the form of (i) expert prior information about good values of the input parameters, and (ii) data from the real system at the end of spin-up, that can be used to calibrate the model. The expert information would ideally take the form of a full, multivariate prior input distribution, and should be elicited from the model developer before any output from the climate model is seen, or taken as a calibration exercise from a previous version of the model.

In reality, only the first few moments of the marginal input distributions, perhaps with some idea about correlation between a few of the more important inputs, can be elicited from a model expert. Due to the nature of model development, it is very likely that output from the model will have been seen by the developer, and therefore a certain amount of 'double counting' will take place in choosing good inputs. The effect can be to produce overconfidence in the set of input distributions, leading to a corresponding overconfidence in predictions of the model. Climate modellers often use a set of uniform prior distributions over fairly arbitrary input ranges as a more 'objective' alternative to a fully Bayesian treatment. This can be a poor choice, as such distributions often do not describe their prior beliefs or knowledge about what constitutes a 'good' model input.

We find that climate model developers often have a good idea of the 'best' value of an input parameter, as well as a firm 'range of applicability' for many inputs, beyond which confidence that the model can accurately describe climate diminishes. We can choose input distributions that are centre weighted, for example triangular or trapezoidal, or even Gaussian. The latter are unbounded, which can be problematic if there is a natural threshold within the input parameter space (e.g. a zero). This problem can be solved by transforming the input space, for example eliciting a distribution for the log of the input, or truncating a Gaussian distribution. We choose to model the expert prior knowledge as a set of independent Beta distributions with $\alpha = 2, \beta = 2$, as these are centre weighted, and probability density can be set to zero at a defined threshold. A priori there is no reason to believe that the 'best' value should be in the centre of the range and an asymmetric Beta might be a better choice – our expert was happier with the symmetric distributions.

16.4.4 Simulations of the MOC in the 21st century

Many model runs show a significant weakening of the MOC by the end of the 21st Century (Figure 16.5). If we are interested in the 'collapse' of the MOC, we need to define what we mean by collapse. The simplest answer would be to define collapse as the reduction of the strength of the overturning to zero; however, a severe reduction in the strength of the overturning would also have a significant impact on the climate. Challenor *et al.* (2006) define collapse as the strength of the overturning being less than 5 Sv at 2100. We can imagine future scenarios where this information does not fully describe the state of the climate, and its impacts. A trajectory which included a sudden collapse at 2030 could have much more profound consequences for society than one where there was a steady decline in overturning through the 21st century. Alternatively, compare the case where the overturning just dips below 5 Sv at 2100 and then recovers,

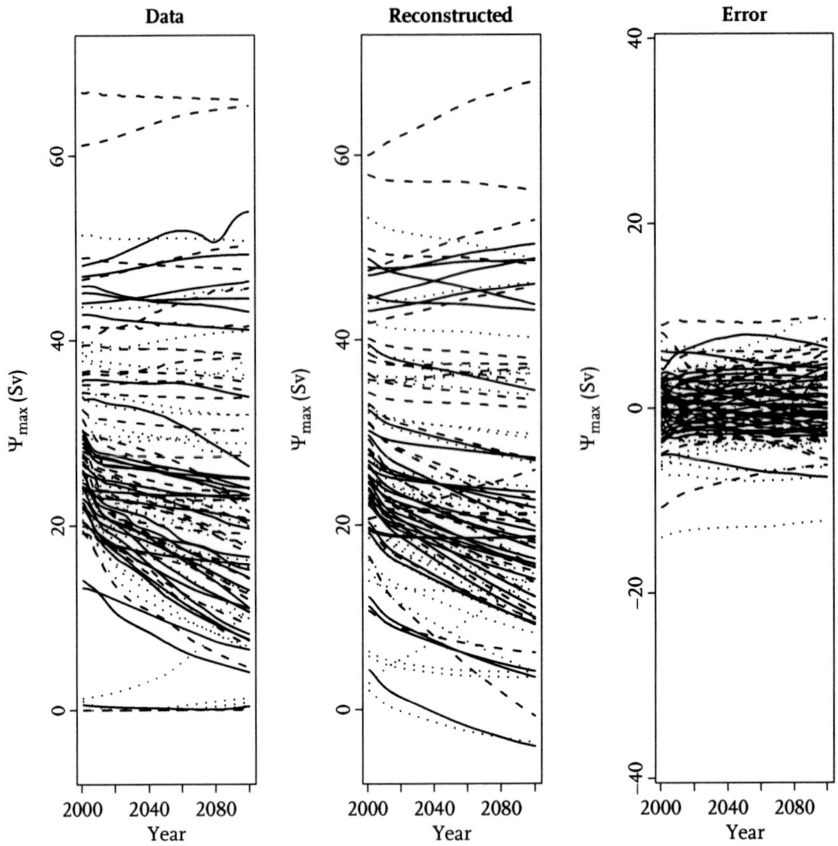

Fig. 16.5 Maximum meridional overturning circulation over the 21st Century, SRES scenario A1B (left). A leave-one-out cross validation reconstruction of the ensemble (centre), and the error, calculated as the difference between the two (right).

with a climate where the strength of the overturning at 2100 is 5.01 Sv but continues to reduce into the future; the latter is a more long term problem. In this paper we therefore reject the idea of defining collapse as simply a measure of the overturning at a specified point in time and instead investigate the concept of climate trajectories. In this study, the training data (ensemble) would appear to rule out a possible abrupt collapse of the system. However, there is an absence of ensemble members with a stable MOC below 10 Sv in the equilibrium spin-up part of the experiment. This suggests that, given continued forcing, an ensemble member that goes below this level is likely to continue to reduce. We suggest then, that 10 Sv may be a 'tipping point', or threshold in the system, and as such we are interested in trajectories that go below this level whichever year that this occurs. Our training data now encompasses the state of the MOC in every ensemble member, *at every year over the entire 21st century* – a dataset with a possible dimensionality of 100. For this reason, we need to construct a multivariate emulator, able to predict climate model output of very high dimension.

16.5 Multivariate and high dimensional emulators

16.5.1 Constructing a high dimensional emulator

We wish to predict the output Y of the climate model g at a previously untried input configuration x. The model output is usually a vector of very high dimension, containing fields and time series of many climate variables. In our case, each output vector contains a length 100 time series of the strength of the MOC through the 21st century. Previous studies using a Gaussian process emulators have been limited to scalar output; choosing a single representative output (Challenor *et al.*, 2006), or summary value of model output, such as a mean (Hankin, 2005). One approach to emulation of high dimensional output would be would be to model each point in the timeseries independently, using time (or location) as a further input to the emulator. Rougier (2007) points out that this would be to ignore the (typically positive) covariances between outputs close to one another and may lead to systematic errors in the emulator, and perhaps larger uncertainty than necessary.

We believe the climate model output (and by implication the climate) lies on a low dimensional manifold embedded within the high dimensional output space. We know that the MOC in any given ensemble member at time $t + 1$ will be very similar to that at time t. This correlation lets us reduce the dimensionality of the output data to q, a manageable number of outputs. We then build q separate Gaussian process emulators and combine them to predict model output of high dimension. Decomposition of high dimensional training data to emulate high dimensional functional output is also used by Higdon *et al.* (2008) and Bayarri *et al.* (2007).

The training data is $\mathbf{Y}$, a centred output matrix of n ensemble members y_i, $i = 1, \ldots, n$ in the rows, each a vector of length p. An ensemble member can be represented

$$y_i = \mu + \sum_{j=1}^{n} \lambda_{ij} v_j \tag{16.3}$$

where μ is the ensemble mean vector, v_j are the columns of $\mathbf{V}$, a set of n basis vectors and λ the reduced dimension *components* of $\mathbf{Y}$. We choose λ and $\mathbf{V}$ so that a few components retain a large portion of the variability of $\mathbf{Y}$ of interest. Truncating $\mathbf{V}$ and λ to the q most important basis vectors and components, the ensemble member can be reconstructed with some error, ϵ. The vector λ_j describes the behaviour of the simulator projected into the space defined by the basis vector v_j. Each vector of components $\lambda_1, \ldots, \lambda_q$ in turn substitutes $\mathbf{Y}$ as training data, so that q separate Gaussian process emulators (see appendix) are built. The emulators are used to estimate $\hat{\lambda}_j$ for an untested proposal input x. High dimensional output at a proposed input can be constructed

$$\hat{y} = \mu + \sum_{j=1}^{q} \hat{\lambda}_j(x)v_j + \epsilon \tag{16.4}$$

where ϵ is an error term and $q < n \ll p$. These components are constrained to be orthogonal, or independent. The error component ϵ can be decomposed into a *truncation error*, ϵ_{trunc}, due to the use of $q < n$ components, and *component error*, ϵ_{comp}, due to imperfect estimation of those components that are used for reconstruction.

$$\epsilon = \epsilon_{\text{comp}} + \epsilon_{\text{trunc}} \tag{16.5}$$

where

$$\epsilon_{\text{comp}} = \sum_{j=1}^{q} (\hat{\lambda}_j(x) - \lambda_j(x))v_j, \tag{16.6}$$

and

$$\epsilon_{\text{trunc}} = \sum_{j=q+1}^{n} \lambda_j(x)v_j. \tag{16.7}$$

It is important to choose enough components (sufficient q) to ensure that the dimension reduction step does not lose too much information about the climate model behaviour, and thus minimize ϵ_{trunc}. It is also important to have enough ensemble members to properly sample the possible model behaviour, build a good emulator and thus minimize ϵ_{comp}. There are several choices for the dimension reduction step; *Independent Components Analysis*, Hyvarinen and Oja (2000), would ensure that the components are both independent and uncorrelated. Bayarri *et al.* (2007) use a wavelet decomposition of functional output. We find that *Principal Components Analysis* (PCA, e.g. Jolliffe, 2002). via the singular value decomposition is quick, and has the advantage that the first few components explain the majority of the variance across the ensemble. A visual inspection of the first few PCs in a scatter plot matrix ensures that the components are approximately independent. However this independence is only guaranteed across the training set, not across all of output space. Because our original design has attempted to span output space we assume that the property of independence can be carried across to the full output space. This assumption is very difficult to test. Performing the calculations with subsets of the training data show that our components are robust within the training set, giving us some confidence that their properties will carry across the whole of output space.

16.6 Uncertainty analysis

We perform a calibrated uncertainty analysis of the MOC through the 21st century in three steps.

1. We calibrate the model (find $Pr(X*)$) using observational data from the 20th century, and a univariate Gaussian process emulator trained on model spin-up data representing the same period.
2. We use the trajectories of maximum MOC, Ψ_{max}, through the 21st century as training data for a high dimensional Gaussian process emulator.
3. We then use the high dimensional emulator to project our calibrated input distribution into model output space corresponding with the MOC through the 21st century.

16.6.1 Calibration of GENIE-1 using MOC data

We use a simple rejection sampling approach, similar to the method of Smith and Gelfand (1992) to generate input samples from the 'best input' posterior distribution $Pr(X^*)$. This approach can be summarized as 'sample from the prior and weight by the likelihood'. We use the prior distributions for the inputs elicited from our climate modeller, and construct a likelihood based on the observational data of the MOC.

The overturning circulation at a latitude of around 25 °N has been estimated by hydrographic section at various times during the last half-century, most recently in 2004 by Bryden *et al.* (2005), who compared their results with previous cruises. Cunningham *et al.* (2007) gives an estimate for the strength of the overturning in 2006 but their methods are different those used by Bryden *et al.* (2005) and are not directly comparable. Bryden *et al.* (2005) found that the overturning had weakened by around 30% during the previous five decades. The data are a sparse timeseries of five measurements of the MOC from 1957 to 2004. The 'best estimate' of each data point is pooled to calculate a mean $m_\psi = 18.4$ Sv, and standard deviation $\sigma_\psi = 3.1$ Sv. (1 Svedrup (Sv) = $10^6\,m^3s^{-1}$). This ignores the stated uncertainty of 6 Sv in the observed data, and any model discrepancy (for example that caused by the difference between the MOC at 26 °N, and the *maximum* MOC, as represented by the model). However, it is reasonable when considering the variability and estimate of the MOC in 2006, made by Cunningham *et al.* (2007), of mean 18.7 Sv, with a standard deviation of 2.8 Sv.

16.6.2 Rejection sampling

We train a Gaussian process emulator, on output from the 92 members of the ensemble that complete their run to the end of the 21st century. The output is Ψ_{max}, a scalar representing the maximum value in Sverdrups of the overturning circulation in the North Atlantic, extracted at the year 1998. This is near the end of the spin-up period, and so the ensemble varies 16 input parameters over the design space $\mathcal{X}_{\mathbf{D}}$.

The posterior input distributions are approximated by Monte Carlo integration, using the following procedure:

1. Take a large sample $x_1, \ldots, x_i$ from the prior input distribution.
2. For each x_i, take a single sample y_i that represents a model output corresponding to historical data from the emulator posterior distribution $Pr(y_i|x_i)$.
3. Calculate the likelihood, based on distance of sample y_i from the historical data, and the model discrepancy, and accept x_i with a probability that is proportional to the likelihood.

For the likelihood function, we use a normal density, based on mean and standard deviation of the observational data. After generating a sample x_i from the prior distribution the output sample y_i is drawn from emulator posterior t distribution, approximated by a normal distribution. We generate a likelihood density for y_i based on the probability density function of the normal distribution, with mean m_ψ and standard deviation σ_ψ:

$$\phi(y_i) = \frac{1}{\sigma_\psi + \sigma_\epsilon\sqrt{2\pi}} \exp\left(-[y_i - (m_\psi + m_\epsilon)]^2 \Big/ \left(2\left[\sigma_\psi^2 + \sigma_\epsilon^2\right)\right]\right). \tag{16.8}$$

Because of missing physical processes and the sheer complexity of the climate system there will be a *discrepancy* between what the model produces and the real world. If we had data on future discrepancies, we could model them as another Gaussian process, as in Kennedy and O'Hagan (2001). However we have no data for climate predictions such as these – our estimates of model discrepancy have to come from our experts. They are unlikely to be able to specify the discrepancy to this level of detail so we may have to model it as a simple uncertainty, increasing the variance of our predictions (although they may be able to say if they believe there will be a bias). We could therefore include a simple model inadequacy term, with mean m_ϵ and standard deviation σ_ϵ. We have no evidence to suspect that the model has a systematic bias, and the aforementioned sparsity of data means we have little with which to estimate a model inadequacy term for the historical behaviour. We have not taken this step here so our inferences are strictly speaking on a calibrated model rather than the real world.

We generate 100,000 samples from the prior distribution for the inputs, and reject around 80% of them to form a sample of approximately 20,000 input configurations, to approximate the posterior distribution of X^*. The 80% of rejected runs are the proportion of climates that have been simulated from the input distributions elicited from the expert are not close enough to the data to be regarded as good representations of the present day MOC. Thus the 20% is a measure of how well the elicited distributions correspond to the data. Once we have a calibrated input distribution, we can project the corresponding model output into the future via an emulator trained on the model future.

Given that we can, fairly accurately, simulate (and emulate) the present day strength of the MOC, does this imply that we can accurately predict the future strength of the MOC and in particular the probability of MOC collapse? The answer is undoubtedly; not necessarily. It would be easy to build a model that reproduced the present day data but failed miserably to predict the future. Our complex simulators are designed to include the physics of those parts of the climate system that our experts believe are important in predicting the future of the MOC. The model we are using is an intermediate complexity model so does not have the fullest and most accurate representation of Earth system processes available. However, our experts believe it should give reasonable answers. In some ways, the intermediate complexity model it is better than many IPCC models (IPCC, 2007) in that it includes a crude model of the melting of the Greenland ice sheet, and has been somewhat validated in experiments to simulate long periods of past climate.

16.6.3 Training and verifying the high dimensional emulator

Transforming the model output through the 21st century, $\mathbf{Y}_f$ via principal components analysis, we reduce the dimensionality of the data from 100 time-steps to two curves plus a mean term (Figure 16.6). The first principal component accounts for 99% of the variance of the output across the ensemble, while the second accounts for just under 1%. After building a high-dimensional Gaussian process emulator, we can measure the its performance using leave-one-out cross-validation. Each member in turn is removed from the the ensemble, and then estimated by an emulator trained on all of the other members. The estimated ensemble, and its difference from the original data can be seen in Figure 16.5. Some estimated ensemble members have a large error, but we find that these are usually from the extremes of the input design space, simulate past climate poorly, and are therefore ruled out by calibration.

16.6.4 Estimating the probability of MOC slowdown

We calculate a calibrated prediction for Y_f, the overturning circulation through the 21st century, by running the emulator at the calibrated input samples. We introduce high and low scenario distributions for the two extra parameters controlling the Greenland melt rate and atmospheric CO_2 removal rate. The uncertainty analysis uses the method of Oakley and O'Hagan (2002), generating 20 *simulated design points* from the emulator posterior distribution. We repeating this sampling 30 times, rebuilding the posterior emulator distribution

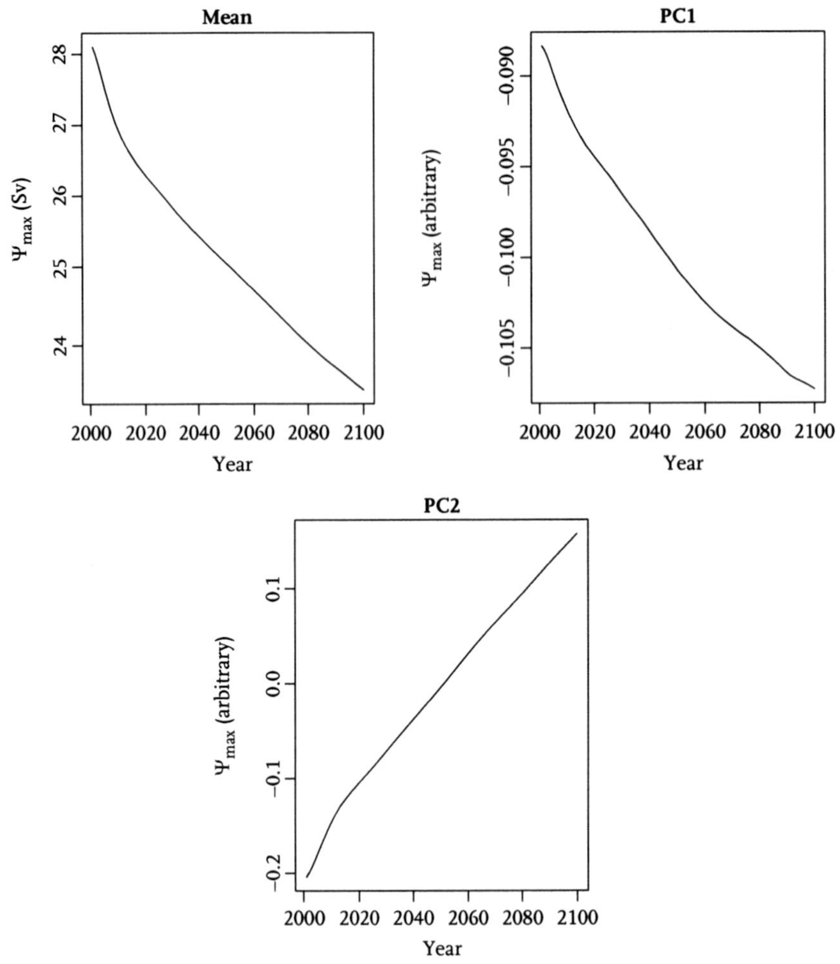

Fig. 16.6 Mean (left), and the first two principal component patterns (centre, right) of the ensemble.

for y_i each time, for every accepted input configuration, to fully account for the code uncertainty in the model output. There are too many trajectories to plot, so we generate a 2D density plot of each, using a kernel density estimator (figure 16.7).

The posterior probability that the MOC lies beneath the 10 Sv threshold at sometime during the 21st century (Table 16.1) is simply the proportion of years where any emulated MOC trajectory lies below that threshold in the scenario.

$$Pr(\Psi_{max} < 10) = \frac{1}{N}\sum_{i=1}^{N} I(y_i) \qquad (16.9)$$

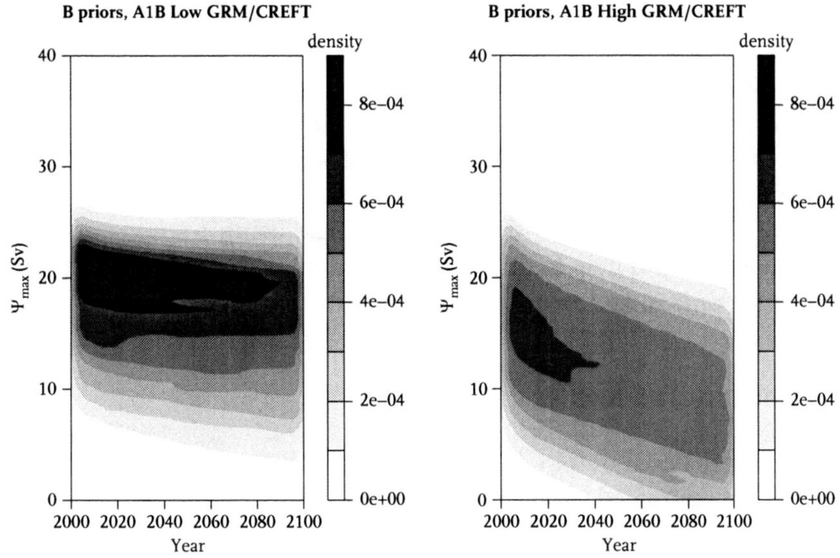

Fig. 16.7 Probability density of the strength of the MOC through the 21st century. The density is conditional on the rate of Greenland melting, CO_2 removal time, expert prior distributions for the model inputs, and a calibration using observed climate data.

where N is the number of simulated data points, or $p \times a$, when we have a accepted samples of $Pr(X^{*})$, and I is the indicator function

$$I = \begin{cases} 1 \text{ if } y_i \leq 10 \\ 0 \text{ if } y_i > 10. \end{cases} \tag{16.10}$$

The probability of ***significant reduction*** (i.e. below 10 Sv) of the MOC is conditional on a low or high rate of Greenland melting, and carbon removal timescale. A low scenario of these parameters generates a probability of crossing the threshold of around 0.1, whereas a high scenario has a probability of above 0.3. The future MOC strength in GENIE-1 is very sensitive to uncertainty in these parameters, implying sensitivity to these poorly understood physical processes.

Table 16.1 Probability of the MOC dropping below a threshold of 10 Sv during the 21st century, for three indicative scenarios.

Scenario	Low melt rate	High melt rate
A1B	0.09	0.35
A2	0.10	0.33
B1	0.10	0.30

As the model does not simulate an 'abrupt' change in the MOC, the estimated probabilities of such a shift are near zero. However, the model does suggest a high probability that the system will cross the emergent threshold of 10 Sv, which may lead to an inevitable shutdown of the MOC over the 21st century. If the threshold in the model is indicative of a real property of the system, this slowdown would have a severe impact on the climate of Northern Europe. A repeat of this experiment with similar forcings, using a spectrum of Earth system models, would help to discover if such a threshold is indeed a property of the system, or is in fact an artefact of this particular model.

16.7 Summary and further discussions

- High dimensional output of the climate model (trajectories of the MOC) can be approximated well using a small number of principal components, and a Gaussian process emulator.
- Expert knowledge in the form of prior distributions for model inputs constrains the model more than the available historical MOC data.
- The model does not simulate an abrupt climate change in our ensemble, but most runs show an increasingly rapidly decreasing MOC. The most feasible runs, as measured by their ability to reproduce past climate data, show a steady decline in MOC strength during the 21st century.
- The model does not show a stable MOC below around 10 Sv, an apparent emergent property of the model. It is unclear if runs that cross this threshold through transient forcing are committed to inevitable shutdown.

These results are conditional on the prior knowledge of the model, as well as the validity of the model itself, the data used in calibration, the scenario of future greenhouse gas emissions, and the statistical framework that was used to combine all of these elements. At every stage, there are subjective decisions being made about what is important to include in the inference. However, this analysis does represent an attempt to comprehensively include uncertainty in a prediction of the future state of the Earth system.

16.7.1 Where next?

Stakeholders such as governments and businesses have an increasing need for accurate climate predictions in order to make informed decisions. This information is always needed at the limit of scientific understanding, and ability to provide; for example future climate impacts are desired at an ever greater spatial resolution. We see no reason to expect that computational power will outstrip the complexity of state-of-the-art climate models, and lead to cheap climate predictions in the foreseeable future. In this case, the development of cheap and accurate climate model emulators should continue to be an area of active research. There is still much work to be done on building efficient high dimensional emulators, especially where the model output consists of many time-steps of field data that may have some correlation and nonlinear interaction; for example, temperature and precipitation. The units used to measure these data, and corresponding variances are very different, and so don't lend themselves well to simultaneous dimension reduction, although some progress in *simultaneous empirical orthogonal functions* has been made in observational climate research. A related problem is in finding sensible ways of dealing with very high dimensional initial conditions – particularly in finding sensible strategies to the sampling the input space for ensemble design. Other difficulties in building emulators include modelling categorical inputs (e.g. discrete switches), or model versions with alternative sections of code. This study uses a single model, which limits our ability to make probability statements about the MOC in the future. A more comprehensive study would use a suite of climate models to predict the future behaviour of the system. The synthesis of information from such a 'grand ensemble' of models is a very active area in climate research, a particular problem being the fact that a collection of climate models does not represent an independent sample from the possible 'model space'. Ultimately, emulators will be used to link a hierarchy, or ensemble of models together, taking into account different approaches to the representation of physical processes, efficiency, and ability to represent past climate.

Appendix

A. Broader context and background

Throughout this paper we use a Gaussian Process (GP) emulator to learn about the model, and perform climate experiments. This type of emulator has been used to model computer codes in, for example, Haylock and O'Hagan (1996); O'Hagan *et al.* (1999); Oakley and O'Hagan (2002) and Oakley (2004). A full demonstration of the GP emulator can be found in Kennedy and O'Hagan (2001), and O'Hagan (2006) provides a good non-technical summary.

The climate model is represented as an unknown function $g(\cdot)$ that relates model inputs X to output Y

$$Y = g(X). \qquad (16.11)$$

The method of learning about $g(\cdot)$ is to draw a sample of inputs, or design $\mathbf{D} = (x_1, \ldots, x_n)^{\mathrm{T}}$ from the input space $\mathcal{X}$ of the model, and run the model to produce output vector $\mathbf{y}$. $\mathbf{D}$ is a matrix of n ensemble members (rows) of d input parameters (columns). The model output is assumed to be smooth, so that we have some knowledge of $g(x')$ for x' close to x. We might a priori believe that $g(x)$ is approximately linear in x, and so in general, the mean of $g(x)$ is given by

$$E\{g(x)|\beta\} = h(x)^{\mathrm{T}}\beta, \tag{16.12}$$

conditional on a vector of coefficients β, and where the vector $h(\cdot)$ consists of q regression functions. The covariance between $g(x')$ and $g(x)$ is given by

$$\mathrm{cov}\{g(x), g(x')|\sigma^2\} = \sigma^2 c(x, x'), \tag{16.13}$$

where $c(x, x')$ is a function that decreases as the distance $|x - x'|$ increases. We choose

$$c(x, x') = \exp\{-(x - x')^T \mathbf{B}(x - x')\} \tag{16.14}$$

where $\mathbf{B}$ is a diagonal matrix of roughness parameters. We use a Gaussian process model to find the distribution of $g(x)$ conditional on σ^2. A Gaussian process can be thought of as an infinite collection of random variables with the property that any subset of these variables has a multivariate normal distribution. For any set of inputs $(x_1, \ldots, x_n)$ then, we can model the outputs $(g(x_1), \ldots, g(x_n))$ as having a multivariate normal distribution.

Using a non-informative prior in the form of $p(\beta, \sigma^2) \propto \sigma^{-2}$, it can be shown that

$$\left.\frac{g(x) - m^*(x)}{\hat{\sigma}\sqrt{c^*(x, x')}}\right| \mathbf{y}, \mathbf{B} \sim t_{r+n} \tag{16.15}$$

where

$$m^*(x) = h(x)^T\hat{\beta} + t(x)^T\mathbf{A}^{-1}(\mathbf{y} - \mathbf{H}\hat{\beta}) \tag{16.16}$$

$$c^*(x, x') = c(x, x') - t(x)^T\mathbf{A}^{-1}t(x') + \{h(x)^T - t(x)^T\mathbf{A}^{-1}\mathbf{H}\}\mathbf{V}^*\{h(x')^T\mathbf{A}^{-1}\mathbf{H}\}^T \tag{16.17}$$

$$t(x) = (c(x, x_1), \ldots, c(x, x_n))^T \tag{16.18}$$

$$\mathbf{H} = (h(x_1), \ldots, h(x))^T \tag{16.19}$$

$$\mathbf{A} = \begin{pmatrix} 1 & c(x_1, x_2) \ldots & c(x_1, x_n) \\ c(x_1, x_n) & 1 & \vdots \\ \vdots & & \ddots \\ c(x_n, x_1) & \ldots & 1 \end{pmatrix} \tag{16.20}$$

$$\hat{\beta} = \mathbf{V}^*(\mathbf{V}^{-1} + \mathbf{H}^T\mathbf{A}^{-1}\mathbf{y}) \tag{16.21}$$

$$\hat{\sigma^2} = \frac{a + z^T\mathbf{V}^{-1}z + \mathbf{y}^T\mathbf{A}^{-1}\mathbf{y} - \hat{\beta^T}(\mathbf{V}^*)^{-1}\hat{\beta}}{n + r - 2} \tag{16.22}$$

$$\mathbf{V}^* = (\mathbf{V}^{-1} + \mathbf{H}^T\mathbf{A}^{-1}\mathbf{H})^{-1} \tag{16.23}$$

$$\mathbf{y} = (g(x_1), \ldots, g(x_n))^T. \tag{16.24}$$

The estimator $m^*(x)$ is the posterior mean of the emulator at x, and from a frequentist perspective can be thought of as the best linear unbiased predictor of $g(x)$. From equation (16.16), we can see it is made up of two parts: the first component $h(x)^T\hat{\beta}$ is our linear prior expectation, with β updated in the light of data $\mathbf{y}$. The second component $t(x)^T\mathbf{A}^{-1}(\mathbf{y} - \mathbf{H}\hat{\beta})$ adjusts the posterior mean so that it passes through all of the observed outputs; there is no uncertainty at locations where we have run the model. The smoothness of the deviation from $h(x)^T\hat{\beta}$ towards the observed output y_i for x close to x_i depends on $\mathbf{B}$. Each element (i, i) of $\mathbf{B}$ describes the 'roughness' of $g(.)$ in the ith input dimension. A smoother $g(.)$ has a higher correlation between $g(x)$ and $g(x')$. As the correlation depends on the distance, the matrix $\mathbf{B}$ determines how close two inputs x and x' need to be for the correlation between $g(x)$ and $g(x')$ to take on a particular value (illustrated in Figure 16.2).

The matrix $\mathbf{B}$ describes the roughness of the function $g(.)$ in each input dimension. Since $g(.)$ is unknown, there is also uncertainty about the elements of $\mathbf{B}$; however, there is no analytical way of dealing with this uncertainty. A simple option is to keep $\mathbf{B}$ fixed, however, since the estimation of $g(x)$ is conditional on $\mathbf{B}$ (e.g. see figure 16.2), this can lead to poor emulator performance. A full Bayesian treatment of $\mathbf{B}$ is possible however, this can be computationally expensive. Oakley (1999) suggests estimating $\mathbf{B}$ from the data, using either cross validation, or via the posterior mode. The density of $\mathbf{B}$ conditional on $\mathbf{y}$ is given by

$$f(\mathbf{B}|\mathbf{y}) \propto (\hat{\sigma}^2)^{-\frac{n-q}{2}}|\mathbf{A}|^{-\frac{1}{2}}|\mathbf{H}^T\mathbf{A}^{-1}\mathbf{H}|^{-\frac{1}{2}}. \tag{16.25}$$

We use optimization methods to find a 'good' set of values for the diagonal matrix $\mathbf{B}$. In practice, both the cross-validation and posterior mode techniques can have problems. With even a moderate number of dimensions, say 10, the estimation of $\mathbf{B}$ via cross validation is computationally expensive, and very slow. Estimation via the posterior mode can also run into problems when the posterior distribution of $\mathbf{B}|\mathbf{y}$ is very flat, or very rough. We find that estimates of $\mathbf{B}$ used to initialize optimization routines, have a large impact on the speed of convergence, and outcome – local minima are common. For the example in this study, we optimise $\mathbf{B}$ for each principal component individually. The first (and most important) principal component required 27,000 iterations of a Nelder – Mead optimization routine to converge, with other components requiring fewer iterations.

Acknowledgements

This work was in part funded by NERC's Rapid Climate Change programme and by the RCUK 'Managing Uncertainty in Complex Models' project. We are grateful to J. Hirschi for providing Figure 16.1.

References

Alley, R., Marotzke, J., Nordhaus, W., Overpeck, J., Peteet, D., Peilke Jr, R., Pierrehumbert, R., Rhines, P., Stocker, T., Talley, L. and Wallace, J., (2003). Abrupt climate change. *Science*, **299**, 2005–2010.

Barber, D. C., Dyke, A., Hillaire-Marcel, C., Jennings, A. E., Andrews, J. T., Kerwin, M. W., Bilodeau, G., McNeely, R., Southon, J., Morehead, M. D. and Gagnon, J.-M. (1999). Forcing of the cold event of 8,200 years ago by catastrophic drainage of laurentide lakes. *Nature*, **400**, 344–348.

Bayarri, M., Berger, J., Cafeo, J., Garcia-Donato, G., Liu, F., Paolmo, J., Parthasarathy, R., Paulo, R., Sack, J. and Walsh, D. (2007). Computer model vaildation with functional output. *The Annals of Statistics*, **35**, 1874–1906.

Broecker, W. (1997). Thermohaline circulation, the Achilles heel of our climate system: Will man-made CO_2 upset the current balance? *Science*, **278**, 1582–1588.

Bryden, H. L., Longworth, H. R. and Cunningham, S. A. (2005). Slowing of the Atlantic meridional overturning circulation at 25 degrees N. *Nature*, **438**, 655–657.

Challenor, P., Hankin, R. and Marsh, R. (2006). Towards the probability of rapid climate change. In *Avoiding Dangerous Climate Change*, (ed. H.J. Schellnhuber, W. Cramer, N. Nakicenoric, T. Wigley and G. Yohe), Chapter 7, pp. 55–63. Cambridge University Press, Cambridge.

Cox, P., Betts, R., Jones, C., Spall, S. A. and Totterdell, I. (2000). Acceleration of global warming due to carbon-cycle feedbacks in a coupled climate model. *Nature*, **408**, 184–187.

Cunningham, S., Kanzow, T., Rayner, D., Baringer, M., Johns, W., Marotzke, J., Longworth, H., Grant, E., Hirschi, J., Beal, L., Meinen, C., and Bryde, H. (2007). Temporal variability of the Atlantic meridional overturning circulation at 26.5 °N. *Science*, **317**, 935–938.

Edwards, N. and Marsh, R. (2005). Uncertainties due to transport-parameter sensitivity in an efficient 3-D ocean-climate model. *Climate Dynamics*, **24**, 415–433.

Etheridge, D., Steele, L., Langenfelds, R., Francey, R., Barnda, J.-M., and Morgan, V. (1998). Historical CO_2 records from the Law Dome DE08, DE08-2, and DSS ice cores. In *Trends: A Compendium of Data on Global Change*, online at Carbon Dioxide Information Analysis Center, Oak Ridge National Laboratory, U.S. Department of Energy, Oak Ridge, Tenn., U.S.A. [`http://cdiac.ornl.gov/trends/co2/lawdome.html`].

Furrer, R., Sain, S. R., Nychka, D. and Meehl, G. R. (2007). Multivariate Bayesian analysis of atmosphere – ocean general circulation models. *Environmental and Ecological Statistics*, **14**, 249–266.

Gregory, J., Dixon, K., Stouffer, R., Weaver, A., Driesschaert, E., Eby, M., Fichefet, T., Hasumi, H., Hu, A., Jungclaus, J., Kamenkovich, I., Levermann, A., Montoya, M., Murakami, S., Nawrath, S., Oka, A., Sokolov, A. and Thorpe, R. (2005). A model intercomparison of changes in the Atlantic thermohaline circulation in response to increasing atmospheric CO_2 concentration. *Geophysical Research Letters*, **32**, L12703.

Grootes, P., Stuiver, M., White, J., Johnsen, S. and Jouzel, J. (1993). Comparison of oxygen isotope records from the GISP2 and GRIP Greenland ice cores. *Nature*, **366**, 552–554.

Hankin, R. (2005). Introducing bacco, an R bundle for Bayesian analysis of computer code output. *Journal of Statistical Software*, **14**, 1–21.

Haylock, R. and O'Hagan, A. (1996). On inference for outputs of computationally expensive algorithms with uncertainty in the inputs. In *Bayesian Statistics 5*, (ed. J. Bernardo, J. Berger, A. Dawid, and A. Smith), pp. 629–637. Oxford University Press, Oxford.

Higdon, D., Gattiker, J., Williams, B. and Rightley, M. (2008). Computer model calibration using high dimensional output. *Journal of the American Statistical Association.*, **103**, 570–583.

Huybrechts, P. and De Wolde, J. (1999). The dynamic response of the Greenland and Antarctic ice sheets to multiple-century climatic warming. *Journal of Climate*, **12**, 2169–2188.

Hyvarinen, A. and Oja, E. (2000). Independent component analysis: algorithms and applications. *Neural Networks*, **13**, 411–430.

IPCC (2007). *Climate Change 2007 – The Physical Basis*. Cambridge University Press, Cambridge.

Jolliffe, I. (2002). *Principal Component Analysis*, (2nd edn.). Springer, New York.

Keeling, C. and Whorf, T. (2005). Atmospheric CO_2 records from sites in the SIO air sampling network. In *Trends: A Compendium of Data on Global Change*, online at Carbon Dioxide Information Analysis Center, Oak Ridge National Laboratory, U.S. Department of Energy, Oak Ridge, Tenn., U.S.A. [`http://cdiac.ornl.gov/trends/co2/sio-mlo.htm`].

Kennedy, M. and O'Hagan, A. (2001). Bayesian calibration of computer models. *Journal of the Royal Statistical Society: B*, **63**, 425–464.

Lenton, T. M., Held, H., Kriegler, E., Hall, J. W., Lucht, W., Rahmstorf, S. and Schellnhuber, H. J. (2008). Tipping elements in the Earth's climate system. *Proceedings of the National Academy of Sciences of the United States of America*, **105**, 1786–1793.

Lorentz, E. (1963). Deterministic nonperiodic flow. *Journal of the Atmospheric Sciences*, **20**, 130–141.

McKay, M., Beckman, R. and Conover, W. (1979). A comparison of three methods for selecting values of input variables in the analysis of output from a computer code. *Technometrics*, **21**, 239–245.

Nakicenovic, N. and Swart, R. (2000). *Special Report on Emissions Scenarios: A Special Report of Working Group III of the Intergovernmental Panel on Climate Change*. Cambridge University Press.

Oakley, J. (1999). Bayesian uncertainty analysis for complex computer codes. Ph.D. thesis, University of Sheffield, Sheffield.

Oakley, J. E. (2004). Estimating percentiles of uncertain computer code outputs. *Applied Statistics*, **53**(Part 1), 83–93.

Oakley, J. E. and O'Hagan, A. (2002). Bayesian inference for the uncertainty distribution of computer model outputs. *Biometrika*, **89**, 769–784.

O'Hagan (2006). Bayesian analysis of computer code output: A tutorial. *Reliability Engineering and System Safety*, **91**, 1290–1300.

O'Hagan, A., Kennedy, M. C. and Oakley, J. E. (1999). Uncertainty Analysis and other inference tools for complex computer codes. In *Bayesian Statistics 6*, (ed. J. Bernardo, J. Berger, A. Dawid, and A. Smith), pp. 525–582. Oxford University Press, Oxford.

Pedlosky, J. (1992). *Geophysical Fluid Dynamics*, (2nd edn.) Springer-Verlag, Berlin.

Rahmstorf, S., Crucifix, M., Ganopolski, A., Goosse, H., Kamenkovich, I., Knutti, R., Lohmann, G., Marsh, R., Mysak, L., Wang, Z. and Weaver, A. (2005). Thermohaline circulation hysteresis: A model comparison. *Geophysical Research Letters*, **32**. L23605, doi:10.1029/2005GL023655.

Rahmstorf, S. and Ganopolski, A. (1999). Long-term global warming scenarios computed with an efficient coupled climate model. *Climatic Change*, **43**, 353–367.

Rohling, E. and Palike, H. (2005). Centennial scale climate cooling with a sudden cold event around 8,200 years ago. *Nature*, **434**, 975–979.

Rougier, J. (2007). Probabilistic inference for future climate using an ensemble of climate model evaluations. *Climatic Change*, **81**, 247–264.

Smith, A. and Gelfand, A. (1992). Bayesian statistics without the tears: A sampling-resampling perspective. *The American Statistician*, **46**, 84–88.

Stommel, H. (1961). Thermohaline convection with two stable regimes of flow. *Tellus*, **13**, 224–230.

Vellinga, M. and Wood, R. (2002). Global climatic impacts of a collapse of the Atlantic thermohaline circulation. *Climatic change*, **54**, 251–267.

·17·

Models for demography of plant populations

James S. Clark, Dave Bell, Michael Dietze, Michelle Hersh, Ines Ibanez, Shannon L. LaDeau, Sean McMahon, Jessica Metcalf, Emily Moran, Luke Pangle and Mike Wolosin

17.1 Introduction

Ecologists seek to understand how demographic rates contribute to species diversity. Birth, growth, and survival together determine population growth. Demographic rates are related to one another, and they depend in complex ways on environmental variables. At a given age, an organism allocates energy in ways that affects current and future growth, fecundity, and survival risk. These relationships change through time as resource availability changes, and organisms develop and age. For the population ecologist challenges include inference not only on specific demographic rates, but also on how they combine to determine population growth. In this chapter we discuss how hierarchical Bayes analysis can help synthesize information from a range of sources to understand how demographic rates relate to one another and might contribute to biodiversity.

The key challenges to demographic inference in ecology include the availability of many sources of incomplete information, often measured at different scales, and the large number of interactions among demographic components. Population growth depends on all demographic rates, birth, growth, death, and migration. Each of these rates can respond to a fluctuating environment, including other organisms. Many factors are observable or only weakly related to factors that can be measured. For example, rarely can the cause of death be determined for organisms in the wild. Without knowledge of cause it is difficult to isolate and model individual risk factors, let alone how they interact. Even where cause might be identified, challenges remain. If death could be attributed to, say, 'drought' (Condit *et al.* 1995, Suarez *et al.* 2004, Nepsted *et al.* 2007, van Mantgem and Stephenson 2007), a series of questions arise: is it the daily, seasonal, or annual average soil moisture that is most important? Is it duration or intensity? Which interactions determine why only a fraction of the population died? We rarely have the information to address such questions, and we often lack understanding of the important scales (e.g. weekly or monthly drought?). These challenges have necessitated a superficial view of demography that is focused on annual rates with limited connection to covariates.

To motivate some of the complexity that follows, consider limitations of current demographic models. Traditionally, most demographic analyses include no predictors beyond age or size (i.e. the standard age-structured model (Leslie matrix) or stage-structured model (see the reviews of Caswell 2001, Gurevitch *et al.* 2002)). Models involving covariates tend to include a demographic rate as a response variable (offspring born per female per year, annual growth rate, fraction of the population making a transition to a different stage of life, survival probability) and, perhaps, a small number of predictors. The conclusions that can be drawn from such models are limited, because it is recognized that they may not accommodate important factors affecting the data, even for simple experiments. For example, seed production of trees is rarely directly observed in natural populations, because it cannot be quantified in forest canopies. The covariates are often represented by crude indices. The standard assumption that error should enter as a stochastic envelope around a deterministic function of predictor variables, such as

$$y_i = x_i'\beta + \varepsilon_i$$

implies that x_i is known much better than y_i. If x_i is a GIS layer, an instrument that records with error, an incubation culture of biological activity, a classification scheme based on unreliable detection, or the interpretation of a fuzzy image, we might better represent the problem by including a model for x. If so, we need to consider how to coherently connect a model for x and a model for y and how to allow for uncertainty and still estimate everything. Growth rates are usually measured with less error than are the environmental covariates used as predictors, including such difficult-to-quantify resources as solar radiation reaching the partially shaded crown, soil moisture, and nutrient supply, having substantial spatiotemporal variation that is never well quantified (Beckage and Clark 2003, Kobe *et al.* 2006, Mohan *et al.* 2007). If mortality is modelled as a function of growth, and observations are not available until after death, the growth rates are rarely available, requiring either annual measurements on trees that both survive and die up until the time of death or increment cores of trees (which are too laborious to obtain on large numbers of individuals). Either way, a model is needed to account for the way in which the covariate data were collected (Kobe *et al.* 1995, Wyckoff and Clark 2000). Moreover, we are now considering multiple demographic rates (growth and survival), connected by virtue of the fact that individuals near death may allocate less to growth.

As ecologists increasingly want comprehensive inference that derives from multiple underlying processes connecting inputs and responses, both of which are partially known, the challenge comes in devising ways to synthesize the sources of information, allow for observation error and uncertainty in the underlying model itself. How can we capture interactions between individual models for growth, fecundity, and mortality risk? Doing so requires that they be fitted simultaneously, as part of an integrated model of demographic change. The model must reflect the uncertainties that enter through both process and data at multiple stages. Useful inference requires that we exploit information coming not only from data, but also from theory and previous observations and experiments.

Multistage models provide a natural framework for organizing how ecologists think about inference. Bayesian techniques provide a natural approach for analysis of such models. Implementation requires that they can be structured to provide transparency regarding assumptions, model behaviour, and parameter estimates. For example, does the process model capture the relationships in realistic ways, and would we recognize failure to do so? Does the model make realistic predictions at all levels, including for state variables and observations at different scales? The application presented here highlights ways to integrate information for demographic inference, connecting models for important components of the problem, each allowing for uncertainty. We model fecundity, dispersal, growth, and survival with covariates that include some of these demographic rates and light availability, a key resource that limits plant growth. We show how the large number of estimates that come from the analysis can be summarized synthetically to provide deeper insight about relationships among demographic rates and how they respond to covariates.

17.2 Demographic data

We use data collected from tree plots $j = 1, \ldots, J$ and covariates to infer demographic rates. Data include observational studies and whole-stand manipulations, which allow us to break up correlations in some of the important covariates. Study areas include $J = 9$ plots of mapped forest stands in the Piedmont (Duke Forest) and southern Appalachian Mountains (Coweeta) of North Carolina, USA. The plots were selected to span a range of topographic, soil moisture, soil types, and elevation characteristics (Table 17.1), supplemented with experimental manipulations. In this study we report observational data come from trees $i = 1, \ldots, I_J$ on plots j in years $t \subset \{t_{ij}, t_{ij} + 1, \ldots, T_{ij}\}$. Plots were established in years $t_j \subset \{1991, \ldots, 2000\}$, when trees were first mapped, identified to species, and measured for diameter. The first observation year for individual ij is the year when plot j was established, t_j, or when the tree grew to sufficient size (2 m in height) to be measured, whichever came first. The last observation year $T_{ij} \leq 2008$ is the last year the individual was observed, at which point it might still be alive or not. Species codes are listed

Table 17.1 Plot characteristics and and number of trees by plot. Species codes are listed in Table 17.3.

Plot	C118	C218	C318	C427	C527	CLG	CUG	DBW	DHW	Total
Elev (m)	780	820	870	1110	410	030	140	70	70	
Lat/Long			35°03′N, 83°27′W					35°58′36″N 79°5′48″W		
Soil type	Typic & humic hapludults, typic dystrochrepts, typic haplumbrepts							Typic & oxyaquic vertic hapludalfs		
First year	1992	1992	1992	1992	1992	2000	2000	2000	1999	
Area (ha)	0.64	0.64	0.64	0.64	0.64	2.75	1.45	4.11	2.40	13.91
acru	496	136	211	257	14	982	608	2761	666	6131
acsa	0	2	1	0	82	35	0	13	0	133
acpe	5	223	15	15	230	451	21	0	0	960
acba	0	0	0	0	0	0	0	4	114	118
acun	0	0	0	0	0	13	0	2	0	15
beal	0	1	2	0	157	0	0	0	0	160
bele	5	38	12	4	66	8	1	0	0	134
beun	0	0	0	0	0	29	38	0	0	67
caca	0	0	0	0	0	0	0	146	439	585
cagl	56	33	41	15	5	36	10	153	31	380
caov	0	0	2	0	0	0	0	0	66	68
cato	1	0	0	0	0	48	3	395	51	498
caun	1	0	1	2	0	162	5	77	73	321
ceca	0	0	0	0	0	0	0	282	39	321
cofl	45	78	27	10	3	118	14	1405	501	2201
fram	0	0	1	4	63	46	0	658	802	1574
list	0	0	0	0	0	0	0	1523	648	2171
litu	11	70	12	9	0	654	6	371	161	1294
nysy	103	30	113	117	0	282	335	361	457	1798
piri	36	0	0	0	0	0	0	0	0	36
pist	2	16	0	0	0	0	0	0	0	18
pita	0	0	0	0	0	0	0	391	175	566
piec	0	0	0	0	0	0	0	76	1	77
pivi	0	0	0	0	0	0	0	25	1	26
qual	12	0	0	0	1	0	0	231	65	309
quco	25	7	1	14	0	51	44	0	0	142
qufa	0	0	0	0	0	0	0	21	5	26
quma	10	0	0	0	0	0	0	20	0	30
quph	0	0	0	0	0	0	0	18	73	91
qupr	91	34	102	101	0	180	173	0	0	681
quru	38	5	21	57	24	139	49	65	0	398
qust	0	0	0	0	0	0	0	41	41	82
quve	45	6	7	3	0	12	14	30	0	117
quun	0	0	0	0	0	18	18	4	0	40
rops	24	6	4	12	0	95	30	1	0	172
tiam	0	5	0	0	87	0	0	0	0	92
tsca	8	27	8	74	3	243	38	0	0	401
ulal	0	0	0	0	0	0	0	719	558	1277
ulam	0	0	0	0	0	0	0	69	147	216
ulru	0	0	0	0	0	0	0	0	64	64
ulun	0	0	0	0	0	0	0	139	86	225
Total trees	1014	717	581	694	735	3602	1407	10001	5264	24015
Seed traps	20	20	20	20	20	73	43	128	66	410
Total seeds	6413	46964	22685	13298	334374	20814	10801	75481	23487	554317

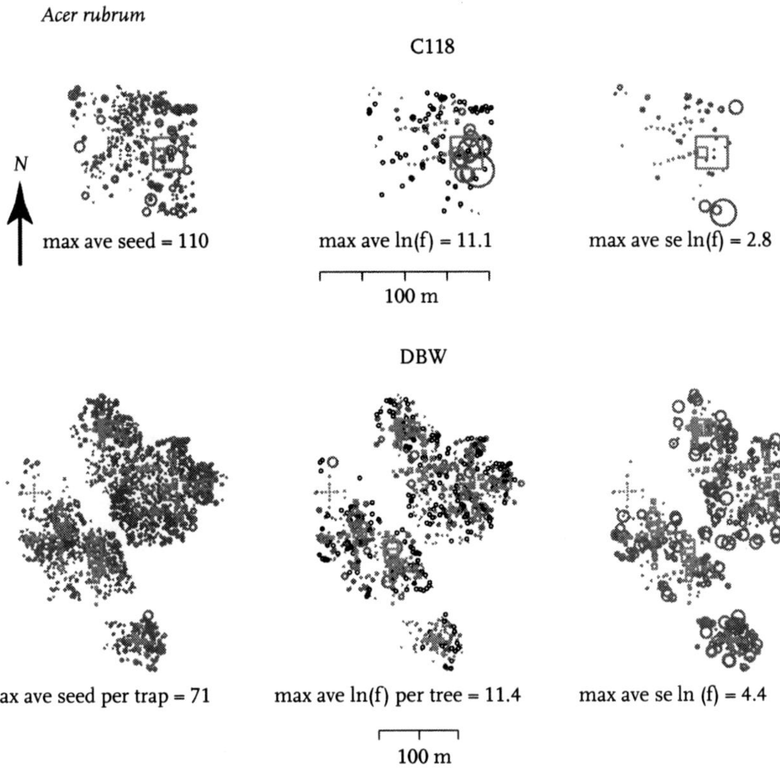

Fig. 17.1 Examples of two mapped stands used for demographic inference, showing stems of a single genus, *Acer*, represented by circles. Boxes are shown at seed trap locations with box size proportional to average annual seed collection for the entire study period. Plot C118 is small and has not been manipulated. Plot DBW is larger (note scale bars), with seed traps clustered in and around the locations of eight canopy gaps created in 2002. From left the series of three maps for both locations have circle size scaled to (1) stem diameter, (2) mean estimate of tree fecundity, and (3) standard error of tree fecundity estimates (not to be confused with the standard deviation in fecundity over time).

in Table 17.2. Example maps of two study plots are shown in Figure 17.1. Environmental data include static variables (elevation, slope, aspect) and variables that fluctuate over time, with additional variation among plots (temperature) and within plots (soil moisture, light penetration). This analysis focuses on how light availability and tree size affects demographic rates, including relationships among them.

A subset of the mapped plots was used for canopy gap experiments. Following collection of several years of pretreatment data, trees were removed from the canopy in 20 m or 40 m wide patches to simulate the small and large treefall gaps that occur when a single tree falls or when groups of trees fall, as during storms. Canopy trees were pulled down using a skidder or bulldozer located outside the plot and left in position, some having snapped, others uprooted.

Table 17.2 Species codes used in other tables and figures.

Code	Species	Code	Species
acru	*Acer rubrum*	piri	*Pinus rigida*
acsa	*Acer saccharum*	pist	*Pinus strobus*
acpe	*Acer pensylvanicum*	pita	*Pinus taeda*
acba	*Acer barbatum*	piec	*Pinus echinata*
acun	*Acer unknown*	pivi	*Pinus virginiana*
beal	*Betula alleghaniensis*	qual	*Quercus alba*
bele	*Betula lenta*	quco	*Quercus coccinea*
beun	*Betula unknown*	qufa	*Quercus falcata*
caca	*Carpinus caroliniana*	quma	*Quercus marilandica*
cagl	*Carya glabra*	quph	*Quercus phellos*
caov	*Carya ovata*	qupr	*Quercus montana*
cato	*Carya tomentosa*	quru	*Quercus rubra*
caun	*Carya unknown*	qust	*Quercus stellata*
ceca	*Cercis canadensis*	quve	*Quercus velutina*
cofl	*Cornus florida*	quun	*Quercus unknown*
fram	*Fraxinus americana*	rops	*Robinia pseudoacacia*
ilde	*Ilex decidua*	tiam	*Tilia americana*
ilop	*Ilex opaca*	tsca	*Tsuga canadensis*
list	*Liquidambar styraciflua*	ulal	*Ulmus alata*
litu	*Liriodendron tulipifera*	ulam	*Ulmus americana*
nysy	*Nyssa sylvatica*	ulru	*Ulmus rubra*
oxar	*Oxydendron arboreum*	ulun	*Ulmus unknown*

Damage to canopy trees and understory trees (snapped or bent by pulled trees) was recorded as the basis for analysis of damage effects on growth (Dietze and Clark 2008). For the analysis presented here, manipulation had the greatest effect on light availability, represented by 'exposed canopy area'.

To allow for inference on demographic interactions, we use a structure that combines important relationships but is constrained by what can be observed or inferred. An individual is characterized by several state variables, some of which are constant, some change over time, some are continuous, and others discrete. State variables differ in terms of how directly each can be observed. Trees are classified according to genus (e.g. *Quercus* for oak) and species (*Q. rubrum* for red oak). The genus to which a tree belongs is known. The species is also taken to be known, in the sense that it will not be inferred, but seeds are confidently identified only to genus. For this reason, trees of the same genus are modeled together. The genera and species are summarized in Table 17.3. There is an unknown species class for trees from several genera, where large individuals could not be confidently ascribed to a particular species (Table 17.3). Because many seeds can only be classified to the level of genus, we infer seed production as the combined contributions from all trees in that genus. For example, the analysis for *Quercus* includes 10 species, because not all acorns could be identified to species level (Clark *et al.* 1998, 2004), whereas the analysis of

Table 17.3 Seed-trap years (same for all taxa) and tree years, grouped by plot and by genus, as analysed here. Species codes are listed in Table 17.3.

Plot	C118	C218	C318	C427	C527	CLG	CUG	DBW	DHW	Total
Trap year	320	320	320	320	320	490	301	1016	342	3749
					Acer					
acru	8432	2312	3587	4369	238	8838	5472	24849	6660	64757
acsa	0	34	17	0	1394	315	0	117	0	1877
acpe	85	3791	255	255	3910	4059	189	0	0	12544
acba	0	0	0	0	0	0	0	36	1140	1176
acun	0	0	0	0	0	117	0	18	0	135
					Betula					
beal	0	17	34	0	2669	0	0	0	0	2720
bele	85	646	204	68	1122	72	9	0	0	2206
beun	0	0	0	0	0	261	342	0	0	603
					Carpinus					
caca	0	0	0	0	0	0	0	1314	4390	5704
					Carya					
cagl	952	561	697	255	85	324	90	1377	310	4651
caov	0	0	34	0	0	0	0	0	660	694
cato	17	0	0	0	0	432	27	3555	510	4541
caun	17	0	17	34	0	1458	45	693	730	2994
					Cercis canadensis					
ceca	0	0	0	0	0	0	0	2538	390	2928
					Cornus florida					
cofl	765	1326	459	170	51	1062	126	12645	5010	21614
					Fraxinus americana					
fram	0	0	17	68	1071	414	0	5922	8020	15512
					Liquidambar styraciflua					
list	0	0	0	0	0	0	0	13707	6480	20187
					Liriodendron tulipifera					
litu	187	1190	204	153	0	5886	54	3339	1610	12623
					Nyssa sylvatica					
nysy	1751	510	1921	1989	0	2538	3015	3249	4570	19543
					Pinus					
piri	612	0	0	0	0	0	0	0	0	612
pist	34	272	0	0	0	0	0	0	0	306
pita	0	0	0	0	0	0	0	3519	1750	5269
piec	0	0	0	0	0	0	0	684	10	694
pivi	0	0	0	0	0	0	0	225	10	235
					Quercus					
qual	204	0	0	0	17	0	0	2079	650	2950
quco	425	119	17	238	0	459	396	0	0	1654
qufa	0	0	0	0	0	0	0	189	50	239
quma	170	0	0	0	0	0	0	180	0	350
quph	0	0	0	0	0	0	0	162	730	892
qupr	1547	578	1734	1717	0	1620	1557	0	0	8753
quru	646	85	357	969	408	1251	441	585	0	4742
qust	0	0	0	0	0	0	0	369	410	779

(cont.)

Table 17.3 (*Continued*).

Plot	C118	C218	C318	C427	C527	CLG	CUG	DBW	DHW	Total
quve	765	102	119	51	0	108	126	270	0	1541
quun	0	0	0	0	0	162	162	36	0	360
					Robinia pseudoacacia					
rops	408	102	68	204	0	855	270	9	0	1916
					Tilia americana					
tiam	0	85	0	0	1479	0	0	0	0	1564
					Tsuga canadensis					
tsca	136	459	136	1258	51	2187	342	0	0	4569
					Ulmus					
ulal	0	0	0	0	0	0	0	6471	5580	12051
ulam	0	0	0	0	0	0	0	621	1470	2091
ulru	0	0	0	0	0	0	0	0	640	640
ulun	0	0	0	0	0	0	0	1251	860	2111

Carpinus includes a single species (Table 17.3). Trees retain the species identity, having some parameters that are species-specific, whereas seeds are modelled as being potentially produced by trees of the entire genus on a probabilistic basis. This approach allows combination of observations at the scale of individual trees and at seed-trap scale (a seed traps accumulate seeds from all trees simultaneously).

In addition to diameter and species, individual level observations include survival, canopy status, and reproductive maturation status. Canopy status involved ordinal classes derived from standard classifications used in forestry. At one to four year intervals, individuals were assigned to:

Class 1: suppressed in the understory, with access limited to sunflecks (e.g. intermittent patches of direct sunlight);

Class 2: intermediate, with not more than 20% of the canopy exposed to some direct sunlight;

Class 3: codominant, with > 20% of the canopy exposed to direct sunlight during part of each day.

A suppressed individual in the understory would be assigned Class 1 status, but could change to Class 3 status if it occupied a canopy gap following loss of the overstory.

Additional information on canopy status comes from low-altitude aerial photo coverage of plots, used to segment and measure canopy areas of trees visible from above. The modelling of canopy exposure based on status and remote sensing observations, combined with allometric models of canopy area is described in Valle *et al.* (2009). Posterior means and variances from that analysis are used as prior means and variances for the analysis presented here.

Maturity status observations were made at irregular years during the flowering season (taxa such as *Ulmus* and *Acer rubrum* have conspicuous flowers before leaf-out in spring), the fruiting season (fruits often visible in the lower canopy include *Cornus florida, Cercis canadensis*), and winter (lack of leaves permits identification reproductive structures of *Liriodendron tulipifera, Liquidambar styraciflua*). Classes of observations are:

uncertain: the largest class, because fruits and/or flowers are difficult to observe;

not mature: if the entire canopy could be clearly observed to have no fruiting structures during the known flowering and/or fruiting season;

flowering: establishes that the individual is mature;

seeds/fruits present: establishes maturity and, for dioecious species, female status;

female or male flowers present: in rare cases individuals of a dioecious species could be assigned gender on the basis of flower structure.

The observations are summarized by maturation status and gender status in Table 17.4, where $\theta_{ij,t} = p(Q_{ij,t} = 1)$ is the probability that individual i on plot j is mature in year t, $v = p(q_{ij,t} = 1 | Q_{ij,t} = 1)$ is the probability that a mature individual will be identified as such, $\phi = p(H_{ij} = 1)$ is the fraction of individuals that are female for the entire population. The probabilities of $Q|q$ and $H|h$ are provided in Table 17.4 primarily for explanatory purposes, because data models discussed in Section 17.3 involve multiple observations. Note that statuses must be modeled only for Table 17.4 entries not containing a zero or a one.

Table 17.4 Indicators and probabilities of maturity and gender conditional on observations.

Observation	Maturity indicator $q_{ij,t}$	Maturity probability[1] $\Pr(Q_{ijt} = 1 \mid q_{ij,t})$	Gender indicator $h_{ij,t}$	Gender probability[1,2] $\Pr(H_{ij} = 1 \mid h_{ij,t})$
no observation	—	$\theta_{ij,t}$	—	ϕ
uncertain	0	$\dfrac{(1-v)\theta_{ij,t}}{1-v\theta_{ij,t}}$	0	ϕ
not mature	−1	0	0	ϕ
flowering	1	1	0	ϕ
seeds/fruits	1	1	1	1
male or female flowers	1	1	0 or 1	0 or 1

[1] The probabilities are shown for the special case that there is a single observation per individual. In fact, there are multiple status observations, modelling of which is discussed in Section 17.3.1. Symbols are θ – probability of being in the mature state, v – probability of detecting mature status, ϕ – female fraction of the population.

[2] Gender status is assumed static (H has subscript *ij*), whereas there are multiple observations of status for each individual (h has subscript *ij,t*).

Unfortunately, most observations are 'uncertain', requiring that most statuses must be modelled.

Seed trap data are the basis for fecundity estimates, and they further contribute to estimates of maturation and gender statuses. Seeds cannot be counted in the dense canopies characteristic of closed forests, but spatiotemporal seed data can be used to model fecundity (Clark *et al.* 1998, 1999, 2004). Fecundity is estimated from seed traps $k = 1, \ldots, K_J$ using models of dispersal (Clark *et al.* 1998, 1999, 2004), where K_J ranged from 20 to 128 seed traps. Seed traps were emptied from two–four times annually, seeds identified to genus or species (some seeds can only be identified to the genus level), and counted. For modelling purposes, seed data were accumulated to total per trap per year, $s_{kj,t}$ (Table 17.3).

Diameter measurements and increment cores provide information on individual tree growth. Because diameter fluctuates with stem moisture content, and diameter measurements have error, we measured diameters $D_{ij,t}$ at 1–4 year intervals. In addition to diameter measurements, increment cores were extracted from a subset of trees, providing a record of past growth $d_{ij,t} = D_{ij,t+1} - D_{ij,t}$ for the individual up to the year in which the core was extracted. Modeling of diameter-census and increment-core data are described by Clark *et al.* (2007). Posterior means and variances for each tree year obtained from that analysis are used as priors for the analysis presented here.

17.3 Models to synthesize data and previous knowledge

Consider a forest containing trees of different species, gender, age, and size, each experiencing the local environment in ways that depend on some factors that can be measured and others that cannot; here we focus on light availability. The responses of interest include growth rate, gender ratio, maturation status, fecundity, seed dispersal, and survival risk. Combinations of these demographic rates within individuals determine the growth rates of populations and, thus, community biodiversity.

An individual's response to the environment produces variation in demographic rates. Resource availability (here we consider light) contributes to overall health. Resources vary at many scales, depending on supply and on competition with neighbours. For example, light levels vary throughout the day and seasonally, and they are reduced by nearby trees that shade one another. Fine scale heterogeneity is most obvious where canopy gaps form, allowing from 10 to 100% of full sunlight to penetrate to the forest floor, depending on gap size. By contrast, the uninterrupted canopy intercepts 95 to 99% of incoming radiation. Due to spatial heterogeneity, individual trees are exposed to conditions that differ from their neighbors, and these differences can change over time with changes in canopy structure and as interannual climate variation moderates the impact of resource supply. Some of these factors can be measured in the environment, but it is important to also allow for variation that cannot be ascribed to measurable factors. Here we describe the model for demographic rates.

17.3.1 Gender and maturation

The gender of a tree remains constant, whereas maturation status changes from immature to mature over time. Maturation status and gender can be confirmed for some trees (Table 17.4). Presence of seeds indicates maturity and (for dioecious species) female status. The presence of flowers indicates maturity, but it does not mean that an individual belonging to a dioecious species is female, unless it can be identified as a female flower. When the entire crown can be observed, lack of flowers or fruits during the flowering/fruiting season is taken to indicate immaturity. In crowded stands, absence of reproductive effort can rarely be confirmed by such observations, so detection is uncertain.

The transition from immature to mature is treated as a hidden Markov process. Modelling of gender and maturation is complicated by the fact that probabilities depend on the entire history of observations on individual ij. Consider an individual that is observed once. If maturation status is uncertain ($q_{ij,t} = 0$ in Table 17.4), then the individual is mature with probability $(1 - v)\theta_{ij,t}/(1 - v\theta_{ij,t})$. But additional observations complicate the model. For example, mature status is more likely for an individual last known to be immature 10 year ago than it is for an individual last known to be immature one year ago. Likewise, an individual is more likely to be mature in year t if it is first known to be mature in year $t + 1$ than if it is first known to be mature in year $t + 10$. Furthermore, an individual observed to be of unknown status once is more likely to be mature than is an individual observed to be of unknown status 10 times. In other words, modelling of status must accommodate not only the probabilities contained in Table 17.4, but also how they must be combined to accommodate the differing observation histories of each individual. Here we discuss these probabilities. We first discuss conditioning on observations listed in Table 17.4, followed by seed data.

The unconditional probability that individual i on plot j is female is termed the female fraction $\Pr(H_{ij} = 1) = \phi$, (Table 17.2) and the probability of being male $\Pr(H_{ij} = 0) = 1 - \phi$. Observations of gender status were obtained at irregular intervals $h_{ij,t}$. If the individual is observed to be female $h_{ij,t} = 1$ or male $h_{ij,t} = 0$ then

$$\Pr(H_{ij} = 1 | h_{ij,t} = 1) = 1$$
$$\Pr(H_{ij} = 0 | h_{ij,t} = 0) = 1.$$

In other words, gender is only assigned if it is certain. If there are no observations for gender, information enters solely though seed rain data. If seed density near an individual is high, then the probability that it is female is large, and vice versa. Thus, fecundity and maturation must be modeled together.

The unconditional probability that the individual is in the mature state $\Pr(Q_{ij,t} = 1) = \theta_{ij,t}$, increases with tree size and canopy exposure (access to sunlight). The probability is parameterized as a logit link to diameter D and canopy exposure λ

$$\theta_{ij,t} = \frac{\exp\left(\beta_0^\theta + \beta_1^\theta D_{ij,t} + \beta_2^\theta \lambda_{ij,t}\right)}{1 + \exp\left(\beta_0^\theta + \beta_1^\theta D_{ij,t} + \beta_2^\theta \lambda_{ij,t}\right)}. \tag{17.1}$$

For computation we require conditional probabilities that derive from this relationship. Two simple examples we detail in the appendix include the probability of making the transition in year t, given previous immaturity and future maturity

$$\delta_{ij,t} = \Pr\left(Q_{ij,t} = 1 \,\middle|\, Q_{ij,t-1} = 0,\ Q_{ij,t+1} = 1\right) = \frac{d\theta_{ij,t}}{d\theta_{ij,t} + d\theta_{ij,t+1}}$$

where

$$d\theta_{ij,t} = \beta_1^\theta d_{ij,t} \theta_{ij,t}\left(1 - \theta_{ij,t}\right) dt$$

and the probability that the transition year τ_{ij} occurred in year t

$$\delta_{ij} = \Pr\left(\tau_{ij} = t \,\middle|\, \tau_{ij}^0 < \tau_{ij} < \tau_{ij}^1\right) = \frac{d\theta_{ij,t}}{\left(\theta_{ij,\tau_{ij}^1} - \theta_{ij,\tau_{ij}^0}\right)}$$

where the lower limit represents that last year the individual was known to be immature and the upper limit represents the first year the individual was known to be mature. These two fundamental relationships are the basis for models that incorporate observations for dioecious and monoecious species, respectively (Appendix: Computation). The first is a Bernoulli probability for each year t; there is a probability for every tree year. The second is a probability associated with an individual and depends on when the maturation event occurred for that individual.

For monoecious species the joint distribution of maturation and fecundity is represented by a discrete mixture that is conditional on the full history of status observations on the individual, a vector $q_{ij} = \{q_{ij,t},\ t = (t_{ij}, \ldots, T_{ij})\}$,

$$\begin{aligned} p&\left(Q_{ij,t}, f_{ij,t} \middle| q_{ij}, x_{ij,t-1}, Q_{ij,t-1}, Q_{ij,t+1}\right) \\ &= \left(1 - \delta_{ij,t}\right)^{1-Q_{ij,t}} \left[\delta_{ij,t} N\left(\ln f_{ij,t} \middle| \mu_{ij,t}^{f|d}, V_{ij,t}^{f|d}\right)\right]^{Q_{ij,t}} \end{aligned} \tag{17.2}$$

where the probability $\delta_{ij,t} = p(Q_{ij,t} = 1 | q_{ij}, Q_{ij,t-1} = 0, Q_{ij,t+1} = 1)$ is based on the history of observations on ij, and the parameters for the log normal distribution are conditional, coming from a bivariate state-space model for fecundity (Appendix: Computation).

For dioecious species, we require gender and the probability for the entire history of maturation Q_{ij} and the probability associated with it, δ_{ij}. This distribution is given by

$$p\left(H_{ij}, Q_{ij}, f_{ij} \left| q_{ij}, x_{ij,t-1}\right.\right) = (1-\phi)^{1-H_{ij}} \delta_{ij} \left\{ \phi \prod_t \left[N\left(\ln f_{ij,t} \left| \mu_{ij,t}^{f|d}, V_{ij,t}^{f|d} \right.\right)\right]^{Q_{ij,t}} \right\}^{H_{ij}}. \quad (17.3a)$$

If an observation establishes an individual as male we have the probability for a maturation history during which no reproduction occurred

$$p\left(Q_{ij}, f_{ij}=0 \left| q_{ij}, x_{ij,t-1}, H_{ij}=0\right.\right) = \delta_{ij} \quad (17.3b)$$

and for a female during which reproduction may or may not have occurred

$$p\left(Q_{ij}, f_{ij} \left| q_{ij}, x_{ij,t-1}, H_{ij}=1\right.\right) = \delta_{ij} \prod_t \left[N\left(\ln f_{ij,t} \left| \mu_{ij,t}^{f|d}, V_{ij,t}^{f|d} \right.\right)\right]^{Q_{ij,t}}. \quad (17.3c)$$

Thus far, we have conditional relationships involving fecundity and maturation. The conditional dependence on seed rain data is discussed in the next section.

17.3.2 Seed data and fecundity

Fecundity (seed production per individual per year) is not directly observed, because seeds cannot be counted in crowded canopies. Like gender and maturation status, indirect information comes from seed trap data, linked by way of a transport model. Individuals that are mature and female can produce seeds. Fecundity is thus zero for immature individuals and all male trees. For mature females fecundity is taken as a continuous, positive variable.

Seeds accumulating in traps located throughout each stand j provide a basis for inverse modeling of fecundity. The likelihood for seeds collected in trap k in stand j in year t is taken to be Poisson

$$Po\left(s_{jk,t} \left| A_{jk} g_{jk}\left(f_{j,t}; u\right)\right.\right) \quad (17.4)$$

where A_{jk} is the area of the seed trap (0.16 m^2 or 0.125 m^2), and $g_{jk,t}$ is the expected density of seed (m^{-2}), including a parameter u. The expected seed density depends on fecundities of all trees in stand k and a dispersal kernel K, added to a crude estimate of small background density of seed that might enter the plot from outside the mapped boundaries, proportional to basal area of the species in stand j, or $BA_{j,t}$.

$$g_{jk}\left(f_{j,t}\right) = c \cdot BA_{j,t} + \sum_i f_{ij,t} K\left(r_{ik}; u\right). \quad (17.5)$$

The dispersal kernel is taken to be a two-dimensional Student's t, previously found to fit seed dispersal data well,

$$K(r_{ik}; u) = \frac{1}{\pi u \left(1 + r_{ik}^2/u\right)^2} \tag{17.6}$$

containing the scale parameter u (Clark *et al.* 1999, 2004). The term in equation (17.5) that includes basal area BA allows for the fact that some small fraction of seed can derive from outside the plot, roughly proportional to the basal area of the species.

In addition to seed data, fecundity depends on covariates. In the appendix we discuss conditional relationships involving the likelihood for seed data (equation 17.3) and a multivariate regression for growth and fecundity, influenced by covariates. We include as covariates tree diameter and light availability, summarized by exposed canopy area. The relationship between covariates and fecundity is described by a bivariate state-space model that additionally includes growth (diameter increment) as a response variable. This bivariate model is described with diameter growth in the next section.

17.3.3 Diameter growth and fecundity

Diameter growth (cm per year) is informed by two sparse data sets and by the state-space model that includes fecundity. Censuses conducted at two to four year intervals, which include measurements of diameter on all trees, provide observations of diameter change over the measurement interval. Increment cores are obtained for some individuals and provide annual rates of growth up until the year in which the core was extracted. Because they are laborious to obtain and they can damage trees, increment cores are not available for many trees. Thus, both types of data are sparse, but in different ways; census data exist for all individuals, but only in a few years, and increment cores were taken from a subset of individuals, but cover all years up until the year the core was collected. In consideration of the multiple data types and sparsity of both, diameter growth was modeled in a two-step process, the first step being a model that assimilates the different types of data and generates posterior estimates of growth for each tree year. This analysis is described in Clark *et al.* (2007) and Metcalf *et al.* (2009). Products of this analysis include estimated means and standard deviations for diameter and diameter increment in each tree-year (Figure 17.2). These are used as priors for the second step, the analysis described here.

The state-space model for growth and fecundity was developed to allow for measurable covariates known to affect demography, random effects at the individual level, year effects, and error. Random effects are included, because many factors could affect individual health that cannot be assigned to observed

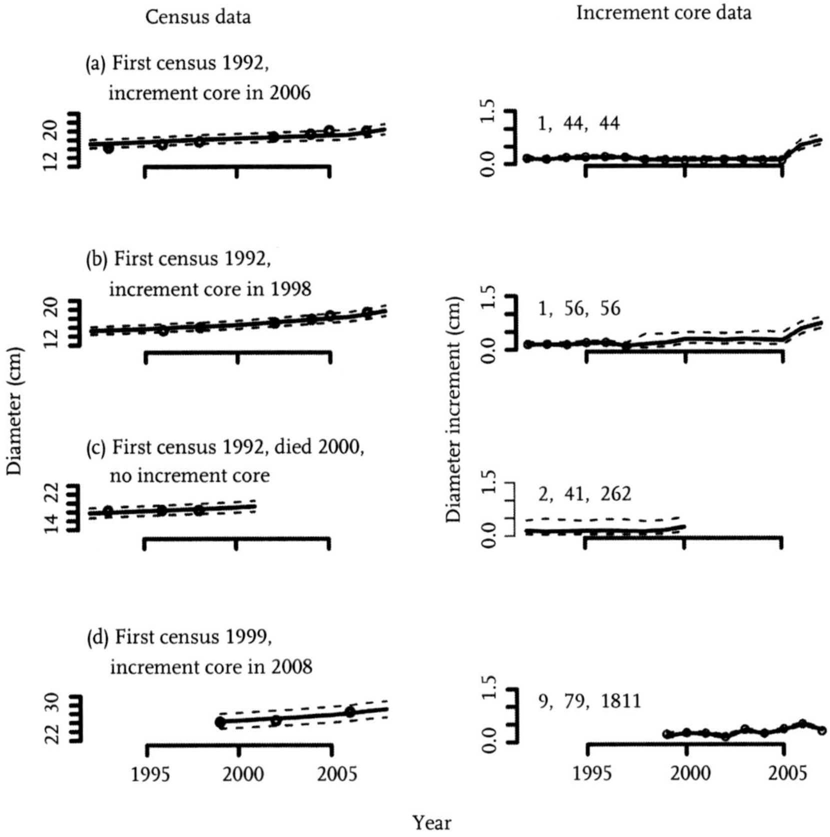

Fig. 17.2 Posteriors for diameter at left and diameter increment at right, represented by median (solid) and 95% CI (dashed) compared with observations (dots). Series vary in length, depending on when observations began and tree survival. Posteriors are generally narrow where increment core data are available, and vice versa. At left are shown corresponding intervals for tree diameter with diameter census data (dots). Numbers at right indicate (individual i, plot j, and a unique order number in the data base). This example is for *Quercus*.

variables. These factors are related to genotype and spatial variation in resources and factors that limit growth. Year effects are included because year-to-year variation in climate is difficult to quantify in ways that might be important for trees. Year effects allow for variation in time that is shared across the entire population. The growth-fecundity submodel is

$$\begin{aligned} y_{ij,t} &= x_{ij,t-1}A + b_t + b_{ij} + \varepsilon_{ij,t} \\ b_{ij} &\sim N_2(0,\ V_b) \\ \varepsilon_{ij,t} &\sim N_2(0,\ \Sigma) \end{aligned} \tag{17.7}$$

with a response vector that includes diameter growth and fecundity

$$y_{ij,t} = \left[\ln\left(d_{ij,t}\right), \ln\left(f_{ij,t}\right)\right]; \tag{17.8}$$

covariates

$$x_{ij,t-1} = \left\lfloor 1, \ln\left(D_{ij,t-1}\right), \ln^2\left(D_{ij,t-1}\right), \ln\left(\lambda_{ij,t-1}\right), \ln\left(d_{ij,t-1}\right) \right\rfloor; \tag{17.9}$$

parameters for fixed covariate effects A, fixed year effects b_t, random individual effects b_{ij}, and error $\varepsilon_{ij,t}$. If the genus includes multiple species, the vector $x_{ij,t}$ includes a fixed effect for each species of the genus. Priors for regression parameters are specified such that $\ln(D)$ term takes up allometric effects of size on fecundity when trees are small, and $\ln^2(D)$ takes up senescence effects on growth and fecundity when trees are large. Growth and fecundity rates are both affected by individual size and by abiotic covariates in any given year. By including allometric relationships in the $\ln(D)$ term, the exposed canopy area λ term estimates the effect of light availability (as opposed to size) on growth and fecundity (Section 17.4).

In addition to the regression for growth and fecundity, both variables conditionally depend on data. For fecundity, conditional dependence for $f_{ij,t}$ includes all seed traps on plot j in year t. Because seed trap count $s_{jk,t}$ depends, in turn, on all trees on plot j, the conditional dependence for all trees on plot j in year t is $f_{j,t} > 0$ involves equations (17.2) and (17.3),

$$p\left(f_{j,t} \middle| s_{j,t}, \ldots\right) \propto \prod_k Po\left(s_{jk,t} \middle| A_{jk} g_{jk}\left(f_{j,t}; u\right)\right) \prod_i p\left(f_{ij,t} \middle| Q_{ij,t}, q_{ij}, x_{ij,t-1}\right). \tag{17.10}$$

The second factor on the right hand side comes from equation (17.2). The conditional means and variances for equation (17.2) are

$$\begin{aligned} \mu_{ij,t}^{f|d} &= \mu_f + \frac{\Sigma_{12}\left(\ln d_{ij,t} - \mu_d\right)}{\Sigma_{11}} \\ V_{ij,t}^{f|d} &= \Sigma_{22} - \Sigma_{12}^2/\Sigma_{11} \end{aligned} \tag{17.11}$$

Σ_{ij} is the ijth element of Σ, and unconditional means are

$$\begin{aligned} \mu_d &= x_{ij,t-1} a_{\bullet d} + b_{d,t} + b_{d,ij} \\ \mu_f &= x_{ij,t-1} a_{\bullet f} + b_{f,t} + b_{f,ij}. \end{aligned}$$

The terms on the right hand side include the columns of A, i.e. $A = [a_{\bullet d}\ a_{\bullet f}]$, and the year and individual effects associated with growth and fecundity. For growth, we have the conditional dependence

$$\begin{aligned} p\left(d_{ij,t}\right) \propto\ & N\left(\ln\left(d_{ij,t}\right) \middle| \mu_{d|f}, V_{d|f}\right) \\ & N\left(\ln\left(d_{ij,t}\right) \middle| \ln\left(d_{ij,t}^{(0)}\right), v_{ij,t}\right) I\left(\left(d_{ij,t}^{(0)} - \sqrt{v_{ij,t}}\right) < \ln\left(d_{ij,t}\right)\right. \\ & \left. < \left(d_{ij,t}^{(0)} + \sqrt{v_{ij,t}}\right)\right) Bernoulli\left(z_{ij,t} \middle| \zeta_{ij,t}\right) \end{aligned} \tag{17.12}$$

where $z_{ij,t}$ is the event that the individual survived interval (t_{-1}, t), $d_{ij,t}^{(0)}$ and $v_{ij,t}$ are the prior mean and variance for log growth rate from the analysis of Clark *et al.* (2007)(Figure 17.2), and conditional means and variances are

$$\mu_{j,t}^{d|f} = \mu_d + \frac{\Sigma_{12}\left(f_{ij,t} - \mu_f\right)}{\Sigma_{22}}$$
$$V_{j,t}^{d|f} = \Sigma_{11} - \Sigma_{12}^2/\Sigma_{22}.$$

Note that the prior for the growth data is truncated to a width of two standard deviations. The standard deviations are large for tree years in which there are census data but no increment data (Figure 17.2). For years lacking increment core data, estimates are more heavily influenced by random effects for individuals and fixed effects for years, thus borrowing information from within the individual over time and from year-to-year variation that is shared by the entire population. The last factor in equation (17.12) comes from survival (Section 17.3.5). The survival probability enters the conditional probability, because it depends on growth rate, through the binned diameter classes (Section 17.3.5).

In addition to equation (17.7) that applies to mature individuals, we fitted an additional model for growth using all trees, regardless of maturation status. This model is univariate, but includes the same covariates as used for mature trees,

$$\begin{aligned} \ln d_{ij,t} &= x_{ij,t-1}a + b_t + b_{ij} + \varepsilon_{ij,t} \\ b_{ij} &\sim N\left(0, V_{b(1,1)}\right) \\ \varepsilon_{ij,t} &\sim N(0, w) \end{aligned} \tag{17.13}$$

where a is the parameter vector (the first column of A in equation 17.7), b_t is the first element of b_t in equation (17.7), and $V_{b(1,1)}$ is the random effects variance for diameter growth (element 1,1 of covariance matrix V_b) from equation (17.7).

17.3.4 Exposed canopy area

Light availability is included as a predictor of growth and fecundity. It is summarized by an index, the area of the crown potentially exposed to sunlight, $\lambda_{ij,t}$. This index has non-zero values, but can be small, such as for the case of a suppressed individual in the understory. It is estimated on the basis of three sources of information in a separate analysis (Wolosin *et al.*, in review). These data sources are (i) low-altitude imagery on which crown area can be measured, (ii) ordinal status classes assigned on the basis of ground observations, and allometric measurements combined with models of solar geometry that yield calculations of light availability throughout the day and over the growing season. A model combines these sources of information. Posterior means and variances enter this analysis as priors, just as described for diameter growth (Section 17.3.3).

17.3.5 Survival

Survival probability is typically modeled as a function of growth rate (see the appendix), which integrates many aspects of tree health (Kobe *et al.* 1995, Clark and Clark 1996, Wyckoff and Clark 2000, Bigler and Bugmann 2004) and of size (Clark and Clark 1996, King *et al.* 2006, Coomes and Allen 2007). A number of functional forms have been used to relate survival to growth rate. The problem with any functional form comes from the facts that (i) this relationship can be strongly nonlinear, changing abruptly at growth rate values close to the lowest range of values typically observed, and (ii) the distribution of data has a large impact on estimates, and individuals close to death may be poorly represented in data sets, because such individuals died disproportionately before the study began. We have developed or modified nonparametric approaches (Wyckoff *et al.* 2000, 2000, Clark *et al.* 2007, Metcalf *et al.*, in review) to describe this relationship and apply one that combines not only growth rate, but also tree size effects on mortality risk (Clark *et al.*, 2007). We include tree size as a predictor of survival, because mortality risk may increase as trees become large and senesce or become susceptible to high winds (Batista *et al.* 1998, Uriarte *et al.* 2004, Rich *et al.* 2007).

Let $z_{ij,t}$ be the event that an individual ij is alive in year t, in which case it survived from year t to $t+1$ with probability

$$\zeta_{ij,t} = 1 - \left(\mu_{d_{ij,t-1}} + \mu_{D_{ij,t-1}} - \mu_{d_{ij,t-1}}\mu_{D_{j,t-1}}\right). \tag{17.14}$$

There are discrete bins for both growth rate and diameter. In year $t-1$ individual ij's growth rate bin is indicated by $\mu_{d_{ij,t-1}}$ and its diameter bin is indicated by $\mu_{D_{ij,t-1}}$. There are monotonicity priors for both sequences, decreasing for growth rate and increasing for diameter (Section 17.4). The likelihood is $Bernoulli(z_{ij,t}|\zeta_{ij,t})$. In the next section we specify priors.

17.4 Prior distributions

The model includes both informative and non-informative prior distributions. Due to the size and complexity of the model, where possible, we used informative priors that are flat but truncated in some fashion, to maximize transparency, i.e. for identification of the contributions of prior versus likelihood. Here we summarize priors and how they were selected to balance information.

The fixed effects in the state-space model (17.7) have flat priors bounded by values either having theoretical justification or sufficiently wide to not impact estimates,

$$vec(A) \sim I(a_1 < vec(A) < a_2) \tag{17.15}$$

where a_1 and a_2 are vectors of minimum and maximum values, respectively. We describe prior values for specific elements of A, using indexing for elements of A that assume a single intercept. Recall that there are separate intercepts for each species included in a given genus (Table 17.3). The actual number of rows of A is p = number of species + 4 (there are four covariates). The first subscript indicates the covariate (equation 17.9) and the second subscript indicates the response (equation 17.8). Truncation points that affect estimates include:

A_{21} – *diameter effect on diameter growth increment constrained to be near zero*: Diameter is included as a predictor, because we expect it to directly affect fecundity – large trees are capable of high seed production. We expect it to also be correlated with the other response variable, diameter growth increment. Because that correlation should be taken up by canopy exposure, rather than by tree diameter directly, we constrain this parameter value to be near zero. From open-grown trees there is no clear evidence for a direct size effect on diameter increment. The correlation between tree size and diameter growth increment is expected to result from the fact that large trees are more likely to have higher light exposure. Because there is no theoretical justification for non-zero values, limits on this parameter are $(-0.02, 0.02)$.

A_{22} – *diameter effect on fecundity constrained between 1.5 and 2.5*: Allometric arguments and empirical evidence suggest that potential fruiting yield should scale with canopy width, which, in turn is roughly proportional to diameter. In fact, this potential should not be realized for trees crowded by neighbours. We allow for this effect of size on potential yield effect due to size with the constraints (1.5, 2.5), as modified by competition, which is reflected in exposed canopy area, i.e. the term including λ.

(A_{31}, A_{32}) – *large diameter effect is negative*: The squared diameter term in equation (17.9) is included to allow for potential senescence, a decline in physiological function with age. Tree data sets rarely have sufficient large (potentially old) individuals to estimate these effects, but we can allow that senescence does eventually occur by specifying that this effect only has impact for especially large individuals. This term is constrained to be negative.

(A_{41}, A_{42}) – *lag-1 effect of growth rate on growth rate and fecundity*: This effect was constrained to be effectively zero for growth rate (A_{41}) but unconstrained for fecundity (A_{42}). We wanted to parametrize the effect on fecundity, so it could be used for predictive modeling of potential tradeoffs in time between growth and fecundity.

The estimates of growth and fecundity represent a balance between contributions from the regression (i.e. the size and light covariates in equation 17.7) and data models for growth rates and seed data. It is not necessarily 'objective' to use a non-informative prior for the state-space error covariance matrix, because there is no objective criterion for balancing information that enters from multiple data types. Data are known to be noisy, particularly seed rain. We used an informative prior on the error covariance matrix Σ to represent a level of variation expected after that taken up by covariates, random effects, and year effects and to assure that covariates were not overwhelmed by noise. The values used for variances on the log growth (cm) and log fecundity (seeds per tree) were 0.05 and 0.2, respectively. These allow for realistic levels of variation on the non-log scale. We used the prior

$$Wishart\left(\Sigma^{-1} \middle| \begin{bmatrix} 0.05 & 0 \\ 0 & 0.2 \end{bmatrix}^{-1}, \quad n_{IJT}\right) \tag{17.16}$$

where $n_{IJT} = \sum_{i,j}(T_{ij} - t_{ij})$ is the number of tree-years in the study. Through extensive sensitivity analysis, this prior was found to provide an acceptable balance of data and regression model, contributing to the conditional posterior approximately twice the weight coming from the regression.

By contrast, priors on random effects and year effects were weak – we wanted data to dominate these estimates. The prior for random effects is

$$Wishart\left(V_b^{-1} \middle| \begin{bmatrix} 0.2 & 0 \\ 0 & 2 \end{bmatrix}^{-1}, \quad \max\left(3, n_{IJ}/100\right)\right) \tag{17.17}$$

where $n_{IJ} = \sum_j I_j$ is the number of trees. The second parameter in equation (17.17) is rounded to an integer value and ranged from 3 to 60 for different species. The contribution to the conditional posterior ranged from about 1/10 to 1/50 of the weight coming from the regression.

The prior for fixed year effects is

$$N_2\left(b_t \middle| [0\ 0]', diag(100, 100)\right) \tag{17.18}$$

and includes a sum-to-zero constraint (intercepts are included in A), implemented directly in the Gibbs sampler.

Because many individuals are not mature, a separate univariate regression is fitted to all tree years, regardless of maturation status (equation 17.13). The covariates are the same as those listed for the multivariate regression given above, and sampling makes use of the univariate distributions corresponding to each of the foregoing multivariate ones. These are Gaussian for fixed (including years) and random effects, and inverse Gamma for variances.

Diameter growth increments have a prior for each tree year taken from the posterior from the analysis of Clark *et al.* (2007) and shown in equation (17.12). Because there are thousands of such densities, the truncation of this posterior at a width of two standard deviations was based on a prior belief that true increments should be within this range. Diagnostics showed that posterior estimates from this analysis did not tend to acccumulate at these truncation values.

As with diameter increment, the Gaussian prior for canopy values is truncated to two standard deviations in width,

$$\ln \lambda_{ij,t} \mid \sim N\left(c_{ij,t}^{(o)}, C_{ij,t}\right) I\left(\left(c_{ij,t}^{(0)} - \sqrt{C_{ij,t}}\right) < \ln\left(\lambda_{ij,t}\right) < \left(c_{ij,t}^{(0)} + \sqrt{C_{ij,t}}\right)\right) \quad (17.19)$$

where $c_{ij,t}^{(0)}$ and $C_{ij,t}$ are the prior mean and variance (log scale), taken from the posterior for the analysis of canopy area (Wolosin *et al.*, in review).

Priors for fecundity, maturation, gender, and missing seed data were either non-informative or derived from previous observation. A flat prior was used for fecundity, truncated at the smallest number of seeds observed for a tree and at values much larger than implied by observation of seed densities,

$$f_{ij,t} \sim unif\left(f_{\min}, f_{\max}\right). \quad (17.20)$$

For instance, when defining $f_{\min}$, we did not expect that a mature individual would produce less seeds than typically contained in a single fruiting structure (e.g. *Pinus, Liriodendron, Liquidambar*). For maximum values, we used parameter estimates similar to those obtained in a simpler model (Clark *et al.* 1999) to 'invert' seed density observations and thus approximate what might constitute unrealistically high seed production for an individual of a given species and size. The model for seed dispersal provides an expected seed density given the spatial locations of trees of different sizes. For example, *Acer rubrum* seeds have been observed at average densities of 10^2 seeds m^{-2} beneath mature trees but not at average densities of 10^3 seeds m^{-2}. This inversion was used not only to set limits on fecundities for individual trees, but also to define the (prior) Poisson means for missing seed data.

The maturation diameter for an individual was assigned a prior that was truncated at values below which we believed that no individuals could be mature and above which we believed all individuals would be mature. These beliefs came from independent observations of trees in similar settings. These limits on maturation diameters translate to limits on maturation year (see appendix). There are prior minimum and maximum diameters, which differ among species. The female fraction was given the prior $Be(\phi|h_1, h_2)$ with $h_1 = h_2 = 4$, having a mean of 0.5 and being dominated by the data. The probability of recognizing a mature individual was assigned the prior $Be(v|v_1, v_2)$ with $v_1 = v_2 = 0.002n_{IJ}$, which has a mean of 0.5 and is dominated by the data.

Parameters for the logit function of maturation equation (17.1) were assigned the prior

$$N\left(\beta^{\theta}\,|b^{\theta},\,V^{\theta}\right)\,I\left(\beta^{\theta}_{2,3}>0\right) \tag{17.21}$$

a truncated normal prior with mean vector $b^{\theta} = [-3,\ 0.1,\ 0.1]$ and covariance $V^{\theta} = diag[10,\ 10,\ 10]$. The positivity constraint on the second and third elements of the vector comes from the prior belief that the relationship between maturation and diameter and light availability is non-negative.

Priors for the seed data model in equation (17.3), including the dispersal parameter and the seed fraction originating outside the map, were

$$p(u,c) = N(u\,|u_0,\,V_u)\,N(c\,|c_0,\,V_c)\,I(u,c>0) \tag{17.22}$$

where parameter values were chosen to be informative. For u they differ among species; we used $c_0 = 0.02$ and $V_c = 0.01$. There is a positivity constraint on u and c.

The monotonicity priors on the parameter sequences μ_d and μ_D in equation (17.14) were designed to allow for uneven distribution of data and strong nonlinearities. Because slow growth is associated with death, the observations of growth rate below a certain threshold are rarely observed. However, this lack of slow growth observations results from the fact that mortality risk increases sharply at slow growth rates. For this reason, our sequence of μ_d values has an intercept at 1. Although zero growth rates do occur in particular years, we used this assumption as a way of approximating the sharp increase in risk that can occur at low growth. This assumption is obviously flexible. In addition to monotonicity, there was an informative prior for values within the sequence μ_D, which was $Be(a_k, b_k)$, where a_k is 0.001 for $k = 1, 2, 3$ and $a_k = 10$ for $k = 4, 5, 6$,

$$b_k = a_k\left(\frac{1}{\mu_k^0} - 1\right)$$

$$\mu_k^0 = [0.00001, 0.00002, 0.00003, 0.00004, 0.00005]\,. \tag{17.23}$$

This prior assures essentially zero values for juvenile trees (bins 1, 2, 3), but is non-informative (but monotonically increasing) for large trees. Thus, the diameter effect only affects large trees. Although small trees grow slowly and thus are at higher mortality risk, this is a growth effect, not a diameter effect. This informative prior allows us to separate the effect of slow growth from that of large size, which could indicate senescence.

17.5 Computation

The posterior was simulated with Gibbs sampling, based on conditional posteriors that are discussed in the appendix, some of the embedded steps being

Metropolis. The simulation was initialized at prior mean parameter values (diameter increments and crown areas), random draws from priors, or MLEs based on simpler models (fecundities for trees were initially estimated without year or individual effects using the approach of Clark *et al.* 1999).

Due to the size of the model, efforts were made to optimize code. Despite the large number of years across many individuals within multiple plots, the main Gibbs loop contains only three loops over years (including one to update maturation/fecundity, one for missing seed data, and another for dispersal and Poisson parameters), and no loops over individuals or plots. Data structures that include pointer arrays were used to rebuild (reorder and restack) matrices of state variables based on the changing gender and maturation statuses of trees and tree-years, respectively.

Convergence was achieved with 10,000 iterations for species with moderate numbers of individuals, but required up to 200,000 iterations for trees with many individuals. There are a large number of parameters, not all of which could be sampled efficiently. The lowest updating rates and highest autocorrelations were obtained for fecundities of dioecious species (*Acer rubrum, A. pennsylvatica, Fraxinus americana,* and *Nyssa sylvatica*), due to the discrete nature of $Q_{ij,t}$ and H_{ij}, and the blocking over all tree-years within a plot. Thus, for fecundity/maturation/gender of dioecious species, we selected for updating at random 30% of the trees for a given iteration and embedded five such iterations within each Gibbs step.

17.6 Diagnostics

From 50,000 to 1,000,000 Gibbs steps were discarded, followed by 50,000 to 100,000 iterations that were retained for analysis. We inspected Gibbs chains for all parameters as well as for samples of individual effects. Experiments involved many parameter initializations; however, results presented here come from single long runs for each taxon group. Acceptance rates for Metropolis steps were generally above 0.2. The exception was for dioecious species, where low acceptance rates were addressed by embedding multiple iterations per Gibbs step (Section 17.5). To help evaluate results we compared priors and posteriors, we considered predictive capacity, in terms of data used to the fit model, and we compared predictive intervals from the model with estimates of latent states that could not be directly observed. Here we discuss some comparisons.

17.6.1 Some prior/posterior comparisons

Some of the estimates for parameters from an example taxon, *Quercus*, are shown in Table 17.5. Estimates for marginal posteriors are accompanied by fitted truncated normal distributions, which would be used in the event that it

Table 17.5 Posterior estimates for some of the main parameters in the model, shown for the taxon *Quercus.*

Symbol (equation)	Parameter p^1	Marginal posterior				Fitted model[2] $N(p\|m, s^2)I(p_1 < p < p_2)$			
		mean	std dev	0.025	0.975	m	s	p_1	p_2
A (17.7)	quun.d	−1.65	0.0271	−1.7	−1.59	−1.65	0.0271	−4	1
	quun.f	4.15	0.283	3.7	4.65	4.15	0.283	−1	10
	diam.d	0.0196	0.000394	0.0186	0.02	0.393	0.0122	−0.02	0.02
	diam^2.d	−0.000155	$5.55e-05$	−0.000301	−0.000101	0.00662	0.000619	−0.2	−1e−04
	cnpy.d	0.0591	0.00478	0.0496	0.0683	0.0591	0.00479	0.01	1
	dlast.d	0.00992	$8.34e-05$	0.00969	0.01	0.0376	0.00163	−0.01	0.01
	qual.d	−1.3	0.0151	−1.33	−1.27	−1.3	0.0151	−4	1
	quco.d	−1.66	0.0168	−1.69	−1.62	−1.66	0.0169	−4	1
	qufa.d	−1.26	0.0324	−1.32	−1.2	−1.26	0.0324	−4	1
	quma.d	−1.61	0.0359	−1.68	−1.54	−1.61	0.0359	−4	1
	quph.d	−1.37	0.0182	−1.41	−1.34	−1.37	0.0182	−4	1
	qupr.d	−1.7	0.0115	−1.72	−1.68	−1.7	0.0115	−4	1
	quru.d	−1.62	0.0123	−1.64	−1.6	−1.62	0.0123	−4	1
	qust.d	−1.5	0.0214	−1.54	−1.46	−1.5	0.0214	−4	1
	quve.d	−1.76	0.0219	−1.8	−1.72	−1.76	0.022	−4	1
	diam.f	1.51	0.0117	1.5	1.54	0.473	0.0724	1.5	2.5
	diam^2.f	−0.183	0.00782	−0.193	−0.159	−0.183	0.00782	−0.25	−1e−04
	cnpy.f	0.131	0.0277	0.0777	0.183	.131	0.0278	0.01	3
	dlast.f	−0.0891	0.0949	−0.236	0.107	−0.0891	0.095	−2	2
	qual.f	4.09	0.16	3.84	4.51	4.09	0.161	−1	10
	quco.f	3.9	0.186	3.6	4.32	3.9	0.186	−1	10
	qufa.f	4.03	0.202	3.72	4.54	4.03	0.202	−1	10
	quma.f	4.06	0.181	3.74	4.45	4.06	0.181	−1	10
	quph.f	4.29	0.155	4.06	4.68	4.29	0.155	−1	10
	qupr.f	3.9	0.18	3.63	4.34	3.9	0.18	−1	10
	quru.f	4.28	0.159	4.03	4.69	4.28	0.16	−1	10
	qust.f	4.19	0.151	3.95	4.58	4.19	0.151	−1	10
	quve.f	4.02	0.187	3.75	4.52	4.02	0.187	−1	10

a (17.13)	diam	0.0185	0.00137	0.015	0.02	0.184	0.016	−0.02	0.02
	diam^2	−0.0538	0.00564	−0.0669	−0.0434	−0.0538	0.00564	−1	−0.001
	cnpy	0.0228	0.00366	0.0155	0.03	0.0228	0.00366	0.005	1
	dlast	0.00987	0.000129	0.00952	0.01	0.0394	0.00211	−0.01	0.01
	qual	−0.76	0.18	−1.1	−0.398	−0.761	0.18	−4	0
	quco	−1.08	0.289	−1.6	−0.477	−1.08	0.289	−4	0
	qufa	−0.554	0.354	−1.36	−0.0301	−0.397	0.465	−4	0
	quma	−0.95	0.464	−1.97	−0.107	−0.907	0.509	−4	0
	quph	−0.67	0.275	−1.21	−0.165	−0.666	0.282	−4	0
	qupr	−1.14	0.0674	−1.27	−1.01	−1.14	0.0675	−4	0
	quru	−1.08	0.114	−1.31	−0.884	−1.08	0.114	−4	0
	qust	−0.736	0.345	−1.44	−0.149	−0.728	0.357	−4	0
	quve	−1.24	0.272	−1.67	−0.57	−1.24	0.272	−4	0
	quun	−1.12	0.471	−2.2	−0.305	−1.12	0.473	−4	0
Σ (17.7)	$\Sigma_{(1,1)}$	0.0511	0.000422	0.0503	0.0519	–	–	–	–
	$\Sigma_{(2,2)}$	0.218	0.00174	0.214	0.221	–	–	–	–
	$\Sigma_{(1,2)}$	$6.35e-05$	0.000593	−0.00111	0.00123	–	–	–	–
β^{γ} (17.4)	intercept	0.757	0.102	0.539	0.935	–	–	–	–
	slope	0.1	0.00714	0.0887	0.118	–	–	–	–
V_b (17.7)	$V_{b\,(1,1)}$	0.105	0.00433	0.0967	0.114	–	–	–	–
	$V_{b\,(2,2)}$	0.34	0.0177	0.307	0.375	–	–	–	–
	$V_{b\,(1,2)}$	0.0205	0.00977	0.000808	0.0387	–	–	–	–
β^{θ} (17.2)	intercept	−1.42	0.0421	−1.5	−1.34	–	–	–	–
	diameter	0.0891	0.00296	0.0826	0.0942	–	–	–	–
	light	0.0197	0.00355	0.0126	0.0264	–	–	–	–
w (17.13)	error variance	0.0415	0.00054	0.0404	0.0425	–	–	–	–
u (17.6)	dispersal	29.9	0.0624	29.8	30	–	–	–	–
v	detection	0.0732	0.00371	0.0662	0.0807	–	–	–	–

[1] For A, parameters are referenced as 'covariate.d' or 'covariate.f'. Thus, 'cnpy.d' is the effect of canopy, or light response, on diameter growth d, and 'quru.f' is the fecundity intercept for *Quercus rubra*. For a, there is a single response variable.
[2] p_1 and p_2 are the lower and upper limits for the truncated normal. For A, they are referenced in equation (17.15).

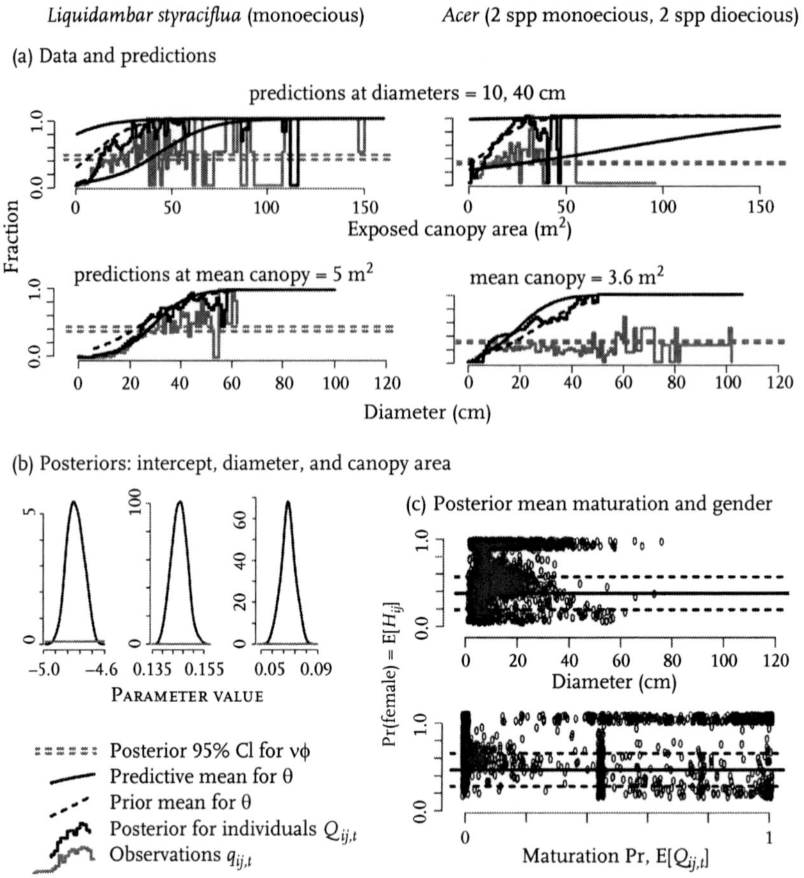

Fig. 17.3 (a) Data and fitted models for maturation and gender. *Liquidambar* (left) is monoecious, *Acer* (right) includes both monoecious and dioecious species. Lower histograms show the faction of observations in diameter bins for which $q_{ij,t} = 1$. Upper histograms show the fraction for which the estimates $Q_{ij,t} = 1$. The function θ is shown for prior (dashed) and posterior (solid) values of β^{θ}. Horizontal dashed lines are 95% CIs for $\upsilon\phi$. (b) Posteriors for β^{θ} for *Liquidambar* compared with priors (flat lines). (c) Posterior means for gender plotted against maturation status.

was desirable to draw samples from it, for purposes of prediction. This would be necessary if one did not have access to the full Gibbs chains. There are two such distributions, one for the parameters of A and second for those of a. The full covariances are not included for space considerations; we have included only standard deviations in Table 17.5. We consider aspects of data, priors, and posteriors, beginning with gender and maturation, followed by growth/fecundity, then survival.

Figure 17.3 provides perspective on how data, priors, and posteriors relate for a monoecious species (*Liquidambar styraciflua* on the left side of Figure 17.3) and the genus *Acer*, which includes both monoecious (*A. barbatum, A. saccharum*) and dioecious (*A. rubrum, A. pensylvanicum*) species (right side of

Figure 17.3). In fact, *A. rubrum* is polygamo-dioecious, having some individuals that are male, some female, and some supporting both male and female flowers. Our female fraction for this species includes both female and monoecious individuals. For the monoecious *Liquidambar*, all mature individuals have female function, so $\phi = 1$. With increasing diameter and canopy area, larger numbers of individuals are observed to be mature (grey histogram in Figure 17.3a) and still more are estimated to be mature (black histogram), because detection probability $\upsilon < 1$ (horizontal dashed grey lines). Note that the posterior 95% credible interval for the estimate of υ roughly averages the red histogram (observations) at sizes where maturation is reached, whereas the black histogram (estimates) approaches 1. This is the expected relationship between observations, detection probability, and the true states. The values approach zero for small diameters, because small trees cannot reproduce. However, values do not approach zero for small exposed canopy areas, because it is possible for trees that are highly shaded to produce at least some fruit.

The estimates for the population-level relationship are given by predictions of θ, shown in Figure 17.3 as predictive means only. These are plotted against exposed canopy area λ (for two values of diameter) and against diameter D (for the mean canopy area). We show prior and posterior means for θ. The estimates of β^{θ}, which are the basis for predictions of θ, are well resolved (Figure 17.3b). They predict maturation at smaller sizes and at lower canopy exposure for *Acer* than they do for *Liquidambar*. The population level predictions (smooth lines) do not appear to run precisely through the histograms of individual level predictions, because the individual level predictions effectively marginalize over diameter and canopy distributions for the entire population, whereas the predictive mean curves are conditional on specific diameter and canopy values.

Capacity to predict gender increases with tree size, because large trees are more likely to be reproductive, and reproduction is the only evidence for gender. The probability of being female tends to zero or one with increasing diameter (Figure 17.3c). As probability of being mature increases, so too does predictability of gender. If data were static, at small diameters, the probability that any individual is female would tend to the posterior estimate of ϕ. This does not occur in Figure 17.3c, because small diameter individuals may later become mature, thus providing evidence of their gender even at small size, i.e. before they were mature. With increasing confidence in maturation status, we see a greater tendency to be female than male. This tendency results from the fact that two of the species in *Acer* are modeled as monoecious and thus will always be counted in the female class.

The influence of truncated priors (equation 17.15) is evident in posteriors for parameters from the growth/fecundity state-space model (Figure 17.4). Due to the size and complexity of the model the flexibility to assign hard boundaries to one or both limits for these parameters and the transparency of prior effects

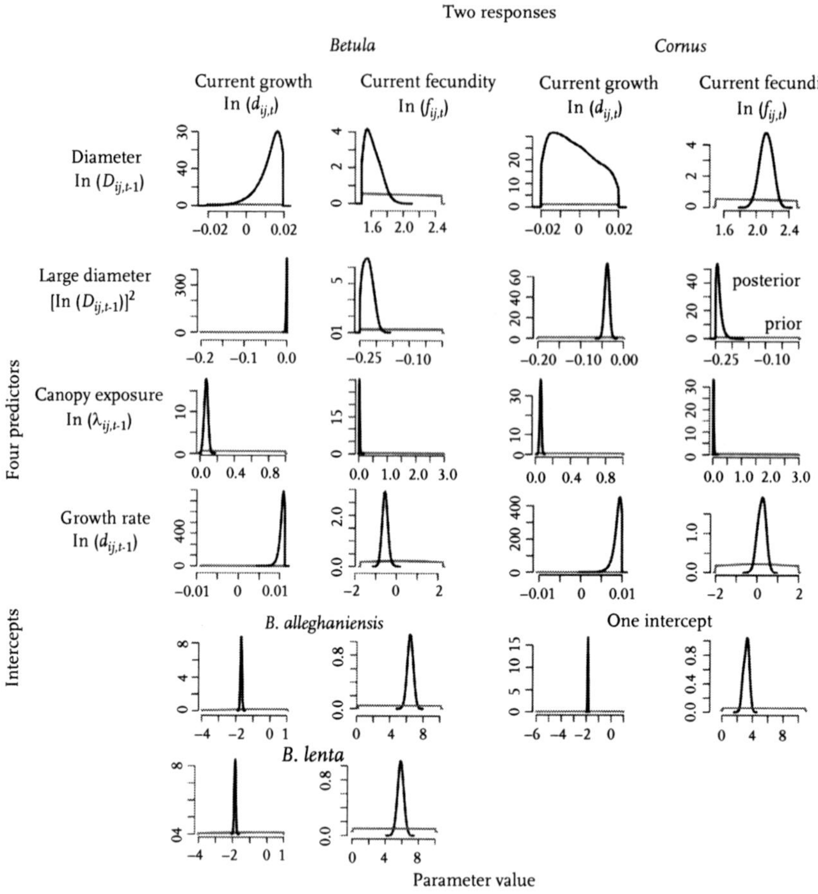

Fig. 17.4 Comparison of priors (flat grey lines) and posteriors (black) for the fixed effects in the matrix A for the state space model (Equation 17.7) for two genera, *Betula* and *Cornus*. *Betula* has two intercepts, one for each of two species. The horizontal axis bounds the prior.

on posteriors was deemed an advantage. We include in this example two shade-tolerant taxa (the canopy exposure effects λ are near zero for both diameter growth and fecundity), one having high fecundity (*Betula*) and another low fecundity (*Cornus*). The fit for *Betula* includes two sets of intercepts, one for the higher fecundity and faster growing *B. alleghaniensis* and one for the lower fecundity and slower growing *B. lenta* (lower panels of Figure 17.4). For this particular fit, we held the diameter effect D on growth rate d to be near zero (there is no prior knowledge to suggest growth rate should respond directly to size until trees become large), but assumed that the effect of D on fecundity should fall between 1.5 and 2.5. Together these assumptions allow for a direct size effect on fecundity that accords with allometric theory, thus allowing that effects of canopy exposure, which can be correlated with size, are more realistic.

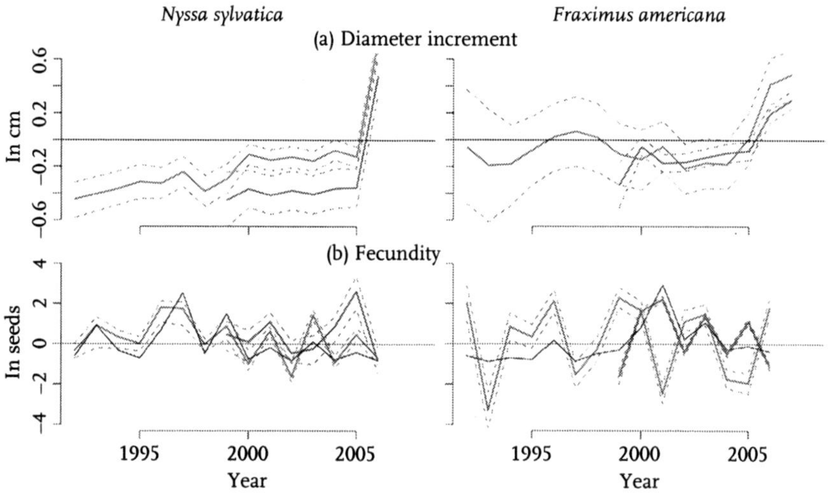

Fig. 17.5 Posterior medians and 95% CIs for year effects b_t for two species. Separate year effects were used for southern Appalachian plots and for Piedmont plots. Solid thin lines in lower panels are proportional to log seed rain averaged over both sites.

To allow for declining physiological function with size, we included the $(\ln D)^2$ term in the model and constrained it to be negative. This term will have increasing influence with size. We did not have large enough trees in these data sets to show clear effects on growth (posteriors clumped at the upper zero boundary), but there was evidence for this negative effect on fecundity for a number of species. Canopy exposure λ has a positive effect on both growth and fecundity. For these shade tolerant species, these estimates were close to zero for both growth and fecundity.

The lagged growth rate effect was constrained to be near zero for growth, because we wanted long-term trends in growth to be taken up by year effects. The tendency for positive correlation was constrained by the upper boundary at 0.01. However, we wanted to explicitly parameterize the lag-1 effect of growth on fecundity, because this could be important for demographic prediction. We obtained a range of values from strongly positive to strongly negative for the lag-1 effect of growth on fecundity.

The different intercepts for genera having more than one species (e.g. *Betula*) allowed us to model fecundities for groups of species having indistinct seeds. Although *B. alleghaniensis* and *B. lenta* have similar life histories, we found substantially higher fecundity for *B. alleghaniensis* (recall that intercepts are on a log scale).

Two sets of fixed year effects b_t were used, one for each of the two regions and shown in Figure 17.5 as longer curves for Coweeta (back to 1992) and shorter curves for Duke Forest (beginning in 1999), both having a sum to zero

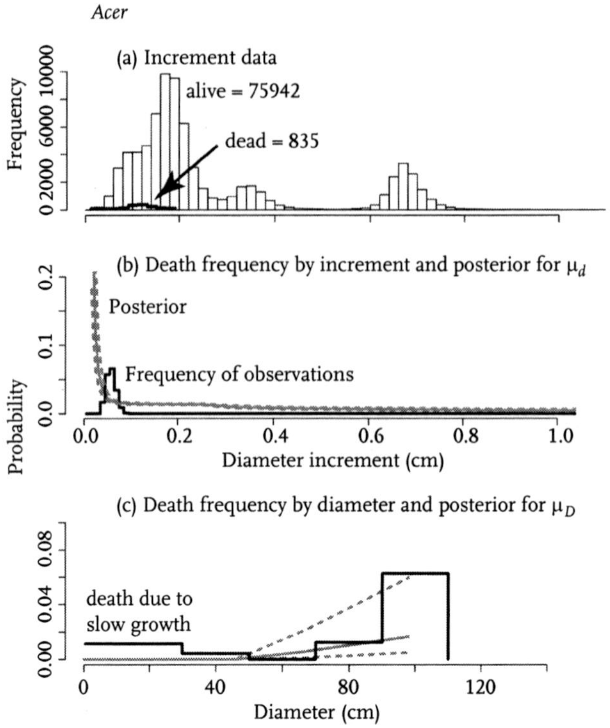

Fig. 17.6 (a) Survival data, (b) relative frequency of deaths (histogram) and posterior median and 95% CI for μ_d, (c) relative frequency of deaths (histogram) and posterior median and 95% CI for μ_D.

constraint. Those for growth show increasing rates in recent years, which could result from several mechanisms. Those for fecundity show a tendency for two-year cycles in *Nyssa*. Note that year effects need not strictly track seed rain trends (solid thin line in Figure 17.5), because other covariates vary from year to year.

The posterior estimates for effects of diameter increment μ_d and diameter μ_D on survival show the effects of the monotonicity assumptions (Figure 17.6). The relationship between growth increment and mortality risk is highly nonlinear close to the limit of increment core data (Figure 17.6a, b). Apparently, trees reach a threshold of low growth, below which mortality risk rises substantially. The histogram of observations in Figure 17.6(c) shows modes not only at the largest sizes, but also the smallest. The latter mode results from the slow growth that results from low light levels in the forest understory. The priors help to discriminate the growth from size effects, by recognizing that mortality risk declines with growth rate, but increases with size. Beyond this relationship already known from previous studies, the prior is weak as to the shapes of these relationships. The prior from equation 17.23 allows the assumption that death

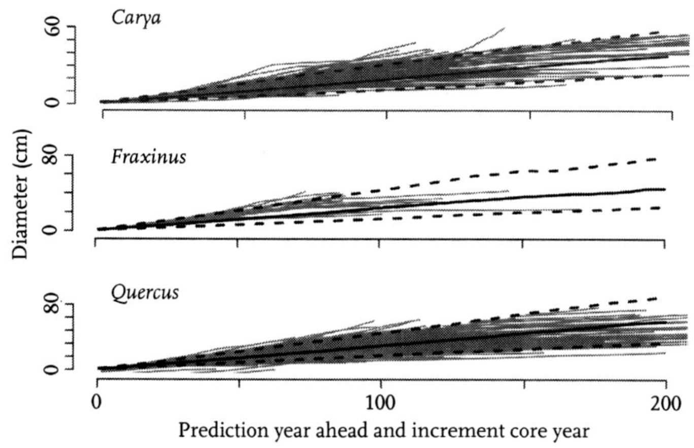

Fig. 17.7 Comparison of increment data from tree ring data not used in fitting the model (thin lines) and predictive distributions of tree diameter from the model (Section 17.6.2).

of these small individuals is not due directly to size, but rather indirectly, due to low growth rates.

17.6.2 Data prediction

To provide further insight into model behaviour we predicted data and latent states. Here we describe some of these predictions and how they compare to data or estimates of latent states.

Predictions of diameter growth were evaluated against an independent data set of growth, obtained from measurements of increment cores spanning decades. The example in Figure 17.7 is typical – we obtain good coverage of size distributions for decade-ahead prediction. It is worth mention that the model contains no explicit age information. And there is no attractor in the model that would necessarily make it converge to a particular diameter value. Moreover, these are not one-step ahead predictions, as is often used to evaluate fits of time series models, but rather 200 year ahead predictions. Here we simply initialized the model and incremented year-by-year predictive distributions, approximating

$$p\left(d_{t+1}, f_{t+1} \middle| X_t^*, X\right) = \int p\left(y_{t+1} \middle| X_t^*, \pi\right) p(\pi|X) d\pi$$

where X is taken to be all data and priors entering the model, π is a vector of parameters, X_t^* is taken to be the current covariates, which include the previously predicted diameter and increment, and a random draw from the distribution of canopy exposure values $\lambda_{ij,t}$ contained in the data. The integrand includes the state-space structure of the model (equation 17.7) and the posterior.

The integral is approximated by drawing at random a row from the iteration-by-parameter matrix of Gibbs sampler output. The tree is initially immature (1 cm diameter) and subject to growth rate in equation (17.13), risk of maturation from equation (17.2), and risk of death from equation (17.14). A draw from V_b determines its random individual effect. If it does not survive an iteration, it is pronounced dead and removed from the simulation. The next growth rate is drawn from a univariate or bivariate normal, depending on maturation status.

A similar approach to prediction was used to evaluate other aspects of the model. We do not observe fecundity, so we cannot compare direct observations of fecundity against model predictions. However, we can check predictions of seed rain. We did this in two ways. Consider that seed rain can be predicted from different levels in the model. The model generates estimates of latent states $f_{ij,t}$. Based on these latent states for all trees on plot j in year t, there is a likelihood for seed rain data at location k in year t. Thus we can consider how well the expected seed rain for all trees at j in year t predict seed rain data at k in year t, or

$$p\left(s_{k,t} \,|\, E\left[f_{j,t}, Q_{j,t}\right], X\right) = \int p\left(s_{k,t} \,|\, E\left[f_{j,t}, Q_{j,t}\right], \pi_1\right) p(\pi_1 \,|X)\, d\pi_1$$

where X represents all data and priors, and the vector $\pi_1 = (u, \beta^\gamma)$. Alternatively, we could predict from a lower level to include the uncertainty in $f_{j,t}$ itself

$$p\left(s_{k,t} \,|\, E\left[X_{j,t-1}\right], X\right) = \int p\left(s_{k,t} \,|f_{j,t}, Q_{j,t}, \pi_1\right) p\left(f_{j,t}, Q_{j,t} \,|\, E\left[X_{j,t-1}\right], \pi_2\right) \times p\left(f_{j,t}, Q_{j,t}, \pi_1, \pi_2 \,|X\right) d\left(f_{j,t}, Q_{j,t}, \pi_1, \pi_2\right)$$

where the vector $\pi_2 = (\beta^\theta, A, a, b_t, \{b_{ij}\}, \Sigma)$. Figure 17.8 compares predictions from these two levels with data (black) and priors for missing data (grey). As expected, the predictions conditional on mean estimates of fecundity (right) have narrow predictive intervals – they include only a subset of the uncertainty, i.e. that contributed by the seed data model assuming known fecundity. Predictions that incorporate the uncertainty in the state-space model itself (left) have broader predictive intervals and provide a more realistic prediction.

We constructed predictive intervals for a population and for individuals within the population, the latter conditioning on known covariates or estimated latent states associated with that individual. In principle, the predictive intervals obtained by methods discussed above should agree with the distributions not only of data (e.g. Figure 17.8), but also of latent states being estimated in the model. To illustrate that this is the case, Figure 17.9 provides an example that includes predictive intervals for growth, fecundity, and survival of *Quercus* where the latent states are represented as grey dots and predictive intervals are in black (a dark understory with $\lambda = 0.1$), grey (an intermediate exposure

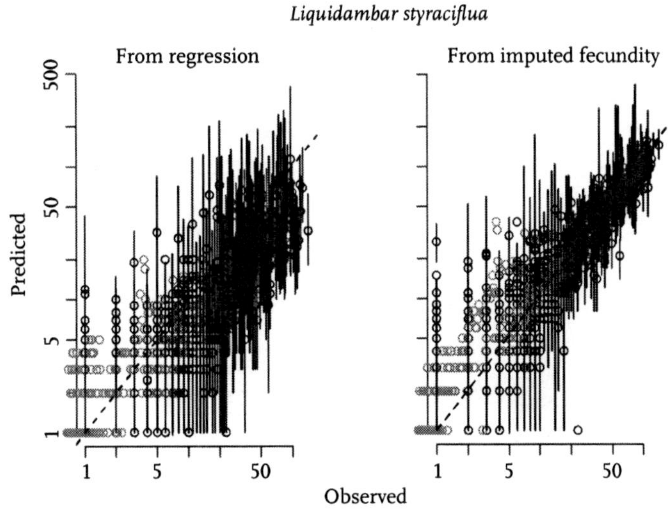

Fig. 17.8 Predictions for seed data conditioned on posterior mean estimates of fecundity (right) and on mean estimates of covariates (left). Predictive intervals are broader on the left, because they integrate not only uncertainty associated with dispersal and sampling, but also in the state-space model of fecundity.

level of $\lambda = 40$), and, for fecundity, dark grey (intermediate exposure, conditional on being mature). For black the sources of uncertainty, from the predictive mean outward are in order: parameter uncertainty (dashed – hardly visible), random individual effects (dotted – in this case small), year-to-year variation (dashed – in this case large), and process error (in this case small). In general we find agreement between estimates of latent states and the predicted variation from the model. The latent states for fecund individuals are covered by the predictive distributions conditional on being mature. The centre plot includes a large number of dots along the bottom of the plot, indicating immature individuals. The black, unconditional fecundity prediction marginalizes over the probability of being mature.

17.7 Summarizing the complexity

The large number of estimates generated by this analysis satisfies the need for a detailed representation of how demographic rates relate to one another, but it produces a new challenge: How do we summarize these results in meaningful ways? Despite the effort needed to simulate the distribution, obtaining the posterior is only the beginning. These results are now the basis for forward simulation experiments to determine how changing environments affect predictions of biodiversity. Here we simply point out the rich set of products that

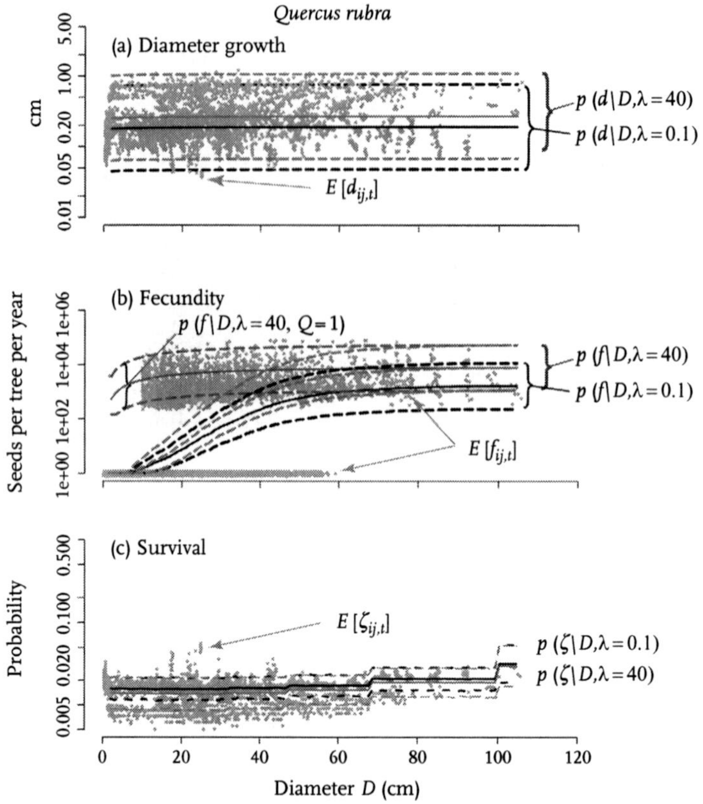

Fig. 17.9 Posterior mean estimates of latent states (dots) and predictive intervals for low ($\lambda = 0.1$; lower set of solid mean and dashed 95%) and high ($\lambda = 40$; upper set of solid mean and dashed 95%) canopy exposure. Included in (b) are predictive intervals for fecundity conditioned on mature status and high canopy exposure ($Q = 1$); zero values are jittered and plotted as one's to make them visible on this log scale.

can be derived from such results. For example, the predictive distributions provide a basis for simulation of interacting populations, a requirement for understanding how competition contributes to species diversity. The simulations in Figure 17.10 are conditioned on a particular assumption of light availability (in this case, a random draw from the data). Competition models generate the light availability based on shading from neighbours (e.g. Pacala *et al.* 1996, Govindarajan *et al.* 2004).

Despite the complexity of this analysis and the large numbers of estimates, predictions can be simple and valuable. For example, the decline in predictive intervals with increasing elevation for fecundity and growth rates in Figure 17.11 can help to explain mechanisms behind species range limits. Correlations among series of rates predicted from the model can be used to identify

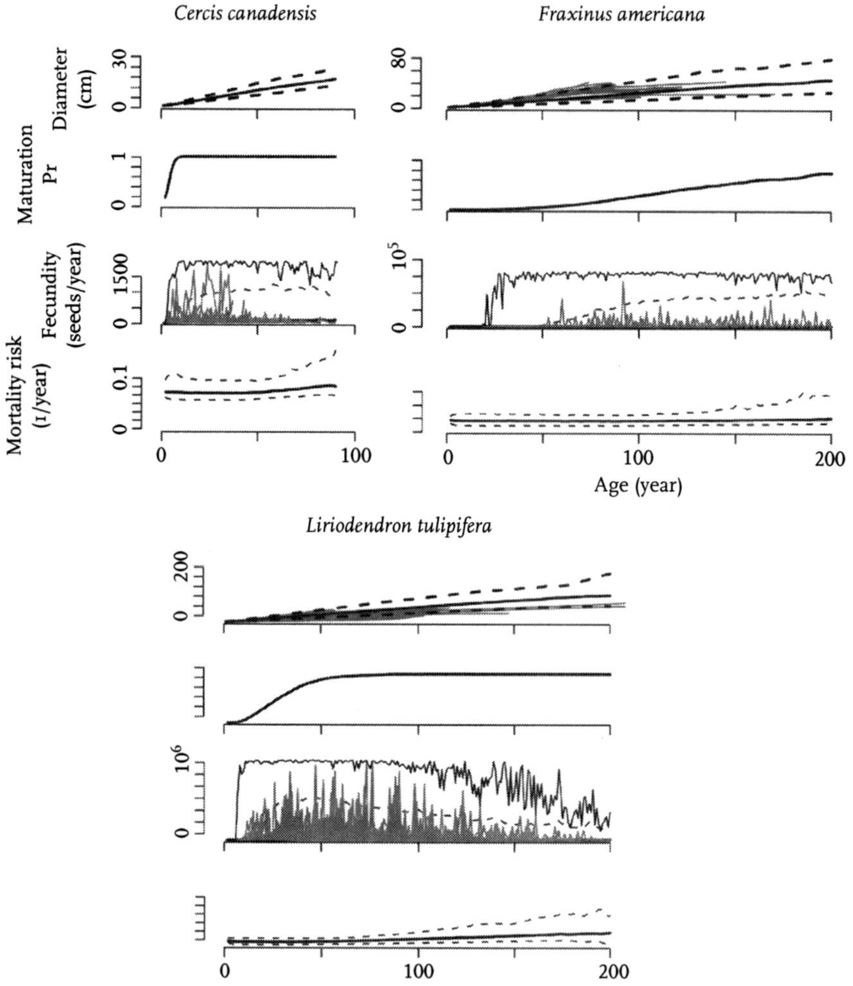

Fig. 17.10 Predictive simulations of demographic rates based on naïve scenarios for light availability, showing inherent differences among species. In each case are shown 95% predictive intervals, which include all sources of uncertainty in the model. For fecundity 30 individual simulations show the range of variability. Note that fecundity has a different scale for each species. For diameter, data are also shown (thin lines).

how lags in growth may affect fecundity (Figure 17.12, left side) and how and when rates of growth and fecundity deteriorate prior to death (Figure 17.12, centre and left). Predictive intervals for specific combinations of demographic rates (Figure 17.13) can be used to test hypotheses about the types of tradeoffs that might be needed for species to coexist. In each of these cases a seemingly incomprehensive number of estimates has been synthesized in ways that allow clear consideration of basic ecological relationships.

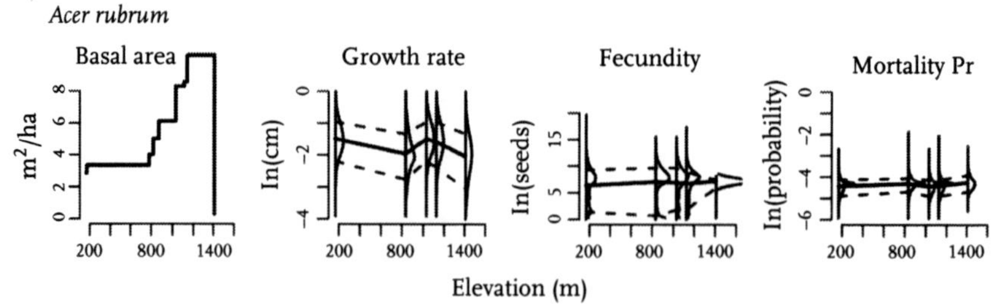

Fig. 17.11 Predictive intervals for demographic rates at different elevations. The predictive densities for different elevations (oriented horizontally) integrate to 1, with solid and dashed lines connecting posterior median and 95% CIs, respectively. The densities include parameter error and individual effects, but they do not include observation error.

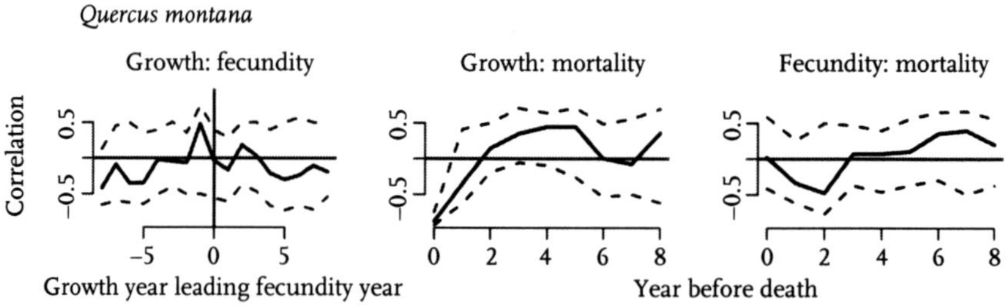

Fig. 17.12 Cross-correlations for demographic rates. The sequences of demographic rates for each individual were detrended and cross-correlated. Shown are bounds for 95% of the individuals (dashed) and medians (solid). In other words, dashed lines bound 95% of the individual cross-correlations.

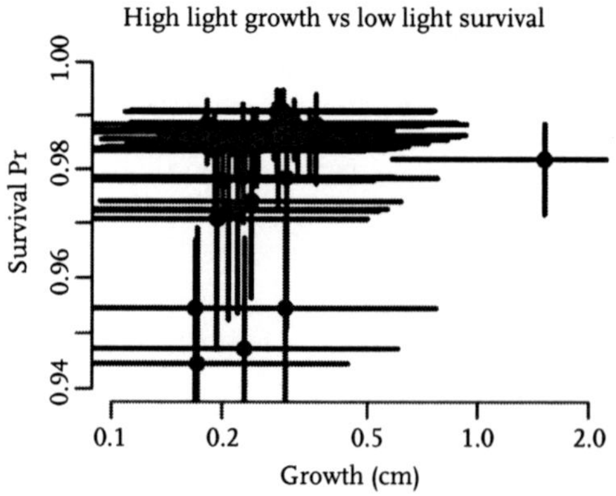

Fig. 17.13 Relationship among species in terms of capacity to grow fast at high light versus survival probability at low light. Predictive intervals are 95% and include only parameter error and variation among individuals.

17.8 Potential

Hierarchical modelling provides a mechanism for synthesis of complex information and interactions. The Bayesian framework is important for incorporation of prior knowledge, the strength of which differs for all parts of the model, including data and theoretical understanding. The advantage it provides for admitting the complexity unavoidable in the real world brings with it the challenge of understanding complex models. Once a large posterior is in hand, predictive distributions of key relationships can help to elucidate patterns of special interest.

Appendix

A. Broader context and background

A.1 Demographic models

Demographic modelling of natural populations has progressed substantially in recent years. Until a decade ago, models for inference on species in the wild involved a single stage, something like response = f(known inputs, stochastic error). The response could be growth rate or fecundity, the inputs being aspects of the individual's state, resources, or other factors that affect health. In fact, the most widely applied methods for inference involved a single predictor, the individual's stage in life, and it assumed that the response was linear. The approach owed its popularity to limited availability of covariate data and to readily available software. For a single response variable, such as fecundity f_i of tree i, linear models dominated. One might apply an allometric function of tree diameter D_i,

$$\begin{aligned} \ln f_i &= \alpha_0 + \alpha_0 \ln D_i + \varepsilon_i \\ \varepsilon_i &\sim N\left(0, \sigma^2\right). \end{aligned}$$

Likewise, tree growth d_i might be modelled as a function of covariates x_i, such as light availability,

$$\begin{aligned} d_i &= f(x_i) + \varepsilon_i \\ \varepsilon_i &\sim N\left(0, \sigma^2\right). \end{aligned}$$

As maximum likelihood has increased in popularity, nonlinear models have been more widely applied. For example, a growth rate might saturate with increasing resource availability. Ecologists have increasing developed code to perform optimizations (e.g. to obtain MLEs) and to simulate distributions (e.g. using a Metropolis algorithm). Mortality risk might be modelled as a function of past annual diameter increment d_i, because growth rate provides some indication of overall tree health,

$$z_i \sim Bernoulli(\zeta_i)$$
$$\zeta_i = f(d_i)$$

using a standard GLM or some alternative form (Kobe *et al.* 1995, Yao *et al.* 2001). Such applications have been tremendously valuable and predispose ecologists to more advanced techniques, such as hierarchical modelling (Clark 2005, Latimer *et al.* 2006, Carlin *et al.* 2007).

The hierarchical application described in this chapter provides examples of many distributional forms and opportunity to mention some background for the hierarchical context. Hierarchical models are now being applied in ways that synthesize complex data to better understand species distributions (Latimer *et al.* 2006), migration (Wikle 2003, Hooten *et al.* 2003), mortality (Metcalf *et al.*, in review), disease spread (LaDeau *et al.* 2007) and environmental variation (Ibanez *et al.* 2007, Ogle *et al.* 2006, Dietze *et al.* 2008), growth (Clark *et al.* 2003, Mohan *et al.* 2007, Ibanez *et al.* 2007), and fecundity and dispersal (Clark *et al.* 2004, HilleRisLambers *et al.* 2006). Our chapter involves a specific application, emphasizing techniques that could be adapted for multiple applications, in part because it models demographic rates together.

A.2 GLMs in a hierarchical setting

Components of our model involve generalized linear models (GLMs) for maturation (Section 17.3.1) and seed data. GLMs involve an underlying linear predictor that is linked to a data distribution, typically binomial or Poisson, by a function that translates the predictor from the real line $(-\infty, \infty)$ to (0,1) for a binomial or $(0, \infty)$ for Poisson by way of a link function. As an example, the Bernoulli example in equation (17.1) shows $\theta_{ij,t}$ as an inverse logit function of the linear predictor $\beta_0^\theta + \beta_1^\theta D_{ij,t} + \beta_2^\theta \lambda_{ij,t}$. A large literature on GLMs dates at least to Nelder and Wedderburn (1972) with applications described for classical and Bayesian settings in many recent texts, including Gelman and Hill (2007). In the Bayesian context, parameters for fixed effects might have Gaussian priors. Our truncated normal priors (equation 17.21) do not importantly complicate the computation.

Hierarchically structured GLMs (or Generalized Linear Mixed Models) typically involve a stochastic specification of one or more terms in the linear predictor. Ecological applications often include random effects associated with location. For example, Latimer *et al.* (2006) include fixed effects as predictors of species occurrence at a geographic location, with a spatial random effect that introduces dependence based on proximity. The prior on this random effect depends on the value for neighbouring cells. This dependence introduces spatial smoothing. Ibanez *et al.* (2008) use random effects for location to absorb some of the differences among stands not accounted for by covariates, in an application where stands are distant from one another. In both of these cases, the variance for random effects requires a prior, introducing an additional stage to the model.

Our application to the binary maturation process in equation (17.1) is applied to a time series for each individual tree, conditionally dependent on previous and future states. However, the underlying predictor is linear, with a logit link. Our approach differs from previous ones in placing the stochasticity not on the coefficients, but rather on the covariates. This structure (parameters as fixed effects assigned to random predictors) was motivated by the fact that tree size is estimated, and canopy exposure estimates are especially crude. The problem is constrained by informative priors on the predictor variables, particularly knowledge that they are bounded within a known range. This range enters as a truncated normal distribution (equations 17.12 and 17.19). Thus, the added stage involves sampling the covariates, rather than the coefficients.

The Poisson distribution for seed density (equation 17.4) differs from a standard GLM in that the expectation is a transport model for seed dispersal. More typically, applications would involve a linear predictor with a log link. Like the binomial, it may have a hierarchical specification. The inclusion of trap area as a coefficient in the Poisson piece of equation (17.3) is standard, accounting for the fact that collecting areas of traps differ.

Related to both the binomial and Poisson is the Zero-Inflated Poisson (*ZIPo*) model, a binomial-Poisson mixture, where the binomial piece can be interpreted as the probability that the Poisson-generating process exists (Lambert 1992, Hall 2000). For example, our fecundity model can be simplified to a simple probability of being mature given tree size (binary) and the expected fecundity, given that the individual is mature (LaDeau and Clark 2001),

$$p(f\,|\,\theta, y) = [1 - \theta + \theta Po(0\,|\,y)]^{1(f=0)}\,[\theta Po(f\,|\,y)]^{1(f>0)}$$

where f is seed production by a tree, 1() is the indicator function, θ is the probability of being mature, and y is the fecundity given that an individual is mature. Marginally we expect $E(f) = \theta y$. Note that the binary part, assigned probability θ, and the conditional Poisson fecundity with mean f are interpreted as different processes. We might include different predictors for θ and f or not (e.g. both depend on tree size and resources). This is a standard interpretation of a *ZIPo* model (e.g. Hall 2000, LaDeau and Clark 2001).

B. Models and computations

Gibbs sampling (Gelfand and Smith 1990) is widely used to simulate posteriors and is described in a number of recent texts (e.g. Carlin and Louis 2000, Gelman *et al.* 1995). In brief, one factors a high-dimensional posterior into a collection of lower-dimensional densities that can be sampled successively, each conditional on others already updated. Conditionals may be sampled directly or indirectly using, for example, a Metropolis step. Here we describe conditional posteriors used for our Gibbs sampling algorithm. The general idea was to embed within the Gibbs sampler Metropolis steps, where direct sampling was not an option, with attention to blocking for efficiency.

B.1 State-space model

For the state space model (equation 17.7), all sampling was direct. The conditional posterior for fixed effect parameters is

$$vec(A)|\, X, Y, \Sigma, \ldots \sim N_{2p}(Vv, (\Sigma + V_b) \otimes V)\; I(a_1 < vec(A) < a_2)$$

where $V^{-1} = X^T\ X$, $v = X^T\ Z$, X is the stacked matrix of $(T_{ij} - t_{ij})$ by p X_{ij} matrices, one for each tree, $Z = [Z_{11}\ Z_{21} \ldots]$ is the stacked matrix $Z_{ij} = Y_{ij} - 1_{ij} b_t$, where 1_{ij} is the length $(T_{ij} - t_{ij})$ vector of 1's, and b_t are taken for the appropriate years in X and Z. The truncated multivariate normal is sampled from the conditional univariate truncated normals.

The error covariance matrix was sampled from an inverse Wishart conditional posterior. Let V_Σ be the parameter matrix for this prior. Then the conditional posterior for Σ^{-1} is

$$\Sigma^{-1} \sim W\left(\left[\sum_{i,j,t} Q_{ij,t}\left(y_{ij,t+1} - x_{ij,t} A - b_t - b_{ij}\right)'\left(y_{ij,t+1} - x_{ij,t} A - b_t - b_{ij}\right) + n_{IJT} V_\Sigma\right]^{-1}, \sum_{i,j,t} Q_{ij,t} + n_{IJT}\right)$$

where n_{IJT} is the total number of tree years (equation 17.16 of the text). The conditional posterior for the random effects variance is

$$\mathbf{V}_b^{-1} \sim \mathrm{W}\left(\left[\sum_{i,j} \max_t (Q_{ij,t})\, b_{ij}' b_{ij} + r_b R_b\right]^{-1}, \sum_{i,j} \max_t (Q_{ij,t}) + r_b\right).$$

Note that individual ij contributes to the conditional posterior only if it is imputed to be mature at some point during the study, in which case $\max_t (Q_{ij,t}) = 1$. The random effects are sampled from

$$b_{ij} \sim N_2(Vv,\ V),$$

where

$$V^{-1} = (T_{ij} - t_{ij})\, \Sigma^{-1} + V_b^{-1},\ v = \Sigma^{-1} \sum_{t=\tau_{ij}}^{T_{ij}} \left(y_{ij,t+1}' - A' x_{ij,t}' - b_t\right),$$

τ_{ij} and T_{ij} are the first and last years during the study in which individual ij is imputed to be mature, and $T_{ij} - \tau_{ij}$ is the number of years mature that ij is mature during the study.

The fixed year effects are sampled from a conditional normal. Let V_t be the covariance matrix for the prior. The conditional posterior is $b_t \sim N_2(Vv, V)$, where $V^{-1} = \Sigma^{-1}\sum_{i,j} Q_{ij,t} + V_t^{-1}$, and $v = \Sigma^{-1}\sum_{i,j} Q_{ij,t}(y_{ij,t+1} - x_{ij,t}\, A - b_{ij})$. This draw was followed by subtraction of the mean for year effect for both $\ln d$ and $\ln f$. Of course b_t and A could be sampled with a single draw from a multivariate normal. A separate step was used due to the large number of b_t and the fact that each tree could have a different subset of total years.

B.2 *Diameter growth*

Diameter growth increments were updated from the conditional posterior given in equation (17.12) using a Gaussian approximation for the third factor, i.e. that corresponding to survival probability. That probability is approximated as $\zeta_{ij,t} \approx 1 - \mu_{d_{ij,t}}$, where the $\mu_{d_{ij,t}}$ sequence contains probabilities for discrete diameter increment bins. The contribution of diameter is omitted, because its contribution to survival probability is small relative to that of growth rate. Then the conditional distribution for the k bins is

$$p\left(\ln d_k \middle| z\right) = \frac{p\left(z \middle| \ln d_k\right) p(\ln d_k)}{\sum_k p\left(z \middle| \ln d_k\right) p(\ln d_k)}$$

where $p(z = 1 | \ln d_k) = 1 - \mu_k$ and $p(z = 0 | \ln d_k) = \mu_k$, z being survival (1) or death (0) in the subsequent year, and $p(\ln\ d_k)$ the distribution of log growth rates. The conditional expectations and variances are

$$\mu_{d|z_{ij,t+1}} \equiv E\left(\ln d \middle| z_{ij,t+1}\right) = \sum_k \ln d_k\, p\left(\ln d_k \middle| z_{ij,t+1}\right)$$

and

$$V_{d|z_{ij,t+1}} \equiv Var\left(\ln d \middle| z_{ij,t+1}\right) = \sum_k (\ln d_k)^2\, p\left(\ln d_k \middle| z_{ij,t+1}\right) - \left[E\left(\ln d \middle| z_{ij,t+1}\right)\right]^2.$$

There is a conditional mean and variance for $z = 0$ and $z = 1$. The log growth rates are sampled from $\ln d_{ij,t} \sim N(Vv, V)$, where

$$V^{-1} = V_{d|f}^{-1} + v_{ij,t}^{-1} + V_{d|z_{ij,t+1}}^{-1}, \quad v = \mu_{d|f}\, V_{d|f}^{-1} + \ln\left(d_{ij,t}^{(0)}\right) v_{ij,t}^{-1} + \mu_{d|z_{ij,t+1}}\, V_{d|z_{ij,t+1}}^{-1},$$

with conditional means and variances contributed by survival are for $z = 0$ or $z = 1$, depending on whether or not the individual survived until the next year.

B.3 *Canopy exposure*

Canopy values are sampled from a conditional posterior that depends on the prior means and variances coming from the analysis that assimilates data and from the state-space model. If the individual is mature in year t, then canopy area is sampled from

$$\ln \lambda_{ij,t} \mid \sim N(Vv, V)\, I\left(\left(c^{(0)}_{ij,t} - \sqrt{C_{ij,t}}\right) < \ln(\lambda_{ij,t}) < \left(c^{(0)}_{ij,t} + \sqrt{C_{ij,t}}\right)\right)$$

where $V^{-1} = A'_{\lambda\bullet}\Sigma^{-1}A_{\lambda\bullet} + C^{-1}_{ij,t}$, and $v = A^T_{\lambda\bullet}\Sigma^{-1}(y_{ij,t} - x^{(-\lambda)}_{ij,t}A_{(-\lambda\bullet)} - b_t - b_{ij}) + c^{(0)}_{ij,t}/C_{ij,t}$. Notation for the first factor follows that from the previous section. If immature in year t, $\ln \lambda_{ij,t}$ is sampled from a normal distribution having

$$V^{-1} = a^2_\lambda / w + C^{-1}_{ij,t}$$

$$v = \frac{\left(\ln\left(d_{ij,t}\right) - x^{(-\lambda)}_{ij,t} a_{(-\lambda)} - b_t - b_{ij}\right) a_\lambda}{w} + \frac{c^{(0)}_{ij,t}}{C_{ij,t}}$$

where $a_{(-\lambda)}$ is the vector a from equation (17.13), but lacking the coefficient for $\ln\lambda$, a_λ is the coefficient for $\ln\lambda$, w is the currently imputed error variance for growth rate regression, having an *IG* prior.

B.4 Fecundity, maturation, gender

Due to their conditional dependence structure, fecundity, maturation, and (for dioecious species) gender are sampled together in a Metropolis step. Here we describe sampling. The basic factoring used for maturation, gender, and fecundity is

$$p\left(f_{ij,t}, Q_{ij,t}, H_{ij}, s_{j,t} \,|q_{ij}, h_{ij}, d_{ij,t-1}, D_{ij,t}, \lambda_{ij,t}\right) = p\left(s_{j,t}\,|\, f_{ij,t}, Q_{ij,t}, H_{ij}\right)$$
$$\times\, p\left(f_{ij,t}, Q_{ij,t}, H_{ij}\,|q_{ij}, h_{ij}, d_{ij,t-1}\, D_{ij,t}, \lambda_{ij,t}\right)$$

where q_{ij} and h_{ij} represent the history of observations on individual ij, both past (before t) and future (after t). The first distribution on the right-hand side is the likelihood for seed trap data, indicating that all seed traps on plot j in year t conditionally depend on every tree i on plot j. The second factor on the right-hand side is the probability of being mature ($Q = 1$), female ($H = 1$), and having fecundity f.

For monoecious species, we use a Metropolis step where maturation status and fecundity are jointly proposed and rejected for all trees in a given stand j in a given year t. For dioecious species we must further sample gender. Because gender applies to an individual across all years, dioecious species are sampled in a different way and are discussed after monoecious species. The blocking differs between these two data types, which we describe here.

Efficient Gibbs sampling requires blocking of variables to facilitate mixing, which is challenging given the ways in which latent variables are linked with the unknown year in which an individual becomes mature τ_{ij}. These linkages include:

(i) the $Q_{ij,t}$ and $f_{ij,t}$ are inherently linked, by virtue of the fact that non-zero fecundity is defined only for mature individuals,

$$f_{ij,t} \left|\left(Q_{ij,t} = 0\right) = 0\right.$$
$$f_{ij,t} \left|\left(Q_{ij,t} = 1\right) > 0;\right.$$

(ii) maturation statuses for an individual over time are mutually dependent according to the one-way transition to maturity in year τ_{ij};
(iii) gender is considered to be fixed; and
(iv) seed trap data conditionally depend on all trees in the plot in a given year.

In light of the conditional relationships involving status and seed production, the choices for blocking are to (1) sample individually every year for every tree (conditioned on other trees for that year and other years for that individual), (2) sample as a block all individuals within a plot for a given year, and (3) sample as a block all trees and years within a plot. The first option has the advantage that high acceptance rates can be achieved, but is computationally slow, entailing loops over plots, individuals, and years, e.g. a Metropolis step for every tree-year in the data set. The third option necessarily results in a high rejection rate, each proposal consisting of $\sum_{i=1}^{n_j} (T_{ij} - t_{ij})$ values. The binary nature of Q and H proposals can make acceptance rates low. Nonetheless, because gender H_{ij} applies to an individual across all years, we use a modification of option 3 for dioecious species. We begin with a description for monoecious species, followed by the description for dioecious species.

Monoecious species – We use the second option for monoecious species, blocking on time and modelling each year successively. The factoring is

$$p\left(f_{j,t}, Q_{j,t} \left| q_{j,t}, x_{j,t-1}, x_{j,t}, Q_{j,t-1}, Q_{j,t+1}, s_{j,t}\right.\right) \propto p\left(s_{j,t} \left| f_{j,t}, Q_{j,t}\right.\right)$$
$$p\left(Q_{j,t}, f_{j,t} \left| q_{j,t}, x_{j,t-1}, Q_{j,t-1}, Q_{j,t+1}, x_{j,t}\right.\right).$$

We propose all values of $\{Q, f\}_{j,t}$ together and accept or reject them as a block. The Markov transition probabilities from t to $t + 1$ are conditioned on observations of status, and they must be combined with probabilities for fecundities and seed trap data. The transition from immature to mature is a hidden Markov process, but only for tree-years in which status is unknown, which is the case after the last year in which immaturity is certain, and before reproduction has been observed. If the status is known through past observations (if previously observed to be mature, then still mature), a current observation (mature or immature), or future observations (if later known to be immature, then immature now), then status $Q_{ij,t}$ is known. This is also the case for imputed statuses. If unknown, status must be modeled as the conditional probability of being in state $Q_{ij,t} = 0$ or 1 given $Q_{ij,t-1}$ and $Q_{ij,t+1}$. These probabilities involve the age-specific rates of making the transition from immature to mature states and can be derived from the cumulative logit probability of being mature given diameter $D_{ij,t}$ and canopy status $\lambda_{ij,t}$ (equation 17.2a). Because blocking is year-by-year, we condition the transition probability on both the foregoing and the following years. Then the trivial probabilities are

$$p\left(Q_{ij,t} = 1 \,\middle|\, Q_{ij,t-1}\right) = 1$$
$$p\left(Q_{ij,t} = 1 \,\middle|\, Q_{ij,t+} = 0\right) = 0.$$

For failure to recognize the reproductive state, we need the additional factor

$$p\left(q_{ij,t} = 0 \,\middle|\, Q_{ij,t} = 1\right) = 1 - v.$$

For Gibbs sampling, we need the year-by-year transition probabilities from immature to mature between $t-1$ and t given that the transition was made between $t-1$ and $t+1$. Let $\delta_{ij,t}$ be the probability of being in the mature state conditional on states in years $t-1,\ t+1$, and on observations. Ignoring observations for the moment, we have

$$\begin{aligned}
\delta_{ij,t} &= p\left(Q_{ij,t} = 1 \,\middle|\, Q_{ij,t-1} = 0,\ Q_{ij,t+1} = 1\right) \\
&= \frac{p\left(Q_{ij,t} = 1 \,\middle|\, Q_{ij,t-1} = 0\right) p\left(Q_{ij,t+1} = 1 \,\middle|\, Q_{ij,t} = 1\right)}{\sum\limits_{k=0,1} p\left(Q_{ij,t} = k \,\middle|\, Q_{ij,t-1} = 0\right) p\left(Q_{ij,t+1} = 1 \,\middle|\, Q_{ij,t} = k\right)} \\
&= \frac{d\theta_{ij,t}/(1-\theta_{ij,t-1})}{d\theta_{ij,t}/(1-\theta_{ij,t-1}) + \left[1 - d\theta_{ij,t}/(1-\theta_{ij,t-1})\right] \times d\theta_{ij,t+1}/(1-\theta_{ij,t})} \\
&= \frac{d\theta_{ij,t}}{d\theta_{ij,t} + \left(\frac{1-\theta_{ij,t-1}-d\theta_{ij,t}}{1-\theta_{ij,t}}\right) d\theta_{ij,t+1}} \\
&= \frac{d\theta_{ij,t}}{d\theta_{ij,t} + d\theta_{ij,t+1}}
\end{aligned}$$

where

$$\begin{aligned}
d\theta_{ij,t} &= \left(\frac{d\theta_{ij,t}}{d\,D_{ij,t}} \times \frac{d\,D_{ij,t}}{dt}\right) dt = \left(\frac{d\theta_{ij,t}}{dD_{ij,t}} \times d_{ij,t}\right) dt \\
&= \beta_1^{\theta} d_{ij,t}\theta_{ij,t}\left(1-\theta_{ij,t}\right) dt.
\end{aligned}$$

Because $\lambda_{ij,t}$ changes much slower than $D_{ij,t}$, we do not include it in the chain rule calculation for the derivative. Because dt is always equal to 1 year, we hereafter omit it.

Observations change the transition probabilities. The previous equation for $d_{ij,t}$ describes the probability of transition in the absence of an observation. If there is an observation in year t and it is 'uncertain' ($q_{ij,t} = 0$: see Table 17.4), then the observer did not identify the tree as mature, and the probability becomes

$$\begin{aligned}\delta_{ij,t} &= \Pr\left(Q_{ij,t} = 1 \,\middle|\, Q_{ij,t-1} = 0,\ Q_{ij,t+1} = 1,\ q_{ij,t} = 0\right) \\ &= \frac{d\theta_{ij,t}(1-v)}{d\theta_{ij,t}(1-v) + d\theta_{ij,t+1}}.\end{aligned}$$

For the first study year, in the absence of an observation (maturation statuses were not obtained on all individuals the first year of the study), we have

$$\delta_{ij,t} = p\left(Q_{ij,t} = 1 \,\middle|\, Q_{ij,t+1} = 1\right) = \frac{\theta_{ij,t}}{\theta_{ij,t} + d\theta_{ij,t+1}}.$$

If there was an observation and that observation was $q_{ij,t} = 0$ (Table 17.4), this becomes

$$\frac{\theta_{ij,t}(1-v)}{\theta_{ij,t}(1-v) + d\theta_{ij,t+1}}.$$

For the last observation year, absent observation,

$$\delta_{ij,T} = p\left(Q_{ij,T} = 1 \,\middle|\, Q_{ij,T-1} = 0\right) = \frac{d\theta_{ij,T}}{\left(1 - \theta_{ij,T}\right)}.$$

If there was an observation, we have

$$\delta_{ij,T} = \frac{d\theta_{ij,T}(1-v)}{\left(1 - \theta_{ij,T}\right)}.$$

The Metropolis steps entail a loop over time (17 year), at each time step proposing values for Q_t and f_t, with the constraints on Q discussed above and $f_{ij,t} = 0$ for all imputed $Q_{ij,t} = 0$. For those imputed to be immature at $t-1$, candidate values come from $Q^*_{ij,t} \sim \textit{Bernoulli}(0.5)$. All others remain mature. For individuals previously imputed to be mature, we propose $\ln f^*_{ij,t} \sim N(\ln f^{(g)}_{ij,t}, 0.1)$, where (g) denotes the current Gibbs step, prior to updating. For individuals previously imputed to be immature, but now mature, we propose from $f^*_{ij,t} | (Q_{ij,t-1} = 0) \sim N(\ln f'_{ij,t}, 0.1)$, where $f'_{ij,t}$ is an auxiliary variable having the value retained from the most recent iteration of the Gibbs sampler in which $Q_{ij,t} = 1$. The acceptance criterion involves the products from equation (17.2a),

$$\begin{aligned}&Q_{j,t}, f_{j,t} \,\middle|\, Q_{j,t-1}, Q_{j,t+1}, q_{j,t}, s_{j,t}, \mu^{f|d}_{j,t}, V^{f|d}_{j,t} \\ &\quad \propto \prod_i \left(1 - \delta_{ij,t}\right)^{1-Q_{ij,t}} \left[\delta_{ij,t} N\left(\ln\left(f_{ij,t}\right) \,\middle|\, \mu^{f|d}_{ij,t}, V^{f|d}_{ij,t}\right)\right]^{Q_{ij,t}} \\ &\quad \times \prod_{k=1} Po\left(s_{jk,t} \,\middle|\, A_{jk} g_{jk}\left(f_{j,t}; u\right)\right).\end{aligned}$$

Note that all individuals imputed to be mature have a conditional density associated with the state-space model. The $\delta_{ij,t}$ are different for each individual and year, as discussed above. All trees contribute to the likelihood for the seed data for plot j in year t, in that by producing seed or not, they affect the parameters of the Poisson sampling distribution for seed. Of course, the set of proposals is accepted or rejected as a block. The sampler is more efficient than it appears, because we can propose statuses and fecundities for all plots simultaneously, and accept/reject them on a plot by plot basis, without actually looping over plots. Once states are updated for time t, we move to $t+1$.

Dioecious species – For dioecious species gender is unchanging over time, so we evaluate the full history of observations for each tree, but still avoiding loops over individual trees. It is efficient to factor the conditional somewhat differently, taking together all trees on plot j over all years,

$$p(f_j, Q_j, H_j \,|q_j, h_j, X_j, s_j) \propto p(s_j \,|f_j, Q_j, H_j)\, p(f_j, Q_j, H_j \,|q_j, h_j, X_j).$$

Because maturation is no longer modeled year-by-year, we require the probability for a history of maturation status, conditioned on observations obtained sporadically over individuals and years.

Let τ_{ij} be the year in which an individual becomes mature, $\tau_{ij}^0 = \max_t(q_{ij,t} = -1)$ be the last year an individual is known to have been immature, and $\tau_{ij}^1 = \min_t(q_{ij,t} = 1)$ be the first year an individual is known to have been mature. Thus, we have the constraint $\tau_{ij}^0 \leq \tau_{ij} \leq \tau_{ij}^1$. The probability assigned to an individual that became mature in year t is,

$$\begin{aligned}\delta_{ij} &= p\left(\tau_{ij} = t \,\middle|\, \tau_{ij}^0 < \tau_{ij} < \tau_{ij}^1\right) = d\theta_{ij,t} \Big/ \left(\theta_{ij,\tau_{ij}^1} - \theta_{ij,\tau_{ij}^0}\right) \\ &= \beta_1^\theta d_{ij,t} \frac{\theta_{ij,t}(1-\theta_{ij,t})}{\left(\theta_{ij,\tau_{ij}^1} - \theta_{ij,\tau_{ij}^0}\right)}.\end{aligned}$$

For individuals imputed to be still immature at the end of the observation period at T_{ij}, the probability is

$$\delta_{ij} = p\left(\tau_{ij} > T_{ij} \,\middle|\, \tau_{ij} > \tau_{ij}^0\right) = 1 - p\left(\tau_{ij} \leq T_{ij} \,\middle|\, \tau_{ij} > \tau_{ij}^0\right) = \frac{1-\theta_{T_{ij}}}{1-\theta_{ij,\tau_{ij}^0}}.$$

For individuals imputed to be already mature before observations began at t_{ij}, the probability is

$$\delta_{ij} = p\left(\tau_{ij} < t_{ij} \,\middle|\, \tau_{ij} < \tau_{ij}^1\right) = \frac{\theta_{t_{ij}}}{\theta_{ij,\tau_{ij}^1}}.$$

We now have a probability for the history of an individual that became mature at time τ_{ij} or remained immature throughout. Our prior specification allows for a minimum and maximum maturation diameter, in which case θ_{ij,τ_{ij}^0} and θ_{ij,τ_{ij}^1} are the values of θ taken at these prior maturation diameter values.

The probability for maturation is combined with observations of status between those years that established it as immature and mature and for gender. Thus far, we have considered observations that definitively establish maturity or immaturity ($q_{ij,t} = -1$ or 1 in Table 17.4). For $q_{ij,t} = 0$, status is uncertain. Status detection is defined as $p(q_{ij,t} = 1 | Q_{ij,t} = 0) = 0$ and $p(q_{ij,t} = 1 | Q_{ij,t} = 1) = v$. The individual has unknown gender if the gender is not observed, the observation is uncertain, or if flowers are observed but not identified to sex, and no observation is available from the fruiting season (Table 17.4). Considering both observations and gender, the probability for individual ij becomes

$$p\left(\tau_{ij}, H_{ij} \middle| \tau_{ij}^0 < \tau_{ij} < \tau_{ij}^1, q_{ij}\right) = \delta_{ij}(1 - v)^{n_{ij}^v}(1 - \phi)^{1-H_{ij}}\phi^{H_{ij}}$$

where n_{ij}^v is defined to be the number of times mature status was 'undetected' during the interval (τ_{ij}, τ_{ij}^1), i.e. the number of times that an individual imputed to be mature in year τ_{ij} was not identified as such. If gender is known, the third factor disappears. The full reproductive history on all individuals has conditional probability

$$p\left(Q_j, H_j, f_j \,|q_j, h_j, s_j, \mu_j^{f|d}, V_j^{f|d}\right) = \prod_i \delta_{ij}(1 - v)^{n_{ij}^v}(1 - \phi)^{1-H_{ij}}$$
$$\times \left[\phi \prod_t N\left(\ln\left(f_{ij,t}\right) \middle| \mu_{ij,t}^{f|d}, V_{ij,t}^{f|d}\right)^{Q_{ij,t}}\right]^{H_{ij}} \prod_t \prod_k Po\left(s_{jk,t} \,| A_{jk} g_{jk}\left(f_{j,t}; u\right)\right).$$

For each individual a maturation diameter is proposed from a uniform distribution

$$\tau_{ij} \sim unif\left(t_{ij}^0, t_{ij}^1\right)$$
$$t_{ij}^0 = \max\left(t_{ij}^{\min}, \tau_{ij}^0\right)$$
$$t_{ij}^1 = \min\left(t_{ij}^{\max}, \tau_{ij}^1\right).$$

The bounds for minimum and maximum maturation diameters are not sooner than the first year in which ij reached the minimum prior diameter for maturation $t_{ij}^{\min} = \max(t | (D_{ij,t} > D_{\min}))$ or it was last known to be immature τ_{ij}^0 and not later than the last year in which ij had not yet reached the prior diameter for certain maturation $t_{ij}^{\max} = \min(t | (D_{ij,t} < D_{\max}))$ or it was known to be mature τ_{ij}^1. There are prior minimum and maximum diameters, which differ among species. The fecundity for an individual proposed to be immature is zero. For individuals currently imputed to be mature, the proposed fecundity is a truncated normal on $(f_{\min}, f_{\max})$ centered on the current estimate. The conditional densities are then the product of Poisson seed data, Gaussian fecundity, and the probability associated with maturation in year t. Because the probability of seed data conditionally depends on all trees in the stand in all years, the ensemble of (f_j, Q_j, H_j) is accepted or rejected as a block.

Recognition error is sampled from

$$Bin\left(v^{(1)} \left| v^{(0)} + v^{(1)}, v\right.\right) Be(v \,|v_1, v_2) = Be\left(v \left| v_1 + v^{(1)}, v_2 + v^{(0)}\right.\right)$$

where the two arguments are sums of prior values and numbers of currently imputed mature individuals for which maturation was recognized as such or not, i.e.

$$v^{(0)} = \sum_{ij,t} I\left(q_{ij,t} = 0, Q_{ij,t} = 1\right)$$

$$v^{(1)} = \sum_{ij,t} I\left(q_{ij,t} = 1, Q_{ij,t} = 1\right).$$

For the female fraction we sample from the conditional posterior

$$Bin\left(h^{(1)} \left| h^{(0)} + h^{(1)}, \phi\right.\right) Be(\phi \,|h_1, h_2) = Be\left(\phi \left| h_1 + h^{(1)}, h_2 + h^{(0)}\right.\right)$$

where the two arguments are sums of prior and currently imputed numbers of the females and males, respectively,

$$h^{(0)} = \sum_{ij} (1 - H_{ij})$$

$$h^{(1)} = \sum_{ij} H_{ij}.$$

Prior values are $h_1 = h_2 = 4$, which has a mean of 0.5 and is dominated by the data.

Parameters for the logit function of maturation equation (17.1) are sampled with Metropolis step. Conditionally we have

$$\prod_{i,j} \delta(\tau_{ij})\; N(\beta^\theta \,|b^\theta, V^\theta)\, I(\beta^\theta_{2,3} > 0)$$

where the first product is the probability associated with maturation years, which depend on β^θ, and the truncated normal prior. A proposal is generated from a multivariate normal truncated at zero for these two parameters.

Parameters for the seed data model are sampled in a single Metropolis step. Conditionally,

$$\prod_{t=t_j}^{T_j} \prod_{k=1}^{K_j} Po\left(s_{jk,t} \left| A_{jk} g_{jk}\left(f_{j,t}; u, c\right)\right.\right) N(u \,|u_0, V_u)\; N(c \,|c_0, V_c)\; I(u, c > 0)\,.$$

Values are proposed from a normal distribution. In the case of missing data, seed counts were replaced with the currently imputed seed value.

Imputation of missing data involved a Metropolis step with proposals of plus or minus 1 from the current value with probability 0.5. The conditional posterior includes a Poisson prior with a mean density as discussed in Section 17.4 multiplied by Poisson density for sample *jk,t*. Proposals were accepted as a block for $s_{j,t}$.

A Metropolis step is used to simultaneously update all of the diameter growth and diameter bins for the nonparametric survival relationship. The growth rates and diameters of all individuals are binned in the sequences μ_d and μ_D are for the all years. For diameter increment there are 31 bins equally spaced with width 0.1 on the $\log_{10}$ scale. For diameter there are six bins, also equally spaced on the $\log_{10}$ scale, with the maximum value chosen to exceed that largest diameter in the data set. Survival from year t to $t+1$ is the event $z_{ij,t}=1$ and death in the subsequent year is $z_{ij,t}=0$. At each Gibbs step new sequences of μ_d and μ_D are proposed each being Gaussian and centred on the currently imputed values, but truncated midway between the current values. For diameter increment the proposal distribution is

$$N\left(\mu_d^* \,|\mu_d,\, V\right)\, I\left(\left(\mu_{d-1}-\mu_d\right)/2 < \mu_d^* < \left(\mu_d-\mu_{d+1}\right)/2\right)$$

where V is a small value (0.1 in this case). In other words, if the currently imputed value for $\mu_{d,k}$ was 0.5 and those for bins $k-1$ and $k+1$ were 0.6 and 0.48, then the proposal would come from the normal centred at 0.5, truncated at 0.55 and 0.49. This procedure allows for any shape subject to monotonic decline. The proposals for the diameter values are done in the same way, with the constraint being monotonic increase and with an informative prior $Be(a_k, b_k)$. All values (both growth rate and diameter) are proposed together and accepted as a block.

References

Batista, W. B., Platt, W. J. and Macchiavelli, R. E. (1998). Demography of a shade-tolerant tree (*Fagus grandifolia*) in a hurricane-disturbed forest. *Ecology*, **79**, 38–53.

Beckage, B. and Clark, J. S. (2003). Seedling survival and growth in southern Appalachian forests: Does spatial heterogeneity maintain species diversity? *Ecology*, **84**, 1849–1861.

Bigler, C. and Bugmann, H. (2004). Predicting the time of tree death using dendrochronological data. *Ecological Applications*, **14**, 902–914.

Carlin, B., Clark, J. S. and Gelfand, A. (2006). Elements of Bayesian Inference. In *Hierarchical Models of the Environment*, (ed. J.S. Clark and A. Gelfand), pp. 3–24. Oxford University Press, Oxford.

Carlin, B. P. and Louis, T. A. (2000). *Bayes and Empirical Bayes Methods for Data Analysis.* Chapman and Hall, Boca Raton, FL.

Caswell, H. (2001). *Matrix Population Models.* Sinauer, Sunderland, MA.

Clark, J. S., LaDeau, S. and Ibanez, I. (2004). Fecundity of trees and the colonization-competition hypothesis, *Ecological Monographs*, **74**, 415–442.

Clark, J. S. (2005). Why environmental scientists are becoming Bayesians. *Ecology Letters*, **8**, 2–14.

Clark, D. B. and Clark, D. A. (1996). Abundance, growth and mortality of very large trees in neotropical lowland rain forest. *Forest Ecology and Management*, **80**, 235–244.

Clark, J. S., Macklin, E. and Wood, L. (1998). Stages and spatial scales of recruitment limitation in southern Appalachian forests. *Ecological Monographs*, **68**, 213–235.

Clark, J. S., Silman, M., Kern, R., Macklin, E. and Hille Ris Lambers, J. (1999). Seed dispersal near and far: Generalized patterns across temperate and tropical forests. *Ecology*, **80**, 1475–1494.

Clark, J. S., Wolosin, M., Dietze, M., Ibanez, I., LaDeau, S., Welsh, M. and Kloeppel, B. (2007). Tree growth inferences and prediction from diameter censuses and ring widths. *Ecological Applications*, **17**, 1942–1953.

Condit, R., Hubbell, S. P. and Foster, R. B. (1995). Mortality rates of 205 neotropical tree and shrub species and the impact of severe drought. *Ecological Monographs*, **65**, 419–439.

Coomes, D. A. and Allen, R. B. (2007). Mortality and tree-size distributions in natural mixed-age forests. *Journal of Ecology*, **95**, 27–40.

Dietze, M., Wolosin, M. and Clark, J. (2008). Capturing diversity and individual variability in allometries: A hierarchical approach. *Forest Ecology and Management*, **256**, 1939–1948.

Govindarajan, S., Dietze, M., Agarwal, P. and Clark, J. S. (2004). A scalable model of forest dynamics. Proceedings of the 20th Symposium on Computational Geometry SCG, pp. 106–115.

Gurevitch, J., Scheiner, S. M. and Fox, G. A. (2002). *The Ecology of Plants*. Sinauer, Sunderland, MA.

Gelfand, A. E. and Smith, A. F. M. (1990). Sampling-based approaches to calculating marginal densities. *Journal of the American Statistical Association*, **85**, 398–409.

Gelman, A. and Hill, J. (2007). *Data Analysis using Regression and Multilevel/Hierarchical Models*. Cambridge University Press, Cambridge.

Gelman, A., Carlin, J. B., Stern, H. S. and Rubin, D. B. (1995). *Bayesian Data Analysis*. Chapman and Hall, London.

Gurevitch, J., Scheiner, S. M. and Fox, G. A. (2002). *The Ecology of Plants*. Sinauer, Sunderland, MA.

Hall, D. B. (2000). Zero-inflated Poisson and binomial regression with random effects: A case study. *Biometrics*, **56**, 1030–1039.

Hooten, M. B., Larsen, D. R. and Wikle, C. K. (2003). Predicting the spatial distribution of ground flora on large domains using a hierarchical Bayesian model. *Landscape Ecology*, **18**, 487–502.

Ibáñez, I., Clark, J. S. and Dietze, M. (2007). Evaluating the sources of potential migrant species: Implications under climate change. *Ecological Applications*, **18**, 1664–1678.

Ibáñez, I., Clark, J. S., LaDeau, S. and Hille Ris Lambers, J. (2007). Exploiting temporal variability to understand tree recruitment response to climate change. *Ecology Monographs*, **77**, 163–177.

King, D. A., Davies, S. J. and Noor, N. S. M. (2006). Growth and mortality are related to adult tree size in a Malaysian mixed dipterocarp forest. *Forenerty and Ecological Management*, **223**, 152–158.

Kobe, R. K., Pacala, S. W., Jr. Silander, J. A. and Canham, C. D. (1995). Juvenile tree survivorship as a component of shade tolerance. *Ecological Applications*, **5**, 517–532.

Kobe, R. K. (2006). Sapling growth as a function of light and landscape-level variation in soil water and foliar N in northern Michigan. *Oecologia*, **147**, 119–133.

LaDeau, S. L. and Clark, J. S. (2006). Elevated CO_2 and tree fecundity: The role of tree size, interannual variability, and population heterogeneity. *Global Change Biology*, **12**, 822–833.

LaDeau, S., Kilpatrick, M. and Marra, P. P. (2007). West Nile virus emergence and large-scale declines of North American bird populations. *Nature*, **447**, 710–714.

Lambert, D. (1992). Zero-inflated Poisson regression, with an application to defects in manufacturing. *Technometrics*, **34**, 1–14.

Latimer, A. M., S. Wu, A. E. Gelfand, and J. A. Silander. (2006). Building statistical models to analyze species distributions, *Ecological Applications*, **16**, 33–50.

Metcalf, C. J. E., Clark, J. S. and McMahon, S. M. (2009). Overcoming data sparseness and parametric constraints in modeling of tree mortality: A new non-parametric Bayesian model. *Canadian Journal of Forest Research*, **39**, 1677–1687.

Mohan, J. E., Clark, J. S. and Schlesinger, W. H. (2007). Long-term CO_2 enrichment of an intact forest ecosystem: Implications for temperate forest regeneration and succession. *Ecological Applications*, **17**, 1198–1212.

Nepstad, D. C., Tohver, I. M., Ray, D., Moutinho, P. and Cardinot, G. (2007). Mortality of large trees and lianas following experimental drought in an Amazon forest. *Ecology*, **88**, 2259–2269.

Nelder, J. A. and Wedderburn, R. W. M. (1972). Generalized linear models. *Journal of the Royal Statistical Society* (Series A), **135**, 370–384.

Ogle, K., Uriarte, M., Thompson, J., Johnstone, J., Jones, A., Lin, Y., McIntire, E., Zimmerman, J. (2006). Implications of vulnerability to hurricane damage for long-term survival of tropical tree species: A Bayesian hierarchical analysis. In *Hierarchical Modeling for the Environmental Sciences: Statistical Methods and Applications*, (ed. J.S. Clark and A.E. Gelfand), Oxford University Press, Oxford.

Rich, R. L., Frelich, L. E. and Reich, P. B. (2007). Wind throw mortality in the southern boreal forest: Effects of species diameter and stand age. *Journal of Ecology*, **95**, 1261–1273.

Suarez, M. L., Ghermandi, L. and Kitzberger, T. (2004). Factors predisposing episodic drought-induced tree mortality in Nothofagus site, climatic sensitivity and growth trends. *Journal of Ecology*, **92**, 954–966.

Thomas, S. C. (1996). Relative size at the onset of maturity in rain forest trees: A comparative analysis of 37 Malaysian species. *Oikos*, **76**, 145–154.

Uriarte, M., Canham, C. D., Thompson, J. and Zimmerman, J. K. (2004). A neighborhood analysis of tree growth and survival in a hurricane-driven tropical forest. *Ecological Monographs*, **74**, 591–614.

Valle, D., Clark, J. S., Dietze, M., Agarwal, P. K., Millette, T. and Schultz, H. (2009). The effect of light competition on growth rate of large trees. In preparation.

van Mantgem, P.J. and Stephenson, N. L. (2007). Apparent climatically induced increase of tree mortality rates in a temperate forest. *Ecology Letters*, **10**, 909–916.

Wyckoff, P. H., and Clark, J. S. (2000). Predicting tree mortality from diameter growth: A comparison of maximum likelihood and Bayesian approaches *Canadian Journal of Forest Research*, **30**, 156–167.

Wikle, C. K. (2003). Hierarchical Bayesian models of predicting the spread of ecological processes. *Ecology*, **84**, 1382–1394.

Yao X., Titus, S. J. and MacDonald, S. E. (2001). A generalized logistic model of individual tree mortality for aspen, white spruce, and lodgepole pine in Alberta mixedwood forests. *Canadian Journal of Forestry Research*, **31**, 283–291.

·18·

Combining monitoring data and computer model output in assessing environmental exposure

Alan E. Gelfand and Sujit K. Sahu

18.1 Introduction

The demand for spatial models to assess regional progress in air quality has grown rapidly over the past decade. For improved environmental decision-making, it is imperative that such models enable spatial prediction to reveal important gradients in air pollution, offer guidance for determining areas in non-attainment with air standards, and provide air quality input to models for determining individual exposure to air pollution. Spatial prediction has the potential to suggest new perspectives in the development of emission control strategies and to provide a credible basis for resource allocation decisions, particularly with regard to network design.

Space-time modelling of pollutants has some history including, e.g. Guttorp *et al.* (1994), Haas (1995) and Carroll *et al.* (1997). In recent years, hierarchical Bayesian approaches for spatial prediction of air pollution have been developed (Brown *et al.*, 1994; Le *et al.*, 1997; Sun *et al.*, 2000). Cressie *et al.* (1999) compared kriging and Markov-random field models in the prediction of PM_{10} (particles with diameters less than $10\,\mu m$) concentrations around the city of Pittsburgh. Zidek *et al.* (2002) developed predictive distributions on non-monitored PM_{10} concentrations in Vancouver, Canada. They noted the under-prediction of extreme values in the pollution field, but their methodology provided useful estimates of uncertainties for large values. Sun *et al.* (2000) developed a spatial predictive distribution for the space-time response of daily ambient PM_{10} in Vancouver. They exploit the temporal correlation structure present in the observed data from several sites to develop a model with two components, a common deterministic trend across all sites plus a stochastic residual. They illustrate the methods by imputing daily PM_{10} fields in Vancouver. Kibria *et al.* (2000) developed a multivariate spatial prediction methodology in a Bayesian context for the prediction of $PM_{2.5}$ in the city of Philadelphia. This approach used both $PM_{2.5}$ and PM_{10} data at monitoring sites with different start-up times. Shaddick and Wakefield (2002) proposeed short term space-time modeling for PM_{10}.

Smith *et al.* (2003) proposed a spatio-temporal model for predicting weekly averages of $PM_{2.5}$ and other derived quantities such as annual averages within three southeastern states in the United States. The $PM_{2.5}$ field is represented as the sum of semiparametric spatial and temporal trends, with a random component that is spatially correlated, but not temporally. These authors apply a variant of the expectation-maximization (EM) algorithm to account for high percentages of missing data. Sahu and Mardia (2005) present a short-term forecasting analysis of $PM_{2.5}$ data in New York City during 2002. Within a Bayesian hierarchical structure, they use principal kriging functions to model the spatial structure and a vector random-walk process to model temporal dependence. Sahu *et al.* (2006) consider modelling of $PM_{2.5}$ through the use of rural and urban process models while Sahu *et al.* (2007) deal with misalignment between ozone data and meteorological information. Sahu *et al.* (2009) develop a hierarchical space-time model for daily eight-hour maximum ozone concentration data covering much of the eastern United States. The model combines observed data and forecast output from a computer model known as the Community Multi-scale Air Quality (CMAQ) Eta forecast model (see below for references) so that the next day forecasts can be computed in real time. They validate the model with a large amount of set-aside data and obtain much improved forecasts of daily O_3 patterns. Berrocal *et al.* (2010) propose a downscaling approach by regressing the observed point level ozone concentration data on grid cell level computer model output with spatially varying regression co-efficients specified through a Gaussian process. Rappold *et al.* (2008) study wet mercury deposition over space and time. Finally, Wikle (2003) provides a nice overview of the role of hierarchical modeling in environmental science. With so much interest in space-time exposure prediction, attention to data fusion models to improve such prediction is not surprising.

18.1.1 Environmental computer models

Computer models are playing an increasing role in our quest to understand complex systems. In this regard, the discussion paper of Kennedy and O'Hagan (2001) reviews prediction and uncertainty analysis for systems which are approximated by complex mathematical models. These models are often implemented as computer codes and typically depend on a number of input parameters which determine the nature of the output. The input parameters are often unknown and are customarily estimated by ad hoc methods such as very crude fitting of the computer model to the observed data. Kennedy and O'Hagan present a Bayesian calibration technique which improves on this usual approach in two respects. First, Bayesian prediction methods allow one to account for all sources of uncertainty including the ones from the estimation of the parameters. Second, any inadequacy in the model specification, even under the best-fitting parameter values, is revealed by discrepancies between the observed data and the model predictions. Illustration is provided using data from a nuclear radiation release at Tomsk and also from a more complex simulated nuclear accident exercise.

Cox *et al.* (2001) describe a statistical procedure for estimation of unknown parameters in a complex computer model from an observational or experimental data base. They develop methods for accuracy assessments of the estimates and illustrate their results in the setting of computer code which models nuclear fusion reactors. Fuentes *et al.* (2003) develop a formal method for evaluation of the performance of numerical models. They apply the method to an air quality model (essentially the CMAQ model) and discuss related issues in the estimation of nonstationary spatial covariance structures.

Turning to environmental computer models, high spatial resolution numerical model output is now widely available for various air pollutants. Our focus here is on the CMAQ forecast model. CMAQ is a modelling system which has been designed to approach air quality as a whole by including capabilities for modelling multiple air quality issues, including tropospheric ozone, fine particles, toxics, acid deposition, and visibility degradation. CMAQ was also designed to have multiscale capabilities so that separate models are not needed for urban and regional scale air quality modelling. The target grid resolutions and domain sizes for CMAQ range spatially and temporally over several orders of magnitude. With the temporal flexibility of the model, simulations can be performed to evaluate longer term (annual to multi-year) pollutant climatologies as well as short term (weeks to months) transport from localized sources. The ability to handle a large range of spatial scales enables CMAQ to be used for urban and regional scale model simulations. See, e.g. `http://www.epa.gov/asmdnerl/CMAQ/`.

It is worth distinguishing the goal of CMAQ which is to provide ambient exposure at high spatial and temporal resolution from computer models that provide individual level exposure. In particular, the former, assimilated with station data, provide the ambient exposures which drive the latter. Again, the contribution of this chapter is to discuss fully model-based implementations of this fusion.

With regard to the latter, Zidek and his co-authors have written a series of papers considering prediction of human exposure to air pollution. In particular, Zidek *et al.* (2007) present a general framework for constructing a predictive distribution for the exposure to an environmental hazard sustained by a randomly selected member of a designated population. The individual's exposure is assumed to arise from random movement through the environment, resulting in a distribution of exposure that can be used for environmental risk analysis. Zidek *et al.* (2005) develop a computing platform, referred to as pCNEM, to produce such distributions. This software is intended for simulating exposures to airborne pollutants. In the paper they illustrate with a model for predicting human exposure to PM_{10}.

Further work along these lines has been an objective of US Environmental Protection Agency (EPA) initiatives. The EPA's National Exposure Research Laboratory (NERL) has developed a population exposure and dose model, particularly for particulate matter (PM), called the Stochastic Human Exposure and Dose Simulation (SHEDS) model (Burke *et al.*, 2003). SHEDS-PM uses a probabilistic approach that incorporates both variability and uncertainty to predict distributions of PM exposure, inhaled dose, and deposited dose for a specified population. SHEDS-PM estimates the contribution of PM from both outdoor and indoor sources (e.g. cigarette smoking, cooking) to total personal PM exposure and dose. In particular, SHEDS-PM generates a simulation population using US Census demographic data for the user-specified population with randomly assigned activity diaries of individuals. Output from the SHEDS-PM model includes distributions of exposure and dose for the specified population, as well as exposure and dose profiles for each simulated individual. It is Bayesian in its conception in the sense that the input parameters (e.g. air exchange rates, penetration rates, cooking and smoking emission rates) are drawn at random from suitable priors.

A similar EPA product, the Air Pollutants Exposure Model (APEX) was developed by the Office of Air Quality and Planning (Richmond *et al.* 2002). It is derived from the probabilistic National Ambient Air Quality Standards (NAAQS) Exposure Model for carbon monoxide (pNEM/CO). APEX serves as the human inhalation exposure model within the Total Risk Integrated Methodology (TRIM) model framework. APEX is intended to be applied at the local, urban, or consolidated metropolitan area scale and currently only addresses inhalation exposures. The model simulates the movement of individuals through time and space and their exposure to the given pollutant in various micro-environments (e.g. outdoors, indoors residence, in-vehicle). Results of the APEX simulations are provided as hourly and summary exposure and/or dose estimates, depending on the application, for each individual included in the simulation as well as summary statistics for the population modelled.

The format of the remainder of this chapter is as follows. In Section 18.2 we review some algorithmic and pseudo-statistical approaches in weather prediction. Section 18.3 provides a review of current state of the art fusion methods for environmental data. We develop a non-dynamic downscaling approach based on our recent work (Sahu, Gelfand, and Holland, 2010) in

Section 18.4. A few summary remarks are provided in Section 18.5. Appendix A contains an introduction to Gaussian processes (GP) and Appendix B outlines the full conditional distributions for the downscaler approach proposed in Section 18.4.

18.2 Algorithmic and pseudo-statistical approaches in weather prediction

A convenient framework within to review algorithmic and pseudo-statistical approaches to data assimilation is in the context of numerical weather prediction. Kalnay (2003) provides a recent development of this material. Such assimilation has a long history dating at least to Charney (1951) who recognized that hand interpolation of available weather observations to a regular grid was too time consuming and that numerical interpolation methods were needed. Earliest work created local polynomial interpolations using quadratic trend surfaces in locations in order to interpolate observed values to grid values. Of course, in the past half century, such polynomial interpolation has come a long way to become a standard device in the statistician's toolkit; we do not detail this literature here.

Instead, we note that what emerged in the meteorology community was the recognition that a first guess (or background field or prior information) was needed (Bergthorsson and Döös, 1955), supplying the *initial conditions.* As short-range forecasts became better and better, their use as a first guess became universal. The climatological intuition here is worth articulating. Over 'data-rich' areas the observational data dominates while in 'data-poor' regions the forecast facilitates transport of information from the data-rich areas. Of course, in the setting of fully-specified models and fully model-based inference we can quantify this adaptation and the associated uncertainty. Indeed, this is the contribution of the following sections of this chapter.

We illustrate several numerical approaches using, illustratively, temperature as the variable of interest. At time t, we let $T_{\text{obs}}(t)$ be an observed measurement, $T_b(t)$ a background level, $T_a(t)$ an assimilated value, and $T_{\text{true}}(t)$ the true value. An early scheme is known as the successive corrections method (SCM) which obtains $T_{i,a}(t)$ iteratively through

$$T_{i,a}^{(r+1)}(t) = T_{i,a}^{(r)}(t) + \left[\sum_k w_{ik}\left\{T_{k,\text{obs}}(t) - T_{k,a}^{(r)}(t)\right\}\right] \Big/ \left(\sum_k w_{ik} + \epsilon^2\right).$$

Here, i indexes the grid cells for the interpolation while k indexes the observed data locations. $T_{k,a}^{(r)}(t)$ is the value of the assimilator at the rth iteration at the observation point k (obtained from interpolating the surrounding grid points). The weights, w_{ik}, can be defined in various ways but usually as a decreasing function of the distance between the grid point and the observation point. In fact, they can vary with iteration, perhaps becoming increasingly local. See, e.g. Cressman (1959) and Bratseth (1986). The analysis reflects the observations more faithfully when ϵ^2 is taken to be zero, see Cressman (1959). Non-zero values of ϵ^2 are assumed when the observations have errors, and the resulting analyses provide some positive weight to the background field.

Another empirical approach is called *nudging* or Newtonian relaxation. Suppose, suppressing location, we think about a differential equation driving temperature, e.g. $dT(t)/dt = a(T(t), t, \theta(t))$. If we write $a(\cdot)$ as an additive form say $a(T(t), t) + \theta(t)$ and let $\theta(t) = (T_{\text{obs}}(t) - T(t))/\tau$ then τ controls the relaxation. Small τ implies that the $\theta(t)$ term dominates while large τ implies that the nudging effect will be negligible.

We next turn to a least squares approach. Again, suppressing location, suppose we assume that $T_{\text{obs}}^{(1)}(t) = T_{\text{true}}(t) + \epsilon_1(t)$ and $T_{\text{obs}}^{(2)}(t) = T_{\text{true}}(t) + \epsilon_2(t)$ where we envision two sources of observational data on the true temperature at t. The ϵ_l have mean 0 and variance $\sigma_l^2, l = 1, 2$. Then, with the variances known, it is a familiar exercise to obtain the best unbiased estimator of $T_{\text{true}}(t)$ based upon these two pieces of information. That is, $T_a(t) = a_1 T_{\text{obs}}^{(1)}(t) + a_2 T_{\text{obs}}^{(2)}(t)$ where $a_1 = \sigma_2^2/(\sigma_1^2 + \sigma_2^2)$ and $a_2 = \sigma_1^2/(\sigma_1^2 + \sigma_2^2)$. Of course, we obtain the same solution as the maximum likelihood estimates (MLE) if we use independent normal likelihoods for the $T_{\text{obs}}^{(l)}(t)$s.

A last idea here is simple sequential assimilation and its connection to the Kalman filter. In the univariate case suppose we write $T_a(t) = T_b(t) + \gamma(T_{\text{obs}}(t) - T_b(t))$. Here, $T_{\text{obs}}(t) - T_b(t)$ is referred to as the observational innovation or observational increment relative to the background. The optimal weight $\gamma = \sigma_{\text{obs}}^2/(\sigma_{\text{obs}}^2 + \sigma_b^2)$, analogous to the previous paragraph. Hence, we only need a prior estimate of the ratio of the observational variance to the background variance in order to obtain $T_a(t)$. To make this scheme dynamic, suppose the background is updated through the assimilation, i.e. $T_b(t + 1) = h(T_a(t))$ where $h(\cdot)$ denotes some choice of forecast model. Then we will also need to create a revised background variance; this is usually taken to be a scalar (>1) multiple of the variance of $T_a(t)$.

Finally, the multivariate assimilation idea is now clear. Now we collect the grid cell variables to vector variables and write $T_a(t) = T_b(t) + W(Y_{\text{obs}}(t) - g(T_b(t)))$. Here, the vector Y_{obs} denotes variables that are different from the ones we seek to interpolate. For temperature, these might be Doppler shifts, radar reflectivities, or satellite radiances. Then, g is the nonlinear operator that converts background temperatures into guesses for these new variables. The dimension of Y is not necessarily the same as that of T. W is the gain matrix that usually appears in the Kalman filter. Finally, we introduce errors in the transitional stage, $T_b(t) = T_{\text{true}}(t) + \epsilon_b(t)$ and $T_a(t) = T_{\text{true}}(t) + \epsilon_a(t)$, as well as errors in the observational stage, i.e. for the $Y_{obs}(t)$. Assuming all errors are Gaussian, the dynamic model is specified and the Kalman filter can be implemented to fit the model.

18.3 Review of data fusion methods for environmental exposure

Recall that our objective is to combine model output and station data to improve assessment of environmental exposure. Such synthesis is referred to as assimilation or fusion. Here we move from the more algorithmic strategies of the previous section to fully model-based approaches. In the next two subsections we review the work of Fuentes and Raftery (2005) which has received considerable attention and the very recent work of McMillan *et al.* (2008). A full development of the approach of Sahu *et al.* (2010) with an example is deferred to the following sections.

18.3.1 Fusion modelling using stochastic integration

The fusion approach proposed by Fuentes and Raftery (2005) builds upon earlier Bayesian melding work in Poole and Raftery (2000). It conceptualizes a true exposure surface and views the monitoring station data as well as the model output data as varying in a suitable way around the true surface. In particular, the average exposure in a grid cell A, denoted by $Z(A)$, differs from the exposure at any particular location s, $Z(s)$. The so-called change of support problem in this context addresses converting the point level $Z(s)$ to the grid level $Z(A)$ through the stochastic integral,

$$Z(A) = \frac{1}{|A_j|} \int_{A_j} Z(s)\, ds, \tag{18.1}$$

where $|A|$ denotes the area of the grid cell A. Fusion modelling, working with *block averaging* as in (18.1) has been considered by, e.g. Fuentes and Raftery (2005).

Let $Y(s)$ denote the true exposure corresponding to $Z(s)$ at a station s. The first model assumption is:

$$Z(s) = Y(s) + \epsilon(s) \tag{18.2}$$

where $\epsilon(s) \sim N(0, \sigma_\epsilon^2)$ represents the measurement error at location s. The true exposure process is assumed to be:

$$Y(s) = \mu(s) + \eta(s) \tag{18.3}$$

where $\mu(s)$ provides the spatial trend often characterized by known functions of the site characteristics such as the components of s, elevation etc. The error term $\eta(s)$ is a spatially coloured process assumed to be the zero mean GP with a specified covariance function. (Appendix A provides an introduction to GPs.) The output of the computer model denoted by $Q(s)$ is often known to be biased and hence this is modelled as:

$$Q(s) = a(s) + b(s)Y(s) + \delta(s) \tag{18.4}$$

where $a(s)$ denotes the additive bias and $b(s)$ denotes the multiplicative bias. The error term, $\delta(s)$, is assumed to be a white noise process given by $N(0, \sigma_\delta^2)$. However, the computer model output is provided in a grid, $A_1, \ldots, A_J$ so the point level process is converted to a grid level one by the stochastic integral (18.1) for the model (18.4), i.e.

$$Q(A_j) = \frac{1}{|A_j|}\left[\int_{A_j} a(s)\,ds + \int_{A_j} b(s)Y(s)\,ds + \int_{A_j} \delta(s)\,ds\right].$$

It is acknowledged that unstable model fitting accrues to the case where we have spatially varying $b(s)$ so $b(s) = b$ is adopted. Spatial prediction at a new location s' is done through the posterior predictive distribution $p(Y(s')|Z, Q)$ where Z denote all the station data and Q denote all the grid-level computer output $Q(A_1), \ldots, Q(A_J)$.

This fusion strategy becomes computationally infeasible in the setting of fusing say, CMAQ data at 12 km^2 grid cells for the eastern United States with station data for this region. We have a very large number of grid cells with a relatively sparse number of monitoring sites. An enormous amount of stochastic integration is required. In this regard, a dynamic implementation over many time periods becomes even more infeasible. Recently Berrocal *et al.* (2010) have shown that the fusion strategy can be outperformed by their proposed downscaling approach both in terms of computing speed and out-of-sample validation.

18.3.2 Fusion modelling by upscaling

While the Fuentes and Raftery (2005) approach models at the point level, the strategy in McMillan *et al.* (2008) scales up to, models at, the grid cell level. In this fashion, computation is simplified and fusion with space-time data is manageable.

In particular, suppose that we have, say, n monitoring stations. As before, let $Q(A_j)$ denote the CMAQ output value for cell A_j while $Z_{A_j}(s_i)$ denotes the station data for site s_i within cell A_j, $i = 1, \ldots, k_j$. Of course, for most of the j's, k_j will be 0 since $n \ll J$. Let $Y(A_j)$ denote the true value for cell A_j.

Then, paralleling (18.2) and (18.4), for each $j = 1, \ldots, J$,

$$Z_{A_j}(s_i) = Y(A_j) + \epsilon_{A_j}(s_i), \quad i = 1, \ldots, k_j \tag{18.5}$$

and

$$Q(A_j) = Y(A_j) + b(A_j) + \gamma(A_j). \tag{18.6}$$

In (18.6), the CMAQ output is modelled as varying around the true value with a bias term, denoted by $b(A_j)$, specified using a B-spline model. Also, the ϵ's are assumed to be independently and identically distributed and so are the γ's, each with a respective variance component. So, the station data and the CMAQ data are conditionally independent given the true surface. Finally, the true surface is modeled analogously to (18.3). But now, the η's are given a CAR specification (see, e.g. Banerjee *et al.*, 2004). For space-time data, McMillan *et al.* (2008) offer a dynamic version of this approach, formulated by assuming a dynamic CAR specification for the η's. They illustrate with a fusion for the year 2001.

18.4 A downscaling approach

Very recently, Sahu *et al.* (2010) proposed a modelling approach that avoids the computationally demanding stochastic integrations required in Fuentes and Raftery (2005) but models at the point rather than the grid cell level as in McMillan *et al.* (2008). In particular, they formalize a latent atmospheric process which is modeled at two different scales, at the point level to align with the station data and at the grid cell level to align with the resolution for the computer model output. The models at these two scales are connected through a measurement error model (MEM). The latent processes are introduced to capture point masses at 0 with regard to chemical deposition while the MEM circumvents the stochastic integration in (18.1). In particular, the point level observed data represent 'ground truth' while gridded CMAQ output are anticipated to be biased. As a result, the MEM enables calibration of the CMAQ model. The opposite problem of disaggregation, i.e. converting the grid level computer output $Q(A_j)$ to point level ones, $Q(s_i)$ is not required. The only assumption is that $Q(A_j)$ is a reasonable surrogate for $Z(s_i)$ if the site s_i is within the grid cell A_j. In this sense, the approach is a downscaler, scaling the grid cell level CMAQ data to the point-level station data.

Sahu *et al.* (2010) model the above fusion approach in a dynamic setting modelling weekly chemical deposition data over a year. They utilize precipitation information to model wet deposition since there can be no deposition without precipitation. They also handle occurrences of zero values in both precipitation and deposition. They introduce a latent space-time atmospheric process which drives both precipitation and deposition as assumed in the mercury deposition modelling of Rappold *et al.* (2008). However, Rappold *et al.* do not address the fusion problem with modeled output. Rather, they used a point level joint process model, specified conditionally for the atmospheric, precipitation and deposition processes. Sahu *et al.* illustrate their methods separately for both wet sulfate and wet nitrate deposition in the eastern United States.

18.4.1 The modelling detail

Here we present detail for the static version of the dynamic spatial model developed in Sahu *et al.* (2010). Let $R(s_i)$ and $Z(s_i)$ denote the observed precipitation and deposition respectively at a site s_i, $i = 1, \ldots, n$. We suppose that $R(s_i)$ and $Z(s_i)$ are driven by a point level latent atmospheric process, denoted by $V(s_i)$, and both take the value zero if $V(s_i) < 0$ to reflect that there is no deposition without precipitation. That is,

$$R(s_i) = \begin{cases} \exp\{U(s_i)\} & \text{if } V(s_i) > 0 \\ 0 & \text{otherwise,} \end{cases} \tag{18.7}$$

and

$$Z(s_i) = \begin{cases} \exp\{Y(s_i)\} & \text{if } V(s_i) > 0 \\ 0 & \text{otherwise.} \end{cases} \tag{18.8}$$

The random variables $U(s_i)$ and $Y(s_i)$ are thus taken as log observed precipitation and deposition respectively when $V(s_i) > 0$. The models described below will specify their values when $V(s_i) \leq 0$ and/or the corresponding $R(s_i)$ or $Z(s_i)$ are missing.

Similar to (18.8) we suppose that the CMAQ model output at grid cell A_j, $Q(A_j)$, is positive if an areal level latent atmospheric process, denoted by $\tilde{V}(A_j)$, is positive,

$$Q(A_j) = \begin{cases} \exp\{X(A_j)\} & \text{if } \tilde{V}(A_j) > 0 \\ 0 & \text{otherwise.} \end{cases} \tag{18.9}$$

The values of $X(A_j)$ when $\tilde{V}(A_j) \leq 0$ will be given by the model described below. As computer model output, there are no missing values in the $Q(A_j)$.

Let R, Z, and Q denote all the precipitation values, wet deposition values and the CMAQ model output, respectively. Similarly define the vectors U, V, and Y collecting all the elements of the corresponding random variable for $i = 1, \ldots, n$. Let X and $\tilde{V}$ denote the vectors collecting the elements $X(A_j)$ and $\tilde{V}(A_j)$, $j = 1, \ldots, J$, respectively.

The first stage likelihood implied by the definitions (18.7), (18.8) and (18.9) is given by:

$$p(R, Z, Q|U, Y, X, V, \tilde{V}) = p(R|U, V) \times p(Z|Y, V) \times p(Q|X, \tilde{V}) \tag{18.10}$$

which takes the form

$$\prod_{i=1}^{n} \left\{ 1_{\exp(u(s_i))} 1_{\exp(y(s_i))} I(v(s_i) > 0) \right\} \prod_{j=1}^{J} \left\{ 1_{\exp(x(A_j))} I(\tilde{v}(A_j) > 0) \right\}$$

where 1_x denotes a degenerate distribution with point mass at x and $I(\cdot)$ is the indicator function.

18.4.2 Second stage specification

In the second stage of modelling we begin by specifying a spatially coloured regression model for log-precipitation based on the latent process $V(s_i)$. In particular, we assume the model:

$$U(s_i) = \alpha_0 + \alpha_1 V(s_i) + \delta(s_i), \quad i = 1, \ldots, n \tag{18.11}$$

where $\delta = (\delta(s_1), \ldots, \delta(s_n))'$ is an independent GP following the $N(0, \Sigma_\delta)$ distribution; Σ_δ has elements $\sigma_\delta(i, j) = \sigma_\delta^2 \exp(-\phi_\delta d_{ij})$, the usual exponential covariance function, where d_{ij} is the geodetic distance between sites s_i and s_j. Using vector notation, the above specification is equivalently written as:

$$U \sim N\left(\alpha_0 + \alpha_1 V, \Sigma_\delta\right).$$

To model $Y(s_i)$, we assume that:

$$Y(s_i) = \beta_0 + \beta_1 U(s_i) + \beta_2 V(s_i) + \left\{b_0 + b(s_i)\right\} X(A_{k_i}) + \eta(s_i) + \epsilon(s_i), \tag{18.12}$$

for $i = 1, \ldots, n$ where, unless otherwise mentioned, A_{k_i} is the grid cell which contains the site s_i.

The error terms $\epsilon(s_i)$ are assumed to follow $N(0, \sigma_\epsilon^2)$ independently, providing the so-called nugget effect. The reasoning for the rest of the specification in (18.12) is as follows. The term $\beta_1 U(s_i)$ is included because of the strong linear relationships between log-deposition and log-precipitation, see Figure 18.3 below. The term $\beta_2 V(s_i)$ captures any direct influence of the atmospheric process $V(s_i)$ on $Y(s_i)$ in the presence of precipitation.

It is anticipated that the relationship between the station data and the CMAQ model output will be roughly linear but that this relationship may vary locally. To specify a rich class of *locally* linear models we may think of a spatially varying slope for the regression of $Y(s_i)$ on log-CMAQ values $X(A_j)$, specified as $\left\{b_0 + b(s_i)\right\} X(A_{k_i})$ in (18.12). Writing $b = (b(s_1), \ldots, b(s_n))'$ we propose a mean 0 GP for b, i.e.

$$b \sim N(0, \Sigma_b)$$

where Σ_b has elements $\sigma_b(i, j) = \sigma_b^2 \exp(-\phi_b d_{ij})$.

The term $\eta(s_i)$ provides a spatially varying intercept which can also be interpreted as a spatial adjustment to the overall intercept parameter β_0. We assume that

$$\eta \sim N(0, \Sigma_\eta),$$

where $\eta = (\eta(s_1), \ldots, \eta(s_n))'$ and Σ_η has elements $\sigma_\eta(i, j) = \sigma_\eta^2 \exp(-\phi_\eta d_{ij})$. The regression model (18.12) is now equivalently written as:

$$Y \sim N\left(\vartheta, \sigma_\epsilon^2 I_n\right)$$

where $Y = (Y(s_1), \ldots, Y(s_n))'$ and $\vartheta = \beta_0 + \beta_1 u + \beta_2 v + b_0 x + X_m b + \eta$ where x is the n-dimensional vector with the ith element given by $x(A_{k_i})$ and X_m is a diagonal matrix whose ith diagonal entry is $x(A_{k_i})$, $i = 1, \ldots, n$ and I_n is the identity matrix of order n.

The CMAQ output $X(A_j)$ is modelled using the latent process $\tilde{V}(A_j)$ as follows:

$$X(A_j) = \gamma_0 + \gamma_1 \tilde{V}(A_j) + \psi(A_j), \quad j = 1, \ldots, J. \tag{18.13}$$

where $\psi(A_j) \sim N(0, \sigma_\psi^2)$ independently for all $j = 1, \ldots, J$, and σ_ψ^2 is unknown. In vector notation, this is given by:

$$X \sim N\left(\gamma_0 + \gamma_1 \tilde{V}, \sigma_\psi^2 I_J\right)$$

where as before, $X = (X(A_1), \ldots X(A_J))'$ and $\tilde{V} = (\tilde{V}(A_1), \ldots \tilde{V}(A_J))'$, see the partitioning of $\tilde{V}$ below equation (18.14) regarding the order of the grid cell indices $1, \ldots, J$.

We now turn to specification of the latent processes $V(s_i)$ and $\tilde{V}(A_j)$. Note that it is possible to have $Z(s_i) > 0$ and $Q(A_{k_i}) = 0$ and vice versa since $Q(A_{k_i})$ is the output of a computer model which has not used the actual observation $Z(s_i)$. This implies that $V(s_i)$ and $\tilde{V}(A_{k_i})$ can be of different signs. To accommodate this flexibility and to distinguish between the point and areal processes we assume the simple measurement error model:

$$V(s_i) \sim N\left(\tilde{V}(A_{k_i}), \sigma_v^2\right), \quad i = 1, \ldots, n \tag{18.14}$$

where σ_v^2 is unknown. Without loss of generality we write $\tilde{V} = (\tilde{V}^{(1)}, \tilde{V}^{(2)})$ where the n-dimensional vector $\tilde{V}^{(1)}$ contains the values for the grid cells where the n observation sites are located and $\tilde{V}^{(2)}$ contains the values for the remaining $J - n$ grid cells. The specification (18.14) can now be written equivalently as

$$V \sim N\left(\tilde{V}^{(1)}, \sigma_v^2 I_n\right).$$

The latent process $\tilde{V}(A_j)$ is assumed to follow a conditionally autoregressive (CAR) process in space (see e.g. Banerjee *et al.*, 2004). That is,

$$\tilde{V}(A_j) \sim N\left(\sum_{i=1}^{J} h_{ji}\tilde{V}(A_i), \frac{\sigma_\zeta^2}{m_j}\right) \tag{18.15}$$

where

$$h_{ji} = \begin{cases} \frac{1}{m_j} & \text{if } i \in \partial_j \\ 0 & \text{otherwise} \end{cases}$$

and ∂_j defines the m_j neighbouring grid cells of the cell A_j. The above improper CAR specification can be written as:

$$p\left(\tilde{V}|\sigma_\zeta^2\right) \propto \exp\left\{-\frac{1}{2}\tilde{V}' D^{-1}(I - H)\tilde{V}\right\} \tag{18.16}$$

where D is diagonal with the jth diagonal entry given by σ_ζ^2/m_j. In summary, the second stage specification is given by:

$$p(Y|U, V, X, \eta, b, \theta) \times p(\eta|\theta) \times p(U|V, \theta) \times p\left(\tilde{V}|\theta\right)$$
$$\times p(V|\theta) \times p\left(X|\tilde{V}, \theta\right) \times p(b|\theta)$$

where θ denote the parameters $\alpha_0, \alpha_1, \beta_0, \beta_1, \beta_2, b_0, \gamma_0, \gamma_1, \rho, \sigma_\delta^2, \sigma_b^2, \sigma_\eta^2, \sigma_\epsilon^2, \sigma_\psi^2$, σ_v^2 and σ_ζ^2. See Appendix B for the prior distributions, the form of the joint posterior distribution and the full conditional distributions needed for Gibbs sampling.

18.4.3 Spatial interpolation at a new location

We can interpolate the deposition surface using the above models as follows. Consider the problem of predicting $Z(s')$ at any new location s' falling on the grid cell A'. The prediction is performed by constructing the posterior predictive distribution of $Z(s')$ which in turn depends on the distribution of $Y(s')$ as specified by equation (18.12) along with the associated $V(s')$. We estimate the posterior predictive distribution by drawing samples from it.

Several cases arise depending on the nature of information available at the new site s'. If precipitation information is available and there is no positive precipitation, i.e. $r(s') = 0$, then we have $Z(s') = 0$ and no further sampling is needed, since there can be no deposition without precipitation. Now suppose that there is positive precipitation, i.e. $r(s') > 0$, then set $u(s') = \log(r(s'))$. We need to generate a sample $Y(s')$. We first generate $V(s') \sim N(\tilde{V}(A'), \sigma_v^2)$ following the measurement error model (18.14). Note that $\tilde{V}(A')$ is already available for any grid cell A' (within the study region) from model fitting, see equation (18.15). Similarly, $X(A')$ is also available either as the log of the

CMAQ output, $\log(Q(A'))$, if $Q(A') > 0$ or from the MCMC imputation when $Q(A') = 0$, see Appendix B. To sample $\eta(s')$ we note that:

$$\begin{pmatrix} \eta(s') \\ \eta \end{pmatrix} \sim N\left[\begin{pmatrix} 0 \\ 0 \end{pmatrix}, \sigma_\eta^2 \begin{pmatrix} 1 & S_{\eta,12} \\ S_{\eta,21} & S_\eta \end{pmatrix}\right],$$

where $S_{\eta,12}$ is $1 \times n$ with the ith entry given by $\exp\{-\phi_\eta d(s_i, s')\}$ and $S_{\eta,21} = S'_{\eta,12}$. Therefore,

$$\eta(s')|\eta, \theta \sim N\left[S_{\eta,12} S_\eta^{-1} \eta, \sigma_\eta^2 \left(1 - S_{\eta,12} S_\eta^{-1} S_{\eta,21}\right)\right]. \qquad (18.17)$$

If the term $b(s)$ is included in the model we need to simulate $b(s')$ conditional on b and model parameters. To do this we note that:

$$\begin{pmatrix} b(s') \\ b \end{pmatrix} \sim N\left[\begin{pmatrix} 0 \\ 0 \end{pmatrix}, \sigma_b^2 \begin{pmatrix} 1 & S_{b,12} \\ S_{b,21} & S_b \end{pmatrix}\right],$$

where $S_{b,12}$ is $1 \times n$ with the ith entry given by $\exp\left\{-\phi_\eta d(s_i, s')\right\}$ and $S_{b,21} = S'_{b,12}$. Therefore,

$$b(s')|b, \theta \sim N\left[S_{b,12} S_b^{-1} b, \sigma_b^2 \left(1 - S_{b,12} S_b^{-1} S_{b,21}\right)\right]. \qquad (18.18)$$

If it is desired to predict $Z(s')$ where $R(s')$ is not available, we proceed as follows. We generate $V(s') \sim N(\tilde{V}(A'), \sigma_v^2)$ following the measurement error model (18.14). If this $V(s') < 0$, then we set both $R(s')$ and $Z(s')$ to zero. If, however, $V(s') > 0$ we need to additionally draw $U(s')$ using the precipitation model (18.11). For this we note that,

$$\begin{pmatrix} U(s') \\ U \end{pmatrix} \sim N\left[\begin{pmatrix} \alpha_0 + \alpha_1 V(s') \\ \alpha_0 + \alpha_1 V \end{pmatrix}, \sigma_\delta^2 \begin{pmatrix} 1 & S_{\delta,12} \\ S_{\delta,21} & S_\delta \end{pmatrix}\right],$$

where $S_{\delta,12}$ is $1 \times n$ with the ith entry given by $\exp\left\{-\phi_\delta d(s_i, s')\right\}$ and $S_{\delta,21} = S'_{\delta,12}$. Therefore,

$$U(s')|U, \theta \sim N\left[\mu(s'), \sigma_\delta^2 \left(1 - S_{\delta,12} S_\delta^{-1} S_{\delta,21}\right)\right], \qquad (18.19)$$

where

$$\mu(s') = \alpha_0 + \alpha_1 V(s') + S_{\delta,12} S_\delta^{-1} (U - \alpha_0 - \alpha_1 v).$$

If $Z(s')$ is not inferred to be zero then we set it to be $\exp\left\{Y(s')\right\}$. If we want the predictions of the smooth deposition surface without the nugget term we simply ignore the nugget term $\epsilon(s')$ in generating $Y(s')$.

18.4.4 An illustration

We illustrate with weekly wet deposition data for 2001 from 120 sites monitored by the National Atmospheric Deposition Program (NADP, `nadp.sws.uiuc.edu`) in the eastern United States, see Figure 18.1. We analyze data

Fig. 18.1 A map of the eastern US with the 120 NADP sites plotted as points.

from the year 2001 since this is the year for which the most recent outputs from the CMAQ model for wet chemical deposition are currently available. These outputs are available for $J = 33,390$ grid cells covering the study region. Our approach is applied separately to the wet sulfate and wet nitrate data. Since there is no need to make any simultaneous inference, a joint model is not required. There is high correlation between the two types of deposition but this is expected since both are driven by precipitation. To facilitate spatial interpolation, we also use weekly precipitation data obtained from a network of 2827 sites located inside the study region.

We model the data separately for the week of January 16–22 and May 22–28 in 2001 to make a comparison between a week in the winter and another one in the summer. Deposition data for these two weeks show significant differences according to classical t-tests, see Figure 18.2. This confirms the fact that the deposition levels are generally higher during the wet summer months and lower during the drier winter months, see e.g. Brook *et al.* (1995). However, in both the weeks there is strong linear relationship between deposition and precipitation (on the log-scale), see Figure 18.3. There is also some, although not very strong, linear relationship between observed NADP data and the CMAQ output for the corresponding grid cell containing the NADP site, see Figure 18.4. Deposition and precipitation values that are 0 are ignored in obtaining the above Figures 18.3 and 18.4.

The spatial interpolation maps are provided in Figure 18.5 for sulfate and Figure 18.7 for nitrate. As seen in Figure 18.2 the model has reconstructed higher levels of both deposition types for May 22–28 than that for January 16–22. Observe that the grey scales are different for the two weeks in each

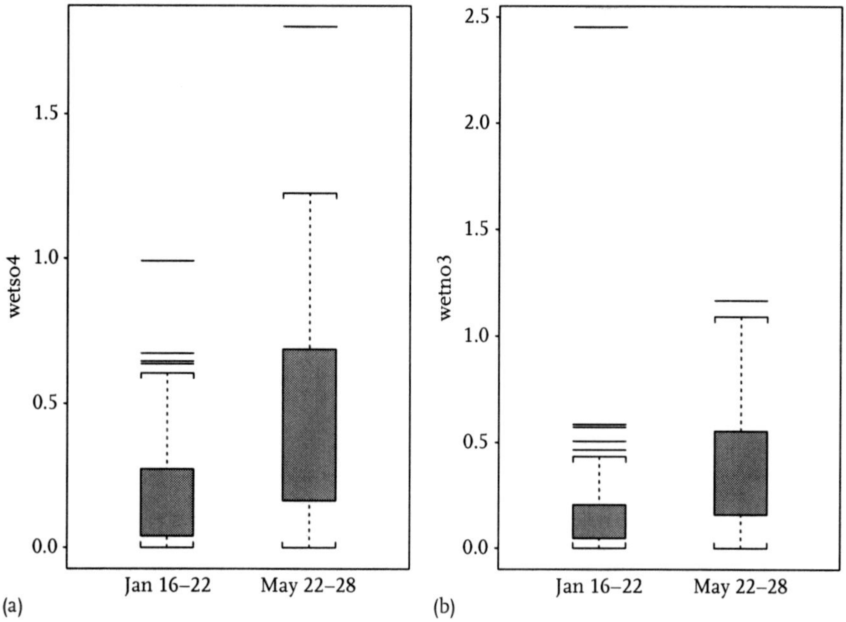

Fig. 18.2 Boxplot of depositions for two weeks in 2001 from 120 NADP sites: (a) wet sulfate and (b) wet nitrate.

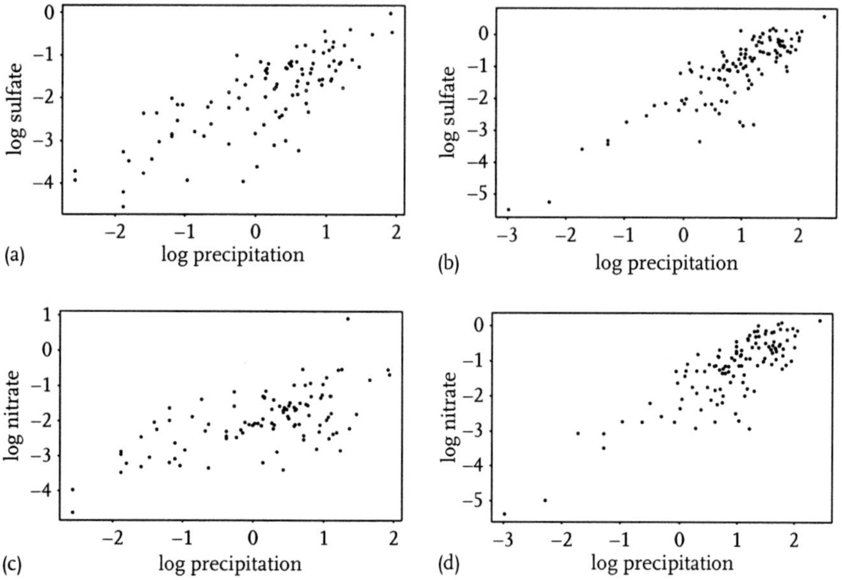

Fig. 18.3 Plot of log deposition against log precipitation in 2001: (a) wet sulfate for January 16–22, (b) wet sulfate for May 22–28, (c) wet nitrate for January 16–22, (d) wet nitrate for May 22–28.

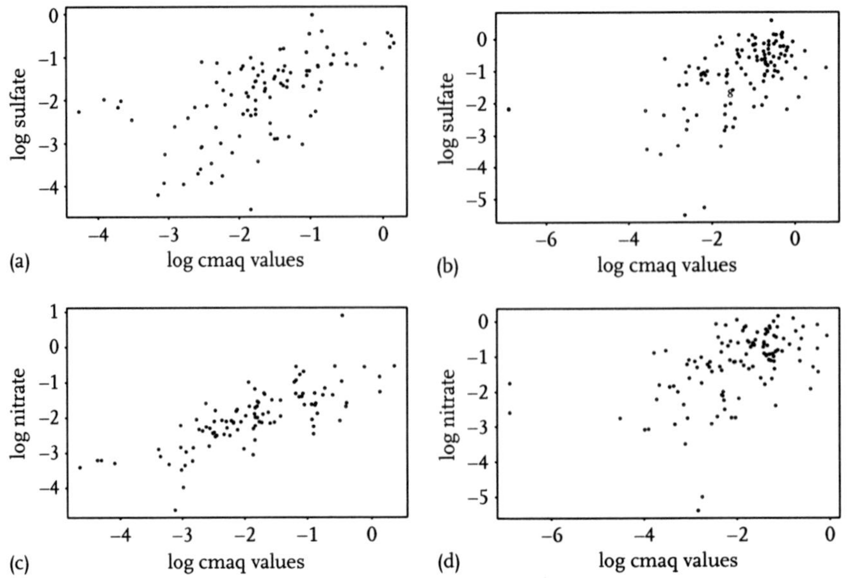

Fig. 18.4 Plot of log deposition against log CMAQ value for the cell containing the corresponding NADP site in 2001: (a) wet sulfate for January 16–22, (b) wet sulfate for May 22–28, (c) wet nitrate for January 16–22, (d) wet nitrate for May 22–28.

of the Figures 18.5 and 18.7. In the corresponding week similar spatial patterns are seen for the sulfate and nitrate deposition values, as expected. Figures 18.6 and 18.8 provide the standard deviation maps for the predictions in Figures 18.5 and 18.7. From these figures we conclude that higher levels of deposition values are predicted with higher levels of uncertainty which is common in this sort of data analysis.

Parameter estimates, model choice analysis, and full spatio-temporal analysis of all the 52 week's dataset with a dynamic version of the foregoing model is presented in the paper of Sahu *et al.* (2010), and hence are not repeated here. They also discuss methods for choosing the decay parameter values ϕ_δ, ϕ_b and ϕ_η. In addition, they validate the models with set aside data and, by suitable aggregation, obtain total annual deposition maps along with their uncertainties.

18.5 Further discussion

A related question of interest is to estimate dry deposition which is defined as the exchange of gases, aerosols, and particles between the atmosphere and earth's surface. Such analysis will enable prediction of total (wet plus dry) sulfur and nitrogen deposition. Using the total predictive surface it will be possible to estimate deposition 'loadings' as the integrated volume of total deposition over ecological regions of interest. If successful, this effort will lead to the first

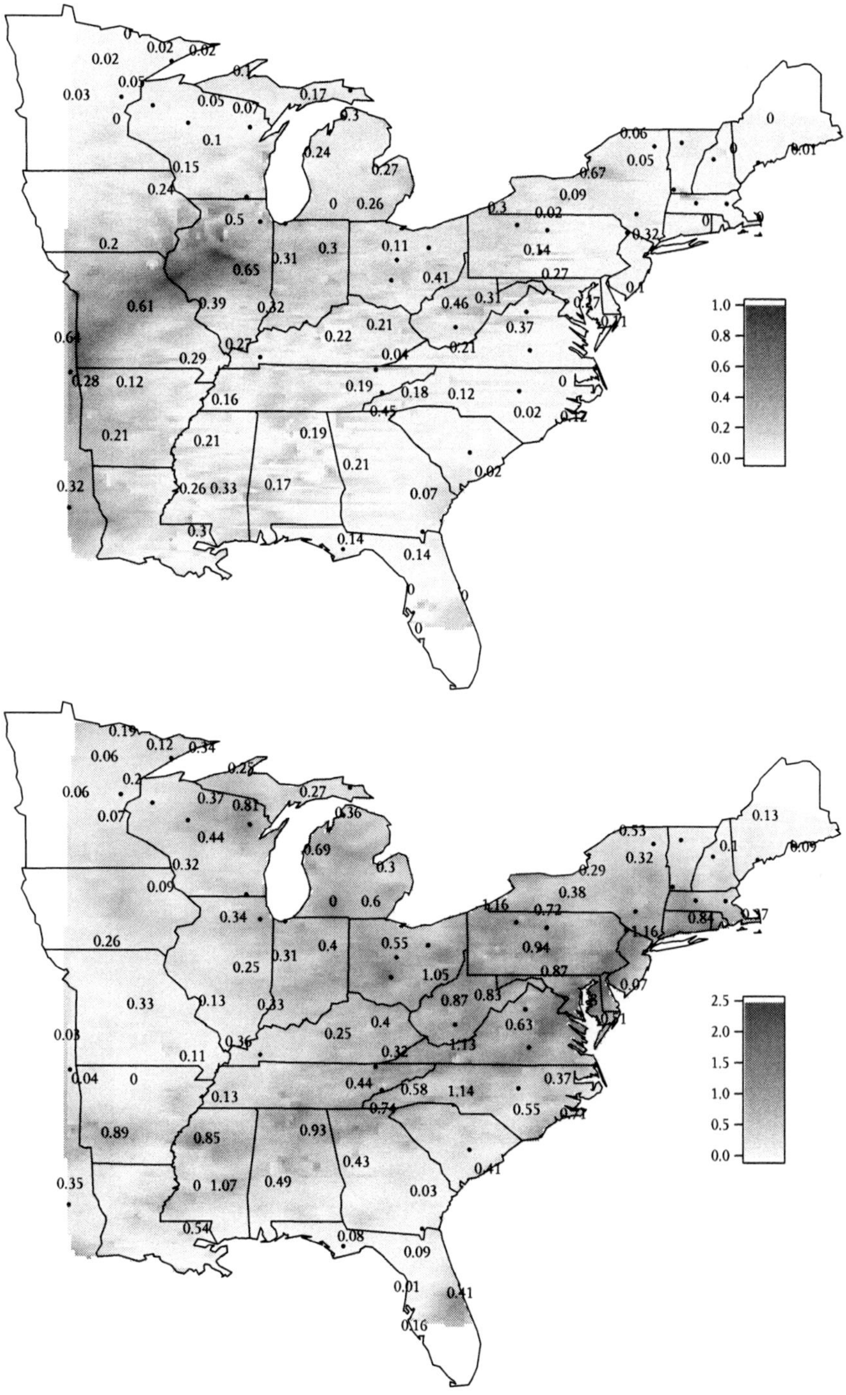

Fig. 18.5 Model interpolated maps for sulfate deposition for two weeks in 2001: top panel for January 16–22 and bottom panel for May 22–28. Observed deposition values from some selected sites are superimposed; the data from the remaining sites are not shown to enhance readability.

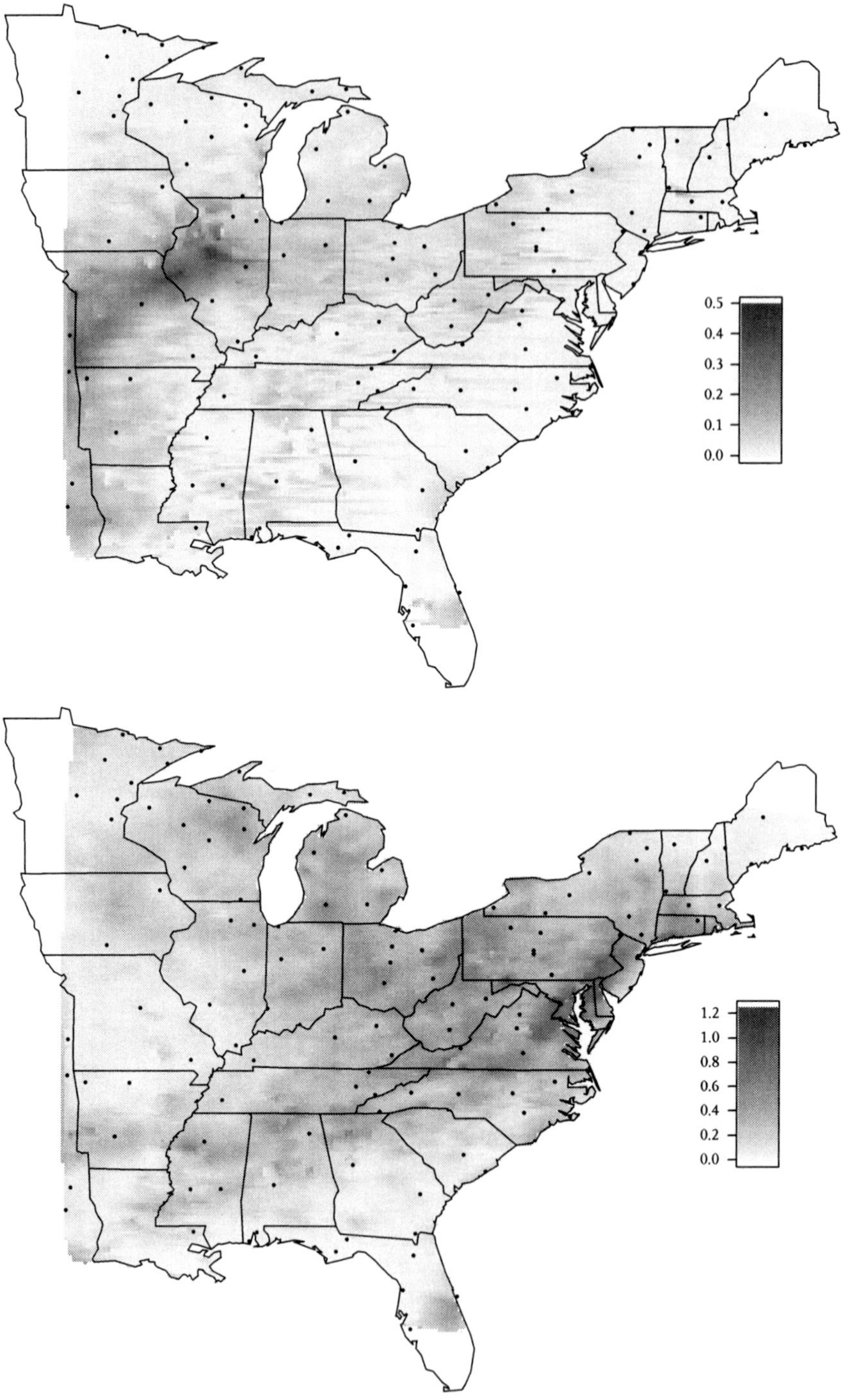

Fig. 18.6 The standard deviation maps for the predictions in Figure 18.5.

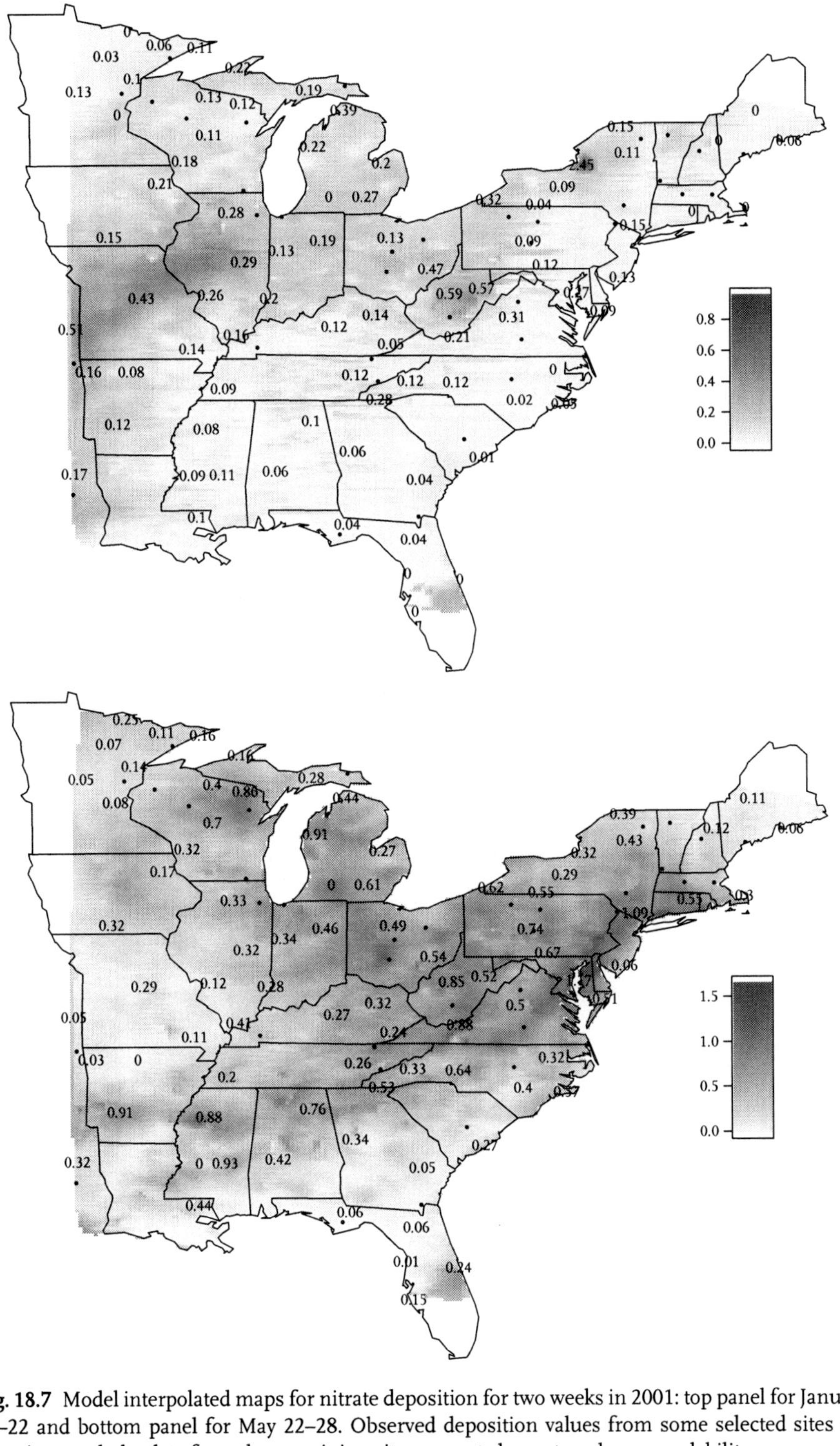

Fig. 18.7 Model interpolated maps for nitrate deposition for two weeks in 2001: top panel for January 16–22 and bottom panel for May 22–28. Observed deposition values from some selected sites are superimposed; the data from the remaining sites are not shown to enhance readability.

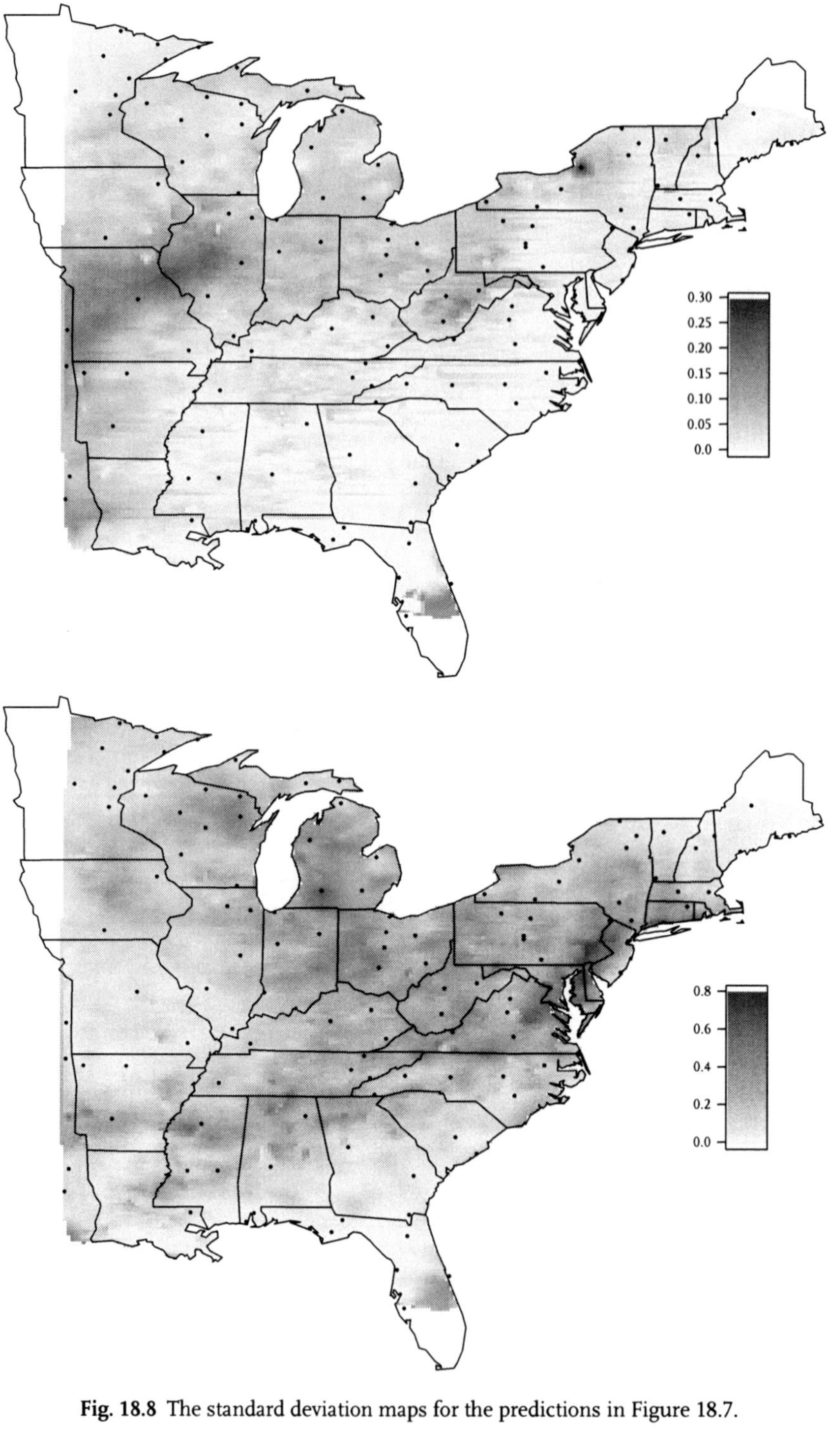

Fig. 18.8 The standard deviation maps for the predictions in Figure 18.7.

ever estimation of total deposition loadings, perhaps the most critical quantity for making ecological assessments. Future work will also address trends in deposition to assess whether regulation has been successful.

Appendix

A. Broader context and background

Gaussian processes play a key role in modeling for spatial and spatio-temporal data. By now, there is an extensive literature on formalizing and characterizing stochastic processes along with analysis their behaviour. However, in a practical setting, ensuring that a stochastic process has been properly defined when the index of the process is over a continuum, say a spatial region, requires care. The primary issue is to guarantee that the joint distribution associated with the entire collection of random variables is consistently defined. The usual strategy is to define the process through its finite dimensional distributions and verify that these distributions satisfy a *consistency condition*. In this regard, the Gaussian process becomes very attractive since its finite dimensional distributions are all multivariate normals. Specification over the set D only requires a mean function, $\mu(s), s \in D$ and a valid covariance function, $C(s, s') = \text{Cov}(Y(s), Y(s'))$, the latter supplying the covariance matrix associated with any finite set of locations.

The convenient conditional distribution theory associated with the multivariate normal distribution is at the heart of kriging (spatial prediction); the convenient marginal distributions facilitate local model specification. Moreover, the only finite dimensional distributions within the class of jointly elliptical distributions that are able to support a stochastic process over a continuum are normals (or scale mixtures of normals).

Additionally, spatial dependence is typically introduced into the modeling in the form of spatial random effects. In general, random effects are modeled as normal variables so a multivariate normal specification for such effects in a spatial setting seems appropriate. In this regard, hierarchical modelling naturally emerges. The spatial random effects are introduced at the second stage of modelling. They appear in the mean (perhaps on a transformed scale if the first stage specification is non-Gaussian as in a spatial generalized linear model such as a binary process where the observation at any location is a 0 or a 1). In this regard, there is practical interest in these random effects. Given their prior specification, the associated revised posterior distributions are of interest with regard to 'seeing' a spatial pattern, again emphasizing the role of Bayesian inference in spatial analysis. While the genesis of spatial modelling for data over a continuum was based primarily on simple least squares theory, modern, fully model-based spatial analysis is almost entirely done within a Bayesian framework.

Specification of a valid covariance function is a separate issue. Such a function must be positive definite, i.e. for any number of and set of spatial locations, the resultant covariance matrix must be positive definite. By now there is a rich literature regarding the choice of such functions in space, enabling isotropy, stationarity, and non-stationarity, and in space time, enabling space-time dependence in association. See, e.g. the book of Banerjee *et al.* (2004) and the recent paper of Stein (2005) and references therein.

Finally, Bayesian computation for space and space-time data analysis is much more demanding than usual analysis. Of course, this is true in general, for fitting hierarchical models but the rewards of full inference will usually justify the effort. Bayesian software to fit spatial data models includes Winbugs (`http://www.mrc-bsu.cam.ac.uk/bugs/`), and two R-packages Geo-R (Ribeiro and Diggle, 1999) and and SpBayes (Finley *et al.*, 2008).

B. Distributions for Gibbs sampling

B.1 Prior and posterior distributions

We complete the Bayesian model specification by assuming prior distributions for all the unknown parameters. We assume that, a priori, each of $\alpha_0, \alpha_1, \beta_0, \beta_1, \beta_2, b_0, \gamma_0, \gamma_1$ is normally distributed with mean 0 and variance 10^3, essentially a flat prior specification. The inverse of the variance components $\frac{1}{\sigma_\delta^2}, \frac{1}{\sigma_b^2}, \frac{1}{\sigma_\eta^2}, \frac{1}{\sigma_\epsilon^2}, \frac{1}{\sigma_\psi^2}, \frac{1}{\sigma_v^2}$, and $\frac{1}{\sigma_\zeta^2}$, are all assumed to follow the Gamma distribution $G(\nu, \lambda)$ having mean ν/λ. In our implementation we take $\nu = 2$ and $\lambda = 1$ implying that these variance components have prior mean 1 and infinite variance.

The log of the likelihood times prior in the second stage specification up to an additive constant is given by:

$$
\begin{aligned}
&-\frac{n}{2}\log\left(\sigma_\epsilon^2\right) - \frac{1}{2\sigma_\epsilon^2}(\gamma - \vartheta)'(\gamma - \vartheta) - \frac{n}{2}\log\left(\sigma_\eta^2\right) - \frac{1}{2\sigma_\eta^2}\eta' S_\eta^{-1}\eta \\
&-\frac{n}{2}\log\left(\sigma_\delta^2\right) - \frac{1}{2\sigma_\delta^2}(u - \alpha_0 - \alpha_1 v)' S_\delta^{-1}(u - \alpha_0 - \alpha_1 v) \\
&-\frac{n}{2}\log\left(\sigma_v^2\right) - \frac{1}{2\sigma_v^2}(v - \tilde{v}^{(1)})'(v - \tilde{v}^{(1)})' \\
&-\frac{J}{2}\log\left(\sigma_\psi^2\right) - \frac{1}{2\sigma_\psi^2}(x - \gamma_0 - \gamma_1\tilde{v})'(x - \gamma_0 - \gamma_1\tilde{v}) \\
&-\frac{J}{2}\log\left(\sigma_\zeta^2\right) - \frac{1}{2}\tilde{v}' D^{-1}(I - H)\tilde{v} \\
&-\frac{n}{2}\log\left(\sigma_b^2\right) - \frac{1}{2\sigma_b^2}b' S_b^{-1}b + \log(p(\theta))
\end{aligned}
$$

where $p(\theta)$ is the prior distribution of θ and $\Sigma_\delta = \sigma_\delta^2 S_\delta$, $\Sigma_b = \sigma_b^2 S_b$, $\Sigma_\eta = \sigma_\eta^2 S_\eta$.

B.2 Handling of the missing values

Note that the transformation equation (18.8) does not define a unique value of $Y(s_i)$ and in addition, there will be missing values corresponding to the missing values in $Z(s_i)$. Any missing value of $Y(s_i)$ is sampled from the model (18.12).

The sampling of the missing $U(s_i)$ for the precipitation process is a bit more involved. The sampling of the missing values must be done using the model (18.11) conditional on all the parameters. Since this model is a spatial model we must use the conditional distribution of $U(s_i)$ given all the $U(s_j)$ values for $j = 1, \ldots, n$ and $j \neq i$. This conditional distribution is obtained using the covariance matrix Σ_δ of δ and is omitted for brevity.

Similarly, equation (18.9) does not define unique values of $X(A_j)$ when $Q(A_j) = 0$. Those values, denoted by $X^*(A_j)$, are sampled using the model equation (18.13), $X^*(A_j)$ is sampled from $N(\gamma_0 + \gamma_1 \tilde{v}(A_j), \sigma_\psi^2)$.

B.3 Conditional posterior distribution of θ

Straightforward calculation yields the following full conditional distributions:

$$\frac{1}{\sigma_\epsilon^2} \sim G\left(\frac{n}{2} + \nu, \lambda + \frac{1}{2}(y - \vartheta)'(y - \vartheta)\right),$$

$$\frac{1}{\sigma_b^2} \sim G\left(\frac{n}{2} + \nu, \lambda + \frac{1}{2} b' S_b^{-1} b\right),$$

$$\frac{1}{\sigma_\eta^2} \sim G\left(\frac{n}{2} + \nu, \lambda + \frac{1}{2} \eta' S_\eta^{-1} \eta\right),$$

$$\frac{1}{\sigma_\delta^2} \sim G\left(\frac{n}{2} + \nu, \lambda + \frac{1}{2}(u - \alpha_0 - \alpha_1 v)' S_\delta^{-1} (u - \alpha_0 - \alpha_1 v)\right),$$

$$\frac{1}{\sigma_\psi^2} \sim G\left(\frac{J}{2} + \nu, \lambda + \frac{1}{2}\left(x - \gamma_0 - \gamma_1 \tilde{v}\right)' \left(x - \gamma_0 - \gamma_1 \tilde{v}\right)\right),$$

$$\frac{1}{\sigma_v^2} \sim G\left(\frac{n}{2} + \nu, \lambda + \frac{1}{2}(v - \tilde{v}^{(1)})'(v - \tilde{v}^{(1)})\right),$$

$$\frac{1}{\sigma_\zeta^2} \sim G\left(\frac{J}{2} + \nu, \lambda + \frac{1}{2} \sum_{j=1}^{J} \left\{m_j (\tilde{V}(A_j) - \mu_j)^2\right\}\right)$$

where $\mu_j = \sum_{i=1}^{J} h_{ji} \tilde{V}(A_i)$.

Let $\beta = (\beta_0, \beta_1, \beta_2)$ and $G = (1, u, v)$ so that G is an $n \times 3$ matrix. The full conditional distribution of β is $N(\Lambda\chi, \Lambda)$ where

$$\Lambda^{-1} = \frac{1}{\sigma_\epsilon^2} G'G + 10^{-3} I_3, \quad \chi = \frac{1}{\sigma_\epsilon^2} G'(y - b_0 x - X_m b - \eta).$$

The full conditional distribution of b_0 is $N(\Lambda\chi,\ \Lambda)$ where

$$\Lambda^{-1} = \frac{1}{\sigma_\epsilon^2}x'x + 10^{-3}, \quad \chi = \frac{1}{\sigma_\epsilon^2}x'(y - \beta_0 - \beta_1 u - \beta_2 v - X_m b - \eta).$$

The full conditional distribution of b is $N(\Lambda\chi,\ \Lambda)$ where

$$\Lambda^{-1} = \frac{1}{\sigma_\epsilon^2}X'X + \Sigma_b^{-1}, \quad \chi = \frac{1}{\sigma_\epsilon^2}X'(y - \beta_0 - \beta_1 u - \beta_2 v - b_0 x - \eta).$$

The full conditional distribution of η is $N(\Lambda\chi,\ \Lambda)$ where

$$\Lambda^{-1} = \frac{I_n}{\sigma_\epsilon^2} + \Sigma_\eta^{-1}, \quad \chi = \frac{1}{\sigma_\epsilon^2}(y - \beta_0 - \beta_1 u - \beta_2 v - b_0 x - X_m b).$$

Let $G = (1, v)$ so that now G is a $n \times 2$ matrix. The full conditional distribution of $\alpha = (\alpha_0, \alpha_1)$ is $N(\Lambda\chi,\ \Lambda)$ where

$$\Lambda^{-1} = G'\Sigma_\delta^{-1}G + 10^{-3}I_2, \chi = G'\Sigma_\delta^{-1}u.$$

Let $G = (1, \tilde{v})$ so that now G is a $J \times 2$ matrix. The full conditional distribution of $\gamma = (\gamma_0, \gamma_1)$ is $N(\Lambda\chi,\ \Lambda)$ where

$$\Lambda^{-1} = \frac{1}{\sigma_\psi^2}G'G + 10^{-3}I_2, \quad \chi = G'x.$$

B.4 Conditional posterior distribution of V

Note that due to the missing and zero precipitation values the full conditional distribution of V will be in a restricted space. First, the unrestricted full conditional distribution of V is $N(\Lambda\chi,\ \Lambda)$ where

$$\Lambda^{-1} = \beta_2^2\frac{I_n}{\sigma_\epsilon^2} + \alpha_1^2\Sigma_\delta^{-1} + \frac{I_n}{\sigma_v^2}, \quad \text{and} \quad \chi = \frac{\beta_2}{\sigma_\epsilon^2}a + \alpha_1\Sigma_\delta^{-1}(u - \alpha_0) + \frac{1}{\sigma_v^2}\tilde{v}^{(1)},$$

where $a = y - \beta_0 - \beta_1 u - b_0 x - X_m b - \eta$. From this n-dimensional joint distribution we obtain the conditional distribution $V(s_i) \sim N(\mu_i, \Xi_i)$, say. If the precipitation value, $r(s_i)$, is missing then there will be no constraint on $V(s_i)$ and we sample $V(s_i)$ unrestricted from $N(\mu_i, \Xi_i)$. If on the other hand the observed precipitation value is zero, $r(s_i) = 0$, we must sample $V(s_i)$ to be negative, i.e. we sample from $N(\mu_i, \Xi_i)I(V(s_i) < 0)$. Corresponding to non-zero precipitation value $r(s_i) > 0$ we sample $V(s_i)$ from $N(\mu_i, \Xi_i)I(V(s_i) > 0)$.

B.5 Conditional posterior distribution of $\tilde{V}$

The full conditional distribution of $\tilde{V} = (\tilde{V}^{(1)}, \tilde{V}^{(2)})$ is $N(\Lambda\chi, \Lambda)$ where

$$\Lambda^{-1} = \begin{pmatrix} \frac{I_n}{\sigma_v^2} & 0 \\ 0 & 0 \end{pmatrix} + \gamma_1^2 \frac{I_J}{\sigma_\psi^2} + D^{-1}(I - H),$$

$$\chi = \begin{pmatrix} \frac{1}{\sigma_v^2} v \\ 0 \end{pmatrix} + \frac{\gamma_1}{\sigma_\psi^2}(x - \gamma_0).$$

Note that this full conditional distribution is a J-variate normal distribution where J is possibly very high (33,390 in our example) and simultaneous update is computationally prohibitive. In addition, we need to incorporate the constraints implied by the first stage likelihood specification (18.10).

The partitioning of $\tilde{V}$, however, suggests an immediate univariate sampling scheme detailed below. First, note that the conditional prior distribution for $\tilde{V}(A_j)$ from the vectorized specification (18.16), as calculated above is given by $N(\xi_j, \omega_j^2)$ where:

$$\omega_j^2 = \sigma_\zeta^2 \frac{1}{m_j} \quad \text{and} \quad \xi_j = \sum_{i=1}^{J} h_{ji} \tilde{v}(A_i).$$

Now for each component $\tilde{V}(A_j)$ of $\tilde{V}^{(1)}$ we extract the full conditional distribution to be viewed as the likelihood contribution from the joint distribution $N(\Lambda_{(1)}\chi_{(1)}, \Lambda_{(1)})$ where

$$\Lambda_{(1)}^{-1} = \frac{I_n}{\sigma_v^2} + \gamma_1^2 \frac{I_n}{\sigma_\psi^2} \quad \text{and} \quad \chi_{(1)} = \frac{1}{\sigma_v^2} v + \frac{\gamma_1}{\sigma_\psi^2}(x^{(1)} - \gamma_0),$$

where $x = (x^{(1)}, x^{(2)})$, partitioned analogusly to $\tilde{V}$. This conditional likelihood contribution is given by $N(\mu_j, \Xi^2)$ where

$$\mu_j = \Xi^2 \left(\tilde{v}(A_j)/\sigma_v^2 + \gamma_1(x(A_j) - \gamma_0)\right)/\sigma_\psi^2, \quad \Xi^2 = 1/\left(1/\sigma_v^2 + 1/\sigma_\psi^2\right).$$

The conditional likelihood contribution for each component of $\tilde{V}^{(2)}$ is the normal distribution $N(\mu_j, \Xi^2)$ where

$$\mu_j = \frac{x(A_j) - \gamma_0}{\gamma_1} \quad \text{and} \quad \Xi^2 = \frac{\sigma_\psi^2}{\gamma_1^2}.$$

Now the unconstrained full conditional distribution of $\tilde{V}(A_j)$, according to the second stage likelihood and prior specification, is obtained by combining the likelihood contribution $N(\mu_j, \Xi^2)$ and the prior conditional distribution $N(\xi_j, \omega_j^2)$ and is given by $N(\Lambda_j\chi_j, \Lambda_j)$ where

$$\Lambda_j^{-1} = \Xi^{-2} + \omega_j^{-2}, \; \chi_j = \Xi^{-2}\mu_j + \omega_j^{-2}\xi_j.$$

In order to respect the constraints implied by the first stage specification we simulate the $\tilde{V}(A_j)$ to be positive if $x(A_j) > 0$ and negative otherwise.

Acknowledgment

The authors thank David Holland, Gary Lear and Norm Possiel of the US EPA for many helpful comments and also for providing the monitoring data and CMAQ model output used in this chapter.

References

Banerjee, S., Carlin, B. P. and Gelfand, A. E. (2004). *Hierarchical Modeling and Analysis for Spatial Data.* Chapman & Hall/CRC, Bora Raton, Florida.

Bergthorsson, P. and Döös, B. (1955). Numerical weather map analysis, *Tellus*, **7**, 329–340.

Berrocal, V. J., Gelfand, A. E. and Holland, D. M. (2010). A spatio-temporal downscaler for output from numerical models. *Journal of Agricultural, Biological and Environmental Statistics*, in press.

Bratseth, A. M. (1986). Statistical interpolation by means of successive corrections. *Tellus*, **38A**, 439–447.

Brook, J. R., Samson, P. J. and Sillman, S. (1995). Aggregation of selected three-day periods to estimate annual and seasonal wet deposition totals for sulfate, nitrate, and acidity. Part I: A synoptic and chemical climatology for eastern North America. *Journal of Applied Meteorology*, **34**, 297–325.

Brown, P. J., Le, N. D. and Zidek, J. V. (1994). Multivariate spatial interpolation and exposure to air pollutants. *The Canadian Journal of Statistics*, **22**, 489–510.

Burke, J. M., Vedantham, R., McCurdy, T. R., Xue, J. and Ozkaynak. A. H. (2003). SHEDS-PM: A population exposure model for predicting distributions of PM exposure and dose from both outdoor and indoor sources. Presented at International Society of Exposure Analysis, Stresa, Italy, September 21–25, 2003.

Carroll, R. J., Chen, R., George, E. I., Li, T. H., Newton, H. J., Schmiediche, H. and Wang, N. (1997). Ozone exposure and population density in Harris County. Texas, *Journal of the American Statistical Association*, **92**, 392–404.

Charney, J. G. (1951). Dynamic forecasting by numerical process, *Compendium of Meteorology*, American Meteorological Society, Boston, MA.

Cox, D. D., Park, J. S. and Singer C. E. (2001). A statistical method for tuning a computer code to a data base. *Computational Statistics and Data Analysis*, **37**, 77–92.

Cressie, N., Kaiser, M. S., Daniels, M. J., Aldworth, J., Lee, J., Lahiri, S. N. and Cox, L. (1999). Spatial analysis of particulate matter in an urban environment. In *GeoEnv II: Geostatistics for Environmental Applications*, (ed. J. Gmez-Hernndez, A. Soares, R. Froidevaux). pp. 41–52. Kluwer, Dordrecht.

Cressman, G. P. (1959). An operational objective analysis system, *Monthly Weather Review*, **87**, 367–374.

Finley, A. O., Banerjee, S. and Carlin, B. P. (2008). Univariate and multivariate spatial modeling. Technical Report, Department of Bio-statistics, University of Minnesota, `http://blue.for.msu.edu/software`.

Fuentes, M., Guttorp, P. and Challenor, P. (2003). Statistical assessment of numerical models. *International Statistical Review*, **71**. 201–221.

Fuentes, M. and Raftery, A. (2005). Model evaluation and spatial interpolation by Bayesian combination of observations with outputs from numerical models. *Biometrics*, **61**, 36–45.

Guttorp, P., Meiring, W. and Sampson, P. D. (1994). A space-time analysis of ground-level ozone data. *Environmetrics*, **5**, 241–254.

Haas, T. C. (1995). Local prediction of a spatio-temporal process with an application to wet sulfate deposition. *Journal of the American Statistical Association*, **90**, 1189–1199.

Kalnay, E. (2003). *Atmospheric Modeling, Data Assimilation and Predictability*. Cambridge University Press, Cambridge.

Kennedy, M. C. and O'Hagan, A. (2001). Bayesian calibration of computer models (with discussion). *Journal of the Royal Statistical Society, Series B*, **63**, 425–464.

Kibria, B. M. G., Sun, L., Zidek, J. V. and Le, N. D. (2002). Bayesian spatial prediction of random space-time fields with application to mapping PM2.5 exposure. *Journal of the American Statistical Association*, **97**, 112–124.

Le, N. D., Sun, W. and Zidek, J. V. (1997). Bayesian multivariate spatial interpolation with data missing by design. *Journal of the Royal Statistical Society, Series B*, **59**, 501–510.

McMillan, N., Holland, D., Morara, M. and Feng, J. (2008). Combining numerical model output and particulate data using Bayesian space-time modeling. *Environmetrics*, DOI:10.1002/env.984.

Poole, D. and Raftery, A. E. (2000). Inference for deterministic simulation models: The Bayesian melding approach. *Journal of the American Statistical Association*, **95**, 1244–1255.

Rappold, A. G., Gelfand, A. E. and Holland, D. M. (2008). Modeling mercury deposition through latent space-time processes. *Journal of the Royal Statistical Society, Series C*, **57**, 187–205.

Ribeoro Jr, P. J. and Diggle, P. J. (1999). geoS: A geostatistical library for S-PLUS. Technical report ST-99–09. Dept of Mathematics and Statistics, Lancaster University, `http://www.leg.ufpr.br/geoR/`.

Richmond, H. M., Palma, T., Langstaff, J., McCurdy, T., Glenn, G. and Smith L. (2002). Further refinements and testing of APEX (3.0): EPA's population exposure model for criteria and air toxic inhalation exposures. Joint Meeting of the Society of Exposure Analysis and International Society of Environmental Epidemiology, Vancouver, Canada.

Sahu, S. K. and Mardia, K. V. (2005). A Bayesian kriged-Kalman model for short-term forecasting of air pollution levels. *Journal of the Royal Statistical Society, Series C*, **54**, 223–244.

Sahu, S. K., Gelfand, A. E. and Holland, D. M. (2006). Spatio-temporal modeling of fine particulate matter. *Journal of Agricultural, Biological and Environmental Statistics*, **11**, 61–86.

Sahu, S. K., Gelfand, A. E. and Holland, D. M. (2007). High resolution space-time ozone modeling for assessing trends. *Journal of the American Statistical Association*, **102**, 1221–1234.

Sahu, S. K., Gelfand, A. E. and Holland, D. M. (2010). Fusing point and areal level space-time data with application to wet deposition. *Journal of the Royal Statistical Society, Series C*, **59**, 77–103.

Sahu, S. K., Yip, S. and Holland, D. M. (2009). Improved space-time forecasting of next day ozone concentrations in the eastern US. *Atmospheric Environment*, doi:10.1016/j.atmosenv.2008.10.028.

Shaddick, G. and Wakefield, J. (2002). Modelling daily multivariate pollutant data at multiple sites. *Journal of the Royal Statistical Society, Series C*, **51**, 351–372.

Smith, R. L., Kolenikov, S. and Cox, L. H. (2003). Spatio-temporal modelling of $PM_{2.5}$ data with missing values. *Journal of Geophysical Research-Atmospheres,* **108**, D249004, doi:10.1029/2002JD002914.

Stein, M. L. (2005). Space-time covariance functions. *Journal of the American Statistical Association 100,* 310–321.

Sun, L., Zidek, J. V., Le, N. D. and Ozkaynak, H. (2000). Interpolating Vancouver's daily ambient PM10 field. *Environmetrics,* **11**, 651–663.

Wikle, C. K. (2003). Hierarchical models in environmental science. *International Statistical Review.* **71**, 181–199.

Zidek, J. V., Shaddick, G., White, R., Meloche, J. and Chatfield, C. (2005). Using a probabilistic model (pCNEM) to estimate personal exposure to air pollution context sensitive links. *Environmetrics,* **16**, 481–493.

Zidek, J. V., Shaddick, G., Meloche, J., Chatfield, C. and White, R. (2007). A framework for predicting personal exposures to environmental hazards. *Environmental and Ecological Statistics,* **14**, 411–431.

Zidek, J. V., Sun, L., Le, N. and Ozkaynak, H. (2002). Contending with space-time interaction in the spatial prediction of pollution: Vancouver's hourly ambient PM_{10} field. *Environmetrics,* **13**, 595–613.

·19·

Indirect elicitation from ecological experts: From methods and software to habitat modelling and rock-wallabies

Samantha Low Choy, Justine Murray, Allan James and Kerrie Mengersen

19.1 Introduction

This work was prompted by a need to model habitat requirements for a threatened Australian species, the brush-tailed rock-wallaby, *Petrogale penicillata*. Modelling wildlife habitat requirements is important for mapping their distribution and therefore informing conservation and management (Guisan and Zimmermann, 2000). For rare and threatened species, it is often difficult to obtain sufficient data coverage over the large spatial and temporal scales required, especially during the early phases of investigation. Moreover there has been considerable debate on appropriate modelling approaches (Austin, 2002; Miller and Franklin, 2002; Guisan and Thuiller, 2005; Elith *et al.*, 2006). This partly explains why expert-defined habitat models are often promoted as best practice in comparison to their statistical data-driven counterparts (e.g. Langhammer *et al.*, 2007).

The Bayesian statistical modelling framework provides a useful 'bridge', from purely expert-defined models, to statistical models allowing survey data and expert knowledge to be 'viewed as complementary, rather than alternative or competing, information sources' (Ferrier *et al.*, 2002). Eliciting an expert-defined prior also clarifies existing knowledge at the outset of modelling (Spiegelhalter *et al.*, 2004).

As summarized in Table 19.1 and discussed in Section A.1.1, quantifying expert opinions for input into a Bayesian statistical framework can be robustly achieved by following statistical design principles: formulate the statistical model, appropriately target and encode expert knowledge, and design an accurate and repeatable elicitation protocol. In particular the desire to achieve elicitation in a transparent, repeatable and robust manner suggests judicious use of technology. Despite the acknowledged benefits of technology during elicitation, there is currently a dearth of elicitation tools available for general application (Leal *et al.*, 2007), with many tools developed for specific problems (Kadane *et al.*, 1980; Kadane and Wolfson, 1998; Denham and Mengersen, 2007;

Table 19.1 A six-component framework for designing elicitation, within a broader framework for implementing a full informative Bayesian analysis (Low Choy *et al.*, 2009).

E1.	**Determine motivation** and purpose for using prior information.
E2.	Specify available prior knowledge from experts or other sources, to define an appropriate and achievable **goal** of elicitation.
E3.	**Formulate a statistical model** representing the ecological conceptual model. Define the likelihood $p(y\|\mu)$ characterizing the data model and the prior $p(\mu)$ reflecting available prior knowledge.
E4.	**Design numerical encoding** (measurement technique) for effective elicitation of prior information and representation as a statistical distribution.
E5.	**Manage uncertainty** for accurate and robust elicitation.
E6.	Design an **elicitation protocol** to manage logistics of implementing elicitation.

Du Mouchel, 1988; Chaloner *et al.*, 1993; O'Hagan, 1997). Nevertheless some exceptions are available (Kynn, 2006; Al-Awadhi and Garthwaite, 2006; Leal *et al.*, 2007).

This chapter has two main aims: to showcase a rigorously designed and implemented expert elicitation for multiple experts; and to describe the use of software to streamline, automate and facilitate an indirect approach to elicitation. For exposition, the elicitation components E1–E6 in Table 19.1 will be used throughout to indicate and consolidate the proposed elicitation framework. Section 19.2 introduces the ecological problem that motivates this work: modelling habitat preferences of a relatively rare rock-wallaby (Murray *et al.*, 2008*b*). Logistic regression is a popular approach for habitat modelling (Guisan and Zimmermann, 2000), and provides a basis for use of the Bayesian framework for elicitation and inference, as outlined in Section 19.3. We propose an extension (Section 19.3.2.2) to the conditional means approach of Bedrick *et al.* (1996), that addresses the practical situation where the number of elicitations exceeds the number of covariates. This new approach has substantial benefits when providing feedback to experts (Sections 19.4,19.5.1). Section 19.4 describes a new software tool *Elicitator* that can be harnessed to assist elicitation for regression within this broader framework of designed elicitation. Results from eliciting expert opinion about preferred habitat of the rock wallaby (Murray *et al.*, 2008*a*) are presented in Section 19.5. We conclude with a discussion in Section 19.6.

19.2 Ecological application: Modelling and mapping habitat of a rock-wallaby

Ecological models of habitat requirements or environmental responses of individual species typically rely on just a few expert opinions (e.g. Kynn, 2005;

Martin *et al.*, 2005; Denham and Mengersen, 2007; O'Leary *et al.*, 2008*b*, 2009). For this study, however, we had access to a substantial number of experts. This could be attributed to several factors: the iconic nature of the species; its threatened status both in New South Wales (NSW) and nationwide; the subsequent interest from conservation managers, community groups and scientists; the profile and experience of the research team as well as funding attracted.

Experts for the study on *Petrogale penicillata* were determined by experience in field observation targeting this species (E6). Field data had recently been collated across two regions but had not yet been reported. Preliminary analysis revealed that rock wallabies used the terrain differently in each region (E6). The first region Q predominantly comprised volcanic rocky areas, where wallabies had often been sighted taking cover amongst rocky outcrops. To avoid predators, wallabies sought the less accessible areas with steeper slopes. The second region N comprised gorge country, therefore of lower elevation, but also containing steep slopes. In these areas wallabies were often observed to take cover amongst vine thicket vegetation as well as rocky outcrops. Hence a comparable number of experts from each region were sought, with five from region N and four from region Q, agreeing to participate (Murray *et al.*, 2008*a*). Expert opinions were considered relatively independent, due to diversity in training, work history as well as rock wallaby expertise (E5).

The elicitor on the project was acknowledged as having the most relevant, comprehensive and up-to-date knowledge of the species' habitat requirements. However, this experience was derived from recent field studies (Murray *et al.*, 2008*b*) which comprised the observed data intended for combining with expert opinion within the Bayesian statistical model. This disqualified the elicitor from actually providing elicited responses (E6). Nevertheless, there were numerous other advantages. Being the acknowledged 'best' expert on the species lent substantial credibility to elicitation and subsequent modelling, attracting involvement from most other experts (E6). Moreover this expert could be interpreted as the 'supra-Bayesian' (Genest and Zidek, 1986) and therefore able to assess and mediate contributions from others (E5).

Most experts worked within the conservation management sector, and were therefore located remotely, relatively close to known rock wallaby colonies. This geographic dispersion, together with funding constraints, meant that elicitation could not employ popular expert consultation approaches that rely on all experts being in the same place at the same time (for instance as promoted by O'Hagan *et al.*, 2006). This led to an imperative for an easily standardised and portable approach to elicitation.

Available field data recorded observed presence/absence at sites across the two regions (E1). As detailed in Murray *et al.* (2008*b*), the sampling design comprised a combination of existing sites supplemented by stratified random sampling to target gaps in existing surveys, with strata based on landscape-scale factors for remnant vegetation, landcover and geology. Gradsect sampling (Austin and Heyligers, 1989) was used to ensure coverage of finer scale variation with respect to elevation, slope and aspect. Landscape scale predictors were sourced from Geographic Information Systems (GIS) map layers (E3). The spatial datasets for geology and remnant vegetation were based on expert delineation of boundaries. Landcover was derived from remote sensing imagery with semi-automated interpretation and ground-truthing based on expert interpretation at field sites. Slope, elevation and aspect were calculated via standard spatial analysis from digital elevation models for topography. Transforming aspect to its cosine provided a measure of 'northerliness' that accounted for the discontinuity between 0° and 360°, of substantial importance for a species believed *a priori* to prefer northerly aspects (Murray *et al.*, 2008*b*).

19.3 Elicitation for regression

The way in which expert knowledge in incorporated as prior information in a Bayesian framework depends on the choice of a specific model. From a practical point of view, logistic regression is a popular choice for habitat modelling for both explanatory and predictive purposes (Guisan and Zimmermann, 2000; Miller and Franklin, 2002). More theoretically, logistic regression is a special case of a generalized linear model (GLM); since two of the encoding methods described below provide the same prior distribution (conditional means and data augmentation), this model is useful for exposition.

We consider presence/absence field data Y_i recorded at independent sites $i = 1 : n$, following a Bern(μ_i) sampling model. The linear predictor η weights covariates x_i by regression coefficients β via $\eta = \sum_{j=1}^{J} \beta_j x_{ij}$. Denote by μ the mean response, here probability of presence, given covariates $\mu_i = \mathrm{E}[y_i | x_i]$. For simplicity we focus on logit link function $g(\mu) = \eta$ such that $g(\mu_i) = \log(\mu_i / (1 - \mu_i))$. The inverse link function is the logistic distribution function, $g^{-1}(\eta) = F(\eta) = (\exp \eta)/(1 + \exp \eta)$ with corresponding probability density $f(\eta) = F(\eta)\,(1 - F(\eta))$. The likelihood for β can therefore be written (Bedrick *et al.*, 1996) as

$$p(y|\beta) = \prod_{i=1}^{n} F\left(x_i^T \beta\right)^{y_i} \left(1 - F\left(x_i^T \beta\right)\right)^{1-y_i}. \tag{19.1}$$

Within a Bayesian framework, prior distributions are placed on the unknown parameters β. Typically a multivariate normal prior is applied $\beta|\phi \sim \mathrm{N}_J(b, \Sigma)$, with Σ a symmetric $J \times J$ matrix with diagonal elements being the variances σ_j^2 and off-diagonal elements for the covariances $\sigma_{jj'}$. A weakly informative prior for β will assign vague distributions favouring neither positive nor negative effects (see for example Spiegelhalter *et al.*, 2003; Robert, 2001; Box and Tiao, 1982). However for an informative Bayesian analysis, prior information can be included but requires careful specification of b and Σ.

19.3.1 Indirect elicitation approaches for regression

An important decision in designing an elicitation method is the choice of a direct or indirect approach (E4). Direct elicitation of regression coefficients from ecological experts is possible for simple regressions with one or two covariates (Appendix, Table 19.5). In this case, however, direct elicitation would pose a substantial cognitive challenge for experts since there are multiple habitat covariates, as discussed in Appendix A.2. Instead we consider an indirect method, particularly useful in ecology (Low Choy *et al.*, 2009), since experts are asked about concrete observable quantities, such as probability of presence, rather than abstract concepts, such as regression coefficients. See Appendix A.2 for further discussion of direct and indirect elicitation approaches.

A useful decomposition of indirect elicitation methods separates elicitation into two components. The first fundamental component elicits expert knowledge together with uncertainty. Standard methods for eliciting distributions are available (Appendix, Table 19.4). The second component comprises elicitation tasks, more unique to the particular indirect method, that take these distributions and encode them into an informative prior for the model. In this section we briefly evaluate options, as outlined in Appendix, Table 19.6.

The posterior predictive approach (Kadane *et al.*, 1980; Kadane and Wolfson, 1998; Denham and Mengersen, 2007) is attractive since it asks experts to predict response at a future hypothetical site, given data observed at previous sites. This approach assumes that experts have already been exposed to observational data, and the posterior predictive distribution is tractable. In this case experts were not exposed to the field data. In addition the effort required to encode the scale parameter Σ can be prohibitive, both for the expert and the modeller (Kadane *et al.*, 1980).

Another option is data augmentation, whereby experts describe a range of possible responses and assign these a prior 'weight' or effective sample size. These approaches have been promoted for log-linear regression in epidemiological contexts (e.g. Greenland, 2006), where priors arise naturally from specifying example cross-tabulations. For logistic regression with multiple covariates this approach was considered too demanding, particularly in this study where several ecological experts had limited statistical experience and where variable selection was still uncertain.

A related approach presented by Chen *et al.* (1999) requires construction of a full initial dataset $\{Z_k\}$, from preliminary data, for specific values of covariates $X_{k=1:K}$. Application of Bayes theorem, with non-informative prior $p(\beta)$, provides an initial posterior:

$$p(\beta|X_{1:K}, Z) \propto p(Z|\beta, X_{1:K})p(\beta). \qquad (19.2)$$

This initial posterior provides a prior estimate for the next cycle of Bayesian learning, when combined with observed data Y elicited at potentially different covariate values $X_{i=1:N}$ with

$$p(\beta|X_{k=1:K,i=1:N}, Y, Z) \propto p(Y|\beta, X_{1:N})p(\beta|Z, X_{1:K}). \qquad (19.3)$$

A key assumption is that the same regression coefficients β relate prior 'data' Z to covariates X in the initial cycle, and observed data Y to covariates X. We may apply this approach in our context, using elicited 'data' as preliminary data $Z_{k=1:K}$, and using a Beta sampling distribution for Z, to enable capture of expert uncertainty, but requiring careful interpretation as the expert's estimate for the probability of presence. Under the Bayesian paradigm, we assume the expert 'data' or opinions are fixed, corresponding to a range of potential expert models represented by β. The prior $p(\beta|Z)$ may instead be defined using a point estimate and standard errors obtained from a regression of Z on X. This assumes that the expert has an underlying conceptual model based on some unknown true value of β, with sampling variation affecting their stated opinions Z.

Alternatively experts may be asked to describe the overall or aggregate relationship between the response and covariates (e.g. Chaloner *et al.*, 1993). Graphically this may involve drawing a response curve of y against covariate x_j (e.g. Willems *et al.*, 2005). This can be difficult in high-dimensional covariate spaces. To simplify the process it is possible to take a conditional approach, based on response curves, by eliciting the relationship between the response and each covariate separately, conditional on all other covariates held fixed, at say their optimum values (Kynn, 2006) or some other reference values (Al-Awadhi and Garthwaite, 2006). Software is already available for the latter curve conditional mean approach, but may appeal to scientists with a more theoretical understanding of the topic, e.g. physiological conceptual understanding that relates biological requirements to species response (Denham and Mengersen, 2007; O'Leary *et al.*, 2008*b*).

The similar case conditional mean prior (CMP) approach of Bedrick *et al.* (1996) asks experts to assess probability of presence μ_k at a number of sites conditioning on the full list of covariate values X_k. For example, in habitat modelling, field-based ecologists often find it easier to conceptualize the probability of presence for known environmental attributes (Denham and Mengersen, 2007). In our study most experts had a predominately field-based background, so we also consider the case CMP approach of Bedrick *et al.* (1996).

19.3.2 Case conditional mean elicitation for regression

19.3.2.1 Case CMP without elicitation error

When the case CMP elicitation approach was first formulated in a general context (Bedrick *et al.*, 1996), an exact computational approach to encoding was proposed, based on eliciting minimal information to solve the regression equations exactly. This exact solution was constrained by an assumption that the number of covariates equals the number of elicitations $J = K$. Following Bedrick *et al.* (1996), the case-specific conditional mean prior relies on eliciting a Beta prior for $\mu_k = \mathrm{E}[y_k|x_k]$ for cases $k = 1 : K$, to induce a prior on β. Estimation of the induced prior on β assumed $K = J$, to ensure existence of X^{-1} and that dimensionality was maintained on changing variable (from μ_k to β). In the 'under-specified' case, this constraint was achieved by conditioning on zero values for excess covariates; in the 'over-specified' case, additional latent variables were introduced to match the number of elicitations. This deterministic approach assumes that each expert directly assesses the probability of presence at each site μ_k, and that no adjustment is required to ensure coherence of the underlying trends across sites.

In the following section we take issue with Bedrick *et al.*'s (1996) assertion that the case $K > J$ is 'unlikely', and argue that in many practical situations, more elicitations than covariates are required to adequately cover covariate space and manage variation. For example in a medical context Landrum and Normand (1999) elicit ratings for 890 covariate sets, whereas covariates numbered less than 100. For this study (Murray *et al.*, 2008*a*) 20 elicitations are obtained on a problem with eight variables. Thus we take a stochastic rather than deterministic approach, positing that elicitations are likely to involve some elicitation or measurement error, which can be 'averaged' out across sites. This is more practical for elicitation compared to a more parameter-rich approach that explicitly adjusts for bias and precision in estimates at each site (in the spirit of Lindley, 1983).

A logistic regression ought to provide a good approximation to the expert's implicit underlying trends in species response: ecological experts are quite familiar with species response curves and their interpretation. However this model, like any other, is only an approximation to the expert's knowledge. The benefit of this case-conditional mean prior approach is that the elicited information does not presume the underlying model is logistic regression. For instance other models could be used on the same elicited information, such as binomial or beta regression with a different link function (probit or complementary log-log), or a regression tree (O'Leary *et al.*, 2008*a*). Indeed just as much effort could be expended to 'reveal' the model that best fits the elicited expert opinions, rather than assuming a particular model form. At the moment, the software *Elicitator* only fits one possible model to the expert's opinion, although more flexibility is intended for future extensions.

19.3.2.2 Case CMP with elicitation error

Denote by Z_k the expert's opinion on the conditional mean, μ_k. This forms the basis of a Beta regression (Branscum *et al.*, 2007) where shape and scale parameters a_k and b_k can be reparameterized in terms of effective prior sample size γ_k and mean μ_k:

$$Z_k \sim \text{Be}(a_k, b_k), \quad \mu_k \equiv a_k/\gamma_k, \quad \gamma_k = a_k + b_k, \quad \text{logit}(\mu_k) = X_k^T \beta. \tag{19.4}$$

This formulation is equivalent to the expert 'data' approach inspired by Chen *et al.* (1999). It permits the expert to have a different level of uncertainty γ_k at each elicited site, rather than enforcing uncertainty γ to be constant across all cases (as in Branscum *et al.*, 2007). It is also possible to elicit a rating of certainty, say on a scale from 0 to 100, about each elicited Z_k. Thus an expert should expect that 90% of their assessments at the 90% level of certainty are correct. These confidence ratings can be normalized, then used to weight the regression of the expert data on the covariates, to reflect unequal precisions in elicitation across sites.

The prior distribution (19.4) obtained in this way establishes an interim result, which is an advantage when collating multiple opinions, or when providing feedback to the expert. Standard regression diagnostics can be used to provide feedback on variation across elicitations (elicitation 'error'). This variation may be dominated by either the expert's epistemic uncertainty in their knowledge or aleatory uncertainty reflected by their difficulty in accurately expressing this knowledge (O'Hagan *et al.*, 2006). Alternatively equation (19.4) can be used within a full Bayesian analysis including the observed data via equation (19.3). Computations are simplified since this prior can be expressed in the same form as the likelihood (Bedrick *et al.*, 1996). For computational convenience we may approximate $Z_k \sim \text{Be}(a_k, b_k)$ by $Z_k^* \sim \text{Bin}\left(\mu_k, \gamma_k^*\right)$ with $\gamma_k^* = \text{Ceiling}[\gamma_k]$ and $Z_k^* = \gamma_k^* Z_k$. Any Bayesian or classical statistical software can be used to fit the full Bayesian model by modifying the observed data to reflect the prior information, through addition of pseudo-observations (Greenland, 2006).

19.3.2.3 Encoding the Beta distribution

Minimally, estimation of the two parameters a_k and b_k requires elicitation of two summary statistics about the distribution (Low Choy *et al.*, 2009). The most obvious starting point is to ask the expert for their 'best estimate' of the probability of presence at a site. Chaloner and Duncan (1983) justify elicitation of the mode since it identifies the point of largest departure from a uniform prior, taking advantage of an expert's tendency to anchor (Tversky and Kahneman, 1981). They argue that the mean is too difficult cognitively, whereas the median is no more difficult than the mode but offers no simplification of arithmetic. Several summary statistic pairs could be used to encode the Beta distribution (see Appendix, Table 19.4). Various methods which average over more than two summary statistics have also been proposed (Chaloner and Duncan, 1983; Hughes and Madden, 2002; Gavasakar, 1988). As shown by Low Choy *et al.* (2008), encoding of skewed distributions, such as the lognormal and the Beta benefits strongly from the use of additional quantities.

For these reasons, in this study and in the software described in Section 19.4, the mode is elicited, together with the upper and lower quartiles, and two more extreme quantiles (E4). We assume that the usual preparations have been made: the protocol has been designed carefully (E6); key terms and quantities to be elicited have been precisely defined (E2); experts have been conditioned to be aware of, or trained to reduce, the main types of biases (E5); and experts have been selected, motivated and quality assessed (E6). The way in which questions are worded strongly impacts on the response and must be tailored to the expertise of the respondents. For example, questions in the habitat modelling context may be worded (E4): 'Let us consider one hundred sites with these habitat characteristics – rainforest, basalt, slope 35°, aspect predominantly north-northeast, and elevation 450 m above sea level. How many of these sites do you believe are occupied by the species? Provide the answer you think is the *most likely*.' Wording in terms of frequencies often leads to more accurate elicitation than probabilities (Kynn, 2008).

It is also advisable to precede this question by a more extreme one: 'What are the absolute minimum and maximum *possible* values for the number of sites....' This provides natural progression to asking the expert to specify the upper and lower bounds (corresponding to a 95% credible interval) on the plausible values for the probability of presence. Asking the expert for these assessments at the outset helps to broaden their thinking initially, and thus helps to avoid over-conservatism arising from anchoring biases (Kynn, 2008; Low Choy *et al.*, 2009). Experts are also asked to estimate another interval, such as the 50% credible interval (CrI), such that the probability of presence has a 25% chance of falling below or above these bounds. In practice we have found that asking for the 95% and 50% CrIs before seeking the mode is also crucial for avoiding representation bias, whereby experts may confuse the 50% CrI for Z_k with the precision of their estimate of its mean or mode.

Exact solutions for estimating a beta distribution given a mode and several quantiles are not available, in the same way that they are for the lognormal distribution (Low Choy *et al.*, 2008) (E4). A simple numerical approach to finding a close-fitting beta distribution would first fix the mode at the elicited value $\widetilde{Z}$, assuming that this is the most accurately elicited quantity. Then we find the pair (a_t, b_t) that provides the best fit between the elicited values $\widetilde{q}(\alpha)$ and the encoded values $Q(\alpha|a_t, b_t)$ over the probability levels α. If only two quantiles are used for fitting, then the remaining two can be used for feedback (Kynn, 2005). A balance is required regarding the number of quantiles that can be accurately and efficiently elicited.

19.4 Software tool for elicitation

Recently prototype software was developed for a geographically assisted approach to elicitation, modifying the posterior predictive approach of Kadane and Wolfson (1998). The software was 'hardwired' for use on two case-studies (Denham and Mengersen, 2007). This section summarizes development of a new software tool, *Elicitator*, inspired by the interfaces of these prototypes, but implemented on a more modern computing platform and based on the new elicitation method (Section 19.3.2.2). The substantial software design and implementation issues are detailed by James *et al.* (2009).

Continually updating visual aids and computations to reflect expert opinions and revisions, including encoding these opinions as prior distributions, can be achieved by hand. However these processes are faster, more accurate and more interactive when supported through software. Other key benefits of a software tool are: bias and error minimization via standardisation of key aspects (E5), inclusion of an elicitation protocol (E6) and underlying statistical design (E5); facilitation of exploration of a wider range of elicitation options *real-time* during elicitation (E4); and provision of a research tool to enable controlled testing of various elicitation options.

In *Elicitator*, we relaxed the requirement of Denham and Mengersen (2007) that covariates have a geographical interpretation, so that the tool would no longer be embedded within a GIS, but instead be loosely coupled to a GIS, being able to communicate via shared filetypes. This also bypassed the difficulty of maintaining version control with commercial GIS packages. The addition of an underlying relational database helped to structure and streamline the collation of information from several elicitations, across multiple phases, for different experts, on various projects (having different elicitation sites). The relational database also facilitates communication with GIS or other databases.

Elicitator's core contains three modules: an *Elicitation* window for managing interaction with the expert to obtain their assessments (Figure 19.1); an *Encoding* module to calculate prior distributions from elicited information; and a *Feedback* window to provide the expert with an opportunity to evaluate their assessments. Input and output to the core is achieved via three additional

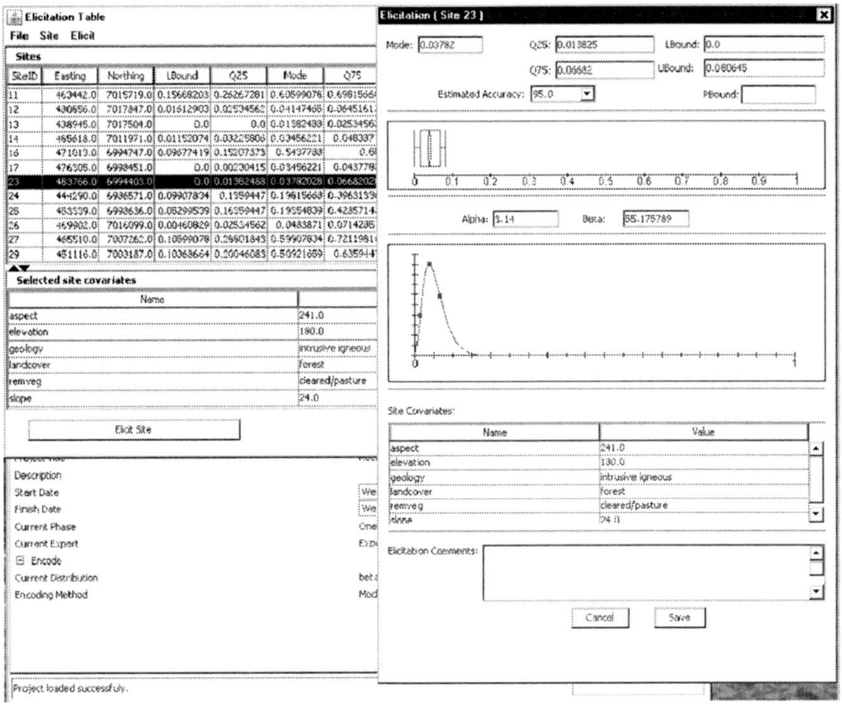

Fig. 19.1 Elicitation window in *Elicitator*. Experts are asked to specify the range of plausible probabilities of presence, and may manipulate the boxplot, beta density plot, or numbers.

modules: a *Setup* window for selecting options defining the elicitation protocol and method; an *Import* window for importing covariate information for each case to be elicited or revisiting previous elicitations; and an *Export* window for exporting and evaluating the prior model. Experts participate in an iterative cycle based on this Elicitation-Encoding-Feedback core. Separating the modules in this way facilitates later extensions, such as: formulating priors for a different GLM; elicitation of a different distribution; different methods of encoding; and feedback tailored to the encoding method and distributional assumptions.

The main windows for elicitation and feedback are separated to streamline functionality. The elicitation window (Figure 19.1; right-hand window) permits experts to specify their estimated probability of success at each elicitation site/case using: numbers for mode and quantiles; a simple graph interactive boxplot to visualize the mode and quantiles; and a more sophisticated interactive beta density plot again depicting mode and quantiles. In addition the encoded parameters of the corresponding beta distribution are shown, for reference. All three representations of the mode and quantiles are updated dynamically when any are edited. Furthermore, the expert may supply a confidence rating for each elicitation case. This window also lists the covariates for the particular elicitation case. The feedback window provides feedback to experts after collating and encoding their elicitations for sufficient cases/sites, and shows univariate response curves and regression diagnostics. Univariate response curves are marginal estimates from predictions formed across covariate profiles taken from the elicitation sites (E4, E5). The regression diagnostics comprise standard GLM diagnostics (Venables and Ripley, 2002), highlighting unusual elicitations (residual versus fitted value plots, standardized residual quantile-quantile plots) or those highly influential in formulating the prior model (Cook's D influence statistic) (E5). The expert may use the mouse to click directly on the feedback graphs, together with numeric identification of cases listed in a table, to select any cases and revisit or review their elicitation.

The inclusion of a project menu enforces a structured filing of elicitation information, first by project defined by a list of experts and a list of elicitation cases/sites each accompanied by covariate information, then by phase within project. This initializes corresponding tables within the database, so that all potential elicitations for each case and expert are appropriately indexed for later retrieval. In addition, the elicitor may use this menu to set database connection details, export elicitation data, set current expert and project phases, and for loading and saving elicitation projects. Encoding can be initiated either from the menu, elicitation windows or feedback windows. This initiates the communication with the R server (Urbanek, 2006), and initialises the data required for the statistical computing environment (covariate information). The database is updated each time elicitations are reviewed and changed. Each time encoding is initiated these details are also uploaded to the statistical computing environment. Results from encoding prior models are also stored in the database, and can be exported as a text file via the menu. Finally the output tab window can also be accessed within the main application window this uses the syntax of the WinBUGS modelling environment (Spiegelhalter *et al.*, 2003) to express the priors for each coefficient for each expert.

To achieve this functionality *Elicitator* links to several open-source libraries in order to manage and dynamically update the graphic user interface windows that link to the underlying data objects, link to a graphing library for graphical representation of elicitations and feedback, link to a relational database package for data persistence and management, import files containing data on elicitation cases/sites and covariates, and link to a statistical computing package for encoding. A modern object-oriented computing platform in Java was selected, for its flexibility, stability, extensibility, portability and popularity, as well as the availability of a wide range of well-developed and thoroughly tested libraries. The dynamic graphing features were achieved using the JFreeChart Java libraries. MySQL was selected as a well-respected and freely available relational database management system. Elicitation information was imported into MySQL from a simple file format that can be output from most commercial or freely available GIS packages. The tool also employs the use of the statistical computing package R (Venables and Ripley, 2002), which is popular (ensuring longevity) and freely available.

19.5 Results

19.5.1 Design of elicitation

Twenty elicitation sites were located within an area, chosen by the elicitor so that it was not familiar to any of the experts consulted, but had good representation of each of the habitat variables (E4), particularly geology. Elicitation sites were selected using stratified random sampling, according to the same strata as data in Murray *et al.* (2008*b*) (E5). Covariates at those sites were determined using GIS, then entered into *Elicitator* (E4). Experts were thus able to view an integrated spatial database, comprising maps showing each habitat characteristic, elicitation site locations, and other useful geographic features. Care was taken to ensure that whilst the expert was able to view maps at scales helpful to interpretation, they were unable to view them at scales allowing identification of the area (E5). Encoding followed the new indirect method (Section 19.3.2.2) (E4). Use of the elicitation tool followed general principles outlined in Section 19.4 (E6). Further details on the design are detailed in Murray *et al.* (2008*a*). A more complete summary of the Bayesian analysis, comparing non-informative priors to expert-informed priors pooled across regions using the moment-averaging approach, is presented in Murray *et al.* (2009).

19.5.2 Encoding

Recall that 4–5 experts from each of two adjacent regions Q and N were consulted. The method of elicitation (described in Sections 19.3 and 19.4) proved sensitive to the diverse range of estimated probabilities of presence provided by individual experts across twenty elicitation sites. For example, Figure 19.2 shows the range in elicited opinion across elicitation sites for one expert from region Q. The raw elicited information is represented by a non-standard boxplot, highlighting the elicited mode, 50% and 98% CrI limits. Some sites are clearly believed to correspond to rock wallaby absence, with narrow encoded beta distributions having mode very close to zero (i.e. sites 6, 12–14, 26, 30). One site had low probability of presence that was definitely non-zero (site 17). Where rock wallaby presence was likely, this expert provided a range of plausible values for the probability of presence. The expert's opinion was fairly vague for one site (site of the esteem of their peers rather than by force of personality).

Experts varied in the consistency of their opinions, across sites, regarding the main effects impacting on the probability of presence for the rock-wallaby.

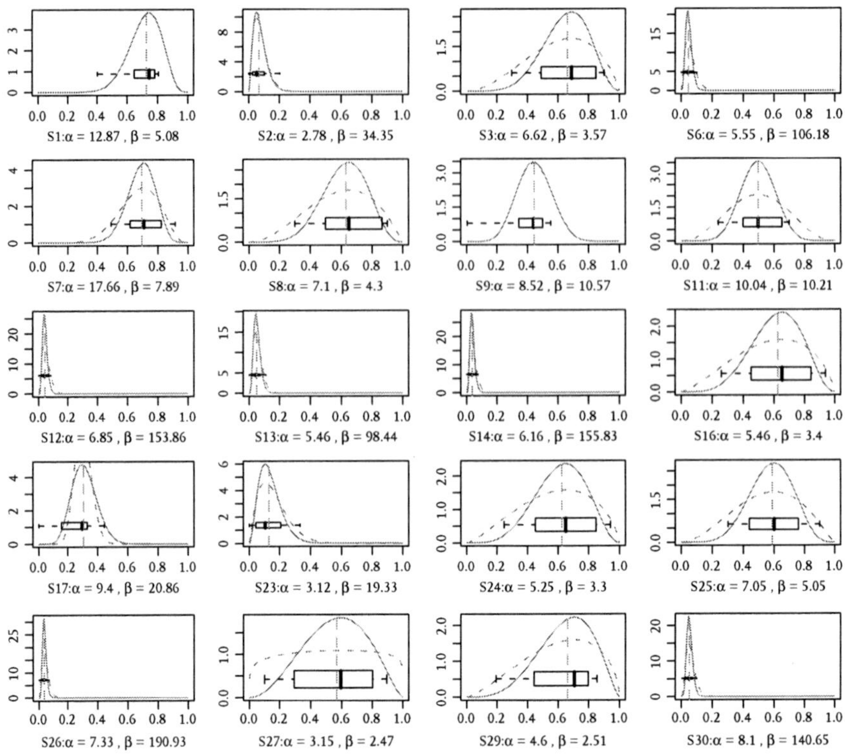

Fig. 19.2 Elicited distribution of the probability of presence for each of twenty sites (each plot), for a single expert. Encoded using mode and upper quartile (dashed line) or using mode, quartiles and 98% credible interval (solid line). Boxplot indicates elicited mode (line), quartiles (box), and 98% credible interval (whiskers).

This consistency is reflected in the regression diagnostics resulting from fitting a main effects model (using Elicitator) to the information elicited from a single expert. Regression diagnostics provided by Elicitator are shown here are standard outputs from R, and two examples are shown (Figures 19.4–19.5). Note these figures show a compressed version of plots produced by Elicitator, redrawn for display here. The elicited information obtained from Expert 3 from region N shows an excellent fit to the logistic regression model, clearly indicating that rock-wallabies prefer higher slopes, forested areas and basalt geology. The accumulation of site-by-site elicitations provided by Expert 3 from region Q show an adequate fit to a logistic regression model, with a moderately strong message that rock-wallabies prefer open forest to other vegetation and forest to other land uses. This expert is less clear on the impact of geology, although confirms a preference for basalt. The effect of aspect estimated by either expert is less clear, with the region N expert suggesting a preference for north-westerly to northerly aspects, and the region Q expert suggesting a preference for northerly to north-easterly aspects. The effect of elevation is

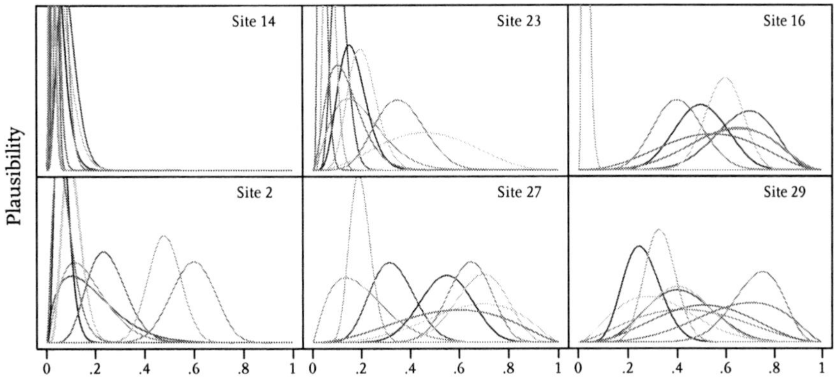

Fig. 19.3 Comparison of elicited probability of presence across experts, for six sites.

indirect and related to rock type and region: in NSW rock-wallabies tend to be found in gorges (leading to inverse relationship between probability of presence and elevation by region N expert); and in QLD rock-wallabies tend to be found in hilly basalt regions (leading to increasing probability of presence with elevation estimated by region Q expert). Furthermore, the site-by-site elicitations performed by region Q expert did not reveal a strong impact by slope. For both experts, there is evidence to suggest that their opinions are subject to more variation as probability of presence increases, confirming the evidence gained across experts (e.g. Figure 19.6).

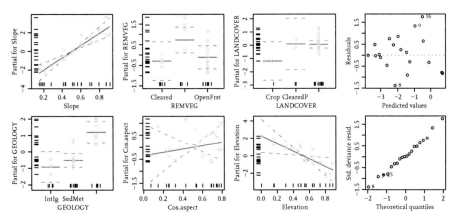

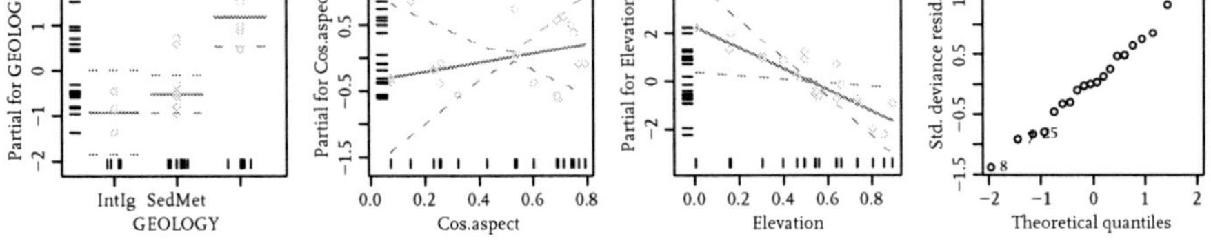

Fig. 19.4 Expert #3 from region N: regression diagnostics from fitting a model to elicited information across sites. The first three plots in the top row show partial residuals and response (with pointwise SE in red) to the effects of: slope, remnant vegetation type (cleared, closed forest, open forest), and landcover (crop, cleared/pasture, forest). The first three plots in the bottom row similarly show partial residuals and response to the effects of: geology (igneous, sedimentary/metamorphic, basalt), cosine of aspect, and elevation. The last plot in the top row shows the pattern of residuals across fitted values, and the last plot in the bottom row shows the quantile-quantile plot of the standardized residuals.

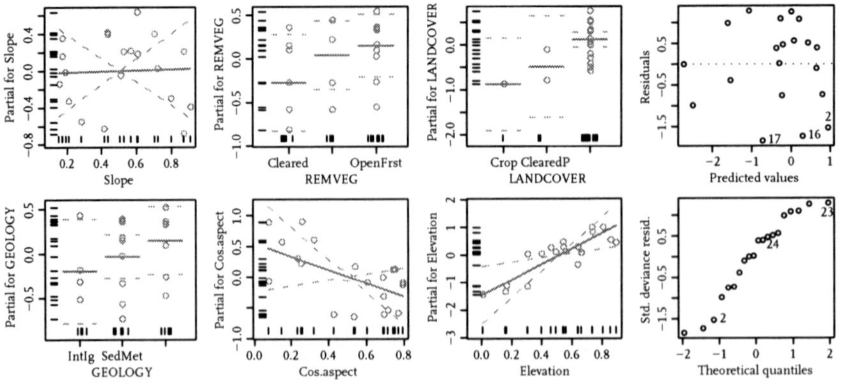

Fig. 19.5 Expert #3 from region Q. Regression diagnostics from fitting a model to elicited information across sites, as detailed in caption to Figure 19.4.

Table 19.2 shows that, after encoding the opinion of expert Q3, we obtained four prior estimates of regression coefficients with high probability of being non-zero. After combining these priors with the observed landscape scale dataset, the posterior probability of being non-zero decreased for two coefficients, increased for one, and became indistinguishable from zero for the fourth (geology). The data provided strong evidence of probability of presence increasing with slope (99% chance that this is a positive effect), overriding the expert's opinion. The posterior estimate of the impact of forest cover increased in magnitude (1.5 instead of 1.3) and had decreased chance of being non-zero (<0.1% instead of 0.6%) with respect to the prior. Posterior correlations between regression coefficients were negligible, mostly 0.09 or below, with the highest being 0.23.

Expert N3 provided case-by-case estimates that corresponded to very low overall prevalence (2%) for baseline habitat. This expert estimated substantial negative impact for increasing elevation, and positive impact of several habitat factors, on probability of presence. On combining with data,

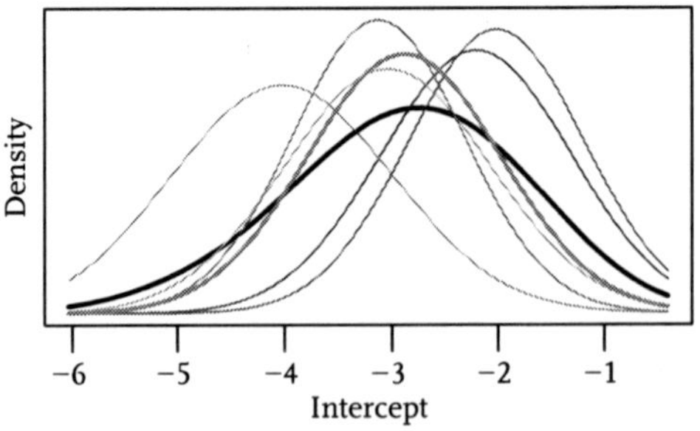

Fig. 19.6 Intercept in habitat model encoded from each expert's 20 site elicitations (coloured lines), pooled across experts using moment averaging (thick grey line) or linear pooling (thick black line).

Table 19.2 The most experienced experts from each region: prior and posterior estimates of regression coefficients β_j on logit scale. Prior was encoded from five elicited quantities; posterior estimated using informative prior. Baseline habitat corresponds to minimum values of habitat covariates: flat ground, eastern or western aspect, elevation at sealevel, intrusive/igneous rock types, forest cleared (of remnant vegetation), and cropping the major landuse.

Covariate j	Prior			Posterior		
	mean	s.e.	$p(\beta_j < 0 \mid x, z)$	mean	s.d.	$p(\beta_j < 0 \mid x, y)$
			Expert Q3			
Intercept	−1.955	0.671	0.998	−1.928	0.629	0.999
Slope	0.068	0.161	0.335	0.326	0.148	0.014
Sedimentary/Metamorphic	0.461	0.283	0.052	0.072	0.252	0.389
Volcanic	0.582	0.401	0.073	−0.282	0.290	0.835
Closed Forest	0.129	0.323	0.344	0.173	0.304	0.285
Open Forest	−0.039	0.293	0.553	0.233	0.251	0.177
Cleared/Pasture use	−0.029	0.499	0.523	−0.063	0.449	0.559
Forest cover	1.309	0.527	0.006	1.449	0.448	0.000
			Expert N3			
Intercept	−4.024	1.027	1.000	−3.899	0.972	1.000
Slope	1.613	0.302	0.000	1.340	0.231	0.000
Sedimentary/Metamorphic	0.237	0.462	0.304	0.656	0.375	0.040
Volcanic	1.979	0.663	0.001	1.598	0.438	0.000
Closed Forest	1.163	0.561	0.019	0.444	0.474	0.174
Open Forest	0.438	0.548	0.212	0.462	0.410	0.133
Cleared/Pasture use	1.569	0.724	0.015	1.412	0.694	0.025
Forest cover	1.060	0.756	0.080	1.458	0.531	0.002
Elevation	−1.035	0.443	0.990	−0.171	0.101	0.9579

sedimentary/metamorphic geology became a clear positive impact, the effect of closed forest became close to zero, and the effect of elevation diminished. The introduction of data appears to have had little impact on the estimated intercept and effect of cleared/pasture landcovers.

In summary this expert from region Q appeared to refer mostly to geology and forest landcover in assessing probability of presence at each site, whereas the expert from region N had a conceptual model for rock wallaby habitat that was clearly driven by strong gradients in at least some categories of all habitat factors considered. This diversity in opinion reflects regional differences as well as the different types of experience held by the two experts. Expert Q3 had broader experience of both regions, from both field-based and desktop GIS-informed studies.

19.5.3 Collating opinions

For exposition we focus on region N, which exhibited less polarity in expressed opinions. For a few experts, sites were eliminated when encoding a particular

Table 19.3 Encoded intercepts.

Expert	Encoded intercept			Sites omitted
	Estimate	SE	p-value	
N1	−2.208	0.893	0.013	24
N2	−3.037	0.961	0.002	12
N3	−4.024	1.027	0.000	
N4	−3.127	0.801	0.000	
N5	−2.018	0.827	0.015	
Q1	−3.504	0.895	0.000	30
Q2	−3.268	0.659	0.000	26, 27
Q3	−1.955	0.671	0.004	
Q4	−6.064	0.799	0.000	12

expert's opinion, only if they were identified as a statistical and conceptual outlier; the site had to be confirmed by the expert to be an unusual site substantially distorting their overall opinion of species response to habitat (see last column in Table 19.3). We apply moment-averaging and linear pooling to combine opinions for the five experts from region N (Figure 19.6). Moment-averaging simply averages, across experts within a region, the means and variances of the Normal priors for each covariate to provide a pooled prior that is also Normally distributed (Martin *et al.*, 2005; Kuhnert *et al.*, 2005). Linear pooling constructs a finite mixture of expert opinions, providing a more complex distribution than a simple Normal, which has more flexibility to reflect diversity and polarity of multiple opinions (Genest and Zidek, 1986; Clemen and Winkler, 1999; Cooke and Goossens, 2004; O'Hagan *et al.*, 2006).

Combining multiple priors on regression coefficients should account for the implicit conditional interpretation of these coefficients. Thus each coefficient indicates the effect of the corresponding variable on the response, conditional on all other variables remaining constant (and in the model).

In region N, priors for the intercept derived from each expert (Table 19.3) are generally overlapping (Figure 19.6). This allows us to avoid the extra complexity of first pooling information on the intercept, then setting this as an offset in each expert's prior model (19.2), before considering pooling information on the habitat covariates. We also ascribe equal weight to each expert, in the absence of any additional information to the contrary, and since experts in this case study derive their knowledge from quite different personal experience.

The moment-averaged estimate (grey) in Figure 19.6 was calculated as $N(-2.88, .9056^2)$, with 95% and 50% credible intervals of $[-4.66, -1.11]$ and $[-3.49, -2.27]$ respectively, and mode equal to the mean -2.88. The linear pooling estimate (black) has 95% and 50% credible intervals of $[-5.18, -0.89]$ and $[-3.64, -2.09]$ respectively, with a mode of -2.75. Although both pooled priors are centred at similar values, the linear pooling estimate reflects two

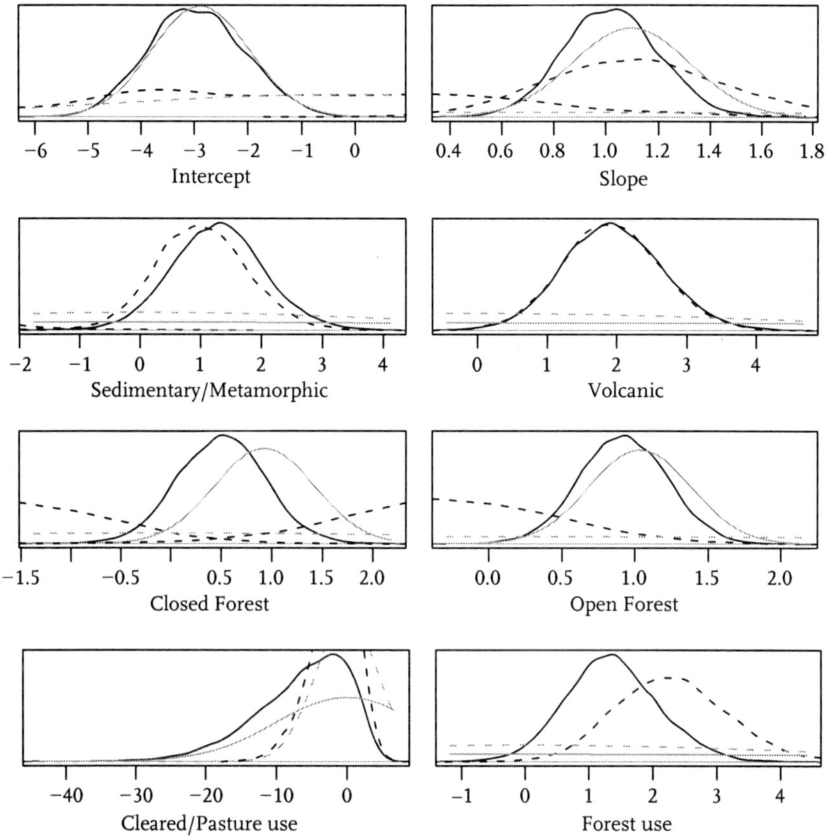

Fig. 19.7 Region N: comparison of posterior (solid line) and prior (dashed line) distributions for Bayesian analysis, with pooled (with voting) expert-informed priors (black) or non-informative priors (grey).

features of the individual opinions not evident in the simple moment-averaged estimate. Firstly, the range of acceptable values has broader range, extending the 95% credible interval by 0.6 and 0.2 units at the lower and upper bounds, respectively. Secondly, the prior is slightly skewed towards values closer to zero.

Posterior estimates are shown for pooling via moment-averaging with voting (Figure 19.7) and by linear pooling (Figure 19.8). In the first case, the posterior is obviously driven by the prior estimated effects of geology, but not influenced by the prior on the intercept. The effects of slope, closed forest and open forest are slightly reduced when accounting for expert knowledge pooled in this way. Analysis of forest use is non-conclusive with non-informative priors, yet clearly indicates a positive impact after accounting for this form of expert knowledge.

Priors derived from linear pooling are generally much more diffuse than corresponding posterior estimates, indicating that the data are re-iterating the

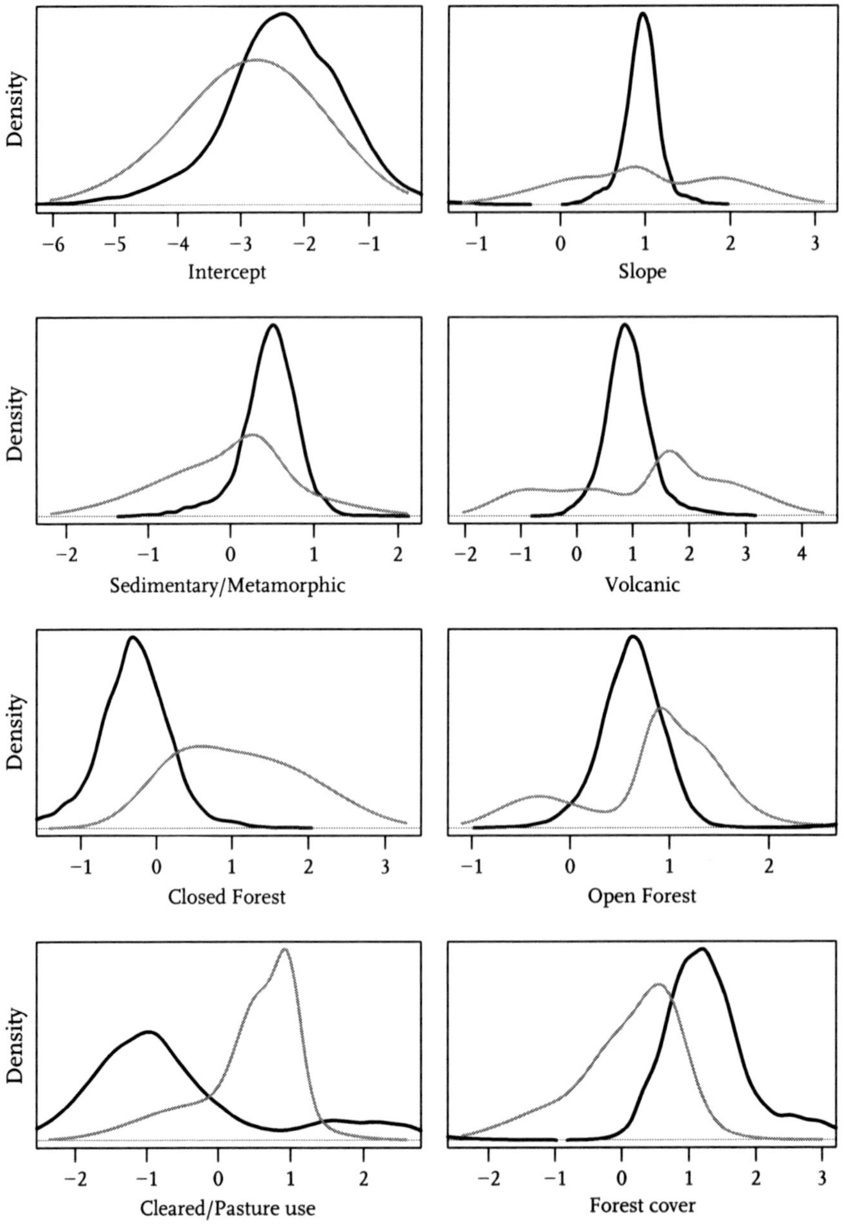

Fig. 19.8 Region N: comparison of posterior (black) and prior (grey) distributions for Bayesian analysis, with linear pooling of expert-informed priors.

opinions of at least some experts. Incorporating information from observed data increases the low level of prevalence of rock-wallabies estimated by experts from a mode under 5% to a mode over 11%, suggesting that expert opinions reflect conservative estimates of prevalence. The positive effects of slope and volcanic rocktype on likely occurrence of rock-wallabies is emphasized through informed posterior analysis, with posteriors more tightly constrained around the mode of more diffuse pooled opinions of experts. The data concur with the more positive estimates of the impact of sedimentary/metamorphic geologies, and the more negative estimates of the impact of closed forest. The posterior estimate of the impact of forest cover is much higher than expected by any of the experts, and is balanced by a much more negative posterior estimate of the impact of cleared land or pasture use. In the posterior estimate of the effect of open forest, the data have dampened the prior estimate of a large positive impact.

19.6 Discussion

The practical outcome of this analysis, is the clear demonstration that, even with a carefully designed empirical dataset obtained from rigorous fieldwork, expert knowledge can contribute to habitat modelling. This is the first case (that we are aware of) where a habitat model where equally rigorous effort has been applied to obtaining empirical field-based data and quantification of expert opinion. Previous use of expert-informed habitat models involving several habitat covariates have been based on elicitation from one or two experts (Kynn, 2005; Al-Awadhi and Garthwaite, 2006; Denham and Mengersen, 2007; O'Leary *et al.*, 2008*b*). Where several experts were used, only a single covariate was considered (Kuhnert *et al.*, 2005; Martin *et al.*, 2005). Elicitation of expert knowledge and empirical data have been used to populate Bayesian networks for defining a habitat model, although in these cases the expert and empirical information are not used to inform estimation of the same model parameters (e.g. Smith *et al.*, 2007). One exception defines a Bayesian net as a prior model, where the prior for each model parameter is defined by model averaging over an expert-defined and data-determined point estimate (Pollino *et al.*, 2007).

This chapter has reviewed the perceived benefits of Bayesian statistical models with informative priors constructed from expert opinion for input into regression models. One advantage of eliciting expert knowledge for input into Bayesian models is the need to quantify precision as well as point estimates in order to encode prior distributions. The issue of assessing the informativeness of priors has been explored in great depth in the context of theoretical construction of reference priors, which seek to maximize the informativeness of the data relative to the prior (e.g. Box and Tiao, 1982). For informative priors, the validity of calibration has been questioned (Kadane and Wolfson, 1998): 'The philosophical argument, and in our opinion, the more compelling [than the mathematical argument] is that what is being elicited is *expert*, not perfect, opinions, and thus they should not be adjusted.' This view is reinforced by Spiegelhalter *et al.* (2004) where the importance of establishing prior opinions before undertaking any analysis is an important undertaking in its own right. A philosophical investigation of the implications of a 'well-calibrated' Bayesian are examined by Dawid (1982) and discussants. He concludes that it is simply too demanding to require that a subjective forecaster is simultaneously coherent and calibrated. Currently we are examining the issue of calibrating and combining expert opinion in more detail.

One of the advantages of the indirect case-by-case approach to elicitation embodied in recent software tools (Denham and Mengersen, 2007; James *et al.*, 2009) is that the information elicited may be used to fit different linear predictors (Bedrick *et al.*, 1996). This arises since elicitation of probability of presence is conditional only on values of covariates, and does not proscribe their transformation or interaction, in contrast with other approaches. This issue has been examined using a standard variable selection approach to indirect elicitation for logistic regression (Chen *et al.*, 1999), or using Bayesian classification trees (O'Leary *et al.*, 2008*a*). Future extensions of *Elicitator* could provide the modeller with more flexibility to specify prior 'models' to be assessed during the encoding phase.

There is little research comparing the use of quite different elicitation techniques, apart from consideration of more subtle differences, such as the number and probabilities of quantiles selected (*e.g.* Kadane and Wolfson, 1998; Garthwaite and O'Hagan, 2000; Garthwaite *et al.*, 2005; Leal *et al.*, 2007). One recent paper that does make these broader comparisons investigates results from applying three approaches to elicitation with two experts, for habitat modelling using regression (O'Leary *et al.*, 2008*b*). This study revealed some large differences in encoded priors and corresponding posteriors, depending on the elicitation technique, and to a lesser extent on the expert. More work is required on comparing and integrating alternative elicitation methods.

The tool *Elicitator* already provides functionality to allow different encoding methods to be used. It is conceivable that different elicitation methods may also be incorporated in later versions. In addition the modes of communication selected by the elicitor – numeric, simple graphical summaries, specialist statistical diagnostic graphs, spatial maps – have led to innovative use of technology. *Elicitator* demonstrates that various communication modes are feasibly incorporated into a single tool, *e.g.* numeric/graphical representations of the elicited distribution of the response variable available in the elicitation window or multiple diagnostics available in the feedback window.

We have endeavoured to illustrate recent technological advances in elicitation by describing one software tool developed as an outcome from our recent experience. Its design has laid the foundations for several extensions, which take advantage of the modern computing environment. The ability to tailor minor aspects of GUIs for specific applications could greatly increase user appeal. Flexibility in terms of model choice provides the modeller with many more choices, by allowing a greater range of GLMs, and potentially even other models. The case-by-case indirect elicitation of expert 'data' may be extended to GLMs by appealing to the very general CMP/DAP framework presented in Bedrick *et al.* (1996). Accuracy of elicitation can depend heavily on the few summary statistics elicited, such as quantiles or which choice of central measure (mean/median/mode), as demonstrated by Low Choy *et al.* (2008). Providing elicitors with flexible choices in terms of summary statistics, and therefore encoding method, enables elicitation to target those summary statistics considered most accurately represented by experts in that field Chaloner *et al.* (1993).

For the brush-tailed rock-wallaby, we found that expert knowledge was informative for modelling habitat in region N (northern NSW). Experts concurred that preferred habitat increases with slope, volcanic rock, open forest and areas with cleared/pasture landuse. Supplementing expert information with empirical data within a Bayesian analysis had least impact on a low estimate of prevalence (5–10%) and confirmed though substantially tightened the experts' estimated positive effect of slope. Inclusion of empirical data slightly changed the experts' estimated little effect of sedimentary/metamorphic rock to be positive, and of negative effect of forest cover to also be positive. In contrast empirical data maintained but decreased the experts' opinions on a positive effect of volcanic rock and of open forest, yet provided sufficient evidence to override the experts' opinions of generally positive impacts of closed forest or of cleared/pasture use to conclude these factors in fact reduced habitat preference. Thus empirical and expert information have varying influence on each habitat factor, so are both essential to the analysis. Full discussion of the ecological interpretation of one way of aggregating expert opinions is provided in Murray *et al.* (2009).

Overall expert and empirical evidence indicates that, in this region, the brush-tailed rock-wallaby clearly prefers sites occurring in landscapes with greater slope, volcanic but less clearly sedimentary/metamorphic rather than intrusive/igneous rocktypes, remnant vegetation that is open forest rather than closed or cleared, and areas that are forested (remnant or regrowth) rather than cleared or used as pasture or cropping. This information will be useful for future work in constructing species distribution maps and assessing spatial connectivity of sites and therefore population dispersal. This work will inform future recovery planning and conservation management of the species, which can refer to rigorous collection and use of both expert knowledge and empirical data. In addition it provides a useful demonstration of an approach and accompanying software that can be applied to any rare and threatened species, particularly where expert knowledge (typically based on experience in the field) is considered valuable.

Applications, including habitat modelling, have provided an important motivation for development and refinement of elicitation techniques for constructing informative priors. The development of software tools to streamline and support elicitation could lead to further improvements in uptake, dissemination as well as improved methodology. We note that tools should be used in conjunction with: training on statistical concepts (Kynn, 2008); conditioning to alert experts to potential biases (Low Choy *et al.*, 2009); an interview proforma, with wording of questions carefully chosen to minimize biases (Chaloner and Duncan, 1983; O'Hagan *et al.*, 2006; Kynn, 2008); within a clearly defined protocol (Low Choy *et al.*, 2009). Where software tools package well-designed elicitation methods together with practical guidelines, they may also promote wider application of good elicitation practice and, consequently, strongly enhance capacity to quantitatively address important practical problems in a wide range of fields.

Appendix

A. Broader context and background

A.1 Designing elicitation

Elicitation, like any other data collection exercise, is more effective, transparent and repeatable when carried out following a designed approach. Five stages during an elicitation (interview) process were proposed by Spetzler and Staël von Holstein (1975) and are still considered relevant (e.g. Renooij, 2001): **motivate** experts to contribute and clarify their motivations, **structure** the problem to focus on quantities of interest, **condition** the expert to follow good practices of probabilistic thinking and to avoid common biases, **elicit** and **encode** their opinions and uncertainty as statistical distributions, and **verify** that the expert is satisfied with their responses. These stages of elicitation reflect the expert's experiences, but rely on significant preparation undertaken *by the statistical modeller before* elicitation to design the process. This is acknowledged by O'Hagan *et al.* (2006) who propose a broader view of the whole elicitation process, considered more from the statistician's perspective. They identify two preliminary phases to (1) establish **background** and undertake **preparation** and to (2) **identify and recruit experts**. These authors also expand the initial motivation stage, combined to some extent with **conditioning**, in a stage of (3) **motivating and training** experts.

Here we focus on expanding the first preparatory phase identified by O'Hagan *et al.* (2006) to detail design of elicitation (Section A.1.1); this phase is of primary concern when designing software to accompany elicitation. Six main components in designing elicitation, distilled from experiences in quantifying knowledge of ecologists, are presented in Table 19.1. These are further expanded to include up to five subcomponents per component, with illustrating examples, in Low Choy *et al.* (2009). Here we provide an indication of what each component involves. When applied in practice, varying levels of emphasis on the six components will be required to tailor the framework to the particular ecological context and ecologists involved.

This framework (Section A.1.1) includes a mathematical step of encoding the expert knowledge into distributions, addressed at its most basic level when encoding a single distribution (Section A.1.2). For more complex situations, indirect approaches to encoding may be more effective (Section A.2).

A.1.1 A framework for designed elicitation At the outset of elicitation it is important to determine how the expert opinion is to be utilized [E19.1]. The end purpose can place constraints on the content and form of information elicited. In some cases expert knowledge is important as an end in itself (Alho *et al.*, 1996) or for designing data collection (Golicher *et al.*, 2006). Expert opinion can be crucial for representing the current state of knowledge before extensive monitoring data is available, either when constructing complex biophysical models (e.g. Boyce *et al.*, 2005; Uusitalo, 2007) or developing environmental guidelines (e.g. Low Choy *et al.*, 2005). Expert knowledge can provide a starting point to be updated with new empirical data within the overall Bayesian learning cycle (e.g. McCarthy and Masters, 2005). In complex process models, expert knowledge may prove valuable for quantifying important input parameters (e.g. Boyce *et al.*, 2005).

A key factor in successful elicitation is determining what experts know, and how they are most comfortable expressing it [E2]. Consider elicitation of preferred habitat for input into logistic regression (O'Leary *et al.*, 2008*b*) or classification trees (O'Leary *et al.*, 2008*a*). Can experts only provide an indication of whether factors increase, decrease or have no effect on the response (O'Leary *et al.*, 2008*b*)? Or are they able to sketch the ecological response to a single factor, with all others held constant (Kynn, 2006)? Alternatively, perhaps they are more comfortable estimating ecological response at a particular site where all the factors are known (Section 19.2; Denham and Mengersen, 2007; Murray *et al.*, 2008*a*). These different forms of elicited information required quite different mathematical encoding techniques.

Model formulation requires understanding the information available, both expert and empirical [E3]. In many cases decomposition of the initial problem may reveal underlying ecological processes, and thus enable the elicitor to better target the expert's knowledge. For example, expert panels are proficient in defining boundaries to delineate bioregions (Neldner, 2006). Alternatively a common data-driven approach clusters sites in a training dataset comprising measurements on various environmental factors, then uses discriminant analysis to allocate new sites to each cluster. The question is how can this expert knowledge define a prior distribution to supplement this type of data within a Bayesian framework? The problem is to define a single model that both identifies and describes the clusters, as well as allocating sites to clusters. One solution is a finite mixture model which focuses on describing the range of values for each environmental factor within each region/cluster, and uses latent variables to allocate sites (training or future) to each region/cluster (Rochester *et al.*, 2004; Accad *et al.*, 2005).

The way in which expert knowledge is translated mathematically into statistical distributions can have substantial impact on results [E4], and is a major concern of this research. An overall decision is whether to use a direct or indirect method of encoding (Section A.2). Other encoding decisions are highly context-specific and may depend on the ecologists involved and resources available, for example: the summary statistics elicited; the ordering of questions; and whether visual feedback is used. Section 19.3 illustrates these concepts in terms of the rock-wallaby application and Section A.1.2 addresses this issue more generally.

Like any measurement technique, the whole process of elicitation is subject to many sources of uncertainty, and it is important to manage these [E5]. For this application, we determine major sources of uncertainty and design a strategy that manages these uncertainties arising from elicitation with a single expert (Section 19.3), as well as variability arising across several experts (Section 19.5.3). See Kynn (2008) for a modern review of the cognitive biases that may inform design and O'Hagan *et al.* (2006, Appendix C) for several practical tips.

Finally an elicitation protocol must address logistical issues and stipulate a repeatable and transparent protocol [E6]. One practical decision concerns delivery. Expert responses can be compiled via questionnaire (Martin *et al.*, 2005), or via interview. Experts can be consulted either individually, perhaps with the aid of a computer (Section 19.4, Kynn, 2005; Denham and Mengersen, 2007), or in groups (Vose, 1996; Garthwaite and O'Hagan, 2000; Low Choy *et al.*, 2005). In some cases experts can be trained in the skill of quantifying their opinions using feedback in the form of scoring rules (Gneiting and Raftery, 2007), or seed variables elicited to help calibrate their responses (Cooke, 1991; Burgman, 2005). An efficient approach with small numbers of experts relies on the elicitor to identify knowledge gaps and address these *in situ* just prior to, or during, elicitation. With a large group of experts, this training can occur during a workshop held prior to elicitation. Sampling issues are paramount in elicitation: the selection of a balanced group of experts, each with sufficient expertise on the topic, greatly affects both the information elicited (particularly in group elicitations) as well as the credibility of the results. Furthermore it is important that each expert's motivating interests (*e.g.* recreation, conservation or development) are clearly stated, which can be related to methods used by the elicitor to motivate their participation. The issues of selecting and motivating experts have been carefully addressed in the environmental risk assessment context (Burgman, 2005, Sections 19.4.1–19.4.2).

A.1.2 Encoding a distribution A plethora of techniques have been developed for encoding a univariate distribution [E4], as outlined in Table 19.4. Moment-matching approaches typically rely on elicitation of a small number of summary statistics. Most often two statistics are elicited to characterize a two-parameter distribution. The most common summary statistic elicited as a 'best estimate' is often taken to represent a measure of location for the distribution (mean, median or mode). Others include quantiles, cumulative probabilities and predictions. Minimally the same number of summary statistics as distributional parameters is required for an exact solution of the distributional parameters. Where these relationships are less algebraically tractable, numerical encoding may be necessary (e.g. Low Choy *et al.*, 2008). Where more statistics are elicited compared to the number of parameters a regression approach can be used (Section 19.3). In some cases a carefully selected series of summary statistics are elicited to fully parametrize a complex variance structure (Kadane *et al.*, 1980; Garthwaite and O'Hagan, 2000).

Table 19.4 Common methods for encoding a distribution (Hughes and Madden, 2002; O'Hagan *et al.*, 2006; Low Choy *et al.*, 2008).

Elicitation method	Description
	Moment-matching approaches
Mean with	A tail probability (Gross, 1971; Weiler, 1965); a quantile (Duran and Booker, 1988); the mode (Léon *et al.*, 2003); or the mean absolute deviation about the mean (Pham-Gia *et al.*, 1992).
Median with	The mean absolute deviation about the median (Pham-Gia *et al.*, 1992), or a quantile (Kynn, 2005; Denham and Mengersen, 2007).
Mode with	The probability of an interval centered on the mode (Fox, 1966); an upper quantile (Gilless and Fried, 2000); several quantiles (Low Choy *et al.*, 2008); 'drop-offs', the odds ratios of the mode ± 1 compared to the mode (Chaloner and Duncan, 1983; Gavasakar, 1988).
	Other approaches
Fractile method	Minimum two quantiles (Duran and Booker, 1988), or more, esp. for skewed distributions (O'Hagan, 1998; Low Choy *et al.*, 2008). Quantiles considered include: quartiles, tertiles and sextiles (Garthwaite and O'Hagan, 2000).
The bisection algorithm	A sequence of quantiles are elicited (O'Hagan *et al.*, 2006) (Leal *et al.*, 2007).
Histogram method	Minimum two intervals (Weiler, 1965), or more (Low Choy *et al.*, 2008).
Hypothetical future samples	Assess future response in light of data to impute prior (Winkler, 1967; O'Hagan *et al.*, 2006); mode with a series of updated estimated responses given hypothetical datasets (Gavasakar, 1988).

A.2 Direct versus *indirect elicitation methods*

Encoding methods [E4] may be categorized as direct (structural) or indirect approaches. Direct approaches simplify statistical encoding, by asking experts to directly describe prior distributions for parameters in the model. For a normal prior on a parameter (such as a regression coefficient), this amounts to asking experts for the mean and standard deviation of the parameter, across all values of other parameters (marginal prior) or for specific values of other parameters (conditional prior). Direct elicitation relies on experts being very familiar with the explanatory side of a model, in particular the interpretation of parameters which are abstract concepts. In contrast, indirect elicitation exploits familiarity of experts with the predictive side of a model, in assessing observables such as expected or predicted responses which are more concrete concepts. Indirect approaches simplify communication with the expert, but incur overheads in statistical encoding to reformulate the model to better target expert knowledge.

In this context, *direct* elicitation (Winkler, 1967) of regression coefficients β and their variability Σ assumes experts have comprehensive understanding of the role of both these parameters in the regression model. Direct methods for encoding regressions are tabulated in Table 19.5. For uncomplicated regressions with a single covariate this may be achievable (e.g. Fleishman *et al.*, 2001; Du Mouchel, 1988). In the more general context with multiple covariates experts are often asked both to select the relevant covariates, as well as specify their 'best' estimate of the coefficients. Indeed this mirrors the approach taken by the popular, non-statistical approach, Multicriteria Decision Analysis (MCDA), often applied to eliciting expert opinion for constructing habitat models in the absence of data (Roloff and Kernohan, 1999).

However, for multiple covariates, direct specification of a covariance matrix can be complicated even with as few as three covariates (Kadane *et al.*, 1980;

Table 19.5 Methods of eliciting information for encoding priors for regression coefficients. Part 1: Direct approaches.

- I. **Simple.** What do you think the effect of the covariate is on the response?
  - I-a. **Moment.** Provide the standard error of your estimated effect (Fleishman *et al.*, 2001).
  - I-b. **Fractile.** Estimate quantiles of the estimated effect (Du Mouchel, 1988; Low Choy *et al.* 2005a).
  - I-c. **Histogram.** Estimate the likelihood of a range of values for the effect (West, 1988).
  - I-d. **Relative precision.** How much information is there in your prior knowledge compared to the observed data (Zellner, 1996).
- II. **Sign.** Do you think the response increases, decreases or is unchanged as the covariate increases (Martin *et al.*, 2005; Kuhnert *et al.*, 2005; O'Leary *et al.*, 2008*b*)?
- III. **Multiple comparisons.** Describe quantiles of a group of effects simultaneously (Du Mouchel, 1988).

Table 19.6 Methods of eliciting information for encoding priors for regression coefficients. Part 2: Main questions defining indirect approaches.

IV. **Predicted response.** Having seen the observed data, what do you predict the response to be for specific cases with known values of all covariates (Kadane *et al.*, 1980; Kadane and Wolfson, 1998)?

IV-a. **Spatial prediction.** As above, replacing 'cases' with 'sites' (Denham and Mengersen, 2007)?

V. **Case Conditional Mean.** What do you expect the response to be, given values of covariates? The number of questions equals the number of theoretical covariates (Oman, 1985; Bedrick *et al.*, 1996).

VI. **Curve Conditional Mean.** What do you expect the response to be, given values of one covariate? All other covariates are fixed at reference values (Willems *et al.*, 2005; Kynn, 2005, 2006; Al-Awadhi and Garthwaite, 2006).

VII. **'Prior' data.** Generate potential responses for specific cases with known values of all covariates (Chen *et al.*, 1999; James *et al.*, 2009).

VIII. **Data Augmentation.** Give examples of expected responses, weighted according to their plausibility, for specific cases with known values of all covariates (Bedrick *et al.*, 1996; Chen *et al.*, 1999; James *et al.*, 2009)?

IX. **Conditional Mean via Measurement Error.** What do you expect the response to be, given values of covariates? Questions may exceed the number of covariates. This new approach (James *et al.*, 2009) extends that of Bedrick *et al.* (1996) inspired by Lindley (1983).

Lindley, 1983; O'Hagan, 1998). One way to simplify its assessment is to reparameterize as $\Sigma = c\sigma^2 X^T X^{-1}$ with $p(\sigma^2) \propto 1/\sigma^2$ (Zellner, 1996). Then c is elicited as the amount of information provided by the prior in comparison to the data. Alternatively assessment of the coefficients can be simplified by eliciting the sign of a sole covariate (Martin *et al.*, 2005; Kuhnert *et al.*, 2005), or several (O'Leary *et al.*, 2008*b*). Experts can be asked whether they believe the response increases, decreases or is insensitive to each covariate. This information can be encoded as an indicator $\beta_j \in \{-1, 0, +1\}$ and then encoded as a single normal distribution representing opinions of several experts (Martin *et al.*, 2005) or as a mixture of normals representing plausibility of each value of the indicator for a single expert, or collated over several experts (O'Leary *et al.*, 2008*b*).

These simplifications will not always be appropriate. In such situations experts may find direct elicitation difficult (Kadane *et al.*, 1980; Denham and Mengersen, 2007; O'Leary *et al.*, 2008*b*). Inaccurate representation of expert knowledge would introduce cognitive and motivational biases (Kynn, 2008). These biases can be reduced by providing experts with a more natural framework for expressing their knowledge, in terms of observables on the same scale as the data, rather than unobservable and abstract model parameters (Kadane and Wolfson, 1998). To this end, elicitation can be restructured (Spetzler and Staël von Holstein, 1975), resulting in indirect elicitation of model parameters (Winkler, 1967), often achieved via decomposition of the statistical model (Low Choy *et al.*, 2009). Table 19.6 lists several indirect approaches to elicitation for regression, with the main question determining how β is encoded. Please see cited references to see how additional questions are required to encode Σ. Several indirect approaches to elicitation have been developed for regression; these are reviewed in Section 19.3.1.

Acknowledgments

We gratefully acknowledge funding support for the first, third and fourth authors under Australian Research Council Discovery Grant DP0667168, and an ARC Linkage Project with Healthy Waterways, Queensland Environmental Protection Agency and Queensland Department of Natural Resources. Initial work by the first author on elicitation was supported by a Queensland University of Technology Postdoctoral research award. Work on collating multiple expert opinions was also seeded by a travel grant from L'Institut Nationale Recherche Agricole (INRA), Paris, France. Work by the third author was supported by the Collaborative Centre for Complex Dynamic Systems & Control.

We thank Robert Denham for useful discussion and access to some of the code from his prototype software, an important inspiration on many levels for *Elicitator*. We thank the organizers of the Environmetrics satellite workshop and the main conference of the International Society for Bayesian Analysis, for an opportunity for initial presentation of aspects of these ideas. The ensuing discussion and comments provided by participants such as Ross McVinish, David Draper and Tony O'Hagan stimulated more careful thought. We also thank Kristen Williams, Mike Austin and Rebecca O'Leary for fruitful discussions about ecological applications of elicitation. The diligence and enthusiasm of ecologists involved in elicitation was pivotal to the project, and greatly appreciated.

References

Accad, A., Low Choy, S., Pullar, D., and Rochester, W. (2005). Bioregional classification via model-based clustering. In MODSIM 2005 International Congress on Modelling and Simulation. Modelling and Simulation Society of Australia and New Zealand.

Al-Awadhi, S. A. and Garthwaite, P. H. (2006). Quantifying expert opinion for modelling fauna habitat distributions. *Computational Statistics*, **21**, 121–140.

Alho, J. M., Kangas, J., and Kolehmainen, O. (1996). Uncertainty in expert predictions of the ecological consequences of forest plans. *Applied Statistics*, **45**, 1–14.

Austin, M. P. (2002). Spatial prediction of species distribution: an interface between ecological theory and statistical modelling. *Ecological Modelling*, **157**, 101–118.

Austin, M. P. and Heyligers, P. C. (1989). Vegetation survey design for conservation – gradsect sampling of forests in northeastern New South Wales. *Biological Conservation*, **50**, 13–32.

Bedrick, E. J., Christensen, R., and Johnson, W. (1996). A new perspective on priors for generalized linear models. *Journal of the American Statistical Association*, **91**, 1450–1460.

Box, G. E. P. and Tiao, G. C. (1982). *Bayesian Inference in Statistical Analysis*. John Wiley, New York.

Boyce, M. S., Irwin, L. L., and Barker, R. (2005). Demographic meta-analysis: synthesizing vital rates for spotted owls. *Journal of Applied Ecology*, **42**, 38–49.

Branscum, A. J., Johnson, W. O., and Thurmond, M. C. (2007). Bayesian Beta regression: applications to household expenditure data and genetic distance between foot-and-mouth disease viruses. *Australian and New Zealand Journal of Statistics*, **49**, 287–301.

Burgman, M. (2005). *Risks and Decisions for Conservation and Environmental Management. Ecology, Biodiversity and Conservation.* Cambridge University Press, Cambridge.

Chaloner, K., Church, T., Louis, T. A., and Matts, J. P. (1993). Graphical elicitation of a prior distribution for a clinical trial. *The Statistician*, **42**, 341–353.

Chaloner, K. M. and Duncan, G. T. (1983). Assessment of a Beta prior distribution: PM elicitation. *The Statistician*, **32**, 174–180.

Chen, M. H., Ibrahim, J. G., and Yiannoutsos, C. (1999). Prior elicitation, variable selection and Bayesian computation for logistic regression models. *Journal of the Royal Statistical Society, Series B (Methodological)*, **61**, 223–242.

Clemen, R. T. and Winkler, R. L. (1999). Combining probability distributions from experts in risk analysis. *Risk Analysis*, **19**, 187–203.

Cooke, R. (1991). *Experts in Uncertainty: Opinion and Subjective Probability in Science.* Oxford University Press, New York.

Cooke, R. M. and Goossens, L. H. J. (2004). Expert judgement elicitation for risk assessments of critical infrastructures. *Journal of Risk Research*, **7**, 643–656.

Dawid, A. P. (1982). The well-calibrated Bayesian (with discussion). *Journal of the American Statistical Association*, **77**, 605–610.

Denham, R. and Mengersen, K. (2007). Geographically assisted elicitation of expert opinion for regression models. *Bayesian Analysis*, **2**, 99–136.

Du Mouchel, W. (1988). A Bayesian model and a graphical elicitation procedure for multiple comparisons. In *Bayesian Statistics 3*, (ed. J. M., Bernardo, M. H., de Groot, D. V., Lindley and A. F. M., Smith), pages 127–145. Oxford University Press, Oxford.

Duran, B. and Booker, J. (1988). A Bayes sensitivity analysis when using Beta distribution as a prior. *IEEE Transactions on Reliability*, **37**, 239–247.

Elith, J., Graham, C. H., Anderson, R. P., Dudík, M., Ferrier, S., Guisan, A., Hijmans, R. J., Huettmann, F., Leathwick, J. R., Lehmann, A., Li, J., Lohmann, L. G., Loiselle, B. A., Manion, G., Moritz, C., Nakamura, M., Nakazawa, Y., Overton, J. M., Peterson, A. T., Phillips, S. J., Richardson, K., Scachetti-Pereira, R., Schapire, R. E., Soberón, J., Williams, S., Wisz, M. S., and Zimmermann, N. E. (2006). Novel methods improve prediction of species' distributions from occurrence data. *Ecography*, **29**, 129–151.

Ferrier, S., Watson, G., Pearce, J., and Drielsma, M. (2002). Extended statistical approaches to modelling spatial pattern in biodiversity in northeast New South Wales. I. Species-level modelling. *Biodiversity and Conservation*, **11**, 2275–2307.

Fleishman, E., Nally, R. M., Fay, J. P., and Murphy, D. D. (2001). Modeling and predicting species occurrence using broad-scale environmental variables: An example with butterflies of the great basin. *Conservation Biology*, **15**, 1674–1685.

Fox, B. (1966). A Bayesian approach to reliability assessment. Technical Report Memorandum RM-5084-NASA, The Rand Corporation, Santa Monica, CA.

Garthwaite, P. H., Kadane, J. B., and O'Hagan, A. (2005). Elicitation. Technical report, University of Sheffield. `http://www.shef.ac.uk/beep`.

Garthwaite, P. H. and O'Hagan, A. (2000). Quantifying expert opinion in the UK water industry: an experimental study. *The Statistician*, **49**, 455–477.

Gavasakar, U. (1988). A comparison of two elicitation methods for a prior distribution for a binomial parameter. *Management Science*, **34**, 784–790.

Genest, C. and Zidek, J. V. (1986). Combining probability distributions: A critique and an annotated bibliography. *Statistical Science*, **1**, 114–135.

Gilless, J. K. and Fried, J. S. (2000). Generating Beta random rate variables from probabilistic estimates of fireline production times. *Annals of Operation Research*, **95**, 205–215.

Gneiting, T. and Raftery, A. E. (2007). Strictly proper scoring rules, prediction, and estimation. *Journal of the American Statistical Association*, **102**, 359–378.

Golicher, D. J., OHara, R. B., Ruiz-Montoya, L., and Cayuela, L. (2006). Lifting a veil on diversity: A Bayesian approach to fitting relative-abundance models. *Ecological Applications*, **16**, 202–212.

Greenland, S. (2006). Bayesian perspectives for epidemiological research: I. Foundations and basic methods. *International Journal of Epidemiology*, **35**, 765–775.

Gross, A. (1971). The application of exponential smoothing to reliability assessment. *Technometrics*, **13**, 877–883.

Guisan, A. and Thuiller, W. (2005). Predicting species distribution: offering more than simple habitat models. *Ecology Letters*, **8**, 993–1009.

Guisan, A. and Zimmermann, N. E. (2000). Predictive habitat distribution models in ecology. *Ecological Modelling*, **135**, 147–186.

Hughes, G. and Madden, L. V. (2002). Some methods for eliciting expert knowledge of plant disease epidemics and their application in cluster sampling for disease incidence. *Crop Protection*, **21**, 203–215.

James, A., Low Choy, S., and Mengersen, K. (2009). *Elicitator*: A software-based expert elicitation tool for regression in ecology. *Environmental Modelling and Software*, **25**, 129–145.

Kadane, J. B., Dickey, J. M., Winkler, R. L., Smith, W. S., and Peters, S. C. (1980). Interactive elicitation of opinion for a normal linear model. *Journal of the American Statistical Association*, **75**, 845–854.

Kadane, J. B. and Wolfson, L. J. (1998). Experiences in elicitation. *The Statistician*, **47**, 3–19.

Kuhnert, P. M., Martin, T. G., Mengersen, K., and Possingham, H. P. (2005). Assessing the impacts of grazing levels on bird density in woodland habitat: A Bayesian approach using expert opinion. *Environmetrics*, **16**, 717–747.

Kynn, M. (2005). Eliciting expert opinion for a Bayesian logistic regression model in natural resources. Ph.D. thesis, School of Mathematical Sciences, Faculty of Science, Queensland University of Technology. `http://adt.library.qut.edu.au/adt-qut/public/adt-QUT20050830.084943`.

Kynn, M. (2006). Designing elicitor: Software to graphically elicit expert priors for logistic regression models in ecology. `http://www.winbugs-development.org.uk/elicitor/files/designing.elicitor.pdf`.

Kynn, M. (2008). The "heuristics and biases" bias in expert elicitation. *Journal of the Royal Statistical Society, Series A*, **171**, 239–264.

Landrum, M. B. and Normand, S.-L. T. (1999). Applying Bayesian ideas to the development of medical guidelines. *Statistics in Medicine*, **18**, 117–137.

Langhammer, P. F., Bakarr, M. I., Bennun, L. A., Brooks, T. M., Clay, R. P., Darwall, W., Silva, N. D., Edgar, G. J., Eken, G., Fishpool, L. D. C., da Fonseca, G. A. B., Foster, M. N., Knox, D. H., Matiku, P., Radford, E. A., Rodrigues, A. S. L., Salaman, P., Sechrest, W., and Tordoff, A. W. (2007). Identification and gap analysis of key biodiversity areas: Targets for comprehensive protected area systems. Best Practice Protected Area Guidelines Series No. 15, IUCN (The World Conservation Union): Gland, Switzerland. `http://www.iucn.org/dbtw-wpd/edocs/PAG-015.pdf`.

Leal, J., Wordsworth, S., Legood, R., and Blair, E. (2007). Eliciting expert opinion for economic models: An applied example. *Value in Health*, **10**, 195–203.

Léon, C. L., Vázquez-Polo, F. J., and González, R. L. (2003). Elicitation of expert opinion in benefit transfer of environmental goods. *Environmental and Resource Economics*, **26**, 199–210.

Lindley, D. (1983). Reconciliation of probability distributions. *Operations Research*, **31**, 866–880.

Low Choy, S., Mengersen, K., and Rousseau, J. (2008). Encoding expert opinion on skewed non-negative distributions. *Journal of Applied Probability and Statistics*, **3**, 1–21.

Low Choy, S., O'Leary, R., and Mengersen, K. (2009). Elicitation by design for ecology: using expert opinion to inform priors for Bayesian statistical models. *Ecology*, **90**, 265–277.

Low Choy, S., Stewart-Koster, B., Eyre, T., Kelly, A., and Mengersen, K. (2005). Benchmarking indicators of vegetation condition: A Bayesian modelling and decision theoretic approach. In MODSIM 2005 International Congress on Modelling and Simulation. Modelling and Simulation Society of Australia and New Zealand.

Martin, T. G., Kuhnert, P. M., Mengersen, K., and Possingham, H. P. (2005). The power of expert opinion in ecological models: A Bayesian approach examining the impact of livestock grazing on birds. *Ecological Applications*, **15**, 266–280.

McCarthy, M. A. and Masters, P. (2005). Profiting from prior information in Bayesian analyses of ecological data. *Journal of Applied Ecology*, **42**, 1012–1019.

Miller, J. and Franklin, J. (2002). Modeling the distribution of four vegetation alliances using generalized linear models and classification trees with spatial dependence. *Ecological Modelling*, **157**, 227–247.

Murray, J., Goldizen, A., O'Leary, R., McAlpine, C., Possingham, H., and Low Choy, S. (2008*a*). How useful is expert opinion for predicting the distribution of a species within and beyond the region of expertise? A case study using brush-tailed rock-wallabies (*Petrogale penicillata*). *Journal of Applied Ecology*, **46**, 842–851.

Murray, J., Low Choy, S., McAlpine, C., Possingham, H., and Goldizen, A. (2008*b*). The importance of ecological scale for wildlife conservation in naturally fragmented environments: A case study of the brush-tailed rock-wallaby (*Petrogale penicillata*). *Biological Conservation*, **141**, 7–22.

Murray, J.V., Goldizen, A.W., O'Leary, R.A., McAlpine, C.A., Possingham, H.P. and Low Choy, S. (2009). How useful is expert opinion for predicting the distribution of a species within and beyond the region of expertise? A case-study using brush-tailed rock wallabies *Petrogale penicillata*. *Journal of Applied Ecology*, **46**, 842–851.

Neldner, J. (2006). Why is vegetation condition important to government? A case study from Queensland. *Ecological Management & Restoration*, **7**, S5–S7.

O'Hagan, A. (1997). The ABLE story: Bayesian asset management in the water industry. In *The Practice of Bayesian Analysis*, (ed. French, S. and Smith, J. Q), pages 173–198. Arnold, London.

O'Hagan, A. (1998). Eliciting expert beliefs in substantial practical applications. *The Statistician*, **47**, 21–35.

O'Hagan, A., Buck, C. E., Daneshkhah, A., Eiser, R., Garthwaite, P., Jenkinson, D., Oakley, J., and Rakow, T. (2006). *Uncertain Judgements: Eliciting Experts' Probabilities*. John Wiley, New York.

O'Leary, R., Murray, J., Low Choy, S., and Mengersen, K. (2008*a*). Expert elicitation for Bayesian classification trees. *Journal of Applied Probability and Statistics*, **3**, 95–106.

O'Leary, R. A., Low Choy, S. J., Fensham, R. J., and Mengersen, K. L. (2009). Improving the performance and interpretation of habitat models: using a two-scale modelling approach to model the envelope and identify excess zeros, with a case study on *stemmacantha australis*. Submitted.

O'Leary, R. A., Low Choy, S. J., Murray, J. V., Kynn, M., Denham, R., Martin, T. G., and Mengersen, K. (2008*b*). Comparison of three expert elicitation methods for logistic regression on predicting the presence of the threatened brush-tailed rock-wallaby *(petrogale penicillata)*. *Environmetrics*, **19**, 1–20.

Oman, S. D. (1985). Specifying a prior distribution in structured regression problems. *Journal of the American Statistical Association*, **80**, 190–195.

Pham-Gia, T., Turkkan, N., and Duong, Q. (1992). Using the mean deviation in the elicitation of the prior distribution. *Statistics and Probability Letters*, **13**, 373–381.

Pollino, C. A., Woodberry, O., Nicholson, A., Korb, K., and Hart, B. T. (2007). Parameterisation and evaluation of a Bayesian network for use in an ecological risk assessment. *Environmental Modelling and Software*, **22**, 1140–1152.

Renooij, S. (2001). Probability elicitation for belief networks: issues to consider. *Knowledge Engineering Reviews*, **16**, 255–269.

Robert, C. P. (2001). *The Bayesian Choice: A Decision-Theoretic Framework*. Springer-Verlag, New York.

Rochester, W., Accad, A., Low Choy, S., Neldner, V., Pullar, D., and Williams, K. (2004). Final report UQ-EPA subregion classification project. Technical report, The University of Queensland, Brisbane, Australia.

Roloff, G. J. and Kernohan, B. J. (1999). Evaluating reliability of habitat suitability index models. *Wildlife Society Bulletin*, **27**, 973–985.

Smith, C. S., Howes, A. L., Price, B., and McAlpine, C. A. (2007). Using a Bayesian belief network to predict suitable habitat of an endangered mammal – the Julia Creek dunnart (*sminthopsis douglasi*). *Biological Conservation*, **139**, 333–347.

Spetzler, C. S. and Staël von Holstein, C.-A. S. (1975). Probability encoding in decision analysis. *Management Science*, **22**, 340–358.

Spiegelhalter, D. J., Adams, K. R., and Myles, J. P. (2004). *Bayesian Approaches to Clinical Trials and Health-Care Evaluation*. John Wiley, Chichester.

Spiegelhalter, D. J., Thomas, A., Best, N. G., and Lunn, D. (2003). WinBUGS version 1.4 user manual. Technical report, MRC Biostatistics Unit, Cambridge.

Tversky, A. and Kahneman, D. (1981). The framing of decisions and the psychology of choice. *Science*, **211**, 453–458.

Urbanek, S. (2006). *RServe*. `http://www.rforge.net/Rserve/`.

Uusitalo, L. (2007). Advantages and challenges of bayesian networks in environmental modelling. *Ecological Modelling*, **203**, 312–318.

Venables, W. N. and Ripley, B. R. (2002). *Modern Applied Statistics using S-Plus* (4th edn). Springer-Verlag, New York.

Vose, D. (1996). *Quantitative Risk Analysis: a Guide to Monte Carlo Simulation Modelling*. John Wiley, Chichester.

Weiler, H. (1965). The use of incomplete Beta functions for prior distributions in binomial sampling. *Technometrics*, **7**, 335–347.

West, M. (1988). Combining expert opinion. In *Bayesian Statistics 3*, (ed. J. M. Bernardo, M. H. de Groot, D. V. Lindley and A. F. M. Smith), pp. 493–508. Oxford University Press, Oxford.

Willems, A., Janssen, M., Verstegen, C., and Bedford, T. (2005). Expert quantification of uncertainties in a risk analysis for an infrastructure project. *Journal of Risk Research*, **8**, 3–17.

Winkler, R. L. (1967). The assessment of prior distributions in Bayesian analysis. *Journal of the American Statistical Association*, **62**, 776–800.

Zellner, A. (1996). Models, prior information, and Bayesian analysis. *Journal of Econometrics*, **75**, 51–68.

·20·

Characterizing the uncertainty of climate change projections using hierarchical models

Claudia Tebaldi and Richard L. Smith

20.1 Climate change and human influences, the current state and future scenarios

There is substantial consensus on many aspects of climate change. It is already with us; we have a large responsibility in many facets of it; and future changes in the absence of significant curbing of greenhouse gas emissions are going to be much more dramatic than what is already experienced, with consequences that will be predominantly detrimental to social and natural systems. Adaptation and mitigation decisions need however detailed information about the reality and the future of climate change, and more often than not at regional, rather than global mean scales. To produce this information, deterministic numerical models are run under different hypothetical scenarios corresponding to alternative greenhouse gas emission pathways into the future. These extremely complex computer models discretize the surface of the Earth, the depths of the oceans and the layers of the atmosphere into regularly spaced grid boxes. Computations through differential equations provide the evolution of a high dimensional state vector representing climate variables, by applying our understanding of climate dynamics and climate interacting processes over land, water, air and sea ice (Washington and Parkinson, 2005). With the evolution of science and technology, more and more processes at increasingly finer scales can be represented explicitly in these simulations, but there still remains the need for approximations, for those processes that act at scales not explicitly represented. It is in these approximations that the source of large uncertainties resides.

20.1.1 Many climate models, all of them right, all of them wrong

Because of the complexities and high dimensional nature of global climate simulations, many alternative solutions to the representation and parametrization of processes are consistent with the state of our scientific understanding. Thus, different models approach the simulation of climate over the Earth with different strategies. It may be as basic as the choice of the resolution used to discretize the globe in its three dimensions (which also affects the time step of the finite difference equations). This choice has important consequences on the range of processes that can be directly simulated and those that have to be represented by parametrization of the subgrid scales.[1] Of course there are less basic choices, like the processes that are included, independently of the computing limitation: is an interactive carbon cycle present? Is vegetation changing with climate? Are urban areas and their effect on climate represented? Even for a given resolution and a given set of processes, the actual computation of the time-evolving series of quantities that make up the climate system may differ because of different values in exogenous parameters of the equations (usually estimated by studying the individual processes in the field) or of different numerical solvers adopted.

Within this large population of model configurations we are hard pressed to find first and second class citizens. There are more than a dozen global climate models (many of which run at different resolutions) which are comparable in terms of overall performance, when validated against the observed record and paleo-climate proxy records. Of course some models are better than others for a given set of metrics (average temperature over North America, precipitation in the monsoon region, the frequency and intensity of the El Niño Southern Oscillation phenomenon) but for any set of 'better than average models' a different set of metrics can be found for which those models will underperform compared to a different set (Gleckler *et al.*, 2007).

Thanks to international efforts like the Intergovernmental Panel on Climate Change periodic assessment reports, the last of which was published in 2007 (IPCC, 2007), modelling centres participate in concerted sets of experiments, running their models under standardized scenarios of external forcings. These are standard greenhouse gas concentration pathways, derived after scientists devised a set of alternative, but equally plausible, future storylines about the social, political, technological and economic future of our world. From these story lines consequences in terms of greenhouse gases and other pollutants' emissions were derived using economic integrated modeling.

[1] Parametrization is performed by estimating the relation between the large scale, resolved processes and the small scale, unrepresented processes that happen within a model grid box. Observational or experimental studies provide the basis for the estimation of the parameters that govern the interaction between them. Uncertainties in observations and in the relations themselves are at the source of the uncertainties in parametrizations, causing a range of values for these parameters to be consistent with their function, but at the same time causing sugnificant differences in the evolution of quantities in the model integrations.

For different families of scenarios (varying from a future of fast technological progress and international collaboration to one of slower integration and fossil-fuel-intensive economies) very different trajectories of greenhouse gas rates of emissions over this century and beyond are specified. These scenarios of greenhouse gas emissions, known as SRES scenarios (Nakicenovic, 2000), have been used by climate modellers as external inputs to run simulations of alternative climate scenarios in the future. By the second half of this century, different emission scenarios cause significantly different climate outcomes, starting from global average temperature change but riverberating in all aspects of Earth's climate. By the end of this century, the uncertainty across scenarios of greenhouse gas increases is larger than the inter-model differences under a specific scenario. However, the climate community feels it inappropriate to attach probabilities to different emission pathways, and we are left with performing our uncertainty analyses conditionally on a given SRES scenario.

After running model experiments under alternative SRES scenarios, the modelling centres are contributing the resulting simulations' output into open access archives. These collections of model simulations have been labelled ensembles of opportunity, i.e. multiple models' collections that are not the result of a statistically designed experiment or random sampling from a population of models, but a post facto collection of what is available, thanks to the voluntary and self-selecting participation of the world's largest and most advanced research centres. The most recent and largest archive of such data sets is maintained by the Program for Climate Model Diagnosis and Intercomparison (PCMDI) at Lawrence Livermore National Laboratory (LLNL), and can be found at `http://www-pcmdi.llnl.gov/ipcc/about_ipcc.php`.

20.1.2 Goals and challenges of analysing ensembles of opportunity

The most direct way to obtain regionally detailed future projections is to process the output of global climate models and determine statistics of the regional climate variables of interest. Determining which model to trust above all others is a daunting task, and one defensible strategy is to utilize all that are available, synthesizing the projections and their uncertainty through a rigorous statistical analysis. This will provide optimal estimates of the changes in store, and will quantify their uncertainty, conditionally on the available information. This kind of representation is of great value to decision makers and stakeholders, notwithstanding the need of communicating the underlying limitations of our current understanding and working hypotheses as embodied by these models, and the assumptions that are at the basis of any statistical representation of the data.

Bayesian statistics has a natural advantage in this particular setting, not only because we are dealing with uncertain events that are not easily framed in a frequentist perspective, but more fundamentally because of its natural framework for incorporating expert judgement, and updating current assessment with the in-flow of additional pieces of information. There are great challenges underlying any statistical analysis of such multimodel ensembles (Tebaldi and Knutti, 2007). They stem from the non-systematic and conversely non-random nature of the sampling of models, which hampers a full representation of the uncertainties at stake; from the dependence among the models in the ensemble, some sharing components, some sharing even their full name, when the same modelling centre contributes output from runs at different resolutions; from the lack of a theory linking model performance for current climate, that we can measure and validate, to model reliability for future projections. A particular difficulty is the impossibility of verifying models' future projections, which sets this problem well apart from weather forecasting, where feedback about the performance of the model can be as immediate as six hours after the forecast is issued.

All these challenges stand in the way of a robust representation of climate change projections, especially at regional scales and especially for variables other than temperature, which is relatively easier to model because of its smoothness. We are using 'robust' here in a more generic sense than is usually understood by the phrase Bayesian robustness, though the latter may also be worth exploring in our setting. Climate scientists, hydrologists and agricultural modelers – among others – are often interested in studying the impacts of climate change and of adaptation measures. The traditional scientists' approach is through scenarios, where alternative futures used to span a range of outcomes are fed through impact models. In contrast with this approach, we argue for a rigorous uncertainty analysis via Bayesian statistics. In fact much of the impact research efforts are veering towards full probabilistic analysis, and probabilistic projection of climate change constitute their stepping stone (Tebaldi and Lobell, 2008). We think it is fair to say that the area is still in a phase of methodological development. We hope that this chapter will offer enough motivation, and open up enough interesting research directions to invite fresh perspectives to this important application of statistical analysis.

20.2 A world of data. Actually, make that 'many worlds'

The most up to date archive of a multimodel ensemble, hosted by PCMDI at LLNL, contains over 35 terabytes of data. It collects output from 23 models, run under a range of SRES emission scenarios. Scores of variables constitute the output of a climate model simulation, as familiar as temperature and precipitation or as esoteric as the mass of water that evaporates over an ice sheet, or metrics of ocean overturning. Many of the variables are archived as daily or monthly means, some are six-hourly, some are yearly averages. The potential for statistical analysis of model output is practically infinite, when we consider that these quantities are produced at each point of grids that discretize the entire Earth surface, and at many layers in the atmosphere and oceans. The median resolution of the climate models in the PCMDI archive is about 2.75 degrees in latitude/longitude, making the standard grid output for each variable and each time step, when vectorized, 8,192 components in length. A typical climate change experiment consists of a simulation that starts from conditions describing the state of the system at a pre-industrial time, chosen often by convention as 1870, and run with only external forcing imposed, otherwise in a self-consistent and isolated manner, until year 2100. External forcings are meant to represent changing greenhouse gas concentrations over time, aerosols, volcano eruptions and solar cycles. The system evolves (at a few minutes' time step) according to the laws of physics known to govern climate dynamics. To the extent permitted by their resolution, climate models represent coastlines and topography, and they impose vegetation types, often evolving along the simulation timeline, in order to represent urbanization, deforestation, switches from natural vegetation to crop growing areas and vice versa. Even for a given experiment (i.e. a given SRES scenario of external forcings) a model is run for a handful of slightly different initial conditions. The trajectories of the members of these single model ensembles give a handle on the characterization of natural internal variability of the system, generated by the intrinsically chaotic nature of weather and, in the long run, climate processes. For an analysis of climate change, its signal has to be isolated and extracted from the background noise of natural variability by averaging the members of the initial-conditions ensemble, and by considering differences between two at least 20-year averages, usually. The idea is that there is initial condition uncertainty to take care of, and there are low frequency natural modes of variability (decadal oscillations and other phenomena like the alternating phases of El Niño/La Niña conditions in the Pacific) that need to be averaged out before starting to talk about anthropogenic (i.e. externally forced by human actions) climate change.

Averaging in time is one way of facilitating the extraction of the signal of change. Averaging over space is its natural counterpart. Local climate is noisy, and climate models, because of their coarse resolution, are not expected to reproduce faithfully the statistics of local climate. Even in the face of a constant push to provide regionally detailed information from these models, it is fair to say that our confidence in their ability to simulate climate is highest when large regions of sub-continental size are concerned, and their climate considered in terms of large area-averages.

20.3 Our simplified datasets

In the following sections of the chapter we present a suite of statistical analyses combining output from ensemble of climate models, from now on referred to as GCMs (General Circulation Models). We will significantly simplify the problem by using summary statistics of their output, regional means of temperature and precipitation, seasonally averaged and aggregated as 10 or 20 year means. We always condition our analysis to a given experiment, defined in terms of the greenhouse gas emission scenario (SRES) used as part of the external forcings by the simulation. For a given scenario a number of GCMs have run simulations covering the period 1870 through 2100 and have archived temperature and precipitation output. We also have observed records (in some region of the world extending as far back) that can be used to gauge the GCM ability to reproduce historic conditions. The model runs take account of changes in forcings, including observed changes in emissions or solar output, but they are not directly calibrated to observational climate data. One consequence of this is that the model outputs include dynamical features such as El Niños, but the El Niños of the model do not correspond in time to the El Niños of the observational climate record. One reason for considering 20-year averages is that over a 20-year time span, such short term fluctuations are likely to average out.

Suppose there are M GCMs, X_j is a simulated value of some current climate variable generated by GCM j, and Y_j a simulated value of some future climate variable generated by GCM j. We also have an observation X_0 of the true current climate, with its associated standard error $\lambda_0^{-1/2}$ that we can estimate from the observations' series and fix in our model. In our typical application, X_j is the mean temperature or precipitation in a particular region for the period 1981–2000, X_0 is the corresponding value calculated from the observational climate record, and Y_j is the corresponding temperature average calculated from the 2081–2100 segment of the GCM simulation.

A modification of this simple setup will involve R regional averages at a time. Accordingly, we add a subscript $i = 1, \ldots, R$ to the variables, and consider $X_{ij}, Y_{ij}, X_{i0}, \lambda_{0i}$.

Finally we will model the joint distribution of two variables, say temperature and precipitation, for a given region and season, and over the entire length of the simulation, as a bivariate time series of decadal averages. Accordingly we will consider X_{jt}, $t = 1, \ldots, 15$ a bivariate vector of temperature and precipitation averages derived from the jth GCM output. Here the time index corresponds to the decades centred at 1955, 1965, . . . , 2005, 2015, . . . , 2095, so that both historical and future periods will be modelled jointly. Similarly, O_t, $t = 1, \ldots 6$ will indicate a two-component vector of observed temperature and precipitation averages. In this case the time index, t, corresponds to the decades centered at 1955, 1965, ..., 2005 (the last one at present is approximated by an estimate based on eight years of data only).

20.4 A hierarchy of statistical models

Our strategy in presenting our approach to multimodel synthesis and uncertainty characterization is to start from a basic representation, highlight its shortcomings and increase the complexity of the statistical treatment gradually to account for as many additional details as possible. This way, we hope to highlight issues, limitations and solutions step by step. Hopefully this will help the reader achieve familiarity with the data and the ultimate goals of the analysis that will engender his or her constructive criticisms and original thinking in the face of this challenging application.

20.4.1 One region at a time

Let us then start from the simplest series of assumptions. We treat each region separately, and we assume the following likelihood model: For $j = 1, \ldots, M$,

$$X_j \sim N\left(\mu, \lambda_j^{-1}\right)$$
$$Y_j \sim N\left(\nu, (\theta\lambda_j)^{-1}\right)$$

and for the observed mean climate variable (from now on we refer to it as temperature, for simplicity),

$$X_0 \sim N\left(\mu, \lambda_0^{-1}\right). \tag{20.1}$$

This likelihood model simply states that each GCM approximates the true mean temperature of the region (μ for current, ν for future climate) with a Gaussian error, whose variance is model specific. The parameter θ allows for the possibility that future precision will be different than current precision, by a factor common to all GCMs (i.e. the likely degradation of the accuracy of future projections affects all GCMs equally). Notice that this is an assumption dictated by our data, which would not permit the identification of a parameter modeling GCM-specific change in precision. In fact the precision parameter λ_j will be estimated on the basis of the minimum necessary number of datapoints, two. Also, the likelihood model assumes no correlation between the error of GCM j in simulating current climate and its error in simulating future climate.

The use of uninformative improper priors for μ and ν, $U(-\infty, +\infty)$, and proper but very diffuse Gamma priors for the precision parameters λ_j and θ, $Ga(a, b)$, with $a = b = 0.01$ completes this basic model. A simple Gibbs sampler can be used to explore the joint posterior distribution of the parameters (see Appendix B1). This approach was first presented in Tebaldi *et al.* (2004, 2005).

Here we highlight a few features of the approximate posterior estimates for the parameters μ, ν and λ_j. The form of the means of μ and ν, given $\lambda_0, \lambda_j, j = 1, \ldots, M$ as well as given the Xs and Ys is respectively,

$$\tilde{\mu} = \frac{\lambda_0 X_0 + \sum \lambda_j X_j}{\lambda_0 + \sum \lambda_j}$$

and

$$\tilde{\nu} = \frac{\sum \lambda_j Y_j}{\sum \lambda_j}.$$

The posterior distribution of λ_j can be approximated by

$$\lambda_j|\text{rest} \sim Ga\left[a + 1, b + \frac{1}{2}(X_j - \mu)^2 + \frac{\theta}{2}\{Y_j - \nu\}^2\right].$$

An easy addition to this model may accommodate the obvious critique that we expect correlation between errors within a single GCM's simulations of current and future climate. We can substitute to the likelihood of Y_j the following:

$$Y_j|X_j \sim N\left(\nu + \beta(X_j - \mu), (\theta\lambda_j)^{-1}\right).$$

By so doing we estimate a possible correlation between X_j and Y_j. Here as before with θ we are forced to assume a common correlation across all GCMs. Perhaps this is too simplistic an assumption, but it is needed for identifiability of the parameter β. The posterior distribution of β will tell us if such assumption is indeed substantiated by the data. The interpretation of the parameter is particularly interesting if we note that $\beta = 0$ corresponds to X_j and Y_j being independent, while $\beta = 1$ corresponds to X_j and $Y_j - X_j$ being independent, conditionally on the other model parameters. We choose an uninformative prior for the correlation parameter by assuming $\beta \sim U(-\infty, +\infty)$, and the Gibbs sampler with an added step still converges to a stable set of posterior estimates for all random variables.

We are being extremely liberal, here, letting each of the parameters have their own diffuse prior. We will see in the application section that the result is indeed a Markov Chain converging to a stationary distribution, but the posterior distribution of temperature change appears extremely sensitive to the actual location of the GCMs' projections in a way that is not easily supported by the scientific understanding behind this application. The difficulty is illustrated by Figure 20.1 where we show some posterior distributions of $\nu - \mu$ based on actual climate models data, in comparison to two alternative approaches introduced later. The current approach (dashed line in Figure 20.1) often shows multimodal posterior distributions, with modes sometimes corresponding to single climate models' projections (see also the original analysis in Tebaldi *et al.* (2004, 2005), which is based on an older set of experiments, prepared

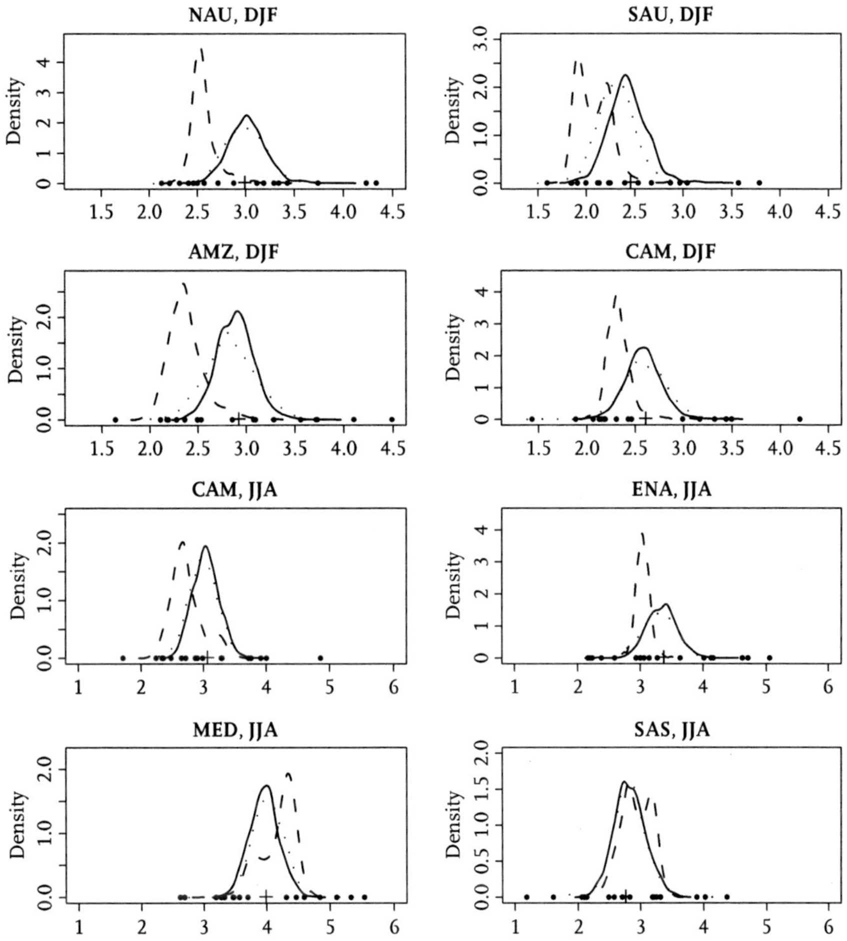

Fig. 20.1 Posterior PDFs of temperature change for a set of four regions and two seasons (December through January, DJF and June through August, JJA). Dashed line: simplest univariate model; dotted line: univariate model with common prior for λ_js; solid line: multivariate model.

for the third assessment report of the IPCC, published in 2001). However, we would expect change in temperature over a given time period to have a smooth distribution. This is the result of dealing with a limited set of climate models, that happen to populate the interval between the extremes of their range very sparsely, and of a statistical model that attributes an uneven set of weights to the participating members of the ensemble, as the λ_j parameters may actually be interpreted according to the form of $\tilde{\mu}$ and $\tilde{\nu}$.

Let us then formalize the fact that we do not expect such an uneven distribution of weights among the different models' precision parameters in this ensemble. The natural way to do that is to use a hierarchical approach, and hypothesize that all λ_js are samples from the same prior distribution with parameters that we estimate in turn, as in

$$\lambda_j \sim Ga(a_\lambda, b_\lambda)$$

with a_λ, b_λ sampled from a Gamma prior, $Ga(a^*, b^*)$ with $a^* = b^* = 0.01$.

This simple extra layer has the effect of smoothing out the estimates of λ_js and, accordingly, the shape of the posterior for $\nu - \mu$. Thus, sensitivity of the estimates to the precise location of X_j and Y_j values is significantly diminished as well, as Figure 20.1 demonstrates by comparing posterior estimates by the first model (dashed lines) to its hierarchical variation (dotted lines), for the same values of the GCMs' $Y_j - X_j$. The only operational consequence is that the joint posterior for this model is no longer estimated by a simple Gibbs sampler, but a Metropolis – Hastings step needs to handle the iterative simulation of a_λ, b_λ (see Appendix B1). This modification of the univariate approach was introduced in Smith *et al.* (2009), but there too the application uses the older data set from IPCC 2001.

20.4.2 Borrowing strength by combining projections over multiple regions

It is sensible to assume that the characteristics of each GCM, which in our development so far were represented by the parameters λ_j, θ and β, could be estimated by gathering information on the model's simulations over a set of regions, rather than just a single one. Consider then X_{ij} and Y_{ij}, which in addition to representing different models $j = 1, \ldots, M$, also represent different regions $i = 1, \ldots, R$. We also have a set of X_{i0}, the current observed mean temperature in region i which is an estimate of the true current mean temperature with standard deviation $\lambda_{0i}^{-1/2}$.

The likelihood model is a natural extension of the univariate case, where

$$X_{i0} \sim N\left[\mu_0 + \zeta_i, \lambda_{0i}^{-1}\right], \quad (\lambda_{0i} \text{ known}), \tag{20.2}$$

$$X_{ij} \sim N\left[\mu_0 + \zeta_i + \alpha_j, (\eta_{ij}\phi_i\lambda_j)^{-1}\right], \tag{20.3}$$

$$Y_{ij}|X_{ij} \sim N\left[\nu_0 + \zeta_i' + \alpha_j' + \beta_i(X_{ij} - \mu_0 - \zeta_i - \alpha_j), (\eta_{ij}\theta_i\lambda_j)^{-1}\right]. \tag{20.4}$$

We choose joint prior densities as in:

$$\mu_0, \nu_0, \beta_i, \beta_0, \zeta_i, \zeta_i' \sim U(-\infty, \infty), \tag{20.5}$$

$$\theta_i, \phi_i, \psi_0, \theta_0, c, a_\lambda, b_\lambda \sim Ga(a, b), \tag{20.6}$$

$$\lambda_j \| a_\lambda, b_\lambda \sim Ga(a_\lambda, b_\lambda) \tag{20.7}$$

$$\eta_{ij} \| c \sim Ga(c, c), \tag{20.8}$$

$$\alpha_j | \psi_0 \sim N\left[0, \psi_0^{-1}\right], \tag{20.9}$$

$$\alpha_j' | \alpha_j, \beta_0, \theta_0, \psi_0 \sim N\left[\beta_0\alpha_j, (\theta_0\psi_0)^{-1}\right], \tag{20.10}$$

all mutually independent unless explicitly indicated otherwise.

This approach is presented in Smith *et al.* (2009). As can be noticed, there are a few substantial changes from the model of Section 20.4.1: new parameters α_j and α'_j, ζ_i and ζ'_i are introduced in the mean components of the likelihood, and the variances in (20.3) and (20.4) have a more complex structure.

The parameters α_j and α'_j represent model biases. They are model-specific quantities, but they are constant across regions, thus introducing correlation between projections from the same model in different regions. By (20.10) we introduce the possibility of a correlation between α_j and α'_j, i.e. between the bias in the current period of the simulation and the future period, using the regression parameter β_0. Similarly, ζ_i and ζ'_i represent region-specific mean components. The idea is that different regions will tend to warm differently, and this region effect will be common to all GCMs. We treat these two sets of mean components in a fundamentally different manner by imposing on the α_j, α'_j Gaussian priors with mean zero, while letting the priors for ζ_i, ζ'_i be improper priors, $U(-\infty, +\infty)$. It is the difference between allowing for significantly different region effects, unconstrained by one another, and imposing a shrinkage effect across model biases according to the expectation that they should cancel each other out.

The variances in (20.3) and (20.4) are modelled as three multiplicative factors, one model-specific, one region-specific, one allowing for an interaction. In the case where $\eta_{ij} \equiv 1$ (a limiting case of (20.8) in which $c \to \infty$) the variance factorizes, with λ_j representing a model reliability and either ϕ_i or θ_i a region reliability. (We note here that two different parameters for current and future simulations of a given region have the same effect of what the model in Section 20.4.1 accomplished by using the parameter θ.) Compared with fitting a separate univariate model to each region, there are many fewer parameters to estimate, so we should get much improved precision. However there is a disadvantage to this approach: if model A has higher reliability than model B for one region, then it will for all regions (and with the same ratio of reliabilities). This is contrary to our experience with climate models, where it is often found that a model's good performance in one region is no guarantee of good performance in another. The parameter η_{ij}, then, may be thought of as an interaction parameter that allows for the relative reliabilities of different models to be different in different regions. As with the other reliability parameters, we assume a prior Gamma distribution, and there is no loss of generality in forcing that Gamma distribution to have mean 1, so we set the shape and scale parameters both equal to some number c that has to be specified. However, if c is too close to 0, then the model is in effect no different from the univariate model, allowing the variances for different models in different regions to be completely unconstrained. Our recommendation is that c should be chosen not too close to either 0 or ∞, to represent some reasonable judgment about the strength of this interaction term. A different solution will be to let c be a random variable, and let the data give us an estimate of it. We can give c a diffuse prior distribution, as with $c \sim Ga(0.01, 0.01)$.

The Gibbs sampler, with a Metropolis – Hastings steps to sample values for c, is described in detail in Appendix B.2. Together with the rest of the discussion of the multivariate approach, it was first detailed in Smith *et al.* (2009).

20.4.3 A bivariate, region-specific model

The statistical treatment described in the preceding section can be applied separately to temperature and precipitation means. The latter does not need to be logarithmically transformed. In our experience, since the data points are seasonal, regional and multidecadal means, the Gaussian likelihood model fits the data without need for transformations. Many studies of impacts of climate change are predicated on the availability of joint projections of temperature and precipitation change. A warmer, wetter future is very different from a warmer, drier one if you are dealing with adaptation of agricultural practices, or water resources management (Groves *et al.*, 2008). Here we propose a joint model for the two climate variables at the individual regional level that we first introduced in Tebaldi and Sansó (2009). Rather than simply extending the set up of Section 20.4.1, we consider decadal means covering the entire observed record and the entire simulation length. These additional data points will help estimate the correlation coefficients. We will model trends underlying these decadal mean time series, and estimate a correlation parameter between temperature and precipitation, once the trend is accounted for.

Here are the new assumptions:

- The vector of observed values O_t is a noisy version of the underlying temperature and precipitation process, with correlated Gaussian noise (we estimate the correlation from the data, through the estimation of the parameter β_{xo}).
- The true process is piecewise linear, for both temperature and precipitation. We fix the 'elbow' at year 2000, which may allow for future trends steeper than the observed ones. Of course a slightly more general model could use a random change point approach, but given the coarse resolution of our time dimension and the limited amount of data at our disposal we choose to fix the change point.
- The model output X_{jt} is a biased and noisy version of the truth. We assume an additive bias and a bivariate Gaussian noise.
- We expect the model biases to be related across the population of models, i.e. we impose a common prior, and we estimate its mean parameter, so that we may determine an overall bias for the ensemble of model simulations, different from zero.

In the notation, superscripts T and P refer to the temperature and precipitation components of the vectors. Thus, the likelihood of the data is:

$$\begin{aligned} O_t^T &\sim N\left(\mu_t^T;\eta^T\right) \quad \text{for } t = 1,\ldots,\tau_0 \\ O_t^P \mid O_t^T &\sim N\left(\mu_t^P + \beta_{xo}\left(O_t^T - \mu_t^T\right);\eta^P\right) \quad \text{for } t = 1,\ldots,\tau_0 \end{aligned} \tag{20.11}$$

where

$$\begin{aligned} \beta_{xo} &\sim N\left(\beta_0, \lambda_o\right), \\ X_{jt}^T &\sim N\left(\mu_t^T + d_j^T;\xi_j^T\right) \quad \text{for } t = 1,\ldots,\tau^* \text{ and } j = 1,\ldots,M \\ X_{jt}^P \mid X_{jt}^T &\sim N\left(\mu_t^P + \beta_{xj}\left(X_{jt}^T - \mu_t^T - d_j^T\right) + d_j^P;\xi_j^P\right) \\ &\qquad\qquad \text{for } t = 1,\ldots,\tau^* \text{ and } j = 1,\ldots,M. \end{aligned}$$

In equations (20.11) we specify bivariate normal distributions for O_t and X_{jt} using conditionality. After accounting for the underlying trends and biases terms, $\beta_{x1},\ldots,\beta_{xM}$ are used to model the correlation between temperature and precipitation in the climate model simulations, while β_{xo} is fixed at the value estimated through the observed record. Also in the likelihood of the observations, η^T and η^P are fixed to their empirical estimates.

The time evolution of the *true climate process* $\mu_t' = (\mu_t^T, \mu_t^P)$, consists of a piecewise linear trend in both components:

$$\begin{pmatrix} \mu_t^T \\ \mu_t^P \end{pmatrix} \equiv \begin{pmatrix} \alpha^T + \beta^T t + \gamma^T (t - \tau_0)\mathcal{I}_{\{t \geq \tau_0\}} \\ \alpha^P + \beta^P t + \gamma^P (t - \tau_0)\mathcal{I}_{\{t \geq \tau_0\}} \end{pmatrix}. \tag{20.12}$$

The priors for the parameters in model (20.11) are specified hierarchically by assuming that

$$\begin{aligned} \beta_{xj} &\sim N\left(\beta_0, \lambda_B\right), \\ d_j^T &\sim N\left(a^T;\lambda_D^T\right), \\ d_j^P &\sim N\left(a^P;\lambda_D^P\right) \end{aligned}$$

for $j = 1,\ldots,M$

$$\xi_j^T \sim Ga\left(a_{\xi^T}, b_{\xi^T}\right)$$

and

$$\xi_j^P \sim Ga\left(a_{\xi^P}, b_{\xi^P}\right).$$

λ_o is fixed to a value estimated on the basis of the observed record.

All the other quantities are assigned uninformative priors:

$$\beta_0, a^T, a^P \sim U(-\infty, +\infty)$$

and

$$\lambda_B, \lambda_D^T, \lambda_D^P, a_{\xi^T}, b_{\xi^T}, a_{\xi^P}, b_{\xi^P} \sim Ga(g, h),$$

where $g = h = 0.01$. Similarly, for the parameters in (20.12), we assume

$$a^T, \beta^T, \gamma^T, a^P, \beta^P, \gamma^P \sim U(-\infty, +\infty).$$

We are assuming that each climate model has its own precision in simulating the true temperature and precipitation time series, but we impose common priors to ξ_j^T and ξ_j^P $\forall j$, whose parameters are in turn estimated by the data. As we discussed in Section 20.4.1, this choice produces more robust estimates of the relative precisions of the different GCMs, not overly sensitive to small perturbations in the GCM trajectories.

The model-specific bias terms d_j^T, d_j^P are assumed constant over the length of the simulation. They model systematic errors in each GCM simulated variable. All the GCM biases for temperature, like all GCM biases for precipitation, are realization from a common Gaussian distribution, whose mean (a^T or a^P), as mentioned, may be different from zero, when the set of model trajectories is distributed around the truth non-symmetrically. We do not expect a systematic behaviour across models when it comes to precipitation versus temperature biases, that is, we do not expect that models having relatively larger temperature biases would show relatively larger precipitation biases, so we do not model a correlation structure between d_j^T, d_j^P. In fact, this correlation structure, if there at all, would not to be identifiable/separable from the correlation modeled through $\beta_{xo}, \beta_{x1}, \ldots, \beta_{xM}$, given the configuration of the present dataset. Notice that the correlation coefficients, β_{xo} and β_{xj} also have a common mean, β_0 possibly different from zero and that will be heavily influenced by the value of the observed correlation coefficient, β_{xo}. All the remaining parameters of the model have non-informative, conjugate distributions. Notice that we use improper priors for the location parameters of the Gaussian distributions and linear regression parameters in the correlation structure and in the trend structure, and proper but diffuse priors for the precision parameters and as hyper-priors of the $\xi_.$ parameters. The likelihood and priors form a conjugate model, and as before a Gibbs sampler can be programmed to explore the posterior distributions for this model, with a Metropolis – Hastings step used to generate sample values for $a_{\xi^T}, b_{\xi^T}, a_{\xi^P}, b_{\xi^P}$. See Appendix B.3 details of the Markov chain Monte Carlo implementation, or the original article Tebaldi and Sansó (2009).

20.5 Validating the statistical models

Climate predictions are a fairly safe business to be in: validation of the forecast comes about no sooner than every 10 or 20 years! We have already commented on how this situation sets climate forecasting well apart from weather forecasting, where statistical calibration of numerical models can be fine tuned continuously, using diagnostics of their performance coming at the steady and frequent rate of every six hours or so. The work by Adrian Raftery, Tilmann Gneiting and colleagues (Raftery *et al.* (2005) and references therein) has set the standard for Bayesian statistical methods brought to bear on ensemble weather forecast calibration and validation. In our application, we are also forecasting conditionally on a specific scenario of greenhouse gases emissions, which is always an idealized scenario. This discussion shows why we are not going to rely on observations to validate our statistical approaches. Rather, we validate our statistical assumptions by performing cross-validation. In all cases but for the simplest univariate model introduced first, we can estimate a posterior predictive distribution for a new model's trajectory or change. We can therefore compare the left-out model projections with their posterior-predictive distribution. In both cases we can do this exercise over all possible models, regions and – separately or jointly – both climate variables, temperature and precipitation averages. Statistical theory provides us with null hypotheses to test about the expected distribution of the values of $P(x*)$ where $x*$ is the left-out value and $P()$ is the posterior predictive cumulative distribution function of that quantity. We expect the values of $P(x*)$ obtained by performing the cross-validation exercise across models and across regions to be a sample from a uniform distribution. In all cases, after conducting the tests, we do not reject the hypothesis more frequently that what would be expected as a result of multiple testing. In the remainder of this section we detail the procedure of the cross-validation exercise, for each of the statistical models proposed. More details and actual results from the cross validation exercises can be found in Smith *et al.* (2009) and Tebaldi and Sansó (2009).

20.5.1 Univariate model with hyperprior on precision parameters

The predictive distribution can be calculated under the assumption that the climate models are exchangeable. Conditionally on the hyperparameters μ, ν, β, θ, a_λ and b_λ, the distribution of $Y_{M+1} - X_{M+1}$ can be derived derived from conditioning on $\lambda_{M+1} \sim Ga(a_\lambda, b_\lambda)$, since, then, $Y_{M+1} - X_{M+1}|\lambda_{M+1} \sim N(\nu - \mu, \{(\beta - 1)^2 + \theta^{-1}\}\lambda_{M+1}^{-1})$.

The conditional distribution should then be convolved with the joint posterior distribution of $(\mu, \nu, \beta, \theta, a_\lambda, b_\lambda)$, to obtain the full posterior predictive distribution. In practice we can carry out this integration within the Gibbs – Metropolis algorithm by

1. Sampling at each step n the hyperparameter values $a_\lambda^{(n)}$, $b_\lambda^{(n)}$, $\nu^{(n)}$, $\mu^{(n)}$, $\beta^{(n)}$, $\theta^{(n)}$, corresponding to one draw from their joint posterior distribution.

A draw of a random $\lambda_{j,n} \sim Ga(a_{\lambda}^{(n)}, b_{\lambda}^{(n)})$ can be generated and the statistic

$$U_j^{(n)} = \Phi\left\{\frac{Y_j - X_j - \nu^{(n)} + \mu^{(n)}}{\sqrt{\left\{\left(\beta_x^{(n)} - 1\right)^2 + \theta^{(n)-1}\right\}(\lambda_{j,n})^{-1}}}\right\}$$

calculated.

2. Over all n iterations we can compute U_j, the mean value of $U_j^{(n)}$, representing an estimate of the predictive distribution function evaluated at the true $Y_j - X_j$. If the statistical model is consistent with the data, U_j should have a uniform distribution on (0, 1).
3. By computing U_j for each region, and each GCM we have a set of test statistics that we can evaluate for discrepancies, applying tests of fit to evaluate the hypothesis that the values are samples from a uniform distribution.

20.5.2 Multivariate model, treating multiple regions at once

The procedure for cross-validation in this model is very similar to what was just described. In this setting we do cross-validation of the variable $Y_{ij} - X_{ij}$, model j's projected change in region i. The full conditional, predictive distribution of this quantity is

$$Y_{ij} - X_{ij}|\text{rest} \sim N\left(\nu_0 - \mu_0 + \zeta_i' - \zeta_i + \alpha_j' - \alpha_j, \frac{1}{\eta_{ij}\lambda_j}\left\{\frac{(\beta_i - 1)^2}{\phi_i} + \frac{1}{\theta_i}\right\}\right).$$

The implementation of the cross-validation exercise within the Gibbs – Metropolis algorithm is as follows:

1. We leave model j out and we run the Gibbs – Metropolis simulation.
2. After burn-in, at every step n, on the basis of the set of parameters currently sampled we generate corresponding values of $\lambda_j^{(n)}$, $\alpha_j^{(n)}$, $\alpha_j'^{(n)}$ and $\eta_{ij}^{(n)}$ as

$$\lambda_j^{(n)} \sim Ga\left(a_{\lambda}^{(n)}, b_{\lambda}^{(n)}\right)$$

$$\alpha_j^{(n)} \sim N\left(0, \frac{1}{\psi_0^{(n)}}\right)$$

$$\alpha_j'^{(n)} \sim N\left(\beta_0^{(n)}\alpha_j^{(n)}, \frac{1}{\psi_0^{(n)}\theta_0^{(n)}}\right),$$

$$\eta_{ij}^{(n)} \sim Ga\left(c^{(n)}, c^{(n)}\right).$$

3. From these values we compute the statistic

$$U_{ij} = \frac{1}{N}\sum_{n=1}^{N}\Phi\left[\frac{Y_{ij} - X_{ij} - \left(\nu_0^{(n)} - \mu_0^{(n)}\right) - \left(\zeta_i'^{(n)} - \zeta_i^{(n)}\right) - \left(\alpha_j'^{(n)} - \alpha_j^{(n)}\right)}{\sqrt{\left(\lambda_j^{(n)}\eta_{ij}^{(n)}\right)^{-1}\left\{\left(\phi_i^{(n)}\right)^{-1}\left(\beta_i^{(n)} - 1\right)^2 + \left(\theta_i^{(n)}\right)^{-1}\right\}}}\right]. \tag{20.13}$$

by generating a sample value at each iteration and averaging them at the completion of the simulation.
4. As with the univariate analysis, we then perform various goodness of fit tests on the statistics U_{ij}. If the model is a good fit, they should be consistent with independent draws from the uniform distribution on [0, 1].

20.5.3 Bivariate model of joint temperature and precipitation projections

After leaving each individual GCM's trajectories out in turn, we simplify the validation of the bivariate time series model by computing three marginal bivariate predictive distributions, one for current climate (defined as the bivariate distribution of average values of temperature and precipitation for the period 1981–2000), one for future climate (defined as the corresponding distribution for average values over the period 2081–2100) and one for climate change (defined as the joint distribution of the temperature and precipitation *differences* between the two same periods). We can then compute the two sets of pairs ($U_1 = P_T(X_*^T = x_*^T)$, $U_2 = P_{P|T}(X_*^P = x_*^P)$) for both current and future time windows, and the pair ($U_1 = P_T(\Delta X_*^T = \Delta x_*^T)$, $U_2 = P_{P|T}(\Delta X_*^P = \Delta x_*^P)$) where the first univariate distribution function is simply the marginal predictive distribution of temperature change, while the second distribution function is the predictive distribution of precipitation change, conditional on the corresponding simulated temperature change. Finally we test the null hypothesis that the pairs of (z_{1j}, z_{2j}) for $j = 1, \ldots, M$ are independent and identically distributed random variates, sampled from uniform distributions on the (0, 1) interval. The form of the posterior predictive and the way we simulate the U statistics is similar to what we described in the previous subsections, with all the parameters being sampled from the estimated value of the hyperparameters within the Gibbs sampler, and the conditional marginal predictive distributions having a manageable Gaussian form.

20.5.4 *U*-statistics and goodness-of-fit tests

Following Smith *et al.* (2009), in all cases described in Sections 20.5.1 through 20.5.3, the sets of U_{ij} (or Z_j) can be tested for goodness of fit with respect to a uniform distribution and independence by traditional tests like

Kolmogorov – Smirnov, Cramér – von Mises and Anderson – Darling. Smith *et al.* (2009) also proposes the use of a correlation test where the ordered values of U_{ij}, $j = 1, \ldots, M$ are correlated to the sequence $1/(M+1), 2/(M+1), \ldots, M/(M+1)$ and large values of the correlation coefficient (or small values of $1 - cor$) indicate a close fit. In most cases the goodness-of-fit tests result in acceptance of the null hypothesis that the U_{ij} or Z_j are independent within each region i.

20.6 Application: The latest model projections, and their synthesis through our Bayesian statistical models

The website of PCMDI provides instructions for the download of model output from all the GCMs that have contributed their simulations to the IPCC-AR4 effort. We choose a set of experiments run under a scenario of greenhouse gas emissions that can be thought of as a 'business-as-usual' scenario, where concentrations of greenhouse gases increase over this century at a rate similar to what is being emitted currently, worldwide. We extract average temperature and precipitation projections from seventeen models. We area-average their output over standard subcontinental regions that have been used by IPCC and many studies in the literature (Giorgi and Francisco, 2000), see Figure 20.2. We consider two seasonal averages, December through February (DJF) and June

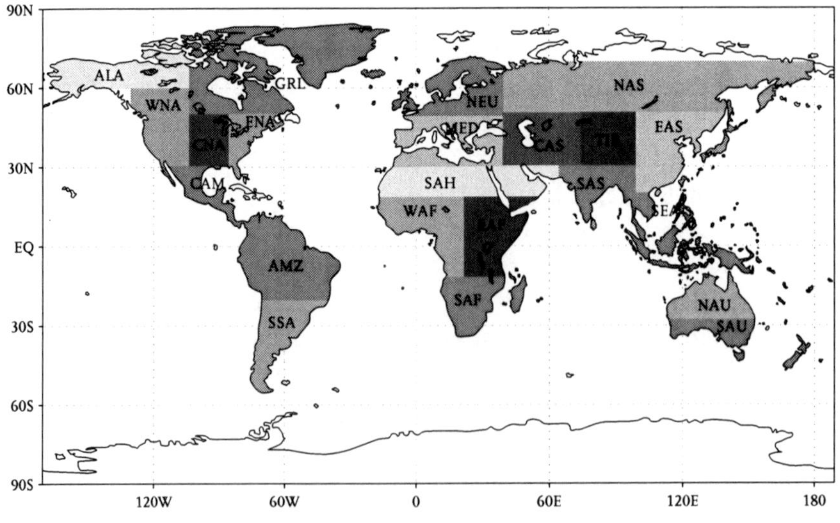

Fig. 20.2 The 22 Giorgi regions. Output of temperature and precipitation for each GCM and observations is area-averaged over each of these large regions, for a given season and (multi-) decadal periods.

through August (JJA). For the models treating temperature or precipitation separately we also average 20-year periods, 1980–1999 as the current climate averages and 2080–2099 as the future. For the model that estimates the joint distribution of temperature and precipitation we consider time series of 15 decadal averages covering the period 1950–2100. We compute the joint posterior distribution of all random parameters according to the univariate (one region at a time, with or without hyperprior over the precision parameters), multivariate (all regions at the same time) and bivariate (temperature and precipitation jointly) models. We then proceed to display and compare results for some quantities of interest. Most importantly, we consider changes in temperature (and precipitation) for all the regions, across the different statistical treatment. We also show posterior distributions of other parameters of interest, like model biases and interaction effects.

20.6.1 Changes in temperature, univariate and multivariate model

The eight panels of Figure 20.1 compare the posterior distribution of temperature change $\nu - \mu$ for the two univariate models (dashed and dotted curves) and $\nu_0 + \zeta_i' - \mu_0 - \zeta_i$ for the multivariate model (solid line) for a group of four regions in DJF and a different group of four regions in JJA. We have already mentioned in Section 20.4.1 that the first model, where the GCM-specific precision parameters are each a sample from a diffuse Gamma prior, suffers from an uneven distribution of 'weight' among GCMs' projections. The multimodality of some of the PDFs in Figure 20.1 is an indication of this behaviour. And even more generally, the fact that these PDFs are often shifted significantly from the location of the ensemble mean (indicated by a cross along the x-axis, whereas the individual GCMs are marked by dots along the basis of the curves) is the effect of some of the GCMs 'stealing the show', i.e. being attributed a much larger precision than others, where we would not expect such difference in relative importance among this family of state-of-the-art climate models. The univariate model with a common prior over the λ_js produces smoother, better centered PDFs that are not significantly different from those produced by the multivariate model (comparing the dashed and solid lines). The two series of boxplots (one for DJF, to the left and one for JJA, to the right) in Figure 20.3 offer a comparison of the three distributions for each region/season combination. For each region, the first boxplot from the bottom shows interquartile range, median and 5th–95th quantiles of the posterior distribution estimated by the unconstrained version of the univariate model, the second and third boxplot show the extent of the posterior for the univariate model with common prior for the λ_js and the multivariate model. These two sets of boxplots confirm the results of Figure 20.1, with the position of the first boxplot in each triplet often shifted away from the other two, which are more similar to one another.

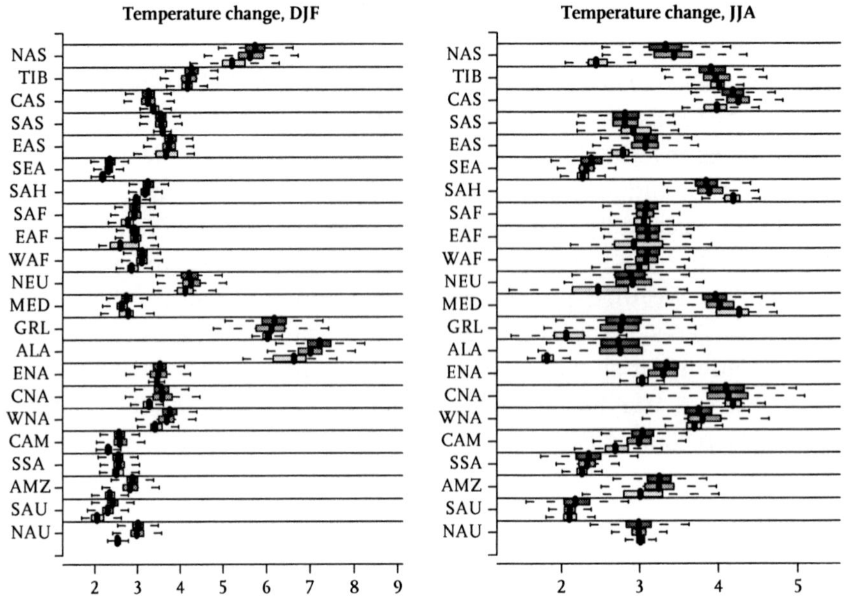

Fig. 20.3 Posterior distributions of temperature change for the 22 regions of Figure 20.2 in DJF (left panel) and JJA (right panel). For each region we show three boxplots corresponding to the posterior distributions derived from the three statistical models represented also in Figure 20.1. The labelling along the vertical axis and the horizontal lines in the plot region identify each region by the acronyms of Figure 20.2. For each region, the lower boxplot, lightest in colour, corresponds to the simplest univariate model. The middle boxplot corresponds to the univariate model with hyperprior on the reliability parameters. The third boxplot, darkest in colour corresponds to the multivariate model.

This display of all the 22 region is also indicative of the large differences in the amount of warming across regions. This justifies the choice of modeling region-specific parameters in the mean component of the likelihood of the multivariate model, ζ_i, ζ_i', with a Uniform prior over the real line that does not have a shrinkage effect towards zero.

20.6.2 Changes in temperature, multivariate and bivariate model

We now take the same eight region/season combinations of Figure 20.1 and compare, in Figure 20.4, the posterior distribution of temperature change for the multivariate model, $\nu_0 + \zeta_i' - \mu_0 - \zeta_i$, (solid line) to the marginal posterior distribution of the temperature change signal from the bivariate model (dotted line). We also show the posterior predictive distribution of a new unbiased GCM's projection of temperature change (dashed line) in the same panels. We note immediately that the posterior distribution from the bivariate model is much narrower than the posterior from the multivariate model. This is no surprise, given that we are using a much richer dataset (six observed decades, 15 simulated decades for each of seventeen GCMs) and we are hypothesizing

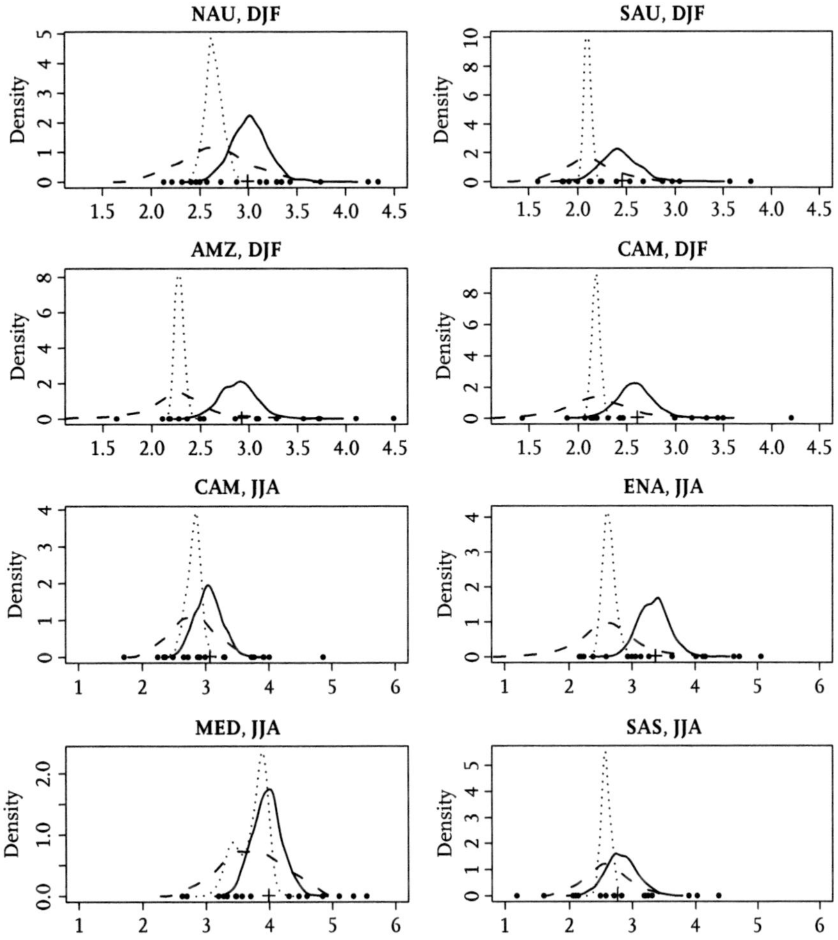

Fig. 20.4 Similar to Figure 20.1, but now the three PDFs are the posterior from the multivariate model (solid line), the posterior from the bivariate model (dotted line) and the posterior predictive from the bivariate model (dashed line). The curves drawn as solid lines are the same as in Figure 20.1.

a piecewise linear function of time as the trajectory of mean temperature, thus we are in fact estimating only three parameters α^T, β^T and γ^T. Note that the joint posterior distribution from the bivariate model is a probabilistic representation of the entire trajectory of temperature (ad precipitation) over the decades (μ_t^T) but the availability of the Gibbs sample makes it straightforward to compute any deterministic function of it (in this case the difference between two bidecadal means). The other obvious feature in the majority of these panels is that the posterior distributions from the bivariate model are significantly shifted from the centre of the multivariate model's PDFs. Remember that the bivariate model is not simply 'smoothing' sets of GCM 'snapshots' in terms of multidecadal averages, but rather fitting a trend to their whole trajectories, anchoring its estimation to the observed series. We are thus comparing two different definitions of 'temperature change', at least in terms of how the statistical model is estimating it. Likely, some of the models produce steeper trends than the underlying trend estimated from the entire ensemble, thus filling the distribution of the 'snapshots' estimate to the right of the trend estimates, and shifting the mean accordingly. This feature may be scenario dependent, with lighter emission scenarios (forcing less of an acceleration in the trends) facilitating a better agreement between the two methods' results. The dashed curves in each figure are posterior predictive distribution of a new GCM's projection. The signal of temperature (change) underlying truth and model simulation is an abstract concept. Even the observations are a noisy representation of this signal. We could be justified then if we thought of a model trajectory as a possible future path for our climate, and accordingly we represented the uncertainty in this future projection by the posterior predictive distribution of a new GCM, whose width is of the same order of magnitude as the range of model projections, rather than being an inverse function of the square root of the number of data points, as the posterior distribution width is. In Figure 20.5 we complete the representation of probabilistic projections of climate change by showing contours of the posterior (tighter set of contours in each panel) and posterior predictive (wider set of contours in each panel) for the same set of region/season pairs of Figures 20.1 and 20.4.

20.6.3 Other uncertain quantities of interest, their full probabilistic characterization

Even the simple univariate treatment, after imposing the regularizing common prior over the λ_js, offers additional information besides the posterior PDFs of temperature change that may be used in applications of impact analysis. Posterior means of λ_js can be regarded as optimal estimates of model reliabilities, and utilized as 'weights' for some further analysis of climate change impacts. For example, Fowler *et al.* (2007) shows how the normalized set of reliabilities can be used for a Monte Carlo simulation of weather scenarios at regional scales by stochastic weather generator models 'driven' by different GCMs. The larger the GCM reliability, the larger the number of weather scenarios one wants to generate according to that GCM, and the total number of Monte Carlo simulations from the weather generator may be divided up proportionally to the normalized weights. In Figure 20.6 we show boxplots of the 16 GCM-specific λ_j's for the eight region/season combinations of Figures 20.1 and 20.4, according to the univariate treatment, displaying them in each panel in increasing order from left to right according to their posterior means. From the figure we note how the distribution of 'weight' among GCMs is very balanced, with no disproportionate attribution to a small subset of models. It is true however

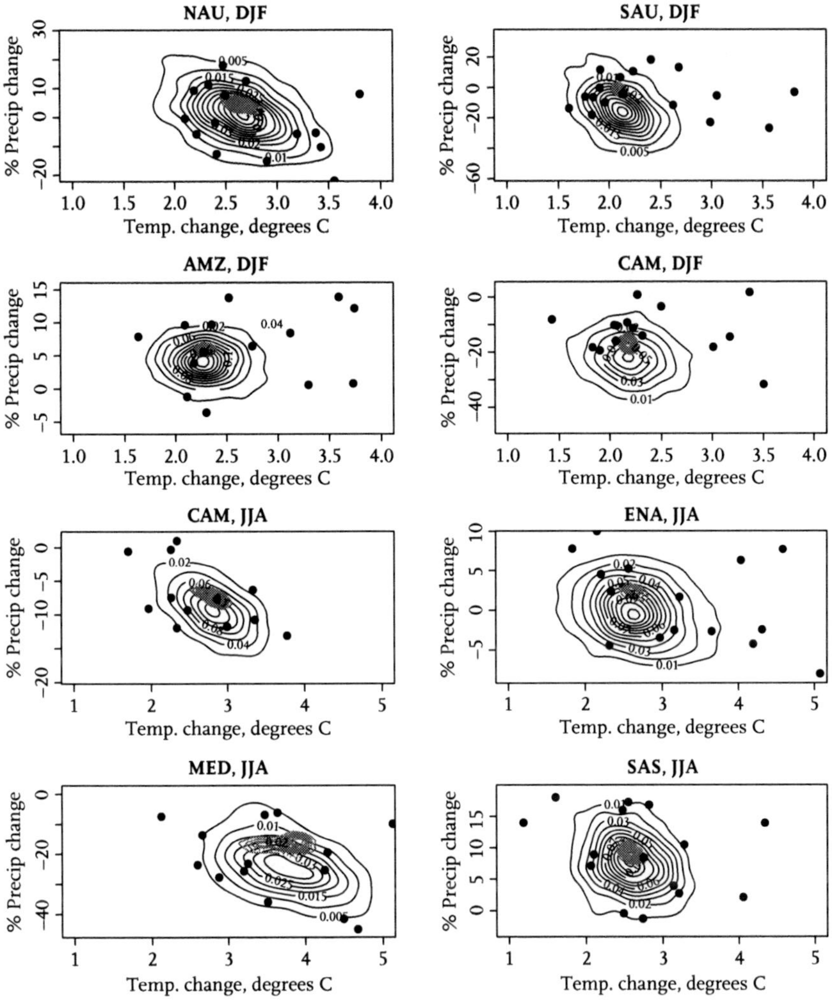

Fig. 20.5 Contours of joint changes in temperature and precipitation (the latter as percentage of current average precipitation) for the same set of regions/seasons as in Figures 20.1 and 20.4. The tight contours correspond to the posterior PDF from the bivariate model. The wider contors represent the posterior predictive PDFs for the projections of a new GCM, unbiased.

that different models gain different ranking depending on the region and season combination. This is consistent with the shared opinion among climate scientists and modelers that no model outperforms all others in every respect, different models showing different strengths in the simulation of regional climates, seasonal processes and so on.

We may be interested in evaluating the posterior estimates of the model-specific mean factors in the multivariate treatment, α_j and α'_j, as they may be interpreted as global average biases that a specific GCM imposes on its simulation of temperature. Figure 20.7 shows boxplots of the two sets of parameters

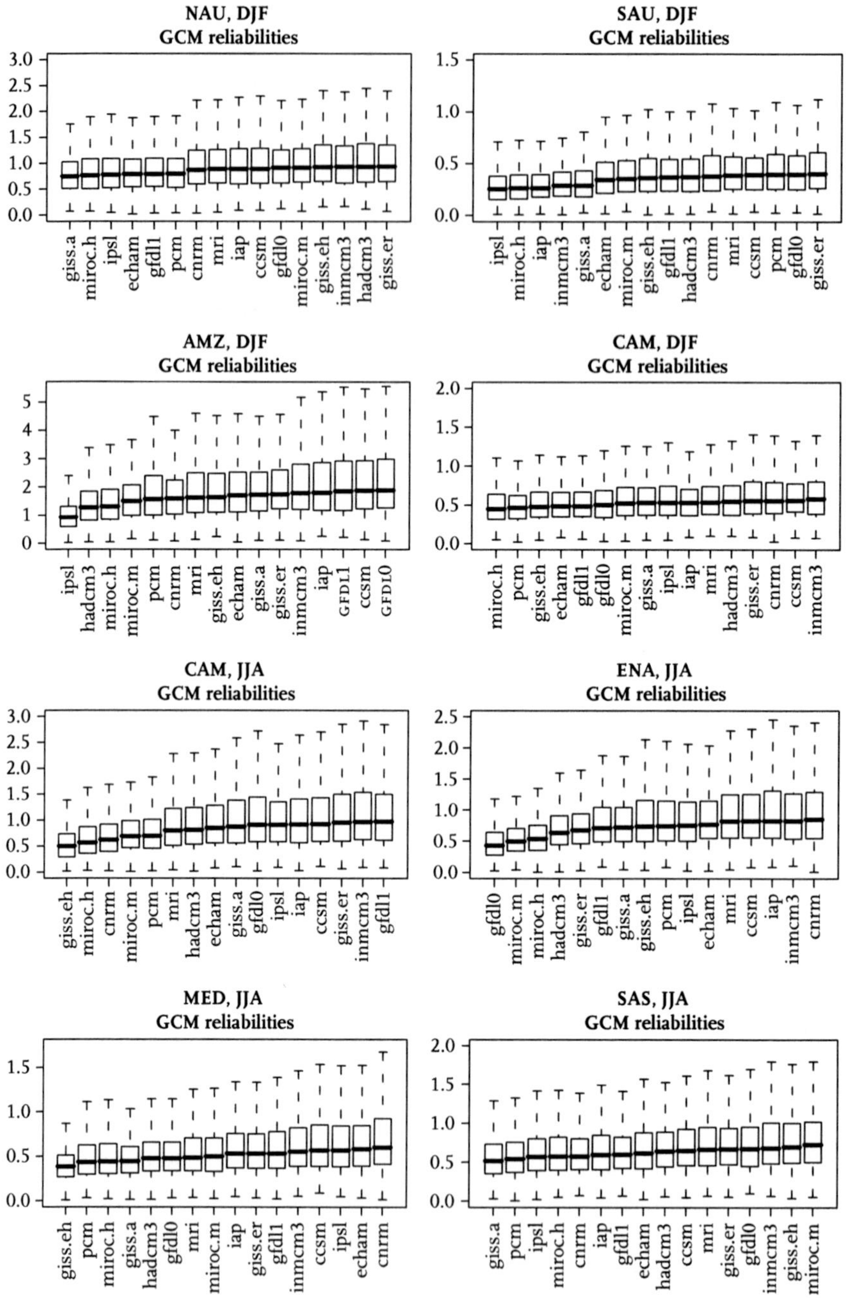

Fig. 20.6 Posterior PDFs for the model reliability parameters λ_js derived from the univariate model with common prior on them. The region/season combinations are the same as in Figures 20.1 and 20.4. Each boxplot in a given panel corresponds to a GCM. Relatively larger values of λ_j are interpretable as better model reliability in the given region/season.

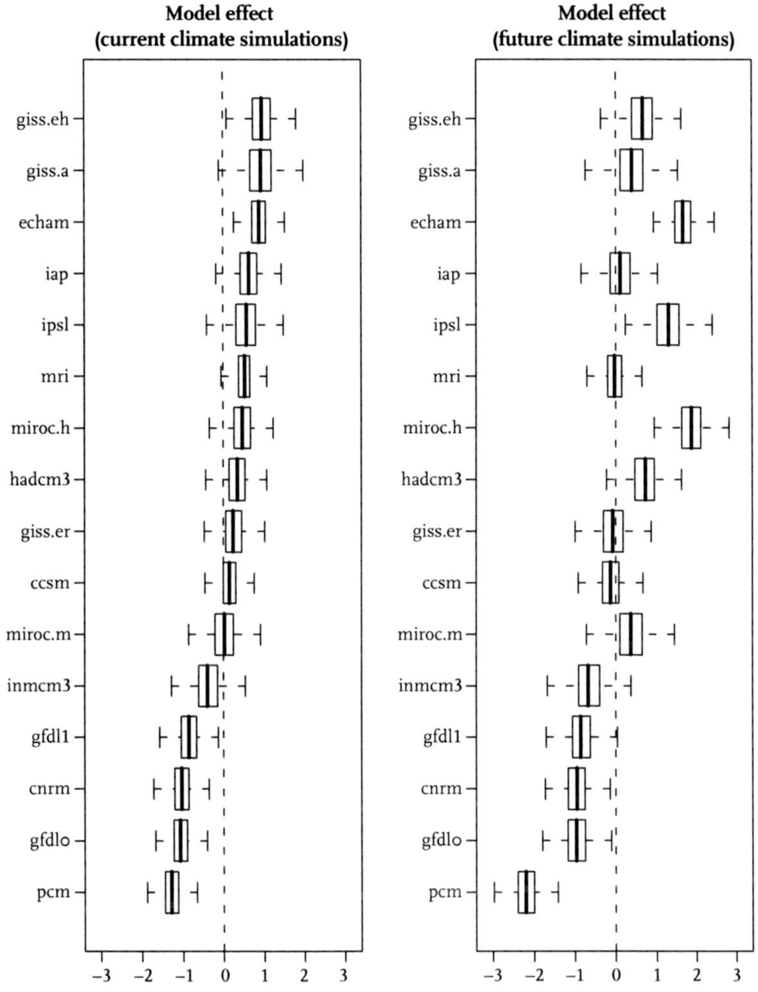

Fig. 20.7 Posterior distributions of model-specific mean effects in the multivariate model. The left panel shows boxplots for the α_js, interpretable as global average biases in each model's current climate simulation. The corresponding distributions of the α'_js, interpretable as future simulation biases, are shown in the panel on the right.

for the simulation of temperature in DJF, as an example. The models have been ordered with respect to the value of the posterior mean of the respective parameter in the current climate simulation (α_j). As can be assessed by the two series of boxplots, there are models with biases that are significantly different from zero, mostly with negative (cold) biases in the left plot. Many more GCMs show significant biases in the future part of the simulation, on both sides of the zero line.

Another interesting question that can be answered through the computation of the marginal posterior distribution of a set of parameters in the multivariate

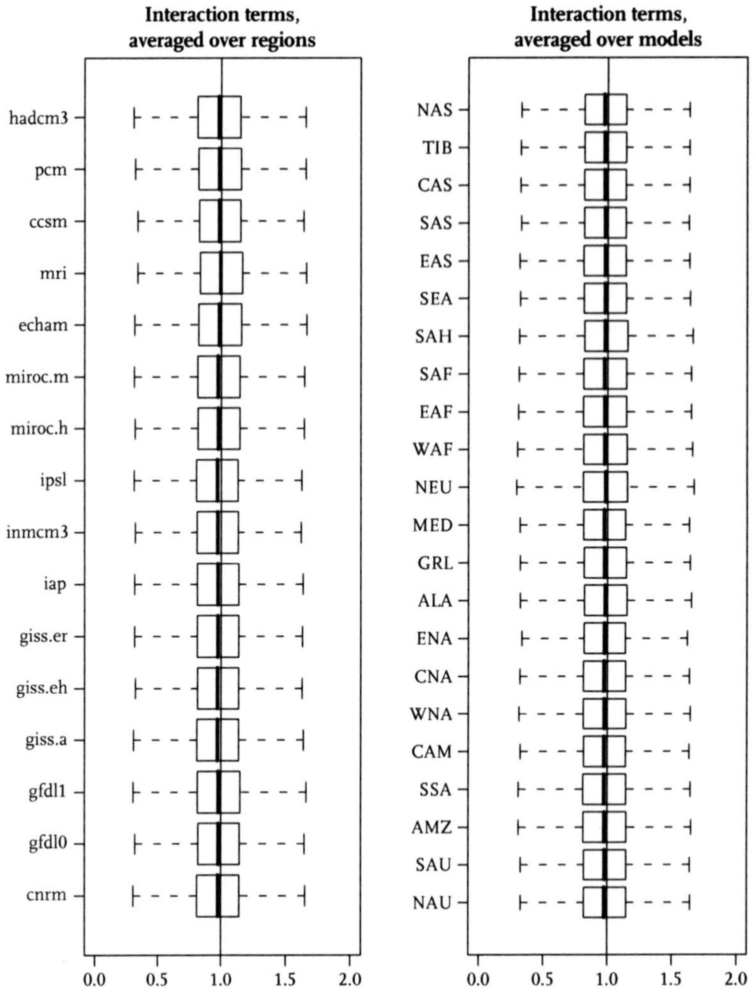

Fig. 20.8 Interaction effect parameters (η_{ij}) of the precisions in the likelihood of the multivariate model. Left panel shows their posterior distributions integrated over regions ($\eta_{.j}$). Right panel shows the distributions integrated over models ($\eta_{i.}$).

model regards the form of the precision (or variance) components in the likelihood of the GCM simulated temperatures. The parameters η_{ij}s are let free to assume values significantly different from one, if an interaction between model-specific and region-specific behaviour is deemed necessary to fit the dataset at hand. As Figure 20.8 demonstrates, however, by showing the distributions of $\eta_{i.}$ and $\eta_{.j}$, i.e. by combining all sample values in the Gibbs simulation across models and across regions, the distribution of these parameters is always centered around the value one, lending support to a model that factorizes the variance component into two terms without interaction (boxplots of all 22*16

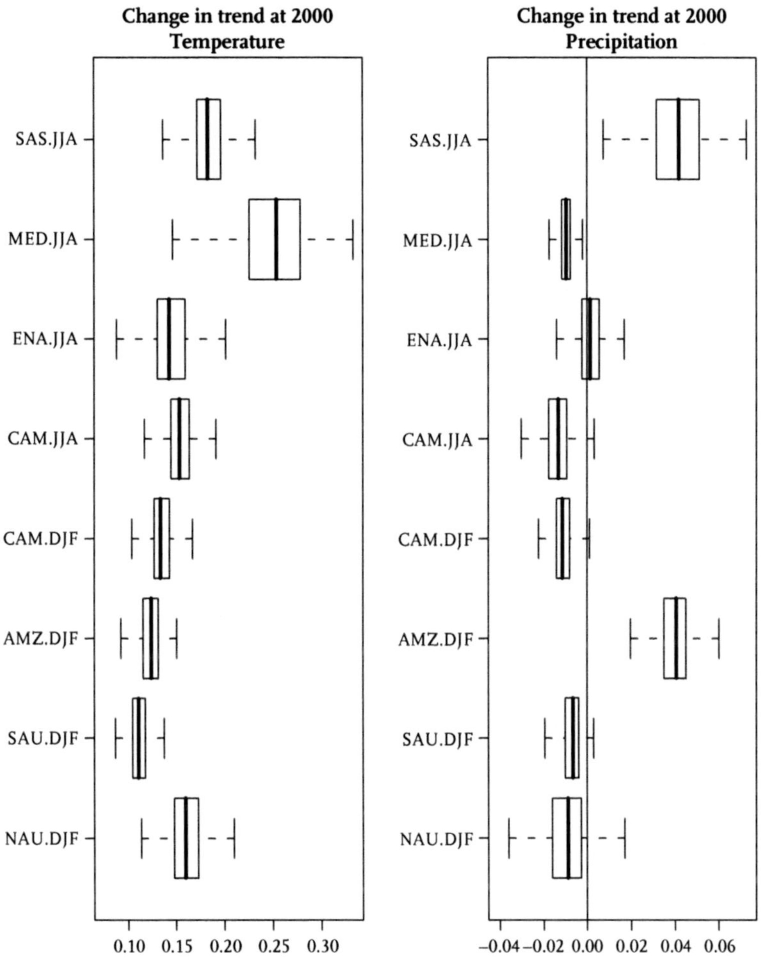

Fig. 20.9 Incremental trend parameters γ^T (temperature time series) and γ^P (precipitation time series) in the bivariate model. When significantly different from zero they estimate a change in the slope of the linear trend after 2000.

region-and-model-specific parameters do not reveal any behavior inconsistent with this averages).

Last, we show posterior PDFs of the γ^T, γ^P trend parameters in the bivariate model. Recall that these parameters are introduced in order to estimate a possible change in the trend underlying the observed and modelled time series, after the year 2000. As can be seen from Figure 20.9 for the usual region/season combinations (listed along the vertical axis), all the parameters of the temperature trends are significantly different from zero (the zero line does not even appear in the plot). The trend parameters estimated for precipitation instead show a whole range of behaviors, with some of the distributions straddling the zero line, some indicating a significant decrease in the rate of change (with the distributions lying for the most part to the left of the zero line), some, at the contrary, estimating an increase in the trend.

20.7 Further discussion

We have presented a series of statistical models aiming at combining multiple climate model simulations. The goal is to arrive at a rigorous characterization of the uncertainty in model projections, quantifying it into probability distributions of climate change. More precisely, we determine PDFs of average temperature and precipitation change at seasonal, multidecadal and regional scales, that can be used for further development as input to impact models in vulnerability and adaptation studies.

The challenge of synthesizing multiple model projections stems from the idiosyncratic nature of the data sample. GCMs are arguably not independent, they do not span the entire range of uncertainty, being fairly conservative guesses of the future trajectory of our climate system, and they may have systematic errors (biases) in their simulation of climate, that do not necessarily cancel out in an overall mean.

Given access to the recently organized, rich archives of GCM data, we are however able to start making sense of these ensembles of simulations. In this chapter we have shown how increasingly complex statistical models can account for features of the data like correlations between current and future simulation errors, biases, differential skill in different periods of the simulation or different regions of the globe by a given model. We have also shown how the difficulty of validating climate predictions may be side-stepped thanks to the calculation of posterior predictive distributions that open the way to cross-validation exercises.

We have dedicated some time to discuss two ways of characterizing the uncertainty of these climate change experiments. One approach cares especially about the common central tendency of these models, and the uncertainty around this estimated central tendency. Thanks to the number of models available and the observed record that 'anchors' our estimates of the true signal of climate, the central tendency is very precisely estimated, within a narrow band of uncertainty. The reality of climate change, however, may look more like one of these model trajectories than like this abstract climate signal. If we subscribe to this view, then a more consistent representation of our uncertain future is naturally found in the posterior predictive distribution, after integrating out all the uncertain parameters of the Bayesian model. The posterior predictive width is of the same order of magnitude as the range of model projections, much wider than the posterior estimate of the climate (change) signal's uncertainty and may look to expert eyes as a more realistic representation of the uncertainty at hands.

There are many ways by which the kind of analyses presented here may be carried forward. Two in particular seem germane to our approach. Furrer *et al.* (2007) chose to model the entire high resolution fields of variables that constitute the original output of these models, after registering them to a common grid. This approach has the advantage of providing a representation of projections that is very familiar to the climate science community, in the form of detailed maps of change, with the added dimension of a probabilistic characterization of their likely ranges. Rougier (2007) uses a set of experiments aimed at perturbing the parametrization of a specific climate model, the Hadley Center GCM. The work therefore represents a natural complement to ours, where intra-model uncertainties are fully characterized. Ideally, it could be incorporated in our framework, in order to treat within model and between model variability in concert.

We think of two main avenues for further development of multimodel analysis. The first is the representation of models' dependencies, which would protect us from the overoptimistic narrowing of the uncertainty with the increasing number of members in these ensembles. It seems clear that not all models add an independent piece of information to the ensemble, while the treatment of each one as independent of the other makes the estimate of the central tendency increasingly more precise with the increasing number of GCMs.

A second direction should lead to the exploration of models' performance within current climate simulations and how it correlates to models' reliability in the future part of their simulation. In our treatment, a coarse metric of performance was implicitly found in the difference between models' average projections in the part of the simulation that overlaps with observed records and the corresponding observed climate average. There are potentially many alternative metrics of performance that could be relevant for a model's ability to simulate future climate reliably, but the question is open, as is that of how to incorporate this information into our statistical models.

Appendix

A. Broader context and background

All of Bayesian statistics derives ultimately from the formula

$$\pi(\theta|x) = \frac{\pi(\theta) f(x|\theta)}{\int_{\Theta} \pi(\theta') f(x|\theta') d\theta'}. \qquad (20.14)$$

Here:

- x is an observation vector, lying in some sample space $\mathcal{X}$;
- θ is a vector of unknown parameters, lying in a parameter space Θ;

- $\pi(\theta)$ represents the prior distribution. If Θ is discrete, this is a probability mass function; if Θ is continuous, this is a probability density function;
- $f(x|\theta)$ is the likelihood function; that is, the conditional distribution of x given θ. This is a probability mass function if $\mathcal{X}$ is discrete and a probability density function if $\mathcal{X}$ is continuous.

In the following discussion, we first review simple applications of (20.14) using *conjugate priors*; then we describe extensions to *hierarchical models*; finally we describe computational methods via the *Gibbs sampler* and the *Hastings – Metropolis sampler*, which the two best known examples of Markov chain Monte Carlo (MCMC) algorithms. We illustrate these concepts with reference to the models for climate change projections that form the bulk of this chapter.

A.1 Conjugate priors

The simplest examples of (20.14) have the following structure: both $\pi(\theta)$ and $f(x|\theta)$ have a parametric structure, which is such that $\pi(\theta|x)$ has the same parametric structure as that of $\pi(\theta)$. In that case, the prior is known as a *conjugate prior*.

A simple example is the Beta prior for a binomial distribution. In this case, $\mathcal{X} = \{0, 1, 2, \ldots, n\}$ for known n, $\Theta = [0, 1]$, and $f(x|\theta) = \frac{n!}{x!(n-x)!}\theta^x(1-\theta)^{n-x}$ (the binomial distribution with n trials and probability of success θ). Consider the prior density

$$\pi(\theta) = \frac{\Gamma(a+b)}{\Gamma(a)\Gamma(b)}\theta^{a-1}(1-\theta)^{b-1}, \quad \theta \in [0, 1], \tag{20.15}$$

the Beta distribution with constants $a > 0,\ b > 0$. Then, after cancelling common terms in the numerator and denominator, (20.14) becomes

$$\pi(\theta|x) = \frac{\theta^{a+x-1}(1-\theta)^{b+n-x-1}}{\int_0^1 (\theta')^{a+x-1}(1-\theta')^{b+n-x-1}d\theta'}. \tag{20.16}$$

But the fact that (20.15) is a proper probability density for all $a > 0,\ b > 0$ implies

$$\int_0^1 (\theta')^{a+x-1}(1-\theta')^{b+n-x-1}d\theta' = \frac{\Gamma(a+x)\Gamma(b+n-x)}{\Gamma(a+b+n)}.$$

Hence (20.16) reduces to

$$\pi(\theta|x) = \frac{\Gamma(a+b+n)}{\Gamma(a+x)\Gamma(b+n-x)}\theta^{a+x-1}(1-\theta)^{b+n-x-1}.$$

Thus, the posterior distribution is of the same parametric form as the prior (20.15), but with a and b replaced by $a + x$ and $b + n - x$. This is the key idea of a conjugate prior.

A second example of a conjugate prior is the *Gamma-normal prior*. Consider the sequence of conditional distributions

$$\lambda \sim Ga(a, b), \tag{20.17}$$

$$\mu | \lambda \sim N[m, (k\lambda)^{-1}], \tag{20.18}$$

$$x_1, \ldots, x_n | \mu, \lambda \sim N[\mu, \lambda^{-1}] \text{ (independent)}. \tag{20.19}$$

In this case the data equation (20.19) consists of normal observations with unknown mean and variance, but for notational convenience, we have written the variance of the normal distribution as λ^{-1} rather than the more conventional σ^2. The parameter λ is often called the precision.

In this case, equations (20.17) and (20.18) together define the joint prior of (λ, μ), with joint density

$$\frac{b^a \lambda^{a-1} e^{-b\lambda}}{\Gamma(a)} \cdot \sqrt{\frac{k\lambda}{2\pi}} \exp\left\{-\frac{k\lambda}{2}(\mu - m)^2\right\} \propto \lambda^{a-1/2} \exp\left[-\lambda\left\{b + \frac{k}{2}(\mu - m)^2\right\}\right]. \tag{20.20}$$

Now consider the prior $\times$ likelihood

$$\frac{b^a \lambda^{a-1} e^{-b\lambda}}{\Gamma(a)} \cdot \sqrt{\frac{k\lambda}{2\pi}} \exp\left\{-\frac{k\lambda}{2}(\mu - m)^2\right\} \cdot \left(\frac{\lambda}{2\pi}\right)^{n/2} \exp\left\{-\frac{\lambda}{2}\sum(x_i - \mu)^2\right\}. \tag{20.21}$$

Noting the side calculation

$$\begin{aligned} k(\mu - m)^2 + \sum(x_i - \mu)^2 = {} & \frac{kn}{k+n}(m - \bar{x})^2 \\ & + \sum(x_i - \bar{x})^2 + (k+n)\left(\mu - \frac{km + n\bar{x}}{k+n}\right)^2 \end{aligned}$$

and defining $\bar{x} = \frac{1}{n}\sum x_i$, $\hat{m} = \frac{km+n\bar{x}}{k+n}$, (20.21) reduces to

$$\begin{aligned} \frac{\sqrt{k} b^a \lambda^{a+(n-1)/2}}{(2\pi)^{(n+1)/2}\Gamma(a)} \exp\Bigg[-\lambda\Bigg\{b + {} & \frac{kn}{2(k+n)}(m - \bar{x})^2 \\ & + \frac{1}{2}\sum(x_i - \bar{x})^2 + \frac{k+n}{2}(\mu - \hat{m})^2\Bigg\}\Bigg]. \end{aligned} \tag{20.22}$$

Comparing (20.22) with (20.20), we see that the posterior distribution is of the same form but with the constants a, b, m, k replaced by $\hat{a}$, $\hat{b}$, $\hat{m}$, $\hat{k}$, where

$$\begin{aligned}
\hat{a} &= a + \frac{n}{2}, \\
\hat{b} &= b + \frac{kn}{2(k+n)}(m - \bar{x})^2 + \frac{1}{2}\sum(x_i - \bar{x})^2, \\
\hat{m} &= \frac{km + n\bar{x}}{k+n}, \\
\hat{k} &= k + n.
\end{aligned}$$

In practice, we generally try to choose the prior to be as uninformative as possible. In (20.20), this is often achieved by setting $k = 0$ (in which case the improper conditional prior for density μ, derived from (20.18), is just a flat prior over $(-\infty, \infty)$) and setting a and b to very small constants, such as $a = b = 0.01$.

A.2 Hierarchical models

Although all Bayesian statistical models may ultimately be written in the form (20.14), this may not be the most convenient form either for conceptual model building or for mathematical solution. Often it is better to impose some structure on the unknown parameters that makes their interdependence clear. The most common way to do this is through a *hierarchical model.*

The simplest structure of hierarchical model consists of three layers:

- a top layer of parameters, that are common to all the individual units of the model,
- a middle layer of parameters that are specific to individual units,
- a bottom layer consisting of observations.

An example is the main univariate model considered in Smith *et al.* (2009), which is defined by the equations

$$\mu, \nu, \beta \sim U(-\infty, \infty), \tag{20.23}$$

$$\theta \sim Ga(a, b), \tag{20.24}$$

$$a_\lambda, b_\lambda \sim Ga(a^*, b^*), \tag{20.25}$$

$$\lambda_1, \ldots, \lambda_M | a_\lambda, b_\lambda \sim Ga(a_\lambda, b_\lambda), \tag{20.26}$$

$$X_0 | \mu \sim N\left[\mu, \lambda_0^{-1}\right], \quad (\lambda_0 \text{ known}) \tag{20.27}$$

$$X_j | \mu, \lambda_j \sim N\left[\mu, \lambda_j^{-1}\right], \tag{20.28}$$

$$Y_j | X_j, \mu, \nu, \theta, \lambda_j, \beta \sim N\left[\nu + \beta(X_j - \mu), (\theta\lambda_j)^{-1}\right]. \tag{20.29}$$

Here the top layer of parameters consists of $(\mu, \nu, \beta, \theta, a_\lambda, b_\lambda)$, with prior distributions specified by (20.23)–(20.25). The middle layer is (20.26), which defines the precisions $\lambda_1, \ldots, \lambda_M$ of the M climate models. The bottom layer is (20.27)–(20.29), which defines the distributions of the data X_0 (present-day observed climate mean) and X_j, Y_j (present-day and future projection under the jth climate model, $1 \le j \le M$). In this case, λ_0 is not considered a parameter because it is assumed known. The constants a, b, a^*, b^* are not unknown parameters of the model but are fixed at the start of the analysis – in practice we set them all equal to 0.01. Sometimes they are called *hyperparameters* to distinguish them from parameters that are actually estimated during the course of the model fitting.

The more general 'multivariate' model of Smith *et al.* (2009) may also be represented in this three-layer hierarchical form, where the model precisions λ_j, having the same prior distribution (20.26), and the interaction parameters η_{ij}, with independent prior distributions $Ga(c, c)$, are in the middle layer of the hierarchy to indicate their dependence on additional unknown parameters a_λ, b_λ and c. For this model, there are many more unknown parameters, but the basic structure of the model is not more complicated than the univariate version of the model. In general, hierarchical models may have more than three layers, or may be subject to more complex dependencies (e.g. the components of the model may be represented as the vertices of a graph, where the edges of the graph represent the dependence structure), but the simple three-layer structure is sufficient for the present application.

A.3 MCMC

Any hierarchical model may be represented conceptually in the form (20.14) by representing all the unknown parameters together as a single long vector θ. However, in virtually all such cases, the model cannot be solved by simple analytic integration to compute the posterior density, while standard numerical methods of integration, such as the trapezoidal rule, are too slow or insufficiently accurate to yield practical results. *Markov chain Monte Carlo* (MCMC) methods are a class of Monte Carlo simulation techniques whose purpose is to simulate a random sample from the posterior distribution of θ. Although these methods are also used in non-hierarchical problems, in cases where a conjugate prior is not available or is inappropriate for a particular application, they really come into their own in solving hierarchical Bayesian problems. The name derives from the fact that they generate a random sample from a Markov chain, which can be proved to converge to the true posterior distribution as the size of the Monte Carlo sample tends to infinity.

The *Gibbs sampler* is based on first partitioning θ into components $\theta_1, \ldots, \theta_p$. Here, p may simply be the total number of unknown parameters in the model, though it is also possible to group parameters together when it is convenient to do so (for example, in the Gamma-normal prior, we are effectively grouping λ and μ together as a single parameter θ: in a hierarchical model, the precision and mean for a single unit could together be one of $\theta_1, \ldots, \theta_p$). The key element is that it must be possible to generate a random variable from any one of $\theta_1, \ldots, \theta_p$ conditional on the other $p-1$ held fixed: this is often achieved in practice through the use of conjugate priors. The algorithm is then as follows:

1. Choose arbitrary starting values $\theta_1, \ldots, \theta_p$, say $\theta_1^{(1)}, \ldots, \theta_p^{(1)}$. Set the counter $b = 1$.
2. Given current values $\theta_1^{(b)}, \ldots, \theta_p^{(b)}$, generate a Monte Carlo random variate from the distribution of θ_1, conditional on $\theta_j = \theta_j^{(b)}$, $j = 2, \ldots, p$. Call the result $\theta_1^{(b+1)}$. Next, generate a Monte Carlo random variate from the distribution of θ_2, conditional on $\theta_1 = \theta_1^{(b+1)}$, $\theta_j = \theta_j^{(b)}$, $j = 3, \ldots, p$. Call the result $\theta_2^{(b+1)}$. Continue with $\theta_3, \theta_4, \ldots$, up to generating a Monte Carlo random variate $\theta_p^{(b+1)}$ from the distribution of θ_p, conditional on $\theta_j = \theta_j^{(b+1)}$, $j = 1, \ldots, p-1$.
3. Set counter $b \rightarrow b+1$ and return to step 2.
4. Continue until $b = B$, the desired number of Monte Carlo iterations. The value of B depends on the complexity of the problem and available computer time, but a value in the range 5000–100,000 is typical.
5. The first $B_0 < B$ samples are discarded as *burn-in* samples, reflecting that the Markov chain has not yet converged to its stationary distribution. Among the last $B - B_0$ iterations, it is usual to retain only every tth iteration, for some suitable $t > 1$. This reflects the fact that successive iterations from the Gibbs sampler are highly correlated, but by thinning in this way, the remaining iterations may be treated as approximately uncorrelated.
6. The resulting $(B - B_0)/t$ values of θ are treated as a random sample from the posterior distribution of θ. Posterior means and standard deviations, predictive distributions, etc., are calculated by averaging over these random samples.

As an example, let us consider the simpler form of (20.23)–(20.29) in which equation (20.25) is omitted and a_λ, b_λ are treated as known constants. As a notational point, θ in (20.24) is a single (scalar) parameter in a multiparameter model, not to be confused with our notation of θ as the vector of all unknown parameters in (20.14). The context should make clear which of the two uses of θ is intended in any particular instance.

In this case, the joint density of all the unknown parameters and observations is proportional to

$$\theta^{a+M/2-1}e^{-b\theta}e^{-\frac{1}{2}\lambda_0(X_0-\mu)^2}\prod_{j=1}^{M}\left[\lambda_j^{a_\lambda}e^{-b_\lambda\lambda_j}\cdot e^{-\frac{1}{2}\lambda_j(X_j-\mu)^2-\frac{1}{2}\theta\lambda_j\{Y_j-\nu-\beta(X_j-\mu)\}^2}\right]. \quad (20.30)$$

It is not possible to integrate (20.30) analytically with respect to all the unknown parameters μ, ν, β, θ, $\lambda_1, \ldots, \lambda_M$. However, for any single one of these parameters, we can integrate using standard forms based on the normal or Gamma densities. For example, as a function of θ alone, conditional on all the others, (20.30) is proportional to

$$\theta^{a+M/2-1}e^{-\theta[b+\frac{1}{2}\sum_j\lambda_j\{Y_j-\nu-\beta(X_j-\mu)\}^2]},$$

which in turn is proportional to a Gamma density with parameters $a+\frac{M}{2}$ and $b+\frac{1}{2}\sum_j\lambda_j\{Y_j-\nu-\beta(X_j-\mu)\}^2$. Therefore the updating step for θ is to generate a random variate with this Gamma distribution, replacing the previous value of θ.

Similarly, the conditional density of λ_j (for fixed $j\in\{1,2,\ldots,M\}$) is Gamma with parameters $a_\lambda+1$ and $b_\lambda+\frac{1}{2}[(X_j-\mu)^2+\theta\{Y_j-\nu-\beta(X_j-\mu)\}^2]$. The remaining parameters μ, ν, β have conditional normal distributions that can be calculating by completing the square in the exponent. As an example, we give the explicit calculation for μ. Expressed as a function of μ, (20.30) is of the form $e^{-Q(\mu)/2}$, where

$$Q(\mu)=\lambda_0(X_0-\mu)^2+\sum_j\lambda_j(X_j-\mu)^2+\theta\sum_j\lambda_j\{Y_j-\nu-\beta(X_j-\mu)\}^2.$$

Completing the square,

$$Q(\mu)=\left(\lambda_0+\sum_j\lambda_j+\theta\beta^2\sum_j\lambda_j\right)\times\left\{\mu-\frac{\lambda_0X_0+\sum_j\lambda_jX_j-\theta\beta\sum_j\lambda_j(Y_j-\nu-\beta X_j)}{\lambda_0+\sum_j\lambda_j+\theta\beta^2\sum_j\lambda_j}\right\}^2+\text{const}$$

where 'const' contains terms that do not depend on μ. But based on this representation, we recognize that $e^{-Q(\mu)/2}$ has the form of a normal density with mean and variance

$$\frac{\lambda_0X_0+\sum_j\lambda_jX_j-\theta\beta\sum_j\lambda_j(Y_j-\nu-\beta X_j)}{\lambda_0+\sum_j\lambda_j+\theta\beta^2\sum_j\lambda_j},\qquad\frac{1}{\lambda_0+\sum_j\lambda_j+\theta\beta^2\sum_j\lambda_j},$$

so the Gibbs sampling step for μ is to generate a random variate from this normal distribution. The calculations for ν and β are similar so we omit the details.

In a typical application of this algorithm, the Gibbs sampler was run for an initial 12,500 iterations as a burn-in, followed by 50,000 iterations, with every 50th iteration being saved. Thus, we ended up with a Monte Carlo sample of size 1000, which are approximately independent from the joint posterior distribution of the unknown parameters.

A.4 The Metropolis – Hastings algorithm

The Gibbs sampler still relies on the assumption that, after partitioning θ (the full vector of unknown parameters) into $\theta_1, \ldots, \theta_p$, the conditional distribution of each θ_j, given all the θ_k, $k \neq j$, may be represented in a sufficiently explicit form that it is possible to generate a Monte Carlo variate directly. In cases where this is not possible, there are by now a wide variety of alternative MCMC algorithms, but the oldest and best known is the Metropolis – Hastings algorithm, which we now describe.

Its general form is as follows. We assume again that θ is an unknown parameter vector lying in some parameter space Θ. Suppose we want to generate Monte Carlo samples from θ, with the probability mass function or probability density function $g(\theta)$. We also assume we have some stochastic mechanism for generating an 'update' θ', given the current value of θ, that is represented by a Markov transition kernel $q(\theta, \theta')$. In other words, for each $\theta \in \Theta$, we assume there exists a probability mass function or probability density function $q(\theta, \cdot)$, that represents the conditional density of θ' given θ. In principle, $q(\theta, \theta')$ is arbitrary subject to only mild restrictions (for example, the Markov chain generated by $q(\theta, \theta')$ must be irreducible, in the sense that samples generated by this Markov chain will eventually cover the whole of Θ). In practice, certain simple forms such as random walks are usually adopted.

The main steps of the algorithm are:

1. Start with an initial trial value of θ, call it $\theta^{(1)}$. Set counter $b = 1$.
2. Given current value $\theta^{(b)}$, generate a new trial θ' from the conditional density $q(\theta^{(b)}, \theta')$.
3. Calculate
$$\alpha = \min\left\{\frac{g(\theta')q(\theta', \theta^{(b)})}{g(\theta^{(b)})q(\theta^{(b)}, \theta')}, 1\right\}.$$
4. With an independent draw from the random number generator, determine whether we 'accept' θ', where the probability of acceptance is α.
5. If the result of step 4 is to accept θ', set $\theta^{(b+1)} = \theta'$. Otherwise, $\theta^{(b+1)} = \theta^{(b)}$.
6. Set counter $b \to b + 1$ and return to step 2.
7. Continue until $b = B$, the desired number of Monte Carlo iterations.

As with the Gibbs sampler, it is usual to discard a large number of initial iterations as 'burn-in', and then to retain only a thinned subset of the remaining iterations, to ensure approximate independence between Monte Carlo variates.

A key feature of the algorithm is that the acceptance probability α depends only on *ratios* of the density g – in other words, it is not necessary to specify g exactly, so long as it is known up to a normalizing constant. This is precisely the situation we face with (20.14), where the numerator $\pi(\theta)f(x|\theta)$ is known explicitly but the denominator requires an integral which is typically intractable. However, with the Metropolis – Hastings algorithm, it is not necessary to evaluate the denominator.

A simpler form of the algorithm arises if the kernel q is symmetric, i.e. $q(\theta',\theta) = q(\theta,\theta')$ for all $\theta \neq \theta'$. In that case, the formula for the acceptance probability reduces to

$$\alpha = \min\left\{\frac{g(\theta')}{g(\theta^{(b)})}, 1\right\}.$$

In this form, the algorithm is equivalent to the Monte Carlo sampling algorithm of Metropolis *et al.* (1953), which was used for statistical physics calculations for decades before the algorithm's rediscovery by Bayesian statisticians. The general form of the algorithm, and its justification in terms of the convergence theory of Markov chains, was due to Hastings (1970).

Although it is possible to apply the Metropolis – Hastings sampler entirely independently of the Gibbs sampler, in practice, the two ideas are often combined, where the Gibbs sampler is used to update all those components of θ for which explicit conditional densities are available in a form that can easily be simulated, and a Metropolis – Hastings step is used for the remaining components. This is exactly the way the algorithm was applied in Smith *et al.* (2009). To be explicit, let us return to the model defined by equations (20.23)–(20.29), where now we treat a_λ, b_λ as unknown parameters with prior density (20.25). Isolating those parts of the joint density that depend just on a_λ, b_λ, we have to generate Monte Carlo samples from a density proportional to

$$g(a_\lambda, b_\lambda) = (a_\lambda b_\lambda)^{a^*-1} e^{-b^*(a_\lambda+b_\lambda)} \prod_{j=1}^{M} \frac{b_\lambda^{a_\lambda} \lambda_j^{a_\lambda} e^{-b_\lambda \lambda_j}}{\Gamma(a_\lambda)}. \tag{20.31}$$

This is not of conjugate prior form so we use the Metropolis – Hastings algorithm to update a_λ and b_λ. The form of the updating rule is that given the current a_λ and b_λ, we define $a'_\lambda = a_\lambda e^{\delta(U_1-\frac{1}{2})}$, $b'_\lambda = b_\lambda e^{\delta(U_2-\frac{1}{2})}$, for random U_1 and U_2 that are uniform on [0, 1] (independent of each other and all other random variates). Both the form of this updating rule and the choice of δ are arbitrary, but we have found it works well in practice, and have used $\delta = 1$ in most of our calculations.

With this rule,

$$q\left\{(a_\lambda, b_\lambda), (a'_\lambda, b'_\lambda)\right\} = \frac{1}{\delta^2 a'_\lambda b'_\lambda}, \quad a'_\lambda \in \left(a_\lambda e^{-\delta/2}, a_\lambda e^{\delta/2}\right), \quad b'_\lambda \in \left(b_\lambda e^{-\delta/2}, b_\lambda e^{\delta/2}\right).$$

Therefore $q\{(a'_\lambda, b'_\lambda), (a_\lambda, b_\lambda)\}/q\{(a_\lambda, b_\lambda), (a'_\lambda, b'_\lambda)\} = a'_\lambda b'_\lambda/(a_\lambda b_\lambda)$ and the acceptance probability α reduces to

$$\min\left\{\frac{g\left(a'_\lambda, b'_\lambda\right) a'_\lambda b'_\lambda}{g(a_\lambda, b_\lambda) a_\lambda b_\lambda}, 1\right\}.$$

A.5 Diagnostics

In this section, we briefly review some of the available diagnostics.

With the Metropolis – Hastings algorithm, the kernel $q(\theta, \theta')$ often contains a tuning parameter (δ in the above example), which controls the size of the jumps. This raises the question of how to choose the tuning parameter. For a specific class of Metropolis sampling rules based on a multivariate normal target, Gelman *et al.* (1996) showed that there is an optimal acceptance probability α, which ranges from 0.44 in one dimension to 0.23 as the dimension tends to ∞. Using this result as a guideline, it is often recommended that for general Metropolis sampling rules, the acceptance probability should be tuned so that it lies between about 0.15 and 0.5 on average. There are many specific proposals for adaptively optimizing this choice; the paper by Pasarica and Gelman (2010) contains one recent proposal and reviews earlier literature.

The question of how many iterations of a MCMC algorithm are required to achieve adequate convergence to the stationary distribution has been the subject of much research. One well-regarded procedure was proposed by Gelman and Rubin (1992) and extended by Brooks and Gelman (1998). The Gelman – Rubin procedure requires running several samplers in parallel, using widely dispersed initial values for the parameters. Then, it calculates 'potential scale reduction factor' R. This can be interpreted as an estimate of the possible reduction in variance of the posterior mean of a particular parameter, if the Markov chain were run to convergence, compared with the current iterations. Since R is itself an unknown parameter estimated from the sample output, it is common to quote both the median and some upper quantile (say, the 0.975 quantile) of the sampling distribution. The ideal value of R is 1; values much above 1 are taken to indicate non-convergence of the MCMC procedure.

Another issue is inference after sampling. Suppose we are trying to estimate the posterior mean of a particular parameter; in our climate application, ν, the estimated mean of future values of a climatic variable, is of particular interest. If we had a Monte Carlo sample of independent estimates of ν, the usual standard error calculation would give an estimate of the sampling variability of the posterior mean of ν. In an MCMC sample, it is desirable to take account of the fact that successive draws from the sample may be autocorrelated. There are many possible procedures; one method, due to Heidelberger and Welch (1981), constructs a nonparametric estimate of the spectral density at low frequencies, and uses this to correct the standard error.

The Gelman – Rubin and Heidelberger – Welch procedures are included, among several others, in the CODA diagnostics package (Plummer *et al.*, 2006), which is available as a downloadable package within R (R Development Core Team, 2007).

A.6 Further reading

There are by now many books on the principles of Bayesian data analysis and MCMC algorithms; a small selection includes Gamerman and Lopes (2006); Gelman *et al.* (2003); Robert and Casella (2004). The book by Robert (2005) is a somewhat more theoretical treatment that makes the link between modern practices in Bayesian statistics and classical decision theory and inference.

B. Computational details

B.1 Computation for the univariate model

We detail here the Gibbs/Metropolis algorithm to sample the posterior distribution of the model in Section 20.4.1, fitting one region at a time, with a hyperprior on the parameters of the prior for the precisions, λ_js. This algorithm is originally described in Smith *et al.* (2009).

Under the region-specific model, the joint density of θ, μ, ν, β, a_λ, b_λ, X_0 and λ_j, X_j, $Y_j (j = 1, \ldots, M)$ is proportional to

$$\theta^{a+M/2-1} e^{-b\theta} e^{-\frac{1}{2}\lambda_0(X_0-\mu)^2} a_\lambda^{a^*-1} e^{-b^* a_\lambda} b_\lambda^{a^*-1} e^{-b^* b_\lambda} \cdot \prod_{j=1}^{M} \left[\lambda_j^{a_\lambda} e^{-b_\lambda \lambda_j} \cdot e^{-\frac{1}{2}\lambda_j (X_j-\mu)^2 - \frac{1}{2}\theta\lambda_j \{Y_j-\nu-\beta(X_j-\mu)\}^2} \right]. \tag{20.32}$$

Define

$$\tilde{\mu} = \frac{\lambda_0 X_0 + \sum \lambda_j X_j - \theta\beta \sum \lambda_j (Y_j - \nu - \beta X_j)}{\lambda_0 + \sum \lambda_j + \theta\beta^2 \sum \lambda_j}, \tag{20.33}$$

$$\tilde{\nu} = \frac{\sum \lambda_j \{Y_j - \beta(X_j - \mu)\}}{\sum \lambda_j}, \tag{20.34}$$

$$\tilde{\beta} = \frac{\sum \lambda_j (Y_j - \nu)(X_j - \mu)}{\sum \lambda_j (X_j - \mu)^2}. \tag{20.35}$$

In a Monte Carlo sampling scheme, all the parameters in (20.32), with the exception of a_λ and b_λ, may be updated through Gibbs sampling steps, as follows:

$$\mu|\text{rest} \sim N\left[\tilde{\mu}, \frac{1}{\lambda_0 + \sum \lambda_j + \theta\beta^2 \sum \lambda_j}\right], \tag{20.36}$$

$$\nu|\text{rest} \sim N\left[\tilde{\nu}, \frac{1}{\theta \sum \lambda_j}\right], \tag{20.37}$$

$$\beta|\text{rest} \sim N\left[\tilde{\beta}, \frac{1}{\theta \sum \lambda_j (X_j - \mu)^2}\right], \tag{20.38}$$

$$\lambda_j|\text{rest} \sim G\left[a + 1, b + \frac{1}{2}(X_j - \mu)^2 + \frac{\theta}{2}\{Y_j - \nu - \beta(X_j - \mu)\}^2\right], \tag{20.39}$$

$$\theta|\text{rest} \sim G\left[a + \frac{M}{2}, b + \frac{1}{2}\sum \lambda_j \{Y_j - \nu - \beta(X_j - \mu)\}^2\right]. \tag{20.40}$$

For the parameters a_λ, b_λ, the following Metropolis updating step is proposed instead:

1. Generate U_1, U_2, U_3, independent uniform on (0, 1).
2. Define new trial values $a'_\lambda = a_\lambda e^{\delta(U_1 - 1/2)}$, $b'_\lambda = b_\lambda e^{\delta(U_2 - 1/2)}$. The value of δ (step length) is arbitrary but $\delta = 1$ seems to work well in practice, and is therefore used here.
3. Compute

$$\begin{aligned} \ell_1 &= M a_\lambda \log b_\lambda - M \log \Gamma(a_\lambda) + (a_\lambda - 1) \sum \log \lambda_j - b_\lambda \\ &\quad \times \sum \lambda_j + a^* \log(a_\lambda b_\lambda) - b^*(a_\lambda + b_\lambda), \\ \ell_2 &= M a'_\lambda \log b'_\lambda - M \log \Gamma(a'_\lambda) + (a'_\lambda - 1) \sum \log \lambda_j - b'_\lambda \\ &\quad \times \sum \lambda_j + a^* \log\left(a'_\lambda b'_\lambda\right) - b^*\left(a'_\lambda + b'_\lambda\right). \end{aligned}$$

   This computes the log likelihood for both (a_λ, b_λ) and (a'_λ, b'_λ), allowing for the prior density and including a Jacobian term to allow for the fact that the updating is on a logarithmic scale.
4. If

$$\log U_3 < \ell_2 - \ell_1$$

   then we accept the new (a_λ, b_λ), otherwise keep the present values for the current iteration, as in a standard Metropolis accept-reject step.

This process is iterated many times to generate a random sample from the joint posterior distribution. In the case where a_λ, b_λ are treated as fixed, the Metropolis steps for these two parameters are omitted and in this case the method is a pure Gibbs sampler. An R program (REA.GM.r) to perform the sampling is available for download from `http://www.image.ucar.edu/~tebaldi/REA`.

B.2 Computation for the multivariate model

The following computations were originally described in Smith *et al.* (2009). Omitting unnecessary constants, the joint density of all the random variables in the model that treats all regions at the same time is

$$
\begin{aligned}
&(ca_\lambda b_\lambda)^{a-1} e^{-b(c+a_\lambda+b_\lambda)} \cdot \left[\prod_{i=0}^{R} \theta_i^{a-1} e^{-b\theta_i}\right] \cdot \left[\prod_{i=1}^{R} \phi_i^{a-1} e^{-b\phi_i}\right] \cdot \left[\prod_{j=1}^{M} \lambda_j^{a_\lambda-1} e^{-b_\lambda \lambda_j} \frac{b_\lambda}{\Gamma(a_\lambda)}\right] \\
&\times \left[\psi_0^{a-1} e^{-b\psi_0}\right] \\
&\times \left[\prod_{i=1}^{R}\prod_{j=1}^{M} \eta_{ij}^{c-1} e^{-c\eta_{ij}} \frac{c^c}{\Gamma(c)}\right] \cdot \left[\prod_{j=1}^{M} \sqrt{\psi_0} e^{-\frac{1}{2}\psi_0 \alpha_j^2}\right] \cdot \left[\prod_{j=1}^{M} \sqrt{\theta_0 \psi_0} e^{-\frac{1}{2}\theta_0 \psi_0 (\alpha_j' - \beta_0 \alpha_j)^2}\right] \cdot \\
&\times \left[\prod_{i=1}^{R} e^{-\frac{1}{2}\lambda_{0i}(X_{i0}-\mu_0-\zeta_i)^2}\right] \cdot \left[\prod_{i=1}^{R}\prod_{j=1}^{M} \sqrt{\eta_{ij}\phi_i\lambda_j} e^{-\frac{1}{2}\eta_{ij}\phi_i\lambda_j (X_{ij}-\mu_0-\zeta_i-\alpha_j)^2}\right] \cdot \\
&\times \left[\prod_{i=1}^{R}\prod_{j=1}^{M} \sqrt{\eta_{ij}\theta_i\lambda_j} e^{-\frac{1}{2}\eta_{ij}\theta_i\lambda_j \left\{Y_{ij}-\nu_0-\zeta_i'-\alpha_j'-\beta_i(X_{ij}-\mu_0-\zeta_i-\alpha_j)\right\}^2}\right]. \qquad (20.41)
\end{aligned}
$$

Define

$$\tilde{\mu}_0 = \frac{\sum_i \lambda_{0i}(X_{i0}-\zeta_i) + \sum_i \phi_i \sum_j \eta_{ij}\lambda_j (X_{ij}-\zeta_i-\alpha_j) - \sum_i \beta_i\theta_i \sum_j \eta_{ij}\lambda_j \left\{Y_{ij}-\nu_0-\zeta_i-\alpha_j'-\beta_i(X_{ij}-\zeta_i-\alpha_j)\right\}}{\sum_i \{\lambda_{0i} + (\phi_i+\beta_i^2\theta_i)\sum_j \eta_{ij}\lambda_j\}}, \qquad (20.42)$$

$$\tilde{\nu}_0 = \frac{\sum_i \theta_i \sum_j \eta_{ij}\lambda_j \left\{Y_{ij} - \zeta_i' - \alpha_j' - \beta_i(X_{ij} - \mu_0 - \zeta_i - \alpha_j)\right\}}{\sum_i \theta_i \{\sum_j \eta_{ij}\lambda_j\}}, \qquad (20.43)$$

$$\tilde{\zeta}_i = \frac{\lambda_{0i}(X_{i0}-\mu_0) + \phi_i \sum_j \eta_{ij}\lambda_j (X_{ij}-\mu_0-\alpha_j) - \beta_i\theta_i \sum_j \eta_{ij}\lambda_j \left\{Y_{ij}-\nu_0-\zeta_i'-\alpha_j'-\beta_i(X_{ij}-\mu_0-\alpha_j)\right\}}{\lambda_{0i} + (\phi_i+\beta_i^2\theta_i)\sum_j \eta_{ij}\lambda_j}, \qquad (20.44)$$

$$\tilde{\zeta}_i' = \frac{\sum_j \eta_{ij}\lambda_j \left\{Y_{ij} - \nu_0 - \alpha_j' - \beta_i(X_{ij} - \mu_0 - \zeta_i - \alpha_j)\right\}}{\sum_j \eta_{ij}\lambda_j}, \qquad (20.45)$$

$$\tilde{\beta}_0 = \frac{\sum_j \alpha_j' \alpha_j}{\sum_j \alpha_j^2}, \qquad (20.46)$$

$$\tilde{\beta}_i = \frac{\sum_j \eta_{ij}\lambda_j \left(Y_{ij} - \nu_0 - \zeta_i' - \alpha_j'\right)(X_{ij} - \mu_0 - \zeta_i - \alpha_j)}{\sum_j \eta_{ij}\lambda_j (X_{ij} - \mu_0 - \zeta_i - \alpha_j)^2}, \quad (i \neq 0), \quad (20.47)$$

$$\tilde{\alpha}_j = \frac{\beta_0\theta_0\psi_0\alpha_j' + \lambda_j \sum_i \eta_{ij}\phi_i(X_{ij} - \mu_0 - \zeta_i) - \lambda_j \sum_i \eta_{ij}\theta_i\beta_i \left\{Y_{ij} - \nu_0 - \zeta_i' - \alpha_j' - \beta_i(X_{ij} - \mu_0 - \zeta_i)\right\}}{\psi_0 + \beta_0^2\theta_0\psi_0 + \lambda_j \sum_i \eta_{ij}\phi_i + \lambda_j \sum_i \eta_{ij}\theta_i\beta_i^2}, \quad (20.48)$$

$$\tilde{\alpha}_j' = \frac{\beta_0\theta_0\psi_0\alpha_j + \lambda_j \sum_i \eta_{ij}\theta_i \left\{Y_{ij} - \nu_0 - \zeta_i' - \beta_i(X_{ij} - \mu_0 - \zeta_i - \alpha_j)\right\}}{\theta_0\psi_0 + \lambda_j \sum_i \eta_{ij}\theta_i}. \quad (20.49)$$

The conditional distributions required for the Gibbs sampler are as follows:

$$\mu_0|\text{rest} \sim N\left[\tilde{\mu}_0, \frac{1}{\sum_i\{\lambda_{0i} + \left(\phi_i + \beta_i^2\theta_i\right)\sum_j \eta_{ij}\lambda_j\}}\right], \quad (20.50)$$

$$\nu_0|\text{rest} \sim N\left[\tilde{\nu}_0, \frac{1}{\sum_i\{\theta_i \sum_j \eta_{ij}\lambda_j\}}\right], \quad (20.51)$$

$$\zeta_i|\text{rest} \sim N\left[\tilde{\zeta}_i, \frac{1}{\lambda_i + \left(\phi_i + \beta_i^2\theta_i\right)\sum_j \eta_{ij}\lambda_j}\right], \quad (20.52)$$

$$\zeta_i'|\text{rest} \sim N\left[\tilde{\zeta}_i', \frac{1}{\theta_i \sum_j \eta_{ij}\lambda_j}\right], \quad (20.53)$$

$$\beta_0|\text{rest} \sim N\left[\tilde{\beta}_0, \frac{1}{\theta_0\psi_0 \sum_j \alpha_j^2}\right], \quad (20.54)$$

$$\beta_i|\text{rest} \sim N\left[\tilde{\beta}_i, \frac{1}{\theta_i \sum_j \eta_{ij}\lambda_j (X_{ij} - \mu_0 - \zeta_i - \alpha_j)^2}\right], \quad (i \neq 0) \quad (20.55)$$

$$\alpha_j|\text{rest} \sim N\left[\tilde{\alpha}_j, \frac{1}{\psi_0 + \beta_0^2\theta_0\psi_0 + \lambda_j \sum_i \eta_{ij}\phi_i + \lambda_j \sum_i \eta_{ij}\theta_i\beta_i^2}\right], \quad (20.56)$$

$$\alpha_j'|\text{rest} \sim N\left[\tilde{\alpha}_j', \frac{1}{\theta_0\psi_0 + \lambda_j \sum_i \eta_{ij}\theta_i},\right], \quad (20.57)$$

$$\theta_0|\text{rest} \sim Gam\left[a + \frac{M}{2}, b + \frac{1}{2}\psi_0 \sum_j \left(\alpha_j' - \beta_0\alpha_j\right)^2\right], \quad (20.58)$$

$$\theta_i|\text{rest} \sim Gam\left[a + \frac{M}{2}, b + \frac{1}{2}\sum_j \eta_{ij}\lambda_j \times \left\{Y_{ij} - \nu_0 - \zeta_i' - \alpha_j' - \beta_i(X_{ij} - \mu_0 - \zeta_i - \alpha_j)\right\}^2\right], \quad (i \neq 0) \quad (20.59)$$

$$\phi_i|\text{rest} \sim Gam\left[a + \frac{M}{2}, b + \frac{1}{2}\sum_j \eta_{ij}\lambda_j(X_{ij} - \mu_0 - \zeta_i - a_j)^2\right], \tag{20.60}$$

$$\lambda_j|\text{rest} \sim Gam\left[a_\lambda + R, b_\lambda + \frac{1}{2}\sum_i \eta_{ij}\phi_i(X_{ij} - \mu_0 - \zeta_i - a_j)^2 \right. \tag{20.61}$$

$$\left. + \frac{1}{2}\sum_i \eta_{ij}\theta_i\left\{Y_{ij} - \nu_0 - \zeta_i' - a_j' - \beta_i(X_{ij} - \mu_0 - \zeta_i - a_j)\right\}^2\right],$$

$$\psi_0|\text{rest} \sim Gam\left[a + M, b + \frac{1}{2}\sum_j a_j^2 + \frac{1}{2}\theta_0\sum_j\left(a_j' - \beta_0 a_j\right)^2\right], \tag{20.62}$$

$$\eta_{ij}|\text{rest} \sim Gam\left[c + 1, c + \frac{1}{2}\phi_i\lambda_j(X_{ij} - \mu_0 - \zeta_i - a_j)^2\right.$$

$$\left. + \frac{1}{2}\theta_i\lambda_j\left\{Y_{ij} - \nu_0 - \zeta_i' - a_j' - \beta_i(X_{ij} - \mu_0 - \zeta_i - a_j)\right\}^2\right]. \tag{20.63}$$

Were a_λ, b_λ and c fixed, as in the univariate analysis, the iteration (20.50)–(20.63) could be repeated many times to generate a random sample from the joint posterior distribution. Having added a layer by making the three parameters random variates, two Metropolis steps are added to the iteration (20.50)–(20.63), as follows.

For the sampling of a_λ and b_λ jointly, define U_1, U_2 two independent random variables distributed uniformly over the interval (0, 1), and the two candidate values $a_\lambda' = a_\lambda e^{(\delta(u_1 - \frac{1}{2}))}$ and $b_\lambda' = b_\lambda e^{(\delta(u_2 - \frac{1}{2}))}$, where δ is an arbitrary increment, chosen as $\delta = 1$ in the implementation to follow. We then compute

$$\begin{aligned} l_1 &= MRa_\lambda \log(b_\lambda) - MR\log(\Gamma(a_\lambda) + R(a_\lambda - 1)\sum_j \log(\lambda_j) \\ &\quad - Rb_\lambda\sum_j \lambda_j + a\log(a_\lambda b_\lambda) - b(a_\lambda + b_\lambda), \end{aligned} \tag{20.64}$$

$$\begin{aligned} l_2 &= MRa_\lambda' \log\left(b_\lambda'\right) - MR\log(\Gamma\left(a_\lambda'\right) + R\left(a_\lambda' - 1\right)\sum_j \log(\lambda_j) \\ &\quad - Rb_\lambda'\sum_j \lambda_j + a\log\left(a_\lambda' b_\lambda'\right) - b\left(a_\lambda' + b_\lambda'\right). \end{aligned} \tag{20.65}$$

In (20.64) and (20.65) we are computing the log likelihoods of (a_λ, b_λ) and (a_λ', b_λ'), allowing for the prior densities and including a Jacobian term, allowing for the fact that the updating is taking place on a logarithmic scale. Then, within each iteration of the Gibbs – Metropolis simulation, the proposed values (a_λ', b_λ') are accepted with probability $e^{l_2 - l_1}$.

Similarly, the updating of c takes place by proposing $c' = ce^{(\delta(u_3 - \frac{1}{2}))}$, where u_3 is a draw from a uniform distribution on (0, 1), and computing

$$\begin{aligned} k_1 &= MRc\log(c) - MR\log(\Gamma(c)) \\ &\quad + (c-1)\sum_i\sum_j \log(\eta_{ij}) - c\sum_i\sum_j \eta_{ij} + a\log(c) - bc \end{aligned} \tag{20.66}$$

$$\begin{aligned} k_2 &= MRc'\log(c') - MR\log(\Gamma(c')) \\ &\quad + (c'-1)\sum_i\sum_j \log(\eta_{ij}) - c'\sum_i\sum_j \eta_{ij} + a\log(c') - bc'. \end{aligned} \tag{20.67}$$

Then, within each iteration of the Gibbs – Metropolis simulation, the proposed value c' is accepted with probability $e^{k_2 - k_1}$.

The method is available as an R program (REAMV.GM.r) from `http://www.image.ucar.edu/~tebaldi/REA/`.

B.3 Computation for the bivariate model

The MCMC algorithm for the bivariate model is taken from Tebaldi and Sansó (2009). Note that in the following, the 'prime' symbol denotes the operation of centring a variable (O_t or X_{jt}) by the respective climate signal $\mu_t = \alpha + \beta t + \gamma t \mathcal{I}_{\{t \geq \tau_0\}}$.

Coefficients of the piecewise linear model:

Define

$$A = \tau_0\eta^T + \tau_0\eta^P\beta_{xo}^2 + \tau^*\sum_j \xi_j^T + \tau^*\sum_j \xi_j^P\beta_{xj}^2$$

and

$$\begin{aligned} B &= \eta^T\sum_{t\leq\tau_0}(O_t^T - \beta^T t) + \eta^P\sum_{t\leq\tau_0}\beta_{xo}^2\left(O_t^T - \beta^T t - \beta_{xo}O_t^{P'}\right) \\ &\quad + \sum_j \xi_j^T \sum_{t\leq\tau_0}(X_{jt}^T - \beta^T t - d_j^T) + \sum_j \xi_j^T\sum_{t>\tau_0}(X_{jt}^T - \beta^T t - \gamma^T(t-\tau_0) - d_j^T) \\ &\quad + \sum_j \xi_j^P\beta_{xj}^2\sum_{t\leq\tau_0}\left(X_{jt}^T - \beta^T t - d_j^T - \beta_{xj}\left(X_{jt}^{P'} - d_j^P\right)\right) \\ &\quad + \sum_j \xi_j^P\beta_{xj}^2\sum_{t>\tau_0}\left(X_{jt}^T - \beta^T t - \gamma^T(t-\tau_0) - d_j^T - \beta_{xj}\left(X_{jt}^{P'} - d_j^P\right)\right). \end{aligned}$$

Then

$$\alpha^T \sim \mathcal{N}\left(\frac{B}{A}, (A)^{-1}\right).$$

Define

$$A = \tau_0 \eta^P + \tau^* \sum_j \xi_j^P$$

and

$$\begin{aligned} B = \eta^P & \sum_{t \le \tau_0} \left(O_t^P - \beta^P t - \beta_{xo} O_t^{T'} \right) \\ & + \sum_j \xi_j^P \sum_{t \le \tau_0} \left(X_{jt}^P - \beta^P t - d_j^P - \beta_{xj} \left(X_{jt}^{T'} - d_j^T \right) \right) \\ & + \sum_j \xi_j^P \sum_{t > \tau_0} \left(X_{jt}^P - \beta^P t - \gamma^P (t - \tau_0) - d_j^P - \beta_{xj} \left(X_{jt}^{T'} - d_j^T \right) \right). \end{aligned}$$

Then

$$\alpha^P \sim \mathcal{N}\left(\frac{B}{A}, (A)^{-1} \right).$$

Define

$$A = \eta^T \sum_{t \le \tau_0} t^2 + \eta^P \beta_{xo}^2 \sum_{t \le \tau_0} t^2 + \sum_j \xi_j^T \sum_{t \le \tau^*} t^2 + \sum_j \xi_j^P \beta_{xj}^2 \sum_{t \le \tau^*} t^2$$

and

$$\begin{aligned} B = \eta^T & \sum_{t \le \tau_0} t \left(O_t^T - \alpha^T \right) + \eta^P \sum_{t \le \tau_0} \beta_{xo}^2 t \left(O_t^T - \alpha^T \right) - \beta_{xo} t\, O_t^{P'}) \\ & + \sum_j \xi_j^T \sum_{t \le \tau_0} t \left(X_{jt}^T - \alpha^T - d_j^T \right) + \sum_j \xi_j^T \sum_{t > \tau_0} t \left(X_{jt}^T - \alpha^T - \gamma^T (t - \tau_0) - d_j^T \right) \\ & + \sum_j \xi_j^P \sum_{t \le \tau_0} t \left(\beta_{xj}^2 \left(X_{jt}^T - \alpha^T - d_j^T \right) - \beta_{xj} \left(X_{jt}^{P'} - d_j^P \right) \right) \\ & + \sum_j \xi_j^P \sum_{t > \tau_0} t \left(\beta_{xj}^2 \left(X_{jt}^T - \alpha^T - \gamma^T (t - \tau_0) - d_j^T \right) - \beta_{xj} \left(X_{jt}^{P'} - d_j^P \right) \right). \end{aligned}$$

Then

$$\beta^T \sim \mathcal{N}\left(\frac{B}{A}, (A)^{-1} \right).$$

Define

$$A = \eta^P \sum_{t \le \tau_0} t^2 + \sum_j \xi_j^P \sum_{t \le \tau^*} t^2$$

and

$$
\begin{aligned}
B = \eta^P \sum_{t \leq \tau_0} t \left(O_t^P - \alpha^P - \beta_{xo} O_t^{T'} \right) \\
+ \sum_j \xi_j^P \sum_{t \leq \tau_0} t \left(X_{jt}^P - \alpha^P - d_j^P - \beta_{xj} \left(X_{jt}^{T'} - d_j^T \right) \right) \\
+ \sum_j \xi_j^P \sum_{t > \tau_0} t \left(X_{jt}^P - \alpha^P - \gamma^P (t - \tau_0) - d_j^P - \beta_{xj} \left(X_{jt}^{T'} - d_j^T \right) \right).
\end{aligned}
$$

Then

$$\beta^P \sim \mathcal{N}\left(\frac{B}{A}, (A)^{-1} \right).$$

Define

$$A = \sum_j \xi_j^T \sum_{t > \tau_0} (t - \tau_0)^2 + \sum_j \xi_j^P \beta_{xj}^2 \sum_{t > \tau_0} (t - \tau_0)^2$$

and

$$
\begin{aligned}
B = \sum_j \xi_j^T \sum_{t > \tau_0} (t - \tau_0) \left(X_{jt}^T - \alpha^T - \beta^T t - d_j^T \right) \\
+ \sum_j \xi_j^P \sum_{t > \tau_0} (t - \tau_0) \left(\beta_{xj}^2 \left(X_{jt}^T - \alpha^T - \beta^T t - d_j^T \right) - \beta_{xj} \left(X_{jt}^{P'} - d_j^P \right) \right).
\end{aligned}
$$

Then

$$\gamma^T \sim \mathcal{N}\left(\frac{B}{A}, (A)^{-1} \right).$$

Define

$$A = \sum_j \xi_j^P \sum_{t > \tau_0} (t - \tau_0)^2$$

and

$$B = \sum_j \xi_j^P \sum_{t > \tau_0} (t - \tau_0) \left(X_{jt}^P - \alpha^P - \beta^P t - d_j^P - \beta_{xj} \left(X_{jt}^{T'} - d_j^T \right) \right).$$

Then

$$\gamma^P \sim \mathcal{N}\left(\frac{B}{A}, (A)^{-1} \right).$$

Bias terms and their priors' parameters:

Define

$$A = \tau^* \xi_j^T + \tau^* \xi_j^P \beta_{xj}^2 + \lambda_D^T$$

and

$$B = \xi_j^T \sum_{t \le \tau^*} X_{jt}^{T'} + \xi_j^P \sum_{t \le \tau^*} \left(\beta_{xj}^2 X_{jt}^{T'} - \beta_{xj}\left(X_{jt}^{P'} - d_j^P\right)\right) + \lambda_D^T a^T.$$

Then

$$d_j^T \sim \mathcal{N}\left(\frac{B}{A}, (A)^{-1}\right).$$

Define

$$A = \tau^* \xi_j^P + \lambda_D^P$$

and

$$B = \xi_j^P \sum_{t \le \tau^*} \left(X_{jt}^{P'} - \beta_{xj}\left(X_{jt}^{T'} - d_j^T\right)\right) + \lambda_D^P a^P.$$

Then

$$d_j^P \sim \mathcal{N}\left(\frac{B}{A}, (A)^{-1}\right).$$

Define $A = M\lambda_D^T$ and $B = \lambda_D^T \sum_j d_j^T$, then

$$a^T \sim \mathcal{N}\left(\frac{B}{A}, (A)^{-1}\right).$$

Define $A = M\lambda_D^P$ and $B = \lambda_D^P \sum_j d_j^P$, then

$$a^P \sim \mathcal{N}\left(\frac{B}{A}, (A)^{-1}\right).$$

$$\lambda_D^T \sim \mathcal{G}\left(1 + \frac{M}{2}; 1 + \frac{\sum_j \left(d_j^T - a^T\right)^2}{2}\right).$$

$$\lambda_D^P \sim \mathcal{G}\left(1 + \frac{M}{2}; 1 + \frac{\sum_j (d_j^P - a^P)^2}{2}\right).$$

The correlation coefficients between temperature and precipitation in the models, and their prior parameters:

Define $A = \xi_j^P \sum_t (X_{jt}^{T'} - d_j^T)^2 + \lambda_B$ and $B = \xi_j^P \sum_t (X_{jt}^{T'} - d_j^T)(X_{jt}^{P'} - d_j^P) + \lambda_B \beta_0$, then

$$\beta_{xj} \sim \mathcal{N}\left(\frac{B}{A}, (A)^{-1}\right).$$

Define $A = M\lambda_B + \lambda_o$ and $B = \lambda_B \sum_{j>0} \beta_{xj} + \lambda_o \beta_{xo}$, then

$$\beta_0 \sim \mathcal{N}\left(\frac{B}{A}, (A)^{-1}\right).$$

$$\lambda_B \sim \mathcal{G}\left(0.01 + \frac{M}{2}; 0.01 + \frac{\sum_j (\beta_{xj} - \beta_0)^2}{2}\right).$$

Precision terms for the models:

$$\xi_j^T \sim \mathcal{G}\left(a_{\xi^T} + \frac{\tau^*}{2}; b_{\xi^T} + \frac{\sum_t (X_{jt}^{T'} - d_j^T)^2}{2}\right).$$

$$\xi_j^P \sim \mathcal{G}\left(a_{\xi^P} + \frac{\tau^*}{2}; b_{\xi^P} + \frac{\sum_t (X_{jt}^{P'} - d_j^P - \beta_{xj}(X_{jt}^{T'} - d_j^T))^2}{2}\right).$$

Only the full conditionals of the hyperparameters $a_{\xi^T}, b_{\xi^T}, a_{\xi^P}, b_{\xi^P}$ cannot be sampled directly, and a Metropolis step is needed. We follow the solution described in Smith *et al.* (2009). The algorithm works identically for the two pairs, and we describe it for a_{ξ^T} and b_{ξ^T} (the sampling is done jointly for the pair). We define U_1, U_2 as independent random variables, uniformly distributed over the interval (0, 1), and we compute two proposal values $a'_{\xi^T} = a_{\xi^T} e^{(\delta(u_1 - \frac{1}{2}))}$ and $b'_{\xi^T} = b_{\xi^T} e^{(\delta(u_2 - \frac{1}{2}))}$, where δ is an arbitrary increment, that we choose as $\delta = 1$. We then compute

$$\ell_1 = Ma_{\xi^T} \log b_{\xi^T} - M \log \Gamma(a_{\xi^T}) + (a_{\xi^T} - 1) \sum_j \log \xi_j^T - b_{\xi^T} \sum_j \xi_j^T$$
$$+ 0.01 \log(a_{\xi^T} b_{\xi^T}) - 0.01(a_{\xi^T} + b_{\xi^T}), \tag{20.68}$$

$$\ell_2 = Ma'_{\xi^T} \log b'_{\xi^T} - M \log \Gamma\left(a'_{\xi^T}\right) + \left(a'_{\xi^T} - 1\right) \sum_j \log \xi_j^T - b'_{\xi^T} \sum_j \xi_j^T$$
$$+ a \log\left(a'_{\xi^T} b'_{\xi^T}\right) - b\left(a'_{\xi^T} + b'_{\xi^T}\right). \tag{20.69}$$

In (20.68) and (20.69) we are computing the log likelihoods of (a_{ξ^T}, b_{ξ^T}) and (a'_{ξ^T}, b'_{ξ^T}). Then, within each iteration of the Gibbs – Metropolis algorithm, the proposed values (a'_{ξ^T}, b'_{ξ^T}) are accepted with probability $e^{\ell_2 - \ell_1}$ if $\ell_2 < \ell_1$, or 1 if $\ell_2 \geq \ell_1$.

An R program that implements this simulation is available as REA.BV.r from `http://www.image.ucar.edu/~tebaldi/REA/`.

Acknowledgements

We thank Linda O. Mearns, Doug Nychka and Bruno Sanso' who have been working with us on various aspects of these analyses. Richard Smith is supported by NOAA grant NA05OAR4310020. Claudia Tebaldi is grateful to the Department of Global Ecology, Carnegie Institution of Washington, Stanford, and the National Center for Atmospheric Research, Boulder for hosting her at the time when this chapter was written.

References

Brooks, S. and Gelman, A. (1998). General methods for monitoring convergence of iterative simulations. *Journal of Computational and Graphical Statistics*, **7**, 434–455.

Furrer, R., Sain, S., Nychka, D. and Meehl, G. (2007). Multivariate Bayesian analysis of atmosphere-ocean general circulation models. *Environmental and Ecological Statistics*, **14**, 249–266.

Gamerman, D. and Lopes, H. (2006). *Markov Chain Monte Carlo: Stochastic Simulation for Bayesian Inference* (2nd edn). Chapman & Hall, Boca Raton, Florida.

Gelman, A., Carlin, J., Stern, H. and Rubin, D. (2003). *Bayesian Data Analysis.* (2nd edn.) Chapman & Hall, CRC Press, Boca Raton, Florida.

Gelman, A., Roberts, G. and Gilks, W. (1996). Efficient Metropolis jumping rules. In *Bayesian Statistics 5*, (ed. J. M. Bernardo, J. Berger, A. P. Dawid and A.F.M. Smith), pp. 599–607. Oxford University Press, Oxford.

Gelman, A. and Rubin, D. (1992). Inference from iterative simulation using multiple sequences. *Statistical Science*, **7**, 457–511.

Giorgi, F. and Francisco, R. (2000). Evaluating uncertainties in the prediction of regional climate change. *Geophysical Research Letters*, **27**, 1295–1298.

Gleckler, P. J., Taylor, K. and Doutriaux, C. (2007). Performance metrics for climate models. *Journal of Geophysical Research*, **113**(D06104). doi:10.1029/2007JD008972.

Groves, D. G., Yates, D. and Tebaldi, C. (2008). Uncertain global climate change projections for regional water management planning. *Water Resources Research*, **44**(W12413). doi:10.1029/2008WR006964.

Hastings, W. (1970). Monte Carlo sampling methods using Markov chains and their applications. *Biometrika*, **57**, 97–109.

Heidelberger, P. and Welch, P. (1981). A spectral method for confidence interval generation and run length control in simulations. *Communications of the ACM*, **24**, 233–245.

IPCC (2007). *Climate Change 2007 – The Physical Science Basis. Contribution of Working Group I to the Fourth Assessment Report of the IPCC. Solomon, S. et al.* (eds.), Cambridge University Press, Cambridge.

Metropolis, N., Rosenbluth, A., Rosenbluth, Teller, A. and Teller, E. (1953). Equation of state calculations by fast computing machines. *Journal of Chemical Physics*, **21**, 1087–1092.

Nakicenovic, N. (2000). *Special Report on Emissions Scenarios: A Special Report of Working Group III of the Intergovernmental Panel on Climate Change.* Cambridge University Press, Cambridge.

Pasarica, C. and Gelman, A. (2010). Adaptively scaling the Metropolis algorithm using expected squared jumped distance. *Statistica Sinica*, **20**, 343–364.

Plummer, M., Best, N., Cowles, K. and Vines, K. (2006). CODA: Convergence Diagnosis and Output Analysis for MCMC. *R NEWS*, **6**, 7–11.

R Development Core Team (2007). *R: A Language and Environment for Statistical Computing.* R Foundation for Statistical Computing. Available from `http://www.R-project.org`.

Raftery, A. E., Gneiting, T., Balabdoui, F. and Polakowski, M. (2005). Using Bayesian model averaging to calibrate forecast ensembles. *Monthly Weather Review*, **133**, 1155–1174.

Robert, C. (2005). *The Bayesian Choice*. (2nd edn). Springer, New York.

Robert, C. and Casella, G. (2004). *Monte Carlo Statistical Methods*. (2nd edn). Springer, New York.

Rougier, J. (2007). Probabilistic inference for future climate using an ensemble of climate model evaluations. *Climatic Change*, **81**, 247–264. doi: 10.1007/s10584-006-9156-9.

Smith, R., Tebaldi, C., Nychka, D. and Mearns, L. (2009). Bayesian modeling of uncertainty in ensembles of climate models. *Journal of the American Statistical Association*, **104**, 97–116.

Tebaldi, C. and Knutti, R. (2007). The use of the multi-model ensemble in probabilistic climate projections. *Philosophical Transactions of the Royal Society, Series A*, **1857**, 2053–2075.

Tebaldi, C. and Lobell, D. (2008). Towards probabilistic projections of climate change impacts on global crop yields. *Geophysical Research Letters*, **35** (L08705). doi:10.1029/2008GL033423.

Tebaldi, C., Mearns, L., Nychka, D. and Smith, R. (2004). Regional probabilities of precipitation change: A Bayesian analysis of multimodel simulations. *Geophysical Research Letters*, **31** (L24213). doi:10.1029/2004GL021276.

Tebaldi, C. and Sansó, B. (2009). Joint projections of temperature and precipitation change from multiple climate models: A hierarchical Bayesian approach. *Journal of the Royal Statistical Society. Series A*, **172**, 83–106.

Tebaldi, C., Smith, R., Nychka, D. and Mearns, L. (2005). Quantifying uncertainty in projections of regional climate change: A Bayesian approach to the analysis of multi-model ensembles. *Journal of Climate*, **18**, 1524–1540.

Washington, W. and Parkinson, C. L. (2005). *Introduction to Three-dimensional Climate Modeling*. University Science Books, London.

PART IV
Policy, Political and Social Sciences

·21·

Volatility in prediction markets: A measure of information flow in political campaigns

Carlos M. Carvalho and Jill Rickershauser

21.1 Introduction

When asked to describe the pivotal moments of the 2004 campaign, Mark Mellman, senior strategist for John Kerry's campaign, first noted that analysts should be careful to:

Avoid what psychologists call fundamental attribution error. Fundamental attribution error consists, as some of you know, of overweighting the significance, the importance of individuals, of personalities, of events, and underweighting the significance, the salience of the structure of the situation of the underlying circumstances.

He went on to outline the factors he believed to have been the most important in determining the vote in the 2004 election: the decline of the Democratic plurality among voters, culture as the primary division in American politics as opposed to class, and Bush's incumbency advantage. While he did list a number of events that the Kerry campaign viewed as important, his central message was that the campaign did the best it could and very well given the political situation in 2004. Matthew Dowd, a Bush strategist, also viewed the campaign as a series of relatively minor ups and downs with a largely expected outcome. Yet when most people think back to the 2004 campaign, some events seem pivotal, such as the Swift Boat Veterans for Truth (SBVT) ads in August and Kerry's comment that he voted for spending $87 billion before he voted against spending $87 billion for emergency funds for troops and reconstruction in Iraq. Did those two events matter much in the final vote tally, or did they not really change the campaign as much as we might think? Analysts look back at campaigns and identify defining moments but, as Popkin (1991) notes, those anecdotes can be simply the story that 'fits' our overall impression of the campaign as opposed to the events that actually affected the election outcome. How can we tell the difference between anecdotal explanations and the events that were actually important? By identifying which campaign events did, in fact, provide new information in a given election year, we can better understand the 'normal' dynamics of campaigns.

In order to investigate what kind of information and events have which kinds of effects on expectations of a candidate's success, we need a measure that reacts quickly to new information. Opinion polls give us a somewhat nuanced view of public opinion during campaigns, but there is a practical problem with surveys: they are taken over several days. A poll taken the day before, the day of, and the day after a presidential debate does measure the effect of a candidate's performance in the debate on vote choice but it also registers the effects of events (e.g. candidate visits or speeches) that occur before and after the debate. If something shocking happened in a debate (would that debate – watchers could be so lucky!), the results from that standard three-day poll would partially reflect it, but we could only surmise what change in candidate preference occurred because of that gaffe and what change was caused by other events.

Prediction markets, like the Iowa Electronic Markets (IEM) and Intrade (previously called Tradesports), offer an attractive alternative to polls or academic panel studies to answer this kind of question because they allow us to clearly analyse people's expectations of political outcomes and the fluctuation in those expectations given new information. The pricing of futures contracts in these markets provides a continuous measure of the probability of a candidate winning the election. The immediacy of online futures markets lets us see exactly how information is incorporated, how quickly, and how expectations about political events stabilize.

This paper uses data from futures on 'Bush wins the popular vote in 2004', or the traded probability, of Bush winning the election, to build a measure of information flow. Our approach builds on theoretical developments in the economics literature, especially the market microstructure literature, that view the variation of asset prices as being driven by the arrival of new information and the process that incorporates that information into market prices. We extend a theoretically justified model that combines returns and volume in creating a measure of information flow to include two regimes of variation. This provides a novel approach to differentiate between the 'normal' campaign occurrences and the noteworthy, important events of particular campaigns.

While the Swift Boat ads have become central to popular accounts of the 2004 presidential race, it appears that Kerry's nomination speech and the debates, particularly the third one, imparted much more information. Other events that were particularly important in the campaign were the CBS story about Bush's National Guard service and the subsequent retraction, the report that explosives went missing in Iraq in late October, and the release of the bin Laden tape a few days before the election.

21.2 Political prediction markets

Prediction markets, which are also called 'information markets', 'idea futures', or 'event futures', pay out returns based on whether or not a specified event occurs. Winner-take-all futures pay \$1 if event X occurs and \$0 if it does not. At one point during the 2004 campaign, a person could buy a 'Bush wins the popular vote' contract for \$0.55. If that person held onto that contract until November 2, then she would receive \$1, making a profit of \$0.45 cents per contract. Investors can resell or buy contracts at any point. For example, she could buy a Bush contract at \$0.55 and then sell it at \$0.58, for a profit of \$0.03 per contract. This paper examines winner-take-all futures markets, though there are several other types of contracts in prediction markets.

Before proceeding to a discussion of information flow in political prediction markets and the evidence about campaign effects in the 2004 election, a few theoretical points about these kinds of markets must be reviewed. In particular, we discuss why prediction markets are able to aggregate information such that the prices of future contracts are reflective of the event's actual probability of occurring.

How do markets aggregate information? When a group of people are asked how many marbles are in a given large jar, very few of them will choose the correct answer. However, the average of the guesses is always a close estimate of the actual number. Similarly, few people can estimate the number of yards between point A and point B, but together the group 'finds' the answer. This occurs even without discussion; opinions are polled privately and then aggregated to determine the distance. This is an intuitive explanation of how a prediction market works. The theoretical explanation rests on the efficient market hypothesis and no arbitrage arguments (Wolfers and Zitzewitz, 2004).

Prediction markets are likely to produce quality information because they provide: (i) incentives to seek information; (ii) incentives for truthful information revelation and (iii) an algorithm for aggregating diverse opinions (Wolfers and Zitzewitz, 2004). These markets force people to *'put their money where their mouths are'*, so their expressed beliefs about a candidate's chances of success are not just cheap talk. It is important to note that the average belief of traders exhibited in the futures price is actually a weighted average, where people with more confidence in their assessment of a candidate's success trade more so that their opinions count more in the price equilibrium. This can be viewed as an advantage over traditional polls: equilibrium prices take into account the strength of people's beliefs about the candidate's chance of electoral success.

Without discussing a candidate's chances, the information held by people all over the country will be combined so that the market estimate of the candidate's probability of winning is accurate. That is, a person who lives in Ohio might trade based on his impression of voters in his town and a person in North Carolina might trade on her impression of the local newspaper coverage. When added together with traders from around the country (and around the world), all the information is combined to produce the accurate value of the contract.

21.2.1 Bounded rationality

The economic theory behind prediction markets suggests that the prices of assets are accurate because people rationally and correctly understand and incorporate information into the market prices. However, behavioral research shows that people are limited information processors and are subject to biases due to their preferences. This is not a major problem because *'all agents need not be informed for a market to be efficient ... survey evidence documenting biased expectations does not necessarily imply that market prices will be similar biased'* (Sauer, 1998). A market is efficient in the sense that it fully reflects all public information because of marginal traders or market makers. In a market of sufficient size and with large enough payoffs to counter transaction costs, arbitrage in prices will occur until all traders, specialists as well as informed and uninformed investors, agree on the appropriate price (Andersen, 1996). At this point in time Tradesports represents the largest political prediction market and appear to have enough participation to alleviate the bias problem.

It is hard to deny that most traders are biased much of the time. Berg and Rietz (2006) notes that *'traders are biased and mistake prone'* and Forsythe *et al.* (1999) concur that *'traders frequently leave money on the table through violations of arbitrage restrictions'*. These conclusions are documented by surveys administered by the Iowa Electronic Market to its traders. The surveys have consistently shown that traders are not in any way representative of the voting population. In particular, traders on the IEM are younger, more educated, more wealthy, and 'more male' than the US population and the voting population. Due to the incentives to seek information when trading, they are also probably more informed than the voting population.

Traders in the IEM tend to hold portfolios that are biased towards their preferred candidates (Forsythe *et al.*, 1992). Forsythe *et al.* (1998) show that investors in the 1993 Canadian elections futures markets held unbalanced portfolios in ways consistent with their preferences. That is, supporters of a party were much more likely to hold more futures of that party than was the market as a whole. Forsythe *et al.* (1999) demonstrate that this bias also exists in the US presidential prediction markets. The trading population of Tradesports is likely somewhat different than the trading population of the IEM due to its larger size and commercial aspect. However, there is still reason to suspect that individual bettors in Tradesports markets are also biased.

These types of biases have been described as 'wish fulfillment' in both the political science literature and the psychological literature. Granberg and Brent (1983) and Uhlaner and Grofman (1986) document the persistence and ubiquity of this phenomenon. Supporters of Carter in 1980 were much more likely to predict that Carter would win than were Reagan supporters (though a few of each candidate's supporters were less biased in their predictions). Whether due to the 'false consensus effect', where supporters of a particular candidate see themselves as more representative of other voters than they actually are, or the 'assimilation-contrast effect', where supporters of a candidate interpret news about that candidate much more positively (Uhlaner and Grofman, 1986), most traders in the IEM exhibit this wishful thinking as evidenced in their portfolio holdings.

21.2.2 Prediction market efficiency

Demonstrations of these biases does not necessarily mean that the prices of candidate futures are incorrect. While there is some evidence that most traders in the IEM are biased in their trading decisions (or at least likely to make mistakes) and it is likely that many, if not most, of the traders in Tradesports markets are similarly behaved, markets are nonetheless accurate because of marginal traders. Though this paper specifically focuses on the effects of information on futures prices as opposed to the markets' predictive power, this section will briefly review the evidence of the accuracy of prices in prediction markets in order to show that analysis of prediction markets' prices is worthwhile.

Many studies of the IEM show that prediction markets are as accurate as or more accurate than opinion polls, both in the week or days before the election and at longer time horizons. Berg *et al.* (2003) note that since 1988, the IEM political prediction markets (conducted in US elections as well as in other countries such as Canada, Austria, and Turkey) have consistently outperformed polls and *'in a few cases (the 1988 and 1992 U.S. Presidential Elections) the market dramatically outperformed polls'*. Not surprisingly, markets with a higher volume of contracts traded and markets with fewer candidates or parties are more successful. US presidential election markets are the most accurate, though the overall market accuracy is also impressive: the average market error in vote-share contracts[1] was 1.49% or 1.58% (depending on the exact measure used) whereas the average poll error was 1.91%. It is more difficult to judge the true probability of an event occurring, but in markets where a winner-take-all contract and a vote-share contract are both traded, evidence suggests

[1] A vote-share contract pays out $1* the percent of the vote won by the candidate or party.

that the prices of the two contracts move together such that the probability expressed in the winner-take-all contract is similarly accurate. The ability to predict the winner is intellectually interesting,[2] though the accuracy of the election markets has been studied so extensively in part because other applications of idea futures markets rely on the markets' ability to accurately forecast events.

If surveys of traders and findings from political science and psychology all show that people invested in the market are biased, how are market predictions themselves accurate? The apparent paradox is resolved by the 'marginal trader' (e.g. Forsythe *et al.*, 1992, 1999). As Forsythe *et al.* (1998) explain, *'as long as there are some traders relatively free of such wish fulfillment biases and with deep enough pockets, they will take advantage of the biases of other traders and in the process bring prices to levels consistent with unbiased expectations'*. These less-biased investors serve as 'market makers', which is to say that they will continue to trade until the price of each futures contract is correct.

Oliven and Rietz (2004) show that the average trader leaves money on the table, either due to bias or mistakes, but that the marginal trader takes advantage of those arbitrage opportunities and adjusts the prices. Other work has shown that market makers, the marginal traders, whose portfolios and trades exhibit no bias, invest much more money than the average trader and are more active in trading. As a result, marginal traders are able to drive prices to efficient levels while profiting from the mistakes of more error-prone traders (Forsythe *et al.*, 1999).

Marginal traders' behavior explains the accuracy of prediction markets, so our starting premise is correct: these markets are worth studying especially in regards to campaign and information effects. While it is true that the equilibrium price in a market with positive information and transaction costs does not summarize some small amount of the information, that fraction is negligible (Grossman and Stiglitz, 1980). These findings suggest that prices of futures are, in fact, reasonable assessment of the 'true' probabilities of a candidate's success given that all information is rationally incorporated into prices, even if individual traders are not always (or even usually) rational.

21.3 Volatility, trading volume and information flow

Studying the changes in the probability of a candidate winning an election is as important as studying the probability itself. In the context of prediction markets, this means that price movements and trading volume must also be

[2] It is also potentially financially rewarding to predict the outcome of elections (see Hartog and Monroe, 2008).

considered in order to understand the dynamics of the campaign. In their study of presidential approval, Gronke and Brehm (2002) write:

> *No one would argue that a statistical distribution can be described solely by its central tendency; nor should our exploration of the patterns and causes of presidential approval look only at the mean. As important in many circumstances is the frequency and sharpness of shifts in public sentiment-volatility.*

Similarly, if we are interested in the question of what information during campaigns is seen as important or consequential (or at least potentially consequential), the movement of futures prices matters. There may be a significant amount of action in a futures price when new information is acquired, though ultimately the price may settle back to the original equilibrium. An analysis of the mean price over one or several days would hide that effect of political information. Therefore, such an analysis would underestimate the perceived news content of specific campaign events.

Prices remain relatively stable when investors learn no new information about the underlying asset. However, when new information such as earnings reports does become public, stock and futures prices can move more dramatically as traders incorporate the arrival of information. This results in prices revealing the full information content through a sequence of trades. This phenomenon is well studied, with a vast literature in economists and finance focusing on the effects of earnings reports and news releases on volatility and volume (see Jennings and Starks, 1985; Brown and Hartzell, 2001; Patell and Wolfson, 1984; Chan, 2003; Dubinsky and Johannes, 2006) as well as the relationship between prices, volatility and volume (e.g. Gallant *et al.* 1992).

In analysing the effects of politics on financial markets, Bernhard and Leblang (2006) pay careful attention to the volatility of asset prices as a measure of uncertainty about a political outcome. Their particular contribution is to highlight the fact that information affects markets conditional on prior expectations, so simple dummy variables in a time period in which some event occurs can miss the actual impact of the event. When they control for expectations about the 2000 presidential election (as measured by poll numbers), they find that exit poll results from battleground states did affect the Standard and Poors and NASDAQ futures and the US dollar-Japanese yen exchange rate. Similarly, this paper tries to identify unexpected events that affected the electoral chances of the candidates in 2004, above and beyond the long-term, underlying electoral landscape.

Volatility can be conceptualized as a measure of the information flow into a market. In the market microstructure literature in economics, the mixture of distributions hypothesis *'posits a joint dependence of returns and volume on an underlying latent event or information flow variable'* (Andersen, 1996). While the theory describes the *'informativeness of market prices, the presence of liquidity traders in the market and the manner in which news is disseminated'* as factors influencing both volatility and volume, it *'points to the rate of information arrival to the market as the primary variable of interest'* (Fleming *et al.*, 2006). The volatility, or the variance of the prices, is *'primarily [caused] by the arrival of new information and the process that incorporates this information into market prices'* (Andersen, 1996).

Information flows into the market and affects both the prices of the assets and the volume of the assets that are traded. When this trading occurs, the price of the asset reaches a new equilibrium state. Andersen (1996) explains:

> *An important finding is that over the course of a (short) period, the sequence of trades reveals the pricing implications of the private signals and subsequently – until new private information arrives – all market participants agree on the value of the traded asset. Thus, private information arrivals induce a dynamic learning process that results in prices fully revealing the content of the private information through the sequence of trades and transaction prices.*

It is new information that starts this process. In the context of political prediction markets, when new information about the campaign – whether polling numbers, a larger ad buy in a battleground state, or an endorsement – becomes known, the price of the candidate's futures contract adjusts accordingly through the increased participation of market makers. Volatility is low despite some new information when rational anticipation of that event was already incorporated into the price. In contrast, high volatility occurs when unexpected information enters the market.

This price adjustment process, as evidenced in the estimated volatility, can be used to characterize information and events. Jennings and Starks (1985) study the effects of high-content earnings announcements as compared with low-content announcements and find that high-content ones induce more volatility for longer periods of time. Studies like that one take the information content as fixed and study the effects on the market, whereas this paper looks at the effects on the market and attempts to make statements about the information content contained in different campaign events. This is done by first estimating a measure of information flow (or volatility) using returns and volume. This measure is then used to evaluate the informativeness of a campaign event. While this approach presents some hurdles to making assertions about causality, it provides a novel and theoretically sound way to identify consequential events, therefore enhancing our understanding of the role of information, campaign events, and other factors on election outcomes.

In this paper we work with a extended version of Andersen (1996), where the underlying information flow process is determined by a Markov-switching stochastic volatility model (Carvalho and Lopes, 2007; So *et al.*, 1998) that allows for occasional shifts in the parameter determining its level. The incorporation of structural changes in the model solves the problem of overestimation of the persistency of volatility in high-frequency data (Lamoureux and Lastrapes, 1990) while providing a straightforward way to identify clusters of high and low information flow in the market. In Andersen's model, returns and trading volume follow a mutually and serially independent bivariate distribution, conditional on the number of information arrivals, where information arrivals are modeled as a stochastic process. This is an empirical representation of a well defined market microstructure model in which both informed and liquidity traders (traders that do not react to news) participate in determining prices in the market and where the volatility, or information flow, process is inferred simultaneously from returns and trading volume.

In detail, let R_t represent the logarithm of returns, V_t a measure of trading volume and I_t the flow of information at time t. The model takes the following form:

$$R_t \sim N(0, I_t) \tag{21.1}$$

$$V_t \sim Po(m_0 + m_1 I_t) \tag{21.2}$$

$$\log(I_t) = \alpha_{S_t} + \beta \log(I_{t-1}) + \epsilon_t \tag{21.3}$$

$$S_t \sim \mathcal{MS}(P) \tag{21.4}$$

where $\epsilon_t \sim N(0, \tau^2)$ and $\mathcal{MS}(P)$ denotes a two-stage first order Markov-switching process with transition probabilities defined by $Pr(S_t = i \mid S_{t-1} = j) = p_{ij}$, for $(i, j) = 1, 2$.

In the model, returns are conditionally normal with variances that reflect the intensity of information arrivals that also characterize how strongly volume fluctuates in response to news. In this context, m_0 represents the participation of liquidity traders in the market whereas the proportion of trades by 'informed' traders (or market makers, see Section 21.2.2) is defined by m_1. Moreover, the information flow I_t (also refereed to as volatility) follows a conditionally stationary ($|\beta| < 1$) *log* AR(1) process where the level α_{S_t} is determined through a two-stage Markovian process. We will refer to S_t as the the level of information (or information state), meaning that when S_t equals 1 there is a reduced level of news arrival. This implies that $\alpha_1 < \alpha_2$ as these parameters define the unconditional mean of the information flow process. It is important to point out that the choice of two stages for S_t has, on the one hand, an applied motivation as we are looking for a way to segregate between periods of low and high information. This device is used by Carvalho and Lopes (2007) in flagging currency crises in emerging markets. On the other hand, evidence in the literature suggests that at least two stages are necessary to properly characterize the volatility of assets (So *et al.*, 1998; Lopes and Carvalho, 2007; Eraker *et al.*, 2003).[3]

Equations (21.1)–(21.4) represent a complex, bivariate dynamic nonlinear system with two latent (non-observable) states, the information flow I_t and the information state S_t. Using standard priors, posterior estimation of this model is carried by a customized Markov chain Monte Carlo algorithm described in the appendix.

We implement the described model and construct a measure of information flow for the 2004 presidential race. The data were acquired from Tradesports. They include all transactions made on the Tradesports 'Bush wins the popular vote' contract from July 1 to Election Day. The transactions were grouped into 3 hour time periods. This choice reflects our hope to work with the finest level of resolution while guaranteing that enough transactions occurred in the period. Observations of the price of the contract at the close of the three hour period and the volume traded in that period were recorded at 3am, 6am, 9am, noon, 3pm, 6pm, 9pm, and 11:59pm. Following the standard in the literature (Andersen, 1996), volume was transformed to the log scale and rounded to the closest integer due to the Poisson model specification. Futures prices, returns and volume are presented in Figure 21.1 and 21.2. Finally, we catalog events from the Drudge Report (`www.drudgereport.com`) in the same 3 hour window. We use the Drudge Report for several reasons. First, to investigate whether the market reacts to campaign events, we need a 'time-stamp' on a news story. Traditionally, media studies have 'timed' events by using the day that the story was on the front page of a major newspaper such as the *New York Times*. Theoretically, we expect the market will react much more quickly, so we need a more fine-grained measure. The Drudge Report provides that; it is constantly updated and is archived every 15 minutes. Second, the Drudge Report is a widely-read political gossip source. In a recent book (Halperin and Harris, 2006), two influential political reporters write:

> *It is a guarantee that most of the reporters, editors, producers, and talk show bookers who serve up the daily national buffet of news recently have checked out his eponymous website, and that* `www.drudgereport.com` *is bookmarked on their computers. That is one reason Drudge is the single most influential purveyor of information about American politics.*

People do use this as a source of information, and that information is on the 'early' side – Drudge routinely reports stories that the major media news outlets have not yet reported or vetted. The most famous example is the Monica Lewinsky story; the Drudge Report was the first place to break most of the news,

[3] We did assess the performance of a model with three regimes and due to the similarity between the levels (a) of of two states we are confident that the model with two regimes is sufficient and appropriate for what we are trying to accomplish here.

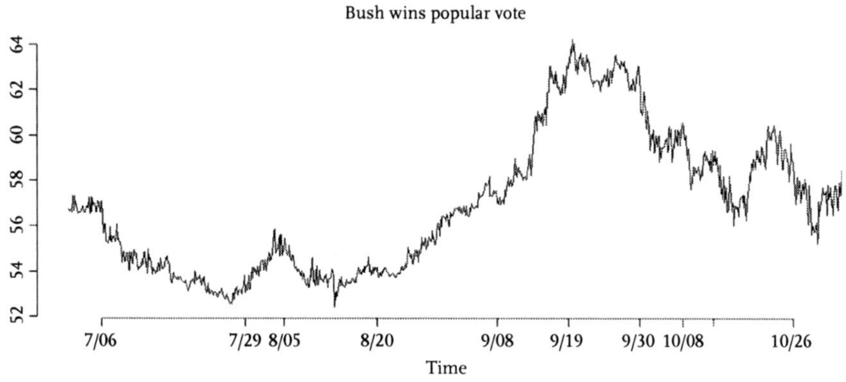

Fig. 21.1 Closing price for 'Bush wins the popular vote' at every three-hour period. Source: Tradesports.com.

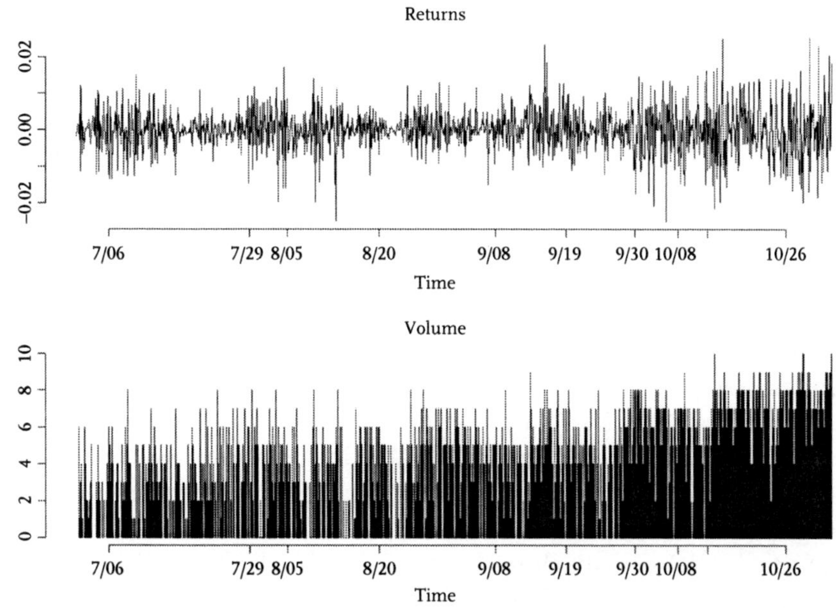

Fig. 21.2 Returns and trading volume.

and covered the story for weeks before the major news outlets. Because it is an agenda-setter with early information and has precise time stamps, we feel that this is the best available source.[4]

[4] To collect the data, we recorded the headlines on the front page of the site at each three hour interval. Major stories remained in the headlines for many periods. In theory, smaller stories could appear and disappear in-between our data collection points. However, if that is the case, we find it unlikely that that could be driving the markets.

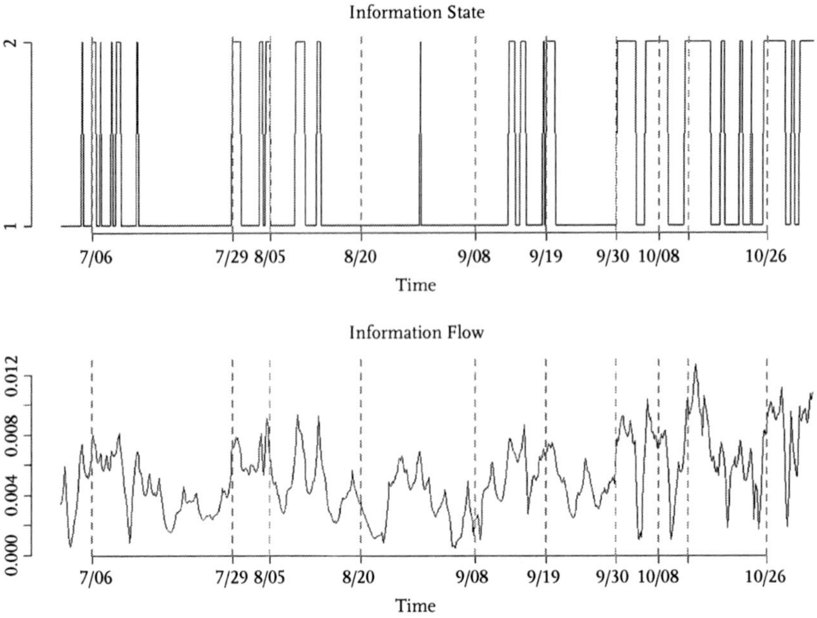

Fig. 21.3 Posterior estimates of $S_{1:T}$ (top) and $\sqrt{I_{1:T}}$ (bottom). We define $\hat{S}_t = 2$ if the posterior $Pr(S_t = 2) > 0.5$. The red lines indicate the key events described in Table 21.2.

21.4 The 2004 presidential election

After fitting our model, we can initially conclude that the flow of information during the campaign was not static. Estimates of the states I_t and S_t are presented in Figure 21.3 and a posterior summary of all parameters appears in Table 21.1. Volatility rose significantly at many points during the campaign, though it remained low for much (but not all) of July, August, and September. The duration of increased volatility also varied. Spikes in volatility were often accompanied by longer stretches of a state of high volatility, indicating greater uncertainty about the effect of new information. However, on many occasions, the volatility increased to a high state and then quickly fell back to the normal level, indicating that the effects of some new information were easily understood and incorporated into the probability of Bush's electoral success. This is also reflected in the estimates of p_{11} and p_{22} given that $p_{11} > p_{22}$.

We are able to assess whether the microstructure assumptions underlying the proposed model are justified by looking at the estimates of m_0 and m_1. If m_1 is close to zero that would indicate that the information arrival does not affect 'informed traders' and the model would be reduced to a simple stochastic volatility model (Jacquier *et al.*, 1994). From the estimates in Table 21.1 we can conclude that the participants in this market do react to new information as, on average, 50% of trades are related to the arrival of news when $S_t = 2$

Table 21.1 Posterior summary.

Parameter	Posterior mean	95% Interval
α_1	−0.369	(−0.395; −0.342)
α_2	−0.306	(−0.327; −0.284)
β	0.967	(0.927;0.983)
τ^2	0.019	(0.018;0.020)
m_0	3.936	(3.81;4.03)
m_1	28370	(25478;32005)
p_{11}	0.983	(0.932;0.991)
p_{22}	0.853	(0.801;0.912)

(high information state). When $S_t = 1$ (low information state) this proportion goes down to 15% with m_0 accounting for the remainder in both cases. These numbers are very similar to the findings in the finance literature as reported by Andersen (1996) and Watanabe (2000).

We would like to underscore an important point: the fact that most campaign events, particularly in mid-August to mid-September, had little or no new information is not saying that the campaign did not have any impact. Rather, it is saying that given the cast of characters and the political and economic context, the campaign itself played out predictably. It is true that during the month of August, there was an elevated state of information on just seven days, for a total of under 70 hours (or 22 periods with our measure), and July had just nine days with elevated information, for a total of just over 70 hours (or 24 periods with our measure). Despite a number of high-profile news stories during August, particularly the Swift Boat Veterans for Truth ads, the investors in the market felt that the underlying context of the election did not change very much. However, information that is not unexpected could, nonetheless, have been crucial to the electoral outcome.

Electoral outcomes can be thought to be composed of three pieces: context (economy, partisanship), standard campaign events, and election year-specific events. First, we will address the events specific to the 2004 presidential election. We will then present evidence regarding the effects of standard campaign events like the nominating conventions. A list of the most influential events appears in Table 21.2 and are marked by the red vertical lines in Figure 21.3.

21.4.1 National security and the 2004 election

Many people have described 2004 as a ‘national security’ election. Opinion polls showed that people were highly concerned about the issue – in an ABC News[5] poll taken in early October, nearly half the respondents reported that either terrorism or the war in Iraq were the most important issues for the

[5] ABC News Poll. October 7, 2004. Data provided by The Roper Center for Public Opinion Research, University of Connecticut. [USABC.100804.R5].

Table 21.2 Key dates in 2004 presidential election.

July 6	Kerry picks Edwards as VP
July 29	Kerry's nomination speech
August 5	First Swift Boat Veterans for Truth advertisement airs
August 20	Second SBVT ad airs
September 8	Bush's National Guard service questioned in CBS story
September 19	CBS retracts the National Guard story and Dan Rather releases statement
September 30	First debate – Bush performs poorly
October 8	Second debate
October 13	Third debate – Kerry makes comment about Cheney's daughter
October 26	Story about missing explosives in Iraq in 2003

next president to deal with, and more importantly, made their vote choice while thinking about that issue (Newsweek, 10/29; Institute of Politics, 2006).[6] Our analysis of the information flow during the campaign corroborates this: a number of national security-related issues significantly influenced the flow of information in the market. We will discuss several of these: the Swift Boat Veterans for Truth advertising campaign, the CBS story about Bush's National Guard service, the news about missing weapons in Iraq, and Osama bin Laden's taped message released just before the election.

Swift Boat Veterans for Truth (SBVT) ads After the 2004 election, a new verb appeared in political discussions: to 'swift boat' someone meant to damage his candidacy with unfair and 'dirty' campaigning. The term itself may be useful, but is it accurate to apportion so much impact to the Swift Boat Veterans for Truth in the 2004 election? Our analysis suggests that while there was an increase in information flow at two points during the height of the story, the influence of the group was actually much less than is conventionally believed.

Though a book had been released months before detailing them, allegations made by the Swift Boat Veterans for Truth's that John Kerry had not earned his medals in the Vietnam War became a major news story on August 4, when SBVT first aired an advertisement. Because SBVT did not have much money to buy air time for their first ad, their strategy was to generate free media through the outlandish claims made in the ad. Needless to say, the strategy worked. Their first ad buy was for $500K, though after they gained notoriety, they significantly increased fundraising and by the end of the campaign, they had spent about $19M on television advertising (Institute of Politics, 2006).

News of the first advertisement regarding Kerry's medals elevated the information state for two days. Several days later, on August 9, the Drudge

[6] Princeton Survey Research International/Newsweek Poll. October 29, 2004. Data provided by The Roper Center for Public Opinion Research, University of Connecticut. [USPSRNEW.103004.R15].

Report and others began questioning whether John Kerry was in Cambodia on Christmas 1969, as he previously claimed. It was later determined (and Kerry confirmed) that he was in Cambodia, but he was there in January 1970, not on Christmas. In retrospect, a rather unimportant point, but the markets reacted: the volatility increased on August 9 and 10. SBVT remained a major news story for most of August and has since made it into the vernacular, but the 'information' from the SBVT allegations was apparently fully incorporated in approximately 36 hours (12 of our periods) over four days. There is another spike in the information flow on August 12 and 13, though there does not appear to be specific information about the SBVT allegations during that time. Even if that increase in volatility is attributed to SBVT, the information was incorporated into the market in just about 50 hours. It should be noted that the debates increased the information flow for a longer period of time than did the SBVT story (see below) and that the story regarding Bush's National Guard service (see below) increased the information flow for about the same amount of time. Additionally, though it is not the point of this paper, it is interesting to note that the likelihood that Bush would win the election did not change much from the beginning of August to the beginning of the Republican convention on August 30; the likelihood went from about 54% to about 56%. At the very least, because the change in the probability of Bush winning reelection was slight and our analysis suggests that the information was incorporated relatively quickly, it seems incorrect to describe Kerry's loss as a result of the 'swift-boating' that took place in August 2004.

CBS story on Bush's national guard story John Kerry's military service was discussed frequently (by his campaign, his supporters, and also his detractors) throughout the primary and general election campaign. However, George W. Bush's military service had largely been ignored until September 8, when CBS News aired a story on '60 Minutes' that Bush had not properly reported for duty while he was in the National Guard. Dan Rather's reporting highlighted documents allegedly written by Bush's commander in the National Guard. The story developed over 12 days, with conservative bloggers immediately questioning the story and the documents and investigating the print on 1970s typewriters. It was several days into the story that the information state increased, on September 13, which coincided with the first reports by major news organizations that the papers were forged. The information flow remained high for most of the time between the 13th and the 20th and peaked on Sunday, September 19, when the '60 Minutes' producer wrote an editorial in the *Washington Post* that stated the story was incorrect. It remained high until Dan Rather released a statement on the evening of September 20 admitting that the story was not properly reported and that the evidence was false. As this story developed, the volatility in the market rose and fell several times, indicating that new information was impacting investors and that it took some time for the potential consequences of the story to become clear.

Missing explosives in Iraq Another national security issue burst into the campaign one week before Election Day. On Monday, October 25, the *New York Times* broke the story that 380 tons of powerful explosives vanished from an Iraq storage depot when the cache was supposed to be secured by American troops. Early that morning, when the Drudge Report initially reported it (presumably after it had appeared on the *New York Times* website), the information state increased immediately. Because this story went along with Senator Kerry's accusations that President Bush had incompetently handled the Iraq war after the initial invasion and the news was completely unexpected, it was viewed as informative. This heightened state of information lasted for three days.

Osama bin Laden videotape The last days of the campaign showed an elevated level of volatility. On the Friday before the election (October 29), Al-Jazeera television showed a taped message from Osama bin Laden. After being authenticated by the FBI, American television stations began showing it. Bin Laden did not directly address the election, but he did say that, *'Your security is not in the hands of Kerry or Bush or al Qaeda. Your security is in your own hands. Any nation that does not attack us will not be attacked'.*[7] When the ad was first played, the information flow increased dramatically; evidently people thought that bin Laden's discussion of 9-11 could influence the election by raising the salience of the issue. The information state remained high throughout the last three days of the campaign, in part because of the tape but presumably also because of new polls and reports of frenzied campaigning.

21.4.2 Effects of standard campaign events

Scholars have found that nominating conventions (Campbell *et al.*, 1992 Holbrook, 1996) and presidential debates (Lanoue, 1992; Benoit *et al.*, 2003; Hillygus and Jackman, 2003) regularly impact voters. This is in part due to the fact that these events are highly publicized. Presidential debates are watched by millions of people; in 2004, 62.5 million viewers tuned into the first debate, 46.7 million watched the second, and 51 million people watched the third.[8] Viewership of nominating conventions has declined, but the candidate's acceptance speeches are still viewed by millions of people. It is debated whether or not the choice of a vice presidential candidate has an influence on the outcome,

[7] 'Bin Laden: *"Your security is in your own hands"*'. CNN, Friday, October 29, 2004 Posted: 10:05 PM EDT. `www.cnn.com/2004/WORLD/meast/10/29/bin.laden.transcript/`, viewed February 17, 2008.

[8] This is in comparison the 15.2 million who watched Fox's coverage of baseball playoffs. Though we find this heartening for democracy, it is nonetheless shocking. We have no idea why anyone would choose a debate over a Red Sox – Yankees game.

but at the least we can agree that the selection is a high-volume news event. Our measure of information flow reveals that the John Edwards' selection as John Kerry's running mate was viewed as important by investors and that the debates contained a significant amount of information. For the most part and contrary to prior findings, we show that the nominating conventions did not alter expectations about the election's outcome.

Vice presidential nomination Political observers begin discussing possible vice presidential candidates early in the nominating process. Though it is unclear whether or not the vice presidential candidate adds much to the party's ticket, the conventional wisdom suggests that it can potentially impact election outcomes by winning additional states or bringing the party together after a nomination battle (Mayer, 2000). Our estimates of information flow suggest John Kerry's announcement that he would pick John Edwards, former senator from North Carolina, was considered to have a potential impact on the election's outcome.

In 2004, after a strong second-place showing in the Iowa caucus, John Edwards was viewed as affable and a charismatic speaker. More importantly, he could perhaps help Kerry win the support of southern voters. On July 6, the day that the Kerry campaign announced that Edwards would be the vice presidential candidate, the information state was heightened. Investors thought that this could affect the outcome of the election. Interestingly, Edwards' selection was incorporated into expectations relatively quickly; the information state remained elevated for only 18 hours.

Nominating conventions Presidential nominating conventions have become highly choreographed media spectacles, though many people do watch parts of the four day conventions. Prior research suggests that party conventions do move voters (e.g. Holbrook, 1996). These findings are based on polls taken just before and just after conventions. Because our information flow variable is continuous and responds immediately to new information, we are able to better investigate the dynamics associated with conventions.

In 2004, the Democratic convention was held in Boston, Mass. on July 26–29 and the Republican convention was held in New York City on August 30–September 2. Our findings suggest that only one small part of one convention affected expectations about the ultimate election outcome: Kerry's acceptance speech on July 29 started an information incorporation period. The information state remained elevated for a full day.[9] At first, the finding that conventions had little effect seems somewhat surprising, though after further consideration, it makes sense: expected information has previously been incorporated

[9] We assume that this elevated information flow was due to Kerry's speech, and not to the failure of the balloons to drop immediately after his speech.

into the market's prices. After four years, there was little that voters could learn about Bush or his policies and nearly all had opinions of him. The fact that volatility did not increase in demonstrates that the Republican convention did not contain any unexpected information.

In contrast, John Kerry was relatively unknown before his presidential bid; before the Democratic convention many people did not know much about Kerry, his background, or his policies. His speech was meant as an introduction to the candidate, and evidently investors thought that that introduction was, in fact, informative. Before the convention, Kerry's chances to win the popular vote were about even, and they remained around 50% after his speech. An analysis of simply the mean price, or probability, of a futures contract would suggest that the convention was completely irrelevant. This volatility measure suggests that despite the fact that there was not a significant change in price, people did learn about Kerry from the convention. The information acquired in this period likely conditioned the effect of later information and was thus indirectly important.

Both campaigns note that a key moment during the campaign occurred during the Republican convention (Institute of Politics, 2006). On September 1, Chechen terrorists invaded an elementary school in the Russian town of Beslan; after the Russian government stormed the school several days later, 200 people, mostly children, were killed. The campaigns argued that this reinforced Bush's message that the world is dangerous. Our analysis finds no such impact. Clearly, this incident could not possibly have been forecast by investors in the market, so what explains this? We can only speculate that because people already placed the issue of terrorism as a high priority for the next president, a certain ceiling effect existed; if the issue was ranked as the most important, it would be hard for the salience to increase very much.

Debates There is a significant amount of evidence suggesting that televised presidential debates impact elections (Geer, 1988; Benoit *et al.*, 2003), though *'their impact has fluctuated, from inconsequential to decisive'* (Graber, 1993). Even if debates do not change the winner of the election, citizens can and do learn from them (Lanoue, 1992; Holbrook, 1994; Miller and MacKuen, 1979). Our research confirms that debates provide a significant amount of information. In 2004, there were three presidential debates: on Thursday, September 30; Friday, October 8; and Wednesday, October 13. Bush was widely viewed as performing poorly in the first debate (e.g. *Los Angeles Times*, 10/2), though the reviews of the second and third debates were mixed. Each debate increased the volatility of the market. In particular, the first and third debates led to an elevated information state for several days after the event. This extended length of volatility fits with findings that post-debate news analysis conditions people's responses to debates (Hwang *et al.*, 2007).

The volatility surrounding the days of the debates highlights the difficulty isolating the cause of that increased information. The information flow increased immediately following the first debate, so it seems reasonable to conclude that that increase was caused by the debate. However, while the debate was the most notable event that occurred from September 30 to October 4 (when the volatility died down), a number of other plausible stories can be told for why the information flow was high three days later, on October 3rd.

Similarly, the level of volatility increased more than a day before the third debate and continued for more than three days after. What information was the market incorporating? In that debate, when asked whether he believed homosexuality was a choice, Kerry replied, *'Dick Cheney's daughter, who is a lesbian, she would tell you that she's being who she was'*. Both liberals and conservatives found the comment awkward and odd and thought he should not have said it. Was it that comment, or the fact that the polls moved only slightly over the course of the three debates, that caused the increased volatility? Or was it some other event? Days after that debate, the popular summary source of political information, ABC's *The Note*, commented:

> *Because the polls moved slightly in Bush's favor after the debates, there must be a reason, and the only two reasons [the media] have been able to come up with are (i) the Mary Cheney remark; and/or (ii) the nation, having considered the totality of the debate round robin, decided it wanted a steady, likeable leader – rather than a voluble debating champion – to be the honcho of the free world. We have no idea if those are the reasons, but that's the best anyone has offered that has smacked against our ears.*

The problem of pinpointing causality in this kind of case seems, to us, intractable. Our method points to the fact that the debates added information to campaign, but we cannot apportion how much is due to comments during the debate, media analysis immediately following the debate, or even to separate events that happen the next day. In July and August, when less is going on in the campaign, it is easier to isolate events. By mid-October, so many polls are being released, so much money is being spent, and so many people are paying attention that it is much more difficult to make any kind of causal statement.

21.5 Concluding remarks

The approach we develop here provides a novel and useful way to investigate the effects of campaigns. The information flow measure allows us to recognize key moments and separate the anticipated, normal dynamics of campaigns from the unexpected occurrences. By using data from prediction markets and a model that attempts to characterize the reactions of market participants to new information, we are able to address an as-yet unresolved question in political science. We build on a vast literature in finance and economics and propose a model with two volatility regimes that simplifies the task of associating events with periods of high information. The development and estimation of these models is greatly facilitated by the use of the Bayesian framework.

We identify a number of events that increased the information flow in the 2004 campaign. Among the standard campaign events, we find that the selection of the vice presidential candidate and the televised debates increased volatility in the prediction market. John Kerry's acceptance speech at the Democratic nominating convention also increased volatility though neither the rest of the convention or any of the Republican convention introduced new information. A number of events specific to the 2004 election were also viewed as informative by investors. In particular, stories related to national security-related issues increased volatility. These include the report that explosives vanished in Iraq under the US's watch, the CBS story about Bush's National Guard service and its subsequent retraction, and the release of the bin Laden tape a few days before the election. We find evidence that the Swift Boat Veterans for Truth ads that were critical of John Kerry's military service record were somewhat informative, but not nearly as central to the outcome of the 2004 campaign as is popularly believed.

We would like to reiterate two important points about this project. First, our goal was to develop a quantitative way to objectively describe the information level over the course of a presidential campaign. The price of the 'Bush wins the popular vote' futures contract chronicles the likelihood of Bush winning in 2004, and clearly whether or not certain events or certain classes of events impact that likelihood is an important question. However, the question of whether an event is informative is related to, but distinct from, the question of what effect a given event has on a particular election. That is, certain events can contain significant information even if they do not alter the outcome in a particular year. For example, if there is an overriding factor in the election like a bad economy then a candidate's acceptance speech at the party convention may not change the likelihood that the incumbent party will lose. Nonetheless, speeches at conventions do contain information, and in a year where the economy was not so salient, that speech could alter the election outcome. Put differently, some information that comes out during the campaign may be trumped by other information or considerations and thus not impact the vote percentage of a candidate. But that event does, in fact, have information in it. In a year with a different underlying political context, that same event may move vote totals significantly. Because our model focuses on the information level itself, we hope our model can be used to develop a more a general theory of campaign event effects over time.

Second, when we find relatively low levels of information in some given period of time, we are not saying that the campaign itself during those periods is irrelevant. We certainly believe that campaigns matter. The expectations about what will happen on election day are dependent on our prior understanding of how campaigns generally progress. When we find low levels of information, we are only arguing that those are periods where people are not learning anything new about the election.

We believe our model is a promising technique for research on campaigns and elections. We will be able to test it and our theory more explicitly in the future because prediction markets grow in popularity in each election cycle. Intrade has had contracts on the party nominees and the winner of the 2008 general election trading with high volume for months. Going forward with this research, we would like to find ways to make stronger causal claims. The question of causality looms large in the social sciences, and this work is no exception. We want to extend our modelling approach in an attempt to more clearly isolate the effects of different events. Addressing that becomes both more difficult and more important in the last weeks of the campaign. The 2008 Democratic primary season provides a particularly useful testing ground for us; because partisanship cannot be the deciding factor between two candidates in a primary, we are likely to see more dramatic movement in the markets during the campaign as the situation progresses.

There certainly is no shortage of political pundits, campaign strategists, and journalists willing to declare 'what the election was about' and highlight the 'turning point' in the campaign, the event or story that pushed the campaign towards its ultimate result. Undoubtedly, pundits and politicos are sometimes correct; but often they are not. When we explain election outcomes with a compact anecdote, we can both overlook events that were actually important and overstate our confidence in our conclusions. A Bayesian modeling approach to this problem, combined with theoretical insight from other fields, gives us greater leverage on a question that is of both academic interest and practical significance – it helps us to understand the dynamics and responsiveness of our democratic institutions.

Appendix

A. Broader context and background

A.1 Posterior computation in state-space models

Consider the general class of state-space models (West and Harrison, 1997):

$$y_t \sim p_y(x_t, \theta) \tag{21.5}$$

$$x_t \sim p_x(x_{t-1}, \theta), \tag{21.6}$$

where x_t represents the states and y_t the vector of observations, both at time t. Observations y_t are treated as independent thought time conditionally on the hidden states x_t and fixed parameters θ. Moreover, the functional form of the densities $p_x(\cdot)$ and $p_y(\cdot)$ is assumed known. These models are often referred to as hidden Markov models (Scott, 2002) and are extensively used in many scientific fields, including economics, biology, engineering, etc.

After observing a sequence $y_{1:T}$, the central element for inference is the joint posterior distribution $p(x_{1:T}, \theta | y_{1:T})$, which, in most cases, cannot be assessed analytically and its exploration requires the development of posterior simulation procedures. Following basic Markov chain Monte Carlo ideas the most direct approach to posterior simulation is to iterate thought the full conditional posterior of individual states $p(x_t | x_{(-t)}, \theta, y_{1:T})$ and fixed parameters $p(\theta | x_{1:T}, y_{1:T})$. This strategy is generally referred to as *single-move* MCMC (Carlin *et al.*, 1992). Given the markovian structure of the model the full posterior for each state can be expressed as

$$p(x_t | x_{(-t)}, \theta, y_{1:T}) \propto p(x_t | x_{t-1}, \theta) p(x_{t+1} | x_t, \theta) p(y_t | x_t, \theta). \tag{21.7}$$

If direct simulation from (21.7) is possible the single-move MCMC becomes a Gibbs sampler, however in many situations draws from (21.7) require a Metropolis – Hastings step.

Although very general, the single-move MCMC can mix very slowly due to the high temporal dependence between states. In some models, the Markov structure can be exploited in the construction of a *block-move* MCMC that iterates through $p(x_{1:T} | \theta, y_{1:T})$ and $p(\theta | x_{1:T}, y_{1:T})$. Drawing the states jointly reduces the the autocorrelation of the Markov chain providing significant efficiency gains in the MCMC sampler. This process begins by writing the joint

$$p(x_{1:T} | \theta, y_{1:T}) = p(x_T | \theta, y_{1:T}) \prod_{t=1}^{T-1} p(x_t | x_{t+1}, \theta, y_{1:t}), \tag{21.8}$$

and noticing that samples from (21.8) can be obtained by first sampling $x_T \sim p(x_T | \theta, y_{1:T})$ followed by, for all t, samples from $x_t \sim p(x_t | x_{t+1}, \theta, y_{1:t})$. This procedure requires the availability of both $p(x_T | \theta, y_{1:T})$ and $p(x_t | x_{t+1}, \theta, y_{1:t})$. This is always possible for situations where the states are discrete (Scott, 2002) or in Dynamic Linear Models (DLM) (Carter and Kohn, 1994; Fruhwirth-Schnatter, 1994) where both $p_x(\cdot)$ and $p_y(\cdot)$ are normal distributions linear in the states. This strategy is also known as *forward filtering backward sampling* algorithms where a forward filter is necessary to assess $p(x_T | \theta, y_{1:T})$ that are then propagated backwards through $p(x_t | x_{t+1}, \theta, y_{1:t})$. Details of the implementation in DLM are presented in Chapters 4 and 15 of West and Harrison (1997) while elements of the discrete case are described below in the drawing of $S_{1:T}$.

Connecting back to the model described in (21.1), we are working with a bivariate, non-normal, state space model with one continuous state λ_t and one discrete state S_t. Posterior simulation for $S_{1:T}$ will be possible through a block-move MCMC whereas a single-move MCMC is necessary for $\lambda_{1:T}$. We complete the model with the following priors: $\alpha_i \sim N(m_{\alpha_i}, C_{\alpha_i})$, $\beta \sim N(m_\beta, C_\beta)\mathbf{I}_{\{-1<\beta<1\}}$, $\tau^2 \sim IG(a_\tau, b_\tau)$, $p_{ii} \sim Be(a_i, b_i)$, $m_0 \sim Ga(a_{m_0}, b_{m_0})$ and $m_1 \sim Ga(a_{m_1}, b_{m_1})$. Let $\theta = \left(m_0, m_1, \alpha_1, \alpha_2, \beta, \tau^2, P\right)$. Based on the following decomposition of the joint posterior distribution

$$p\left(I_{1:T}, S_{1:T}, \theta | R_{1:T}, V_{1:T}\right) \propto p\left(R_{1:T} | I_{1:T}\right) p\left(V_{1:T} | I_{1:T}, \theta\right) p\left(I_{1:T} | S_{1:T}, \theta\right) p\left(S_{1:T} | \theta\right) p\left(\theta\right) \tag{21.9}$$

we can generate posterior samples thought a MCMC schemes defined by the full conditionals described below.

A.2 *Drawing the states $I_{1:T}$ and $S_{1:T}$*

The full conditional for $S_{1:T}$ can be written as:

$$\begin{aligned} p\left(S_{1:t} | \cdot\right) &= p\left(S_T | \lambda_{1:T}, \theta\right) \prod_{t=T-1}^{1} p\left(S_t | S_{t+1}, \lambda_{1:T}, \theta\right) \\ &\propto p\left(S_T | \lambda_{1:T}, \theta\right) \prod_{t=T-1}^{1} p\left(S_{t+1} | S_t\right) p\left(S_t | \lambda_t, \theta\right). \end{aligned} \tag{21.10}$$

This allows for a block sampling of $S_{1:T}$ through a forward filtering backward sampling algorithm as described above (?)see also][]Kim99. In detail, assume that at time t we have available the set of posterior probabilities $p\left(S_t = i | \lambda_t, \theta\right)$, for $i = 1, 2$ (clearly, this can be generalized to k discrete states). Prior probabilities for time $t + 1$ can be obtained via

$$p\left(S_{t+1} = j | \lambda_t, \theta\right) = \sum_{i=1}^{2} p\left(S_{t+1} = j | S_t = i, \theta\right) p\left(S_t = i | \lambda_t, \theta\right). \tag{21.11}$$

This probabilities are then updated via

$$p\left(S_{t+1} = j | \lambda_{t+1}\right) \propto p\left(\lambda_{t+1} | S_{t+1} = j, \theta\right) p\left(S_{t+1} = j | \lambda_t, \theta\right). \tag{21.12}$$

The above steps provide all the necessary distributions for the generation of a joint draw of $S_{1:T}$.

Sampling of I_t is the only 'hard' step in the MCMC. The full conditional $p(I_t | I_{-t}, \cdot)$ takes the following form

$$p\left(I_t | I_{-t}, \cdot\right) \propto \exp\left(-\frac{1}{2 I_t} R_t^2\right) \exp\left(m_0 + m_1 I_t\right) \left(m_0 + m_1 I_t\right)^{V_t} \tag{21.13}$$

$$\times \exp\left\{-\frac{1}{2\tau^2} \left(\lambda_t - \alpha_{S_t} - \beta\lambda_{t-1}\right)^2\right\} \exp\left\{-\frac{1}{2\tau^2} \left(\lambda_{t+1} - \alpha_{S_{t+1}} - \beta\lambda_t\right)^2\right\}. \tag{21.14}$$

Joint samples from the full conditional of $I_{1:T}$ cannot be generated without approximations such as de Jong and Shephard (1995). In this work, we implement a single-move MCMC and generate draws from $p(I_t|I_{-t},\cdot)$. Random-walk metropolis was used. It is our experience that this simple approach performed well enough and our MCMC presented satisfactory convergence diagnostics.

A.3 Drawing θ

Full conditionals for θ can be easily obtained as follows:

$$(\alpha_i|\cdot) \sim N\left\{\left(\frac{T_i}{\tau^2}+\frac{1}{C_{\alpha_i}}\right)\left(\frac{\sum_{t:S_t=i}\lambda_t-\beta\lambda_{t-1}}{\tau^2}+\frac{m_{\alpha_i}}{C_{\alpha_i}}\right),\left(\frac{T_i}{\tau^2}+\frac{1}{C_{\alpha_i}}\right)^{-1}\right\}; \quad (21.15)$$

$$(\beta|\cdot) \sim N\left\{\left(\frac{\sum_{t=1}^{T}\lambda_{t-1}^2}{\tau^2}+\frac{1}{C_\beta}\right)\left(\frac{\sum_{t=1}^{T}\left(\lambda_t-\alpha_{S_t}\right)\lambda_{t-1}}{\tau^2}+\frac{m_\beta}{C_\beta}\right),\right.$$

$$\left.\left(\frac{\sum_{t=1}^{T}\lambda_{t-1}^2}{\tau^2}+\frac{1}{C_\beta}\right)^{-1}\right\}; \quad (21.16)$$

$$(\tau^2|\cdot) \sim IG\left(a_\tau+\frac{T}{2},\quad b_\tau+\frac{1}{2}\sum_{t=1}^{T}\lambda_t-\alpha_{S_t}-\beta\lambda_{t-1}\right); \quad (21.17)$$

$$(p_{11}|\cdot) \sim Be(a_1+n_{11}, b_1+n_{12}); \quad (21.18)$$

$$(p_{22}|\cdot) \sim Be(a_2+n_{22}, b_2+n_{21}); \quad (21.19)$$

$$p(m_0|\cdot) \propto \exp\left[-(b_{m_0}+T)\,m_0+\sum_{t=1}^{T}\log\left\{m_0^{a_{m_0}-1}(m_0+m_1 I_t)^{V_t}\right\}\right]; \quad (21.20)$$

$$p(m_1|\cdot) \propto \exp\left[-\left(b_{m_1}+\sum_{t=1}^{T}I_t\right)m_1+\sum_{t=1}^{T}\log\left\{m_1^{a_{m_1}-1}(m_0+m_1 I_t)^{V_t}\right\}\right], \quad (21.21)$$

where $\lambda_t = \log(I_t)$, T_i represents the number of observations allocated to the ith state and n_{ij} are counts of transitions from state i to j. Sampling from the above distributions is straightforward. The only minor complication is the Metropolis step necessary for sampling m_0 and m_1. We use a random-walk Metropolis with normal proposals. Initial values where defined based on the MLE of both parameters in a Poisson regression where $I_{1:T}$ was fixed at the posterior mean from a standard two stage Markov-switching stochastic volatility model where an efficient Gibbs sampler can be implemented. These estimates of the information flow are also used as initial values for $I_{1:T}$.

References

Andersen, T. (1996). Return volatility and trading volume: An information flow interpretation of stochastic volatility. *Journal of Finance*, **51**, 169–204.

Benoit, W. L., Hansen, G. and Verser, R. (2003). A meta-analysis of the effects of viewing us presidential debates. *Communication Monographs*, **70**, 335–350.

Berg, J., Nelson, F. and Rietz, T. (2003). Accuracy and forecast standard error of prediction markets. Discussion paper, Tippie College of Business, University of Iowa.

Berg, J. and Rietz, T. (2006). The Iowa electronic market: Lessons learned and answers yearned. In *Information Markets: A New Way of Making Decisions*, (ed. R. Hahn and P. Tetlock), AEI Press, Washington, DC.

Bernhard, W. and Leblang, D. (2006). *Democratic Processes and Financial Markets: Pricing Politics*. Cambridge University Press, New York.

Brown, G. and Hartzell, J. (2001). Market reaction to public information: The atypical case of the Boston celtics. *Journal of Financial Economics*, **60**, 333–370.

Campbell, J., Cherry, L. and Wink, K. (1992). The convention bump. *American Politics Research*, **20**, 287–307.

Carlin, B., Polson, N. and Stoffer, D. (1992). A Monte Carlo approach to nonnormal and nonlinear state-space modeling. *Journal of the American Statistical Association*, **87**, 493–500.

Carter, C. and Kohn, R. (1994). On Gibbs sampling for state space models. *Biometrika*, **81**, 541–553.

Carvalho, C. and Lopes, H. (2007). Simulation-based sequential analysis of Markov switching stochastic volatility models. *Computational Statistics and Data Analysis*, **51**, 4526–4542.

Chan, W. (2003). Stock price reaction to news and no-news: drift and reversal after headlines. *Journal of Financial Economics*, **70**, 223–260.

de Jong, P. and Shephard, N. (1995). The simulation smoother for time series models. *Biometrika*, **82**, 339–350.

Dubinsky, A. and Johannes, M. (2006). Earnings annoucement and equity options. Working paper. University of Columbia Business School. Available at SSRN: http://ssrn.com/abstract=600593.

Eraker, B., Johannes, M. and Polson, N. (2003). The impact of jumps in volatility and returns. *The Journal of Finance*, **58**, 1269–1300.

Fleming, J., Kirby, C. and Ostdiek, B. (2006). Stochastic volatility, trading volume, and the daily flow of information. *The Journal of Business*, **79**, 1551–1590.

Forsythe, R., Frank, M., Krishnamurthy, V. and Ross, T. (1998). Markets as predictors of election outcomes: Campaign events and judgement bias in the 1993 UBC election stock market. *Canadian Public Policy*, **24**, 329–351.

Forsythe, R., Nelson, F., Neumann, G. and Wright, J. (1992). Anatomy of an experimental political stock market. *American Economic Review*, **82**, 1142–1161.

Forsythe, R., Rietz, T. and Ross, T. (1999). Wishes, expectations and actions: a survey on price formation in election stock markets. *Journal of Economic Behavior and Organization*, **39**, 83–110.

Fruhwirth-Schnatter, S. (1994). Data augmentation and dynamic linear models. *Journal of Time Series Analysis*, **15**, 183–202.

Gallant, A., Rossi, P. and Tauchen, G. (1992). Stock prices and volume. *Review of Financial Studies*, **5**, 199–242.

Geer, J. G. (1988). The effects of presidential debates on the electorate's preferences for candidates. *American Politics Quarterly*, **16**, 486–501.

Graber, D. (1993). *Mass Media and American Politics*, (4th edn). Congressional Quarterly Press, Washington.

Granberg, D. and Brent, E. (1983). When prophecy bends: The preference–expectation link in US presidential elections, 1952–1980. *Journal of Personality and Social Psychology*, **45**, 477–491.

Gronke, P. and Brehm, J. (2002). History, heterogeneity, and presidential approval: a modified arch approach. *Electoral Studies*, **21**, 425–452.

Grossman, S. and Stiglitz, J. (1980). On the impossibility of informationally efficient markets. *American Economic Review*, **70**, 393–408.

Halperin, M. and Harris, J. F. (2006). *The Way to Win: Taking the White House in 2008*. Random House, New York.

Hartog, D. and Monroe, N. (2008). The value of majority status: The effect of Jeffords's switch on asset prices of republican and democratic firms, *Legislative Studies Quarterly*, **33**, 63–84.

Hillygus, D. and Jackman, S. (2003). Voter decision making in election 2000: Campaign effects, partisan activation, and the Clinton legacy. *American Journal of Political Science*, **47**, 583–596.

Holbrook, T. (1994). The behavioral consequences of vice-presidential debates: Does the undercard have any punch? *American Politics Research*, **22**, 469–482.

Holbrook, T. (1996). *Do Campaigns Matter?* Sage Publications, London.

Hwang, H., Gotlieb, M., Nah, S. and McLeod, D. (2007). Applying a cognitive-processing model to presidential debate effects: Postdebate news analysis and primed reflection. *Journal of Communication*, **57**, 40–59.

Jacquier, E., Polson, N. and Rossi, P. (1994). Bayesian analysis of stochastic volatility models. *Journal of Business and Economic Statistics*, **12**, 371–389.

Jennings, R. and Starks, L. (1985). Information content and the speed of stock price adjustment. *Journal of Accounting Research*, **23**, 336–350.

Kim, C. and Nelson, C. (1999). *State-Space Models with Regime Switching: Classical and Gibbs-Sampling Approaches with Applications*. The MIT Press, Cambridge, Massachusetts.

Lamoureux, C. and Lastrapes, W. (1990). Persistence in variance, structural change and the garch model. *Journal of Business and Economic Statistics*, **8**, 225–234.

Lanoue, D. (1992). One that made a difference: Cognitive consistency, political knowledge, and the 1980 presidential debate. *Public Opinion Quarterly*, **56**, 168–184.

Lopes, H. and Carvalho, C. (2007). Factor stochastic volatility with time varying loadings and markov switching regimes. *Journal of Statistical Planning and Inference*, **137**, 3082–3091.

Mayer, W. G. (2000). *In Pursuit of the White House 2000: How we Choose our Presidential Nominees*. Chatham House Publishers, New York.

Miller, A. and MacKuen, M. (1979). Learning about the candidates: The 1976 presidential debates. *Public Opinion Quarterly*, **43**, 326–346.

Oliven, K. and Rietz, T. (2004). Suckers are born but markets are made: Individual rationality, arbitrage, and market efficiency on an electronic futures market. *Management Science*, **50**, 336–351.

Patell, J. and Wolfson, M. (1984). The intraday speed of adjustment of stock prices to earnings and dividend announcements. *Journal of Financial Economics*, **13**, 223–252.

Popkin, S. (1991). *The Reasoning Voter: Communication and Persuasion in Presidential Campaigns*. University of Chicago Press, Chicago.

Sauer, R. (1998). The economics of wagering markets. *Journal of Economic Literature*, **36**, 2021–2064.

Scott, S. (2002). Bayesian methods for hidden Markov models: Recursive computing in the 21st century. *Journal of the American Statistical Association*, **97**, 337–351.

So, M., Lam, K. and Li, W. (1998). A stochastic volatility model with Markov switching. *Journal of Business and Economic Statistics*, **16**, 244–253.

Uhlaner, C. and Grofman, B. (1986). The race may be close but my horse is going to win: Wish fulfillment in the 1980 presidential election. *Political Behavior*, **8**, 101–129.

Watanabe, T. (2000). Bayesian analysis of dynamic bivariate mixture models: Can they explain the behavior of returns and trading volume? *Journal of Business and Economic Statistics*, **18**, 199–210.

West, M. and Harrison, P. (1997). *Bayesian Forecasting and Dynamic Models*. Springer-Verlag, New York.

Wolfers, J. and Zitzewitz, E. (2004). Interpreting prediction market prices as probabilities. NBER Working Paper No. 12200, National Bureau of Economic Research, USA. Available at `www.nber.org/papers/w12200`.

·22·

Bayesian analysis in item response theory applied to a large-scale educational assessment

Dani Gamerman, Tufi M. Soares and Flávio B. Gonçalves

22.1 Introduction

The OECD's[1] Programme for International Student Assessment (PISA) collects information, every three years, about 15-year-old students in several countries around the world. It examines how well students are prepared to meet the challenges of the future, rather than how well they master particular curricula. The data collected in PISA surveys contains valuable information for researchers, policy makers, educators, parents and students. It allows countries to assess how well their 15-year-old students are prepared for life in a large context and to compare themselves to other countries. As it is mentioned in PISA 2003 Technical Report[2], it is now recognized that the future economic and social well-being of countries is closely linked to the knowledge and skills of their population.

The first PISA survey was in 2000 and 43 countries participated in the assessment. The others were in 2003 with 41 countries, in 2006 with 57 countries and in 2009 with 67 countries. At least one country from each continent is present in the surveys, the majority being from Europe. A list with the countries participating in 2003 is presented in Appendix B.

Since PISA is applied to a large number of countries, many differences are found among the students participating, from cultural to curricular ones. These differences may influence the characteristics of some items in each country. For example, if an item is strongly related to some cultural aspect of an specific country, it is expected that this item would be easier for students from this country than for students from another one where such cultural aspect is not present. If this difference is not taken into account, such item is not good to assess the abilities of students from these two countries. Such phenomenon is called differential item functioning (DIF) and has been heavily studied in the

[1] OECD – Organisation for Economic Co-Operation and Development.

[2] PISA 2003 Technical Report. `http://www.pisa.oecd.org/dataoecd/49/60/35188570.pdf`.

psychometric literature. In general, one group is fixed as the *reference group* and the other ones as the *focal groups*, and the item functioning in the latter ones is compared to the item functioning in the reference group.

DIF analysis can be separated in two different stages. The first one is the detection of items with DIF and the second one is the quantification and explanation of the DIF detected in the first stage. Normally, the latter one is based on regression structures to identify features of the items that present differential functioning. Usually, the first stage is divided in two substages: the estimation of the students abilities or a matching of the students using groups, where students in the same group are assumed to have the same ability; and the DIF detection using these estimated abilities. Both substages are linked because, in general, it is impossible to generate abilities which are totally free of DIF. Therefore, the most natural approach would be to treat them together.

However, along with the integrated modelling of the DIF, comes the problem of model identifiability, which reflects the conceptual difficulty of the problem. For this reason, additional hypotheses on the DIF parameters of the model are assumed in order to guarantee identifiability. Nevertheless, such hypotheses may be too restrictive.

Situations like this, where, besides the information from the data, initial hypotheses on the parameters are needed, are very suitable for Bayesian approaches. The use of appropriate prior distributions, elicited in a Bayesian framework and based on previous knowledge on the items' DIF, can be less restrictive than usual hypotheses, but still enough to guarantee identifiability.

Preparation of a large test like PISA somehow screens the DIF to avoid such items to compose the test. It is then not expected that items with large DIF, that have a great influence on the proficiencies, are found. On the other hand, it is impossible to eliminate all DIF in the tests, specially in such large-scale tests.

This chapter presents a DIF analysis of the Mathematics test in PISA 2003, for the English speaking countries, using an integrated Bayesian model. Such model allows items to present DIF without affecting the quality of the estimated proficiencies. Moreover, the DIF analysis may be useful to detect educational differences among the countries.

22.2 Programme for International Student Assessment (PISA)[3]

22.2.1 What does PISA assess?

PISA is based on a dynamic learning model that takes into account the fact that knowledge and ability must be continuously acquired to have a successful

[3] The information presented in this section is based on the PISA 2003 Technical Report. `http://www.pisa.oecd.org/dataoecd/49/60/35188570.pdf`.

adaptation to a constant changing world. Therefore, PISA aims to assess how much knowledge and ability, essential for an effective participation in society, is acquired by students near the end of compulsory education. Differently from previous international assessments (IEA, TIMMS, OREALC, etc.),[4] PISA does not focus only in curricular contents. It also emphasizes the knowledge required in modern life. Besides that, data about the student's study habits and their motivation and preferences for different types of leaning situation is also collected. The word Literacy is used in the sequel to show the amplitude of the knowledge, abilities and competencies being assessed, and it covers:

- contents or structures of the knowledge the students have to acquire in each domain;
- processes to be executed;
- the context in which these knowledge and abilities are applied.

In Mathematics, for example, the competence is assessed in items that cover from basic operations to high order abilities involving reasoning and mathematical discoveries. The Literacy in Maths is assessed in three dimensions:

- the content of Mathematics – firstly defined in terms of wide mathematical concepts and then related to branches of the discipline;
- the process of Mathematics – general mathematical competence in use of mathematical language, selection of models and procedures, and abilities to solve problems;
- situations where Mathematics is used – varying from specific contexts to the ones related to wider scientific and public issues.

In PISA 2003, four subject domains were tested, with mathematics as the major domain, and reading, science and problem solving as minor domains. It was a paper-and-pencil test. The tests were composed by three item formats: multiple-choice response, closed-constructed response and open-constructed response. Multiple choice items were either standard multiple choice, where the students had to choose the best answer out of a limited number of options (usually four), or complex multiple choice where the students had to choose one of several possible responses (true/false, correct/incorrect, etc.) for each of the statements presented. In closed-constructed response items, a wider range of responses was possible. Open-constructed response items required more extensive writing, or showing a calculation, and frequently included some explanation or justification. Pencils, erasers, rulers, and in some cases calculators, were provided. The decision on providing or not calculators was made

[4] IEA – International Association for Evaluation of Educational Achievement, TIMSS – Trends in International Mathematics and Science Study, OREALC/UNESCO – Oficina Regional de Educación para América Latina e Caribe.

by the national centres and was based on standard national practice. No items in the pool required a calculator but, in some items, its use could facilitate computation. Since the model used here to analyse the data is for standard multiple-choice items, the items which do not have this format were converted into dichotomous items. Most of them were already dichotomized by PISA. The remaining items were dichotomized by putting 1 for full credit and 0 for partial or no credit.

The complex probabilistic sample, involving stratification and clustering, was obtained by randomly selecting the schools. Inside the selected schools, the students who were from 15 years and 3 months to 16 years and 2 months old were also randomly chosen. The students had to be in the 7th or 8th year of Basic School or in High School. The sample was stratified by the schools' location (urban or rural). Besides, information on physical infrastructure of the school, geographic region, type of school (private or public) and number of enroled students were also used as variables in the implicit stratification.

22.2.2 The structure of PISA 2003 Maths test

More than a quarter of a million students, representing almost 30 million 15-year-old students enrolled in the schools of the 41 participating countries, were assessed in 2003.

The Maths test was composed by 85 items, from which 20 were retained from PISA 2000 for linking purposes. These items were selected from a previous set of 217 items by expert groups based on several criteria. Some important characteristics of the 85 selected items are presented in Tables 22.1, 22.2 and 22.3. The characteristics presented in these tables will be considered in the DIF analysis performed here.

22.3 Differential item functioning (DIF)

Differential item functioning (DIF) is the phenomenon where the characteristics of an item differ among groups of individuals. Such characteristics may

Table 22.1 Mathematics main study items (item format by competence cluster).

Item format	Competence cluster			
	Reproduction	Connections	Reflection	Total
Multiple-choice response	7	14	7	28
Closed-constructed response	7	4	2	13
Open-constructed response	12	22	10	44
Total	26	40	19	85

Table 22.2 Mathematics main study items (content category by competence cluster).

Content category	Competence cluster			
	Reproduction	Connections	Reflection	Total
Space and shape	5	12	3	20
Quantity	9	11	3	23
Change and relationships	7	8	7	22
Uncertainty	5	9	6	20
Total	26(31%)	40(47%)	19(22%)	85

include discrimination and difficulty. For example, students with same level of knowledge and ability but different cultural background may lead to different probabilities of correctly answering an item with DIF.

The concern about DIF arose with the desire of creating test items that were not affected by cultural and ethnic characteristics of the individuals taking the tests (cf. Cole, 1993). If one does not consider DIF when it exists, one may be led to wrong conclusions regarding, especially, the students knowledge and abilities.

Studies conducted by the Educational Testing Service (ETS) (see Stricker and Emmerich, 1999), in the USA, indicate that DIF may exist due to three factors in a large-scale assessment context: the familiarity to the item's content, which can be associated to the exposure to the theme or to a cultural factor; the personal interest in the content; and a negative emotional reaction caused by the item's content.

It seems reasonable to think that items with DIF should be avoided since these would favor some group(s) of students, besides having some technical and statistical implications and other ethical issues. Nevertheless, DIF items may be very useful to study social and cultural differences that are not easily noticed. In particular, in educational assessment, DIF items can help detecting contents that are treated differently among the groups and may point out in which groups the instruction of such contents should change.

Table 22.3 Mathematics main study items (content category by item format).

Content category	Item format			
	Multiple-choice response	Closed-constructed response	Open-constructed response	Total
Space and shape	8	6	6	20
Quantity	6	2	15	23
Change and relationships	3	4	15	22
Uncertainty	11	1	8	20
Total	28	13	44	85

Explanation of DIF is a hard task. Besides that, the pedagogical and technical structure of a large-scale assessment like PISA aims to create high quality items that do not in general present differential functioning. However, it is known that the characteristics of a country and its level of economic development have great influence in the social and cultural life of its population, which reflects in education. For this reason, different countries are expected to organize the curricula in different ways, some countries may give more importance to some themes and explore more some contents. The challenge in explaining the DIF in PISA is to notice patterns in the items that present DIF. In order to do so, it is important to have a large number of items that are quite different.

22.4 Bayesian model for DIF analysis

The most common approach to undertake a statistical analysis of educational assessment data is by using item response theory (IRT). It is a psychometric theory extensively used in education assessment and cognitive psychology to analyse data arising from answers given to items belonging to a test, questionnaire, etc., and it is very useful to produce scales.

The IRT arose, formally, in the work of Lord (1952) and Rasch (1960). The basic idea of the theory is to apply models, generally parametric, where the parameters represent important features of the items and of the subjects answering the items. Some common item parameters are discrimination, difficulty and guessing. The subject parameters are individual characteristics, for example, a type of ability or some other latent trait possibly associated to his/her psychological condition.

An item can be dichotomous, if the answer is either correct or not; polytomous, if the answer can be classified in more than two categories; and also continuous, if the answer is classified in a continuous scale. There are IRT models for all those situations, but only models for dichotomic items will be considered here.

The increasing use of IRT in educational assessment and the concerns about differential item functioning are leading researchers to propose new models that take DIF into account. In this context, Soares *et al.* (2009) propose a new IRT Bayesian model that is a generalization of the three-parameter logistic model. The proposed model incorporates the detection, quantification and explanation of DIF in an integrated approach.

Typically, in educational assessment, a test is formed by I items, but a student j only answers a subset $I(j)$ of these items. Let Y_{ij}, $j = 1, \ldots, J$, be the score attributed to the answer given by the student j to the item $i \in I(j) \subset \{1, \ldots, I\}$. In the case where i is a dichotomous item, $Y_{ij} = 1$ if the answer is correct and $Y_{ij} = 0$ if the answer is wrong.

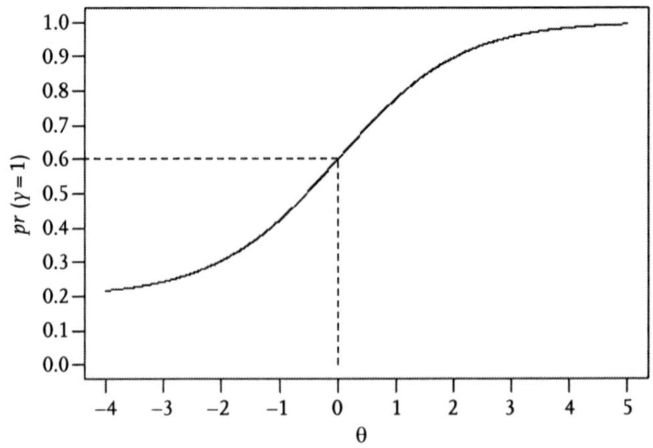

Fig. 22.1 Item's characteristic curve for the three-parameter logistic model with $a = 0.5$, $b = 0$ and $c = 0.2$.

Define $Pr(Y_{ij} = 1|\theta_j, a_i, b_i) = \pi_{ij}$, and consider $logit(\pi_{ij}) = \ln(\pi_{ij}/1 - \pi_{ij})$. If $logit(\pi_{ij}) = \Delta_{ij}$, then $\pi_{ij} = logit^{-1}(\Delta_{ij}) = 1/1 + e^{-\Delta_{ij}}$. One of the most used models in IRT is the three parameters logistic model proposed by Birnbaum (1968), which models the probabilities π_{ij} as

$$\begin{aligned}\pi_{ij} &= c_i + (1 - c_i)logit^{-1}(\Delta_{ij}) \\ \Delta_{ij} &= Da_i(\theta_j - b_i)\end{aligned} \tag{22.1}$$

where a_i, b_i and c_i are the discrimination, difficulty and guessing of item i, respectively, $\forall i$. θ_j is the ability, proficiency or knowledge of student j, $\forall j$, and D is a scale factor designed to approximate the logistic link to the normal one (and set here to 1.7).

A good way to interpret the three-parameter logistic model is by analysing the item's characteristic curve generated by the model, presented in Figure 22.1. Note from the model that the greater a_i is, the greater will be the slope of the curve. This means that, the more discriminant an item is, the larger is the difference in the probability of correct answer for students with different proficiencies, that is, the larger is the capability of the item to discriminate the students. Moreover, the maximum slope is attained at $\theta = b$, as can be seen in the figure.

The difficulty parameter b_i interferes in the location of the curve. If the value of b_i is increased, the curve moves right and the probability of correct answer is reduced (the item becomes harder). Alternatively, if b_i is decreased, the curve moves left and the probability of correct answer is increased (the item becomes easier).

Furthermore, note that $\lim_{\theta_j \to -\infty} Pr(Y_{ij} = 1|\theta_j, a_i, b_i, c_i) = c_i$. Hence, c_i is the minimum probability that a student has to correctly answer item i. c_i is then called guessing parameter because, since it is a multiple choice item, there is always a probability that the student answers it correctly by guessing. Lord (1952) noticed that, in general, the percentage of correct answers for an item, in very low levels of proficiency, was smaller than the inverse of the number of options in the item. Experts in the area report that they have observed varied behaviours for those percentages.

In general, there can be different types of DIF (see Hanson (1998) for a wider characterization). For the three parameters model, the types of DIF can be characterized according to the difficulty, discrimination and guessing. The model proposed here does not consider the possibility of DIF in the guessing parameter. Although it is possible, the applicability of this case is substantially limited by the known difficulties in the estimation of this parameter and by practical restrictions.

Suppose that the students are grouped in G groups. The integrated Bayesian DIF model used in this chapter associates the student's answer to his/her ability via (22.1) with

$$\Delta_{ij} = Da_{ig}(\theta_j - b_{ig}),$$

where a_{ig} is the discrimination parameter for item i and group g, b_{ig} is the difficulty parameter for item i and group g, c_i is the guessing parameter for item i, for $i = 1, \ldots, I$, $j = 1, \ldots, J$ and $g = 1, \ldots, G$.

Since the item difficulty is a location parameter, it is natural to think about its DIF in the additive form and thus it is set as $b_{ig} = b_i - d^b_{ig}$. Analogously, since the discrimination is a scale parameter, it is natural to think about its DIF in the multiplicative form and thus it is set as $a_{ig} = e^{d^a_{ig}} a_i$ (>0). Thus, d^b_{ig} (with $d^b_{i1} = 0$) represents the DIF related to the difficulty of the item in each group and d^a_{ig} (with $d^a_{i1} = 0$) represents the DIF related to the discrimination of the item in each group. Use of the exponential term in the discrimination places the DIF parameter over the line and combines naturally with a normal regression equation setting. Alternatively, the DIF parameter can be specified directly without the exponential form. This leads to a log-normal regression model. The two forms are equivalent but the former was preferred here.

It is assumed, a priori, that $\theta_j|\lambda_{g(j)} \sim N(\mu_{g(j)}, \sigma^2_{g(j)})$, where $g(j)$ means the group which the student j belongs to and $\lambda_g = (\mu_{g(j)}, \sigma^2_{g(j)})$, $j = 1, \ldots, J$. It is admitted that $\lambda_1 = (\mu_1, \sigma^2_1) = (0, 1)$ to guarantee the identification of the model. On the other hand, $\lambda_g = (\mu_g, \sigma^2_g)$, $g = 2, \ldots, G$, is unknown and must be estimated along with the other parameters.

The model is completed with specifications of the prior distributions for the parameters. Let N be the Normal distribution, LN the Log-Normal distribution, Be the Beta distribution and IG the Inverse-Gamma distribution, so, the prior distributions assumed for the structural parameters are:

$$a_i \sim LN\left(\mu_{a_i}, \sigma^2_{a_i}\right), b_i \sim N\left(\mu_{b_i}, \sigma^2_{b_i}\right) \text{ and } c_i \sim Be\left(\alpha_{c_i}, \beta_{c_i}\right), \text{ for } i = 1, \ldots, I.$$

The prior distributions for the parameters of the abilities' distributions are:

$$\mu_g \sim N\left(\mu_{0g}, \sigma^2_{0g}\right) \text{ and } \sigma^2_g \sim IG\left(\alpha_g, \beta_g\right), \quad \forall\, g = 2, \ldots, G.$$

The set of anchor items (items for which $d^a_{ig} = d^b_{ig} = 0,\ \forall\, g = 1, \ldots, G$) is represented by $I_A \subset \{1, \ldots, I\}$. The set of items for which the parameters vary between the groups is represented by $I_{dif} = \{1, \ldots, I\} - I_A$. Moreover, $I^a_{dif} \subset I_{dif}$ is the set of items with DIF in the discrimination and $I^b_{dif} \subset I_{dif}$ is the set corresponding to the DIF in the difficulty. Naturally $I_{dif} = I^a_{dif} \cup I^b_{dif}$. Notice that if an item belongs to I_{dif}, it does not necessarily mean that this item has DIF in the usual meaning of the term. It means that it is not an anchor item and it can potentially have DIF. Besides that, this can be used as admissible information for the explanatory structure imposed to the DIF, which cannot be performed with the anchor items.

Let Z^h_{ig}, be the DIF indicator variable of item i in group g, for parameter h, for $h = a, b$. Therefore, $Z^h_{ig} = 1$ if parameter h of item i has DIF in group g, and $Z^h_{ig} = 0$, otherwise. Two possibilities may be considered: one where Z^h_{ig} is known, which means that the anchor items are known a priori and the DIF analysis considers all the other items; and another one where Z^h_{ig} is unknown and must be identified. In other words, it is not known a priori if the item has or does not have DIF. The latter one will be used in the analysis of PISA.

Finally, a regression structure is considered for d^h_{ig} in the DIF explanation as follows

$$d^h_{ig} = \gamma^h_{0g} + \sum_{k=1}^{K^h} \gamma^h_{kg} W^h_{ik} + \eta^h_{ig}, \quad \text{if } Z^h_{ig} = 1. \tag{22.2}$$

γ^h_{kg} are the fixed parameters of the DIF model, W^h_{ik} are the explanatory variables associated to the items and η^h_{ig} is the item specific random factor for each group. It is also assumed for modelling simplification that random factors are independent with $\eta^h_{ig} \sim N(0, (\tau^h_g)^2)$, for $g = 2, \ldots, G$.

The regression structure is imposed for all items but the anchor ones. Set $W^h_i = (1, W^h_{i1}, \ldots, W^h_{iK^h})$ and $\gamma^h_g = (\gamma^h_{0g}, \ldots, \gamma^h_{K^h g})'$. When $Z^h_{ig} = 1$, the conditional distribution of d^h_{ig} is given by $(d^h_{ig} | \gamma^h_g, W^h_i, (\tau^h_g)^2) \sim N(W^h_i \gamma^h_g, (\tau^h_g)^2)$. When $Z^h_{ig} = 0$, d^h_{ig} will be assumed to have a reduced variance normal distribution $(d^h_{ig} | (\tau^h_g)^2, Z^h_{ig} = 0) \sim N(0, s^2(\tau^h_g)^2)$, where s^2 is chosen to be small enough to ensure that d^h_{ig} is tightly concentrated around (but not degenerated at) 0. This strategy was proposed for variable selection in a regression model by George and McCulloch (1993).

The distribution of $d_{ig}^h | \gamma_g^h, W_i^h, Z_{ig}^h, (\tau_g^h)^2$ can then be written as follows

$$\left(d_{ig}^h | \gamma_g^h, W_i^h, Z_{ig}^h, \left(\tau_g^h\right)^2\right) \sim N\left(\left(W_i^h \gamma_g^h\right) Z_{ig}^h\,,\ [s^2]^{1-Z_{ig}^h} \left(\tau_g^h\right)^2\right).$$

Suitable prior distributions are $\gamma_g^h \sim N(\gamma_0^h, S_0^h)$, $(\tau_g^h)^2 \sim IG(\alpha_g^h, \beta_g^h)$ and $Z_{ig}^h \sim Ber(\pi_{ig}^h)$, where *Ber* is the Bernoulli distribution.

The model proposed here is very general. Apart from the usual parameters, it also has a DIF indicator variable Z_{ig}^h which can either be estimated along with all the other parameters of the model or be fixed a priori. The items for which Z_{ig}^h is not fixed at 0 include additional variables to the model. These variables are used for the DIF explanation and constitute a regression model which may or not have covariates.

Prior distributions play a very important role in a Bayesian model. In this particular model, they are very important in the selection of the anchor items, as it will be seen. If one wishes to set an item as an anchor one, one simply sets $\pi_{ig}^h = 0,\ \forall\, g = 2, \ldots, G,\ \forall\, h = a, b$. Naturally, π_{ig}^h may be set as zero for some but not all groups. In the same way, if one wishes to include an item in the DIF analysis, independent of the DIF's magnitude, one simply sets $\pi_{ig}^h = 1,\ \forall\, g = 2, \ldots, G$, for some h. However, the most interesting use of this prior distributions is to consider previous information and beliefs about the items' functioning to identify parameters more precisely and effectively. The estimation of the parameters is performed by using MCMC methods to obtain a sample from the joint posterior distribution of all the parameters. Details on these methods are presented in Appendix B.

The level of generality introduced by this model aims to represent the complexity of the problem. However, along with the integrated modelling of the DIF, comes the problem of model identifiability. Such problems are described and studied in Soares *et al.* (2009). The authors show that identifiability is achieved by either imposing informative prior distributions for some Z_{ig}^h parameters or fixing some items not to have DIF. Nevertheless, in many cases, the model is identifiable without any of these two actions because of the informative priors attributed to the other parameters.

22.5 DIF analysis of PISA 2003

The Bayesian model presented in Section 22.4 is now used to perform a DIF analysis of the PISA 2003 Maths test in English speaking countries. The database obtained had 84 of the 85 items of the test. The estimates of such missing item's parameters are not presented in the PISA Technical Report 2003 either. Great Britain is defined as the reference group (group 1). Table 22.4 shows the other groups.

Table 22.4 English speaking countries used in the DIF analysis.

Country	Group
Great Britain	1
Canada	2
Australia	3
Ireland	4
USA	5
New Zealand	6

22.5.1 Prior distribution

The following prior distributions are used:

$$a_i \sim LN(0, 2), b_i \sim N(0, 1), c_i \sim Be(5, 17), \mu_g \sim N(0, 1), \sigma_g^2 \sim IG(0.1, 0.1),$$
$$\gamma_g^h \sim N(0, I)\,,\ \left(\tau_g^h\right)^2 \sim IG(0.5, 0.5) \quad \text{and} \quad Z_{ig}^h \sim \text{Ber }(0.5),$$
$$\text{for} \quad i = 1, \ldots, 84, g = 2, \ldots, 6, h = a, b.$$

The value chosen for parameter s^2 is 1/200,000.

The prior distributions of the item parameters are chosen according to what is expected, in general, when the proficiencies follow a standard Normal distribution, which is the case of the reference group. The discrimination parameters are expected to vary mostly between 0 and 3, and the $LN(0, 2)$ has probability 0.70 of being smaller than 3. Most of the difficulty parameters are expected to lie between -2 and 2, and a few of them to have absolute value greater than 2. The standard Normal distribution is then a suitable prior to describe this behavior. The guessing parameter is expected to be low (less than 0.3), since many items were not multiple choice ones, and then should have a low probability of correct guessing. The $Be(5, 17)$ distribution gives 0.8 probability to values smaller than 0.3.

The mean of the proficiencies in the focal groups are not expected to be very far (more than 1 unit) from the mean of the reference group. For this reason, a standard Normal distribution is used for these parameters. For the variance of the proficiencies, a prior distribution with large variance was preferred.

The DIF parameters are expected to have absolute value smaller than 0.5 and, for a few items, between 0.5 and 1. For this reason, the effect of a binary covariate is also expected to be around these values. Therefore, a standard Normal distribution is used for the coefficients in the regression analysis. For the variance of the regression error, a prior distribution with large variance is adopted.

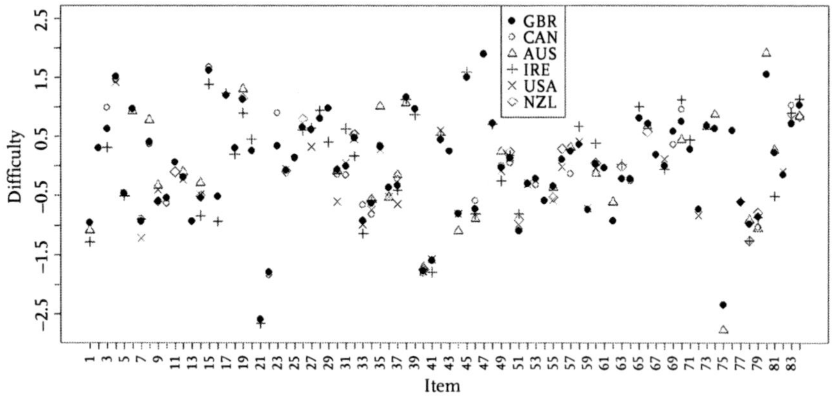

Fig. 22.2 Estimates of difficulty parameters in the reference group and in the groups where they had a posterior probability of DIF greater than 0.5.

Since there is no prior information about how likely each item is to present DIF, a symmetric prior distribution Ber(0.5) is used for the DIF indicator variables Z_{ig}^{h}.

22.5.2 DIF detection

Figures 22.2, 22.3 and 22.4 show some results for the item parameters. For the discrimination and difficulty parameters, the estimates in the reference group (GBR) and in the groups where the item is likely to have DIF a posteriori are presented.

A nice feature of the model is that it incorporates the uncertainty about DIF in the estimates. This means that the model does not have to 'decide' if an item has DIF. It outputs a posterior distribution on the parameters that describes

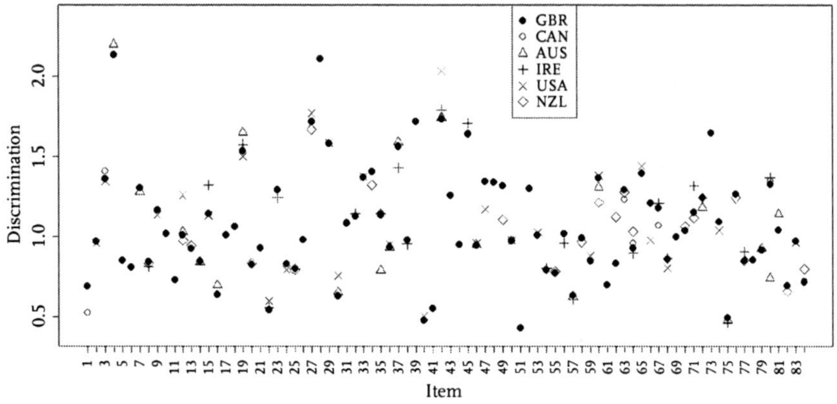

Fig. 22.3 Estimates of discrimination parameters in the reference group and in the groups where they had a posterior probability of DIF greater than 0.5.

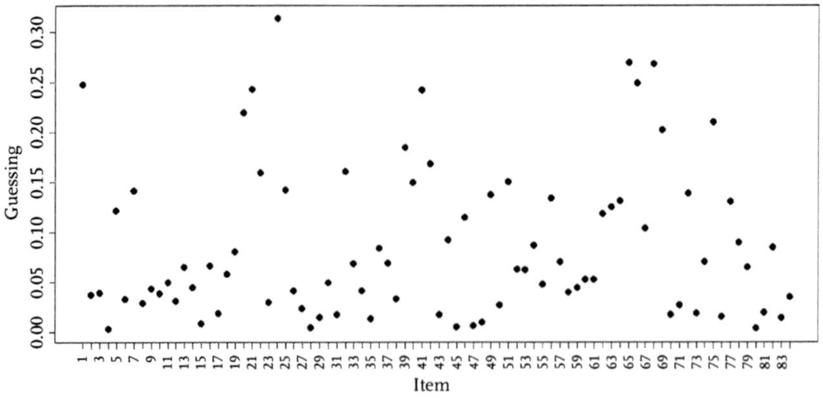

Fig. 22.4 Estimates of guessing parameter.

the uncertainty about that, particularly, for the posterior distribution of the DIF parameters. It is up to the researcher to analyse this uncertainty and draw conclusions from it.

If an item i is likely to have DIF a posteriori in group g and parameter h, say $P(Z^h_{ig} = 1|Y) > 0.5$, the posterior distribution of d^h_{ig} will probably be bimodal with one mode around 0. In order words, this posterior distribution is a mixture of a distribution with mean around 0 and a very small variance and another distribution. The former one is the distribution of $(d^h_{ig}|Z^h_{ig} = 0, Y)$ and the latter is the distribution of $(d^h_{ig}|Z^h_{ig} = 1, Y)$. The more likely the item is to have DIF a posteriori, the greater is the weight of the second component of the mixture.

The decision on an item presenting or not DIF is based on the posterior distribution of the respective Z^h_{ig}. If one item is assumed to have DIF, the estimation of this DIF can be made by the distribution of $(d^h_{ig}|Z^h_{ig} = 1, Y)$. For this reason, the estimates of the item parameters in the focal groups, when the item has a posterior probability of DIF greater than 0.5, presented in Figures 22.2 and 22.3, are the posterior means of $(d^h_{ig}|Z^h_{ig} = 1, Y)$.

Concerning difficulty, 20 items have posterior probability of DIF smaller than 0.5 in all groups, that is, they are more likely not to present any noticeable DIF. No item has this DIF probability greater than 0.5 for more than three countries. Forty-five DIF parameters have absolute value greater than 0.3, which is a considerable magnitude for DIF, and nine have this value greater than 0.5, which is a large value for DIF.

Considering discrimination, seven items have posterior probability of DIF smaller than 0.5 in all groups. Nine items have this DIF probability greater than 0.5 for four countries and no item has this probability greater than 0.5 for all the countries. On the other hand, only eight DIF parameters are greater than

0.3 (which increases the discrimination by 35% if it is positive and decreases by 26% if it is negative). Only one DIF parameter is greater than 0.5; this value increases the discrimination by 65% if it is positive and decreases by 40% if it is negative.

Finally, note that most of the guessing parameters estimates are less than 0.15. This was expected since most of the items are not multiple choice ones. It would be reasonable to fix $c = 0$ for these items. However, this was not imposed, in order to enable the comparison of the results obtained with the integrated Bayesian model with the ones from Bilog-mg software (see Thissen, 2001) without DIF. Bilog-mg uses the same scale for the proficiencies and fits a three parameters logistic model.

22.5.3 DIF explanation

Four covariates were chosen to explain the DIF in the English speaking countries. The first one is the Content category shown in Table 22.2. It is a categorical variable with four categories: Space and shape; Quantity; Change and relationships; and Uncertainty. Three dummy variables are used to introduce this covariate to the regression model. Quantity is the base category, the first dummy variable refers to Space and shape, the second one to Change and relationships, and the last one to Uncertainty.

The second covariate used in the DIF explanation is Competence cluster shown in Table 22.1. It is also a categorical variable and has three categories: Reproduction; Connections; and Reflections. Two dummy variables represent the covariate where Connections is the base category, the first dummy variable refers to Reproduction and the second one represents Reflection.

The third covariate is a binary variable and indicates if the item has support of graphical resource. Finally, the fourth covariate represents the size of the question, measured by the number x of words and standardized using the rule $(x - 75)/100$.

Although the complete analysis can be performed in an integrated way, incorporating all model uncertainty, the DIF explanation is performed here in a separate step. The DIF magnitude among the six countries is not large, since few items with large DIF were detected in the previous section. This way, it would be difficult to highlight possible effects of covariates in the DIF.

The analysis is performed by fixing the items that had a small probability of presenting DIF in all groups in the first analysis as anchor items (13, 21, 43, 61, 67 and 74). All the other items are fixed to have DIF, that is $\pi_{ig}^{h} = 1$. This way, the analysis presented in this section is designed only to identify possible effects of covariates in the DIF.

Table 22.5 *Results* of the DIF explanation regression model for DIF in difficulty. * – significant at 90% and ** – significant at 95%. Significant at α% means that the α% posterior credibility interval does not include 0.

Covariates	Coefficients				
	Canada	Australia	Ireland	USA	New Zealand
Model 1					
Intercept	0.03	−0.09	0.001	0.01	−0.02
Space and shape	−0.11	0.02	−0.27**	−0.13	0.11*
Change	−0.13*	−0.01	−0.06	−0.15	0.01
Uncertainty	−0.14**	−0.15**	−0.15	−0.33*	−0.07
Model 2					
Intercept	−0.03	−0.05	0.08	−0.03	0.03
Space and shape	−0.04	0.06	−0.24**	−0.07	0.10*
Uncertainty	−0.08	−0.15**	−0.11*	−0.29**	−0.10*
Reproduction	0.04	0.003	−0.09	−0.003	0.009
Reflection	−0.02	−0.03	−0.10	0.08	0.002
Model 3					
Intercept	−0.03	−0.19*	0.03	0.02	−0.04
Space and shape	−0.05	0.02	−0.25**	−0.03	0.10*
Uncertainty	−0.08	−0.12*	−0.13*	−0.27**	−0.09*
Graphical support	0.003	0.11	−0.02	−0.09	0.01
Model 4					
Intercept	0.02	−0.03	0.03	−0.01	0.02
Space and shape	−0.04	0.05	−0.24**	−0.02	0.11*
Uncertainty	−0.08	−0.15*	−0.13*	−0.23**	−0.08
Question size	0.03	−0.03	−0.01	0.07	0.002

Several models were fitted for the DIF explanation. The first model for DIF in difficulty has only the covariates to indicate the Content category. In each of the following steps, a new covariate was included and the covariates that were not significant at 95% in all group in the previous model were removed. For the discrimination, the same models used for difficulty were fitted.

The results, presented in Tables 22.5 and 22.6, show that Space and shape items are more difficult for students from Ireland compared to the other countries. These items are also somewhat easier for students from New Zealand. Moreover, items related to Uncertainty are harder for students from the USA compared to the other five countries. They are also slightly harder for students from Australia and Ireland than for the ones from the other three countries.

Regarding the discrimination of the items, the results show that, in general, items with DIF in discrimination are more discriminant for students from Great Britain, followed by New Zealand and Ireland, Canada and Australia, and they are less discriminant for students from the USA. On the other hand,

Table 22.6 *Results* of the DIF explanation regression model for DIF in discrimination. * – significant at 90% and ** – significant at 95%. Significant at α% means that the α% posterior credibility interval does not include 0.

Covariates	Coefficients				
	Canada	Australia	Ireland	USA	New Zealand
Model 1					
Intercept	−0.14**	−0.14**	−0.12*	−0.27**	−0.05
Space and shape	0.06	−0.006	0.12	0.09	0.008
Change	−0.01	−0.02	0.02	−0.01	−0.02
Uncertainty	−0.006	0.05	0.07	0.15	0.03
Model 2					
Intercept	−0.16**	−0.17**	−0.06	−0.26**	−0.11*
Space and shape	0.08	0.008	0.10	0.09	0.02
Uncertainty	−0.003	0.04	0.05	0.15*	0.03
Reproduction	−0.01	0.01	−0.08	−0.02	−0.02
Reflection	0.02	0.10	−0.03	0.07	0.05
Model 3					
Intercept	−0.19**	−0.16**	−0.16**	−0.27**	−0.09*
Space and shape	0.06	0.02	0.09	0.09	0.05
Uncertainty	0.01	0.07	0.07	0.17*	0.03
Graphical support	0.04	0.02	0.05	0.04	−0.05
Model 4					
Intercept	−0.17**	−0.16**	−0.12**	−0.22**	−0.09*
Space and shape	0.08	0.01	0.08	0.11	0.02
Uncertainty	−0.02	0.05	0.01	0.16*	0.01
Question size	−0.06	−0.09	−0.20**	−0.08	−0.13*

if only items related to Uncertainty are considered, they are, on average, as discriminant for students from the USA as they are for students from Great Britain and they are more discriminant in these two countries than in the other four countries. Furthermore, the question size has an influence on the DIF in discrimination in Ireland and New Zealand. Larger questions make the items less discriminant in these countries, specially in Ireland.

22.5.4 Analysis of the proficiencies

The results obtained for the distribution of the proficiencies in the DIF analysis without covariates is presented. They are compared with the results from the original analysis of PISA and from the analysis with the Bilog-mg software.

Bilog-mg also allows the existence of DIF, but only in the difficulty. The scale in Bilog-mg is defined by assuming a standard normal distribution for the proficiencies from the reference group. The same is used for the Bayesian model proposed in this chapter.

Table 22.7 *Estimates* of the mean and standard deviation of the proficiencies in each country. * refers to the modified scales.

Country	Parameter	PISA	Bilog-mg*	IBM*	Bilog-mg	IBM
Australia	Mean	524.11	522.95	522.15	0.1602	0.1515
N = 235,486	Std. Dev.	95.60	94.40	93.78	1.0209	1.1042
Canada	Mean	532.70	529.43	528.77	0.2303	0.2231
N = 330,098	Std. Dev.	87.33	90.00	85.26	0.9734	0.9221
Great Britain	Mean	508.14	508.14	508.14	0	0
N = 696,215	Std. Dev.	92.47	92.47	92.47	1	1
Ireland	Mean	503.52	502.87	508.10	−0.0570	−0.0005
N = 54,838	Std. Dev.	85.32	85.62	82.37	0.9259	0.8908
New Zealand	Mean	524.17	521.56	523.68	0.1452	0.1681
N = 48,606	Std. Dev.	98.17	96.72	92.71	1.0459	1.0026
U.S.A.	Mean	483.64	484.85	474.99	−0.2518	−0.3585
N = 3,140,301	Std. Dev.	95.37	91.40	97.52	0.9884	1.0546
Total	Mean	493.81	494.33	487.45	−0.1494	−0.2238
N = 4,505,544	Std. Dev.	95.72	92.88	97.46	1.0045	1.0540

Table 22.7 shows the mean and variance of the distributions of the proficiencies in each country considering:

- the original PISA proficiency (pv1math), which does not consider DIF;
- the results obtained with Bilog-mg without DIF;
- the results obtained with Bilog-mg without DIF in a modified scale;
- the results obtained with the integrated Bayesian model (IBM) proposed in this chapter;
- the results obtained with the integrated Bayesian model (IBM) in a modified scale.

The modified scale referred to above consists in transforming the estimates in order to make the mean and variance of the reference group (GBR) the same as in the PISA scale.

PISA uses the Rasch model and a partial credit model and fixes the mean and standard deviation of all the proficiencies to be 500 and 100, respectively. For this reason, the proficiencies obtained with the integrated Bayesian model can not be directly compared to the ones from the original PISA results. They should be compared to the results from Bilog-mg, with GBR as the reference group. The transformed scales of the IBM and Bilog-mg are presented just to give an idea about the differences among the countries, compared with the original results from PISA.

Table 22.7 shows that the results are very similar with and without DIF in four of the six countries. Differences are only found in Ireland, where the mean increases when DIF is considered, and in the USA, where the mean decreases if DIF is accounted. It is important to mention that the data was weighted by the sampling weights.

22.6 Conclusions

The analysis presented here shows the importance of appropriately accounting for all sources of heterogeneity present in educational testing. Incorporation of differentiation in the education pattern of countries allows the possibility for explanation of possible causes for it. This can lead the way for improvement in the schooling systems. The use of the Bayesian paradigm provides a number of advantages ranging from inclusion of relevant background information to allowance for model identification.

In the context of the specific application considered, a host of indicators differentiating the educational systems of the English-speaking countries were identified. These may help to understand the nature and possible origins of the difference between them and lead the way for incorporation of beneficial practices in the currently available systems.

Appendix

A. Broader context and background

A.1 Markov chain Monte Carlo methods

The use of Bayesian statistics to handle statistical inference problems has grown considerably since the 1990s. This is due, mainly, to the advances in the field of stochastic simulation, particulary *Markov chain Monte Carlo* (MCMC) methods.

Bayesian inference is mainly based on the *posterior distribution* of the quantities of interest (generally parameters). But in most practical situations this distribution is not fully available. In most of the cases it is known apart from a constant. Bayes theorem states that the posterior distribution of a quantity θ, which is the distribution of θ given some observed data X, is given by:

$$p(\theta|X) = \frac{p(X|\theta)\, p(\theta)}{\int_{\Theta} p(X|\theta)\, p(\theta)\, d\theta}. \tag{22.3}$$

The posterior distribution of θ is proportional to the product of the likelihood function and the prior distribution of θ as it is shown in the numerator of (22.3). The denominator of (22.3) is a constant with respect to θ and is generally not analytically available due to the complexity of the integral.

Since such integral cannot be solved analytically, it is not possible to compute expectations with respect to the posterior distribution of θ. Examples include the expectation and variance of θ or any probability like $P(\theta \in A|X)$ for a given subset A of the state space of θ. Therefore, specially when θ is multidimensional, knowing only the numerator in (22.3) is not enough to have useful posterior information on θ.

However, if it is possible to have a sample from this posterior distribution, one could obtain the Monte Carlo (MC) estimator of the desired expectations. The Monte Carlo estimator of $E[g(\theta)|X]$ is given by

$$\hat{E}[g(\theta)\,|X] = \frac{1}{n}\sum_{i=1}^{n} g\left(\theta^{(i)}\right) \tag{22.4}$$

where $\theta^{(1)}, \ldots, \theta^{(n)}$ is a sample from the posterior distribution of θ.

Monte Carlo estimators have some nice properties. First of all, they are unbiased estimators. Secondly, if $g^2(\theta)$ has finite expectation under $p(\theta|X)$, the variance of the MC estimator is $O(1/n)$. Moreover, they are strongly consistent estimators, since $\hat{E}[g(\theta)|X]$ converges almost surely to $E[g(\theta)|X]$ by the strong law of large numbers.

Markov chain Monte Carlo methods consist of obtaining a sample from the posterior distribution of θ through iterative simulation and using this sample to calculate estimates of expectations. The idea is to construct a Markov chain for θ which has $p(\theta|X)$ as invariant distribution. This means that, $p(\theta^{(i)}) \xrightarrow{i\to\infty} p(\theta|X)$, for any arbitrary initial value $\theta^{(0)}$ of the chain. Thus, for an arbitrarily large iteration i, $\theta^{(i)} \sim p(\theta|X)$ approximately, and consequently $\theta^{(i+1)} \sim p(\theta|X)$. This way, starting from an arbitrary seed $\theta^{(0)}$, values simulated from such Markov chain after a large number of iterations should come from a distribution very close to the posterior distribution of θ.

In practice, a number N of iterations is chosen in a way that the chain is believed to be in equilibrium after this iteration and a sample $\theta^{(N+1)}, \ldots, \theta^{(N+M)}$ is used to obtain estimates of expectations. These first N iterations of the chain are called *burn-in* period and the estimators used are

$$\hat{E}(g(\theta)|X) = \frac{1}{M}\sum_{i=N+1}^{N+M} g(\theta^{(i)}). \tag{22.5}$$

Note that the sample obtained is not independent since it consists of a realisation of a Markov chain. However, it is still reasonable to use the estimator in (22.5) since the ergodic theorem states that, if the chain is ergodic (see Gamerman and Lopes, 2006) and $E[g(\theta)|X] < \infty$, then

$$\frac{1}{n}\sum_{i=1}^{n} g(\theta^{(i)}) \xrightarrow{a.s.} E[g(\theta)|X], \quad \text{as } n \to \infty. \tag{22.6}$$

This is a Markov chain equivalent of the law of large numbers and states that averages of values from the Markov chain are strongly consistent estimators of expectations w.r.t. to the invariant distribution, despite the dependence structure imposed by the chain.

The most used MCMC methods in Bayesian inference are the Gibbs sampling and the Metropolis – Hastings algorithms. The former consists in defining the transition distribution of the chain as the full conditional distributions.

Suppose that $\theta = (\theta_1, \ldots, \theta_d)$, where each component θ_i can be a scalar, a vector or a matrix, and that the full conditional distributions $p(\theta_i|\theta_{-i}, X)$ are known and it is possible to simulate from them, where θ_{-i} is the vector θ without the i-th component.

It can be shown that a Markov chain with transition distributions given by the full conditional distributions has $p(\theta|X)$ as invariant distribution. This way, the Gibbs sampling algorithm is given by:

1. Initialize the iteration counter of the chain $j = 1$ and set initial values $\theta^{(0)} = (\theta_1^{(0)}, \ldots, \theta_d^{(0)})$.
2. Obtain a new value $\theta^{(j)} = (\theta_1^{(j)}, \ldots, \theta_d^{(j)})$ from $\theta^{(j-1)}$ through successive generation of values

$$\theta_1^{(j)} \sim p\left(\theta_1|\theta_2^{(j-1)}, \ldots, \theta_d^{(j-1)}\right)$$

$$\theta_2^{(j)} \sim p\left(\theta_2|\theta_1^{(j)}, \theta_3^{(j-1)}, \ldots, \theta_d^{(j-1)}\right)$$

$$\vdots$$

$$\theta_d^{(j)} \sim p\left(\theta_1|\theta_1^{(j)}, \ldots, \theta_{d-1}^{(j)}\right).$$

3. Change counter j to $j+1$ and return to step 2 until convergence is reached.

In the case where it is not feasible to simulate directly from the full conditional distributions, the Metropolis – Hastings algorithm can be used. The idea is to choose a transition distribution $p(\psi, \phi)$ in a way that it constitutes a Markov chain that has $p(\theta|X)$ as invariant distribution. A sufficient condition for that is to have a reversible chain, that is

$$\pi(\psi)\, p(\psi, \phi) = \pi(\phi)\, p(\phi, \psi), \quad \forall (\psi, \phi), \tag{22.7}$$

where π represents the posterior distribution.

The distribution $p(\psi, \phi)$ consists of two elements: an arbitrary transition distribution $q(\psi, \phi)$ and a probability $\alpha(\psi, \phi)$ such that

$$p(\psi, \phi) = q(\psi, \phi)\alpha(\psi, \phi), \quad if\ \psi \neq \phi. \tag{22.8}$$

This transition kernel defines a density $p(\psi, \phi)$ for every possible value of θ different from ψ. Therefore, there is a positive probability left for the chain to remain in ψ given by

$$p(\psi, \psi) = 1 - \int q(\psi, \phi)\alpha(\psi, \phi)d\phi. \tag{22.9}$$

Hastings (1970) proposed to define the acceptance probability in a way that it defines a reversible chain when combined with the arbitrary transition distribution. Such probability is given by

$$\alpha(\psi, \phi) = \min\left\{1, \frac{\pi(\phi)q(\phi, \psi)}{\pi(\psi)q(\psi, \phi)}\right\}. \tag{22.10}$$

Note that it is not necessary to know the constant in the denominator of (22.3), since it cancels in the expression of the acceptance probability. The Metropolis – Hastings algorithm may be detailed as follows:

1. Initialize the iteration counter of the chain $j = 1$ and set initial values $\theta^{(0)} = \left(\theta_1^{(0)}, \ldots, \theta_d^{(0)}\right)$.
2. Move the chain to a new value ϕ generated from the density $q(\theta^{(j-1)}, \cdot)$.
3. Evaluate the acceptance probability of the move $\alpha(\theta^{(j-1)}, \phi)$ given by (22.10).
4. If the move is accepted, $\theta^{(j)} = \phi$. If it is not accepted, $\theta^{(j)} = \theta^{(j-1)}$ and the chain does not move.
5. Change counter j to $j + 1$ and return to step 2 until convergence is reached.

Step 4 is performed by generating a value u from a unit uniform distribution. If $u \leq \alpha$, the move is accepted, if $u > \alpha$ the chain does not move.

In the case where some (but not all) the full conditional distributions are known, the algorithm sometimes called Metropolis-within-Gibbs can be used. This is the algorithm used in this chapter to draw samples from the joint posterior distribution of all the parameters.

The idea of the Metropolis-within-Gibbs algorithm is to construct a Gibbs sampler and, for the components of θ that cannot be directly sampled from the full conditional distribution, a Metropolis – Hastings step is used. That is, such components are drawn from an arbitrary transition distribution and have the moves accepted with an acceptance probability that constitutes a chain that has the correspondent full conditional as invariant distribution. In other words, on the $(j + 1)$-th iteration, a component θ_i which cannot be directly drawn from its full conditional distribution is drawn from an arbitrary transition distribution

$q\left(\theta_i^{(j)}, \phi\right)$ and has the move accepted with probability

$$\alpha(\theta_i^{(j)}, \phi) = \min\left\{1, \frac{\pi_i(\phi) q\left(\phi, \theta_i^{(j)}\right)}{\pi_i\left(\theta_i^{(j)}\right) q\left(\theta_i^{(j)}, \phi\right)}\right\}$$

where π_i is the full conditional distribution of the i-th component of θ.

Such transition distributions constitute a Markov chain with invariant distribution given by the posterior distribution of θ. For more details on the MCMC methods described here and other issues on stochastic simulation in Bayesian inference see Gamerman and Lopes (2006).

A.2 MCMC details for our DIF model

The estimation of the parameters of the model presented in this chapter is performed by using MCMC methods to obtain a sample from the joint posterior distribution of these parameters. The method used is Gibbs sampling with Metropolis – Hastings steps.

Basically, all the parameters that appear explicitly in the likelihood are drawn using a Metropolis – Hastings step with a suitable random walk as the proposal distribution since it is not possible to directly draw from their full conditional distributions. These parameters are the proficiencies, the item parameters and the DIF parameters. For all the other parameters, it is possible to directly draw from their full conditional distributions.

The parameters are simulated as follows:

- Abilities.
  Samples from $p(\theta|\beta, d, \lambda, \gamma, T, Y, W, Z)$, are drawn from

$$\begin{aligned}
p(\theta_j|\theta_{\neq j}, \beta, d, \lambda, \gamma, T, Y, W, Z) &= p\left(\theta_j|\beta_{I(j)}, d_{I(j)g(j)}, \lambda_{g(j)}, Y_j\right) \\
&\propto p\left(Y_j|\theta_j, \beta_{I(j)}, d_{I(j)g(j)}, \lambda_{g(j)}\right) \\
&\quad\; p\left(\theta_j|\beta_{I(j)}, d_{I(j)g(j)}, \lambda_{g(j)}\right) \\
&= p\left(Y_j|\theta_j, \beta_{I(j)}, d_{I(j)g(j)}\right) p\left(\theta_j|\lambda_{g(j)}\right) \\
&= \prod_{i\in I(j)} p\left(Y_{ij}|\theta_j, \beta_i, d_{ig(j)}\right) p\left(\theta_j|\lambda_{g(j)}\right), \\
&\quad\; \forall j = 1, \ldots, J.
\end{aligned}$$

  The calculations above are obtained by assuming independence between the students' abilities, and between the answers Y_{ij} when conditioned to the abilities and to the items' parameters. It is not easy to directly draw from the distribution above because of its complexity. So, the Metropolis – Hastings algorithm is used. A normal transition kernel is used and the proposal for the new state is

$$\theta_j^l \sim q\left(\theta_j|\theta_j^{l-1}\right) = N\left(\theta_j^{l-1}, c_\theta^2\right).$$

The tuning parameter is set as $c_\theta = 0.1$, chosen after a pilot study to assure an appropriate acceptance rate of the chain.

- Parameters of the distributions of the groups' abilities.
  It is possible to directly draw from the full conditional distributions of the means and variances of the abilities' distribution since conjugate prior distributions were chosen.
  - Mean of the distributions of the groups' abilities.
    If a prior distribution $\mu_g \sim N(\mu_{0g}, \sigma_{0g}^2)$ is chosen, the following full conditional distribution is obtained:

$$\begin{aligned} p(\mu_g|\cdot) &= p\left(\mu_g|\theta_{J(g)}, \sigma_g^2\right) \propto p\left(\theta_{J(g)}|\mu_g, \sigma_g^2\right) p\left(\mu_g|\sigma_g^2\right) \\ &= \prod_{j\in J(g)} p\left(\theta_j|\mu_g, \sigma_g^2\right) p(\mu_g) \end{aligned}$$

so:

$$(\mu_g|\cdot) \sim N\left(m_g, s_g^2\right), \quad \text{where:}$$

$$m_g = \frac{\sum_{j\in J(g)} \theta_j \sigma_{0g}^2 + \mu_{0g}\sigma_g^2}{n_g\sigma_{0g}^2 + \sigma_g^2} \quad \text{and} \quad s_g = \frac{\sigma_g\sigma_{0g}}{\sqrt{n_g\sigma_{0g}^2 + \sigma_g^2}}.$$

$J(g)$ represents the set of the students in group g and n_g is the number of students in group g, for $g = 2, \ldots, G$.

  - Variance of the distributions of the groups' abilities.

$$p\left(\sigma_g^2|\cdot\right) = p\left(\sigma_g^2|\theta_{J(g)}, \mu_g\right) \propto \prod_{j\in J(g)} p\left(\theta_j|\mu_g, \sigma_g^2\right) p\left(\sigma_g^2\right).$$

If a prior distribution $\sigma_g^2 \sim IG(\alpha_g, \beta_g)$ is chosen, the following full conditional distribution is obtained:

$$\left(\sigma_g^2|\cdot\right) \sim IG\left(\alpha_g + \frac{n_g}{2}, \frac{\sum_{j\in J(g)}(\theta_j - \mu_g)^2) + 2\beta_g}{2}\right), \quad g = 2, \ldots, G.$$

- Structural parameters β.
  Under the hypotheses of local independence of the items, samples of $p(\beta|\cdot)$ are drawn from:

$$p\left(\beta_i|\theta_{J(i)}, d_i, Y_{J(i)}\right) \propto p\left(Y_{J(i)}|\theta_{J(i)}, \beta_i, d_i\right) p\left(\beta_i|\theta_{J(i)}, d_i\right)$$
$$= \prod_{j\in J(i)} p\left(Y_{ij}|\theta_j, \beta_i, d_{ig(j)}\right) p(a_i)p(b_i)p(c_i),$$
$$\forall i = 1, \ldots, I.$$

The last equality comes from the prior independence between item parameters. The chosen prior distributions of the parameters are:

$$a_i \sim LN\left(\mu_{a_i}, \sigma^2_{a_i}\right); \quad b_i \sim N\left(\mu_{b_i}, \sigma^2_{b_i}\right) \quad \text{and} \quad c_i \sim Beta\left(\alpha_{c_i}, \beta_{c_i}\right).$$

In general, $\mu_{a_i} = 0,\ \sigma^2_{a_i} = 2,\ \mu_{b_i} = 0,\ \sigma^2_{b_i} = 1,\ \alpha_{c_i} = 5,\ \beta_{c_i} = 17$. These values are used, for example, as default values in the software Bilog-mg. In some cases, these values have to be modified to assure a better fit of current data or to add past information about the parameters.
Once again, it is not possible to directly draw from the full conditional distribution and the Metropolis – Hastings algorithm is used assuming the following transition kernels:

$$a^l_i \sim LN\left(\ln\left(a^{l-1}_i\right), c_a\right),\ \ b^l_i \sim N\left(b^{l-1}_i, c^2_b\right)\ \text{ e }\ c^l_i \sim U\left[c^{l-1}_i - \delta, c^{l-1}_i + \delta\right].$$

In general, the values $c_a = 0.02,\ c_b = 0.1,\ \delta = 0.05$ are used.

- Structural DIF parameters.
  To draw samples from $p(d^h|\cdot)$, $h = a, b$, samples are independently drawn from:

$$p\left(d^h_{ig}|\cdot\right) = p\left(d^h_{ig}|d^h_{\neq i,g}, d^{\neq h}_g, Z^h, \theta_{J(i,g)}, \beta_i, \gamma_g, T, Y_{J(i,g)}, W\right)$$
$$\propto p\left(Y_{J(i)}|\theta_{J(i,g)}, \beta_i, d^h_{ig}\right) p\left(d^h_{ig}|d^h_{\neq i,g}, d^{\neq h}_g, W, \gamma_g, T, Z^h_{ig}\right)$$
$$= \prod_{j\in J(i,g)} p\left(Y_{ij}|\theta_j, \beta_i, d^h_{ig}\right) p\left(d^h_{ig}|W^h_i, \gamma^h_g, \tau^h_g, Z^h_{ig}\right),$$
$$\forall i \in I^h_{dif},\ g = 2, \ldots, G.$$

In the last equality, it is assumed that $T^h = \left(\tau^h_g\right)^2 I$, where I is the identity matrix $nid_h \times nid_h$, and nid_h is the number of items for which DIF in parameter h is assumed, $h = a, b$. The conditional prior distribution of the DIF parameters is $(d^h_{ig}|W^h_i, \gamma^h_g, \tau^h_g, Z^h_{ig}) \sim N(W^h_i\gamma^h_g, (\tau^h_g)^2)$ if $Z^h_{ig} = 1$, and the transition kernel used in the Metropolis – Hastings algorithm is the following:

$$\left(d_{ig}^{h}\right)^{l+1} \sim N\left(\left(d_{ig}^{h}\right)^{l}, 0.1\right), \ \forall i.$$

On the other hand, $(d_{ig}^{h} | W_i^h, \gamma_g^h, \tau_g^h, Z_{ig}^h) \sim N(0, s_i^2(\tau_g^h)^2)$ if $Z_{ig}^h = 0$. In practical situations, since s_i is very small, $d_{ig}^h \cong 0$ if $Z_{ig}^h = 0$.

- Parameters of the DIF regression structure.
For the parameters γ_g, g = 2, ..., G, samples are drawn from:

$$p\left(\gamma_g^h|\cdot\right) = p\left(\gamma_g^h|d_g^h, T_g^h, W^h, Z^h\right) \propto p\left(d_g^h|\gamma_g^h, W^h, T_g^h, Z^h\right) p\left(\gamma_g^h\right).$$

If a prior distribution $\gamma_g^h \sim N\left(\gamma_0^h, S_0^h\right)$ is assumed, the following full conditional distribution is obtained:

$$\left(\gamma_g^h|d_g^h, T_g^h, W^h, Z^h\right) \sim N(H, L), \quad \text{where:}$$

$$L = \left[\left(W^h_{I\left(Z_{ig}^h=1\right)}\right)^T \left(T_g^h\right)^{-1} W^h_{I\left(Z_{ig}^h=1\right)} + \left(S_0^h\right)^{-1}\right]^{-1} \quad \text{and}$$

$$H = L\left[\left(W^h_{I\left(Z_{ig}^h=1\right)}\right)^T \left(T_g^h\right)^{-1} d^h_{I\left(Z_{ig}^h=1\right),g} + \left(S_0^h\right)^{-1} \gamma_0^h\right].$$

Samples of $\left(\tau_g^h\right)^2$ are drawn from:

$$p\left(\left(\tau_g^h\right)^2|\cdot\right) = p\left(\left(\tau_g^h\right)^2|d_g^h, \gamma_g^h, W^h_{I\left(Z_{ig}^h=1\right)}, Z^h\right)$$
$$\propto p\left(d_g^h|\left(\tau_g^h\right)^2, \gamma_g^h, W^h_{I\left(Z_{ig}^h=1\right)}, Z^h\right) p\left(\left(\tau_g^h\right)^2\right).$$

If a prior distribution $(\tau_g^h)^2 \sim IG(\alpha_{\tau_g^h}, \beta_{\tau_g^h})$ is assumed, the following full conditional distribution is obtained:

$$\left(\left(\tau_g^h\right)^2|\cdot\right) \sim IG\left(\alpha_{\tau_g^h} + \frac{\sum Z_{ig}^h}{2}, \left[\frac{1}{2}\left(d_g^h - W^h_{I\left(Z_{ig}^h=1\right)}\gamma_g\right)^T \left(d_g^h - W^h_{I\left(Z_{ig}^h=1\right)}\gamma_g\right) + \beta_{\tau_g^h}\right]\right),$$

for g = 2,...,G.

- DIF indicator variable.

$$p\left(Z_{ig}^h|\cdot\right) = p\left(Z_{ig}^h|d_{ig}^h, T^h, W_i^h, \gamma^h\right) = p\left(d_{ig}^h|Z_{ig}^h, T^h, W_i^h, \gamma^h\right) p\left(Z_{ig}^h\right)$$

If a prior distribution $Z_{ig}^h \sim Ber(\pi_{ig}^h)$ is assumed, the full conditional distribution $(Z_{ig}^h|\cdot) \sim Ber(\omega_{ig}^h)$ is obtained, where:

$$\omega_{ig}^h = \frac{1}{c_\omega}\pi_{ig}^h \exp\left(-\frac{1}{2\left(\tau_g^2\right)}\left(d_{ig} - W_i^h\gamma_g^h\right)^2\right), \text{ and:}$$

$$c_\omega = \pi_{ig}^h \exp\left(\frac{-1}{2\left(\tau_g^2\right)}\left(d_{ig} - W_i^h\gamma_g^h\right)^2\right) + \left(1 - \pi_{ig}^h\right)\exp\left(\frac{-1}{2\left(s_i^2\tau_g^2\right)}\left(d_{ig}^h\right)^2\right),$$

for $g = 2, \ldots, G$.

A.3 *Bayesian methods in item response theory*

Since the late 1960s, knowledge about populations of students has been used in order to provide improved estimates for individual abilities. In the early works of Birnbaum (1969) and Owen (1975), Bayes estimates of ability parameters were obtained in item response models under the assumption that the item parameters are known. Novick *et al.* (1972) and Rubin (1980) proposed Bayesian and empirical Bayesian solutions, respectively, to predict the performance of students from a Law school using information from other Law schools. Also, the most used traditional Bayesian methods for ability estimation were proposed in the early 1980s: the Expected A-Posteriori (EAP; Bock and Mislevy 1982) and Modal A-Posteriori (MAP; Samejima 1980).

Bayesian methods have been used in IRT for the estimation of structural parameters since the early 1980s. Applying the hierarchical Bayesian approach suggested in Lindley and Smith (1972), Swaminathan and Gifford (1982), Swaminathan and Gifford (1985) and Swaminathan and Gifford (1986) proposed Bayesian procedures for simultaneous estimation of the item and ability parameters in the Rasch, two-parameter and three parameter models, respectively. The authors showed that the non-convergence of the estimates of the discrimination and guessing parameters, which is common in joint maximum likelihood methods, can be controlled by appropriate specification of prior distributions for all parameters. Mislev (1986) and Tsutakawa and Lin (1986) proposed to use the EM algorithm (Dempster *et al.* 1977) to obtain a modal marginal posterior distribution estimator for the item parameters, as it is proposed by Bock and Aitkin (1981) for maximum marginal likelihood estimation.

The importance of Bayesian approaches in item response theory have been steadily growing since the 1980s, with incorporation of new models. Such models try to include the effect of covariates or grouping in the estimation of latent abilities. Multidimensional structures and local dependence of the items have also been proposed. This way, as the complexity of the models increase, also does the difficulty on the estimation of the parameters of the models.

Classical methods, like maximization of marginal likelihood, and even some Bayesian methods, like maximum posterior distribution, use the EM algorithm or some of its variations and, basically, treat the latent variables as 'missing data'.

However, the use of the EM algorithm becomes harder as the complexity increases. For this reason, the use of methods based on stochastic simulation has grown since the 1990s, specially MCMC methods. In a pioneering study, Albert (1992) applied the Gibbs sampler to the two-parameter model using the normal distribution function as the link function. Béguin and Glas (2001) proposed a Gibbs sampler for the three-parameter and the multidimensional models. Fox and Glass (2001, 2003) studied an IRT model with a hierarchical regression structure on the ability parameters, with both latent and observed covariates. The latent covariates are jointly estimated with the parameters of the model. All of those works use normal distributions for the latent variables and estimate the abilities using augmented data techniches (cf. Tanner and Wong, 1987).

Patz and Junker (1999*b*) applied the Gibbs sampler to estimate parameters in a two-parameter model. Differently from the works cited above, that use augmented data, the authors use the Metropolis-within-Gibbs algorithm. Patz and Junker (1999*a*) extended this work for the three-parameter model, polytomic models and models with missing data.

DIF analysis is an appropriate environment for a genuine Bayesian formulation due to the complex structure of the models and the subjective decision features involved, which can be formulated through Bayesian arguments. For example, Zwick *et al.* (1999, 2000) and Zwick and Thayer (2002) considered a formulation where the MH D-DIF statistic is represented by a normal model where the mean is equal to the real DIF parameter. These authors use Empirical Bayes (EB) for the posterior estimation of the parameters. Sinharay *et al.* (2006) considered the same formulation and proposed informative prior distributions based on past information. They showed that the full Bayes (FB) method leads to improvements if compared to the two other approaches, specially in small samples. Wang *et al.* (2008) proposed a Bayesian approach to study DIF based on testlet response theory (cf. Wainer *et al.*, 2007).

B. Countries that participated in PISA 2003

OECD countries:

Australia, Austria, Belgium, Canada, Czech Republic, Denmark, Finland, France, Germany, Greece, Hungary, Iceland, Ireland, Italy, Japan, Korea, Luxembourg, Mexico, Netherlands, New Zealand, Norway, Poland, Portugal, Slovak Republic, Spain, Sweden, Switzerland, Turkey, United Kingdom, Scotland, United States.

Partner countries:

Brazil, Hong Kong-China, Indonesia, Latvia, Liechtenstein, Macao-China, Russian Federation, Serbia and Montenegro, Thailand, Tunisia, Uruguay.

Acknowledgments

The first author acknowledges grants from CNPq-Brazil and FAPERJ-Brazil.

References

Albert, J. H. (1992). Bayesian estimation of normal ogive item response models using Gibbs sampling. *Journal of Educational Statistics*, **17**, 251–269.

Béguin, A. A. and Glass, C. A. W. (2001). MCMC estimation and some model-fit analysis of multidimensional IRT models. *Psychometrika*, **66**, 541–562.

Birnbaum, A. (1968). Some latent traits models and their use in inferring examinee's ability. In reissue of *Statistical Theories of Mental Test Scores* (ed. F. M. Lord and M. R. Novick). Addison-Wesley, Reading, MA.

Birnbaum, A. (1969). Statistical theory for logistic mental test models with a prior distribution of ability. *Journal of Mathematical Psychology*, **6**, 258–276.

Bock, R. D. and Aitkin, M. (1981). Marginal maximum likelihood estimator of item parameters: an application of an EM algorithm. *Psychometrika*, **46**, 443–459.

Bock, R. D. and Mislevy, R. J. (1982). Adaptive EAP estimation of ability in a microcomputer environment. *Applied Psychological Measurement*, **6**, 431–444.

Cole, N. S. (1993). *Differential Item Functioning*, Chapter on History and development of DIF. Lawrence Erlbaum, Hillsdale, NJ.

Dempster, A. P., Laird, N. M. and Rubin, D. B. (1977). Maximum likelihood from incomplete data via the EM algorithm (with discussion). *Journal of the Royal Statistical Society B*, **39**, 1–38.

Fox, J. P. and Glass, C. A. W. (2001). Bayesian estimation of a mutilevel model using Gibbs sampling. *Psychometrika*, **66**, 271–288.

Fox, J. P. and Glass, C. A. W. (2003). Bayesian modeling of measurement error in predictor variables using IRT. *Psychometrika*, **68**, 169–191.

Gamerman, D. and Lopes, H. F. (2006). *Markov Chain Monte Carlo: Stochastic Simulation for Bayesian Inference*, (2nd edn). Chapman & Hall/CRC, New York.

George, E. I. and McCulloch, R. E. (1993). Variable selection via Gibbs sampling. *Journal of American Statistical Association*, **85**, 398–409.

Hastings, W. K. (1970). Monte Carlo sampling methods using Markov chains and their applications. *Biometrika*, **57**, 97–109.

Lindley, D. V. and Smith, A. F. (1972). Bayesian estimates for the linear model (with discussion). *Journal of the Royal Statistical Society B*, **34**, 1–41.

Lord, F. M. (1952). A theory of test scores. *Psychometric Monograph*, **7**. Psychometric society.

Mislev, R. J. (1986). Bayes modal estimation in item response models. *Psychometrika*, **51**, 177–195.

Novick, M. R., Jackson, P. H., Thayer, D. T. and Cole, N. S. (1972). Estimating multiple regression in M-groups: a cross-validation study. *British Journal of Mathematical and Statistical Psychology*, **5**, 33–50.

Owen, R. J. (1975). A Bayesian sequential procedure for quantal response in the context of adaptive mental testing. *Journal of the American Statistical Association*, **70**, 351–360.

Patz, R. J. and Junker, B. W. (1999a). Applications and extensions of MCMC in IRT. *Journal of Educational and Behavioral Statistics*, **24**, 342–366.

Patz, R. J. and Junker, B. W. (1999b). A straightforward approach to Markov chain Monte Carlo methods for item response models. *Journal of Educational and Behavioral Statistics*, **24**, 146–178.

Rasch (1960). *Probabilistic Models for Some Intelligence and Attainment Tests*. Copenhagen: Institute for Educational Research.

Rubin, D. B. (1980). Using empirical Bayes techniques in the law school validity studies. *Journal of the American Statistical Association*, **73**, 801–827.

Samejima, F. (1980). Is Bayesian estimation proper for estimating the individual's ability? Research Report 80–3, University of Tennessee, Department of Psychology, Knoxville, TN.

Sinharay, S., Dorans, N. J., Grant, M. C., Blew, E. O. and Knorr, C. M. (2006). Using past data to enhance small-sample DIF estimation: a Bayesian approach. Technical Report ETS RR-06-09, Educational Testing Service, Princeton, NJ.

Soares, T. M., Gonçalves, F. B. and Gamerman, D. (2009). An integrated Bayesian model for DIF analysis. *Journal of Educational and Behavioral Statistics*, **34**, 348–377.

Stricker, L. J. and Emmerich, W. (1999). Possible determinants of differential item functioning: familiarity, interest and emotional reaction. *Journal of Educational Measurement*, **36**, 347–366.

Swaminathan, H. and Gifford, J. A. (1982). Bayesian estimation in the rasch model. *Journal of Educational Statistics*, **7**, 175–191.

Swaminathan, H. and Gifford, J. A. (1985). Bayesian estimation in the two parameter logistic model. *Psychometrika*, **50**, 349–364.

Swaminathan, H. and Gifford, J. A. (1986). Bayesian estimation in the three parameter logistic model. *Psychometrika*, **51**, 589–601.

Tanner, M. A. and Wong, W. H. (1987). The calculation of posterior distribution by data augmentation. *Journal of the American Statistical Association*, **82**, 528–550.

Thissen, D. (2001). IRTLRDIF v.2.0.b: Software for the Computation of the Statistics Involved in Item Response Theory Likelihood-Ratio Tests for Differential Item Functioning. `http://www.unc.edu/ dthissen/dl.html`: Scientific Software International.

Tsutakawa, R. K. and Lin, H. Y. (1986). Bayesian estimation of item response curves. *Psychometrika*, **51**, 251–267.

Wainer, H., Bradlow, E. T. and Wang, X. (2007). *Testlet Response Theory*. Cambridge University Press, New York.

Wang, X., Bradlow, E. T., Wainer, H. and Muller, E. S. (2008). A Bayesian method for studying DIF: a cautionary tale filled with surprises and delights. *Journal of Educational and Behavioral Statistics*, **33**, 363–384.

Zwick, R. and Thayer, D. T. (2002). Application of an empirical Bayes enhancement of mantel-haenszel DIF analysis to a computerized adaptive test. *Applied Psychological Measurement*, **26**, 57–76.

Zwick, R., Thayer, D. T. and Lewis, C. (1999). An empirical Bayes approach to Mantel–Haenszel DIF analysis. *Journal of Educational Measurement*, **36**, 1–28.

Zwick, R., Thayer, D. T. and Lewis, C. (2000). Using loss functions for DIF detection: an empirical Bayes approach. *Journal of Educational and Behavioral Statistics*, **25**, 225–247.

·23·

Sequential multilocation auditing and the New York food stamps program

Karl W. Heiner, Marc C. Kennedy and Anthony O'Hagan

23.1 Introduction

23.1.1 The food stamps program

In the United States, the Department of Agriculture's Food and Nutrition Services (FNS) Division oversees a programme designed to assist needy families with the purchase of groceries, the food stamps program. This programme is administered by the states who share in the cost of the programme, although in some larger states, e.g. New York State (NYS), administration is actually handled by the counties within the state. The programme allows low-income families to buy nutritious food with coupons and Electronic Benefits Transfer cards. To be eligible for this programme, a household must have a gross monthly income below a level that depends on household size. For example, in 2004 a one-person household must have had a gross monthly income no more than 960 US dollars while an eight-person household could earn no more than 3296 dollars. Furthermore, there are limits on resources, e.g. bank accounts. This limit is 2000 dollars unless one member of the household is at least 60 years old, in which case the resource limit is 3000 dollars. A household's monthly food stamps benefit is a function of household size and income. Because resources, income and even household size are likely to fluctuate in time, benefits determination is prone to error.

As part of its oversight, FNS directs each state to sample and audit approximately 1000 transactions each year. State auditors review each transaction and categorize them as correct, ineligible, overpaid, or underpaid. When there is an error, the amount in error is also determined. Federal auditors review a sample of the state's findings. Based on the results of this quality control audit, the Department of Agriculture imposes rather large sanctions on states whose ratio of error dollars to total payments is excessive. This excessive ratio is based on the average of the error rates for the states and a target error rate of 6%. Should a state's error rate exceed both 6% and the average error rate of all states, a large penalty is imposed on the state by FNS. From 1998 to 2002, the proportion of audited transactions found to be in error in NYSs food stamps quality control audits were 0.124, 0.101, 0.121, 0.084, and 0.074, which suggests an improving trend. However, in view of the 6% error rate limit, the uncertainty over whether this trend would be maintained and the uncertainty associated with the average error rate among all states, NYS officials were rightly concerned about the possibility that their food stamps program would soon face large federally-imposed fiscal sanctions.

In NYS, the counties administer the food stamps program and the state's portion of the food stamps program's cost is shared equally between the state and each county. Should there be a sanction, state officials wish to share the sanction with the counties, in particular those contributing most to the error rate, but with slightly more than one thousand randomly audited cases statewide, sample sizes in many counties are very small. It would be hard to justify penalising a county with a high error rate when that rate was observed in only a small sample number of transactions. A statistical method was required to estimate error rates at the county level that would be perceived as fair.

23.1.2 Auditing concepts and terminology

Statistical methods are widely used in auditing all but very small organizations, because of the impracticality of checking every single transaction. Sampling may be by means of a random sample of all transactions, but more often a form of 'sampling proportional to size' known as *monetary unit sampling* is used.

The key terminology in auditing distinguishes between the *book value*, which is the monetary value of a transaction recorded in the accounts being audited, and the *audit value*, which is the value determined by the auditor as the correct value for that transaction. Book values are known for all transactions, or can be found readily from the organization's account data. Audit values are only known for the sample of transactions that have been audited. When the audit value differs from the book value, there is an error. If we denote the book value of transaction k by b_k and its audit value by a_k, then the error is $e_k = b_k - a_k$. A positive error denotes an overpayment, while a negative error is an underpayment.

Although it is natural to be interested in the total error $\sum_k e_k$, auditors may focus on the total absolute error $\sum_k |e_k|$ since this better represents the quality of the accounts. This is the case in audits conducted by the US Department of Agriculture for states' food stamps determination and distribution.

It is usual to express errors proportionately to the book value; the *taint* in transaction k is e_k/b_k. Book values and audit values are non-negative,[1] so the

[1] Accounts for payments and receipts are audited separately. A negative value would indicate a transaction that should be in the other account. Whilst this can happen, it would trigger a specific investigation rather than being handled within the statistical analysis.

taint is less than or equal to 1, but is unbounded below. When we are concerned with absolute error, taint is bounded below by zero, but not bounded above. A taint of 1 is sometimes referred to as a *bogus* transaction, since it means that no payment should have been made. In the food stamps program audit, another term for such a transaction is 'ineligible.'

The food stamps program is characterized also by multilocation audits. Large organizations very often operate at a number of locations. Accounts for the transactions at a given location may be located there, making simple random sampling of transactions impractical since it is necessary to visit the location of each sampled transaction. Sampling may then be stratified (if every location is audited) or two-stage (if only a sample of locations is audited). There is then interest in estimating the total (absolute) error at each location as well as for the organization as a whole. This is obviously challenging in the case of two-stage sampling, but is not straightforward even if all locations have been sampled. To identify a location as having a particularly high error rate on the basis of a small sample may be contentious, and it is natural to think in this case of estimation methods that involve shrinkage.

It was in this context that the National Audit Office (UK) commissioned the development of Bayesian methods for the analysis of multilocation audit data, leading to the models and techniques described by Laws and O'Hagan (2000, 2002). This work represented the first systematic approach for multilocation sampling that allowed the incorporation of the auditor's non-sample based knowledge of the organization, as well as allowing the total error in each location to be estimated. A key feature of this analysis was a hierarchical model that linked each location's error rate and taint profile to underlying common error rates and taint distributions, and so facilitated the shrinkage of small sample estimates and inference about unsampled locations.

Whereas this method was clearly well suited to the context of the NYS food stamps program audit, where the sample sizes in individual counties (locations) were often small, it was necessary also to address another complication. The food stamps program audit demanded extensions to allow for repeated annual audits and for the possibility of trends over time in the error rates. In general, an organization is nearly always audited periodically, usually at least annually. There are therefore data available from previous years which may provide useful information about current error rates. There is some previous work suggesting methods to incorporate such information sequentially (Heiner 1999, Heiner and Laws 1999). but nothing that combines both multilocation and sequential elements. The method of Laws and O'Hagan (2002) allows knowledge from previous periods to be incorporated informally through the auditor's prior distribution. We present here a model that allows more formal and open incorporation of such data.

23.1.3 Review of previous work

Several features of the audit problem make it a challenging one for statistical analysis. The first of these is that errors are comparatively rare in most audits and yet the audit samples are usually rather small. The great majority of transactions yield zero error but the few transactions that are in error can have large positive or negative errors. The distributions of errors are often referred to in the field as *non-standard*. With samples typically comprising only a few hundred transactions and containing only a handful of errors, inference about such distributions is not straightforward. There has been extensive study of statistical methods to address the problem of estimating the total error in the population.

Neter and Loebbecke (1977) empirically study the precision and reliability of various classical approaches (e.g. mean per unit, ratio, etc.) to estimating total error amounts. They concluded that these classical methods did not provide good coverage, and although they did not note that in positively skewed audit distributions which are typical, both ends of mean per unit confidence intervals are too low, their tabulated results clearly demonstrate this point. Cochran (1977) had observed this fact which is particularly useful in adversarial situations since judges and hearing officers frequently rely on lower confidence limits when determining damages (Heiner, Wagner and Fried 1984; Heiner and Whitby 1980). Smith (1979) discusses statistical sampling to estimate total errors in an audit, while Frost and Tamura (1982) suggest the use of a jackknife ratio estimator for obtaining reliable interval estimates of the true value of an audit population total. Because financial audit sample results are typically a mixture of mainly zero findings and various non-zero findings, Neter and Godfrey (1985) discuss using Bayesian mixture models in monetary unit sampling to estimate total overstatement. Their mixture distribution combines the likelihood of an overstated error and the density of the magnitude of the overstated errors. Kvanli, Shen and Deng (1998) propose constructing interval estimates by inverting likelihood ratio tests when data are assumed to come from a mixture of a Bernoulli distribution and a normal distribution or an exponential distribution. Their simulation demonstrated that when error distributions were Gaussian or Exponential, the coverage probabilities were better than the traditional methods. Lower confidence limits for their likelihood profile method were almost always greater than the lower control limit of the traditional method. In a 1989 review of statistical models and analysis in auditing, the Panel on Nonstandard Mixtures of Distributions (1989) discusses various approaches used in statistical auditing and provides an annotated bibliography on statistical practices in auditing. Tsui, Matsumura and Tsui (1985) provide a method of constructing bounds on the proportion of amounts that are overstated using a Bayesian multinomial Dirichlet model applied to monetary unit sampling and Moors and Janssens (1989) consider using Bayesian models to find an upper bound on total overstatement error. van Batenburg, O'Hagan and Veenstra (1994) describe a Bayesian hierarchical model that may be used by auditors to establish compliance with a prescribed minimum standard of error prevalence in financial statements that uses the auditor's experience with a firm. Cohen and Sackrowitz (1996) develop lower confidence bounds for total overpayments in insurance claims using pilot samples to decide whether or not a more extensive audit is necessary and then adjusts the confidence limits obtained by the more extensive sample.

We describe a Bayesian model suited for audits when within the audit sampling frame there are multiple locations and when these audits are conducted periodically. Using this model, we analyse several years of data from audits of the food stamps program in NYS.

23.1.4 Notation and outline of the paper

Laws and O'Hagan (2002) (L&O'H) model the probability distribution of the error in any given transaction in location i (for $i = 1, 2, \ldots, L$, where L is the number of locations in the organization) using four steps.

- The transaction is in error with probability θ_i, and so is correct (zero error) with probability $1 - \theta_i$. We refer to θ_i as the error rate in location i.
- If the transaction is in error, the error falls into error class j with probability ψ_{ij}, for $j = 1, 2, \ldots, p$, where p is the number of error classes. Error classes can be flexibly defined with reference to the range of possible taints. L&O'H primarily used $p = 3$ classes – bogus transactions (taint 1), overpayments (taint greater than 0 but less than 1) and underpayments (taint less than 0). The vector of class error probabilities for location i is denoted by $\psi_i = (\psi_{i1}, \psi_{i2}, \ldots, \psi_{ip})$.
- If the transaction falls into error class j, and if that class covers more than a single taint value, then it has a taint sampled from a taint distribution appropriate to that class.
- Finally, having determined the taint, the error arises from multiplying the taint by the book value. Book values are usually known exactly, although in some situations might be uncertain.

The model presented here extends L&O'H by modelling evolution over time of the θ_is and the ψ_is. The modelling of the taint distribution within error classes is as in L&O'H, and these distributions are assumed to be constant over time. Apart from a small change detailed in Section 23.5.1 the treatment of book values is also as in L&O'H, and the reader is referred to that source for details of both these aspects of the modelling and analysis.

We denote the values of θ_i and ψ_i at time period t by θ_i^t and $\psi_i^t = (\psi_{i1}^t, \psi_{i2}^t, \ldots, \psi_{ip}^t)$, respectively. L&O'H introduce hyperparameters θ_0 and ψ_0 by modelling the location-specific parameters hierarchically, and these will also have values θ_0^t and $\psi_0^t = (\psi_{01}^t, \psi_{02}^t, \ldots, \psi_{0p}^t)$ at time t. Furthermore, to model the evolution over time we introduce a trend parameter that takes value τ_0^t at time t. The set of parameters at time t is therefore

$$\beta^t = \left(\theta_0^t, \theta_1^t, \ldots, \theta_L^t, \tau_0^t, \psi_0^t, \psi_1^t, \ldots, \psi_L^t\right).$$

Finally, we denote the information up to time t by D^t. The sequential analysis can then be considered as a cycle in two stages.

1. *Update.* Given the prior distribution for β^t, based on D^{t-1}, and given the audit data obtained in period t, derive the posterior distribution for β^t.
2. *Project.* Given the posterior distribution of β^t, based on D^t, construct a distribution for β^{t+1}, also based on D^t. This then becomes the prior distribution for the next update step.

In Section 23.2 we present the form of the hierarchical prior distribution for β^t based on D^{t-1}, which is the same as in L&O'H except for the addition of the trend parameter. The update step is described in Section 23.3. Section 23.4 presents the project step, which embodies the modelling of the evolution of the system over time. In Section 23.5 we return to the food stamps problem which originally motivated the development of this model, and present the resulting analysis of the New York food stamps data. We conclude with a brief discussion in Section 23.6.

23.2 Modelling of error rates and error classes

23.2.1 Modelling relationships between parameters

We introduce here a technique that we will use repeatedly to relate one parameter, or vector of parameters, to another. Suppose first that p and p' are random variables lying in [0.1] such that we view p' as a perturbation of p. We represent this via the Beta conditional distribution

$$p' \mid p \sim Be(\omega g(p, r), \omega\{1 - g(p, r)\}), \tag{23.1}$$

where

$$g(p, r) = \frac{p \exp(r)}{1 + p\{\exp(r) - 1\}}. \tag{23.2}$$

The interpretation of this modelling is as follows. First note from (23.1) that $E(p' \mid p) = g(p, r)$, and from (23.2) that

$$r = \log\left[\frac{g(p, r)}{1 - g(p, r)} \Big/ \frac{p}{1 - p}\right].$$

Therefore r is the log-odds ratio of the expected value of p' versus p. So r represents an expected shift of p' away from p. If $r = 0$, $g(r, p) = p$ and there is no systematic shift of p' from p. If r is positive/negative, then there is a positive/negative shift such that p' is expected to be larger/smaller than p. The second parameter in this model is ω, which is a precision parameter that controls the variability of the perturbation:

$$V(p' \mid p) = g(p, r)\{1 - g(p, r)\}/(\omega + 1)\,. \tag{23.3}$$

Now consider a generalization of this idea to vectors $\boldsymbol{p} = (p_1, p_2, \ldots, p_d)$ and $\boldsymbol{p}' = (p'_1, p'_2, \ldots, p'_d)$ taking values in the $(d - 1)$-dimensional simplex. That is, each p_i and p'_i lies in $[0, 1]$ and $\sum_{i=1}^d p_i = \sum_{i=1}^d p'_i = 1$. We similarly represent $\boldsymbol{p}'$ as a perturbation of $\boldsymbol{p}$ via the Dirichlet conditional distribution

$$\boldsymbol{p}' \mid \boldsymbol{p} \sim Di(\omega \boldsymbol{h}(\boldsymbol{p}, \boldsymbol{r}))\,, \tag{23.4}$$

where the vector function $\boldsymbol{h}$ has elements

$$h_i(\boldsymbol{p}, \boldsymbol{r}) = \frac{p_i \exp(r_i)}{\sum_{k=1}^d p_k \exp(r_k)}, \qquad i = 1, 2, \ldots, d. \tag{23.5}$$

The Beta model is the special case $d = 2$ of the Dirichlet model where $\boldsymbol{p} = (p, 1 - p)$, and with the g function corresponding to the $\boldsymbol{h}$ function by setting $\boldsymbol{r} = (r, 0)$. In the Dirichlet case we can without loss of generality constrain r_d to equal zero.

23.2.2 Error rates

Following L&O'H, we model the location error rates θ_i^t at any time t hierarchically, introducing a 'typical' error rate θ_0^t. Each θ_i^t is linked to θ_0^t through a log-odds ratio parameter ρ_i^t and a precision parameter ω_i^t:

$$\theta_i^t \mid \theta_0^t, D^{t-1} \sim Be\left(\omega_i^t g\left(\theta_0^t, \rho_i^t\right), \omega_i^t \left\{1 - g\left(\theta_0^t, \rho_i^t\right)\right\}\right).$$

Note that L&O'H modelled this relationship with an odds ratio parameter δ that corresponds to the exponential of our ρ.

At the first time point, the information D^0 is the prior information before any audit data become available. The initial values ρ_i^1 and ω_i^1 are then set as in L&O'H to reflect the auditor's prior knowledge. The ρ_i^1 can represent the auditor's belief that location i is more or less error-prone than the typical. In the absence of any special beliefs about differing error rates at location i, we could set $\rho_i^1 = 0$, but positive or negative values can reflect the auditor's expectation of a higher or lower error rate than the typical. The meaning of a 'typical' location is determined by these initial choices of the auditor, since a typical location is one having $\rho_i^1 = 0$. The location specific precision parameter ω_i^t allows the auditor to express different levels of prior information concerning each location.

The values of ρ_i^t and ω_i^t at subsequent time points are then determined by the update–project cycle as described in subsequent sections. Although in later times there may be no location with $\rho_i^t = 0$, the meaning of θ_0 as an underlying error rate to which the error rates in other locations are referred is preserved, and the hierarchical structure induces posterior shrinkage at each time point.

At the next stage of the hierarchy we follow L&O'H by assuming a Beta prior distribution for the typical error rate θ_0^t, but this is now conditional on the trend τ_0^t. The trend parameter works as another log-odds ratio parameter in

$$\theta_0^t \mid \tau_0^t, D^{t-1} \sim Be\left(\omega_0^t g\left(m_0^t, \tau_0^t\right), \omega_0^t \left\{1 - g\left(m_0^t, \tau_0^t\right)\right\}\right), \tag{23.6}$$

where ω_0^t determines precision and m_0^t expresses the expectation of θ_0^t if there is no trend, i.e. if $\tau_0^t = 0$. However, if $\tau_0^t > 0$ there is an increasing trend and the expectation of θ_0^t will be higher, whereas if $\tau_0^t < 0$ it will be lower. The initial values ω_0^1 and m_0^1 will be based on the auditor's initial prior information, as in L&O'H. Then subsequent values are determined by the update–project cycle.

The prior model for error rates is completed by a normal distribution for τ_0^t,

$$\tau_0^t \mid D^{t-1} \sim N\left(d_0^t, v_0^t\right) . \tag{23.7}$$

Again, the initial values d_0^1 and v_0^1 will represent the auditor's prior information and subsequent values will be determined by the update–project cycle.

23.2.3 Error classes

We again follow L&O'H and use a Dirichlet hierarchical model to link $\boldsymbol{\psi}_i^t$ to a baseline vector of typical category probabilities $\boldsymbol{\psi}_0^t = (\psi_{01}^t, \psi_{02}^t \ldots, \psi_{0p}^t)$.

$$\boldsymbol{\psi}_i^t \mid \boldsymbol{\psi}_0^t, D^{t-1} \sim Dir\left(\gamma_i^t \boldsymbol{h}(\boldsymbol{\psi}_0^t, \boldsymbol{\rho}_i^t)\right) ,$$

where the distribution is determined by the precision parameter γ_i^t (an effective sample size) and the $\boldsymbol{\rho}_i^t$ vector. Without loss of generality, we let $\rho_{ip}^t = 0$. The initial values γ_i^1 and $\boldsymbol{\rho}_i^1$ define the auditor's prior knowledge about how $\boldsymbol{\psi}_i^t$ relates to the typical $\boldsymbol{\psi}_0^t$, and thereafter they are determined by the update–project cycle.

The prior distribution for $\boldsymbol{\psi}_0^t$ is also Dirichlet, but we do not introduce a trend component:

$$\boldsymbol{\psi}_0^t \mid D^{t-1} \sim Dir\left(\gamma_0^t \boldsymbol{h}_0^t\right), \tag{23.8}$$

where γ_0^t is a scalar precision parameter and $\boldsymbol{h}_0^t$ is the mean vector and satisfies $\sum_{j=1}^{p} h_{0j}^t = 1$. Initially, γ_0^1 and $\boldsymbol{h}_0^1$ define the auditor's prior information, and the update–project cycle determines subsequent values.

23.3 Updating

At time t, the auditor visits some or all of the locations, takes samples of transactions at the visited locations, and observes which transactions are in error, which error class those transactions fall into, and the size of the taint. All of this information is added to D^{t-1} to form D^t.

Consider first the posterior distribution for the error rates. Suppose that the auditor discovers r_i^t errors in a sample of size n_i^t from location i, $i = 1, 2, \ldots, L$. For any location which is not visited, $r_i^t = n_i^t = 0$. Following the development in L&O'H, the posterior distributions of the θ_i^ts conditional on θ_0^t are independent Beta distributions, i.e. the Beta posterior distribution whose parameters are the parameters of the Beta prior updated by adding the number of errors and number of correct cases, respectively:

$$\theta_i^t \mid \theta_0^t, D^t \sim Be\left(\omega_i^t g(\theta_0^t, \rho_i^t) + r_i^t, \omega_i^t\{1 - g(\theta_0^t, \rho_i^t)\} + n_i^t - r_i^t\right) . \tag{23.9}$$

After integrating out these parameters, the joint posterior distribution of θ_0^t and τ_0^t is found to be

$$p(\theta_0^t, \tau_0^t \mid D^t) \propto \frac{(\theta_0^t)^{a_1(\tau_0^t)-1}(1-\theta_0^t)^{a_2(\tau_0^t)-1}}{B(a_1(\tau_0^t), a_2(\tau_0^t))} \exp\left\{-\frac{1}{2v_0^t}(\tau_0^t - d_0^t)^2\right\}$$
$$\times \prod_{i=1}^{L} \frac{B\left(\omega_i^t g(\theta_0^t, \rho_i^t) + r_i^t, \omega_i^t\{1 - g(\theta_0^t, \rho_i^t)\} + n_i^t - r_i^t\right)}{B\left(\omega_i^t g(\theta_0^t, \rho_i^t), \omega_i^t\{1 - g(\theta_0^t, \rho_i^t)\}\right)}, \tag{23.10}$$

where $a_1(\tau_0^t) = \omega_0^t g(m_0^t, \tau_0^t)$ and $a_2(\tau_0^t) = \omega_0^t\{1 - g(m_0^t, \tau_0^t)\}$.

Now consider the posterior distribution for the error class probabilities. The auditor observes the vector $\boldsymbol{c}_i^t = (c_{i1}^t, c_{i2}^t, \ldots, c_{ip}^t)$ of error class counts at location i, $i = 1, 2, \ldots, L$, where $\sum_{j=1}^{p} c_{ij}^t = r_i^t$. The posterior distributions of the ψ_i^ts conditional on ψ_0^t are independent Dirichlet distributions:

$$\psi_i^t \mid \psi_0^t, D^t \sim Dir\left(\gamma_i^t \boldsymbol{h}(\psi_0^t, \rho_i^t) + \boldsymbol{c}_i^t\right) .$$

After integrating out these parameters, the posterior distribution of ψ_0^t is given by

$$p(\psi_0^t \mid D^t) \propto \left(\prod_{j=1}^{p} (\psi_{0j}^t)^{\gamma_0^t h_{0j}^t - 1}\right) \prod_{i=1}^{L} \frac{D\left(\gamma_i^t \boldsymbol{h}(\psi_0^t, \rho_i^t) + \boldsymbol{c}_i^t\right)}{D\left(\gamma_i^t \boldsymbol{h}(\psi_0^t, \rho_i^t)\right)}, \tag{23.11}$$

where if $\boldsymbol{x}$ has elements $x_1, x_2, \ldots, x_p$ then $D(\boldsymbol{x}) = \prod_{i=1}^{p} \Gamma(x_i) / \Gamma(\sum_{i=1}^{p} x_i)$.

Finally, the auditor will also learn the taint values for the transactions in error, and may also learn some book value information, but these are handled in exactly the same way as in L&O'H. In particular, since we assume that taint distributions within error classes are constant over time, the posterior distribution for these at time t is based on the simple accumulation of all taint values observed up to time t combined with the auditor's original prior distribution.

The posterior distributions (23.10) and (23.11) are generally intractable, but low-dimensional. Hence we can derive posterior inference by numerical integration of these distributions. This is discussed in L&O'H, particularly with reference to using importance sampling for ψ_0. Through their conditional distributions, we can make inference also about θ_i^t and ψ_i^t. L&O'H discuss doing this by Monte Carlo sampling but we can also evaluate moments directly by numerical integration with respect to (23.10) and (23.11); for example $E\left(\theta_i^t \mid D^t\right)$ can be evaluated as the expectation of

$$\frac{\omega_i^t g\left(\theta_0^t, \rho_i^t\right) + r_i^t}{\omega_i^t + n_i^t}$$

with respect to (23.10). Note that the structure of our model means that we can use this simple method based on Monte Carlo sampling, and so do not need to employ the much more intricate and computationally intensive Markov chain Monte Carlo (MCMC) approach that is widely used for deriving inferences in many other Bayesian applications.

23.4 Projection

23.4.1 Models for projection

In the projection step we formalize the evolution over time of the parameters from β^t to β^{t+1}. Based on the information D^t available at time t, we model beliefs about the parameters at time $t+1$. In order to keep the sequential analysis of audit data manageable, we need to express the projection so that the distribution of β^{t+1} based on data D^t has the same form as that of β^t based on data D^{t-1}, given in Section 23.2. For instance, if we consider the typical error rate parameter θ_0 we have a Beta distribution $Be(\omega_0^t g(m_0^t, \tau_0^t), \omega_0^t\{1 - g(m_0^t, \tau_0^t)\})$ for θ_0^t given τ_0^t and data D^{t-1}. We therefore need to express the distribution of θ_0^{t+1} given τ_0^{t+1} and data D^t as $Be(\omega_0^{t+1} g(m_0^{t+1}, \tau_0^{t+1}), \omega_0^{t+1}\{1 - g(m_0^{t+1}, \tau_0^{t+1})\})$. The projection step needs to determine how the parameters ω_0^{t+1} and τ_0^{t+1} are to be constructed based on the current information about θ_0^t given by its posterior distribution (23.10).

To motivate the projection in the audit model, consider the following simple process model for a vector $\boldsymbol{p}^t$ of error rates at time t. We suppose that rates evolve by the model

$$p^{t+1} \mid p^t \sim Di(kp^t), \tag{23.12}$$

so that the expectation of p^{t+1} given p^t is p^t, and the evolution precision parameter k determines how far it might be from p^t. Hence k represents the time volatility of error rates or the level of information decay, with smaller k implying more variability from t to $t + 1$.

Now suppose that the distribution of p^t given information at time t is $Di(\phi m^t)$. The marginal distribution of p^{t+1} resulting from this formulation is not Dirichlet, but we can find its mean and variance and approximate as a $Di(\omega m^t)$ where the precision parameter is

$$\omega = \frac{k\phi}{k + \phi + 1}. \tag{23.13}$$

Our formulation of the projection step will use (23.13) as a model for the way that uncertainty is increased by the evolution of error rates from time t to time $t + 1$. However, a variety of considerations are needed to specify the evolution mean.

23.4.2 Error rates

First consider the distribution of θ_i^{t+1} conditional on θ_0^{t+1} and information D^t. As discussed in Section 23.4.1, this needs to match the prior model in Section 23.2. So it should be of the form $Be(\omega_i^{t+1}\, g(\theta_0^{t+1}, \rho_i^{t+1}), \omega_0^{t+1}\, \{1 - g(\theta_0^{t+1}, \rho_0^{t+1})\})$. The projection step needs to determine ω_i^{t+1} and ρ_i^{t+1}.

The posterior conditional distribution (23.9) is Beta with precision parameter $\omega_i^t + n_i^t$. Following the above model and equation (23.13), we let

$$\omega_i^{t+1} = \frac{k_i^t\left(\omega_i^t + n_i^t\right)}{k_i^t + \omega_i^t + n_i^t + 1}, \tag{23.14}$$

for some specified evolution decay parameter k_i^t. We do not suppose any trend in the location-specific error rates conditional on the typical error rate (the trend being expressed in the typical error rate θ_0^t), so we should let the conditional expectation of θ_i^{t+1} be the same as the posterior conditional mean of θ_i^t, but this is not of the required form $g(\theta_0^{t+1}, \rho_i^{t+1})$ for any ρ_i^{t+1}. We therefore refer back to the idea of ρ as a log-odds ratio and define

$$\rho_i^{t+1} = \log\left[\frac{E\left(\theta_i^t \mid D^t\right)}{1 - E\left(\theta_i^t \mid D^t\right)} \Big/ \frac{E\left(\theta_0^t \mid D^t\right)}{1 - E\left(\theta_0^t \mid D^t\right)}\right], \tag{23.15}$$

where both expectations in this equation need to be evaluated numerically from the posterior distribution (23.10). In particular, the expectation of θ_i^t is evaluated as described at the end of Section 23.3.

Next consider the distribution of θ_0^{t+1} conditional on the trend τ_0^{t+1}. As discussed above, this needs to be of the form $Be(\omega_0^{t+1} g(m_0^{t+1}, \tau_0^{t+1}), \omega_0^{t+1}\{1 - g(m_0^{t+1}, \tau_0^{t+1})\})$. The projection is defined by identifying appropriate formulae for ω_0^{t+1} and m_0^{t+1}. From the way the trend is modelled, we set

$$m_0^{t+1} = E\left(\theta_0^t \mid D^t\right) , \tag{23.16}$$

where this expectation is evaluated numerically from the posterior distribution (23.10). To set the precision parameter ω_0^{t+1} we use again the model of Section 23.4.1. The posterior distribution of θ_0^t from (23.10) is not Beta, but we infer a precision value

$$\phi_0^t = \frac{E\left(\theta_0^t \mid D^t\right)\left\{1 - E\left(\theta_0^t \mid D^t\right)\right\}}{V\left(\theta_0^t \mid D^t\right)} - 1 , \tag{23.17}$$

where $V\left(\theta_0^t \mid D^t\right)$ is also evaluated numerically from (23.10). Then we have

$$\omega_0^{t+1} = \frac{k_0^t \phi_0^t}{k_0^t + \phi_0^t + 1} , \tag{23.18}$$

for some specified evolution decay parameter k_0^t.

We now need a projection for the trend parameter. The distribution of τ_0^{t+1} given data D^t needs to be of the same form as that of τ_0^t given data D^{t-1} in Section 23.2, i.e. $N(d_0^{t+1}, v_0^{t+1})$. We need to define d_0^{t+1} and v_0^{t+1}, which we do as follows. First

$$d_0^{t+1} = r^t E\left(\tau_0^t \mid D^t\right) , \tag{23.19}$$

where the trend shrinkage parameter r^t represents a belief that any trend can peter out over time. If $r^t = 1$ the prior expectation of τ_0^{t+1} is the same as the posterior expectation of τ_0^t, while if $r^t = 0$ the prior expectation of τ_0^{t+1} is 0. So at one extreme $r^t = 1$ represents an expectation that a trend at time t will be sustained at time $t+1$, whereas at the other extreme $r^t = 0$ means that we don't believe in trends. For the variance evolution we let

$$v_0^{t+1} = (r^t)^2 V\left(\tau_0^t \mid D^t\right) + w^t , \tag{23.20}$$

where the role of w^t is as an evolution variance, which represents the loss of information over time. Both $E\left(\tau_0^t \mid D^t\right)$ and $V\left(\tau_0^t \mid D^t\right)$ are evaluated numerically from (23.10).

23.4.3 Error classes

The projection for ψ_i^{t+1} given ψ_0^{t+1} proceeds as for θ_i^{t+1} given θ_0^{t+1}. Thus we first have the new precision parameter

$$\gamma_i^{t+1} = \frac{l_i^t \left(\gamma_i^t + r_i^t\right)}{l_i^t + \gamma_i^t + r_i^t + 1} ,$$

where l_i^t is a specified evolution decay parameter. The log-odds parameters are then defined by solving the equations

$$E\left(\psi_i^t \mid D^t\right) = h_i\left(E\left(\psi_0^t \mid D^t\right), \rho_i^{t+1}\right), \quad i = 1, 2, \ldots, p,$$

where the expectations are evaluated with respect to the posterior distribution (23.11).

Finally, we need a projection for ψ_0^{t+1}, which should have a Dirichlet distribution $Di(\gamma_0^{t+1}\boldsymbol{h}_0^{t+1})$. We do not model any trend in these parameters over time, and hence we set

$$\boldsymbol{h}_0^{t+1} = E\left(\psi_0^t \mid D^t\right).$$

The choice of γ_0^{t+1} entails a compromise because we could define it by reference to the posterior variance of any one element of ψ_0^t. We do so conservatively by first letting

$$\chi_0^t = \min_i \frac{E\left(\psi_{0i}^t \mid D^t\right)\left\{1 - E\left(\psi_{0i}^t \mid D^t\right)\right\}}{V\left(\psi_{0i}^t \mid D^t\right)} - 1,$$

and then

$$\gamma_0^{t+1} = \frac{l_0^t \chi_0^t}{l_0^t + \chi_0^t + 1},$$

for some specified evolution decay parameter l_0^t.

23.4.4 Choosing the evolution parameters

The remaining question is how to specify the evolution parameters for the projection step. We have parameters $\{k_0^t, k_1^t, \ldots, k_p^t\}$ and $\{l_0^t, l_1^t, \ldots, l_p^t\}$ which all describe the volatility of the error and class rates over time. Small/large values of these parameters imply high/low variability from one time point to the next. These evolution decay parameters should be set to reflect how much we think error rates might change over time. We would generally expect less volatility in the overall baseline error rates than in individual locations, in which case k_i^t and l_i^t should be less than k_0^t and l_0^t, for instance.

There may be reasons for supposing that some locations might change more from time t to time $t + 1$ than others, and this would lead us to use different k_i^t values for different locations. A location that has been warned to put its house in order, for example, could change more and we could give it a lower k_i^t.

We also need to specify the trend persistence parameter r^t and the trend evolution variance parameter w^t, which should reflect beliefs about the consistency and longevity of trends in the underlying mean error rate.

We discuss these parameters further in the next section when we return to the food stamps program, and also in Section 23.6.

23.5 Application to New York food stamps audit

23.5.1 Modelling

To apply the general model the locations are the $L = 57$ upstate counties (i.e. excluding New York city) of New York state. We report here an analysis for the fiscal years 1998 to 2003, so $t = 1$ corresponds to 1998 and $t = 6$ to 2003. Because the statewide sample is random and county populations are extremely varied, data in some counties for some years are sparse or even nonexistent.

We first define error classes. Whereas previous applications of the L&O'H model used three classes, we chose to subdivide the understatements class. Understatements for which the absolute magnitude of the taint was greater than 5 were treated separately from those with less extreme taint values because the transactions were seen to have quite different book values from the other classes. Our model assumes that book values and taints are independent within each error class. The data clearly showed a small set of negative taint values associated with the minimum benefit payouts. Since the values of these taints were larger, in absolute value, than the other transactions, it was simple to create a new error category for large understatements. We therefore had $p = 4$ error classes: bogus transactions (taint 1), overstatements (taint greater than 0 and less than 1), understatements (taint less than 0 but greater than -5), large understatements (taint less than or equal to -5).

Evolution parameters were set as follows. We set $k_0^t = l_0^t = 99$ for all time periods t, and $k_i^t = l_i^t = 24$ for all counties i and time periods t. The value 99 says that if the baseline rate for some error category is 0.1, then the standard deviation of its change in one time period is 0.03, which we consider to be a suitably small value. Similarly, if the baseline rate for some category is 0.04, the standard deviation of its one period change is 0.02. The value of 24 for individual county evolutions implies a standard deviation of the one period change being 0.06 if the current rate is 0.1, or 0.04 if the current rate is 0.04.

Evolution parameters for the trend were taken as $r^t = 0.7$ and $w^t = 2.0$, implying a belief in any trend being reasonably persistent but highly variable.

Finally, we used deliberately vague priors in order to see how the trend and rates would adapt to the data. Specifically, we assumed $m_0^1 = 0.5$, $\omega_0^1 = 2$ for the prior distribution of the typical error rate; $d_0^1 = 0$, $v_0^1 = 100$ for the trend; $\rho_i^1 = \rho_{ij}^1 = 0$, for all locations and classes (i.e. we regard all counties as 'typical', having no prior information to expect any given county to be more error prone than any other); equal county specific precisions $\gamma_i^1 = 2$, $\omega_i^1 = 2$ for all i; and $\gamma_0^1 = 4$ and $h_{01}^1 = \ldots = h_{04}^1 = 0.25$ for the typical error class rates. Table 23.1 illustrates how trend and rates adapt to data.

In Section 23.5.3 some of these values are modified as part of a sensitivity study.

Table 23.1 Means (variances) of typical error rate θ_0 and trend τ_0.

Year	θ_0		τ_0	
1998	0.217	(9.07×10^{-4})	0.143	(0.359)
1999	0.177	(7.19×10^{-4})	−0.276	(0.132)
2000	0.131	(5.25×10^{-4})	−0.399	(0.170)
2001	0.130	(5.70×10^{-4})	−0.108	(0.192)
2002	0.126	(5.11×10^{-4})	−0.082	(0.193)
2003	0.128	(5.32×10^{-4})	−0.036	(0.196)

23.5.2 Results

Figure 23.1 plots the posterior mean for the typical overall error rate θ_0^t with its uncertainty, together with posterior mean location specific error rates for two example counties, Genessee and Monroe. Figure 23.2 plots in the same way the posterior mean and 95% credible intervals for the typical proportion of errors ψ_{ij}^t in each of the error classes and, for comparison, the posterior mean proportions for Genessee and Monroe counties. In 2002 and 2003 the proportions of understatements and overstatements in Genessee have deviated significantly from the typical proportions, but it is clear from Figure 23.1 that error information generally is poor for these years within Genessee. This can be explained by the fact that sampled transactions consist of only one understatement and one correct transaction for 2002, and three correct transactions for 2003. Sampled transactions from earlier years contain no errors, so the rates are almost equal to the typical rates. The error class proportions for Monroe deviate much more from the typical, as we have more sampled error transactions of each type from this county.

Figure 23.3 plots posterior mean error rates for all counties in NYS. Each panel in this trellis display shows a county's estimated error rate for each year. Panels are ordered from highest error rate (lower left) to lowest error rate (upper right) in the most recent year. This allows one to observe the trend for each county and each county's pattern compared to the other counties.

Since simulation draws are made from the posterior distributions, it is easy to estimate the probability that a county error rate will exceed any threshold and to develop the distribution of ranks. Figure 23.4 shows the posterior densities of error rates for nine example counties. The vertical line at 0.06 is of particular interest because of the Federally defined threshold having to do with sanctions. Figure 23.5 graphs cumulative posterior distribution functions (CDF) for four counties while Figure 23.6 displays the posterior probability that the error rate is less than 6% for each county. The posterior means and credible intervals for the ranks are shown in Figure 23.7. State and county officials have found these figures useful in managing the quality of the program.

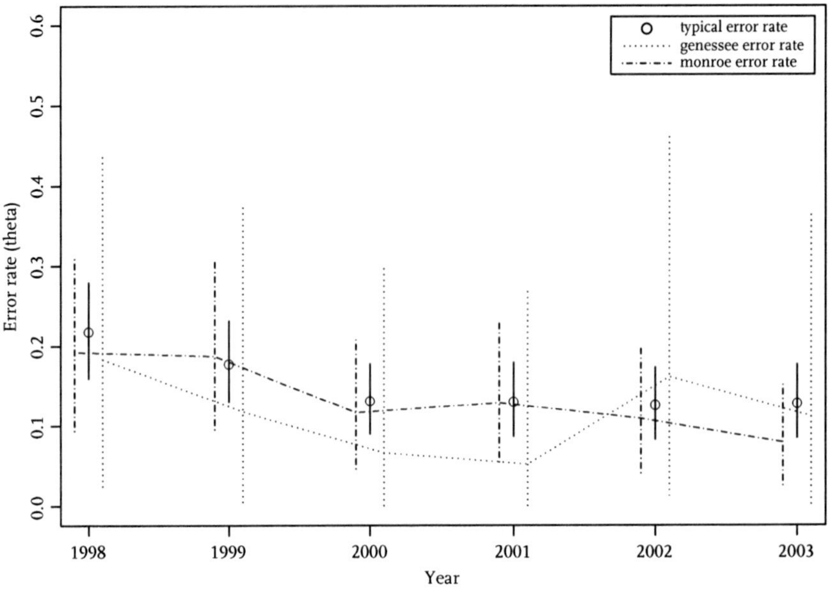

Fig. 23.1 A selection of error rates, with 95% posterior probability intervals.

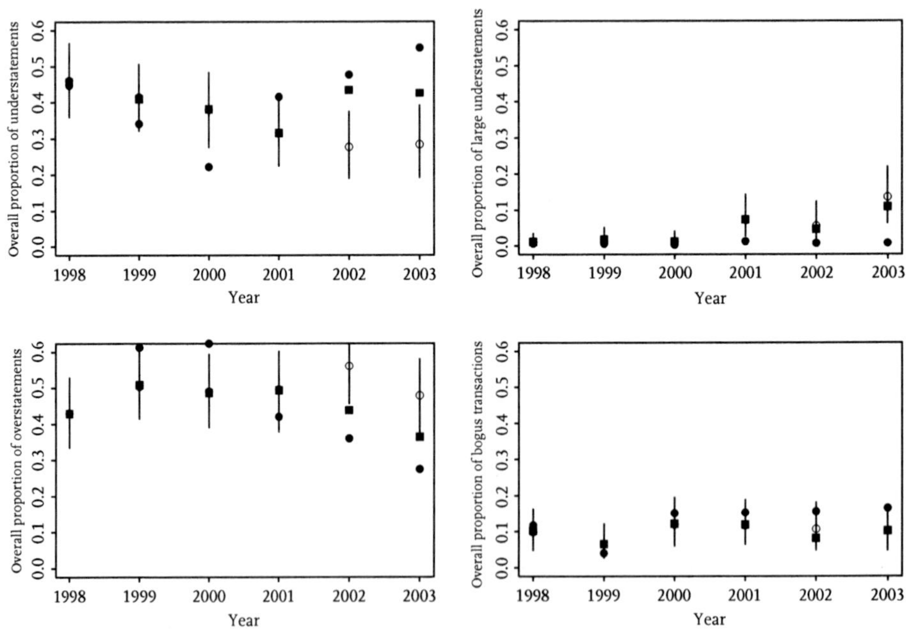

Fig. 23.2 Typical error class proportions, showing the means (circles) and 95% posterior probability intervals. Corresponding proportions for Genessee county (filled squares) and Monroe county (filled circles) are also shown.

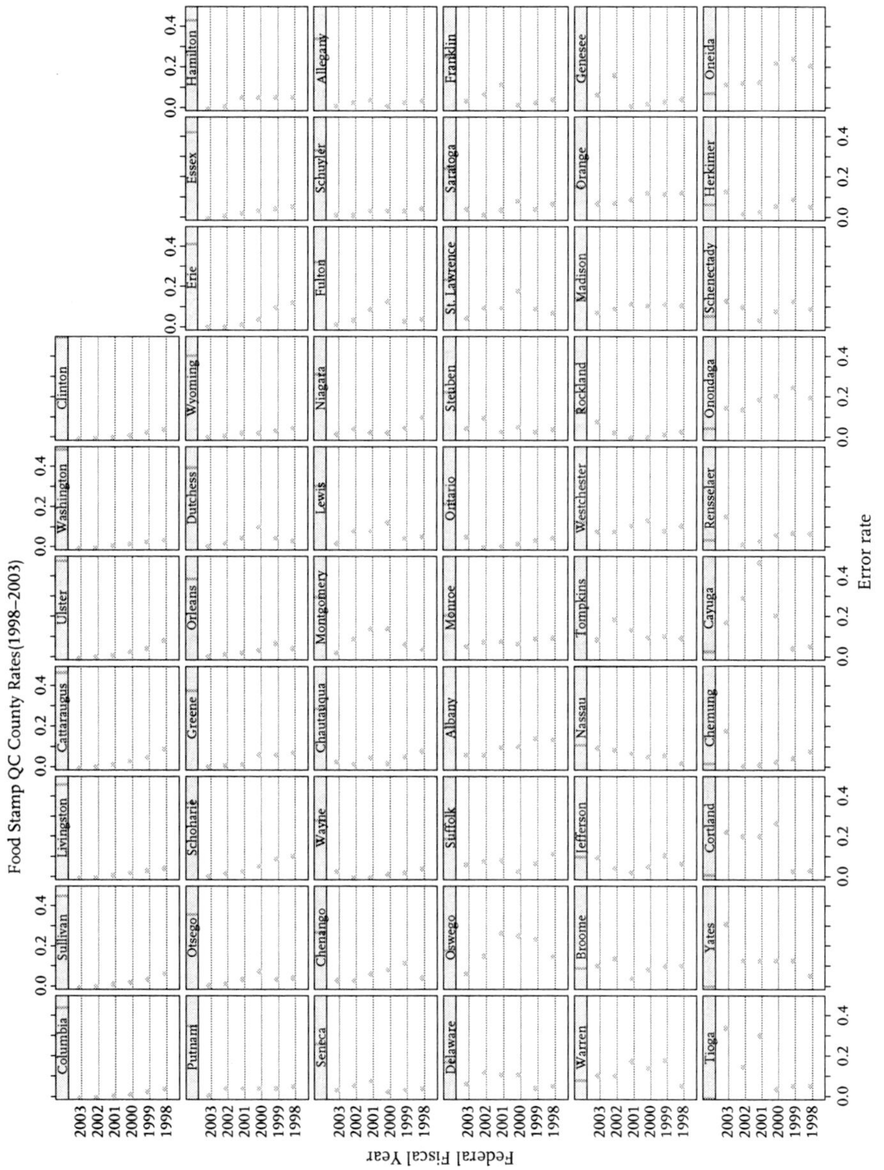

Fig. 23.3 Trellis display of food stamp quality control error rates by county and year (1998–2003).

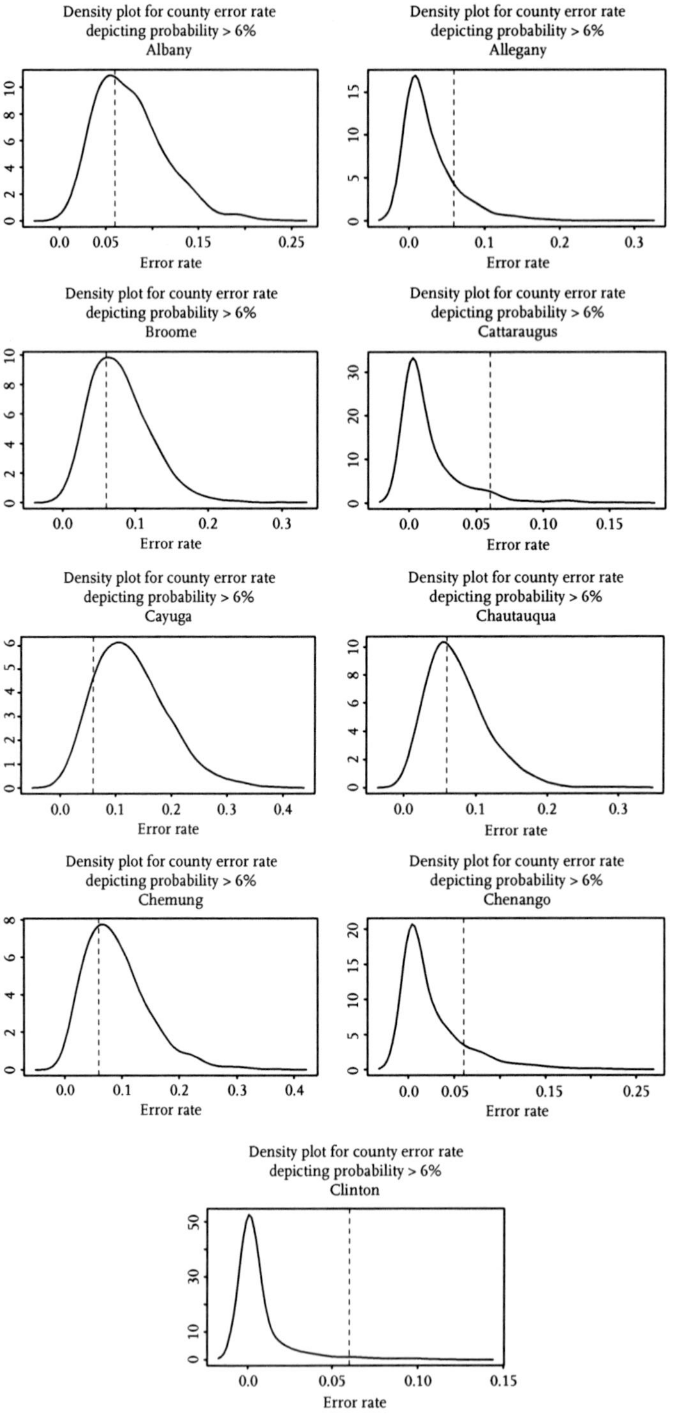

Fig. 23.4 Densities of error rates for some example counties.

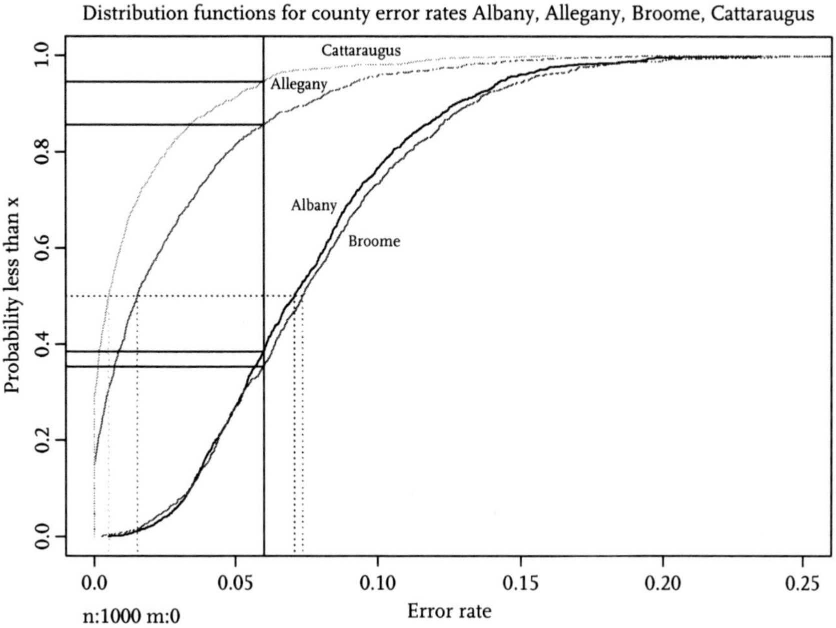

Fig. 23.5 CDFs of error rate for four example counties.

23.5.3 Sensitivity study

We tried four variations on the model and considered the effect on the evolution of θ_0 and τ_0. The first three modify the evolution parameters. The final one turns off trend learning altogether. The tables should be compared with Table 23.1.

1. Set $v^e = 10$ (increases loss of information about trend) (Table 23.2).
2. Set $r^e = 0.3$ (trend more likely to peter out) (Table 23.3).
3. Set $k_0^e = 30$ and $k_i^e = 10$ (reduced precision of error rate evolutions) (Table 23.4).
4. Set $v^e = 1 \times 10^{-8}$ and $\tau_0 \sim N(0, 1 \times 10^{-8})$ (effectively turns trend modelling off) (Table 23.5).

Table 23.2 Means (variances) of typical error rate θ_0 and trend τ_0 Variation 1.

Year	θ_0		τ_0	
1998	0.217	(9.07×10^{-4})	0.143	(0.359)
1999	0.176	(6.50×10^{-4})	−0.324	(0.143)
2000	0.131	(5.18×10^{-4})	−0.411	(0.195)
2001	0.129	(5.49×10^{-4})	−0.104	(0.205)
2002	0.123	(4.96×10^{-4})	−0.104	(0.210)
2003	0.125	(5.44×10^{-4})	−0.064	(0.203)

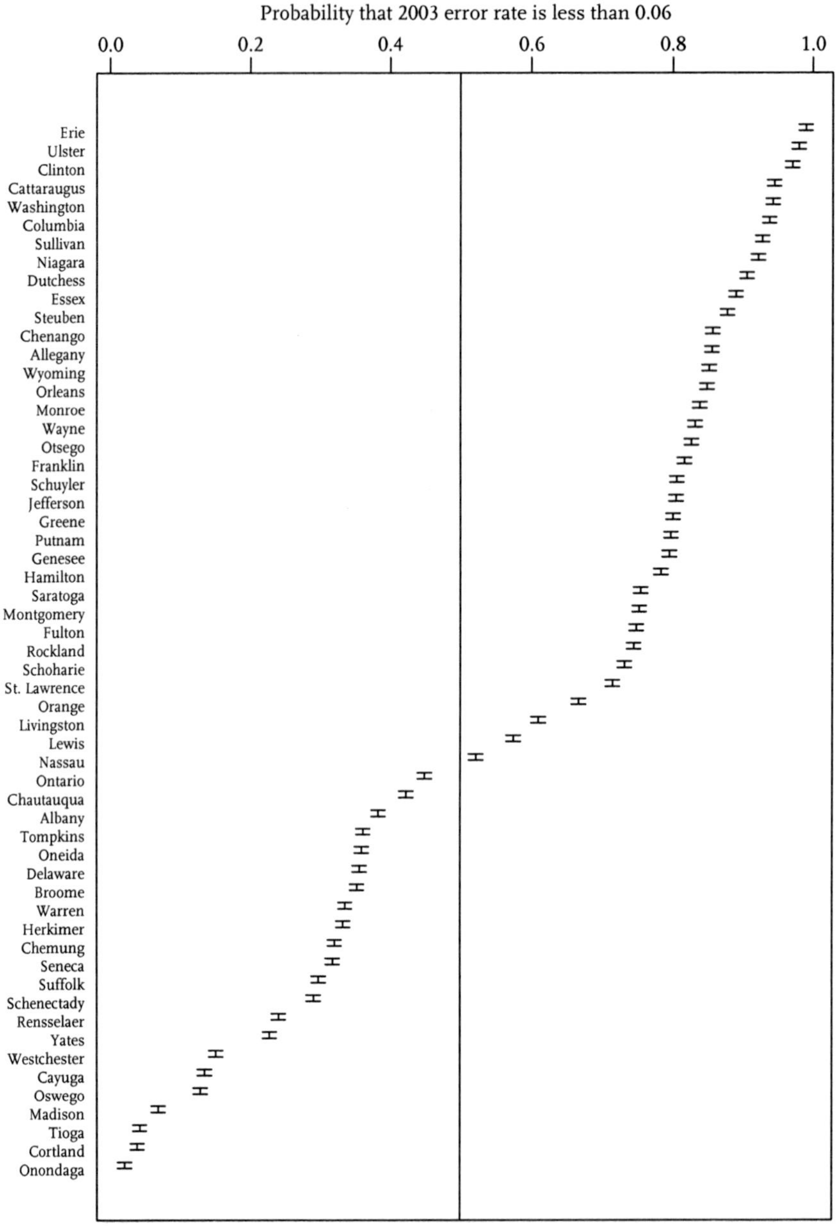

Fig. 23.6 Probability that the error rate is less than 6% for each county.

The first two variations are seen to have little effect on the posterior distributions for the typical error rate and underlying trend. In the third variation, the increased volatility assumed for all the error rates seems to obscure the learning about trend, with the result that the underlying trend in the data is not picked up so well by the model. Similarly, the fourth variation removes the

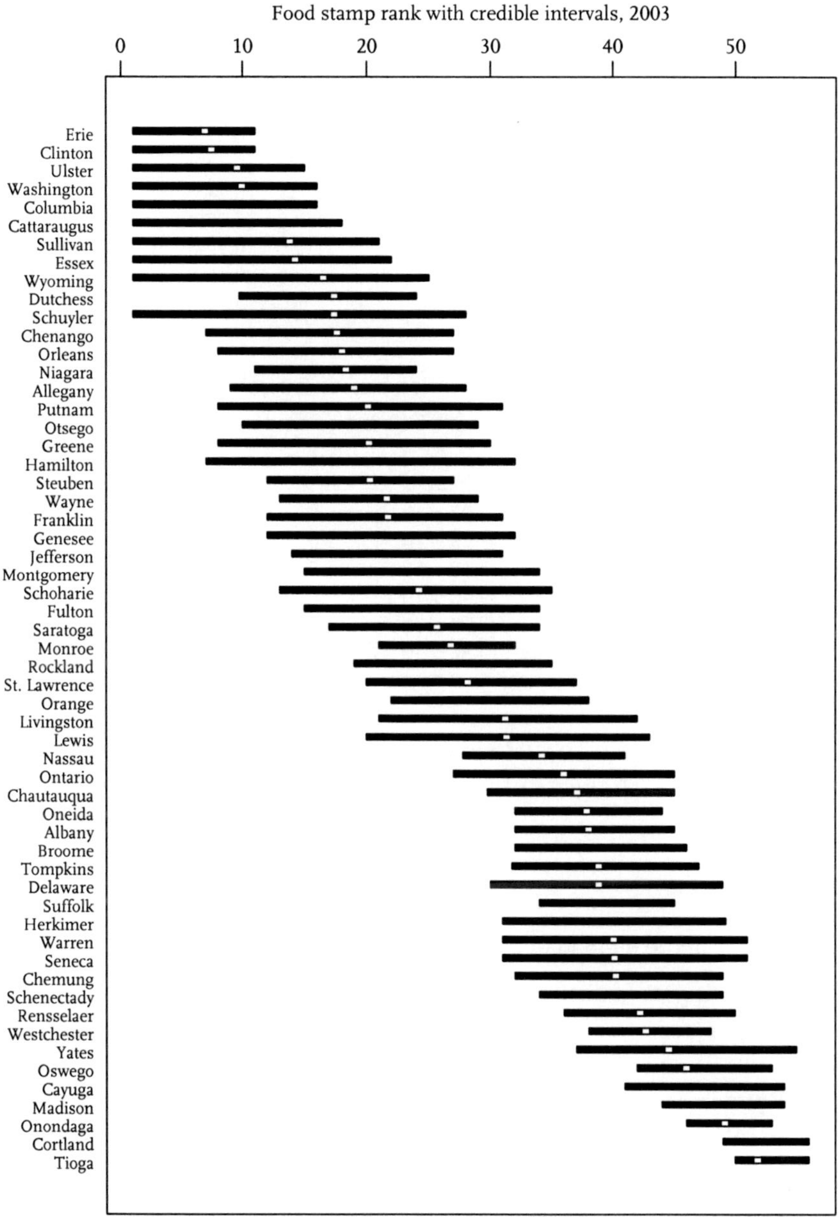

Fig. 23.7 County 2003 food stamp error rate ranks with credible intervals.

trend component from the model, and also results in the posterior estimates not showing the full trend towards lower overall error rates that was seen in the data. To the extent that we can learn from this simple sensitivity analysis, it seems that the trend component is important in the model, and its posterior distribution may be quite robust to the trend evolution parameters. Also, if there

Table 23.3 Means (variances) of typical error rate θ_0 and trend τ_0 Variation 2.

Year	θ_0		τ_0	
1998	0.217	(9.07 $\times 10^{-4}$)	0.143	(0.359)
1999	0.175	(6.52 $\times 10^{-4}$)	−0.302	(0.132)
2000	0.132	(5.07 $\times 10^{-4}$)	−0.351	(0.145)
2001	0.130	(6.06 $\times 10^{-4}$)	−0.063	(0.179)
2002	0.126	(5.11 $\times 10^{-4}$)	−0.086	(0.178)
2003	0.126	(5.35 $\times 10^{-4}$)	−0.059	(0.180)

Table 23.4 Means (variances) of typical error rate θ_0 and trend τ_0 Variation 3.

Year	θ_0		τ_0	
1998	0.217	(9.07 $\times 10^{-4}$)	0.143	(0.359)
1999	0.180	(7.78 $\times 10^{-4}$)	−0.258	(0.285)
2000	0.147	(7.18 $\times 10^{-4}$)	−0.304	(0.342)
2001	0.150	(9.40 $\times 10^{-4}$)	−0.093	(0.374)
2002	0.153	(7.98 $\times 10^{-4}$)	−0.078	(0.371)
2003	0.165	(9.86 $\times 10^{-4}$)	−0.005	(0.342)

is (assumed to be) much annual volatility in error rates (as expressed in the ρ parameters), then learning about underlying trend is reduced.

23.6 Discussion

23.6.1 Success of the method in New York state

In 2003, reporting of county error rates was initiated in New York state. Counties with high error rates objected, but it was pointed out that the hierarchical nature of the model shrunk county specific error rates toward the statewide rate, resulting in modeled rates being less than empirical rates in counties with high error rates. Some counties reported that they reviewed and revised their

Table 23.5 Means (variances) of typical error rate θ_0 Variation 4.

Year	θ_0	
1998	0.212	(8.48 $\times 10^{-4}$)
1999	0.180	(5.28 $\times 10^{-4}$)
2000	0.144	(4.15 $\times 10^{-4}$)
2001	0.138	(4.41 $\times 10^{-4}$)
2002	0.132	(4.46 $\times 10^{-4}$)
2003	0.134	(4.40 $\times 10^{-4}$)

administrative approach despite believing that their systems were not prone to error. Subsequently, a number of counties improved and the statewide error rates for 2003 through 2006 were 0.0588, 0.0574, 0.0723, and 0.0456, respectively, causing the federal government to award the State several million dollars as part of an incentive programme to promote cost savings. The incentive was then distributed among the counties based in part on the rates resulting from the sequential multicentre model.

During this period, at a meeting of county officials, a state official was asked by a county official to explain how his county could have a published error rate when hardly any of the statewide sample came from his county. The state official could not explain and announced that the practice would be abolished. After a short moratorium, at the request of some state-level programme managers and some county officials, modelling county level error rates was reinstated, although the rates from 2007 have been used internally but have not been published.

23.6.2 Wider applicability

In our applied experience we have encountered several audit situations where the organization being audited is distributed over several locations. Furthermore, a number of these multilocation audits occur periodically. The example that we cite, food stamp quality control sampling across counties within a state, is not atypical. Many governmental programmes are audited in a similar fashion.

In the United States, health care is frequently provided by large Health Maintenance Organizations (HMOs) that consist of a health care insurance mechanism and several health care providers, e.g. hospitals, physicians, etc. In fact, individual hospitals are composed of several units, each responsible for their own record keeping. At universities and in research organizations, each principal investigator manages a budget, but there is collective responsibility and therefore audit requirements that dictate sampling over several locations. Of course, large corporations almost invariably have multiple locations and accounts.

When these multilocation organizations are sampled, the sample size in many locations is small and sometimes zero. The hierarchical nature of the Bayesian approach provides useful shrinkage allowing information to flow from the whole organization to the subunits or locations and from period to period. This allows statements to be made about each location and precision to be reported.

The question of the importance of the information from previous periods and from the whole organization to its parts requires careful consideration and sensitivity analysis is highly recommended. Research into this question is needed.

The sequential multilocation auditing model allows for prior information from expert opinion. Elicitation in this setting needs study.

We have found the Bayesian sequential multilocation model appropriate in a number of audits. Results have been useful and accepted by managers who have successfully used the results in managing their programmes.

Appendix

A. Broader context and background

A.1 Prior information

In any Bayesian analysis it is interesting to see the extent to which the prior distribution is based on genuine prior information. This chapter extends an earlier approach to multilocation auditing, Laws and O'Hagan (2002), in which there was substantial emphasis on real prior information. In that approach, auditors were required to formulate quite complex prior information. We argued that they generally have such information, based in part on the fact that auditors typically audit the same organizations on an annual basis. The present work explores the basis for using previous audit results by explicitly modelling the evolution of error rates over time, rather than requiring the auditor to accommodate earlier results implicitly in his/her prior specification.

The auditor is still required to make subjective judgements, but now in a much simpler way. He/she is required only to specify a few evolution parameters to describe, for instance, how rapidly the error rate in a given location may change, and how consistent any trend will be. It is possible, however, to reintroduce some of the subjective judgement of the earlier model, because in practice the auditor's information prior to each audit comprises more than the results of previous audits. This includes some examination of the financial systems and controls that are in place, knowledge of any changes in key personnel, and so on. It will be possible for the auditor to use the prior distribution coming from our new dynamic model, and to modify it to reflect any other factors that he/she feels represent relevant prior information.

Similarly, in the context of Bayesian forecasting, West and Harrison (1997) have stressed the possibility of intervening to modify the prior distribution at any time point. In any Bayesian analysis, there should be scope for using genuine prior information.

A.2 Dynamic models

The modelling and analysis here also has similarities to Bayesian forecasting using dynamic linear models. Such models have an observation equation to relate observations to the underlying system state, plus a system equation that expresses how that state evolves over time. We have a set of parameters θ^t defining the system state at time t, and the current distribution of these parameters given observations up to time $t-1$ is $p(\theta^t \mid D^{t-1})$. Using the observation equation, we update this distribution by Bayes' theorem to $p(\theta^t \mid D^t)$, where the information D^t comprises the prior information D^{t-1} plus the new observation at time t. We then project using the system equation to $p(\theta^{t+1} \mid D^t)$, which then becomes the prior distribution for the next time step. In the usual Gaussian form of this model, all of these distributions are (multivariate) normal. The updating and projection steps then comprise the well-known Kalman filter (see Chapter 1 of West and Harrison 1997).

If we make the underlying distributions non-Gaussian, or the equations nonlinear in the state parameters, then it is no longer possible to retain the simplicity of the Kalman filter without making approximations. Exact analysis would result in increasingly complex distributions $p(\theta^t|D^t)$ and $p(\theta^{t+1}|D^t)$ after each updating and projection step, and the computations rapidly become unmanageable. One approach to this problem is to use particle filtering (see Doucet and Gordon 2001, and `http://www-sigproc.eng.cam.ac.uk/smc/papers.html`), while another is to employ approximations at each step (Chapters 13 and 14 of West and Harrison 1997).

In our dynamic auditing model, we employ the approximation approach. Our equivalent of the state vector θ^t is made up of θ_0^t, τ_0^t, θ_i^t for each location i, ψ_0^t and ψ_i^t for each location. After Bayesian updating with the results of the audit at time t, the posterior distribution of these parameters is complex and no longer of the same form as the prior. Hence we apply a simplified approximation during the projection step to reinstate a consistent form of prior distribution for the parameters at the next time period. Thus θ_0^{t+1} and θ_i^{t+1} have beta distributions, ψ_0^{t+1} and ψ_i^{t+1} have Dirichlet distributions, and τ_0^t has a normal distribution, given data D_t.

Acknowledgements

The work presented here was begun with the collaboration of David Laws, and owes much to his contributions. Sadly, David was already suffering from Motor Neurone Disease, a dreadful progressive condition for which there is no cure. David died during the preparation of this chapter, in 2008.

The authors are also grateful to Arilee Bagley and Shannon Smosarski for their help with the preparation of this chapter.

References

Cochran, W. G. (1977). *Sampling Techniques.* John Wiley, New York.

Cohen, A. and Sackrowitz, H. B. (1996). Lower confidence bounds using pilot samples with an application to auditing. *Journal of the American Statistical Association*, **91**, 338–342.

Doucet, A. de Freitas, N. and Gordon, N. (2001). *Sequential Monte Carlo Methods in Practice.* Springer Verlag, New York.

Frost, P. A. and Tamura, H. (1982). Jackknifed ratio estimation in statistical auditing. *Journal of Accounting Research*, **20**, 103–120.

Heiner, K. and Laws, D. (1999). Simulation based inference for auditing. In *Computing Science and Statistics. Models, Predictions, and Computing*. (ed. K. Berk and M. Pourahmadi), pp. 64–68. Proceedings of the 31st Symposium on the Interface, Interface Foundation of North America.

Heiner, K. W. (1999). Issues in statistical auditing and some Bayesian solutions. *In* ASA Proceedings of the Business and Economic Statistics Section, American Statistical Association, pp. 18–24.

Heiner, K. W., Wagner, N. L. and Fried, A. C. (1984). Successfully using statistics to determine damages in fiscal audits. *Jurimetrics*, **24**, 273–281.

Heiner, K. W. and Whitby, O. (1980). Maximizing restitution for erroneous medical payments when auditing samples. *Interfaces (INFORMS)*, **10**, 46–53.

Kvanli, A. H., Shen, Y. K. and Deng, L. Y. (1998). Construction of confidence intervals for the mean of a population containing many zero values. *Journal of Business and Economic Statistics*, **16**, 362–368.

Laws, D. J. and O'Hagan, A. (2000). Bayesian inference for rare errors in populations with unequal unit sizes. *Journal of the Royal Statistical Society, Series C: Applied Statistics*, **49**, 577–590.

Laws, D. J. and O'Hagan, A. (2002). A hierarchical Bayes model for multilocation auditing. *Journal of the Royal Statistical Society, Series D: The Statistician*, **51**, 431–450.

Moors, J. J. A. and Janssens, M. J. B. T. (1989). Exact distributions of Bayesian Cox–Snell bounds in auditing. *Journal of Accounting Research*, **27**, 135–144.

Neter, J. and Godfrey, J. (1985). Robust Bayesian bounds for monetary-unit sampling in auditing. *Applied Statistics*, **34**, 157–168.

Neter, J. and Loebbecke, J. K. (1977). On the behavior of statistical estimators when sampling accounting populations. *Journal of the American Statistical Association*, **72**, 501–507.

Smith, T. M. F. (1979). Statistical sampling in auditing: A statistician's viewpoint. *The Statistician*, **28**, 267–280.

The Panel on Nonstandard Mixtures of Distributions (1989). Statistical models and analysis in auditing. *Statistical Science*, **4**, 2–33.

Tsui, K.-W., Matsumura, E. M. and Tsui, K.-L. (1985). Multinomial-Dirichlet bounds for dollar-unit sampling in auditing. *Accounting Review*, **60**, 76–96.

van Batenburg, P. C., O'Hagan, A. and Veenstra, R. H. (1994). Bayesian discovery sampling in financial auditing: A hierarchical prior model for substantive test sample sizes. *The Statistician*, **43**, 99–110.

West, M. and Harrison, P. (1997). *Bayesian Forecasting and Dynamic Models*, (2nd edn), Springer-Verlag, New York.

·24·

Bayesian causal inference: Approaches to estimating the effect of treating hospital type on cancer survival in Sweden using principal stratification

Donald B. Rubin, Xiaoqin Wang, Li Yin, and Elizabeth R. Zell[1]

24.1 Introduction

This chapter concerns Bayesian causal inference, in particular, using the posterior predictive approach of the 'Rubin causal model', originally outlined in Rubin (1975) and developed in numerous articles by him and coauthors, as referenced in this chapter. Here we apply this approach to a difficult and important problem in medicine, deciding which type of hospital, large volume versus small volume, is superior for treating certain serious conditions, where volume refers to the number of patients with that condition treated there. The data we use came from central and northern Sweden, which arose in a study at the Karolinska Institute in Stockholm. More precisely, the situation we have is the following.

If one has a serious medical condition, which is diagnosed in a small volume hospital, is it be better to be transferred to a major medical centre for treatment, where the medical staff are relatively more experienced with the condition and possibly better equipped to handle it, or is it better to be treated in the local hospital where personal attention is more likely, including from family and friends? A version of this question also confronts governmental agencies when making budgetary decisions concerning maintaining smaller volume treatment facilities in local hospitals versus just maintaining treatment facilities in the larger volume hospitals, and transferring all patients who are diagnosed at small volume hospitals to larger volume hospitals for treatment. If data suggest that patient outcomes are better when treated at large-volume hospitals, then there is no reason to maintain treatment facilities at small volume hospitals, thereby saving resources.

[1] The authors' names are in alphabetic order.

The findings and conclusions in this chapter are those of the authors and do not necessarily represent the views of their respective institutions.

Ideally, one would have data from randomized experiments, where patients with specific conditions are randomized to be treated in large versus small volume hospitals, but such data are not available, and so to address this question, observational data must be relied upon. In some cases, such as ours, medical judgment may be that the assignment of the diagnosing (home) hospital type (large versus small) may reasonably be considered ignorable, at least with a rich enough set of covariates available. Thus, the estimation of the causal effect of large versus small volume hospital type is standard. However, the estimation of the causal effects of treating hospital type (large versus small) is complicated by the existence of transfers between hospital types for treatment, typically from small home hospital types to large treating hospital types, which implies that the assignment of treating hospital type is nonignorable because the reasons for the transfers undoubtedly involved unmeasured variables, such as the patients' willingness to undergo invasive medical operations.

We address this problem using principal stratification (Frangakis and Rubin, 2002), where we view the transfers between hospital types as a form of noncompliance with assigned type of hospital. We apply our formulation to the data from Sweden on cardia and stomach cancer. This formulation leads to a Bayesain version of the classical 'instrumental variables' method from economics (Angrist, Imbens and Rubin, 1996), as initially described in Imbens and Rubin (1997). Critical assumptions are discussed at length: the monotonicity assumption and exclusion restriction. The Bayesian analyses suggest an interesting and intuitively reasonable conclusion: the positive effect of large treating hospital type on cardia cancer, which is a rare and difficult to treat condition, but no such effect for the more common stomach cancer. Our analytic approach should have many applications for addressing similar questions in other contexts.

The remainder of the chapter has two main parts. The first part summarizes the general Bayesian approach to causal inference, primarily with ignorable treatment assignment. The second part describes our application in more detail, motivates our approach using simple method-of-moments summary statistics, explicates our Bayesian model, and analyzes our data.

24.2 Bayesian causal inference – General framework

The framework for Bayesian causal inference that we use is now commonly referred to as 'Rubin's causal model' (RCM, Holland, 1986), for a series of articles written in the 1970s (Rubin, 1974–1980). The full RCM framework has three parts. The first part defines causal effects of treatments on units through potential outcomes and is developed in Section 24.2.1. The second part of this framework concerns the assignment mechanism, as described in Section 24.2.2. The third part of the RCM is the use of Bayesian posterior predictive inference for causal effects as initially developed in Rubin (1975, 1978), and this is presented in Section 24.2.3. These descriptions in Section 24.2 borrow in some places from a related chapter on general causal inference in Rubin (2008).

24.2.1 Units, treatments, potential outcomes, and SUTVA

For causal inference, there are several primitives – concepts, that are basic and on which we must build. A 'unit' is a physical object, e.g. a person, at a particular point in time. A 'treatment' is an action that can be applied or withheld from that unit. We focus on the case of two treatments, although the extension to more than two treatments is simple in principle although not necessarily so with real data. Associated with each unit are two 'potential outcomes': the value of an outcome variable Y at a point in time after the active treatment is applied and the value of that outcome variable at the same point in time when the active treatment is withheld. The objective is to learn about the causal effect of the application of the active treatment relative to the control treatment (the control treatment implies that the active treatment is withheld) on the variable Y.

For example, the unit could be 'you now' with your headache, the active treatment could be taking aspirin for your headache, and the control treatment could be not taking aspirin (as in Rubin, 1974). The outcome Y could be the intensity of your headache pain in two hours, with the potential outcomes being the headache intensity if you take aspirin now and if you do not take aspirin now. 'You now' is a different unit from 'you later'.

Notationally, let W indicate which treatment the unit, you, received: $W = 1$ for the active treatment, $W = 0$ for the control treatment. Also let $Y(1)$ be the value of the potential outcome if the unit received the active version, and $Y(0)$ the value if the unit received the control version. The causal effect of the active treatment relative to its control version is the comparison of $Y(1)$ and $Y(0)$ – typically the difference, $Y(1)$ minus $Y(0)$, or perhaps the difference in logs, $\log[Y(1)]$ minus $\log[Y(0)]$, or some other comparison, possibly the ratio.

We can observe the value of $Y(W)$ as indicated by W. The fundamental problem facing causal inference (Rubin, 1978: p. 38, Holland, 1986) is that, for any individual unit, we observe only the value of the potential outcome under one of the possible treatments, namely the treatment actually assigned, and the potential outcome under the other treatment is missing. Thus, inference for causal effects is a missing-data problem – the 'other' value is missing. For example, your reduction in blood pressure one week after taking a drug is a change in time, in particular, a change from before taking the drug to after taking the drug on you, and so is not a causal effect without additional assumptions. The comparison of your blood pressure after taking the drug with what it would have been at the same point in time without taking the drug is a causal effect.

We learn about causal effects using replication, more units. The way we personally learn from our own experience is replication involving the same physical object (e.g. you) at more points in time, thereby generating more experimental units (e.g. you at the various points in time). That is, if I want to learn about the effect of taking aspirin on headaches for me, I learn from replications in time when I do and do not take aspirin to relieve my headache, thereby having some observations of $Y(1)$ and some of $Y(0)$. When we want to generalize to units other than ourselves, we typically use more physical objects; that is what is done in epidemiology and medical experiments.

Suppose instead of only one unit we have two, and use subscripts 1 and 2 on Y to denote their potential outcomes. Now in general we have at least four potential outcomes for each unit: the outcome for unit 1 if both unit 1 and unit 2 received control, $Y_1(0, 0)$; the outcome for unit 1 if both units received the active treatment, $Y_1(1, 1)$; the outcome for unit 1 if unit 1 received control and unit 2 active, $Y_1(0, 1)$, and the outcome for unit 1 if unit 1 received active and unit 2 received control, $Y_1(1, 0)$; and analogously for unit 2 with potential outcomes $Y_2(1, 1)$, $Y_2(0, 0)$, $Y_2(1, 0)$, and $Y_2(0, 1)$. In fact, generally there are even more potential outcomes because there have to be at least two 'doses' of the active treatment available to contemplate all assignments, and it could make a difference which doses was taken. For example, for the aspirin case, one tablet could be very effective and the other quite ineffective.

Clearly, replication does not help unless we can restrict the explosion of potential outcomes. As in all theoretical work with applied value, simplifying assumptions are crucial. The most straightforward assumption to make is the 'stable unit treatment value assumption' (SUTVA, Rubin, 1980; 1990*a*, 1990*b*) under which the potential outcomes for the ith unit just depend on the treatment the ith unit received. That is, there is 'no interference between units' (Cox, 1958) and there are 'no unrepresented treatments for any unit'. Then, all potential outcomes for N units with two possible treatments can be represented by an array with N rows and two columns, the ith unit having a row with two potential outcomes, $Y_i(0)$ and $Y_i(1)$.

Obviously, SUTVA is a major assumption. But there is no assumption-free causal inference, and nothing is wrong with this. It is the quality of the assumptions that matters, not their existence or even their absolute correctness. Good researchers attempt to make such assumptions plausible by the design of their studies. For example, SUTVA becomes more plausible when units are isolated from each other. For example, when studying an intervention such as a smoking prevention programme in schools (e.g. see Peterson *et al.*, 2000), the units were defined to be intact schools that were geographically separated rather than individual students or classes in the schools.

The stability assumption (SUTVA) is very commonly made, even though it is not always appropriate. For example, consider a study of the effect of vaccination on a contagious disease. The greater the proportion of the population that gets vaccinated, the less any unit's chance of contracting the disease, even if the unit is not vaccinated – an example of interference. Throughout this chapter, we assume SUTVA, although there are other assumptions that could be made to restrict the exploding number of potential outcomes with replication and no assumptions.

In general, some of the N units may receive neither the active treatment $W_i = 1$ nor the control treatment $W_i = 0$. For example, some of the units may be in the future, as when we want to generalize to a future population. Then formally W_i must take on a third value, $W_i = *$ representing neither 1 nor 0; we often avoid this extra notation in this chapter.

In addition to (a) the vector indicator of the treatment for each unit in the study, W, (b) the array of potential outcomes when exposed to the treatment, $Y(1) = \{Y_i(1)\}$, and (c) the array of potential outcomes when exposed to the control, $Y(0) = \{Y_i(0)\}$, we have (d) an array of covariates $X = \{X_i\}$, which are, by definition, unaffected by treatment exposure, such as age or pretreatment baseline blood pressure. All causal estimands involve comparisons of $Y_i(0)$ and $Y_i(1)$, on either all N units, or a common subset of units; for example, the average causal effect across all units that are female as indicated by their X_i, or the median causal effect for units with $Y_i(0)$ values indicating unacceptably high blood pressure after exposure to the control treatment.

Thus, under SUTVA, all causal estimands can be calculated from the matrix of 'scientific values' with ith row: $(X_i, Y_i(0), Y_i(1))$. By definition, all relevant information is encoded in X_i, $Y_i(0)$, $Y_i(1)$ and so the labelling of the N rows is a random permutation of $1, \ldots, N$. In other words, the N-row array

$$(X, Y(0), Y(1)) = \begin{bmatrix} X_1 & Y_1(0) & Y_1(1) \\ & \vdots & \\ X_i & Y_i(0) & Y_i(1) \\ & \vdots & \\ X_N & Y_N(0) & Y_N(1) \end{bmatrix}$$

is row exchangeable. We call this array 'the science' because its values are beyond our control; by changing treatments, we get to change which values are actually observed, but not the values themselves. That is, the observed values of Y are $Y_{\text{obs}} = \{Y_{\text{obs},i}\}$, where $Y_{\text{obs},i} = (1 - W_i)Y_i(0) + W_i Y_i(1)$. In Section 24.2, we consider X fully observed. However, the model we use in Section 24.3 has some unobserved covariates that define 'principal strata', as defined in Frangakis and Rubin (2002).

Covariates (such as age, race and sex) play a particularly important role in observational studies for causal effects. In some studies, the units exposed to the active treatment differ on their distribution of covariates in important ways from the units not exposed. To see how this issue influences our formal framework, we must define the 'assignment mechanism', the probabilistic mechanism that determines which units get the active version of the treatment and which units get the control version. The assignment mechanism is the topic of Section 24.2.2.

24.2.2 Treatment assignment mechanism

A model for the assignment mechanism is needed for all forms of statistical inference for causal effects. The assignment mechanism gives the conditional probability of each vector of assignments for all the units given the covariates and potential outcomes:

$$\Pr(W \mid X, Y(0), Y(1)). \tag{24.1}$$

Here W is an $N \times 1$ vector and X, $Y(1)$ and $Y(0)$ are all matrices with N rows. A specific example of an assignment mechanism is a completely randomized experiment with N units, where $n < N$ are assigned to the active treatment, and $N - n$ are assigned to the control treatment:

$$\Pr(W \mid X, Y(0), Y(1)) = \begin{cases} 1/C_n^N & \text{if} \sum W_i = n \\ 0 & \text{otherwise.} \end{cases} \tag{24.2}$$

The assignment mechanism is fundamental to causal inference because it tells us how we got to see what we saw. As stated earlier, causal inference is basically a missing data problem with at least half of the potential outcomes missing. When we have no understanding of the process that creates missing data, we have no hope of inferring anything about the missing values. That is, without a stochastic model for how treatments are assigned to individuals, formal causal inference, at least using probabilistic statements, is impossible. This statement does not mean that we need to *know* the assignment mechanism, but rather that without positing one, we cannot make any statistical claims about causal effects, such as the coverage of Bayesian posterior intervals, defined later.

A special class of assignment mechanisms is particularly important to Bayesian inference: ignorable assignment mechanisms (Rubin, 1978). Ignorable assignment mechanisms are defined by their freedom from dependence on any missing potential outcomes:

$$\Pr(W \mid X, Y(0), Y(1)) = \Pr(W \mid X, Y_{\text{obs}}). \tag{24.3}$$

In our application in Section 24.3, the assignment mechanism is 'latently' ignorable (Frangakis and Rubin, 1999) in the sense that we posit an assignment mechanism such that one of the 'covariates' needed to make the assignment mechanism ignorable is partially missing, meaning if it were fully observed, the assignment mechanism would be ignorable.

24.2.3 Posterior predictive causal inference

Bayesian inference for causal effects requires a model for the underlying science, $\Pr(X, Y(0), Y(1))$. A virtue of the RCM framework is that it separates science – a model for the underlying science – from what we do to learn about the science – the assignment mechanism, $\Pr(W|X, Y(0), Y(1))$. Notice that together, these two models specify a joint distribution for all observables, the approach commonly called Bayesian.

Bayesian inference for causal effects directly and explicitly confronts the missing potential outcomes, $Y_{\text{mis}} = \{Y_{\text{mis},i}\}$, where $Y_{\text{mis},i} = W_i Y_i(0) + (1 - W_i) Y_i(1)$. The perspective takes the specification for the assignment mechanism and the specification for the underlying science and derives the 'posterior predictive' distribution of Y_{mis}, that is, the distribution of Y_{mis} given all observed values:

$$\Pr(Y_{\text{mis}}|X, Y_{\text{obs}}, W). \tag{24.4}$$

This distribution is 'posterior' because it is conditional on all observed values (X, Y_{obs}, W) and 'predictive' because it predicts (stochastically) the missing potential outcomes. From (a) this distribution, (b) the observed values of the potential outcomes, Y_{obs}, (c) the observed assignments, W, and (d) the observed covariates, X, the posterior distribution of any causal effect can, in principle, be calculated.

This conclusion is immediate if we view the posterior predictive distribution in equation (24.4) as specifying how to take a random draw of Y_{mis}, because once a value of Y_{mis} is drawn using equation (24.4), any causal effect can be directly calculated from the drawn value of Y_{mis} and the observed values of X and Y_{obs}. For example, the median causal effect for males: $\text{med}\{Y_i(1) - Y_i(0)|X_i \text{indicate males}\}$. Repeatedly drawing values of Y_{mis} and calculating the causal estimand for each draw generates the posterior distribution of the causal estimand. Thus, we can view causal inference entirely as a missing data problem (Little and Rubin, 1987, 2002), where we multiply-impute (Rubin, 1987, 2004) the missing potential outcomes to generate a posterior distribution for the causal effects, and thereby point and interval estimates for these effects.

We now describe how to generate these imputations, that is the posterior predictive distribution of Y_{mis} under ignorable treatment assignment. Under non-ignorable assignment mechanisms, the situation is more complicated, and although the general case is discussed in Rubin (1978), real progress depends on the particular situation, for example, the case of Section 24.3.

In general:

$$\Pr(Y_{\text{mis}}|X, Y_{\text{obs}}, W) = \frac{\Pr(X, Y(0), Y(1)) \Pr(W|X, Y(0), Y(1))}{\int \Pr(X, Y(0), Y(1)) \Pr(W|X, Y(0), Y(1)) d Y_{\text{mis}}}. \quad (24.5)$$

With ignorable treatment assignment, equation (24.5) becomes:

$$\Pr(Y_{\text{mis}}|X, Y_{\text{obs}}, W) = \frac{\Pr(X, Y(0), Y(1))}{\int \Pr(X, Y(0), Y(1)) d Y_{\text{mis}}}. \quad (24.6)$$

Equation (24.6) reveals that, under ignorability, all that we need to model is the science, $\Pr(X, Y(0), Y(1))$. With latent ignorability, as in our application, we assume that the partially observed X is actually known, and then multiply-impute the missing values in this partially observed variable, and then carry out the remaining analysis assuming ignorable treatment assignment for each such imputation. Details are given in the context of our application, where a model also needs to be specified for the partially missing X given the fully observed X's.

Because all information is in the underlying data, the unit labels are effectively just random numbers, and hence the array $(X, Y(0), Y(1))$ is row exchangeable. With essentially no loss of generality as in Rubin (1978), therefore, by de Finetti's (1963) theorem, we have that the distribution of $(X, Y(0), Y(1))$ may be taken to be iid (independent and identically distributed) given some parameter θ, with prior distribution $p(\theta)$:

$$\Pr(X, Y(0), Y(1)) = \int \left[\prod_{i=1}^{N} f(X_i, Y_i(0), Y_i(1)|\theta) \right] p(\theta) d\theta. \quad (24.7)$$

Equation (24.7) provides the bridge between fundamental theory and the common practice of using iid models. Of course, there remains the critical point that the distributions $f(\cdot|\theta)$ and $p(\theta)$ are rarely, if ever known, and this limitation haunts Bayesian inference despite its great flexibility.

Without loss of generality, we can factor $f(X_i, Y_i(0), Y_i(1)|\theta)$ into:

$$f(Y_i(0), Y_i(1)|X_i, \theta_{y\cdot x}) f(X_i|\theta_x),$$

where $\theta_{y\cdot x} = \theta_{y\cdot x}(\theta)$ is the parameter governing the conditional distribution of $Y_i(0), Y_i(1)$ given X_i, and analogously, $\theta_x = \theta_x(\theta)$ is the parameter governing the marginal distribution of X, where both parameters are functions of θ. The reason for doing this factorization is that we are assuming X is fully observed, and so we wish to predict the missing potential outcomes Y_{mis} from X and the observed potential outcomes, Y_{obs}, and therefore must use $f(Y_i(0), Y_i(1)|X_i, \theta_{y\cdot x})$ as the basic distribution.

To do this, we factor $f(Y_i(0), Y_i(1)|X_i, \theta_{y\cdot x})$ into either

$$f(Y_i(0)|X_i, Y_i(1), \theta_{0\cdot x1})f(Y_i(1)|X_i, \theta_{1\cdot x})$$

when $Y_i(1)$ is observed, indexed by the set $S_1 = \{i|W_i = 1\}$, or

$$f(Y_i(1)|X_i, Y_i(0), \theta_{1\cdot x0})f(Y_i(0)|X_i, \theta_{0\cdot x})$$

when $Y_i(0)$ is observed, indexed by the set $S_0 = \{i|W_i = 0\}$, where the various subscripted θ's are all functions of θ governing the appropriate distributions in an obvious notation.

These factorizations allow us to write (24.7) as

$$\int \prod_{i\in S_1} f(Y_i(0)|X_i, Y_i(1), \theta_{0\cdot x1}) \prod_{i\in S_1} f(Y_i(1)|X_i, \theta_{1\cdot x}) \tag{24.8a}$$

$$\times \prod_{i\in S_0} f(Y_i(1)|X_i, Y_i(0), \theta_{1\cdot x0}) \prod_{i\in S_0} f(Y_i(0)|X_i, \theta_{0\cdot x}) \tag{24.8b}$$

$$\times \prod_{i\in S_*} f(Y_i(0), Y_i(1)|X_i, \theta_{y\cdot x}) \tag{24.8c}$$

$$\times \prod_{i\in S} f(X_i|\theta_x)p(\theta)d\theta, \tag{24.8d}$$

where $S_* = \{i|W_i =^*\}$.

Notice that the first factor in (24.8a), times the first factor in (24.8b), times (24.8c) is proportional to the posterior predictive distribution of Y_{mis} given θ (i.e. given X, Y_{obs} and θ), $\Pr(Y_{\text{mis}}|X, Y_{\text{obs}}, \theta)$. Also notice that the remaining factors in (24.8), that is the second factor in (24.8a) times the second factor in (24.8b) times (24.8d) is proportional to the posterior distribution of θ, $\Pr(\theta|X, Y_{\text{obs}})$, which is equal to the likelihood of θ, $L(\theta|X, Y_{\text{obs}})$, times the prior distribution of θ, $p(\theta)$.

24.2.4 Assumptions on the science

Let us now assume that $\theta_{y\cdot x}$ and θ_x are a priori independent:

$$p(\theta) = p(\theta_{y\cdot x})p(\theta_x). \tag{24.9}$$

This assumption is not innocuous, but it is useful and is standard in many prediction environments. For an example of a situation where it might not be reasonable, suppose X includes many base-line measurements of cholesterol going back many years; the relationships among previous X values may provide useful information for predicting $Y(0)$, (i.e. Y without intervention), from X, using for example, a time-series model (e.g. Box and Jenkins, 1970).

For simplicity, here we make the assumption implied by equation (24.9), although, as with all such assumptions, it should be carefully considered. Then the integral over θ_x in (24.8) passes through all the products in (24.8a), (24.8b) and (24.8c), and we are left with the integral over (24.8d); (24.8d) after this integration is proportional to $p(\theta_{y\cdot x})d\theta_{y\cdot x}$.

As a consequence, (24.8) becomes

$$\int \Pr(Y_{\text{mis}}|X, Y_{\text{obs}}, \theta_{y\cdot x}) \Pr(\theta_{y\cdot x}|X, Y_{obs})d\theta_{y\cdot x}, \tag{24.10}$$

where the second factor in (24.10) is proportional to the product of the second factors in (24.8a) and (24.8b), and the first factor in (24.10) is, as before, proportional to the product of the first factors of (24.8a) and (24.8b) times (24.8c).

We now assume that entirely separate activities are to be used to impute the missing $Y_i(0)$ and the missing $Y_i(1)$. This is accomplished with two formal assumptions:

$$f(Y_i(0), Y_i(1)|X_i, \theta_{y\cdot x}) = f(Y_i(0)|X_i, \theta_{0\cdot x}) f(Y_i(1)|X_i, \theta_{1\cdot x}) \tag{24.11}$$

and

$$p(\theta_{y\cdot x}) = p(\theta_{0\cdot x}) p(\theta_{1\cdot x}). \tag{24.12}$$

Thus, in (24.11) we assume $Y_i(0)$ and $Y_i(1)$ are conditionally independent given X_i and $\theta_{y\cdot x}$, and in (24.12), that the parameters governing these conditional distributions are a priori independent. We make this assumption in our application in Section 24.3. Consequently, $f(Y_i(0)|X_i, Y_i(1), \theta_{0\cdot x1}) = f(Y_i(0)|X_i, \theta_{0\cdot x})$, and $f(Y_i(1)|X_i, Y_i(0), \theta_{1\cdot x0}) = f(Y_i(1)|X_i, \theta_{1\cdot x})$.

Thus, equations (24.8) or (24.10) can be described in four distinct parts with associated activities as follows.

1. Using the control units, obtain the posterior distribution of $\theta_{0\cdot x}$, the parameters governing the distribution of the control potential outcomes given the covariates:

$$\Pr(\theta_{0\cdot x}|X, Y_{\text{obs}}) \propto L(\theta_{0\cdot x}|X, Y_{\text{obs}}) p(\theta_{0\cdot x}) \propto \prod_{i\in S_0} f(Y_i(0)|X_i, \theta_{0\cdot x}) p(\theta_{0\cdot x}).$$

2. Using $\theta_{0\cdot x}$, obtain the conditional posterior predictive distribution of the missing control potential outcomes, $Y_i(0)$:

$$\prod_{i\in S_1\cup S_*} f(Y_i(0)|X_i, \theta_{0\cdot x}).$$

3. Using the treated units, obtain the posterior distribution of $\theta_{1\cdot x}$, the parameters governing the distribution of the treatment potential outcomes given the covariates.

$$\Pr(\theta_{1\cdot x}|X, Y_{\text{obs}}) \propto L(\theta_{1\cdot x}|X, Y_{\text{obs}}) p(\theta_{1\cdot x}) \propto \prod_{i\in S_1} f(Y_i(1)|X_i, \theta_{1\cdot x}) p(\theta_{1\cdot x}).$$

4. Using $\theta_{1\cdot x}$, obtain the conditional posterior predictive distribution of the missing treatment potential outcomes, $Y_i(1)$:

$$\prod_{i \in S_0 \cup S_*} f(Y_i(1)|X_i, \theta_{1\cdot x}).$$

For simulation, perform steps 1–4 repeatedly with random draws, thereby multiply imputing Y_{mis}. As stated earlier, with latently ignorable treatment assignment, a model relating the partially observed X to the fully observed X's must be specified, and then the partially observed covariates missing values can be multiply imputed. This process is described in Section 24.3 that follows.

24.3 Bayesian inference for the causal effect of large versus small treating hospitals

The task of many researchers in the medical and social sciences is to estimate the causal effects of treatments (e.g. medical interventions, social policies) on outcomes (e.g. cancer survival, unemployment). This task is often attempted from observational studies where the treatment assignment can be non-ignorable because of unmeasured background covariates, which makes the estimation of causal effects using simple methods of data analysis inappropriate. Here we illustrate how principal stratification (Frangakis and Rubin, 2002) and Bayesian causal inference can be used to address such a problem in a study of the effect of hospital type, defined by cancer treatment volume – large versus small – on the survival of cardia cancer and stomach cancer patients in central and northern Sweden, an example where the assumptions underlying conventional methods are not medically plausible, but where the assumptions underlying our approach are quite plausible. The most critical issue concerns whether the treatment centres at small volume hospitals can be closed without having a deleterious effect on patient survival. If so, resources could be saved because patients diagnosed at small volume hospitals could be transferred to large treating hospitals.

24.3.1 Assignment of home and treating hospital type

According to investigators at the Karolinska Institute, the assignment of 'home hospital type', where patients were first diagnosed with their medical condition, can reasonably be considered ignorable, at least after conditioning on known demographic covariates describing patients, such as age, male/female and urban/rural. Assuming this ignorability allows us to estimate the causal effect of the home hospital type using conventional methods, but this approximates the effect of treating hospital type only when transfers between hospital types are rare. However, in this data set, transfers between hospital types for treatment are frequent, and although the transfers are well documented, the reasons for the transfers often appear to involve factors that are unmeasured, such as patients' willingness to tolerate surgical interventions and to be treated in locations distant from friends and family. Consequently, it would be inappropriate to estimate the causal effect of treating hospital type assuming ignorability of the assignment of treating hospital type given home hospital type and observed covariates.

Fortunately, however, our situation with transfers between hospital types has strong analogies to randomized trials with non-compliance. The assignment to home hospital type is considered ignorable, and the transfers to a different treating hospital type can be thought of as non-compliance with the assigned home hospital type. For such situations, Angrist, Imbens and Rubin (1996) showed that the econometric 'instrumental variables' method can be interpreted, under certain assumptions, as estimating the causal effect for the 'true' compliers, who are defined to be those who would comply with their assigned treatment no matter what their assignment, and Imbens and Rubin (1997) have addressed such non-compliance using fully Bayesian methods from the RCM perspective, which provides better estimation than the conventional instrumental variables approach. Little and Yau (1998) and Hirano *et al.* (2000) have extended the approach to allow for the presence of covariates. More recently, Frangakis and Rubin (2002) extended the idea behind instrumental variables to a general framework for a class of causal inference problems where treatment comparisons are validly adjusted for post-treatment variables.

In Section 24.3.2 we introduce our data sets and their medical importance, and present some descriptive statistics. Section 24.3.3 discusses home (= diagnosing) hospital type and treating hospital type, and the conditions for valid causal inference for each type, as well as defines 'principal stratification' and its application to our problem. Section 24.3.4 introduces the 'monotonicity' assumption and how to obtain rough estimates of the sizes of principal strata in our problem under it. Section 24.3.5 discusses general problems with two commonly used estimators, the 'as-treated' and 'per-protocol' estimators. Section 24.3.6 introduces the 'exclusion restriction' and shows how it leads to the 'instrumental variables' estimate of the 'complier average causal effect' of treating hospital on survival. The specific Bayesian models that we use, including their prior distributions, is the topic of Section 24.3.7, and Section 24.3.8 discusses the MCMC computation. Section 24.3.9 presents the resulting posterior distributions and discusses practical implications.

24.3.2 Data sets, notation and basic results

Our two data sets involve, first, 150 cardia cancer patients, and second, 853 stomach cancer patients between ages 35 and 85, who were diagnosed between 1988 and 1995 at hospitals located in central and northern Sweden. For each patient, we have measurements of the following variables: home hospital, treating hospital, survival time from date of diagnosis with cardia or stomach cancer, and demographic background variables: age (old/young), male/female, and area type of residence (urban/rural). The home and treating hospitals are categorized into two types: large volume, defined for the first data set as treating more than 10 patients for cardia cancer during the period 1988–1995, and for the second data set defined as treating more than 60 patients for stomach cancer during the period 1988–1995; the remaining hospitals for each data set are considered small volume. If our analyses of these data sets are considered medically revealing, extensions of them may be applied to more recent data sets to inform decisions on the status of small volume treatment centers.

Throughout the analysis, we make the stable-unit-treatment-value-assumption – SUTVA discussed in Section 24.2.2, so that there is no interference between units, and there are no hidden versions of treatments available to any patient. SUTVA appears entirely plausible under the Swedish medical system where medical care is free with reasonably sufficient resources, so there is no competition for treating hospital type among patients, and there is only one plausible home hospital for each patent.

Denote the home hospital type by h, which takes the value ℓ when assigned large hospital type and s when assigned small home hospital type. Similarly, let T denote treating hospital type, which takes the value L when the treating hospital is large, and takes the value S when treating hospital is small. We observe two types of transfers between home and treating hospital types: large volume to small volume, $\ell \rightarrow S$ and small volume to large volume, $s \rightarrow L$. Columns (1)–(3) of Table 24.1 show the observed transfers for cardia cancers. Columns (1)–(3) of Table 24.2 show the corresponding results for stomach cancers. Both tables show, not surprisingly, that patients mostly transferred to large-volume hospitals. Comparing Tables 24.1 and 24.2, we see that cardia cancer patients transferred proportionally more often than stomach cancer patients.

We assume that h is ignorably assigned (conditionally given age category, male/female, and urban versus rural residence of the patient), which is considered reasonable by Swedish medical experts. There are two types of outcomes: (1) the treating hospital type, denoted by $T = L$ or S, which will be considered a partially missing covariate that defines the principal strata; and (2) the survival time since diagnosis, Y, which, for medical reasons, takes the values 5+ for survival for more than five years, 1–5 for survival between one and five years, and 0 for death within one year. Under the RCM, each patient has potential outcomes $T(\ell)$, $T(s)$ and $Y(\ell)$, $Y(s)$ when assigned home hospital type large and small, respectively. The observed assignment for patient i is denoted by h_i, and the observed potential outcomes for patient i are denoted by $T_{\text{obs},i}$ and $Y_{\text{obs},i}$, where the fully observed background variables

Table 24.1 Cardia cancer: Observed counts in observed groups and approximate counts in principal strata.

					Conventional comparisons			Generally valid for ITT for LS		
Observed groups				N for observed groups	Valid for home hospital type	Generally invalid for any causal effect				
'Assigned'/ randomized home hospital type			Treating hospital type T	N	Intent to treat	As treated	Per protocol	Under-lying principal strata*	Approximate proportion in population in principal strata	Approximate N in LS principal stratum
	H	N			Ignores T	Ignores h	Ignores observed non-compliers			
	(1)		(2)	(3)	(4)	(5)	(6)	(7)	(8)	(9)
(1)	ℓ	75	L	73	Trt = 73	Trt = 73	Trt = 73	L L L S	33/75 40/75**	Trt = 40***
(2)			S	2	Trt = 2	Ctl = 2	—	S S	2/75	
(3)	s	75	L	33	Ctl = 33	Trt = 33	—	L L	33/75	
(4)			S	42	Ctl = 42	Ctl = 42	Ctl = 42	S S L S	2/75 40/75**	Ctl = 40***

* Assuming monotonicity, i.e. no defiers, no SL principal stratum; first letter equals treating hospital with $h = \ell$; second letter equals treating hospital when $h = s$.

** $1 - (\%SS + \%LL) = 1 - (\frac{2}{75} + \frac{33}{75}) = \frac{40}{75} \approx 53.3\%$; $\%LL \approx 44.0\%$; $\%SS \approx 2.7\%$.

*** Number assigned to home hospital type × proportion that are true compliers (i.e. in the LS principal stratum).

Table 24.2 Stomach cancer: Observed counts in observed groups and approximate counts in principal strata.

					Conventional comparisons			Generally valid for ITT for *LS*		
	Observed groups			*N* for observed groups	Valid for home hospital type	Generally invalid for any causal effect				
	'Assigned'/ randomized home hospital type		Treating hospital type *T*	*N*	Intent to treat	As treated	Per protocol	Underlying principal strata*	Approximate proportion in population in principal strata	Approximate *N* in *LS* principal stratum
	H	N			Ignores *T*	Ignores *h*	Ignores observed non-compliers			
	(1)		(2)	(3)	(4)	(5)	(6)	(7)	(8)	(9)
(1)	ℓ	506	L	502	Trt = 502	Trt = 502	Trt = 502	L L L S	44/347 ~86.5%**	Trt = 438***
(2)			S	4	Trt = 4	Ctl = 4	—	S S	4/506	
(3)	*s*	347	L	44	Ctl = 44	Trt = 44	—	L L	44/347	
(4)			S	303	Ctl = 303	Ctl = 303	Ctl = 303	S S L S	4/506 ~86.5%**	Ctl = 300***

* Assuming monotonicity, i.e. no defiers, no *SL* principal stratum; first letter equals treating hospital with $h = \ell$; second letter equals treating hospital when $h = s$.

** $1 - (\%SS + \%LL) = 1 - \left(\frac{4}{506} + \frac{44}{347}\right) \approx 86.5\%$; $\%LL \approx 12.7\%$; $\%SS \approx 0.8\%$.

*** Number assigned to home hospital type × proportion that are true compliers (i.e. in the *LS* principal stratum).

for patients are denoted by a three component vector, x_i, where each component of x_i is dichotomous. The partially missing covariate for person i is the bivariate $G_i \equiv (T_i(\ell), T_i(s))$, whose values will be denoted by LL, LS, SL, or SS (where for simplicity LL means the same as (L, L), etc.). These define four possible 'principal strata' (Frangakis and Rubin, 2002), as developed in Section 24.3.3.

The average causal effect of home hospital type on survival is the comparison of the potential survival outcomes of all N patients under $h = \ell$ and under $h = s$,

$$\text{ITT} = \frac{1}{N} \sum_{i=1}^{N} [Y_i(\ell) - Y_i(s)], \tag{24.13}$$

where ITT is the Intention-To-Treat (ITT) effect in randomized trials. Under ignorable treatment assignment, we are able to estimate ITT by, for example, using x_i to define eight strata of patients, then taking the average observed difference in Y for treated (i.e. large volume hospitals) and control (i.e. small volume hospitals) patients within each stratum, and finally averaging across these stratum-specific estimates using the stratum sizes as weights. The fourth columns of Tables 24.1 and 24.2 show which groups of patients are to be considered treated (Trt) and control (Ctl) for estimating ITT for cardia cancer and stomach cancer, respectively.

In contrast to the home hospital type, the treating hospital type cannot be considered to be ignorably assigned given the observed background covariates and home hospital type. Invalid methods of estimating the causal effect of treating hospital type involve directly comparing patients from the treating hospital types, a so-called 'as-treated' analysis – indicated in column (5) in Tables 24.1 and 24.2. Also invalid is removing all observed transfer patients from the analysis, a so-called 'per-protocol' analysis displayed in column (6) in Tables 24.1 and 24.2. The general invalidity of the as-treated and per-protocol comparisons is well known (e.g. Lee *et al.*, 1991). Principal stratification leads to valid inference for causal effects by estimating the ITT effect within each principal stratum.

24.3.3 Causal effects of hospital type within principal strata

The idea of principal stratification (Frangakis and Rubin, 2002) is to stratify units according to the potential outcomes of an 'intermediate' post-treatment variable, which here is the treating hospital type indicated by G_i. Such stratification is legitimate because the values of post-treatment potential outcomes are *not* affected by assignment of home hospital type – which values are observed *is* affected by treatment assignment, but not the values themselves, and therefore G_i is, formally, a partially observed covariate.

As defined in Section 24.3.2, there are four possible principal strata, which are labeled *LL, LS, SL, SS*. *LS* can be thought of as the stratum of compliers, i.e. non-transfer patients; the *LL* and *SS* strata can be thought of as non-compliers who will always be treated at the same home hospital type no matter where assigned (diagnosed), and *SL* can be thought of as defiers, who will transfer no matter where assigned (diagnosed). Because the assignment of home hospital type is ignorable given $\mathbf{x}_i$, it is ignorable given both $\mathbf{x}_i$ and G_i.

When $G_i = LS$, the home hospital type equals the treating hospital type, i.e. $h_i = T_i$. The causal effect of home hospital type when $G_i = LS$ is defined to be

$$\text{ITT}_{LS} = \frac{1}{N_{LS}} \sum_{i \in LS} (Y_i(L) - Y_i(S)), \tag{24.14}$$

where N_{LS} is the number of *LS* patients; ITT_{LS} is easily estimated once we identify the individuals in the *LS* stratum, but explicit assumptions are needed to make this estimation straightforward. Notice that ITT_{LS} can be interpreted as either the intention-to-treat effect of home hospital type for complying patients or the intention-to-treat effect of treating hospital type for complying patients, because for the *LS* principal stratum, $h_i = T_i$. The *LS* and *SL* principal strata are the only strata of patients where we can learn about the causal effects of treating hospital types because the patients in the other principal strata, *LL* and *SS*, will always be exposed to the same treating hospital type. Imbens and Rubin (1997) call ITT_{LS} 'CACE', for 'Complier Average Causal Effect'.

24.3.4 The monotonicity assumption and the approximate size of the *LS* group

First, we consider what is called the 'monotonicity' assumption or the 'no-defier' assumption – that is, the *SL* principal stratum is empty. Specifically this assumption can be viewed as having two parts. First, if patient i is the type of patient who would be treated in a large volume hospital when assigned to a small volume hospital, patient i would be treated in a large volume hospital when assigned a large volume hospital. Second, if patient i is the type of patient who would be treated in a small volume hospital when assigned to a large volume hospital, patient i would be treated in a small volume hospital if assigned to a small volume hospital. In our setting, this assumption is very plausible, and because it excludes the *SL* principal stratum, we have only three possible principal strata: *LL, LS*, and *SS*.

The possible principal strata for each *observed* combination of home hospital type and treating hospital type (i.e. each of the four rows of Tables 24.1 and 24.2) are shown in the seventh columns of Table 24.1 and 24.2. The observed $\ell \rightarrow S$ group (the second row in Tables 24.1 and 24.2) must be composed of *SS* patients because they can be neither *LL* nor *SL* patients, respectively – because they were assigned ℓ but treated in S and therefore not LL patients, and there are no SL patients by the monotonicity assumption. Similarly, the observed $s \rightarrow L$ group (the third row of Tables 24.1 and 24.2) must be LL patients because they were assigned s but were treated in L.

In contrast, the observed $\ell \rightarrow L$ subgroup (the first row of Tables 24.1 and 24.2) could be compliers, and so be in LS, or non-compliers who are members of the LL principal stratum (who were assigned to home hospital type L, and to which they would have transferred for their treating hospital type if they were assigned to a small home hospital type). Hence, we split row 1 into two subrows in columns (7) of Tables 24.1 and 24.2. Similarly, the observed $s \rightarrow S$ subgroups (the fourth row of Tables 24.1 and 24.2) could be compliers, and so be in LS, or non-compliers who are members of the SS principal stratum, and so is also split into two subrows.

We begin with a simple method-of-moments analysis to convey the essential idea of the estimation of the size of the LS group before considering our fully Bayesian analysis. Ignoring the covariates and thinking of the assignment of small home hospital type as assigned completely at random, we can approximate the size of each principal stratum, as shown in the eighth columns of Tables 24.1 and 24.2. More explicitly, from the second row of Table 24.1, columns (1) and (3), we see that 2/75 are observed to be $\ell \rightarrow S$. Because of the assumed random assignment into ℓ and s, we have that approximately 2/75 of the patients belong to the principal stratum SS for cardia cancer, as shown in column (8) in Table 24.1. Analogously, we see from Table 24.2 that approximately 4/506 of the patients belong to principal stratum SS for stomach cancer, as shown in column (8). Similarly, from the third row of Table 24.1, columns (1) and (3), we infer that approximately 33/75 of patients belong to principal stratum LL for cardia cancer, and approximately 44/347 of patients belong to principal stratum LL for stomach cancer, as shown in columns (8) of Tables 24.1 and 24.2, respectively.

Thus we can approximate the fraction of patients in the SS principal stratum, 2/75 for cardia cancer and 4/506 ($\approx 0.8\%$) for stomach cancer, and we can approximate the fraction of patients in the LL principal stratum, 33/75 ($\approx$44.0%) for cardia cancer and 44/347 ($\approx 12.7\%$) for stomach cancer. Hence, we can approximate the fraction of compliers, the LS principal stratum, by simple subtraction: $1 - (\frac{2}{75} + \frac{33}{75}) = \frac{44}{75} \approx 53.3\%$ for cardia cancer; and $1 - (\frac{4}{506} + \frac{44}{347}) \approx 86.5\%$ for stomach cancer. These results are displayed in the first and fourth rows of columns (7) and (8) of Tables 24.1 and 24.2, and summarized in the first three rows of Table 24.3. The ninth columns in Tables 24.1 and 24.2 indicate the approximate number of LS patients in each of the four rows of observed patients, and the approximate number to be considered assigned to an L large volume hospital (labelled Trt for treated) and a S small volume hospital (labelled Ctl for controls).

Table 24.3 Method-of-moments estimates ignoring covariates.

% in principal strata	Cardia cancer			Stomach cancer		
π_{LL}		44.0			12.7	
π_{SS}		2.7			0.8	
π_{LS}		53.3			86.5	
$\pi_{LL}/(\pi_{LL}+\pi_{LS})$		94.2			94.1	
% survival - σ	1 year	5 year	5-1 year	1 year	5 year	5-1 year
σ_{LL}	60.6[1]	18.2[1]	30.0	50.0[2]	25.0[2]	50.0
σ_{SS}	50.0[3]	50.0[3]	100.0	75.0[4]	25.0[4]	33.3
σ_{ℓ}	34.7[5]	13.3[5]	38.5	43.0[6]	17.4[6]	40.4
σ_s	37.3[7]	9.3[7]	25.0	45.5[8]	17.6[8]	38.6
$\sigma_{LS\bullet\ell}$	12.5[9]	7.4[9]	43.0[9]	41.7[9]	16.2[9]	39.1[9]
$\sigma_{LS\bullet s}$	17.4[10]	0.0[10]	17.6[10]	44.6[10]	16.4[10]	37.0[10]
CACE = ITT_{LS}	−4.9[11]	7.5[11]		−2.3[11]	−0.2[11]	

Estimated y survival among:
[1] 33 $s \rightarrow L$ in Table 24.1 (20,6); [2] 44 $s \rightarrow L$ in Table 24.2 (22,11); [3] 2 $\ell \rightarrow s$ in Table 24.1 (1,1); [4] 4 $\ell \rightarrow s$ in Table 24.2 (3,1); [5] 75 assigned ℓ in Table 24.1 (26,10); [6] 506 assigned ℓ in Table 24.2 (218,88); [7] 75 assigned s in Table 24.1 (28,7); [8] 347 assigned s in Table 24.2 (158,61); [9] $[\sigma_\ell - \pi_{LL}\sigma_{LL} - \pi_{SS}\sigma_{SS}]/\pi_{LS}$; [10] $[\sigma_s - \pi_{LL}\sigma_{LL} - \pi_{SS}\sigma_{SS}]/\pi_{LS}$; [11] $[\sigma_\ell - \sigma_s]/\pi_{LS}$.

24.3.5 Problems with the as-treated and per-protocol analyses

We can also see why the as-treated [column (5)] and per-protocol [column (6)] estimators generally do not estimate actual causal effects. For example, consider the as-treated analysis for cardia cancer. The 'Treatment' group in these analyses, in rows one and three of Table 24.1, has 73 + 33 = 106 patients. But from column (9), this group of 106 patients has approximately 40 *LS* patients, and therefore the remaining 66 must be *LL* patients. In contrast, the 'Control' group in these analyses, in rows two and four, has 2 + 42 = 44 patients, but from column (9) has approximately 40 *LS* patients, and therefore the remaining four patients must be *SS*. Even assuming complete randomization to home hospital type, the as-treated analysis compares the observed values of potential outcomes for overlapping but not identical types of patients, in contrast to doing what is valid: comparing observed values of potential outcomes on a randomly divided common set of units, here, those in the *LS* principal stratum, who comprise less than half (only about 40 of 106) of those treated in the large volume hospitals, but a great majority (about 40 of the 44) of those treated in the small volume hospitals.

We will use the stomach cancer data to illustrate a similar problem with the per-protocol estimator [column (6)]. Here the 'Treatment' group in row one of Table 24.2 has 502 patients, but from column (9) has approximately 438 *LS* patients, and therefore the remaining 64 must be *LL* patients. In contrast, the 'Control' group in row four has 303 patients, but from column (9) has approximately 300 *LS* patients, and the remaining three must be *SS* patients. Even assuming complete randomization to home hospital type, we are comparing the potential outcomes for subsets of non-identical types of patients, in contrast to doing what is valid: comparing potential outcomes on a randomly divided common set of units, here, those in the *LS* principal stratum, who comprise most (about 438 of 502) of those randomized to and treated in the large volume hospitals, but nearly all (about 300 of 303) of those randomized to and treated in the small volume hospitals. For the stomach cancer data, this is not a terribly serious problem because the SS principal stratum is very small, and the LL principal stratum is relatively small.

24.3.6 The exclusion restriction and the instrumental variables estimate of CACE

Notice that we have not yet identified any particular member of the $\ell \rightarrow L$ or $s \rightarrow S$ rows (rows one and four) in Tables 24.1 and 24.2 as being in the *LS* principal stratum, and so we cannot yet compare average outcomes in this stratum. Nevertheless, we can find a unique method-of-moments estimate of the causal effect of assigned (=home) hospital type within the *LS* principal stratum under, what are considered, medically very justifiable assumptions, which in general are called 'exclusion restrictions'. This estimator of CACE is known as the 'instrumental variable estimate' (Angrist, Imbens and Rubin, 1996). We follow this simple analysis with a Bayesian model for estimating this causal effect.

The first exclusion restriction is for patients in the *LL* principal stratum. It states that, for all $i \in LL$, $Y_i(\ell) = Y_i(s)$; that is, there is no effect on potential outcomes Y of being assigned to a large (ℓ) or small (s) home hospital type for patient $i \in LL$. The medical justification for this restriction is that patient i would be treated in a large hospital type (L) under either assignment, and one's medical outcome is considered a result of where one is treated not where one is diagnosed. The exclusion restriction for patients in the *SS* principal stratum is analogous; for all $i \in SS$, $Y_i(\ell) = Y_i(s)$; that is, for those patients who would be treated in a small hospital type (S) whether assigned to l or s, there is no effect of assignment on the Y potential outcomes.

Table 24.3 presents simple method-of-moments estimates to convey essential ideas. The first three rows have already been discussed, and they present the relative sizes of the principal strata for both the cardia and the stomach cancer data sets, as estimated in Tables 24.1 and 24.2. Row 4 presents the approximate relative sizes of the *LL* group among noncompliers (*LL* or *SS*) for both cancers. Row 5 gives one-year and five-year survival rates, for both kinds of cancers, for the *LL* group, as estimated by the known *LL* groups (i.e. group of 33 patients with cardia cancer and the group of 44 patients with stomach cancer who were assigned to small hospital type but treated in a large hospital type). For these, by the exclusion restriction, their survival rate would be the same if assigned to a large hospital type. Row 6 gives the analogous results for the SS group, now based on those observed to be $\ell \rightarrow S$. Row 7 gives the survival rate for all patients assigned ℓ, and row 8 gives the survival rate for all patients assigned s. Because of the assumed randomization to home hospital type (ℓ versus s), the survival rates among those assigned ℓ, σ_ℓ, is the weighted combination of the survival rates across the principal strata when assigned to ℓ,

$$\sigma_\ell = \pi_{LL}\sigma_{LL} + \pi_{SS}\sigma_{SS} + \pi_{LS}\sigma_{LS\bullet\ell},$$

where by exclusion, σ_{LL} and σ_{SS} are unaffected by assignment to ℓ or s. Thus, the survival rate within the LS principal stratum when assigned to ℓ is,

$$\sigma_{LS\bullet\ell} = \left(\sigma_\ell - \pi_{LL}\sigma_{LL} - \pi_{SS}\sigma_{SS}\right)/\pi_{LS},$$

which is displayed in row 9 for both cancers and both survival outcomes. Analogously, the survival rate for the LS principal stratum when assigned to s is:

$$\sigma_{LS\bullet s} = \left(\sigma_s - \pi_{LL}\sigma_{LL} - \pi_{SS}\sigma_{SS}\right)/\pi_{LS},$$

which is displayed in row 10. Now $\text{ITT}_{LS} \equiv \text{CACE}$ is the difference,

$$\text{CACE} = \sigma_{LS\bullet\ell} - \sigma_{LS\bullet s} = \left(\sigma_\ell - \sigma_s\right)/\pi_{LS},$$

which is displayed in row 11 for both cancers and both survival outcomes.

Another, more direct derivation of CACE, which we believe is less insightful, is as follows. ITT for all patients can be written as

$$\text{ITT} = \pi_{LS}\text{ITT}_{LS} + \pi_{SS}\text{ITT}_{SS} + \pi_{LL}\text{ITT}_{LL} \tag{24.15}$$

where ITT_{LS}, ITT_{SS}, and ITT_{LL} are the intention-to-treat effects in the LS, SS, and LL strata, respectively. Because the exclusion restrictions force ITT_{SS} and ITT_{LL} to be identically zero, equation (24.15) becomes

$$\text{ITT} = \pi_{LS}\text{ITT}_{LS}. \tag{24.16}$$

The instrumental variables estimate of the ITT effect of treating hospital on one-year survival is $-2.6\%/53.3\% = -4.9\%$, and on five-year survival is $4\%/53.3\% = 7.5\%$. The corresponding estimates for stomach cancer are both approximately 0% because there appears to be no evidence of an ITT effect of home hospital type for stomach cancer survival rates. However, these estimates are not good ones for two reasons. First, they ignore important differences between patients in diagnosing hospital types in the distributions of observed covariates, and second, because they are inefficient method of moment estimates rather than the preferable Bayesian estimates.

24.3.7 Specific Bayesian models

We now proceed to describe the specific model that follows the general framework of Section 24.2. First, we assume latently ignorable treatment assignment; that is, we assume equation (24.3) holds when X_i is defined to include the three fully observed background variables, x_i, *and* the three level indicator G_i of the principal stratum of patient i, $G_i \in \{LS, LL, SS\}$. Thus, no further modelling is needed for the assignment mechanism. What is needed, however, is a model for the principal strata given x_i, which is described later in this section.

Also in parallel with equation (24.9), we assume that the parameters governing the marginal distribution of the fully observed covariates, x_i, are a priori independent of all other parameters. All additional model specifications use distributions that are independent for each distinct value of x_i. Thus, because x_i defines a $2 \times 2 \times 2$ table, there are eight entirely independent models. This allows the following descriptions to drop the repeated conditioning on x_i. We first describe the distribution of G_i governed by parameter θ_g, and then the distribution of $(Y_i(\ell), Y_i(s))$ given G_i.

The distribution of G_i is specified by two Bernoulli probabilities, two because G_i has three levels. The first parameter, γ, gives the probability that $G_i \in LS$ (i.e. that patient i is a complier). The second parameter, η, gives the probability, if patient i is a noncomplier, that $G_i = LL$ versus $G_i = SS$. The reason for using this parametrization with $\theta_g = (\gamma, \eta)$ is that interest focuses on the compliers and CACE, the causal effect of the treating hospital where it can be estimated. These two Bernoulli parameters, γ and η, are a priori independent, with prior distributions for each created by appending one fake observation to the data. This one observation for γ is split according to the crude estimates of the marginal proportion of compliers among all patients given in row 3 of Table 24.3. More specifically, for cardia cancer the proportions of compliers is estimated to be about 53.3%, and for stomach cancer it is about 86.5%, so, the fake observation for γ is split according to these fractions. The one fake observation for η is split between LL and SS according to the marginal proportion of LL versus SS among all patients for cardia cancer; this split is estimated to be 94.2% versus 5.8%, and for stomach cancer about 94.1% versus 5.9%. The idea to use such prior distribution was first proposed in Rubin (1984; reprinted in Rubin 2004, Appendix II). It was explored further in Rubin and Schenker (1987) and Clogg *et al.* (1991).

We complete the model specification by providing the conditional distribution of the potential outcomes given $X_i = (G_i, x_i)$ and $\theta_{y \bullet g}$, described ignoring x_i because the models are specified independently for each of the eight values of x_i. First, we make the two assumptions implied by equations (24.11) and (24.12): the potential outcomes $Y_i(s)$ and $Y_i(\ell)$ are conditionally independent given G_i and $\theta_{y \bullet g}$, and their associated parameters are also a priori independent. Second, for medically relevant reasons, we summarize Y_i by two indicator variables. The first, Y_i^{+1}, for lived at least a year $(Y_i^{+1} = 1)$ versus died within a year after the cancer diagnosis $(Y_i^{+1} = 0)$, and the second for, given $Y_i^{+1} = 1$, lived at least five years $(Y_i^{+5} = 1)$ versus died within five years $(Y_i^{+5} = 0)$. In the *LS* principal stratum there is one distribution of $Y_i(\ell)$ to be estimated and one distribution of $Y_i(s)$ to be independently estimated. By the exclusion restrictions, within the *LL* principal stratum there is one distribution for $Y_i(\ell) \equiv Y_i(s)$, and in the *SS* principal stratum there is one distribution for $Y_i(\ell) \equiv Y_i(s)$. Each of these four distributions of Y_i is modelled in the same way as the G_i; that is, each of the four will be modelled as two independent Bernoulli distributions, and thus will have associated Beta posterior distributions.

In *LS*, there is one Bernoulli distribution for $Y_i^{1+}(\ell)$ with parameter ${\phi^\ell}_{LS}$ and one Bernoulli distribution for $Y_i^{5+}(\ell)$ given $Y_i^{1+}(\ell) = 1$ with parameter ${\psi^\ell}_{LS}$; and analogously for $Y_i^{1+}(s)$ and $Y_i^{5+}(s)$ with Bernoulli distribution parameters ${\phi^s}_{LS}$ and ${\psi^s}_{LS}$. The parameters for the distribution of $Y_i(\ell) = Y_i(s)$ in the *LL* principal stratum and the distribution of $Y_i(\ell) = Y_i(s)$ in the *SS* principal stratum are analogously ϕ_{LL}, ψ_{LL}, ϕ_{SS}, and ψ_{SS}. Thus, the total number of Bernoulli parameters being estimated for the conditional distribution of the potential outcomes is eight, and over all eight values of x_i, the number of scalar parameters is $8 \times 8 = 64$. So θ has $64 + 16 = 80$ a priori independent Bernoulli parameters, because there is a γ and an η at each value of x_i. The prior distribution on these is created by appending one fake observation to each of the 80 subdata sets, split according to the crude estimates of the marginal (across all eight values of x_i) probabilities of 'success' and 'failure', implied by the point estimates in Table 24.3 of one-year survival for ϕ and of five-year survival given one-year survival for ψ. Other prior distributions could, of course, be used: better ones would be based on carefully considered medical knowledge.

24.3.8 Computation

Our plan is to use DA (data augmentation, Tanner and Wong 1987), which is a stochastic version of the EM algorithm (Demster, Laird, Rubin 1977) to implement the computation. Many references now exist for this Markov Chain Monte Carlo (MCMC method) e.g. Gelman, *et al.* (1995) and Tierney (1994).

We begin with a description of how we create a starting 'completed' data set, with all missing G_i and missing Y_i randomly imputed. Specifically, use the method of moments estimates of the probabilities in Table 24.3 to draw, first the missing values of G_i, initially, *LS* versus *LL* or *SS*, and then if the latter is drawn, *LL* versus *SS*; now knowing all the values of G_i, we draw the survival outcomes, first one-year survival versus one-year death, and then if one-year survival is drawn, we draw five-year survival using the drawn data set. In order to use this completed dataset to create an overdispersed starting distribution for drawing parameters, we reduced the sample size by a factor of 60%. All subsequent iterations are done independently across all values of x.

Each iteration of the MCMC has two main steps. The first step draws the parameters from their complete data posterior distribution, assuming all imputed values are real. The first step consists of five independent substeps: (1) one for the parameters governing the principal strata; (2) one for the parameters governing the parameters of the $Y(1)$ potential outcomes in LS; (3) one for the parameters governing the parameters of the $Y(0)$ potential outcomes in LS; (4) one for the parameters governing the $Y(1)$ potential outcomes in LL; and (5) one for the parameters governing the $Y(0)$ potential outcomes in SS. Each of these substeps comprises two sub-substeps: (i) one based on all the data at that value of x (for drawing being a complier or not for substep 1), and the other (ii) conditional on the being a noncomplier (for the first substep) or surviving at least one year (for the other four substeps). Each of these 10 sub-substeps draws the parameter from a Beta posterior distribution governed by: the prior distribution (fake observation) plus the data that has been observed plus the data that has been imputed.

The second step of the MCMC considers the values of the parameters drawn in the first step to be the correct values and draws the missing principal strata and missing potential outcomes in five substeps: (1) independently draw the missing principal strata for all the patients using the parameters drawn in step 1, substep 1; (2) independently draw the missing $Y(1)$ potential outcomes for the patients in LS who were assigned 0 (control = Small) using the parameters drawn in step 1, substep 2; (3) independently draw the missing $Y(0)$ for the patients in LS who were assigned to 1 (treated=Large), using the parameters drawn in step 1, substep 3; and analogously for substeps (4) and (5). Each of these substeps has two parts, corresponding to the sub-substeps of step 1. Each of the 10 sub-substeps draws missing data from a Bernoulli distribution.

So in all, within each stratum defined by x, there are 10 draws from Beta posterior distributions. and 10 sets of draws from Bernoulli sampling distributions. Three such independent chains were drawn at each value of x, and allowed to 'burn in' until approximate convergence. Convergence of the logits of all 80 Bernoulli parameters was monitored using the GR statistic (Gelman and Rubin, 1992*a*,*b*). All GR statistics were very close to 1.0 after only 100 iterations. Thereafter 100,000 iterations were run for each chain with every tenth iteration saved as a draw from the posterior distribution.

24.3.9 Results and discussion

The results of our MCMC simulation are partially summarized in Table 24.4, which gives the posterior medians and central 95% posterior intervals for several important estimands. We first discuss the results in Table 24.4 for cardia cancer and then for stomach cancer.

Table 24.4 Posterior medians (and 95% intervals) for one- and five-year survival adjusted for background variables.

Principal strata	% in strata	When assigned	% Survival rates	
			One year survival	Five year survival
		Cardia cancer		
LS	52.0 (42.7, 60.0)	Large home hospital Small home hospital	33.3 (19.4, 48.8) 21.1 (12.0, 31.4)	13.0 (3.7, 25.3) 4.7 (0.0, 11.5)
LL	44.0 (36.7, 51.3)	No effect*	52.1 (43.9, 60.6)	16.9 (12.2, 21.7)
SS	4.0 (1.3, 8.7)	No effect*	33.3 (12.5, 71.4)	22.2 (9.1, 50.0)
All	100	Large home hospital Small home hospital	42.0(36.0, 48.0) 35.3(30.0, 41.3)	14.7(11.3, 20.7) 10.7(8.0, 14.7)
		Stomach cancer		
LS	85.9 (83.0, 88.4)	Large home hospital Small home hospital	44.1 (40.6, 47.7) 44.7 (40.5, 49.2)	16.0 (13.2, 18.8) 16.3 (13.4, 19.8)
LL	12.8 (10.4, 15.5)	No effect*	50.9 (41.4, 60.6)	25.0 (17.5, 33.3)
SS	1.3 (0.7, 2.3)	No effect*	66.7 (40.0, 90.0)	26.7 (8.3, 50.0)
All	100	Large home hospital Small home hospital	42.3(42.7, 47.8) 45.8 (42.0, 49.9)	17.2 (15.6, 19.5) 17.6 (14.9, 20.6)

* By the exclusion restriction

Notice first we estimate that only about 50% of the cardia cancer patients are compliers in the sense that they would be treated in their diagnosing (home) hospital type. We estimate that roughly 45% would be treated in a large volume hospital no matter where diagnosed, and only roughly 5% would be treated in a small volume hospital no mater where diagnosed. Notice that the non-compliers appear to be healthier than the compliers, no matter where the compliers are treated with better one-year or five-year survival. This may make medical sense: LL consists of patients who may be willing to undergo more aggressive treatment provided by large volume centres because their prognosis is relatively good, and SS may consist of patients whose cancers were caught early and thus also have a reasonably good long-term prognosis and thus choose to be closer to family or friends for treatment. These results are not causal but describe the type of cardia cancer patient in the three principal strata; all these descriptive statements are subject to substantial posterior uncertainty due to the small sample sizes involved.

With respect to the results in Table 24.4 for the much more common stomach cancer, we see that the vast majority, roughly 86%, are estimated to be compliers, in the sense of being treated in the same hospital type as where they were diagnosed, and nearly all the rest, roughly 13%, are estimated to be in LL; only 1% are estimated to be in SS. But as with the cardia cancer patients,

Table 24.5 'Causal' answers from four Bayesian analyses of cardia and stomach cancers: Posterior medians (and 50% and 95% intervals) for effects of large versus small hospital types on one- and five-year survival rates, adjusted for background covariates.

	One year survival	Five year survival
	% (Large – Small)	% (Large – Small)
	Cardia cancer	
CACE = ITT_{LS}	12.2 (−4.8, 6.3, 18.3, 29.6)	8.1 (2.8, 3.9, 12.6, 21.5)
ITT	4.0 (−7.3, 0.7, 8.0, 15.3)	7.3 (−0.7, 4.7, 10.7, 16.7)
	Stomach cancer	
CACE = ITT_{LS}	−0.7 (−5.4, −2.4, 0.9, 4.2)	−0.4 (−4.7, −1.8, 1.1, 3.7)
ITT	−2.9 (−7.7, −4.6, −1.3, 1.9)	−0.1 (−3.9, −1.4, 1.1, 3.4)

the non-complying stomach cancer patients appear to have somewhat better survival rates than the complying stomach cancer patients, no matter where treated, presumably for similar reasons as for the cardia cancer patients. In fact, it appears that the survival rates for the complying patients are very similar no matter where treated.

Table 24.5 directly addresses the primary causal questions by providing posterior medians and central 50% and central 95% posterior intervals for CACE = ITT_{LS} and for ITT. For stomach cancer, there appears to be no evidence for any advantage to being treated in one type of hospital or the other: even the 50% intervals for CACE comfortably cover zero. The story for cardia cancer appears different, however. The posterior medians for the causal CACE effects are in favor of being treated in a large-volume hospital, approximately a 12% advantage for one-year survival – 33% versus 21%, a factor of about 1.5, and approximately an 8% advantage for five-year survival – 13% versus 5%, a factor of nearly 3. Both 50% posterior intervals for CACE easily exclude zero, and even the 95% interval for CACE for five-year survival is entirely positive. Therefore, it is more likely than not that being treated for cardia cancer in a large-volume hospital versus a small volume hospital causes better survival outcomes, both short term and long term, and very likely that it causes better long-term survival. Using ITT instead of CACE just makes conclusions weaker and far more ambiguous.

The implications for patient choices and for policy seem clear. For patients with stomach cancer, there is essentially no difference on survival whether treated in a large-volume or small-volume hospital. In contrast, for patients with the much rarer cardia cancer, there appears to be a relatively clear survival advantage to being treated in a large-volume rather than a small-volume hospital, and therefore it would appear prudent for patients diagnosed in small-volume hospitals to be transferred to large-volume hospitals.

Sharper, that is, more precise, estimation could possibly be obtained by including medical knowledge to modify prior distributions, either by restricting parameters or by assuming some parameters are a priori dependent across some values of x and principal strata. However, there is something very 'clean' and agnostic about the Bayesian models that we have used. And notice that this sort of fitting could not have been accomplished using standard frequency tools at each value of x, such as the method-of-moments estimates used to provide the motivation for Tables 24.1–24.3.

Appendix

A. Broader context and background

Here we provide additional background about the RCM in the context of observational studies for causal effects. A critical feature of the RCM is that assumptions must be made explicit using full probability models (Rubin, 1978, 2008).

First consider the simpler case in which Bayesian causal inference can be based on one regression model relating the observed value of the outcome variable to the treatment indicator and background covariates. Under the RCM, the model for the treatment assignment mechanism is distinct from the model for the data, which include the potential outcomes and covariates. Given a rich enough set of covariates X, the treatment assignment W is assumed to be ignorable

$$\Pr(W \mid X, Y(0), Y(1)) = \Pr(W \mid X, Y_{obs}),$$

where $Y(0)$ and $Y(1)$ are the potential outcomes under two treatments indicated by $W = 0$ and $W = 1$, and Y_{obs} is the observed outcome. Under this assumption and the stability assumption (SUTVA), Bayesian causal inference can be based on a single regression model using as predictors the treatment indicator and the covariates X. In observational studies, where there does not exist a known mechanism for treatment assignment, one posits an assignment mechanism and then assesses the propriety of the assumption of ignorable treatment assignment.

In this chapter, the assumption of the ignorable assignment of home hospital type h given the observed background covariates X was plausible because of the Swedish health care system and Sweden's social economic structure. Although we could not assess all aspects of this assumption, we did assess some aspects, such as its implications for weight loss, an important prognostic correlate, and the results suggested that ignorability was reasonable. The ignorability assumption for the assignment of treating hospital type, however, was considered unreasonable. If we had observed covariates, such as the patients' true cancer stage and their willingness to undergo invasive medical operations, then the assumption of ignorable assignment of the treating hospital type could have been plausible. If so, then Bayesian causal inference for the effect of treating hospital type could have been based on a single regression model predicting the observed outcome from an indicator for treating hospital type, an indicator for home hospital type, and the covariates.

Because we had to allow for a non-ignorable treatment assignment mechanism, we had to consider the case where Bayesian causal inference could not be based on just one single regression model. Our solution is to consider those patients who are treated in the same hospital type as their home hospital type to be compliers with their original assignment, and the other patients to be non-compliers, and seek causal inference for the compliers – where home hospital is the same as treating hospital, and the former is ignorably assigned. Even though the true compliers cannot be exactly identified from the data, under medically very plausible assumptions in our setting, called 'monotonicity' and 'exclusion restrictions,' estimation of the complier average causal effect (CACE) can be accomplished. The critical idea is to conduct Bayesian inference in what economists call instrumental variables models (Imbens and Rubin, 1997). The resulting analyses are related to those in Hirano *et al.* (2000), and are special cases of more general analyses based on the concept of 'principal stratification' (Frangakis and Rubin, 2002).

Acknowledgements

We are grateful to Olof Nyren for stimulating discussions and advice on medical aspects of the problem, without which the work would be impossible. Donald B. Rubin's work was partially supported by grants from the US National Science Foundation (Grant SES 0550887) and from the US National Institutes of Health (Grant R01 DA023879-01).

References

Angrist, J.D., Imbens, G.W. and Rubin, D.B. (1996). Identification of causal effects using instrumental variables (with discussion). *Journal of the American Statistical Association*, **91**, 444–455.

Box, G.E.P. and Jenkins, G.M. (1970). *Time Series Analysis: Forecasting and Controls.* Holden-Day, San Francisco.

Carlin, B.P. and Louis, T.A. (2000). *Bayes and Empirical Bayes Methods for Data Analysis*, (2nd edn). Chapman & Hall, London.

Clogg, C.C., Rubin, D.B., Schenker, N., Schultz, B. and Weidman, L. (1991). Multiple imputation of industry and occupation codes in census public-use samples using Bayesian logistic regression. *Journal of the American Statistical Association*, **86**, 68–78.

Cox, D.R. (1958). *The Planning of Experiments.* John Wiley, New York.

de Finetti, B. (1963). Foresight: Its logical laws, its subjective sources. In *Studies in Subjective Probability*, (ed. H.E. Kyburg and H.E. Smokler). John Wiley, New York.

Dempster, A.P., Laird, N. and Rubin, D.B. (1977). Maximum likelihood estimation from incomplete data using the EM algorithm (with discussion). *Journal of the Royal Statistical Society, Series B*, **39**, 1–38.

Frangakis, C.E. and Rubin, D.B. (1999). Addressing complications of intention-to-treat analysis in the combined presence of all-or-none treatment-noncompliance and subsequent missing outcomes. *Biometrika*, **86**, 365–379.

Frangakis, C.E. and Rubin, D.B. (2002). Principal stratification in causal inference. *Biometrics*, **58**, 21–29.

Gelman, A. and Rubin, D.B. (1992*a*). A single sequence from the Gibbs sampler gives a false sense of security. In *Bayesian Data Analysis 4*, (ed. J.M. Bernardo, J.O. Berger, A.P. Dawid, and A.F.M. Smith), pp. 623–631. Oxford University Press, New York.

Gelman, A. and Rubin, D.B. (1992*b*). Inference from iterative simulation using sequence (with discussion). *Statistical Science*, **7**, 457–511.

Gelman, A., Carlin, J.B., Stern, H.S. and Rubin, D.B. (1995). *Bayesian Data Analysis*, Chapter 11. Chapman & Hall/CRC, Boca Raton.

Gelman, A., Carlin, J. Rubin, D.B. and Stern, H. (2003). *Bayesian Data Analysis* (2nd edn). CRC Press, New York.

Geman, S. and Geman, D. (1984). Stochastic relaxation, Gibbs distributions and the Bayesian restoration of images. *IEEE Transactions on Pattern Analysis and Machine Intelligence*, **6** 721–741.

Imbens, G.W. and Rubin, D.B. (1997). Bayesian inference for causal effects in randomized experiment with non-compliance. *Annals of Statistics*, **25**, 305–327.

Hastings, W. (1970). Monte Carlo sampling methods using Markov chains and their application. *Biometrika*, **57**, 97–109.

Hirano, K., Imbens, G.W., Rubin, D.B. and Zhou, X. (2000). Assessing the effect of an influenza vaccine in an encouragement design. *Biostatistics*, **1**, 69–88.

Holland, P. (1986). Statistics and causal inference. *Journal of the American Statistical Association*, **81**, 945–970.

Lee, Y.J., Ellenberg, J.H., Hirtz, D.G. and Nelson, K.B. (1991). Analysis of clinical trials by treatment actually received: Is it really an option? *Statistics in Medicine*, **10**, 1595–1605.

Little, R.J.A. and Rubin, D.B. (1987). *Statistical Analysis with Missing Data.* John Wiley, New York.

Little, R.J.A. and Rubin, D.B. (2002). *Statistical Analysis with Missing Data* (2nd edn.). John Wiley, New York.

Little, R. and Yau, L. (1998). Statistical technique for analyzing data from prevention trials, treatment of no-shows using Rubin's causal model. *Psychological Methods*, **3**, 147–159.

Metropolis, N. and Ulam, S. (1949). The Monte Carlo method. *Journal of the American Statistical Association*, **44**, 335–341.

Peterson, A.V. Jr., Kealey, K.A., Mann, S.L., Marek, P.M. and Sarason, I.G. (2000). Hutchinson smoking prevention project: long-term randomized trial in school-based tobacco use prevention-results on smoking. *Journal of the National Cancer Institute*, **92**, 1979–1991.

Robert, C. and Casella, G. (2008). A History of Markov Chain Monte Carlo – Subjective Recollections from Incomplete Data. Technical Report, Department of Statistics, University of Florida; `http://www.stat.ufl.edu/~casella/Papers/MCMCHistory.pdf`

Rubin, D.B. (1974). Estimating causal effects of treatments in randomized and nonrandomized studies. *Journal of Educational Psychology*, **66**, 688–701.

Rubin, D.B. (1975). Bayesian inference for causality: The importance of randomization. In *The Proceedings of the Social Statistics Section of the American Statistical Association*, pp. 233–239. American Statistical Association, Alexandria, VA.

Rubin, D.B. (1976). Inference and missing data. *Biometrika*, **63**, 581–592.

Rubin, D.B. (1977). Assignment to treatment group on the basis of a covariate. *Journal of Educational Statistics*, **2**, 1–26. Printer's correction note 3, p. 384.

Rubin, D.B. (1978). Bayesian inference for causal effects: The role of randomization. *The Annals of Statistics*, **6**, 34–58.

Rubin, D.B. (1979). Discussion of 'Conditional Independence in Statistical Theory', by A.P. Dawid. *The Journal of the Royal Statistical Society, Series B*, **41**, 27–28.

Rubin, D.B. (1980). Discussion of 'Randomization analysis of experimental data in the Fisher randomization test', by Basu. *Journal of the American Statistical Association*, **75**, 591–593.

Rubin, D.B. (1984). Progress Report on Project for Multiple Imputation of 1980 Codes. Report for the U.S. Bureau of the Census.

Rubin, D.B. (1987). *Multiple Imputation for Nonresponse in Surveys*. John Wiley, New York.

Rubin, D.B. (1990*a*). Formal modes of statistical inference for causal effects. *Journal of Statistical Planning and Inference*, **25**, 279–292.

Rubin, D.B. (1990*b*). Comment: Neyman (1923) and causal inference in experiments and observational studies. *Statistical Science*, **5**, 472–480.

Rubin, D.B. (2004). *Multiple Imputation for Nonresponse in Surveys*. Reprinted with appendices as a "Wiley Classics." John Wiley, New York.

Rubin (2008). Statistical inference for causal effects, with emphasis on applications in epidemiology and medical statistics. In: *Handbook of Statistics: Epidemiology and Medical Statistics*. (ed. C.R. Rao, J.P. Mill and D.C. Rao). Elsevier, Amsterdam.

Rubin, D.B. and Schenker, N. (1987). Logit-based interval estimation for binomial data using the Jeffreys prior. *Sociological Methodology*, **17**, 131–144.

Tanner, M. and Wong, W. (1987). The calculation of posterior distribution by data augmentation (with discussion). *Journal of the American Statistical Association*, **82**, 528–550.

Tierney, L. (1994). Markov chains for exploring posterior distributions (with discussion). *The Annals of Statistics*, **22**, 1701–1762.

PART V
Natural and Engineering Sciences

·25·

Bayesian statistical methods for audio and music processing

A. Taylan Cemgil, Simon J. Godsill, Paul Peeling and Nick Whiteley

25.1 Introduction

Computer-based music composition and sound synthesis date back to the first days of digital computation. However, despite recent technological advances in synthesis, compression, processing and distribution of digital audio, it has not yet been possible to construct machines that can simulate the effectiveness of human listening – for example, an expert human listener can accurately write down a fairly complex musical score based solely on listening to the audio. Statistical methodolgies are now migrating into human – computer interaction, computer games and electronic entertainment computing. Here, one ambitious research goal focuses on computational techniques to equip computers with musical listening and interaction capabilities. This is essential for the construction of intelligent music systems and virtual musical instruments that can listen, imitate and autonomously interact with humans. For flexible interaction it is essential that music systems are aware of the semantic content of the music, are able to extract structure and can organize information directly from acoustic input. For generating convincing performances, they need to be able to analyse and mimic master musicians. These outstanding technological challenges motivate this research, in which fundamental modelling principles are applied to gain as much information as possible from ambiguous audio data.

Musical audio processing is a rather broad field and the research is driven by both scientific and technological motivations – two related but distinct goals. For technological needs, the primary motivation is to develop practical engineering solutions to enhance classification, denoising, source separation or score transcription. The ultimate goal here is to construct computer systems that display aspects of human, or super-human, performance levels in an automated fashion. In the second, the goal is to aid the scientific understanding of cognitive processes behind the human auditory system (Moore 1997) and the physical sound generation process of musical instruments or voices (Fletcher and Rossing 1998).

The starting point in this chapter is that in both contexts, scientific and technological, Bayesian statistical methods provide a sound formalism for making progress. This is achieved via models which quantify prior knowledge about the physical properties and semantics of sound, combined with powerful computational methodology. The key equation, then, is Bayes' theorem and in the context of audio processing it can be stated as

$$p(\text{Structure}|\text{Audio Data}) \propto p(\text{Audio Data}|\text{Structure})\, p(\text{Structure}).$$

Thus inference is made from the posterior distribution for the hidden structure given observed audio data. One of the strengths of this simple and intuitive view of audio processing is that it unifies a variety of tasks such as source tracking, enhancement, transcription, separation, identification or resynthesis into a single Bayesian inference framework. The approach also inherits the benefit common to all applications of Bayesian statistical methods that the problem formulation and computational solution strategy are well separated. This is in contrast with many of the more heuristic and ad hoc approaches to audio processing. Popular aproaches here involve the design of custom-built algorithms for solving specific tasks, and in which the problem formulation and computational solution are blended together, taking account of practical and pragmatic considerations only. These techniques potentially miss out on the generality and accuracy afforded by a well-defined Bayesian model and associated estimation algorithms.

We firstly consider mainstream applications of audio signal processing, give a very brief introduction to the properties of musical audio, and then proceed to pose the principal challenges as Bayesian inference tasks.

25.1.1 Applications

A fundamental task that will be a focus of this paper is music-to-score transcription (Cemgil 2004; Klapuri and Davy 2006). This involves the analysis of raw audio signals to produce a musical score representation. This is one of the most challenging and comprehensive tasks facing us in computational music analysis, and one that is certainly ill-defined, since there are many possible written scores corresponding to one performance. An expert human listener could transcribe a relatively complex piece of musical audio but the score produced would be dissimilar in many respects to that of the composer. However, it would be reasonable to hope that the transcriber could generate a score having similar pitches and durations to those of the composer. The subtask of generating a pitch-and-duration map of the music is the main aim of many so-called 'transcription' systems. Others have considered the task of score generation from this point on and software is available commercially for this highly subjective part of the process – we will not consider it further here.

Applications that require the transcription task include analysis of ethnomusicological recordings, transcription of jazz and other improvised forms for analysis or publication of performance versions, and transcriptions of rare or historical pieces which are no longer available in the form of a printed score. Apart from applications which directly require the full transcription there are many applications, for example those below, which are fully or partially solved as a result of a solution to the transcription problem.

Signal separation is a second fundamental challenge (Hyvärinen, Karhunen, and Oja 2001; Virtanen 2006b) – here we attempt to separate out individual instruments or notes from a polyphonic (many-note) mixture. This finds application in many areas from sound remastering in the recording studio through to *karaoke* (extraction of a principal vocal line from a source, leaving just the accompaniment). Source separation finds much wider application of course in non-musical audio, especially in hearing aids, see below. Instrument classification is a further important component of musical analysis systems, i.e. the task of recognizing which instruments are playing at any given time in a piece. A related concept is timbre determination – extraction of the tonal character of a pitched musical note (in coarse terms, is it harsh, sweet, bright, etc.; Herrera-Boyer, Klapuri, and Davy 2006).

Finally, at the signal level, audio restoration and enhancement (Godsill and Rayner 1998) form another key area. In this application the quality of an audio source is enhanced, for example by reduction of background noise. This task comes as a by-product of many model-based analysis tasks, such as source separation above, since a noise-reduced version of the input signal will often be available as one of the possible inferences from the Bayesian posterior distribution.

The fundamental tasks above will find use in many varied acoustical applications. For example, with vast amounts of audio data available digitally in on-line repositories, it is not unreasonable to predict that almost all audio material will be available digitally in the near future. This has rendered automated processing of audio for sorting and choice of musical content an important and central information processing task, affecting literally millions of end users. For flexible interaction it is essential that systems are able to extract structure and organize information from the audio signal directly. Our view is that the associated fundamental computational problems require both a fresh look at existing signal processing techniques and development of novel statistical methodologies.

25.1.2 Introduction to musical audio

The following discussion gives a basic introduction to some of the properties of musical audio signals, following closely that of Godsill (2004). Musical audio is highly *structured*, both in the time domain and in the frequency domain.

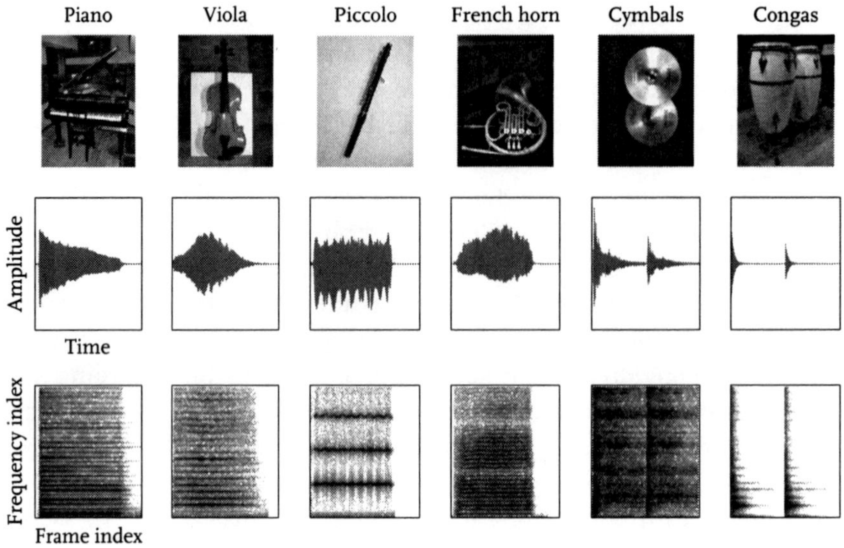

Fig. 25.1 Some acoustical instruments, examples of typical time series and corresponding spectrograms (time varying magnitude spectra – modulus of short time Fourier transform) computed with FFT. (Audio data and images from RWCP Instrument samples database.)

In the time domain, *tempo* and *beat* specify the range of likely times where note transitions occur. In the frequency domain, two levels of structure can be considered. First, each note is composed of a fundamental frequency (related to the 'pitch' of the note) and partials whose relative amplitudes determine the timbre of the note. This frequency-domain description can be regarded as an empirical approximation to the true process, which is in reality a complex nonlinear time-domain system (McIntyre, Schumacher, and Woodhouse 1983; Fletcher and Rossing 1998). The frequencies of the partials are approximately integer multiples of the fundamental frequency, although this clearly doesn't apply for instruments such as bells and tuned percussion. See Figure 25.1 for examples of pitched and percussive musical instruments and typical time series. Second, several notes played at the same time form chords, or polyphony (Figure 25.2). The fundamental frequencies of each note comprising a chord are typically related by simple multiplicative rules. For example, a C major chord may be composed of the frequencies 523 Hz, 659 Hz $\approx 5/4\times523$ Hz and 785 Hz $\approx 3/2\times523$ Hz. Figure 25.3 shows the waveform for a simple monophonic (single note) flute recording and Figure 25.4 shows the corresponding time – frequency spectrogram analysis (this may be auditioned at `www-sigproc.eng.cam.ac.uk/~sjg/haba`, where other extracts used in this paper may also be found). In this both the temporal segmentation and the frequency-domain structure are clearly visible on the plot. Focusing on a single localized time frame, at around 2 s in the same extract, we can clearly see

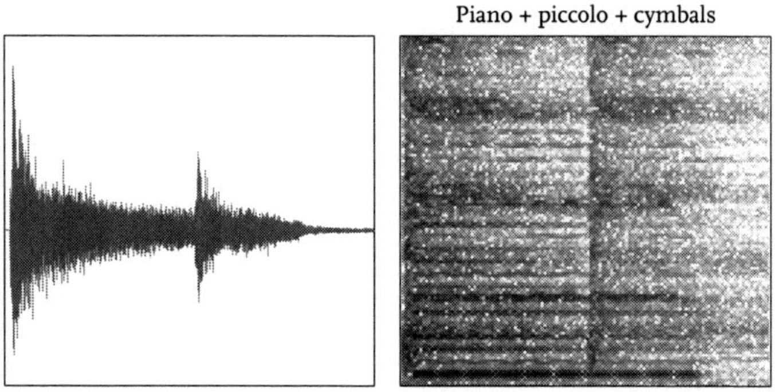

Fig. 25.2 Superposition. The time series and the magnitude spectrogram of the resulting signal when some of the instruments play concurrently.

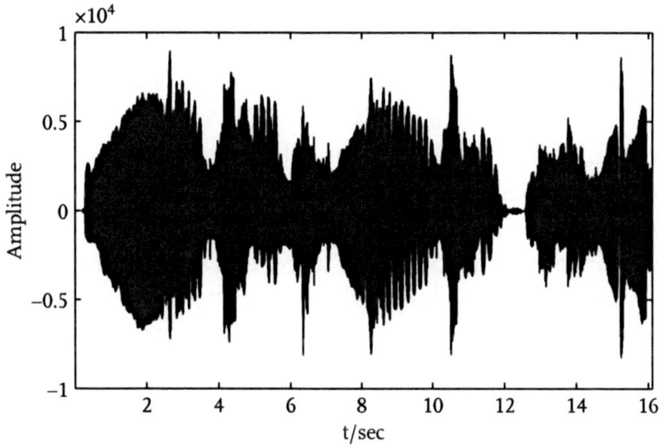

Fig. 25.3 Time-domain waveform for a solo flute extract.

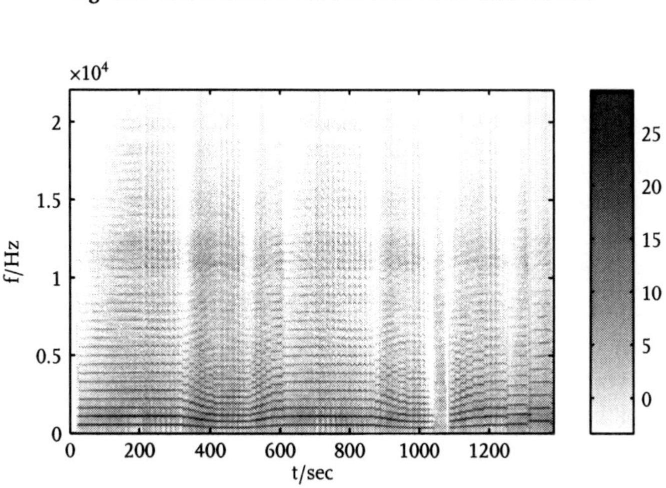

Fig. 25.4 Time – frequency spectrogram representation for the flute recording.

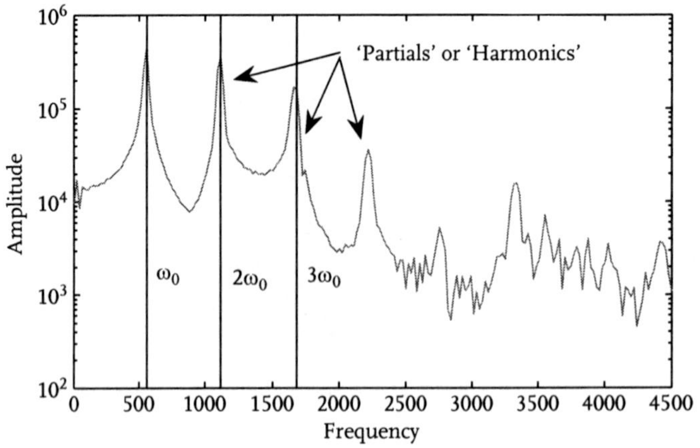

Fig. 25.5 Short-time Fourier analysis of a single frame of data from the flute extract.

the fundamental frequency component, labelled ω_0, and the partial stucture, at frequencies $2\omega_0$, $3\omega_0, \ldots$ of a single musical note in Figure 25.5. It is clear from spectra such as Figure 25.5 that it will be possible to estimate the pitch from single-note data that is well segmented in time (so that there is not significant overlap between more than one separate musical note within any single segment). We will refer to pitch interchangeably with fundamental frequency ω_0, although it should be noted that *perceived* pitch is a more complex function of the fundamental and amplitudes and number of its harmonics. There are many ways to achieve pitch detection, based on sample autocorrelation functions, spectral peak locations, etc. Of course, real musical extracts don't usually arrive in conveniently segmented single-note form or extracts, and much more complex structures need to be considered, as detailed in the sections below.

25.1.3 Superposition and the Bayesian approach

In applications that involve acoustical and computational modelling of sound, a fundamental obstacle is *superposition*, i.e. concurrent sound events (music, speech or environmental sounds) are mixed and modified due to reverberation and noise present in the acoustic environment. This situation is of primary importance in polyphonic music, in which several instruments sound simultaneously and one of the many possible processing goals is to separate or identify the individual voices. In domains such as these, information about individual sources cannot be directly extracted, owing to the superposition effect, and significant focus is given in the literature to source separation (Hyvärinen, Karhunen, and Oja 2001), deconvolution and perceptual organization of sound (Wang and Brown 2006).

25.1.4 Fundamental audio processing tasks

From the above discussion of the challenges facing audio processing, some fundamental tasks can be identified for treatment by Bayesian techniques. Firstly, we can hope to address the superposition task in a model-based fashion by posing models that capture the behaviour of superimposed signals. These are similar in flavour to the latent factors analysed in some statistical modelling problems. A generic model for observed data Y, under a linear superposition assumption, will then be:

$$Y = \sum_{i=1}^{I} s_i \tag{25.1}$$

where the s_i represent each of the I individual audio sources present. We pose this very basic model here as a single-channel observation model, although it is straightforward to extend the model to the multichannel case, in which case it will be usual to include also channel-specific mixing coefficients. The sources and data will typically be audio time series but can also represent expansion coefficients of the audio in some other domain such as the Fourier or wavelet domain, as will be made clear in context later. We may render the model a little more sophisticated by making the data a stochastic function of the sources, and in this case we will specify some non-degenerate likelihood function $p(Y|\sum_{i=1}^{I} s_i)$ that models an additive noise component in addition to the desired signals.

We typically assume that the individual sources s_i are independent a priori. They are parametrized by θ_i, which represent information about the sound generation process for that particular source, including perhaps its pitch and other characteristics (number of partials, etc.), encoded through a conditional distribution and prior distribution for each source:

$$p(s_i, \theta_i) = p(s_i|\theta_i)p(\theta_i).$$

Dependence between the θ_i, for example to model the harmonic relationships of notes within a chord, can of course be included as desired when considering the joint distribution of sources and parameters. To this model we can add unknown hyperparameters Λ with prior $p(\Lambda)$ in the usual way, and incorporate model uncertainty through an additional prior distribution on the number of components I. The specification of suitable source models $p(s_i|\theta_i)$ and $p(\theta_i)$, as well as the form of likelihood function $p(Y|\sum_{i=1}^{I} s_i)$, will form a substantial part of the remainder of the paper.

Several fundamental inference tasks can then be identified from this generic model, including the source separation and polyphonic music transcription tasks previously identified.

25.1.4.1 Source separation

In source separation the task is to infer the *source signals* s_i themselves, given the *observed signal* Y. Collecting the sources together as $S = \{s_i\}_{i=1}^{I}$ and the parameters as $\Theta = \{\theta_i\}_{i=1}^{I}$, the Bayesian formulation of the problem can be stated, under a fixed number of sources I, as (see for example Mohammad-Djafari 1997; Knuth 1998; Rowe 2003; Févotte and Godsill 2006; Cemgil, Fevotte, and Godsill 2007)

$$p(S|Y) = \frac{1}{P(Y)} \int p(Y|S, \Lambda)p(S|\Theta, \Lambda)p(\Lambda)p(\Theta)d\Lambda d\Theta \qquad (25.2)$$

where, under our deterministic model above in equation (25.1), the likelihood function $p(Y|S, \Lambda)$ will be degenerate. The marginal likelihood $P(Y)$ plays a key role when model order uncertainty is to be incorporated into the problem, for example when the number of sources N is unknown and needs to be estimated (Miskin and Mackay 2001). Additional considerations which may additionally be included in the above framework include convolutive (filtered) and non-stationary mixing of the sources – both scenarios are of practical interest and still pose significant computational challenges. Once the posterior distribution is computed by evaluating the integral, point estimates of the sources can be obtained using suitable estimation criteria, such as marginal MAP or posterior mean estimation, although in both cases one has to be especially careful with the interpretation of expectations in models where likelihoods and priors are invariant to source permutations.

25.1.4.2 Polyphonic music transcription

Music transcription refers to extraction of a human readable and interpretable description from a recording of a music performance, see Figure 25.6. In cases where more than a single musical note plays at a given time instant, we term this task *polyphonic music transcription* and we are once again in the superposition regime. The general task of interest is to infer automatically a musical notation, such as the traditional western music notation, listing the pitch values of notes, corresponding timestamps and other expressive information in a given performance. These quantities will be encoded in the above model through the parameters θ_i of each note present at a given time. Simple models will encode only the pitch of the note in θ_i while more complex models can include expressive information, instrument-specific characteristics and timbre, etc.

Apart from being an interesting modelling and computational problem in its own right, automated extraction of a score-like description is potentially very useful in a broad spectrum of applications such as interactive music performance systems, music information retrieval and musicological analysis

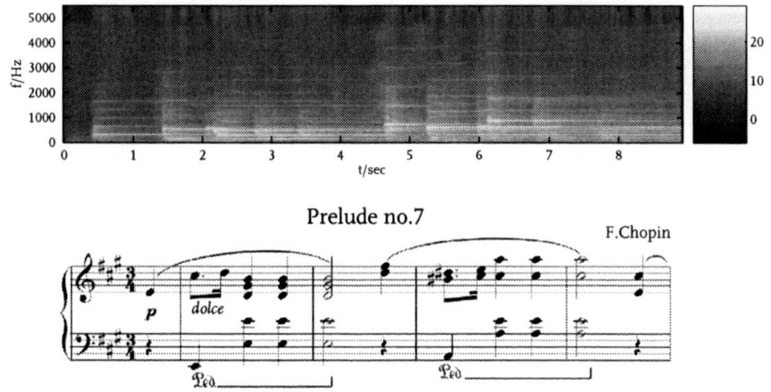

Fig. 25.6 Polyphonic music transcription. The task is to generate a human readable score as shown below, given the acoustic input. The computational problem here is to infer pitch, number of notes, rhythm, tempo, meter, time signature. The inference can be achieved online (filtering) or offline (smoothing), depending upon requirements.

of musical performances, not to mention as an aid to the source separation task identified above. However, in its most unconstrained form, i.e. when operating on an arbitrary acoustical input, music transcription remains a very challenging problem, owing to the wide variation in acoustical conditions and characteristics of musical instruments. In spite of these difficulties, a practical engineering solution is possible by careful incorporation of prior knowledge from cognitive science, musicology, musical acoustics, and by use of computational techniques from statistics and digital signal processing.

Music transcription is an inference problem in which we wish to find a musical score that is consistent with the encoded music. In this context, a score can be contemplated as a collection of 'musical objects' (e.g. note events) that are rendered by a performer to generate the observed signal. The term 'musical object' comes directly from an analogy to visual scene analysis where a scene is 'explained' by a list of objects along with a description of their intrinsic properties such as shape, colour or relative position. We view music transcription from the same perspective, where we wish to 'explain' individual samples of a music signal in terms of a collection of musical objects and where each object has a set of intrinsic properties such as pitch, tempo, loudness, duration or score position. It is in this respect that a score is a high level description of music.

Musical signals have a very rich temporal structure, and it is natural to think of them as being organized in a hierarchical way. At the highest level of this organization, which we may call as the cognitive (symbolic) level, we have a score of the piece, as, for instance, intended by a composer.[1] The performers add their interpretation to music and render the score into a collection of

[1] In reality the music may be improvised and there may be actually not a written score. In this case we replace the generative model with the intentions of the performer, which can still be expressed in our framework as a 'virtual' musical score.

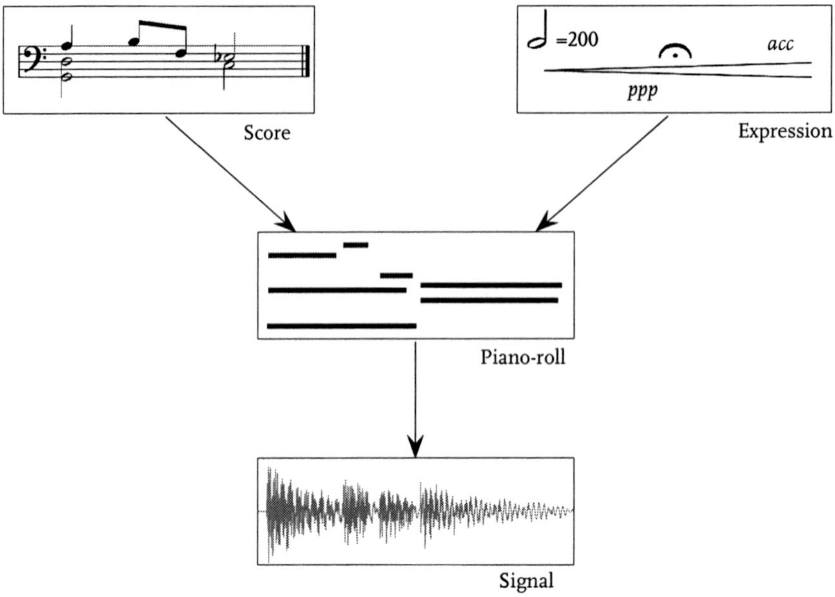

Fig. 25.7 A hierarchical generative model for music transcription. In this model, an unknown score is rendered by a performer into a 'piano roll'. The performer introduces expressive timing deviations and tempo fluctuations. The piano roll is rendered into audio by a synthesis model. The piano roll can be viewed as a symbolic representation, analogous to a sequence of MIDI events. Given the observations, transcription can be viewed as Bayesian inference of the score. Somewhat simplified, the techniques described in this chapter can be viewed as inference techniques as applied to subgraphs of this graphical model.

'control signals'. Further down at the physical level, the control signals trigger various musical instruments that synthesize the observed sound signal. We illustrate these generative processes using a hierarchical graphical model (see Figure 25.7), where the arcs represent generative links.

In describing music, we are usually interested in a symbolic representation and not so much in the 'details' of the actual waveform. To abstract away from the signal details we define an intermediate layer that represents the control signals. This layer, that we call a 'piano roll', forms the interface between a symbolic process and the actual signal process. Roughly, the symbolic process describes how a piece is composed and performed. Conditioned on the piano roll, the signal process describes how the actual waveform is synthesized. Conceptually, the transcription task is then to 'invert' this generative model and recover back the original score. As an intermediate and but still very challenging task, we may try and invert back only as far as the piano roll.

25.1.5 Organization of the chapter

In Section 25.2, signal models for audio are developed in the time domain, including some examples of their inference for a musical acoustics problem.

Section 25.3 describes models in the frequency transform domain that lead to greater computational tractability. In particular, we describe new dependence structures across time and frequency that allow for very accurate prior modelling for the audio. A final conclusion section is followed by appendices covering some basic methods and technical detail.

25.2 Time-domain models for audio

We begin by describing some basic note and chord models for musical audio, based in the time domain. As already discussed, a basic property of most non-percussive musical sounds is a set of oscillations at frequencies related to the fundamental frequency ω_0. Consider for the moment a short-time frame of musical audio data, denoted $y(\tau)$, in which note transitions do not occur. This would correspond, for example, to the analysis of a single musical chord. Throughout, we assume that the continuous time audio waveform $y(\tau)$ has been discretised with a sampling frequency ω_s rad s^{-1}, so that discrete time observations are obtained as $y_t = y(2\pi t/\omega_s)$, $t = 0, 1, 2, \ldots, N-1$. We assume that $y(\tau)$ is bandlimited to $\omega_s/2$ rad s^{-1}, or equivalently that it has been prefiltered with an ideal low-pass filter having cut-off frequency $\omega_s/2$ rad s^{-1}. We will not consider for the moment the time evolution of one chord to the next, or of note changes in a melody. This critical issue is treated in later sections.

The following model for, say, the ith note out of a chord comprising I notes in total can be written as

$$s_{i,t} = \sum_{m=1}^{M_i} \alpha_{m,i} \cos(m\omega_{0,i}t) + \beta_{m,i} \sin(m\omega_{0,i}t) \tag{25.3}$$

for $t \in \{0, \ldots, N-1\}$. Here, $M_i > 0$ is the number of partials present in note i, $\sqrt{\alpha_{m,i}^2 + \beta_{m,i}^2}$ gives the amplitude of a partial and $\tan^{-1}(\beta_{m,i}/\alpha_{m,i})$ gives the phase of that partial. Note that $\omega_{0,i} \in (0, \pi)$ is here scaled for convenience – its actual frequency is $\frac{\omega_{0,i}}{2\pi}\omega_s$. The unknown parameters for each note are thus $\omega_{0,i}$, the fundamental frequency, M_i, the number of partials and $\alpha_{m,i}$, $\beta_{m,i}$, which determine the amplitude and phase of each partial.

The extension to the multiple note case is then straightforwardly obtained by linear superposition of a number of notes:

$$y_t = \sum_{i=1}^{I} s_{i,t} + v_t$$

where v_t is a random background noise component (compare this with the deterministic mixture in equation 25.1). In this model v_t will also have to model any residual transient noise from the musical instruments themselves. We now have in addition an unknown parameter I, the number of notes present, plus any unknown statistics of the background noise process.

Such a model is a reasonable approximation for many steady musical sounds and has considerable analytical tractability, especially if a Gaussian form is assumed for v_t and for the priors on amplitudes α and β. Nevertheless, the posterior distribution is highly non-Gaussian and multimodal, and sophisticated computational tools are required to infer accurately from this model. This was precisely the topic of the work in Walmsley, Godsill, and Rayner (1998, 1999), where a reversible jump sampler was developed for such a model under the above-mentioned Gaussian prior assumptions.

The basic form above is, however, over-idealized in a number of ways: principally from the assumption of constant amplitudes α and β over time, and in the fixed integer relationships between partials, i.e. partial m in note i lies exactly at frequency $m\omega_{0,i}$. The modification of the basic model to remove these assumptions was the topic of our later work (Davy and Godsill 2002; Godsill and Davy 2002; Davy, Godsill, and Idier 2006; Godsill and Davy 2005), still within a reversible jump Monte Carlo framework. In particular, it is fairly straightforward to modify the model so that the partial amplitudes α and β may vary with time,

$$s_{i,t} = \sum_{m=1}^{M_i} \alpha_{m,i,t} \cos(m\omega_{0,i}t) + \beta_{m,i,t} \sin(m\omega_{0,i}t) \tag{25.4}$$

and we typically expand $\alpha_{m,i,t}$ and $\beta_{m,i,t}$ on a finite set of smooth basis functions $\psi_{i,t}$ with expansion coefficients a_i and b_i:

$$\alpha_{m,i,t} = \sum_{j=1}^{J} a_i\psi_{i,t}, \quad \beta_{m,i,t} = \sum_{j=1}^{J} b_i\psi_{i,t}.$$

In our work we have adopted 50%-overlapped Hamming windows for the basis functions, see Figure 25.8, with support either chosen a priori by the user or treated as a Bayesian random variable (Godsill and Davy 2005).

Alternative more general representations allow a fully stochastic variation of $\alpha_{m,i,t}$ in the state-space formulation. Further idealisations in these models include the assumption of constant fundamental frequencies with time and the Gaussian prior and noise assumptions, but in principle all can be addressed in a principled Bayesian fashion.

25.2.1 A prior distribution for musical notes

Under the above basic time-domain model we need to assign prior distributions over the unknown parameters for a single note in the mix, currently $\{\omega_{0,i}, M_i, \boldsymbol{\alpha}_i, \boldsymbol{\beta}_i\}$, where $\boldsymbol{\alpha}_i, \boldsymbol{\beta}_i$ are the vectors of parameters $\alpha_{m,i}, \beta_{m,i}$,

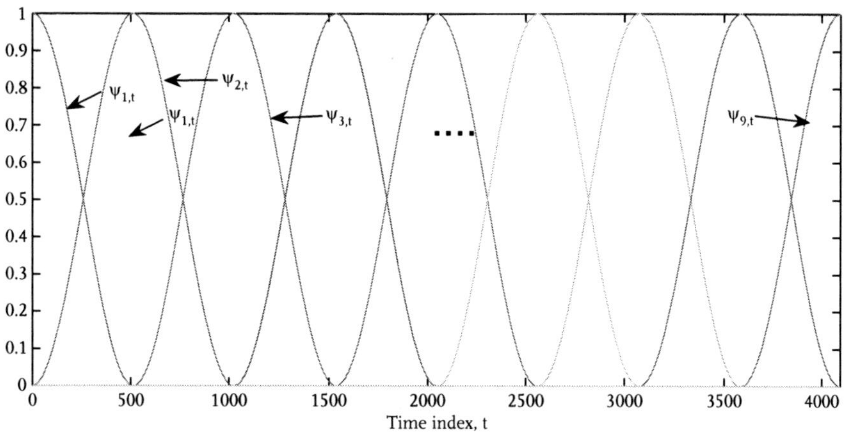

Fig. 25.8 Basis functions $\psi_{i,t}$, $I = 9$, 50% overlapped Hamming windows.

$m = 1, 2, \ldots, M_i$. Under an assumed note system such as an equally tempered Western note system, we can augment this with a note number index n_i. A suitable scheme is the MIDI note numbering system[2] which labels middle C (or 'C4') as note number 60, and all other notes as integers relative to this – the A below this would be 57, for example, and the A above middle C (usually at 440 Hz in modern Western tuning systems) would be note number 69. Other non-Western systems could also be encoded within variants of such a scheme. The fundamental frequency would then be expected to lie 'close' to the expected frequency for a particular note number, allowing for performance and tuning deviations from the ideal. Thus a prior for the observed fundamental frequency $\omega_{0,i}$ can be constructed fairly straightforwardly. We adopt here a truncated log-normal distribution for the note's fundamental frequency:

$$p(\log(\omega_{0,i})|n_i) \propto \begin{cases} N\left(\mu(n_i), \sigma_\omega^2\right), & \log(\omega_{0,i}) \in [(\mu(n_i-1)+\mu(n_i))/2, (\mu(n_i)+\mu(n_i+1))/2)] \\ 0, & \text{otherwise} \end{cases}$$

where $\mu(n)$ computes the expected log-frequency of note number n, i.e. when we are dealing with music in the equally tempered Western system,

$$\mu(n) = (n-69)/12 \log(2) + \log(440/\omega_s) \tag{25.5}$$

where once again ω_s rad s^{-1} is the sampling frequency of the data. Assuming $p(n)$ is uniform for now, the resulting prior $p(\omega_{0,i})$ is plotted in Figure 25.9, capturing the expected clustering of note frequencies at semitone spacings relative to A440.

[2] See for example `www.harmony-central.com/MIDI/doc/table2`.

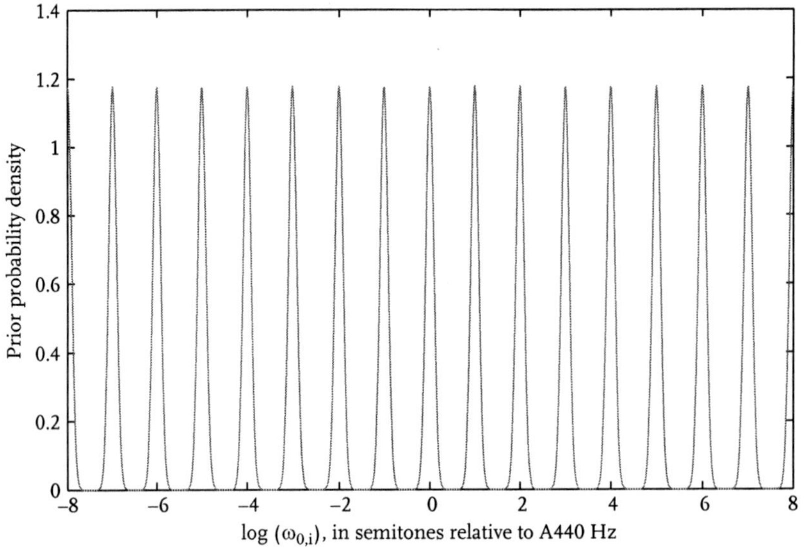

Fig. 25.9 Prior for fundamental frequency $p(\omega_{0,i})$.

The prior model for a note is completed with two components. Firstly, a prior for the number of partials, $p(M_i|\omega_{0,i})$, is specified as uniform over the range $\{M_{\min}, \ldots, M_{\max}\}$, with limits truncated to prevent partials at frequencies greater than $\omega_s/2$, the Nyquist rate. Secondly, a prior for the amplitude parameters $\boldsymbol{\alpha}_i, \boldsymbol{\beta}_i$ must be specified. This turns out to be quite crucial to the modelling performance and here we initially proposed a Gaussian form. It is expected however that partials at high frequencies will have lower energy than those at lower frequencies, generally following a low-pass filter shape in the frequency domain. Coefficents $\alpha_{m,i}$ and $\beta_{m,i}$ are then assigned independent Gaussian prior distributions such that their amplitudes are assumed to decay with increasing frequency of the partial number m. The general form of this is

$$p(\alpha_{m,i}, \beta_{m,i}) = N\left(\beta_{m,i}|0, g_i^2 k_m\right) N\left(\alpha_{m,i}|0, g_i^2 k_m\right).$$

Here g_i is a scaling factor common to all partials in a note and k_m is a frequency-dependent scaling factor to allow for the expected decay with increasing frequency for partial amplitudes. Following Godsill and Davy (2005) the amplitudes are assumed to decay as follows:

$$k_m = 1/(1 + (Tm)^\nu)$$

where ν is a decay constant and T determines the cut-off frequency. Such a model is based on empirical observations of the partial amplitudes in many real instrument recordings, and essentially just encodes a low pass filter with unknown cut-off frequency and decay rate. See for example the family of curves

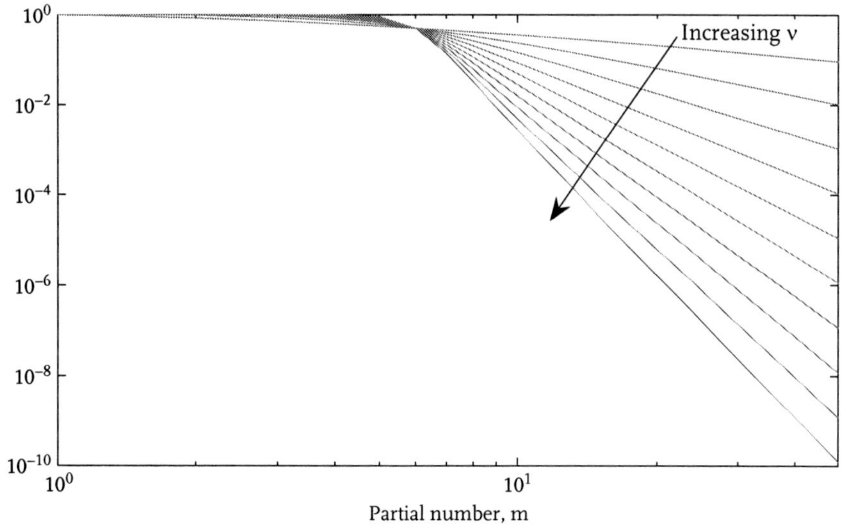

Fig. 25.10 Family of k_m curves (log-log plot), $T = 5$, $\nu = 1, \ldots, 10$.

with $T = 5$, $\nu = 1, 2, \ldots, 10$, Figure 25.10. It is worth pointing out that this model does not impose very stringent constraints on the precise amplitude of the partials: the Gaussian distribution will allow for significant departures from the $k_m = 1/(1 + (Tm)^\nu)$ rule, as dictated by the data, but it does impose a generally low-pass shape to the harmonics across frequency. It is possible to keep these parameters as unknowns in the MCMC scheme (see Godsill and Davy 2005), although in the examples presented here we fix these to appropriately chosen values for the sake of computational simplicity. g_i, which can be regarded as the overall 'volume' parameter for a note, is treated as an additional random variable, assigned an inverted Gamma distribution for its prior. The Gaussian prior structure outlined here for the α and β parameters is readily extended to the time-varying amplitude case of equation (25.4), in which case similar Gaussian priors are applied directly to the expansion coefficients a and b, see Davy, Godsill, and Idier (2006).

In the simplest case, a polyphonic model is then built by taking an independent prior over the individual notes and the number of notes present:

$$p(\Theta) = p(I) \prod_{i=1}^{I} p(\theta_i)$$

where

$$\theta_i = \{n_i, \omega_{0,i}, M_i, \boldsymbol{\alpha}_i, \boldsymbol{\beta}_i, g_i\}.$$

This model can be explored using MCMC methods, in particular the reversible jump MCMC method (Green 1995), and results from this and related models

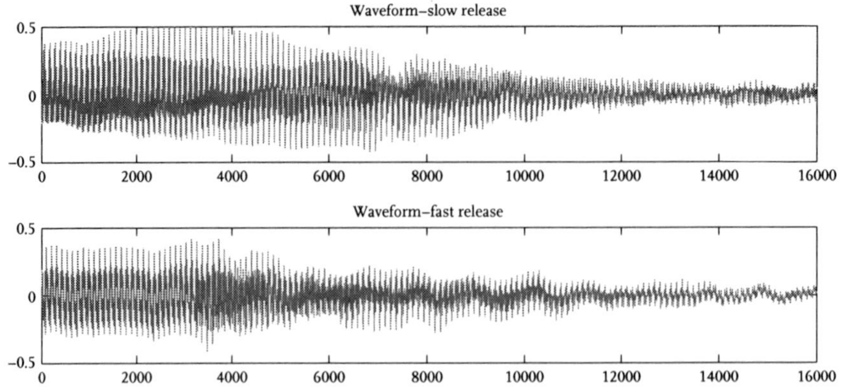

Fig. 25.11 Waveforms for release transient on pipe organ. Top: slow release; bottom: fast release.

can be found in Godsill and Davy (2005) and Davy, Godsill, and Idier (2006). In later sections, however, we discuss simple modifications to the generative model in the frequency domain which render the computations much more feasible for large polyphonic mixtures of sounds.

The models of this section provide a quite accurate time-domain description of many musical sounds. The inclusion of additional effects such as inharmonicity and time-varying partial amplitudes (Godsill and Davy 2005; Davy, Godsill, and Idier 2006) makes for additional realism.

25.2.2 Example: Musical transient analysis with the harmonic model

A useful case in point is the analysis of musical transients, i.e. the start or end of a musical note, when we can expect rapid variation in partial amplitudes with time. Here we take as an example a pipe organ transient, analysed under different playing conditions: one involving a rapid release at the end of the note, and the other involving a slow release, see Figure 25.11. There is some visible (and audible) difference between the two waveforms, and we seek to analyse what is being changed in the structure of the note by the release mode. Such questions are of interest to acousticians and instrument builders, for example.

We analyse these datasets using the prior distribution of the previous section and the model of equation (25.4). A fixed length Hamming window of duration 0.093 s was used for the basis functions. The resulting MCMC output can be used in many ways. For example, examination of the expansion coefficients α_i and β_i allows an analysis of how the partials vary with time under each playing condition. In both cases the reversible jump MCMC identifies nine significant partials in the data. In Figures 25.12 and 25.13 we plot the first five ($m = 1, \ldots, 5$) partial energies $a_{m,i}^2 + b_{m,i}^2$ as a function of time.

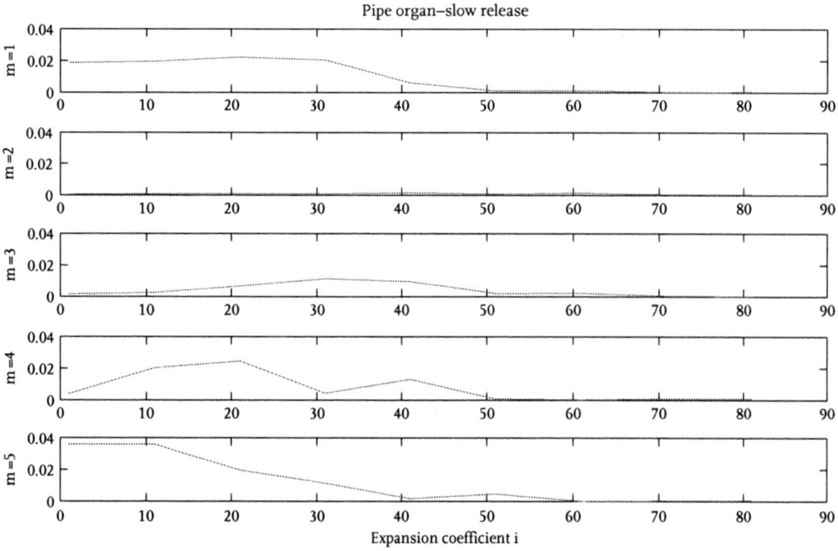

Fig. 25.12 Magnitudes of partials with time: slow release.

Examining the behaviour from the MCMC output we can see that the third partial is substantially elevated during the slow release mode, between coefficients $i = 30$ to 40. Also, in the slow release mode, the fundamental frequency ($m = 1$) decays at a much later stage relative to, say, the fifth partial, which itself decays more slowly in that mode. One can also use the model output to perform signal modification; for example time stretching or pitch shifting of the transient are readily achieved by reconstructing the signal using the MCMC-estimated parameters but modifying the Hamming window basis

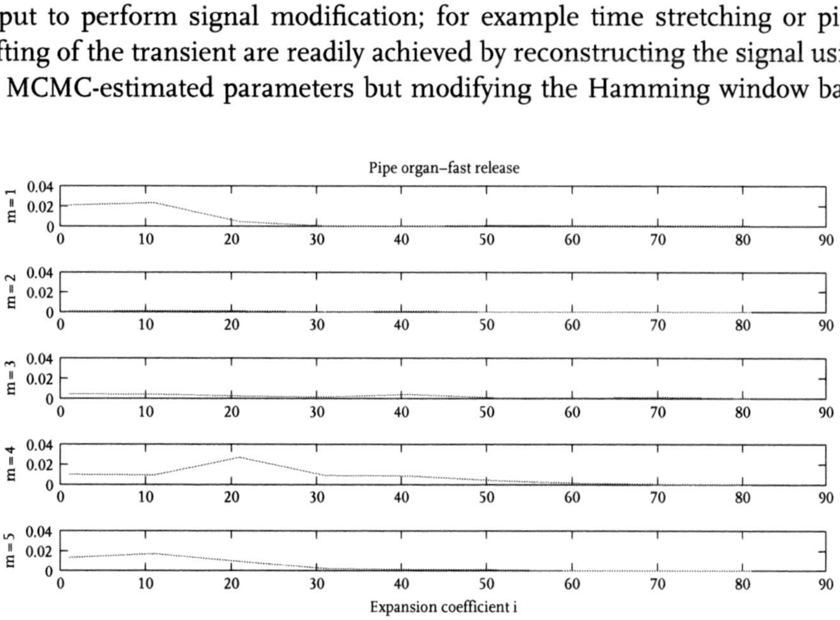

Fig. 25.13 Magnitudes of partials with time: fast release.

function length (for time-stretching) or reconstructing with modified fundamental frequency ω_0, see `www-sigproc.eng.cam.ac.uk/~sjg/haba`. The details of our reversible jump MCMC scheme are quite complex, involving a combination of specially designed independence Metropolis – Hastings proposals and random-walk-style proposals for the note frequency variables. In the frequency-domain models described in Section 25.3 we use essentially the same MCMC scheme, with simpler likelihood functions – some more details of the proposals used are given there.

25.2.3 State-space models

A more general and potentially more realistic modelling of audio in the time domain is given by the state-space formulation – essentially extending the sinusoidal models so far considered to allow for dynamic evolution with time. Specifically these models are readily amenable to inclusion of note change-points, stochastic amplitude/frequency variations and polyphonic music. For space reasons we do not include any detailed discussion here but the interested reader is referred to Cemgil, Kappen, and Barber (2006) and Cemgil (2007). Such state-space models are quite accurate for many examples of audio, although they show some non-robust properties in the case of signals which are far from steady-state oscillation and for instruments which do not closely obey the laws described above. Perhaps more critically, for large polyphonic mixes of many notes, each having potentially many partials, the computations – in particular the calculation of marginal likelihood terms in the presence of many Gaussian components $\boldsymbol{\alpha}_i$ and $\boldsymbol{\beta}_i$ – can become very expensive. Computing the marginal likelihood is costly as this requires computation of Kalman filtering equations for a large state space (that scales with the number of tracked harmonics) and for very long time series (as typical audio signals are sampled at 44.1 kHz). Hence, either efficient approximations need to be developed or simplified models need to be constructed. The latter approach is taken by frequency domain models which we will review in the following section.

25.3 Frequency-domain models

The preceding sections described various time-domain models for musical audio based on sinusoidal modelling. In this section we at least partially bypass the computational issues of the time-domain models by working with approximate models in the frequency domain. These allow for direct likelihood calculations without resorting to expensive matrix inversions and determinant calculations. Later in the chapter these models will be elaborated further to give sophisitcated Bayesian non-negative matrix factorization algorithms which are capable of learning the structure of the audio events in a semi-blind fashion. Here initially, though, we work with simple model-based structures in the frequency-domain that are analogous to the time-domain priors of the Section 25.2. There are several routes to a frequency-domain representation, including multiresolution transforms, wavelets, etc., though here we use a simple windowed discrete Fourier transform as examplar. We now propose two versions of a frequency-domain likelihood model, both of which bypass the main computational burden of the high-dimensional time-domain Gaussian models.

25.3.1 Gaussian frequency-domain model

The first model proposed is once again a Gaussian model. In the frequency domain we will have typically complex-valued expansion coefficients of the data on a one-dimensional lattice of frequency values $\nu \in N$, i.e. a set of spectrum values y_ν. The assumption is that the contribution of each musical source term to the expansion coefficients is as independent zero-mean (complex) Gaussians, with variance determined by the parameters of the musical note:

$$s_{i,\nu} \sim N_C(0, \lambda_\nu(\theta_i))$$

where $\theta_i = \{n_i, \omega_{0,i}, M_i, g_i\}$ has the same interpretation as for the earlier time-domain model, but now we can neglect the α and β coefficients since the random behaviour is now directly modelled by $s_{i,\nu}$. This is a very natural formulation for generation of polyphonic models since we can add a number of sources together to make a single complex Gaussian data model:

$$y_\nu \sim N_C(0, S_{v,\nu} + \sum_{i=1}^{I} \lambda_\nu(\theta_i)).$$

Here, $S_{v,\nu} > 0$ models a Gaussian background noise component in a manner analogous to the time-domain formulation's v_t and it then remains to design the positive-valued 'template' functions λ. Once again, Figure 25.5 gives some guidance as to the general characteristics required. We then model the template using a sum of positive valued pulse waveforms ϕ_ν, shifted to be centred at the expected partial position, and whose amplitude decays with increasing partial number:

$$\lambda_\nu(\theta_i) = \sum_{m=1}^{M_i} g_i^2 k_m \phi_{\nu - m\omega_{0,i}} \tag{25.6}$$

where k_m, g_i and M_i have exactly the same interpretation as in the time-domain model. An example template construction is shown in Figure 25.14, in which a Gaussian pulse shape has been utilized.

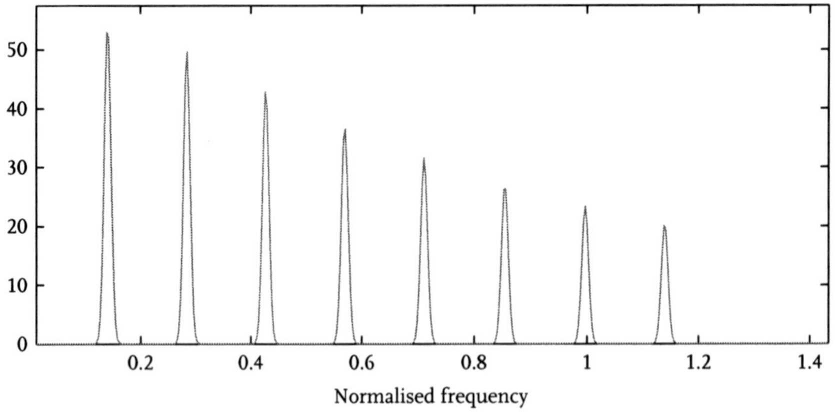

Fig. 25.14 Template function $\lambda_\nu(\theta_i)$ with $M_i = 8$, $\omega_{0,i} = 0.71$, Gaussian pulse shape.

25.3.2 Point process frequency-domain model

The Gaussian frequency-domain model requires a knowledge of the conditional distribution for the whole range of spectrum values. However, the salient features in terms of pitch estimation appear to be the *peaks* of the spectrum (see Figure 25.5). Hence a more parsimonious likelihood model might work only with the peaks detected from the Fourier magnitude spectrum. Thus we propose, as an alternative to the Gaussian spectral model, a point process model for the peaks in the spectrum. Specifically, if the peaks in the spectrum of an individual note are assumed to be drawn from a one-dimensional inhomogeneous Poisson point process having intensity function $\lambda_\nu(\theta_i)$ (considered as a function of continuous frequency ν), then the combined set of peaks from many notes may be combined, under an independence assumption, to give a Poisson point process whose intensity function is the sum of the individual intensities (Grimmett and Stirzaker 2001). Suppose we detect a set of peaks in the magnitude spectrum $\{p_j\}_{j=1}^J$, $\nu_{\min} < p_j < \nu_{\max}$. Then the likelihood may be readily computed using:

$$p\left(\{p_j\}_{j=1}^J, J \mid \Theta\right) = \text{Po}(J \mid Z(\Theta)) \prod_{j=1}^{J} \frac{\left(S_{v,p_j} + \sum_{i=1}^{I} \lambda_{p_j}(\theta_i)\right)}{Z(\Theta)}$$

where $Z(\Theta) = \int_{\nu_{\min}}^{\nu_{\max}} (S_{v,\nu} + \sum_{i=1}^{I} \lambda_\nu(\theta_i))d\nu$ is the normalizing constant for the overall intensity function. Here once again we include a background intensity function $S_{v,\nu}$ which models 'false detections', i.e. detected peaks that belong to no existing musical note. The form of the template functions λ can be very similar to that in the Gaussian frequency model, equation (25.6). A modified form of this likelihood function was successfully applied for chord detection problems by Peeling, Li, and Godsill (2007).

25.3.3 Example: Inference in the frequency-domain models

The frequency-domain models provide a substantially faster likelihood calculation than the earlier time-domain models, allowing for rapid inference in the presence of significantly larger chords and tone complexes. Here we present example results for a tone complex containing many different notes, played on a pipe organ. Analysis is performed on a very short segment of 4096 data points, sampled at a rate of $\omega_s = 2\pi \times 44,100$ rad s^{-1} – hence just under 0.1 s of data, see Figure 25.15. From the score of the music we know that there are four notes simultaneously playing: C5, F♯5, B5, and D6, or MIDI note numbers 72, 78, 83 and 86. However, the mix is complicated by the addition of pipes one octave below and one or more octaves above the principal pitch, and hence we have at least 12 notes present in the complex, MIDI notes 60, 66, 71, 72, 74, 78, 83, 84, 86, 90, 95, and 98. Since the upper octaves share all of their partials with notes from one or more octaves below, it is not clear whether the models will be able to distinguish all of the sounds as separate notes. We run the frequency-domain models using the prior framework of Section 25.2.1 and a reversible jump MCMC scheme of the same form as that used in the previous transient analysis example. Firstly, using the Gaussian frequency-domain model of Section 25.3.1, the MCMC burn-in for the note number vector $n = [n_1, n_2, \ldots, n_I]$ is shown in Figure 25.16. This is a variable-dimension vector under the reversible jump MCMC and we can see notes entering or leaving the vector as iterations proceed. We can also see large moves of an octave ($\pm$12 notes) or a fifth (+7 or -5 notes), corresponding to specialized Metropolis – Hastings moves which centre their proposals on the octave or fifth as well as the locality of the current note. As is typical of these models, the MCMC becomes slow-moving once converged to a good mode of the distribution and further large moves only occur occasionally. There is a good case here for using adaptive or population MCMC schemes to improve the properties of the MCMC. Nevertheless, convergence is much

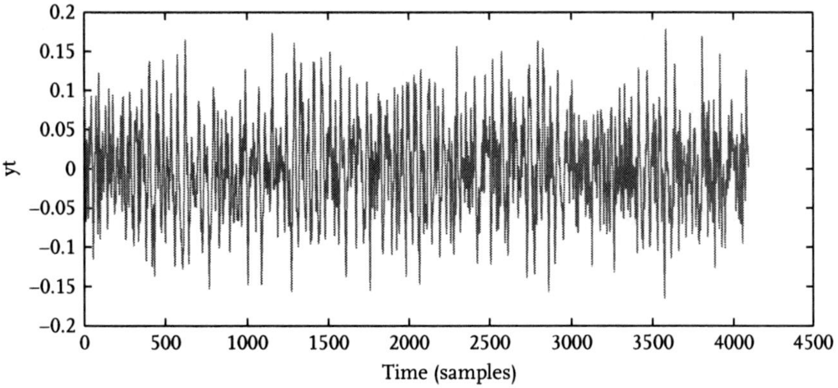

Fig. 25.15 Audio waveform – single chord data.

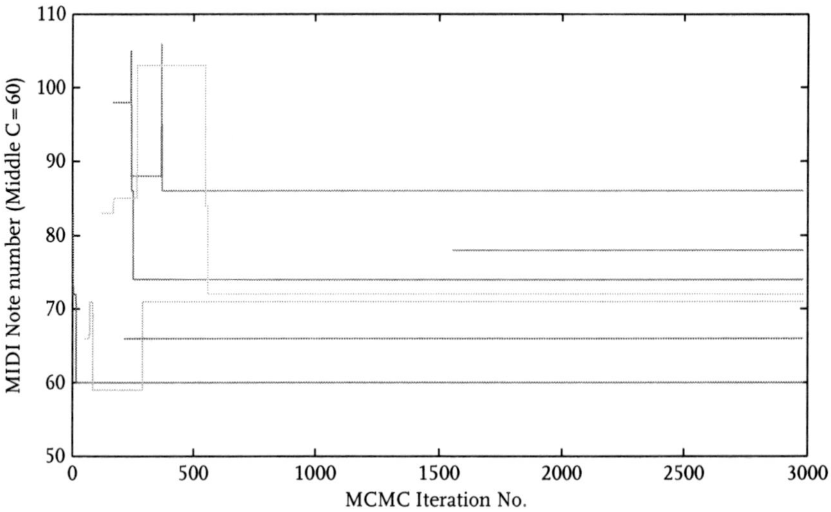

Fig. 25.16 Evolution of the note number vector with iteration number – single chord data. Gaussian frequency-domain model.

faster than for the earlier proposed time-domain models, particularly in terms of the model order sampling, which was here initialized at $I = 1$, i.e. one single note present at the start of the chain. Specialized independence proposals have also been devised, based on simple pitch estimation methods applied to the raw data. These are largely responsible for the initiation of new notes in the MCMC chain. In this instance the MCMC has identified correctly seven out of the (at least) 12 possible pitches present in the music: 60, 66, 71, 72, 74, 78, 86. The remaining five unidentified pitches share all of their partials with lower pitches estimated by the algorithm, and hence it is reasonable that they remain unestimated. Examination of the discrete Fourier magnitude spectrum (Figure 25.17) shows that the higher pitches (with the possible exception of $n_7 = 83$, whose harmonics are modelled by $n_3 = 71$) are generally buried at very low amplitude in the spectrum and can easily be absorbed into the model for pitches one or more octaves lower in pitch.

We can compare these results with those obtained using the Poisson model of Section 25.3.2. The MCMC was run under identical conditions to the Gaussian model and we plot the equivalent note index output in Figure 25.18. Here we see that fewer notes are estimated, since the basic point process model takes no account of the amplitudes of the peaks in the spectrum, and hence is happy to assign all harmonics to the lowest possible fundamental pitch. The four predominant pitches estimated are the four lowest fundamentals: 60, 66, 71 and 74. The sampler is, however, generally more mobile and we see a better and more rapid exploration of the posterior.

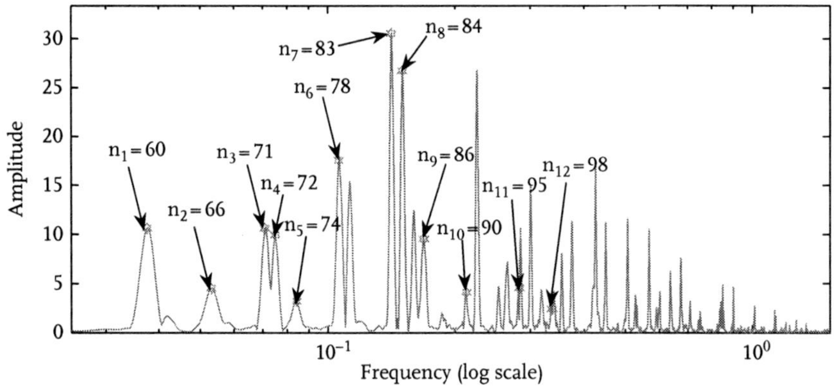

Fig. 25.17 Discrete Fourier magnitude spectrum for 12-note chord. True note positions marked with a pentagram.

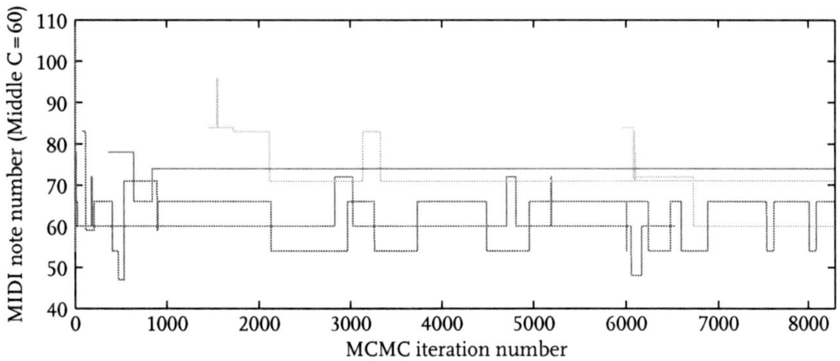

Fig. 25.18 Evolution of the note number vector with iteration number – single chord data. Poisson frequency-domain model.

25.3.4 Further prior structures for transform domain representations

In audio processing, the energy content of a signal across frequencies is time-varying and hence it is natural to model audio as an evolving process with a time-varying power spectral density in the time-frequency plane and several prior structures are proposed in the literature for modelling the expansion coefficients (Reyes-Gomez, Jojic, and Ellis 2005; Wolfe, Godsill, and Ng 2004; Févotte, Daudet, Godsill, and Torrésani 2006). The central idea is to choose a latent variance model varying over time and frequency bins

$$s_{\nu,k}|q_{\nu,k} \sim \mathcal{N}(s_{\nu,k}; 0, q_{\nu,k})$$

where the normal is interpreted either as complex Gaussian or real Gaussian depending on the transform used – the Fourier representation is complex, the discrete sine/cosine representation is real. In Wolfe, Godsill, and Ng (2004), the following structure is proposed under the name *Gabor regression*. The variance parameters $q_{\nu,k}$ are treated as independent conditional upon a lattice of activity variables $r_{\nu,k}$ which are modelled as dependent using Markov chains and Markov random fields:

$$q_{\nu,k}|r_{\nu,k} \sim [r_{\nu,k} = \text{on}]\, IGa(q_{\nu,k}; a, b/a) + [r_{\nu,k} = \text{off}]\, \delta(q_{\nu,k}).$$

Moreover, the joint distribution over the latent indicators $r = r_{0:W-1,0:K-1}$ is taken as a pairwise Markov random field (MRF) where u denotes a double index $u = (\nu, k)$

$$p(r) \propto \prod_{(u,u')\in\mathcal{E}} \phi(r_u, r_{u'}).$$

Several MRF constructions are considered, including Markov chains across time or frequency and Ising-type models.

25.3.5 Gamma chains and fields

An alternative model is introduced by Cemgil and Dikmen (2007) and Cemgil *et al.* (2007), where a Markov random field is directly placed on the variance terms as

$$p(q) = \int d\lambda p(q, \lambda)$$

using a so-called Gamma field.

To understand the construction of a Gamma field, it is instructive to look first at a chain, where we have an alternating sequence of Gamma and inverse Gamma random variables

$$q_u|\lambda_u \sim IGa(q_u; a_q, a_q\lambda) \qquad\qquad \lambda_{u+1}|q_u \sim Ga(\lambda_{u+1}; a_\lambda, q_u/a_\lambda).$$

Note that this construction leads to conditionally conjugate Markov blankets that are given as

$$p(q_u|\lambda_u, \lambda_{u+1}) \propto IGa(q_u; a_q + a_\lambda, a_q\lambda_u + a_\lambda\lambda_{u+1})$$
$$p(\lambda_u|q_{u-1}, q_u) \propto Ga\left(\lambda_u; a_\lambda + a_q, a_\lambda q_{u-1}^{-1} + a_q q_u^{-1}\right).$$

Moreover it can be shown that any pair of variables q_i and q_j are positively correlated, and q_i and λ_k are negatively correlated. Note that this is a particular type of *stochastic volatility* model useful for characterization of non-stationary behaviour observed in, for example, financial time series (Shepard 2005).

We can represent a chain by a graphical model where the edge set is $\mathcal{E} = \{(u, u)\} \cup \{(u, u + 1)\}$. Considering the Markov structure of the chain, we define a Gamma field $p(q, \lambda)$ as a bipartite undirected graphical model consisting of the vertex set $\mathcal{V} = \mathcal{V}_\lambda \cup \mathcal{V}_q$, where partitions $\mathcal{V}_\lambda$ and $\mathcal{V}_q$ denotes the collection of variables λ and q that are conditionally distributed Ga and IGa respectively.

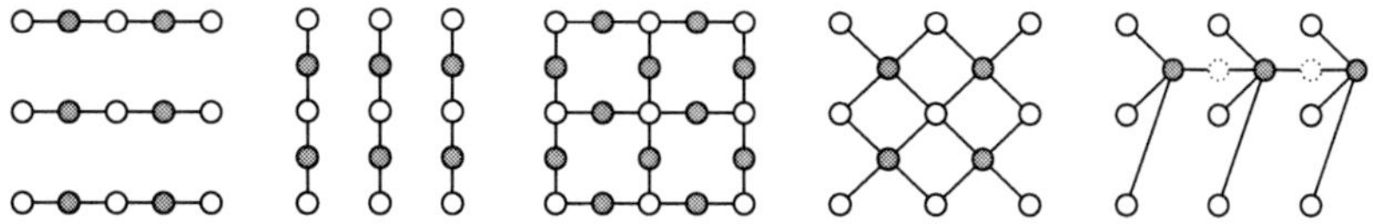

Fig. 25.19 Possible model topologies for gamma fields. White and grey nodes correspond to $\mathcal{V}_q$ and $\mathcal{V}_\lambda$ nodes respectively. The horizontal and vertical axis corresponds to frequency ν and frame index k. Each model describes how the prior variances are coupled as a function of time – frequency index. For example, the first model from the left corresponds to a source model with 'spectral continuity', the energy content of a given frequency band changes only slowly. The second model is useful for modelling impulsive sources where energy is concentrated in time but spread across frequencies.

We define an edge set $\mathcal{E}$ where an edge $(u, u') \in \mathcal{E}$ such that $\lambda_u \in \mathcal{V}_\lambda$ and $q_{u'} \in \mathcal{V}_q$, if the joint distribution admits the following factorization

$$p(\lambda, q) \propto \left(\prod_{u \in \mathcal{V}_\lambda} \lambda_u^{(\sum_{u'} a_{u,u'} - 1)} \right) \left(\prod_{u' \in \mathcal{V}_q} q_u^{-(\sum_u a_{u,u'} + 1)} \right) \left(\prod_{(u,u') \in \mathcal{E}} \exp\left(-a_{u,u'} \frac{\lambda_u}{q_{u'}} \right) \right).$$

Here, the shape parameters play the role of coupling strengths; when $a_{u,u'}$ is large, adjacent nodes are correlated. Given, this construction, various signal models can be developed – see Figure 25.19.

25.3.6 Models based on latent variance/intensity factorization

The various Markov random field priors of the previous section introduced couplings between the latent variances $q_{\nu,k}$. Another alternative and powerful approach is to decompose the latent variances as a product. We define the following hierarchical model (see Figure 25.21)

$$\begin{aligned} s_{\nu,k} &\sim N(s_{\nu,k}; 0, q_{\nu,k}) & q_{\nu,k} &= t_\nu v_k \\ t_\nu &\sim IGa\left(t_\nu; a_\nu^t, a_\nu^t b_\nu^t\right) & v_k &\sim IGa\left(v_k; a_k^v, a_k^v b_k^v\right). \end{aligned} \tag{25.7}$$

Such models are also particularly useful for modelling acoustic instruments. Here, the t_ν variables can be interpreted as average expected energy template as a function of frequency bin. At each time index this template is modulated by v_ν, to adjust the overall volume. An example is given in Figure 25.20 to represent a piano sound. The template gives the harmonic structure of the pitch and the excitation characterises the time varying energy.

A simple factorial model that uses the Gamma chain prior models introduced in Section 25.3.5 is constructed as follows:

$$x_{\nu,k} = \sum_i s_{\nu,i,k} \quad s_{\nu,i,k} \sim N(s_{\nu,i,k}; 0, q_{\nu,i,k}) \quad Q = \{q_{\nu,i,k}\} \sim p(Q|\Theta^t). \tag{25.8}$$

The computational advantage of this class of models is the conditional independence of the latent sources given the latent variance variables. Given the

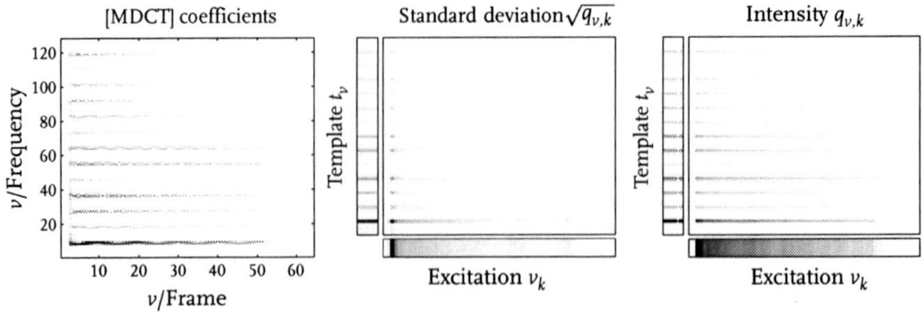

Fig. 25.20 (Left) the spectrogram of a piano $|s_{\nu,k}|^2$. (Middle) estimated templates and excitations using the conditionally Gaussian model defined in equation (25.7), where $q_{\nu,k}$ is the latent variance. (Right) estimated templates and excitations using the conditionally Poisson model defined in the next section.

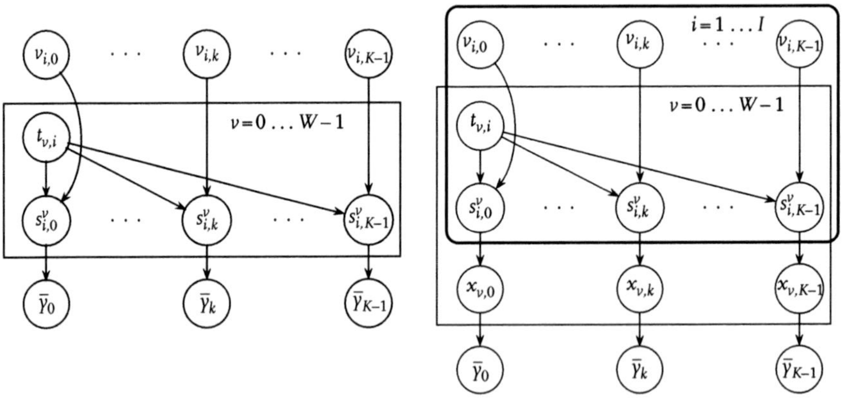

Fig. 25.21 (Left) latent variance/intensity models in product form (equation 25.7). Hyperparameters are not shown. (Right) factorial version of the same model, used for polyphonic estimation as used in Section 25.3.7.1.

latent variances and data, the posterior of the sources is a product of Gaussian distributions. In particular, the individual marginals are given in closed form as

$$p(s_{\nu,i,k}|X, Q) = \mathcal{N}(s_{\nu,i,k}; \kappa_{\nu,i,k}x_{\nu,k}, q_{\nu,i,k}(1 - \kappa_{\nu,i,k}))$$
$$\kappa_{\nu,i,k} = q_{\nu,i,k} \Big/ \sum_{i'} q_{\nu,i',k}.$$

This means that if the latent variances can be estimated, source separation can be easily accomplished. The choice of prior structures on the latent variances $p(Q|\cdot)$ is key here.

Below we illustrate this approach in single channel source separation for transient/harmonic decomposition. Here, we assume that there are two sources $i = 1, 2$. The prior variances of the first source $i = 1$ are tied across time frames using a Gamma chain and aims to model a source with harmonic continuity. The prior has the form $\prod_\nu p(q_{\nu,i=1,1:K})$. This model simply assumes

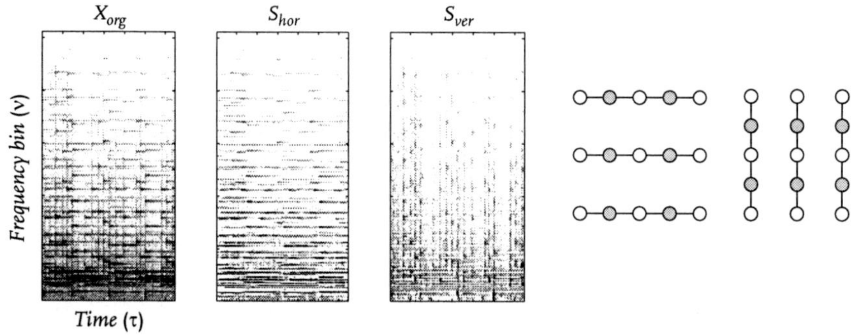

Fig. 25.22 Single channel source separation example, left to right, log-MDCT coefficients of the original signal and reconstruction with horizontal and vertical IGMRF models.

that for a given source the amount of energy in a frequency band stays roughly constant. The second source $i = 2$ is tied across frequency bands and has the form $\prod_k p(q_{1:W,i=2,k})$; this model tries to capture impulsive/percusive structure (for example compare the piano and conga examples in Figure 25.1). The model aims to separate the sources based on harmonic continuity and impulsive structure.

We illustrate this approach to separate a piano sound into its constituent components and drum separation. We assume that $J = 2$ components are generated independently by two gamma chain models with vertical and horizontal topology. In Figure 25.22, we observe that the model is able to separate transients and harmonic components. The sound files of these results can be downloaded and listened at the following url: `http://www-sigproc.eng.cam.ac.uk/sjg/haba`, which is perhaps the best way assess the sound quality.

The variance/intensity factorization models described in equation 25.7 have also straightforward factorial extensions

$$x_{\nu,k} = \sum_i s_{\nu,i,k}$$

$$s_{\nu,i,k} \sim N(s_{\nu,i,k}; 0, q_{\nu,i,k}) \qquad q_{\nu,i,k} = t_{\nu,i} v_{i,k} \tag{25.9}$$

$$T = \{t_{\nu,i}\} \sim p(T|\Theta^t) \qquad V = \{v_{i,k}\} \sim p(V|\Theta^v). \tag{25.10}$$

If we integrate out the latent sources, the marginal is given as

$$x_{\nu,k} \sim N\left(x_{\nu,k}; 0, \sum_i t_{\nu,i} v_{i,k}\right).$$

Note that, as $\sum_i t_{\nu,i} v_{i,k} = [TV]_{\nu,k}$, the variance 'field' Q is given compactly as the matrix product $Q = TV$. This resembles closely a matrix factorisation and is used extensively in audio modelling. In the next section, we discuss models of this type.

25.3.7 Non-negative matrix factorization models

Up to this point we have described conditionally Gaussian models. Recently, a popular branch of source separation and analysis of musical audio literature has focused on non-negativity of the magnitude spectrogram $X = \{x_{\nu,\tau}\}$ with $x_{\nu,\tau} \equiv \|s_{\nu,k}\|_2^{1/2}$, where $s_{\nu,k}$ are expansion coefficients obtained from a time – frequency expansion. The basic idea of NMF is representing a spectrogram by enforcing a factorization as $X \approx TV$ where both T and V are matrices with positive entries (Smaragdis and Brown 2003; Abdallah and Plumbley 2006; Virtanen 2006a; Kameoka 2007; Bertin, Badeau, and Richard 2007; Vincent, Bertin, and Badeau 2008). In music signal analysis, T can be interpreted as a codebook of templates, corresponding to spectral shapes of individual notes and V is the matrix of activations, somewhat analogous to a musical score. Often, the following objective is minimized:

$$(T, V)^* = \min_{T,V} D(X||TV) \tag{25.11}$$

where D is the information (Kullback – Leibler) divergence, given by

$$D(X||\Lambda) = \sum_{\nu,\tau} \left(x_{\nu,\tau} \log \frac{x_{\nu,\tau}}{\lambda_{\nu,\tau}} - x_{\nu,\tau} + \lambda_{\nu,\tau} \right). \tag{25.12}$$

Using Jensen's inequality (Cover and Thomas 1991) and concavity of $\log x$, it can be shown that $D(\cdot)$ is non-negative and $D(X||\Lambda) = 0$ if and only if $X = \Lambda$. The objective in (25.11) could be minimized by any suitable optimization algorithm. Lee and Seung (2000) have proposed an efficient variational bound minimization algorithm that has attractive convergence properties. that has been since successfully applied to various applications in signal analysis and source separation.

It can also be shown that the minimization algorithm is in fact an EM algorithm with data augmentation (Cemgil 2008). More precisely, it can be shown that minimizing D w.r.t. T and V is equivalent finding the ML solution of the following hierarchical model

$$\begin{aligned}
x_{\nu,k} &= \sum_i s_{\nu,i,k} \\
s_{\nu,i,k} &\sim \mathcal{P}o(s_{\nu,i,k}; 0, \lambda_{\nu,i,k}) & \lambda_{\nu,i,k} &= t_{\nu,i} v_{i,k} & (25.13)\\
t_{\nu,i} &\sim \mathcal{G}a\left(t_{\nu,i}; a^t_{\nu,i}, b^t_{\nu,i}/a^t_{\nu,i}\right) & v_{i,k} &\sim \mathcal{G}a\left(v_{i,k}; a^v_{i,k}, b^v_{i,k}/a^v_{i,k}\right). & (25.14)
\end{aligned}$$

Note that this model is quite distinct from the Poisson point model used in Section 25.3.2 since it models each time – frequency coefficient as a Poisson random variable, while the previous approach models detected peaks in the spectrum as a spatial point process.

The computational advantage of this model is the conditional independence of the latent sources given the variance variables. In particular, we have

$$p(s_{\nu,i,k}|X, T, V) = \mathcal{B}i(s_{\nu,i,k}; x_{\nu,k}, \kappa_{\nu,i,k})$$

$$\kappa_{\nu,i,k} = \lambda_{\nu,i,k} / \sum_{i'} \lambda_{\nu,i',k}$$

This means that if the latent variances can be estimated somehow, source separation can be easily accomplished as $\mathsf{E}(s)_{\mathcal{B}i(s;x,\kappa)} = \kappa x$. It is also possible to estimate the marginal likelihood $p(X)$ by integrating out all of the templates and excitations. This can be done via Gibbs sampling or more efficiently using a variational approach that we outline in Appendix A.

25.3.7.1 Example: Polyphonic pitch estimation

In this section, we illustrate Bayesian NMF for polyphonic pitch detection. The approach consists of two stages:

1. Estimation of hyperparameters given a corpus of piano notes.
2. Estimation of templates and excitations given new polyphonic data and fixed hyperparameters.

In the first stage, we estimate the hyperparameters $a^t_{\nu,i} = a^t_i$ and $b^t_{\nu,i}$ (see equation 25.14), via maximization of the variational bound given in equation 25.20. Here, the observations are matrices X_i; a spectrogram computed given each note $i = 1 \ldots I$. In Figure 25.23, we show the estimated scale parameters $b^t_{\nu,i}$ as

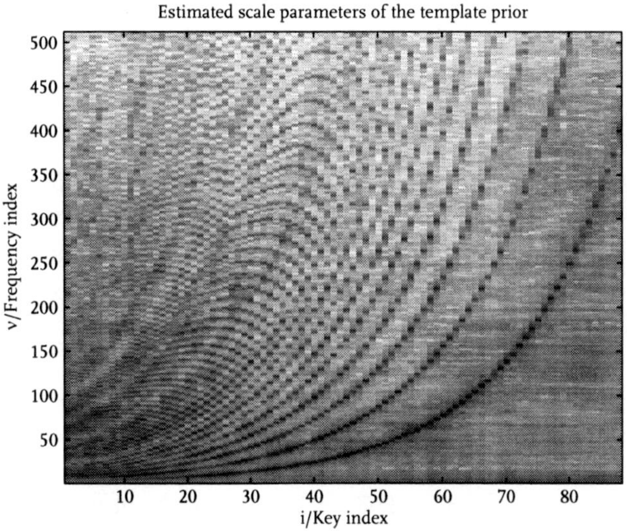

Fig. 25.23 Estimated template hyperparameters $b^t_{\nu,i}$.

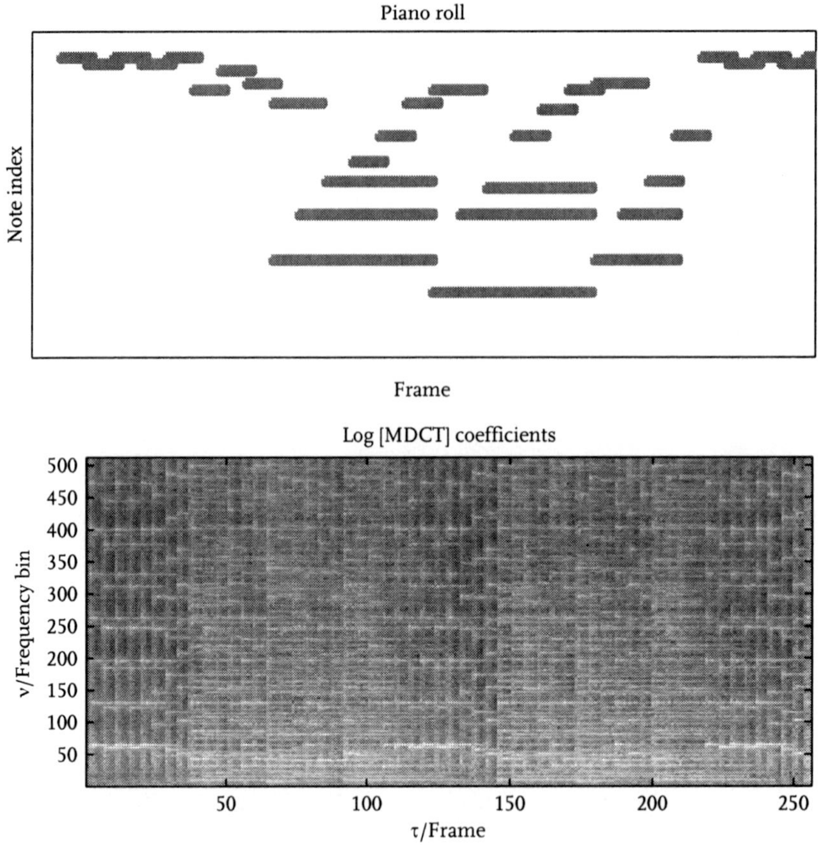

Fig. 25.24 The ground truth piano roll and the spectrum of the polyphonic data.

a function of frequency band ν and note index i. The harmonic structure of each note is clearly visible.

To test the approach, we synthesize a music piece (here, a short segment from the beginning of *Für Elise* by Beethoven), given a MIDI piano roll and recordings of isolated notes from a piano by simply appropriately shifting each time series and adding. The piano roll and the the spectrogram of the synthesized audio are shown in Figure 25.24. The pitch detection task is infering the excitations given the hyperparameters and the spectrogram.

The results are shown in Figure 25.25. The top figure shows the excitations estimated give the prior shown in equation 25.14. The notes are visible here but there are some artifacts. The middle figure shows results from a model where excitations are tied across time using a Gamma chain introduced in Section 25.3.5. This prior is highly effective here and we are able to get a more clearer picture. The bottom figure displays results

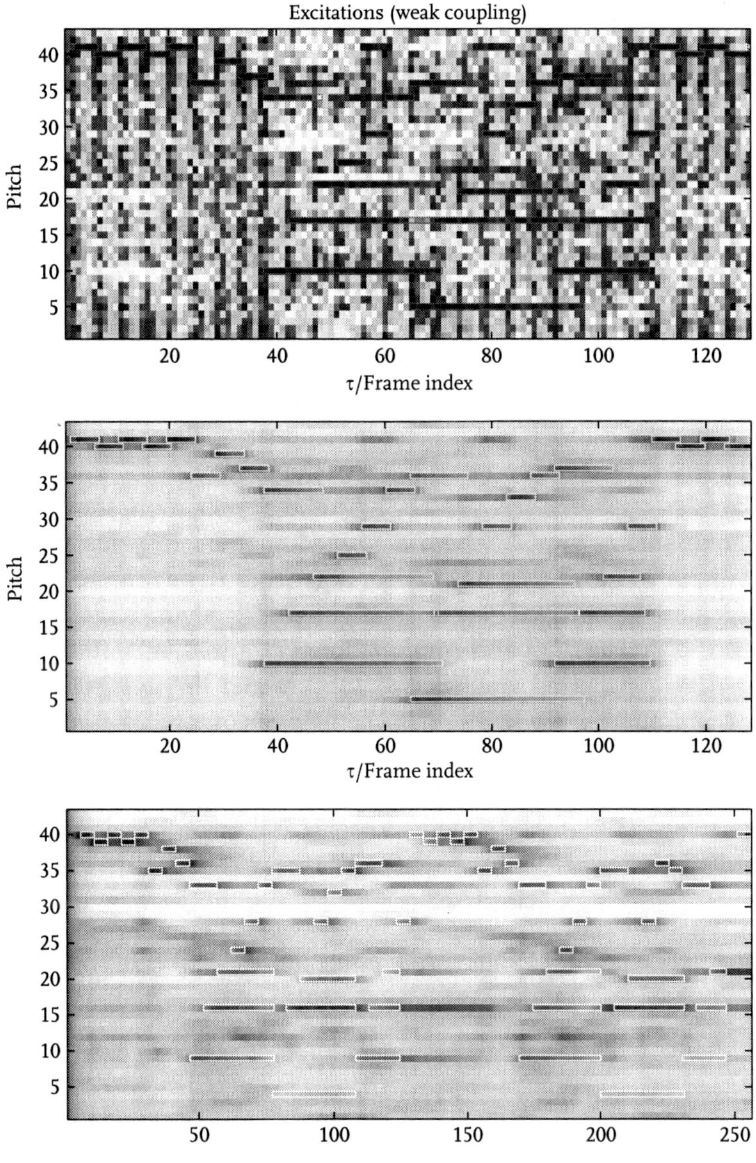

Fig. 25.25 Polyphonic pitch detection. Estimated expected excitations. (Top) uncoupled excitations. (Middle) tied excitations using a Gamma chain, ground truth shown in white. (Bottom) excitations estimated from a guitar using the hyperparameters estimated from a piano – ground truth shown in black.

obtained from a real recording of *Für Elise*, performed on electric guitar. Interestingly, whilst we are still using the hyperparameters estimated from a piano, the inferred excitations show significant overlap with the original score.

25.4 Conclusions

In this chapter we have described recently proposed Bayesian methods for analysis of audio signals. The Bayesian models exhibit complex statistical structure and in practice, highly adaptive and powerful computational techniques are needed to perform inference. We have reviewed and developed some of these statistical models and described how various problems in audio and music processing can be cast into the Bayesian inference framework. We have also illustrated inference methods based on Monte Carlo simulation or other deterministic techniques (such as mean field, variational Bayes) originating in statistical physics to tackle computational problems posed by inference in these models. We described models in both the time domain and transform domains, the latter typically offering greater computational tractability and modelling flexibility at the expense of some accuracy in the models.

The Bayesian approach has two key advantages over more traditional engineering solutions: it provides both a unified methodology for probabilistic model construction and a framework for algorithm development. Apart from the pedagogical advantages (such as highlighting algorithmic similarities, convergence characteristics and computational requirements), the framework facilitates development of sophisticated models and the automation of code generation procedures. We believe that the field of computer hearing, which is still in its infancy compared to topics such as computer vision and speech recognition, has great potential for advancement in coming years, with the advent of powerful Bayesian inference methodologies and accompanying increases in computational power.

Appendix

A. Broader context and background

Audio processing applications require efficient inference in fairly complex hierarchical Bayesian models. In statistics, the fundamental computational tools to such high dimensional integrals are based on Markov chain Monte Carlo strategies such as the Gibbs sampler (Gilks, Richardson, and Spiegelhalter 1996). The main advantage of MCMC is its generality, robustness and attractive theoretical properties. However, the method comes at the price of heavy computational burden which may render it impractical for data intensive applications.

An alternative approach for computing the required integrals is based on deterministic fixed point iterations (Variational Bayes – Structured Mean field; Ghahramani and Beal 2000; Wainwright and Jordan 2003; Bishop 2006). This set of methods have direct links with the well-known expectation-maximization (EM) type of algorithms. Variational methods have been extensively applied to various models for source separation by a number of authors (Attias 1999; Valpola 2000; Girolami 2001; Miskin and Mackay 2001; Hojen-Sorensen, Winther, and Hansen 2002; Winther and Petersen 2006).

From an algorithmic point of view, the VB method can be viewed as a 'deterministic' counterpart of the Gibbs sampler. Especially for models where a Gibbs sampler is easy to construct (e.g. in models with conjugate priors leading to known full conditionals) the VB method is equally easy to apply. Like the Gibbs sampler, the framework facilitates generalization to more complex models and to automation of code generation procedure. Moreover, the method directly provides an approximation (a lower bound) to the marginal likelihood. Although in general not much is known about how tight the bound is, there is empirical evidence that for many models the bound can provide a good approximation to an estimate obtained from Gibbs sampling via Chib's method (Chib 1995).

A.1 *Bounding marginal likelihood via variational Bayes*

We sketch here the Variational Bayes (VB) (Ghahramani and Beal 2000; Bishop 2006) as a method to bound the marginal loglikelihood

$$\mathcal{L}_X(\Theta) \equiv \log p(X|\Theta) = \log \int dT dV p(X, T, V|\Theta). \tag{25.15}$$

We first introduce an instrumental distribution $q(T, V)$.

$$\mathcal{L}_X(\Theta) \geq \int dT, dVq \log \frac{p(X, T, V|\Theta)}{q} \tag{25.16}$$

$$= \mathsf{E}(\log p(X, V, T|\Theta))_q + H[q] \equiv \mathcal{B}_{VB}[q]. \tag{25.17}$$

Here, $H[q]$ denotes the entropy of q. From the general theory of EM we know that the bound is tight for the exact posterior $q(T, V) = p(T, V|X, \Theta)$. The VB idea is to assume a simpler form for the instrumental distribution by ignoring some of the couplings present in the exact posterior. A natural candidate is a factorized distribution

$$q(T, V) = q(T)q(V) \equiv \prod_{\alpha \in C} q_\alpha.$$

In the last equation, we have formally written the q distribution as a product over variables from disjoint clusters $\alpha \in C$ and $C = \{\{T\}, \{V\}\}$ denotes the set of disjoint clusters. Since in general the family of q distributions won't include the exact posterior density, we are no longer guaranteed to attain the exact marginal likelihood $\mathcal{L}_X(\Theta)$. Yet, the bound property is preserved and the strategy of VB is to optimize the bound. Although the best q distribution respecting the factorization is not available in closed form, it turns out that a local optimum can be attained by the following fixed point iteration:

$$q_\alpha^{(n+1)} \propto \exp\left(\mathsf{E}(\log p(X, T, V|\Theta))_{q_{\neg\alpha}^{(n)}}\right) \tag{25.18}$$

where $q_{\neg\alpha} = q/q_\alpha$. This iteration monotonically improves the individual factors of the q distribution, i.e. $\mathcal{B}[q^{(n)}] \leq \mathcal{B}[q^{(n+1)}]$ for $n = 1, 2, \ldots$ given an initialisation $q^{(0)}$. The order is not important for convergence – one could visit blocks in arbitrary order. However, in general, the attained fixed point depends upon the order of the updates as well as the starting point $q^{(0)}(\cdot)$. This approach is computationally rather attractive and is very easy to implement (Cemgil 2008).

B. Variational Bayesian NMF

In this section we derive a variational Bayes algorithm for the NMF model described in equations (25.13) and (25.14). The marginal likelihood is given as

$$\mathcal{L}_X(\Theta) \equiv \log p(X|\Theta) \geq \sum_S \int d(T, V) q \log \frac{p(X, S, T, V|\Theta)}{q} \tag{25.19}$$

$$= \mathsf{E}(\log p(X, S, V, T|\Theta))_q + H[q] \equiv \mathcal{B}_{VB}[q] \tag{25.20}$$

where q is defined as

$$q(S, T, V) = q(S)q(T)q(V)$$

$$= \left(\prod_{\nu,\tau} q(\nu, 1:I, \tau)\right)\left(\prod_{\nu,i} q(t_{\nu,i})\right)\left(\prod_{i,\tau} q(v_{i,\tau})\right) \equiv \prod_{\alpha\in\mathcal{C}} q_\alpha.$$

Here, $\alpha \in \mathcal{C} = \{\{S\}, \{T\}, \{V\}\}$ denotes a set of disjoint clusters. A local optimum can be attained by the following fixed point iteration:

$$q_\alpha^{(n+1)} \propto \exp\left(\mathsf{E}(\log p(X, S, T, V|\Theta))_{q_{\neg\alpha}^{(n)}}\right) \tag{25.21}$$

where $q_{\neg\alpha} = q/q_\alpha$.

The expectations of $\mathsf{E}(\log p(X, S, T, V|\Theta))$ are functions of the sufficient statistics of q. The fixed point iteration for the latent sources S (where $m_{\nu,\tau} = 1$), and excitations V leads to the following

$$q(\nu, 1:I, \tau) = \mathcal{M}(\nu, 1:I, \tau; x_{\nu,\tau}, p_{\nu,1:I,\tau}) \tag{25.22}$$

$$p_{\nu,i,\tau} = \exp(\mathsf{E}(\log t_{\nu,i}) + \mathsf{E}(\log v_{i,\tau})) \Big/ \sum_i \exp(\mathsf{E}(\log t_{\nu,i}) + \mathsf{E}(\log v_{i,\tau})) \tag{25.23}$$

$$q(v_{i,\tau}) = \mathcal{G}a\left(v_{i,\tau}; \alpha_{i,\tau}^v, \beta_{i,\tau}^v\right) \tag{25.24}$$

$$\alpha_{i,\tau}^v = a_{i,\tau}^v + \sum_\nu m_{\nu,\tau}\mathsf{E}(\nu, i, \tau) \quad \beta_{i,\tau}^v = \left(\frac{a_{i,\tau}^v}{b_{i,\tau}^v} + \sum_\nu m_{\nu,\tau}\mathsf{E}(t_{\nu,i})\right)^{-1}. \tag{25.25}$$

The variational parameters of $q(t_{\nu,i}) = \mathcal{G}a\left(t_{\nu,i}; \alpha^t_{\nu,i}, \beta^t_{\nu,i}\right)$ are found similarly. The hyperparameters can be optimized by maximizing the variational bound $\mathcal{B}_{VB}[q]$. While this does not guarantee to increase the true marginal likelihood, it leads in this application to quite practical and fast algorithms and is very easy to implement (Cemgil 2008).

For the same model, it is also straightforward to implement a Gibbs sampler. A comparison showed that both algorithms give qualitatively very similar results, both for inference as well as model order selection (Cemgil 2008). We find the variational approach somewhat more practical as it can be expressed as simple matrix operations, where both the fixed point equations as well as the bound can be compactly and efficiently implemented using matrix computation software. In contrast, our Gibbs sampler is computationally more demanding and the calculation of marginal likelihood is somewhat more tricky. With our implementation of both algorithms the variational method is faster by a factor of around 13.

In terms of computational requirements, the variational procedure has several advantages. First, one circumvents sampling from multinomial variables, which is the main computational bottleneck with a Gibbs sampler in this model. Whilst efficient algorithms are developed for multinomial sampling (Davis 1993), the procedure is time consuming when the number of latent sources I is large. In contrast, the variational method computes the expected sufficient statistics via elementary matrix operations. Another advantage is hyperparameter estimation. In principle, it is possible to maximize the marginal likelihood via a Monte Carlo EM procedure (Tanner 1996; Quintana, Liu, and del Pino 1999), yet this potentially requires many more iterations of the Gibbs sampler. In contrast, the evaluation of the derivatives of the lower bound is straightforward and can be implemented without much additional computational cost.

Acknowledgements

We would like to thank Andrew Feldhaus for carefully proofreading the manuscript.

References

Abdallah, S. A. and Plumbley, M. D. (2006). Unsupervised analysis of polyphonic music using sparse coding. *IEEE Transactions on Neural Networks*, **17**, 179–196.

Attias, H. (1999). Independent factor analysis. *Neural Computation*, **11**, 803–851.

Bertin, N., Badeau, R. and Richard, G. (2007). Blind signal decompositions for automatic transcription of polyphonic music: NMF and K-SVD on the benchmark. In Proceedings of the International Conference on Audio, Speech and Signal Processing (ICASSP), Honolulu.

Bishop, C. M. (2006). *Pattern Recognition and Machine Learning*. Springer, New York.

Cemgil, A. T. (2004). Bayesian music transcription. Ph.D. thesis, Radboud University of Nijmegen.

Cemgil, A. T. (2007). Strategies for sequential inference in factorial switching state space models. In Proceedings of the IEEE International Conference on Acoustics, Speech and Signal Processing (ICASSP 07), Honolulu, Hawaii, pp. 513–516.

Cemgil, A. T. (2008). Bayesian inference in non-negative matrix factorisation models. Technical Report CUED/F-INFENG/TR.609, University of Cambridge.

Cemgil, A. T. and Dikmen, O. (2007). Conjugate Gamma Markov random fields for modelling nonstationary sources. In ICA 2007, 7th International Conference on Independent Component Analysis and Signal Separation, London, UK.

Cemgil, A. T., Fevotte, C. and Godsill, S. J. (2007). Variational and stochastic inference for Bayesian source separation. *Digital Signal Processing*, **17**, 891–913. Special Issue on Bayesian source separation.

Cemgil, A. T., Kappen, H. J. and Barber, D. (2006). A generative model for music transcription. *IEEE Transactions on Audio, Speech and Language Processing*, **14**, 679–694.

Cemgil, A. T., Peeling, P., Dikmen, O. and Godsill, S. J. (2007). Prior structures for time-frequency energy distributions. In Proceedings of IEEE Workshop on Applications of Signal Processing to Audio and Acoustics, New Paltz, NY.

Chib, S. (1995). Marginal likelihood from the gibbs output. *Journal of the Acoustical Society of America*, **90**, 1313–1321.

Cover, T. M. and Thomas, J. A. (1991). *Elements of Information Theory*. John Wiley, New York.

Davis, C. S. (1993). The computer generation of multinomial random variates. *Computational Statistics and Data Analysis*, **16**, 205–217.

Davy, M., Godsill, S. and Idier, J. (2006). Bayesian analysis of polyphonic Western tonal music. *Journal of the Acoustical Society of America*, **119**, 2498–2517.

Davy, M. and Godsill, S. J. (2002). Detection of abrupt spectral changes using support vector machines. An application to audio signal segmentation. In Proceedings IEEE International Conference on Acoustics, Speech and Signal Processing, Orlando, FL.

Févotte, C., Daudet, L., Godsill, S. J. and Torrésani, B. (2006). Sparse regression with structured priors: Application to audio denoising. In Proceedings ICASSP, Toulouse, France.

Févotte, C. and Godsill, S. (2006). A Bayesian approach for blind separation of sparse sources. *IEEE Transactions on Audio, Speech and Language Processing*, **14**, 2174–2188.

Fletcher, N. H. and Rossing, T. (1998). *The Physics of Musical Instruments*. Springer, Berlin.

Ghahramani, Z. and Beal, M. (2001). Propagation algorithms for variational Bayesian learning. In *Neural Information Processing Systems*, (ed. T. Leen, T. Dietterich and V. Tresp, V.), Vol. **13**, 507–513. The MIT Press, Cambridge, Massachusetts.

Gilks, W. R., Richardson, S. and Spiegelhalter, D. J. (eds.) (1996). *Markov Chain Monte Carlo in Practice*. CRC Press, London.

Girolami, M. (2001). A variational method for learning sparse and overcomplete representations. *Neural Computation*, **13**, 2517–2532.

Godsill, S. (2004). Computational modeling of musical signals. *Chance Magazine*, **17**, 23–29.

Godsill, S. and Davy, M. (2005). Bayesian computational models for inharmonicity in musical instruments. In Proceedings of IEEE Workshop on Applications of Signal Processing to Audio and Acoustics, New Paltz, NY.

Godsill, S. J. and Davy, M. (2002). Bayesian harmonic models for musical pitch estimation and analysis. In Proceedings IEEE International Conference on Acoustics, Speech and Signal Processing. Orlando, FL.

Godsill, S. J. and Rayner, P. J. W. (1998). *Digital Audio Restoration: A Statistical Model-Based Approach*. Springer, Berlin.

Green, P. J. (1995). Reversible jump Markov-chain Monte Carlo computation and Bayesian model determination. *Biometrika*, **82**, 711–732.

Grimmett, G. and Stirzaker, D. (2001). *Probability and Random Processes*, (3rd edn). Oxford University Press, Oxford.

Herrera-Boyer, P., Klapuri, A. and Davy, M. (2006). Automatic classification of pitched musical instrument sounds. See Klapuri, and Davy (2006), pp. 163–200.

Hojen-Sorensen, P., Winther, O. and Hansen, L. K. (2002). Mean-field approaches to independent component analysis. *Neural Computation*, **14**, 889–918.

Hyvärinen, A., Karhunen, J. and Oja, E. (2001). *Independent Component Analysis*. John Wiley, New York.

Kameoka, H. (2007). Statistical approach to multipitch analysis. Ph.D. thesis, University of Tokyo.

Klapuri, A. and Davy, M. (Eds.) (2006). *Signal Processing Methods for Music Transcription*. Springer, New York.

Knuth, K. H. (1998). Bayesian source separation and localization. In SPIE'98: Bayesian Inference for Inverse Problems, San Diego, pp. 147–158.

Lee, D. D. and Seung, H. S. (2000). Algorithms for non-negative matrix factorization. In *Advances in Neural Information Processing Systems* (NIPS), Volume 12, pp. 556–562. MIT Press, Cambridge, MA.

McIntyre, M. E., Schumacher, R. T. and Woodhouse, J. (1983). On the oscillations of musical instruments. *Journal of the Acoustical Society of America*, **74**, 1325–1345.

Miskin, J. and Mackay, D. (2001). Ensemble learning for blind source separation. In *Independent Component Analysis*, (ed. S. J. Roberts and R. M. Everson), pp. 209–233. Cambridge University Press, Cambridge.

Mohammad-Djafari, A. (1997). A Bayesian estimation method for detection, localisation and estimation of superposed sources in remote sensing. In SPIE'97, San Diego.

Moore, B. (1997). *An Introduction to the Psychology of Hearing*, (4th edn). Academic Press, New York.

Peeling, P. H., Li, C. and Godsill, S. J. (2007). Poisson point process modeling for polyphonic music transcription. *Journal of the Acoustical Society of America Express Letters*, **121**, EL168–EL175.

Quintana, F. A., Liu, J. S. and del Pino, G. E. (1999). Monte Carlo EM with importance reweighting and its applications in random effects models. *Computational Statistics and Data Analysis*, **29**, 429–444.

Reyes-Gomez, M., Jojic, N. and Ellis, D. (2005). Deformable spectrograms. In AI and Statistics Conference, Barbados.

Rowe, D. B. (2003). *Multivariate Bayesian Statistics: Models for Source Separation and Signal Unmixing*. Chapman & Hall/CRC, Boca Raton, Florida.

Shepard, N. (ed.) (2005). *Stochastic Volatility, Selected Readings*. Oxford University Press, Oxford.

Smaragdis, P. and Brown, J. (2003). Non-negative matrix factorization for polyphonic music transcription. In IEEE Workshop on Applications of Signal Processing to Audio and Acoustics (WASPAA). New Paltz, NY. IEEE.

Tanner, M. A. (1996). *Tools for Statistical Inference: Methods for the Exploration of Posterior Distributions and Likelihood Functions* (3rd edn). Springer, New York.

Valpola, H. (2000). Nonlinear independent component analysis using ensemble learning: Theory. In Proceedings of the Second International Workshop on Independent Component Analysis and Blind Signal Separation, ICA 2000, Helsinki, Finland, pp. 251–256.

Vincent, E., Bertin, N. and Badeau, R. (2008). Harmonic and inharmonic nonnegative matrix factorization for polyphonic pitch transcription. In Proceedings IEEE

International Conference on Acoustics, Speech and Signal Processing (ICASSP), Las Vegas. IEEE.

Virtanen, T. (2006*a*). Sound source separation in monaural music signals. Ph.D. thesis, Tampere University of Technology.

Virtanen, T. (2006*b*). Unsupervised learning methods for source separation in monaural music signals. See Klapuri and Davy (2006), pp. 267–296.

Wainwright, M. and Jordan, M. I. (2003). Graphical models, exponential families, and variational inference. Technical Report 649, Department of Statistics, UC Berkeley.

Walmsley, P., Godsill, S. J. and Rayner, P. J. W. (1998). Multidimensional optimisation of harmonic signals. In Proceedings European Conference on Signal Processing. Rhodes, Greece.

Walmsley, P. J., Godsill, S. J. and Rayner, P. J. W. (1999). Polyphonic pitch tracking using joint Bayesian estimation of multiple frame parameters. In Proceedings IEEE Workshop on Audio and Acoustics, Mohonk, NY State, Mohonk, NY State.

Wang, D. and Brown, G. J. (eds.) (2006). *Computational Auditory Scene Analysis: Principles, Algorithms, Applications.* John Wiley, New York.

Winther, O. and Petersen, K. B. (2006). Flexible and efficient implementations of Bayesian Independent Component Analysis. *Neuro Computing*, **71**, 221–233.

Wolfe, P. J., Godsill, S. J. and Ng, W. (2004). Bayesian variable selection and regularisation for time-frequency surface estimation (with discussion). *Journal of the Royal Statistical Society, B*, **66**, 575–589.

·26·

Combining simulations and physical observations to estimate cosmological parameters

Dave Higdon, Katrin Heitmann, Charles Nakhleh and Salman Habib

26.1 Introduction

Over the last three decades observational cosmology has made extraordinary progress in determining the make-up of the universe and its evolution. Precision measurements from all-sky COsmic Background Explorer (COBE) (Smoot *et al.*, 1992) and, more recently, the Wilkinson Microwave Anisotropy Probe (WMAP) (Spergel *et al.*, 2007) have given very detailed measurements of the cosmic microwave background. In addition, large galaxy surveys such as the Sloan Digital Sky Survey (SDSS) (Adelman-McCarthy *et al.*, 2006) advance our understanding of structure formation and yield complementary information to the CMB to determine the make-up of the universe.

The Λ-cold dark matter (ΛCDM) model is the simplest cosmological model in agreement with the CMB and large scale structure measurements. This model is determined by a small number of parameters which control the composition, expansion and fluctuations of the universe. The precision measurements from different cosmological probes reveal a highly unexpected result: roughly 70% of the universe is made up of a mysterious dark energy that accelerates the recent expansion of the universe.

In this chapter we combine computationally intensive simulation results with measurements from the SDSS to infer about a subset of the parameters that control the ΛCDM model. We also describe a statistical framework adapted from Kennedy and O'Hagan (2001) and Higdon *et al.* (2008) to determine a posterior distribution for these cosmological parameters given the simulation output and the physical observations. We then go on to demonstrate how this formulation can be extended to combine information from simulations and observations corresponding to both the large scale structure of the universe and the CMB.

26.1.1 Large scale structure of the universe

The SDSS maps out the spatial location of galaxies around our own Milky Way Galaxy (Figure 26.1, right). Note that this spatial distribution of galaxies exhibits a combination of voids and high density 'filaments' of matter. This peculiar spatial distribution is due to the cumulative effects of gravity (and other forces) acting on slight matter density fluctuations present shortly after the big bang, as evidenced by the CMB.

Predicting the spatial distribution of matter at our current time in the universe, given the parameters of the ΛCDM model, requires a substantial computing effort. For a given parameter setting, a very large-scale N-body simulation is carried out that evolves tracer dark matter particles according to gravity and other forces from an initial setting based on the CMB to our current time in the universe. The result of one such simulation is shown in the left frame of Figure 26.1. Under different cosmologies (i.e. cosmological parameter settings) the spatial structure of the resulting simulation output differs. We would like to determine which cosmologies are consistent with physical observations of our universe – such as those from the SDSS shown in the right hand frame of Figure 26.1.

Direct comparison between the simulation output and the SDSS data is not possible. The simulations evolve an idealized cube of dark matter particles, while the SDSS data give a censored, local snapshot of the large scale structure of the universe. Instead, we summarize the simulation output and physical observations by their dark matter power spectra which describe the spatial distribution of matter density at a wide range of length scales. Note that the wave number k on the x-axis of these spectra is given in h Mpc^{-1}. Mpc is a

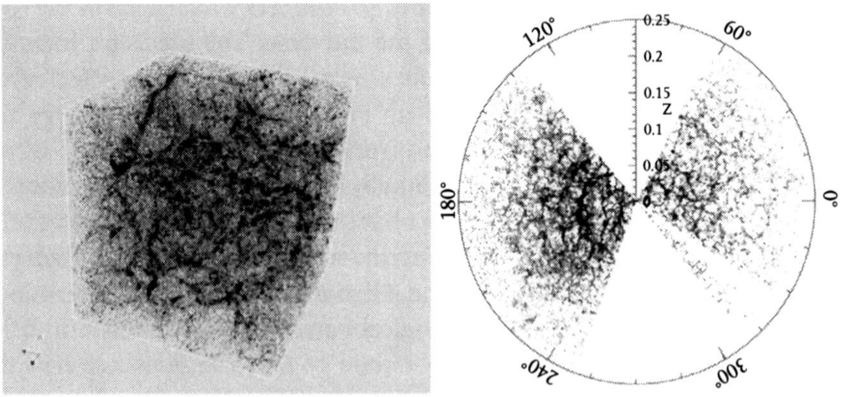

Fig. 26.1 Left: Simulation results from a large scale N-body simulation that evolves particles from an early time in the universe to now. The large scale structure in the output depends on the cosmological parameters θ^* under which the simulation was carried out. The goal of this analysis is to determine which cosmologies are consistent with observations, such as the right hand figure from the Sloan Digital Sky Survey (Credit: Sloan Digital Sky Survey).

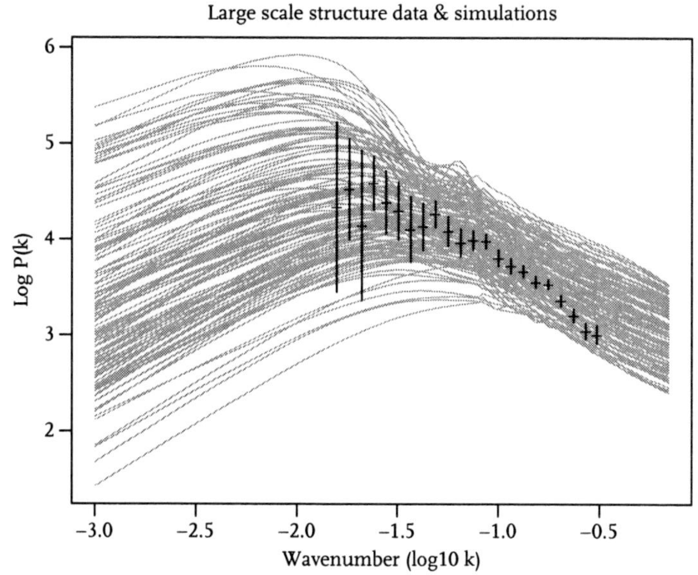

Fig. 26.2 Twenty-two data points for the dark matter power spectrum derived by Tegmark *et al.* (2004) from the Sloan Digital Sky Survey. The error bars denote two standard deviations. The grey lines show the 128 simulated power spectra, discussed in Section 26.2.1.

length scale; two galaxies are separated by about 1 Mpc on average. The grey lines in Figure 26.2 show a number of matter power spectra produced by carrying out simulations using different cosmological parameter settings.

Computing the matter power spectrum is trivial for the simulation output since it is defined on a periodic, cubic lattice. In contrast, determining matter power spectrum from the SDSS data is a far more challenging task since one must account for the many difficulties that accompany observational data: non-standard survey geometry, redshift space distortions, luminosity bias and noise, just to name a few. Because of these challenges, we use the published data and likelihood of Tegmark *et al.* (2004) which is summarized in Figure 26.2. The resulting data correspond to 22 pairs (y_i, k_i). The data vector $y = (y_1, \ldots, y_{22})'$ has a diagonal covariance Σ_y; the two standard deviation bars are shown in Figure 26.2 for each data point.

For the N-body simulations, we consider five ΛCDM parameters

$$\theta = (n, h, \sigma_8, \Omega_{\mathrm{CDM}}, \Omega_{\mathrm{B}}),$$

where

n	is the spectral index,
h	is the Hubble constant,
σ_8	is the galaxy fluctuation amplitude,
Ω_{CDM}	is the density of dark matter, and
Ω_{B}	is the density of baryonic matter.

Since we assume a flat universe and a constant dark energy equation of state, we expect that any variation in the remaining ΛCDM parameters will not affect the resulting matter power spectra.

The framework we use for carrying out this analysis is given in Higdon *et al.* (2008), but gentler introductions to the topic of combining simulations and field data can be found in Kennedy and O'Hagan (2001) or Higdon *et al.* (2004). The framework integrates the following concepts:

- simulation design – the determination of the parameter settings at which to carry out the simulations;
- emulation – given simulation output at a fixed set of input parameter settings, how to estimate the output at new, untried settings;
- uncertainty and sensitivity analysis – determining the variations in simulation output due to uncertainty or changes in the input parameters;
- calibration – combining observations (with known errors) and simulations to estimate parameter values consistent with the observations;
- prediction – given parameter uncertainties and uncertainties in other nuisance parameters, predict system behaviour with uncertainty.

In this paper we discuss the framework methodology in detail through a specific application: Estimation of five parameters from dark matter structure formation simulations and SDSS measurements of the matter power spectrum (Figure 26.2) in Section 26.2. We extend this framework in Section 26.3 to include measurements of the CMB temperature power spectrum derived from the WMAP data release. We perform a combined large scale structure and CMB analysis and demonstrate how to extend our framework to include data from diverse data sources.

26.2 The statistical framework

In this section we describe the statistical methodology to combine physical observations with output from a simulation model to infer about unknown model parameters. We use observations y from the matter power spectrum (Figure 26.2) and matter power spectra derived from physical simulations.

Generally, the simulation model requires p_θ-vector θ^* of input parameter settings in order to produce a matter power spectrum $\eta(\theta^*)$. The simplest possible model one might postulate is that the vector of observations y is a noisy version of the simulated spectrum $\eta(\theta)$ at the true setting θ

$$y = \eta(\theta) + \epsilon,$$

where the observation error vector is normal, with mean 0 and variance Σ_y. Given a prior distribution $\pi(\theta)$ for the true parameter vector θ, the resulting posterior distribution $\pi(\theta|y)$ for θ is given by

$$\pi(\theta|y) \propto L(y|\eta(\theta)) \cdot \pi(\theta),$$

where $L(y|\eta(\theta))$ comes from the normal sampling model for the data

$$L(y|\eta(\theta)) = \exp\left\{\frac{1}{2}(y - \eta(\theta))'\Sigma_y^{-1}(y - \eta(\theta))\right\}.$$

In principle, this posterior distribution could be explored via MCMC. However, if a single evaluation of $\eta(\theta)$ requires hours (or days) of computation, a direct MCMC-based approach is infeasible.

Our approach deals with this computational bottleneck by treating $\eta(\cdot)$ as an unknown function to be estimated from a fixed collection of simulations $\eta(\theta_1^*), \ldots, \eta(\theta_m^*)$ carried out at input settings $\theta_1^*, \ldots, \theta_m^*$. This approach requires a prior distribution for the unknown function $\eta(\cdot)$, and treats the simulation output $\eta^* = (\eta(\theta_1^*), \ldots, \eta(\theta_m^*))'$ as additional data to be conditioned on for the analysis. Hence there is an additional component of the likelihood obtained from the sampling model for η^* by $L(\eta^*|\eta(\cdot))$.

For this case, the resulting posterior distribution has the general form

$$\pi(\theta, \eta(\cdot)|y, \eta^*) \propto L(y|\eta(\theta)) \cdot L(\eta^*|\eta(\cdot)) \cdot \pi(\eta(\cdot)) \cdot \pi(\theta), \qquad (26.1)$$

which has traded direct evaluations of the simulator model for a more complicated form which depends strongly on the prior model for the function $\eta(\cdot)$. Note that the marginal distribution for the cosmological parameters θ will be affected by uncertainty regarding $\eta(\cdot)$.

In the following subsections, we describe in detail a particular formulation of equation (26.1) in the context of this large scale structure application. This formulation has proven fruitful in a variety of physics and engineering applications which combine field observations with detailed simulation models for inference. In particular we cover approaches for choosing the m parameter settings at which to run the simulation model, and a (prior) model – or emulator – which describes how $\eta(\cdot)$ is modeled at untried parameter settings. Section 26.2.3 describes how the observed data is combined with the simulations and the emulator to give the posterior distribution. In the following section we will demonstrate how this formulation can be extended to combine information from different data sets from galaxy surveys and cosmic microwave background measurements.

26.2.1 The simulation design

The dark matter simulations are quite demanding since they must compute the force interactions for over two million interacting particles. Simulation accuracy is particularly important for the smaller length scales ($k \geq 0.2h$ Mpc^{-1}), where the gravitational effects become strongly nonlinear. For this application, we use $m = 128$ simulations. Future investigations which will make use of next generation of large scale structure surveys will require far more resolution in the simulations. With such large numbers of particles, we expect to be able to carry out no more than 100 simulation runs.

Generally, the design specifies m input settings which vary over predefined ranges for each of the p_θ input parameters:

$$\begin{pmatrix} \theta_1^* \\ \vdots \\ \theta_m^* \end{pmatrix} = \begin{pmatrix} \theta_{11}^* & \cdots & \theta_{1p_\theta}^* \\ \vdots & \vdots & \vdots \\ \theta_{m1}^* & \cdots & \theta_{mp_\theta}^* \end{pmatrix}. \tag{26.2}$$

We use θ^* to differentiate the design input settings from the true value of the parameter vector θ which is what we are trying to estimate.

For our purposes, we would like a design that leads to an accurate, Gaussian process (GP)-based emulator which is described in Section 26.2.2. Space-filling Latin hypercube (LH) designs have proven to be well suited for building GP models to estimate simulator output at untried settings (Sacks *et al.*, 1989; Currin *et al.*, 1991). In particular, we have used orthogonal array-based LH designs (Tang, 1993) as well as symmetric LH designs (Ye *et al.*, 2000). Figure 26.3 shows the $m = 128$ point design over $p_\theta = 5$ dimensions used in this analysis. This design was constructed by perturbing a five-level orthogonal array design so that each one-dimensional projection gives an equally spaced set of points along the standardized parameter range [0,1]. See Santner *et al.* (2003), Chapters 5 and 6, for a recent survey of the area.

The actual parameter ranges used for the $m = 128$ simulations are

$$\begin{aligned} 0.8 \le &\ n \le 1.4, \\ 0.5 \le &\ h \le 1.1, \\ 0.6 \le &\ \sigma_8 \le 1.6, \\ 0.0 \le &\ \Omega_{\text{CDM}} \le 0.6, \\ 0.02 \le &\ \Omega_{\text{B}} \le 0.12. \end{aligned} \tag{26.3}$$

These ranges are standardized to $[0, 1]^5$ by shifting and scaling each interval.

26.2.2 Emulating simulator output

For a given input θ, the simulator produces a matter power spectrum, as shown in Figure 26.2. Each spectrum is a n_η-vector of values. A key component of our analysis is a probability model to describe this functional simulator output at untried settings θ. To do this, we use the output of the simulation runs to construct a GP model that 'emulates' the simulator at arbitrary input settings over the (standardized) input space $[0, 1]^{p_\theta}$. The emulator models the simulation

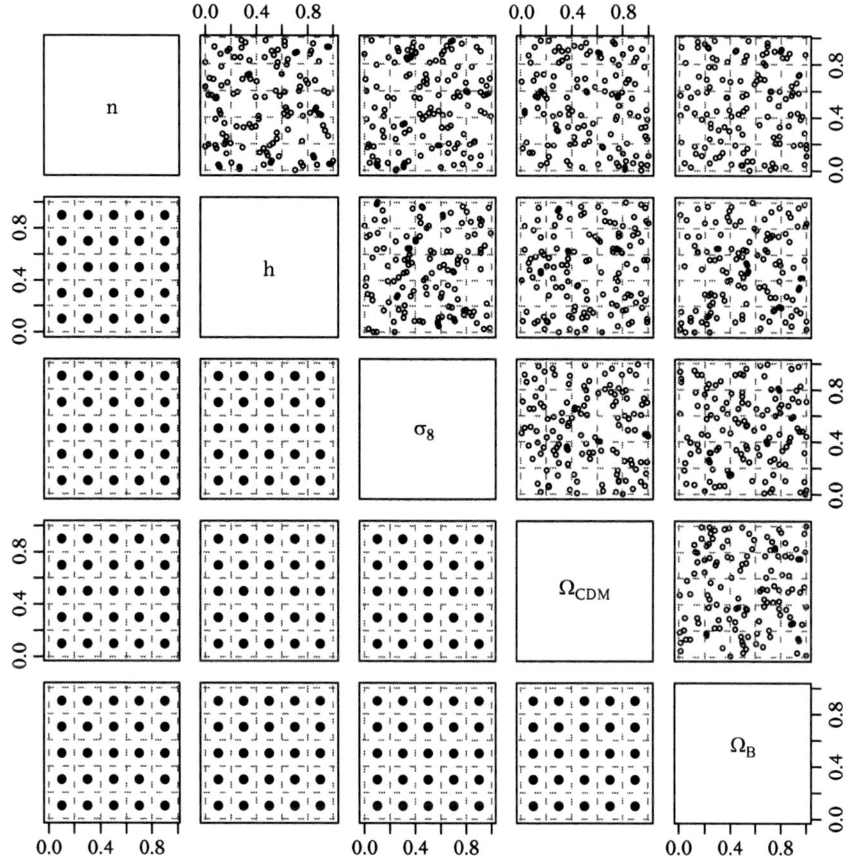

Fig. 26.3 Lower triangle of plots: two-dimensional projections of a $m = 128$ point, five-level, OA design. Upper triangle of plots: an OA-based LH design obtained by spreading out the five-level OA design so that each one-dimensional projection gives an equally spaced set of points along [0,1].

output using a p_η-dimensional basis representation:

$$\eta(\theta) = \sum_{i=1}^{p_\eta} \phi_i w_i(\theta) + \epsilon, \;\; \theta \in [0, 1]^{p_\theta}, \tag{26.4}$$

where $\{\phi_1, \ldots, \phi_{p_\eta}\}$ is a collection of orthogonal, n_η-dimensional basis vectors, the $w_i(\theta)$s are GPs over the input space, and ϵ is an n_η-dimensional error term. This type of formulation reduces the problem of building an emulator that maps $[0, 1]^{p_\theta}$ to R^{n_η} to building p_η independent, univariate GP models for each $w_i(\theta)$. The details of this model specification are given below.

Output from each of the m simulation runs prescribed by the design results in n_η-dimensional vectors which we denote by $\eta_1, \ldots, \eta_m$. Since the simulations give complete output, the simulation output can be efficiently represented via principal components (Ramsay and Silverman, 1997). We first centre the simulations about 0 by subtracting the mean $(\frac{1}{m}\sum_{j=1}^{m}\eta_j)$ from each output vector. We note that, depending on the application, some alternative standardization may be preferred. Whatever the choice of the standardization, the same standardization is also applied to the experimental data.

We define Ξ to be the $n_\eta \times m$ matrix obtained by column-binding the (standardized) output vectors from the simulations

$$\Xi = [\eta_1; \cdots; \eta_m].$$

The size of a given simulation output n_η is much larger than the number of simulations carried out m. We apply the singular value decomposition (SVD) to the simulation output matrix Ξ giving

$$\Xi = UDV',$$

where U is a $n_\eta \times m$ orthogonal matrix, D is a diagonal $m \times m$ matrix holding the singular values, and V is a $m \times m$ orthonormal matrix. To construct a p_η-dimensional representation of the simulation output, we define the principal component (PC) basis matrix Φ_η to be the first p_η columns of $[\frac{1}{\sqrt{m}}UD]$. The resulting principal component loadings or weights is then given by $[\sqrt{m}V]$.

For the matter power spectrum application we take $p_\eta = 5$ so that $\Phi_\eta = [\phi_1; \phi_2; \phi_3; \phi_4; \phi_5]$; the basis functions ϕ_1, ϕ_2, ϕ_3, ϕ_4 and ϕ_5 are shown in Figure 26.4. Note that the ϕ_is are functions of log wave number.

We use the basis representation of equation (26.4) to model the n_η-dimensional simulator output over the input space. Each basis weight $w_i(\theta)$, $i = 1, \ldots, p_\eta$, is then modeled as a mean 0 GP

$$w_i(\theta) \sim \mathrm{GP}\left(0, \lambda_{wi}^{-1} R(\theta, \theta'; \rho_{wi})\right),$$

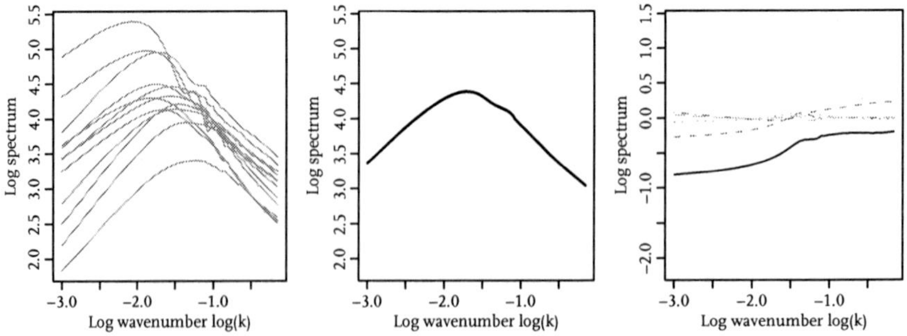

Fig. 26.4 Simulations (left), mean (centre), and the first five principal component bases (right) derived from the simulation output.

where λ_{wi} is the marginal precision of the process and the correlation function is given by

$$R(\theta, \theta'; \rho_{wi}) = \prod_{k=1}^{p_\theta} \rho_{wik}^{4(\theta_k - \theta'_k)^2}. \tag{26.5}$$

This is the Gaussian covariance function, which gives very smooth realizations, and has been used previously by Kennedy and O'Hagan (2001) and Sacks *et al.* (1989) to model computer simulation output. An advantage of this product form is that only a single additional parameter is required per additional input dimension, while the fitted GP response still allows for rather general interactions between inputs. We use this Gaussian form for the covariance function because the simulators we work with tend to respond very smoothly to changes in the inputs. The parameter ρ_{wik} controls the spatial range for the kth input dimension of the process $w_i(\theta)$. Under this parametrization, ρ_{wik} gives the correlation between $w_i(\theta)$ and $w_i(\theta')$ when the input conditions θ and θ' are identical, except for a difference of 0.5 in the kth component. Note that this interpretation makes use of the standardization of the input space to $[0, 1]^{p_\theta}$.

Restricting to the m input design settings given in (26.2), we define the m-vector w_i to be

$$w_i = \left(w_i\left(\theta_1^*\right), \ldots, w_i\left(\theta_m^*\right)\right)', \quad i = 1, \ldots, p_\eta.$$

In addition we define $R(\theta^*; \rho_{wi})$ to be the $m \times m$ correlation matrix resulting from applying equation (26.5) to each pair of input settings in the design. The p_θ-vector ρ_{wi} gives the correlation distances for each of the input dimensions.

At the m simulation input settings, the mp_η-vector $w = (w_1', \ldots, w_{p_\eta}')'$ then has prior distribution

$$\begin{pmatrix} w_1 \\ \vdots \\ w_{p_\eta} \end{pmatrix} \sim N\left(\begin{pmatrix} 0 \\ \vdots \\ 0 \end{pmatrix}, \begin{pmatrix} \lambda_{w1}^{-1} R(\theta^*; \rho_{w1}) & 0 & 0 \\ 0 & \ddots & 0 \\ 0 & 0 & \lambda_{wp_\eta}^{-1} R(\theta^*; \rho_{wp_\eta}) \end{pmatrix}\right), \tag{26.6}$$

which is controlled by p_η precision parameters held in λ_w and $p_\eta \cdot p_\theta$ spatial correlation parameters held in ρ_w. The prior above can be written more compactly as

$$w \sim N(0, \Sigma_w),$$

where Σ_w, controlled by parameter vectors λ_w and ρ_w, is given in equation (26.6).

We specify independent $Ga(a_w, b_w)$ priors for each λ_{wi} and independent $Be(a_{\rho_w}, b_{\rho_w})$ priors for the ρ_{wik}s.

$$\pi(\lambda_{wi}) \propto \lambda_{wi}^{a_w-1} e^{-b_w \lambda_{wi}}, \quad i = 1, \ldots, p_\eta,$$

$$\pi(\rho_{wik}) \propto \rho_{wik}^{a_{\rho w}-1}(1-\rho_{wik})^{b_{\rho w}-1}, \quad i = 1, \ldots, p_\eta, \ k = 1, \ldots, p_\theta.$$

We expect the marginal variance for each $w_i(\cdot)$ process to be close to one due to the scaling of the basis functions. For this reason we specify that $a_w = b_w = 5$. In addition, this informative prior helps stabilize the resulting posterior distribution for the correlation parameters which can trade off with the marginal precision parameter.

Because we expect only a subset of the inputs to influence the simulator response, our prior for the correlation parameters reflects this expectation of 'effect sparcity.' Under the parametrization in equation (26.5), input k is inactive for PC i if $\rho_{wik} = 1$. Choosing $a_{\rho_w} = 1$ and $0 < b_{\rho_w} < 1$ will give a density with substantial prior mass near 1. We take $b_{\rho_w} = 0.1$, which makes $\Pr(\rho_{wik} < 0.98) \approx \frac{1}{3}$ a priori. In general, the selection of these hyperparameters should depend on how many of the p_θ inputs are expected to be active. Alternatively, the prior could be specified to have some point mass at 1 as in Linkletter *et al.* (2006).

If we take the error vector in the basis representation of equation (26.4) to be i.i.d. normal, we can then develop the sampling model, or likelihood, for the simulator output. We define the $n_\eta m$-vector η to be the concatenation of all m simulation output vectors

$$\eta = \text{vec}(\Xi) = \text{vec}\left(\left[\eta\left(\theta_1^*\right); \cdots ; \eta\left(\theta_m^*\right)\right]\right),$$

where vec(Ξ) produces a vector by stacking the columns of matrix Ξ. Given precision λ_η of the errors the likelihood is then

$$L(\eta|w, \lambda_\eta) \propto \lambda_\eta^{\frac{mn_\eta}{2}} \exp\left\{-\frac{1}{2}\lambda_\eta(\eta - \Phi w)'(\eta - \Phi w)\right\},$$

where the $n_\eta \times mp_\eta$ matrix Φ is given by

$$\Phi = [I_m \otimes \phi_1; \cdots ; I_m \otimes \phi_{p_\eta}],$$

and the ϕ_is are the p_η basis vectors previously computed via SVD. A $Ga(a_\eta, b_\eta)$ prior is specified for the error precision λ_η.

Since the likelihood factors as shown below

$$L(\eta|w, \lambda_\eta) \propto \lambda_\eta^{\frac{mp_\eta}{2}} \exp\left\{-\frac{1}{2}\lambda_\eta(w - \hat{w})'(\Phi'\Phi)(w - \hat{w})\right\}$$

$$\times \lambda_\eta^{\frac{m(n_\eta - p_\eta)}{2}} \exp\left\{-\frac{1}{2}\lambda_\eta \eta'(I - \Phi(\Phi'\Phi)^{-1}\Phi')\eta\right\},$$

the formulation can be equivalently represented with a dimension reduced likelihood and a modified $Ga(a_\eta^*, b_\eta^*)$ prior for λ_η:

$$L(\hat{w}|w, \lambda_\eta) \propto \lambda_\eta^{\frac{mp_\eta}{2}} \exp\left\{-\frac{1}{2}\lambda_\eta(\hat{w} - w)'(\Phi'\Phi)(\hat{w} - w)\right\},$$

where

$$a_\eta^* = a_\eta + \frac{m(n_\eta - p_\eta)}{2},$$
$$b_\eta^* = b_\eta + \frac{1}{2}\eta'(I - \Phi(\Phi'\Phi)^{-1}\Phi')\eta, \quad \text{and}$$
$$\hat{w} = (\Phi'\Phi)^{-1}\Phi'\eta. \tag{26.7}$$

Thus the Normal-Gamma model

$$\eta|w, \lambda_\eta \sim N\left(\Phi w, \lambda_\eta^{-1} I_{n_\eta}\right), \ \lambda_\eta \sim Ga(a_\eta, b_\eta)$$

is equivalent to the reduced form

$$\hat{w}|w, \lambda_\eta \sim N(w, (\lambda_\eta\Phi'\Phi)^{-1}), \ \lambda_\eta \sim Ga\left(a_\eta^*, b_\eta^*\right)$$

since

$$L(\eta|w, \lambda_\eta) \times \pi(\lambda_\eta; a_\eta, b_\eta) \propto L(\hat{w}|w, \lambda_\eta) \times \pi\left(\lambda_\eta; a_\eta^*, b_\eta^*\right). \tag{26.8}$$

The likelihood depends on the simulations only through the computed PC weights $\hat{w}$. After integrating out w, the posterior distribution becomes

$$\pi(\lambda_\eta, \lambda_w, \rho_w|\hat{w}) \propto \tag{26.9}$$
$$\left|(\lambda_\eta\Phi'\Phi)^{-1} + \Sigma_w\right|^{-\frac{1}{2}} \exp\left\{-\frac{1}{2}\hat{w}'([\lambda_\eta\Phi'\Phi]^{-1} + \Sigma_w)^{-1}\hat{w}\right\}$$
$$\times \lambda_\eta^{a_\eta^*-1} e^{-b_\eta^*\lambda_\eta} \times \prod_{i=1}^{p_\eta} \lambda_{wi}^{a_w-1} e^{-b_w\lambda_{wi}} \times \prod_{i=1}^{p_\eta}\prod_{j=1}^{p_\theta}(1 - \rho_{wij})^{b_\rho-1}.$$

This posterior distribution is a milepost on the way to the complete formulation, which also incorporates experimental data. However, it is worth considering this intermediate posterior distribution for the simulator response. It can be explored via MCMC using standard Metropolis updates and we can view a number of posterior quantities to illuminate features of the simulator. In Oakley and O'Hagan (2004) the posterior of the simulator response is used to investigate formal sensitivity measures of a univariate simulator; in Sacks *et al.* (1989) this is done from a non-Bayesian perspective. For example, Figure 26.5 shows boxplots of the posterior distributions for the components of ρ_w. From this figure it is apparent that PCs 1 and 2 are most influenced by σ_8 and Ω_{CDM}.

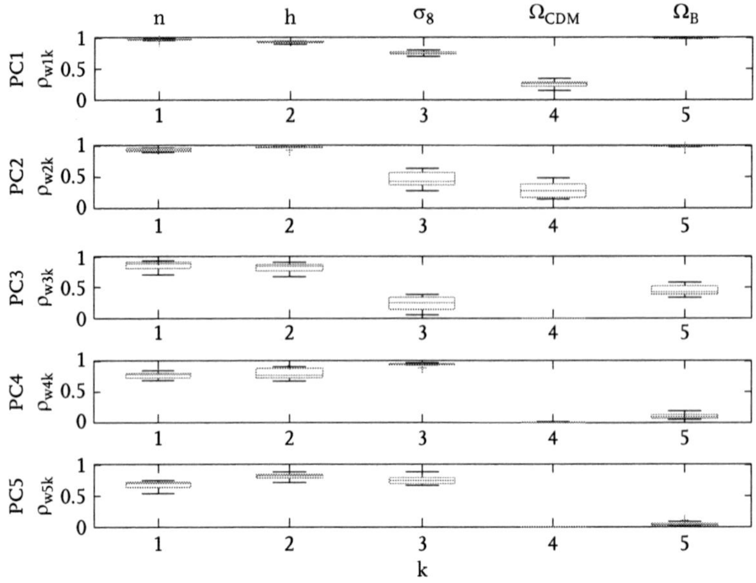

Fig. 26.5 Boxplots of posterior samples for each ρ_{wik} for the large scale structure application.

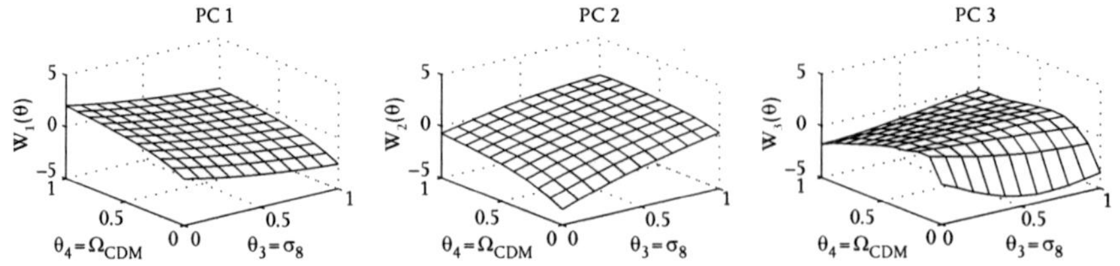

Fig. 26.6 Posterior mean surfaces for $w_i(\theta)$, $i = 1, 2, 3$. Here the other three parameters were held at their midpoints as σ_8 and Ω_{CDM} vary over the design range.

Figure 26.6 shows the resulting posterior mean surfaces for $w_1(\theta)$, $w_2(\theta)$ and $w_3(\theta)$ as a function of σ_8 and Ω_{CDM}.

Given the posterior realizations from equation (26.9), one can generate realizations from the process $\eta(\theta)$ at any input setting θ^*. Since

$$\eta(\theta^*) = \sum_{i=1}^{p_\eta} \phi_i w_i(\theta^*),$$

realizations from the $w_i(\theta^*)$ processes need to be drawn given the MCMC output. For a given draw $(\lambda_\eta, \lambda_w, \rho_w)$ a draw of $w^* = (w_1(\theta^*), \ldots, w_{p_\eta}(\theta^*))'$ can be produced by making use of the fact

$$\begin{pmatrix} \hat{w} \\ w^* \end{pmatrix} \sim N\left(\begin{pmatrix} 0 \\ 0 \end{pmatrix}, \left[\begin{pmatrix} (\lambda_\eta \Phi'\Phi)^{-1} & 0 \\ 0 & 0 \end{pmatrix} + \Sigma_{w,w^*}(\lambda_w, \rho_w)\right]\right),$$

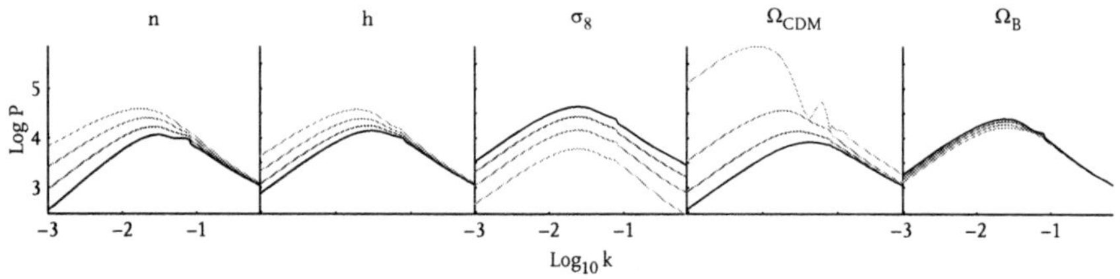

Fig. 26.7 Changes to the posterior mean simulator predictions obtained by varying one input, while holding others at their central values, i.e. at the midpoint of their range. The light to dark lines correspond to the smallest parameter setting to the biggest, for each parameter.

where Σ_{w,w^*} is obtained by applying the prior covariance rule to the augmented input settings that include the original design and the new input setting (θ^*). Recall $\hat{w}$ is defined in equation (26.7). Application of the conditional normal rules then gives

$$w^*|\hat{w} \sim N\left(V_{21}V_{11}^{-1}\hat{w},\, V_{22} - V_{21}V_{11}^{-1}V_{12}\right),$$

where

$$V = \begin{pmatrix} V_{11} & V_{12} \\ V_{21} & V_{22} \end{pmatrix} = \left[\begin{pmatrix} (\lambda_\eta \Phi'\Phi)^{-1} & 0 \\ 0 & 0 \end{pmatrix} + \Sigma_{w,w^*}(\lambda_w, \rho_w)\right]$$

is a function of the parameters produced by the MCMC output. Hence, for each posterior realization of $(\lambda_\eta, \lambda_w, \rho_w)$, a realization of w^* can be produced. The above recipe easily generalizes to give predictions over many input settings at once.

Figure 26.7 shows posterior means for the simulator response η where each of the inputs is varied over its prior (standardized) range of [0, 1] while the other four inputs are held at their nominal setting of 0.5. The posterior mean response conveys an idea of how the different parameters affect the highly multivariate simulation output. Other marginal functionals of the simulation response can also be calculated such as sensitivity indicies or estimates of the Sobol decomposition (Sacks *et al.*, 1989; Oakley and O'Hagan, 2004). Note that a simplified emulator can be constructed by taking plug in estimates for $(\lambda_\eta, \lambda_w, \rho_w)$.

26.2.2.1 *Assessing emulator fit*

The effectiveness of the GP emulator fit depends in part on how smoothly the output changes as inputs vary over their prior range, the effectiveness of the PC basis representation, the number of 'active' parameter inputs, and the complexity of the simulator response. We assess the accuracy of this emulator by predicting the simulated power spectrum over a 64 run holdout design. This 64 run holdout design is also an OA-based LH design over the same

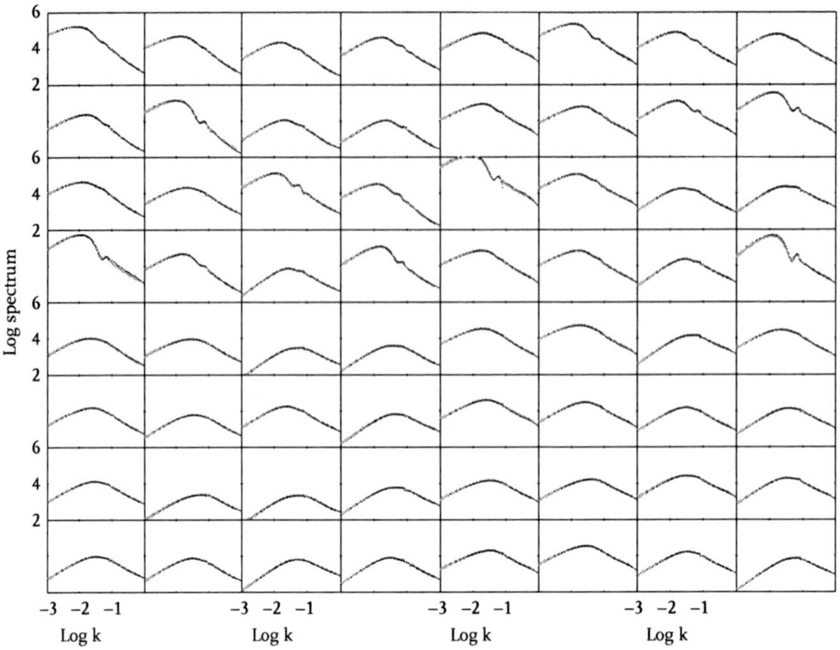

Fig. 26.8 Posterior mean estimates (light lines) and actual simulated spectra (black dots) for a 64 run design that was not used to train the response surface model.

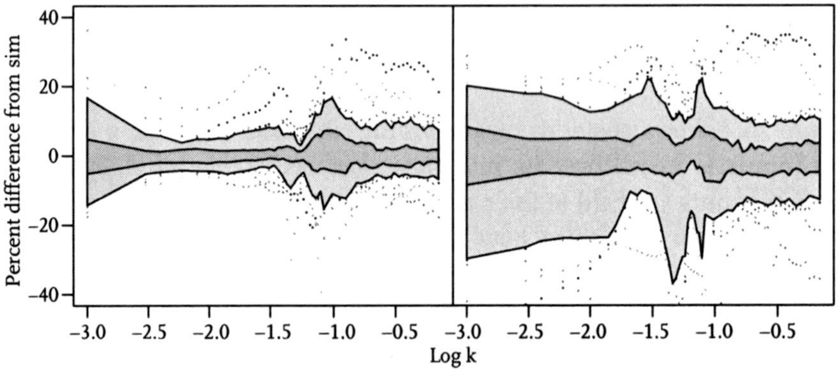

Fig. 26.9 Emulator tests on a 64 run design. Left: emulator based on 128 runs, right: emulator based on 32 runs. The central grey region contains the middle 50% of the residuals, the wider light grey region, the middle 90%. The residuals outside the 90% bands are shown as dots.

parameter ranges. The resulting simulated and predicted power spectra are shown in Figure 26.8. The green lines show the emulator predictions and the black dots show the actual simulation output for the 64 runs. Overall, the emulator performance is sufficient for the task of estimating the cosmological parameters. A more detailed look at the emulator performance is shown in Figure 26.9, left frame. Here we display the residuals of the emulator prediction compared to the simulation runs. The dark gray band contains the middle 50% of the residuals, the light gray band the middle 90%. The overall accuracy of our emulator is over a wide range better than 5%. Only on the edges of the parameter ranges investigated is the quality slightly worse.

Next we investigate how the accuracy of the emulator changes with number of simulations. This is important since large scale simulations are very costly, therefore we'd like to know what the minimal number of simulations required might be. To this end, we create a design for 32 simulations using an orthogonal Latin hypercube sampling design and test it on the same 64 design run we have used for testing the 128 run emulator. The results for this limited run emulator are shown in Figure 26.9, right frame. The overall quality of the emulator is still sufficient, with accuracy at the 10% level. Compared with the larger design emulator, the predictions for the medium k ranges are not as good. Such studies help us assess the relative tradeoff between a sparse set of high accuracy simulations and a larger set of less accurate simulations.

26.2.3 Full statistical formulation

Given the model specifications for the simulator $\eta(\theta)$, we can now consider the sampling model for the experimentally observed data. The data are contained in an n-vector y. For the matter power spectrum application $n = 22$, corresponding to different wave numbers as shown in Figure 26.2. As stated in Section 26.1 the data are modelled as a noisy version of the simulated spectrum $\eta(\theta)$ run at the true, but unknown, parameter setting θ. Thus

$$y = \eta(\theta) + \epsilon,$$

where the errors are assumed to be $N(0, \Sigma_y)$. For notational convenience we represent the precision Σ_y^{-1} as $\lambda_y W_y$, leaving open the option to estimate a scaling of the error covariance with λ_y^{-1}. Using the basis representation for the simulator this equation becomes

$$y = K_y w(\theta) + \epsilon$$

where $w(\theta)$ is the p_η-vector $(w_1(\theta), \ldots, w_{p_\eta}(\theta))'$. Because the wave number support of y is not necessarily contained in the support of the simulation output, the basis vectors in K_y may have to be interpolated over wave number from the columns of K_η. Since the simulation output over wave number is quite dense, this interpolation is straightforward.

We specify a $Ga(a_y, b_y)$ prior for the precision parameter λ_y resulting in a Normal-Gamma form for the data model

$$y|w(\theta), \lambda_y \sim N(K_y w(\theta), (\lambda_y W_y)^{-1}), \; \lambda_y \sim Ga(a_y, b_y). \tag{26.10}$$

The observation precision W_y is fairly well-known for the SDSS data. Hence we use an informative prior with $a_y = b_y = 5$, encouraging λ_y to be near one.

Equivalently, equation (26.10) can be represented in terms of the basis weights

$$\hat{w}_y|w(\theta), \lambda_y \sim N\left(w(\theta), \left(\lambda_y K_y' W_y K_y\right)^{-1}\right), \; \lambda_y \sim Ga\left(a_y^*, b_y^*\right),$$

with

$$\begin{aligned}
\hat{w}_y &= \left(K_y' W_y K_y\right)^{-1} K_y' W_y y, \\
a_y^* &= a_y + \frac{1}{2}(n - p_\eta), \qquad \text{and} \\
b_y^* &= b_y + \frac{1}{2}(y - K_y \hat{w}_y)' W_y (y - K_y \hat{w}_y).
\end{aligned}$$

This equivalency follows from equation (26.8) given in Section 26.2.2.

The (marginal) distribution for the combined, reduced data obtained from the experiments and simulations given the covariance parameters has the form

$$\begin{pmatrix} \hat{w}_y \\ \hat{w} \end{pmatrix} \sim \mathrm{N}\left(\begin{pmatrix} 0 \\ 0 \end{pmatrix}, \begin{pmatrix} \Lambda_y^{-1} & 0 \\ 0 & \Lambda_\eta^{-1} \end{pmatrix} + \begin{pmatrix} I_{p_\eta} & \Sigma_{w_y w} \\ \Sigma_{w_y w}' & \Sigma_w \end{pmatrix} \right), \tag{26.11}$$

where Σ_w is defined in (26.6),

$$\begin{aligned}
\Lambda_y &= \lambda_y K_y' W_y K_y, \\
\Lambda_\eta &= \lambda_\eta K' K, \\
I_{p_\eta} &= p_\eta \times p_\eta \text{ identity matrix}, \\
\Sigma_{w_y w} &= \begin{pmatrix} \lambda_{w1}^{-1} R(\theta, \theta^*; \rho_{w1}) & 0 & 0 \\ 0 & \ddots & 0 \\ 0 & 0 & \lambda_{wp_\eta}^{-1} R(\theta, \theta^*; \rho_{wp_\eta}) \end{pmatrix}.
\end{aligned}$$

Above, $R(\theta, \theta^*; \rho_{wi})$ denotes the $1 \times m$ correlation submatrix obtained by applying equation (26.5) to the observational setting θ crossed with the m simulator input settings $\theta_1^*, \ldots, \theta_m^*$.

26.2.3.1 Posterior distribution

If we take $\hat{z}$ to denote the reduced data $(\hat{w}_y', \hat{w}')'$, and $\Sigma_{\hat{z}}$ to be the covariance matrix given in equation (26.11), the posterior distribution has the form

$$\pi(\lambda_\eta, \lambda_w, \rho_w, \lambda_y, \theta|\hat{z}) \propto |\Sigma_{\hat{z}}|^{-\frac{1}{2}} \exp\left\{-\frac{1}{2}\hat{z}'\Sigma_{\hat{z}}^{-1}\hat{z}\right\} \times \lambda_\eta^{a_\eta^*-1} e^{-b_\eta^*\lambda_\eta} \times \prod_{i=1}^{p_\eta} \lambda_{wi}^{a_w-1} e^{-b_w\lambda_{wi}}$$

$$\times \prod_{i=1}^{p_\eta}\prod_{k=1}^{p_\theta} \rho_{wik}^{a_{\rho w}-1}(1-\rho_{wik})^{b_{\rho w}-1} \times \lambda_y^{a_y^*-1} e^{-b_y^*\lambda_y} \times I[\theta \in C],$$

where C denotes the p_θ-dimensional rectangle defined in (26.3).

Realizations from the posterior distribution are produced using standard, single site MCMC. Metropolis updates (Metropolis *et al.*, 1953) are used for the components of ρ_w and θ with a uniform proposal distribution centred at the current value of the parameter. The precision parameters λ_η, λ_w and λ_y are sampled using Hastings updates (Hastings, 1970). Here the proposals are uniform draws, centred at the current parameter values, with a width that is proportional to the current parameter value. In a given application the candidate proposal width can be tuned for optimal performance.

The resulting posterior distribution estimate for θ is shown in Figure 26.10 on the original scale. These estimates are consistent with the current best estimates of these parameters.

26.3 Combined CMB and large scale structure analysis

So far we have focused our analysis on the matter power spectrum. A more complete analysis of cosmological data should also include additional data sources, such as the WMAP. Figure 26.11 shows a reconstruction of the temperature field for the CMB produced by the WMAP team. A reconstruction is required since the observations from WMAP (and other sources) do not give a complete picture of the cosmic sky. From these incomplete measurements, estimates of the temperature spectrum (called the TT spectrum) are made at different multipole moments ℓ which index the spherical harmonics.

In this section we extend our analysis to include the TT spectrum data given by the WMAP five-year data release. We use the CAMB (Lewis *et al.*, 2000) code to produce a TT power spectrum given the ΛCDM parameters. For modeling the TT spectrum, we consider an additional parameter τ which controls the optical depth to reionization. Hence each CAMB simulation is determined by the six-dimensional parameter vector

$$\theta = (n, h, \sigma_8, \Omega_{\text{CDM}}, \Omega_{\text{B}}, \tau).$$

Different groups analysing different data sets (Spergel *et al.*, 2007; Tegmark *et al.*, 2006) found that the model specified by these six parameters consistently

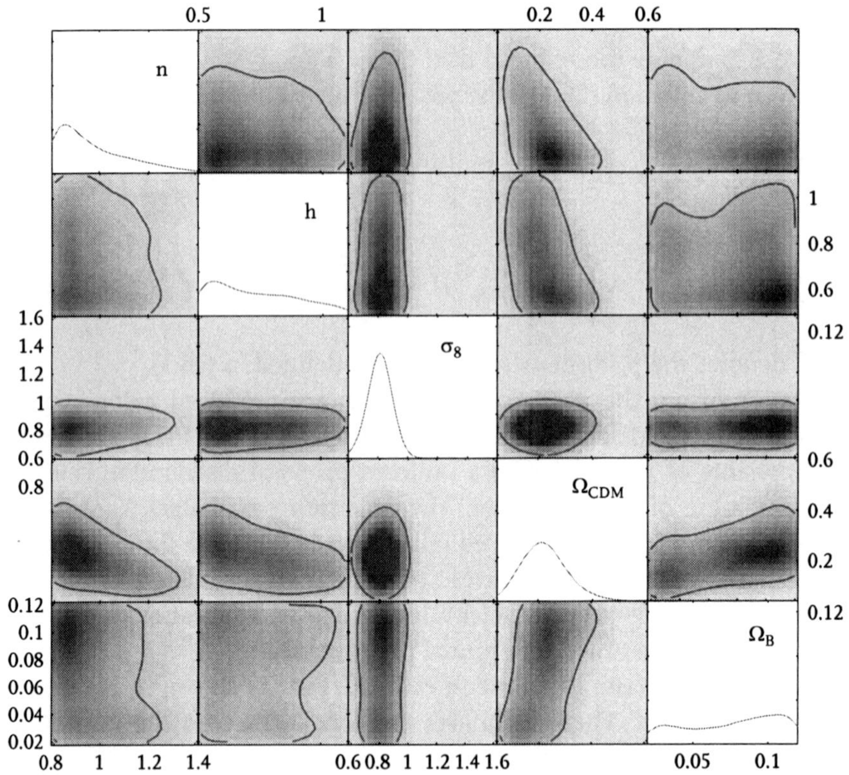

Fig. 26.10 Estimated posterior distribution of the parameters $\theta = (n, h, \sigma_8, \Omega_{CDM}, \Omega_B)$. The diagonal shows the estimated marginal posterior pdf for each parameter; the off-diagonal images give estimates of bivariate marginals; the contour lines show estimated 90% hpd regions.

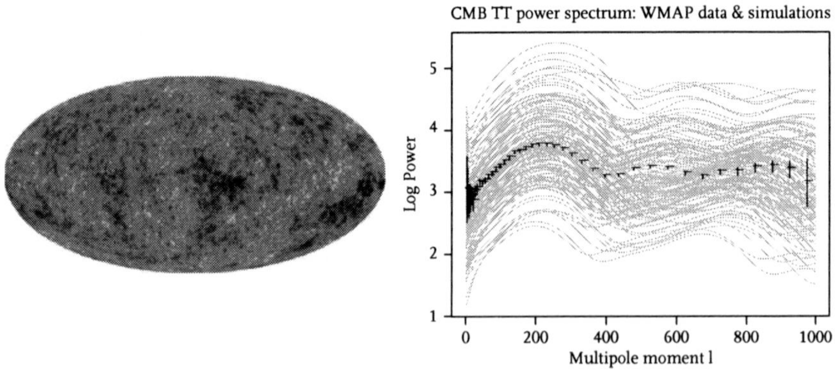

Fig. 26.11 The TT temperature spectrum of the CMB as measured by WMAP. Left: a reconstruction of the spherical CMB temperature field from WMAP observations (Credit: NASA/WMAP Science Team). Right: simulated and measured temperature spectra for the CMB using spherical harmonics. The black vertical lines show two standard deviation uncertainties determined from the WMAP five-year data release.

fits all currently available data. We allow a prior range for τ of (0, 0.3) in our analysis below.

26.3.1 Constraints from the cosmic microwave background

Before we carry out a combined analysis of large scale structure and CMB data, we investigate how well our framework works on the TT power spectrum alone. We use a Gaussian approximation to the likelihood supplied by the Legacy Archive for Microwave Background Data Analysis (LAMBDA) resulting in a 999-dimensional observed vector y, and error covariance matrix Σ_y. Figure 26.11 shows the data vector averaged over local bins, along with corresponding uncertainties.

As with the matter power spectrum analysis, we create a design for 128 runs, this time for a six-parameter space. In the right hand frame of Figure 26.11 the grey lines show the TT power spectra produced by the 128 CAMB runs, and the black +s show the data points with two standard deviation uncertainties derived from the likelihood. Note that the data here do have a slight amount correlation which is encoded in the error matrix Σ_y.

In Figure 26.12 we show the analog to Figure 26.9, demonstrating that the TT emulator predicts 90% of the holdout runs to better than 10% and 50% of the runs to better than 5%. This accuracy is impressive considering the dynamic range and complexity of the TT spectra. However, this complexity requires more

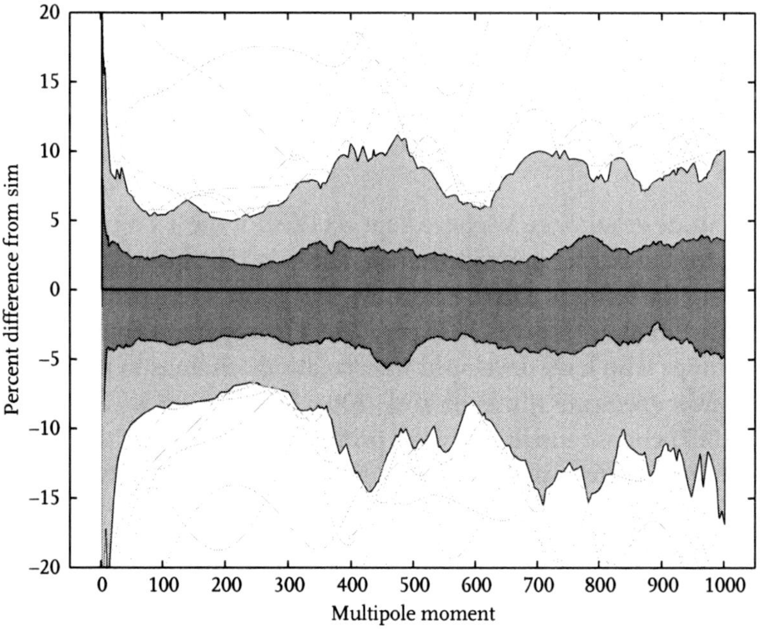

Fig. 26.12 Emulator performance on holdout test. The central grey region contains the middle 50% of the residuals, the wider light grey region, the middle 90%. The lines show residuals extending beyond the 90% bounds.

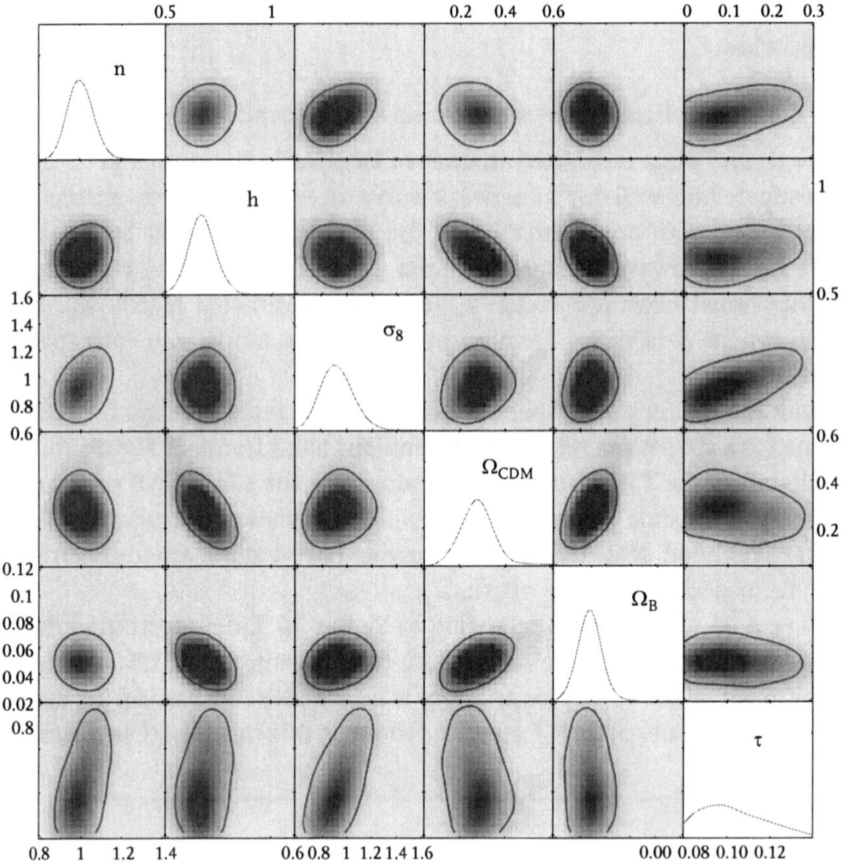

Fig. 26.13 Posterior distribution for the cosmological parameters using just the data from the WMAP TT power spectrum. The TT analysis uses the additional parameter τ.

PCs for accurate emulation. We have kept six PCs for the TT analysis compared to five PCs for the matter power spectrum analysis. The bivariate marginal plot summarizing the inference in the cosmological parameters taking into account the TT spectrum alone is given in Figure 26.13 (compare to Figure 26.10).

Other groups who have developed interpolation schemes to predict the temperature power spectrum (Jimenez *et al.*, 2004; Kosowsky *et al.*, 2002; Fendt and Wandelt, 2007) choose much narrower priors than we have in this application. The GP emulator affords us comparable accuracy over a much broader range of parameter values.

26.3.2 Combined constraints

Now we combine the information contained in the matter power spectrum with the TT spectrum. Given two sets of observed data, y_1 and y_2 – corresponding to the SDSS and WMAP observations respectively – that inform on a common

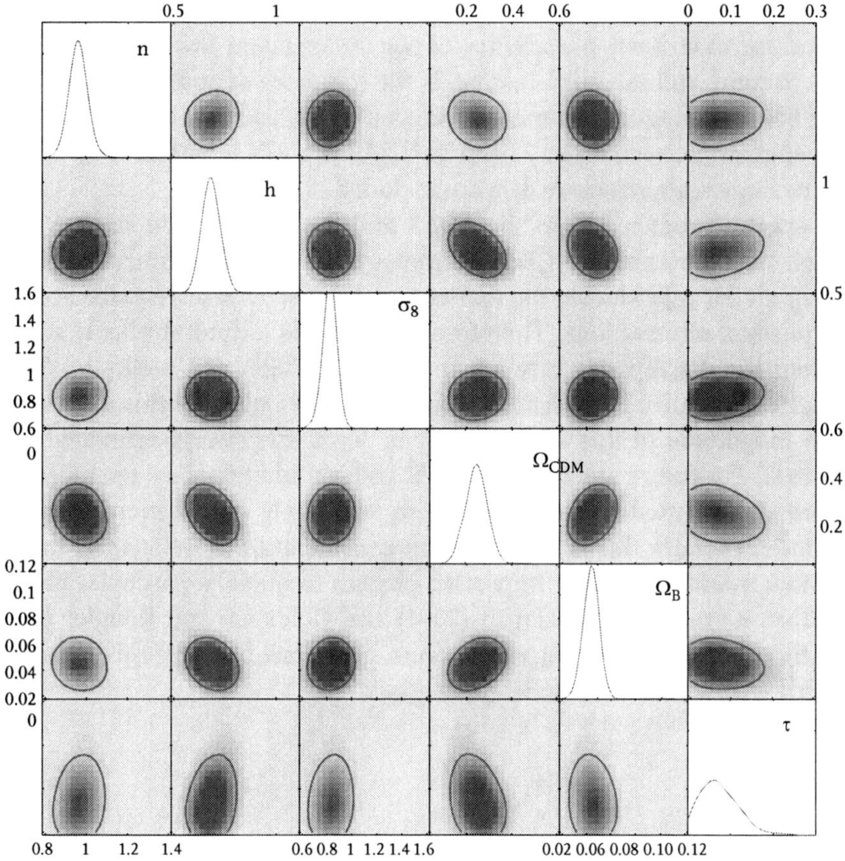

Fig. 26.14 Posterior distribution for the cosmological parameters using data from the matter power spectrum and the CMB TT power spectrum.

set of cosmological parameters θ and statistical parameters ξ, the posterior density is

$$\pi(\theta, \xi|y_1, y_2) \propto L(y_1, y_2|\theta) \cdot \pi(\xi) \cdot \pi(\theta).$$

Assuming the two datasets are independent given the true cosmology θ, the likelihood factors into

$$\pi(\theta, \xi|y_1, y_2) \propto L(y_1|\theta, \xi_1) \cdot \pi(\xi_1) \cdot L(y_2|\theta, \xi_2) \cdot \pi(\xi_2) \cdot \pi(\theta).$$

Our analysis code simply takes the product of the two separate posteriors, while ensuring that the prior information for the common parameter vector θ is counted only once.

The payoff from including both sets of observational data is illustrated in Figure 26.14. The posterior volume of the cosmological parameter space is significantly reduced by the inclusion of the large scale structure information.

There are two main reasons for this volumetric reduction. First is the expected statistical increase due to the addition of two independent and consistent pieces of data. Second, and more interesting, is the influence of posterior correlations among the cosmological parameters induced by the two datasets. For example, the degeneracy between n and τ from the CMB analysis is significantly reduced when the large scale structure data are included.

The spectra produced from the SDSS and the WMAP data sources agree very well with a common ΛCDM cosmology. We feel the simulation models are adequate for reproducing the matter and TT spectra relative to the accuracy of the physical observations. Therefore our analysis did not explicitly account for systematic discrepancy between the simulator $\eta(\theta)$ and reality y. We did investigate alternative formulations which included a term for this discrepancy, but the magnitude of this additional error term was always estimated to be quite small. For future studies which will include additional data sources (and simulators), this model discrepancy term will likely play a more important role. More generally, this is an important consideration if we suspect that the simulation model is missing important physics to model a particular physical observable. Kennedy and O'Hagan (2001) and Goldstein and Rougier (2008), along with their accompanying discussions, give some background on the topic of dealing with simulation model inadequacy.

Appendix

A. Broader context and background

A.1 GP-based calibration of computer models

In many settings, a model for physical reality is replaced by a computer model which typically has a number of unknown, tunable parameters. The goal is to use observations of the physical system to constrain, or calibrate, these parameters. Given the resulting posterior distribution of these parameters, the computer model can now be used to make predictions of the physical system at new, untried conditions.

At various initial conditions x, observations $y(x)$ are made of the physical system $\zeta(x)$

$$y(x_i) = \zeta(x_i) + \epsilon_i,\ i = 1, \ldots, n,$$

where $\zeta(x_i)$ denotes the physical system at initial conditions x_i, and ϵ_i denotes the error in the ith physical observation. The n observations are modeled statistically using the computer model $\eta(x, \theta)$ where θ denotes the unknown calibration parameters. The physical system is commonly modeled as $\zeta(x) = \eta(x, \theta) + \delta(x)$, so that $\delta(x)$ accounts for inadequacies in the computer model (Kennedy and O'Hagan, 2001; Higdon *et al.*, 2004; Bayarri *et al.*, 2007*b*). Thus for the n observations we have

$$y(x_i) = \eta(x_i, \theta) + \delta(x_i) + \epsilon_i, \; i = 1, \ldots, n.$$

This model discrepancy term can account for numerical error (Kaipio and Somersalo, 2007) as well as missing physics. The discrepancy term is most commonly modeled as a GP.

When the computer model can be evaluated very quickly, the resulting posterior can be sampled directly using MCMC as is typically done in the Bayesian solution to inverse problems (Kaipio and Somersalo, 2004; Higdon *et al.*, 2003). In applications where the evaluation of the computer code is limited, an *emulator* of the computer is required. Most commonly, a GP model is used to model $\eta(x, \theta)$ at untried input settings, as is done in this chapter. Figure 26.15 shows the posterior decomposition of the various model elements in an example where both inputs x and θ are univariate. In this case, the discrepancy $\delta(x)$ is given a GP prior which ensures that $\delta(x)$ changes smoothly with x.

This simple example highlights the complications that arise when the discrepancy term is clearly non-zero. First, the posterior distribution for the calibration parameters θ is typically not well determined, and is sensitive to the prior for $\delta(x)$. Second, the mere existence of the discrepancy means that our computer model is not physical reality (even at the best θ). Hence one must be careful in attributing any physical meaning to the posterior for θ. In general, this indeterminacy does not vanish with additional simulations or physical observations.

Predictions of the physical system $\zeta(x)$ at 'nearby' xs are less affected by this indeterminacy resulting from the inclusion of model discrepancy. Intuitively this makes sense since even an empirical model will be quite accurate given sufficient numbers of physical observations. However, the allure of the physics-based computer model is to make extrapolative predictions, at initial conditions x that may be far away from our observational experience. For extrapolations, the prior model for discrepancy is quite important. For example, we may trust our computational model to predict accurately for extrapolations in initial temperature, but not extrapolations initial pressure. Ideally, such concepts should be incorporated in the prior specification of the model discrepancy. Exactly how to utilize the various sources of available information to specify this model discrepancy is an important area of ongoing investigation. Kennedy and O'Hagan (2001), Goldstein and Rougier (2008) and their discussions are a good starting points for the interested reader.

Finally we note that utilizing low-fidelity simulations may be very helpful in predicting the output of a high-fidelity computer model run. In emulator-free applications the low-fidelity simulators can be used to construct auxiliary formulations which speed up the MCMC sampling. See Higdon *et al.* (2002)

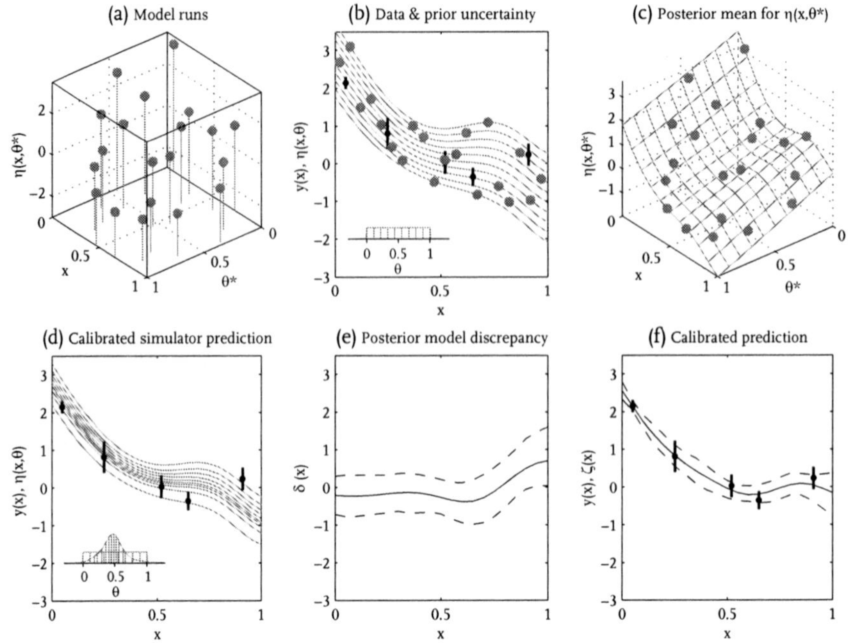

Fig. 26.15 Basic computer model calibration framework. (a) An initial set of simulation runs are carried out over the input settings (x_j^*, θ_j^*), $j = 1, \ldots, m$. (b) Experimental data are collected at $n = 5$ initial conditions; data are given by the black dots; 90% uncertainties are given by the black lines. The light circles correspond to the $m = 20$ simulation output values. (c) Posterior mean estimate for the simulator output $\eta(x, \theta^*)$. (d) Posterior distribution for the calibration parameter θ and the resulting simulator-based predictions (lines). (e) Posterior mean estimate and pointwise 90% prediction intervals for the model discrepancy term $\delta(x)$. (f) Posterior mean estimate and pointwise 90% prediction intervals for the physical system $\zeta(x)$ which incorporate parameter uncertainty, observation error, emulator uncertainty and model discrepancy.

for such an example. An alternative is to use the delayed acceptance scheme of Christen and Fox (2005). Both approaches use low-fidelity simulations to speed up the exploration of the posterior based on the high-fidelity simulator. In settings where an emulator is required, the work of Kennedy and O'Hagan (2000) can be employed which constructs an emulator over multiple model resolutions, resulting in improved accuracy at the highest resolution.

A.2 Emulating multivariate computer model output

When the output from the computer model is highly multivariate, the computational demands of fitting a GP model to the output becomes burdensome. For the CMB TT spectrum output in this chapter's application, the output is a 999-vector. If the output were simply treated as just another dimension in the GP model, the resulting covariance matrix in the likelihood would be over $10^5 \times 10^5$. A number of approaches have been developed to deal with the computational bottleneck produced by multivariate output.

If one treats the output support dimension ($\log k$) as just one more dimension in the covariance model in (26.5), then the resulting covariance matrix for the GP has a kronecker form. Thus one only needs to compute two Cholesky decompositions to evaluate the likelihood – one of a 128×128 matrix and one of a 999×999 matrix – instead of a single decomposition of a $10^5 \times 10^5$ matrix. Taking advantage of this Kronecker structure in the full posterior for the calibration problem (26.12) is a bit more involved. See Williams *et al.* (2006) or Bayarri *et al.* (2007b) for two examples. A detailed description of a multivariate GP emulator which also exploits the kronecker structure of the mean function can be found in Rougier (2007).

While very efficient, the kronecker approach restricts every element of the simulation output to have the same covariance model over the input space. An alternative is to represent the multivariate output with a basis decomposition as in (26.4) of this chapter. These basis approaches allow separate GP models to control each basis element. In this chapter's application, the basis was constructed using a principal component basis derived from the 128 simulation results. While this has worked well for a number of physics-based applications, other approaches for constructing bases have also met with success. For example Efendiev *et al.* (2009) use a Karhunen – Loeve basis for an oil reservoir representation and Bayarri *et al.* (2007*a*) use wavelets to represent one dimensional time history of force.

Acknowledgments

We thank Adrian Pope for useful discussions and insight regarding the matter power spectrum. Also, we would like to thank Antony Lewis for help with CAMB and Licia Verde for clarification on the WMAP-III data sets. A special acknowledgment is due to supercomputing time awarded to us under the LANL Institutional Computing Initiative. This research is supported by the DOE under contract W-7405-ENG-36, and with support from the LDRD program at Los Alamos National Laboratory.

References

Adelman-McCarthy, J., Agueros, M., Allam, S., Anderson, K., Anderson, S., Annis, J., Bahcall, N., Baldry, I., Barentine, J., Berlind, A., *et al.* (2006). The fourth data release of the Sloan Digital Sky Survey. *The Astrophysical Journal Supplement Series*, **162**, 38–48.

Bayarri, M., Berger, J., Cafeo, J., Garcia-Donato, G., Liu, F., Sacks, J., and Walsh, D. (2007*a*). Computer model validation with functional output. *Annals of Statistics*, **35**, 1874–1906.

Bayarri, M., Berger, J., Paulo, R., Sacks, J., Cafeo, J., Cavendish, J., Lin, C., and Tu, J. (2007*b*). A framework for validation of computer models. *Technometrics*, **49**, 138–154.

Christen, J. and Fox, C. (2005). Markov chain Monte Carlo using an approximation. *Journal of Computational and Graphical Statistics*, **14**, 795–810.

Currin, C., Mitchell, T., Morris, M., and Ylvisaker, D. (1991). Bayesian prediction of deterministic functions, with applications to the design and analysis of computer experiments. *Journal of the American Statistical Association*, **86**, 953–963.

Efendiev, Y., Datta-Gupta, A., Ginting, V., Ma, X., and Mallick, B. (2009). An efficient two-stage Markov chain Monte Carlo method for dynamic data integration. To appear in *Water Resources Research*.

Fendt, W. and Wandelt, B. (2007). Pico: Parameters for the impatient cosmologist. *The Astrophysical Journal*, **654**, 2–11.

Frieman, J., Turner, M., and Huterer, D. (2008). Dark energy and the accelerating universe. Arxiv preprint arXiv:0803.0982.

Goldstein, M. and Rougier, J. C. (2008). Reified Bayesian modelling and inference for physical systems (with discussion). *Journal of Statistical Planning and Inference*, **139**, 1221–1239.

Hastings, W. K. (1970). Monte Carlo sampling methods using Markov chains and their applications. *Biometrika*, **57**, 97–109.

Higdon, D., Gattiker, J. R., and Williams, B. J. (2008). Computer model calibration using high dimensional output. *Journal of the American Statistical Association*, **103**, 570–583.

Higdon, D., Kennedy, M., Cavendish, J., Cafeo, J., and Ryne, R. D. (2004). Combining field observations and simulations for calibration and prediction. *SIAM Journal of Scientific Computing*, **26**, 448–466.

Higdon, D., Lee, H., and Bi, Z. (2002). A Bayesian approach to characterizing uncertainty in inverse problems using coarse and fine scale information. *IEEE Transactions in Signal Processing*, **50**, 389–399.

Higdon, D. M., Lee, H., and Holloman, C. (2003). Markov chain Monte Carlo-based approaches for inference in computationally intensive inverse problems. In *Bayesian Statistics 7* (eds. J. M. Bernardo, M. J. Bayarri, J. O. Berger, A. P. Dawid, D. Heckerman, A. F. M. Smith, and M. West), pp. 181–197. Oxford University Press, Oxford.

Jimenez, R., Verde, L., Peiris, H., and Kosowsky, A. (2004). Fast cosmological parameter estimation from microwave background temperature and polarization power spectra. *Physical Review D*, **70**, 23005.

Kaipio, J. and Somersalo, E. (2007). Statistical inverse problems: Discretization, model reduction and inverse crimes. *Journal of Computational and Applied Mathematics*, **198**, 493–504.

Kaipio, J. P. and Somersalo, E. (2004). *Statistical and Computational Inverse Problems*. Springer, New York.

Kennedy, M. and O'Hagan, A. (2000). Predicting the output from a complex computer code when fast approximations are available. *Biometrika*, **87**, 1–13.

Kennedy, M. and O'Hagan, A. (2001). Bayesian calibration of computer models (with discussion). *Journal of the Royal Statistical Society (Series B)*, **68**, 425–464.

Kosowsky, A., Milosavljevic, M., and Jimenez, R. (2002). Efficient cosmological parameter estimation from microwave background anisotropies. *Physical Review D*, **66**, 63007.

Lewis, A., Challinor, A., and Lasenby, A. (2000). Efficient computation of cosmic microwave background anisotropies in closed Friedmann–Robertson–Walker models. *The Astrophysical Journal*, **538**, 473–476.

Linkletter, C., Bingham, D., Hengartner, N., Higdon, D., and Ye, K. (2006). Variable selection for Gaussian process models in computer experiments. *Technometrics*, **48**, 478–490.

Metropolis, N., Rosenbluth, A., Rosenbluth, M., Teller, A., and Teller, E. (1953). Equations of state calculations by fast computing machines. *Journal of Chemical Physics*, **21**, 1087–1091.

Oakley, J. and O'Hagan, A. (2004). Probabilistic sensitivity analysis of complex models. *Journal of the Royal Statistical Society (Series B)*, **66**, 751–769.

Ramsay, J. O. and Silverman, B. W. (1997). *Functional Data Analysis*. Springer, New York.

Rougier, J. (2007). Efficient emulators for multivariate deterministic functions. *Journal of Computational and Graphical Studies*, **17**, 827–843.

Sacks, J., Welch, W. J., Mitchell, T. J., and Wynn, H. P. (1989). Design and analysis of computer experiments (with discussion). *Statistical Science*, **4**, 409–423.

Sahni, V. and Starobinsky, A. (2000). The case for a positive cosmological Lambda-term (2000). *International Journal of Modern Physics D*, **9**, 373.

Santner, T. J., Williams, B. J., and Notz, W. I. (2003). *Design and Analysis of Computer Experiments*. Springer, New York.

Smoot, G., Bennett, C., Kogut, A., Wright, E., Aymon, J., Boggess, N., Cheng, E., de Amici, G., Gulkis, S., Hauser, M., *et al.* (1992). Structure in the COBE differential microwave radiometer first-year maps. *Astrophysical Journal*, **396**, 1.

Spergel, D., Bean, R., Dore, O., Nolta, M., Bennett, C., Dunkley, J., Hinshaw, G., Jarosik, N., Komatsu, E., Page, L., *et al.* (2007). Three-year Wilkinson Microwave Anisotropy Probe (WMAP) observations: Implications for cosmology. *The Astrophysical Journal Supplement Series*, **170**, 377–408.

Tang, B. (1993). Orthogonal array-based Latin hypercubes. *Journal of the American Statistical Association*, **88**, 1392–1397.

Tegmark, M., Blanton, M., Strauss, M., Hoyle, F., Schlegel, D., Scoccimarro, R., Vogeley, M., Weinberg, D., Zehavi, I., Berlind, A., *et al.* (2004). The three-dimensional power spectrum of galaxies from the Sloan Digital Sky Survey. *The Astrophysical Journal*, **606**, 702–740.

Tegmark, M., Eisenstein, D., Strauss, M., Weinberg, D., Blanton, M., Frieman, J., Fukugita, M., Gunn, J., Hamilton, A., Knapp, G., *et al.* (2006). Cosmological constraints from the SDSS luminous red galaxies. *Physical Review D*, **74**, 123507.

Williams, B., Higdon, D., Gattiker, J., Moore, L., McKay, M., and Keller-McNulty, S. (2006). Combining experimental data and computer simulations, with an application to flyer plate experiments. *Bayesian Analysis*, **1**, 765–792.

Ye, K. Q., Li, W., and Sudjianto, A. (2000). Algorithmic construction of optimal symmetric Latin hypercube designs. *Journal of Statistical Planning and Inference*, **90**, 145–159.

·27·

Probabilistic grammars and hierarchical Dirichlet processes

Percy Liang, Michael I. Jordan and Dan Klein

27.1 Introduction

The field of natural language processing (NLP) aims to develop algorithms that allow computers to understand and generate natural language. The field emerged from computational linguistics, a field whose early history was shaped in part by a rejection of statistical approaches to language, where 'statistical' at the time generally referred to simplistic Markovian models on observed sequences of words. Despite this unfavorable historical context, statistical approaches to NLP have been in ascendancy over the past decade (Manning and Schütze, 1999), driven in part by the availability of large corpora of text and other linguistic resources on the Internet, and driven in part by a growth in sophistication among NLP researchers regarding the scope of statistical modelling, particularly latent-variable modelling. The phenomenon of language itself is also responsible: language is replete with ambiguity, so it is inevitable that formal inferential methods should play a significant role in managing this ambiguity.

The majority of the work in statistical NLP has been non-Bayesian, but there is reason to believe that this is a historical accident. Indeed, despite the large corpora, sparse data problems abound to which hierarchical Bayesian methods seem well suited. Also, the conditional perspective of Bayesian statistics seems particularly appropriate for natural language – conditioning on the current sentence and the current context can provide precise inference despite the high degree of ambiguity.

In the current chapter, we discuss a Bayesian approach to the problem of *syntactic parsing* and the underlying problems of *grammar induction* and *grammar refinement*. The central object of study is the *parse tree*, an example of which is shown in Figure 27.1. A substantial amount of the syntactic structure and relational semantics of natural language sentences can be described using parse trees. These trees play a central role in a range of activities in modern NLP, including machine translation (Galley *et al.*, 2004), semantic role extraction (Gildea and Jurafsky, 2002), and question answering (Hermjakob, 2001),

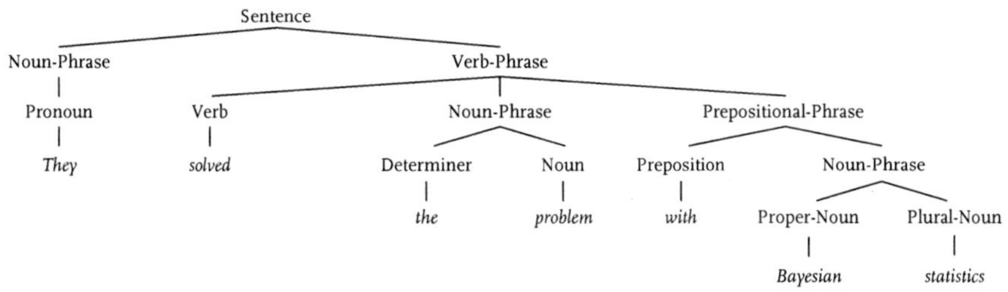

Fig. 27.1 A parse tree for the sentence *They solved the problem with Bayesian statistics.*

just to name a few. From a statistical perspective, parse trees are an extremely rich class of objects, and our approach to capturing this class probabilistically will be to make use of tools from Bayesian nonparametric statistics.

It seems reasonable enough to model parse trees using context-free grammars (CFGs); indeed, this goal was the original motivation behind the development of the CFG formalism (Chomsky, 1956), and it remains a major focus of research on parsing to this day. Early work on NLP parsing concentrated on efficient algorithms for computing the set of all parses for a sentence under a given CFG. Unfortunately, as we have alluded to, natural language is highly ambiguous. In fact, the number of parses for a sentence grows exponentially with its length. As a result, systems which enumerated all possibilities were not useful in practice. Modern work on parsing has therefore turned to probabilistic models which place distributions over parse trees and probabilistic inference methods which focus on likely trees (Lari and Young, 1990).

The workhorse model family for probabilistic parsing is the family of *probabilistic context-free grammars* (PCFGs),[1] which are probabilistic generalizations of CFGs and structural generalizations of hidden Markov models (HMMs). A PCFG (described formally in Section 27.1.1) is a branching process in which nodes iteratively rewrite from top to bottom, eventually terminating in dedicated *lexical* items, i.e. words. Each node is rewritten *independently* according to a multinomial distribution specific to that node's symbol. For example, a noun phrase frequently rewrites as a determiner followed by a noun (e.g. *the problem*).

Early work focused on *grammar induction* (also known as grammatical inference): estimating grammars directly from raw sentences without any other type of supervision (Carroll and Charniak, 1992). Grammar induction is an important scientific problem connecting cognitive science, linguistics, statistics, and even philosophy. A successful grammar induction system would have important implications for human language learning, and it would also be a valuable asset for being able to parse sentences with little human

[1] Also known as stochastic context-free grammars (SCFGs).

effort. However, a combination of model misspecification and local optima issues with the EM algorithm stymied these initial attempts. It turned out that it was necessary to impose more constraints on the tree structure (Pereira and Shabes, 1992).

Only with the advent of *treebanks* (Marcus *et al.*, 1993) – hand-labelled collections of parse trees – were NLP researchers able to develop the first successful broad-coverage natural language parsers, but it still took a decade before the performance of the best parsers started to level off. In this supervised setting, maximum likelihood estimates for PCFGs have a simple closed-form solution: the rule probabilities of the PCFG are proportional to the counts of the associated grammar productions across all of the trees in the treebank (Charniak, 1996). However, such statistical grammars do not perform well for parsing. The problem is that treebanks contain only a handful of very coarse symbols, such as NP (noun phrase) and VP (verb phrase), so the conditional independences assumed by the PCFG over these coarse symbols are unrealistic. The true syntactic process is vastly more complex. For example, noun phrases can be subjects or objects, singular or plural, definite or indefinite, and so on. Similarly, verb phrases can be active or passive, transitive or intransitive, past or present, and so on. For a PCFG to adequately capture the true syntactic process, finer-grained grammar symbols are required. Much of the past decade of NLP parsing work can be seen as trying to optimally learn such fine-grained grammars from coarse treebanks.

Grammars can be refined in many ways. Some refinements are syntactically motivated. For example, if we augment each symbol with the symbol of its parent in the tree, we get symbols such as NP-VP, which represents direct object noun phrases, distinct from NP-S, which represents subject ones. This strategy is called *parent annotation* (Johnson, 1998) and can be extended (Klein and Manning, 2003). For the parse tree in Figure 27.1, if each node's symbol is augmented with the symbols of its parent and grandparent, a maximum likelihood grammar would only allow *they* to be produced under a noun phrase in subject position, disallowing ungrammatical sentences such as *The problem solved they with Bayesian statistics.*

Other refinements are semantically motivated. In many linguistic theories, each phrase is identified with a *head word*, which characterizes many of the important properties of the phrase. By augmenting the symbol of each node with the head word of the phrase under that node, we allow some degree of semantic plausibility to be captured by the parse tree. This process is called *lexicalization*, and was the basis for the first generation of practical treebank parsers (Collins, 1999; Charniak, 2000). Consider the sentence in Figure 27.1. It is actually ambiguous: did they use Bayesian statistics to solve the problem at hand (A) or did Bayesian statistics itself have a fundamental flaw which they resolved (B)? Though both are perfectly valid syntactically, (B) is implausible semantically, and we would like our model to prefer (A) over (B). If we lexicalize the verb phrase with *solved* (replace VP with VP-*solved*) and the preposition phrase with *statistics* (replace PP with PP-*statistics*),[2] we allow semantics to interact through the tree, yielding a better model of language that could prefer (A) over (B). Lexicalization produces a very rich model at the cost of multiplying the number of parameters by millions. In order to cope with the resulting problems of high-dimensionality, elaborate smoothing and parameter-tying methods were employed (Collins, 1999; Charniak, 2000), a recurrent theme in NLP.

Both parent annotation and lexicalization kept the grammar learning problem fully observed: given the coarse trees in the treebank, the potential uncertainty resides in the choice of head words or parents, and these were typically propagated up the tree in deterministic ways. The only inferential problem remaining was to fit the grammar parameters, which reduces (in the point estimation setting generally adopted in statistical NLP) to counting and smoothing.

More recently, latent-variable models have been successfully employed for automatically refining treebank grammars (Matsuzaki *et al.*, 2005; Petrov *et al.*, 2006). In such approaches, each symbol is augmented with a latent cluster indicator variable and the marginal likelihood of the model is optimized. Rather than manually specifying the refinements, these latent-variable models let the data speak, and, empirically, the resulting refinements turn out to encode a mixture of syntactic and semantic information not captured by the coarse treebank symbols. The latent clusters allow for the modeling of long-range dependencies while keeping the number of parameters modest.

In this chapter, we address a fundamental question which underlies all of the previous work on parsing: what priors over PCFGs are appropriate? In particular, we know that we must trade off grammar complexity (the number of grammar symbols) against the amount of data present. As we get more data, more of the underlying grammatical processes can be adequately modelled. Past approaches to complexity control have considered minimum description length (Stolcke and Omohundro, 1994) and procedures for growing the grammar size heuristically (Petrov *et al.*, 2006). While these procedures can be quite effective, we would like to pursue a Bayesian nonparametric approach to PCFGs so that we can state our assumptions about the problem of grammar growth in a coherent way. In particular, we define a nonparametric prior over PCFGs which allocates an unbounded number of symbols to the grammar through a Dirichlet process (Ferguson, 1973, 1974), then shares those symbols throughout the grammar using a Bayesian hierarchy of Dirichlet processes

[2] According to standard NLP head conventions, the preposition *with* would be the head word of a prepositional phrase, but here we use a non-standard alternative to focus on the semantic properties of the phrase.

(Teh *et al.*, 2006). We call the resulting model a *hierarchical Dirichlet process probabilistic context-free grammar* (HDP-PCFG). In this chapter, we present the formal probabilistic specification of the HDP-PCFG, algorithms for posterior inference under the HDP-PCFG, and experiments on grammar learning from large-scale corpora.

27.1.1 Probabilistic context-free grammars (PCFGs)

Our HDP-PCFG model is based on probabilistic context-free grammars (PCFGs), which have been a core modelling technique for many aspects of syntactic structure (Charniak, 1996; Collins, 1999) as well as for problems in domains outside of natural language processing, including computer vision (Zhu and Mumford, 2006) and computational biology (Sakakibara, 2005; Dyrka and Nebel, 2007).

Formally, a PCFG is specified by the following:

- a set of *terminal* symbols Σ (the words in the sentence),
- a set of *nonterminal* symbols S,
- a designated *root* nonterminal symbol $\textsc{Root} \in S$, and
- *rule probabilities* $\phi = (\phi_s(\gamma) : s \in S, \gamma \in \Sigma \cup (S \times S))$ with $\phi_s(\gamma) \geq 0$ and $\sum_\gamma \phi_s(\gamma) = 1$.

We restrict ourselves to rules $s \to \gamma$ that produce a right-hand side γ which is either a single terminal symbol ($\gamma \in \Sigma$) or a pair of non-terminal symbols ($\gamma \in S \times S$). Such a PCFG is said to be in *Chomsky normal form*. The restriction to Chomsky normal form is made without loss of generality; it is straightforward to convert a rule with multiple children into a structure (e.g. a right-branching chain) in which each rule has at most two children.

A PCFG defines a distribution over sentences and parse trees via the following generative process: start at a root node with $s = \textsc{Root}$ and choose to apply rule $s \to \gamma$ with probability $\phi_s(\gamma)$; γ specifies the symbols of the children. For children with nonterminal symbols, recursively generate their subtrees. The process stops when all the leaves of the tree are terminals. Call the sequence of terminals the *yield*. More formally, a *parse tree* has a set of nonterminal nodes N along with the symbols corresponding to these nodes $s = (s_i \in S : i \in N)$. Let N_E denote the nodes having one terminal child and N_B denote the nodes having two nonterminal children. The tree structure is represented by $c = (c_j(i) : i \in N_B, j = 1, 2)$, where $c_j(i) \in N$ is the j-th child node of i from left to right. Let $z = (N, s, c)$ denote the parse tree and $x = (x_i : i \in N_E)$ denote the yield.

The joint probability of a parse tree z and its yield x is given by

$$p(x, z|\phi) = \prod_{i \in N_B} \phi_{s_i}\left(s_{c_1(i)}, s_{c_2(i)}\right) \prod_{i \in N_E} \phi_{s_i}(x_i). \tag{27.1}$$

PCFGs are similar to hidden Markov models (HMMs) and these similarities will guide our development of the HDP-PCFG. It is important to note at the outset, however, an important qualitative difference between HMMs and PCFGs. While the HMM can be represented as a graphical model (a Markovian graph in which the pattern of missing edges corresponds to assertions of conditional independence), the PCFG cannot. Conditioned on the structure of the parse tree (N, c), we have a graphical model over the symbols s, but the structure itself is a random object – we must run an algorithm to compute a probability distribution over these structures. As in the case of the forward-backward algorithm for HMMs, this algorithm is an efficient dynamic programming algorithm – it is referred to as the 'inside-outside algorithm' and it runs in time cubic in length of the yield (Lari and Young, 1990). The inside-outside algorithm will play an important role in the inner loop of our posterior inference algorithm for the HDP-PCFG. Indeed, we will find it essential to design our model such that it can exploit the inside-outside algorithm.

Traditionally, PCFGs are defined with a fixed, finite number of non-terminals S, where the parameters ϕ are fit using (smoothed) maximum likelihood. The focus of this chapter is on developing a nonparametric version of the PCFG which allows S to be countably infinite and which performs approximate posterior inference over the set of non-terminal symbols and the set of parse trees. To define the HDP-PCFG and develop effective posterior inference algorithms for the HDP-PCFG, we need to bring several ingredients together – most notably the ability to generate new symbols and to tie together multiple usages of the same symbol on a parse tree (provided by the HDP), and the ability to efficiently compute probability distributions over parse trees (provided by the PCFG). Thus, the HDP-PCFG is a Bayesian nonparametric generalization of the PCFG, but it can also be viewed as a generalization along the Chomsky hierarchy, taking a nonparametric Markovian model (the HDP-HMM) to a nonparametric probabilistic grammar (the HDP-PCFG).

The rest of this chapter is organized as follows. Section 27.2 provides the probabilistic specification of the HDP-PCFG for grammar induction, and Section 27.3 extends this specification to an architecture appropriate for grammar refinement (the HDP-PCFG-GR). Section 27.4 describes an efficient variational method for approximate Bayesian inference in these models. Section 27.5 presents experiments: using the HDP-PCFG to induce a small grammar from raw text, and using the HDP-PCFG-GR to parse the *Wall Street Journal*, a standard large-scale dataset. For supplementary information, see the appendix, where we review DP-based models related to and leading up to the HDP-PCFG (Appendix A.1), discuss general issues related to approximate inference for these types of models (Appendix A.2), and provide some empirical intuition regarding the interaction between model and inference (Appendix A.3). The details of the variational inference algorithm are given in Appendix B.

27.2 The hierarchical Dirichlet process PCFG (HDP-PCFG)

At the core of the HDP-PCFG sits the Dirichlet process (DP) mixture model (Antoniak, 1974), a building block for a wide variety of nonparametric Bayesian models. The DP mixture model captures the basic notion of clustering which underlies symbol formation in the HDP-PCFG. From there, the HDP-PCFG involves several structural extensions of the DP mixture model. As a first stepping stone, consider hidden Markov models (HMMs), a dynamic generalization of mixture models, where clusters are linked structurally according to a Markov chain. To turn the HMM into a nonparametric Bayesian model, we use the hierarchical Dirichlet process (HDP), yielding the HDP-HMM, an HMM with a countably infinite state space. The HDP-PCFG differs from the HDP-HMM in that, roughly speaking, the HDP-HMM is a chain-structured model while HDP-PCFG is a tree-structured model. But it is important to remember that the tree structure in the HDP-PCFG is a random object over which inferences must be made. Another distinction is that rules can rewrite to two nonterminal symbols jointly. For this, we need to define DPs with base measures which are products of DPs, thereby adding another degree of complexity to the HDP machinery. For a gradual introduction, see Appendix A.1, where we walk through the intermediate steps leading up to the HDP-PCFG – the Bayesian finite mixture model (Appendix A.1.1), the DP mixture model (Appendix A.1.2), and the HDP-HMM (Appendix A.1.3).

27.2.1 Model definition

Figure 27.2 defines the generative process for the HDP-PCFG, which consists of two stages: we first generate the grammar (which includes the rule probabilities) specified by (β, ϕ); then we generate a parse tree and its yield (z, x) using that grammar. To generate the grammar, we first draw a countably infinite set of stick-breaking probabilities, $\beta \sim \mathrm{GEM}(\alpha)$, which provides us with a base distribution over grammar symbols, represented by the positive integers (Figure 27.3a; see Appendix A.1.2 for a formal definition of the stick-breaking distribution). Next, for each symbol $z = 1, 2, \ldots$, we generate the probabilities of the rules of the form $z \rightarrow \gamma$, i.e. those which condition on z as the left-hand side. The emission probabilities ϕ_z^E are drawn from a Dirichlet distribution, and provide multinomial distributions over terminal symbols Σ. For the binary production probabilities, we first form the product distribution $\beta\beta^\top$ (Figure 27.3b), represented as a doubly-infinite matrix. The binary production probabilities ϕ_z^B are then drawn from a Dirichlet process with base distribution $\beta\beta^\top$ – this provides multinomial distributions over pairs of nonterminal symbols (Figure 27.3c). The Bayesian hierarchy ties these distributions together through the base distribution over symbols, so that the grammar effectively has a globally shared

HDP-PCFG

$\beta \sim \text{GEM}(\alpha)$ [draw top-level symbol probabilities]
For each grammar symbol $z \in \{1,2,\ldots\}$:
  $\phi_z^T \sim \text{Dir}(\alpha^T)$ [draw rule type parameters]
  $\phi_z^E \sim \text{Dir}(\alpha^E)$ [draw emission parameters]
  $\phi_z^B \sim \text{DP}(\alpha^B, \beta\beta^{\top})$ [draw binary production parameters]
For each node i in the parse tree:
  $t_i \sim \text{Mult}(\phi_{z_i}^T)$ [choose rule type]
  If t_i = EMISSION:
    $x_i \sim \text{Mult}(\phi_{z_i}^E)$ [emit terminal symbol]
  If t_i = BINARY-PRODUCTION:
    $(z_{c_1(i)}, z_{c_2(i)}) \sim \text{Mult}(\phi_{z_i}^B)$ [generate child symbols]

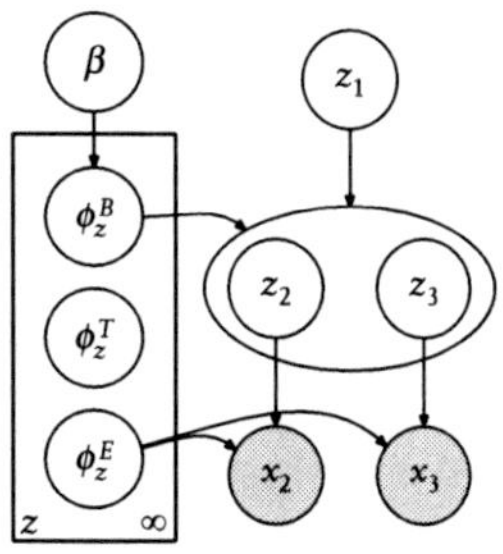

Fig. 27.2 The probabilistic specification of the HDP-PCFG. We also present a graphical representation of the model. Drawing this graphical model assumes that the parse is known, which is *not* our assumption, so this representation should be viewed as simply suggestive of the HDP-PCFG. In particular, we show a simple fixed tree in which node z_1 has two children (z_2 and z_3), each of which has one observed terminal child.

inventory of symbols. Note that the HDP-PCFG hierarchy ties distributions over symbol pairs via distributions over single symbols, in contrast to the hierarchy in the standard HDP-HMM, where the distributions being tied are defined over the same space as that of the base distribution. Finally, we generate a 'switching' distribution ϕ_z^T over the two rule types {Emission, Binary-Production} from a symmetric Dirichlet. The shapes of all the Dirichlet distributions and the Dirichlet processes in our models are governed by concentration hyperparameters: α, α^T, α^B, α^E.

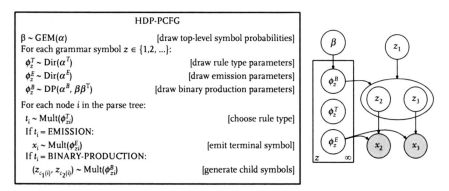

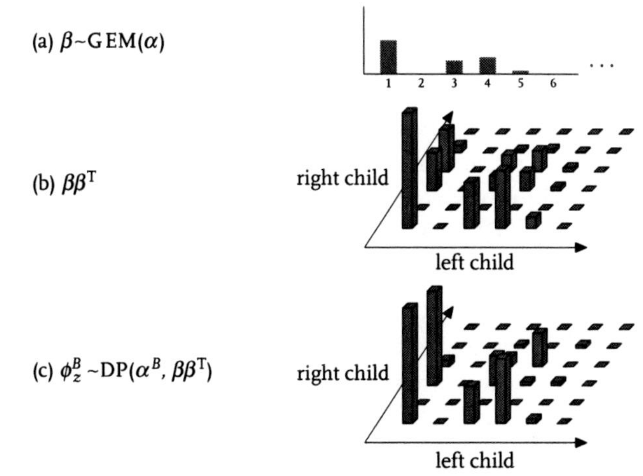

Fig. 27.3 The generation of binary production probabilities given the top-level symbol probabilities β. First, β is drawn from the stick-breaking prior, as in any DP-based model (a). Next, the outer-product $\beta\beta^{\top}$ is formed, resulting in a doubly-infinite matrix (b). We use this as the base distribution for generating the binary production distribution from a DP centered on $\beta\beta^{\top}$ (c).

Given a grammar, we generate a parse tree and sentence in the same manner as for an ordinary PCFG: start with the root node having the designated root symbol. For each non-terminal node i, we first choose a rule type t_i using $\phi_{z_i}^T$ and then choose a rule with that type – either an emission from Mult $(\phi_{z_i}^E)$ producing x_i or a binary production from Mult $(\phi_{z_i}^T)$ producing $(z_{c_1(i)}, z_{c_2(i)})$. We recursively apply the procedure on any non-terminal children.

27.2.2 Partitioning the preterminal and constituent symbols

It is common to partition non-terminal grammar symbols into those that emit only terminals (*preterminal symbols*) and those that produce only two child grammar symbols (*constituent symbols*). An easy way to accomplish this is to let $\alpha^T = (0, 0)$. The resulting Dir(0, 0) prior on the rule type probabilities forces any draw $\phi_z^T \sim \text{Dir}(\alpha^T)$ to put mass on only one rule type. Despite the simplicity of this approach, is not suitable for our inference method, so let us make the partitioning more explicit.

Define two disjoint inventories, one for preterminal symbols ($\beta^P \sim \text{GEM}(\alpha)$) and one for constituent symbols ($\beta^C \sim \text{GEM}(\alpha)$). Then, we can define the following four rule types: preterminal-preterminal productions with rule probabilities $\phi_z^{PP} \sim \text{DP}\left(\alpha^{PP}, \beta^P(\beta^P)^\top\right)$, preterminal-constituent productions with $\phi_z^{PC} \sim \text{DP}\left(\alpha^{PC}, \beta^P(\beta^C)^\top\right)$, constituent-preterminal productions with $\phi_z^{CP} \sim \text{DP}\left(\alpha^{CP}, \beta^C(\beta^P)^\top\right)$, and constituent-constituent productions with $\phi_z^{CC} \sim \text{DP}\left(\alpha^{CC}, \beta^C(\beta^C)^\top\right)$. Each node has a symbol which is either a preterminal (P, z) or a constituent (C, z). In the former case, a terminal is produced from ϕ_z^E. In the latter case, one of the four rule types is first chosen given ϕ_z^T and then a rule of that type is chosen.

27.2.3 Other nonparametric grammars

An alternative definition of an HDP-PCFG would be as follows: for each symbol z, draw a distribution over left child symbols $\phi_z^{B_1} \sim \text{DP}(\alpha, \beta)$ and an independent distribution over right child symbols $\phi_z^{B_2} \sim \text{DP}(\alpha, \beta)$. Then define the binary production distribution as the product $\phi_z^B = \phi_z^{B_1}\phi_z^{B_2 \top}$. This also yields a distribution over symbol pairs and defines a different type of nonparametric PCFG. This model is simpler than the HDP-PCFG and does not require any additional machinery beyond the HDP-HMM. However, the modelling assumptions imposed by this alternative are unappealing as they assume the left child and right child are independent given the parent, which is certainly not the case in natural language.

Other alternatives to the HDP-PCFG include the *adaptor grammar* framework of Johnson *et al.* (2006) and the *infinite tree* framework of Finkel *et al.* (2007). Both of these Bayesian nonparametric models have been developed using the Chinese restaurant process rather than the stick-breaking representation. In an adaptor grammar, the grammar symbols are divided into *adapted* and *non-adapted* symbols. Non-adapted symbols behave like parametric PCFG symbols, while adapted symbols are associated with a Bayesian nonparametric prior over subtrees rather than over pairs of child symbols as in PCFG. While this gives adaptor grammars the ability to capture more global properties of a sentence, recursion is not allowed on adaptor symbols (that is, coexistence of rules of the form $A \rightarrow BC$ and $B \rightarrow DA$ is disallowed if A is an adaptor symbol). Applications of adaptor grammars have mostly focused on the modelling of word segmentation and collocation rather than full constituency syntax. The infinite tree framework is more closely related to the HDP-PCFG, differing principally in that it has been developed for dependency parsing rather than constituency parsing.

27.3 The HDP-PCFG for Grammar Refinement (HDP-PCFG-GR)

While the HDP-PCFG is suitable for grammar induction, where we try to infer a full grammar from raw sentences, in practice we often have access to *treebanks*, which are collections of tens of thousands of sentences, each hand-labelled with a syntactic parse tree. Not only do these observed trees help constrain the model, they also declare the coarse symbol inventories which are assumed by subsequent linguistic processing. As discussed in Section 27.1, these treebank trees represent only the coarse syntactic structure. In order to obtain accurate parsing performance, we need to learn a more refined grammar.

We introduce an extension of the HDP-PCFG appropriate for grammar refinement (the HDP-PCFG-GR). The HDP-PCFG-GR differs from the basic HDP-PCFG in that the coarse symbols are fixed and only their subsymbols are modelled and controlled via the HDP machinery. This formulation makes explicit the level of coarse, observed symbols and provides implicit control over model complexity within those symbols using a single, unified probabilistic model.

27.3.1 Model definition

The essential difference in the grammar refinement setting is that now we have a collection of HDP-PCFG models, one for each symbol $s \in S$, and each HDP-PCFG operates at the subsymbol level. In practical applications we also need to allow unary productions – a symbol producing exactly one child symbol – the PCFG counterpart of transitions in HMMs. Finally, each non-terminal node $i \in N$ in the parse tree has a symbol-subsymbol pair (s_i, z_i), so each subsymbol needs to specify a distribution over both child symbols and subsymbols. The former is handled through a finite Dirichlet distribution since the finite set of symbols is known. The latter is handled with the HDP machinery, since the (possibly unbounded) set of subsymbols is unknown.

Despite the apparent complexity of the HDP-PCFG-GR model, it is fundamentally a PCFG where the symbols are (s, z) pairs with $s \in S$ and $z \in$

HDP-PCFG for grammar refinement (HDP-PCFG-GR)

For each symbol $s \in S$:
  $\beta_s \sim \mathrm{GEM}(\alpha)$ [draw subsymbol probabilities]
  For each subsymbol $z \in \{1,2,\ldots\}$:
    $\phi^T_{sz} \sim \mathrm{Dir}(\alpha^T)$ [draw rule type parameters]
    $\phi^E_{sz} \sim \mathrm{Dir}(\alpha^E(s))$ [draw emission parameters]
    $\phi^u_{sz} \sim \mathrm{Dir}(\alpha^u(s))$ [draw unary symbol production parameters]
    For each child symbol $s' \in S$:
      $\phi^U_{szs'} \sim \mathrm{DP}(\alpha^U, \beta_{s'})$ [draw unary subsymbol production parameters]
    $\phi^b_{sz} \sim \mathrm{Dir}(\alpha^b(s))$ [draw binary symbol production parameters]
    For each pair of child symbols $(s', s'') \in S \times S$:
      $\phi^B_{szs's''} \sim \mathrm{DP}(\alpha^B, \beta_{s'}\beta^T_{s''})$ [draw binary subsymbol production parameters]

For each node i in the parse tree:
  $t_i \sim \mathrm{Mult}(\phi^T_{s_iz_i})$ [choose rule type]
  If t_i = EMISSION:
    $x_i \sim \mathrm{Mult}(\phi^E_{s_iz_i})$ [emit terminal symbol]
  If t_i = UNARY-PRODUCTION:
    $s_{c_1}(i) \sim \mathrm{Mult}(\phi^u_{s_iz_i})$ [generate child symbol]
    $z_{c_1}(i) \sim \mathrm{Mult}(\phi^U_{s_iz_isc_1(i)})$ [generate child subsymbol]
  If t_i = BINARY-PRODUCTION:
    $(s_{c_1(i)}, s_{c_2(i)}) \sim \mathrm{Mult}(\phi_{s_iz_i})$ [generate child symbols]
    $(z_{c_1(i)}, z_{c_2(i)}) \sim \mathrm{Mult}(\phi^B_{s_iz_isc_1(i)sc_2(i)})$ [generate child subsymbols]

Fig. 27.4 The definition of the HDP-PCFG for grammar refinement (HDP-PCFG-GR).

$\{1, 2, \ldots\}$. The rules of this PCFG take one of three forms: $(s, z) \rightarrow x$ for some terminal symbol $x \in \Sigma$, unary productions $(s, z) \rightarrow (s', z')$ and binary productions $(s, z) \rightarrow (s', z')(s'', z'')$. We can think of having a single (infinite) multinomial distribution over all right-hand sides given (s, z). Interestingly, the prior distribution of these rule probabilities is no longer a Dirichlet or a Dirichlet process, but rather a more general Pólya tree over distributions on $\{\Sigma \cup (S \times Z) \cup (S \times Z \times S \times Z)\}$.

Figure 27.4 provides the complete description of the generative process. As in the HDP-PCFG, we first generate the grammar: For each (coarse) symbol s, we generate a distribution over its subsymbols β_s, and then for each subsymbol z, we generate a distribution over right-hand sides via a set of conditional distributions. Next, we generate a distribution ϕ^T_{sz} over the three rule types {EMISSION, UNARY-PRODUCTION, BINARY-PRODUCTION}. The emission distribution ϕ^T_{sz} is drawn from a Dirichlet as before. For the binary rules, we generate from a Dirichlet a distribution ϕ^b_{sz} over pairs of coarse symbols and then for each coarse symbol $s' \in S$, we generate from a Dirichlet process a distribution $\phi^b_{szs's''}$ over pairs of subsymbols (z', z''). The unary probabilities are generated analogously. Given the grammar, we generate a parse tree and a sentence via a recursive process similar to the one for the HDP-PCFG; the main difference is that at each node, we first generate the coarse symbols of its children and then the child subsymbols conditioned on the coarse symbols.

The HDP-PCFG-GR is a generalization of many of the DP-based models in the literature. When there is exactly one symbol ($|S| = 1$), it reduces to the HDP-PCFG, where the subsymbols in the HDP-PCFG-GR play the role of symbols in the HDP-PCFG. Suppose we have two distinct symbols A and B and that all trees are three node chains of the form $A \rightarrow B \rightarrow x$. Then the HDP-PCFG-GR is equivalent to the nested Dirichlet process (Rodriguez *et al.*, 2008). If the subsymbols of A are observed for all data points, we have a plain hierarchical Dirichlet process (Teh *et al.*, 2006), where the subsymbol of A corresponds to the group of data point x.

27.4 Bayesian inference

In this section, we describe an approximate posterior inference algorithm for the HDP-PCFG based on variational inference. We present an algorithm only for the HDP-PCFG; the extension to the HDP-PCFG-GR is straightforward, requiring only additional bookkeeping. For the basic HDP-PCFG, Section 27.4.1 describes mean-field approximations of the posterior. Section 27.4.2 discusses the optimization of this approximation. Finally, Section 27.4.3 addresses the use of this approximation for predicting parse trees on new sentences. For a further discussion of the variational approach, see Appendix A.2.1. Detailed derivations are presented in Appendix B.

27.4.1 Structured mean-field approximation

The random variables of interest in our model are the parameters $\theta = (\beta, \phi)$, the parse tree z, and the yield x (which we observe). Our goal is thus to compute the posterior $p(\theta, z \mid x)$. We can express this posterior variationally as the solution to an optimization problem:

$$\underset{q \in \mathcal{Q}}{\operatorname{argmin}} \operatorname{KL}(q(\theta, z) \,||\, p(\theta, z \mid x)). \tag{27.2}$$

Indeed, if we let $\mathcal{Q}$ be the family of all distributions over (θ, z), the solution to the optimization problem is the exact posterior $p(\theta, z \mid x)$, since KL divergence is minimized exactly when its two arguments are equal. Of course, solving this optimization problem is just as intractable as directly computing the posterior. Though it may appear that no progress has been made, having a variational formulation allows us to consider tractable choices of $\mathcal{Q}$ in order to obtain principled approximate solutions.

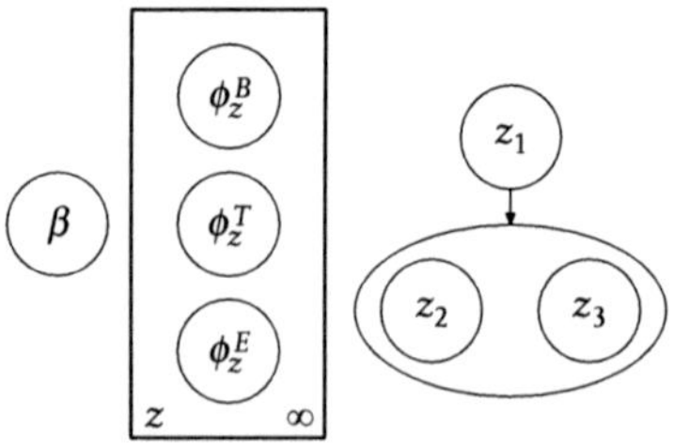

Fig. 27.5 We approximate the posterior over parameters and parse trees using structured mean-field. Note that the posterior over parameters is completely factorized, but the posterior over parse trees is not constrained.

For the HDP-PCFG, we define the set of approximate distributions $\mathcal{Q}$ to be those that factor as follows:

$$\mathcal{Q} = \left\{q : \left[q(\beta) \prod_{z=1}^{K} q\left(\phi_z^T\right) q\left(\phi_z^E\right) q\left(\phi_z^B\right)\right] q(z)\right\}. \quad (27.3)$$

Figure 27.5 shows the graphical model corresponding to the family of approximate distributions we consider. Furthermore, we impose additional parametric forms on the factors as follows:

- $q(\beta)$ is degenerate $(q(\beta) = \delta_{\beta^*}(\beta)$ for some $\beta^*)$ and truncated $(\beta_z^* = 0 \text{ for } z > K)$.
  The truncation is typical of inference algorithms for DP mixtures that are formulated using the stick-breaking representation (Blei and Jordan, 2005). It can be justified by the fact that a truncated stick-breaking distribution well approximates the original stick-breaking distribution in the following sense: Let $G = \sum_{z=1}^{\infty} \beta_z \delta_{\phi_z}$ and let $G^K = \sum_{z=1}^{K} \beta'_z \delta_{\phi_z}$ denote the truncated version, where $\beta'_z = \beta_z$ for $z < K$, $\beta'_K = 1 - \sum_{z=1}^{K-1} \beta_z$ and $\beta'_z = 0$ for $z > K$. The variational distance between G and G^K decreases exponentially as a function of the truncation level K (Ishwaran and James, 2001).
  We also require $q(\beta)$ to be degenerate to avoid the computational difficulties due to the nonconjugacy of β and ϕ_z^B.
- $q\left(\phi_z^T\right)$, $q\left(\phi_z^E\right)$, and $q\left(\phi_z^B\right)$ are Dirichlet distributions. Although the binary production parameters ϕ_z^B specify a distribution over $\{1, 2, \ldots\}^2$, the K-component truncation on β forces ϕ_z^B to assign zero probability to all but a finite $K \times K$ subset. Therefore, only a finite Dirichlet distribution and not a general DP is needed to characterize the distribution over ϕ_z^B.
- $q(z)$ is any multinomial distribution (this encompasses all possible distributions over the discrete space of parse trees). Though the number of parse trees is exponential in the sentence length, it will turn out that the

optimal $q(z)$ always has a factored form which can be found efficiently using dynamic programming.

Note that if we had restricted all the parameter distributions to be degenerate ($q(\phi) = \delta_{\phi^*}(\phi)$ for some ϕ^*) and fixed β^* to be uniform, then we would get the objective function optimized by the EM algorithm. If we further restrict $q(z) = \delta_{z^*}(z)$ for some z^*, we would obtain the objective optimized by Viterbi EM.

27.4.2 Coordinate ascent

We now present an algorithm for solving the optimization problem described in (27.2)–(27.3). Unfortunately, the optimization problem is non-convex, and it is intractable to find the global optimum. However, we can use a simple coordinate ascent algorithm to find a local optimum. The algorithm optimizes one factor in the mean-field approximation of the posterior at a time while fixing all other factors. Optimizing $q(z)$ is the analog of the E-step in fitting an ordinary PCFG, and optimizing $q(\phi)$ is the analog of the M-step. Optimizing $q(\beta)$ has no analog in EM. Note that Kurihara and Sato (2004) proposed a structured mean-field algorithm for Bayesian PCFGs; like ours, their algorithm optimizes $q(z)$ and $q(\phi)$, but it does not optimize $q(\beta)$.

Although mathematically our algorithm performs coordinate ascent on the mean-field factors shown in (27.3), the actual implementation of the algorithm does not actually require storing each factor explicitly; it stores only summaries sufficient for updating the other factors. For example, to update $q(\phi)$, we only need the expected counts derived from $q(z)$; to update $q(z)$, we only need the multinomial weights derived from $q(\phi)$ (see below for details).

27.4.2.1 Updating posterior over parse trees $q(z)$ ('E-step')

A sentence is a sequence of terminals $x = (x_1, \ldots, x_n)$; conditioned on the sentence length n, a parse tree can represented by the following set of variables: $z = \{z_{[i,j]} : 1 \leq i \leq j \leq n\}$, where the variable $z_{[i,j]}$ indicates whether there is a node in the parse tree whose terminal span is $(x_i, \ldots, x_j)$, and, if so, what the grammar symbol at that node is. In the first case, the value of $z_{[i,j]}$ is a grammar symbol, in which case we call $[i, j]$ a *constituent*; otherwise, $z_{[i,j]}$ takes on a special NON-NODE value. In order for z to specify a valid parse tree, two conditions must hold: (1) $[1, n]$ is a constituent, and (2) for each constituent $[i, j]$ with $i < j$, there exists exactly one $k \in [i, j-1]$ for which $[i, k]$ and $[k+1, j]$ are constituents.

Before deriving the update for $q(z)$, we introduce some notation. Let $B(z) = \{[i, j] : i < j \text{ and } z_{[i,j]} \neq \text{NON-NODE}\}$ be the set of binary constituents. Let

$c_1([i, j]) = [i, k]$ and $c_2([i, j]) = [k + 1, j]$, where k is the unique integer such that $[i, k]$ and $[k + 1, j]$ are constituents.

The conditional distribution over the parse tree is given by the following:

$$\begin{aligned} p(z \mid \theta, x) &\propto p(x, z \mid \theta) \\ &= \prod_{i=1}^{n} \phi^{T}_{z[i,i]}(\mathrm{E})\, \phi^{E}_{z_{[i,i]}}(x_i) \prod_{[i,j] \in B(z)} \phi^{T}_{z[i,j]}(\mathrm{B})\, \phi^{B}_{z_{[i,j]}}(z_{c_1([i,j])} z_{c_2([i,j])}), \end{aligned} \tag{27.4}$$

where we have abbreviated the rule types: E = Emission and B = Binary-Production. The first product is over the emission probabilities of generating each terminal symbol x_i, and the second product is over the binary production probabilities used to produce the parse tree. Using this conditional distribution, we can form the mean-field update by applying the general mean-field update rule (see (27.35) in Appendix B.1):

$$q(z) \propto \prod_{i=1}^{n} W^{T}_{z_{[i,i]}}(\mathrm{E})\, W^{E}_{z_{[i,i]}}(x_i) \prod_{[i,j] \in B(z)} W^{T}_{z_{[i,j]}}(\mathrm{B})\, W^{B}_{z_{[i,j]}}(z_{c_1([i,j])}, z_{c_2([i,j])}), \tag{27.5}$$

where the multinomial weights are defined as follows:

$$W^{E}_{z}(x) \stackrel{\text{def}}{=} \exp\left\{E_{q(\phi)} \log \phi^{E}_{z}(x)\right\}, \tag{27.6}$$

$$W^{B}_{z}(z', z'') \stackrel{\text{def}}{=} \exp\left\{E_{q(\phi)} \log \phi^{B}_{z}(z'z'')\right\}, \tag{27.7}$$

$$W^{T}_{z}(t) \stackrel{\text{def}}{=} \exp\left\{E_{q(\phi)} \log \phi^{T}_{z}(t)\right\}. \tag{27.8}$$

These multinomial weights for a d-dimensional Dirichlet are simply d numbers that 'summarize' that distribution. Note that if $q(\phi)$ were degenerate, then W^T, W^E, W^B would be equal to their non-random counterparts ϕ^T, ϕ^E, ϕ^B. In this case, the mean-field update (27.5) is the same as conditional distribution (27.4). Even when $q(\phi)$ is not degenerate, note that (27.4) and (27.5) factor in the same way, which has important implications for computational efficiency.

To compute the normalization constant of (27.5), we can use using dynamic programming. In the context of PCFGs, this amounts to using the standard inside-outside algorithm (Lari and Young, 1990); see Manning and Schütze (1999) for details. This algorithm runs in $O(n^3 K^3)$ time, where n is the length of the sentence and K is the truncation level. However, if the tree structure is fixed as in the case of grammar refinement, then we can use a variant of the forward-backward algorithm which runs in just $O(nK^3)$ time (Matsuzaki *et al.*, 2005; Petrov *et al.*, 2006).

A common perception is that Bayesian inference is slow because one needs to compute integrals. Our mean-field inference algorithm is a counterexample: because we can represent uncertainty over rule probabilities with single numbers, much of the existing PCFG machinery based on EM can be imported into the Bayesian framework in a modular way. In Section A.3.2, we give an empirical interpretation of the multinomial weights, showing how they take into account the uncertainty of the random rule probabilities they represent.

27.4.2.2 Updating posterior over rule probabilities $q(\phi)$ ('M-step')

The mean-field update for $q(\phi)$ can be broken up into independent updates for the rule type parameters $q(\phi_z^T)$ the emission parameters $q(\phi_z^E)$ and the binary production parameters $q(\phi_z^B)$ for each symbol $z \in S$.

Just as optimizing $q(z)$ only required the multinomial weights of $q(\phi)$, optimizing $q(\phi)$ only requires a 'summary' of $q(z)$. In particular, this summary consists of the expected counts of emissions, rule types, and binary productions, which can be computed by the inside-outside algorithm:

$$C^E(z, x) \stackrel{\text{def}}{=} E_{q(z)} \sum_{1 \le i \le n} \mathbb{I}[z_{[i,i]} = z, x_i = x], \tag{27.9}$$

$$C^B(z, z' z'') \stackrel{\text{def}}{=} E_{q(z)} \sum_{1 \le i \le k < j \le n} \mathbb{I}[z_{[i,j]} = z, z_{[i,k]} = z', z_{[k+1,j]} = z''], \tag{27.10}$$

$$C^T(z, t) \stackrel{\text{def}}{=} \sum_{\gamma} C^t(z, \gamma). \tag{27.11}$$

Applying (27.35), we obtain the following updates (see Appendix B.3 for a derivation):

$$q(\phi_z^E) = \text{Dir}(\phi_z^E; \alpha^E + C^E(z)),$$
$$q(\phi_z^B) = \text{Dir}(\phi_z^B; \alpha^B + C^B(z)),$$
$$q(\phi_z^T) = \text{Dir}(\phi_z^T; \alpha^T + C^T(z)).$$

27.4.2.3 Updating the top-level component weights $q(\beta)$

Finally, we need to update the parameters in the top level of our Bayesian hierarchy $q(\beta) = \delta_{\beta^*}(\beta)$. Unlike the other updates, there is no closed-form expression for the optimal β^*. In fact, the objective function (27.2) is not even convex in β^*. Nonetheless, we can use a projected gradient algorithm (Bertsekas, 1999) to improve β^* to a local optimum. The optimization problem is as follows:

$$\min_{\beta^*} \mathrm{KL}(q(\theta, z) \| p(\theta, z \mid x)) \tag{27.12}$$

$$= \max_{\beta^*} E_q \log p(\theta, z \mid x) + H(q(\theta, z)) \tag{27.13}$$

$$= \max_{\beta^*} \log \mathrm{GEM}(\beta^*; \alpha) + \sum_{z=1}^{K} E_q \log \mathrm{Dir}\left(\phi_z^B; \alpha^B \beta^* \beta^{*T}\right) + \text{constant} \tag{27.14}$$

$$\stackrel{\text{def}}{=} \max_{\beta^*} L(\beta^*) + \text{constant}. \tag{27.15}$$

We have absorbed all terms that do not depend on β^* into the constant, including the entropy, since δ_{β^*} is always degenerate. See Appendix B.4 for the details of the gradient projection algorithm and the derivation of $L(\beta)$ and $\nabla L(\beta)$.

27.4.3 Prediction: Parsing new sentences

After having found an approximate posterior over parameters of the HDP-PCFG, we would like to be able to use it to parse new sentences, that is, predict their parse trees. Given a loss function $\ell(z, z')$ between parse trees, the Bayes optimal parse tree for a new sentence x_{new} is given as follows:

$$z^*_{\text{new}} = \operatorname{argmin}_{z'_{\text{new}}} E_{p(z_{\text{new}} \mid x_{\text{new}}, x, z)} \ell\left(z_{\text{new}}, z'_{\text{new}}\right) \tag{27.16}$$

$$= \operatorname{argmin}_{z'_{\text{new}}} E_{p(z_{\text{new}} \mid \theta, x_{\text{new}}) p(\theta, z \mid x)} \ell\left(z_{\text{new}}, z'_{\text{new}}\right). \tag{27.17}$$

If we use the 0-1 loss $\ell\left(z_{\text{new}}, z'_{\text{new}}\right) = 1 - \mathbb{I}\left[z_{\text{new}} = z'_{\text{new}}\right]$, then (27.16) is equivalent to finding the maximum marginal likelihood parse, integrating out the parameters θ:

$$z^*_{\text{new}} = \operatorname*{argmax}_{z_{\text{new}}} E_{p(\theta, z \mid x)} p(z_{\text{new}} \mid \theta, x_{\text{new}}). \tag{27.18}$$

We can substitute in place of the true posterior $p(\theta, z \mid x)$ our approximate posterior $q(\theta, z) = q(\theta) q(z)$ to get an approximate solution:

$$z^*_{\text{new}} = \operatorname*{argmax}_{z_{\text{new}}} E_{q(\theta)} p(z_{\text{new}} \mid \theta, x_{\text{new}}). \tag{27.19}$$

If $q(\theta)$ were degenerate, then we could evaluate the argmax efficiently using dynamic programming (the Viterbi version of the inside algorithm). However, even for a fully-factorized $q(\theta)$, the argmax cannot be computed efficiently, as noted by MacKay (1997) in the context of variational HMMs. The reason behind this difficulty is that the integration over that rule probabilities couples distant parts of the parse tree which use the same rule, thus destroying the Markov property necessary for dynamic programming. We are thus left with the following options:

1. Maximize the expected log probability instead of the expected probability:

$$z^*_{\text{new}} = \operatorname*{argmax}_{z_{\text{new}}} E_{q(\theta)} \log p(z_{\text{new}} \mid x_{\text{new}}, \theta). \tag{27.20}$$

   Concretely, this amounts to running the Viterbi algorithm with the same multinomial weights that were used while fitting the HDP-PCFG.
2. Extract the mode $\theta^* = q(\theta)$ and parse using θ^* with dynamic programming. One problem with this approach is that the mode is not always well defined, for example, when the Dirichlet posteriors have concentration parameters less than 1.
3. Use options 1 or 2 to obtain a list of good candidates, and then choose the best candidate according to the true objective (27.19).

In practice, we used the first option.

27.4.3.1 *Grammar refinement*

Given a new sentence x_{new}, recall that the HDP-PCFG-GR (Section 27.3.1) defines a joint distribution over a coarse tree s_{new} of symbols and a refinement z_{new} described by subsymbols. Finding the best refined tree $(s_{\text{new}}, z_{\text{new}})$ can be carried out using the same methods as for the HDP-PCFG described above. However, in practical parsing applications, we are interested in predicting coarse trees for use in subsequent processing. For example, we may wish to minimize the 0–1 loss with respect to the coarse tree $\left(1 - \mathbb{I}\left[s_{\text{new}} = s'_{\text{new}}\right]\right)$. In this case, we need to integrate out z_{new}:

$$s^*_{\text{new}} = \operatorname*{argmax}_{s_{\text{new}}} E_{q(\theta)} \sum_{z_{\text{new}}} p(s_{\text{new}}, z_{\text{new}} \mid \theta, x_{\text{new}}). \tag{27.21}$$

Note that even if $q(\theta)$ is degenerate, this expression is difficult to compute because the sum over z_{new} induces long-range dependencies (which is the whole point of grammar refinement), and as a result, $p(s_{\text{new}} \mid x_{\text{new}}, \theta)$ does not decompose and cannot be handled via dynamic programming.

Instead, we adopt the following two-stage strategy, which works quite well in practice (cf. Matsuzaki *et al.*, 2005). We first construct an approximate distribution $\tilde{p}(s \mid x)$ which does decompose and then find the best coarse tree with respect to $\tilde{p}(s \mid x)$ using the Viterbi algorithm:

$$s^*_{\text{new}} = \operatorname*{argmax}_{s} \tilde{p}(s \mid x). \tag{27.22}$$

We consider only tractable distributions $\tilde{p}$ which permit dynamic programming. These distributions can be interpreted as PCFGs where the new symbols $S \times \{(i, j) : 1 \le i \le j \le n\}$ are annotated with the span. The new rule probabilities must be consistent with the spans; that is, the only rules allowed to have non-zero probability are of the form $(a, [i, j]) \to (b, [i, k]) \, (c, [k+1, j])$ (binary) and $(a, [i, j]) \to (b, [i, j])$ (unary), where $a, b, c \in S$ and $x \in \Sigma$. The 'best' such distribution is found according to a KL-projection:

$$\tilde{p}(s \mid x) = \underset{\tilde{p}' \text{ tractable}}{\operatorname{argmin}} \ \mathrm{KL}\left(\exp E_{q(\theta)} \log p(s_{\text{new}} \mid x_{\text{new}}, \theta) \| \tilde{p}'(s \mid x)\right). \quad (27.23)$$

This KL-projection can be done by simple moment-matching, where the moments we need to compute are of the form $\mathbb{I}[(s_{[i,j]} = a, s_{[i,k]} = b, s_{[k+1,j]} = c)]$ and $\mathbb{I}[s_{[i,j]} = a, s_{[i,j]} = b]$ for $a, b, c \in S$, which can be computed using the dynamic programming algorithm described in Section 27.4.2.1.

Petrov and Klein (2007) discuss several other alternatives, but show that the two-step strategy described above performs the best.

27.5 Experiments

We now present an empirical evaluation of the HDP-PCFG and HDP-PCFG-GR models for grammar induction and grammar refinement, respectively. We first show that the HDP-PCFG and HDP-PCFG-GR are able to recover a known grammar more accurately than a standard PCFG estimated with maximum likelihood (Section 27.5.1–Section 27.5.2). We then present results on a large-scale parsing task (Section 27.5.3).

27.5.1 Inducing a synthetic grammar using the HDP-PCFG

In the first experiment, the goal was to recover a PCFG grammar given only sentences generated from that grammar. Consider the leftmost grammar in Figure 27.6; this is the 'true grammar' that we wish to recover. It has a total of eight nonterminal symbols, four of which are left-hand sides of only emission rules (these are the preterminal symbols) and four of which are left-hand sides of only binary production rules (constituent symbols). The probability of each rule is given above the appropriate arrow. Though very simple, this grammar still captures some of the basic phenomena of natural language such as noun phrases (NPs), verb phrases (VPs), determiners (DTs), adjectives (JJs), and so on.

From this grammar, we sampled 1000 sentences. Then, from these sentences alone, we attempted to recover the grammar using the standard PCFG and the HDP-PCFG. For the standard PCFG, we allocated 20 latent symbols, 10 for preterminal symbols and 10 for constituent symbols. For the HDP-PCFG (using the version described in Section 27.2.2), we set the stick-breaking truncation level to $K = 10$ for the preterminal and constituent symbols. All Dirichlet hyperparameters were set to 0.01. We ran 300 iterations of EM for the standard PCFG and variational inference for the HDP-PCFG. Since both algorithms are prone to local optima, we ran the algorithms 30 times with different random initializations. We also encouraged the use of constituent symbols by

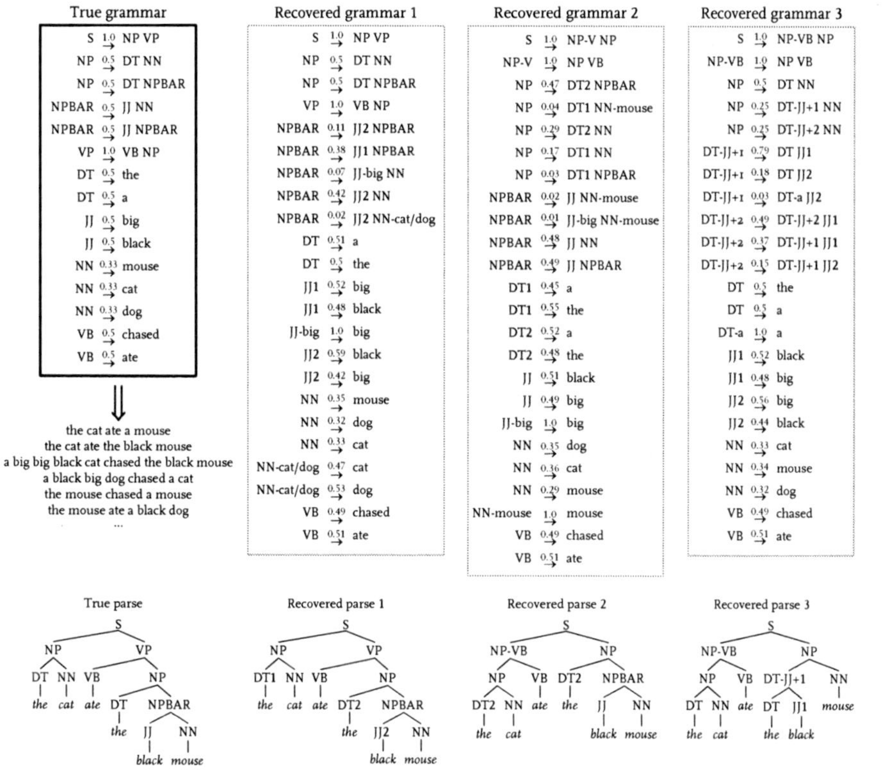

Fig. 27.6 We generated 1000 sentences from the true grammar. The HDP-PCFG model recovered the three grammars (obtained from different initializations of the variational inference algorithm). The parse trees of a sentence under the various grammars are also shown. The first is essentially the same as that of the true grammar, while the second and third contain various left-branching alternatives.

placing slightly more prior weight on rule type probabilities corresponding to production-production rules.

In none of the 30 trials did the standard PCFG manage to recover the true grammar. We say a rule is *active* if its multinomial weight (see Section 27.4.2.1) is at least 10^{-6} and its left-hand side has total posterior probability also at least 10^{-6}. In general, rules with weight smaller than 10^{-6} can be safely ignored without affect parsing results. In a typical run of the standard PCFG, all 20 symbols were used and about 150 of the rules were active (in contrast, there are only 15 rules in the true grammar). The HDP-PCFG managed to do much better. Figure 27.6 shows three grammars (of the 30 trials) which had the highest variational objective values. The symbols are numbered 1 to K in the model, but we have manually labeled them with to suggest their actual role in the syntax. Each grammar has around 4–5 constituent symbols (there were four in the true grammar) and 6–7 preterminal symbols (four in the true grammar), which yields around 25 active rules (15 in the true grammar).

While the HDP-PCFG did not recover the exact grammar, it was able to produce grammars that were sensible. Interestingly, each of the three grammars used a slightly different type of syntactic structure to model the data. The first one is the closest to the true grammar and differs only in that it allocated three subsymbols for JJ and two for NN instead of one. Generally, extra subsymbols provide a better fit of the noisy observations, and here, the sparsity-inducing DP prior was only partially successful in producing a parsimonious model.

The second grammar differs fundamentally from the true grammar in that the subject and the verb are grouped together as a constituent (NP-VB) rather than the verb and the object (VP). This is a problem with non-identifiability; in this simple example, both grammars describe the data equally well and have the same grammar complexity. One way to break this symmetry is to use an informative prior – e.g., that natural language structures tend to be right-branching.

The third grammar differs from the true grammar in one additional way, namely that noun phrases are also left-branching; that is, for *the black mouse*, *the* and *black* are grouped together rather than *black* and *mouse*. Intuitively, this grammar suggests the determiner as the head word of the phrase rather than the noun. This head choice is not entirely unreasonable; indeed there is an ongoing debate in the linguistics community about whether the determiner (the DP hypothesis) or the noun (the NP hypothesis) should be the head (though even in a DP analysis the determiner and adjective should not be grouped).

Given that our variational algorithm converged to various modes depending on initialization, it is evident that the true Bayesian posterior over grammars is multimodal. A fully Bayesian inference procedure would explore and accurately represent these modes, but this is computationally intractable. Starting our variational algorithm from various initializations provides a cheap, if imperfect, way of exploring some of the modes.

Inducing grammars from raw text alone is an extremely difficult problem. Statistical approaches to grammar induction have been studied since the early 1990s (Carroll and Charniak, 1992). As these early experiments were discouraging, people turned to alternative algorithms (Stolcke and Omohundro, 1994) and models (Klein and Manning, 2004; Smith and Eisner, 2005). Though we have demonstrated partial success with the HDP-PCFG on synthetic examples, it is unlikely that this method alone will solve grammar induction problems for large-scale corpora. Thus, for the remainder of this section, we turn to the more practical problem of grammar refinement, where the learning of symbols is constrained by a coarse treebank.

27.5.2 Refining a synthetic grammar with the HDP-PCFG-GR

We first conduct a simple experiment, similar in spirit to the grammar induction experiment, to show that the HDP-PCFG-GR can recover a simple grammar while a standard PCFG-GR (a PCFG adapted for grammar refinement) cannot. From the grammar in Figure 27.7(a), we generated 2000 trees of the form shown in Figure 27.7(b). We then replaced all X_is with X in the training data to yield a coarse grammar. Note that the two terminal symbols always have the same subscript, a correlation not captured by their parent X. We trained both the standard PCFG-GR and the HDP-PCFG-GR using the modified trees as the input data, hoping to estimate the proper refinement of X. We used a truncation of $K = 20$ for both S and X, set all hyperparameters to 1, and ran EM and the variational algorithm for 100 iterations.

Figure 27.8 shows the posterior probabilities of the subsymbols in the grammars produced by the PCFG-GR (a) and HDP-PCFG-GR (b). The PCFG-GR used all 20 subsymbols of both S and X to fit the noisy co-occurrence statistics of left and right terminals, resulting in 8320 active rules (with multinomial weight larger than 10^{-6}). On the other hand, for the HDP-PCFG-GR, only four subsymbols of X and one subsymbol of S had non-negligible posterior mass; 68 rules were active. If the threshold is relaxed from 10^{-6} to 10^{-3}, then only 20 rules are active, which corresponds exactly to the true grammar.

$$
\begin{aligned}
S &\rightarrow X_1X_1 \mid X_2X_2 \mid X_3X_3 \mid X_4X_4 \\
X_1 &\rightarrow a_1 \mid b_1 \mid c_1 \mid d_1 \\
X_2 &\rightarrow a_2 \mid b_2 \mid c_2 \mid d_2 \\
X_3 &\rightarrow a_3 \mid b_3 \mid c_3 \mid d_3 \\
X_4 &\rightarrow a_4 \mid b_4 \mid c_4 \mid d_4
\end{aligned}
$$

(a)

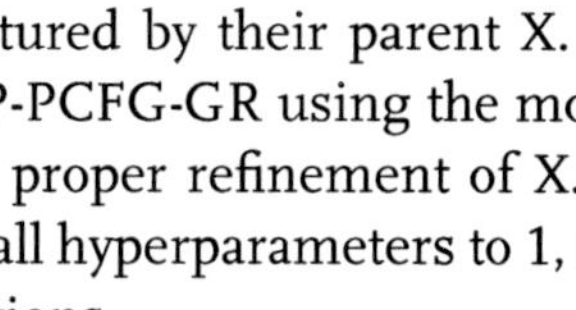

(b)

Fig. 27.7 (a) A synthetic grammar with a uniform distribution over rules. (b) The grammar generates trees of the form shown on the right.

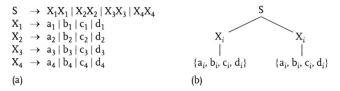

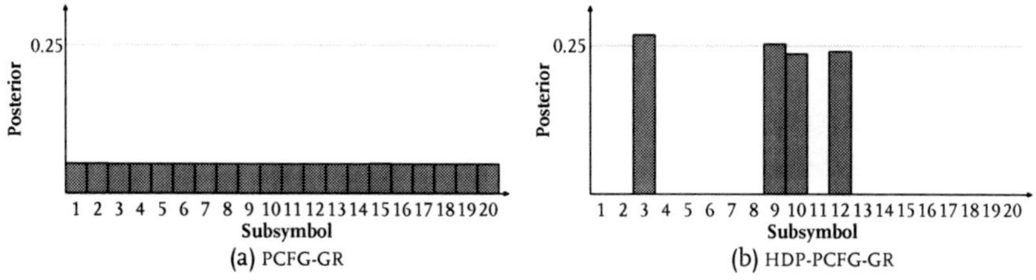

Fig. 27.8 The posterior probabilities over the subsymbols of the grammar produced by the PCFG-GR are roughly uniform, whereas the posteriors for the HDP-PCFG-GR are concentrated on four subsymbols, the true number in the original grammar.

27.5.3 Parsing the Penn Treebank

In this section, we show that the variational HDP-PCFG-GR can scale up to real-world datasets. We truncated the HDP-PCFG-GR at K subsymbols, and compared its performance with a standard PCFG-GR with K subsymbols estimated using maximum likelihood (Matsuzaki *et al.*, 2005).

27.5.3.1 Dataset and preprocessing

We ran experiments on the *Wall Street Journal* (WSJ) portion of the Penn Treebank, a standard dataset used in the natural language processing community for evaluating constituency parsers. The dataset is divided into 24 sections and consists of approximately 40,000 sentences. As is standard, we fit a model on sections 2–21, used section 24 for tuning the hyperparameters of our model, and evaluated parsing performance on section 22.

The HDP-PCFG-GR is defined only for grammars with unary and binary production rules, but the Penn Treebank contains trees in which a node has more than two children. Therefore, we use a standard *binarization* procedure to transform our trees. Specifically, for each non-terminal node with symbol X, we introduce a right-branching cascade of new nodes with symbol $\overline{X}$.

Another issue that arises when dealing with a large-scale dataset of this kind is that words have a Zipfian distribution and in parsing new sentences we invariably encounter new words that did not appear in the training data. We could let the HDP-PCFG-GR's generic Dirichlet prior manage this uncertainty, but would like to use prior knowledge such as the fact that a capitalized word that we have not seen before is most likely a proper noun. We use a simple method to inject this prior knowledge: replace any word appearing fewer than five times in the training set with one of 50 special 'unknown word' tokens, which are added to Σ. These tokens can be thought of as representing manually constructed word clusters.

We evaluated our predicted parse trees using F_1 score, defined as follows. Given a (coarse) parse tree s, let the labelled brackets be defined as

$$\mathrm{LB}(s) = \{(s_{[i,j]}, [i, j]) : s_{[i,j]} \neq \textsc{Non-Node}, 1 \leq i \leq j \leq n\}.$$

Given the correct parse tree s and a predicted parse tree s', the precision, recall and F_1 scores are defined as follows:

$$\mathrm{Precision}(s, s') = \frac{|\mathrm{LB}(s) \cap \mathrm{LB}(s')|}{|\mathrm{LB}(s)|} \quad \mathrm{Recall}(s, s') = \frac{|\mathrm{LB}(s) \cap \mathrm{LB}(s')|}{|\mathrm{LB}(s')|} \tag{27.24}$$

$$\mathrm{F}_1(s, s')^{-1} = \frac{1}{2}\left(\mathrm{Precision}(s, s')^{-1} + \mathrm{Recall}(s, s')^{-1}\right). \tag{27.25}$$

Table 27.1 For each truncation level, we report the α^B that yielded the highest F_1 score on the development set.

truncation K	2	4	8	12	16	20
best α^B	16	12	20	28	48	80
uniform α^B	4	16	64	144	256	400

27.5.3.2 Hyperparameters

There are six hyperparameters in the HDP-PCFG-GR model, which we set in the following manner: $\alpha = 1$, $\alpha^T = 1$ (uniform distribution over unaries versus binaries), $\alpha^E = 1$ (uniform distribution over terminal words), $\alpha^u(s) = \alpha^b(s) = \frac{1}{N(s)}$, where $N(s)$ is the number of different unary (binary) right-hand sides of rules with left-hand side s in the treebank grammar. The two most important hyperparameters are α^U and α^B, which govern the sparsity of the right-hand side for unary and binary rules. We set $\alpha^U = \alpha^B$ although greater accuracy could probably be gained by tuning these individually. It turns out that there is not a single α^B that works for all truncation levels, as shown in Table 27.1. If α^B is too small, the grammars are overly sparse, but if α^B is too large, the extra smoothing makes the grammars too uniform (see Figure 27.9).

If the top-level distribution β is uniform, the value of α^B corresponding to a uniform prior over pairs of child subsymbols is K^2. Interestingly, the best α^B appears to be superlinear but subquadratic in K. We used these best values of α^B in the following experiments.

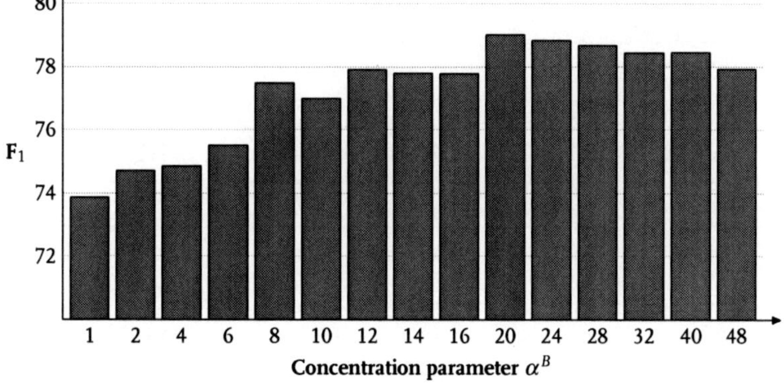

Fig. 27.9 Development F_1 performance on the development set for various values of α^B, training on only section 2 with truncation $K = 8$.

27.5.3.3 Results

The regime in which Bayesian inference is most important is when the number of training data points is small relative to the complexity of the model. To explore this regime, we conducted a first experiment where we trained the PCFG-GR and HDP-PCFG-GR on only section 2 of the Penn Treebank.

Table 27.2 shows the results of this experiment. Without smoothing, the PCFG-GR trained using EM improves as K increases but starts to overfit around $K = 4$. If we smooth the PCFG-GR by adding 0.01 pseudo-counts (which corresponds using a Dirichlet prior with concentration parameters $1.01, \ldots, 1.01$), we see that the performance does not degrade even as K increases to 20, but the number of active grammar rules needed is substantially more. Finally, we see that the HDP-PCFG-GR yields performance comparable to the smoothed PCFG-GR, but the number of grammar rules needed is smaller.

We also conducted a second experiment to demonstrate that our methods can scale up to realistically large corpora. In particular, in this experiment we trained on all of sections 2–21. When using a truncation level of $K = 16$, the standard PCFG-GR with smoothing yielded an F_1 score of 88.36 using 706,157 active rules. The HDP-PCFG-GR yielded an F_1 score of 87.08 using 428,375 active rules. We thus see that the HDP-PCFG-GR achieves broadly comparable performance compared to existing state-of-the-art parsers, while requiring a substantially smaller number of rules. Finally, note that $K = 16$ is a relatively stringent truncation and we would expect to see improved performance from the HDP-PCFG-GR with a larger truncation level.

Table 27.2 Shows the development F_1 scores and grammar sizes (the number of active rules) as we increase the truncation K. The ordinary PCFG-GR overfits around $K = 4$. Smoothing with 0.01 pseudo-counts prevents this, but produces much larger grammars. The HDP-PCFG-GR attains comparable performance with smaller grammars.

K	PCFG-GR		PCFG-GR (smoothed)		HDP-PCFG-GR	
	F_1	Size	F_1	Size	F_1	Size
1	60.47	2558	60.36	2597	60.50	2557
2	69.53	3788	69.38	4614	71.08	4264
4	75.98	3141	77.11	12436	77.17	9710
8	74.32	4262	79.26	120598	79.15	50629
12	70.99	7297	78.80	160403	78.94	86386
16	66.99	19616	79.20	261444	78.24	131377
20	64.44	27593	79.27	369699	77.81	202767

27.6 Discussion

The HDP-PCFG represents a marriage of grammars and Bayesian nonparametrics. We view this marriage as a particularly natural one, given the need for syntactic models to have large, open-ended numbers of grammar symbols. Moreover, given that the most successful methods that have been developed for grammar refinement have been based on clustering of grammar symbols, we view the Dirichlet process as providing a natural starting place for approaching this problem from a Bayesian point of view. The main problem that we have had to face has been that of tying of clusters across parse trees, and this problem is readily solved via the hierarchical Dirichlet process.

We have also presented an efficient variational inference algorithm to approximate the Bayesian posterior over grammars. Although more work is needed to demonstrate the capabilities and limitations of the variational approach, it is important to emphasize the need for fast inference algorithms in the domain of natural language processing. A slow parser is unlikely to make inroads in the applied NLP community. Note also that the mean-field variational algorithm that we presented is closely linked to the EM algorithm, and the familiarity of the latter in the NLP community may help engender interest in the Bayesian approach.

The NLP community is constantly exploring new problem domains that provide fodder for Bayesian modeling. Currently, one very active direction of research is the modeling of multilingual data – for example, synchronous grammars which jointly model the parse trees of pairs of sentences which are translations of each other (Wu, 1997). These models have also received a initial Bayesian treatment (Blunsom *et al.*, 2009). Also, many of the models used in machine translation are essentially already nonparametric, as the parameters of these models are defined on large subtrees rather than individual rules as in the case of the PCFG. Here, having a Bayesian model that integrates over uncertainty can provide more accurate results (DeNero *et al.*, 2008). A major open challenge for these models is again computational; Bayesian inference must be fast if its application to NLP is likely to succeed.

Appendix

A. Broader context and background

In this section, we present some of the background in Bayesian nonparametric modelling and inference needed for understanding the HDP-PCFG. Appendix A.1 describes various simpler models that lead up to the HDP-PCFG. Appendix A.2 overviews some general issues in variational inference and

Appendix A.3 provides a more detailed discussion of the properties of mean-field variational inference in the nonparametric setting.

A.1 DP-based models

We first review the Dirichlet process (DP) mixture model (Antoniak, 1974), a building block for a wide variety of nonparametric Bayesian models, including the HDP-PCFG. The DP mixture model captures the basic notion of clustering which underlies symbol formation in the HDP-PCFG. We then consider the generalization of mixture models to hidden Markov models (HMMs), where clusters are linked structurally according to a Markov chain. To turn the HMM into a nonparametric Bayesian model, we introduce the hierarchical Dirichlet process (HDP). Combining the HDP with the Markov chain structure yields the HDP-HMM, an HMM with a countably infinite state space. This provides the background and context for the HDP-PCFG (Section 27.2).

A.1.1 Bayesian finite mixture models We begin our tour with the familiar Bayesian finite mixture model in order to establish notation which will carry over to more complex models. The model structure is summarized in two ways in Figure 27.10: symbolically in the diagram on the left side of the figure and graphically in the diagram on the right side of the figure. As shown in the figure, we consider a finite mixture with K *mixture components*, where each component $z \in \{1, \ldots, K\}$ is associated with a parameter vector ϕ_z drawn from some prior distribution G_0 on a parameter space Φ. We let the vector $\beta = (\beta_1, \ldots, \beta_K)$ denote the mixing proportions; thus the probability of choosing the mixture component indexed by ϕ_z is given by β_z. This vector is assumed to be drawn from a symmetric Dirichlet distribution with hyperparameter α.

Given the parameters β and $\phi = (\phi_z : z = 1, \ldots, K)$, the data points $x = (x_1, \ldots, x_n)$ are assumed to be generated conditionally i.i.d. as follows: first choose a component z_i with probabilities given by the multinomial probability vector β and then choose x_i from the distribution indexed by z_i; i.e. from $F(\cdot; \phi_{z_i})$.

Bayesian finite mixture model

$\beta \sim \mathrm{Dir}(\alpha, \ldots, \alpha)$ [draw component probabilities]
For each component $z \in \{1, \ldots, K\}$:
  $\phi_z \sim G_0$ [draw component parameters]

For each data point $i \in \{1, \ldots, n\}$:
  $z_i \sim \mathrm{Mult}(\beta)$ [choose component]
  $x_i \sim F(\cdot; \phi_{z_i})$ [generate data point]

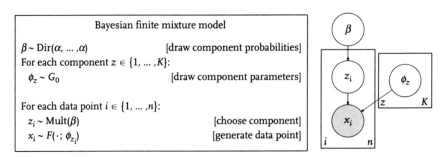

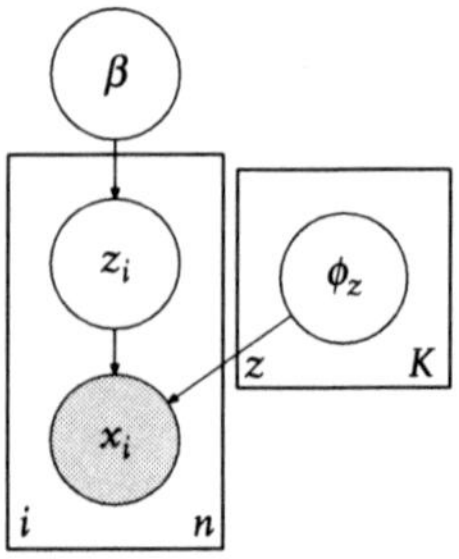

Fig. 27.10 The definition and graphical model of the Bayesian finite mixture model.

There is another way to write the finite mixture model that will be helpful in developing Bayesian nonparametric extensions. In particular, instead of expressing the choice of a mixture component as a choice of a label from a multinomial distribution, let us instead combine the mixing proportions and the parameters associated with the mixture components into a single object – a *random measure* G – and let the choice of a mixture component be expressed as a draw from G. In particular, write

$$G = \sum_{z=1}^{K} \beta_z \delta_{\phi_z}, \tag{27.26}$$

where δ_{ϕ_z} is a delta function at location ϕ_z. Clearly G is random, because both the coefficients $\{\beta_z\}$ and the locations $\{\phi_z\}$ are random. It is also a measure, assigning a nonnegative real number to any Borel subset B of Φ: $G(B) = \sum_{z:\phi_z \in B} \beta_z$. In particular, as desired, G assigns probability β_z to the atom ϕ_z. To summarize, we can equivalently express the finite mixture model in Figure 27.10 as the draw of a random measure G (from a stochastic process that has hyperparameters G_0 and α), followed by n conditionally independent draws of parameter vectors from G, one for each data point. Data points are then drawn from the mixture component indexed by the corresponding parameter vector.

A.1.2 Dirichlet process mixture models The Dirichlet process (DP) mixture model extends the Bayesian finite mixture model to a mixture model having a countably infinite number of mixture components. Since we have an infinite number of mixture components, it no longer makes sense to consider a symmetric prior over the component probabilities as we did in the finite case; the prior over component probabilities must decay in some way. This is achieved via a so-called *stick-breaking distribution* (Sethuraman, 1994).

The stick-breaking distribution that underlies the DP mixture is defined as follows. First define a countably infinite collection of *stick-breaking proportions* $V_1, V_2, \ldots,$ where the V_z are drawn independently from a one-parameter Beta distribution: $V_z \sim \text{Beta}(1, \alpha)$, where $\alpha > 0$. Then define an infinite random sequence β as follows:

$$\beta_z = V_z \prod_{z'<z} (1 - V_{z'}). \tag{27.27}$$

As shown in Figure 27.11, the values of β_z defined this procedure can be interpreted as portions of a unit-length stick. In particular, the product $\prod_{z'<z}(1 - V_{z'})$ is the currently remaining portion of the stick and multiplication by V_z breaks off a proportion of the remaining stick length. It is not difficult to show that the values β_z sum to one (with probability one) and thus β can be viewed as an infinite-dimensional random probability vector.

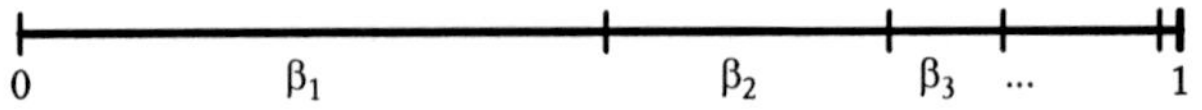

Fig. 27.11 A sample $\beta \sim \text{GEM}(\alpha)$ from the stick-breaking distribution with $\alpha = 1$.

DP mixture model

$\beta \sim \text{GEM}(\alpha)$ [draw component stick probabilities]
For each component $z \in \{1,2, \ldots\}$:
  $\phi_z \sim G_0$ [draw component parameters]

For each data point $i \in \{1, \ldots, n\}$:
  $z_i \sim \text{Mult}(\beta)$ [choose component]
  $x_i \sim F(\cdot; \phi_{z_i})$ [generate data point]

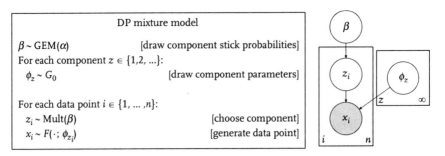

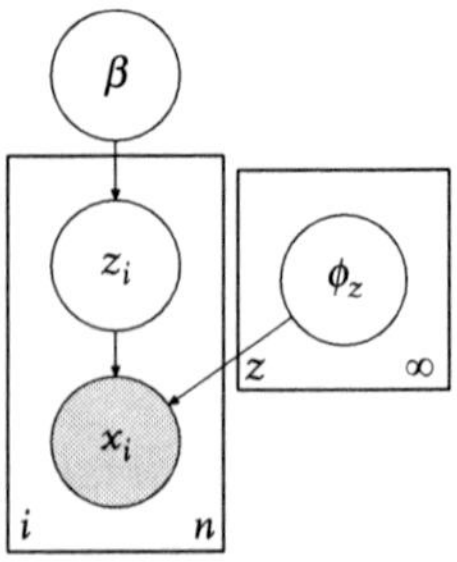

Fig. 27.12 The definition and graphical model of the DP mixture model.

We write $\beta \sim \text{GEM}(\alpha)$ to mean that $\beta = (\beta_1, \beta_2, \ldots)$ is distributed according to the stick-breaking distribution. The hyperparameter α determines the rate of decay of the β_z; a larger value of α implies a slower rate of decay.

We present a full specification of the DP mixture model in Figure 27.12. This specification emphasizes the similarity with the Bayesian finite mixture; all that has changed is that the vectors β and ϕ are infinite dimensional (note that 'Mult' in Figure 27.12 is a multinomial distribution in an extended sense, its meaning is simply that index z is chosen with probability β_z).

Let us also consider the alternative specification of the DP mixture model using random measures. Proceeding by analogy to the finite mixture model, we define a random measure G as follows:

$$G = \sum_{z=1}^{\infty} \beta_z \delta_{\phi_z}, \qquad (27.28)$$

where δ_{ϕ_z} is a delta function at location ϕ_z. As before, the randomness has two sources: the random choice of the ϕ_z (which are drawn independently from G_0) and the random choice of the coefficients $\{\beta_z\}$, which are drawn from $\text{GEM}(\alpha)$. We say that such a random measure G is a draw from a *Dirichlet process*, denoted $G \sim \text{DP}(\alpha, G_0)$. This random measure G plays the same role in the DP mixture as the corresponding G played in the finite mixture; in particular, given G the parameter vectors are independent draws from G, one for each data point.

The term 'Dirichlet process' is appropriate because the random measure G turns out to have finite-dimensional Dirichlet marginals: for an arbitrary partition $(B_1, B_2, \ldots, B_r)$ of the parameter space Φ (for an arbitrary integer r), Sethuraman (1994) showed that

$$\Big(G(B_1), G(B_2), \ldots, G(B_r)\Big) \sim \text{Dir}\Big(\alpha G_0(B_1), \alpha G_0(B_2), \ldots, \alpha G_0(B_r)\Big).$$

A.1.3 Hierarchical Dirichlet Process Hidden Markov Models (HDP-HMMs) The hidden Markov model (HMM) can be viewed as a dynamic version of a finite mixture model. As in the finite mixture model, an HMM has a set of K mixture components, referred to as *states* in the HMM context. Associated to each state, $z \in \{1, 2, \ldots, K\}$, is a parameter vector ϕ_z^E; this vector parametrizes a family of *emission distributions*, $F\left(\cdot; \phi_z^E\right)$, from which data points are drawn. The HMM differs from the finite mixture in that states are not selected independently, but are linked according to a Markov chain. In particular, the parametrization for the HMM includes a *transition matrix*, whose rows ϕ_z^T are the conditional probabilities of transitioning to a next state z' given that the current state is z. Bayesian versions of the HMM place priors on the parameter vectors $\left\{\phi_z^E, \phi_z^T\right\}$. Without loss of generality, assume that the initial state distribution is fixed and degenerate.

A nonparametric version of the HMM can be developed by analogy to the extension of the Bayesian finite mixture to the DP mixture. This has been done by Teh *et al.* (2006), following on from earlier work by Beal *et al.* (2002). The resulting model is referred to as the *hierarchical Dirichlet process HMM* (HDP-HMM). In this section we review the HDP-HMM, focusing on the new ingredient, which is the need for a *hierarchical* DP.

Recall that in our presentation of the DP mixture, we showed that the choice of the mixture component (i.e. the state) could be conceived in terms of a draw from a random measure G. Recall also that the HMM generalizes the mixture model by making the choice of the state conditional on the previous state. This suggests that in extending the HMM to the nonparametric setting we should consider a *set* of random measures, $\{G_z\}$, one measure for each value of the current state.

A difficulty arises, however, if we simply proceed by letting each G_z be drawn independently from a DP. In this case, the atoms forming G_z and those forming $G_{z'}$, for $z \neq z'$, will be distinct with probability one (assuming that the distribution G_0 is continuous). This means that the set of next states available from z will be entirely disjoint from the set of states available from z'.

To develop a nonparametric version of the HMM, we thus require a notion of *hierarchical Dirichlet process* (HDP), in which the random measures $\{G_z\}$ are tied. The HDP of Teh *et al.* (2006) does this by making a single global choice of the atoms underlying each of the random measures. Each of the individual measures G_z then weights these atoms differently. The general framework of the HDP achieves this as follows:

$$G_0 \sim \mathrm{DP}(\gamma, H) \tag{27.29}$$

$$G_z \sim \mathrm{DP}(\alpha, G_0), \quad z = 1, \ldots, K, \tag{27.30}$$

where γ and α are concentration hyperparameters, where H is a measure and where K is the number of measures in the collection $\{G_z\}$. The key to this hierarchy is the presence of G_0 as the base measure used to draw the random measures $\{G_z\}$. Because G_0 is discrete, only the atoms in G_0 can be chosen as atoms in the $\{G_z\}$. Moreover, because the stick-breaking weights in G_0 decay, it is only a subset of the atoms – the highly weighted atoms – that will tend to occur frequently in the measures $\{G_z\}$. Thus, as desired, we share atoms among the $\{G_z\}$.

To apply these ideas to the HMM and to the PCFG, it will prove helpful to streamline our notation. Note in particular that in the HDP specification, the same atoms are used for all of the random measures. What changes among the measures G_0 and the $\{G_z\}$ is not the atoms but the stick-breaking weights. Thus we can replace with the atoms with the positive integers and focus only on the stick-breaking weights. In particular, we re-express the HDP as follows:

$$\beta \sim \text{GEM}(\gamma) \tag{27.31}$$

$$\phi_z^T \sim \text{DP}(\alpha, \beta), \quad z = 1, \ldots, K. \tag{27.32}$$

In writing the hierarchy this way, we are abusing notation. In (27.31), the vector β is a vector with an infinite number of components, but in (27.32), the symbol β refers to the measure that has atoms at the integers with weights given by the components of the vector. Similarly, the symbol ϕ_z^T refers technically to a measure on the integers, but we will also abuse notation and refer to ϕ_z^T as a vector.

It is important not to lose sight of the simple idea that is expressed by (27.32): the stick-breaking weights corresponding to the measures $\{G_z\}$ are reweightings of the global stick-breaking weights given by β. (An explicit formula relating ϕ_z^T and β can be found in Teh *et al.* (2006).)

Returning to the HDP-HMM, the basic idea is to use (27.31) and (27.32) to fill the rows of an infinite-dimensional transition matrix. In this application of the HDP, K is equal to infinity, and the vectors generated by (27.31) and (27.32) form the rows of a transition matrix for an HMM with a countably infinite state space.

We provide a full probabilistic specification of the HDP-HMM in Figure 27.13. Each state z is associated with transition parameters ϕ_z^T and emission parameters ϕ_z^E. Note that we have specialized to multinomial observations in this figure, so that ϕ_z^E is a finite-dimensional vector that we endow with a Dirichlet distribution. Given the parameters $\{\phi_z^T, \phi_z^E\}$, a state sequence $(z_1, \ldots, z_n)$ and an associated observation sequence $(x_1, \ldots, x_n)$ are generated as in the classical HMM. For simplicity we assume that z_1 is always fixed to a designated START state. Given the state z_t at time t, the next state z_{t+1} is obtained by drawing

HDP-HMM

$\beta \sim \mathrm{GEM}(\gamma)$ [draw top-level state probabilities]

For each state $z \in \{1, 2, \ldots\}$:

$\phi_z^E \sim \mathrm{Dir}(\gamma)$ [draw emission parameters]

$\phi_z^T \sim \mathrm{DP}(\alpha, \beta)$ [draw transition parameters]

For each time step $i \in \{1, \ldots, n\}$:

$x_i \sim \mathrm{Mult}(\phi_{z_i}^E)$ [emit current observation]

$z_{i+1} \sim \mathrm{Mult}(\phi_{z_i}^T)$ [choose next state]

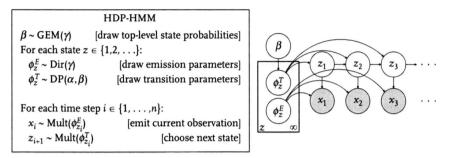

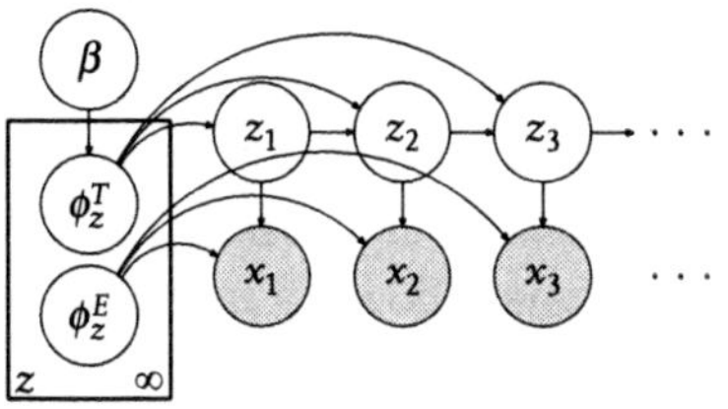

Fig. 27.13 The definition and graphical model of the HDP-HMM.

an integer from $\phi_{z_t}^T$, and the observed data point at time t is obtained by a draw from $\phi_{z_t}^E$.

A.2 Bayesian inference

We would like to compute the full Bayesian posterior $p(\theta, z \mid x)$ over the HDP-PCFG grammar θ and latent parse trees z given observed sentences x. Exact computation of this posterior is intractable, so we must resort to an approximate inference algorithm.

A.2.1 Sampling versus variational inference The two major classes of methodology available for posterior inference in the HDP-PCFG are Markov chain Monte Carlo (MCMC) sampling (Robert and Casella, 2004) or variational inference (Wainwright and Jordan, 2008). MCMC sampling is based on forming a Markov chain that has the posterior as its stationary distribution, while variational inference is based on treating the posterior distribution as the solution to an optimization problem and then relaxing that optimization problem. The two methods have complementary strengths: under appropriate conditions MCMC is guaranteed to converge to a sample from the true posterior, and its stochastic nature makes it less prone to local optima than deterministic approaches. Moreover, the sampling paradigm also provides substantial flexibility; a variety of sampling moves can be combined via Metropolis – Hastings. On the other hand, variational inference can provide a computationally efficient approximation to the posterior. While the simplest variational methods can have substantial bias, improved approximations can be obtained (at increased computational cost) via improved relaxations. Moreover, storing and manipulating a variational approximation to a posterior can be easier than working with a collection of samples.

In our development of the HDP-PCFG, we have chosen to work within the variational inference paradigm, in part because computationally efficient inference is essential in parsing applications, and in part because of the familiarity of variational inference ideas within NLP. Indeed, the EM algorithm can be viewed as coordinate ascent on a variational approximation that is based on a degenerate posterior, and the EM algorithm (the inside-outside algorithm in the case of PCFGs; Lari and Young, 1990) has proven to be quite effective in NLP applications to finite non-Bayesian versions of the HMM and PCFG. Our variational algorithm generalizes EM by using non-degenerate posterior distributions, allowing us to capture uncertainty in the parameters. Also, the variational algorithm is able to incorporate the DP prior, while EM cannot in a meaningful way (see Section A.3.3 for further discussion). On the other hand, the variational algorithm is procedurally very similar to EM and incurs very little additional computational overhead.

A.2.2 Representation used by the inference algorithm Variational inference is based on an approximate representation of the posterior distribution. For DP mixture models (Figure 27.12), there are several possible choices of representation. In particular, one can use a stick-breaking representation (Blei and Jordan, 2005) or a *collapsed* representation based on the Chinese restaurant process, where the parameters β and ϕ have been marginalized out (Kurihara *et al.*, 2007).

Algorithms that work in the collapsed representation have the advantage that they only need to work in the finite space of clusterings rather than the infinite-dimensional parameter space. They have also been observed to perform slightly better in some applications (Teh *et al.*, 2007). In the stick-breaking representation, one must deal with the infinite-dimensional parameters by either truncating the stick (Ishwaran and James, 2001), introducing auxiliary variables that effectively provide an adaptive truncation (Walker, 2004), or adaptively allocating more memory for new components (Papaspiliopoulos and Roberts, 2008).

While collapsed samplers have been effective for the DP mixture model, there is a major drawback to using them for structured models such as the HDP-HMM and HDP-PCFG. Consider the HDP-HMM. Conditioned on the parameters, the computation of the posterior probability z can be done efficiently using the forward-backward algorithm, a dynamic program that exploits the Markov structure of the sequence. However, when parameters have been marginalizing out, the hidden states z sequence are coupled, making dynamic programming impossible.[3] As a result, collapsed samplers for the HDP-HMM generally end up sampling one state z_i at a time conditioned on the rest. This sacrifice can be especially problematic when there are strong dependencies along the Markov chain, which we would expect in natural language. For example, in an HMM model for part-of-speech tagging where the hidden states represent parts-of-speech, consider a sentence containing a two-word fragment *heads turn*. Two

[3] However, it is still possible to sample in the collapsed representation and still exploit dynamic programming via Metropolis – Hastings (Johnson *et al.*, 2007).

possible tag sequences might be NOUN VERB or VERB NOUN. However, in order for the sampler to go from one to the other, it would have to go through NOUN NOUN or VERB VERB, both of which are very low probability configurations.

An additional advantage of the non-collapsed representation is that conditioned on the parameters, inference on the parse trees decouples and can be easily parallelized, which is convenient for large datasets. Thus, we chose to use the stick-breaking representation for inference.

A.3 Mean-field variational inference for DP-based models

Since mean-field inference is only an approximation to the HDP-PCFG posterior, it is important to check that the approximation is a sensible one – that we have not entirely lost the benefits of having a nonparametric Bayesian model. For example, EM, which approximates the posterior over parameters with a single point estimate, is a poor approximation, which lacks the model selection capabilities of mean-field variational inference, as we will see in Section A.3.3.

In this section, we evaluate the mean-field approximation with some illustrative examples of simple mixture models. Section A.3.1 discusses the qualitative nature of the approximated posterior, Section A.3.2 shows how the DP prior manifests itself in the mean-field update equations, and Section A.3.3 discusses the long term effect of the DP prior over multiple mean-field iterations.

A.3.1 Mean-field approximation of the true posterior Consider simple mixture model in Figure 27.14. Suppose we observe $n = 3$ data points drawn from the model: (1, 4), (1, 4), and (4, 1), where each data point consists of five binomial trials. Figure 27.15 shows how the true posterior over parameters compares with EM and mean-field approximations. We make two remarks:

1. The true posterior is symmetrically bimodal, which reflects the non-identifiability of the mixture components: ϕ_1 and ϕ_2 can be interchanged without affecting the posterior probability. The mean-field approximation can approximate only one of those modes, but does so quite well in this simple example. In general, the mean-field posterior tends to underestimate the variance of the true posterior.

Two-component mixture model	
$\phi_1 \sim \text{Beta}(1,1)$	[draw parameter for component 1]
$\phi_2 \sim \text{Beta}(1,1)$	[draw parameter for component 2]
For each data point $i \in \{1, \ldots, n\}$:	
$z_i \sim \text{Mult}(\frac{1}{2}, \frac{1}{2})$	[choose component]
$x_i \sim \text{Bin}(5, \phi_{z_i})$	[generate data point]

Fig. 27.14 A two-component mixture model.

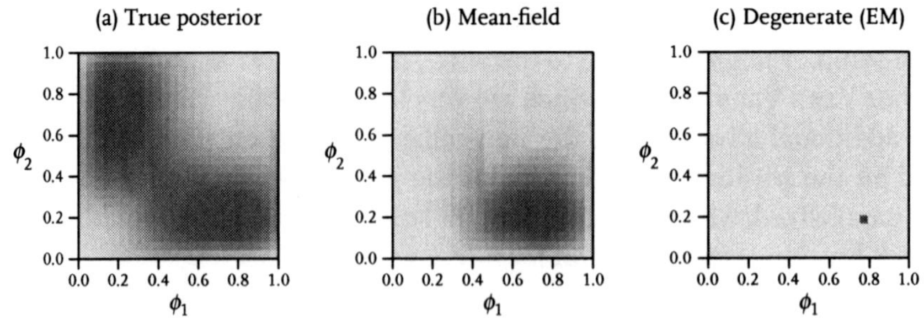

Fig. 27.15 Shows (a) the true posterior $p(\phi_1, \phi_2 \mid x)$, (b) the optimal mean-field posterior, and (c) the optimal degenerate posterior (found by EM).

2. The mode found by mean-field has higher variance in the ϕ_1 coordinate than the ϕ_2. This is caused by the fact that there component 2 has two data points ((1, 4) and (1, 4)) supporting the estimates whereas component 1 has only one ((4, 1)). EM (using MAP estimation) represents the posterior as a single point at the mode, which ignores the varying amounts of uncertainty in the parameters.

A.3.2 EM and mean-field updates in the E-step In this section we explore the difference between EM and mean-field updates as reflected in the E-step. Recall that in the E-step, we optimize a multinomial distribution $q(z)$ over the latent discrete variables z; for the HDP-PCFG, the optimal multinomial distribution is given by (27.5) and is proportional to a product of multinomial weights (27.36). The only difference between EM and mean-field is that EM uses the maximum likelihood estimate of the multinomial parameters whereas mean-field uses the multinomial weights. Unlike the maximum likelihood solution, the multinomial weights do not have to sum to one; this provides mean-field with an additional degree of freedom that the algorithm can use to capture some of the uncertainty. This difference is manifested in two ways, via a local tradeoff and a global tradeoff between components.

Local tradeoff Suppose that $\beta \sim \text{Dir}(\alpha, \ldots, \alpha)$, and we observe counts $(c_1, \ldots, c_K) \sim \text{Mult}(\beta)$. Think of β as a distribution over the K mixture components and $c_1, \ldots, c_K$ as the expected counts computed in the E-step. In the M-step, we compute the posterior over β, and compute the multinomial weights needed for the next E-step.

For MAP estimation (equivalent to assuming a degenerate $q(\beta)$), the multinomial weights would be

$$W_i = \frac{c_i + \alpha - 1}{\sum_{j=1}^{K}(c_j + \alpha - 1)}.$$

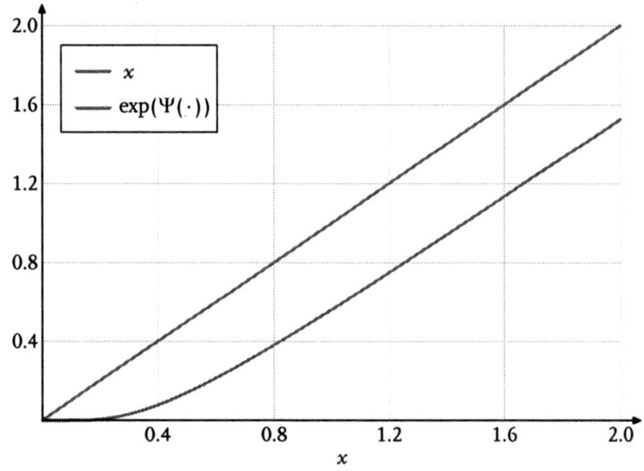

Fig. 27.16 The $\exp(\Psi(\cdot))$ function, which is used in computing the multinomial weights for mean-field inference. It has the effect of adversely impacting small counts more than large counts.

When $\alpha = 1$, maximum likelihood is equivalent to MAP, which corresponds to simply normalizing the counts ($W_i \propto c_i$).

Using a mean-field approximation for $q(\beta)$ yields the following multinomial weights:

$$W_i = \exp\{E_{\beta \sim \mathrm{Dir}(c_1+\alpha,\ldots,c_K+\alpha)} \log \beta_i\} = \frac{\exp(\Psi(c_i+\alpha))}{\exp\left(\Psi\left(\sum_{j=1}^{K}(c_j+\alpha)\right)\right)}. \tag{27.33}$$

Let $\alpha = \alpha'/K$ and recall from that for large K, $\mathrm{Dir}(\alpha'/K, \ldots, \alpha'/K)$ behaves approximately like a Dirichlet process prior with concentration α' (Theorem 2 of Ishwaran and Zarepour 2002).

When K is large, $\alpha = \alpha'/K \approx 0$, so, crudely, $W_i \propto \exp(\Psi(c_i))$ (although the weights need not sum to 1). The $\exp(\Psi(\cdot))$ function is shown in Figure 27.16. We see that $\exp(\Psi(\cdot))$ has the effect of reducing the weights; for example, when $c > \frac{1}{2}$, $\exp(\Psi(c)) \approx c - \frac{1}{2}$. However, the relative reduction $1 - \exp(\Psi(c_i))/c_i$ is much more for small values of c_i than for large values of c_i.

This induces a rich-gets-richer effect characteristic of Dirichlet processes, where larger counts get further boosted and small counts get further diminished. As c increases, however, the relative reduction tends to zero. This asymptotic behavior is in agreement with the general fact that the Bayesian posterior over parameters converges to a degenerate distribution at the maximum likelihood estimate.

Global tradeoff Thus far we have considered the numerator of the multinomial weight (27.33), which determines how the various components of the multinomial are traded off *locally*. In addition, there is a *global* tradeoff that occurs between different multinomial distributions due to the denominator. Consider the two-component mixture model from Section A.3.1. Suppose that after the E-step, the expected sufficient statistics are (20, 20) for component 1 and (0.5, 0.2) for component 2. In the M-step, we compute the multinomial weights W_{zj} for components $z = 1, 2$ and dimensions $j = 1, 2$. For another observation $(1, 0) \sim \text{Bin}^1(\phi_z)$, where $z \sim \text{Mult}\left(\frac{1}{2}, \frac{1}{2}\right)$, the optimal posterior $q(z)$ is proportional to the multinomial weights W_{z1}.

For MAP ($q(\phi_1, \phi_2)$ is degenerate), these weights are

$$W_{11} = \frac{20}{20 + 20} = 0.5 \quad \text{and} \quad W_{21} = \frac{0.5}{0.5 + 0.2} \approx 0.714,$$

and thus component 2 has higher probability. For mean-field ($q(\phi_1, \phi_2)$ is a product of two Dirichlets), these weights are

$$W_{11} = \frac{e^{\Psi 20+1}}{e^{\Psi 20+20+1}} \approx 0.494 \quad \text{and} \quad W_{21} = \frac{e^{\Psi 0.5+1}}{e^{\Psi 0.5+0.2+1}} \approx 0.468,$$

and thus component 1 has higher probability. Note that mean-field is sensitive to the uncertainty over parameters: even though the mode of component 2 favors the observation (1, 0), component 2 has sufficiently higher uncertainty that the optimal posterior $q(z)$ will favor component 1.

This global tradeoff is another way in which mean-field penalizes the use of many components. Components with more data supporting its parameters will be preferred over their meager counterparts.

A.3.3 Interaction between prior and inference In this section we consider some of the interactions between the choice of prior (Dirichlet or Dirichlet process) and the choice of inference algorithm (EM or mean-field). Consider a K-component mixture model with component probabilities β with a 'null' data-generating distribution, meaning that $F(\cdot; \cdot) \equiv 1$ (see Section A.1.1). Note that this is not the same as having zero data points, since each data point still has an unknown assignment variable z_i. Because of this, the approximate posterior $q(\beta)$ does not converge to the prior $p(\beta)$.

Figure 27.17 shows the impact of three priors using both EM and mean-field. Each plot shows how the posterior over components probabilities change over time. In general, the component probabilities would be governed by both the influence of the prior shown here and the influence of a likelihood.

With a Dir(1, ..., 1) prior and EM, the component probabilities do not change over iterations since the uniform prior over component probabilities does not prefer one component over the other (a). If we use mean-field with the same prior, the weights converge to the uniform distribution since $\alpha = 1$ roughly has the effect of adding 0.5 pseudo-counts (d) (see Section A.3.2). The same asymptotic behaviour can be obtained by using Dir(1.1, ..., 1.1) and EM, which is equivalent to adding 0.1 pseudo-counts (b).

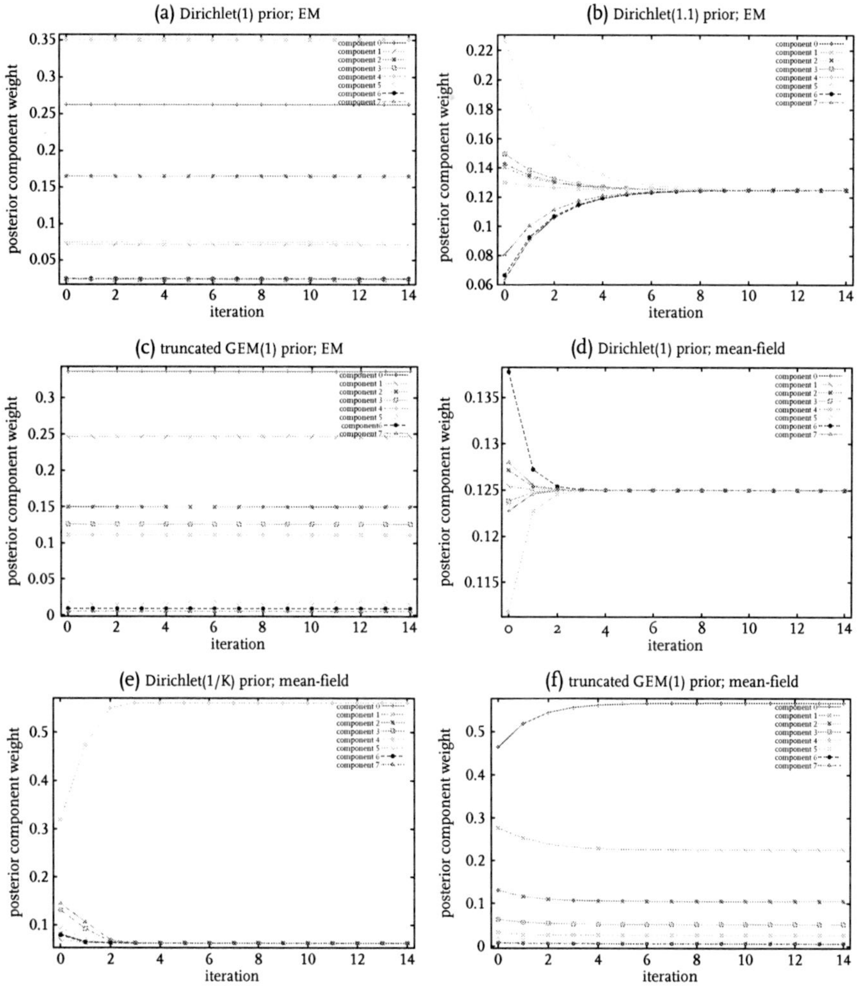

Fig. 27.17 The evolution of the posterior mean of the component probabilities under the null data-generating distribution for three priors $p(\beta)$ using either EM or mean-field.

Since the data-generating distribution is the same for all components, one component should be sufficient and desirable, but so far, none of these priors encourage sparsity in the components. The Dir$(1/K, \ldots, 1/K)$ prior (e), which approximates a DP prior (Ishwaran and Zarepour, 2002), does provide this sparsity if we use mean-field. With $\alpha = 1/K$ near zero, components with more counts are favoured. Therefore, the probability of the largest initial component increases and all others are driven close to zero. Note that EM is not well defined because the mode is not unique.

Finally, plots (c) and (f) provide two important insights:

1. The Dirichlet process is defined by using the GEM prior (Section A.1.2), but if we use EM for inference the prior does not produce the desired sparsity in the posterior (c). This is an instance where the inference algorithm is too weak to leverage the prior. The reason behind this failure is that the truncation of GEM(1) places a uniform distribution on each of the stick-breaking proportions, so there is no pressure to move away from the initial set of component probabilities.
   Note that while the truncation of GEM(1) places a uniform distribution over stick-breaking proportions, it does not induce a uniform distribution over the stick-breaking probabilities. If we were to compute the MAP stick-breaking probabilities under the induced distribution, we would obtain a different result. The reason because this discrepancy is that although the likelihood is independent of the parametrization, the prior is not. Note that full Bayesian posterior inference is independent of the parametrization.
2. If we perform approximate Bayesian inference using mean-field (f), then we can avoid the stagnation problem that we had with EM. However, the effect is not as dramatic as in (e). This shows that although the $\mathrm{Dir}(1/K, \ldots, 1/K)$ prior and the truncated GEM(1) prior both converge to the DP, their effect in the context of mean-field is quite different. It is perhaps surprising that the stick-breaking prior, which places emphasis on decreasing component sizes, actually does not result in a mean-field posterior that decays as quickly as the posterior produced by the symmetric Dirichlet.

B. Derivations

This section derives the update equations for the coordinate-wise ascent algorithm presented in Section 27.4. We begin by deriving the form of the general update (Section B.1). Then we apply this result to multinomial distributions (Section B.2) for the posterior over parse trees $q(z)$ and Dirichlet distributions (Section B.3) for the posterior over rule probabilities $q(\phi)$. Finally, we describe the optimization of the degenerate posterior over the top-level component weights $q(\beta)$ (Section B.4).

B.1 General updates

In a general setting, we have a fixed distribution $p(y)$ over the variables $y = (y_1, \ldots, y_n)$, which can be computed up to a normalization constant. (For the HDP-PCFG, $p(y) = p(z, \theta \mid x)$.) We wish to choose the best mean-field approximation $q(y) = \prod_{i=1}^{n} q(y_i)$, where the $q(y_i)$ are arbitrary distributions. Recall that the objective of mean-field variational inference (27.2) is to minimize the KL-divergence between $q(y)$ and $p(y)$. Suppose we fix $q(y_{-i}) \stackrel{\text{def}}{=} \prod_{j \neq i} q(y_j)$ and wish to optimize $q(\gamma_i)$. We rewrite the variational objective with the intent of optimizing it with respect to $q(\gamma_i)$:

$$\begin{aligned}
\mathrm{KL}(q(\gamma)||p(\gamma)) &= -E_{q(\gamma)}[\log p(\gamma)] - H(q(\gamma)) \\
&= -E_{q(\gamma)}[\log p(\gamma_{-i})p(\gamma_i \mid \gamma_{-i})] - \sum_{j=1}^{n} H(q(\gamma_j)) \\
&= \Big(-E_{q(\gamma_i)}[E_{q(\gamma_{-i})}\log p(\gamma_i \mid \gamma_{-i})] - H(q(\gamma_i))\Big) + \\
&\qquad \Big(-E_{q(\gamma_{-i})}\log p(\gamma_{-i}) - \sum_{j\neq i} H(q(\gamma_j))\Big) \\
&= \mathrm{KL}(q(\gamma_i)||\exp\{E_{q(\gamma_{-i})}\log p(\gamma_i \mid \gamma_{-i})\}) + \text{constant}, \qquad (27.34)
\end{aligned}$$

where the constant does not depend on $q(\gamma_i)$. We have exploited the fact that the entropy $H(q(\gamma))$ decomposes into a sum of individual entropies because q is fully factorized.

The KL-divergence is minimized when its two arguments are equal, so if there are no constraints on $q(\gamma_i)$, the optimal update is given by

$$q(\gamma_i) \propto \exp\{E_{q(\gamma_{-i})}\log p(\gamma_i \mid \gamma_{-i})\}. \qquad (27.35)$$

In fact, we only need to compute $p(\gamma_i \mid \gamma_{-i})$ up to a normalization constant, because $q(\gamma_i)$ must be normalized anyway.

From (27.35), we can see a strong connection between Gibbs sampling and mean-field. Gibbs sampling draws $\gamma_i \sim p(\gamma_i \mid \gamma_{-i})$. Both algorithms iteratively update the approximation to the posterior distribution one variable at a time using its conditional distribution.

B.2 Multinomial updates

This general update takes us from (27.4) to (27.5) in Section 27.4.2.1, but we must still compute the multinomial weights defined in (27.6)–(27.7). We will show that the multinomial weights for a general Dirichlet are as follows:

$$W_i \stackrel{\text{def}}{=} \exp\{E_{\phi\sim\mathrm{Dir}(\gamma)}\log\phi_i\} = \frac{\exp\{\Psi(\gamma_i)\}}{\exp\{\Psi(\sum_j \gamma_j)\}}, \qquad (27.36)$$

where $\Psi(x) = \frac{d}{dx}\log\Gamma(x)$ is the digamma function.

The core of this computation is the computation of the mean value of $\log\phi_i$ with respect to a Dirichlet distribution. Write the Dirichlet distribution in exponential family form:

$$p(\phi \mid \gamma) = \exp\Big\{\sum_i \gamma_i \underbrace{\log\phi_i}_{\substack{\text{sufficient}\\ \text{statistics}}} - \underbrace{\Big[\sum_i \log\Gamma(\gamma_i) - \log\Gamma\Big(\sum_i \gamma_i\Big)\Big]}_{\text{log-partition function } A(\gamma)}\Big\}.$$

Since the expected sufficient statistics of an exponential family are equal to the derivatives of the cumulant function, we have

$$E \log \phi_i = \frac{\partial A(\gamma)}{\partial \gamma_i} = \Psi(\gamma_i) - \Psi\Big(\sum_j \gamma_j\Big).$$

Exponentiating both sides yields (27.36).

Finally, in the context of the HDP-PCFG, (27.36) can be applied with Dir(γ) as $q(\phi^E)$, $q(\phi^B)$, or $q(\phi^T)$.

B.3 Dirichlet updates

Now we consider updating the Dirichlet components of the mean-field approximation, which include $q(\phi^E)$, $q(\phi^B)$, and $q(\phi^T)$. We will only derive $q(\phi^E)$ since the others are analogous. The general mean-field update (27.35) gives us

$$q\left(\phi_z^E\right) \propto \exp\left\{ E_{q(z)} \log p\left(\phi_z^E \mid \theta \backslash \phi_z^E, z, x\right) \right\} \tag{27.37}$$

$$\propto \exp\left\{ E_{q(z)} \log \prod_{x \in \Sigma} \phi_z^E(x)^{\alpha^E(x)} \prod_{x \in \Sigma} \phi_z^E(x)^{\sum_{i=1}^n \mathbb{I}[z_{[i,i]}=z, x_i=x]} \right\} \tag{27.38}$$

$$= \exp\left\{ \sum_{x \in \Sigma} E_{q(z)}\left(\alpha^E(x) + \sum_{i=1}^n \mathbb{I}[z_{[i,i]} = z, x_i = x]\right) \log \phi_z^E(x) \right\} \tag{27.39}$$

$$= \exp\left\{ \sum_{x \in \Sigma} (\alpha^E(x) + C^E(z, x)) \log \phi_z^E(x) \right\} \tag{27.40}$$

$$= \prod_{x \in \Sigma} \phi_z^E(x)^{\alpha^E(x) + C^E(z,x)} \tag{27.41}$$

$$\propto \text{Dir}\left(\phi_z^E; \alpha^E(\cdot) + C^E(z, \cdot)\right). \tag{27.42}$$

B.4 Updating top-level component weights

In this section, we discuss the computation of the objective function $L(\beta)$ and its gradient $\nabla L(\beta)$, which are required for updating the top-level component weights (Section 27.4.2.3).

We adapt the projected gradient algorithm (Bertsekas, 1999). At each iteration t, given the current point $\beta^{(t)}$, we compute the gradient $\nabla L(\beta^{(t)})$ and update the parameters as follows:

$$\beta^{(t+1)} \leftarrow \Pi(\beta^{(t)} + \eta_t \nabla L(\beta^{(t)})), \tag{27.43}$$

where η_t is the step size, which is chosen based on approximate line search, and Π projects its argument onto the K-dimensional simplex, where K is the truncation level.

Although β specifies a distribution over the positive integers, due to truncation, $\beta_z = 0$ for $z > K$ and $\beta_K = 1 - \sum_{z=1}^{K-1} \beta_z$. Thus, slightly abusing notation, we write $\beta = (\beta_1, \ldots, \beta_{K-1})$ as the optimization variables, which must reside in the set

$$\mathcal{T} = \left\{(\beta_1, \ldots, \beta_{K-1}) : \forall 1 \leq z < K, \beta_z \geq 0 \text{ and } \sum_{z=1}^{K-1} \beta_z \leq 1\right\}. \tag{27.44}$$

The objective function is the sum of two terms, which can be handled separately:

$$L(\beta) = \underbrace{\log \text{GEM}(\beta; \alpha)}_{\text{'prior term' } L_{\text{prior}}} + \underbrace{\sum_z E_q \log \text{Dir}\left(\phi_z^B; \alpha_B \beta\beta^\top\right)}_{\text{'rules term' } L_{\text{rules}}}. \tag{27.45}$$

Prior term Let $\pi : [0, 1]^{K-1} \mapsto \mathcal{T}$ be the mapping from stick-breaking proportions $\mathbf{u} = (u_1, \ldots, u_{K-1})$ to stick-breaking weights $\beta = (\beta_1, \ldots, \beta_{K-1})$ as defined in (27.27). The inverse map is given by

$$\pi^{-1}(\beta) = (\beta_1/T_1, \beta_2/T_2, \ldots, \beta_{K-1}/T_{K-1})^\top,$$

where

$$T_z \stackrel{\text{def}}{=} 1 - \sum_{z'=1}^{z-1} \beta_{z'}$$

are the tail sums of the stick-breaking weights.

The density for the GEM prior is naturally defined in terms of the densities of the independent Beta-distributed stick-breaking proportions $\mathbf{u}$. However, to produce a density on $\mathcal{T}$, we need to perform a change of variables. First, we compute the Jacobian of π^{-1}:

$$\nabla \pi^{-1} = \begin{pmatrix} 1/T_1 & 0 & 0 & \cdots & 0 \\ * & 1/T_2 & 0 & \cdots & 0 \\ * & * & 1/T_3 & \cdots & 0 \\ \vdots & \vdots & \vdots & \ddots & \vdots \\ * & * & * & * & 1/T_{K-1} \end{pmatrix}. \tag{27.46}$$

The determinant of a lower triangular matrix is the product of the diagonal entries, so we have that

$$\log \det \nabla \pi^{-1}(\beta) = -\sum_{z=1}^{K-1} \log T_z, \tag{27.47}$$

which yields

$$
\begin{aligned}
L_{\text{prior}}(\beta) &= \log \text{GEM}(\beta; \alpha) \\
&= \log \prod_{z=1}^{K-1} \text{Beta}(\beta_z / T_z; 1, \alpha) + \log \det \nabla \pi^{-1}(\beta) \\
&= \log \prod_{z=1}^{K-1} \frac{\Gamma(\alpha+1)}{\Gamma(\alpha)\Gamma(1)} (1 - \beta_z / T_z)^{\alpha-1} + \log \det \nabla \pi^{-1}(\beta) \\
&= (K-1) \log \alpha + \sum_{z=1}^{K-1} (\alpha - 1) \log(T_{z+1} / T_z) + \log \det \nabla \pi^{-1}(\beta) \\
&= (\alpha - 1) \log T_K - \sum_{z=1}^{K-1} \log T_z + \text{constant}.
\end{aligned}
$$

Note that the Beta prior is only applied to the first $K-1$ stick-breaking proportions, since the K-th proportion is always fixed to 1. The terms from the log densities reduce to $(\alpha - 1) \log T_K$ via a telescoping sum, leaving the terms from the Jacobian as the main source of regularization.

Observing that $\frac{\partial T_z}{\partial \beta_k} = -\mathbb{I}[z > k]$, we can differentiate the prior term with respect to β_k:

$$
\frac{\partial L_{\text{prior}}(\beta)}{\partial \beta_k} = -\frac{\alpha - 1}{T_K} - \sum_{z=1}^{K-1} \frac{-\mathbb{I}[z > k]}{T_z} \tag{27.48}
$$

$$
= \sum_{z=k+1}^{K-1} \frac{1}{T_z} - \frac{\alpha - 1}{T_K}. \tag{27.49}
$$

To compute the effect of β_k on the objective, we only need to look at tail sums of the component weights that follow it.

For comparison, let us compute the derivative when using a Dirichlet instead of a GEM prior:

$$
\tilde{L}_{\text{prior}}(\beta) = \log \text{Dir}(\beta; \alpha) = (\alpha - 1) \sum_{z=1}^{K} \log \beta_k + \text{constant}. \tag{27.50}
$$

Differentiating with respect to β_k for $k = 1, \ldots, K-1$ (β_K is a deterministic function of $\beta_1, \ldots, \beta_{K-1}$) yields:

$$
\frac{\partial \tilde{L}_{\text{prior}}(\beta)}{\partial \beta_k} = (\alpha - 1)\left(\frac{1}{\beta_k} - \frac{1}{\beta_K}\right) = \frac{\alpha - 1}{\beta_k} - \frac{\alpha - 1}{T_K}. \tag{27.51}
$$

Rules term We compute the remaining term in the objective function:

$$L_{\text{rules}}(\beta) = \sum_{z=1}^{K} E_{q(\phi)} \log \text{Dir}\left(\phi_z^B; \alpha_B \beta\beta^\top\right) \tag{27.52}$$

$$= \sum_{z=1}^{K} \left(\log \Gamma(\alpha_B) - \sum_{i=1}^{K}\sum_{j=1}^{K} \log \Gamma(\alpha_B \beta_i \beta_j) + \sum_{i=1}^{K}\sum_{j=1}^{K} (\alpha_B \beta_i \beta_j - 1) E_{q(\phi)} \log \phi_z^B(i, j) \right).$$

Before differentiating $L_{\text{rules}}(\beta)$, let us first differentiate a related function $L_{\text{rules}-K}(\beta, \beta_K)$, defined on K arguments instead of $K-1$:

$$\frac{\partial L_{\text{rules}-K}(\beta, \beta_K)}{\partial \beta_k} = \sum_{z=1}^{K} \left(-2 \sum_{i=1}^{K} \alpha_B \beta_i \Psi(\alpha_B \beta_i \beta_k) + \sum_{i=1}^{K} \alpha_B \beta_i E_{q(\phi)} \log \phi^B(k, i) \phi_z^B(i, k) \right) \tag{27.53}$$

$$= \alpha_B \sum_{z=1}^{K}\sum_{i=1}^{K} \beta_i \left(E_{q(\phi)} \log \phi_z^B(k, i) \phi_z^B(i, k) - 2\Psi(\alpha_B \beta_i \beta_k) \right).$$

Now we can apply the chain rule using the fact that $\frac{\partial \beta_K}{\partial \beta_k} = -1$ for $k = 1, \ldots, K-1$:

$$\frac{\partial L_{\text{rules}}(\beta)}{\partial \beta_k} = \frac{\partial L_{\text{rules}-K}(\beta, \beta_K)}{\partial \beta_k} + \frac{\partial L_{\text{rules}-K}(\beta, \beta_K)}{\partial \beta_K} \cdot \frac{\partial \beta_K}{\partial \beta_k}$$

$$= \alpha_B \sum_{z=1}^{K}\sum_{i=1}^{K} \beta_i (E_{q(\phi)} \log \phi_z^B(k, i) \phi_z^B(i, k) - 2\Psi(\alpha_B \beta_i \beta_k)) -$$

$$\alpha_B \sum_{z=1}^{K}\sum_{i=1}^{K} \beta_i (E_{q(\phi)} \log \phi_z^B(K, i) \phi_z^B(i, K) - 2\Psi(\alpha_B \beta_i \beta_K)). \tag{27.54}$$

References

Antoniak, C. E. (1974). Mixtures of Dirichlet processes with applications to Bayesian nonparametric problems. *Annals of Statistics*, **2**, 1152–1174.

Beal, M., Ghahramani, Z. and Rasmussen, C. (2002). The infinite hidden Markov model. In *Advances in Neural Information Processing Systems (NIPS)*, Volume 14, pp. 577–584. MIT Press, Cambridge, MA.

Bertsekas, D. (1999). *Nonlinear Programming*. Athena Scientific, Belmont, MA.
Blei, D. and Jordan, M. I. (2005). Variational inference for Dirichlet process mixtures. *Bayesian Analysis*, **1**, 121–144.
Blunsom, P., Cohn, T. and Osborne, M. (2009). Bayesian synchronous grammar induction. In *Advances in Neural Information Processing Systems (NIPS)*, Volume 21, pp. 161–168. MIT Press, Cambridge, MA.
Carroll, G. and Charniak, E. (1992). Two experiments on learning probabilistic dependency grammars from corpora. In Workshop Notes for Statistically-Based NLP Techniques, pp. 1–13. American Association for Artificial Intelligence (AAAI).
Charniak, E. (1996). Tree-bank grammars. In *Proceedings of the Thirteenth National Conference on Artificial Intelligence*, American Association for Artificial Intelligence (AAAI), pp. 1031–1036. MIT Press, Cambridge, MA.
Charniak, E. (2000). A maximum-entropy-inspired parser. In Applied Natural Language Processing and North American Association for Computational Linguistics (ANLP/NAACL), Seattle, Washington, pp. 132–139. Association for Computational Linguistics.
Chomsky, N. (1956). Three models for the description of language. *IRE Transactions on Information Theory*, **2**, 113–124.
Collins, M. (1999). Head-driven statistical models for natural language parsing. Ph.D. thesis, University of Pennsylvania.
DeNero, J., Bouchard-Côté, A. and Klein, D. (2008). Sampling alignment structure under a Bayesian translation model. In Empirical Methods in Natural Language Processing (EMNLP), Honolulu, HI, pp. 314–323.
Dyrka, W. and Nebel, J. (2007). A probabilistic context-free grammar for the detection of binding sites from a protein sequence. *Systems Biology, Bioinformatics and Synthetic Biology*, **1**, 78–79.
Ferguson, T. S. (1973). A Bayesian analysis of some nonparametric problems. *Annals of Statistics*, **1**, 209–230.
Ferguson, T. S. (1974). Prior distributions on spaces of probability measures. *Annals of Statistics*, **2**, 615–629.
Finkel, J. R., Grenager, T. and Manning, C. (2007). The infinite tree. In Association for Computational Linguistics (ACL), Prague, Czech Republic, pp. 272–279. Association for Computational Linguistics.
Galley, M., Hopkins, M., Knight, K. and Marcu, D. (2004). What's in a translation rule? In Human Language Technology and North American Association for Computational Linguistics (HLT/NAACL), Boston, MA, pp. 273–280.
Gildea, D. and Jurafsky, D. (2002). Automatic labeling of semantic roles. *Computational Linguistics*, **28**, 245–288.
Hermjakob, U. (2001). Parsing and question classification for question answering. In Workshop on Open-domain Question Answering, ACL, Toulouse, France, pp. 1–6.
Ishwaran, H. and James, L. F. (2001). Gibbs sampling methods for stick-breaking priors. *Journal of the American Statistical Association*, **96**, 161–173.
Ishwaran, H. and Zarepour, M. (2002). Exact and approximate sum-representations for the Dirichlet process. *Canadian Journal of Statististics*, **30**, 269–284.
Johnson, M. (1998). PCFG models of linguistic tree representations. *Computational Linguistics*, **24**, 613–632.
Johnson, M., Griffiths, T. and Goldwater, S. (2006). Adaptor grammars: A framework for specifying compositional nonparametric Bayesian models. In *Advances in Neural Information Processing Systems (NIPS)*, Volume 19, pp. 641–648. MIT Press, Cambridge, MA.

Johnson, M., Griffiths, T. and Goldwater, S. (2007). Bayesian inference for PCFGs via Markov chain Monte Carlo. In Human Language Technology and North American Association for Computational Linguistics (HLT/NAACL), Rochester, New York, pp. 139–146.

Klein, D. and Manning, C. (2003). Accurate unlexicalized parsing. In Association for Computational Linguistics (ACL), Sapporo, Japan, pp. 423–430. Association for Computational Linguistics.

Klein, D. and Manning, C. D. (2004). Corpus-based induction of syntactic structure: Models of dependency and constituency. In Association for Computational Linguistics (ACL), Barcelona, Spain, pp. 478–485. Association for Computational Linguistics.

Kurihara, K. and Sato, T. (2004). An application of the variational Bayesian approach to probabilistic context-free grammars. In International Joint Conference on Natural Language Processing Workshop Beyond Shallow Analyses, Japan.

Kurihara, K., Welling, M. and Teh, Y. W. (2007). Collapsed variational Dirichlet process mixture models. In International Joint Conference on Artificial Intelligence (IJCAI), Hyderabad, India.

Lari, K. and Young, S. J. (1990). The estimation of stochastic context-free grammars using the inside-outside algorithm. *Computer Speech and Language*, **4**, 35–56.

MacKay, D. (1997). Ensemble learning for hidden Markov models. Technical report, University of Cambridge.

Manning, C. and Schütze, H. (1999). *Foundations of Statistical Natural Language Processing*. MIT Press, Cambridge, MA.

Marcus, M. P., Marcinkiewicz, M. A. and Santorini, B. (1993). Building a large annotated corpus of English: the Penn Treebank. *Computational Linguistics*, **19**, 313–330.

Matsuzaki, T., Miyao, Y. and Tsujii, J. (2005). Probabilistic CFG with latent annotations. In Association for Computational Linguistics (ACL), Ann Arbor, Michigan, pp. 75–82. Association for Computational Linguistics.

Papaspiliopoulos, O. and Roberts, G. O. (2008). Retrospective MCMC for Dirichlet process hierarchical models. *Biometrika*, **95**, 169–186.

Pereira, F. and Shabes, Y. (1992). Inside-outside reestimation from partially bracketed corpora. In Association for Computational Linguistics (ACL), Newark, Delaware, pp. 128–135. Association for Computational Linguistics.

Petrov, S., Barrett, L., Thibaux, R. and Klein, D. (2006). Learning accurate, compact, and interpretable tree annotation. In International Conference on Computational Linguistics and Association for Computational Linguistics (COLING/ACL), pp. 433–440. Association for Computational Linguistics.

Petrov, S. and Klein, D. (2007). Learning and inference for hierarchically split PCFGs. In Human Language Technology and North American Association for Computational Linguistics (HLT/NAACL), Rochester, New York, pp. 404–411.

Robert, C. P. and Casella, G. (2004). *Monte Carlo Statistical Methods*. Springer, New York.

Rodriguez, A., Dunson, D. B. and Gelfand, A. E. (2008). The nested Dirichlet process. *Journal of the American Statistical Association*, **103**, 1131–1144.

Sakakibara, Y. (2005). Grammatical inference in bioinformatics. *IEEE Transactions on Pattern Analysis and Machine Intelligence*, **27**, 1051–1062.

Sethuraman, J. (1994). A constructive definition of Dirichlet priors. *Statistica Sinica*, **4**, 639–650.

Smith, N. and Eisner, J. (2005). Contrastive estimation: Training log-linear models on unlabeled data. In Association for Computational Linguistics (ACL), Ann Arbor, Michigan, pp. 354–362. Association for Computational Linguistics.

Stolcke, A. and Omohundro, S. (1994). Inducing probabilistic grammars by Bayesian model merging. In International Colloquium on Grammatical Inference and Applications, London, UK, pp. 106–118. Springer-Verlag, Berlin.

Teh, Y. W., Jordan, M. I., Beal, M. and Blei, D. (2006). Hierarchical Dirichlet processes. *Journal of the American Statistical Association*, **101**, 1566–1581.

Teh, Y. W., Newman, D. and Welling, M. (2007). A collapsed variational Bayesian inference algorithm for latent Dirichlet allocation. In *Advances in Neural Information Processing Systems (NIPS)*, Volume 19, pp. 1353–1360. MIT Press, Cambridge, MA.

Wainwright, M. and Jordan, M. I. (2008). Graphical models, exponential families, and variational inference. *Foundations and Trends in Machine Learning*, **1**, 1–307.

Walker, S. G. (2004). Sampling the Dirichlet mixture model with slices. *Communications in Statistics – Simulation and Computation*, **36**, 45–54.

Wu, D. (1997). Stochastic inversion transduction grammars and bilingual parsing of parallel corpora. *Computational Linguistics*, **23**, 377–404.

Zhu, S. C. and Mumford, D. (2006). A stochastic grammar of images. *Foundations and Trends in Computer Graphics and Vision*, **2**, 259–362.

·28·

Designing and analysing a circuit device experiment using treed Gaussian processes

Herbert K. H. Lee, Matthew Taddy, Robert B. Gramacy and Genetha A. Gray

28.1 Introduction

This chapter describes work in the development of a new circuit device, and is part of a collaboration with scientists at Sandia National Laboratories. Circuit devices need to be tested for various capabilities during the development phase, in order to eventually create a device that will be effective in a range of operational environments. Both physical and computer simulation experiments are used. Our work here focuses on two parts of this process: creation of the design for the physical experiment, and optimization during the calibration of the computer model. Our key statistical tool is the treed Gaussian process.

The goal of our collaborators was to both 'calibrate' and 'validate' computer models of the electrical circuit devices. These models can then be used for the next stage of the design process. In this context, calibration uses data from physical experiments to inform upon uncertain input parameters of a computer simulation model (Kennedy and O'Hagan, 2001, for example). Model parameters are tuned so that the code calculations in the simulator most closely match the observed behaviour of the experimental data. Accurate calibration both improves predictive capabilities and minimizes the information lost by using a numerical model instead of the actual system.

In contrast, the validation process is applied in order to quantify the degree to which a computational model is an accurate representation of the real world phenomena it seeks to represent (Oberkampf *et al.*, 2003). Validation is critical to ensure some level of predictive capability of the simulation codes so that these codes can be subsequently used to study or predict situations for which experimental data are unavailable due to environmental or economic limitations. Model validation is a major effort regularly undertaken by numerous government agencies and private companies. Our role as statisticians focused on the calibration aspects, and did not deal directly with validation.

28.1.1 Circuit experiments

The circuit devices under study are bipolar junction transistors (bjt), which are used to amplify electrical current. They exist in both PNP and NPN constructions (Sedra and Smith, 1997). Twenty different such devices were studied; in this chapter we present results only for a few of them. For example, we will consider the *bft92a*, which is a PNP, and the *bfs17a* which is an NPN. The primary interest is in understanding current output as a function of the intensity of a pulse of gamma radiation applied to the devices, where the current output is characterized by the peak amplitude reached during the experiment. Because of the physical setup of the experiment, a particular testing facility is capable of only a rather limited range of possible radiation doses, so experiments were run at three different facilities to span a broader range of possible dose rates. Although in principle the relationship between dose rate and peak amplitude should not depend on the facility, some of the results do appear to show a facility effect. Experimental runs were done at three different temperature settings. The scientists do allow for the results to depend on temperature, and they have separate computer simulation models for each temperature. Further details on the physical experiment are available in Gray *et al.* (2007).

The physical experiment is accompanied by a collection of computer simulation models. These radiation-aware models are built on a Xyce implementation of the Tor Fjeldy photocurrent model for the bjt. Xyce is an electrical circuit simulator developed at Sandia National Laboratories (Keiter, 2004), and the Tor Fjeldy photocurrent model is described in detail in Fjeldy *et al.* (1997). The model input is a radiation pulse expressed as a dose rate over time. The corresponding output is a curve of current value over time which reflects the response of the electrical device. This curve is summarized by its maximum value. The model also involves 38 user-defined tuning parameters. It is these parameters that need to be calibrated using the physical data so that the simulator output is as close as possible to the results from the physical experiments. All bjts share a basic underlying model and the same computer simulator can be used to approximate their behavior, with only changes to the parameters required.

28.1.2 Treed Gaussian processes

Treed Gaussian processes are a highly flexible and computationally efficient model for spatially correlated data, as well as for more general functions. They combine standard Gaussian processes with treed partition models, producing an effective semi-parametric non-stationary model. In this section, we start with a brief review of Gaussian processes, cover the basics of treed Gaussian processes, then discuss their use for adaptive sampling in computer experiments and for optimization. Additional details are available in the appendix.

Gaussian processes are the standard model for creating a statistical emulator of the output of a computer simulator (Sacks *et al.*, 1989; Kennedy and O'Hagan, 2001; Santner *et al.*, 2003). A Gaussian process (GP) specifies that the set of responses $z(x_1, \ldots, x_m)$ at any finite collection of locations x (which could be spatial locations or multivariate inputs to a computer model) has a multivariate Gaussian distribution. In general, we can write $z(x) = \mu(x) + w(x)$ where μ is a mean function and $w(x)$ is a zero mean random process with covariance $C(x_j, x_k)$. Typically we will assume stationarity for the correlation structure, and thus we have $C(x_j, x_k) = \sigma^2 K(x_j, x_k)$ where K is a correlation matrix whose entries depend only on the difference vectors $x_j - x_k$. Sometimes a further assumption of isotropy is made, and then the correlations depend only on the distance between x_j and x_k, typically specified with a simple parametric form (Abrahamsen, 1997). We take the mean function to be linear in the inputs, $\mu(x) = \beta' x$. We also use a small nugget to allow for noisy data, for smoothing, or for numerical stability. For more details on GPs, we refer the reader to Cressie (1993) or Stein (1999). Herein we use the separable power family for the correlation structure, with a separate range parameter d_i in each dimension ($i = 1, \ldots, m_X$), and power $p_0 = 2$ to give smooth realizations:

$$K(x_j, x_k | d) = \exp\left\{ -\sum_{i=1}^{m_X} \frac{|x_{ij} - x_{ik}|^{p_0}}{d_i} \right\}. \tag{28.1}$$

Standard GPs have three drawbacks that affect us here. First, they are computationally intensive to fit, with effort growing with the cube of the sample size due to the need to invert the covariance matrix. Second, GP models usually assume stationarity, in that the same covariance structure is used throughout the entire input space, which may be too strong of a modelling assumption, or else they are fully nonstationary and too computationally expensive to be used for more than a relatively small number of datapoints. Third, the estimated predictive error of a stationary model does not directly depend on the locally observed response values. Rather, the predictive error at a point depends only on the locations of the nearby observations and on a global measure of error that uses all of the discrepancies between observations and predictions without regard for their distance from the point of interest (because of the stationarity assumption). Thus there is no ready local measure of lack-of-fit or uncertainty, which will be needed for sequential experimental design (adaptive sampling).

All of these issues can be addressed by combining the stationary GP model with treed partitioning, resulting in what we call a treed Gaussian process (TGP). The input space is partitioned into disjoint regions, with an independent stationary GP fit in each region. This approach provides a computationally efficient way of creating a non-stationary model. It reduces the overall computational demands by fitting separate GP models to smaller data sets (the individual partitions). The partitioning approach is based on that of Chipman *et al.* (1998, 2002), who used it to develop the Bayesian classification and regression trees. Reversible jump Markov chain Monte Carlo (Green, 1995) with tree proposal operations (prune, grow, swap, change, and rotate) allows simultaneous fitting of the tree and the parameters of the individual GP models. In this way, all parts of the model can be learned automatically from the data, and Bayesian model averaging through reversible jump allows for explicit estimation of predictive uncertainty. It also provides a smooth mean fit when appropriate (as the partition locations are also averaged over, so the mean function does not exhibit discontinuities unless the data call for such a fit). We provide more details in the appendix, and also point the reader to Gramacy and Lee (2008*a*). Software is available in the `tgp` library for the statistical package R at `http://www.cran.r-project.org/web/packages/tgp/index.html` (Gramacy and Taddy, 2008; Gramacy, 2007).

28.1.2.1 Adaptive sampling

When each datapoint is difficult or expensive to collect, either because a physical experiment is involved or because the computer simulator takes a long time to run, it is important to choose the sample points with great care. One needs to select the set of design points that will provide maximum information about the problem. Traditional methods for experimental design involve starting with a specific model then creating a design of all of the planned runs. If a simple enough model is chosen, or enough assumptions about parameter values are made, then optimal designs can be found in closed form. For more complex cases, optimal designs are found numerically. Various optimality criteria lead to designs such as *A*-optimal, *D*-optimal, maximin, maximum entropy, and Latin hypercube designs. A good review of the Bayesian literature on optimal designs is provided by Chaloner and Verdinelli (1995). However, such an approach is not readily amenable to learning about the process as the data are collected. The concept of sequential experimental design is less well-developed, but work has progressed in some areas, such as computer experiments (Sacks *et al.*, 1989; Santner *et al.*, 2003). By updating the design after learning from each new observation (or each new batch), one can better deal with lack of knowledge about model parameters, or even about the model itself. Here we are concerned about learning both the tree structure and the GP parameters during the experiment, and so direct application of optimality criteria is not feasible. Some additional background on experimental design is in the appendix.

In selecting design points sequentially, we want those which will provide the most additional information, conditional on the data we have already collected. Because we are fitting a fully Bayesian model, we can consider the predictive variance as a measure of our uncertainty. Thus one way to define the amount of information we expect to learn from a new datapoint is to look at the expected reduction in squared error averaged over the input space. This approach was developed in the active learning literature in computer science by Cohn (1996). This expected reduction in global error for a proposed new sample $\tilde{x}$ under a GP model can be expressed as:

$$\Delta\hat{\sigma}^2(\tilde{x}) = \int \Delta\hat{\sigma}^2_{\tilde{x}}(y)dy \equiv \int \hat{\sigma}^2(y) - \hat{\sigma}^2_{\tilde{x}}(y)dy = \int \frac{\sigma^2\left[q'(y)C^{-1}q(x) - \kappa(x, y)\right]^2}{\kappa(x, x) - q'(x)C^{-1}q(x)}dy \tag{28.2}$$

where y represents the input space, σ^2 is the overall variance, $\hat{\sigma}^2_{\tilde{x}}(y)$ is the estimated (posterior) predictive variance at y when $\tilde{x}$ is added into the design, and

$$\begin{aligned} C^{-1} &= (K + \tau^2 FWF')^{-1} \\ q(x) &= k(x) + \tau^2 FWf(x) \\ \kappa(x, y) &= K(x, y) + \tau^2 f'(x)Wf(y) \end{aligned}$$

with $f'(x) = (1, x')$, and $k(x)$ an n-vector with $k_{\nu,j}(x) = K(x, x_j)$, for all $x_j \in X$. Refer to the appendix for definitions of the other quantities, e.g. W and K. For a treed GP, this expression is evaluated within partitions each MCMC iteration and the result averaged across the whole MCMC run. For points in separate partitions, there is no change in predictive variance. More details on this derivation are given by Gramacy and Lee (2009). In practice the integral is evaluated as a sum over a grid of locations.

Searching for the optimal design point over a multidimensional continuous space is quite difficult and computationally expensive, so we restrict our attention to a smaller set of candidate points that are well-spaced out relative to each other, and then choose the best point from our candidate set. To obtain a well-spaced set, we return to standard optimal designs, such as maximum entropy designs, maximin designs, and/or Latin hypercubes. This approach is then easily extensible to choosing more than one point, either because a physical experiment is being done in batches or because an asynchronous parallel computing environment is being used for a computer experiment. Since the points are spread apart, a simple approach to choosing n_b design points for the next batch is to just select the n_b points in the candidate set with the highest $\Delta\hat{\sigma}^2(\tilde{x})$ as per equation (28.2). A better approach, if computational resources allow, is to choose the first design point as the one with the highest $\Delta\hat{\sigma}^2(\tilde{x})$, then to add a pseudo-datapoint at this chosen location with value equal to its predictive mean value and to recompute equation (28.2), choosing the second point as the one that maximizes this among the remaining points in the candidate set. This approximation has the effect of reducing uncertainty in the local region of the first point, so that the second point will typically be selected from a different part of the input space. This process is iterated until n_b design points are obtained.

28.1.2.2 *Sensitivity analysis*

Global sensitivity analysis (SA; not to be confused with local derivative based analysis) is a resolving of the sources of output variability by apportioning elements of this variation to different sets of input variables (Saltelli *et al.*, 2000). In large engineering problems there can be a huge number of input variables over which the objective is to be optimized, but only a small subset will be influential within the confines of their uncertainty distribution. Thus global SA is an important (but often overlooked) aspect of efficient optimization and it may be performed, at relatively little additional cost, on the basis of a statistical model fit to the initial sample. Variance-based SA methods decompose the variance of the objective function, with respect to an uncertainty distribution placed on the inputs, into variances of conditional expectations. These provide a natural measure of the output association with specific sets of variables and provide a basis upon which the importance of individual inputs may be judged.

We will concentrate on two influential sensitivity indices: the first order for the jth input variable, $S_j = \text{var}\left(\text{E}\left[f(\mathbf{x})|x_j\right]\right)/\text{var}(f(\mathbf{x}))$, and the total sensitivity for input j, $T_j = \text{E}\left[\text{var}\left(f(\mathbf{x})|\mathbf{x}_{-j}\right)\right]/\text{var}(f(\mathbf{x}))$. Here, f denotes the objective function, $\mathbf{x}_{-j}$ is the input vector excluding the jth input, and all expectations and variances are taken with respect to the uncertainty distribution placed on $\mathbf{x}$. The uncertainty distribution may be any probability function defined over the input space, but we will assume that it consists of independent uniform distributions over each (bounded) dimension of the input space. The first order indices measure the portion of variability that is due to variation in the main effects for each input variable, while the total effect indices measure the portion of variability that is due to total variation in each input. Thus, the difference between T_j and S_j provides a measure of the variability in the objective function due to interaction between input j and the other input variables. A large difference may lead the investigator to consider other sensitivity indices to determine where this interaction is most influential, and this is often a key aspect of the dimension-reduction that SA provides for optimization problems. Refer to Sobol' (1993) and Homma and Saltelli (1996) for a complete discussion of the properties and derivation of variance-based SA indices.

The influential paper by Oakley and O'Hagan (2004) describes an empirical Bayes estimation procedure for the sensitivity indices; however, some variability in the indices is lost due to plug-in estimation of GP model parameters and, more worryingly, the variance ratios are only possible in the form of a ratio of expected values. Likelihood based approaches are proposed by Welch *et al.* (1992) and in Morris *et al.* (2008). The technique proposed here is, in contrast, fully Bayesian and provides a complete accounting of the uncertainty involved. Briefly, at each iteration of an MCMC sampler that is taking draws from the TGP posterior, output is predicted over a large (carefully chosen) set of input locations. Conditional on this predicted output, the sensitivity indices can be calculated via Monte Carlo integration. In particular, Saltelli (2002) describes an efficient LHS based scheme for estimation of both first order and total effect indices in such situations, and we follow this technique exactly. That is, the locations chosen for TGP prediction are precisely those prescribed in Saltelli's approach. At each MCMC iteration, after calculating Monte Carlo estimates of the integrals involved conditional on the TGP predicted response, we obtain a posterior realization of the variance indices. The resultant full posterior sample then incorporates variability from both the integral estimation and uncertainty about the function output.

Apart from the variance-related quantities, another common component of global SA is an accounting of the main effects for each input variable, $\mathrm{E}[f(\mathbf{x})|x_j]$ as a function of x_j. These can easily be obtained as a byproduct of the above variance analysis procedure, again through Monte Carlo integration conditional upon the TGP predicted response.

28.1.2.3 Optimization

In Section 28.1.2.1, TGP prediction was used to guide the intelligent collection of data, with the goal being to minimize the predictive variance. An alternative goal would be one of optimization – it is not the entire response surface which is of interest, but rather only the minimum (or maximum) response point on this surface. In this case, although TGP prediction still forms the backbone of our inference, we need a different objective function and a different sampling approach.

Statistical methods are useful in optimization problems, particularly where the function being optimized is best treated as an expensive black-box, i.e. each function evaluation is relatively costly to obtain (physically or in computing time), and no additional information about the function (such as parametric form or gradient evaluations) is available. Creation of a statistical surrogate model, or emulator, of the black-box then allows optimization on the cheaper statistical model, reducing the need for expensive full evaluations (Booker *et al.* 1999). The most direct application of this approach would have statistical prediction fully determine the search for an optimum input configuration, and the most prominent example of this strategy is the Expected Global Optimizer (EGO) algorithm of Jones *et al.* (1998). The method is designed to search the input space and converge towards the global minimum. At each iteration, a GP is fit to the current set of function evaluations and a new location for data collection is chosen based upon the GP posterior expectation for the improvement statistics:

$$I(x) = \max\{(f_{\text{best}} - f(x)), 0\}, \tag{28.3}$$

where f_{best} is response corresponding to the current best point in the search. The location with maximum expected improvement is chosen for evaluation, and following evaluation the GP is then fit anew to the data set augmented by these results. Schonlau *et al.* (1998) provide an extensive discussion of improvement statistics. The key to success for this algorithm is that, in the expectation, candidate locations are rewarded both for a near-optimal mean predicted response as well as for high response uncertainty (indicating a poorly explored region of the input space). Hence, the posterior expectation for improvement provides an ideal statistic to inform intelligent optimization.

For this particular project, our collaborators were interested in not simple global point optimization, but rather robust local optimization. Cost and time constraints often make it infeasible to execute a search of the magnitude required to guarantee global convergence (such as EGO). On the other hand, purely local optimization algorithms will fail on highly multimodal problems where one can easily get stuck in poor local optima. Furthermore, in engineering applications (such as that discussed herein) it is essential to avoid local solutions on a *knife's edge* portion of the response surface, where small changes in the input lead to large changes in the response.

Thus, the problem at hand may be characterized as robust local optimization – we need to find a solution with a response that is close to the global optimum, while using many fewer iterations than a truly global search would require. The proposed solution is to combine existing local optimization methods, for quick convergence, with TGP statistical prediction, to give the algorithm a global scope. Briefly, a local optimization search pattern is periodically augmented with locations chosen to maximize the TGP predicted expected improvement.

The TGP generated search pattern consists of m locations that maximize (over a discrete candidate set chosen through some space-filling design) the expected multilocation improvement, $E\left[I(x_1, \ldots, x_m)\right]$, where

$$I(x_1, \ldots, x_m) = \max\{(f_{\text{best}} - f(x_1)), \ldots, (f_{\text{best}} - f(x_m)), 0\} \tag{28.4}$$

(Schonlau *et al.*, 1998). Taking a fully Bayesian approach, the improvement $I(\tilde{x})$ is drawn for each $\tilde{x}$ in the candidate set at each iteration of MCMC sampling from the TGP posterior. This offers an improvement over competing algorithms that use only point estimates of the parameters governing the probability distribution around the response surface. Our Bayesian analysis results in a full posterior predictive distribution for the response (and, hence, the improvement) at any desired location in the input space. This full posterior predictive sample is essential to the maximization of the multivariate expected improvement in (28.4): locations x_1 through x_m may be chosen iteratively, such that each x_i maximizes the expected i-location improvement conditional on x_1 through x_{i-1} already having been selected. Full posterior samples for the improvement statistics at each $\tilde{x}$ are required to recalculate the expected improvement conditional on the iteratively expanding list of selected locations. This simplifies what would have itself been a complex optimization problem, and has the added benefit of defining an order to the list of m search locations.

Finally, although our general hybrid optimization approach will work with any local pattern search algorithm, the local optimization scheme used here is the asynchronous parallel pattern search (APPS) (Kolda, 2005; Gray and Kolda, 2006) developed at Sandia National Laboratories. APPS is a derivative-free optimization method that works by evaluating points in a search pattern of decreasing radius around the present best point. Software is publicly available at `http://software.sandia.gov/appspack/`. The primary stopping criterion is based on step length, and APPS is locally convergent under mild conditions. In addition, APPS is an efficient method for finding a local optimum when already in the neighbourhood of this optimum. By combining it with a TGP emulator, we can more quickly find the correct neighbourhood. Thus we use more points chosen by TGP early in the optimization process, and more points chosen by APPS later in the process. More details of optimization through posterior expected improvement via TGP, as well as the hybrid optimization algorithm with APPS and a parallel implementation, are provided by Taddy *et al.* (2009).

28.2 Experimental design

Calibration of the circuit devices under study consists of a minimization of a loss function for the distance between simulated current amplitude curves and experimental data. Before turning to the optimization problem in the following section, it is necessary to physically test the devices. The goal of this calibration is to obtain a single simulator parametrization for each device that will provide accurate current amplitude curve predictions under a variety of different physical situations. In particular, simulation will be required at different temperatures, for different radiation dosages, and at different DC voltage bias levels for the relevant complex system composed of the circuit devices. Thus the experiment design should provide as full a picture as possible of the device performance over the space defined by these three physical variables. Although device performance is completely characterized as a current amplitude curve, the peak amplitude provided a univariate output which the experimentalists felt would be representative of the device behaviour at different input variable values.

For each of 20 circuit devices, historical testing had yielded a bank (approximately 50 observations per device) of existing data for device current amplitude behaviour at various levels of temperature, radiation dosage, and bias. The experimentalists requested a list of 150 additional variable location vectors for testing. Although, in each case, the underlying input variables are continuous (and unbounded over the region of interest), the limitations of physical experimentation reduced the possible input values to three temperature levels, six dosage levels, and five bias levels. Hence, the space of potential experiment input locations consisted of a $3 \times 6 \times 5 = 90$ point grid. These factor levels correspond only to the set-up configuration; actual temperature, dosage, and bias amounts for each experiment are measurable and will be in the neighbourhood of the specified input configuration, but will not be exactly the same as designed.

The experiments are performed in batches, with five circuit devices grouped together on each of four different boards. Due to the difficulties inherent in finding a design that is optimal for each batch of five circuits, one device was chosen to be representative of each board and the experiment was designed around this single device. There is random noise in the results, and the data already include replication at individual input vectors. The list for additional testing locations should include replication where necessary to reduce the overall variance of output (peak amplitude) predictions.

Our approach to design is an iterative application of the adaptive sampling procedure outlined in Section 28.1.2.1. The expected reduction in global error shown in (28.2) is not easily extended to an analogous criterion for combinations of multiple locations. This issue is resolved through the implementation of a greedy algorithm which repeatedly chooses a single new location for testing to be added to an existing list. A first new input vector is chosen exactly as proposed above in Section 28.1.2.1, through adaptive sampling based on the TGP prior for peak amplitude conditional on temperature, dosage, and usage. A value for realized peak amplitude at this input location is then drawn from the conditional posterior predictive distribution, and this value is used as the imputed output corresponding to a future test at this location. Thus in searching for the second location, we treat the existing data as the combination of the observed data and the one new imputed point, and now look to maximize the expected reduction in global error of this updated dataset. The process is repeated, and at each iteration the treed GP model is fit to the existing data augmented by randomly imputed output values at all of the locations already chosen for future testing. All of the existing imputed output values are redrawn from their conditional posterior predictive distributions at each iteration. This additional variability helps to account for the variability in the physical observations (in contrast to the typically deterministic behaviour of a computer simulator). This iterative adaptive sampling algorithm is used to provide an ordered list of 150 locations (including repetition) for additional testing of each of the 20 devices. The prioritization implied by the ordering of the list is especially valuable in the motivating example, as the expense of individual experiments is unknown in advance and experimentation is terminated once the study has reached a predetermined budget constraint. At some point, a batch of experimental data becomes available, and all of the placeholder values are replaced by the new real data, and the iterative process continues for the next round of physical experiments.

Results for two of the devices, *bft92a* and *bfs17a*, are shown in Figure 28.1. The predictions shown in the left hand figure expose a discontinuity in the log peak amplitude surface which occurs for log dosages between 20 and 21. Although the exhibited results are conditional on a temperature of 75 celsius and a bias of zero, similar behaviour was discovered at other parameter configurations. The posterior mean for error reduction, $\Delta\hat{\sigma}^2(x)$ as defined in (28.2), from additional testing at a single new point x, is plotted in the right hand panel. We see that $\Delta\hat{\sigma}^2(x)$ is substantially reduced following the additional testing, with significant room for variance reduction only for x in the zone of discontinuity and at the boundaries. Finally, we note that the complex surfaces shown in the left hand figure mean that significant modeling gains are available by using treed Gaussian processes instead of more traditional stationary models.

28.3 Calibrating the computer model

The second part of this project focused on calibration and validation of a computer simulation model for the circuits. Numerical simulation is increasingly used because of advances in computing capabilities and the rising costs associated with physical experiments. In this context the computer models are often treated as an objective function to be optimized, and this is how they were treated by our collaborators. The challenges inherent in implementing this optimization are characterized by an inability to calculate derivatives and by the expense of obtaining a realization from the objective function. Fast convergence

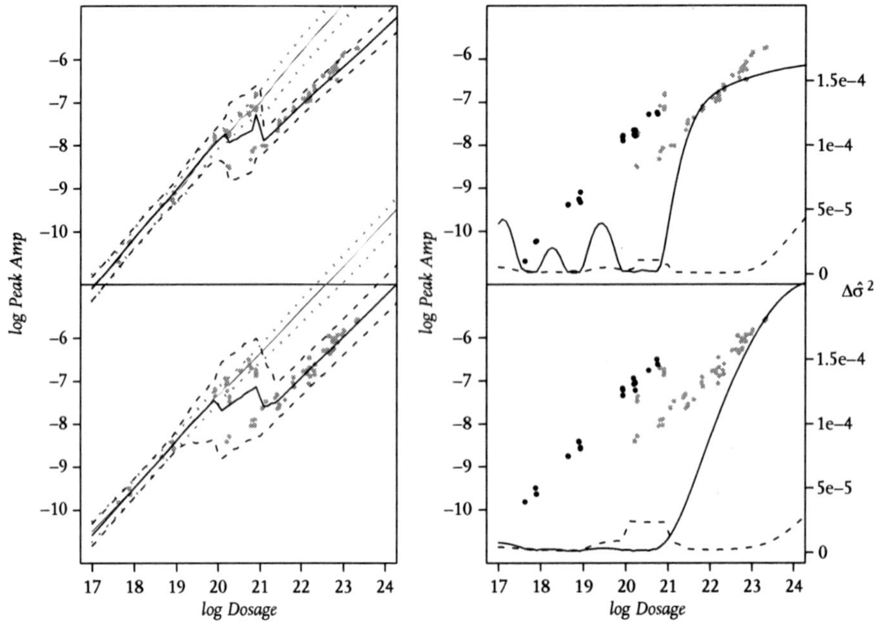

Fig. 28.1 Experiment results for *bft92a* on the top and *bfs17a* on the bottom, given a temperature of 75 celsius and zero bias. The left hand figures show mean (solid lines) and 90% interval predicted log peak amplitude; grey lines correspond to a TGP fit for the original observations, and black corresponds to a TGP fit to all observations. The right hand figure shows the original data in black and the data obtained through additional testing in grey. Plotted over the data we see the expected posterior reduction in error variance, $\Delta\hat{\sigma}^2$ for this temperature and bias, as a function of log dosage. The solid line refers to the posterior expectation after the original testing, and the dashed line corresponds to posterior expectation conditional on the completed dataset.

of the optimizer is needed because of the cost of simulation, and a search of the magnitude required to guarantee global convergence is not feasible. However, it is important to be aware that these large engineering problems typically have multimodal objective functions, so we want to avoid converging to low quality solutions. Thus we combine the local optimization method APPS for quick convergence with TGP to help provide a more robust solution, as described in Section 28.1.2.3.

The goal here is to find tuning parameter values for the Xyce simulator that lead to predicted current output that is as close as possible to the results obtained in the physical experiments. Here 'close' is defined through a squared-error objective function:

$$f(\mathbf{x}) = \sum_{i=1}^{N} \frac{1}{T_i} \sum_{t=1}^{T_i} \left[(S_i(t;\mathbf{x}) - E_i(t))^2 \right], \tag{28.5}$$

where N is the number of physical experimental runs (each corresponding to a unique radiation pulse), T_i is the total number of time observations for experiment i, $E_i(t)$ is the amount of electrical current observed during experiment i at time t, and $S_i(t;\mathbf{x})$ is the amount of electrical current at time t as computed by the simulator with tuning parameters $\mathbf{x}$ and radiation pulse corresponding to experiment i. Since each physical experiment may result in a different number of usable time observations, the weights of the errors are standardized so that each experimental run is counted equivalently. We note that the traditional statistical approach would also include a term for model discrepancy. However, our collaborators did not want such a term, as the mathematical modellers want to know what the best fit is for their model, and then they intend to address any remaining discrepancies by updating the physics in their model.

Because of the need to do both calibration and validation, only six experimental runs were used for each calibration, with the remaining datapoints saved for the validation stage. For each circuit device and each temperature setting, the six points were chosen to be representative of the whole set of data collected, and were selected by fitting a six-component mixture model. The selection details are not needed herein, we just treat the six points as the available data, but if the reader is interested, the details are given by Lee *et al.* (2009).

The simulator involves 38 user-defined tuning parameters for modeling current output as a function of radiation pulse input. Through discussion with experimentalists and researchers familiar with the simulator, 30 of the tuning parameters were fixed in advance to values either well known in the semiconductor industry or determined through analysis of the device construction. The semiconductor engineers also provided informative bounds for the remaining eight parameters. It is these eight parameters which are the inputs for our objective function (28.5). These parameters include those that are believed to have both a large overall effect on the output of the model and a high level of uncertainty with respect to their ideal values. Figure 28.2 shows the results of an MCMC sensitivity analysis, as described in Section 28.1.2.2, based on a TGP model fit to an initial Latin Hypercube Sample (LHS) of 160 input locations and with respect to a uniform uncertainty distribution over the bounded parameter space. All of the eight parameters appear to have significant effect on the objective function variability, although the main effect and first-order plots indicate that some variables are only effective in interaction with the other inputs. These higher-order interactions create challenges for optimization. In addition, the posterior mean main effect plots alerted the researchers to the possibility of optimal solutions on the boundaries of the input space (especially for x_2 and x_7) and will provide valuable guidance for checking the validity of the calibrated simulator.

The objective function (28.5) was optimized using both APPS by itself and the hybrid algorithm TGP-APPS. In the case of the hybrid algorithm, a LHS of 160 points was used to provide and initial fit for the TGP. The wall clock

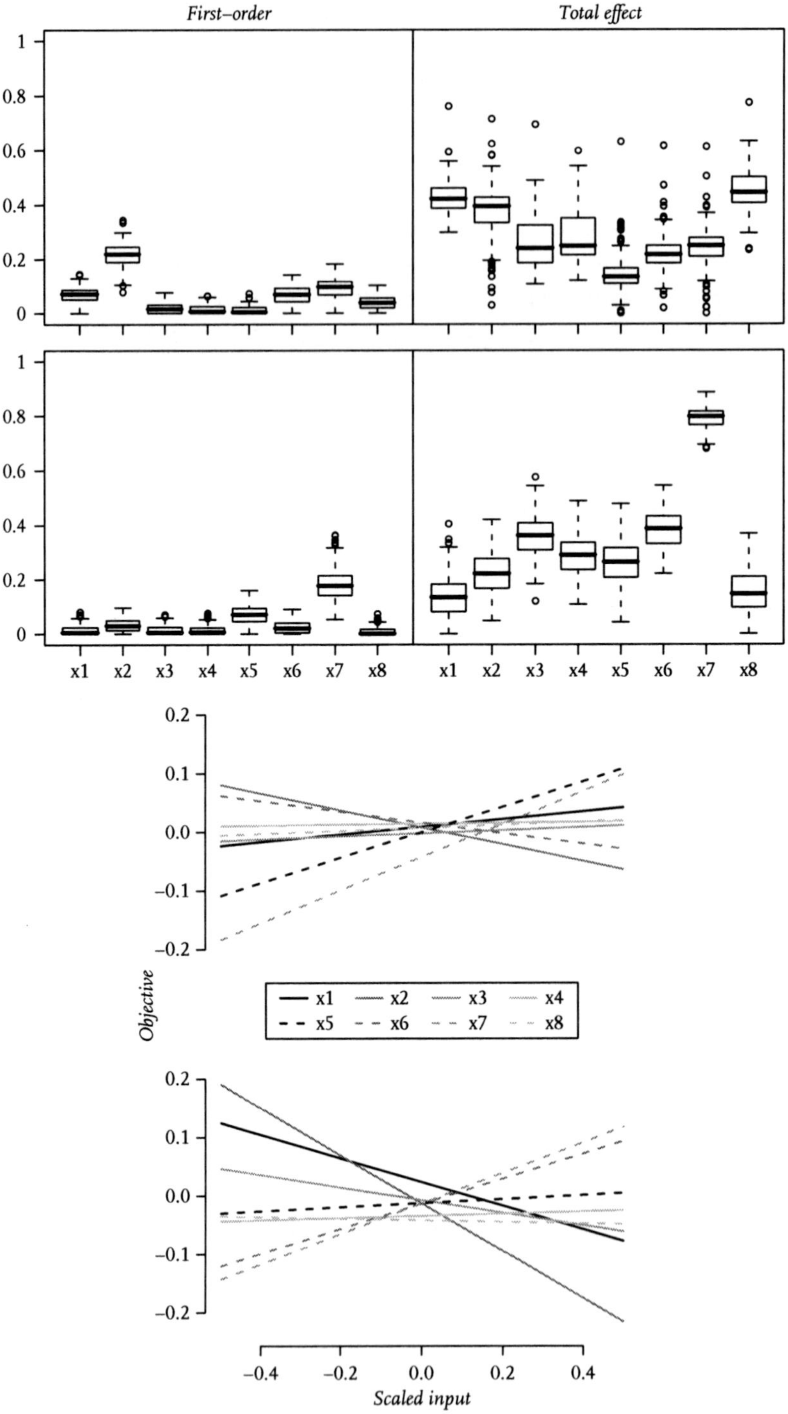

Fig. 28.2 Sensitivity analysis for *bft92a* (*top*) and *bfs17a* (*bottom*) optimization objective functions, summarized by posterior distributions for the first order (*left*), total (middle) sensitivity indices, and posterior mean main effects (*right*).

Table 28.1 For each bjt device and each optimization algorithm, the number of objective function evaluations and total wall clock time required to find a solution.

Device	Method	Evaluations	Time (hours)
bft92a	APPS	6823	94.8
bft92a	APPS-TGP	962	13.8
bfs17a	APPS	811	10.3
bfs17a	APPS-TGP	1389	18.1

time and the number of objective function evaluations corresponding to each device and each optimization algorithm are shown in Table 28.1. Figure 28.3 shows simulated current response curves corresponding to each solution and to the initial guess for tuning parameter values, as well as the data, for a single radiation pulse input to each device. Results for the other radiation pulse input values exhibit similar properties.

In the case of *bft92a*, the solutions produced by the two optimization algorithms are practically indistinguishable (they appear on top of each other in the figure). However, the APPS solution required over seven times as may functional evaluations, leading to much additional computational expense and elapsed time. The gain of the hybrid algorithm here is in its ability to move the search pattern quickly into better areas of the input space. We note that even if we started with the hybrid algorithm without an initial LHS (and starting

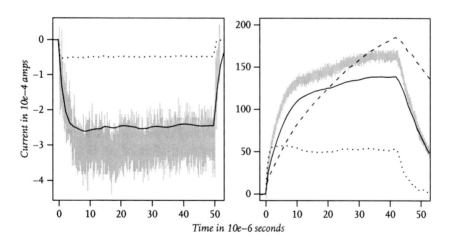

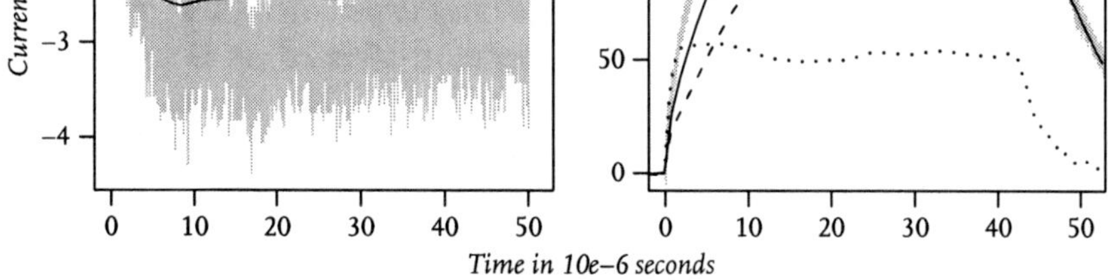

Fig. 28.3 Simulated response current curves for the *bft92a* (*left*) and *bfs17a* (*right*) devices. The solid line shows the response for parameters found using TGP-APPS, the dashed line for parameters found through APPS alone, and the dotted line for the initial parameter vector guess. The experimental current response curves for the radiation impulse used in these simulations are shown in grey.

from the same initial parameter vector as for APPS), it only takes about two more hours (a total of 15.8 hours) to obtain an equivalent solution through TGP-APPS – still a huge gain over APPS alone.

For the *bfs17a* device, the difference in the resulting response curves is striking and illustrates the desirable robustness of our hybrid algorithm. The response curve created using the parameter values obtained by APPS alone differs significantly from the data in overall shape. In contrast, the curve resulting from the parameters found by TGP-APPS is a reasonable match to the experimental data. These results suggest that the APPS algorithm was unable to overcome a weak local minimum while the inclusion of TGP allowed for a more comprehensive search of the design space. Note that the time results support this, as they show that the APPS algorithm converged relatively quickly. The extra computational cost of TGP-APPS is well justified by the improvement in fit. Of course, it is still clear that the simulator is not completely matching the data, and at this point we suspect that there is an inherent bias in the simulator. A complete statistical calibration would thus require the modelling of a bias term, as in the work of Kennedy and O'Hagan (2001). However, for our purposes in this project on optimum control, both our collaborating modellers and experimentalists were quite happy with the robust solution with respect to minimization of the provided objective function that TGP-APPS was able to find.

28.4 Further discussion

We have illustrated how the treed Gaussian process (TGP) model can be useful for spatial data and semiparametric regression in the context of a computer experiment for designing a circuit device. We have seen how the model can be used towards sequential design of (computer) experiments (via Bayesian adaptive sampling), sequential robust local optimization (with the help of APPS), validation, calibration, and sensitivity analysis all by simply sampling from the posterior distribution. In both optimization and experiment design, full posterior sampling combined with recursive iteration allowed us to use univariate prediction to optimize multivariate criteria.

Some of the models and methods described herein have also been used to design a computational fluid dynamics computer experiment for a rocket booster at NASA (Gramacy and Lee, 2009), and have been validated as competitive regression and spatial models on numerous synthetic and real data sets (Gramacy and Lee, 2008*a*,*b*; Gramacy, 2007). TGP is indeed a flexible model with many potential applications. However, one limitation is that the current methodology only supports real-valued inputs and responses. An adaptation of these methods to support categorical inputs and outputs promises to be a fruitful future direction of research.

Allowing categorical inputs will widen the scope of regression and design applications that can be addressed by the model. While the GP part of the model can easily handle binary-encoded categorical inputs on its own, it represents a sort of overkill. For example, a separable correlation function with width parameter d_i will tend to infinity if the output does not vary with binary input x_i, and will tend to zero if it does. Clearly, this functionality is more parsimoniously served by partitioning, e.g. using a tree. However, in a TGP implementation the tree will never partition on the binary inputs because doing so would cause the resulting design matrices at the leaves to be rank deficient. So without special care, any benefits of a divide-and-conquer approach (e.g. speed) to a nonparametric (TGP) regression with categorical inputs are lost. Once a careful implementation has been realized, one can imagine many further extensions. For example, including latent variable categorical inputs could enable the model to be used for clustering.

Extending the methodology to handle categorical responses will allow TGP to be applied to problems in classification. Separately, treed models and GP models have enjoyed great success in classification problems. Adapting treed models for classification (e.g. CART) is straightforward, whereas adapting GP models is a bit more complicated, requiring the introduction of $O(nk)$ latent variables where k is the number of classes. A combined modelling approach via TGP has the potential to be as fruitful for classification as it is for regression. It will be exciting to see how this extension develops, as well as accompanying methods for adaptive sampling, optimization, validation, calibration, and sensitivity that can be developed alongside.

Appendix

A. Broader context and background

A.1 Treed Gaussian processes

Here we provide more background on treed Gaussian processes, with further details available in Gramacy and Lee (2008*a*). First, the structure of the tree. We partition the input space using a tree, along the lines of models such as CART Breiman *et al.* (1984). Such a tree is constructed with a series of binary recursive partitions. For example, in two dimensions, one might consider an input space on [0,1] X [0,1]. The first split could be at $X_1 = .4$ separating the space into two rectangles, [0,0.4] X [0,1] and [0.4,1] X [0,1]. The second split might be at $X_2 = .3$ in the first partition, thus creating a third partition. Note that this second split does not affect the [0.4,1] X [0,1] region. By allowing multiple splits on the same variable, any arbitrary axis-aligned partitioning structure can be achieved. The restriction of axis-alignment allows us to build models in a computationally efficient manner, without losing too much modeling flexibility. Arbitrary partitions would require significantly more computing resources. We denote the whole tree structure by $\mathcal{T}$ and the leaf nodes by $\eta \in \mathcal{T}$, each of which represents a region of the input space. A prior for the tree is defined through a growth process. We start with a null tree (no partitions, all of the data is together in a single leaf node). Each leaf node η splits with probability $a(1+q_\eta)^{-b}$, where q_η is the depth of $\eta \in \mathcal{T}$ and a and b are parameters chosen to give an appropriate size and spread to the distribution of trees. Further details are available in Chipman *et al.* (1998, 2002). Here we use the default values of $a = 0.5$ and $b = 2$ from the **R** package (Gramacy and Taddy, 2008). The tree recursively partitions the input space into into R non-overlapping regions $\{r_\nu\}_{\nu=1}^R$. Each region r_ν contains data $D_\nu = \{x_\nu, Z_\nu\}$, consisting of n_ν observations.

Let $m \equiv m_X + 1$ be number of covariates in the design (input) matrix X plus an intercept. For each region r_ν, the hierarchical generative GP model is

$$\begin{aligned} Z_\nu|\beta_\nu, \sigma_\nu^2, K_\nu &\sim N_{n_\nu}\left(f_\nu\beta_\nu, \sigma_\nu^2 K_\nu\right) & \beta_0 &\sim N_m(\mu, B) \\ \beta_\nu|\sigma_\nu^2, \tau_\nu^2, W, \beta_0 &\sim N_m\left(\beta_0, \sigma_\nu^2\tau_\nu^2 W\right) & \tau_\nu^2 &\sim IG(\alpha_\tau/2, q_\tau/2), \\ \sigma_\nu^2 &\sim IG(\alpha_\sigma/2, q_\sigma/2) & W^{-1} &\sim W((\rho V)^{-1}, \rho) \end{aligned}$$

with $F_\nu = (1, X_\nu)$, and W is an $m \times m$ matrix. The N, IG, and W are the (Multivariate) Normal, Inverse-Gamma, and Wishart distributions, respectively. Hyperparameters μ, B, V, ρ, $\alpha_{\hat{\sigma}}$, q_σ, α_τ, q_τ are treated as known, and we use the default values in the **R** package. This model (28.4) specifies a multivariate normal likelihood with linear trend coefficients β_ν, which are also modeled hierarchically. Each region is fit independently (conditional on the hierarchical structure), which gives this approach some similarity to change-point models.

The GP correlation structure within each region is given by $K_\nu(x_j, x_k) = K_\nu^*(x_j, x_k|d) + g_\nu\delta_{j,k}$, where K_ν^* is the separable power family given in equation (28.1) and g is the nugget term. Our choice of priors encodes a preference for a model with a nonstationary global covariance structure, giving roughly equal mass to small d representing a population of GP parameterizations for wavy surfaces, and a separate population for those which are quite smooth or approximately linear:

$$p(d, g) = p(d) \times p(g) = p(g) \times \frac{1}{2}[Ga(d|\alpha = 1, \beta = 20) + Ga(d|\alpha = 10, \beta = 10)]. \tag{28.6}$$

We take the prior for g to be exponential with rate λ.

In some cases, a full GP may not be needed within a partition; instead a simple linear model may suffice. Because of the linear mean function in our implementation of the GP, the standard linear model can be seen as a limiting case. The linear model is more parsimonious, as well as much more computationally efficient. We augment the parameter space with indicator variables $b = \{b\}_{i=1}^{m_X} \in \{0, 1\}^{m_X}$. The boolean b_i selects either the GP ($b_i = 1$) or its limiting linear model for the ith dimension. The prior for b_i specifies that smoother GPs (those larger range parameters d_i) are more likely to jump to the limiting linear model:

$$p_{\gamma,\theta_1,\theta_2}(b_i = 0|d_i) = \theta_1 + (\theta_2 - \theta_1)/(1 + \exp\{-\gamma(d_i - 0.5)\})$$

and we use the R package default values of $(\gamma, \theta_1, \theta_2) = (10, 0.2, 0.95)$. More details are available in Gramacy and Lee (2008*b*).

A.2 Experimental design

The basic ideas for experimental design in computer experiments follow those of standard experimental design, i.e. one wants a relatively small set of points that are expected to provide maximal information about parameters for a particular choice of model. For a GP, as with most models, one generally wants to spread out the points, as each observation gives a fair amount of local information because of the smoothness properties of the GP model. Approaches include maximin distance, Latin hypercube, D-optimal, maximum entropy, and orthogonal array designs (McKay *et al.*, 1979; Santner *et al.*, 2003). In general, no replications are planned when the computer simulator is deterministic.

With computer experiments, it is natural to move on to sequential collection of data, i.e. Sequential Design of Experiments (DOE) or Sequential Design and Analysis of Computer Experiments (SDACE) (Sacks *et al.*, 1989; Currin *et al.*, 1991; Welch *et al.*, 1992). Depending on whether the goal of the experiment is inference or prediction, the choice of utility function will lead to different algorithms for obtaining optimal designs (Shewry and Wynn, 1987; Santner *et al.*, 2003).

In the machine learning literature, sequential design of experiments is often referred to as active learning. Two approaches applied to Gaussian processes are that of Cohn (1996) described in the main text (maximizing the expected reduction in average squared error) and that of MacKay (1992), which chooses the new point as that with the largest standard deviation in predicted output. While the Mackay approach is simpler, it is also more localized, and less useful in the presence of heteroskedasticity.

B. Computations

We fit these models using the `tgp` package in R (Gramacy and Taddy, 2008). A tutorial is provided by Gramacy (2007). The core of the package is based on C++

code that employs reversible jump Markov chain Monte Carlo to fit both the tree structure and the GPs in each of the partitions. By averaging across the Markov chain realizations, estimates of the posterior mean and of predictive intervals are obtained. This averaging includes the tree structures, and as a result, we typically obtain smooth posterior mean fits because of mixing over the location of partitions.

Conditional on a tree structure, most parameters can be updated via Gibbs sampling. The linear regression parameters β_ν and their prior mean β_0 all have multivariate normal full conditionals. The data variance parameter σ^2 and the linear variance parameter τ^2 are both conditionally inverse-Gamma, and the linear model covariance matrix W is conditionally inverse-Wishart. Correlation parameters d and g require Metropolis – Hastings updates. The tree structure itself is updated with reversible jump steps: grow, prune, change, swap, and rotate. The first two require care in accounting for the change of dimension, while the latter three are straightforward Metropolis – Hastings steps. More details on estimation and prediction are available in Gramacy and Lee (2008*a*). It can be helpful to standardize the data before running the **R** code, so that the default parameter values are reasonable.

Acknowledgements

This work was partially supported by NASA awards 08008-002-011-000 and SC 2003028 NAS2-03144, Sandia National Laboratories grants 496420 and 673400, and National Science Foundation grants DMS 0233710 and 0504851. Any opinions, findings, and conclusions expressed in this material are those of the authors and do not necessarily reflect the views of the funding organizations.

References

Abrahamsen, P. (1997). A review of Gaussian random fields and correlation functions. Technical Report 917, Norwegian Computing Center, Box 114 Blindern, N-0314 Oslo, Norway.

Booker, A. J., Dennis, J. E. Frank, P. D., Serafini, D. B., Torczon, V. and Trosset, M. W. (1999). A rigorous framework for optimization of expensive functions by surrogates. *Structural and Multidisciplinary Optimization*, **17**, 1–13.

Breiman, L., Friedman, J. H., Olshen, R. and Stone, C. (1984). *Classification and Regression Trees.* Wadsworth, Belmont, CA.

Chaloner, K. and Verdinelli, I. (1995). Bayesian experimental design, a review. *Statistical Science*, **10**, 273–304.

Chipman, H., George, E. and McCulloch, R. (1998). Bayesian CART model search (with discussion). *Journal of the American Statistical Association*, **93**, 935–960.

Chipman, H., George, E. and McCulloch, R. (2002). Bayesian treed models. *Machine Learning*, **48**, 303–324.

Cohn, D. A. (1996). Neural network exploration using optimal experimental design. In *Advances in Neural Information Processing Systems (NIPS)*, Volume 6(9), pp. 679–686. Morgan Kaufmann Publishers. San Mateo, CA.

Cressie, N. A. C. (1993). *Statistics for Spatial Data*, revised edition. John Wiley, New York.

Currin, C., Mitchell, T., Morris, M. and Ylvisaker, D. (1991). Bayesian prediction of deterministic functions, with applications to the design and analysis of computer experiments. *Journal of the American Statistical Association*, **86**, 953–963.

Fjeldy, T. A., Ytterdal, T. and Shur, M. S. (1997). *Introduction to Device Modeling and Circuit Simulation*. Wiley-Interscience, New York..

Gramacy, R. B. (2007). `tgp`: An **R** package for Bayesian nonstationary, semiparametric nonlinear regression and design by treed gaussian process models. *Journal of Statistical Software*, **19**, 1–46.

Gramacy, R. B. and Lee, H. K. H. (2008*a*). Bayesian treed Gaussian process models with an application to computer modeling. *Journal of the American Statistical Association*, **103**, 1119–1130.

Gramacy, R. B. and Lee, H. K. H. (2008*b*). Gaussian processes and limiting linear models. *Computational Statistics and Data Analysis*, **53**, 123–136.

Gramacy, R. B. and Lee, H. K. H. (2009). Adaptive design and analysis of supercomputer experiments. *Technometrics*, **51**, 130–145.

Gramacy, R. B. and Taddy, M. A. (2008). tgp: Bayesian treed Gaussian process models. R package version 2.1-2. `http://www.ams.ucsc.edu/~rbgramacy/tgp.html`.

Gray, G. A. and Kolda, T. G. (2006). Algorithm 856: APPSPACK 4.0: Asynchronous parallel pattern search for derivative-free optimization. *ACM Transactions on Mathematical Software*, **32**, 485–507.

Gray, G. A., Martinez-Canales, M., Lam, C., Owens, B. E., Hembree, C., Beutler, D. and Coverdale, C. (2007). Designing dedicated experiments to support validation and calibration activities for the qualification of weapons electronics. In Proceedings of the 14th NECDC. Also available as Sandia National Labs Technical Report SAND2007-0553C.

Green, P. (1995). Reversible jump Markov chain monte carlo computation and Bayesian model determination. *Biometrika*, **82**, 711–732.

Homma, T. and Saltelli, A. (1996). Importance measures in global sensitivity analysis of nonlinear models. *Reliability Engineering and System Safety*, **52**, 1–17.

Jones, D., Schonlau, M. and Welch, W. J. (1998). Efficient global optimization of expensive black box functions. *Journal of Global Optimization*, **13**, 455–492.

Keiter, E. R. (2004). Xyce parallel elctronic simulator design: mathematical formulation. Technical Report SAND2004-2283, Sandia National Labs, Albuquerque, NM.

Kennedy, M. and O'Hagan, A. (2001). Bayesian calibration of computer models (with discussion). *Journal of the Royal Statistical Society, Series B*, **63**, 425–464.

Kolda, T. G. (2005). Revisiting asynchronous parallel pattern search for nonlinear optimization. *SIAM Journal of Optimization*, **16**, 563–586.

Lee, H. K. H., Taddy, M. and Gray, G. A. (2009). Selection of a representative sample. *Journal of Classification*. To appear.

MacKay, D. J. C. (1992). Information-based objective functions for active data selection. *Neural Computation*, **4**, 589–603.

McKay, M. D., Conover, W. J. and Beckman, R. J. (1979). A comparison of three methods for selecting values of input variables in the analysis of output from a computer code. *Technometrics*, **21**, 239–245.

Morris, R. D., Kottas, A., Taddy, M., Furfaro, R. and Ganapol, B. (2008). A statistical framework for the sensitivity analysis of radiative transfer models used in remote sensed data product generation. *IEEE Transactions on Geoscience and Remote Sensing*, **12**, 4062–4074.

Oakley, J. and O'Hagan, A. (2004). Probabilistic sensitivity analysis of complex models: a Bayesian approach. *Journal of the Royal Statistical Society, Series B*, **66**, 751–769.

Oberkampf, W. L., Trucano, T. G. and Hirsch, C. (2003). Verification, validation, and predictive capability. Technical Report SAND2003-3769, Sandia National Labs, Albuquerque, NM.

Sacks, J., Welch, W. J., Mitchell, T. J. and Wynn, H. P. (1989). Design and analysis of computer experiments. *Statistical Science*, **4**, 409–435.

Saltelli, A. (2002). Making best use of model evaluations to compute sensitivity indices. *Computer Physics Communications*, **145**, 280–297.

Saltelli, A., Chan, K. and Scott, E. (eds.) (2000). *Sensitivity Analysis*. John Wiley, New York.

Santner, T. J., Williams, B. J. and Notz, W. I. (2003). *The Design and Analysis of Computer Experiments*. Springer-Verlag, New York.

Schonlau, M., Jones, D. and Welch, W. (1998). Global versus local search in constrained optimization of computer models. In *New Developments and Applications in Experimental Design*, (ed. N. Flournoy, W. F. Rosenberger and W. K. Wong) IMS Lecture Notes – Monograph Series, Vol. 34, pp. 11–25. Institute of Mathematical Statistics, USA.

Sedra, A. S. and Smith, K. C. (1997). *Microelectronic Circuits*, (4th edn). Oxford University Press, Oxford.

Shewry, M. and Wynn, H. (1987). Maximum entropy sampling. *Journal of Applied Statistics*, **14**, 165–170.

Sobol, I. M. (1993). Sensitivity analysis for nonlinear mathematical models. *Mathematical Modeling and Computational Experiment*, **1**, 407–414.

Stein, M. L. (1999). *Interpolation of Spatial Data*. Springer, New York.

Taddy, M., Lee, H. K. H., Gray, G. A. and Griffin, J. D. (2009). Bayesian guided pattern search for robust local optimization. *Technometrics*, **51**, 389–401.

Welch, W. J., Buck, R. J., Sacks, J., Wynn, H. P., Mitchell, T. and Morris, M. D. (1992). Screening, predicting, and computer experiments. *Technometrics*, **34**, 15–25.

·29·

Multistate models for mental fatigue

Raquel Prado

29.1 Goals and challenges in the analysis of brain signals: The EEG case

Electroencephalographic signals or EEGs are recordings of waves that represent electrical activity in the brain over time. An EEG measures changes of voltage levels at the scalp surface in cycles per second (Hz). In clinical settings, the analysis of EEG signals has been useful in the diagnostic of seizures and other neurological disorders (e.g. Le Van Quyen *et al.* 2001), and in assessing the efficacy of treatments for major depression such as electroconvulsive therapy or ECT (Krystal, Prado, and West 1999). In non-clinical settings EEGs have been used in the characterization of cognitive fatigue (Trejo *et al.* 2007) and in automatic classification of cognitive overload (Trejo, Matthews, and Allison 2007).

Here, we present analyses of EEG data recorded on a subject who performed continuous arithmetic operations for a period of three hours. The most relevant aspects of the experimental setting that led to these electroencephalogram recordings are described below. Other analyses of these data, as well as analyses of EEG data from other subjects who participated in the same experiment, appear in Trejo *et al.* (2006) and Prado (2009). The main objectives of EEG monitoring in this area include detection, monitoring and prediction of mental fatigue in awake subjects, which is defined as the willingness of alert, motivated subjects to continue performing mental work (Trejo *et al.* 2007). More specific goals and some of the challenges of dealing with these signals are also discussed.

EEG monitoring of brain activity usually involves the recording of EEG traces at multiple channels located over a subject's scalp. For instance, in the analysis of EEG signals measured after ECT was administered to a patient, Prado, West, and Krystal (2001) studied traces from 19 channels recorded at a sampling rate of 256 Hz. The monitoring time depended on the subject and typically varied from 1 to 3 minutes. For the patient considered in Prado, West, and Krystal (2001), the recording time was a bit longer than one and a half minutes and so, the full data set consisted of 19 time series with about 26,000 observations per series. In the cognitive fatigue EEG application presented below, we have traces recorded at 30 channels at a sampling rate of 128 Hz. The recording time in this study was quite long, ranging from 1 to 3 hours depending on whether or not a given subject completed the experiment. For subjects who performed continuous arithmetic operations for 180 minutes, the EEG traces consist of 1,382,400 observations per channel. Consequently, one of the difficulties faced in an analysis of this type of signals is the size of the data sets.

The EEG can be broken up into four main frequency bands referred to as the delta (0–4 Hz), theta (4–8 Hz), alpha (8–13 Hz) and Beta (above 13 Hz) bands. Various types of activity/behaviour may induce the appearance of brain waves in one or more of these frequency bands. For example, the normal resting EEG usually consists of alpha and beta rhythms (Dyro 1989), while anesthesia effects in electrically induced seizures may be characterized by a mixture of slow and fast frequency activity (Weiner, Coffey, and Krystal 1991). In the area of monitoring and detection of mental fatigue, prior EEG studies have suggested an association of mental fatigue with an increase in the theta band power at channels located in mid-line frontal areas, and with an increase in the alpha band power at one or more parietal locations (Trejo *et al.* 2006). The fact that different mental states can lead to changes in the spectral characteristics of the EEGs over time results in signals that are typically non-stationary when relatively long recording time periods are considered – and often also during very short intervals due to the nature of the process being measured. From the modelling viewpoint this represents a challenge, as many of the time domain and frequency domain methods used in time series analysis are based on the assumption of stationarity. Extending such methods to deal with non-stationary time series can be a difficult task analytically and computationally, as it is often hard to develop efficient algorithms for inference and prediction given and that the models considered may be non-linear and/or non-Gaussian, and the data sets are large. In addition, in many EEG applications, such as the one considered here, it is necessary to perform on-line inference and prediction, which further complicates the analysis.

We focus on the study of EEG signals for characterization of cognitive fatigue, however, we emphasize that many of the models and methods discussed here can also be applied to other kinds of EEG signals and, in general, to time series that can be described by means of multiple autoregressive processes such as some biomedical signals, speech signals and seismic recordings. In Section 29.1.1, we describe the data and briefly explain the experimental setting that led to such data. In Section 29.2.1 we discuss analyses, based on autoregressive models and related time series decompositions, of EEG signals recorded in one of the subjects who participated in the experiment. The results obtained with these analyses motivate the general class of mixtures of autoregressive models, or multi-AR(p) processes, proposed in Section 29.2.2. Due to the computational complexity of achieving on-line inference within this general model class, we consider approximate inference in low order multi-AR processes (with $p = 1$ or $p = 2$). In spite of their limitations, these low order models are useful for on-line characterization of cognitive fatigue as shown in Section 29.2.4. Finally, Section 29.3 discusses current and future trends in the development of methods for on-line analysis and classification of multichannel EEG signals.

29.1.1 Data description

The EEG data analysed here were recorded at the NASA Ames Research Center in Moffett Field, California by L. Trejo and collaborators. We present a study of EEG traces recorded in one of 16 subjects who participated in the experiment. We refer to this subject as **skh**. A detailed description of the experiment and previous data analyses can be found in Trejo *et al.* (2006, 2007). We now summarize the key aspects of the experimental setting and describe the format of the data.

The subjects who participated in this study were asked to solve simple arithmetic operations continuously for up to 3 hours. More specifically, the participants sat in front of a computer with their right hand resting on a key pad, and were asked to solve the summation problems that appeared on the screen as quickly as possible without sacrificing accuracy. The summations consisted on four randomly generated single digits, three operators and a target sum, e.g. $5 + 2 - 1 + 1 = 8$. The participants had to solve each problem and decide whether the summation on the left was less than, equal to, or greater than the target sum on the right by pressing the appropriate button on the keypad (<,=,>). Once an answer was received, the monitor went blank for one second and after this a new summation appeared on the screen. Each participant performed the task until 180 minutes had elapsed or they quit from exhaustion.

Electroencephalograms were recorded from each of the participants at 30 channels locations using 32 Ag/AgCl electrodes (two of the electrodes were used to record vertical and horizontal electrooculograms). A schematic representation with the approximate location of the channels over the scalp is shown in Figure 29.1. The EEGs were amplified, digitized and submitted to algorithms for elimination of eye-artifacts, as well as visually inspected. Blocks of data containing artifacts were removed as detailed in Trejo *et al.* (2006). The EEGs were also epoched around the times of the stimuli, which were the times at which the summations appeared on the computer screen. In particular, the data of subject **skh** consist of 864 consecutive epochs per channel. Each of these epochs corresponds to the EEG recording that goes from 5 seconds prior to a given stimulus, to 8 seconds after such stimulus. As mentioned before, the sampling rate was 128 Hz, and so, each epoch has a total number of 1,664 observations.

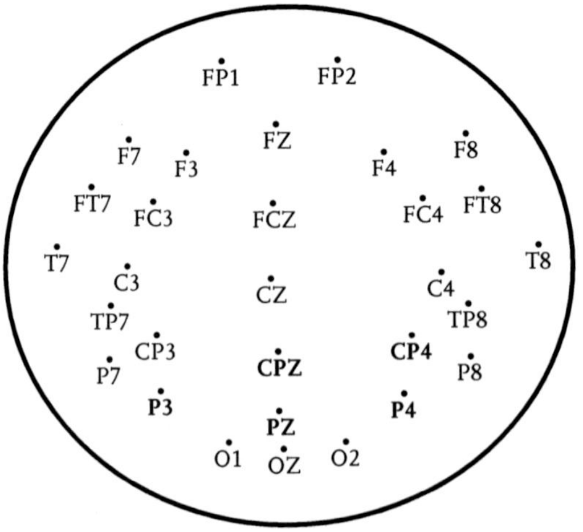

Fig. 29.1 Approximate location of the 30 EEG channels over the scalp. The channels in bold are those for which it is possible to discriminate epochs recorded at the beginning of the experiment – when the individual is alert – from those recorded towards the end – when the individual is fatigued – based on a multi-AR(1) analysis.

The ultimate goal of this study is the development of automatic methods for detection and prediction of cognitive fatigue. In this chapter we focus on models for characterizing fatigue in terms of the changes in the spectral content of some of the latent processes underlying the EEG signals over time. Modelling these signals is a balancing act, as we need to use models that are sophisticated enough to capture the possible – and often very subtle – changes in the signals over time, and yet such models must be simple enough so that on-line posterior estimation and prediction are feasible. It is also known from preliminary analyses that the changes in the EEG spectral characteristics that may be associated to cognitive fatigue vary across subjects and that, within a given subject, such effects may also vary across channels. Therefore, the ideal models would be those that allow us to incorporate relevant subject and channel-specific prior information about the mental states – based for instance on previous experiments involving the same subject or other subjects – and can also adapt over time.

29.2 Modelling and data analysis

29.2.1 Discovering EEG features associated with mental fatigue via AR models

We begin by showing how autoregressions (AR models) and related time series decompositions can be used to discover spectral features in the EEG signals that may be associated with cognitive fatigue. This approach was followed in Trejo *et al.* (2007) and Prado (2009) to estimate EEG frequencies in the alpha range, as well as their associated peaks, in the power spectra for epochs recorded during the first 15 min. and the last 15 min. of the experiment.

Let $y_{t,q,j}$ be the t-th observation of epoch q for channel j, with $t = 1 : T$, $q = 1 : Q$, and $j = 1 : J$. For subject **skh** we have that $T = 1,664$, $Q = 864$ and $J = 30$. Following Trejo *et al.* (2006), the AR-based approach presented in Trejo *et al.* (2007) and Prado (2009) divides the EEG recordings into intervals, each interval corresponding to 15 minutes of recording. Data from the first and the last intervals are then used to determine if there are changes in the spectral characteristics of the EEG signals that allow us to discriminate epochs recorded when the individual was supposed to be alert – e.g., during the first 15 min. of the experiment – from those recorded when the individual was supposed to be fatigued – e.g., during the last 15 min. of the experiment. This was done by fitting AR models to the epochs from the first and last intervals, and then looking at their estimated AR-based time series decompositions as detailed below.

For subject **skh** we have 70 epochs in the first interval and 92 epochs in the last interval. An AR(p) model is fitted to each of these epochs and each of the 30 channels, i.e.

$$y_{t,q,j} = \sum_{i=1}^{p} \theta_{i,q,j} y_{t-i,q,j} + \epsilon_{t,q,j}, \quad \epsilon_{t,q,j}, \sim N(0, v_{q,j}). \tag{29.1}$$

Bayesian inference is achieved assuming a normal prior on the AR coefficients and a Gamma prior on $\phi_{q,j} = 1/v_{q,j}$, as detailed in Appendix A. For each channel, inference can be summarized in terms of the posterior means of the AR coefficients for all the epochs in the first and last intervals, this is, $\hat{\theta}_{q,j} = (\hat{\theta}_{1,q,j}, \ldots, \hat{\theta}_{p,q,j})'$ for $q = 1 : 70$ and $q = 773 : 864$. Using these posterior estimates, AR-based decompositions of each time series (or EEG epoch) can be computed following the results of West (1997) (see Appendix A for details), and so we can write

$$y_{t,q,j} = \sum_{i=1}^{n^c_{q,j}} z_{i,t,q,j} + \sum_{i=1}^{n^r_{q,j}} x_{i,t,q,j}, \tag{29.2}$$

where the $z_{i,t,q,j}$s are real process associated with the $n^c_{q,j}$ pairs of complex reciprocal roots of the AR characteristic polynomial $\hat{\Theta}_{q,j}(u) = 1 - \hat{\theta}_{1,q,j}u - \cdots - \hat{\theta}_{p,q,j}u^p$, where u is a complex number, and the $x_{i,t,q,j}$s are real processes associated with the $n^r_{q,j}$ real reciprocal roots of the same AR characteristic polynomial. More specifically, if the pairs $(r_{l,q,j}, \lambda_{l,q,j})$ denote the moduli and periods of the complex reciprocal roots of $\hat{\Theta}_{q,j}(u)$, for $l = 1 : n^c_{q,j}$, and $r_{l,q,j}$ for $l = \left(2n^c_{q,j} + 1\right) : p$ are the moduli of the real reciprocal roots, then, each $z_{i,t,q,j}$ is a quasi-periodic AR(2, 1) process with modulus $r_{i,q,j}$, and period

$\lambda_{i,q,j}$, while each $x_{i,t,q,j}$ is an AR(1) process with modulus $r_{2n^c_{q,j}+i,q,j}$. The numbers of real and complex pairs of reciprocal roots, $n^r_{q,j}$ and $n^c_{q,j}$ with $p = n^r_{q,j} + 2n^c_{q,j}$, do vary slightly across epochs and channels for a given p. However, for the AR(10)-based analyses discussed here we found that most epochs in subject **skh** displayed exactly four pairs of complex reciprocal roots and two distinct real reciprocal roots. The moduli of these roots were always below one, indicating stationarity within each epoch, however, some of the roots were rather persistent: typically one of the real roots and at least one pair of complex roots had estimated moduli higher than 0.8, and very often close to one.

In Prado (2009) models with the same order p were considered for all the epochs and all the channels. The model order was chosen as follows. For each channel, AR(p) models with $8 \leq p \leq 20$ were fitted to the epochs in the first and last intervals. Then, the model order with the best average 'leave-one-epoch-out' predictive performance, averaging across channels, was chosen as the optimal order. The predictive performance was measured as the percentage of the epochs that were correctly classified as epochs from the first or the last intervals. For subject **skh** such model order was found to be $p = 10$. The details of this approach will not be discussed here. Other criteria that have been used for choosing optimal model orders in EEG analyses include BIC and AIC. When these criteria were used, model orders $p = 9, 10$ and $p = 11$ were found to be optimal for the various epochs and channels.

More general model classes can also be considered to describe the epochs, such as time-varying AR models (TVARs). TVARs were successfully used in the analyses of EEG traces presented in West, Prado, and Krystal (1999). TVAR models have the form given in (29.1) but with AR coefficients changing over time. West, Prado, and Krystal (1999) considered models in which the AR coefficients vary according to a random walk. In such models, the smoothness of the changes in time is controlled by a parameter δ, with $\delta \in (0, 1]$, referred to as a discount factor (West and Harrison, 1997). Low values of δ are related to rapid changes in the AR parameters over time, while high values are consistent with slow changes. The case of $\delta = 1$ corresponds to the standard AR model, since no changes of the AR coefficients over time are allowed when $\delta = 1$. Optimal values of p and δ can be chosen by maximum likelihood as explained in West, Prado, and Krystal (1999). We considered the class of TVAR models in the analysis of the single epoch EEG data of subject **skh**. We found that, for almost all the epochs and all the model orders considered, the optimal value of δ was $\delta = 1$, indicating that there is no advantage in using TVAR models, instead of simpler AR models, to describe individual epochs.

The AR(10)-based analyses presented in Prado (2009) show that some of the channels in subject **skh** display significant differences in the alpha frequency band between the epochs recorded in the first 15 min. of the experiment,

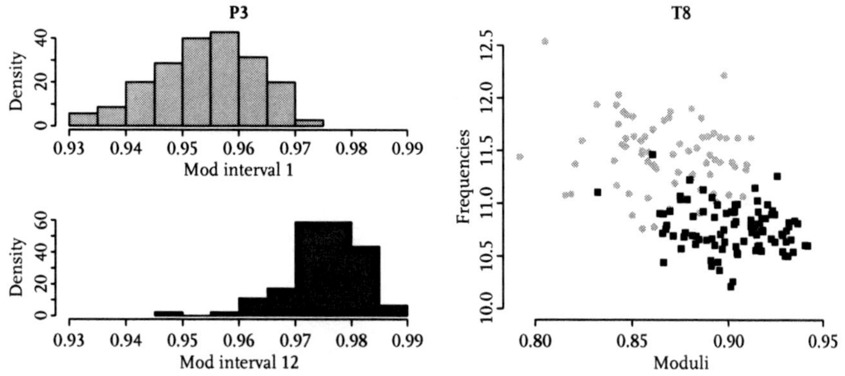

Fig. 29.2 Left panel: histograms of the real reciprocal roots with the highest estimated modulus for epochs recorded in the first 15 min. of the experiment (top graph) and those for epochs recorded in the last 15 min (bottom graph). Right panel: Estimated frequencies and corresponding moduli in the alpha band in channel T_8 for epochs in the first interval (light points) and epochs in the last intervals (dark points).

and those recorded in the last 15 min. In particular, it was found that channels O_z, T_8, FT_8 and F_4 showed lower estimated frequency values and higher estimated moduli values of the component in the alpha frequency band for most (but not all) the epochs in the last interval. Other latent processes in the decomposition also show differences in their spectral characteristics. More specifically, for some of the channels, the estimated values of the moduli of the real characteristic roots with the highest moduli are larger for epochs recorded in the last 15 min. of the experiment than those for epochs recorded during the first 15 min. Some of these findings are illustrated in Figure 29.2. The left panel depicts histograms of the estimated real characteristic roots with the highest moduli (as mentioned before, there are typically two real roots for each epoch) in channel P_3. The top histogram shows the estimated moduli for epochs in the first interval, while the histogram at the bottom corresponds to epochs recorded in the last 15 min. The right panel in Figure 29.2 shows that most epochs in the last interval in channel T_8 have estimated frequency values – in the alpha band – that are lower than the estimated frequency values (in the same band) for epochs in the first interval, while the corresponding estimated moduli are higher for epochs in the last intervals than those for early epochs.

The AR-based results summarized here motivate the class of multi-AR(p) models with structured priors developed in the next section. The idea behind these models is to describe the mental process as a mixture of states such that, at any given time, we can compute the probability that an individual is in a particular mental state, e.g. alert or fatigue. Each of the states is characterized in terms of the spectral content of the EEG signals. For instance, fatigue may be defined as a state for which the AR component that lies in the alpha band has a modulus that is above a certain threshold, say x_f, while for the alert state the modulus of such component is below some other threshold x_a, with $x_a \leq x_f$. Our approach assumes that the number of states is known a priori. The models allow us to incorporate prior information in a structured and interpretable manner, and so, EEG data from previous studies can be used to build the priors on the model parameters that characterize each mental state.

29.2.2 Multi-AR models: General case and computational difficulties

In the equations below we drop the index j, that indicates the specific EEG channel, in order to simplify the notation as much as possible.

For each epoch q, consider a collection of K models $\{\mathcal{M}_q(1), \ldots, \mathcal{M}_q(K)\}$, representing K different states, with

$$\mathcal{M}_q(k): \quad y_q = X_q \theta^{(k)} + \epsilon_q^{(k)}, \quad \epsilon_q^{(k)} \sim N(0, \phi^{-1} I), \tag{29.3}$$

for $k = 1 : K$. Here $y_q = (y_{p+1,q}, \ldots, y_{T,q})'$ is a vector of dimension $n = (T - p)$, $\theta^{(k)}$ is a p-dimensional parameter vector $\theta^{(k)} = \left(\theta_1^{(k)}, \ldots, \theta_p^{(k)}\right)'$, $\epsilon_q^{(k)}$ is a vector of innovations of dimension n, ϕ^{-1} is the precision and

$$X_q = \begin{pmatrix} y_{p,q} & \cdots & y_{1,q} \\ y_{p+1,q} & \cdots & y_{2,q} \\ \vdots & \vdots & \vdots \\ y_{T-1,q} & \cdots & y_{T-p,q} \end{pmatrix}.$$

We assume that for each epoch q, the observed series y_q follows model $\mathcal{M}_q(k)$ with some probability. Each of these models is an AR(p) with AR coefficients $\theta^{(k)}$. In order to specify the mechanisms by which the models are chosen, we follow a probabilistic procedure that provides a multiprocess based on discrete probability mixtures of AR models, as detailed later in Section 29.2.2.2.

A key aspect of this modeling approach is the choice of the prior distributions. We propose the use of structured priors on the AR coefficients of each model, following the developments of Huerta and West (1999) for standard AR models. The main reason for using structured priors in the context of EEG monitoring and characterization of cognitive fatigue, is that they allow us to incorporate relevant information on the mental states in terms of interpretable parameters, such as those defining the spectral features of the EEG signals.

We describe the priors for the general case in Section 29.2.2.1 and discuss the computational difficulties for posterior inference under these priors in Section 29.2.2.2. In Section 29.2.3 we consider the specific cases of multiprocesses for which all the models are either AR(2)s or AR(1)s, and summarize methods for approximate posterior inference in this framework. Finally, in Section 29.2.4, we illustrate the use of multi-AR(1) and multi-AR(2) processes in the characterization of cognitive fatigue for subject **skh**.

29.2.2.1 Prior structure

Let $\alpha^{(k)} = \left(\alpha_1^{(k)}, \ldots, \alpha_p^{(k)}\right)$ denote the reciprocal roots of the AR(p) characteristic polynomial given by

$$\Theta^{(k)}(u) = 1 - \theta_1^{(k)} u - \theta_2^{(k)} u^2 - \ldots - \theta_p^{(k)} u^p, \tag{29.4}$$

where u is a complex number. Therefore, model $\mathcal{M}_q(k)$ in (29.3) is stationary if the roots of (29.4) lie outside the unit circle, or equivalently, if $|\alpha_l^{(k)}| < 1$ for all $l = 1 : p$. Some of these reciprocal roots will be real and the rest will appear in complex conjugate pairs. Assume that for all models we have n_c pairs of complex reciprocal roots, each of them defined in terms of its modulus and wavelength, i.e., $\alpha_{2j-1}^{(k)} = r_j^{(k)} \exp\left(-2\pi i/\lambda_j^{(k)}\right)$ and $\alpha_{2j}^{(k)} = r_j^{(k)} \exp\left(+2\pi i/\lambda_j^{(k)}\right)$, for $j = 1 : n_c$, and n_r real reciprocal roots $\alpha_j^{(k)} = r_j^{(k)}$ for $j = (2n_c + 1) : p$, where the $r_j^{(k)}$s denote the moduli and the $\lambda_j^{(k)}$s are the wavelengths or periods. The assumption of having the same values of n_c and n_r across q is a reasonable one for the EEG signals recorded during the cognitive fatigue experiment. Analyses of EEG traces in some of the subjects who participated in this experiment indicate that the numbers of complex reciprocal pairs and real reciprocal roots generally remain unchanged over the course of the experiment, and so, we assume that the AR models $\mathcal{M}_q(k)$ have the same model order $p = 2n_c + n_r$ for $q = 1 : Q$ and $k = 1 : K$.

We propose the following prior structure on the moduli and periods of the characteristic reciprocal roots of each of the K models. For the complex reciprocal roots we have that

$$\left(r_j^{(k)}|\mathcal{D}_0\right) \sim f_{j,k}\left(r_j^{(k)}\right), \quad \text{and} \quad \left(\lambda_j^{(k)}|\mathcal{D}_0\right) \sim g_{j,k}\left(\lambda_j^{(k)}\right), \quad j = 1 : n_c,$$

with $f_{j,k}\left(r_j^{(k)}\right)$ a continuous distribution on $l_{j,k}^c(1) \leq r_j^{(k)} \leq u_{j,k}^c(1)$, with $0 < l_{j,r}^c(1) \leq u_{j,k}^c(1) < 1$, and $g_{j,k}\left(\lambda_j^{(k)}\right)$ a continuous distribution on $l_{j,k}^c(2) \leq \lambda_j^{(k)} \leq u_{j,k}^c(2)$, with $2 \leq l_{j,k}(2) \leq u_{j,k}(2) \leq \lambda_{j,k}^*$ for some fixed value $\lambda_{j,k}^* \geq 2$. In addition, for the real reciprocal roots we have that

$$\left(r_j^{(k)}|\mathcal{D}_0\right) \sim h_{j,k}\left(r_j^{(k)}\right), \quad j = (2n_c + 1) : p,$$

with $h_{j,k}\left(r_j^{(k)}\right)$ a continuous distribution on $l_{j,k}^r \leq r_j^{(k)} \leq u_{j,k}^r$, with $-1 < l_{j,k}^r \leq u_{j,k}^r < 1$. $\mathcal{D}_0$ denotes all the information available initially.

Finally, if the precision parameter ϕ is unknown, a Gamma prior distribution is used, with $(\phi|\mathcal{D}_0) \sim Ga(n_0/2, d_0/2)$. Some examples follow. In these examples we assume that ϕ is known. We illustrate how truncated normal priors can be useful in specifying which spectral characteristics of the signals define a particular mental state.

Example 29.1 (Multi-AR(1) process with truncated normal priors) Assume that we have two possible states and so, $K = 2$. Suppose that it is known that each of these states can be represented by an AR(1) such that, when $k = 1$ we have a stationary but very persistent process, with $\left(r^{(1)}|\mathcal{D}_0\right) \sim TN(0.95, 0.001, \mathcal{R}^{(1)})$, and $\mathcal{R}^{(1)} = (0.8, 1.0)$. Here, $TN(x|\mu, \sigma^2, \mathcal{R})$ denotes a truncated normal distribution on x with location parameter μ, scale parameter σ and truncation region $\mathcal{R}$. In addition, when $k = 2$, the process is much closer to white noise and $\left(r^{(2)}|\mathcal{D}_0\right) \sim TN(0.1, 0.01, \mathcal{R}^{(2)})$, where $\mathcal{R}^{(2)} = (0, 0.3)$. In this multi-AR(1) setting we have that, before observing any data, $E\left(r^{(1)}|\mathcal{D}_0\right) = 0.946$, $E\left(r^{(2)}|\mathcal{D}_0\right) = 0.123$, $V\left(r^{(1)}|\mathcal{D}_0\right) = 0.0008$ and $V\left(r^{(2)}|\mathcal{D}_0\right) = 0.0052$. As data y_q arrives, the distributions of $\left(r^{(k)}|\mathcal{D}_q\right)$ need to be sequentially updated, as well as $Pr(\mathcal{M}_q(k)|\mathcal{D}_q)$.

Example 29.2 (Multi-AR(2) processes) Take $K = 2,\ p = 2$ and assume that the two states can be characterized a priori by

$$\left(r^{(1)}|\mathcal{D}_0\right) \sim TN\left(0.95, 0.001, \mathcal{R}_1^{(1)}\right), \quad \left(\lambda^{(1)}|\mathcal{D}_0\right) \sim TN\left(10, 4, \mathcal{R}_2^{(1)}\right),$$

$$\left(r^{(2)}|\mathcal{D}_0\right) \sim TN\left(0.95, 0.001, \mathcal{R}_1^{(2)}\right), \quad \left(\lambda^{(2)}|\mathcal{D}_0\right) \sim TN\left(17, 4, \mathcal{R}_2^{(2)}\right),$$

with $\mathcal{R}_1^{(1)} = \mathcal{R}_1^{(2)} = (0.8, 1)$, $\mathcal{R}_2^{(1)} = (8, 12)$ and $\mathcal{R}_2^{(2)} = (14, 20)$. Here, $E\left(\lambda^{(1)}|\mathcal{D}_0\right) = 10$, $E\left(\lambda^{(2)}|\mathcal{D}_0\right) = 17$, and $V\left(\lambda^{(1)}|\mathcal{D}_0\right) = 1.1645$ and $V\left(\lambda^{(2)}|\mathcal{D}_0\right) = 2.206$. So, both states have the same prior distribution on the moduli of the complex reciprocal roots that characterize each process in the mixture. In particular, these roots are assumed to be rather persistent, given that their moduli are constrained to be above 0.8 by the prior structure. State $k = 1$ has a period in the (8, 12) interval, while the second state has larger period in the (14, 20) interval. Then, via the prior structure, the two states are differentiated in terms of their frequency content.

29.2.2.2 Posterior inference

In this section we follow the multiprocess notation of West and Harrison (1997, Chapter 12). Specifically, let $\pi_q(k) = Pr(\mathcal{M}_q(k)|\mathcal{D}_{q-1})$, be the prior probability of model $\mathcal{M}_q(k)$ before observing the data of epoch q, where $\mathcal{D}_{q-1}$ denotes all the information available up to $q-1$. We assume that $\mathcal{D}_0 = \{y_{1,1}, \ldots, y_{p,1}\}$ and $\mathcal{D}_q = \{\mathcal{D}_{q-1}, y_q\}$. Similarly, let $p_q(k) = Pr(\mathcal{M}_q(k)|\mathcal{D}_q)$ be the posterior probability of model $\mathcal{M}_q(k)$ after epoch q has been observed. In addition, a first-order Markov assumption relates each $p_{q-1}(i)$, for $i = 1 : K$, with $\pi_q(k)$ via fixed and known transition probabilities denoted by

$$\pi(k|i) = Pr\left(\mathcal{M}_q(k)|\mathcal{M}_{q-1}(i), \mathcal{D}_{q-1}\right) = Pr\left(\mathcal{M}_q(k)|\mathcal{M}_{q-1}(i), \mathcal{D}_0\right),$$

for all q and so, $\pi_q(k) = \sum_{i=1}^{K} \pi(k|i) p_{q-1}(i)$. Now, for each q and h such that $0 \leq h < q$, define

$$p_q(k_q, k_{q-1}, \ldots, k_{q-h}) = Pr\left(\mathcal{M}_q(k_q), \ldots, \mathcal{M}_{q-h}(k_{q-h})|\mathcal{D}_q\right),$$

and consequently, the posterior probability of model $\mathcal{M}_q(k_q)$ after observing the data of epoch q is $p_q(k_q)$.

In our EEG application, we want to update $p\left(\theta^{(1)}, \ldots, \theta^{(K)}, \phi|\mathcal{D}_q\right)$ and $p_q(k_q)$ sequentially as the epoched EEG data arrives. We begin with $q = 1$, and so,

$$p\left(\theta^{(1)}, \ldots, \theta^{(K)}, \phi|\mathcal{D}_1\right) = \sum_{k=1}^{K} p\left(\theta^{(1)}, \ldots, \theta^{(K)}, \phi|\mathcal{M}_1(k_1), \mathcal{D}_1\right) p_1(k_1), \quad (29.5)$$

where

$$\begin{aligned} p\left(\theta^{(1)}, \ldots, \theta^{(K)}, \phi|\mathcal{M}_1(k_1), \mathcal{D}_1\right) \\ = \frac{p\left(y_1|\mathcal{M}_1(k_1), \theta^{(1)}, \ldots, \theta^{(K)}, \phi\right) \times p\left(\theta^{(1)}, \ldots, \theta^{(K)}, \phi|\mathcal{D}_0\right)}{p(y_1|\mathcal{M}_1(k_1), \mathcal{D}_0)}, \end{aligned} \quad (29.6)$$

and

$$\begin{aligned} p(y_1|\mathcal{M}_1(k_1), \mathcal{D}_0) = \int \cdots \int \frac{1}{(2\pi|\phi^{-1} I|)^{1/2}} \exp\left\{-\frac{\phi\left(y_1 - X_1\theta^{(k_1)}\right)'\left(y_1 - X_1\theta^{(k_1)}\right)}{2}\right\} \\ \times p\left(\theta^{(1)}, \ldots, \theta^{(K)}, \phi|\mathcal{D}_0\right) d\theta^{(1)} \cdots d\theta^{(K)} d\phi. \end{aligned} \quad (29.7)$$

In addition, $p_1(k_1)$ in equation (29.5) is given by

$$p_1(k_1) = Pr(\mathcal{M}_1(k_1)|\mathcal{D}_1) \propto p(y_1|\mathcal{M}_1(k_1), \mathcal{D}_0) Pr(\mathcal{M}_1(k_1)|\mathcal{D}_0), \quad (29.8)$$

with $p_1(k_1)$ normalized such that $\sum_{k_1=1}^{K} p_1(k_1) = 1$.

The first difficulty in computing (29.5) arises because the priors proposed in Section 29.2.2.1 do not lead to closed form expressions of the posterior. Furthermore, the expression in (29.7) is generally not available analytically. Huerta and West (1999) develop a Markov chain Monte Carlo approach to compute the posterior distribution of the parameters of a single AR model when structured priors on the characteristic reciprocal roots are used. Such approach is not useful in this setting since it does not allow us to directly compute (29.7) for each k_1 with $k_1 = 1 : K$. More importantly, an MCMC scheme is not computationally efficient to achieve on-line posterior inference, which is a requirement in our EEG application. Even if normal-Gamma prior distributions were set on $(\theta^{(k)}|\phi)$ and ϕ, computing the posterior distribution at steps $q > 1$ would be highly computationally demanding, particularly for K large, as it would involve a mixture of K^q components. This is

$$p\left(\theta^{(1)}, \ldots, \theta^{(K)}, \phi|\mathcal{D}_q\right) = \sum_{k_q=1}^{K} \cdots \sum_{k_1=1}^{K} p\left(\theta^{(1)}, \ldots, \theta^{(K)}, \phi|\mathcal{M}_q(k_q), \ldots, \mathcal{M}_1(k_1), \mathcal{D}_q\right) \times p_q(k_q, \ldots, k_1). \quad (29.9)$$

West and Harrison (1997) obtain approximate posterior inference for multiprocess models of class II – i.e. mixture models in which each model component has a dynamic linear model structure with conjugate normal-Gamma priors. Such inference is obtained by approximating the posterior distribution at each time period by a fixed number of mixtures, say K^{h+1} for some h, with h typically small. We follow a similar approach here for the specific cases of multi-AR(1) and multi-AR(2) models with structured priors, as detailed below. The approximations used in these low order mixture models are not helpful in obtaining approximate posterior inference for multi-AR models with $p > 2$ if structured priors are considered. An alternative solution for achieving real-time inference in higher order multi-AR(p) processes with structured priors is to consider sequential Monte Carlo methods. We revisit this topic in Section 29.3 when discussing current and future research directions.

29.2.3 The multi-AR(1) and multi-AR(2) cases: Approximate posterior inference

We summarize the methodology for approximate posterior inference in multi-AR(1) and multi-AR(2) models. We present results for cases in which the prior distributions on the AR coefficients – implied by impossing the structured priors on the AR characteristic reciprocal roots proposed in Section 29.2.2.1 – are truncated normal priors, or can be approximated by truncated normal priors.

29.2.3.1 The multi-AR(1) case

We discuss some features of the methodology to approximate the posteriors sequentially when ϕ is known. Details on approximations for multi-AR(1) models in which ϕ is unknown are given in Appendix B.

When $p = 1$ we have K AR coefficients – one per mixture component – or equivalently, K real reciprocal roots, each with modulus $r^{(k)}$ and so, $\theta^{(k)} = r^{(k)}$, for $k = 1 : K$. Assume that $f_k(r^{(k)}) = TN\left(r^{(k)}|m_0(k), C_0(k), \mathcal{R}^{(k)}\right)$. Then,

$$\begin{aligned} p\left(\theta^{(1:K)}|\mathcal{D}_q\right) &= \sum_{k_q=1}^{K}\sum_{k_{q-1}=1}^{K}\cdots\sum_{k_1=1}^{K} p\left(\theta^{(1:K)}|\mathcal{M}_q(k_q), \mathcal{M}_{q-1}(k_q), \ldots, \mathcal{M}_1(k_1), \mathcal{D}_q\right) \\ &\quad \times p_q(k_q, k_{q-1}, \ldots, k_1) \\ &\approx \sum_{k_q=1}^{K}\sum_{k_{q-1}=1}^{K} p\left(\theta^{(1:K)}|\mathcal{M}_q(k_q), \mathcal{M}_{q-1}(k_{q-1}), \mathcal{D}_q\right) \times p_q(k_q, k_{q-1}). \end{aligned} \tag{29.10}$$

In other words, we are using the approximation

$$\begin{aligned} &p\left(\theta^{(1:K)}|\mathcal{M}_q(k_q), \mathcal{M}_{q-1}(k_{q-1}), \ldots, \mathcal{M}_1(k_1), \mathcal{D}_q\right) \\ &\quad \approx p\left(\theta^{(1:K)}|\mathcal{M}_q(k_q), \mathcal{M}_{q-1}(k_{q-1}), \mathcal{D}_q\right), \end{aligned}$$

suggested in West and Harrison (1997) in the context of multiprocess models of class II. This implies that the number of components in the mixture at any given time period will be at most K^2. Then, when epoch q arrives, we only need to consider the models for the previous epoch to perform the analysis, instead of looking at all the models for all the epochs, starting from the previous one down to the first one. As mentioned before, approximations can use $K^{(h+1)}$ components in the mixture for h any integer with $h \geq 1$. In such cases we would need to consider the models in epochs $(q-h) : (q-1)$ when performing the analysis for epoch q.

Using the approximation with $h = 1$ it can be shown that, if $\left(\theta^{(k)}|\mathcal{M}_{q-1}(k_{q-1}), \mathcal{D}_{q-1}\right)$ is approximated by a single truncated normal, the distribution of $\left(\theta^{(k)}|\mathcal{M}_q(k_q), \mathcal{M}_{q-1}(k_{q-1}), \mathcal{D}_q\right)$ can be approximated by

$$\begin{aligned} &\left(\theta^{(k)}|\mathcal{M}_q(k_q), \mathcal{M}_{q-1}(k_{q-1}), \mathcal{D}_q\right) \\ &\quad \approx TN\left(\theta^{(k)}|m_q^{(k)}(k_q, k_{q-1}),\; C_q^{(k)}(k_q, k_{q-1}), \mathcal{R}^{(k)}\right). \end{aligned} \tag{29.11}$$

The expressions to compute $m_q^{(k)}(\cdot, \cdot)$, and $C_q^{(k)}(\cdot, \cdot)$ in (29.11), as well as $p_q(\cdot, \cdot)$ in (29.10) are given in Appendix B.

Finally, the distribution of $\left(\theta^{(k)}|\mathcal{M}_q(k_q), \mathcal{D}_q\right)$, which can be written as a mixture of K components, can be approximated by a single truncated normal distribution, i.e.

$$\left(\theta^{(k)}|\mathcal{M}_q(k_q), \mathcal{D}_q\right) \approx TN\left(\theta^{(k)}\,|m_q^{(k)}(k_q), C_q^{(k)}(k_q), \mathcal{R}^{(k)}\right), \tag{29.12}$$

by further collapsing the K components in such mixture. Then, we can proceed with approximate on-line inference by sequentially computing the moments of the distributions in (29.11) and (29.12), as well as $p_q(k_q, k_{q-1})$ and $p_q(k_q)$, with

$p_q(k_q) = \sum_{k_{q-1}=1}^{K} p_q(k_q, k_{q-1})$. Once again, the expressions to compute $m_q^{(k)}(\cdot)$ and $C_q^{(k)}(\cdot)$ are given in Appendix B.

When ϕ is unknown, a Gamma prior distribution $(\phi|\mathcal{D}_0) \sim Ga(n_0/2, d_0/2)$ is assumed and so, the posterior distributions in (29.11) and (29.12) can be approximated by truncated Student-t distributions instead of truncated normals, while the distribution of $(\phi|\mathcal{M}_q(k_q), \mathcal{D}_q)$ is approximated by a Gamma distribution (see Appendix B).

29.2.3.2 *The multi-AR(2) case*

For $p = 2$ we can either have two real reciprocal roots per model, or one pair of complex reciprocal roots. We describe the later case as it is relevant in the EEG framework.

Let $r^{(k)}$ and $\lambda^{(k)}$ denote the modulus and wavelength of the k-th AR(2) component in the mixture. In AR(2) models, the relationship between the AR coefficients and the reciprocal roots of the characteristic polynomial is given by

$$\theta_2^{(k)} = -\left(r^{(k)}\right)^2, \quad \text{and} \quad \theta_1^{(k)} = 2r^{(k)}\cos\left(2\pi/\lambda^{(k)}\right).$$

Then, if $r^{(k)} \in (l_k(1), u_k(1))$ and $\lambda^{(k)} \in (l_k(2), u_k(2))$, with $l_k(\cdot)$ and $u_k(\cdot)$ such that $0 < l_k(1) \leq u_k(1) < 1$ and $2 < l_k(2) \leq u_k(2) < \lambda_k^*$, we have that

$$\theta_2^{(k)} \in \left(-\,(u_k(1))^2, -(l_k(1))^2\right) \quad \text{and}$$

$$\theta_1^{(k)} \in \left(2\sqrt{-\theta_2^{(k)}}\cos(2\pi/l_k(2)), 2\sqrt{-\theta_2^{(k)}}\cos(2\pi/u_k(2))\right). \quad (29.13)$$

Now, assume that the prior structure on $r^{(k)}$ and $\lambda^{(k)}$ is such that $\left(r^{(k)}|\mathcal{D}_0\right) \sim f_k(r^{(k)})$ and $\left(\lambda^{(k)}|\mathcal{D}_0\right) \sim g_k(\lambda^{(k)})$, with

$$f_k(r^{(k)}) = TN\left(r^{(k)}|m_{0,1}(k), C_{0,1}(k), \mathcal{R}_1^{(k)}\right), \quad \text{and}$$

$$g_k(\lambda^{(k)}) = TN\left(\lambda^{(k)}|m_{0,2}(k), C_{0,2}(k), \mathcal{R}_2^{(k)}\right), \quad (29.14)$$

where $\mathcal{R}_1^{(k)} = (l_k(1), u_k(1))$ and $\mathcal{R}_2^{(k)} = (l_k(2), u_k(2))$. Such prior imposes the restrictions (29.13) on the prior (and the posterior) of $\theta_1^{(k)}$ and $\theta_2^{(k)}$ and, when combined with (29.3) for $p = 2$, it does not lead to closed form posterior distributions on $\theta_1^{(k)}$ and $\theta_2^{(k)}$. In order to perform approximate on-line inference in multi-AR(2) models, we propose a prior on $\theta^{(k)}$ of the form

$$\left(\theta^{(k)}|\mathcal{D}_0\right) \sim TN\left(\theta^{(k)}|m_0(k), C_0(k), \mathcal{R}^{(k)}\right), \quad (29.15)$$

with $\mathcal{R}^{(k)} = (a_1(k), b_1(k)) \times (a_2(k), b_2(k))$, where $a_2(k) = -(u_k(1))^2$, $b_2(k) = -(l_k(1))^2$, $a_1(k) = 2l_k(1)\cos(2\pi/l_k(2))$ and $b_1(k) = 2u_k(1)\cos(2\pi/u_k(2))$. This prior structure guarantees that $r^{(k)} \in (l_k(1), u_k(1))$, but its support includes regions of the parameter space that lead to $\lambda^{(k)}$ values that are not in $(l_k(2), u_k(2))$. This could be a problem in situations where we want to discriminate between two mental states characterized only in terms of the frequency of a given latent component of the EEG signal – instead of the frequency and the modulus of such component – and the frequency values for those two states are relatively close. In our EEG application we choose $m_0(k)$ and $C_0(k)$ in (29.15) that result in a prior structure as similar as possible to the implied prior of $\theta^{(k)}$ that would be obtained if a structure of the form (29.14) was used on $r^{(k)}$ and $\lambda^{(k)}$. We illustrate this issue in the example below.

The priors (29.15) lead to the following approximate results, when ϕ is known

$$\left(\theta^{(k)}|\mathcal{M}_q(k_q), \mathcal{M}_q(k_{q-1}), \mathcal{D}_q\right) \approx TN\left(\theta^{(k)}\left|m_q^{(k)}(k_q, k_{q-1}), C_q^{(k)}(k_q, k_{q-1}), \mathcal{R}^{(k)}\right.\right),$$

$$\left(\theta^{(k)}|\mathcal{M}_q(k_q), \mathcal{D}_q\right) \approx TN\left(\theta^{(k)}\left|m_q^{(k)}(k_q), C_q^{(k)}(k_q), \mathcal{R}^{(k)}\right.\right). \tag{29.16}$$

Then, as in the multi-AR(1) case, approximate on-line posterior inference is achieved by computing the moments of the distributions in (29.16), as well as $p_q(k_q, k_{q-1})$ and $p_q(k_q)$. In addition, when ϕ is unknown and modelled with a Gamma prior, the distributions used in the approximations (29.16) will be truncated Student-t distributions (see Appendix B for details).

Before proceeding with the EEG analysis, we show how the approximate inference works in the analysis of data simulated from two AR(2) processes, each with a pair of complex conjugate characteristic roots with similar moduli values but different periods.

Example 29.3 (Simulated data) A time series with 9,000 data points was simulated from two AR(2) processes as follows. The first 2,000 observations were simulated, in batches of 100 observations, from an AR(2) process with modulus $r_1 = 0.95$ and wavelength $\lambda_1 = 16$. The following 3,000 data points were simulated from an AR(2) process with modulus $r_2 = 0.99$ and wavelength $\lambda_2 = 6$ (also in batches of 100 observations). Then, the next 2,000 observations were again simulated from the AR(2) process with reciprocal roots $r_1 e^{\pm 2\pi i/\lambda_1}$ and finally, the last 2,000 observations were simulated from the AR(2) model with reciprocal roots $r_2 e^{\pm 2\pi i/\lambda_2}$. The innovations for both types of processes followed independent Gaussian distributions centered at zero with variance $v = 1/\phi = 100$. In order to mimic the structure of our EEG data, we assume that an epoch consists of 100 observations and so, the simulated data has a total number of 90 epochs.

We fitted multi-AR(2) models with $K = 2$, and the following prior structure

$$\begin{aligned}
(\theta^{(1)}|\mathcal{D}_0) &\sim TN\left(m_0(1), C_0(1), \mathcal{R}^{(1)}\right),\\
\mathcal{R}^{(1)} &= \left(1.4\cos\left(\frac{2\pi}{4}\right), 2\cos\left(\frac{2\pi}{10}\right)\right) \times (-1^2, -0.7^2),\\
(\theta^{(2)}|\mathcal{D}_0) &\sim TN\left(m_0(2), C_0(2), \mathcal{R}^{(2)}\right),\\
\mathcal{R}^{(2)} &= \left(1.4\cos\left(\frac{2\pi}{12}\right), 2\cos\left(\frac{2\pi}{20}\right)\right) \times (-1^2, -0.7^2), \qquad (29.17)\\
(\phi|\mathcal{D}_0) &\sim Ga(1/2, 10/2).
\end{aligned}$$

The values of $m_0(k)$ and $C_0(k)$ for $k = 1, 2$ were chosen to approximate a prior with the following structure on $r^{(k)}$ and $\lambda^{(k)}$

$$\begin{aligned}
r^{(1)} &\sim TN\left(r^{(1)}|0.8, 1, \mathcal{R}_1^{(1)}\right), \quad \lambda^{(1)} \sim TN\left(\lambda^{(1)}|5, 1, \mathcal{R}_2^{(1)}\right), \qquad (29.18)\\
r^{(2)} &\sim TN\left(r^{(2)}|0.8, 1, \mathcal{R}_1^{(2)}\right), \quad \lambda^{(2)} \sim TN\left(\lambda^{(2)}|13, 1, \mathcal{R}_2^{(2)}\right),
\end{aligned}$$

and with $\mathcal{R}_1^{(1)} = \mathcal{R}_1^{(2)} = (0.7, 1)$, $\mathcal{R}_2^{(1)} = (4, 10)$ and $\mathcal{R}_2^{(2)} = (12, 20)$. The graphs (a) and (c) in Figure 29.3 show 1,000 simulated values from the implied prior on $\theta^{(k)} = \left(\theta_1^{(k)}, \theta_2^{(k)}\right)'$ if the priors in (29.18) are chosen for $r^{(k)}$ and $\lambda^{(k)}$, while the graphs (b) and (d) correspond to 1,000 values simulated from the prior in (29.17). For these graphs we used $m_0(1) = (1.6\cos(2\pi/5), -0.8^2)$, $m_0(2) = (1.6\cos(2\pi/13), -0.8^3)$ and $C_0(1)$ and $C_0(2)$ such that $C_0(1)[1, 1] = C_0(1)[2, 2] = 1$, $C_0(1)[1, 2] = C_0(1)[2, 1] = -0.4$, $C_0(2)[1, 1] = C_0(2)[1, 1] = 1$ and $C_0(2)[1, 2] = C_0(2)[2, 1] = -0.9$. As mentioned before, the priors in (29.17) and (29.18) are not the same, with (29.17) giving non-zero probability to regions of $\theta^{(k)}$ that result in values of $\lambda^{(k)}$ that do not lie in $\mathcal{R}_2^{(k)}$. This may be a problem in models for which different states are characterized by the same modulus and distinct but very close wavelength values. In such cases, as expected, it will be typically hard to discriminate between states. In the EEG data for subject **skh** channels that show discrepancies in the EEG signals recorded at the beginning of the experiment, and those recorded towards the end of the experiment are typically characterized by relatively different moduli values and often by different wavelength values as well.

Finally, to complete the model structure in this analysis we set $\pi(1|1) = \pi(2|2) = 0.9$ and $\pi_0(1) = 0.9$. The left panel in Figure 29.4 shows the approximate values of $p_q(1) = Pr(\mathcal{M}_q(1)|\mathcal{D}_q)$ for $q = 1:90$. We have that $\hat{p}_1(1) \approx 0.2$, and then $\hat{p}_q(1) = 0$ for $q = 2:20$ and $q = 51:70$, while $\hat{p}_q(1) = 1$ for $q = 21:50$ and $q = 71:90$. This corresponds precisely to the structure used to simulate the data. The right panel shows approximate values of $E\left(\phi^{-1}|\mathcal{D}_q\right)$ for $q = 1:90$, indicating that the approximations also work well in terms of the posterior

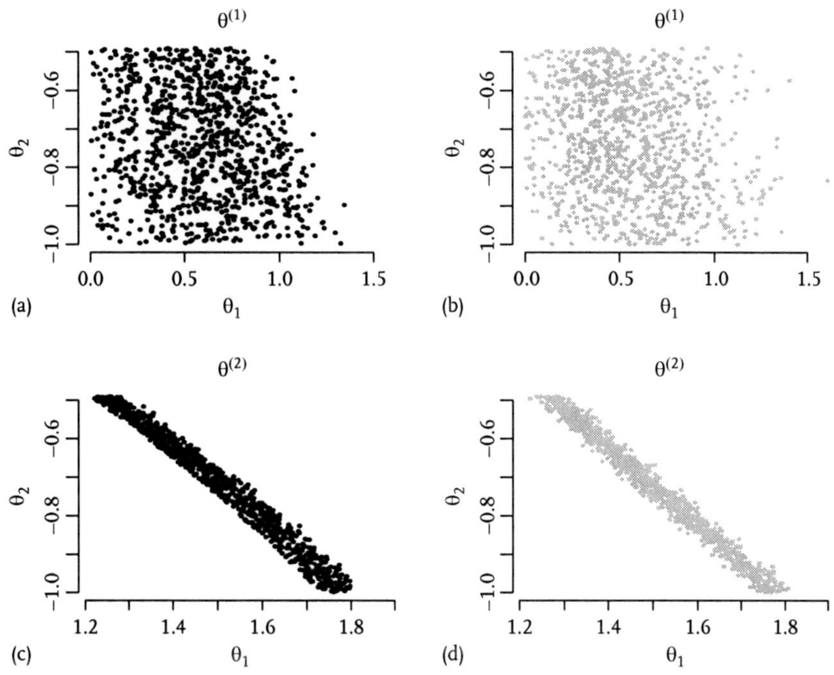

Fig. 29.3 (a) Simulated values from the prior (29.18) for $\theta^{(1)}$; (b) simulated values from the prior (29.17) for $\theta^{(1)}$; (c) simulated values from the prior (29.18) for $\theta^{(2)}$; (d) simulated values from the prior (29.17) for $\theta^{(1)}$.

posterior inference for ϕ. In addition, we obtain that $E(r^{(1)}|\mathcal{D}_{90}) \approx 0.9909$, $E(r^{(2)}|\mathcal{D}_{90}) \approx 0.9557$, $E(\lambda^{(1)}|\mathcal{D}_{90}) \approx 5.9919$, and $E(\lambda^{(2)}|\mathcal{D}_{90}) \approx 16.1103$. Then, $\mathcal{M}_q(1)$ captures the AR(2) structure whose characteristic roots have moduli $r_2 = 0.99$ and period $\lambda_2 = 6$, while $\mathcal{M}_q(2)$ captures the AR(2) structure whose characteristic roots have moduli $r_1 = 0.95$ and period $\lambda_1 = 16$.

29.2.4 Analysis of EEG data via multi-AR(1) and multi-AR(2)

We use multi-AR(1) and multi-AR(2) models to study changes in the latent components of the EEG epochs over time. We begin by computing the decompositions in (29.2) for each channel and each epoch and then extract some of the latent processes as follows. For each channel j, we fit AR(10) models to each of the 864 epochs and estimate the AR reciprocal roots based on the posterior means of the AR coefficients. Most epochs display four pairs of complex roots and two distinct real roots. We order the reciprocal roots by moduli and extract the latent process with the highest modulus for each epoch and each channel. Therefore, for each channel we obtain a collection of 864 time series (one per epoch). For subject **skh**, each of these latent processes is associated with a real reciprocal root for all the epochs and all the channels. We denote each time

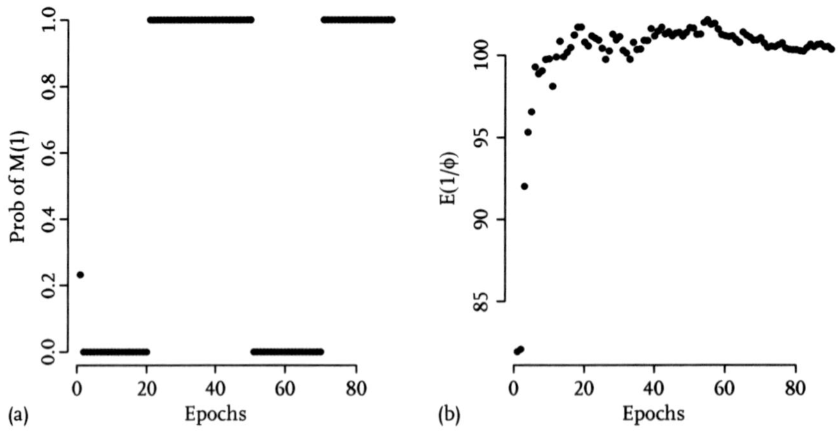

Fig. 29.4 (a) Approximate values of $p_q(1) = Pr(\mathcal{M}_q(1)|\mathcal{D}_q)$, with $q = 1:90$ for the simulated data; (b) $E\left(\phi^{-1}|\mathcal{D}_q\right)$ based on approximations.

series in this collection by $x_{1:T,q,j}$. Similarly, we extract the quasi-periodic latent components that lie in the alpha frequency band for all the epochs and all channels. Each of those latent processes is a time series, denoted by $z_{1:T,1:Q,j}$, with an ARMA(2, 1) structure. Then, for each channel j we proceed to analyse the collection of time series $x_{1:T,1:Q,j}$ with multi-AR(1) models, and $z_{1:T,1:Q,j}$ with multi-AR(2) models.

Based on analyses of EEG data for another subject who participated in the same study (labelled as **ess**), we use a model with two states to describe $x_{1:T,1:Q,j}$. One of the two states is characterized by a very persistent modulus in (0.975, 1.0), while the other state is characterized by a modulus below 0.975. Specifically, for each channel j we model the series $x_{1:T,q,j}$ for $q = 1:864$ as a mixture of two AR(1) process with the following prior structure

$$\left(\theta^{(1)}|\mathcal{D}_0\right) \sim TN\left(\theta^{(1)}|0.94,\, C_0(1),\, \mathcal{R}^{(1)}\right),\quad \left(\theta^{(2)}|\mathcal{D}_0\right) \sim TN\left(\theta^{(2)}|0.98,\, C_0(2),\, \mathcal{R}^{(2)}\right),$$
$$(\phi|\mathcal{D}_0) \sim Ga(1/2,\, 10/2),$$

with $\mathcal{R}^{(1)} = (0.9, 0.975)$ and $\mathcal{R}^{(2)} = (0.975, 1.0)$. Values of $C_0(1) = C_0(2) = c$, with $c = 0.01, 1, 10$ were used, leading to similar results in terms of the posterior inference. In addition, we set $\pi(1|1) = \pi(2|2) = 0.999$ and $\pi_0(1) = 0.9$.

Figure 29.5 displays the estimated values of $p_q(1) = Pr(\mathcal{M}_q(1)|\mathcal{D}_q)$ (light dots) and $p_q(2) = Pr(\mathcal{M}_q(2)|\mathcal{D}_q)$ (dark squares) for channels Pz, P_4, CP_4, P_3 and CPz. For these channels, the process in the mixture with AR coefficient restricted to (0.9, 0.975) dominates the first epochs of the experiment, while that with AR coefficient above 0.975 dominates the last epochs of the experiment. There is a fair amount of variability across channels. Channels Pz, P_4 and CP_4 behave similarly, displaying large values of $p_q(1)$ for most epochs before $q = 600$ (i.e. before approx. 106 min.), and large values of $p_q(2)$ for

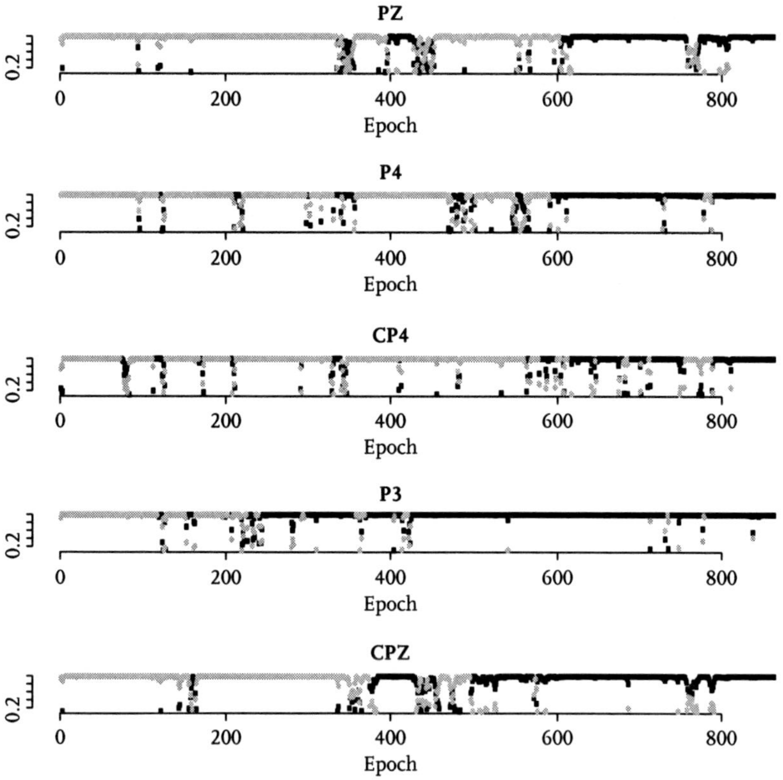

Fig. 29.5 Estimates of $p_q(1)$ (light dots) and $p_q(2)$ (dark dots) for the latent processes with the highest moduli in channels Pz, P_4, CP_4, P_3 and CPz.

most epochs after $q = 600$. These three channels are located close to each other in left-parietal and left-middle areas, as shown in Figure 29.1. Channel P_3 displays large values of $p_q(2)$ for most epochs after about $q = 250$ (i.e. after ± 45 min. had elapsed). Pictures for the remaining channels (not displayed) showed $p_q(1) \approx 1$ for almost all the epochs in the experiment, and so, no multistate evidence for the EEG latent process with the largest modulus was found in such channels.

The same type of analyses were carried out to describe the quasi-periodic latent processes $z_{1:T,q,j}$, for each j and all q with $q = 1 : 864$. Figure 29.6 shows the estimates of $p_q(1)$ (light dots) and $p_q(2)$ (dark squares) for channel T_8, based on a multi-AR(2) model with $K = 2$ and truncated normal priors on $\theta^{(1)}$ and $\theta^{(2)}$ with truncation regions

$$\mathcal{R}^{(1)} = (1.2\cos(2\pi/9.5), 1.6\cos(2\pi/11.5)), \times (-0.85^2, -0.6^2),$$

$$\mathcal{R}^{(2)} = (1.6\cos(2\pi/11.5), 1.9\cos(2\pi/13.5)) \times (-1.0^2, -0.85^2).$$

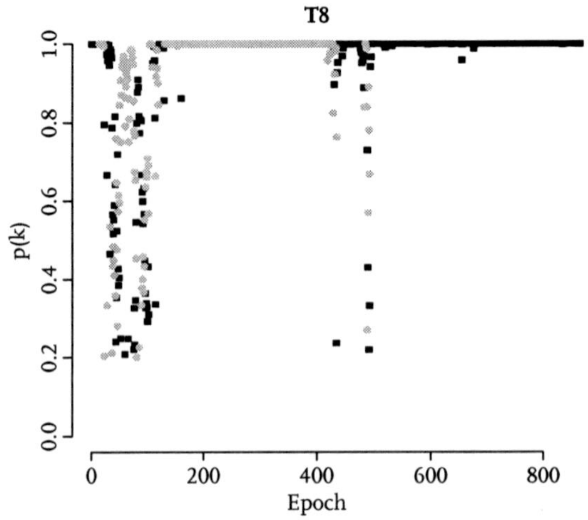

Fig. 29.6 Estimates of $p_q(1)$ (light dots) and $p_q(2)$ (dark squares) for channel T_8 based on a multi-AR(2) analysis.

Different values of $m_0(1) \in \mathcal{R}^{(1)}$ and $m_0(2) \in \mathcal{R}^{(2)}$ and $C_0(k)$ for $k = 1, 2$ were chosen, leading to similar posterior estimates. Given that the sampling rate in the data is 128 Hz, the prior for $k = 1$ aims to restrict the frequency approximately to the 11–13.5 Hz range, and with corresponding modulus in the 0.6–0.85 range while for $k = 2$, the range of the frequency is approximately (9.5, 11) Hz and its corresponding modulus is in the 0.85–1.0 range. We fitted this model to the collection of series $z_{1:T,q,j}$ for $q = 1 : 864$ and for each channel. We found evidence in favor of a process with two states as defined above in channels T_8 and FT_8. For channel T_8, $p_q(2) \approx 1$ for all the epochs after $q \approx 500$ ($\approx$ 80 min.) as seen in Figure 29.6. Channel FT_8 has $p_q(2) \approx 1$ for most (but not all) epochs after $q = 650$ (not shown). For the remaining channels $p_q(1) \approx 1$ for almost all the epochs, indicating no evidence in support of more than one state based on this two-state multi-AR(2) process.

29.3 Further discussion

In this chapter we discuss the challenges of developing models for EEG data when the main goal is automatic detection and prediction of cognitive fatigue. We first show how AR models and related decompositions can be used as a descriptive tool for discovering EEG features that may be associated with one or more mental states. We have used these models in the analyses of EEG epochs from subjects who participated in the experiment described in Section 29.1.1 (Trejo *et al.* 2007). For some of these individuals, such as subject **skh**, the AR-based inference points towards differences in the spectral characteristics of the signals recorded during the first 15 min. of the experiment, and those recorded during the last 15 min. Some of these findings are consistent with the results shown in Trejo *et al.* (2006) obtained using a very different modelling approach. Given that the individuals were rested prior to the experiment and clearly fatigued after the experiment ended – as confirmed by measures of subject performance and pre and post-task moods (Trejo *et al.* 2006) – it is possible to hypothesize that the observed differences in the spectral characteristics of the signals recorded at the beginning of the experiment, and those recorded at the end, are associated with cognitive fatigue.

If a particular mental state can be characterized in terms of the frequencies and their corresponding peaks in the power spectra of the EEG signals for different channels – as seems to be feasible based on the results obtained using AR models in the cognitive fatigue study – the class of multi-AR(p) processes provides a modeling framework that allows practicioners to perform on-line detection of multiple mental states in a probabilistic fashion, as illustrated with the multi-AR(1) and multi-AR(2) analyses shown in Section 29.2.4. We emphasize that our intention with the analyses presented here was not to provide a precise definition of cognitive fatigue in terms EEG features, but to show that, if hypotheses about which EEG spectral features define a particular cognitive state are available, multi-AR processes can be used to determine if and when an individual enters such mental state.

Due to the computational difficulties for on-line posterior inference in multi-AR models, we present an approximate analysis for models with $p = 1$ and $p = 2$. Because the EEG signals are composed by several latent processes, most of which are quasi-periodic, low order models cannot be directly applied to the actual data and so, we first extracted relevant latent processes from the signals using the AR decompositions and then analyzed such processes with low order multi-AR models. In spite of their limitations, these low order multi-AR models provide useful insights in the description of the EEG data of subject **skh**. Approximate posterior inference can be achieved in a computationally efficient manner, allowing practioners to use these methods descriptively to study EEG data and hypothesize about which EEG features are associated with a particular mental state.

Current and future research directions include developing efficient algorithms for on-line posterior estimation in general multi-AR processes with structured priors for cases with $p \gg 2$. We are currently exploring sequential Monte Carlo (SMC) approaches for this purpose. In particular, we are considering SMC algorithms such as those proposed in Liu and West (2001) and Carvalho *et al.* (2008). One of the challenges in connection with the use of sequential MC approaches in this framework is dealing with relatively large parameter spaces, specially when K is large. Other areas of future research include developing models that can simultaneously handle data from multiple channels. Mixtures of vector autoregressions and factor models are two of the model classes that will be considered in the analysis of mulvariate EEG data.

Appendix

A. Broader context and background

A.1 Model specification for TVARs and ARs

A time-varying autoregressive model of order p, or TVAR(p), is a model of the form

$$y_t = \sum_{i=1}^{p} \theta_{t,i} y_{t-i} + \epsilon_t, \tag{29.19}$$

where $\theta_t = (\theta_{t,1}, \ldots, \theta_{t,p})'$ is the vector or AR coefficients at time t and ϵ_t is a zero-mean innovation, typically assumed Gaussian with variance v_t (West, Prado, and Krystal 1999). Two additional equations are used to describe the evolution of the model parameters over time. One of such equations describes the changes in θ_t while the other one models the variance v_t. Typically, a random walk is used to model θ_t and the multiplicative version of a random walk equation is used to describe the changes in v_t. This is,

$$\theta_t = \theta_{t-1} + \psi_t, \quad \psi_t \sim N(0, W_t), \tag{29.20}$$

and

$$v_t = v_{t-1}(\beta/\eta_t), \quad \eta_t \sim Be(a_t, b_t), \tag{29.21}$$

where the stochastic terms ψ_t and η_t are independent and mutually independent, as well as independent of ϵ_t. The matrix W_t can be defined using a discount factor $\delta \in (0, 1]$, as described in West and Harrison (1997). Low values of δ are consistent with rapid changes in θ_t over time, while large values describe smooth changes. A value of $\delta = 1$ leads to the standard autoregressive model, AR(p), i.e.

$$y_t = \sum_{i=1}^{p} \theta_i y_{t-i} + \epsilon_t, \tag{29.22}$$

when $\delta = 1$. Similarly, in equation (29.21), β acts as a discount factor taking values in $(0, 1]$. $\beta = 1$ leads to a model for which $v_t = v$ for all t. The model defined by equations (29.19), (29.20) and (29.21) is completed by specifying a prior distribution on $(\theta_1, v_1|\mathcal{D}_0)$, where $\mathcal{D}_0$ denotes all the information available initially. Conjugate normal-inverse-Gamma distributions are generally used, this is, $(\theta_1|v_1, \mathcal{D}_0) \sim N(\theta_1|m_0, v_1 C_0)$ and $\left(v_1^{-1}|\mathcal{D}_0\right) \sim Ga(n_0/2, d_0/2)$.

A.2 Time series decompositions

West (1997) and West, Prado, and Krystal (1999) provide the basic methodology for obtaining the time series decompositions for AR and TVAR models. Here we summarize such results.

Consider the following stucture

$$y_t = F'\theta_t, \quad \theta_t = G_t\theta_{t-1} + \eta_t, \qquad \eta_t \sim N(0, W_t).$$

Assume that at each time t the eigenvalues of G_t are all distinct, with c pairs of complex eigenvalues denoted by $r_{t,j}\exp(\pm 2\pi i/\lambda_{t,j})$ for $j = 1:c$, and r real eigenvalues $r_{t,j}$ with $j = 1:r$. Then, it is possible to show that

$$y_t = \sum_{j=1}^{c} z_{t,j} + \sum_{j=1}^{r} x_{t,j},$$

where $z_{t,j}$ is a real process associated to $r_{t,j}\exp(\pm 2\pi i/\lambda_{t,j})$ and $x_{t,j}$ is a real process associated to r_j. In the particular case of TVAR(p) models, $F = (1, 0, \dots, 0)'$, $\eta_t = (\epsilon_t, 0, \dots, 0)'$ and G_t is given by

$$\begin{pmatrix} \theta_{t,1} & \theta_{t,2} & \cdots & \theta_{t-1,p} & \theta_{t,p} \\ 1 & 0 & \cdots & 0 & 0 \\ 0 & 1 & \cdots & 0 & 0 \\ \vdots & & \ddots & \cdots & \vdots \\ 0 & 0 & \cdots & 1 & 0 \end{pmatrix},$$

and its eigenvalues correspond precisely to the reciprocal roots of the characteristic polynomial $\theta_t(u) = 1 - \theta_{t,1}u - \ldots - \theta_{t,p}u^p$. Each of the $z_{t,j}$ processes follows approximately a TVARMA(2, 1), with modulus $r_{t,j}$ and wavelength $\lambda_{t,j}$. Similarly, each of the $x_{t,j}$ processes follows approximately a TVAR(1), with modulus $r_{t,j}$. In the case of AR(p) models $\theta_t = \theta$ for all t and so, $z_{t,j}$ and $x_{t,j}$ follow ARMA(2, 1) and AR(1) processes, respectively.

B. Computation

B.1 Bayesian inference in AR and TVAR models

Inference in the TVAR models described above can be easily achieved using standard Dynamic Linear Model (DLM) theory, as detailed in West and Harrison (1997) for general DLMs, and in West, Prado, and Krystal (1999) for the particular case of TVAR models with the structure described above.

The software `tvar`, available on-line at `www.stat.duke.edu`, performs Bayesian inference and computes the time series decompositions for TVAR and AR models with conjugate normal-inverse-Gamma priors on θ_t and v_t. This software allows the user to choose optimal values for the model order

p and the discount factors δ and β. In particular, when $\delta = 1$ and $\beta = 1$ the `tvar` software provides posterior inference for standard AR models based on conjugate normal-Gamma priors on $\theta|\phi$ and $\phi = 1/v$.

A related software for Bayesian inference in AR models is `arcomp`, also available on-line at `www.stat.duke.edu`. This software implements the AR models with structured priors developed in Huerta and West (1999). There are two interesting features in the models implemented in `arcomp`. One of them is that it allows for unit roots. The other one is that it handles model order uncertainty.

B.2 Approximate posterior inference in multi-AR models

We refer the reader to West and Harrison (1997) for general theory of multi-process models. Such models are mixtures in which each model component has a dynamic linear model structure. West and Harrison (1997) develop methodology for approximate on-line posterior updating when normal-Gamma priors are used on the parameters of each DLM considered in the mixture. We summarize the steps to obtain approximate on-line posterior inference in multi-AR processes below, assuming that a truncated normal prior is set on the AR coefficients of each model component, and a Gamma prior is set on ϕ. In our multi-AR setting the AR coefficients do not change over time, and our priors are not conjugate within each mixture component. Therefore, the steps needed to sequentially update the approximate posterior distributions are not exactly the same as those used in the DLM multi-process framework, however they are very similar, and so, the reader would benefit from reading Chapter 12 of West and Harrison (1997).

Case with ϕ known

(a) Updating the posteriors at $q = 1$. After observing the data of the first epoch, we have that

$$p\left(\theta^{(1)}, \ldots, \theta^{(K)}|\mathcal{D}_1\right) = \sum_{k_1=1}^{K} p\left(\theta^{(1)}, \ldots, \theta^{(K)}|\mathcal{M}_1(k_1), \mathcal{D}_1\right) p_1(k_1),$$

where

$$p\left(\theta^{(1)}, \ldots, \theta^{(K)}|\mathcal{M}_1(k_1), \mathcal{D}_1\right) = \frac{p\left(y_1|\mathcal{M}_1(k_1), \theta^{(1)}, \ldots, \theta^{(K)}\right) p\left(\theta^{(1)}, \ldots, \theta^{(K)}|\mathcal{D}_0\right)}{p(y_1|\mathcal{M}_1(k_1), \mathcal{D}_0)},$$

and with

$$p(y_1|\mathcal{M}_1(k_1), \mathcal{D}_0) = \int p(y_1|\mathcal{M}_1(k_1), \theta^{(1)}, \ldots, \theta^{(K)}) \, p\left(\theta^{(1)}, \ldots, \theta^{(K)}|\mathcal{D}_0\right) d\theta^{(1)} \ldots d\theta^{(K)}.$$

In addition,

$$p_1(k_1) \propto p(y_1|\mathcal{M}_1(k_1), \mathcal{D}_0)\pi_1(k_1).$$

Now, if $(\theta^{(k)}|\mathcal{D}_0) \sim TN\left(m_0(k), C_0(k), \mathcal{R}^{(k)}\right)$, we have that

$$\left(\theta^{(k)}|\mathcal{M}_1(k_1), \mathcal{D}_1\right) = TN\left(\theta^{(k)}\left|m_1^{(k)}(k_1), C_1^{(k)}(k_1), \mathcal{R}^{(k)}\right.\right),$$

and

$$p_1(k_1) \propto \frac{\kappa_1(k_1)\pi_1(k_1)\left|C_1^{(k_1)}(k_1)\right|^{1/2}}{\kappa_0^*(k_1)\left|C_0(k_1)\right|^{1/2}} \exp\left\{-\frac{1}{2}\left(\phi y_1' y_1 + [m_0(k_1)]'[C_0(k_1)]^{-1}[m_0(k_1)]\right)\right\}$$
$$\times \exp\left\{\frac{1}{2}\left[m_1^{(k_1)}(k_1)\right]'\left[C_1^{(k_1)}(k_1)\right]^{-1}\left[m_1^{(k_1)}(k_1)\right]\right\},$$

with

$$C_1^{(k)}(k_1) = \begin{cases} C_0(k) & \text{if} \quad k \neq k_1, \\ \left(C_0^{-1}(k) + \phi X_1' X_1\right)^{-1} & \text{if} \quad k = k_1, \end{cases}$$

and

$$m_1^{(k)}(k_1) = \begin{cases} m_0(k) & \text{if} \quad k \neq k_1, \\ C_1^{(k_1)}(k_1)\left(C_0(k)^{-1} m_0(k) + \phi X_1' y_1\right) & \text{if} \quad k = k_1. \end{cases}$$

In addition,

$$\kappa_0^*(k_1) = \int_{\mathcal{R}^{(k_1)}} N\left(\theta^{(k_1)}|m_0(k_1), C_0(k_1)\right) d\theta^{(k_1)},$$
$$\kappa_1(k_1) = \int_{\mathcal{R}^{(k_1)}} N\left(\theta^{(k_1)}\left|m_1^{(k_1)}(k_1), C_1^{(k_1)}(k_1)\right.\right) d\theta^{(k_1)}.$$

(b) Updating the posteriors when $q > 1$. Assume that the distribution of $\left(\theta^{(k)}|\mathcal{M}_{q-1}(k_{q-1}), \mathcal{D}_{q-1}\right)$ can be approximated by a truncated normal distribution and that approximations for $p_{q-1}(k_{q-1})$ with $k_{q-1} = 1 : K$ are also available. Then, after observing the data of epoch q for $q > 1$, we have

$$\left(\theta^{(k)}|\mathcal{M}_q(k_q), \mathcal{M}_{q-1}(k_q), \mathcal{D}_q\right) \approx TN\left(\theta^{(k)}\left|m_q^{(k)}(k_q, k_{q-1}), C_q^{(k)}(k_q, k_{q-1})\right.\right),$$

and

$$p_q(k_q, k_{q-1}) \propto \frac{\pi(k_q|k_{q-1})p_{q-1}(k_{q-1})\kappa_q(k_q, k_{q-1})\left|C_q^{(k_q)}(k_q, k_{q-1})\right|^{1/2}}{\kappa^*_{q-1}(k_q, k_{q-1})\left|C_{q-1}^{(k_q)}(k_{q-1})\right|^{1/2}}$$
$$\times \exp\left\{-\frac{1}{2}\left(\phi y_q' y_q + \left[m_{q-1}^{(k_q)}(k_{q-1})\right]'\left[C_{q-1}^{(k_q)}(k_{q-1})\right]^{-1}\left[m_{q-1}^{(k_q)}(k_{q-1})\right]\right)\right\}$$
$$\times \exp\left\{\frac{1}{2}\left[m_q^{(k_q)}(k_q, k_{q-1})\right]'\left[C_q^{(k_q)}(k_q, k_{q-1})\right]^{-1}\left[m_q^{(k_q)}(k_q, k_{q-1})\right]\right\},$$

with $p_q(k_q, k_{q-1})$ normalized such that $\sum_{k_q=1}^K \sum_{k_{q-1}=1}^K p_q(k_q, k_{q-1}) = 1$ and with

$$C_q^{(k)}(k_q, k_{q-1}) = \begin{cases} C_{q-1}^{(k)}(k_{q-1}) & \text{if } k \neq k_q, \\ \left(\left[C_{q-1}^{(k)}(k_{q-1})\right]^{-1} + \phi X_q' X_q\right)^{-1} & \text{if } k = k_q, \end{cases}$$

$$m_q^{(k_q)}(k_q, k_{q-1}) = \begin{cases} m_{q-1}^{(k)}(k_{q-1}) & \text{if } k \neq k_q, \\ C_q^{(k)}(k_q, k_{q-1})\left(\left[C_{q-1}^{(k)}(k_{q-1})\right]^{-1} m_{q-1}^{(k)}(k_{q-1}) + \phi X_q' y_q\right) & \text{if } k = k_q. \end{cases}$$

In addition,

$$\kappa^*_{q-1}(k_q, k_{q-1}) = \int_{\mathcal{R}^{(k_q)}} N\left(\theta^{(k_q)}\left|m_{q-1}^{(k_q)}(k_{q-1}), C_{q-1}^{(k_q)}(k_{q-1})\right.\right) d\theta^{(k_q)}$$
$$\kappa_q(k_q, k_{q-1}) = \int_{\mathcal{R}^{(k_q)}} N\left(\theta^{(k_q)}\left|m_q^{(k_q)}(k_q, k_{q-1}), C_q^{(k_q)}(k_q, k_{q-1})\right.\right) d\theta^{(k_q)}.$$

Now, taking $p_q(k_q) = \sum_{k_{q-1}=1}^K p_q(k_q, k_{q-1})$, we have that, if $p_q(k_q) \neq 0$, $p(\theta^{(k)}|\mathcal{M}_q(k_q), \mathcal{D}_q)$ can be written as follows

$$p(\theta^{(k)}|\mathcal{M}_q(k_q), \mathcal{D}_q) = \sum_{k_{q-1}=1}^{K} p(\theta^{(k)}|\mathcal{M}_q(k_q), \mathcal{M}_{q-1}(k_{q-1}), \mathcal{D}_q)\ p_q(k_q, k_{q-1})/p_q(k_q),$$

and so, we can approximate this mixture as

$$(\theta^{(k)}|\mathcal{M}_q(k_q), \mathcal{D}_q) \approx TN\left(\theta^{(k)}\left|m_q^{(k)}(k_q), C_q^{(k)}(k_q), \mathcal{R}^{(k)}\right.\right),$$

with

$$m_q^{(k)}(k_q) = \sum_{k_{q-1}=1}^{K} m_q^{(k)}(k_q, k_{q-1})p_q(k_q, k_{q-1})/p_q(k_q),$$

and

$$C_q^{(k)}(k_q) = \sum_{k_{q-1}=1}^{K} \Big\{ \Big[C_q^{(k)}(k_q, k_{q-1}) + \big(m_q^{(k)}(k_q) - m_q^{(k)}(k_q, k_{q-1})\big)' \big(m_q^{(k)}(k_q) - m_q^{(k)}(k_q, k_{q-1})\big)\Big] \times \frac{p_q(k_q, k_{q-1})}{p_q(k_q)} \Big\}.$$

Case with ϕ unknown

(a) Updating the posteriors at $q = 1$. When ϕ is unknown, having a prior of the form

$$p(\theta^{(1)}, \ldots, \theta^{(k)}, \phi | \mathcal{D}_0) = \prod_{k=1}^{K} TN\big(\theta^{(k)} | m_0(k), \phi^{-1} C_0^*(k), \mathcal{R}^{(k)}\big) \times Ga(\phi | n_0/2, d_0/2),$$

leads to the following results

$$\begin{aligned}(\theta^{(k)} | \mathcal{M}_1(k_1), \mathcal{D}_1, \phi) &= TN\Big(\theta^{(k)} \Big| m_1^{(k)}(k_1), \phi^{-1} C_1^{*,(k)}(k_1), \mathcal{R}^{(k)}\Big),\\ (\phi | \mathcal{M}_1(k_1), \mathcal{D}_1) &\approx Ga(n_1/2, d_1(k_1)/2),\end{aligned}$$

with

$$C_1^{*,(k)}(k_1) = \begin{cases} C_0^*(k) & \text{if } \; k \neq k_1, \\ \big(C_0^*(k)^{-1} + X_1' X_1\big)^{-1} & \text{if } \; k = k_1, \end{cases}$$

$$m_1^{(k)}(k_1) = \begin{cases} m_0(k) & \text{if } \; k \neq k_1, \\ C_1^{*,(k_1)}(k_1)\big(C_0^*(k_1)^{-1} m_0(k_1) + X_1' y_1\big) & \text{if } \; k = k_1, \end{cases}$$

$n_1 = n_0 + n$, where n is the dimension of the y_q vectors, and

$$d_1(k_1) = d_0 + (y_1 - X_1 m_0(k_1))' \big(Q_1^*(k_1)\big)^{-1} (y_1 - X_1 m_0(k_1)),$$

where $Q_1^*(k_1) = \big(X_1 C_0^*(k_1) X_1' + I\big)$. We can also obtain the approximation

$$\big(\theta^{(k)} | \mathcal{M}_1(k_1), \mathcal{D}_1\big) \approx TT_{n_q}\Big(\theta^{(k)} \Big| m_1^{(k)}(k_1), C_1^{(k)}(k_1), \mathcal{R}^{(k)}\Big),$$

where $TT_\nu(\cdot | m, C, \mathcal{R})$ denotes a truncated Student-t distribution with ν degrees of freedom, location m, scale C and truncation region $\mathcal{R}$. The value of $C_1^{(k)}(k_1)$ in the equation above is given by $C_1^{(k)}(k_1) = C_1^{*,(j)}(k_1) S_1(k_1)$, with $S_1(k_1) = d_1(k_1)/n_1$. In addition, writing

$$\tilde{p}_1(k_1) \propto \frac{\kappa_1(k_1)\pi_1(k_1)}{\kappa_0^*(k_1)} \times \frac{\big|C_1^{(k)}(k_1)\big|^{1/2} d_0^{(n_0/2)} \Gamma(n_1/2)}{|C_0(k_1)|^{1/2} \Gamma(n_0/2) [d_1(k_1)]^{n_1/2}},$$

where the $\tilde{p}_1(\cdot)$s are normalized such that $\sum_{k_1=1}^{K} \tilde{p}_1(k_1) = 1$, we have that $p_1(k_1) \approx \tilde{p}_1(k_1)$. Finally, $\kappa_0^*(\cdot)$ and $\kappa_1(\cdot)$ are defined as

$$\kappa_0^*(k_1) = \int_{\mathcal{R}^{(k_1)}} TT_{n_0}\left(\theta^{(k_1)} | m_0(k_1), C_0(k_1)\right) d\theta^{(k_1)},$$

$$\kappa_1(k_1) = \int_{\mathcal{R}^{(k_1)}} TT_{n_1}\left(\theta^{(k_1)} \middle| m_1^{(k_1)}(k_1), C_1^{(k_1)}(k_1)\right) d\theta^{(k_1)}.$$

(b) Updating the posteriors when $q > 1$. After y_q is observed, we have the following approximations

$$\left(\theta^{(k)} | \mathcal{M}_q(k_q), \mathcal{M}_{q-1}(k_{q-1}), \mathcal{D}_q, \phi\right) \approx TN\left(\theta^{(k)} \middle| m_q^{(k)}(k_q, k_{q-1}), C_q^{*,(k)}(k_q, k_{q-1})/\phi, \mathcal{R}^{(k)}\right),$$

with

$$C_q^{*,(k)}(k_q, k_{q-1}) = \begin{cases} C_{q-1}^{*,(k)}(k_{q-1}) & \text{if } k \neq k_q, \\ \left(\left[C_q^{*,(k)}(k_q, k_{q-1})\right]^{-1} + X_q' X_q\right)^{-1} & \text{if } k = k_q, \end{cases}$$

$$m_q^{(k)}(k_q, k_{q-1}) = \begin{cases} m_{q-1}^{(k)}(k_{q-1}) & \text{if } k \neq k_q, \\ C_q^{*,(k)}(k_q, k_{q-1})\left(\left[C_{q-1}^{*,(k)}(k_{q-1})\right]^{-1} m_q^{(k)}(k_{q-1}) + X_q' y_q\right) & \text{if } k = k_q, \end{cases}$$

and

$$(\phi | \mathcal{M}_q(k_q), \mathcal{M}_{q-1}(k_{q-1}), \mathcal{D}_q) \approx Ga(n_q/2, d_q(k_q, k_{q-1})/2),$$

with $n_q = n_{q-1} + n$, and

$$d_q(k_q, k_{q-1}) = d_{q-1}(k_{q-1}) + \left(y_q - X_q m_{q-1}^{(k_q)}(k_{q-1})\right)' \left[Q_q^*(k_q, k_{q-1})\right]^{-1} \left(y_q - X_q m_{q-1}^{(k_q)}(k_{q-1})\right),$$

where $Q_q^*(k_q, k_{q-1}) = \left(X_q C_{q-1}^{(k_q)}(k_{q-1}) X_q' + I\right)$. In addition, we approximate $p_q(k_q, k_{q-1})$ by $\tilde{p}_q(k_q, k_{q-1})$, where

$$\tilde{p}_q(k_q, k_{q-1}) \propto \frac{\pi(k_q | k_{q-1}) \kappa_q(k_q, k_{q-1})}{\kappa_{q-1}^*(k_q, k_{q-1})} \times \frac{\left|C_q^{(k_q)}(k_q, k_{q-1})\right|^{1/2} (d_{q-1}(k_{q-1}))^{n_{q-1}/2} \Gamma(n_q/2)}{\left|C_{q-1}^{(k_q)}(k_{q-1})\right|^{1/2} (d_q(k_q, k_{q-1}))^{n_q/2} \Gamma(n_{q-1}/2)},$$

with $\tilde{p}_q(k_q, k_{q-1})$ normalized such that $\sum_{k_q=1}^{K} \sum_{k_{q-1}=1}^{K} p_q(k_q, k_{q-1}) = 1$ and

$$\kappa_{q-1}^{*}(k_q, k_{q-1}) = \int_{\mathcal{R}^{(k_q)}} T_{n_{q-1}} \left(\theta^{(k_q)} \middle| m_{q-1}^{(k_q)}(k_{q-1}), C_{q-1}^{*,(k_q)}(k_{q-1}) \frac{d_{q-1}(k_{q-1})}{n_{q-1}} \right) d\theta^{(k_q)}$$

$$\kappa_q(k_q, k_{q-1}) = \int_{\mathcal{R}^{(k_q)}} T_{n_q} \left(\theta^{(k_q)} \middle| m_q^{(k_q)}(k_q, k_{q-1}), C_{q-1}^{*,(k_q)}(k_q, k_{q-1}) \frac{d_q(k_q)}{n_q} \right) d\theta^{(k_q)}.$$

Then, we approximate $p_q(k_q, k_{q-1})$ by $\tilde{p}_q(k_q, k_{q-1})$. Finally, using the fact that $p(\theta^{(k)} | \mathcal{M}_q(k_q), \mathcal{D}_q, \phi)$ can be written as a mixture of K components, we can obtain the following approximations by collapsing the K components into a single component (see West and Harrison, 1997, Chapter 12), and so

- $(\phi | \mathcal{M}_q(k_q), \mathcal{D}_q) \approx Ga(n_q/2, d_q(k_q)/2)$, with $d_q(k_q) = n_q S_q(k_q)$ and

$$S_q^{-1}(k_q) = \sum_{k_{q-1}=1}^{K} S_q^{-1}(k_q, k_{q-1}) p_q(k_q, k_{q-1}) / p_q(k_q).$$

- $\left(\theta^{(k)} | \mathcal{M}_q(k_q), \mathcal{D}_q, \phi^{-1}\right) \approx TN\left(\theta^{(k)} \middle| m_q^{(k)}(k_q), C_q^{*,(k)}(k_q)/\phi, \mathcal{R}^{(k)}\right)$, with $C_q^{*,(j)}(k_q) = \frac{C_q^{(k)}(k_q)}{S_q(k_q)}$. The values of $m_q^{(k)}(k_q)$ and $C_q^{(k)}(k_q)$ are computed as follows

$$m_q^{(k)}(k_q) = \sum_{k_{q-1}=1}^{K} m_q^{(k)}(k_q, k_{q-1}) p_q^{*}(k_q, k_{q-1}),$$

$$C_q^{(k)}(k_q) = \sum_{k_{q-1}=1}^{K} \Big\{ \Big[C_q^{(k)}(k_q, k_{q-1}) + \left(m_q^{(k)}(k_q) - m_q^{(k)}(k_q, k_{q-1}) \right)' \left(m_q^{(k)}(k_q) - m_q^{(k)}(k_q, k_{q-1}) \right) \Big] \times p_q^{*}(k_q, k_{q-1}) \Big\},$$

where $p_q^{*}(k_q, k_{q-1}) = S_q(k_q) S_q^{-1}(k_q, k_{q-1}) p_q(k_q, k_{q-1}) / p_q(k_q)$, normalized such that, for all k_q, $\sum_{k_{q-1}=1}^{K} p_q^{*}(k_q, k_{q-1}) = 1$.

Acknowledgements

We are grateful to L. Trejo for useful discussions and for providing access to the data.

References

Carvalho, C., Johannes, M., Lopes, H. F. and Polson, N. (2008). Particle learning and smoothing. Technical report, Graduate School of Business, University of Chicago.

Dyro, F. (1989). *The EEG Handbook.* Little, Brown and Company, Boston, Massachusetts.

Huerta, G. and West, M. (1999). Priors and component structures in autoregressive time series models. *Journal of the Royal Statistical Society B*, **61**, 881–899.

Krystal, A., Prado, R. and West, M. (1999). New methods of time series analysis of non-stationary EEG data: eigenstructure decompositions of time-varying autoregressions. *Clinical Neurophysiology*, **110**, 2197–2206.

Le Van Quyen, M., Martinerie, J., Navarro, V., Boon, P., D'Have, M., Adam, C., Renault, B., Varela, F. and Baulac, M. (2001). Anticipation of epileptic seizures from standard EEG recordings. *Lancet*, **357**, 183–8.

Liu, J. and West, M. (2001). Combined parameter and state estimation in simulation-based filtering. In *Sequential Monte Carlo Methods in Practice*, (ed. A. Doucet, N. de Freitas, and N. Gordon), pp. 197–223. Springer, New York.

Prado, R. (2009). Characterization of latent structure in brain signals. In *Statistical Methods for Modeling Human Dynamics* (ed. S. Chow, E. Ferrer, and F. Hsieh). Psychology Press, Taylor and Francis Group, New York.

Prado, R., West, M. and Krystal, A. (2001). Multi-channel EEG analyses via dynamic regression models with time-varying lag/lead structure. *Journal of the Royal Statistical Society, Series C (Applied Statistics)*, **50**, 95–109.

Trejo, L., Knuth, K., Prado, R., Rosipal, R., Kubitz, K., Kochavi, R., Matthews, B. and Zhang, Y. (2007). EEG-based estimation of mental fatigue: Convergent evidence for a three-state model. In *Augmented Cognition, HCII 2007, LNAI 4565*, (ed. D. Schmorrow and L. Reeves), New York, pp. 201–211. Springer, New York.

Trejo, L., Kochavi, R., Kubitz, K., Montgomery, L., Rosipal, R. and Matthews, B. (2006). EEG-based estimation of mental fatigue. Technical report, Available at `http://publications.neurodia.com/Trejo-et-al-EEG-Fatigue2006-Manuscript.pdf`.

Trejo, L., Matthews, R. and Allison, B. (2007). Experimental design and testing of a multimodal cognitive overload classifier. In *Foundations of Augmented Cognition* (4th edn), (ed. D. Schmorrow, D. Nicholson, J. Drexler, and L. Reeves), pp. 13–22. Strategic Analysis, Inc, Arlington, VA.

Weiner, R. D., Coffey, E. and Krystal, A. D. (1991). The monitoring and management of electrically induced seizures. *Psychiatric Clinics of North America*, **14**, 845–69.

West, M. (1997). Time series decompositions. *Biometrika*, **84**, 489–94.

West, M. and Harrison, J. (1997). *Bayesian Forecasting and Dynamic Models*, (2nd edn). Springer-Verlag, New York.

West, M., Prado, R. and Krystal, A. (1999). Evaluation and comparison of EEG traces: Latent structure in nonstationary time series. *Journal of the American Statistical Association*, **94**, 1083–1095.

Index

The index entries appear in letter-by-letter alphabetical order.
Index entries displayed in bold text indicate entries from 'A' Appendices.

Lightning Source UK Ltd.
Milton Keynes UK
UKOW02n1329310114

225652UK00001B/3/P